Lecture Notes in Civil Engineering

Volume 211

Lecture Notes in Civil Engineering (LNCE) publishes the latest developments in Civil Engineering—quickly, informally and in top quality. Though original research reported in proceedings and post-proceedings represents the core of LNCE, edited volumes of exceptionally high quality and interest may also be considered for publication. Volumes published in LNCE embrace all aspects and subfields of, as well as new challenges in, Civil Engineering. Topics in the series include:

- Construction and Structural Mechanics
- Building Materials
- Concrete, Steel and Timber Structures
- Geotechnical Engineering
- Earthquake Engineering
- Coastal Engineering
- Ocean and Offshore Engineering; Ships and Floating Structures
- Hydraulics, Hydrology and Water Resources Engineering
- Environmental Engineering and Sustainability
- Structural Health and Monitoring
- Surveying and Geographical Information Systems
- Indoor Environments
- Transportation and Traffic
- Risk Analysis
- Safety and Security

To submit a proposal or request further information, please contact the appropriate Springer Editor:

- Pierpaolo Riva at pierpaolo.riva@springer.com (Europe and Americas);
- Swati Meherishi at swati.meherishi@springer.com (Asia - except China, and Australia, New Zealand);
- Wayne Hu at wayne.hu@springer.com (China).

All books in the series now indexed by Scopus and EI Compendex database!

More information about this series at https://link.springer.com/bookseries/15087

Wei Guo · Kai Qian
Editors

Proceedings of the 2022 International Conference on Green Building, Civil Engineering and Smart City

Springer

Editors
Wei Guo
Central South University
Changsha, China

Kai Qian
Guilin University of Technology
Guilin, China

ISSN 2366-2557 ISSN 2366-2565 (electronic)
Lecture Notes in Civil Engineering
ISBN 978-981-19-5216-6 ISBN 978-981-19-5217-3 (eBook)
https://doi.org/10.1007/978-981-19-5217-3

This Springer imprint is published by the registered company Springer Nature Singapore Pte Ltd.
The registered company address is: 152 Beach Road, #21-01/04 Gateway East, Singapore 189721, Singapore

Preface

On behalf of the Conference Committee, we welcome you to the International Conference on Green Building, Civil Engineering and Smart City, held in Guilin, China, during July 24–27, 2022. The conference aims to provide an international forum for researchers and industry practitioners to share their new ideas, original research results, and practical development experiences from related areas, including green building, civil engineering and smart city and exchange technical knowledge, discuss innovative and effective solutions and address challenges in the fields of green building, civil engineering and smart city.

The conference featured keynote speeches which were presented by Prof. Kai Qian from Guilin University of Technology, Prof. Wei Guo from Central South University, Prof. Marco Cimillo from Xi'an Jiaotong-Liverpool University, Prof. Hazem Samih Mohamed from Southwest Petroleum University, and Prof. Liu Jin from Beijing University of Technology. Invited speeches were delivered by Prof. JunWang from Wenzhou University, Prof. Yifei Hao from Hebei University of Technology and Dr. Chee-Ming Chan from Tun Hussein Onn University of Malaysia. Furthermore, two parallel sessions were held successfully. The first session was themed on advanced civil engineering technology, and the second one was regarding smart city and green buildings. The above presentations were shown by the face-to-face presentation and live broadcast via VooV Meeting. The location and affiliation of participants are various, including Guilin University of Technology, Central South University, Nanyang Technological University, University of Calabria, Beijing University of Technology, Southwest Petroleum University, Dalian University of Technology, Southeast University, Changsha University of Science and Technology, Tongji University, Chongqing University, Zhejiang University and Huazhong University of Science and Technology, etc.

Most of the submissions came from the workshops which were proposed by workshop chairs. All manuscripts need to go through an Initial Check by assistant editors to check if it is relevant to the theme of the conference. Once a manuscript passes the initial checks by assistant editors, it will be assigned to at least two independent experts for peer review. Guest editors would be responsible for all rebuttals and review all materials. Eventually, 129 submissions were accepted and recommended to be published. This proceeding was divided into three volumes, among which, 40 submissions were about green building, 76 submissions were related to civil engineering, and 13 submissions were categorized as smart city. We would like to thank all conference committee members and reviewers, who had offered a high-quality program with diligence on GBCESC 2022.

We would also like to express our special thanks for the following affiliations' support:

- Financial sponsor: Guilin University of Technology
- Technical sponsor: Central South University
- Co-sponsors: Dalian University of Technology, Zhejiang University and Wenzhou University, University of Saint Joseph, Soochow University, etc.
- Academic supporters: ISHMII, China Green Building Council of CSUS, Committee for the Promotion of Green Building Investment of China Association of Building Energy Efficiency(CPGBI), American Concrete Institute, China-Portugal Joint Laboratory of Cultural Heritage Conservation Science supported by the Belt and Road Initiative, Laboratory of Multifunctional Shaking Table High-Speed Railway of Central South University, High-Speed Railway of Construction Technology of Central South University Railway Campus of National Engineering Laboratory of China, State Key Laboratory of Geomechanics and Geotechnical Engineering of China of Institute of Rock and Soil Mechanics of Chinese Academy of Sciences, and Jiangsu Key Laboratory of Engineering Mechanics.

We wish to sincerely thank all conference committee members, reviewers and supporters for their invaluable efforts in ensuring a timely, fair, and highly effective GBCESC 2022 program.

In the end, we sincerely thank LNCE and Springer publishing team for their dedicated services and support.

Wei Guo
Kai Qian

Organization

General Chairs

Kai Qian	Guilin University of Technology, China
Wei Guo	Central South University, China
H. S. Mohamed	Southwest Petroleum University, China
Liu Jin	Beijing University of Technology, China

Co-chairs

Zijian Guo	Dalian University of Technology, China
Lei He	Southeast University, China
Wenbing Wu	China University of Geosciences, China

Honorary Advisors

Stephen Morgan	University of Saint Joseph, Macao
Sean Monkman	CarbonCure Technologies, Canada
Shana Kelley	KPFF, USA

Technical Chairs

Zhen Guo	Zhejiang University, China
Jiang He	Gunagxi University, China
Zhiqi He	Southeast University, China

Technical Program Committees

Shishun Zhang	Huazhong University of Science and Technology, China
Lizhong Jiang	Central South University, China
Bai Yu	Monash University, Australia

Yao Wu	Soochow University, China
Kaiming Bi	Kurtin University, Australia
Xuhui He	Central South University, China
Elżbieta Maciorzek	Silesian University of Technology, Poland
Yan Han	Changsha University of Science and Technology, China
Huseyin Seker	Northumbria University, UK
S. V. Aruna Kumar	University of Beira Interior, Portugal
Ugur Albayrak	Department of Civil Engineering, Turkey
Feng Chen	Tongji University, China
Li Chen	Southeast University, China
Zhigang Zhang	Chongqing University, China
Jianlin Liu	China University of Petroleum, China
Sheng Li	Lanzhou Jiaotong University, China
Supaporn Chairongsee	Walailak University, Thailand
Florence Jean Talirangan	Misamis University, Philippines
Jinyang Fu	Central South University, China
Yuancheng Guo	Zhengzhou University, China
Lin Zhao	Tongji University, China
Linh Truang-Hong	University College Dublin, Ireland
Caihua Shen	Hohai University, China
Miguel Ángel Campeno Laborda	Universidad de Sevilla, Spain
Xusheng Wan	Southwest Petroleum University, China
Yonghui Li	University of Sydney, Australia
Mohamed Rihan	Menoufia University, The Arab Republic of Egypt
Georgios Theodoropoulos	Durham University, Britain
Bo Li	Lanzhou University, China
Zhiwu Yu	Central South University, China
Yifei Hao	Hebei University of Technology, China
Qian Fang	Beijing Jiaotong University, China
Hui Liu	Central South University, China
Tony Yang	University of British Columbia, UK
Junxian Zhao	South China University of Technology, China
Hongjing Li	Nanjing Tech University, China
Jiandong Zhang	Nanjing Tech University, China
Jun Wang	Wenzhou University, China
Shuang Li	Harbin Institute of Technology, China
Rong Zhu	Jiangnan University, China
Biao Wei	Central South University, China
Yaoting Zhang	Huazhong University of Science and Technology, China
Xiuxin Wang	Southeast University, China
Xin Cai	Central South University, China

Decheng Feng	Southeast University, China
Liangjiu Jia	Tongji University, China
Liping Wang	Central South University, China
Zhi Li	Guilin University of Technology, China
Yonghong Wang	Qingdao University of Technology, China
Guoming Liu	Shandong University of Science and Technology, China
Bing Liu	Guilin University of Technology, China
Yi Zhang	Guilin University of Technology, China

Contents

Green Building

Numerical Verification of Offline Hybrid Testing for the Seismic Response of Energy Dissipation Structure 3
Haodong Liu, Zhenyun Tang, and Fukang Gao

State-of-the-Art of BRBs in Reinforced Concrete Structures: BRB Units, Connections, Seismic Design, and Performance Evaluation 14
Huiming Chen and Jiulin Bai

Design Strategies for Green Retrofitting of a University Student Dormitory in the Hot-humid and Less-windy Climate 22
Binhua Li, Xuexiu Zhao, and Yanwen Luo

Analysis of Urban Dual-Carbon Targets and Countermeasures Based on GeoSOS-FLUS Simulation–A Case of Hangzhou City, Zhejiang Province 32
Weiwu Wang, Huan Chen, Shiyuan Cao, and Biyan Wang

Research on Macao Cultural Heritage Protection and Sustainable Development——A Case Study of Ruins of St. Paul's 44
Yishan Mao, Yao Wu, and Yonglan Wu

3D Printing of Fast Setting Basalt Fiber Reinforced Cement-Based Materials 57
Weihong Li, Xuhao Chen, and Nan Li

Using Nighttime Light Data to Achieve Precise Estimation of Residents' Living Carbon Emissions and Landscape Pattern Characteristics 65
Huan Chen, Biyan Wang, Weiwu Wang, and Shan Wang

Research on the Kang in Cold Rural Area Based on Spatial Adaptation and Technological Improvement 78
Yiming Li, Hongpeng Xu, Zirui Fang, Chengyuan Bao, and Jianmei Wu

Constructing the Mode and Model of Spatial Appeal, Spatial Response for Conservation and Utilization of Chen Hongmou Xiangguo Mansion Site in Qing Dynasty of China Based on Value Evaluation ... 92
Zhengmin Wen, Rongsong Quan, Hui Wang, Bing Li, and Zhenqiang Li

Air Distribution Analysis and Optimization Layout of In-Row Air Conditioning in Data Center ... 108
Jianqiang Xu and Zhen Liang

Intelligent Protection and Inheritance Method of the Historic Centre of Macao Based on BIM+ Concept —— A Case Study of Mandarin's House ... 122
Yao Wu and Hui Zhu

Recycled Concrete Powder Supported Nano-TiO_2 Composites: Efficient Utilization of RCP and Excellent Photocatalytic Performance . 133
Gang Liao and Wu Yao

A Survey on Life Cycle Sustainability Assessment for Triple Bottom Line of Buildings ... 141
Kun Lu, Xueyuan Deng, and Baoquan Cheng

Spatial and Temporal Prediction of Carbon Peaking Goals and Zonal Governance Approaches to Achieve Carbon neutrality—A Case Study from Zhejiang Province, China ... 151
Weiwu Wang, Huan Chen, Ronglu Yang, Biyan Wang, and Yuxin Yang

Peak Projections and Solutions for Carbon Neutrality and Carbon Peaking in the Construction Industry ... 166
Baoxi Zhang, Zhiqiang Kang, and Tongxin Jin

Synthesis and Application of Tackifying Resin in Silane Modified Polyether Adhesive for Prefabricated Building ... 176
Yangqing Chen, Yuliang Qi, Hengchao Huang, Jianjun Chen, and Keke Huang

Analysis on the Regeneration Strategy of Shikumen Residential Building ... 185
Ranran Sun

Research on the Transformation and Development of Modern Western-Style Villas in Suzhou Under the Concept of Sustainable Development ... 190
Yonglan Wu, Yao Wu, and Yishan Mao

A Multi-objective Optimization Approach of Green Building Performance Based on LGBM and AGE-MOEA ... 202
Yuxuan Shen and Yue Pan

Carbon Emissions of Passive Building During Production and Operation in Cold Region of China 211
Zhiqiang Kang, Bo He, and Xiaodong Xia

Strength of an Improved Connection for Modular Concrete Structures Without Onsite Casting 221
Chengguang Xu, Xudong Qian, Ran Tao, and Rong Wang

Research on Application Effect of Built-In Louver Insulating Glass Window 230
Aofeng Shi and Guoqing Yu

Micro-scale Regional Spatial Low-Carbon Planning System Framework Based on Scenario Analysis 239
Zhexuan Chen and Junting Chen

The Optimization of High-Density Campus Transportation Based on Green Transport 248
Ming Liu

Design of Wave Energy Capture Structure and Research on Mooring System 255
Mingjun Li, Xinhui Duan, Zheng Huang, and Yan Wei

The Concept and Strategy of Territorial Spatial Planning System for Construction Waste Reduction and Resource Utilization 263
Shuaibo Li and Yingying Zhou

The Study on the Development of Contemporary International Timber Heritage Conservation 276
Xi Chen and Mei Huang

Analytical Evaluation for Optimization of Landside Traffic System Planning at a Hub Airport 288
Hongjing Ge and Jiang He

Study on the Protection and Development of Farming-Type Traditional Village Landscape Under the Concept of Sustainability – A Case Study of Western Henan 299
Sai Yin

Surrogate Based Multi-objective Optimization for Energy-Saving Building Design 311
Yuchen Chen and Xuhua Shi

Research on the Development of Modern Mosaic Wall Decoration Materials Based on Nostalgic Memory—Taking Wumi Congee (Congee Shop) in Cantonese Area of China as an Example 319
Qiang Tang, Yile Chen, Junzhang Chen, and Liang Zheng

Exploration and Practice of the Application of Natural Lighting and Shading Technology in Building Energy Conservation 329
Jinghua Shen, Jie Zhang, Zhaoyang Xue, and Jie Wu

Analysis of the Effect of Stress Intensity Factors on Temperature-Shrinkage Reflective Cracks 337
Qinshou Huang

Exploration of the Emergency Renovation of Sports Venues Resilience Space Based on Green Building Technology Under the Normal Epidemic Situation 347
Jie Liu and Tong Yang

Research on Boundary Space Design of Urban Residential Community in the Pearl River Delta 356
Junrui Cao and Yichenfei Guo

Distributed Ledger for Construction Material Providers to Control Carbon Emissions, a Case Study of Glulam Material 365
Wenhao Liu and Zhan Shu

Dynamic Response Analysis of Stress Intensity Factor of Reflective Cracking in Asphalt Pavement Addition 373
Qinshou Huang

Study on the Evaluation Index System of Reinforcement and Renovation of Existing Campus Buildings: Take Yueyi Building as an Example 385
Xiaoxue Tang, Jie Wu, Youqing Zhu, Ruijie Xu, Shunuo Li, and Qianqian Yan

In-Process Reinforcement Method for 3D Concrete Printing: Status, Potentials and Challenges 394
Billy Kazadi, Liang Yao, and Li Wang

Systematic Framework of Economic KPIs for Prefabricated Building Construction 403
Bingyan Wang

Civil Engineering

Numerical Simulation of Resists Progressive Collapse of the Elevated Station 415
Xiaolan Yuan, Weizhuo Huang, Guangtao Li, Zhi Li, and Xiaofang Deng

Numerical Simulation Method of Dynamic Coupling of Complicated Mountains and the Wake of Multiple Wind Turbines 427
Chi Ma, Shuanglong Fan, and Zhenqing Liu

Numerical Simulation of Substructure Shake Table Testing for Base-Isolated Structure Using Model-Based Integration Algorithm 436
Bo Fu, Futai Zhang, and Jin Chen

Effect of Ceramic Waste Powder on Mechanical Properties of Recycled Coarse Aggregate Concrete 445
Ronghui Yang, Xin Min, Feixiang Chen, and Gang Li

Construction Technology Innovation of Light-Weight Steel-Concrete Composite Cable-Stayed Bridge 455
Xueyun Kang, Qi You, Gang Li, and Huijun Shen

Numerical Simulations of the Effects of the Radiant Floor Combined with the Displacement Ventilation of the Spread of Exhaled Contaminants in the Confined Space 465
Jie Wu, Liang Xu, Jing Hua Shen, António Candeias, and Wei Zhang

Research on Impact Collapse Resistance of RC Beams After Fire 474
Zhi Li, Kaiqian Huang, and Xiaolan Yuan

Method Research on Traffic Volume Forecasting Based on Bio-LSTM 484
Xiao Li, Chaoyang Li, Tao Wang, Yi Zhang, and Changqin Xi

Application of BIM in Automatic Typesetting of Universal Segments in Shield Tunnel 496
Liujie Zhu, Yuxian Zhang, Jixun Zhang, and Zichang Ma

Rating Techniques Research on Operation Performance of Continuous Box Beam with Common Span in China High Speed Railway 504
Wei Wang, Penghui Liu, Xin Meng, Zhensheng Dong, Jing Yin, and Yigan Wang

Static Simple Shear Tests on Mechanical Behavior of Interface Between Coral Sand and Concrete 519
Chunyong Jiang, Xuanming Ding, and Gang Yang

Concrete and Steel Structure of High-Rise Modular Buildings: A Systematic Review 527
Zhen Yu

Dynamic Properties of Hydrophobic Sand as a Filling Material in Road Subgrade 536
Gege Huang and Hongjie Lin

Effects of Redispersible Latex Powder on the Performance of 3D Printing Cementitious Materials 542
Huan Wang, Xinchen Tian, Huixian Wei, Xueyan Dai, Yujun Che, and Huashan Yang

Study on the Deterioration Law of Sleeve Grouting Defects on the Lateral Bearing Capacity of Assembled Shear Walls 552
Qizhou Liu, Yuhan Lin, Jinzhou Li, and Min Zhang

Monitoring Technology of Runway Pavement with Holes Based on Vertical Shear Strain 562
Jing Cai, Hanxue Wang, Yue Li, and Zhiqiang Du

Research on Vibration Control for Transformers of Off-Shore Wind Turbine 571
Xianqing Guo, Chengyuan Wang, and Hongju Wu

Hybrid Simulation Platform for Onshore Wind Turbine Under Coupled Wind-Earthquake Load 579
Hao Hu, Zhu Mei, Yang Liu, and Kao-shan Dai

Risk-Based Inspection Planning for Support Structure of Offshore Wind Turbine 591
Weiheng Zhang, Dagang Lu, Jianjun Qin, Min Liu, and Michael Havbro Faber

Upper Bound Limit Analysis for the Stability of Elliptical Tunnels Based on the Node-Based Smoothed Finite Element Method and Second Order Cone Programming 599
Feng-Tao Liu, Xi-Wen Zhou, and Yan Zhang

Response of Soil-Structure with Respect to Different Shapes of Pile Cross-section Under Cyclic Loads in the Context of CO_2 Emission 607
Linfeng Li and Jianjun Qin

Simplified Calculation of Waist Hole of Straddle Monorail Pin Bearing 616
Hui Wang, Shuyi Duan, and Yanqiang Ji

Experimental and Numerical Studies on the Seismic Performance of Crawler-Type Connection Node 629
Yixin Zeng, Wei Lin, Jing Li, Yixi Hu, Zhilong Jiang, and Peixing Xie

Experimental Study on the Lateral Performance of Traditional Wooden Structure with Bamboo-Woven Mud Infill 649
Yunyang Qi and Hao Huang

Seismic Response Analysis of Shallow Tunnel Under Simultaneous Oblique Incidence of P and SV Wave 660
Dong-Dong Zhang, Yang Liu, Feng Xiong, and Zhu Mei

Experimental and Numerical Investigations on the Structural Behaviour of the Joints in Aluminium Alloy Frames Supported Glass Curtain Wall 672
Mengyu Li, Chunyin Ma, Zhongxing Wang, Guoxing Li, and Lulu Fan

Discussion on Key Problems of 3D Printing Concrete in Coastal Environment 679
Xinyue Zhang, Weijiu Cui, Chengping Qu, Sheng Wang, and Qi Yu

Experimental and Numerical Investigations on the Compression Behavior of Calcareous Sand-Rubber Mixture 689
Ziye Zhou, Hongjie Lin, and Jiankun Liu

Stress Analysis of Narrow Steel Box Continuous Composite Beam Based on ABAQUS 696
Jing Huang, Shuai Liu, Yan Zheng, and Xiaobei Liu

Optimization Design and Simulation of TMD for a Wind Turbine Based on SIMPACK 708
Yang Li and Chao Wang

Measurement and Analysis of Flow Velocity in a Large Air Duct Elbow System 716
Yi Tang, Sen Yu, Xiaocheng Cao, and Haidong Wang

Modal Identification of Underwater Precast Self-centering Segmental CFST Bridge Under Horizontal or Vertical Ground Motion Based on Covariance Driven Random Subspace Identification 724
Mo Zhang, Yunfei Wang, Dan Zhang, Junfei Zhang, and Ning Li

Earthquake Damages to Substation Buildings 734
Weixian Che, Yanfeng Wang, Baichong Pan, Xinghua Wang, Wenwei Zhu, Jingen Guo, and Hantuo Dong

Automatic Object Detection of Construction Workers and Machinery Based on Improved YOLOv5 741
Yuanzhi Xiang, Jiayue Zhao, Wenjing Wu, Caifeng Wen, and Yunzhong Cao

Study on Comparison and Selection of Grouting Pre Reinforcement Schemes for Underground Excavation Land Section of River Crossing Tunnel – Taking Rongjiang Fourth Road River Crossing Tunnel as an Example 750
Hongwei Zhang, Helin Fu, Hongyu Guo, Guijun Luo, and Kaiyuan Zhang

Study on Impact Resistance of Corroded Concrete Beams 762
Xiaofang Deng, Longmei Jin, and Zhi Li

GPU-Accelerated and Machine-Learning-Based Wind Turbine Damper Optimization 771
Shi Liu, Yize Wang, and Zhenqing Liu

Influence of Openings on Fluctuating Wind Pressure Characteristics of Large-Span Cantilevered Roofs 779
Min Zhang, Pailin Fang, Qizhou Liu, Dan Lu, and Dan Pan

Influence of Scour Depth on Pile Group with Complex Configuration in Clay 789
Zengliang Wang, Hang Zhou, and Longyong Tong

Finite Element Simulation Analysis of Double-Wall Steel Boxed Cofferdam for MaLuKou Zishui River Bridge on Guanzhuang-Xinhua Expressway 798
Lu Yi

Defect Detection and Damage Evaluation of Steel Rods Based on Ultrasonic Guided Waves 809
Han Zhu and Jing Zhang

Study on Enhancing the Performance of Unglazed Solar Air Collector by Attached Jet 816
Jian Jin, Jianjun Hu, Gui Hu, and Shenghua Du

Experimental Investigation on Stress-Strain Curve of Phosphogypsum Under Uniaxial Compression 826
Li Zhou, Song Hu, Chao Yin, Yifeng Xu, and Qingyu Zou

Numerical Simulation on Progressive Collapse Resistance of RC Beam-Slab-Column Substructures 836
Yi Zhang, Yirong Gao, Xiaolan Yuan, and Bing Liu

Effect of Vertical Load on Lateral Response of Single Piles in Coral Sand 851
Chunyan Wang and Xuanming Ding

Study on the Effect of Pier Explosion on the Performance of RC Curved Girder Bridge with Double-Column Pier 860
Guohua Song, Sandeep Gharti Magar, Mingming Yang, and Jian Su

A Comparison Between PDEM-Based Approach and Linear-Regression-Based Approach in Seismic Fragility Assessment: Application into Low-Rise Frame Buildings 869
Xuyang Cao, Jigang Xu, and Shizhi Chen

Study on the Fissure Propagation of Jointed Rock Slope Under Unloading Condition 876
Sheng Li and Chen Yang

Determining and Estimating the Fracture Risk of Diaphragm Wall from Observed Lateral Deflection 884
Xiong Wang, Yue Pan, Mingguang Li, and Jinjian Chen

Hysteretic Performance of Self-centering Glulam Beam-To-Column Connections with Steel Core Panel Zone 893
Zheng Li, Fei Chen, Xiaofeng Sun, Zhan Shu, and Minjuan He

Behavior of CFRP-Confined Basalt Fibre Reinforced Recycled Aggregate Concrete 901
Yi Xu, Jiangfeng Dong, and Qingyuan Wang

Influence of Hammer Geometry on Impact Performance of RC Beams 910
Tianqi Xue, Xiaolan Yuan, Xin Zhang, Zhi Li, and Xiaofang Deng

Component-Based Simulation on Progressive Collapse of a 2D-Framed Steel Structure Under Debris Impact 918
Yufang Yuan, Kang Chen, and Bo Yang

Analysis of Mechanical Properties of Turbogenerator Foundation Based on ANSYS 926
Mingyi Tang, E Wenqi, and Cong Zeng

Mechanical Properties of Spray-Based 3D Printed Micro-cable Reinforced Concrete 934
Ji Zhou, Guanyu Hou, Xiongfei Liu, Qi Li, Li Wang, and Jixiang Li

The Distributed Hybrid Test and Validation of Multi-substructural Space Steel-Braced Frame Models Based on OpenFresco 943
Tengfei Li

Kalman Filter Based Adaptive Control for Real-Time Hybrid Simulation with Nonlinear Substructure 951
Wei Huang and Xizhan Ning

Comparative Analysis of Tunnel Structures with Different Sections in Open Excavation and Buried Section of Urban River Passage 960
Hongyu Guo, Helin Fu, Guijun Luo, Hongwei Zhang, and Kaiyuan Zhang

Study on Flexural Behavior of Coral Concrete Beams Reinforced with BFRP Bars 974
Yi Zhang, Baikun Chen, Xiaolan Yuan, and Bing Liu

Stochastic Model of Typhoons Along the China Coast and Its Application into Risk Assessment of Sea-Crossing Bridge Under Multiple Hazards 984
Kai Wei, Zhonghui Shen, Xi Zhong, and Zhiyu Bai

Comparison of Stability of Four Numerical Integration Methods Under Negative Stiffness 992
Fantao Meng, Jianfeng Zhao, and Xingqun Ruan

Study on Phase Change Process of Heat Source for a Water Source Heat Pump 1001
Junhua Wu and Nuo Wang

Bidirectional Hybrid Simulation of a Six-Story Full-Scale Frame Specimen with Incomplete Boundary Conditions 1009
Shangzhang Wang, Ge Yang, Liang Tang, Zhen Wang, Bin Wu, Yanfang Liang, Zhen Tian, and Jiajun Xiao

Main-Aftershock Modeling According to Measured Records 1022
Zixin Liu, Yunmu Jiang, Zhangjun Liu, and Xinxin Ruan

Study on Mechanical Performance of Column Base Joint with Slip-Friction Arc Endplates 1030
Chengyu Li and Zimeng Lai

Effect of Tung Oil on Mechanical Behavior of Silica Sand 1039
Ke Chen and Sérgio D. N. Lourenço

A Study on Wind Resistance of Steel Arch Pedestrian Bridge 1045
Jiahua Xian, Junfeng He, Shaocai Zhu, and Minting Zhong

Research on Impact Resistance of H-shaped Steel Beam 1053
Yunhao Weng, Shihong Li, Xiaolan Yuan, and Tianqi Xue

Segmental Prefabrication and Assembly Technology of Composite Box Girder with Corrugated Steel Webs 1062
Huijun Shen, Hehui Zheng, Fei Tian, and Hao Dai

Correlation of Spectral Shapes of Horizontal and Vertical Ground Motions at the Same Period Based on Copula Function 1070
Xiaolei Wang, Zixu Zhao, and Ruiguang Zhu

Hydrophobicity Assessment of Sands Treated with Tung Oil 1084
Xin Xing and Sérgio D. N. Lourenço

Study on Essential Maintenance of Long-Span Bridge in Railway 1091
Xin Jiang, Zhouyu Zhang, Peng Wang, Xinlin Ban, Min Huang, and Jichao Guo

Load-Path Method for Design of Disturbed Regions in Concrete Structures 1099
Jianlin Wang, Zhiqiang Wang, and Zhiqi He

A Study on the Design Strategy of Living Space Renovation Under the Background of "Aged Caring at Home" 1107
Feng Wang, Xiangyun Wang, and Jie Liu

Experimental Study on Bending Toughness of Engineered Cementitious Composites 1116
Huiqing Xue

Near-Fault Pulse Ground Motion Decomposition Based on Improved Wavelet Multi-scale Analysis Method 1125
Longlong Zhou, Huiguo Chen, Yixin Lei, and Xiaorui Yang

Smart City

A Comparative Study of Point Cloud Mapping Algorithms Towards Heterogeneous Traffic Scenarios 1139
Jiazhong Zhang, Shuai Wang, Xiaojun Tan, and Minghao Chen

Assessment of Urban Bikeability with Structural Equation Model 1151
Yijun Liu, Aitong Pang, and Gege Jiang

Long Short-Term Memory Network and Ordinary Kriging Method for Prediction of PM2.5 Concentration 1158
Junyou Liu, Bohong Zheng, and Jinyu Fan

The Influence of the Characteristics of the Elderly on the Landscape Design of Residential Area Suitable for Aging 1170
Xiangyun Wang and Feng Wang

Travel Mode Recognition Using Mobile Phone Signaling Data 1179
Fanghao Fu, Jiemin Xie, Shuqi Zhong, and Ming Cai

Construction and Study of Automotive Working Conditions Based on Improved K_means Clustering 1188
Wei Ding and Songwen Tian

Research on the Innovative Design Mode of Guangzhou Vernacular Architecture Under the Background of Carbon Neutrality 1201
Qiang Tang, Yile Chen, Liang Zheng, and Junzhang Chen

The Thermal Comfort Characteristics of PCS Applied in EVs 1209
Jingjing Wu and Jingde Zhao

Mode Choice Behavior of Ride-Sourcing Passengers Under Travel Time Variability 1218
Aitong Pang, Xin Zhang, and Gege Jiang

Intelligent Control of Shield Tunneling from the Perspective of Complex Network 1226
Xiaojing Zhou, Yue Pan, and Jianjun Qin

A Systematic Review of Urban Design and Computer Modelling Methods to Support Smart City Development in a Post-COVID Era . . . 1234
Liu Yang, Michiyo Iwami, Yishan Chen, Mingbo Wu, and Koen H. van Dam

Impact Estimation of Traffic Accident Duration Based on Survival Analysis by Using Field Urban Traffic Condition 1247
Duanya Lyu and Yongjie Lin

An Integrated Framework for Population Synthesis at Fine-Grained Spatial Scales 1255
Yinni Huang, Meng Zhou, Runqin Deng, Zeteng Huang, and Linlin You

Author Index 1263

Green Building

Numerical Verification of Offline Hybrid Testing for the Seismic Response of Energy Dissipation Structure

Haodong Liu, Zhenyun Tang(✉), and Fukang Gao

The Key Laboratory of Urban Security and Disaster Engineering of Ministry of Education, Beijing University of Technology, Beijing 100124, China
tzy@bjut.edu.cn

Abstract. Real-time hybrid testing (RTHT) has been developed rapidly for seismic testing. When energy dissipation structure is tested by RTHT, the damper is usually adopted as physical substructure and the structure is adopted as numerical substructure. In order to minimize the seismic response of energy dissipation structure, many dampers installed at different positions are often necessary. The dynamic response of dampers at different positions is different in seismic action. Therefore, more than one damper needs to be tested physically in RTHT, and multiple dampers loaded by multiple actuators in real time will increase control difficulty. To solve this issue, an offline hybrid testing method for energy dissipation structure is proposed based on machine learning, which avoids the challenge of control and stability caused by real-time loading of multiple actuators in RTHT. The feasibility of this method is proved by numerical simulation, which provides a new idea for seismic testing of energy dissipation structure.

Keywords: Energy dissipation structure · Real-time hybrid testing · Multi-interface loading · Machine learning · Offline hybrid testing

1 Introduction

The resilience of structure subjected to earthquake has become the focus of earthquake engineering. Seismic mitigation [1] is an important measure to improve seismic performance of structure, among which damper is a widely used damping device. Due to the strong nonlinearity of damper, it is usually required shaking table testing [2] to evaluate the seismic response characteristics of damper-installed structure. Shaking table testing often require large scale reduction, which makes the damper difficult to machine and its mechanical properties are distorted. RTHT [4] divides structure into two parts. One part with strong nonlinearity or difficult to model is taken as physical substructures for physical testing; the rest part is taken as numerical substructures for numerical simulation. As this method only requires the physical testing of the part of the tested structure, the size of the specimen can be enlarged using the same loading facilities. And then damper can be tested in full size.

W. Guo and K. Qian (Eds.): GBCESC 2022, LNCE 211, pp. 3–13, 2023.
https://doi.org/10.1007/978-981-19-5217-3_1

Real-time hybrid testing requires real-time calculation and real-time loading, which leads to the higher standard for numerical solution and physical loading. In order to ensure the smooth progress of RTHT, the numerical integral solution needs to meet three requirements [5]: stability, accuracy and high efficiency. For example, the central difference method has high computational efficiency with stability boundary but cannot be used for numerical substructure solving with large integration steps or high-frequency characteristics. In order to ensure the stability and accuracy, many scholars studied the unconditional stable algorithm. Chang [6] proposed an unconditional stable integral algorithm with second order accuracy, the algorithm is unconditionally stable, and the error propagation is smaller than central difference method and Newmark-β, but velocity is hidden. An unconditionally stable explicit integral algorithm proposed by Chen and Ricles [7], the displacement and velocity with only first-order accuracy. Kolay [8] proposed the KR-method, a discrete integral algorithm that controls energy dissipation, which can attach numerical damping to the numerical integration with unconditionally stable properties.

In addition, the real-time loading accuracy is also a key factor affecting the stability of RTHT. In the early stages of RTHT development, Horiuchi [9, 10] regarded the loading error of loading system as time delay, and proposed a time-delay compensation method based on polynomial interpolation. The actuator time delay is also associated with the dynamic properties of specimen as shown by Darby [11].To fully compensate for the systematic time-varying delay, Darby [11], Ahmadizadeh [12] developed an adaptive estimation and compensation methods. In RTHT with high test quality or requiring high-frequency performance, the loading system has both magnitude error and phase lag. To enhance the control accuracy further, Spencer [13] and Tang [14] used model-based inverse compensation to improve the control accuracy of the loading system. At present, many control methods have been developed for the stability and control accuracy of a single loading system, which has achieved good results and has promoted the development and application of RTHT. However, the existing control method does not consider the collaborative compensation control of the multi-loading system, and the collaborative loading hybrid test of multiple control systems cannot be realized.

For real-time hybrid testing of energy dissipation structures, the building structure is usually used as the numerical substructure, and the damper is used as the physical substructure. However, the dynamic response of dampers at different locations is different. In order to ensure the testing accuracy, multiple dampers need to be tested physically in real-time hybrid testing. In this case, multiple actuators are necessary for real time loading. This leads greater challenges for the implementation of RTHT. This work proposed an offline hybrid testing method for testing energy dissipation structure, which avoids the challenge of control and stability caused by real-time loading of multiple actuators in RTHT.

2 Principle of Offline Hybrid Testing

The offline hybrid testing method suitable for the energy dissipation structure is shown in Fig. 1. The dampers are taken as physical substructures and the building structure is taken as numerical substructure. The dynamic response characteristics of the damper are obtained by the dynamic testing of the damper in advance. And the neural network is used to learn the dynamic characteristics of the damper. The hybrid dynamic analysis is carried out between the neural network model of damper and the numerical substructure model of building structure. And obtain the dynamic response of the whole damped structure under seismic excitation.

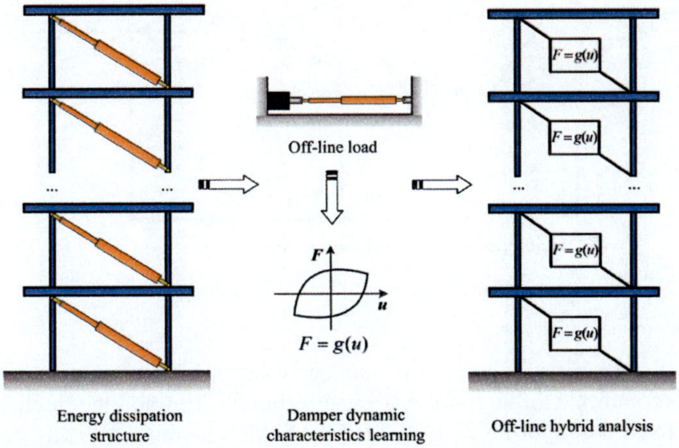

Fig. 1. Schematic diagram of offline hybrid testing for energy dissipation structure

The dynamic response equation for the energy dissipation structure is shown in Eq. (1):

$$[M]\{a\} + [C]\{v\} + [K]\{u\} + \{F\} = -[M]a_g \tag{1}$$

$[M]$, $[C]$, $[K]$ are the mass matrix, damping matrix, stiffness matrix. $\{a\}$, $\{v\}$, $\{u\}$ are the acceleration, velocity, and displacement vector of structure. a_g is the acceleration of ground motion, and $\{F\}$ is the reaction force vector of dampers.

The relationship $g(\cdot)$ between the displacement and the reaction force of damper is learned by neural networks:

$$F = g(\tilde{u}) \tag{2}$$

$\tilde{u}$ is the displacement of damper.

The dynamic analysis by neural networks is shown in Eq. (3):

$$[M]\{a\} + [C]\{v\} + [K]\{u\} + \{g(\tilde{u})\} = -[M]a_g \tag{3}$$

3 Neural Network Model

Neural networks are widely used in the field of civil engineering, including health monitoring [15], constitutive model [16], etc. In this work, the BP neural network [17] is used to learn the dynamic property of damper, with the topology is shown in Fig. 2.

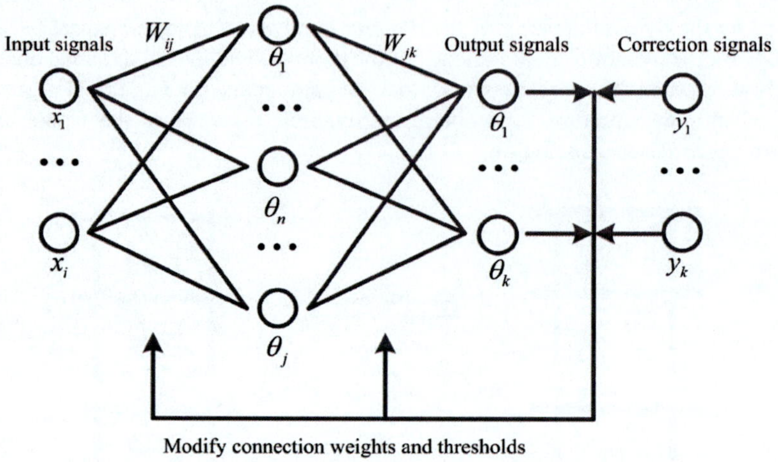

Fig. 2. BP neural network topology

BP neural network structure consists of input layer, hidden layer and output layer. Each layer is connected through weights and thresholds. The training sample information is inputted from the input layer, processed by hidden layer, and finally outputted the prediction results. Comparing the error of prediction and correction information, when the error is bigger than expected, the error signal is back propagated, the corresponding weights and thresholds in the network connection are adjusted according to the learning algorithm, the training sample is inputted into the network with adjusted parameters the above steps keep repeating until the error accuracy meets the requirements.

In the process of error back-propagation, the training algorithm for the original BP neural network is the gradient descent method. The parameters move in the opposite direction of the error gradient until the minimum value is achieved. However, the linear convergence rate of this algorithm is slower. The Neural network learning algorithm selects the Levenberg-Marquardt (LM) [18] algorithm in this paper. The LM algorithm belongs to the approximate second-order algorithm, with the combination of gradient descent method and Gaussian-Newtonian method. In contrast to the gradient descent method, its convergence rate is faster. The function has less probability of falling into local minima. The correction rule of weight and threshold is shown in Eq. (4):

$$W_{k+1} = W_k - \left[J^T J + \mu I\right]^{-1} J^T e \tag{4}$$

W_{k+1} is the weight correction vector for step $k + 1$, e is the error vector, J is the Jacobian matrix derived from the error vector e for the weights, I is the unit matrix, μ is the positive parameter. When μ is big, the LM algorithm approaches the gradient descent method, when μ is small, the LM algorithm approaches the Gaussian-Newton method. Multi-layer feedforward neural networks can theoretically approximate any nonlinear function.

In this work, the input feature vector [16] of neural network model is 6 variables, which are the deformation of the damper at the current moment and the previous moment u_i, u_{i-1}, the reaction force of the damper at the previous moment f_{i-1}, the product of the reaction force and the deformation at the previous moment $f_{i-1} * u_{i-1}$, the product of the reaction force and the sign function of the deformation's increment $F_{i-1} = f_{i-1} * sign(u_i - u_{i-1})$, and the energy expenditure in a time step $e_{i-1} = (f_{i-1} + f_{i-2}) * (u_{i-1} - u_{i-2})/2$, where the subscript i represents the number of time steps, and one output feature variable is the reaction force f_i at the current moment. The equation of neural network is shown in Eq. (5):

$$f = h(w \ * \cdots g(w * x + \theta) + \theta) \tag{5}$$

x is the input feature vector, at the moment i, $x = [u_i, u_{i-1}, f_{i-1}, f_{i-1} * u_{i-1}, F_{i-1}, e_{i-1}]$, w, θ are the weights and thresholds of each hidden and output layer. $g(\cdot)$ is the activation function of hidden layer, $h(\cdot)$ is the activation function of output layer.

4 Simulation Validation

4.1 Simulation Process

To verify the feasibility of the offline hybrid testing, this section performs the offline hybrid simulation as shown in Fig. 3. The overall model is a ten-layer frame structure with the damper arranged each layer. The frame structure is taken as numerical substructure and the damper is taken as physical substructure. The frame structure parameters are shown in Table 1. Bouc-Wen model is used to represent the dynamics of damper. The formula of Bouc-Wen model is shown in Eq. (6), f is reaction force, k is initial stiffness, d and v are the system input displacement and velocity, z is state displacement, α is the second stiffness coefficient. n, β, γ are the parameters which controlling the shape of the hysteresis loop, when n tends to infinity, the Bouc-Wen model will be equivalent to bilinear model. The used parameters [19] are $k = 8 \times 10^4$ kN/m, $\alpha = 0.02$, $\beta = 60$, $n = 1$, $\gamma = 50$.

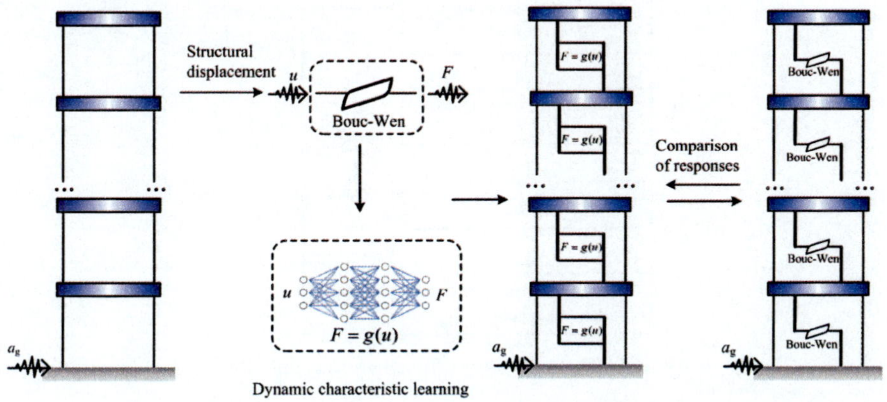

Fig. 3. Simulation process of offline hybrid testing

Table 1. Structural parameters of frame structure model

Floor	1	2	3	4	5	6	7	8	9	10
Weight (t)	563	552	552	552	552	552	552	552	552	552
Damp (e^3 kN/(m/s))	5.23	4.99	4.72	4.57	4.49	4.40	4.30	4.21	4.12	4.01
Rigidity (e^5 kN/m)	3.17	3.06	2.89	2.81	2.75	2.7	2.65	2.58	2.53	2.46

$$\begin{aligned} f &= \alpha k d + (1-\alpha) k z \\ \dot{z} &= v - \beta |v| |z|^{n-1} z - \gamma v |z| \end{aligned} \tag{6}$$

First, the seismic response of frame structure without dampers is simulated to obtain the time-history of displacement response at singular floor. And then taking the simulated displacement response as the incentive of dampers at corresponding locations, the time-history of reaction force can be obtained through Bouc-Wen model. The above displacement and reaction force are used to train the dynamic model of damper by neural network. Finally, the trained neural network replaces the original damper, conducting hybrid simulation with the frame structure, and comparing with the response of the original energy dissipation structure under the same excitation. The process of offline hybrid simulation is shown in Fig. 4, at the moment i, the frame structure is affected by seismic excitation ag_i and damper reaction force f_i, and the interface response u_{i+1} is obtained by integral solution. Then the interface response is transmitted to the physical substructure and the neural network predicts damper reaction force f_{i+1}, then transmitted to the numerical substructure for the next step. The above steps are repeated until the end of the simulation. The BP model used in this work has 2 hidden layers and 15 neuronal nodes in each layer. The hidden layer activation function is the logsig, and the output layer activation function is linear function.

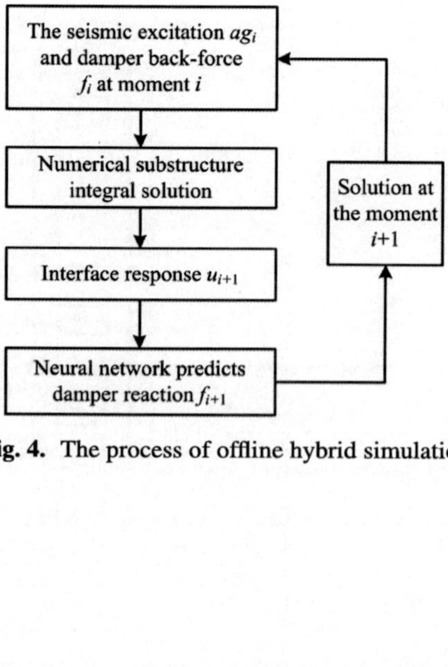

Fig. 4. The process of offline hybrid simulation

4.2 Accuracy Evaluation

For evaluating the accuracy of simulation results, the Peak Error (PE) and the Root Mean Square Deviation (RMSD) are selected as the index. The PE is determined by the error at the peak of the dynamic response, and its expression is shown in Eq. (7). The RMSD is determined by the time history, the expression is shown in Eq. (8).

$$PE = \frac{F_m^{true} - F_m^k}{F_m^{true}} \tag{7}$$

$$RMSD = \sqrt{\frac{\sum_{i=1}^{t} (F_i^{true} - F_i^{k})^2}{\sum_{i=1}^{t} (F_i^{true})^2}} \tag{8}$$

F^{true} represents the real value of verification, F^k represents the predicted value of the offline hybrid simulation, i represents the corresponding time step, and m represents the time of the peak point.

4.3 Simulation Result

First, the training accuracy of the neural network was verified. And the different excitation were given to the trained neural network to obtain the corresponding reaction force. The results are shown in Fig. 5. The predicted error is very small, the RMSD is 1.34% and PE is 0.1%. It shows that the neural network has a good ability for nonlinear learning.

Hybrid simulation between the trained neural network and frame structure can gives the seismic response of the damped structure. Meanwhile, when the dampers are modelled by Bouc-Wen model, the seismic response of the damped structure also can be obtained by the integrated numerical simulation. The El-centro wave is used as the external excitation. The dynamic responses are shown in Fig. 6. In Fig. 6, the solid line is integrated numerical simulation, and the dash line represents the result of offline hybrid simulation. Figure 6 demonstrated that offline hybrid simulation gave the same accuracy with the integrated simulation.

The errors of dynamic response for each floor are shown in Fig. 7. The PE of dynamic response with each layer is within 0.2% and the RMSD is within 0.5%. The simulation verification shows that offline hybrid testing has good potential for seismic testing of energy dissipation structure.

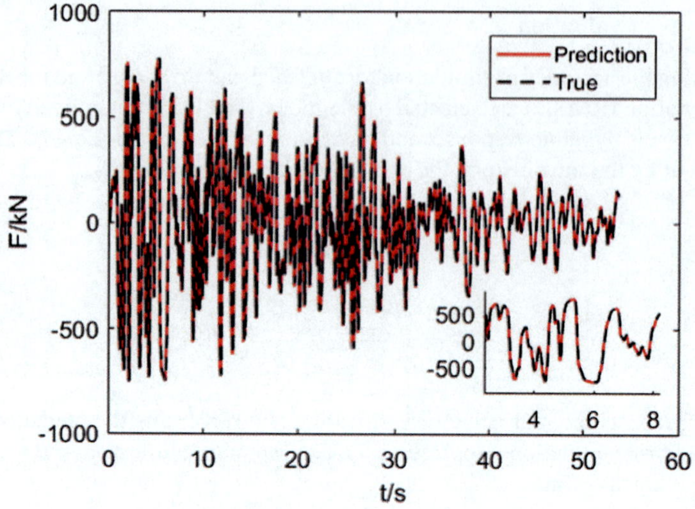

(a) Time-history of restoring force

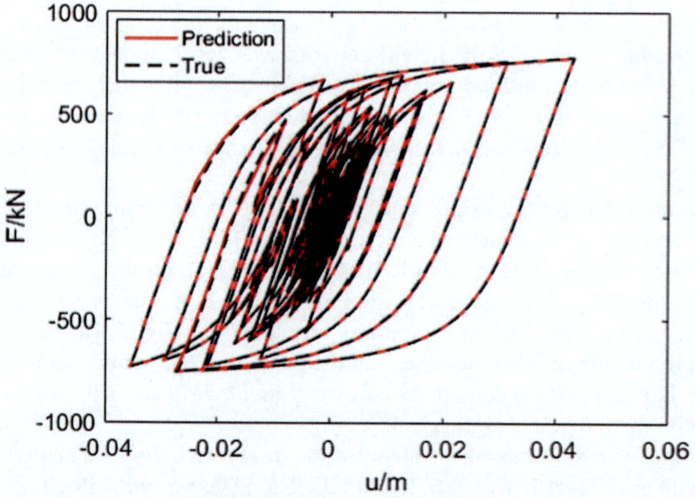

(b) Force-displacement curve

Fig. 5. Learning accuracy of damper dynamic characteristics

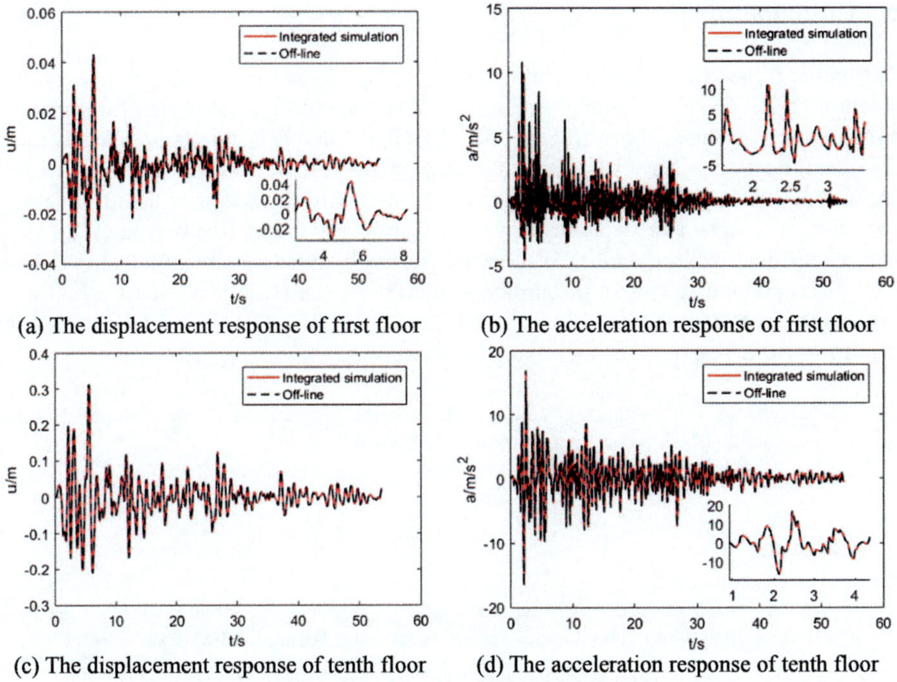

(a) The displacement response of first floor
(b) The acceleration response of first floor
(c) The displacement response of tenth floor
(d) The acceleration response of tenth floor

Fig. 6. The dynamic response of offline hybrid simulation

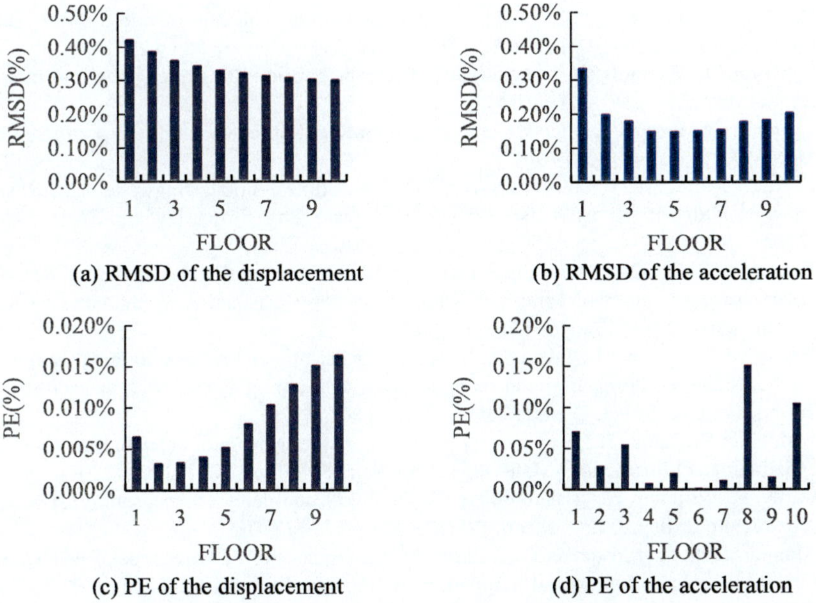

(a) RMSD of the displacement
(b) RMSD of the acceleration
(c) PE of the displacement
(d) PE of the acceleration

Fig. 7. Error of dynamic response with each layer

5 Conclusion

This paper proposes an offline hybrid testing method for seismic testing of energy dissipation structure. The building structure is taken as numerical substructure and the damper taken as physical substructure. Through the dynamic test of damper in advance, the neural network is used to learn its dynamic characteristics. Then the learned neural network is used to replace the damper and carry out hybrid simulation with the numerical substructure to obtain the dynamic response of the energy dissipation structure. The simulation verification shows that the method effectively avoids the insufficient real-time control accuracy and system instability in RTHT. At the same time, the method also avoids the needs of real-time numerical solution required by RTHT, and provides a new idea for seismic testing of energy dissipation structure.

Acknowledgement. The authors gratefully acknowledge the support of the National Natural Science Foundation of China under grant number 51978016.

References

1. Kwag, S., Eem, S., Kwak, J.: Mitigation of seismic responses of actual nuclear piping by a newly developed tuned mass damper device. Nucl. Eng. Technol. **53**(8), 2728–2745 (2021)
2. Jian, S., Kaiming, B., Kun, X., Qiang, H., Xiuli, D.: Seismic responses of adjacent bridge structures coupled by tuned inerter damper. Eng. Struct. **243**, 112654 (2021)
3. Nguyen, V., Seo, J., Huh, J., Ahn, J., Haldar, A.: Seismic response investigation of 1/20 scale container crane through shake table test and finite element analysis. Ocean Eng. **234**, 109266 (2021)
4. Nakashima, M., Kato, H., Takaoka, E.: Development of real-time pseudo dynamic testing. Earthquake Eng. Struc. Dynam. **21**(1), 79–92 (1992)
5. McCrum, D., Williams, M.: An overview of seismic hybrid testing of engineering structures. Eng. Struct. **118**(1), 40–261 (2016)
6. Chang, S.-Y.: Explicit pseudodynamic algorithm with improved stability properties. J. Eng. Mech. **136**(5), 599–612 (2010)
7. Chen, C., Ricles, M.: Stability analysis of SDOF real-time hybrid testing systems with explicit integration algorithms and actuator delay. Earthquake Eng. Struct. Dynam. **37**(4), 597–613 (2008)
8. Kolay, C., Ricles, M.: Development of a family of unconditionally stable explicit direct integration algorithms with controllable numerical energy dissipation. Earthquake Eng. Struct. Dynam. **43**(9), 1361–1380 (2014)
9. Toshihiko, H., Inoue, M., Takao, K., Namita, Y.: Real-time hybrid experimental system with actuator delay compensation and its application to a piping system with energy absorber. Earthquake Eng. Struct. Dynam. **28**(10), 1121–1141 (1999)
10. Toshihiko, H., Takao, K.: A new method for compensating actuator delay in real-time hybrid experiments. Philos. Trans.: Math. Phys. Eng. Sci. **359**(1786), 1893–1909 (2001)
11. Darby, P., Williams, S., Blakeborough, A.: Stability and delay compensation for real-time substructure testing. J. Eng. Mech. **128**(12), 1276–1284 (2002)
12. Ahmadizadeh, M., Mosqueda, G., Reihorn, M.: Compensation of actuator delay and dynamics for real-time hybrid structural simulation. Earthquake Eng. Struct. Dynam. **37**(1), 21–42 (2008)

13. Juan, C., Spencer, J., Brain, P.: Real-time hybrid simulation for structural control performance assessment. Earthquake Eng. Eng. Vib. **8**(4), 481–492 (2009)
14. Zhenyun, T., Dietz, M., Yue, H., Zhenbao, L.: Performance extension of shaking table-based real-time dynamic hybrid testing through full state control via simulation. Struct. Control. Health Monit. **27**(10), e2611 (2020)
15. Shanwu, L., Suchao, L., Shujin, L., Hui, L.: Data-driven modeling of bridge buffeting in the time domain using long short-term memory network based on structural health monitoring. Struct. Control Health Moni. **28**(8), 1–13 (2021)
16. Yun, J., Ghaboussi, J., Elnashai, S.: A new neural network-based model for hysteretic behavior of materials. Int. J. Numer. Meth. Eng. **73**(4), 447–469 (2008)
17. Rumelhart, E., Hinton, E., Williams, J.: Learning representations by back-propagating errors. Nature **323**(6088), 533–536 (1986)
18. Smith, S., Wu, B., Wilamowski, M.: Neural network training with levenberg-marquardt and adaptable weight compression. IEEE Trans. Neural Networks Learn. Syst. **30**(2), 580–587 (2019)
19. Fan, C., Yurong, G.: Adaptive unscented kalman filter based off-line model updating hybrid simulation of frame structures. Earthquake Eng. Eng. Dynam. **40**(1), 162–170 (2020)

State-of-the-Art of BRBs in Reinforced Concrete Structures: BRB Units, Connections, Seismic Design, and Performance Evaluation

Huiming Chen[1,2] and Jiulin Bai[1,2](✉)

[1] Key Laboratory of New Technology for Construction of Cities in Mountain Area (Chongqing University), Ministry of Education, Chongqing 400045, China
baijiulin@cqu.edu.cn

[2] School of Civil Engineering, Chongqing University, Chongqing 400045, China

Abstract. Owing to the stable hysteresis behavior and favorable costs, the buckling-restrained brace (BRB) is a high-performance damper in seismic areas to protect the building structures from severe damage, especially for steel frames. However, the combination of BRBs and reinforced concrete (RC) structures is complicated, due to the material difference between BRBs and the RC. This study briefly offers state-of-the-art views on the key issues of the combination of BRBs in RC frames (RCFs), such as BRB units, BRB-to-frame connections, seismic design approaches, and performance evaluation procedures. Despite excellent seismic behaviors can be ensured in many BRBs and gusset plates, more high-resilience components require further research. The BBR-to-frame interaction behavior, story shear force distribution relationship, strength and stiffness requirement, and BRB fatigue damage should be taken into account in the design of BRB-RCF systems. This paper aims to provide some recommendations for the development and future prospects of BRBs in RCFs.

Keywords: BRB · RC structures · BRB-to-frame connections · Structural design · Performance evaluation

1 Introduction

The first practical buckling-restrained brace (BRB) was designed in 1988, and has been quickly accepted for structural seismic protection in seismic areas [1, 2]. A typical BRB consists of an axial force-carrying steel core sandwiched in a mortar/concrete filled tube (see Fig. 1). Due to the external restraining members, BRBs can yield both in tension as well as in compression, and thus exhibiting stable hysteretic behavior without buckling under strong earthquakes [3, 4]. After the 2008 M_w 7.9 WenChuan earthquake, BRBs are now widely employed to improve the seismic resilience of reinforced concrete frames (RCFs), which are the most common type of construction in China and seriously damaged/destroyed during strong seismic excitation. The combination of BRBs in RCFs can make the dual structural system (BRB-RCF) achieve excellent seismic behaviors. The BRBs significantly contribute to resistant lateral forces in the elastic stage. Under

W. Guo and K. Qian (Eds.): GBCESC 2022, LNCE 211, pp. 14–21, 2023.
https://doi.org/10.1007/978-981-19-5217-3_2

strong earthquakes, BRBs yield before the frame system and are called upon to dissipate seismic energy, and thus the RC components are protected from severe damage. As a result, the structural seismic responses (e.g., maximum and residual drifts) can be effectively mitigated. To achieve these objectives, the resilience of BRB units, BRB-to-frame connections, seismic design approaches, and performance evaluation procedures should be ensured.

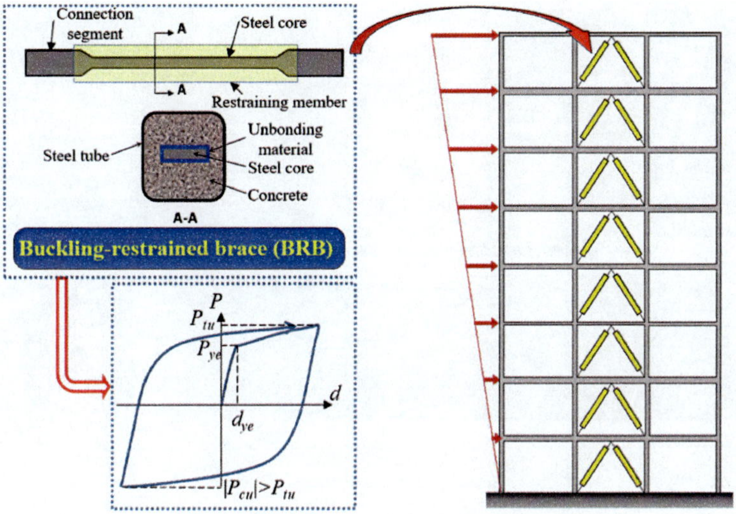

Fig. 1. The application of BRBs in RCFs.

2 BRB Units

Although a significant number of novel BRBs have been investigated to date, their working mechanism is consistent and robust. The steel core is allowed to slide freely and resist axial forces by a separation unit, and the restraining members are called upon to limit the core buckling. The test results usually demonstrate that BRBs possess stable hysteresis curves, excellent ductility, high low-cycle fatigue properties, and so on. However, it should be pointed out that, as a structural damper, the BRB should not only focus on enhancing the mechanical property, but also meet the architectural requirements (e.g., lightweight, low-cost, convenient for assembly and replacement, easy to post-earthquake detection, and so on).

In comparison to the concrete-filled restraining tube, the all-steel assembled BRBs have the merits of more lightweight and convenient fabrication. Moreover, it is easy for BRBs to assemble by high-strength bolts or welding [5]. Dusicka et al. [6] used the fiber-reinforced polymers for restraining units to develop an ultra-lightweight aluminum-core BRB. Moreover, for the rapid inspection and replacement, the BRBs with partially restraining units and spacers were proposed [7, 8]. In order to immediately examine the core damage after seismic excitations, Bai et al. [9] developed a novel BRB (FTA-BRB) using four concrete filled tubes. As shown in Fig. 2, the FTA-BRB restraining units are composed of U-shaped hoops. After strong earthquakes, disassembling the restraining hoops and thus the core damage can be easily observed. In addition, the reutilization of some undamaged units (such as steel square tubes) can reduce the maintenance and repair costs.

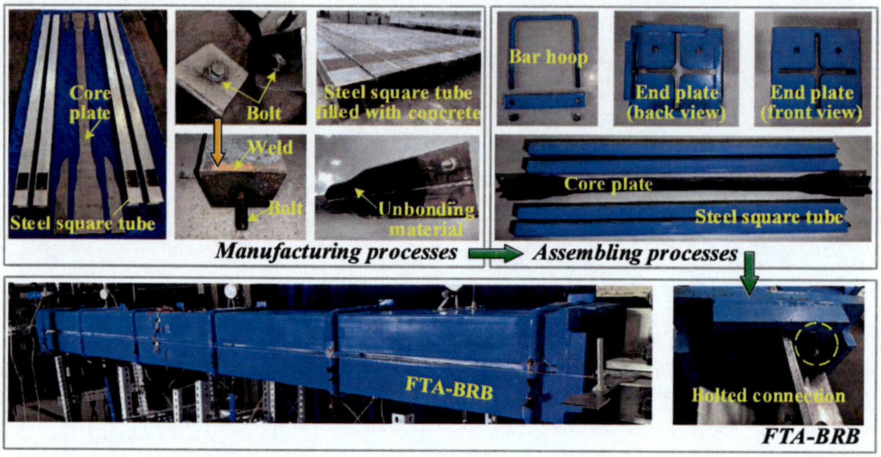

Fig. 2. Configuration of the FTA-BRB [9].

3 BRB Connections

The robustness of BRB-RCFs is dependent on the BRB-to-frame connections, where the gusset plates shall exhibit stable behaviors during strong earthquakes. The uniform force method and equivalent strut model were widely employed to calculate the internal force demands (BRB action and frame action) for the gusset design, respectively [10]. Moreover, various novel gusset connections have also been performed to eliminate the negative effect of frame actions. To release gusset-to-frame deformation, Zhao et al. [11] performed a BRB connection with a sliding configuration. Moreover, Berman et al. [12] developed the unconstrained gusset connection to entirely eliminate the frame action. Due to the material difference between the BRB and concrete, the connection of BRBs in RCFs is complex. For this reason, Bai et al. [13] constructed a new joint to connect BRBs and RC frames using shear connectors (see Fig. 3). The research showed that the novel BRB connection is stable and robust under strong earthquakes.

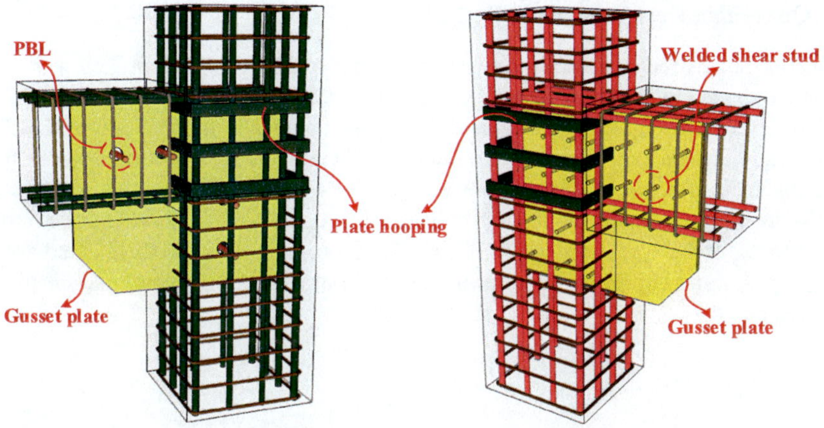

Fig. 3. Shear connector BRB-to-RC joint connections [13].

4 Structural Seismic Design

4.1 Interaction in BRB-RCF Systems

In the dual structural system, the overall behavior of BRB-RCFs comprises the contribution of BRBs and frames. It should be pointed out that the coupled work of BRB and RCFs is complicated especially under strong earthquakes, because of their different mechanical properties. Bai et al. [14] decomposed the dual system into the bare frame system and BRB system (see Fig. 4) to quantify the interaction. On the basis of the pushover and cyclic analysis, the internal force redistribution can be found in the inelastic stage. The BRB-RCF system exhibit better seismic performance compared with the bare RC frame not only due to the improvement of BRBs, but also caused by their interaction behaviors.

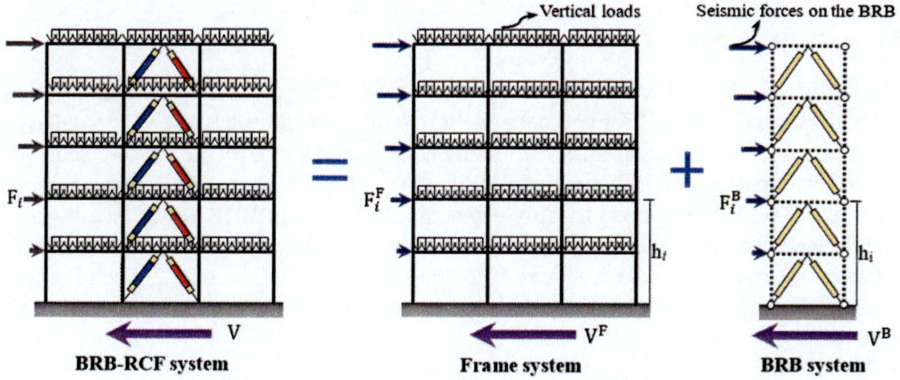

Fig. 4. Decomposition of the dual structural system [14].

4.2 Quantified Design for BRB-RCFs

To obtain the desired structural behavior, a new quantified design approach was developed by Bai and Ou [15] for BRB-RCFs (see Fig. 5). By quantifying the structural internal force and deformation demands, and the energy balance relationship, the BRB-RCF system can perform expected failure modes: BRB yield and strong-column weak-beam mechanisms. Furthermore, the responses of a series of BRB-RC frames were systematically investigated under far-fault and near-fault records [16, 17], where the number of stories, story shear resisted by BRBs, and BRB configurations were selected as variables in prototype buildings. As a result, the optimum structural design parameter was recommended.

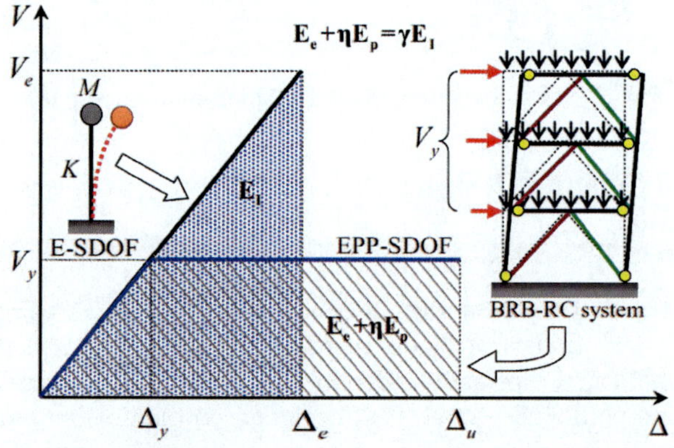

Fig. 5. The energy balance relationship for BRB-RCF [15].

5 Seismic Performance Evaluation

5.1 Loading Protocol

For performance quantification of BRBs, quasi-static testing is the most commonly used method. However, a significant number of BRB loading histories have been stipulated in many countries and regions. They are mainly constructed based on the seismic analyses of BRBFs, experimental investigations, judgments of experts, and so on. For this reason, Chen and Bai [18] developed loading protocols for braces in BRB-RCFs (see Fig. 6). The control parameters were determined on the basis of the behaviors of the most critical BRBs under earthquakes. Three types of loading histories were recommended to reflect the seismic responses of BRBs under maximum considered earthquake hazard level input.

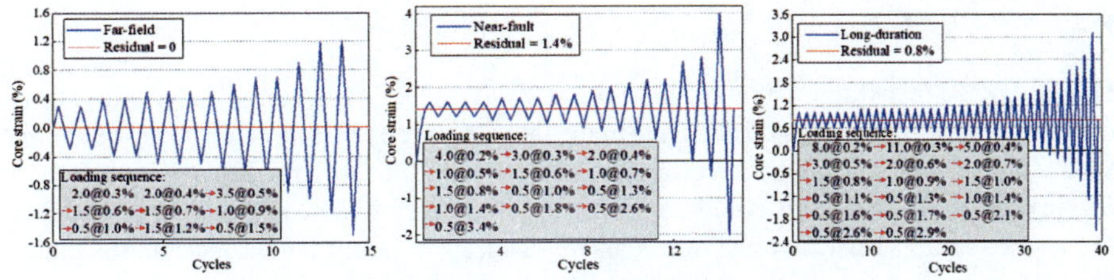

Fig. 6. Loading protocols for BRBs in BRB-RCFs [18].

5.2 Low-Cycle Fatigue Damage

Damage is an important index for BRB performance quantification, which has an impact on the behaviors of BRB-RCFs during strong earthquakes [19]. The BRB fatigue damage is mainly dependent on the core strain. Therefore, the BRB strain-time history during seismic excitation should be quantified using the rainflow counting method (see Fig. 7). Moreover, the Coffin-Manson equation [20] can calculate fatigue damage caused by the strain:

$$\varepsilon_i = \varepsilon_0 (N_\mathrm{f})^\mathrm{c} \tag{1}$$

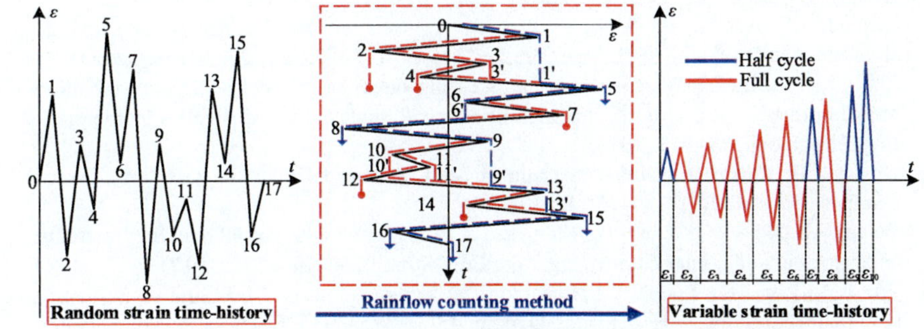

Fig. 7. Rainflow counting method [19].

where N_f = failure cycle; ε_i = strain amplitude; ε_0 and c are material parameters. Furthermore, the BRB accumulated damage during earthquakes can be determined by the Miner's rule [21]:

$$D = \sum \frac{n_i}{N_{\mathrm{f}i}} \tag{2}$$

where D is the damage quantified parameter; $D = 0$ and 1 means the core without any damage and fracture, respectively; n_i is strain cycles. According to the quantified strain amplitudes, the BRB damage D can be determined by Eqs. (1) and (2).

6 Conclusions

Unlike BRB-steel frames, the application of BRBs in RCFs is complicated, due to the difference between BRBs and concrete materials. This study briefly offers state-of-the-art views on the key issues of the development of BRBs in RCFs, such as BRB units, BRB-to-frame connections, seismic design approaches, and performance evaluation procedures. Despite excellent seismic behaviors can be ensured in many BRBs and gusset plates, more high-resilience components require further research. Moreover, the design of the BRB should not only focus on enhancing the mechanical property, but also meet the architectural requirements. The BBR-to-frame interaction behavior, story shear force distribution relationship, strength and stiffness requirement, and BRB fatigue damage shall be considered in the design of dual structural systems.

References

1. Watanabe, A., Hitomi, Y., Saeki, E., Wada, A., Fujimoto, M.: Properties of brace encased in buckling-restraining concrete and steel tube. In: Proceedings of ninth world conference on earthquake engineering, WCEE, pp. 719–724. Tokyo, Japan (1988)
2. Uang, C.M., Nakashima, M., Tsai, K.C.: Research and application of buckling-restrained braced frames. Int. J. Steel Struct. **4**(4), 301–313 (2004)
3. Tan, Q., Wu, B., Shi, P., Xu, G., Wang, Z., Sun, J., Lehman, D.E.: Experimental performance of a full-scale spatial RC frame with buckling-restrained braces subjected to bidirectional loading. J. Struct. Eng. **147**(3), 04020352 (2021)
4. Castaldo, P., Tubaldi, E., Selvi, F., Gioiella, L.: Seismic performance of an existing RC structure retrofitted with buckling restrained braces. J. Build. Eng. **33**, 101688 (2021)
5. Zhao, J., Wu, B., Li, W., Ou, J.: Local buckling behavior of steel angle core members in buckling-restrained braces: cyclic tests, theoretical analysis, and design recommendations. Eng. Struct. **66**, 129–145 (2014)
6. Dusicka, P., Tinker, J.: Global restraint in ultra-lightweight buckling-restrained braces. J. Compos. Constr. **17**(1), 139–150 (2013)
7. Wang, C.L., Chen, Q., Zeng, B., Meng, S.: A novel brace with partial buckling restraint: an experimental and numerical investigation. Eng. Struct. **150**, 190–202 (2017)
8. Wu, A.C., Lin, P.C., Tsai, K.C.: High-mode buckling responses of buckling-restrained brace core plates. Earthquake Eng. Struct. Dynam. **43**(3), 375–393 (2014)
9. Bai, J.L., Chen, H.M., Gao, M., Duan, L.: Development of a four-tube-assembled buckling-restrained brace for convenient post-earthquake damage examination and replacement. J. Build. Eng. **50**, 104209 (2022)
10. Tsai, C.Y., Tsai, K.C., Chen, L.W., Wu, A.C.: Seismic performance analysis of BRBs and gussets in a full-scale 2-story BRB-RCF specimen. Earthquake Eng. Struct. Dynam. **47**(12), 2366–2389 (2018)
11. Zhao, J., Chen, R., Wang, Z., Pan, Y.: Sliding corner gusset connections for improved buckling-restrained braced steel frame seismic performance: subassemblage tests. Eng. Struct. **172**, 644–662 (2018)
12. Berman, J.W., Bruneau, M.: Cyclic testing of a buckling restrained braced frame with unconstrained gusset connections. J. Struct. Eng. **135**(12), 1499–1510 (2009)
13. Bai, J., Chen, H., Zhao, J., Liu, M., Jin, S.: Seismic design and subassemblage tests of buckling-restrained braced RC frames with shear connector gusset connections. Eng. Struct. **234**, 112018 (2021)

14. Bai, J., Chen, H., Jin, S.: Investigation on the interaction between BRBs and the RC frame in BRB-RCF systems. Eng. Struct. **243**, 112685 (2021)
15. Bai, J., Ou, J.: Earthquake-resistant design of buckling-restrained braced RC moment frames using performance-based plastic design method. Eng. Struct. **107**, 66–79 (2016)
16. Bai, J., Cheng, F., Jin, S., Ou, J.: Assessing and quantifying the earthquake response of reinforced concrete buckling-restrained brace frame structures. Bull. Earthq. Eng. **17**(7), 3847–3871 (2019)
17. Du, K., Cheng, F., Bai, J., Jin, S.: Seismic performance quantification of buckling-restrained braced RC frame structures under near-fault ground motions. Eng. Struct. **211**, 110447 (2020)
18. Chen, H., Bai, J.: Loading protocols for seismic performance evaluation of buckling-restrained braces in RC frames. J. Build. Eng. **45**, 103522 (2022)
19. Chen, H., Bai, J.: Seismic performance evaluation of buckling-restrained braced RC frames considering stiffness and strength requirements and low-cycle fatigue behaviors. Eng. Struct. **239**, 112359 (2021)
20. Uriz, P.: Towards Earthquake Resistant Design of Concentrically Braced Steel Structures. University of California, Berkeley (2005)
21. Fisher, J.W., Kulak, G.L., Smith, I.F.: A Fatigue Primer for Structural Engineers. National Steel Bridge Alliance. American Institute of Steel Construction. Lehigh University, Bethlehem (1998)

Check for updates

Design Strategies for Green Retrofitting of a University Student Dormitory in the Hot-humid and Less-windy Climate

Binhua Li[1], Xuexiu Zhao[1], and Yanwen Luo[2](✉)

[1] Guangxi University, Guangxi 530004, People's Republic of China
[2] Guangxi Arts University, Guangxi 530007, People's Republic of China
20100081@gxu.edu.cn

Abstract. Green retrofitting of old buildings is an effective approach for the sustainable urban development. It is different from new green building design and lays much emphasis on the locality and adaptability including climate, function and technology. This study uses a case study approach to demonstrate how to conduct green retrofitting of an old university student dormitory building in a southern city of China, Nanning. The target dormitory building was investigated through field measurement and improvement strategies were proposed based on the investigated results. The improvement effects of the proposed strategies were also quantitatively analyzed using numerical simulation. As a result, it was found that wind speeds around the dormitory are increased by about 0.4 m/s; the average noise level is reduced by about 6 dB; the average outdoor air temperature is lowered by about 5 °C; the daylighting hours inside the dormitory are up by above 1 h; the standard-reaching rate of daylighting is improved from 58.34% to 79.64% and the energy saving rate of building air-conditioning load is raised up by about 0.6%. The findings described in this paper can provide design guidance for green retrofitting of university student dormitories in the hot-humid and less-windy climate.

Keywords: Green retrofitting · University dormitory · Filed investigation · Improvement strategy · Numerical simulation

1 Introduction

With the development of the construction industry in the world, building energy consumption is increasing, having accounted for about 36% of total global energy consumption in recent years [1]. Facing the energy crisis, it is important to reduce building energy consumption for ecological construction, energy planning and sustainable urban development, which has become a global consensus [2]. For this, the Chinese government proposed a 'two-carbon' goal (carbon emission peaking and carbon neutralization) towards the world in September 2020 [3]. The development of green buildings is just one of the most important steps to this goal.

Construction of green buildings has played an important role in sustainable society creation and urban development [4, 5]. It mainly manifested in adding the concept of

W. Guo and K. Qian (Eds.): GBCESC 2022, LNCE 211, pp. 22–31, 2023.
https://doi.org/10.1007/978-981-19-5217-3_3

green buildings to design requirements for new buildings and a series of evaluation standards for green buildings has been constituted. On the other hand, green reconstruction of old buildings is also important as found in many countries [6, 7]. In other words, green retrofitting of old buildings also takes an important role in the development of green buildings. Retrofitting design of old buildings is different from new green building design, in which building characteristics and embodied problems are considered so as to optimize design proposals with retrofitting strategies suitable for local climate, locality and required functions. Thus, green retrofitting of old buildings should be carried out case by case. It is also necessary to make sure whether green retrofitting of an old building is suitable for local climate [8].

Green retrofitting strategies are different from different buildings and climates as described in many previous studies. One previous study by Amirhosein Ghaffarianhoseini et al. [9] revealed that thermal comfort in a courtyard for the hot-humid climate can be improved by increasing vegetation coverage around the courtyard, and the duration of thermal discomfort can be shortened from 9 h to 3 h. Another study by Edward Halawa et al. [10] found that building energy consumption can be reduced through greening on the wall façade. The study conducted by Yingli Lou et al. [11] shown that the same retrofitting measures have different effects on CO_2 emission reduction under different climatic conditions.

A university student dormitory is the main living place of students and its energy consumption is four times as much as that of an ordinary residential building [12–14]. Therefore, it is of great importance to conduct researches on green retrofitting of old university student dormitories [15]. As reported in a study [16] related to retrofitting of university dormitories, building energy consumption in the dormitory can be greatly reduced by strengthening thermal insulation of building envelope and using solar photovoltaic generation. The study by Sertac Gorgulu et al. [17] reported a finding that lighting energy consumption can be reduced after using PV panels to provide lighting power. Margarita-Niki Assimakopoulos et al. [18] investigated the improvement effects of light shelves on energy, lighting and comfort in a refurbished student dormitory. Nevertheless, there still is a lack of the studies covering building energy and built environments including wind, sound and daylight in green retrofitting of old university student dormitories, especially in hot and humid climates.

For filling the above research scarcity, this study conducted a case study on green retrofitting of a university student dormitory in a southern city of China (Nanning) with hot-humid and less-windy climate. This dormitory building was investigated through field measurement and improvement measures were proposed based on the investigated results. The improvement effects of the proposed measures were also analyzed using numerical simulation.

2 Methodology

2.1 Field Investigation

The investigated building is a 6-storey student dormitory at Guangxi University, and there are 66 rooms on each floor from the second to sixth floor. Figure 1 is a bird view of this building and its surroundings. The first floor is open and used as commercial service areas. The outdoor units of air conditioners are mounted on the building facades below the windows and there are a few trees between dormitory buildings (Fig. 2). The field investigation found that the daylighting is poor in the indoor space and corridors, and there is high noise around the dormitory during lunch break hours. In addition, the building is located in Nanning where is in a hot-humid and less-windy climate. There are serious problems such as thermal discomfort and poor ventilation during the hot-humid and long summer in Nanning. The hot air from the outdoor fans of the air-conditioners is a big heat source to deteriorate the thermal environment around the building.

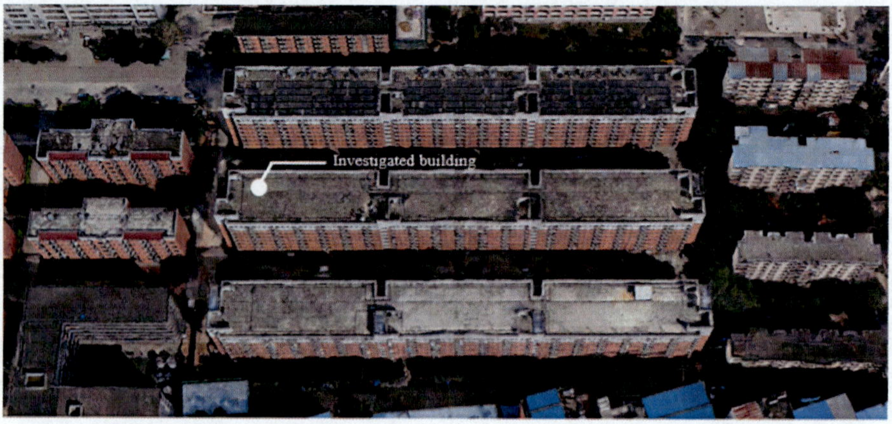

Fig. 1. Bird view of the study dormitory building and its surroundings

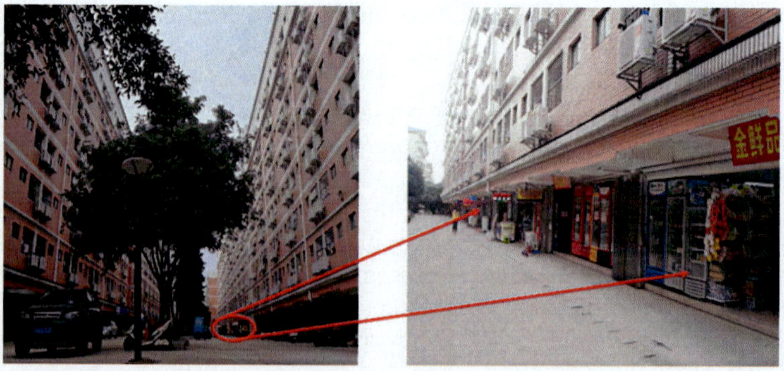

Fig. 2. A view between two dormitories where the right building is the investigated

To address the above problems, improvement measures were proposed from the four aspects of space renewing, energy-saving strategies, natural lighting and vegetation increase. As shown in Fig. 3a, space renewing is to open the enclosed space on the first floor to form a ventilation corridor. Increased ventilation through the first floor can be expected. Besides, green screens are added between dormitories to reduce noise.

The following energy-saving technologies are used as illustrated in Fig. 3b: PV panel installation on the roof, thermal reflective coatings to the walls and replacing the original glass with double-layer hollow glass. Two dormitory rooms are reconstructed to a public space on the 2nd to 6th floors (Fig. 3c). The public space overhangs 1500mm outward to improve daylighting in the corridor. Light pipes are designed to improve daylighting in the rooms on 2nd to 6th floors, as shown in Fig. 3d. As seen in Fig. 3e, vegetation around the dormitory, vertical greening and rooftop greening are increased. Figure 4. presents the retrofitted dormitory building with the proposed design strategies.

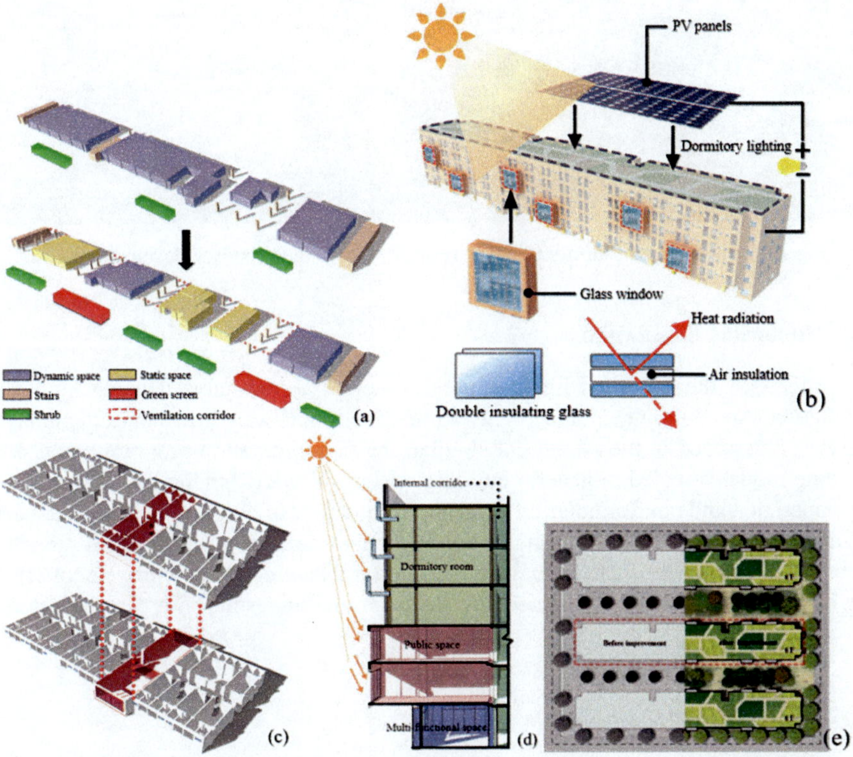

Fig. 3. The proposed improvement measures: (a) space renewing; (b) energy-saving technology; (c) two dormitory rooms are reconstructed to be a public space; (d) installation of light pipes; (e) increased vegetation

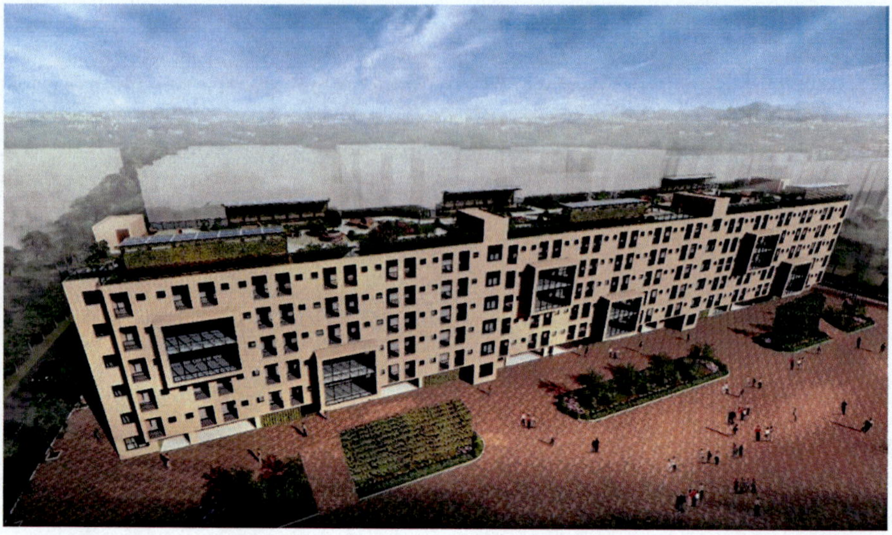

Fig. 4. View of the dormitory building retrofitted with the proposed design strategies.

2.2 Numerical Simulation

The effects of the proposed improvement strategies were analyzed using numerical simulation. An environment and energy simulation tool (Sware green building analysis software) was used in the numerical simulation. This simulation program can use the building models created in general building design software, having the advantages of easy operation and rapid simulation [19]. The simulations of wind environment, acoustic environment, luminous environment, thermal environment and air-conditioning load were carried out under meteorological condition in Nanning. The improvement effects can be quantified through comparison between simulation results before and after green retrofitting.

3 Simulation Results

3.1 Wind Environment

Figure 5 shows simulation results of air velocity distribution around the dormitory before and after retrofitting under the summer wind condition. It can be seen that airflow around the dormitory has been improved after opening a ventilation corridor on the first floor. Air velocity in the ventilation corridor was increased from 1.49 m/s to 1.86 m/s. There are windless areas and vortex zones before retrofitting, and they almost disappear after retrofitting. Moreover, ventilation around the building reached the standard in the summer, winter and transitional season after retrofitting.

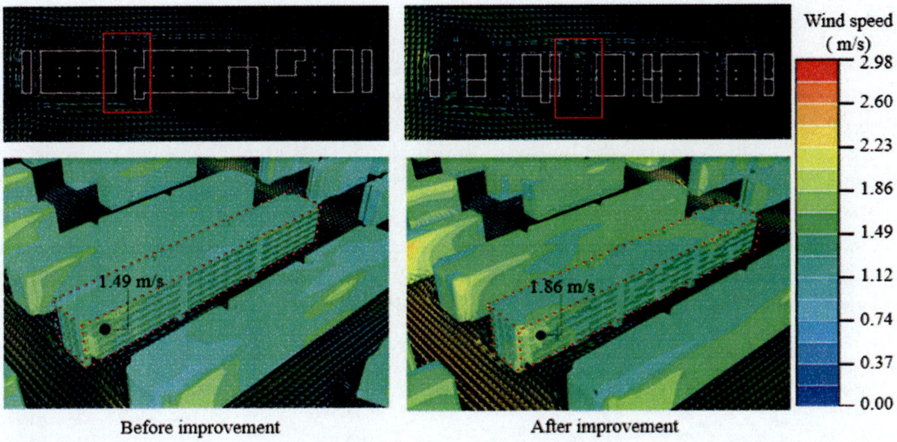

Fig. 5. Simulation results of air velocity distribution around the building before and after retrofitting.

3.2 Acoustic Environment

Figure 6 is acoustic pressure distribution around the dormitory before and after retrofitting. The measure for noise reduction is to design green screens between the dormitories. It can be found that acoustic pressure around the dormitory was reduced after retrofitting, as shown in Fig. 6. The maximum noise decibel at a height of 1.5 m around the dormitory was down from 56 dB to 50 dB during the daytime, being up to the Green Building Evaluation Standard.

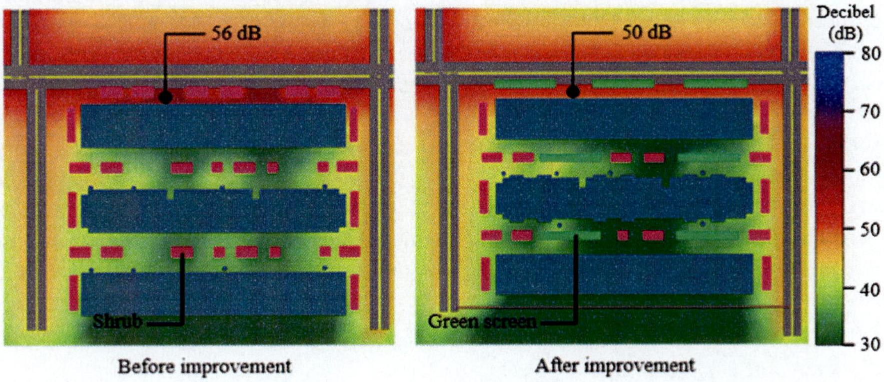

Fig. 6. Simulation results of acoustic pressure distribution before and after retrofitting.

3.3 Heading and Cooling Load

Annual heating and cooling load in the building before and after retrofitting are given in Fig. 7. The measures proposed in retrofitting are to install PV panels on the building roof and coat a thermal reflective material on the walls, use double insulating glass and increase vegetation around the building. Simulation results show that there is almost no difference in the heating load before and after retrofitting, and there is a great difference in the cooling load before and after retrofitting. The peak cooling load is in July, being reduced from 900 kW to 500 kW by 44.44% after retrofitting, and the peak latent-heat cooling load was reduced from 700 kW to 400 kW by 42.86%. Furthermore, the basic energy saving rate of the building load increased from 51.71% to 52.31%.

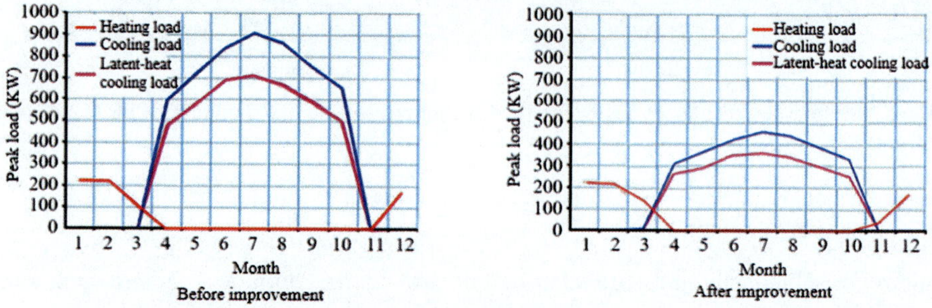

Fig. 7. Simulation results of heating and cooling load before and after retrofitting.

3.4 Luminous Environment

Figure 8 presents distribution of daylighting hours on the second floor in the building before and after retrofitting. Because the first floor is for commercial use and the second to sixth floors are for residence, the second floor was selected as an example for daylighting analysis below. The improvement measures are to reconstruct two dormitory rooms to a public space on each floor from the 2nd to 6th floor and the reconstructed rooms overhang 1500 mm outward. In addition, light pipes are installed in each dormitory room. As seen in Fig. 8, daylighting hours in the dormitory corridor and dormitory rooms were about 0.5 h before retrofitting, and the daylighting compliance rate was 58.34%. After retrofitting, the daylighting hours in the dormitory corridor and dormitory rooms increased up to 2 h and 1 h respectively, and the daylighting compliance rate rose to 79.64%.

3.5 Thermal Environment

The thermal improvement measures are to increase vegetation around the dormitory and use thermal reflective coatings on the walls. Surface and air temperatures around the dormitory on the summer solstice were simulated before and after retrofitting. Air temperatures in the analyzed area are calculated by Eq. (1):

$$t_a(\tau) = t_{a \cdot TMD} + \Delta t_{sol}(\tau) - \Delta t_{lw}(\tau) - \Delta t_{lat}(\tau) \tag{1}$$

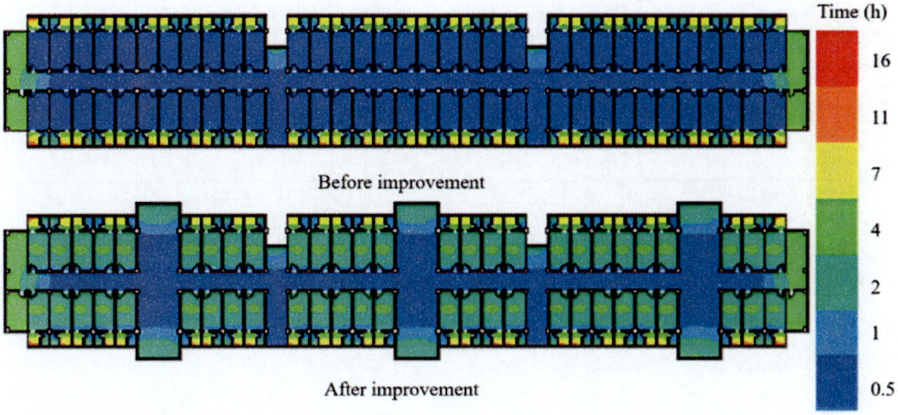

Fig. 8. Simulation results of daylighting on the second floor before and after retrofitting

where, $t_{a \cdot TMD}$ is the air temperature on a typical meteorological day, $\Delta t_{sol}(\tau)$ is the change of air temperature by solar radiation, $\Delta t_{lw}(\tau)$ is the change of air temperature by long wave radiation, and $\Delta t_{lat}(\tau)$ is the change of air temperature by evaporation, τ is time.

Figure 9 and Fig. 10 give the simulated results of ground surface temperature distribution at 14:00 and average air temperatures around the dormitory, respectively. The ground at a surface temperature of 38 °C before retrofitting was lowered to about 34 °C after retrofitting, as revealed in Fig. 9. As seen in Fig. 10, the average air temperatures around the dormitory dropped down by 5 °C from the noontime to the nighttime after retrofitting. The vegetation coverage on the building roof increased from 0% to 92% after retrofitting, and the shading coverage in the outdoor activity areas around the dormitory was from 4% up to 59%. Both the two indexes meet the requirements of "Design Standard for Thermal Environment of Urban Residential Areas".

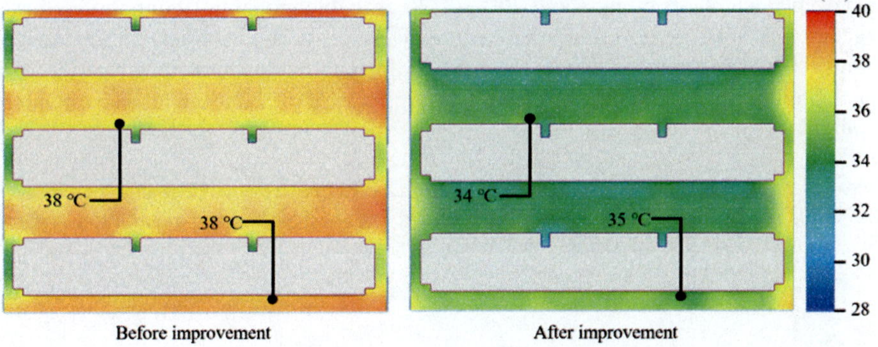

Fig. 9. Ground surface temperatures around the dormitory on the summer solstice at 14:00.

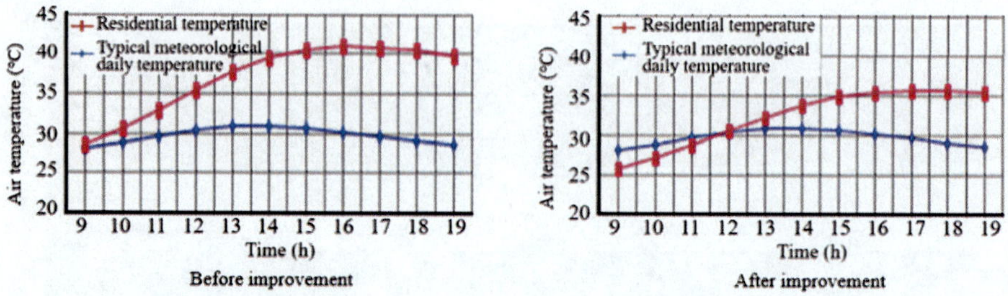

Fig. 10. Average outdoor air temperatures around the dormitory on the summer solstice.

4 Conclusion

This paper presented a case study for green retrofitting of a university student dormitory building in a southern city of China (Nanning). The improvement measures were proposed under consideration of the local hot-humid climate and regional characteristics, being summarized as follows: (1) a ventilation corridor is designed on the first floor; (2) green screens are added between the buildings to reduce noise; (3) two dormitory rooms on each floor from the 2nd to 6th floor are reconstructed to a public space with overhanging outwards, and light pipes are installed in each room to improve indoor daylighting; (4) increase of vegetation on the roof and walls as well as around the building; (5) thermal reflective coatings are used on the outside walls to reduce the absorbed solar heat on the walls; (6) double-layer hollow glass is used to replace single glass; (7) solar photovoltaic panels are newly mounted to reduce HVAC energy consumption in the dormitory.

Through comparison of simulation results for wind environment, acoustic environment, luminous environment, thermal environment and air-conditioning load, the following improvement effects of the proposed measures were found. (1) The newly-opened ventilation corridor can increase airflow by about 0.4 m/s, and improve ventilation around the building. (2) The green screens can reduce noise by about 6 dB. (3) The daylighting hours in the dormitory corridor and rooms can be raised by about 1 h and the daylighting compliance rate is up by about 20%. (4) Reasonable greening layout and increased vegetation around the dormitory can averagely lower the ambient air temperatures by about 5 °C on the summer solstice. (5) The building cooling load can be reduced by 44% at a maximum value, and the basic energy saving rate of the building load can be up by about 0.6%.

Acknowledgement. This work was supported from the Natural Science Foundation of Guangxi (No. 2021GXNSFBA220030), Guangxi Science and Technology Base and Talent Project (Grant No. AD21075050) and the Special Project of Guangxi Educational Science '14th Five-Year' Plan (No. 2021ZJY1548).

References

1. He, C., Hou, Y., Ding, L., Li, P.: Visualized literature review on sustainable building renovation. J. Build. Eng. **44**, 102622 (2021)
2. Zhang, Y., Wang, J., Hu, F., Wang, Y.: Comparison of evaluation standards for green building in China, Britain, United States. Renew. Sustain. Energy Rev. **68**, 262–271 (2017)
3. Yanhua, L., Yuhang, L., Wentao, W.: Challenges, opportunities and actions for China to achieve the targets of carbon peak and carbon neutrality. China Popul. Resour. Environ. **31**(9), 1–5 (2021)
4. Ge, J., Zhao, Y., Luo, X., Lin, M.: Study on the suitability of green building technology for affordable housing: a case study on Zhejiang Province. J. Cleaner Prod. **275**, 122685 (2020)
5. Yang, Z., Chen, H., Mi, L., Li, P., Qi, K.: Green building technologies adoption process in China: How environmental policies are reshaping the decision-making among alliance-based construction enterprises. Sustain. Cities Soc. **73**, 103122 (2021)
6. Awadh, O.: Sustainability and green building rating systems: LEED, BREEAM, GSAS. J. Build. Eng. **11**, 25–29 (2017)
7. Liu, Y., Chen, H., Wang, X.-J.: Research on green renovations of existing public buildings based on a cloud model –TOPSIS method. J. Build. Eng. **34**, 101930 (2021)
8. Ma, G., Liu, T., Shang, S.: Improving the climate adaptability of building green retrofitting in different regions: a weight correction system for Chinese national standard. J. Build. Eng. **69**, 102843 (2021)
9. Ghaffarianhoseini, A., Berardi, U., Ghaffarianhoseini, A.: Thermal performance characteristics of unshaded courtyards in hot and humid climates. Build. Env. **87**, 154–168 (2015)
10. Halawa, E., et al.: A review on energy conscious designs of building façades in hot and humid climates: Lessons for (and from) Kuala Lumpur and Darwin. Renew. Sustain. Energy Rev. **82**, 2147–2161 (2018)
11. Lou, Y., Yang, Y., Ye, Y., Zuo, W., Wang, J.: The effect of building retrofit measures on CO_2 emission reduction–a case study with U.S. medium office buildings. Energy Build. **253**, 111514 (2021)
12. Ding, Y., Zhang, Z., Zhang, Q., Lv, W., Yang, Z., Zhu, N.: Benchmark analysis of electricity consumption for complex campus buildings in China. Appl. Thermal Eng. **131**, 428–436 (2018)
13. Zhou, Y., Zhou, M., Xia, Q., Hong, W.-C.: Construction of EMD-SVR-QGA model for electricity consumption: case of university dormitory. Mathematics **7**(12), 1188 (2017)
14. Zhou, Y., Sun, L., Hu, X., Ma, L.: Clustering and statistical analyses of electricity consumption for university dormitories: a case study from China. Energy Build. **245**, 110862 (2021)
15. Chunsong, W., Zongyi, S., Jiuling, N.: Energy consumption analysis and energy saving measures of university buildings. Dist. Heat. **4**(8), 28–49 (2017)
16. Petidis, I., Aryblia, M., Daras, T., Tsoutsos, T.: Energy saving and thermal comfort interventions based on occupants' needs: a students' residence building case. Energy Build. **174**, 347–364 (2018)
17. Gorgulu, S., Kocabey, S.: An energy saving potential analysis of lighting retrofit scenarios in outdoor lighting systems: a case study for a university campus. J. Cleaner Prod. **260**, 121060 (2020)
18. Margarita-Niki, A., et al.: Application of light shelves in a refurbished student dormitory. Energy Reports **7**, 253–258 (2021)
19. Qiuyue, C., Long, C., Zijing, Z.: Research on energy-saving renovation of old residential area under Sware software. Hous. Real Estate **2**, 50–51 (2020)

Analysis of Urban Dual-Carbon Targets and Countermeasures Based on GeoSOS-FLUS Simulation–A Case of Hangzhou City, Zhejiang Province

Weiwu Wang[1,2](✉), Huan Chen[1,2], Shiyuan Cao[1], and Biyan Wang[1]

[1] College of Civil Engineering and Architecture, Zhejiang University, Hangzhou 310058, China
weiwuwang@zju.edu.cn

[2] Center for Balance Architecture, Zhejiang University, Hangzhou 310058, China

Abstract. Revealing the evolution characteristics of land-use carbon emissions in urban areas is helpful to understand the relationship between human activities, urbanization and their influencing factors. It can also provide a scientific basis for the current national land spatial planning and realizing the "dual carbon" goal. Based on the land use data and carbon emission driver data in 2010, 2015, and 2020, this study uses the GeoSOS-FLUS model to simulate the land use situation in 2030, aiming to pass the carbon emission standards and land-use types of different land-use types. The carbon emission in the conversion process is estimated to be the characteristics of Hangzhou's carbon emission zoning in 2030, and the future carbon governance methods are discussed. After the Kappa and FoM value test, the simulation accuracy of the land use prediction system reaches 95.547%, and the reliability is high. The results show that (1) Hangzhou's carbon emissions in 2030 will show an upward trend, and the estimated carbon emissions will reach 9706352.272t. (2) The spatial difference of carbon emissions in Hangzhou in 2030 is quite significant. Among them, the carbon footprint pressure index and carbon emission risk index of Binjiang District are the highest, 3082.21 and 27427.467 tons per square kilometre, respectively. The study results can provide evidence that Hangzhou needs to take measures such as establishing a carbon trading market, constructing a carbon monitoring system, and controlling the direction of urban development to achieve carbon governance.

Keywords: Land use prediction · GeoSOS-FLUS · Carbon footprint pressure · Carbon risk index · Carbon governance

1 Introduction

In 2020, General Secretary Xi Jinping put forward the goal of "Carbon Peak and Carbon Neutral" in the General Debate of the Seventy-fifth UN General Assembly to solve climate change and its impacts caused by excessive carbon dioxide emissions. Revealing the evolution characteristics of carbon emission from land-use helps clarify its relationship with the influencing factors of nature and human activities and provides a basis for the current land spatial planning and double carbon construction.

W. Guo and K. Qian (Eds.): GBCESC 2022, LNCE 211, pp. 32–43, 2023.
https://doi.org/10.1007/978-981-19-5217-3_4

The GeoSOS-FLUS model can simulate and analyze land-use change and future land use well. It has further development and inheritance based on its predecessor, the geographic simulation and optimization system GeoSOS. The GeoSOS-FLUS model is a "bottom-up" dynamic simulation tool, which can make up for the insufficiency of GIS technology in complex spatiotemporal dynamic changes and simulate the spatiotemporal evolution process of complex systems. First of all, the software interface adopts a cross-platform C++ graphical user interface application development framework, which can display the land-use change process in the simulated area in real-time; simultaneously, the neural network algorithm it adopts can quickly obtain the suitability probability of various land distributions. Therefore, the model is highly applicable to this case study and studies in other cities. Carbon governance planning and decision-making need to be forward-looking, and the use of land use prediction models can provide more scientific and rational data support. Li Xia et al. coupled geographic simulation and spatial optimization model. They established GeoSOS-FLUS software, a geographic simulation-optimization system, to provide a technical support platform for the guidance, regulation and optimization of spatial control and land use in urbanized areas [1]. The Automata Model (CA), jointly developed with Stanford University, TNC and WWF, is an evaluation model in InVEST that can quantify various ecosystem services. Scholars have widely used it at home and abroad in ecological change and climate research [2].

2 Data Sources and Research Methods

2.1 Data Sources and Processing Methods

Overview of the Study Area. Hangzhou City is located at 118°21′E–120°30′E, 29°11′W–30°33′W in the north of the southeast coast of China. In the north of Zhejiang Province, it is adjacent to Hangzhou Bay in the east, adjacent to Shaoxing City, adjacent to Quzhou City in the southwest, adjacent to Huzhou City and Jiaxing City in the north, bordering Huangshan City in Anhui Province in the southwest, and Xuancheng City in Anhui Province in the northwest. The whole city is hilly and mountainous. It accounts for 65.6% of the total area, 26.4% for plains, and 8% for rivers, lakes, and reservoirs. The world's longest artificial canal, the Beijing-Hangzhou Grand Canal, and the Qiantang River, famous for its large tides, pass through it. The total area is 16,596 km^2, including 8,000 km^2 of municipal districts, 10 municipal districts, 2 counties, and 1 county-level city.

Data Sources and Preprocessing. Land use data (2000, 2010, and 2020) and vector data such as railways, highways, and administrative divisions were obtained from the Chinese Academy of Sciences Resource and Environmental Science and Data Center. The ground elevation data comes from Geospatial Data Cloud. According to the needs of this research, the land use in the study area is divided into 6 types according to the LUCC China Land Use and Land Cover Remote Sensing Detection Data Classification System, namely cultivated land, construction land, forest land, water area, grassland and unused land. Data such as slope, aspect, and road distance were extracted using surface analysis tools in ArcGIS, and the population and GDP data in social drivers came from the 2019 Hangzhou Statistical Yearbook.

The driving force factors are divided into traffic location factors, natural location factors and socioeconomic factors. The interconversion between land-use types is caused by the interaction of natural, social and economic aspects in different time scales [3]. Based on the principles of data availability, consistency and comprehensiveness. The traffic location factors mainly include the distance from the city centre (located at the seat of the Hangzhou Municipal Government), the distance from the expressway (including the Hangzhou expressway), the distance from the national road, the distance from the provincial road, and the distance from the county road. Natural location factors mainly include water system distance, DEM digital elevation model, slope analysis, and slope aspect analysis. The main socioeconomic factors are regional population and regional GDP. The limiting factors of land-use types are selected from the large water surface of Hangzhou (including Qiantang River, Qingdao Lake, etc.), the main tourist areas and scenic spots in Hangzhou, and the large area of forests and nature reserves in Hangzhou. These types of limiting factors are obtained by binarisation.

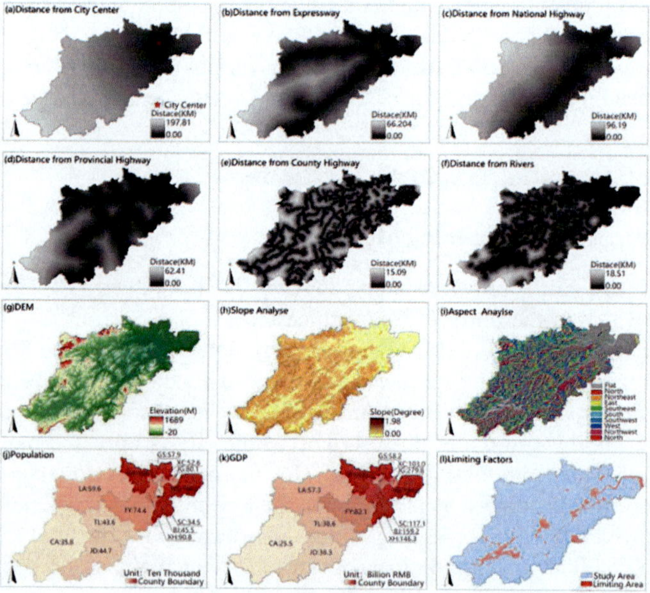

Fig. 1. Driving force factors.

2.2 Prediction Method

Calculation of Driving Probability Based on Artificial Neural Network (ANN). The artificial neural network is an intelligent algorithm designed to imitate the structure of biological brain neurons, and it is mainly used to estimate nonlinear functions with various inputs [4]. By inputting the driving force factors (elevation, slope, aspect, etc.)

of land use type transformation, the competition and interaction between different land-use types are simulated. The occurrence probability of different land types based on a neural network is calculated. The land-use type prediction result is obtained using probability as the input factor.

CA Simulation Based on Adaptive Inertial Mechanism. GeoSOS-FLUS software builds adaptive inertia and competition mechanism in a cellular automata (CA) model for handling complex competition and interactions between different land-use types. Based on the suitability probability of neural network and the restrictive factors constraining land-use change, land use prediction results are obtained using domain factors and iteration.

Using Kappa and FoM Values to Predict the Accuracy of the Results. The Kappa coefficient is widely used in evaluating the classification accuracy of remote sensing images. The classification accuracy of remote sensing images can be more accurately verified by counting the classified images and reference images pixel by pixel and establishing an error matrix [5]. Pontius [6] and others believe that the quality factor FoM can be used for model accuracy verification. The GeoSOS-FLUS model combines the two verification methods to obtain a more reliable model accuracy test.

Calculation of Direct Carbon Emissions. According to the research results of et al. [7], considering the actual situation of Hangzhou, the calculation formula for direct carbon emissions is determined as follows:

$$E_m = \sum e_i = \sum T_i \times \delta_i \tag{1}$$

Among them, E_m represents the direct carbon emissions; i represents the land type, i = 1, 2, 3, 4 and 5 correspond to cultivated land, forest land, grassland, water land and unused land respectively; e_i represents the carbon emissions from the i-th type of land; T_i indicates the area of the i-th type of land; δ_i indicates the carbon emission (absorption) coefficient of the i-th type of land. The carbon emission coefficients (kg C/(m^2·a) of cultivated land, forest land, grassland, water area and unused land were 0.0422, −0.0644, −0.0021, −0.0253, and 0.0005, respectively.

Calculation of Indirect Carbon Emissions. We mainly select 8 kinds of energy, including raw coal, cleaned coal, other washed coal, coke, gasoline, kerosene, diesel oil, and fuel oil for calculation. The energy consumption data of each district and county is indirectly calculated by the ratio of the GDP of the district and county to the GDP of Hangzhou City. The calculation formula is as follows:

$$E_{\mathrm{t}} = \sum_{\mathrm{i}=1}^{8} E_{\mathrm{ti}} = \sum_{\mathrm{i}=1}^{8} (E_{\mathrm{ni}} \times \theta_{\mathrm{i}} \times \varphi_{\mathrm{i}}) \tag{2}$$

In the formula: E_t indicates the total amount of carbon emissions from the final consumption of fossil energy; Eti indicates the carbon emission of the i-th type of fossil energy terminal consumption; Eni indicates the terminal consumption of various types of fossil energy; θ_i indicates the converted standard coal coefficient of various types of fossil energy; ρ_i indicates the carbon emission coefficient of various energy sources. The converted standard coal coefficient is from the "China Energy Statistical Yearbook" (2016), and the carbon emission coefficient is converted from the default value of carbon content in the "IPCC Guidelines for National Greenhouse Gas Inventories" (Table 1).

Table 1. Conversion coefficient of standard coal and carbon emission coefficient.

Energy name	Standard coal conversion factor(kgce/kg)	Carbon emission factor (tC/t)
Raw coal	0.7143	0.7559
Cleaned coal	0.9000	0.7559
Other washed coal	0.2857	0.7559
Coke	0.9714	0.8550
Gasoline	1.4174	0.5538
Kerosene	1.4174	0.5714
Diesel oil	1.4571	0.5921
Fuel oil	1.4286	0.6185

Calculation Method of Carbon Footprint. Based on the concept and calculation method of ecological footprint proposed by Wackernage [8], and combined with the carbon footprint calculation method proposed by Zhao Rongqin et al. [9], this study defines carbon footprint as The biologically productive land area required to absorb the greenhouse gases emitted by human economic activities. It is used to measure the ecological carrying capacity of Hangzhou and the impact of human economic activities on nature.

$$C_p = C_m / C_s \tag{3}$$

In the formula, C_p is the carbon footprint pressure index; C_m is the carbon source of different land use patterns, C_s is the carbon sink of different land use patterns.

Calculation of Carbon Emission Risk Index. To more clearly reflect the carbon emission risk of various districts and counties in Hangzhou, the carbon emission risk index is selected to measure it. The calculation formula is as follows:

$$C_{RI} = \sum_{i=1}^{n} (A_i \times P_i)/A \tag{4}$$

In the formula: C_{RI} represents the carbon emission risk index, A_i represents the area of the ith land use type, A represents the total land area of the region, and P_i represents the carbon emission coefficient of the ith land use type.

3 Results and Analysis

3.1 Results of Simulation Predictions

Calculation Results of Driving Probability Based on Artificial Neural Network (ANN). The land use data is taken from the 2020 TIFF image, and the above driving factors are input. The sampling mode of the neural network to obtain training samples uses the uniform sampling mode. According to expert experience, the number of sampling points is 2% of the effective pixel points. The number of layers is set to 12, and the normalization processing module with the model is used to normalize the input driving force factors. After the calculation, the land use type driving probability map is finally obtained (Fig. 2).

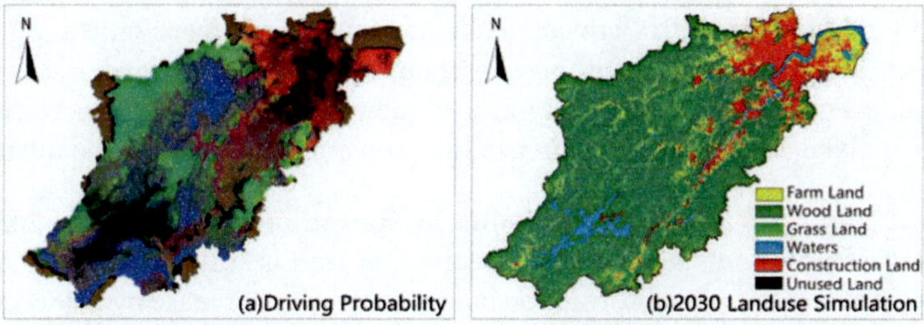

Fig. 2. Driving probability analysis and 2030 land use simulation.

Verification of Simulation Accurac. The verification takes the land use data of Hangzhou in 2000 as the initial state. The land-use situation in Hangzhou in 2010 was obtained through the FLUS model simulation. Compared with the actual land-use situation in Hangzhou in 2010, the urban land use simulation results using the FLUS model were generally consistent with the distribution of the actual data urban land. To verify the simulation accuracy of the FLUS model, the Kappa coefficient and From coefficient are introduced to evaluate the model accuracy. After calculation, the Kappa value is 0.894, the overall accuracy is 94.733%, and the Form value is 0.0758. The results show that the simulation accuracy of the FLUS model is high. It can better present the urban development form close to the real. Therefore, the reliability of applying the model to predict the future development of urban land is also high (Fig. 1) (Fig. 3).

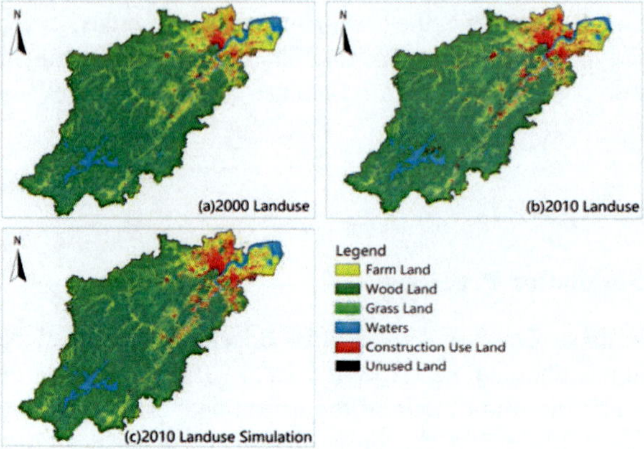

Fig. 3. The Verification of stimulation accuracy.

CA Simulation Results Based on Adaptive Inertia Mechanism. The land use data took 2020 TIFF images as input neural training-driven probability data while inputting land-use constraints. Referring to Wu Xinxin et al. [10], parameters such as iterative layers and land transfer matrix are set. They are using the dimensionless value of TA changes to provide a reference for the neighbourhood weight parameter setting of the FLUS model. For the simulation of land use quantity in 2030, SPSS performs curve regression analysis to obtain each land use function curve and predict the quantity.

Analysis of Land use Simulation Results in Hangzhou from 2020 to 2030 From 2020 to 2030, Hangzhou's main urban construction land is still concentrated along the Qiantang River, and the total construction land is forecast to increase by 289.6 km^2, from 1218.61 Km^2 to 1507.42 km^2. Compared with 2010–2020, the growth rate has slowed down, and the main growth points are located in Xiaoshan District and Yuhang District on the south bank of Qiantang River.

Analysis of Land Use Simulation Results in Hangzhou from 2020 to 2030. From 2020 to 2030, Hangzhou's main urban construction land is still concentrated in the area along the Qiantang River, and the total construction land is forecast to increase by 289.6 km^2, from 1218.61 Km^2 to 1507.42 km^2. Compared with 2010–2020, the growth rate has slowed down, and the main growth points are located in Xiaoshan District and Yuhang District on the south bank of Qiantang River. The reduction of cultivated land and forest land is relatively significant. It is predicted that cultivated land will change from 2938.44 km^2 to 2792.64 km^2, a decrease of about 145.8 km^2. The woodland 11410.792 km^2 becomes 11286.59 km^2, a decrease of about 124.203 km^2. From 2020 to 2030, the reduced forest land range is mainly located in the construction land of villages and towns in the piedmont valley area of Lin'an District in the west. The reduced cultivated land area is mainly located in the area occupied by urban construction land in the northern part of Yuhang District.

The water area is reduced by 1.1457 km^2, and the forecast shows that the area of Xiaoshan District on the south bank of Qiantang River and the area of Xixi Wetland is reduced. The grassland and unused land areas ranged from approximately 393.05 km^2 to 394.19 km^2 and from 6.07 km^2 to 5.74 km^2. The area remains unchanged.

From 2020 to 2030, the spatial change characteristics of Hangzhou will mainly extend outward from the existing urban construction land, forming a development trend of spire-shaped land.

3.2 Calculation of Carbon Emissions and Carbon Sinks from Land Use in Hangzhou

Estimation of Carbon Emissions in Hangzhou

See Table 2

Table 2. Estimation of carbon emissions in each district of Hangzhou (t).

2030 carbon emissions														
		Binjiang district	Chun'an county	Fuyang district	Jiande city	Gongshu district	Lin'an district	Linping district	Shangcheng district	Qiantang district	Xiaoshan district	Xihu district	Tonglu county	Yuhang district
Carbon source	Arable land	207.48474	9951.9838	9166.7684	8805.1144	136.76598	12346.876	3719.11554	296.35794	5186.9708	47230.907	2148.45264	7620.476	10732.13
	Unused	0.00585	0.2853	0	0.02835	0.02835	0.03555	0.0441	0	0	0.0648	0	0	0.85545
	Construction land	1485727.331	204769.4311	691160.6607	333537.5729	1534494.116	510987.3916	1038860.176	1014570.402	932526.4573	1862669.895	1390236.349	320260.7519	1558290.264
	Total	1485934.822	214721.7002	700327.4291	342342.7156	1534630.91	523334.3042	1042579.336	1014866.76	937713.4281	1909900.867	1392384.802	327881.2279	1569023.255
Carbon sink	Woodland	275.25204	169341.088	61267.584	88420.556	363.9888	127519.728	741.0186	311.36112	172.08324	12884.73984	5403.9006	70383.404	19174.134
	Grassland	0	233.2722	84.27132	152.55891	0.64638	81.39285	3.69873	0	4.24494	7.20279	2.78208	46.69245	7.19901
	Waters	207.16146	8869.674	938.69325	1035.26082	19.87821	340.68474	121.97889	555.63354	984.36987	2362.22811	564.67323	698.72022	432.42507
	Total	482.4135	178444.0342	62290.54857	89608.37573	384.51339	127941.8056	866.69622	866.99466	1160.69805	15254.17074	5971.35591	71128.81667	19613.75
	Net carbon emissions	1485452.408	36277.66597	638036.8805	252734.3399	1534246.397	395392.4976	1041712.639	1013999.765	936552.7301	1894646.696	1386413.446	256752.4112	1549409.497

The Carbon Footprint Pressure and Carbon Risk Index of Each District in Hangzhou

See Table 3

Table 3. Carbon footprint pressure and carbon emission risk index of districts and counties in Hangzhou.

Districts and counties	Carbon footprint pressure index	Carbon emission risk index (Ton/km^2)
Binjiang district	3082.21	27427.467
Chun'an county	1.20	10.819
Fuyang district	11.24	48.506
Gongshu district	3990.74	20539.764
Jiande city	3.82	144.651
Lin'an district	4.09	167.394
Linping district	1202.94	4887.985
Qiantang district	807.89	3872.210
Shangcheng district	1170.56	9548.101
Tonglu county	4.61	184.108
West lake district	233.18	5863.086
Yuhang district	80.00	2165.740
Xiaoshan strict	125.21	2035.066

3.3 Discussion on the Carbon Control Measures in Hangzhou

It can be seen from the above that the spatial difference of carbon emissions in Hangzhou is pronounced. Due to the differences in economic development and natural environment factors, the spatial distribution of carbon emissions in different districts and counties is also quite different.

The overall carbon emission characteristics show the difference of "high in the east and low in the west" and "high in the river and low in the mountains". Binjiang District and Gongshu District have the highest carbon emission values, and the pressure index and risk index are also significantly higher than other districts and counties. The five districts and counties of Chun'an, Tonglu, Fuyang, Jiande, and Lin'an are significantly lower than the main urban areas in terms of carbon emissions and related indicators due to their low level of economic development and their ideal natural environment with mountainous forests and waters. According to the unique characteristics of the carbon emission pattern in Hangzhou, the following regional carbon governance measures should be proposed according to the carbon emission governance goals.

Adjustment of Cultivated Land Spatial Structure. From the simulation results, the general cultivated land area in Hangzhou will decrease by about 145.8 km^2 in 2030. The reduced area is mainly located in the northern part of Yuhang District and Gongshu District. It is mainly transformed into urban construction land, and the area and scope of cultivated land that has a more significant impact on the forest carbon sink system in the forest area of Lin'an District have not been effectively controlled. Achieving the 2030 "carbon peak" goal still requires a certain degree of conversion of farmland to forests in mountainous areas to improve the ecological carbon sink system in mountainous areas.

Control the Development Direction of Construction Land. The well-preserved large-area woodlands in the west and south of Hangzhou provide most of the carbon sinks for the city. In the land-use simulation scenario, villages and towns in the mountains, forests and valleys tend to develop towards the surrounding woodlands. It is necessary to control construction land development and delineate the ecological red line and the urban development boundary in combination with the relevant content of national land space planning.

Pay attention to the Risk of Carbon Emission in the Main Urban Area. The actual calculation results of the carbon footprint pressure index and carbon risk index show that the carbon risk index of the central urban districts and counties such as Binjiang District, Gongshu District and Shangcheng District is relatively high. In future carbon emission control, a regional detection system should be established in the main urban area for the carbon risk index and carbon footprint pressure, effectively reflecting the urban carbon emission risk and the urban ecological carrying capacity and improving urban development.

Constructing a Carbon Emission Trading System Between Peripheral Counties and Central Urban Areas. Through the flow of resources and capital between the main urban area and the peripheral districts and counties, the overall net carbon emissions of the city will be reduced, the development of the surrounding districts and counties of the central city will be driven, and preliminary carbon neutralization will be carried out within the city to reduce the overall carbon emission level of the city.

4 Discussion and Policy Orientation

In this case, study, based on analyzing the land use pattern changes in Hangzhou in 2000, 2010 and 2020, combined with the driving factors of nature, traffic location and socio-economic factors, the GeoSOS-FLUS software was used to analyze the changes in Hangzhou 2030. Based on the prediction and analysis of the land use pattern in 2030, combined with statistical data and corresponding index coefficients, the spatial pattern of carbon emissions and related indicators in Hangzhou in 2030 was calculated and analyzed. According to the spatial and quantitative characteristics of carbon emissions, corresponding carbon governance methods are proposed.

Numerous studies have shown that land-use change and land management can directly or indirectly affect the greenhouse gas exchange and carbon cycle processes between terrestrial ecosystems and the atmosphere [11]. In the process of achieving carbon peaking and carbon neutrality goals, the carbon emissions and corresponding changes caused by land use and its type changes should be given sufficient attention. The research shows that using the GeoSOS-FLUS prediction model is feasible to predict and analyze the land-use carbon emission pattern. It can provide data and direction guidance for decision-making related to the dual carbon goals.

Although the driving force factors of land use simulation selected in this paper cover three aspects of nature, traffic location and social economy, they are still not comprehensive enough from good simulation. In follow-up research, the accuracy of model prediction results can be improved by inputting more raw data. At the same time, the selection of limiting factors is relatively simple. Perhaps land use policy, planning, and territorial spatial planning control can provide a higher reference opinion on the limiting factors for the prediction model. Finally, the parameters in the FLUS model are set differently, and the results will be different. In calculating carbon emissions and related indicators, more comprehensive data such as energy consumption can also be used to provide more reference opinions for urban carbon governance.

References

1. Li, X., et al.: The Implementation and application of geographical simulation and optimization systems(GeoSOS). Acta Sci. Naturalium Univ. SunYatSen **49**(04), 1–5+15 (2010)
2. Crescenzi, A., Palermo, A., Trimboli, P.: Cancer prevalence in the subcategories of the indeterminate class III (AUS/FLUS) of the Bethesda system for thyroid cytology: a meta-analysis. J. Endocrinol. Invest. **44**(7), 1343–1351 (2021)
3. Yang, L., Xie, Y., Zong, L., Qiu, T., Jiao, J.: Land use optimization configuration based on multi- objective genetic algorithm and FLUS model of agro- pastoral ecotone in Northwest China. J. Geo-inform. Sci. **22**(3), 568–579 (2020)
4. Xiaoping, L., et al.: A future land use simulation model (FLUS) for simulating multiple land use scenarios by coupling human and natural effects. Landsc. Urban Plan. **168**, 94–116 (2017)
5. Wenning, X., Pengxin, W., Ping, H., Tailai, Y., Shuyu, Z.: Application of Kappa coefficient to accuracy assessments of drought forecasting model:a case study of Guanzhong Plain. J. Nat. Disasters **20**(06), 81–86 (2011)
6. Pontius, R., et al.: Comparing the input, output, and validation maps for several models of land change. The Ann. Regional Sci. **42**(1), 11–37 (2008)
7. He, S., Hong, M.L., Xue, L.C., Qing, C.C., Yun, T.: Land use patterns on carbon emission and spatial association in China. Econ. Geogr. **35**(03), 154–162 (2015)
8. Wackernagel, M., et al.: National natural capital accounting with the ecological footprint concept. Ecol. Econ. **29**(3), 375–390 (1999)
9. Zhao, R., Huang, X., Zhong, T., Peng, J.: Analysis on carbon emission intensity and carbon footprint of different industrial spaces in China. Acta Geogr. Sinica **65**(09), 1048–1057 (2010)
10. Wu, X., Liu, X., Liang, X., Chen, G.: Multi-scenarios simulation of urban growth boundaries in Pearl River Delta based on FLUS-UGB. J. Geo-inform. Sci. **20**(4), 532–542 (2018)
11. Ke, R.L.: Land use change, net greenhouse gas emissions and terrestrial ecosystem carbon cycle. China Meteorological Press, Beijing (2002)

Research on Macao Cultural Heritage Protection and Sustainable Development——A Case Study of Ruins of St. Paul's

Yishan Mao, Yao Wu(✉), and Yonglan Wu

Soochow University, 199 Renai Road Suzhou, Suzhou, China
wuyaoseu@foxmail.com

Abstract. From the perspective of cultural heritage protection and sustainable development, the paper combines the conference's concerns on cultural heritage protection, effective development and environmental improvement to ensure the survival, inheritance and development of heritage and make heritage a truly sustainable social activity. Cultural heritage landscape is a renewal model of urban space, which can form a transition space from cultural landscape to modern urban landscape. The information dissemination, spatial renewal and unique culture brought by cultural heritage landscape help to enrich the use function of urban historical and cultural space, and ultimately promote the conservation and development of cultural heritage and tourism development. Ruins of St. Paul's is one of Macao's most iconic tourist attractions. The paper takes Ruins of St. Paul's as a carrier for research, analyzes the landscape status and restrictive factors in the region, and puts forward the landscape control and landscape reconstruction strategy of Ruins of St. Paul's, which further drives the protection and sustainable development of Macao's cultural heritage.

Keywords: Cultural heritage protection · Cultural heritage landscape · Ruins of St. Paul's · Style restoring

1 Macao Cultural Heritage Protection and Sustainable Development Needs

1.1 Planning with Urban Cultural Space Characteristics

Macao's diverse cultural background provides unique cultural characteristics for the renewal of urban public space, which can not only be presented in the urban space, but also enable people to have a deep understanding of the city and deepen their impression of the city in the unique cultural atmosphere. It is necessary to construct cultural heritage landscape with cultural heritage as the carrier, because it contains the rich connotation of the interaction between man and culture, man and city. We propose that the conservation and development of Macao's cultural heritage should first identify the characteristics of different cultural Spaces. Different spatial characteristics create different cultural values, such as historical and cultural space, industrial and cultural space, religious

W. Guo and K. Qian (Eds.): GBCESC 2022, LNCE 211, pp. 44–56, 2023.
https://doi.org/10.1007/978-981-19-5217-3_5

and cultural space, etc. These Spaces all contain different cultural characteristics, thus creating different cultural landscape space.

From the perspective of cultural heritage protection and sustainable development, the protection and development of cultural heritage needs to be planned in combination with the characteristics of cultural space. However, the cultural landscape constructed with cultural heritage as the carrier is not simply attached to the cultural space, and we also need to avoid the loss of landscape characteristics caused by homogenous design. Secondly, let cultural heritage space form a good landscape environment and an effective landscape network in the process of Macao's rapid urban development, and give play to the value of landscape environment. In addition, landscape green space has special functions. It can form a buffer area for urban activities and cultural heritage space, which can not only beautify the cultural heritage landscape space, but also alleviate the direct or indirect impact and damage caused by urban activities on the cultural protection area, forming a certain protection effect. Therefore, in the conservation and development of cultural heritage, it is necessary not only to build cultural landscape space that makes people happy, but also to set up buffer zone with environmental function and protection of cultural heritage space.

1.2 The Protection, Development and Utilization of Local Cultural Heritage

The cultural heritage in the historic city of Macao reflects the history and cultural precipitation of a city. These heritage Spaces bear the past history of the city and witness the development process of the city in different times. The tourism resources with cultural heritages are the unique advantages of Macao, which drive the development of Macao tourism. The urban renewal and development of Macao requires the protection, development and utilization of cultural heritage. Only by formulating the planning and design consistent with the connotation of cultural heritage space and constantly improving the environmental quality of cultural heritage space can Macao be loved by people. The renewal, conservation, repair, development and utilization of cultural heritage involves different types of cultural heritage space, which presents different cultural characteristics, environmental conditions, distribution regions, economic conditions and so on. The conservation work should be based on the assessment after analysis and statistics, so that the cultural heritage space can be integrated according to different characteristics, so that the cultural heritage conforms to the location of the space. At the same time, the conservation of cultural heritage should be carried out systematically and hierarchically. First, according to the local planning and design of cultural heritage space, a distinctive cultural landscape is formed. Then, different forms and sizes of urban cultural heritage space are planned as a whole to form a connected landscape ecological network and further improve the quality of urban space. Conservation measures from micro to macro make cultural heritage better integrated with other urban landscape areas.

2 Continue the Urban Context and Promote Tourism Development

2.1 Improve the Understanding of Urban Culture Among Residents and Tourists

Cultural cognition is regional. Such regionalism has different understandings of different types of audiences, and local residents have a deeper understanding of these cultures, which reflects their deep understandings and recognition of culture from living customs to public activities. For tourists, the culture of a city is usually a macro and overall impression, and the cultural experience of a city mainly comes from the intuitive symbols and the impression of well-known places, so it is inevitable to have a rough understanding of it due to the short experience time. Different cities contain different cultural deposits, people's cultural cognition will be different, and their impression of urban experience will also be affected. Macao, for example, gives people a cultural cognitive impression of exotic customs and entertainment. Of course, cultural cognition will also be reflected in many other aspects of the city, which will be affected by regional and crowd factors. Moreover, the establishment and dissemination of cultural cognition also depend on regional areas, so that people can experience diverse urban cultures. The author believes that targeted cultural experience can be established through analysis and combined with corresponding tourism development projects to improve people's cognition and experience of urban culture.

The construction of cultural heritage landscape is closely related to urban cultural space. In the form of planning and design, it is combined with the urban context and adapted to the characteristics of cultural space. The construction of cultural heritage landscape combined with urban context can not only spread through the cultural information carried by cultural heritage landscape, but also make it become the carrier of urban context. Urban culture is presented through this carrier, which is conducive to deepening the public's experience of cultural heritage landscape. Cultural heritage landscape not only has the function of protection and display but also has the function of popular science. Popularizing science by echoing with cultural heritage space helps people understand the original urban cultural space.

2.2 The Role of Promoting Tourism Development

As mentioned above, the protection and development of cultural heritage and construction of cultural heritage landscape can improve the landscape environment of urban cultural space and highlight the value of cultural heritage. Forming landscape buffer zone for some cultural heritage space can also alleviate the destruction of cultural space by urban public activities, and play a role of protecting urban context and promoting tourism development. The protection of urban cultural space by cultural landscape can not only protect the material context in the space, but also form a transition space from cultural landscape to modern urban landscape, so that cultural landscape plays a role of connecting the past and the future [1]. The valuable architectural style, urban texture, historical and cultural atmosphere and unique landscape features in Macao's cultural heritage are the focus of attracting tourists. Cultural heritage conservation and development can use the form of cultural landscape to express the connotation of space and present the characteristics of space. After cultural heritage becomes a carrier of cultural space, the information conveyed can constantly improve the popularity of cultural space, and also make people who use the space receive information more quickly and accurately. The renewal of the historical and cultural space is conducive to improving the ecological environment of the historical and cultural space, enriching the use function of the space and attracting people to stay. The development and utilization of cultural heritage is an important means of urban historical district renewal, and the renewal of cultural heritage is conducive to promoting tourism development. Such development and utilization for the purpose of tourism development can increase the attraction and tourism experience of cultural heritage to tourists [2]. However, the cultural heritage landscape does not fully reflect the relevant cultural information. It emphasizes the characteristics of the cultural space and should make the characteristics of the original cultural space more prominent to deepen the experience of tourists. The information dissemination, spatial renewal and unique culture brought by cultural heritage landscape greatly enhance the attraction of cultural heritage, form the effect of drainage, and promote the conservation and development of cultural heritage and tourism development.

3 The Inner Drive of Macau's Urban Tourism Planning

Macao's unique tourism resources make it different from other cities as a tourist resort, forming a tourism mode represented by cultural heritage landscape. Macao cultural heritage landscape has unique regional characteristics, which is the result of the interaction between human beings and the natural and social environment of a specific region. Therefore, the protection and development of cultural heritage is not carried out at a single level, but needs to be carried out together with government policies, urban areas and cultural conservation. In 2020, the Macao SAR government promulgated Draft Consultation Text of the Master Plan of the Macao Special Administrative Region (2020–2040), through which relevant policies and measures were adopted to build cities and protect cultural heritage, which played an intrinsic role in driving tourism. Article 23, 24 and 26 of "Guiding Principles" in Chapter 6 of the draft respectively guide the protection of cultural heritage of historic urban areas, the reconstruction of old areas and the construction of urban landscape, including: "1. Protect the prominent universal value of historic urban areas and the style of the community... To create a tourist city with rich cultural heritage and unique charm; 2. Optimize regional carrying capacity by setting restrictions or constraints on adjacent buildings of cultural heritage, so as to reduce the impact of adjacent development on cultural heritage. 3.... Dedicated to preserving the traditional and distinctive urban fabric and original architectural style; 4. Strengthen the urban landscape...... To strike a balance between conservation and development..." [3]. Ruins of St. Paul's is an architectural heritage with a unique cultural identity and an important cultural heritage in Macao. From the perspective of cultural heritage protection and sustainable development, it is of practical significance to reconstruct the Ruins of St. Paul's attractions and the surrounding tourist areas in the "Central District −2" area of the Macao Peninsula.

3.1 Macao Tourism Planning Driven by Cultural Heritage Landscape Construction

For a long time, the image of Macao's gaming tourism has been deeply rooted in people's hearts. Although it has brought huge economic benefits to Macao, most tourists only stay for a short time and it is difficult to form the requirement of continuous promotion, which greatly weakens the unique cultural advantage of Macao. In addition, with the improvement of consumption level, people tend to pay attention to cultural experience and high-quality cultural tourism enjoyment. In view of this situation, the author thinks that Macao should make full use of its unique advantages of diverse cultural gathering, reasonably plan cultural landscape and cultural heritage resources, and combine cultural heritage, museums, old streets and historic urban areas. From the planning level to get rid of the single "gambling city" image to transform into a historical and cultural city, improve the popularity of cultural tourism. The construction of Macao famous cultural and tourism city can be carried out from the following aspects: 1. Find out the factors that stimulate the interactive development of urban space; 2. Further promote the construction of complex landscape space and urban cultural landscape; 3. Strengthen cultural tourism supporting services and publicity and promotion activities; 4. Macao tourism planning is driven by cultural heritage landscape construction.

Because landscape is not a single existence, but a complete system (or process), all landscape elements interact and restrict each other. In the process of cultural heritage landscape construction, we should explore the structural, functional and procedural aspects of urban cultural landscape, form urban cultural heritage landscape network through landscape layout from point to surface and from far to near, and improve the cultural atmosphere of Macao city. At the present stage, the urban space of Macao is fragmented, resulting in poor connection of cultural heritage landscape and isolated landscape. Therefore, the integration and connection of cultural heritage landscape should be restored by means of landscape construction. As a "soft power" to promote Macao's urban development and a strong support for urban renewal, multi-culture can combine natural landscape and cultural heritage landscape to create a complex landscape space, so as to enrich landscape features and enhance cultural connotation, and effectively drive the development of tourism.

Macao, with its unique natural landscape and coastal city characteristics, has a strong attraction to people. The combination of natural landscape and cultural heritage landscape can create a vivid city tourism card, such as Nice in France, Barcelona in Spain and other urban construction cases give us a good reference for construction guidance. Therefore, the overall and vivid spatial structure of urban landscape can be formed through strategies such as strengthening the construction of natural landscape structure, maintaining the characteristics of human landscape and promoting the connection between natural and human landscapes [4].

4 Ruins of St. Paul's Cultural Heritage Landscape Control and Spatial Remodeling

4.1 Ruins of St. Paul's Cultural Heritage Landscape

Ruins of St. Paul's and its surrounding areas are the epitome of Macao's political, economic and cultural changes. They are also important cultural nodes in Macao, and Macao people have a special sense of identity and belonging to them. Ruins of St. Paul's and its adjacent Mount Fortress area are restricted due to cultural preservation, so there are restrictions on building height, but Patane to the north has no definite planning restrictions. In recent years, due to the influence of real estate construction, several 100-m-tall buildings have been erected to the north of Ruins of St. Paul's, which has seriously affected the landscape view and skyline shape of Ruins of St. Paul's and destroyed the landscape order [5]. In addition, some historical streets and architectural cultural heritage are also submerged in the modern development of the city, and the modern urban landscape hinders the historical landscape with cultural cognition. At present, the public Spaces of Ruins of St. Paul's surrounding streets are in disorder. Some buildings of historical value are falling into disrepair. Moreover, private construction is common, which seriously affects the spatial characteristics of streets in historic cities. Therefore, the author believes that corresponding strategies should be adopted to rebuild the order of cultural heritage landscape, protect the sight of Macau's urban landscape, and improve the overall image and style control of the historic town around Ruins of St. Paul's.

4.2 Cultural Heritage Landscape Style Control and Reconstruction Strategy

In the past, based on the population density and land area of Macao, the "one-size-fits-all" building height limit measures were adopted, which inevitably affected the development and utilization of land resources, hindered social and economic development and even led to the rise of building prices. Therefore, it is necessary to coordinate the relationship between cultural heritage landscape conservation and socio-economic needs. After field investigation, the author thinks that the height buffer zone can be established by using stepped regional building height according to the different regional distance. With the goal of preserving the landscape skyline and landscape of the Ruins of St. Paul's tourist area, it is proposed to set the height of buildings according to the distance between different areas and the Ruins of St. Paul's Tourist area. Set height limits for buildings within the Ruins of St. Paul's axis. Based on the human view in front of Ruins of St. Paul's, it is connected to the top of the building within the central axis to form a baseline. The new buildings that are close to Ruins of St. Paul's are made lower, and those that are further away can be increased. Create a visual buffer between the line of sight and the construction of the building, and propose local solutions to protect the central axis landscape of Ruins of St. Paul's (see Fig. 1).

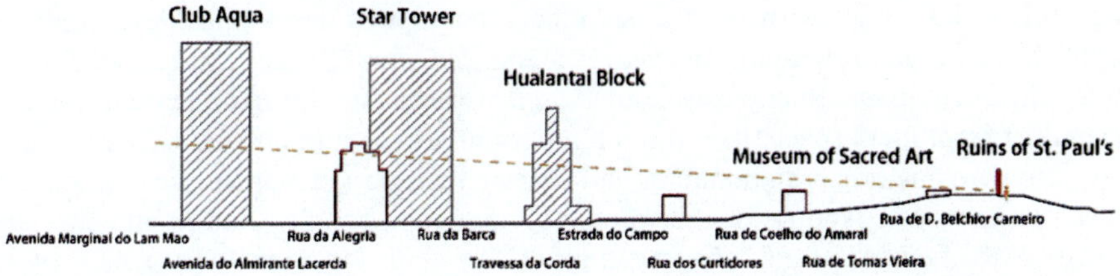

Fig. 1. Construction strategy of Macao famous cultural tourism city (**Source:** Drawn by the author)

On the other hand, the historic streets of Ruins of St. Paul's are located in the tourist center and have strong elements of the city's historical context. The reconstruction of historical streets and alleys also means the revitalization of the construction of the city's historic urban areas. On the premise of not destroying the overall architectural form and appearance, the renovation of historical streets and alleys can be carried out "from outside to inside". It provides possibilities for the increase of spatial functions. It focuses on the transformation of street and lane spatial functions, extends the existing plot and strengthens the connection between blocks, integrates the broken space to expand the usable area of space and forms continuity and orientation (see Fig. 2). In addition, from the perspective of residents and tourists, the idle areas of streets and lanes should be used to build more humanized space, provide tourism-oriented functions, promote the vitality regeneration of streets and lanes, and create new areas with both leisure and entertainment activities.

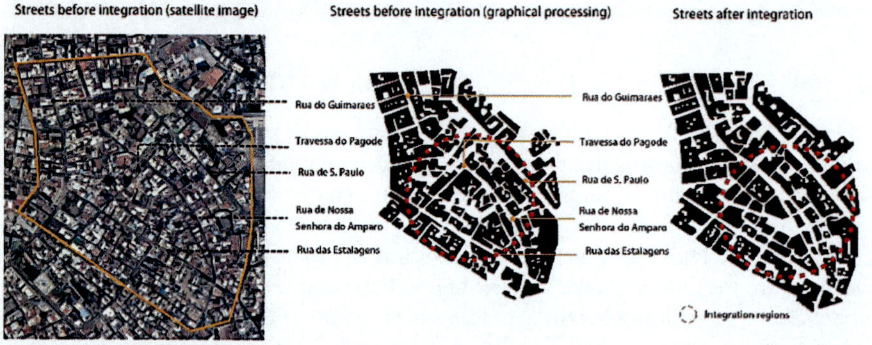

Fig. 2. Comparison of historical streets and alleys before and after space integration (**Source:** Drawn by the author)

As landscape urbanism emphasizes to solve urban problems from the perspective of "landscape", landscape is not only a plot between buildings, but also a broad and inclusive urban surface. Meanwhile, urban linear landscape can be divided into different node Spaces, and each node space has a specific structure and function to serve the city [6]. Therefore, the author believes that the streets and alleys in Macao's historic urban areas can be integrated and updated through the spatial method of "node and linear". Firstly, the illegal private buildings that affect the historical landscape of streets and alleys should be demolished and their functions replaced. For example, these Spaces can be transformed into public open Spaces integrating Chinese and Western cultures to increase the area of rest and entertainment space. In addition, important architectural nodes were renovated to restore their historical authenticity and further tourism development activities. For linear space, the key is to rationalize the first and second facades, adding integrity and richness to the facade in terms of appearance. The first facade refers to the facade of the building, which requires the renovation and conservation of the facade of the building with historical value, so as to change the facade image of the streets and lanes, so as to form a continuous cultural landscape. The second facade refers to the overall facade of the buildings in the street. Due to the narrow streets, insufficient lighting, and lack of activity space and green space, the second facade requires the establishment of transition space at the entrance of the streets to enhance visual identification.

The rebuilt Ruins of St. Paul's is a linear landscape that connects historic and cultural blocks with commercial, cultural and tourist areas and heritage landscape Spaces, creating an attractive tourist route that effectively perpetuates the deep cultural heritage of historic streets. At the same time, the measures of building height limit prevent the historical city from being submerged in the modern urban construction environment, on the one hand protect the diverse cultural advantages of Macao, on the other hand also protect the skyline landscape of Macao, and promote the protection and development of cultural heritage landscape.

4.3 Three-Dimensional and Multi-dimensional Space System Design

Ruins of St. Paul's and the surrounding area are the focal point of Macao's World Heritage, It brings together Ruins of St. Paul's, Na Tcha Temple, Catholic Art Museum, Macau Museum, Mount Fortress and other famous historical and cultural heritage sites. It is a tourist area with three main areas: historical and cultural landscape, museum landscape and business service area. It attracts a large number of tourists every day. According to the preliminary investigation, this area is affected by the difference of terrain, dense flow of people and spatial scale, which makes it difficult to move in tourist peak season. Taking Ruins of St. Paul's as an example, this study discusses the "three-dimensional" design and the establishment of multi-dimensional spatial system in the region, and proposes the intervention of "thickened ground" and urban green infrastructure in the spatial structure of the city [7]. Therefore, "thickened ground", as a layered structure, can perform functions through interconnection, interaction and sharing [8]. The most typical way of expression is the establishment of three-dimensional multi-level space system, which is the multi-dimensional landscape system of the city, integrating the functions of recreation and transportation. The establishment of Ruins of St. Paul's three-dimensional and multi-faceted space system can be used to divert tourists. Specific visiting routes are set up to enhance the guidance of the site and facilitate the realization of transportation, sightseeing and landscape services.

Enhance the Connectivity of Pedestrian Transportation System and Spatial Guidance. The area around ruins of St. Paul's has obvious terrain differences, with Ruins of St. Paul's and the Macau Museum as the highest points. Due to the rugged terrain and uneven roads in the tourist area, tourists mainly rely on maps and walking to visit. At present, the tourist roads in the region have no substantial guidance and poor visual guidance (see Fig. 3).

Now, the construction of infrastructure is no longer a simple large-scale construction separated from the environment, but integrates the environment and facilities, making road traffic become the container of people's life, thus generating a new public area [9]. Using Kelly Shannon's ideas, the author believes that the existing infrastructure and terrain can be used to create new public areas and optimize the pedestrian system, Connect Ruins of St. Paul's, Na Tcha Temple, Catholic Art Museum, Macau Museum, Mount Fortress and other attractions, and make the road traffic become a container for

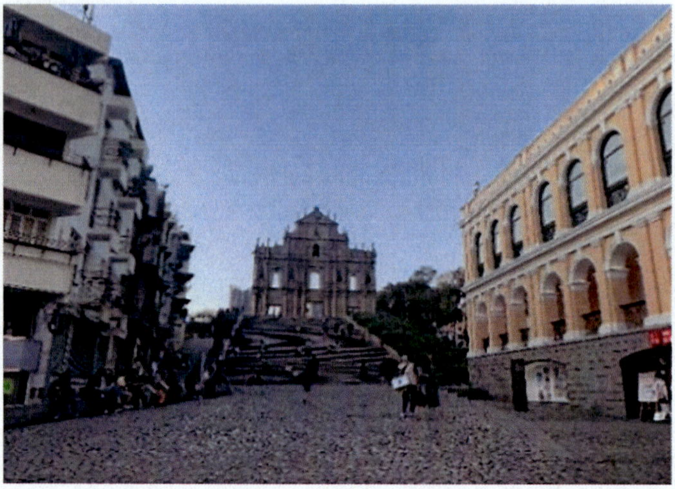

Fig. 3. Obvious terrain differences of Ruins of St. Paul's (**Source.** Photo by the author)

people's life. In this way, it is not only convenient for tourists to go sightseeing and realize the rapid evacuation of people flow in the scenic spot, but also can connect the cultural heritage landscape inside the scenic spot to improve the connection and integrity(see Fig. 4).

Ruins of St. Paul's tourist area has many cultural heritage landscape categories, and the road walking system is not well targeted. In the past, tourists did not have clear guidance on the visiting route, so it was easy to cause confusion on the visiting route.

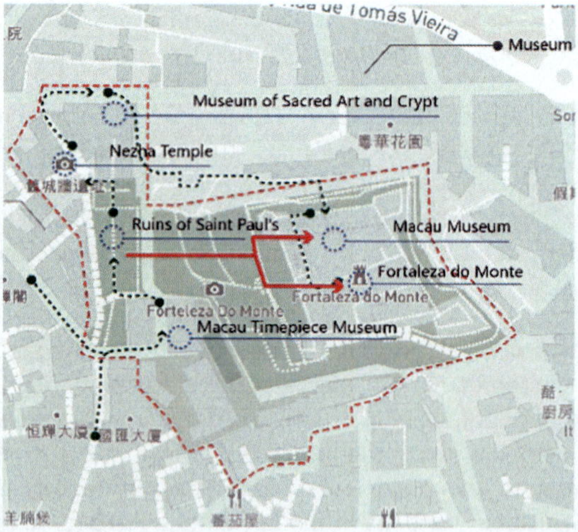

Fig. 4. Plane diagram of walking system (**Source:** Drawn by the author)

Therefore, the buildings in the tourist area can be classified or the corresponding functional areas can be set up according to the purpose of tourists' behavior. For example, the category of historical and cultural landscape area, museum landscape area and commercial landscape block is set, or it can be divided according to the behavioral purposes of tourists, such as visiting, commercial shopping, food tour and so on. In the design, different materials, lights or colors are used to divide the pedestrian road system, so as to connect different types of cultural heritage landscape and form a new visual guidance system and "three-dimensional" traffic system (see Fig. 5). Compared with traditional roads, visual signs as a guide system have stronger guiding function, which can attract tourists to use according to the instructions and play a role in dredging the flow of people. Therefore, it can be used as the main road connecting the historical and cultural landscape in the tourist area, enhancing the spatial guidance and connectivity of the pedestrian traffic system.

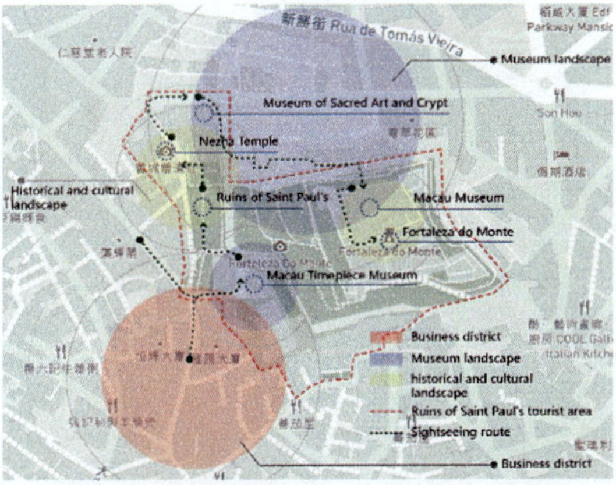

Fig. 5. Schematic diagram of different landscape areas and tour routes in Ruins of St. Paul's tourist area (**Source:** Drawn by the author)

Enrich the Service Functions of Landscape Nodes and Optimize the Multi-level Function System. At present, the back of Ruins of St. Paul's has only a simple steel structure to support the building and serve as an observation deck. Most of the remaining space is left vacant (see Fig. 6). As the core of Ruins of St. Paul's, the Ruins of St. Paul's tourist area often attracts a large number of tourists. Recreational facilities can be added on both sides of the area for tourists to stay for a short time. It can also be used as a tourist transit area, which helps to disperse tourists and buffer the flow of people in the area. At the same time, related display props and facilities can be set up to increase the historical and cultural introduction and information display function of the tourist area. Such measures can enable tourists to know the history and culture of Macao more directly and quickly, and improve the cultural image display of Macao city (see Fig. 7).

From the perspective of cultural heritage protection and sustainable development, relevant methods and strategies are proposed to intervene in the planning and updating of tourism areas, and further control the cultural heritage landscape of Ruins of St. Paul's tourism area. The author thinks that it can effectively change the isolated situation of architectural cultural heritage and historical city, and enhance the connection between architecture and region, region and region. At the same time, landscape corridors should be added in important spatial nodes to connect various important scenic spots and build a three-dimensional and multi-dimensional spatial system to facilitate the development of tourism activities and promote regional interaction, and ultimately drive the development of cultural tourism in Macao.

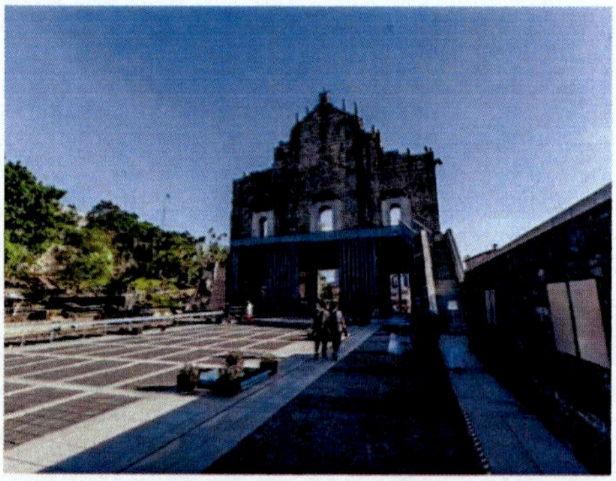

Fig. 6. The simple rear support structure of Ruins of St. Paul's (**Source:** www.sohu.com/a/278244383, 2019)

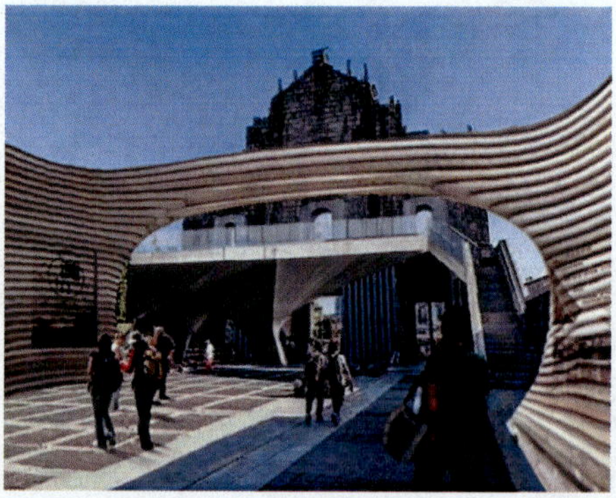

Fig. 7. Increase the leisure area for tourists (**Source:** The author repainted from Fig. 6)

4.4 Strengthen the Social Management Mechanism

To sum up, a series of rectification measures integrate the urban spatial form. Both regional landscape and spatial system have been greatly improved. On the one hand, this effect reflects the "interaction" between urban regional space and real activities; on the other hand, it also accords with the "sustainability" of cultural heritage protection and urban renewal planning, and builds the city with the operation mode of underlying infrastructure. However, in order to further protect the existing achievements, this study believes that measures to protect Ruins of St. Paul's tourism area can be put forward at the level of government policies and regulations, so as to continue to strengthen the government and social management mechanism and protect them "top-down". This kind of protection mode is the guarantee basis for urban cultural heritage protection and renewal activities, which can not only provide guidance for the development of protection activities, but also provide basis for the implementation of sustainable development strategy. Therefore, the incorporation of strategies and methods to strengthen the social management mechanism is practical and effective, and is also a favorable guarantee for the research on cultural heritage protection and sustainable development strategy, which reflects the standardization and professionalism of the research on cultural heritage protection and sustainable development in Macao.

5 Conclusion

Since The Historic Centre of Macao was designated as a "World Cultural Heritage", relevant departments of The Macao SAR government have focused on the conservation and development of existing cultural heritage, paid attention to the protection of historical and cultural heritage in urban renewal, and established a conservation mechanism for cultural heritage. The cultural characteristics and diversity of Macao's historical and cultural heritage greatly enhance the attraction and place spirit of the region, and the conservation and development of cultural heritage will help residents and tourists to enhance their understanding of the city's culture, promote the continuation of the city's cultural context and drive tourism planning. Based on the perspective of cultural heritage protection and sustainable development, this paper takes Ruins of St. Paul's tourism area as an important node to plan corresponding tourism strategies to promote the tourism development of Macao. 1. Proposed the control and reconstruction strategy of cultural landscape of Ruins of St. Paul's tourism area. By setting height limits on buildings adjacent to Ruins of St. Paul's, a visual buffer protects the central axis landscape dominated by Ruins of St. Paul's; 2. Rebuilding the street space around Ruins of St. Paul's tourist Area Through "from the outside to the inside" renovation and construction of idle space, promote the continuation of urban context and vitality regeneration of street space; 3. Develop three-dimensional and multi-dimensional space system design. By enhancing the connectivity of the pedestrian transportation system, different cultural landscape Spaces are connected and the guidance of space sightseeing is enhanced to facilitate the development of tourism activities; 4. We will strengthen social management mechanisms. To promote cultural heritage protection and sustainable development research through policy and institutional management and strengthening its rectification effect.

Acknowledgement. The writing and research of this paper are obtained from National Key R&D Program of China (2021YFE0200100) and 2021 Policy Directed Program of Jiangsu Province (BZ2021015).

References

1. Rong, Z.: Research on the Protection and Management of The World Cultural Heritage of Macao. Social Sciences Dedicated to publishing house, Beijing (2015)
2. Yao, W., Yishan, M.: The rostoration, conservation and tourism development of the cultural heritage of residential architecture in macau: taking the mandarin's house as an example. Art J. **123**(06), 026–032 (2020)
3. Government of Macau Special Administrative Region: Draft Consultation Text of the Master Plan of the Macao Special Administrative Region (2020–2040). Land, Public Works and Transport Bureau, Macau (2020)
4. Jorgensen, A.: Beyond the view: future directions in landscape aesthetics research. Landsc. Urban Plan. **100**(4), 353–355 (2011)
5. Yaohong, L.: Study on Landscape Control in the Historic Centre of Macau. Unpublished Master's Thesis. Huaqiao University, Xiamen, pp. 79 (2013)
6. Waldheim, C.: The Landscape Urbanism Reader. Princeton Architectural Press, New York (2006)
7. Waldheim, C.: Landscape as Urbanism: A General Theory. Princeton University Press, New York (2016)
8. Corner, J.: Recovering Landscape: Essays in Contemporary Landscape Architecture. Princeton Architectural Press, New York (2008)
9. Shannon, K., Smets, M.: The Landscape of Contemporary Infrastructure. NAI Publishers, Rotterdam (2010)

Check for updates

3D Printing of Fast Setting Basalt Fiber Reinforced Cement-Based Materials

Weihong Li(✉), Xuhao Chen, and Nan Li

College of Architectural Engineering, Dalian University, Dalian 116000, China
liweihong714@163.com

Abstract. In this study, the traditional 3D printing process was enhanced, the mixing and pumping processes were removed, and the concrete 3D printer with simultaneous mixing and extrusion of materials was independently developed. The printing of rapid-setting cement-based materials was successfully realized, and the printing performance of rapid-setting basalt fiber reinforced cement-based materials were tested. The results show that the mixing-extrusion integrated concrete 3D printer can successfully print the rapid solidification cement-based composite material with the final setting time of less than 5 min. Under the requirements of printing performance, the dosage range of basalt fiber is 0–2.0%, and the printing performance is the best when the dosage is 1.5%.

Keywords: 3D Printing · Mixing-extrusion integrated printer · Printability

1 Introduction

In recent years, 3D printing concrete technology has developed rapidly in the construction industry. The "additive manufacturing" technology of stacking concrete materials layer by layer to manufacture the required three-dimensional objects has the advantages of intelligence, rapid construction, and cost-saving [1–4]. Among them, one of the key technologies of 3D printing is that the printing material is required to adapt to the 3D printer. The printing material has good printing properties such as fluidity, extrusion, and construction, to ensure the continuous, smooth, non-collapse of the printing process and the stable formation of the printing components [5]. Most of the existing printing equipment is composed of four parts: storage system, pumping system, discharge system, and printing system. The stirred concrete slurry needs to be put into the storage bin, and the concrete is transported to the printer head through the pumping system. Finally, the printer head squeezes out the printing material to achieve 3D printing. This requires that the setting time of printing materials used cannot be too short, and it is difficult to print fast-setting cement-based materials. And due to the existence of time loss, resulting in material printing performance and late structure and seismic performance are affected, failed to give full play to the advantages of 3D printing technology rapid construction.

To realize the printing of quick-setting cement-based materials, this paper updates and improves the existing traditional 3D printing equipment, and independently develops a concrete 3D printer with simultaneous mixing and extrusion of materials. The

W. Guo and K. Qian (Eds.): GBCESC 2022, LNCE 211, pp. 57–64, 2023.
https://doi.org/10.1007/978-981-19-5217-3_6

quick-setting cement-based materials with a setting time of less than 5 min are prepared with sulphoaluminate cement, and basalt fiber is added to improve its mechanical properties and crack resistance. The printing performance is tested by the self-developed mixing-extrusion integrated 3D concrete printer, and the proportion meeting the printing performance is selected, which provides a reference for the subsequent practical engineering application [6].

2 Materials and Experimental Methods

2.1 Materials

In this experiment, composite sulphoaluminate cement (SAC), grade I fly ash, and silica fume are used as cementitious materials. The main chemical composition is shown in Table 1 Aggregate: quartz sand, with particle sizes of 20–40 meshes, 40–70 mesh, and 70–120 meshes, respectively. The main indexes are shown in Table 2 Water reducing agent with water reducing rate >30% and volume stabilizer (thickener and thixotropic agent, etc.) were used.

Table 1. Chemical composition of cementitious materials

Component	Average percentage/%		
	SAC	FA	SF
SiO_2	6.13	51.97	92.46
MgO	1.24	1.71	
Cao	42.54	6.27	
Fe_2O_3	1.53	5.27	
SO_3	10.79	1.46	
Al_2O_3	35.17	28.77	
TiO_2		1.66	
Loss	2.10	2.89	2.2

Table 2. Main physical and mechanical properties of aggregate

Refractoriness/°C	Uniformity/%	Hardness	Water content/%	Powder in sand/%
1700	90	6.7	1	0.4

Five cement-based materials with different basalt fiber volume fractions were prepared. The volume ratios of BF were 0%, 0.5%, 1.0%, 1.5%, 1.5% and 2.0%, respectively, and the test setting time was within 5 min. Table 3 shows the mix ratios of the materials. Using 6 mm basalt fiber, the main physical and mechanical properties are shown in Table 4.

Table 3. Mix proportion of 3D print fiber reinforced cement base material

Group	SAC/g	FA/g	SF/g	Sand/g	Water/g	SP/g	HPMC/g	VMA/g	Fiber/%
B0	400	100	40	600	160	2	0.4	4	0
B1	400	100	40	600	160	2	0.4	4	0.5
B2	400	100	40	600	160	2	0.4	4	1.0
B3	400	100	40	600	160	2	0.4	4	1.5
B4	400	100	40	600	160	2	0.4	4	2.0

Table 4. Main physical and mechanical properties of basalt fiber

Length/mm	Diameter/μm	Tensile strength/MPa	Elongation/%	GPA	Density/(g cm^{-3})
6	11	3900	2.7	100	2.64

3 The Mixing-Extrusion Integrated 3D Concrete Printer

The design of a 3D printer will directly affect the 3D printing process. In this study, the printability of composite sulphoaluminate cement-based materials was evaluated by using a concrete 3D printer (shown in Fig. 1) with material mixing and extrusion simultaneously. The printer adopts the way of mixing dry powder material and water in the printer head, stirring and extrusion at the same time, cancels the traditional pumping pipeline, and avoids the influence of transmission pipeline on the time of concrete material. When printing, only the dry powder material mixed by the ratio is loaded into the storage bin in advance, and transmitted into the print head through the dry powder circulation system. The dry powder is mixed with the water of the inlet pipe and stirred by the stirring plate group. The mixed wet material is deposited at the bottom of the stirring plate group, and then the printing head is extruded by the extrusion cavity spiral through the extrusion nozzle.

Through software programming, the printer sets printing parameters such as feed speed, layer height, and trajectory mode, converts the target model into the G-CODE code file identified by the machine tool, and the transmits it to the machine tool system through DNC for printing. The effective stroke of the gantry frame is 1.5 m (L) × 1.0 m (W) × 0.6 m (H), the rotation speed of the printer head is 60 r/min, and the extrusion diameter is 18 mm. The print system and print head structure are shown in Fig. 2.

The rotary printing head adopts the planetary wheel group structure. The outer cylinder wall rotates clockwise and the stirring plate group rotates counterclockwise, which not only enhances the stirring intensity of the material but also cooperates with the inlet

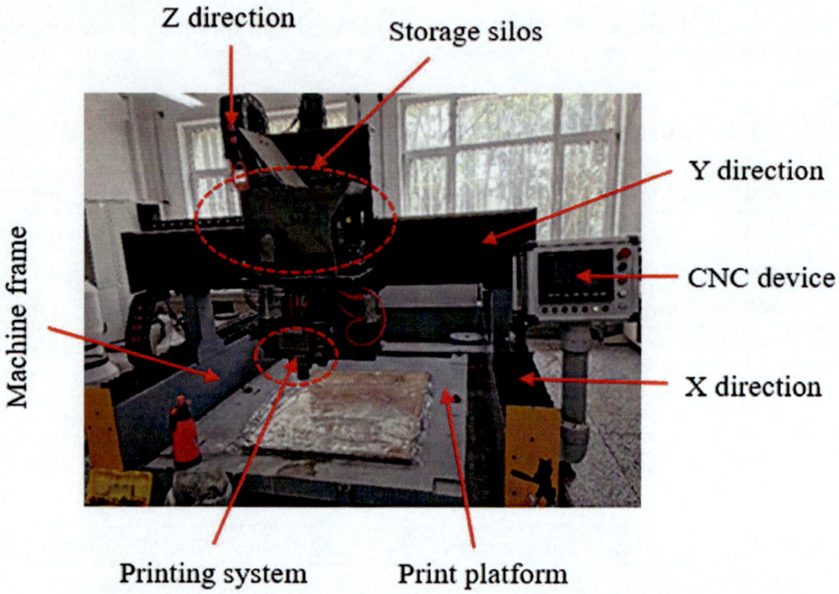

Fig. 1. The gantry mixing-extrusion integrated 3D concrete printer

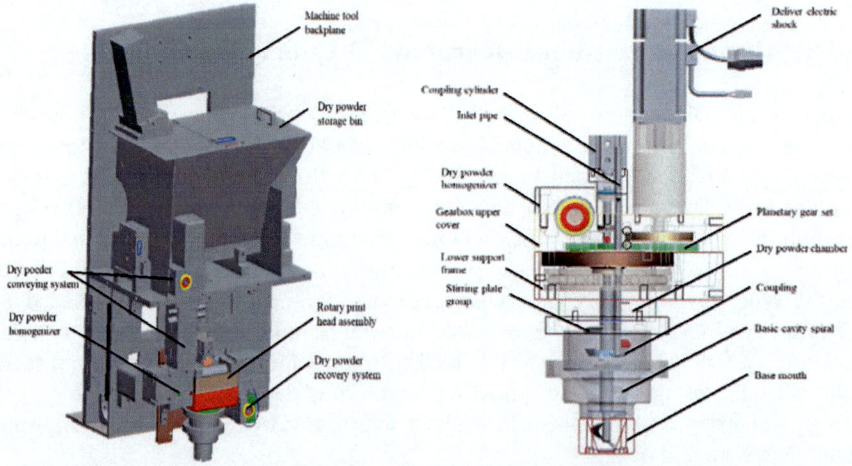

(a) Printing system structure diagram **(b)** Printhead head structure diagram

Fig. 2. Printing system and print head structure schematic

pipe wall on both sides to scrape the rapid solidification material adhered to the inner cylinder wall and keep the inner cylinder wall clean, to ensure the continuous printing. At the same time, since the wet materials are mainly concentrated around the extrusion cavity spiral, the dry powder is above the inlet pipe outlet is unbonded, and after printing, only the extrusion cavity spiral section needs to be removed and cleaned.

The printer effectively solves the problem of uneven stirring caused by short stirring time due to the use of 'water in powder falling plate'. The mixing, stirring and extrusion process of dry powder materials and water is only 3–5 s, which has unique advantages for printing quick-setting cement-based materials within 3–5 min of setting time. Combined with the characteristics of high hydration heat of quick-setting materials, 3D printing of buildings in extreme cold and other special environments can also be realized.

4 Printing Results and Discussion

4.1 Buildability Testing

Buildability refers to the ability to maintain stability without excessive deformation and collapse in the process of stacking layers [7, 8]. In this study, a hollow ring with a diameter of 300 mm, a layer height of 15 mm, and 20 layers were printed as shown in Fig. 3, and its construction was evaluated by observing whether it collapsed and deformed after molding.

After printing, the total height of the molded component was 296 mm, the theoretical height was 300 mm, and the change rate of the total height was 1.33%. Compared with the bottom diameter, the upper diameter has no change, the overall composition has no collapse phenomenon, and the construction is good.

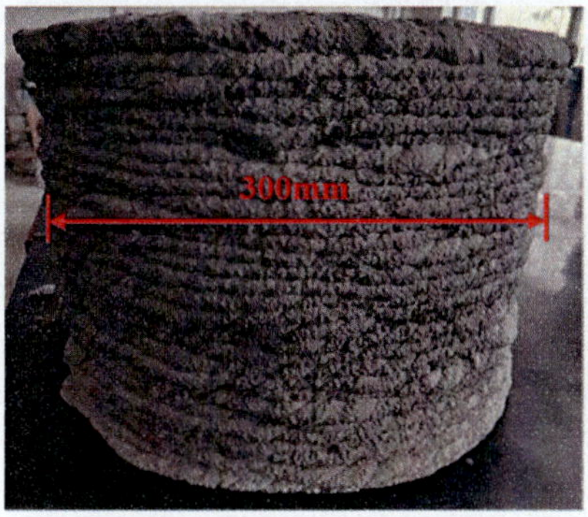

Fig. 3. Buildability testing

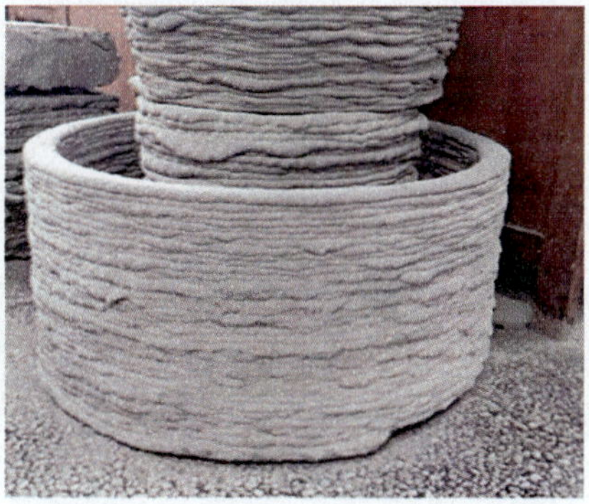

Fig. 4. Solid printing circular components

Figure 4 shows a hollow ring after continuous printing for 1 h. Due to the use of sulphoaluminate cement-based printing material, the water requirement of the material is large, the hydration reaction speed is fast, the hydration heat is large, and the water evaporation is too fast, resulting in atomization on the surface of the component. Therefore, this study adopts the method of water spray maintenance while printing in the printing process to effectively prevent the dehydration and cracking of the later specimens. It is also because of this feature that the material and printer can be used for printing in cold and underwater environments.

4.2 Effect of Fiber Content on Printability

In this study, the rectangular models of 800 mm × 200 mm × 48 mm in size and 12 mm in layer height were printed by five groups of material ratios shown in Table 3. The first layer was printed within 4 min, and the overall model was completed within 18 min. The printing process is shown in Fig. 5.

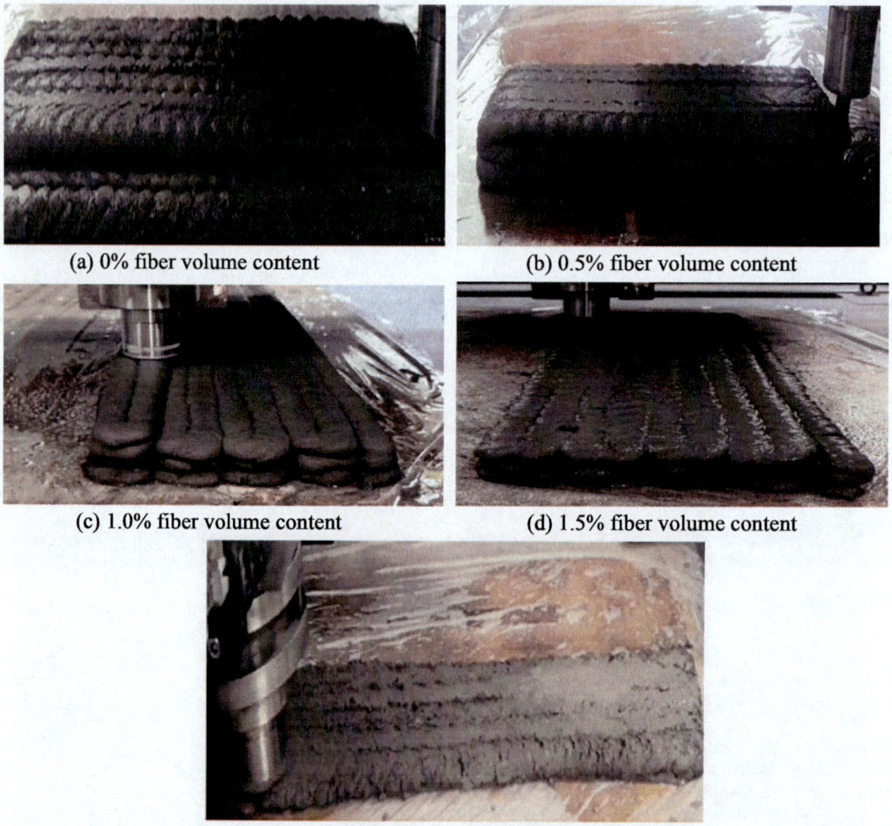

(a) 0% fiber volume content

(b) 0.5% fiber volume content

(c) 1.0% fiber volume content

(d) 1.5% fiber volume content

(e) 2.0% fiber volume content

Fig. 5. 3D printing of fiber volume content

It can be seen from Fig. 5 (a), (b) and (c) that the extrusion function of the rotary printer head can be realized by setting the height of the layer. When the layer height is high, because the rotary printer head adopts the spiral model, the component surface will be corrugated, which can enhance the interlayer bonding force. On the contrary, when the layer height is low, the rotary printer head has an extrusion effect on the printing material, and the printing layer can be flattened. It can be seen from Fig. 5 (d) that the edge of the component of the rotary printer head is rough in the printing process, and the leveling device will be prepared next to the rotary printer head to solve this problem.

In addition, it can be seen from Fig. 5 that with the increase of fiber content, the extrusion of the material decreases. As shown in Fig. 5 (e), when the fiber content reaches 2.0%, discontinuous fracture occurs when printing the first layer, failing the target size component. The reason for this phenomenon is that with the increase of fiber volume content, the working performance of printing materials decreases, and there will be intermittent and tearing phenomena in the extrusion process. At the same time, the fluidity of printing materials becomes worse, which increases the risk of printer blockage. In addition, adding too much fiber into cement-based materials will increase the porosity of the internal structure of the material and micro-cracks appear on the surface.

In summary, the printer can print the fiber volume content of 0%–2.0%, and the printing performance is the best when the basalt fiber content is 1.5%.

5 Conclusion

In this paper, five kinds of rapid setting fiber-reinforced cement-based printing materials with different basalt fiber contents were prepared based on the self-developed mixing-extrusion integrated 3D concrete printer, and the printability of rapid setting 3D basalt fiber reinforced cement-based materials were tested. The results show that:

The concrete 3D printer with mixing and extrusion of the material can realize the printing of quick setting basalt fiber reinforced cement-based materials with a final setting time of less than 5 min. With the increase of fiber volume content, the extrusion of cement-based composites decreases, and micro-cracks appear on the surface of components. Excessive fiber content will lead to printer blockage and printing interruption. When the volume content of basalt fiber is 0–2.0%, and the volume content of basalt fiber is 1.5%, the printing performance is the best. (2) When the basalt fiber volume content is 1.5%, the compressive strength of 3D printing composite cement-based material is the largest, and the flexural strength decreases slightly compared with the fiber volume content of 1.0%, and the decrease is less than 5.0%. Therefore, as fiber-reinforced rapid solidification 3D printing cement-based material used in this study with the mixing-extrusion integrated 3D concrete printer, the optimal volume content of basalt fiber is 1.5%.

In addition, using this printer, magnesium phosphate cement printing with light burned magnesia as raw material, without retarder and setting time less than 1 min was successfully realized (related research results, discussed in another paper), which solved the problem that rapid setting magnesium phosphate cement cannot be applied to 3D printing.

Acknowledgment. The authors would like to acknowledge the support by the National Nature Foundation Major Fund of China (52032011); Scientific Research Project of Liaoning Provincial Department of Education (LJKZ1177).

References

1. Cesaretti, G., et al.: Building components for an outpost on the lunar soil using a novel 3D printing technology. Acta Astronaut. **2014**(93), 430–450 (2014)

2. Pegna, J.: Exploratory investigation of solid freeform construction. Autom. Constr. **5**(5), 427–437 (1997)
3. Wolfs, R.J.M., Bos, F.P., Salet, T.A.M.: Early age mechanical behaviour of 3D printed concrete: numerical modelling and experimental testing. Cem. Concr. Res **106**, 103–116 (2018)
4. Bos, F., Wolfs, R., Ahmed, Z., Salet, T.: Additive manufacturing of concrete in construction: potentials and challenges of 3D concrete printing. Virtual Phys. Prototyp. **11**(3), 209–225 (2016)
5. Ma, G., Li, Z., Wang, L., Fang, W., Jay, S.: Mechanical anisotropy of aligned fiber reinforced composite for extrusion-based 3D printing. Constr. Build. Mater. **202**, 770–783 (2019). https://doi.org/10.1016/j.conbuildmat.2019.01.008
6. Ranjbar, N., Zhang, M.: Fiber-reinforced geopolymer composites: a review. Cem. Concr. Compos. **107**, 103498 (2020). https://doi.org/10.1016/j.cemconcomp.2019.103498
7. Wang, L., Hua, M., Zhijian, L., Guowei, M.: Cementitious composites blending with high belite sulfoaluminate and medium-heat Portland cements for largescale 3D printing. Addit. Manuf. **46**, 102189 (2021)
8. Thanh, L., et al.: Mix design and fresh properties for high-performance printing concrete. Mater. Struct. **45**(8), 1221–1232 (2012)

Using Nighttime Light Data to Achieve Precise Estimation of Residents' Living Carbon Emissions and Landscape Pattern Characteristics

Huan Chen[1,2](✉), Biyan Wang[1], Weiwu Wang[1,2], and Shan Wang[1,2]

[1] College of Civil Engineering and Architecture, Zhejiang University, Hangzhou 310058, China
chenhuan01@zju.edu.cn

[2] Center for Balance Architecture, Zhejiang University, Hangzhou 310058, China

Abstract. A method for estimating residents' living carbon emissions and analyzing the spatiotemporal changes of carbon emissions landscape patterns using multi-source data at the township or street scale is proposed. After overlaying the registration of nighttime light data with the registration of Hangzhou township and street-level administrative divisions, the distribution of carbon emissions and carbon emission intensity divisions of the township or street residents in Hangzhou were obtained. Fragstats software was used to analyze the landscape pattern characteristics of living carbon emissions of residents in townships and streets in Hangzhou in 2010, 2015, 2020. The results show that: (1) There is a high Power-law correlation between the living carbon emissions of the township or street residents in Hangzhou and the population distribution, showing a three-layer circle structure. (2) High-intensity areas are concentrated in the northeastern part of the main urban area of Hangzhou, and continue to expand to the northeast, showing the transfer of low-value carbon emission intensity areas and the transfer of high-value carbon emission intensity areas. (3) The landscape pattern of carbon emission intensity has gradually shifted from single-patch control to multi-patch control, among which low-value carbon emission intensity landscape types still play a significant role in controlling urban areas, accounting for more than 70%, and the degree of fragmentation is increasing year by year.

Keywords: Nighttime light data · Township or street scale · Urban residents' living energy consumption carbon emissions · Landscape pattern characteristics

1 Introduction

1.1 A Subsection Sample

Global warming and its impact on social production activities have aroused widespread concern in the international community, and just like that, in recent years, countries have advocated a low-carbon economy and green development. September 2020, President Xi Jinping announced at the 75th United Nations General Assembly that China's carbon

W. Guo and K. Qian (Eds.): GBCESC 2022, LNCE 211, pp. 65–77, 2023.
https://doi.org/10.1007/978-981-19-5217-3_7

dioxide emissions will peak by 2030 and strive to achieve carbon neutrality by 2060. Many carbon emissions are attributed to the burning of energy in human economic activities [1]. In 2019, fossil fuels associated with human activities resulted in global carbon emissions reaching 34.36 billion tons. In 2020, despite the impact of the COVID-19, global carbon emissions will still get 32.28 billion tons, of which the global carbon emissions caused by natural resources, power generation and heating account for 43% [2]. It can be seen that analyzing the interaction between human economic activities and carbon dioxide emissions is a benign interaction mechanism for the coordinated development of people and the environment. It is also the fundamental choice for the "dual-carbon goal" of future scientific planning for green cities.

Human economic activities are closely related to nighttime lights, and the scale of nighttime lights can be used as an essential factor to reflect the intensity of human economic activities. Luminous images can capture information such as night town lights, fishing boats at night, natural gas burning, forest fires and other information and are widely used in economic growth assessment, power consumption simulation, etc.At present, most of the research on carbon emission analysis using night light data at home and abroad takes countries, provinces and regional cities as the research objects for calculation and analysis. On the global scale, Doll et al. (2000) [3] selected countries in different development stages as research objects and found that the regression coefficient of determination between the carbon emissions and the logarithm of the night lighting area in all countries in the world reached more than 0.84 and based on this Using the DMSP/OLS night light area, a linear distribution model was used to draw a $1^\circ \times 1^\circ$ resolution grid map of global carbon emissions in 1995. At the national scale, Zhang Q et al. [4] used nighttime light data to map the urbanization dynamics of India, China, Japan and the United States. The differences in urbanization trajectories from 1992 to 2008 explained the internal mechanism affecting urbanization dynamics; Ma Zhongyu et al. [5] (2017) constructed a carbon emission simulation model for China's provinces based on satellite nighttime light data and the overall, annual, local. Annual and provincial carbon emissions estimation results the model runs Works well. At the regional scale, Lv Qian et al. [6] (2020) used a four-step method to fuse and correct DMSP/OLS and NPP/VIIRS nighttime light data and combined the provincial energy consumption statistical carbon emission data to construct a multi-scale carbon emission estimate in the Yellow River Basin Model; Sun Guiyan et al. [7] (2020) based on night light images, estimated the carbon emissions of energy consumption in the upper reaches of the Yangtze River from 2000 to 2017, and studied the spatial characteristics and influencing factors of carbon emissions from energy consumption. Judging from the current research progress, most of the calculation of carbon emission data is based on the macro scale. The current energy consumption statistics are only perfected to the prefecture-level city level. In contrast, the lack of energy consumption statistics at the township and county-level administrative divisions and the different statistical calibres have seriously hindered the research process of small-scale carbon emission quantitative calculation.

2 Study Area Overview and Data Sources

2.1 Study Area

The research site is located in Hangzhou City, Zhejiang Province, China (Fig. 1), between 118°21′–120°30′ east longitude and 29°11′–30°33′ north latitude, located in the southern wing of the Yangtze River Delta in China and the western end of Hangzhou Bay, the lower reaches of the Qiantang River, the southern end of the Beijing-Hangzhou Grand Canal, with a total area of 16,850 km^2. It is an important central city in the Yangtze River Delta and a transportation hub in southeastern China. The study area is divided into two levels, namely the city-wide (including 10 municipal districts, 2 counties and 1 county-level city) and the urban area (including 10 municipal districts).

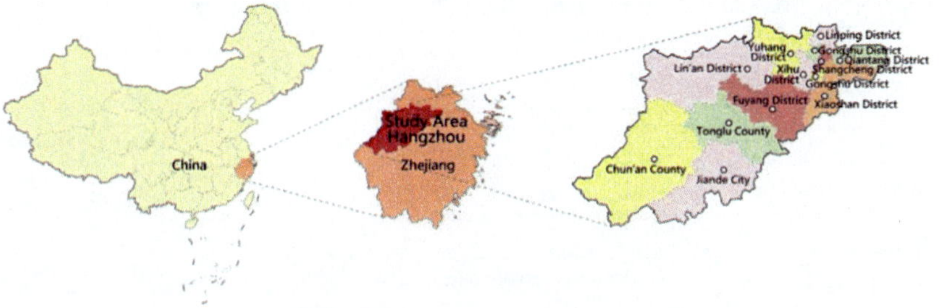

Fig. 1. Study area overview.

2.2 Data Sources

The experimental data includes night light data (2010 is DMSP/OLS data, 2015, 2020 is NPP/VIIRS data, data preprocessing such as noise reduction, correction, and desaturation is performed on the luminous data), Hangzhou township street data (subject to the 2020 Hangzhou administrative division), Hangzhou's resident population data at the township level and the electricity consumption data of urban and rural residents.

Since the precision of the annual raster data of the night lights is 1000 m, which is too large for the research scale of this paper, the raster data is refined in this paper, and the raster is resampled into 100 m precision units. The township and street data and sub-regional resident population data in Hangzhou are based on the national census data, supplemented by the statistical yearbooks of Hangzhou and its subordinate districts and counties and the statistical bulletin of national economic and social development. Collect 196 township-level regions. Resident population. According to the 2011 and 2016 Statistical Yearbooks of Hangzhou and the 2020 Statistical Bulletin of National Economic and Social Development, the electricity consumption of urban and rural residents in Hangzhou was 6.345 billion kWh in 2010, and the electricity consumption of urban and rural residents in Hangzhou in 2015 was 90.26 TWh. In 2020, Hangzhou's urban and rural residents will consume 14.4 billion kWh of electricity.

3 Research Methods

The technical route of this study is shown in Fig. 2. The first is to invert and calculate the carbon emissions of urban and rural residents in the streets and towns of Hangzhou [8]; the second is to classify the carbon emission intensity of residents in the urban area of Hangzhou; the last is to analyze the urban area (10 districts of Hangzhou) from the perspective of landscape pattern index.) landscape pattern change characteristics of residents' carbon emissions [9]. Due to the lack of the consumption structure and specific statistical data of residents' energy consumption at the township and street level, electricity consumption accounts for a large part of the residents' energy consumption. It is increasing year by year [10]. Therefore, this paper uses electricity consumption of urban and rural residents to replace energy consumption to calculate the carbon emissions of urban and rural residents at the scale of townships and streets in Hangzhou. Still, the substitution of this data does not affect the adaptability of this paper.

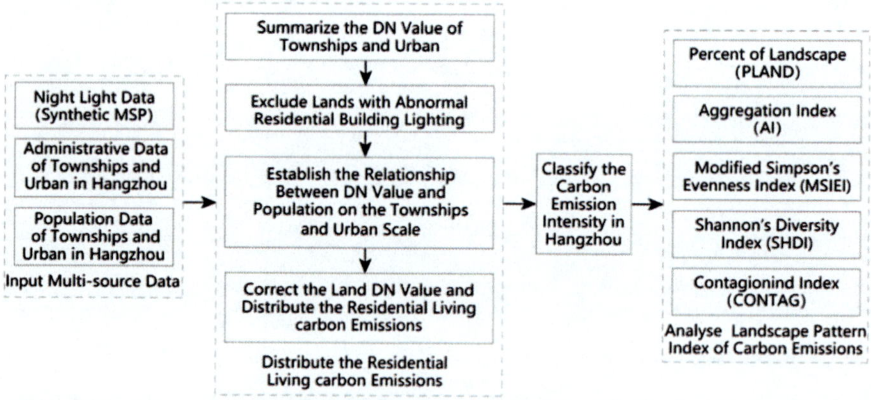

Fig. 2. Research methods and technical routes.

3.1 Street-Scale Residential Carbon Emissions Allocation Method

It is assumed that the weight of the light value of the plot corrected by the population of the plot to the total light value represents the weight of the energy consumption of the plot to the total energy consumption so that the total energy consumption of urban residents is allocated to each plot.

Fitting of Light DN Total Value to Population. First of all, manual interpretation should be carried out. The plots with abnormal lighting of residential buildings should be eliminated or appropriately reduced, including plots with intensive commercial activities, mainly lakes, sparsely populated and those with transportation hubs. However, due to the lack of data on the time of the study and the nature of plot planning, no relevant studies have proposed a method for determining the reduction factor. Therefore, the manual interpretation method in this study is relatively rough, that is, excluding the plots containing large business circles and transportation hubs (mainly the streets under the jurisdiction of Hangzhou urban area), and excluding the plots containing a large number of mountains, lakes and industrial groups (mainly Hangzhou Suburbs govern townships).

The correlation between the population data and the total light DN value of the township streets in Hangzhou was tested, the correlation between the population and the total light DN value was discussed, and a suitable fitting model was selected to correct the light value of the township streets. The three correlations are power-law correlation, linear correlation, and logarithmic correlation. The dependent variable is the resident population data of the 117 township-level census areas in the seventh Hangzhou City after manual interpretation. The independent variable y is the total DN value of lights in all 117 township-level administrative districts in Hangzhou City.

In Excel, the power-law correlation, linear correlation and logarithmic correlation were explored for the above two groups of variables, respectively. The fitting relationship between the population of Hangzhou and the total value of lighting DN was obtained. At the township street scale, it is found that in 2020, the power-law fitting accuracy R^2 of the population and light DN values in Hangzhou township streets is 0.7838 (Fig. 3a), the fitting accuracy R^2 of the linear correlation is 0.4551 (Fig. 3b), the fitting accuracy R^2 of the logarithmic correlation is 0.4839 (Fig. 3c). In the existing research, Wang Meiling et al. [11] (2021) conducted a fitting study on the light value and population space of the NPP/VIIRS 4 districts, and the fitting accuracy was 0.714–0.968; Jia Tao et al. [12] (2020). The power-law relationship between the population of the plot and the light intensity was fitted, and the fitting accuracy R^2 was 0.35; Zhang Qiuyuan et al. [13] (2019) obtained the population distribution of Guizhou Province based on the DMSP/OLS nighttime light data, and obtained the total light intensity and the resident population. The correlation coefficient of 0.881.

Therefore, it can be considered that in this study, in 2020, there is an excellent power-law fitting relationship between the population and lighting of township streets in Hangzhou, and the appropriate model is as shown in Eq. (2). It is worth noting that the fitting relationship between the population of towns and streets in Hangzhou and the DN value of lights is different each year. In 2015, there was an excellent power-law relationship between the population and lighting of township streets in Hangzhou. The fitting accuracy R^2 was 0.8103, and the fitting Equation was as shown in Eq. (3). In the power-law relationship, the fitting accuracy R^2 is 0.6883, and the fitting Equation is shown in Eq. (4).

$$DN_i' = 0.0942 * P_i^{0.8606} \tag{1}$$

$$DN_i' = 0.0789 * P_i^{0.9166} \tag{2}$$

$$DN_i' = 0.0643 * P_i^{0.881} \tag{3}$$

where: DN_i' is the total value of pixel grey level in the ith administrative division after resampling, P_i is the resident population of each township and street.

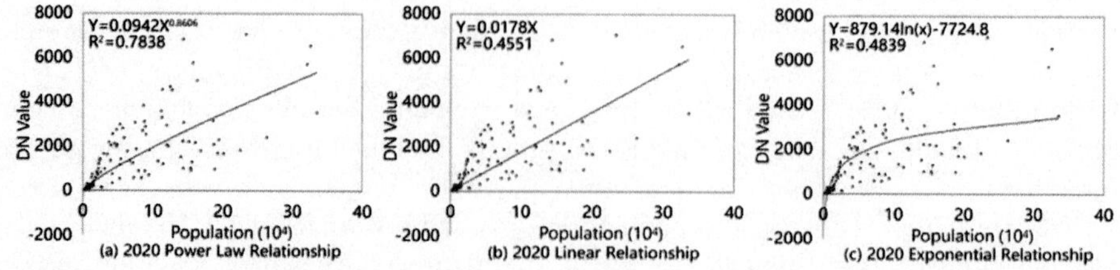

Fig. 3. Fitting of three correlations between the population of towns and streets in Hangzhou and the DN value of lights in 2020.

Distribution of Carbon Emission and Carbon Emission Intensity Division of Residents' Living at the Township and Street Scale The distribution method of residential carbon emissions at the township and street level is as follows: 1) Overlay nighttime light images and administrative division data, and calculate the total DN of nighttime lights in Hangzhou and its subordinate townships and streets according to the spatial distribution. 2) The corrected light values of all township streets are calculated using the population-night light DN value fitting model (such as formula (2), the state of the fitting model in different years is different); 3) According to the corrected light values of township streets, the proportion of Hangzhou city The weight of the total light value is used to allocate the electric energy consumption of urban and rural residents to each plot, as shown in formula (5); the carbon emissions of each plot are obtained according to the electric energy carbon emission conversion coefficient 28912 KG/TJ provided by IPCC.

$$f_i = E * EF * DN_i' / \sum DN_i' \tag{4}$$

where: f_i is the living carbon emissions of residents in the ith township street; E is the electric energy consumption of urban and rural residents; EF is the carbon emission conversion coefficient of electric energy consumption.

Based on the distribution method of residents' living carbon emissions at the township and street scale, the spatial distribution of residents' living carbon emissions in the towns and streets of Hangzhou in 2010, 2015 and 2020 was inverted. Then compare the carbon emissions of townships and streets with their administrative area to get the carbon emissions intensity per unit area. According to the carbon emission intensity of residents' life, the study area is divided into 5-grade areas (Table 1). According to this level, the distribution of Hangzhou residents' carbon emission intensity level in different periods is obtained.

Table 1. Residential life carbon emission intensity grade division.

Grade	Font size and style
low carbon emission zone	0 kg/ha ≤ carbon emission intensity < 2000 kg/ha
Low-medium carbon emission area	2000 kg/ha ≤ carbon emission intensity < 4000 kg/ha
Medium carbon emission area	4000 kg/ha ≤ carbon emission intensity < 6000 kg/ha
Medium-high carbon emission areas	6000 kg/ha ≤ carbon emission intensity < 8000 kg/ha
High carbon emission area	8000 kg/ha ≤ carbon emission intensity

3.2 Analysis on the Landscape Pattern Characteristics of Residents' Living Carbon Emissions in Hangzhou

Landscape pattern analysis is an analysis method to explore the spatial configuration and changes between different landscape elements by calculating pattern indices. From the landscape pattern index perspective, this study analyzes the temporal and spatial changes of the landscape pattern of urban residents' carbon emissions in Hangzhou. Reference related research, this paper selects five landscape indices: Landscape Area Ratio (PLAND), Aggregation Index (AI), Landscape Aggregation Index (CONTAG), Shannon Diversity (SHDI) and Modified Simpson Evenness Index (MSIEI) to explore The spatial and temporal evolution characteristics of carbon emissions of residents of all levels in the study area, such as fragmentation degree, area ratio, aggregation degree and diversity.

4 Results Analysis

4.1 Analysis of the Residential Carbon Emission Distribution in Hangzhou

It is known that in 2020, the urban and rural residents of Hangzhou will consume 14.4 billion kWh of electricity, an increase of 4.7%. According to Eq. (3), the total carbon emission of residents in Hangzhou is allocated to each township and street to obtain the carbon emission of residents at the scale of township and street (Fig. 4).

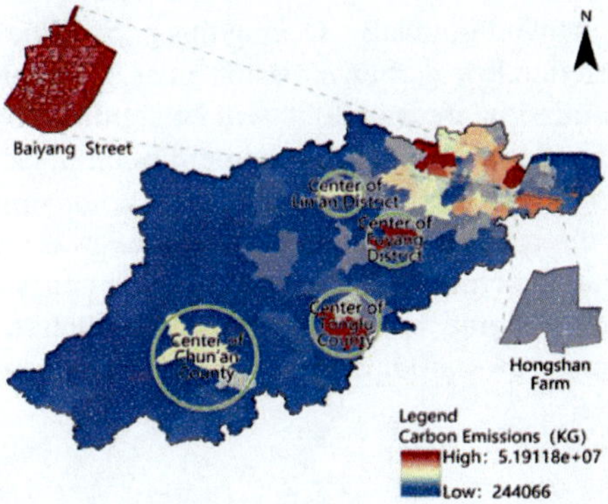

Fig. 4. Spatial distribution of living carbon emissions of residents in Hangzhou.

As can be seen from Fig. 4, the living carbon emissions of residents in towns and streets in Hangzhou in 2020 have the following characteristics: 1) The overall structure is a three-layer circle, and the central layer is the area east of the West Lake Scenic Spot in the old city and north of the Qianjiang River. (West Lake Street is the most typical), affected by factors such as West Lake's control of the surrounding building height and plot ratio and the urban functions undertaken by the street, it is a low-value area for residents' living carbon emissions; the central layer is Yuhang District, Lin Hangzhou New City, surrounded by Ping District, Qiantang District, and Xiaoshan District, is an emerging development area in Hangzhou, which is a high-value area for residents' living carbon emissions, and showed a decreasing trend from the northeast township to the southwest. There were high-value areas of residential carbon emissions in the central areas of Lin'an District, Fuyang District, Tonglu County and Chun'an County. 2) Township streets with dense residential buildings have higher carbon emissions, such as Baiyang Street, while townships and streets with sparse residential buildings have low carbon emissions, such as Hongshan Farm. 3) The townships and streets with high carbon emissions on the southwestern edge of Hangzhou are often due to the large land area. They contain many people, so the light value is high, and the significant carbon emissions.

4.2 Temporal and Spatial Variation Characteristics of Residential Carbon Emission Intensity in Hangzhou City

From 2010 to 2020, the temporal and spatial changes of residents' residential carbon emission intensity in Hangzhou can be divided into two trends (Fig. 5). One is that residents' carbon emission intensity o\ in non-urban areas is steady-state, mainly due to the low-value carbon emission intensity in the central area of Tonglu County. They are converted to medium and low carbon emission intensity. Second, the carbon emission intensity of residents' living in urban areas has undergone substantial changes, reflected explicitly in the following contents.

The first is that the number of residential carbon emissions of residents with different intensity levels in the urban area of Hangzhou has changed dramatically in different periods (Table 2). Compared with 2010, the low-carbon emission areas in 2015 increased significantly by 6.38%, and the high carbon emission areas increased by 6.38%. The proportion of the area decreased by 6.38%, and the proportion of low-, medium-, and high-carbon emission areas changed gradually. During this period, the high-carbon emission areas were mainly turned into low-carbon emission areas. Compared with 2015, the proportion of low-carbon emission areas in 2020 will be significantly reduced by 32.62%. The proportion of medium-low and high-carbon emission areas will increase significantly by 6.38% and 24.82%. The medium- and high-carbon emission areas increased steadily, and the transition from low-carbon emission areas was the mainstay during this period. It can be seen that the changes in the living carbon emissions of urban residents in Hangzhou are relatively significant, and the overall situation is unstable. Still, the main changes are reducing low-carbon emission areas and the increase of high-carbon emission areas.

Then, the median carbon emission intensity area, the medium and high-value carbon emission intensity area and the high-value carbon emission intensity area are mainly concentrated in the northeast of Hangzhou urban area and expand to the northeast. Among them, the distribution of medium and high value carbon emission intensity areas There are apparent changes: in 2010, the distribution area was small and relatively concentrated, mainly in Gongshu District, the southern part of Shangcheng District, the northeastern part of Xihu District, the eastern part of Binjiang District and the southwest part of Yuhang District; Extending on both sides, it is mainly distributed in the west of Gongshu District, the south of Shangcheng District, a small part of the north of Xihu District, Qiantang District along the Qiantang River, the west and east of Binjiang District and a small part of the east of Lin'an District; 2020 The annual distribution area has expanded significantly and re-concentrated to the southeast central area compared to 2015, mainly in Gongshu District, Shangcheng District, Binjiang District, the north side of Xihu District, the central part of Xiaoshan District near Binjiang District and the Qiantang River in the middle of Fuyang District. Nearby areas. Based on the spatial distribution of urban residents' carbon emission intensity in Hangzhou, it can be seen that Gongshu District, Shangcheng District and the northern part of Xihu District have always been in the areas of medium and high carbon emission intensity. Construction, commercial facilities, etc., with frequent economic activities, resulting in many carbon emissions from residents' lives, forming a medium-to-high-value carbon emission intensity area.

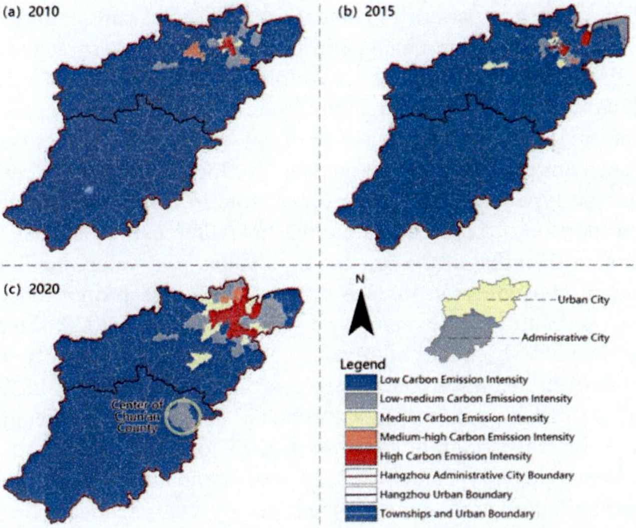

Fig. 5. Zoning map of residential carbon emission intensity level of Hangzhou residents

Table 2. Statistical table of the grading of carbon emission intensity of residents in Hangzhou urban area.

Residential carbon emission level area	2010		2015		2020	
	Number	%	Number	%	Number	%
Low carbon emission zone	94	66.67	103	73.05	57	40.43
Medium-low carbon emission area	14	9.93	12	8.51	21	14.89
Medium carbon emission area	3	2.13	6	4.26	10	7.09
Medium-high carbon emission area	6	4.26	5	3.55	3	2.13
High carbon emission area	24	17.02	15	10.64	50	35.46

Note: The statistical scope of this table covers 141 townships and streets under the jurisdiction of 10 urban districts in Hangzhou.

4.3 Analysis of the Results of the Carbon Emission Distribution of Hangzhou Residents

Hangzhou residents' living carbon emission intensity shows that the residents' living carbon emission intensity changes are concentrated in the urban area of Hangzhou, while the non-urban residents' living carbon emission intensity of Hangzhou is in a steady state. Therefore, it focuses on analyzing the characteristics of the urban residents' living carbon emission landscape pattern in Hangzhou. The urban carbon emission intensity of Hangzhou City has apparent changes in the landscape pattern at the type level in different periods (Table 3). The details are as follows:

First, the landscape pattern of residential carbon emission in urban areas of Hangzhou has gradually shifted from a single patch type of low-value carbon emission intensity to a landscape pattern controlled by multiple patches, in which the low-value carbon emission intensity landscape type still plays a significant role in controlling the urban area. In terms of the proportion of landscape area, the low-value carbon emission intensity is significantly higher than that of the low-medium value, mid-value, medium-high value and high-value carbon emission intensity, with an average proportion of more than 70%; however, on the time scale, compared with In 2010 and 2020, the proportion of low-, medium-, medium-high and high-value carbon emission intensity in urban areas will increase relatively. Low-carbon emission areas will decrease from 90% to about 75%. The revised Simpson evenness index value did not change significantly from 2010 to 2015. Still, it increased significantly from 2015 to 2020, indicating that the area distribution between patch types before 2015 was dominated by one patch type. After 2015, the area distribution among the plaque types gradually tended to be uniform.

Secondly, the fragmentation degree of the landscape pattern of residential carbon emission intensity in the urban areas of Hangzhou gradually increased, forming a dense pattern with multiple elements. From the characteristics of the changes in the aggregation of landscape patches, it can be seen that the overall landscape aggregation index shows a downward trend over time, from 90% in 2010 to about 70% in 2020. From the perspective of landscape heterogeneity, its Shannon The overall diversity value showed a trend of increasing with the increase of the year, and the large span increased significantly from

2015 to 2020, indicating that the type of plaques increased significantly, the degree of fragmentation increased. The information content of its uncertainty also gradually increased.

Table 3. Landscape pattern index of urban residents' living carbon emissions in Hangzhou.

Residential carbon emission level area	2010 (%)	2015 (%)	2020 (%)
PLAND (low carbon emission zone)	92.55	92.09	73.92
PLAND (medium-low carbon emission area)	4.40	2.28	12.47
PLAND (medium carbon emission area)	0.49	0.70	5.62
PLAND (medium-high carbon emission area)	1.68	2.91	1.10
PLAND (high carbon emission area)	0.87	2.00	6.89
SHDI	34.51	37.80	87.87
MSIEI	9.45	10.11	34.93
CONTAG	88.55	87.50	71.46
AI	99.51	99.51	99.30

5 Conclusion

The research results show that: (1) At the township and street scale, the residential carbon emissions of Hangzhou residents have a high power-law correlation with the population distribution. At the township and street scale, there is a three-layer circle structure. The volume is small, and the living carbon emissions of residents in the new urban area are massive, while the living carbon emissions of residents in non-urban areas are the lowest. (2) From 2010 to 2020, the residential carbon emission intensity of non-urban residents in Hangzhou was in a steady state; while the residential carbon emission intensity of urban residents changed significantly, and the median, medium-high and high-value carbon emission intensity areas were mainly concentrated in the northeastern part of Hangzhou urban area. It expands to the northeast, mainly manifested in the transfer out of low-value carbon emission intensity areas and the inflow of high-value carbon emission intensity areas. (3) The characteristics of changes in the landscape pattern of urban residents' life carbon emission intensity in Hangzhou. First, the landscape pattern of carbon emission intensity has gradually shifted from single-patch type control to multi-patch control. The low-value carbon emission intensity landscape type is still in the central control of the urban area. Second, the fragmentation degree of the landscape pattern of carbon emission intensity gradually increased, forming a dense pattern with multiple elements.

6 Discussion

This paper superimposes the DMSP/OLS and NPP/VIIRS nighttime light data after manual interpretation, fitting and correction, and the data of Hangzhou administrative divisions and provides a feasible method for using nighttime light data to invert residential energy consumption and carbon emissions. The research results can provide a basis for management decisions such as carbon emission control, resource allocation and low-carbon economic model at the township scale in Hangzhou. However, on the one hand, the residential space carbon emission data used in this study is relatively simple and cannot entirely reflect the actual development of residential space in Hangzhou. For example, the lighting data, traffic, industry, and other factors to the residential carbon emission should be included.; On the other hand, due to the unavailability of energy consumption data of individual residential buildings, this study uses the energy consumption data of all urban residents to carry out the inversion calculation of carbon emissions, in which the plot-level energy consumption using night light data The accuracy of the assignment model is not high and needs further improvement. Therefore, the next step will be to improve further the essential morphological factors, socioeconomic factors and plot planning factors of buildings and use data such as land types to study the advanced inversion, spatiotemporal analysis, and texture interpretation of residential buildings energy consumption in Hangzhou. And through a sampling survey, the actual energy consumption data of residential buildings and the consumption habits of residents and other factors that indirectly affect energy consumption are verified to improve the calculation accuracy and applicability of the model.

References

1. Shaojian, W., Guangdong, L., Chuanglin, F.: Urbanization, economic growth, energy consumption, and CO2 emissions: empirical evidence from countries with different income levels. Renew. Sustain. Energy Rev. **81**(pt. 2), 2144–2159 (2018)
2. BP: World Energy Statistical Yearbook, 2021 Edition. BP, London (2021)
3. Doll, C., et al.: Night-time imagery as a tool for global mapping of socioeconomic parameters and greenhouse gas emissions. AMBIO—A Journal of the Human Environment **29**(3), 157–162 (2000)
4. Qingling, Z., Karn, C.S.: Mapping urbanization dynamics at regional and global scales using multi-temporal DMSP/OLS nighttime light data. Remote Sens. Environ. **115**(9), 2320–2329 (2011)
5. Zhongyu, M., Hongwei, X.: Space-time simulation of carbon emissions in Chinese provinces based on satellite nighttime light data. China Population, Re-sources and Environment **27**(09), 143–150 (2017)
6. Qian, L., Haibin, L.: Multi-scale analysis of spatiotemporal evolution of carbon emissions from energy consumption in the Yellow River Basin based on nighttime light data. Econ. Geogr. **40**(12), 10 (2020)
7. Guiyan, S., Sheng, W., Lei, X.: Research on carbon emissions and influencing factors of energy consumption in the upper reaches of the Yangtze river based on night light data. Regional Research and Development **39**(4), 5 (2020)
8. Tao, J., Shihao, Y., Xin, L.: Inversion calculation and spatiotemporal analysis of carbon emissions from residential buildings in Wuhan. J. Earth Inf. Sci. **22**(5), 10 (2020)
9. Ting, P., Caige, S., Yongdong, Z., Fenglei, F.: Temporal and spatial variation of anthropogenic heat emission landscape patterns in the central city of Guangzhou. Journal of South China Normal University (Natural Science Edition) **53**(05), 92–102 (2021)

10. Jinzhao, S., Jiayao, Z., Xiaoping, W., Zhen, H.: Research on the potential of urban residential carbon emission reduction based on system dynamics: taking Xi'an as an example. Journal of Chongqing University of Technology (Natural Science) **33**(08), 213–221 (2019)
11. Meiling, W., Hesheng, Z.: Research on population spatialization based on Luojia No. 1 night light data. Geospatial Information **19**(09), 53–56+7 (2021)
12. Tao, J.: Inversion calculation and spatiotemporal analysis of carbon emissions from residential buildings in Wuhan. Journal of Geoinformation Science **22**(05), 1063–1072 (2020)
13. Qiuyuan, Z., Mingchun, P., Chongyun, W., Changwei, Z.: Analysis of population distribution and influencing factors in Guizhou province based on DMSP/OLS nighttime light data. Journal of Yunnan University (Natural Science Edition) **41**(05), 992–1000 (2019)

Research on the Kang in Cold Rural Area Based on Spatial Adaptation and Technological Improvement

Yiming Li[1], Hongpeng Xu[1,2], Zirui Fang[1], Chengyuan Bao[1,2], and Jianmei Wu[1,2](✉)

[1] School of Architecture, Harbin Institute of Technology, Harbin, China
wjmtutu@126.com

[2] Key Laboratory of Cold Region Urban and Rural Human Settlement Environment Science and Technology, Ministry of Industry and Information Technology, Harbin 150006, China

Abstract. Low technology, low cost, high thermal efficiency and other characteristics make cold rural inhabitants reluctant to give up the Kang heating method. The unique Kang technology in some cases has been research hotspots, however how to improve the Kang technology and adapt to the contemporary life is meaningful. Through the on-site investigation and questionnaire survey, it is found that the modern spatial feature is the key constraint of the Kang technology improvement. This paper developed an inheritance pattern of Kang integrated both spatial adaptation and heating effect into account. The numerical simulation was used to prove the heated kang compatible with the modern rural lifestyle, where two conditions were examined: 1)the temperature of the main room was increased by 6% and the temperature fluctuation was reduced, 2)the temperature of the secondary room was rised by 14%, achieving better thermal efficiency.

Keywords: Kang · Spatial adaptation · Technological improvement · Thermal efficiency

1 Introduction

In the past two thousand years, the kang has been one of the main heating methods in cold rural areas in China. According to statistics in 2004, 67 million kang was widely used by 1.75 million people [1]. In 2006, Tsinghua University also found that the utilization rate of kang reached 73.3% in North China, 86.3% in Northwest China and 96.2% in Northeast China [2]. On the basis of the data of 2019, the building area of rural heating in northern China was 7 billion square meters, where the proportion of heating with kang and earth heating used to be high [3]. Meanwhile, the kang also had an important role in reducing the burden of energy consumption and environmental pressure in China. As the social center of rural housing, kang also showed unique cultural values. In particular,

W. Guo and K. Qian (Eds.): GBCESC 2022, LNCE 211, pp. 78–91, 2023.
https://doi.org/10.1007/978-981-19-5217-3_8

kang was the place where many family activities took place. However, in recent years, with the improvement of rural living standards, kang began to conflict with the use of modern space. Thus, it is the key to put forward the improvement of traditional kang combining spatial adaptation and technical improvement, for the future renewal and transformation of heating mode of rural buildings in northern China.

Since the end of the 1970s, kang have been widely researched by many domestic scholars. On the one hand, the existing form of kang was optimized, in order to improve their thermal performance to improve the thermal environment quality of rural houses. Such as the smoke flow inside the kang was simulated and analyzed by computer software, and the optimal design of the internal structure was put forward to improve the kang surface temperature (Jie Zhang et al., 2019) [4]. On the other hand, a new form of kang was studied to improve the utilization rate.

For example, a solar kang heating system has been designed, which could provide domestic hot water in summer and heating in winter (Feng Jing et al., 2006) [5]. Additionally, some researches was mainly focused on measurements and evaluations, i.e., the thermal performance of heated wall and kang was evaluated and tested by experiments, to provide basic data for its popularization and application (Lin Duanmu et al., 2016) [6]. However, most of studies are mainly correlated with thermal performance of kang, combination of kang technology and modern rural housing pattern. Nevertheless, there are no research for behavioral and functional space that is affected heating equipment. Therefore, the purpose of this work is to investigate the inheritance mode of rural kang in cold areas, based on the perspective of the combination of spatial adaptation and technical improvement. To address this need, through fieldwork and questionnaire research, we put forward the improvement strategies of space and technology to adapt to modern rural life, and then carried out the simulation and verification of the thermal environment of the typical scheme.

2 Research on Needs and Issues

In order to sort out the actual needs of the inheritance of heating with Kang, we conducted a survey of Shijianfang in Shijian Village from December 2021 to January 2022. Shijian Village has 174 households including 435 people, the average household is about 2–3. We were based on 30 households in two types of family structures. Among the 174 houses, 50.6% of the buildings built before 2000, and the number of floors of traditional rural houses is mainly one (Table 1).

Table 1. Statistics on the age and number of floors

The age of the house construction				
	Before 1990	1990–2000	2000–2021	total
Number of households	34	54	86	174
Percentage	19.5	31	49.4	100
Statistics on the number of residential floors				
	Bungalow	Second floor	Three floors	total
Number of households	167	4	3	174
Percentage	96	2.3	1.7	100

2.1 Heated Kang is Still Popular

Within the 30 houses surveyed, 25 used traditional heated kangs and stoves, three added heated walls on this basis, other two used radiator and air conditioning. We got a whole year's heating time through research (Fig. 1-a). among 2 households started heating in September, 16 households started heating after the autumn harvest in October, and the other 12 households started heating in November. In February of the next year, 5 households stopped heating, 30 surveyed households finally stopped heating by early May. We also investigated the heating time of the day (Fig. 1-b), 28 households heated twice a day. The heating time was mostly 7:00 a.m. and 15:00 p.m. The indoor temperature of houses with traditional kang and stove was as low as 4 °C in the early morning, and can be maintained at 12 °C–16 °C after heating. In addition, the indoor temperature of residential buildings using kang and heating was 9 °C in the early morning, the temperature was mostly maintained at 18 °C after heating.

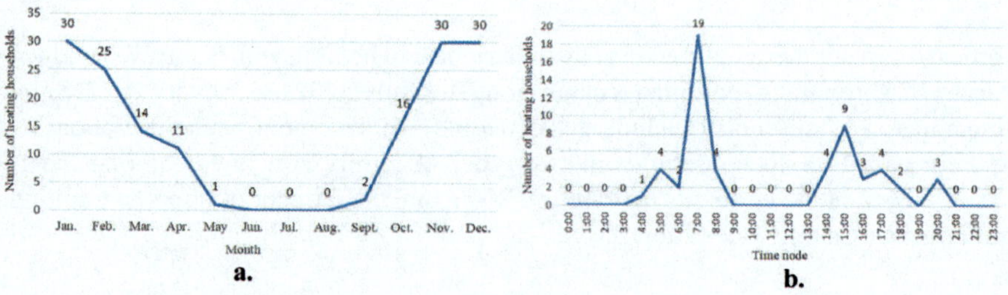

Fig. 1. -a. Annual heating schedule. -b. Daily heating schedule

The survey results show that the kang and heated wall are still used as a heating facility commonly used in rural areas in cold areas. Residents' winter daily activities are also mainly concentrated around the heated kang and heated wall, which possess the advantages that other heating facilities cannot surpass.

2.2 Traditional Heating Technology Leads to Serious Uneven Distribution of Spatial Heat

The kang and heated wall can be inherited because of their low cost and economical heating materials. However, there are many disadvantages, such as uneven indoor temperature, overheating near the heating equipment more than 70 °C. If people rely on it for a long time, it will lead to burns, but the temperature in the area far away from the heating equipment is low. Therefore, rural residents can only carry out daily life around heating equipment, which also leads to a contradiction with the production and life style of modern rural residents.

In addition, the residential area of more than 80 m^2 accounts for 63.3% of the total number of households (Table 2). This shows that the area of rural housing has increased. However, due to the low room temperature in rooms far away from heating equipment, many residents choose to add heated walls and kang in such rooms or directly use them as storage rooms.

Table 2. Residential floor area statistics

Residential floor area				
	$AA \leq 80\ m^2$	$80\ m^2 < AA \leq 100\ m^2$	$100\ m^2 < AA \leq 150\ m^2$	total
Number of households	11	13	6	30
Percentage	36.7	43.3	20	100

2.3 The Heating Method of the Kang and the Modern Rural Lifestyle Form a Spatial Contradiction

According to functional partitions, the houses in Shijian Village can be divided into main space (living room, bedroom) and secondary space (kitchen, storage room). With the improvement of living standards, villagers' living patterns are gradually enriched. However, there are few types of traditional residential space, and the functional zoning is not clear, so the life behavior is cross mixed. As a heating space equipment integrating sleep, leisure and catering in the lives of residents, the layout form of kang directly plays a role in organizing the overall space of living [7]. Consequently, the combination of kang and modern rural lifestyle is crucial.

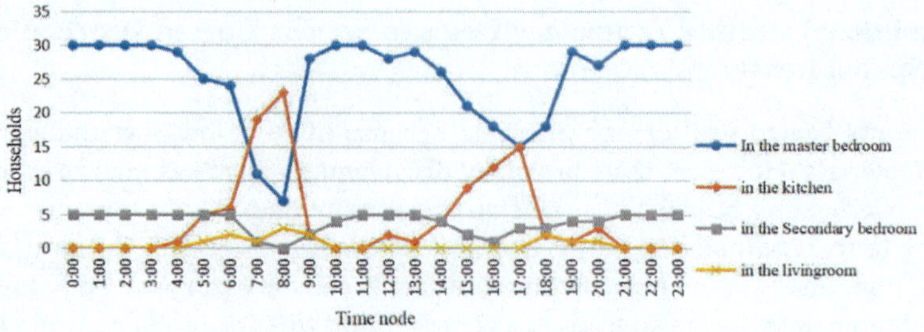

Fig. 2. Time-by-hour occupancy rate

Of the 30 dwellings surveyed, 7 had a living room, of which 6 used the living room to receive guests, and the other 1 had a low frequency due to the low temperature of the living room.

The functions of other houses are not ideal, the mode is relatively single, and there are few changes. Bedrooms are mainly used for daily life, leisure and entertainment. As shown in Fig. 2, From 20:00 p.m. to 6:00 a.m., about half of the villagers were concentrated in the master bedroom and the second bedroom in winter. villagers mainly stay in the kitchen from 6:00 a.m. to 8:00 a.m. and from 14:00 p.m. to 17:00 p.m. Among them, 4 households ate in the living room, 2 in the dining room, and the rest ate in the master bedroom due to heating problems.

After the interview, most residents expressed the need to use other types of rooms, such as living rooms and restaurants, and gradually developed the awareness of separating other life behaviors from the bedroom. Furthermore, in the traditional residence, the remaining space of the bedroom is large and the space utilization rate is low, resulting in the poor integrity of indoor thermal comfort [8]. Therefore, residential transformation is meaningful. After the transformation, we should also ensure the heating needs of different rooms. At the same time, rooms with temporary use requirements need to be combined with new heating equipment to meet their heating demands, so as to solve the problem that heating traditional heating facilities requires more time.

3 The Proposal of the Inheritance Model

From the above survey, the main problem lies in the spatial contradiction and the uneven distribution of spatial heat [9]. Therefore, the spatial transformation strategy and the technical improvement method are proposed, which can not only satisfy the modern lifestyle, but also improve the thermal efficiency of kang.

3.1 Adaptive Transformation of Space

The following transformation strategies are proposed for the lifestyle, increasing the efficiency of the heated kang, and adapting to the needs of modern living space.

Heating Facilities are Reasonably Arranged According to the Primary and Secondary Space. According to the frequency in different spaces, the primary space, secondary space and auxiliary space is distinguished. By using the kang and heated wall to

Table 3. Selection and transformation of typical types

plan	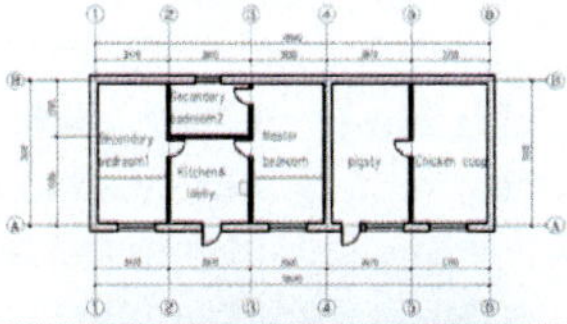	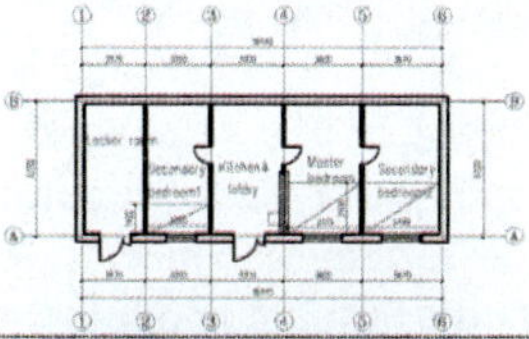
Outdoor photos	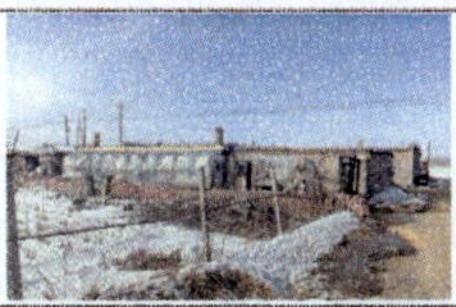	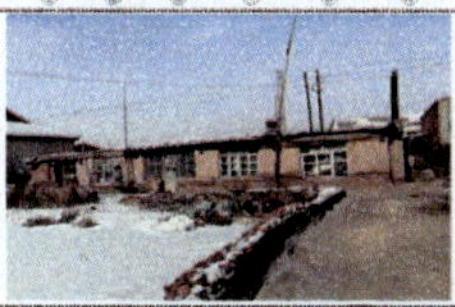
Interior photos	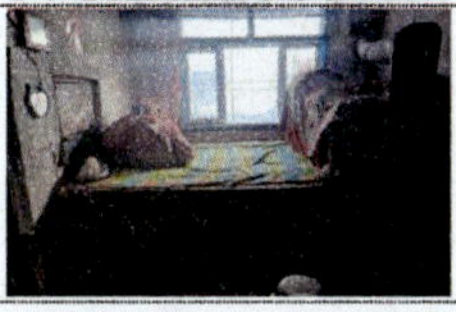	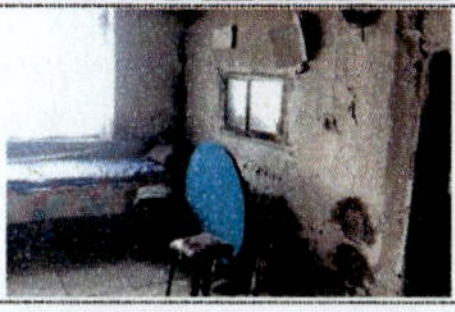
Life behavior	The main range of activities is in the master bedroom, relatives and friends eat in the master bedroom during the day, live in the second bedroom 1 at night.	The main range of activities is in the master bedroom and secondary bedroom,the secondary bedroom possesses simultaneously the function of the restaurant, the bathroom is located outdoors.
Functional evaluation	1. The master bedroom has poor privacy and low space utilization 2. Lack of toilets indoors 3. The secondary bedroom 2 faces the north side, there is no heating equipment inside, the used frequency is low	1. There is a lack of public activity space 2. Poor privacy in the master bedroom 3. The heated wall is mainly heated in the master bedroom, secondary bedrooms possess poor thermal comfort.
Floor plan after the renovation	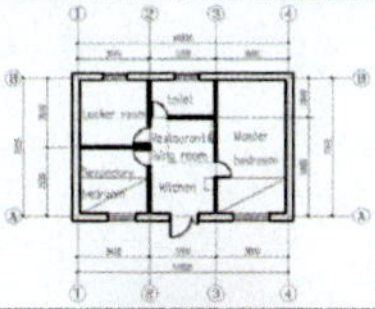	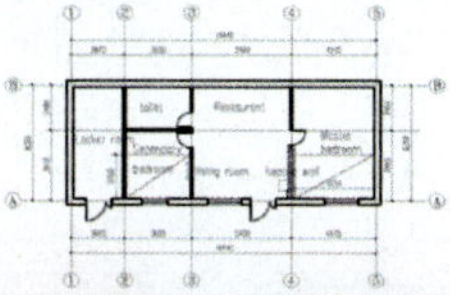
Spatial transformation	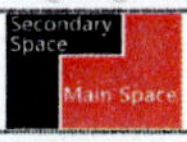 The location of the main and secondary space has been not transformed, thus the functional composition of the main space has increased the living room.	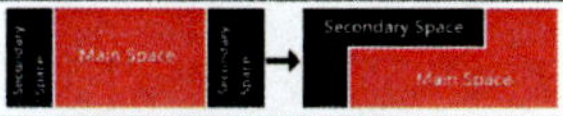 The main and secondary space is adjusted from dispersion to a combined layout,it is convenient for heating equipment to be placed in the main space, auxiliary heating to be placed in the secondary space.
Functional evaluation after the transformation	1. Heating facilities are arranged in the center,they can satisfy the heating needs of living and dining spaces 2. There are fewer family members, the shared functions are arranged close 3. Reserve the secondary bedroom for children overnight, with instant heating facilities	1. In daily life,the master bedroom and secondary bedroom are used in high frequency, the heating equipment is arranged around the master bedroom, secondary bedroom and living room according to needs 2. Reduce the intersection of life behaviors,avoid sharing between the living room and the dining room

satisfy the heating needs of the main space. Meanwhile, the radiators are used to ensure the short-term heating demand of other space, maximizing the thermal performance of different heating facilities.

Heating Facilities are Arranged in the Center to Reduce Energy Consumption. Different from the layout of traditional heating facilities close to the external wall, this adaptive transformation concentrate the heating facilities in the center of the residence, it can satisfy the farmers' modern lifestyle needs while minimizing energy consumption.

According to the individuals in households, two typical family structures are selected, the types select one-storey houses with north-south facing brick and stone structures. The shared characters are built in early and in low cost, the heating equipment includes heated wall and kang.

3.2 Improvement of Heated Kang Technology

This research focuses on the impact of the internal valve position, flue design and its location under the heated wall and kang. In addition, starting from the new form of heated wall and kang to improve the utilization rate of fuel, as an example, Junliang. Z et al. combined the heated wall type heated kang with hot water heating system [13], Haofang Z. et al. using solar kang to heat the indoor temperature [14], and Yunbo. Z et al. used flue gas waste heat for tunnel-type heating.

Table 4. Technical improvements of traditional heating equipment

	The improved technology of heated kang	The improved technology of heated wall
Schematic of the heating system	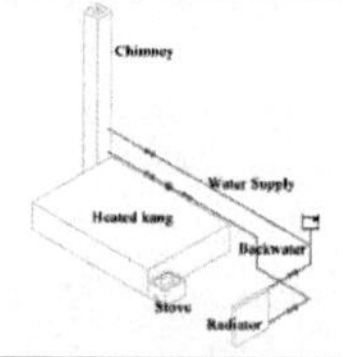	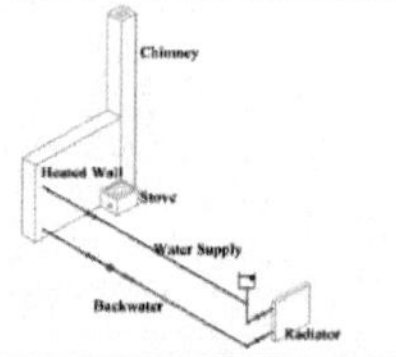
Structural drawing of the heating system	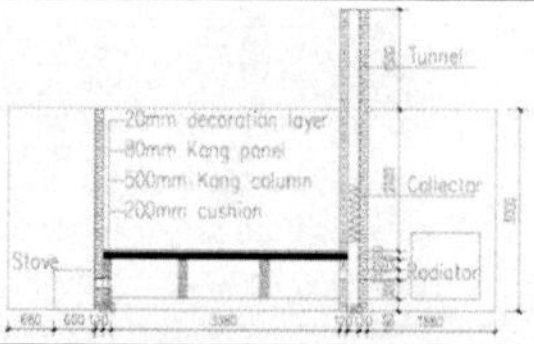	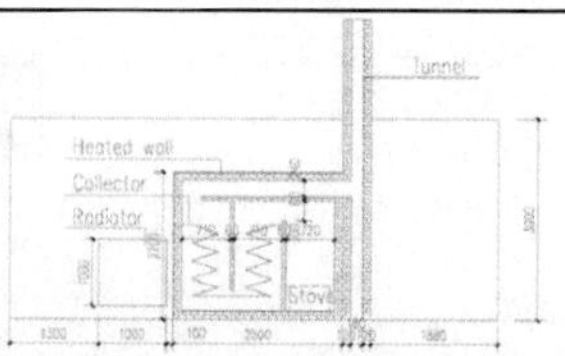
Technical characteristics	The water-filled collector is placed near the heat source, in the heated kang system, the collector is placed in the tunnel, the flue waste heat is used to heat the radiator.	Copper pipes are added inside the heated wall, the flue gas heats the water in the spiral copper pipe when the heated wall is working, the hot water is stored in the top tank. Place the radiator in the room that requires intermittent heating, open the valve when the room has heating demand.

Since this study is based on improving indoor heating equipment and minimizing the cost, the relatively high cost of solar kang and tunnel heating is abandoned.The flue waste heat is used in the kang system, a collector is added inside the heated wall, the technical improvement is carried out with the hot water system.

4 Numerical Simulation

Inability to meet the modern lifestyle and uneven indoor temperature is the main problem, consequently two strategies are proposed around heating facilities, meanwhile, making technical improvements to heating equipment. Based on this, in order to verify the improved strategy, it is necessary to simulate indoor temperature, in order to compare to transformational situation.

4.1 Software Selection and Reasons

Literature [10–13] use Fluent to simulate the performance of different heated kang, among which Geyu F. et al. simulate and analyze the impact of three kinds of heating equipment in different location [10], then Zhaojun W. et al. used Fluent to analyze the influencing factors of indoor thermal comfort, comparing the temperature of test point in the simulation with the measured temperature, the relative error did not exceed 10% [11], the simulation results were verified within a reasonable error range. Since the heating equipment involved in this study is simple and does not transform the building structure, materials and area, the temperature distribution transforms can be obtained by using Fluent to select the steady-state condition.

4.2 Simulation Setup

The data used in the following simulation was taken from surveys of typical house types. For the indoor spatial distribution of the physical model was established and meshed, the air was used as the working medium, a steady-state simulation was carried out. Moreover, by calculating the Grashof numbers, it can be found that the indoor air flow is laminar flow under working conditions, consequently the laminar flow model is used for calculation. The Coupled algorithm was used to decouple the pressure and velocity to avoid the occurrence of unreasonable pressure and velocity fields. Determine the thermal boundary conditions for each wall surface by setting its surface data of temperature. The temperature of each exterior wall surface was set to 10 °C, the interior wall was set to the thermal boundary condition, the surface temperature of the heated wall was set to 60 °C, the surface temperature of the kang and the radiator was 40 °C. In addition, the form was treated according to the wall.

By simulating the working state of the heated wall, the kang and the radiator, the distribution of the indoor temperature before and after transformation is compared. The dimensions of the plan are shown in Table 3, further the technical renovation of heated wall and kang in Table 4. The building is a flat roof with a height of 3000mm, in order to compare the distribution of indoor temperature, the performance of the heated wall is simulated under the condition that other parameters are the same.

5 Results and Discussion

5.1 Thermal Performance in the Main Room

According to the layout of the house type 1, the plane temperature distribution at Z = 1.7m was obtained by the three-dimensional simulation. Before the transformation (Fig. 3-a.) of indoor temperature fluctuations are large, the heat is released in the bedroom mainly. Moreover, the main space temperature fluctuates between 15–18 °C before the hot water system is added. Thus, the heat distribution of the main space is uneven, the secondary space temperature is simultaneously too low; after the transformation (Fig. 3-b.), the flat temperature fluctuation is small, relatively stable heat source is obtained with the bedroom and living room. Due to the use of waste heat and the addition of hot water system, the indoor temperature reaches uniform, the main space temperature is stable at 18 °C, the flue gas heat is transferred to other rooms through the hot water system. In addition, the secondary bedroom is higher than the master bedroom temperature under steady conditions, which is due to the fact that the indoor temperature is also affected by area. Comparing Fig. 4-a and Fig. 4-b, it can be found that the indoor temperature range is increased from 13–18 °C to 16–19 °C after adding the hot water system, the main room temperature is increased by 6%, the indoor temperature distribution is relatively concentrated.

When the use of secondary bedroom is frequent, the temperature of the master bedroom can be increased by reducing the area of master bedroom and increasing the heat storage of wall;In contrast, the frequency of use of the secondary bedroom is low, according to the heating needs of different types of secondary bedrooms, adjust the length of the copper pipe disc in tunnel, or appropriately increase the volume of the water tank and the size of the radiator; In addition, when the secondary bedroom is not used for a long time, the radiator can be set up the master bedroom, thus the secondary bedroom radiator can be closed in daily to concentrate heat in main room.

5.2 Thermal Performance in the Secondary Room

The layout type 2 based on two-dimensional simulation obtains temperature distribution at Z = 1.7m. In the plane before the renovation (Fig. 5-a.), it is observable that the area around the heated wall is warmer, the average temperature of the master bedroom and foyer reach 20 °C. Thus, the occupancy and residence time of the foyer are less, the energy consumption is lost. In addition, the houses on both sides of the house cannot satisfy needs for thermal comfort.

After the renovation (Fig. 5-b.), the layout of function has been adjusted, the main and secondary spaces are combined and arranged, the heating equipment is placed in the main space, the auxiliary is placed in the secondary space. Consequently the functions on both sides of the heated wall are transformed to the living room and the master bedroom in frequency, so that the temperature of the master bedroom and the living room is maintained at approximately 20 °C to avoid wasting energy. Meanwhile, the radiator is added in the secondary room to increase the temperature to 15 °C-20 °C. Comparing Fig. 6-a with Fig. 6-b, it can be found that the area in 10–15 °C is reduced from 48.0 m^2 to 37.4 m^2 after adding the hot water system, the area in 15–20 °C increases by 13.4 m^2, the temperature of the secondary room is increased by 14%. In addition, the temperature gradient of the area near the heated wall does not transform much, in contrast, the secondary space temperature increases more.

When the use of the secondary bedroom is frequent, the surface area of the radiator or the length of the copper tube disc can be appropriately increased. In general, the heat system of dissipation generally accounts for 30% [13]. When the collector is located inside the heated wall, radiators can be set more; In contrast, the collector is located inside the flue, its heat dissipation only bears the needs of one second bedroom. In addition, when the radiator position is far away from the heating center of the heating stove, it is necessary to increase the height difference between the heating center of the heating furnace and the radiator, in order to avoid the poor circulation of insufficient pressure.

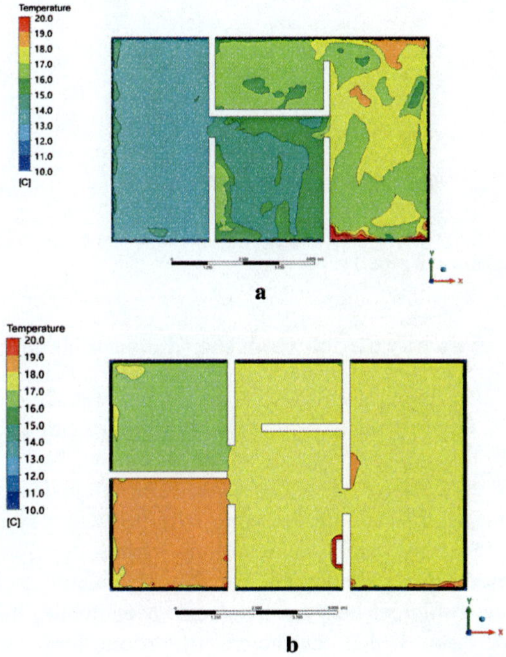

Fig. 3. -a Temperature field before the renovation of type 1. -b Temperature field after the transformation of type 1

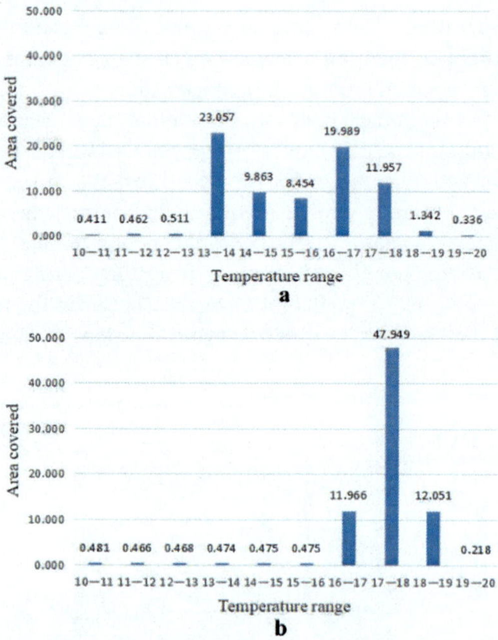

Fig. 4. -a Temperature range and area before the renovation of type 1. -b Temperature range and area after the transformation of type 1

5.3 The Influence of Family Structure on the Choice of Heating Equipment

Two-individual families only need kang and radiator to satisfy requirements in temperature, and large-scale apartments with three or more individuals need to add heated walls.

In the two-individual house type, the indoor area is small, the functional space is single and centralized, thus the heated kang with hot water system can satisfy the indoor temperature. In the three-individual and more house type, the area is larger, the functional type is more, the space layout is scattered, thus the heated hot-wall kang is required to provide sufficient heat for the room. Meanwhile, the radiator is set up in the low-frequency room, which can increase the room temperature to over 15 °C. The heated wall directly affects the indoor temperature gradient, the main factor is the ratio of the surface area of the heated wall to volume, the ratio is too large so that the temperature near the heated wall is higher than 30 °C, in contrast, the ratio is too small will make the room temperature unable to satisfy requirements. The ratio of the surface area of the heated wall to the room (including living room and master bedroom) after the renovation of the type 2 is 0.12, the value is within the comfort range of the indoor thermal environment. According to the difference in household structure and behavior, heating equipment is set up, similarly the household type is adjusted.

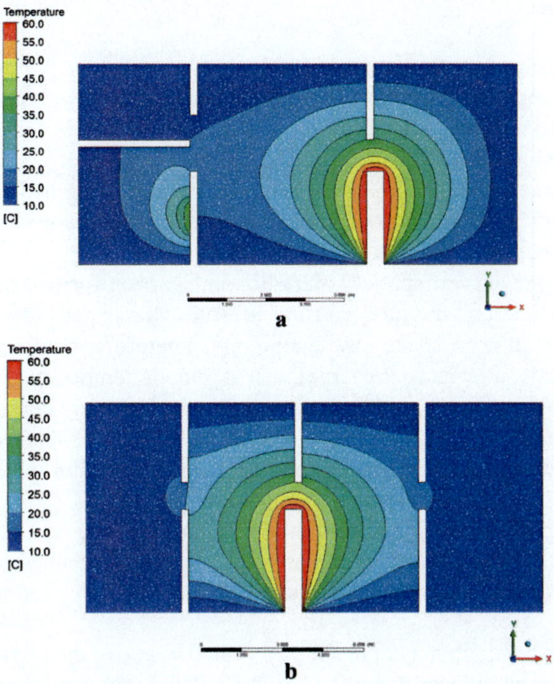

Fig. 5. -a Temperature field before the renovation of type 2. -b Temperature field after the transformation of type 2

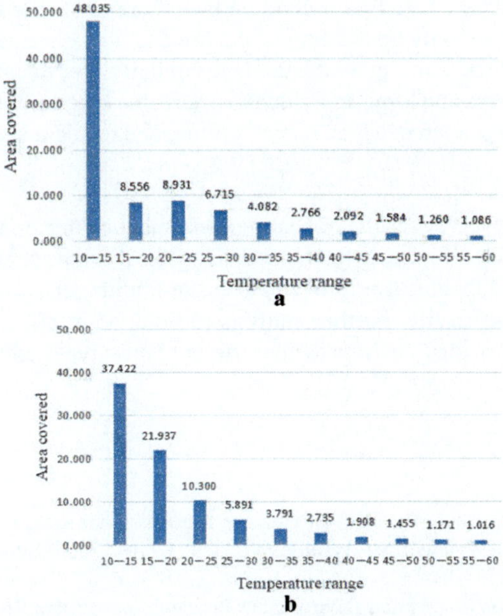

Fig. 6. -a Temperature range and area before the renovation of type 2. -b Temperature range and area after the renovation of type 2

6 Conclusion

In this study, the behavior of villagers in Shijian was investigated. Afterwards, the space and heating equipment were improved. Through the simulation verification of the indoor temperature field in winter. These findings of the research have led the author to the following conclusion.

Through field research, 93.3% of households are still using heated wall and kang, the mode can only be carried out around heating equipment is no longer suitable for people's modern needs.

According to the survey, spatial contradiction has been formed between the traditional heating method and the modern rural lifestyle. this paper proposed the mode of adaptive transformation in space and technology. Therefore, the two-individual family structure mainly alleviates the problems of indoor temperature fluctuations and low temperature, large-area house with three-individual or more can reach indoor heat distribution transfer.

In the empirical scheme and simulation, the following results are obtained.

- After adjusting the house type and adding radiators in the heated kang room, the temperature of the main room is increased by 6% and the temperature fluctuation is reduced. In addition, the room temperature of the secondary room is increased by 1–5 °C.
- After adjusting the function near the heated wall, the ratio of the surface area of heated wall to the room is reduced from 0.24 to 0.12. The secondary room is transformed from no heat to radiator in heating, which temperature increases by 14%. Meanwhile, the transform of temperature gradient near the heated wall is small.
- For small house type where two-individual live, there are fewer functions, the space are dense. Therefore, only the heating of the fire and water heating system is needed, the range of 16–19 °C can satisfy the demands of daily life; In contrast, for the house type with a large area and three-individual or more, the types of function are more, the space is relatively scattered. Thus, it is recommended to add a heated wall to provide sufficient heat.

According to the survey results of heating equipment layout strategies to improve of heating equipment. By simulating the typical schemes, it can be shown that the mode of space adaptation and technical improvement make traditional heating equipment play a greater role in modern life. Further study is to be done in effecting of human thermal comfort, simultaneously in improving different house types and heating equipment according to PMV indicators [15, 16] .

References

1. Zhuang, Z., Li, Y., Chen, B., Guo, J.: Chinese Kang as a domestic heating system in rural northern China—a review. Energy Build. **41**(1), 111–119 (2009). https://doi.org/10.1016/j.enbuild.2008.07.013
2. Summary Report on Energy and Environment Research in Northern Rural Areas of Tsinghua University in 2006. Tsinghua University, Beijing (2007)

3. Jianguo, Z.: Thoughts on pathways of promoting low-carbon tanstormation of heating under the background of carbon peak and carbon neutrality. Energy Industry **43**(09), 32–37 (2021)
4. Jie, Z., Lizhu, C., Yun, X.: Thermal comfort of large complexity Kang based on improving Kang hole structural form. Sci. Technol. Eng. **19**(08), 262–267 (2019)
5. Feng, J., Jing, W., Xu, Z.: Solar Kang heating system for cold areas. Low Temperature Architecture Technology **03**, 113–115 (2006)
6. Ling, D.M., Pengli, Y., Ce, X., Zongshan, W.: Thermal performance and heating capacity of hot-wall Kang. J. Therm. Sci. Technol. **15**(01), 59–66 (2016). https://doi.org/10.13738/j.issn.1671-8097.2016.01.010
7. Jing, W., Shan, L., Haihong, L.: A regional study on the "Heated Kang" of the traditonal residential houses in the south of Hebei province. Huazhong Architecture **37**(07), 112–115 (2019). https://doi.org/10.13942/j.cnki.hzjz.2019.07.026
8. Ma, L., et al.: A study on indoor thermal environment of rural residence in south Liaoning province. Procedia Engineering. Elsevier Ltd, Dalian, China, pp. 366–374(2016)
9. Xiping, Z., Peng, Z., Tiegang, Z.: Influences of hot-wall on rural residential thermal environment in severe cold areas. Building Energy Efficiency **44**(03), 1–5 (2016)
10. Geyu, F., Kaiying, J., Bozhi, L.: Numerical Simulation of Suspended Kang, Huoqiang, Radiator Heat-room Under the Two Layouts[J]. Building Energy Efficiency **37**(01), 33–35 (2009)
11. Zhaojun, W., Yushan, W., Yuchen, J., Teng, Z., Chang, L.: Analysis of the thermal comfort temperature with heated Kang houses and its influencing factors. Build. Sci. **37**(06), 122–129 (2021). https://doi.org/10.13614/j.cnki.11-1962/tu.2021.06.16
12. Yushan, W.: Study on thermal environment and thermal comfort in winter at rural houses with Kang in severe cold region. Harbin Inst. Technol. (2019). https://doi.org/10.27061/d.cnki.ghgdu.2019.003248
13. Junliang, Z., Zongshan, W., Lin, D.M., Jing, L.: Heating performance of coupled hot-water and hot-wall heating system based on Chinese Kang. Heating Ventilating and Air Conditioning **42**(02), 85–91 (2012)
14. Haofang, Z., Jie, J., Chao, G., Dongsheng, Z., Wei, W.: Experimental study of active and passive compound heating house based on solar Kang. Acta Energ. Sol. Sin. **39**(04), 940–945 (2018)
15. Guohui, J., Peng, Y., Nafei, L., Weifeng, L.: Thermal environment analysis and heating mode of grassland dwellings in west of inner Mongolia. Journal of University of Jinan (Science and Technology) **34**(05), 504–509+520 2020. https://doi.org/10.13349/j.cnki.jdxbn.20200513.001
16. Tao, L.: Study on differentiated thermal requirements in temporal and spatial dimensions and the optimization of heating system in rural residential buildings in northwest China. Xi'an Univ. Archit. Technol. (2020). https://doi.org/10.27393/d.cnki.gxazu.2020.000001

Constructing the Mode and Model of Spatial Appeal, Spatial Response for Conservation and Utilization of Chen Hongmou Xiangguo Mansion Site in Qing Dynasty of China Based on Value Evaluation

Zhengmin Wen[1](✉), Rongsong Quan[1], Hui Wang[1], Bing Li[2], and Zhenqiang Li[1]

[1] Guilin University of Technology, Guilin 541004, China
wzm812475701@163.com
[2] Guilin Institute of Architectural Design and Research, Guilin 541002, China

Abstract. In this paper, the status of Chen Hongmou's Xiangguo Mansion Site has been found out basically by Linear space survey, and the Value Evaluation Index System of the Site has been established, through comparing the Site with the Value Evaluation System of Great Site established by Hongwei Liu, then adjusting and supplementing. After evaluating the Site's value conclusions are drawn as follows: 1) There has a high value of noumenon and environment and it can provide the basis to declare key cultural relics protection units for Xiangguo Mansion Site. 2) There exist three kinds of demands including whole space protection, ontological space protection, and dynamic space change. 3) The thesis formulates three aspects of the response method which include "protection plus overall subarea", "protection plus ontological perfection" and "protection plus dynamic adaptation". 4) According to the summary of Visual experience, a height control model is constructed to control the building height by zone. The research results have a good guiding effect on the scientific, reasonable and orderly protection and utilization of the Xiangguo Mansion sites, and provide a mode and model for the protection and utilization of the mansion sites in the Qing Dynasty in Lingnan area of China, and can also provide a reference mode and model for the study of similar mansion protection and utilization in China and abroad.

Keywords: Construction of the mode and model · Spatial appeals · Spatial responses · Value evaluation · Xiangguo mansion site

1 Introduction

Mansion, originated in Ming and Qing Dynasties, refers to the residence of nobles with more than one grade of official residence, such as the first-class Duke, the sixth Duke, Bachelor (prime minister or cabinet chief assistant) according to the Qing Collected Statutes. It is a special type of residence in ancient architecture and the most precious

W. Guo and K. Qian (Eds.): GBCESC 2022, LNCE 211, pp. 92–107, 2023.
https://doi.org/10.1007/978-981-19-5217-3_9

and unique historical resource for the origin and development of history and culture. Compared with the north, the Lingnan mansion site is less, but its geomantic omen pattern, site environment, and site layout are unique, which has important protection and research significance. Chen Hongmou (1696–1771), a famous minister in the Qing Dynasty, his Xiangguo Mansion is located in Hengshan village, Sitang Town, Lingui District, Guilin City, Guangxi Province, which is a national traditional village. It is a typical representative of Lingnan mansion site, the birthplace of "Guangxi spirit" [1].

At present, Chinese scholars have made some achievements in the protection and utilization of the northern mansion sites. From the point of view of the value evaluation and protection of the Palace site, Li Chunqing used historical map comparison method, on-site research method, spatial syntax and other methods to study the layout structure and architectural regulation of important Mansion buildings in Beijing such as Prince Fu's Mansion, Prince Alashan's Mansion, South Mansion of Chun Prince, Princess Gulun Hejing's Mansion and Prince Xun's Mansion and summarized the value of the northern Mansion buildings from the three aspects of historical and cultural value, artistic value and available value [2–6]; Li Pengpeng analyzed the artistic achievements of Palace Garden from the aspects of its composition, garden-making concept and artistic technique, and expounded the value of its research and protection [7]; From the perspective of spatial form, Xu Junhui and others analyzed the spatial relationship between Xiangyang City and King's palace in the Ming Dynasty by studying the construction process and location of the base site of Xiangyang City and King's palace [8].

There is relatively little research literature of Lingnan Mansion. Lai Yanni and so on, from the point of structure characteristics of Prince Jingjiang's palace, analyzed the overall layout of palace and the way of palace's utilization of the surrounding environmental elements [9]. From the analysis of the architectural space pattern of Jingjiang Palace and the completed protective construction activities, Ha Ying pointed out that the Palace site should adopt a dynamic protection method to continue the vitality of cultural relics and monuments [10]. The relevant literature research from the perspective of spatial appeal and spatial response of the "site" itself has not yet been seen.

Therefore, it is of great significance to study the protection and utilization of "ruins" from the perspective of spatial appeal and spatial response.

The article is divided into five parts. 1) The basic situation of the "site" was found out using linear space survey, visiting the elderly, and consulting materials; 2) Based on the value evaluation system of the Liu Weihong site, the comprehensive value evaluation index system of the site was constructed by comparison, adjustment, supplement; 3) By Delphi method, the comprehensive value evaluation index system of the site was established "Site" ontology and environmental value; 4) Summarize the space demands of "site" protection and utilization; 5) Put forward three response methods of "protection + overall zoning, ontology improvement, dynamic adaptation", and build a building height control model of zoning.

2 Current Situation of "Site"

Three times, authors visited the site for a linear space survey, interviewed nearly 10 old people who are familiar with the historical background of Hengshan village, and consulted a large number of materials, and found out the relevant situation of the site.

2.1 Overall Space Status

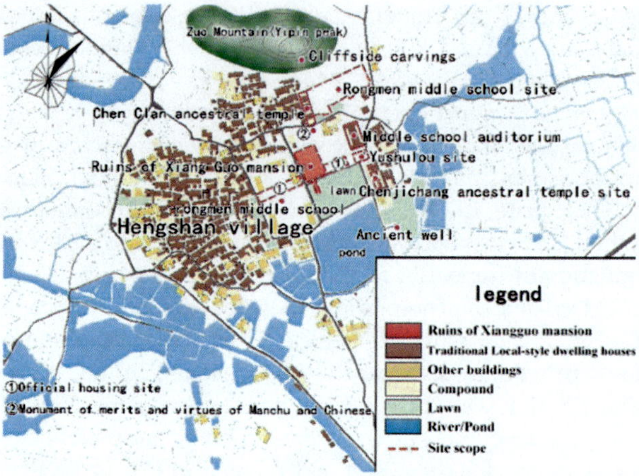

Fig. 1. Current distribution of mansion complex. The topographic map is provided by the housing and Construction Bureau of Lingui district and the text labels are drawn by the authors.

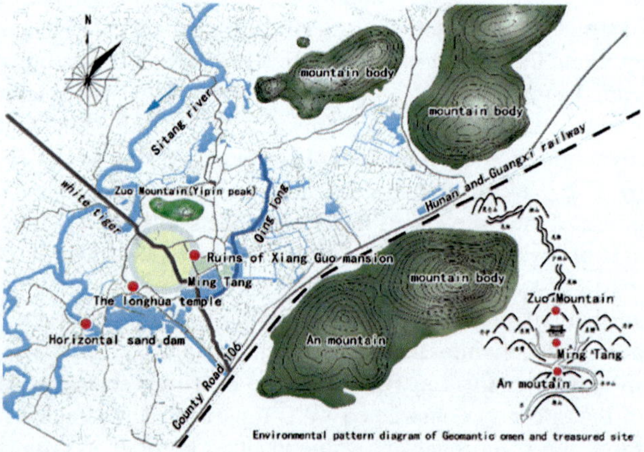

Fig. 2. Site body environment. The map comes from Baidu and the text labels are drawn by the authors.

Chen Hongmou's Xiang Guo mansion complex, including the site of the main body and the affiliated buildings Rongmen middle school and auditorium, the site of the official residence, the site of the imperial library, the site of ancestral temple of Chen Jichang, who successively won the first place in the ancient imperial examination three times, the ancient well and the cultural relics protection unit of Guangxi District, Chen's ancestral temple and Hengshan stone carvings (Fig. 1).

Hengshan stone carvings include 41 stone carvings in Chen clan ancestral temple, Monument of merits and virtues of Manchu and Chinese in front of the ancestral temple and three cliff stone carvings in the south foot of Yipin Peak; Rongmen middle school was initiated by Li Zongren, a famous person in Guangxi, and it was founded in 1943 on Chen's Jiashu site in memory of Chen Hongmou, existing site and middle school auditorium; there are 9 sites of Chen's official residence in Qing Dynasty on the left and right sides of the mansion. The site selection of the main building is based on the Yipin Peak (Zuo Mountain) in the north; the open lawns and pond form the "Ming Tang" in the south; "An mountain" is the "Zhao An"; the stream in the east is the "Qing long", and the road in the west is the "White Tiger"; the site is surrounded by mountains and water, with hundreds of hectares of fertile land; thus forming the "U" type semi-enclosed space, which conforms to the tradition Site selection requirements of "sitting north facing south, moderate in the middle, riding on the gas, and surrounded by mountains and water" (Fig. 2).

In the West and north of the site, there are many traditional houses, streets and lanes, as well as the registered cultural relics Longhua ancient temple and Hengshan Dayan (Fig. 1).

2.2 Current Situation of Ontology Space

The main body of the building was built in the Qing Dynasty. Because of the war and man-made destruction, the main body of the building has been destroyed. Only one pair of stone lions stand in the grass, and the foundation is buried underground.

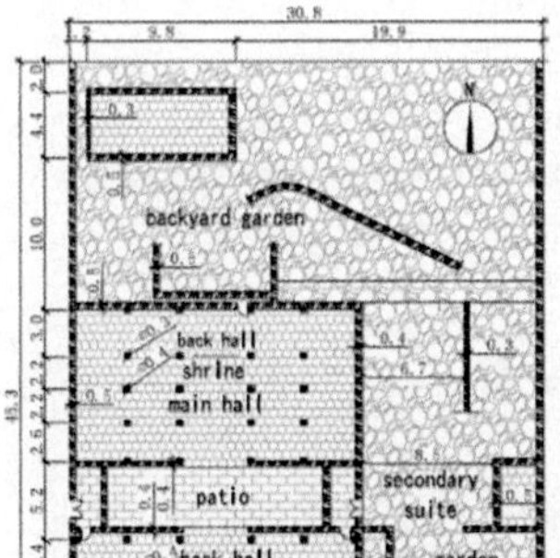

Fig. 3. Site plan of the main body. On-site drawing by the authors

Around 2015, part of the platform base was discovered. The body site is 45.3 m long from south to north, 30.4 m wide from east to west, 0.45 m high platform base, and covers an area of 1377 m^2. It is composed of three parts: the main building, the back garden, and the attached house. The main body of the building is 18.7 m wide and 29.3 m deep, with a layout of 2-in-depth and 5-bay, and the front and back two bays are patios; the back garden is 30.4 m × 16.0 m, and the attached rooms are 11.7 m × 29.3 m (Fig. 3). At present, the main site has not been declared as a cultural relic's protection unit.

3 Comprehensive Value Index System and Value Evaluation of "Site"

3.1 Construction of Comprehensive Value Evaluation Index System of "Site"

Combining the characteristics of the "site", it is adjusted and supplemented based on the value evaluation system of Liu Weihong great site [11], as the basis of the value evaluation of the "site", and it has become the basis for the excavation of its protection and utilization of space demands and the formulation of space response methods. The value index system of "site"-A can be divided into site ontology value-B1 and site derived value-B2. Among them, B1 includes historical value-C1, scientific value-C2 and artistic value-C3 while B2 includes social value-C4, cultural value-C5, commercial value-C6, environmental value-C7 and contextual value-C8. C1 includes site age-D1, site scale-D2, site level-D3, the information value of relevance to historical events or figures-D4 and site integrity-D5. C2 includes site selection-D6, Site planning ideas-D7, planning layout-D8, specific technology-D9 and the scientific value of unique remains-D10. C3 includes aesthetic value-D11, ornamental value-D12 and value of art historical materials-D13. C4 includes social reference-D14, cultural identity or emotional identity-D15, cultural identity or emotional identity-D15, popular science education-D16, sightseeing-D17 and village development-D18. C5 includes material culture-D19 and intangible culture-D20.C6 includes business value-D21 and resource value-D22. C7 includes natural environment-D23 and human environment-D24. C8 includes continuity-D25 and regionalism-D26.

Table 1. List of "site" value composition

B	C	Site characteristics
B1	C1	① It has a long history: it was founded in Zhengde period of the Ming Dynasty and has a history of more than 500 years, which is a typical case study of social migration history [12]; ② historical figures have a great influence: Chen Hongmou is a clean official in Kang Qian period, a pragmatic science courtier, known as Lingnan Confucianism [13], which has a typical impact on political and ideological circle [14]. After Chen Hongmou, there are many talents. The most outstanding representative is "Sanyuanjidi" Chen Jichang who successively won first place in the ancient imperial examination three times (Chen Hongmou's great-great-grandson)

(*continued*)

Table 1. (*continued*)

B	C	Site characteristics
	C2	① The site selection and layout are in line with the layout concept of geomantic omen that "the mountains, the water, and the atmosphere are the same"; ② the combination mode of typical "architecture in northern Guangxi"
	C3	The planning and layout of the site is ingenious and harmonious with the environment
B2	C4	① It is of great significance to the research of popular science, education, and related context; ② Cultural Origin and Emotional Attribution Space
	C5	The birthplace of cultural achievements and ideological extension, such as Chen Hongmou's *Five Heritage Codes*
	C6	Develop the tertiary industry, optimize the industrial structure and improve the living standard of villagers
	C7	Protect and continue the overall space and environmental pattern
	C8	Preservation and continuation of historical memory

Source: Self-made by the authors

The value composition of the research object is detailed in Table 1.

3.2 Comprehensive Value Evaluation of "Site"

The value evaluation system of "site" consists of four levels: general target level A, sub-target level B, evaluation index level C, and evaluation factor level D. According to the objective judgment, the relative importance of two elements in each level is expressed quantitatively. At the same time, the Delphi method is used to determine the weight coefficient of C-D indicators at all levels [15], and the 5-point scoring system is used (the highest 5 points, the lowest 1 point). A total of 10 experts and villagers are invited to score.

The indicators at all levels, and finally, the value of the "site" is determined. There is 1 expert in urban and rural planning, 1 expert in landscape architecture, 1 expert in architecture, 1 expert in history and 1 expert in human geography, 3 experts in cultural heritage, and 2 villagers who are familiar with the historical background of Hengshan village.

The evaluation of the value of "site" can directly reflect the score pros and cons of the evaluation index level and factor level of the site, as well as the differences between different values. Table 2 shows that the value of location selection, environmental layout, and plane layout of Xiang Guo mansion and the related derivative value is outstanding; the overall value score is 4.58 (5 out of 5), which is quite high. Therefore, the space demands for protection and utilization of "site" are quite strong in the three aspects of site selection pattern, site body, and site space dynamic change. Accordingly, it should respond positively in space.

Table 2. List of "site" value evaluation results

Evaluation index level		Factor level of evaluation elements			
First level index C	Weight coefficient	Weighted calculation score	Two level index D	Weight coefficient	Expert rating (maximum 5 points, minimum 1 point)
		Xiang Guo Fu site			Xiang Guo Fu site
C1	0.2	0.76	D1	0.04	5 ● 3 2 1
			D2	0.04	5 ● 3 2 1
			D3	0.04	5 ● 3 2 1
			D4	0.04	● 4 3 2 1
			D5	0.04	5 4 3 ●.1
C2	0.1	0.46	D6	0.02	● 4 3 2 1
			D7	0.02	● 4 3 2 1
			D8	0.02	● 4 3 2 1
			D9	0.02	5 ● 3 2 1
			D10	0.02	5 ● 3 2 1
C3	0.1	0.41	D11	0.04	● 4 3 2 1
			D12	0.03	5 4. ● 2.1
			D13	0.03	5 ● 3 2 1
C4	0.2	1.00	D14	0.04	● 4 3 2 1
			D15	0.04	● 4 3 2 1
			D16	0.04	● 4 3 2 1
			D17	0.04	● 4 3 2 1
			D18	0.04	● 4 3 2 1
C5	0.1	0.45	D19	0.05	5 ● 3 2 1
			D20	0.05	● 4 3 2 1
C6	0.1	0.50	D21	0.05	● 4 3 2 1
			D22	0.05	● 4 3 2 1
C7	0.1	0.50	D23	0.05	● 4 3 2 1
			D24	0.05	● 4 3 2 1
C8	0.1	0.50	D25	0.05	● 4 3 2 1
			D26	0.05	● 4 3 2 1
Xiang Guo Fu site value A		4.58	Comprehensive evaluation		5 ● 3 2 1

Source: Self-made by the authors

4 Space Demands Protection and Utilization of "Sites"

4.1 Site Selection and Overall Pattern Protection and Zoning Protection Demands

The overall space not only refers to the site itself and the surrounding environment but also includes the geomantic pattern where the site is located. It is a coordinated and indivisible organism [16]. As mentioned before, the site itself is located in a typical geomantic omen spatial pattern with mountains and water behind the mountains and reservoirs of wind and gas. To keep the geomantic omen pattern of "ruins" completely can not only increase the sense of hierarchy of the overall pattern landscape and realize the environmental view of "harmony between human and nature", but also provide a realistic basis for the scientific nature and artistry of modern human settlements environmental planning. From the point of view of the relationship between the site itself and the environment, the site itself and the affiliated buildings, the surrounding traditional houses, farmland, water area, and mountains form a "U" semi-closed space, which has a unified historical environment, culture, and overall style, and is an inseparable whole. Therefore, there is an urgent need to protect the overall space of the site.

4.2 Ontology Space Protection and Derivative Demands

The protection of ontology space demands the inheritance and expression of culture, the reconstruction, and the deduction of the architectural system. The site has a long history, which is the end of Chen's family's historical migration and evolution. It is also the origin of Chen Hongmou's philosophy of governing the world, the way of self-cultivation and other thoughts. It has an important value of protection, inheritance, and expression. At the same time, the layout of the site is in line with the system of "below Duke Hou, above the third grade, two feet high platform base, five wide rooms and nine columns", which is of great value to the reconstruction and display of Xiang Guo mansion, and also provides basis and reference for the study of mansion buildings.

4.3 Space Dynamic Change Demands

The protection and utilization of sites are characterized by uncertainty and dynamic change [17]. Combined with the current situation and value evaluation of the site, it has dynamic demands of protection methods, management mechanisms, planning strategies, and support systems. It is of great significance for the protection and healthy development of "ruins" to establish flexible planning ideas and dynamic management mechanisms.

5 Spatial Responses to the Protection and Utilization of "Ruins"

5.1 Response Mode of "Protection + location and Pattern"

Through the division of three types, the relationship between the whole site and the village is coordinated while the protection of the site and various elements is implemented.

Analysis of Defining Factors of the Site Protection Area

(1) Core protection scope. The purpose of defining the core protection scope is to protect the authenticity of the site itself and the integrity of the site environment. The definition of core protected areas is mainly carried out from seven aspects, and the specific definition factors are shown in Table 3.

(2) Construction control zone. With the purpose of "buffer and transition" and the principle of resisting the impact of the changes of the external environment on the core protection scope, the delineation of the construction control area mainly involves four factors: Relevant national and local regulations, Spatial texture characteristics, Integrity of land space, Business development needs and Aesthetic requirements. Relevant standards refer to provincial and national special standards, general standards and Guilin local standards.

Table 3. Analysis list of factors defining the core protection scope of "site"

Research object	Defining factors	Specific definition
Core protection scope	Site boundary	According to the existing site boundary or historical original boundary of the site
	Site value	According to different types of site value, the core protection scope of the site is defined
	Pattern and spatial form	The relationship between site pattern, spatial form and surrounding spatial pattern should be considered in delimiting the scope of protection
	Historical function, role and significance	According to the historical function, role and significance of the site, the scope of protection shall be determined
	Site protection level	The site level can be divided into national, provincial, municipal and county cultural relic protection units, cultural relic protection points and non-cultural relic protection units. If the cultural relic has not been evaluated shortly after its excavation, the corresponding level of protection can be formulated according to its value degree

(continued)

Table 3. (*continued*)

Research object	Defining factors	Specific definition
	Relevant national and local regulations	1) Special standards (provincial and national): ① The scope of underground immovable sites can be appropriately expanded according to the actual situation and protection needs ② Landmark sites or sites with great landscape value can be expanded according to the actual situation ③ As for sites in cities and towns, under the premise of being able to resist external damage factors, the protection scope shall be reduced according to actual needs, but not less than 20 m 2) General standard (city, county): 200 m outside the ancient site is the red line, 300 m outside the red line is the purple line 3) Guilin standard: Based on the existing boundary or historical boundary of cultural relics
	Relationship between historical context and environment	Include related buildings (structures) and environment with important value, function and significance to the site

Source: Self-made by the authors

Among them, the special standard stipulates: First of all, for sites with poor resistance, the construction control range can be appropriately increased according to actual needs, and the height limit cannot exceed 8 times the height of the site. The second is a site with high landscape value. According to the value requirements and the actual situation, the scope of control can be appropriately expanded, and corresponding requirements are put forward. The height limit should be below the site body, and the architectural style and color should be coordinated with the cultural relics body. In addition, according to the actual situation, the control range of landmark places can be divided into different levels and areas, which can extend up to 500 m. The general standard of cities and counties stipulates: Based on the core protection scope, the new buildings within the scope shall be 50–100 m away from the site, and the height limit shall not be higher than the cultural relics. In order to meet the local requirements and combined with Guilin standards: Construction control zone can be divided into two types of zones: Class I zone (non-construction area), a control boundary is set at 10 m around the cultural relics (except Jingjiang King City, which is 16 m), and the control area can only be used for greening and construction of fire passages. A Class II zone is an area 20 m from a Class I area. New buildings have a height limit of 10 m. The nature, form, style, volume, color, etc. of the building all need to be coordinated with the cultural relics.

In addition to meeting relevant national and local regulations, the construction control area also needs to follow the texture characteristics of traditional streets, plots, passages, buildings and their layout. In the definition of the specific control area, it is first necessary to consider the nature, scale, function, courtyard, boundary and other factors of the land to maintain the integrity of the land space, secondly, it is necessary to consider the local industrial structure to reserve enough space for industrial development, and finally It is also necessary to meet the coordination and overall planning requirements of aesthetic design elements such as humanities and natural landscapes.

(3) Environmental coordination area. The method and intensity of construction and utilization of this area need to be coordinated with the overall pattern, overall style, colour, and height of the core protection area and construction control zone, and the overall geomantic omen pattern should be protected in integrity. It is mainly composed of a building environment coordination area and a natural environment coordination area, of which the building environment coordination area includes the construction environment area related to the core protection area of the site in addition to the construction control area.

Regional Protection Response to the Overall Pattern of "Site"

1) Partition response. Core protection scope. It is divided into two parts. One is the protection scope of the site body. Based on the boundary of the historical site, it is shifted outwards by 10 m, and the integrity of the surrounding land space is taken into consideration. The local area is adjusted according to the boundary of the adjacent plot, with a division area of 1.89 hm^2. No engineering construction is allowed within the scope. The second is the scope of environmental protection of the site. From the perspective of the historical context -- Chen Hongmou's mainline, the site's affiliated buildings, and the sites for Zuo Mountain, Ming tang and Qing long are included in the overall protection, with a division area of 5.07 hm^2. In this area, the Residential house and Rongmen Middle School need to be replaced with the site of Chen Jichang ancestral temple and the residence of the official eunuch on the left side of the mansion; without disturbing the authenticity and damaging the environment of the related sites, the scope can be used for site environment and landscape protection activities, and the overall pattern should be coordinated with the original spatial scale and style (Fig. 4). If there are special needs, it should be reported to the government or departments according to the procedures Approval. The core protection area is 6.96 hm^2.

Construction of control zone. Based on the core protection scope, it is shifted outward by 20 m, taking into account the factors of aesthetics, landscape value, business. development. needs, and surrounding environment relations, and combining with the

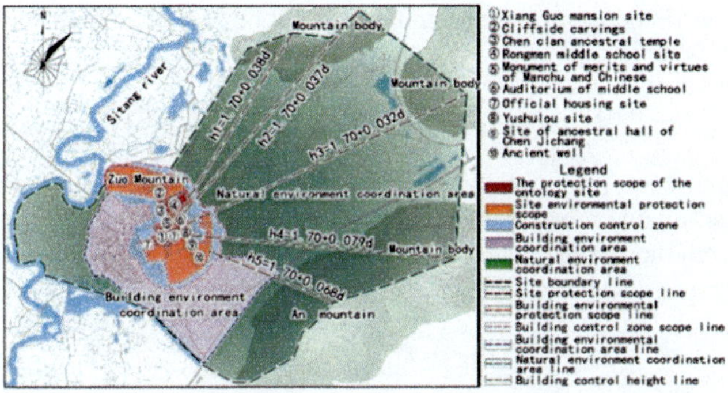

Fig. 4. Overall protection division of "site". The topographic map is provided by the housing and Construction Bureau of Lingui district and the text labels are drawn by the authors.

spatial integrity of land use and the spatial texture of villages, the scope line of the construction control zone is adjusted, with a zoning area of 4.90 hm^2. Under the condition of not affecting the site environment and the overall traditional style and features of the village, Under the condition of not affecting the environment of the site and the overall traditional style of the village, the protection method of "small-scale, small-scale" gradual renovation and transformation should be adopted to prevent the impact on the core protection scope (Fig. 4).

Environment coordination area. With an area of 127.97 hm^2, it is divided into two parts: one is the construction environment coordination area, with an area of 23.70 hm^2, including the constructed land on the west side of the construction control zone and the farmland and sporadic residential land on the south side; the other is the natural environment coordination area, with an area of 104.27 hm^2, including the farmland on the west side of the construction control zone and the mountains on the south side and the countryside and mountains on the east side. The environmental coordination zone adopts the method of "ecological buffer + coordination", mainly through coordinating the nature, colour, form, and height of the land with the construction control zone and the core protection scope, and establishing the ecological buffer zone to resist the invasion and destruction of the village by the external material environment (Fig. 4).

2) Partition Height Control Response Model. According to the description of the different partitions above, the construction control area and the environmental coordination area need to be controlled for building height. According to relevant regulations of Guilin City, the height of new buildings in the former is controlled by 10 m. But it only applies to building height control on the west side. It is necessary to establish a height control model to control the height of these two zones, to protect the overall spatial pattern of the site.

To achieve this goal, it is necessary to take the site as the centre and keep an open space for viewing the surrounding mountains. Visual experience shows that the exposed part of the mountain should not be less than half of the mountain, so the height of the new building at the bottom of the mountain should not exceed 1/2 of the height of the mountain, that is, 0.5.

$$H = h_{person} + 0.5(h_{mountain} - h_{land})d/D \quad (1)$$

In the formula, h is the height of the new building, h person is the constant, the value is 1.70 m, h mountain is the mountain elevation, h land is the site floor elevation, D is the distance from the site to the positive projection of the mountain top, d is the distance from the site to the new building (Fig. 5). After calculation, in Fig. 4:

$$h_1 = 1.70 + 0.038d \quad (2)$$

$$h_2 = 1.70 + 0.037d \quad (3)$$

$$h_3 = 1.70 + 0.032d \quad (4)$$

$$h_4 = 1.70 + 0.079d \quad (5)$$

$$h_5 = 1.70 + 0.068d \quad (6)$$

To sum up, the building height of the construction control zone on the west side of the site is controlled by 10 m; the height of the new construction control zone and the environmental coordination zone on the northeast, east and south sides are controlled by h1, h2, h3, h4 and h5, which form a space sector height control surface (Fig. 4).

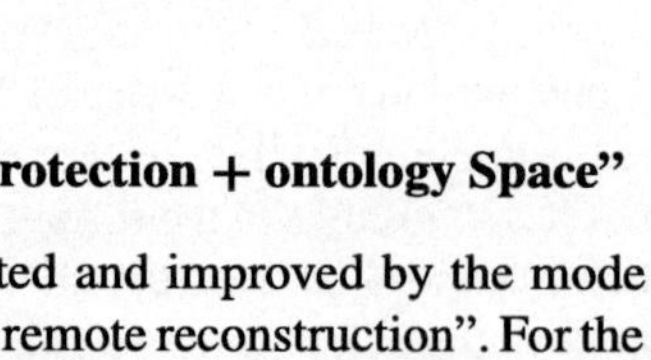

Fig. 5. Area height control sight space. Self-made by the authors.

5.2 Response Mode of "Protection + ontology Space"

"Ruins" Ontology is protected and improved by the mode of "complete protection in place + real scene display + remote reconstruction". For the protection of the site layout, the integrity and authenticity of the original site are protected by the use of automatic shrinkage organic transparent glass, and the chemical and physical methods are used to reinforce and repair the existing site layout and components, to achieve the purpose of "authenticity" real-world protection and display, and also provide a platform for the research and reconstruction of relevant institutions. At the same time, a "mansion" cultural theme landscape park was built on the periphery of the original site to express and display the cultural and organizational relationship of the mansion site through the landscape wall, sketch, and landscape elements, as well as provide leisure and entertainment places for villagers and tourists.

For the cultural expression and inheritance of the site itself, the Xiangguo Mansion can be rebuilt in the Architectural Environment Coordination Zone to the south of the site as the "World Chen Hongmou" cultural museum to display and interpret Chen's life, deeds and historical culture. At the same time, we should make full use of the site of Rongmen middle school, establish a "Chen Hongmou" cultural exhibition base integrating culture, science popularization and high-end meetings, and attract more fans of Chen Hongmou culture to study here. At the same time, apply to hold related cultural heritage seminars or summit meetings, so that more people can know and learn about Chen Hongmou's culture, to revive and continue the "Guangxi Spirit" and shape and promote the "Chen Hongmou" cultural brand.

5.3 Response Mode of "Protection + Space Dynamic"

The protection and utilization of the site can adopt the method of "live protection + sound mechanism + perfect system" to adapt to the impact of the external environment at different stages on the site. According to the problems in the actual development process, the protection methods of the site can be analysed and fed back, then the solutions can be put forward quickly and effectively, and the planning can be repeatedly modified and dynamically maintained to establish a "reversible" dynamic protection method. The management mechanism of the site should follow the principle of "government organization, experts leading, public participation, and social cooperation". Under the leadership of the Lingui district government, a dynamic and transparent supervision and management mechanism should be established to effectively supervise and manage the protection and utilization of the site. The construction of the ruins protection support system should speed up the information collection of the ruins, and formulate the corresponding "rolling + circular" local regulations and documents to ensure that the protection and utilization of the ruins are carried out in a reasonable, legal, and orderly manner. At the same time, restructuring the financing mechanism means establishing a third-party special fund management platform, implementing a diversified financing method of "government-led + villagers' equity + social financing" to protect the vital interests and lives of villagers, and effectively support the protection and utilization of the site.

To sum up, from the four aspects of the geomantic pattern, overall environment, site ontology and spatial dynamics of the "site", the protection and utilization appeal body of the "site" in the overall space-ontology space-dynamic space will be deeply excavated to formulate the corresponding protection and utilization spatial response mode. The mode preserves the integrity, and authenticity of the "site" and develop and inherit it. At the same time, it also provides a model for the protection and utilization of the mansion sites in Lingnan area.

6 Conclusions

The contributions of this paper are as follows: 1) the overall spatial pattern of Chen Hongmou's Xiang Guo mansion site selection, the spatial layout of the main building, and the layout of the mansion complex are restored in the field investigation, which provides the data basis for Chinese and foreign scholars to further study the "site"; 2) the value evaluation index system and evaluation method of the "site" are constructed, which provides a feasible evaluation for the value evaluation of the Lingnan mansion site Reference mode; 3) the overall value of the "site" is quite high, which provides a reliable basis for the local government to apply for cultural relics protection units; 4) the space demand and space response mode suitable for the protection and utilization of Lingnan mansion site is proposed, and the general model of building height control in building construction control zone and environmental coordination zone in the overall spatial pattern zoning protection is constructed. The mode and model can provide a reference for the scientific, reasonable, and orderly protection and utilization of Lingnan mansion site, as well as other similar mansion studies in China and abroad.

Acknowledgement. The authors acknowledge Zhuge Ya, the stationmaster of the village construction station in Sitang Town, Lingui District, for providing a lot of convenience for the field investigation of this article.

References

1. Liang-Hong, L.: The summary of Chen Hongmou research. J. Lit. Hist. Expo (Theory) **10**, 9–12 (2008)
2. Chun-Qing, L., Li-Ying, K.: Research on building regulation and typical value of prince Fu's mansion in Beijing of Qing Dynasty. J. Architect. Cult. **10**, 208–209 (2018)
3. Chun-Qing, L., Ting-Sheng, L.: Study on historical evolution and architectural characteristics of Mongolian prince Alashan's mansion in Beijing in Qing dynasty. J. Study Nat. Cult. Heritage **4**(05), 100–106 (2019)
4. Chun-Qing, L., Yi-Tong, L.: Research on architecture of the south mansion of Chun prince in Qing dynasty. J. Study Nat. Cult. Heritage **3**(04), 103–108 (2018)
5. Chun-Qing, L., Bo-Xuan, L., Ze-Yang, Z., Fan, F.: Research on historical evolution and architectural layout characteristics of princess Gulun Hejing's mansion building in Beijing in Qing dynasty. J. Beijing Univ. Civil Eng. Archit. **37**(01), 41–50 (2021)
6. Chun-Qing, L., Guang-Zhe, W., Zi-Yi, Z.: Research on architectural regulation and spatial characteristics of prince Xun's mansion building in Beijing in Qing dynasty. J. Beijing Univ. Civil Eng. Archit. **37**(02), 1–8 (2021)
7. Peng-Peng, L.: The gardening art characteristics in prince's mansions of Beijing. J. Study Nat. Cult. Heritage **3**(02), 87–90 (2018)
8. Jun-Hui, X., Long-Yue, W.: A preliminary research on the spatial form of Xiangyang City and the King's palace. J. Huazhong Archit. **32**(07), 134–137 (2014)
9. Yan-Ni, L., Fang, Z., Qiu-Tong, C., Zi-Liu, Q.: Analysis on the characteristics of princess Jingjiang residence Jing Wangfu garden in Guilin. J. Green Sci. Technol. (15), 21–22+27 (2019)
10. Ying, H.: The architectural space pattern of Guilin Jingjiang palace and its protection and utilization. J. Dongguan Univ. Technol. **24**(01), 81–86 (2017)

11. Wei-Hong, L.: Construction of qualitative evaluation system of complex site conservation. J. Northwest Univ. (Nat. Sci. Ed.) **41**(05), 907–912 (2011)
12. Xiuming, K.: Chen Hongmou's research in Japan. In: Zhu, F. (ed.) Chen Hongmou's Research Collection. Guangxi Normal University Press, Guilin (1998)
13. Li-Sha, Z.: The Value and reference significance of Chen Hongmou's historical wisdom. J. Party Sch. CPC Guilin Municipal Committee **19**(01), 33–36 (2019)
14. Yu-Lin, F.: On the stlyles of work of Chen Hongmou. J. North Univ. China (Soc. Sci. Ed.), **34**(02), 121–124+128 (2018)
15. Chun-Hui, W., Dong-Bo, Y., Fang, D.: Research on value evaluation and tourism development of historic districts. J. Lantai World (08), 33–34(2011)
16. Ji-Xiang, S.: Thoughts on the overall protection of large ancient city ruins. J. Archaeol. (05), 3–14+2 (2006)
17. Wen-Liang, C.: The application of elasticity planning strategy in large sites protection-based on the Yong cite site research. J. Urban Dev. Stud. **16**(08), 77–82+90 (2009)

Air Distribution Analysis and Optimization Layout of In-Row Air Conditioning in Data Center

Jianqiang Xu and Zhen Liang(✉)

Institute of Heating, Ventilation and Air Conditioning,
College of Environmental Science and Engineering,
Donghua University, Shanghai 201620, China
liangzhen@dhu.edu.cn

Abstract. In order to study the energy-saving and consumption reduction of the data center, taking a practical engineering data center as the object, the physical model of the machine room is established by simulation software, and the airflow organization in the data room is simulated. It is found that there are serious airflow mixing and local hot issues in the machine room. Through the simulation of two kinds of air supply forms of the closed cold aisle and closed hot aisle, the temperature field, airflow field, SHI, and RTI thermal environment evaluation indexes at different heights are compared and analyzed. It is found that the closed hot aisle has better airflow uniformity and temperature stability than the traditional closed cold aisle. At the same time, It is proposed to change the airflow organization of in-row air conditioning from the front supply and the rear return to the side supply and the rear return. The research shows that the airflow organization of the side supply and the rear return has better cooling effects. Through further research, it is found that the air distribution optimization effect is the best when the air conditioning body is pulled out by 30%.

Keywords: In-row air conditioning · Airflow organization optimization · Closed aisle · Numerical analysis

1 Introduction

With the rapid development of information technology, the power of a single server in the data center room has already exceeded 10 kW, even up to 20–30 kW. At the same time, the increase of cabinet capacity also makes the server energy consumption more and more. In order to solve the problem of high energy consumption in data centers, many air conditioning refrigeration schemes have emerged on the market, among which the in-row air conditioning is a special scheme.

In-row air conditioning is a precision refrigeration system for high heat density racks. The cold air is sent into the cold aisle by the in-row air conditioning from the front end.

W. Guo and K. Qian (Eds.): GBCESC 2022, LNCE 211, pp. 108–121, 2023.
https://doi.org/10.1007/978-981-19-5217-3_10

After absorbing the heat dissipation of the server, it becomes hot air and is discharged from the cabinet. At this time, the hot air is inhaled again from the back of the air conditioning under the negative pressure of the in-row air conditioning fan and then sent into the cold aisle again after being cooled in the air conditioning for the next cycle. In this way, the cycle is repeated to achieve the refrigeration of the data center.

Airflow organization optimization is one of the main energy-saving technologies of air conditioning systems in data centers [1]. QUN Z [2] found that a reasonable airflow organization scheme could reduce the energy consumption of the air conditioning system. XUE JIAO Z et al. [3] found that in-row air supply had better energy utilization rate and airflow organization form when the machine room environment and various parameters of air conditioning system were fixed by comparing the three air supply modes of upper air supply, underfloor air supply, and in-row air supply. SHI JIE Y [4] compared the temperature and velocity fields of three different air conditioning arrangements between rows and found that different air conditioning arrangements between rows had a certain impact on the energy consumption of the machine room. ZHI GANG W et al. [5] found through simulation that the diagonal layout of in-row air conditioning can make the distribution of airflow organization more uniform than the positive layout, and reduce the occurrence of local hot spots on the server. Based on the problems of airflow confusion, the insufficient air volume in some areas and local hot spots in the data center of in-row air conditioning, FANG L et al. [6] put forward the scheme of improving air supply, changing the layout of air conditioning and closing cold aisle in the data center.

Referring to the related research of in-row air conditioning in data centers, this paper takes the existing air distribution problems in a bank data center as the research content.

2 Model Verification

With the experimental data of a scaled in-row air conditioning data center constructed by Abbas et al. [7] in the laboratory, the reliability of the established model is verified. The data center is a 45 × 42 × 33 cm size, using a length ratio of 1/6 of the standard proportional model, by the blower into 22.5 °C cold air, take hot and cold aisle closure technology. The scaling model theory has been confirmed by Fernando et al. [8], and the effectiveness of thermal similarity has been proved [9–12]. The average temperature of hot and cold aisles at different heights was measured by a T-type thermocouple. In this paper, the CFD simulation software is used to establish the model with the same size and boundary conditions as the above experimental model. The temperature measurement results obtained after numerical calculation are compared with the experimental data, as shown in Fig. 1.

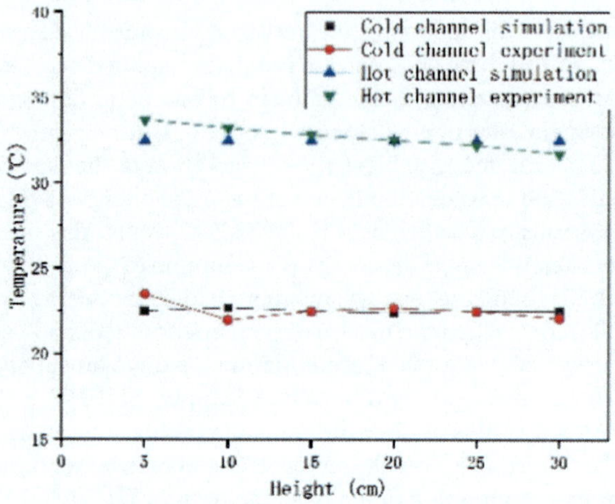

Fig. 1. Comparison of simulation results with experimental data

As shown in the figure, the average temperature in the cold aisle is 22.5 °C, while that in the hot aisle is 32.7 °C. It can be seen from the figure that the experimental values of the cold and hot aisles are consistent with the simulation variation, and the relative error is controlled within 5%. Although the temperature error of the measuring point at the lower part of the machine room is slightly large (which may be related to the simplification of the machine room model) because of the overall error of less than 5%, it is considered that the data center model used in this paper is reliable has reference significance for similar situations.

3 Model Establishment of Data Center Machine Room

3.1 Project Profile

The physical model of the machine room is shown in Fig. 2. The data room area is 263.835 m^2, and the floor height is 3.7 m. A 600 mm × 1200 mm × 2200 mm cabinet is selected in the machine room. Single cabinet power consumption 5 kW, internal filling 4 U × 1600 W server. There are 4 cold aisles and 5 hot aisles in the room. There are 32 air conditioning units between cabinet and cabinet, a single cooling capacity of 20 kW, the air volume of 5000 m^3/h. The cabinets are arranged from top to bottom and from left to right according to A–H and 1–12 (such as the upper left cabinet is A1).

3.2 Solution to Control

In this paper, CFD simulation software is used to simulate the data center room. In order to ensure the safety and stability of the solution process, it is necessary to simplify the established model. The simulation process is as follows:

(1) Data center wall insulation is good. So the wall is treated as adiabatic;

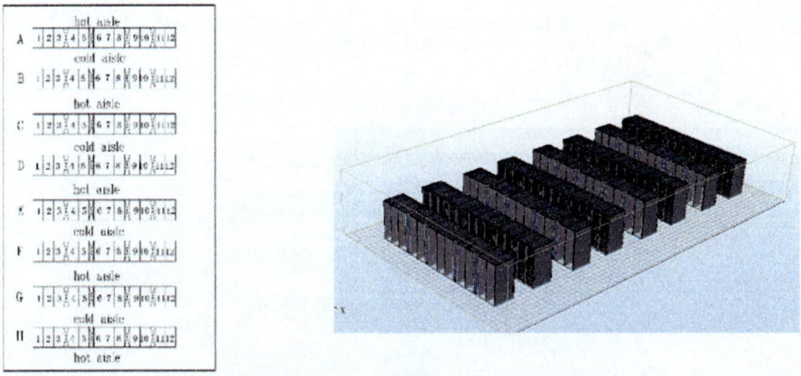

(a) Machine room equipment plan (b) 3D drawing of machine room equipment

Fig. 2. Physical model of the machine room

(2) Airflow is assumed to be incompressible fluid;
(3) The standard $k - \varepsilon$ turbulence model is adopted to solve the calculation, and the convergence accuracy of the simulation residual tends to 1;
(4) Air conditioning return air temperature is 32 °C.

4 Analysis of Simulation Results of Air Distribution in the Machine Room

4.1 Cabinet Temperature Analysis

The temperature state of the cabinet largely reflects the airflow organization in the machine room, and it is also a common indicator of the energy consumption utilization rate of the data center. The average temperature of the cabinet obtained by simulation is shown in Fig. 3. It is found that many cabinets in the machine room do not meet the ASHRAE temperature standard, and the temperature is high in many places. The larger the temperature difference between the inlet and outlet of the cabinet is, the more uniform the air supply of the air conditioner is, the more reasonable the layout of the room is, and the more sufficient the utilization of the cooling capacity is. The maximum temperature at the inlet and outlet of each cabinet is simulated as shown in Fig. 4. Figure 4 shows that the maximum air inlet temperature of the cabinet in the machine room is 21.0–28.4 °C, and the exhaust air temperature is 35.4–41.2 °C. The inlet/outlet temperature of A row and 12 rows of cabinets is higher, and there are local hotspots.

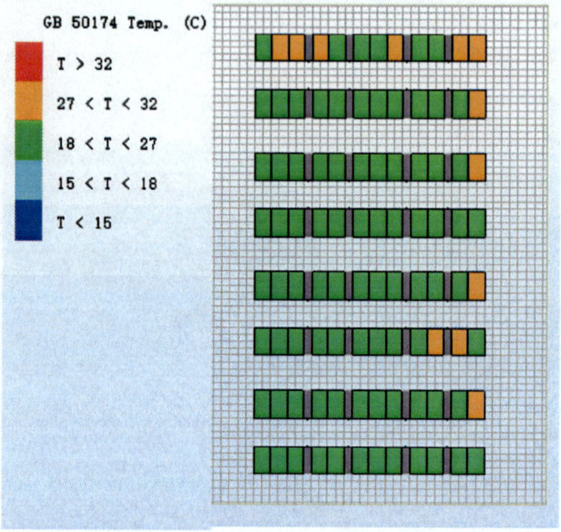

Fig. 3. Status of the cabinet under ASHRAE temperature standard

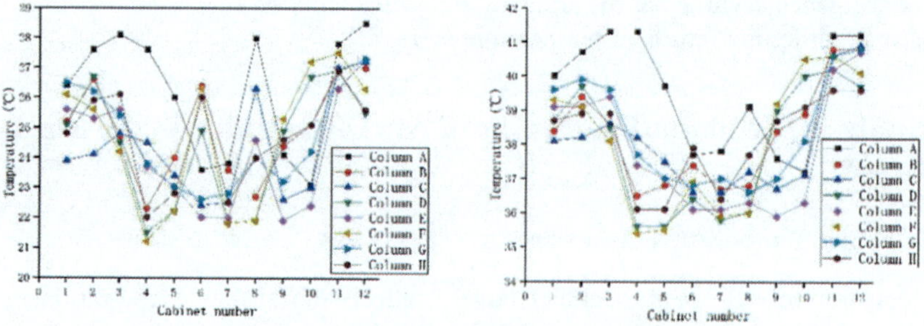

Fig. 4. Maximum temperature of air inlet and outlet of cabinet

4.2 Analysis of the Temperature Field in the Machine Room

The temperature distribution in the machine room is the key to ensuring the safe operation of the server and optimizing the airflow organization. The temperature sections of the bottom (Y = 0.3 m), the middle (Y = 1.0 m), and the top (Y = 2.1 m) of the cabinet are selected as Fig. 5.

Comprehensive analysis of the temperature fields at the bottom, middle, and top of the cabinet can reflect the thermal environment of the machine room to a certain extent. It can be seen from Fig. 5 that at the height of 0.3 m from the ground, the average outlet temperature of the two sides of the cabinet close to the air conditioner is high, and the utilization rate of cold energy is low. There are hot air recirculation and cold air bypass at both ends of the cold aisle of the cabinet, especially at the rightmost end of the aisle, and the maximum temperature of the cross-section is 40.9 °C; at the height of 1.0 m from the ground, the mixing of cold and hot air at both ends of the hot aisle is more obvious, and the temperature distribution is not uniform; at the height of 2.1 m from the

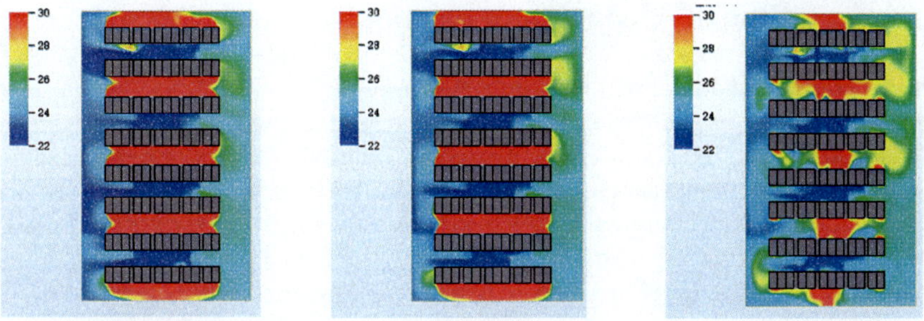

(a) Y=0.3m (b) Y=1.0m (c) Y=2.1m

Fig. 5. Temperature distribution of each section of the original machine room

ground, the airflow in the cold and hot aisles is further disordered, and the air supply and exhaust of the server are mixed to varying degrees, resulting in local overheating. The maximum temperature of the section is 38.5 °C.

4.3 Flow Field Analysis of Machine Room

In order to further understand the airflow organization of the machine room, the airflow organization streamlines diagram in the machine room is simulated, as shown in Fig. 6. The airflow streamline diagram can have a more intuitive understanding of the actual airflow direction, flow velocity, and dissipation movement. It can be seen from Fig. 6 that there is a self-vortex in the airflow near the middle cabinet of the room. This part of the airflow does not enter the cabinet to cool the server, which will greatly increase the loss of cooling capacity and air volume. In addition, the bypass of cold air occurs at both ends of the cold aisle of the machine room, and the reflux of hot air flow occurs in some cabinets. At the same time, the doping of these cold and hot airflows will lead to the attenuation of cooling capacity and local overheating.

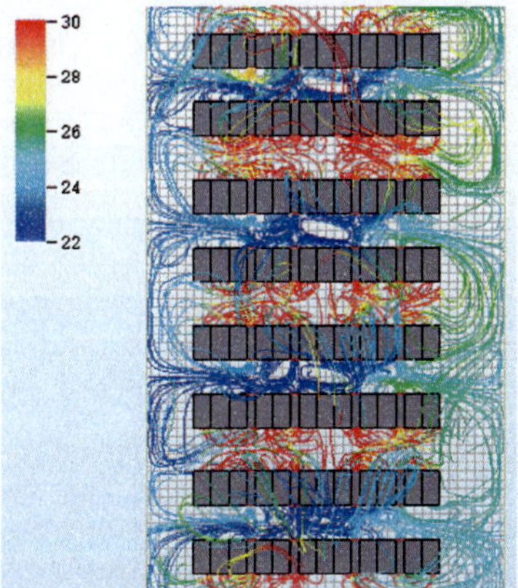

Fig. 6. Flow line diagram of the machine room

5 Optimization of Airflow Organization

If the server is in a bad thermal environment for a long time, the reliability of the data center will be reduced and the service life of the equipment will be shortened, and even downtime will occur in serious cases. Therefore, improving the airflow organization of the data center is an important means of energy saving. Given the cold and hot air mixing and local hot spots in the room, the closed aisle method and the change of the traditional air conditioning airflow organization form are proposed. And through a large number of simulations to find the best placement of the air conditioning body.

5.1 Comparative analysis of Closed Aisle Optimization

Based on the original machine room, scheme A: closed cold aisle and scheme B: closed hot aisle are proposed. The physical models under the two schemes are shown in Fig. 7.

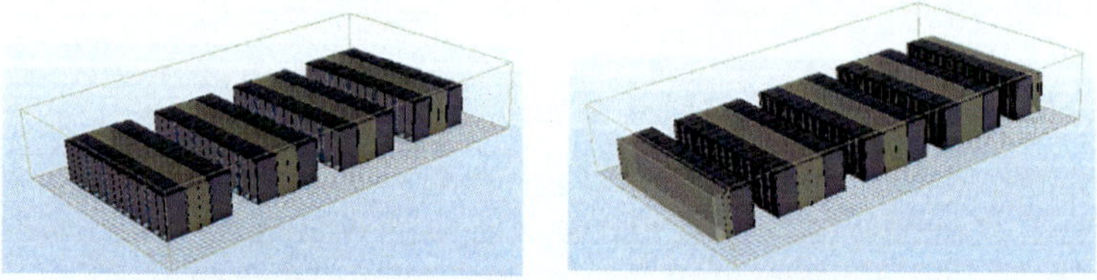

(a) Scheme A (b) Scheme B

Fig. 7. Physical 3D model under the closed aisle

Temperature Field Analysis. Compared with the original machine room, regardless of the closed cold/hot aisle, the model results show that the cabinet meets the ASHRAE/GB50174 temperature standard, and the temperature distribution is between 18 °C and 27 °C, without local hot spots. Analysis of temperature field at typical section Y = 0.3 m, 1.0 m, and 2.1 m.

Scheme A found that the height Y = 0.3, 1.0 and 2.1 m, the average temperature of the three horizontal planes were 28.0, 28.1, and 28.3 °C, respectively, which were 1.0, 1.4, and 0.8 °C lower than the corresponding temperature of the original machine room. At the same time, the maximum inlet temperature of the cabinet in the scheme was 23 °C, which was 5.5 °C lower than that of the original machine room. Figure 8(a)(b) (c) also show that the accumulation of high-temperature airflow in the machine room is improved, and the distribution of airflow organization is more uniform, but the phenomenon of high temperature at the rightmost end has not been completely solved.

Scheme B found that the height Y = 0.3, 1.0, and 2.1 m, the average temperature of the three horizontal planes were 25.9, 26.2, and 26.5 °C, respectively, which were 3.1, 3.3, and 2.6 °C lower than the corresponding temperature of the original machine room. At the same time, the maximum inlet temperature of the cabinet in the scheme was 22.6 °C, which was 5.9 °C lower than that of the original machine room. The temperature distribution of the non-closed room after optimization is between 20 °C and 22 °C, which

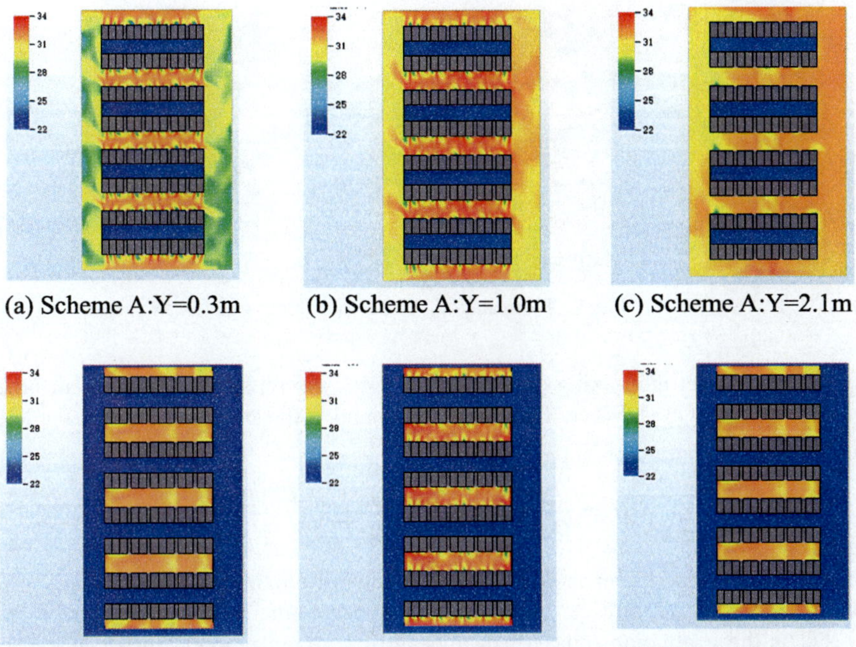

(a) Scheme A:Y=0.3m (b) Scheme A:Y=1.0m (c) Scheme A:Y=2.1m

(d) Scheme B:Y=0.3m (e) Scheme B:Y=1.0m (f) Scheme B:Y=2.1m

Fig. 8. Temperature distribution of each section of the machine room after optimization

has a rising space for the thermal environment of personnel activities. On this basis, a large space is reserved for further reducing energy consumption.

Flow Field Analysis. The simulation results of airflow organization in the data center room under the two schemes are shown in Fig. 9. Compared with the original air supply mode in Fig. 6, it can be seen that regardless of the closed cold aisle or hot aisle, the mixing phenomenon of cold and hot airflow is greatly reduced, and the airflow flowing into the cabinet is more uniform, so as to improve the utilization rate of cooling capacity. At the same time, Fig. 9 (a) shows that under Scheme A, the hot airflow in the upper left corner of the machine room is large, which is prone to heat accumulation, resulting in the risk of local overheating in some cabinets. Moreover, because it is a closed cold aisle, the cold air flows through the cabinet into a hot air flow to an open hot aisle, resulting in high temperature in the whole room. In addition, as shown in Fig. 9 (b), the cold airflow direction of the closed hot aisle is clear, which reduces the supply air velocity, reduces the circulating air resistance, and reduces the power consumption of the fan.

Thermal Evaluation Index. This paper selects two widely used evaluation indexes to analyze the results: SHI (Supply Heat Index) and RTI(Return Temperature Index).

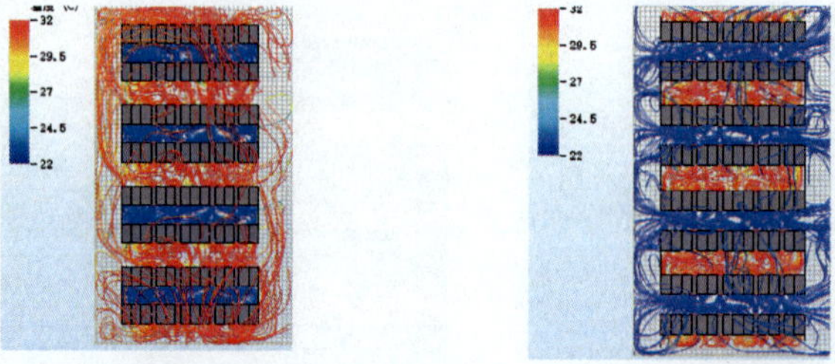

(a) Scheme A (b) Scheme B

Fig. 9. Flow diagram of the machine room

SHI can reflect the mixing degree of cold airflow with surrounding hot air before cooling the server [13]. When SHI is between 0 and 1, the smaller the value, the lower the loss.

$$SHI = \frac{\delta Q}{Q + \delta Q} = \frac{\sum_j \sum_i ((T^r_{in})_{i.j} - T_{ref})}{\sum_j \sum_i ((T^r_{out})_{i.j} - T_{ref})} \tag{1}$$

where T_{ref} is the outlet air temperature of in-row air conditioning, T_{in} is the average intake temperature of the rack, and T_{out} is the average outlet temperature of the rack.

RTI is the evaluation index of cabinet airflow organization, which can evaluate whether the system air supply is excessive or insufficient [13]. RTI is closer to 100% more ideal.

$$RTI = \left[\frac{T_{re} - T_{su}}{\Delta T_{eq}}\right] \times 100\% \tag{2}$$

where T_{re} is the return air temperature of air conditioners, T_{su} is the supply air temperature, ΔT_{eq} is the difference of the average intake air temperature and exhaust air temperature.

The SHI and RTI values calculated according to the simulation of the original machine room and various schemes are shown in Fig. 10. It can be seen from Fig. 10 that, compared with the original airflow organization model of the machine room, RTI under the two closure methods is improved, indicating that no matter the cold/hot aisle closure method, the cold air short circuit problem has been improved to a certain extent. At the same time, the heating index SHI also drops by more than an order of magnitude, and the SHI of Scheme A is 5.6×10^{-4}, and that of Scheme B is 6.8×10^{-4}, with little difference between the two.

Considering the temperature field, flow field, and thermal evaluation index, the data center has the best airflow organization effect under the closed thermal aisle mode without changing the layout of the room equipment.

5.2 Air Conditioning Supply and Return Air Form Optimization

The traditional in-row air conditioning is placed between the rack rows and columns, using the front air supply and rear air return as Fig. 11 (a). The cold air is fed into the

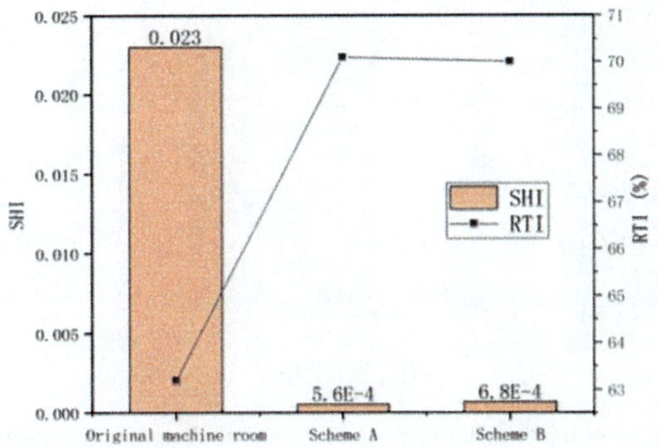

Fig. 10. Thermal evaluation index of the air distribution scheme

cold aisle from the air conditioning and flows through the cabinet. After heat transfer, the air is returned from the hot aisle to complete the refrigeration cycle. But it is difficult to take care of the cabinets on both sides of the air conditioner, reducing the flow of cold air into the server. Therefore, the air distribution form of the data room is changed to the side air supply and rear air return (the air conditioning body pulls out 25%) as shown in Fig. 11 (b), and the optimized room is compared with the original room.

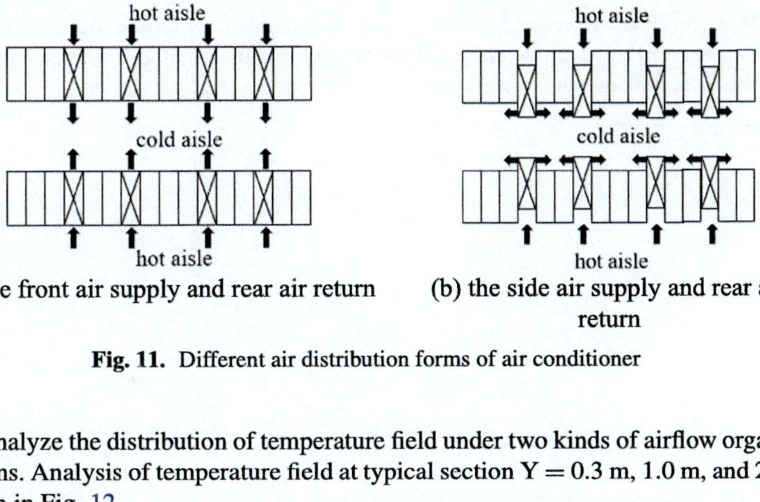

(a) the front air supply and rear air return (b) the side air supply and rear air return

Fig. 11. Different air distribution forms of air conditioner

To analyze the distribution of temperature field under two kinds of airflow organization forms. Analysis of temperature field at typical section Y = 0.3 m, 1.0 m, and 2.1 m, as shown in Fig. 12.

Figure 12 shows that the section height of the machine room Y = 0.3, 1.0 and 2.1 m, the average temperature of the three horizontal planes is 28.4, 28.7 and 28.4 °C, respectively, which is 0.6, 0.8, and 0.7 °C lower than the corresponding temperature of the original machine room, and the maximum inlet temperature of the room cabinet is 26.3 °C, which is 2.2 °C lower than that of the original room. By comparing Fig. 5 and Fig. 12, it can be seen that optimizing the air supply mode of air conditioning into side supply and back return can reduce the mixing degree of cold and hot air in the machine

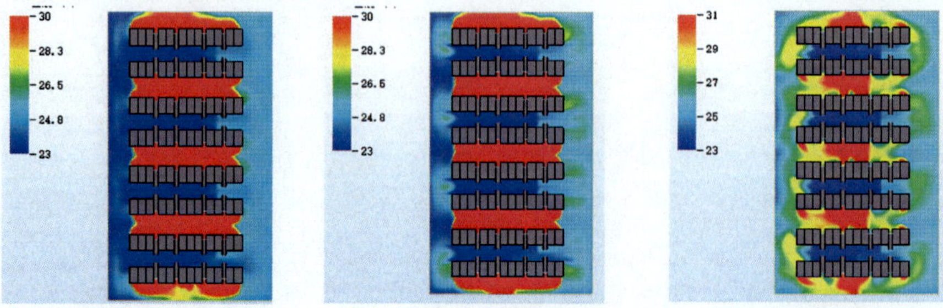

(a) Y=0.3m (b) Y=1.0m (c) Y=2.1m

Fig. 12. Temperature distribution map of each section of air conditioning room

room, avoid the short-circuit phenomenon of cold aisle air supply, reduce local hot spots and improve the utilization rate of cooling capacity.

5.3 Air Conditioning Placement Optimization

To further explore the influence of the air conditioning location on the optimization effect of this air supply mode, based on the optimization scheme of the side air supply and rear air return, the pull-out value of the air conditioning body is assumed to be in the range of 12.5%–40%, and 12 machine room models are established with the density of 2.5% interval for simulation research. The maximum inlet temperature of the engine room cabinet and the average temperature of the typical section of the cabinet from the ground 1 m plane are selected as the comparison parameters, as shown in Fig. 13.

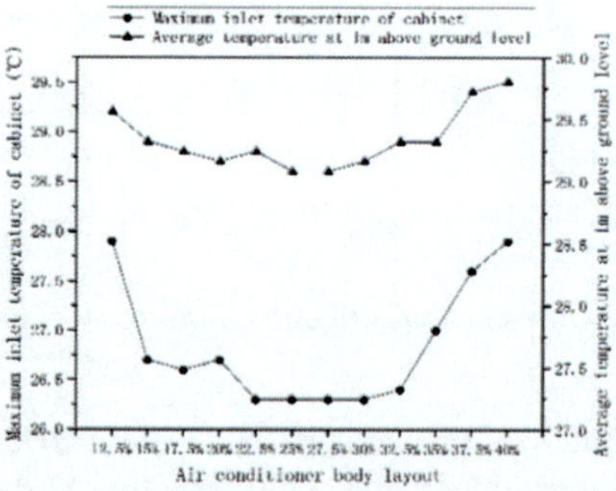

Fig. 13. Temperature comparison chart under different air conditioning body arrangement

Figure 13 shows that no matter the maximum inlet temperature of the cabinet or the average temperature of the Y = 1m level, the pull-out of 12.5%–40% of the air conditioning body decreases first and then increases, indicating that the pull-out of 22.5%–32.5% of the air conditioning body can reduce the disturbance between the cold airflow and improve the cooling effect of the server.

In addition, the RTI and SHI values of the 12 models are calculated according to the server vent temperature and air conditioning return air temperature in the simulation results, as shown in Fig. 14. It can be seen from Fig. 14 that under different models, the return air temperature index RTI remains at about 63.4%, with little change, indicating that the arrangement of air conditioning units after side air supply has little effect on the air recirculation and air bypass in the machine room. However, the SHI value of the heating index changes parabolically with the increase of the pull-out ratio of the air conditioning body, which is the lowest at about 30%. This shows that in this interval, the cooling loss of air conditioning supply before entering IT equipment is the smallest, and the air distribution is the most uniform.

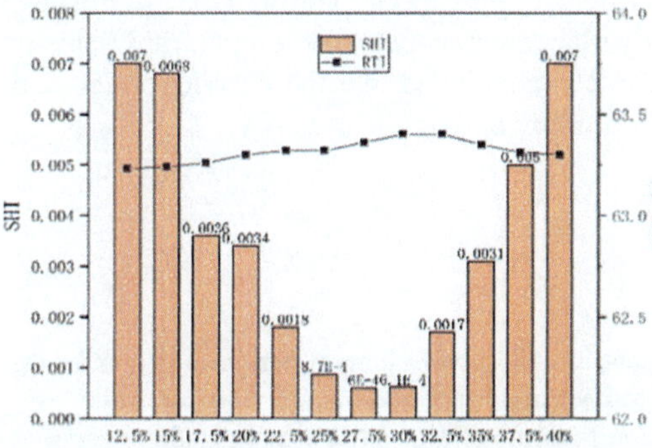

Fig. 14. Thermal evaluation index values under different models

Considering the influence factors of the above studies, it is found that under the mode of side air supply and rear air return, the airflow optimization effect of 30% pulled by the air conditioning body is the best.

6 Conclusion

(1) Only using in-row air conditioning, the cold and hot airflow in the original room with open cold/hot aisle is seriously doped and the high-temperature airflow accumulation occurs frequently. The local hot spot temperature is up to 41.3 °C, which is not conducive to the safe and efficient operation of servers and other equipment. The cold/hot aisle closure makes the cold and hot air flow orderly enter and exit the server equipment, so as to avoid the short-circuit phenomenon of the cold aisle air supply flow and improve the utilization rate of cooling capacity.

(2) The stability of the flow field and temperature field of the machine room model under the hot aisle closure technology is better than that under the cold aisle closure technology. The cold aisle closure technology has the problem of high overall temperature in the room. The cabinet near the wall side may produce heat accumulation. Instead, The closed hot aisle technology is to isolate the hot airflow, so that the room is dominated by cold aisles. This technology also makes the air conditioning return air temperature significantly higher than that of the closed cold aisle,

increases the supply and return air temperature difference of the air conditioning system, and reduces the energy consumption of the data center.

(3) By optimizing the traditional front air supply and rear air return mode of the in-row air conditioning in the data center to the side supply and rear air return mode, the air conditioning can directly send the airflow organization to the vicinity of the suction outlet of the server, which avoids the problems of large air supply resistance and small air supply on both sides of the air conditioning, reduces the airflow short circuit and mixing phenomenon existing in the air conditioning of the head and tail columns, and improves the utilization rate of cold energy.

(4) Based on the optimization scheme of the in-row air-conditioning side air supply and rear air return mode, it is found that the position of the air conditioner will affect the cooling effect of this air supply mode. Through the establishment of 12 different air conditioning body placement models, the temperature field and thermal evaluation index are studied. It is concluded that the cooling effect of the server is the best when the air conditioning body pulls out 30%.

References

1. Xiuming, C., Jiuchen, Z.: Air conditioning system design and air distribution optimization for the data center room. Fluid Mach. **42**(11), 79–82+86 (2014)
2. Qun, Z.: Research on the influence of airflow organization design on the cooling effect of data room. Telecommun. Eng. Technol. Stand. **34**(02), 16–21 (2021)
3. Xuejiao, Z., Junli, Z., Qinli, I.D.: Comparative study of three air supply modes in data centers. Architectural Sci. **35**(02), 102–107+15 (2019)
4. Shijie, Y.: Research on the optimal layout scheme of air conditioning between columns in data center. Telecommun. Eng. Technol. Stand. **30**(05), 82–86 (2017)
5. Zhigang, W., Dan, H.: Research on the influence of different airflow organization forms on the thermal environment of row-level data center. HVAC **1**(S1), 170–175 (2021)
6. Fang, L., Zhigang, W.: Simulation study on indoor air conditioning airflow organization in a data center. Build. Energ. Effi. **44**(10), 11–17 (2016)
7. Abbas, A.M., Huzayyin, A.S., Mouneer, T.A., Nada, S.A.: Thermal management and performance enhancement of data centers architectures using aligned/staggered in-row cooling arrangements. Case Studies Thermal Eng. **24**, 100884 (2021)
8. Fernando, H., Siriwardana, J., Halgamuge, S.: Can a data center heat-flow model be scaled down? In: Proceedings of the IEEE Information and Automation for Sustainability ICIAfS, pp. 273–278. IEEE (2012)
9. Nada, S.A., Elfeky, K.E., Attia, A.M.A., Alshaer, W.G.: Experimental parametric study of servers cooling management in data centers buildings. Heat Mass Transf. **53**(6), 2083–2097 (2017)
10. Nada, S.A., Elfeky, K.E.: Experimental investigations of thermal managements solutions in data centers buildings for different arrangements of cold aisles containments. J. Build. Eng. **5**, 41–49 (2016)

11. Meng, X., Zhou, J., Zhang, X., Luo, Z., Gong, H., Gan, T.: Optimization of the thermal environment of a small-scale data center in China. Energy **196**, 117080 (2020)
12. Nada, S.A., Attia, A.M.A., Elfeky, K.E.: Experimental study of solving thermal heterogeneity problem of data center servers. Appl. Therm. Eng. **109**, 466–474 (2016)
13. Nada, S.A., Said, M.A., Rady, M.A.: CFD investigations of data centers' thermal performance for different configurations of CRACs units and aisles separation. Alex. Eng. J. **55**(2), 959–971 (2016)

Intelligent Protection and Inheritance Method of the Historic Centre of Macao Based on BIM+ Concept —— A Case Study of Mandarin's House

Yao Wu and Hui Zhu(✉)

Soochow University, 199 Renai Road, Suzhou, China
jan_000@sina.com

Abstract. The intervention of new technology in the protection of cultural heritage has become the consensus and common goal of heritage protection in all countries. Mandarin's House is one of the trials conducted by the cultural Bureau of Macau Special Administrative Region in the intelligent protection and utilization of cultural heritage and has made some progress. However, in the process of field investigation and literature research, it can be found that there are still some problems in its protection and utilization. These problems are not only reflected in the protection and utilization of Mandarin's House, but also in the process of protection and reuse of other cultural heritage in The Historic Centre of Macao. Therefore, this study takes Mandarin's House as the main research object, explores the problems it is facing at the present stage through field research, case analysis and literature survey, and proposes a solution strategy for intelligent protection and inheritance based on BIM, combined with 3D GIS, Extended Reality, Internet of Things and other technologies. For reference to Cultural heritage protection in Mandarin's House and other Historic Centre of Macao.

Keywords: The Historic Centre of Macao · Mandarin's house · Intelligent protection and inheritance · Historical building information modeling

1 The Introduction

Since The Historic Centre of Macao successfully applied for world Cultural Heritage in 2005, Macao has made great efforts in traditional heritage protection. However, due to the special geographical location, history and culture of Macao, the protection and inheritance of The Historic Centre of Macao still face some problems. Mandarin's House, as an important part of The Historic Centre of Macao, faces the same problems as other cultural heritages. Intelligent conservation is the trend of international cultural heritage conservation. In 2009, Murphy first mentioned the new concept of HBIM in Historic Building Information Modeling (HBIM) [1]. In China, the Institute of Architecture and Cultural Heritage Conservation of Tsinghua University also proposed the "Heritage/Historical Building Information Modeling (HBIM)" technology in 2009 based on BIM technology

W. Guo and K. Qian (Eds.): GBCESC 2022, LNCE 211, pp. 122–132, 2023.
https://doi.org/10.1007/978-981-19-5217-3_11

and aiming at the particularity of Historical buildings. Historical Building Information Modeling (HBIM), which originated from Murphy and Tsinghua University, in essence, is a historical building life-cycle information management system based on BIM technology, which collects data from the early stage and generates 3D models for historical buildings and integrates them into BIM platform. By combining 3D GIS, Extended Reality, Internet of Things and other technologies, this Historical Building Information Modeling can promote the protection and utilization of Historical buildings in various stages such as the protection, management, reuse design, construction and so on.

2 An Overview of Macao World Cultural Heritage Protection

2.1 The Historic Centre of Macao

The Historic Centre of Macao (see Fig. 1) is located in Freguesia de Santo António, Sé Freguesias and Freguesia de S. Lourenço from southwest to central Macao Peninsula. The Historic Centre of Macao extends east to west from Colina da Guia to Avenida de Almeida Ribeiro, near Terminal Marítimo de Passageiros do Porto Interior, and north to south from Jardim de Luís de Camões e Cemitério Protestante toColina da Barra.

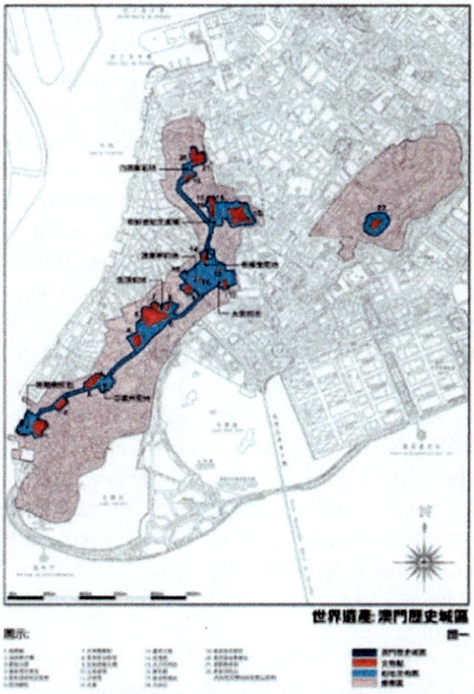

Fig. 1. The Historic Centre of Macao (Picture Source: https://www.culturalheritage.mo/mwhinfo05/)

The Historic Centre of Macao consists of 22 buildings and eight "Praça" in a narrow terrain. The 22 buildings are dominated by Western architecture, interspersed with Chinese architecture, blending Chinese and Western architecture: A Chinese lion is seen in the cathedral building, and A Western ship is seen in the A-Ma Temple. These buildings reflect the characteristics of Macao as a seaport city and a place where Chinese and Western cultures blend for more than 400 years: cultural diversity and symbiosis, aesthetic tolerance and harmony. "Praça" is the Portuguese word for square. It is located in front of churches and municipal buildings. The paving style is mostly Portuguese gravel, made of small black and white stones with various patterns that reflect the colorful buildings surrounding the front site.

2.2 The Development of Cultural Heritage Protection in Macao

Macao's cultural heritage protection can be divided into two periods before and after Macao's return to China.

The preservation of Macau's cultural heritage started earlier, dating back to 1953 when a commission of inquiry was appointed by Macau governor Joaquim Marques Esparteiro. The Commission is responsible for investigating and identifying buildings of cultural value in Macao. During The term of office of The Macao Portuguese Government, Macao has also followed The Portuguese way of protecting cultural heritage, and the protection work has been mainly led by the government. With two large migrations in the 20th century, much land in Macao has been used for living. Later, The Macao Portuguese Government and other developers built and rebuilt houses wantonly, destroying many historical buildings and the urban context in Macao. Until the 1970s, Macao made new progress in cultural heritage protection policy. Decree No. 34/76/M, issued in 1976, was Macao's first law on the protection of cultural relics. The law stipulates the standard of cultural relics evaluation in Macao, lists the buildings to be preserved in Macao one by one, and provides detailed provisions on the rights and obligations of members of the Committee for the Maintenance of Macao Urban Scenery and Cultural Property and the law on the protection of cultural relics [2]. Decree No. 56/84/M on the Protection of Architectural Landscapes and Cultural Property, issued in 1984, established the Commission for the Protection of Architectural landscapes and Cultural Property, replacing the Commission for the Maintenance of Macau Urban Landscapes and Cultural Property established in 1976, and redefined the functions and powers of the Commission. The Act also defines standards for tangible cultural property and regulations on the protection, utilization, transfer, maintenance and taxation of designated monuments, designated combinations, designated places and protected areas [3].

After Macao returned to China in 1999, the Cultural Affairs Bureau of Macao was in charge of the protection of Macao's cultural heritage. In 2005, The Historic Centre of Macao was added to The World Cultural Heritage List, becoming China's 31st World Heritage site. In 2013, the Legislative Council of the Macao Special Administrative Region enacted law No. 11/2013 on the Protection of Cultural Heritage, which is a policy for the protection of Macao's historical, paleontological, archaeological, architectural, linguistic, documentary, artistic, ethnological, scientific, social, industrial or technological properties. The policy has provided legal support for Macao's cultural heritage protection and utilization from various aspects. In particular, it has defined

"The Historic Centre of Macao", clarified its scope, and formulated overall and partial protection and comprehensive management plans [4]. After Macao returned to China, with the opening of gambling rights, the success of applying for the World Heritage List, the construction of Guangdong-Hong Kong-Macao Greater Bay Area and other events, the protection and utilization of Macao's cultural heritage have attracted more attention and accelerated the protection of cultural heritage.

Regarding the digital protection of Cultural heritage in Macao, in terms of law, Macao has no relevant laws and standards for cultural heritage and historical urban areas in terms of digital or intelligent intervention. In terms of theory, Zhao Qiang discussed the problems, countermeasures and prospects of scientific and technological protection of Macao cultural heritage in his paper "Problems, Countermeasures and Prospects of Technological Protection of Macao Cultural Heritage". He believes that the team of scientific research talents should be strengthened first [5]. In terms of practice, the Macao Foundation launched the cultural and historical website "Macao Memory" in 2019 to collect and integrate historical and cultural materials of Macao and display them in words, images, sounds and impressions, etc., and output them through different online ports such as web pages and social software combined with offline exhibitions [6]. As for the establishment of The Historic Centre of Macao exhibition platform, The Cultural Bureau of Macao SPECIAL Administrative Region established The Macao Cultural Heritage website and the mobile phone app "WH Macau" (see Fig. 2). WH Macau contains a brief introduction, general location, news and activity information, photos, loopback and other information about The Historic Centre of Macao for people to have a general understanding of The Historic Centre of Macao. In 2022, the Macao Cultural Bureau launched VR and AR virtual Tours based on photos of individual scenic spots. However, a more systematic and refined information system is needed for in-depth understanding, protection and relevant research of The Historic Centre of Macao.

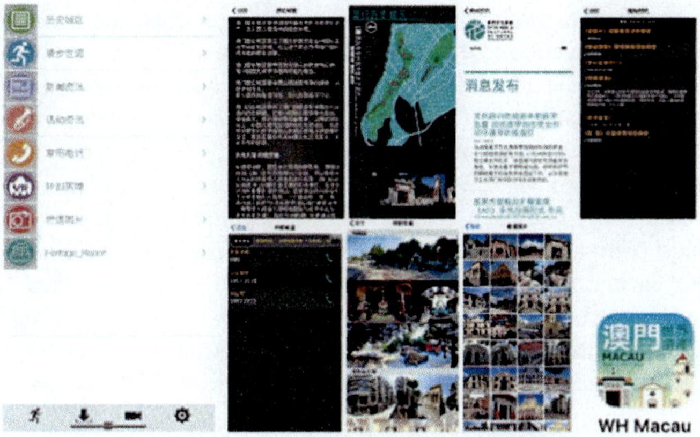

Fig. 2. WH Macau (Picture Source: APP "WH Macau")

3 Preservation Process and Existing Problems of Mandarin's House

3.1 Preservation Process of Mandarin's House

Mandarin's House is an important part of The Historic Centre of Macao. In terms of architectural style, although Mandarin's House is a Lingnan Style Chinese building because it is located in Macao, a city where Chinese and Western cultures blend, western architectural elements can be seen everywhere in this Chinese building, such as the western clay sculpture relief decoration on the large garden window and the ceiling inside the House. A large number of brick arch structures, a relatively rare load-bearing structure in Chinese architecture, can also be seen in the building.

In terms of history and culture, Mandarin's House was built in 1869 by the father of Zheng Guanying, a modern Chinese thinker. Mandarin's House has been built for more than 150 years, during which it has experienced prosperity and decline. When it was first built, Mandarin's House was one of the largest Chinese residences in Macau. Later, due to the migration of descendants of the Zheng family to other places and the housing shortage caused by the influx of a large number of immigrants into Macao, Mandarin's House was rented out, and the original courtyard of a powerful family became a " Tenement" where seventy-two families from all over the world live together. Tenants have built Mandarin's House according to their personal needs. Although this kind of behavior destroys the original style of the building, it is the epitome of a period of Macao's history and has the same protection value.

In terms of data collection, repair and protection, the Government of Macau Special Administrative Region acquired the property right of Mandarin's House in 2001. Began an eight-year restoration of Mandarin's House. In the process of renovation, the Cultural Bureau of Macau Special Administrative Region collected historical materials of Mandarin's House from all walks of life and invited experts in Lingnan architectural protection to conduct research, which provided a large number of materials and laid a foundation for the renovation and research of Mandarin's House.

In terms of activation and utilization, Mandarin's House was officially opened to the public in 2010. It is usually used for visiting, and can also be used as a venue for exhibitions and activities. The protection mode has changed from protection to simultaneous protection and utilization. In 2021, the online VR virtual tour function of Mandarin's House will be officially launched. Visitors can visit Mandarin's House online and learn about it through panorama, photos of architectural details, text and voice introduction. The online VR virtual tour has promoted the digital utilization of Macao cultural heritage and strengthened the promotion and education of The Historic Centre of Macao. In early 2022, the Cultural Bureau of Macau Special Administrative Region launched Mandarin's House augmented Reality (AR) mobile application. When tourists visit Mandarin's House on the spot, they can explore the interactive scenes in Mandarin's House and make movies through mobile phones, and also understand the appearance of Mandarin's House in different periods. Augmented Reality (AR) mobile application has realized simultaneous cooperation between the offline visit of Mandarin's House and online digital display, further promoting the intelligent protection and inheritance of The Historic Centre of Macao.

3.2 Existing Problems of Mandarin's House

Mandarin's House is an important cultural heritage in Macao with its unique charm. In the process of preservation and inheritance of Mandarin's House, certain progress has been made in data collection, repair and restoration, and intelligent display. However, in the process of field investigation and literature research, it can be found that the current protection and utilization of Mandarin's House still have some common problems with other cultural heritage protection in Macao:

The Integration of Chinese and Western Architectural Styles. Cultural diversity and complexity lead to the integration of Chinese and western architectural styles. The most striking feature of Macao culture is that Chinese and western cultures have gathered on this land since ancient times. Through the baptism of time, these cultures have kept their characteristics while blending in Macao. Therefore, Macao is a city in which foreign cultures join local history and culture to form their own unique and distinct personality. From the architectural style and development history of Mandarin's House, it can be seen that Mandarin's House preserves the historical memory of the city, reflects the characteristic culture of the city, and witnesses the historical changes and cultural development of the city. However, the collection of different cultures leads to different architectural styles, materials, functions and protection methods. Therefore, an information platform is needed to integrate all information and prevent information from being isolated from each other and forming information islands.

Disaster Prevention and Post-disaster Repair. Mandarin's House may face three main types of disasters: wind disaster, fire and erosion of buildings caused by Macao's climate characteristics. First of all, Macao is often hit by typhoons. In typhoon season every year, Mandarin's House not only needs to resist strong winds but also faces the crisis of being damaged by surrounding trees and public facilities. Some cultural heritage sites in low-lying areas, they may be damaged by seawater flooding. Secondly, according to Macao Data 2021, written by the Government of Macao Special Administrative Region Statistics and Census Service, the overall population density of Macao is 208,000 people/square kilometer, while the population density of The Macao Peninsula is 583,000 people/square kilometer [7]. Under the pressure of high-density population and housing, the streets and lanes in Macao are also very narrow, with historical buildings and former sites distributed among them. Mandarin's House is such a cultural heritage surrounded by residential buildings and narrow streets (Fig. 3). Relevant units need to timely predict and avoid risks before the fire, timely notify the fire when it happens, evacuate people and clear roads for fire vehicles, to ensure the smooth launch of rescue work. Finally, Macau Peninsula, where Mandarin's House is located, is surrounded by the sea on three sides. It has abundant rainfall and is windy. The average humidity in 2020 is 82%. When the relative humidity exceeds 70%, mold breeds, so the exterior of historic buildings is often eroded. In addition to humidity, acid rain caused by air pollution and daily wind and rain erosion can also cause damage to cultural heritage, which requires long-term testing and protection.

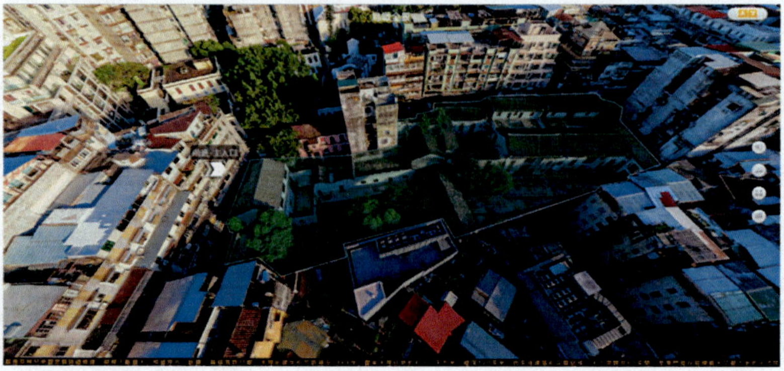

Fig. 3. An aerial view of Mandarin's House and surrounding area (Picture Source: https://vr.icm.gov.mo/wh/mandarinhouse/)

The Suppression of Gambling. As can be seen from Fig. 4, before the outbreak of COVID-19, lucky gaming revenue had been the main source of total revenue for the Macao government. In 2018, for example, the total revenue of the government was 14.1313 billion MOP, and the revenue of lucky gambling was 11.3512 billion MOP, accounting for 80% of the total revenue of the government. From the perspective of industrial structure, Macao's gaming and gaming intermediary in 2018 also accounted for 50.7%. After the outbreak of COVID-19 in China at the end of 2019, the total government revenue shrank significantly to 9.47 billion MOP in 2020, among which the revenue of lucky gambling was 2.9808 billion MOP, accounting for 31.5% of the total government revenue. With the rapid development of the gaming industry, Macao has always been regarded as a "gambling city" with a very simple industrial structure. The outbreak of the epidemic has severely hit Macao's economy. Therefore, the economic development mode supported by the single gambling industry makes the economy unable

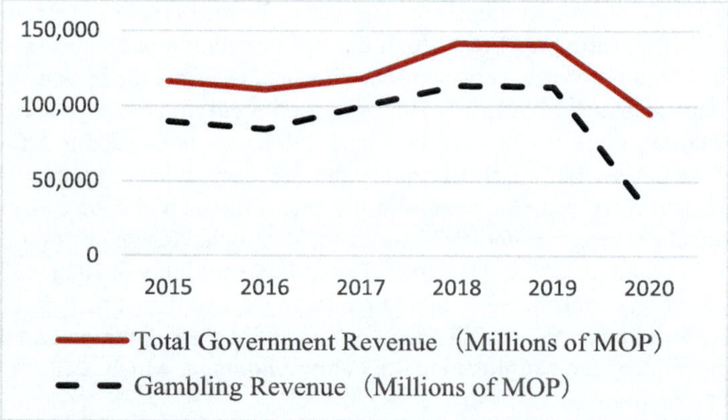

Fig. 4. Macau Government Revenue & Gambling Revenue (Picture Source: Macao Data 2021)

to develop sustainably and healthily, and also makes people ignore the cultural heritage tourism of Macao. In February 2019, the Outline Development Plan for the Guangdong-Hong Kong-Macao Greater Bay Area defined Macao as the "World Tourism and Leisure Center". Therefore, we should make reasonable use of The Historic Centre of Macao's rich cultural heritage resources, promote the protection and reuse of Macao's cultural heritage, and promote Macao's cultural heritage tourism.

The Impact of COVID-19. After being affected by COVID-19 at the end of 2019, the number of tourists in Macao plummeted from 39.4062 million in 2019 to 5.8968 million, reducing the number of inbound tourists by nearly 85% [8]. For Macao's cultural heritage, its protection and inheritance should not only be limited to architecture, space or a specific object but also pay attention to the dissemination of culture and the inheritance of cultural context. The epidemic has sharply reduced the number of tourists in Macao, blocking a major transmission route for Macao's culture and cultural heritage. Therefore, actively developing new cultural and cultural heritage publicity paths can help establish cultural communication paths and lay a foundation for the sustainable development of cultural heritage in Macao.

4 Intelligent Protection and Inheritance Strategy of Mandarin s House Based on BIM+ Concept

Building Information Modeling (BIM) is a concept to create a building information base that can be integrated throughout the life cycle of a building. At the same time, it also established a building design, construction, management, operation, use and other stages of the personnel sharing information platform. In 1975, Professor Chunk Eastman of Georgia Tech University first proposed the concept of the "Building Description System", which is the origin of the concept of BIM. In 2002, Autodesk developed BIM software, followed by Bentley and Graph iSOFT, two software developers. BIM technology began to develop rapidly, and has been recognized and promoted worldwide. BIM technology in China started late, but it is also in the stage of rapid development. It has the support of policies and regulations such as The Guidance on Promoting the Coordinated Development of Intelligent Construction and Building Industrialization and Several Opinions on Accelerating the Industrialization developments of New Buildings. It has applied BIM technology in many projects such as Shanghai Tower, the National Aquatics Center in Beijing and Shanghai Bund SOHO.

According to the characteristics and existing problems of Mandarin's House, based on the concept of BIM + and the existing protection and inheritance of Mandarin's House, the protection and inheritance strategy of Mandarin's House can be put forward, including the collection of cultural heritage information, the establishment of the information base and the future application mode (see Fig. 5):

The first is the collection and recording of the architectural information of Mandarin's House. Mandarin's House is a mixture of architectural styles, so it is necessary to record the history, style, material, construction method and detailed size of the building. This is not only convenient for information exchange between relevant personnel in the process of cultural heritage protection and utilization but also convenient for the later restoration of cultural heritage based on the detailed information of the building to select appropriate protection methods. Based on existing texts, historical photos and drawings, historical information about historical buildings and former places can be collected and integrated, and elements such as area, orientation, height, color, material and facade of Mandarin's House from ancient times can be collected and recorded in combination with 3D scanning and field surveying and mapping survey. At the same time, aiming at the disaster prevention problem in Macao, the historical buildings are equipped with temperature and humidity sensors, water level sensors, wind speed sensors, etc., and the building environment information is collected through the Internet of things technology.

Secondly, the architectural components of Mandarin's House were constructed and divided to form internal libraries or databases. Although the architectural style of Mandarin's House combines Chinese and Western characteristics, if the architectural components are separated, they can still be distinguished according to the style. Therefore, it is possible to classify and model the architectural components of Mandarin's House according to their functions and styles, and create internal libraries or databases for historical architectural components. After the establishment of internal libraries or databases, the information collected at the earlier stage can be integrated and input into BIM software to establish the full life cycle model of Mandarin's House.

After the establishment of the information base and the whole life cycle model, the model of Mandarin's House and its surrounding areas is established in combination with 3D GIS technology, which can be applied in digital restoration, disaster prevention and collaborative display of Mandarin's House. In terms of digital restoration, information collection in the early stage and internal libraries or databases of historical buildings provide the basis for building restoration, and BIM technology is used to establish the 3D model of historical buildings. This is convenient for cultural heritage protection units to inquire about information, in the process of restoration and activation design, and provide complete information to assist with design. At the same time, the design scheme can be modeled in advance. On the one hand, the effect after construction can be foreseen and analyzed in the overall environment of the block to prevent adverse effects on the surrounding environment. On the other hand, problems should be found in the process of model building, and prevention and avoidance schemes should be provided in advance to prevent damage to the architectural heritage itself in the process of transformation. In terms of disaster prevention, according to the timely collection and analysis of data, disasters can be given an early warning, and the erosion and destruction of buildings can be reduced by geographical calculation and building a disease simulation system for prediction and prevention in advance. GIS can also be used for analysis before and after the disaster to obtain evacuation plans for personnel and vehicles, which is convenient for efficient and rapid development of rescue work. In terms of collaborative display, Mandarin's House has established its simulated atlas, VR virtual tour and AUGMENTED reality (AR) mobile application. But the way of presentation is mainly text, photos, audio and video. Based on the 3D model established by BIM software, the virtual roaming scene of Mandarin's House is established by combining extended reality technology and game Engine (Unity3D, Unreal Engine 4, etc.). Create a more immersive heritage tour experience by improving the verisimilitude of the Mandarin's House display. At The same time, it will also bring a new path for The Historic Centre of Macao and Macao culture to spread, and improve The situation that The gambling industry has suppressed cultural heritage tourism and Macao's tourism industry has been hit by The COVID-19.

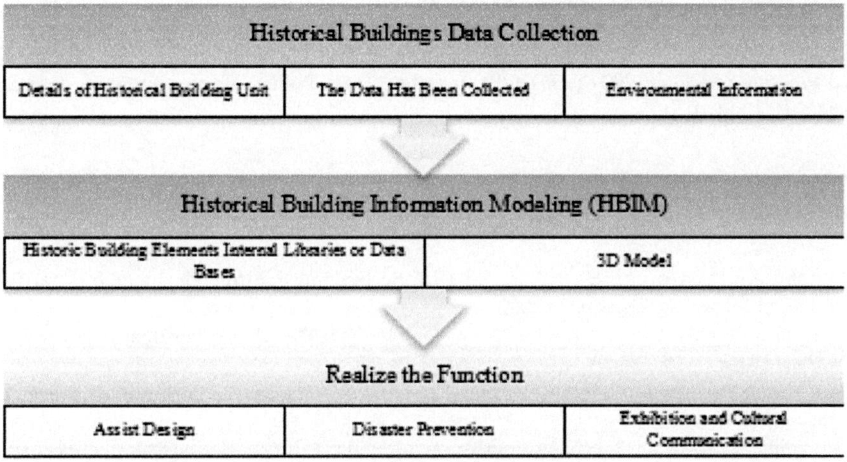

Fig. 5. Flow Charts for HBIM Using

5 Conclusion

Today, with the intervention of intelligent research and practice in various fields, heritage protection in all countries in the world has entered a new stage of transformation from intelligent protection. The problems faced by The Historic Centre of Macao at the present stage also need to be improved through the intervention of new technological means.

From a geographical perspective, located in southern China and surrounded by the sea on three sides, Macao has been an important port city since ancient times. It has attracted visitors from all over the world and created conditions for Macao's multicultural development. From a historical perspective, the integration of Chinese and western cultures in Macao has a history of several hundred years. The ancient buildings and Praça in the historic city tell people about the history and cultural development of Macao, and the time-honored history and culture have been preserved in the process of The Times. The central government's policy support for the protection and development of Macao's culture after Macao's return has helped the protection of Macao's cultural heritage and the development of its urban culture.

Mandarin's House, as a world cultural heritage in The Historic Centre of Macao, has the characteristics of other World cultural heritage in Macao. However, as one of the pilots of Intelligent protection and Inheritance of Macao's cultural heritage, it has its unique characteristics in terms of protection and utilization. Therefore, taking Mandarin's House as an example, Analyze the problems in the process of intelligent protection and inheritance. Based on BIM and combined with 3D GIS, Extended Reality,

Internet of Things and other technologies, a solution strategy of intelligent protection and inheritance including cultural heritage information collection, information database establishment and future application is proposed. It not only provides a new way for the protection and inheritance of Mandarin's House but also lays a foundation for Macao to build itself into a "world tourism and leisure center".

Acknowledgement. The writing and research of this paper are obtained from the National Key R&D Program of China (2021YFE0200100) and the 2021 Policy Directed Program of Jiangsu Province (BZ2021015).

References

1. Murphy, M., McGovern, E., PaVía, S.: Historic building information modelling (HBIM). Structural Survey **3**(27), 11–27 (2009
2. The Macao Portuguese government: Law No. 34/76/M. The Macao Portuguese government, Macao (1976)
3. The Macao Portuguese government: Law No. 56/84/M. The Macao Portuguese government, Macao (1984)
4. Government of Macau Special Administrative Region: Cultural Heritage Protection Law of Macao Law No. 11/2013. Legislative Council of the Macao Special Administrative Region, Macao (2013)
5. Qiang, Z.: Problems, countermeasures and prospects of technological protection of macao cultural heritage. Public Communication Science Technol. **13**(24), 128–130 (2021)
6. GOV. MO https://www.gov.mo/zh-hans/news/258845/ Accessed 8 Feb 2022
7. Wenda, W.: A Story of Macau. Macao Education Press, Macao (2003)
8. Government of Macau Special Administrative Region: Macao Data 2021. Government of Macao Special Administrative Region Statistics and Census Service (2021)

Recycled Concrete Powder Supported Nano-TiO_2 Composites: Efficient Utilization of RCP and Excellent Photocatalytic Performance

Gang Liao[1] and Wu Yao[1,2](✉)

[1] School of Materials Science and Engineering, Tongji University, Shanghai 201804, China
yaowuk@tongji.edu.cn

[2] Key Laboratory of Advanced Civil Engineering Materials, Ministry of Education, Tongji University, Shanghai 201804, China

Abstract. In this paper, in order to promote the application of recycled concrete powder (RCP), RCP was used to synthesize RCP/TiO_2 composites of high value. The microstructure of RCP/TiO_2 composites were investigated in detail, and photocatalytic degradation of RhB under UV-light was carried out to assess the photocatalytic performance of RCP/TiO_2 composites. The result showed that nano-TiO_2 was evenly dispersed on the surface of RCP and firmly combined with RCP by chemical bonds. The RCP/TiO_2 composite with only 40% addition of TiO_2 exhibited the same photocatalytic performance as equal mass of pure TiO_2 within 60 min. There was a threshold RhB concentration for per RCP/TiO_2 composite under which an excellent photocatalytic performance can be achieved. RCP/TiO_2 composites showed high stability in cyclic tests. The added value of RCP is enhanced by the form of RCP/TiO_2 composites. Our work provides a feasible way to promote the upcycling of RCP by preparing value-added products.

Keywords: Recycled concrete powder · Solid waste · Tio_2 · Photocatalytic degradation · Environment remediation

1 Introduction

In China, a large amount of concrete is demolished every year, resulting mass of waste concrete, mainly containing recycled aggregate (RA) and recycled concrete powder (RCP). At present, RA can be well used in producing recycled concrete [1], however, RCP is difficult to reuse as building material in large scale due to some limitations, such as low activity, high water absorption, and variable composition [2]. Most RCP is mainly disposed by landfilling, and thus the accumulation of RCP will occupy a lot of land and cause severe environment pollution [3]. Discard of RCP causes such a waste in resource [4]. Therefore, it is extremely urgent to improve the value of RCP and expand the application scope of RCP.

TiO_2 is a promising photocatalytic material in environment remediation, but its price is relatively high. Recently, studies showed that immobilization of nano-TiO_2

W. Guo and K. Qian (Eds.): GBCESC 2022, LNCE 211, pp. 133–140, 2023.
https://doi.org/10.1007/978-981-19-5217-3_12

on carriers could not only reduce the consumption of nano-TiO_2 but also enhance its photocatalytic activity [5–8]. Liao [7] reported that zeolite has large specific surface area and strong adsorption ability, thus leading to a high concentration of pollutants around the active sites of TiO_2 and enhancing the photocatalytic efficiency greatly. Hence, finding sustainable and cost-effective carriers is essential for promoting the application of TiO_2. Inspired by the design of natural mineral/TiO_2 composites, RCP can replace natural minerals to prepare supported TiO_2 composites, and RCP shows great advantages in cost and environmental protection.

In this study, in order to increase the value of RCP and improve its utilization rate, we used RCP as carriers to synthesize RCP/TiO_2 composite and then put it into use of the purification of dye waste water. Specifically, a wet mechanical grinding method was used to combine TiO_2 with RCP. Photocatalytic degradation of RhB solution was applied to evaluate the photocatalytic performance of as-prepared RCP/TiO_2 composite. We focused on the relationship between the photocatalytic efficiency and the microstructure of RCP/TiO_2 composite. The reusability of RCP/TiO_2 composite including was also tested. Our work may provide a new way to efficiently utilize RCP and reduce the cost of TiO_2 in curbing environment pollution.

2 Experimental Procedure

2.1 Materials

The nano-TiO_2 was provided by Shanghai Titan Technology Co., Ltd. RCP was obtained from Shanghai Youhong Environmental Protection Technology Co., Ltd. Analytically pure Rhodamine B (RhB) was also used in this experiment.

2.2 Sample Preparation

RCP/TiO_2 composite was prepared by a wet grinding method. Firstly, TiO_2 and 1% sodium dodecyl benzene sulfonate (SDBS) by mass of TiO_2 were ultrasonicated in water for 30 min. Then the well-dispersed TiO_2 slurry and RCP slurry were stirred for 30 min. The solid to water ratio was fixed at 25%. Secondly, the mixed slurry was ground by a planetary ball mill (QM-3SP2, China) at a speed of 900r/min for 30 min. Finally, the composite slurry was filtered and dried at 105 °C for 12 h (KSL-1100X-S, China), obtaining RCP/TiO_2 composites. According the content of TiO_2 in the composite, the sample was labeled as RCPTn (n% stood for the content of TiO_2, $n = 0, 10, 20, 30, 40$), for instance, RCPT10 meant that the content of TiO_2 in this sample was 10%. The pure TiO_2 was named as P25.

2.3 Characterization

The phase composition analysis was carried out on an X-ray diffractometer (D/max2550, RIGAKU). The morphology of as-prepared RCP/TiO_2 composites were observed via a scanning electron microscope equipped with an energy dispersive spectroscope (TM4000 Plus, HITACHI). The BET specific surface area and mesopore size distribution were measured by a BET automatic specific surface and porosity analyzer (ASAP2460, MICROMERITICS, America). The Fourier-transform infrared (FT-IR) spectra were recorded on an infrared spectroscope (EQUINOX55, BRUKER).

2.4 Photocatalytic Performance Evaluation

A 175 W high-pressure mercury lamp with a wavelength of 365 nm was the light source. A typical test procedure is as follow. Firstly, 50 mg RCP/TiO_2 composite was mixed with 50 mL RhB solution (10 mg/L) in a petri dish and stirred for 30 min without illumination until adsorption equilibrium. Secondly, the lamp was switched on and 3 mL RhB solution was sampled at an interval of 10 min. The irradiation intensity of the UV-light over the liquid was about 3 mW/cm^2. Finally, the sampled RhB solution was centrifuged. The absorbance of the RhB solution was measured by a UV-VIS spectrophotometer (U-3310, HITACHI). And the sediments were washed by ethanol and dried at 80 °C for cyclic tests. The pseudo first order reaction model was used to characterize the photocatalytic reaction kinetics as shown in Eq. (1).

$$\mathrm{Ln}(c/c_0) = -k_{app} \cdot t \quad (1)$$

where, C_0 was the initial concentration of RhB, and C was the concentration of RhB at time t, and k_{app} was the apparent reaction rate constant.

3 Results and Discussion

3.1 Microstructure of RCP/TiO_2 Composites

The X-ray diffraction patterns are shown in Fig. 1(a). The main phases of RCPT0 were quartz and calcite. The diffraction pattern of P25 indicated that the nano-TiO_2 was a co-crystalline of anatase and rutile. For RCPT20 and RCPT40, the diffraction intensity of anatase, quartz, and calcite decreased with TiO_2 content. Therefore, it can be inferred that the combination of TiO_2 and RCP only occurred on the surface of RCP, whether this combination was caused by chemical or physical effect. As shown in Fig. 1(b),

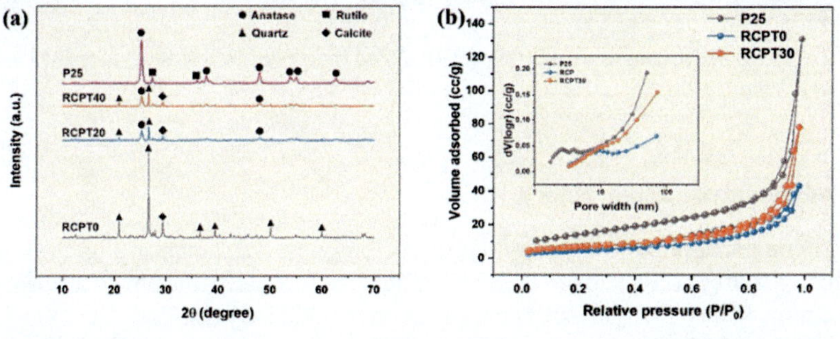

Fig. 1. (a) X-ray diffraction patterns of RCP0, RCPT20, RCPT40, and P25; (b) BET surface area of P25, RCPT0, and RCPT30

the N_2 adsorption-desorption isotherms of P25 and RCPT30 were a hybrid of type III and type V according to the classification of IUPAC, and small hysteresis loops can be observed, indicating that mesopore structure existed in these two samples. The pore size distribution curves confirmed that the mesopore content of RCPT0 was significantly lower than that of P25 and RCPT30, and BTE surface area of RCPT0 (18.7427 m^2/g) was also lower than others, suggesting that RCPT0 had poor adsorption ability. After modified by nano-TiO_2, BET surface area of RCPT30 was enhanced remarkably. It can be reasoned that nano-TiO_2 distributed well on the surface of RCP and formed a new porous structure on RCP surface, which was conductive to accelerate the photocatalytic reaction [5].

Figure 2 presents the micro-morphologies of RCP0 and RCPT30. As shown in Fig. 2(a), it can be seen that the particle size of RCP0 was uniform after grinding. As depicted in Fig. 2(b), the RCP surface was rough and porous, which can provide abundant caves for accommodating nano-TiO_2 particles and contribute to anchor TiO_2 [9]. After loaded with TiO_2, the surface of RCP became flocculent as shown in Fig. 2(c), and the magnified image in Fig. 2(d) revealed the detail of nano-TiO_2 particles as the yellow arrows indicated, and nano-TiO_2 particles were firmly grafted on the surface of RCP. These flocculent substances were proved to be TiO_2 by EDS results given in Fig. 2(e) and (f).

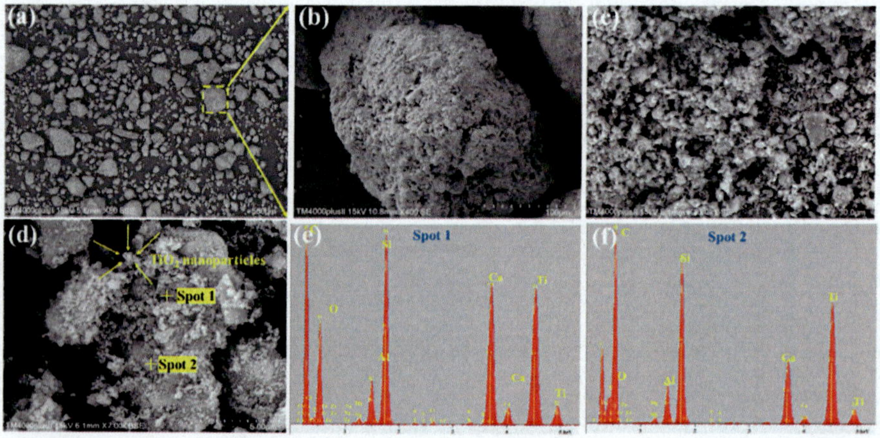

Fig. 2. (a)-(b) SEM images of RCPT0; (c)-(d) SEM images of RCPT30; (e)-(f) EDS result

3.2 Photocatalytic Performance

Figure 3 compares photocatalytic efficiency of samples. As shown in Fig. 3(a), before illumination, the concentration of RhB had a slight decline due to the physical adsorption. After turning on the UV-light, RhB concentration immediately decreased, but the decline rate differed with TiO_2 content. RCP0 exhibited the lowest degradation efficiency, and only about 10% of RhB was removed until 80 min, which indicated that RCP had little photocatalytic activity but a certain adsorption ability. P25 and RCPT40 held the highest photodegradation rate, and RhB was completely removed within 80 min. However, the apparent reaction rate constant of P25 (0.4686 min^{-1}) was 1.18 times higher than that of RCPT40 (0.03951 min^{-1}) as shown in Fig. 3(b). RCPT20 and RCPT30 showed the second highest photocatalytic degradation rate, and 86% of RhB was removed within 80 min. The reaction rate of RCPT10 (0.00513 min^{-1}) was much lower than that of RCPT20 (0.01593 min^{-1}). RCPT40 with only 40% addition of TiO_2 can achieve the equal photocatalytic degradation rate with pure TiO_2 within 60 min. This may be related to the balance between the quantity of reactive oxygen species (ROS) and the concentration of RhB. It can be concluded that RCP has the enhanced effect on improving the photocatalytic performance [10].

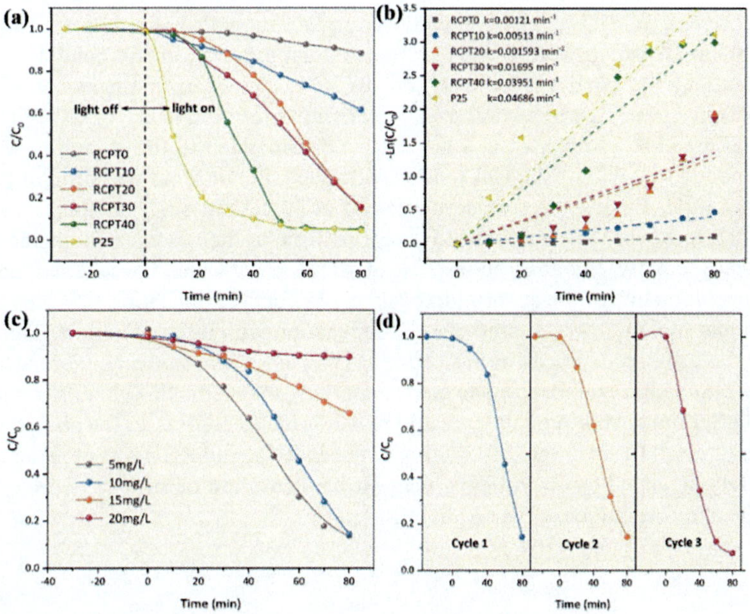

Fig. 3. (a) Photocatalytic degradation of RhB; (b) Linear fitting of -lnc/c_0 and time; (c) Effect of RhB concentration on photocatalytic efficiency of RCPT30; (d) Cyclic degradation curves over RCPT30

Previous studies reported that the concentration of RhB would affect the light transmittance [11]. Figure 3(c) shows the effect of RhB concentration on photocatalytic efficiency of RCPT30. Apparently, the degradation rate declined with the concentration of RhB. When RhB concentration was at a high level, the photocatalytic degradation was controlled by the amount of ROS. And when RhB concentration decreased, more ROS were produced due to stronger irradiation, leading to a higher degradation rate, which was consistent with the aforementioned analysis. For RCPT30, when RhB concentration was lower than 10 mg/L, the photocatalytic performance was excellent within 80 min whatever the RhB concentration. Therefore, it can be inferred that there is a threshold of RhB concentration for specific RCP/TiO_2 composite. If RhB concentration exceeds this threshold, the photocatalytic performance of RCP/TiO_2 composite would noticeably decrease. Figure 3(d) shows the photocatalytic performance of RCPT30 in the cyclic test. It can be observed that the photocatalytic performance of RCPT30 was stable as there was little decline in degradation rate of RhB even after tested for 3 times. Therefore, it is indicated that RCP/TiO_2 composites prepared in this study held excellent stability.

3.3 Binding Mechanism

The binding force between nano-TiO_2 and RCP is crucial for the durability of RCP/TiO_2 composites. The infrared spectra of RCP/TiO_2 composites were tested in Fig. 4. For P25, a wide absorption band was observed in the range of 400–700 cm^{-1}, which was attributed to the stretching vibration of Ti-O-Ti bonds and regarded as the characteristic absorption peaks of TiO_2 [12]. Meanwhile, two absorption peaks appeared at 3422 cm^{-1} and 1634 cm^{-1}, which were caused by the stretching vibration and bending vibration of the surface O-H bond, respectively. For RCPT0, the absorption peaks at 873 cm^{-1} and 1447 cm^{-1} were ascribed to the characteristic absorption peaks of CO_3^{2-} in calcite [13]. And the absorption peak at 1084 cm^{-1} corresponded to the stretching vibration of Si-O bond [14]. After RCP was loaded with TiO_2, the infrared spectrum of RCPT40 changed a little. Firstly, the surface hydroxyls at 32503750 cm^{-1} in RCPT0 widened and shifted to lower wavenumber [15], suggesting that the association degree of O-H increased. Secondly, the absorption peak of CO_3^{2-} at 873 cm^{-1} weakened, and there was also a blue-shift of the absorption peak at 1447 cm^{-1}, indicating that the chemical environment of CO_3^{2-} was altered due to the reaction with other materials. Thirdly, the characteristic absorption peak of Si-O bond at 1084 cm^{-1} was noticeably broadened and the absorption edge extended to 985 cm^{-1}, implying that some Si-O bonds were broken and Ti-O-Si bonds were probably created in situ [16]. According to the analysis above, it can be inferred that a chemical bonding between TiO_2 and RCP was generated which was mainly achieved by two effects, namely, condensation of surface hydroxyl group on calcite and formation of Ti-O-Si bond.

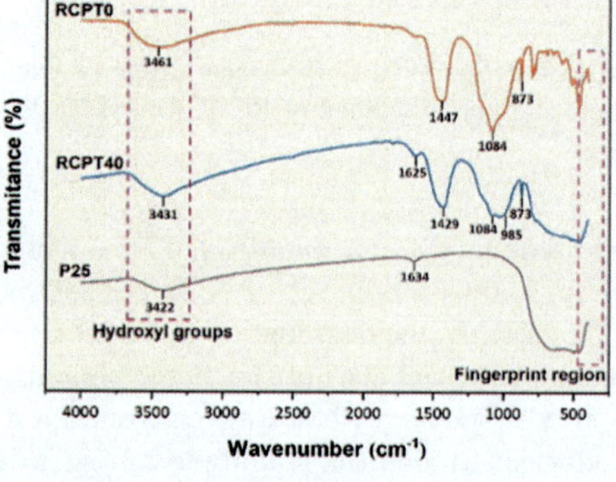

Fig. 4. Fourier-transform infrared spectra of RCP0, RCPT40, and P25

4 Conclusions

In summary, RCP/TiO_2 composites were successfully synthesized by a wet grinding method and exhibited high photocatalytic degradation performance towards RhB. RCP was a suitable carrier for TiO_2 due to its stability and rough surface texture. Nano-TiO_2 particles were dispersed homogenously on the surface of RCP and firmly combined with RCP by chemical bonds. The photocatalytic efficiency of RCP/TiO_2 composites increased with the content of TiO_2 and decreased with the concentration of RhB. RCP/TiO_2 composite with 40% addition of TiO_2 can achieve the same photocatalytic efficiency as pure TiO_2 within 60 min. There was a threshold concentration of RhB for RCP/TiO_2 composite with specific addition of TiO_2 under which high photocatalytic efficiency can be obtained. The cyclic photocatalytic degradation performance was excellent. Our current work proved that RCP is an ideal carrier for nano-TiO_2 and provided a new way to efficiently utilize RCP.

Acknowledgement. This work was supported by the National Key R&D Program of China (grant no. 2019YFC1906203).

References

1. Caijun, S., Yake, L., Jiake, Z., Wengui, L., Linlin, C., Zhaobin, X.: Performance enhancement of recycled concrete aggregate-a review. Journal of Cleaner Production **112**, 466-472 (2016)
2. Yong, J.K., Yun, W.C.: Utilization of waste concrete powder as a substitution material for cement. Constr. Build. Mater. **30**, 500–504 (2012)
3. Olugbenga, O.A., et al.: Designing out construction waste using BIM technology: Stakeholders' expectations for industry deployment. J. Clean. Prod. **180**, 375–385 (2018)
4. Qin, T., Zhiming, M., Huixia, W., Wan, W.: The utilization of eco-friendly recycled powder from concrete and brick waste in new concrete: a critical review. Cement Concr. Compos. **114**, 103807 (2020)
5. Youhei, N., Shuji, F., Taku, F.: Thermodynamics of removing crotamiton and its transformation byproducts from water by a rotating advanced oxidation contactor with zeolite/TiO_2 composite sheets. Chem. Eng. J. **380**, 122479 (2020)
6. Michaela, T., Jonáš, T., Kristina, Č, Vlastimil, M., Kateřina, M.K., Jana, S.: The stability of photoactive kaolinite/TiO_2 composite. Compos. B Eng. **67**, 262–269 (2014)
7. Gang, L., Wei, H., Yuming, H.: Investigation of microstructure and photocatalytic performance of a modified zeolite supported nanocrystal TiO_2 composite. Catalysts **9**(6), 502 (2019)
8. Maria, V., Luminita, A., Anca, D.: Fly ash-TiO_2 nanocomposite material for multi-pollutants wastewater treatment. J. Environ. Manage. **150**, 336–343 (2015)
9. Fazhou, W., Lu, Y., Guoxin, S., Lingyue, G., Shuguang, H.: The hierarchical porous structure of substrate enhanced photocatalytic activity of TiO_2/cementitious materials. Constr. Build. Mater. **64**, 488–495 (2014)
10. Jie, W., Sijia, S., Lei, P., Zhuoqun, X., Hao, D., Wei, L.: Preparation and Properties of $CaCO_3$-Supported Nano-TiO_2 Composite with Improved Photocatalytic Performance. Materials**12**(20), 3369 (2019)

11. Youji, L., Shuguo, S., Mingyuan, M., Yuzhu, O., Wenbin, Y.: Kinetic study and model of the photocatalytic degradation of rhodamine B (RhB) by a TiO_2-coated activated carbon catalyst: Effects of initial RhB content, light intensity and TiO_2 content in the catalyst. Chemical Eng. J. **142**(2), 147–155 (2008)
12. Tamara, B., Tetiana, G., Galyna, P., Valentyna, S., Jan, B., Michal, M.: Spectroscopic study of TiO_2 (rutile)–benzophenone heterogeneous systems. J. Mol. Struct. **614**(1), 315–324 (2002)
13. Jincai, C., Lan, X.: Controllable synthesis of calcium carbonate polymorphs at different temperatures. Powder Technol. **189**(1), 64–69 (2009)
14. Jie, W., Sijia, S., Hao, D., Wanting, C., Yu, L.: Preparation of a composite photocatalyst with enhanced photocatalytic activity: smaller TiO_2 carried on SiO_2 microsphere. Appl. Surf. Sci. **493**, 146–156 (2019)
15. Tamara, B., Galyna, P., Valentyna, S., Jan, B., Henryk, R.: IR-analysis of H-bonded H_2O on the pure TiO_2 surface. J. Mol. Struct. **700**(1), 175–181 (2004)
16. Vernon, A.Z., Charles, A.B.: The infrared spectra of some Ti-O-Si, Ti-O-Ti and Si-O-Si compounds. J. Phys. Chem. **61**(9), 1174–1177 (1957)

A Survey on Life Cycle Sustainability Assessment for Triple Bottom Line of Buildings

Kun Lu[1], Xueyuan Deng[1,2](✉), and Baoquan Cheng[3]

[1] School of Naval Architecture, Ocean and Civil Engineering, Shanghai Jiao Tong University, Shanghai 200240, China
dengxy@sjtu.edu.cn

[2] Shanghai Key Laboratory for Digital Maintenance of Buildings and Infrastructure, School of Naval Architecture, Ocean and Civil Engineering, Shanghai Jiao Tong University, Shanghai 200240, China

[3] School of Civil Engineering, Central South University, Changsha 410083, China

Abstract. The idea of life cycle sustainability assessment (LCSA) has recently been embraced in the building field, and academics and practitioners are paying more attention to it. However, the applications of LCSA for triple bottom line (TBL) are still not clear. Therefore, this survey tries to close this research problem by reviewing the peer-reviewed literature papers involved. After illustrating the definition and component of LCSA and TBL, 20 academic publications obtained from the Scopus database were chosen for further analysis and discussion. By summarizing these studies, economic, environmental, social indicators for LCSA are presented. Then, a methodological framework for LCSA consists of four steps: 1) system boundary definition; 2) inventory; 3) impact assessment; 4) interpretation. Finally, the implementations of building information modelling, artificial intelligence, and internet of things will be future directions. Findings from this review are thought to be valuable for academics and practitioners.

Keywords: Life cycle sustainability assessment (LCSA) · Literature review · Triple bottom line (TBL) · Sustainable construction

1 Introduction

Human benefit from the building industry because it enhances their quality of life and satisfies their socioeconomic requirements. Nevertheless, it generates a few of adverse environmental impacts and caused environmental degradation. The construction industry is generally regarded as a non-sustainable sector due to its substantial environmental effects, social risks, and low productivity. This low sustainability performance of the building industry presents a unique chance to improve its performance, thus enhancing the world-wide sustainability.

Sustainable development was described as meeting current demands without jeopardizing the ability of future generations to meet their own needs. For improving the sustainability of building field, the triple bottom line (TBL) has been included into the

W. Guo and K. Qian (Eds.): GBCESC 2022, LNCE 211, pp. 141–150, 2023.
https://doi.org/10.1007/978-981-19-5217-3_13

life cycle sustainability assessment (LCSA). Using LCSA in the building industry could be a good strategy to increase long-term sustainability. As a result, Goh et al. revisited TBL for sustainable construction [1]. Backes & Traverso reviewed the application of LCSA in the construction sector [2]. Llatas et al. reviewed the integration of LCSA and building information modelling during design stages [3]. However, the status of applications of LCSA for TBL in the building field is unclear, and a critical review of that is scarce. As a result, the goal of this research is to fill that gap by evaluating the relevant peer-reviewed publications.

Excluding the introduction, Sect. 2 delivers a concise conceptual underpinning for LCSA and TBL; Sect. 3 explains the research methodology of this paper; Sect. 4 summarizes existing research; Sect. 5 reviews the indicators and framework for LCSA of buildings, and discusses the intelligent technology for future works. Section 6 brings the major findings to a close.

2 Literature Review

2.1 Life Cycle Sustainability Assessment

Life cycle sustainability assessment (LCSA) consists of: 1) life cycle costing (LCC), also called as economic life cycle assessment; 2) life cycle assessment (LCA), also known as environmental LCA; 3) social life cycle assessment (S-LCA). LCSA aims to address the three dimensions of economic, environmental, and social performances for sustainable assessment. In other words, LCC, LCA, S-LCA are the methodology of economic assessment, environmental assessment, and social assessment, as shown in Fig. 1.

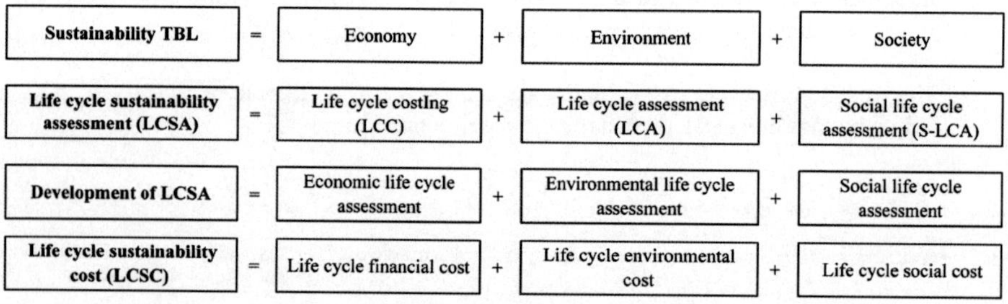

Fig. 1. Life cycle sustainability assessment and its relative information

2.2 Triple Bottom Line

What has been widely accepted are three dimensions framed around economic, environmental, and social responsibilities, which are called sustainability triple bottom line (TBL). There are three stakeholders around the sustainable TBL: profit, planet, and people, which is known as the 3P triple bottom line. More information about TBL is shown in Fig. 2.

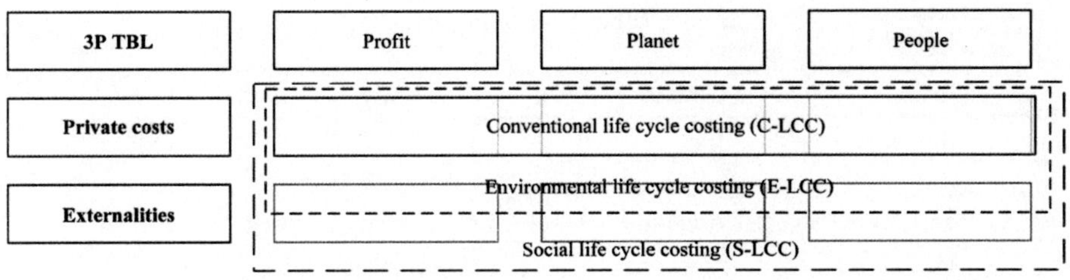

Fig. 2. Triple bottom line and its relative information

3 Research Methodology

This study follows four steps: 1) Conducting a search on a search engine; 2) identifying relevant articles; and 3) presenting the results and discussing them. Scopus is utilized in this study to search research articles, and its advanced search is carried out using the following code:

*TITLE-ABS-KEY(("life cycle sustainability assessment" OR "life cycle sustainable assessment" OR LCSA OR ("life cycle" AND "triple bottom *line*" OR TBL)) AND *building*) AND SRCTYPE (j) AND LANGUAGE (English) AND DOCTYPE (ar OR re).*

A total of 80 articles have been retrieved by this code until December 2021. Through the title, abstract, keywords, and conclusion of these articles, publications not related to this research were further excluded. Consequently, 20 articles regarding the LCSA of buildings for TBL were found (listed in Table 1).

4 Result

The research materials connected to LCSA for TBL in the building field are included in Table 1. *International Journal of Life Cycle Assessment* has the most significant publication number among all the articles. Three levels are identified for the research interests or applications of LCSA for TBL, including: (1) building component; (2) building project; (3) building industry. Among them, the application of building project level is the most, involving residential buildings, educational buildings, office buildings, heritages, and so on.

Table 1. Existing researches of LCSA of building.

Ref	Journal	Country	Case study	Level
[4]	J. Build. Eng.	Spain	A residential building	Project
[5]	Sust. Cities Soc.	India	A mass housing project	Project

(*continued*)

Table 1. (*continued*)

Ref	Journal	Country	Case study	Level
[6]	Buildings	Iran	Two frames for multi-story buildings	Project
[7]	Sust. Cities Soc.	Australia	A university building	Project
[8]	Int. J. Life Cycle Assess.	Malaysia	Concrete construction techniques for a residential building	Project
[9]	Build. Environ.	Brazil	A residential building	Project
[10]	Environ. Sustain. Indic.	Australia	Detached houses	Project
[11]	J. Constr. Eng. Manage.	the United States	Fugitive dust control	Component
[12]	Build. Environ.	the United States	Window-to-wall ratios of office buildings	Component
[13]	J. Build. Eng.	Malaysia	Window renovations in a school building	Component
[14]	J. Environ. Manage.	Australia	A conventional building	Project
[15]	Sustain. Dev.	Singapore	Three structural designs for a student residential hall	Project
[16]	Int. J. Life Cycle Assess.	Spain	Restoration of a medieval fortress with cultural heritage	Project
[17]	J. Build. Eng.	Malaysia	Different timber structures for a residential building	Project
[18]	Int. J. Life Cycle Assess.	Hong Kong, China	A public rental house	Project
[19]	Eng. Struct.	the US	Reinforced concrete buildings in seismic regions	Project
[20]	J. Environ. Plan. Manag.	Canada	Timber frame and concrete frame buildings	Project
[21]	Clean Technol. Environ. Policy	-	Net-zero buildings	Project

(*continued*)

Table 1. (*continued*)

Ref	Journal	Country	Case study	Level
[22]	Int. J. Life Cycle Assess.	the US	US buildings	Industry
[23]	Int. J. Life Cycle Assess.	France, Sweden, Netherlands	Refurbishment of three different buildings	Project

5 Discussion

5.1 Indicators and Impacts

Through a summary of existing literature, a list of impact categories and indicators is shown in Table 2. As shown in Table 2, there is a lack of uniform results of TBL indicators for buildings.

Economic Indicators. Economic indicators of building projects include cost, benefit, and profit. Additionally, some studies added carbon tax saving to economic indicators [5, 10, 14]. It is worth noting that for LCSA in building industry level, foreign purchase and Gross Domestic Product (GDP) also included [22].

Environmental Indicators. Environmental indicators include Global Warming Potential (GWP) (also called greenhouse gas emission), Fossil Depletion Potential (FDP) (also called energy consumption), Acidification Potential (AP), Eutrophication Potential (EP), Ozone Depletion Potential (ODP), Smog Formation Potential (SFP), Water Consumption (WC), Land Use (LU), Human toxicity (HT), Biodiversity depletion (BD), material waste depletion (MWD), Ecotoxicity potential (ETP), Smog formation potential (SFP), Photochemical smog potential (PSP), Ionizing radiation (IR), Indoor air (IA) and so on.

Social Indicators. Social indicators can be divided into main five stakeholders, including: 1) Workers: Associational and collective bargaining rights, Occupational safety and health, Living/non-poverty wages, Use of forced labor, Discrimination. 2) Local community: Job creation, Land use, Protection of cultural heritage, Levels of industrial water use, Presence of complaints by local community, Extraction of material resources, Pollution level, Use of local labor, Waste generation. 3) Society: Installation difficulty, GDP contribution, Functionality and appeal, Durability, Maintenance difficulty. 4) Occupant: Consumer complaints, Sound barrier. 5) Value chain: On-time payment, Monopoly and anti-competitive behavior [8, 13].

The classification of these above indicators is controversial sometimes. For example, noise is classified as an environmental indicator by Tokede et al. [7], but as a social indicator by Janjua et al. [10] and Janjua et al. [14]. In addition, in the study of building industry level by Onat et al. [22], its indicators are difference with the project level.

Table 2. Indicators of LCSA for TBL of buildings.

Ref	Economic	Environmental	Social
[4]	Cost	GWP	Working hours
[5]	Cost, Social Cost of Carbon	GWP, WC, LU, FDP	Accessibility, Job Generation Potential
[6]	Cost	GWP	Safety and health, Fair salary, Local employment, Local community acceptance, Extraction of material resources
[7]	Cost	IR, HT, AP, BD, LU, GWP, FDP, Noise, WC, MWD, IA	Workers, Occupant, Local community, Society, Value chain
[8]	Cost	GWP, ODP, HT, FDP, ETP	Workers, Occupant, Local community, Society, Value chain
[9]	Cost	GWP, AP, EP	Fair Wage Potential
[10]	Cost, Potential Savings, Benefit-Cost ratio, Net benefit, Carbon Tax Saving	GWP, AP, EP, LU, BD, WC, FDP, MWD	Affordability, Living conditions, Thermal comfort, Noise, Local material sourcing
[11]	Cost	FDP WC GWP AP, EP, ETP, ODP, SFP	Social cost
[12]	Cost	AP, EP, GWP, ODP, SFP	Occupant satisfaction
[13]	Cost	GWP	Workers, Local community, Society, Occupant, Value chain
[14]	Cost, Potential savings, Benefit-cost ratio, Net benefit, Carbon tax saving	GWP, AP, EP. LU, BD, WC FDP MWD	Affordability, Living conditions, Thermal comfort, Noise, Local material sourcing, Energy conservation
[15]	Cost	SFP, BD, MWD	Workers, Occupant, Local community, Society
[16]	-	MWD, FDP, WC, GWP, EP	Workers, Consumers, Local communities, Society
[17]	Cost, Present Value	GWP, FDP, AP, EP HT	Foreman wage, Job creation

(*continued*)

Table 2. (*continued*)

Ref	Economic	Environmental	Social
[18]	Cost	HT, GWP, PSP, IR, SP, LU EP, ETP, FDP, MWD	Workers, Local community, Society
[19]	Cost, Downtime	Environmental performance score	Fatalities
[20]	Cost	FDP, GWP, AP, SP, MWD, EP, PSP, ODP, ETP, IA, LU	Occupant comfort, Safety, Seismic resistance, Fire resistance, Affordability
[21]	Cost	WC, FDP, MWD, GWP	-
[22]	Foreign purchase, Business profit, GDP	GWP, FDP, WC, MWD	Income, Government tax, Injuries
[23]	Cost	ReCiPe indicate	-

5.2 Framework and Methodology

By summarizing these 20 studies, a general LCSA methodological framework is summarized for TBL in building field, as shown in Fig. 3. This framework is comprised of four steps:

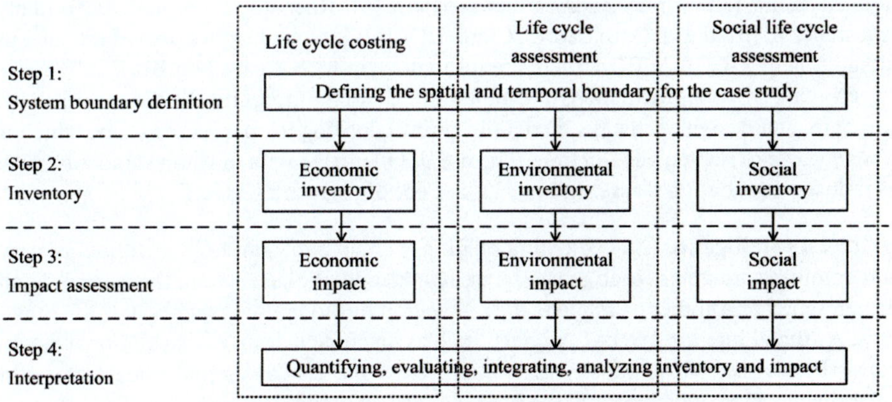

Fig. 3. The general LCSA framework for TBL in building field.

Step 1: system boundary definition

The system boundary defines shall be considered in the LCC, LCA, and S-LCA. The research scope encompasses the entire whole life of buildings, including material production stage, construction stage, operation stage, maintenance stage, and demolition stage. Then, interpreting the target audience and determining the functional unit are also important aspects of this phase.

Step 2: inventory
Bills of quantities will be used to gather project information, covering resource and material consumption, and other detailed data, which is used to establish an input and output inventory for LCC and LCA. Furthermore, additional project data can be gathered through questionnaire surveys of project managers or residents, which is required for S-LCA.

Step 3: impact assessment
The findings are acquired via the measurement of the flow, which is then totaled to create an overall effect, which includes economic, environmental, and social consequences.

Step 4: interpretation
This phase involves a more in-depth examination of the data obtained from the three phases. The findings of the inventory and impact are measured and analyzed according to the identified goal and scope. Then, Improvement measures are proposed to optimize sustainable TBL performance, as well as presenting conclusions and recommendations.

5.3 Intelligent Technology

Building Information Modeling (BIM). It is challenging to get a comprehensive LCSA result for construction projects since assessing all aspects is a complicated task in building's life cycle. The inventory analysis process for buildings is more difficult and complex compared with other products because different types of machines, various kinds of construction materials, and chaotic construction techniques. Building information modeling (BIM) may address this problem [3]. Recently, there have been many researches focused to BIM-based LCC and LCA [24]. However, because of the lack of integration of S-LCA in BIM, there are difficulties in LCSA based on BIM.

Existing BIM-based methods of evaluating TBL of buildings rely on the specific BIM software, and designers always modeling again according to specialized tools, which is a huge waste of manpower and time. The use of OpenBIM with uniform standards, such as industry foundation classes, allows LCSA out of software vendors.

Artificial Intelligence (AI). Existing methods cannot carry out LCSA without precise and complete drawings, such as, in the early conceptual phase where there is no detail design scheme. Another example is the renovation of old buildings without CAD drawings. Artificial intelligence (AI) models, such as machine learning, convolutional neural network, or artificial neural network, can predict the environmental impact and cost performance [25], which provides a potential way to evaluate building LCSA.

In addition, using big data and AI, owners or consultants can quickly predict the TBL performance of new buildings through the accumulation of historical LCSA data.

Internet of Things (IoT). The current research mainly focused on the simulation of LCSA results, but the research is still scarce about monitoring the actual value of LCSA using Internet of things (IoT). The IoT technology can carry out real-time perception and dynamic collection of a large amount of information generated in the life cycle of buildings, realizing the real-time acquisition and summary of all kinds of information and data. The real TBL data by IoT can be compared with the simulated TBL results, thereby detecting the implementation of the LCSA and facilitating the audit of the TBL results. Although IoT has been widely used in other industries [26], its application in the LCSA of building industry is still scarce.

6 Conclusion

Following a systematic review of 20 researches particularly related to LCSA for TBL of buildings, the main conclusions include: TBL indicators are involved in buildings' LCSA, including economic (such as cost, benefit, and profit), environmental (such as GWP, FDP, AP, EP, ODP), and social (five main stakeholders: worker, occupant, local community, society, and value chain). Then, a general framework is summarized for LCSA of buildings, consisting of: 1) system boundary definition; 2) inventory; 3) impact assessment; 4) interpretation.

In the future, intelligent methods could be combined with LCSA, including: 1) BIM offers a way to quickly forecast the TBL performances, and OpenBIM will break the limitations of specific LCSA tools; 2) artificial intelligence can predict TBL result through historic LCSA data; 3) internet of things will be an important tool for monitoring and auditing LCSA results.

The main limitation of this survey is that the number of 20 papers is limited, because the scope of this review only includes peer-reviewed journal papers.

References

1. Goh, C.S., et al.: Revisiting triple bottom line within the context of sustainable construction: a systematic review. J. Cleaner Production **252**, 119884 (2020)
2. Backes, J.G., Traverso, M.: Application of life cycle sustainability assessment in the construction sector: a systematic literature review. Processes, **9**(7), 1248 (2021)
3. Llatas, C., Soust-Verdaguer, B., Passer, A.: Implementing life cycle sustainability assessment during design stages in building information modelling: from systematic literature review to a methodological approach. Building and Environment **182**, 107164 (2020)
4. Soust-Verdaguer, B., et al.: How to conduct consistent environmental, economic, and social assessment during the building design process. a BIM-based life cycle sustainability assessment method. J. Building Eng. **45**, 103516 (2022)
5. Francis, A., Thomas, A.: A framework for dynamic life cycle sustainability assessment and policy analysis of built environment through a system dynamics approach. Sustainable Cities and Society **76**, 103521 (2022)
6. Oladazimi, A., et al.: Sustainability identification of steel and concrete construction frames with respect to triple bottom line. Buildings **11**(11), 565 (2021)
7. Tokede, O.O., Roetzel, A., Ruge, G.: A holistic life cycle sustainability evaluation of a building project. Sustainable Cities and Society **73**, 103107 (2021)
8. Balasbaneh, A.T., Sher, W.: Life cycle sustainability assessment analysis of different concrete construction techniques for residential building in Malaysia. Int. J. Life Cycle Assessment **26**(7), 1301–1318 (2021). https://doi.org/10.1007/s11367-021-01938-6
9. Figueiredo, K., et al.: Sustainable material choice for construction projects: a life cycle sustainability assessment framework based on BIM and Fuzzy-AHP. Building and Environment **196**, 107805 (2021)

10. Janjua, S.Y., Sarker, P.K., Biswas, W.K.: Sustainability implications of service life on residential buildings – an application of life cycle sustainability assessment framework. Environmental and Sustainability Indicators **10**, 100109 (2021)
11. Raymond, A.J., et al.: Life cycle sustainability assessment of fugitive dust control methods. J. Construction Eng. Manage. **147**(3), 04020181 (2021)
12. Phillips, R., et al.: Triple bottom line sustainability assessment of window-to-wall ratio in US office buildings. Building and Environment **182**, 107057 (2020)
13. Balasbaneh, A.T., Yeoh, D., Zainal Abidin, A.R.: Life cycle sustainability assessment of window renovations in schools against noise pollution in tropical climates. J. Building Eng. **32**, 101784 (2020)
14. Janjua, S.Y., Sarker, P.K., Biswas, W.K.: Development of triple bottom line indicators for life cycle sustainability assessment of residential bulidings. J. Environmental Manage. **264**, 110476 (2020)
15. Liu, S., Qian, S.: Towards sustainability-oriented decision making: Model development and its validation via a comparative case study on building construction methods. Sustain. Dev. **27**(5), 860–872 (2019)
16. Mohaddes Khorassani, S., Ferrari, A.M., Pini, M., Settembre Blundo, D., García Muiña, F.E., García, J.F.: Environmental and social impact assessment of cultural heritage restoration and its application to the Uncastillo Fortress. Int. J. Life Cycle Assessment **24**(7), 1297–1318 (2018). https://doi.org/10.1007/s11367-018-1493-1
17. Balasbaneh, A.T., Marsono, A.K.B., Khaleghi, S.J.: Sustainability choice of different hybrid timber structure for low medium cost single-story residential building: environmental, economic and social assessment. J. Building Eng. **20**, 235–247 (2018)
18. Dong, Y.H., Ng, S.T.: A modeling framework to evaluate sustainability of building construction based on LCSA. Int. J. Life Cycle Assessment **21**(4), 555–568 (2016). https://doi.org/10.1007/s11367-016-1044-6
19. Gencturk, B., Hossain, K., Lahourpour, S.: Life cycle sustainability assessment of RC buildings in seismic regions. Eng. Struct. **110**, 347–362 (2016)
20. Hossaini, N., et al.: AHP based life cycle sustainability assessment (LCSA) framework: a case study of six storey wood frame and concrete frame buildings in Vancouver. J. Environ. Planning Manage. **58**(7), 1217–1241 (2015)
21. Hossaini, N., Hewage, K., Sadiq, R.: Spatial life cycle sustainability assessment: A conceptual framework for net-zero buildings. Clean Technol. Environ. Policy **17**(8), 2243–2253 (2015)
22. Onat, N.C., Kucukvar, M., Tatari, O.: Integrating triple bottom line input-output analysis into life cycle sustainability assessment framework: the case for US buildings. Int. J. Life Cycle Assess. **19**(8), 1488–1505 (2014)
23. Ostermeyer, Y., Wallbaum, H., Reuter, F.: Multidimensional Pareto optimization as an approach for site-specific building refurbishment solutions applicable for life cycle sustainability assessment. Int. J. Life Cycle Assessment **18**(9), 1762–1779 (2013). https://doi.org/10.1007/s11367-013-0548-6
24. Lu, K., et al.: Integration of life cycle assessment and life cycle cost using building information modeling: a critical review. J. Cleaner Production **285**, 125438 (2021)
25. Elmousalami, H.H.: Artificial intelligence and parametric construction cost estimate modeling: state-of-the-art review. J. Construction Eng. Manage. **146**(1), 03119008 (2020)
26. Gu, F., et al.: Internet of things and Big Data as potential solutions to the problems in waste electrical and electronic equipment management: An exploratory study. Waste Manage. **68**, 434–448 (2017)

Spatial and Temporal Prediction of Carbon Peaking Goals and Zonal Governance Approaches to Achieve Carbon neutrality—A Case Study from Zhejiang Province, China

Weiwu Wang[1,2](✉), Huan Chen[1,2], Ronglu Yang[1], Biyan Wang[1], and Yuxin Yang[1,2]

[1] College of Civil Engineering and Architecture, Zhejiang University, Hangzhou 310058, China
weiwuwang@zju.edu.cn

[2] Center for Balance Architecture, Zhejiang University, Hangzhou 310058, China

Abstract. With global warming, the United Nations General Assembly once again emphasized the control of greenhouse gas emissions. In the General Assembly, China proposed the ambitious goal of achieving carbon peaking by 2030 and carbon neutrality by 2060. In order to study the potential and policy direction of Zhejiang Province in achieving the goal of carbon neutrality, based on the STIRPAT model and combined with ridge regression analysis, two models were constructed to predict the carbon emissions of Zhejiang Province by using scenario analysis. The time to carbon peak was determined by numerical calculation. According to the spatial differences of carbon emission driving factors of low-level cities in Zhejiang Province, the timetable and roadmap for achieving the "dual carbon" goal in Zhejiang Province are put forward. This study also explored the establishment of a regional governance mechanism for carbon peaks in Zhejiang Province. According to carbon peaks' spatial differential distribution characteristics, a differential low-carbon optimization strategy for land space is proposed.

Keywords: Carbon peak · Carbon neutrality · Scenario simulation · STIRPAT model · Path choice · Subarea management

1 Introduction

How to achieve carbon peak and promote economic growth and green and low-carbon development are the dual tasks that China is currently focusing on. It is an important prerequisite for China to achieve carbon peaking and carbon neutrality goals [1]. At present, significant differences in the carbon emission situation of various provinces in China have also emerged and will continue in the long run. Therefore, each region in China must choose a suitable path to achieve the dual-carbon target according to local conditions. Zhejiang Province, as the birthplace of General Secretary Xi Jinping's "Two

W. Guo and K. Qian (Eds.): GBCESC 2022, LNCE 211, pp. 151–165, 2023.
https://doi.org/10.1007/978-981-19-5217-3_14

Mountains" concept, has taken the lead in tackling climate change and has achieved good results. However, under the pressure of rapid population growth and insufficient power for the transformation of traditional industries, the development of a low-carbon economy in Zhejiang Province is under great pressure as a whole [2]. In order to better promote carbon peaking and carbon neutrality, it is necessary to explore the path to achieve carbon peaking in Zhejiang Province and formulate a development plan for carbon peaking so as to provide an effective reference for the implementation of total carbon emission control in Zhejiang Province and the formulation of related policies.

In this context, many scholars have carried out a lot of research on the current status of carbon emissions and the goal of "carbon peaking", covering the driving factors of carbon emissions, carbon dioxide peak forecasting and many other fields. In terms of the driving factors of carbon emissions, Ma Zheng and Cui Yuhong [3] used the structural decomposition model to decompose the changes in China's carbon emissions and mainly carried out research on the structural effect, technology effect, demand effect, and total amount effect. Lin Boqiang [4], Xu Guoquan [5], etc., respectively established a factor decomposition model of China's carbon emissions, quantifying factors such as per capita GDP, urbanization rate, energy structure.

In terms of carbon dioxide peak prediction methods, most scholars use a combination of model prediction and scenario analysis. The STIRPAT model proposed by York et al. [6] is based on the IPAT model, which is more scalable and random and has been widely used in energy emission prediction and evaluation research. The combination of the STIRPAT model and statistical method realizes the prediction function of the model, which is mainly the application of ridge regression and time series analysis. Yan Wei [7] predicted the changes and trends of total carbon emissions in the future by setting scenarios. Huang Rui et al. [8] obtained a multivariate linear model of energy consumption carbon emissions and population, per capita GDP, energy intensity, etc.

This paper mainly uses the emission coefficient method to measure the carbon emissions of fossil energy in Zhejiang Province from 2010 to 2020. Using the STIRPAT expansion model and regression analysis combined with the scenario analysis method to simulate the carbon emissions of Zhejiang Province from 2021 to 2040, the carbon peak time and carbon emissions peak are predicted and combined with the prediction results to explore the "carbon peak" realization path, in order to achieve the carbon dioxide emission control goal of Zhejiang Province, improve the selection of emission reduction measures during the "14th Five-Year Plan", improve the ability to deal with climate change, and promote the sustainable economic and social development of Zhejiang Province, it provides a specific decision-making basis.

2 Data Sources and Research Methods

2.1 Overview of the Study Area

Zhejiang Province is located in the southeastern coastal area of China and has 11 prefecture-level cities under its jurisdiction (including two sub-provincial cities). Zhejiang Province is a small resource province in China, but it is a big economic province (Fig. 1).

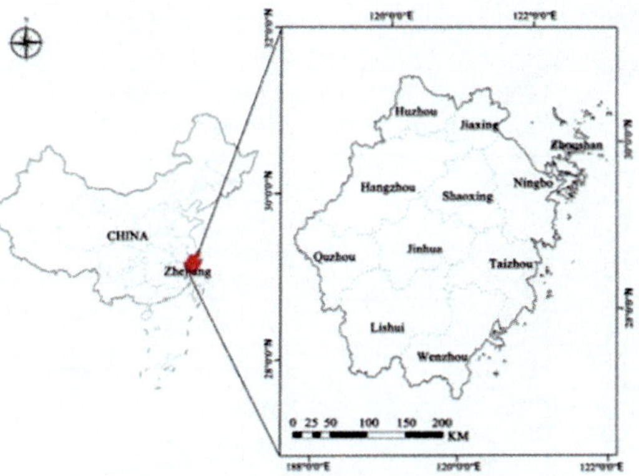

Fig. 1. Location map of Zhejiang Province

2.2 Data Sources

This study takes 2021 as the base period and 2021 to 2040 as the forecast period. The economic and population data involved are from the "Zhejiang Statistical Yearbook" (2010–2020) and the statistical bulletin of the national economic and social development of prefecture-level cities. The energy consumption data comes from the China Energy Statistical Yearbook (2010–2020) and the Zhejiang Statistical Yearbook. The energy structure is calculated based on the proportion of coal consumption to total energy consumption. The energy intensity is calculated based on the ratio of total energy consumption to GDP. The ratio was calculated, and the individual missing data were filled by interpolation. The selection of scene-setting parameters mainly refers to the literature [9].

2.3 Research Methods

Calculation of Carbon Dioxide Emissions. At present, the estimation method of carbon dioxide emissions is mainly based on the estimation of energy carbon sources, and the most widely used method is the carbon emission coefficient method [10]. According to IPCC guidelines, carbon dioxide emissions are measured based on primary energy consumption. The calculation formula is as follows:

$$I = \sum_{i=1}^{n} F_i * E_i = \sum_{i=1}^{n} \frac{O_i * NCV_i * CC_i}{CF_i} * E_i * \frac{44}{12} \quad (1)$$

In the formula: I is the carbon dioxide emission, i is the type of primary energy, Fi is the carbon emission coefficient of energy i, Ei is the consumption of energy i, Oi is the carbon oxidation rate of energy i, NCVi is the average low-level calorific value of energy i, CCi is the carbon content per unit calorific value of energy i, and CFi is the conversion coefficient of energy i. The carbon emission coefficients of various energy sources in Eq. (1) are from references [11, 12] (Table 1).

Table 1. Carbon emission factor for various energy sources.

Energy type	Carbon emission coefficient	Energy type	Carbon emission coefficient
Raw coal	0.76	Kerosene	0.57
Charred coal	0.86	Diesel oil	0.59
Crude oil	0.59	Fuel oil	0.62
Gasoline	0.55	Natural gas	0.45
Wash the fine coal	0.76	Liquefied petroleum gas	0.50
Coal oven gas	0.35	Other gas	0.35
Other coking products	0.64	Other petroleum products	0.59

Model Selection. The STIRPAT model is currently the most widely used and highly recognized in the study of carbon emission peaks. The model can be used to quantitatively study the relationship between carbon emissions and various influencing factors. It has good flexibility and certain room for expansion. The basic expression of the model is:

$$I = aP^bA^cT^d\mathrm{e} \tag{2}$$

In the formula: I represent the environmental pressure, P, A, and T represent the population size, affluence, and technological level, respectively, a is the model coefficient, b, c, and d represent the elasticity coefficients of the corresponding variables, respectively, and e is the model error term.

According to existing research, six factors, population, per capita GDP, urbanization rate, industrial structure, energy intensity and energy structure, are widely used in carbon emission-related research. All of them can significantly affect carbon dioxide emissions [13]. The above factors are introduced into the model, and finally, the model is further extended to:

$$lnI = lna + blnP + clnA + dlnU + f2lnIS + glnEI + hlnES + lne \tag{3}$$

where I is carbon dioxide emissions, P is population size, A is GDP per capita, U is urbanization rate, IS is industrial structure, ES is energy structure, EI is energy intensity, a is model coefficient, b, c, d, f, g, and h represent the elastic coefficients of the corresponding variables, respectively, and e is the model error term. The specific variable descriptions are shown in Table 2.

Table 2. Key variables and descriptions.

Variable	Explanation	Unit
CO_2 Emissions (I)	Total carbon dioxide emissions in that year	10^6t
Population size (P)	Resident population at the end of the year	Thousands of people
GDP per capita (A)	GDP / year-end resident population	Ten thousand yuan, / person
Urbanization rate (U)	The ratio of the urban resident population to the total population	Natural gas
Industrial structure (IS)	The ratio of the secondary industry to GDP	%
Energy intensity (EI)	Energy consumption per unit of GDP	t standard coal / ten thousand yuan
Energy structure (ES)	The ratio of coal consumption to total energy consumption	%

Development Scenario Settings. The "scenario analysis method" can be used to predict certain trends in future carbon emissions changes, and the accuracy of the prediction can be controlled by setting parameters. In order to explore the prospect of "carbon peaking" in Zhejiang Province, it is necessary to model the change rates of six factors, including future population size, per capita GDP, urbanization rate, industrial structure, energy structure and energy intensity. The scenario simulation takes 2020 as the base year for the forecast value in Zhejiang Province, and the forecast period is set from 2021 to 2040. In accordance with China's national "five-year" plan, the rate of change of different scenarios is adjusted in a five-year cycle. When setting the scenario conditions, the influence of the current situation of economic development in Zhejiang Province and the impact of the new crown pneumonia epidemic on the changes of various factors was comprehensively considered. The specific settings are as follows:

Scenario Setting of Population Size (P). The United Nations (2009) judged that China's population in the future would reach its maximum value in the 2030s, and the population will gradually begin to grow negatively after entering the 2040s. Zhejiang Province, as an economically developed area along the eastern coast, will also show a gradually decreasing trend in its population growth rate. Therefore, the population development of Zhejiang Province is set with reference to the United Nations (2009) forecast of China's population in the future, and the average annual population growth rate of Zhejiang Province is set at a medium speed. That is, from 2021 to 2030, the average annual growth rate of the population is 0.30%. From 2031 to 2040, the average annual growth rate of the population is 0.25%.

Scenario Setting of Per Capita GDP (a). In 2020, the per capita GDP of Zhejiang Province had reached 101,000 yuan per person, and the per capita GDP exceeded the development goal of 130,000 yuan per person during the "14th Five-Year Plan" period. Based on this, this paper sets the growth rate of per capita GDP in Zhejiang Province is at a medium speed. That is, from 2021 to 2030, the per capita GDP growth rate is 5.25%. From 2030 to 2040, the per capita GDP growth rate is 4.25%.

Scenario Setting for the Urbanization Rate (U). At present, the urbanization rate of the permanent population in Zhejiang Province has exceeded 70%, and the overall urbanization has entered a mature development period. In 2020, the urbanization rate will reach 72.2%, and it is proposed that "by 2025, the urbanization rate of the province's permanent population will reach 75%. By 2035, the urbanization rate of the province's permanent population will reach about 80%". In the past ten years, the growth rate of urbanization in Zhejiang Province has shown a trend of slowing down. In the future, the rural population in Zhejiang Province and the population outside the province will further gather in cities, but the urbanization rate will not increase significantly, and it is expected to drop to less than one percentage point per year. Based on this, this paper sets the average annual growth rate of urbanization in Zhejiang Province at a medium speed. That is, from 2021 to 2030, the average annual growth rate is 1.00%. From 2030 to 2040, the average annual growth rate is 0.72%.

Scenario Setting of Industrial Structure (iS). The energy consumed by the secondary industry is the main source of carbon dioxide. In 2019, Zhejiang's secondary industry accounted for about 45.50% of GDP. Based on this, this paper sets the rate of change of industrial structure in Zhejiang Province at a medium speed. That is to say, Zhejiang Province is in a period of rapid development from 2021 to 2040, and the proportion of the secondary industry will drop to 35% by 2040, with an average annual decline rate of 1.08%.

Scenario Setting of Energy Intensity (EI). By 2025, the province's energy consumption per unit of GDP will be reduced by 15%, with an average annual reduction of 3.2%". From 2021 to 2030, the average annual decline rate is 3.30%. From 2030 to 2040, the average annual decline rate is 3.00%.

Scenario Setting of Energy Structure (ES). The main source of carbon dioxide emissions in Zhejiang Province is the combustion of fossil fuels such as coal, oil, and natural gas, and the consumption of coal produces the largest amount of carbon dioxide. As of 2019, coal accounted for 45.30% of the energy consumption structure in Zhejiang Province, and coal has become the basic energy supporting the economic development of Zhejiang Province. According to the development requirements of low-carbon cities in Zhejiang Province, the burning of coal-based fossil energy will be reduced in the future, and the use of low-carbon and clean energy will be increased. Based on this, this paper assumes that by 2040, the proportion of coal consumption in Zhejiang Province will drop to 35%, with an average annual decline rate of 0.6%. The details of the setting of the change rate of each factor are shown in Table 3.

Table 3. Rate of change set for each factor.

Rate of change	Period	P	A	U	IS	EI	ES
Medium speed	2020–2025	0.32%	5.50%	1.10%	−1.28%	−3.30%	−0.70%
	2026–2030	0.28%	5.00%	0.90%	−1.18%	−3.30%	−0.60%
	2031–2035	0.26%	4.50%	0.82%	−1.08%	−3.10%	−0.60%
	2036–2040	0.24%	4.00%	0.62%	−0.78%	−2.90%	−0.50%
Fast speed	2020–2025	0.34%	6.00%	1.20%	−1.38%	−3.40%	−0.80%
	2026–2030	0.30%	5.50%	1.00%	−1.28%	−3.30%	−0.70%
	2031–2035	0.28%	5.00%	0.92%	−1.18%	−3.20%	−0.70%
	2036–2040	0.26%	4.50%	0.72%	−0.88%	−3.10%	−0.60%

Due to the tightening of resource constraints and the decline of environmental carrying capacity in Zhejiang Province, Zhejiang Province intends to achieve sustainable development by reversing the extensive growth pattern of high consumption and high carbon emissions in terms of social and economic development strategy. Zhejiang Province's "Thirteenth Five-Year Plan" Implementation Plan for Controlling Greenhouse Gas Emissions points out that green and low-carbon development should be regarded as a major strategy for social and economic development and an important way to build ecological civilization. In 2021, Zhejiang Province issued the "Zhejiang Carbon Peak Carbon Neutral Science and Technology Innovation Action Plan", which is the first carbon peak carbon-neutral science and technology innovation action plan introduced in the country. The plan proposes to closely follow the energy planning of Zhejiang Province and the green transformation needs of key industries and systematically promote the innovation and development of green and low-carbon technologies.

According to the social strategic layout of Zhejiang Province and the setting of the change rate of each influencing factor in the previous article, this paper designs two different carbon emission development models for Zhejiang Province in the next 20 years. The details of scenario combination settings are shown in Table 4.

Table 4. Carbon emission scenarios in Zhejiang Province.

Scenario combination	Scenario setting					
	P	A	U	IS	EI	ES
Baseline Scenario (S1)	medium	medium	medium	medium	medium	medium
Green Development Scenario (S2)	medium	medium	medium	fast	fast	fast

Baseline Scenario (S1): The rate of change of each influencing factor is set at a medium speed, aiming to explore the possible future trend of carbon emission changes based on the historical development law and the development goals of the 14th Five-Year

Plan. At the same time, the carbon peaking prospects of Zhejiang Province are forecasted under the condition of maintaining the existing policies.

Green Development Scenario (S2): The rate of change of industrial structure, energy intensity and energy structure are set to be fast, and the rate of change of other factors is set to be medium. In this scenario, it is assumed that Zhejiang Province actively implements the national low-carbon economy awareness, the government vigorously advocates new energy consumption, and enterprises increase investment in energy conservation and emission reduction. The actual energy consumption structure has been further optimized based on the medium and long-term plan of Zhejiang Province in 2030, which can fully take into account the needs of future social, economic and environmental development and ultimately achieve green development.

3 Results and Analysis

3.1 Prediction of Carbon Peak Period

STIRPAT Model Regression Analysis. Based on the relevant statistical data of Zhejiang Province from 2010 to 2020, the STIRPAT model was used to fit the carbon dioxide emissions and various influencing factors. In order to avoid the possibility of multicollinearity between variables, the ridge regression method was used for analysis [15, 16], and it was finally determined that the model tends to be stable when $k = 0.03$, the overall fitting degree is relatively good, and it can pass the significant sex test. The carbon emission prediction model of Zhejiang Province was constructed. The detailed regression process is shown in Table 5. The model equation is as follows:

$$lnI = 4.194 + 0.305lnP + 0.198lnA + 1.079lnU - 0.172lnIS + 0.073lnEI + 0.076lnES \quad (4)$$

Table 5. Regression fitting results of carbon emission prediction model of Zhejiang Province.

	LnP	lnA	lnU 1	nIS	lnEI	lnES	Cons	k	R^2
Coef	0.305	0.198	1.079	-0.172	0.073	0.076	4.194	0.03	0.964
t, price	0.362	3.586	3.231	-1.613	1.802	1.326	0.617		
Standard error	0.844	0.055	0.334	0.106	0.041	0.057	6.796		

After calculation, it is found that the influence degree of carbon emission influencing factors in Zhejiang Province varies. The significant impacts are ranked as urbanization rate, population size, per capita GDP, industrial structure, energy structure and energy intensity in order of effect.

Prediction Results of Peak Carbon Emissions. Based on the setting of the future development model, the data of the relevant scenarios are substituted into the formula (4), and the carbon emissions in the future years are predicted (Table 6).

Table 6. Predicted value, peak value and peak time of future carbon emissions under different scenarios.

	In 2025	In 2030	In 2035	In 2040	Peak time	Peak value
S1	530.33	552.80	544.83	538.78	2030	552.80
S2	503.31	497.31	482.40	473.28	2025	503.31

By analyzing the results of different combinations of scenarios, it can be found that under the baseline scenario (S1), the peak time of carbon emissions in Zhejiang Province will occur in 2030, and the peak amount is 552.80 million tons. Under the green development scenario (S2), the carbon emissions of Zhejiang Province remained basically stable. Suppose Zhejiang Province improves the planning goals of various environmental protection and energy resources on the existing basis and strengthens the implementation of policies. At the same time, adopting the "three-pronged" emission reduction means of industrial upgrading, energy-saving development and clean development can effectively reduce the total amount of carbon dioxide emissions while ensuring stable economic development and achieving "carbon peaking" in 2025. The peak is expected to be around 503.31 million tonnes. The green development scenario can be regarded as the best choice for Zhejiang Province to achieve "carbon peak" (Fig. 2).

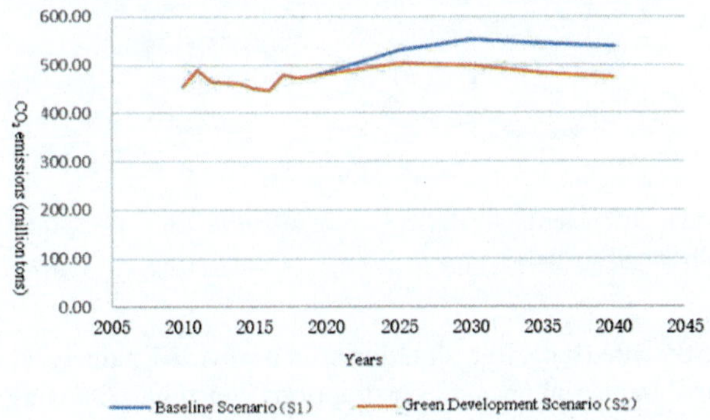

Fig. 2. Scenario Prediction of carbon emission in Zhejiang Province

Spatially Differentiated Distribution of Carbon Peaks. The predicted data are the industrial energy consumption and urbanization rate of 11 prefecture-level cities in Zhejiang Province (Table 7). According to the fitting data analysis of the urbanization rate (divided into four levels from high to low) and total industrial energy consumption (divided into four levels from low to high), the relevant data of Zhejiang Province in 2020 were selected respectively. Plot a two-line radar chart (Fig. 3). In the figure, the urbanization rate is getting higher and higher from the circumference to the center of the circle, the lower the total industrial energy consumption is from the circumference to the center of the circle, and the lower the total industrial energy consumption is, the closer it is to the carbon peak. Therefore, the higher the fit of the relative values of the two curves, the faster the carbon peaks [14].

Table 7. Industrial energy consumption and urbanization rate of cities in Zhejiang Province.

	Hangzhou	Ningbo	Wenzhou	Jiaxing	Huzhou	Shaoxing	Jinhua	Quzhou	Zhoushan	Taizhou	Lishui
Total industrial energy consumption in 2020(10,000 tons of standard coal)	1809	2854	871	1543	822	1165	673	758	261	1492	305
Urbanization rate (%)	83.29	78.00	72.16	71.34	65.64	71.02	68.19	57.57	71.89	61.98	61.82

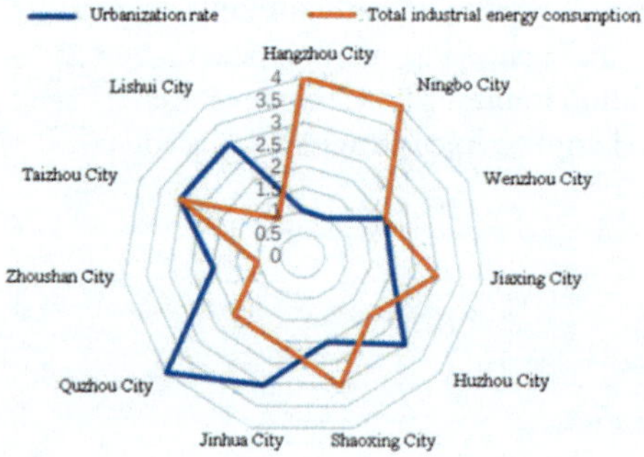

Fig. 3. Fitting spatial difference analysis diagram of urbanization rate and total industrial energy consumption in Zhejiang Province

According to the fitted data, four situations can be divided, namely: (1) high urbanization rate - low total industrial energy consumption: Wenzhou, Shaoxing, Zhoushan; (2) low urbanization rate, total industrial energy consumption High consumption: Taizhou; (3) High urbanization rate, high total industrial energy consumption: Ningbo, Hangzhou, Jiaxing; (4) Low urbanization rate, low total industrial energy consumption: Huzhou, Jinhua, Quzhou City, Lishui City. According to the 2025 peak target. Corresponding to the degree of fit is divided into four types of partitions. That is to reach the peak first, possibly to reach the peak overdue, to reach the peak on schedule (weak), and to reach the peak on schedule (strong). Among them, the peak on schedule is divided into two categories. When the total industrial energy consumption is low, it is easy to achieve carbon peaking. When the total industrial energy consumption is high, it is more difficult to achieve carbon peaking. Therefore, the former situation is classified as scheduled peaking (strong), and the latter is scheduled peaking. (weak) (Fig. 4).

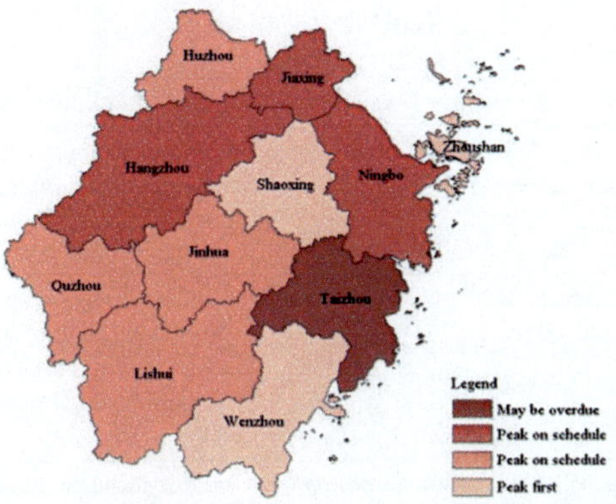

Fig. 4. Spatial Differential Distribution of peak carbon dioxide emissions in Zhejiang Province

4 Exploration on the Regional Control of Carbon Peaking in Zhejiang Province

4.1 Zhejiang Province's "Double-Carbon" Goal Realization Route

First of all, according to the "Zhejiang Province Carbon Peak Carbon Neutral Science and Technology Innovation Action Plan" (hereinafter referred to as the "Action Plan"), it should be clear that Zhejiang Province will achieve carbon peaking in 2025 and carbon neutrality in 2030. According to the predicted carbon peak time, referring to the growth law of fast, slow, and stable growth, the dual-carbon target development plan is subdivided according to the development speed of the 5-year period. At the same time, combined with the progress forecast of carbon emission reduction and carbon absorption and the relevant policy documents such as the Action Plan, a timetable and roadmap for the realization of the dual carbon goals in Zhejiang Province should be proposed (Table 8).

Table 8. Five-year development roadmap of carbon neutrality in peak carbon dioxide emissions, Zhejiang Province.

Year	Specific "two-carbon" development goals	Policies and Initiatives
2021	Strengthen the "two-carbon" basic frontier innovation leadership, and strengthen the collaborative innovation of applied basic research	Carry out "double carbon" basic frontier research and talent cultivation
2025	Zhejiang Province has fully realized the carbon peak, started to carry out the carbon neutralization plan, and shared Zhejiang's experience with the whole China Zhejiang Province has initially established a green and low-carbon technology innovation system	Innovation of key core technologies for carbon emission reduction Transformation of advanced technological achievements in carbon emission reduction Utilizing Supply and Demand to Reduce Emissions

(continued)

Table 8. (*continued*)

Year	Specific "two-carbon" development goals	Policies and Initiatives
2030	China's carbon peaks, Zhejiang province achieves carbon neutrality	Promote the construction of sustainable development demonstration zones and lead the green and low-carbon development of high-tech zones
2035	Zhejiang Province begins to comprehensively promote carbon constraints	Implement low-carbon technology open cooperation (domestic and domestic scientific and technological cooperation and regional innovation cooperation in the Yangtze River Delta)
2040	Prosperous carbon industry development	
2050	Most of China has achieved carbon neutrality	Provide technical assistance outside Zhejiang Province
2060	Achieving carbon neutrality across China	

4.2 The Regional Governance Approach of Zhejiang Province Under the Background of the Implementation of China's Territorial Space Planning

The above analysis shows that due to the different influences of resource endowments, environmental characteristics and development orientations in different regions, the corresponding carbon emission reduction potentials and carbon neutralization capabilities are also different. In order to effectively achieve the "dual carbon" goal, it is very necessary to carry out the regional governance of carbon peaks in all prefecture-level cities in Zhejiang Province. Due to the regional differences in carbon emission reduction potential, Zhejiang Province needs to carry out different carbon emission control according to the above four types of regions (Table 9).

Focusing on the overall goal of "reducing carbon emissions and increasing carbon sinks", carbon emission control measures need to focus on key areas such as spatial pattern, energy, resource utilization, and ecological environment.

For the areas that reach the peak first and reach the peak on schedule (strong), it is also necessary to establish a new leap-forward development mechanism led by the creation of a "carbon neutral" pioneer area. According to the characteristics of carbon emissions and carbon sinks in ecological resource-rich areas, a mechanism for creating a "carbon neutral" pilot zone can be established and improved at the national level. It is used to support pioneering exploration in key areas such as clean energy supply, carbon sink trading, cultivation of low-carbon and high-efficiency industries, green and low-carbon life, the transformation of ecological value, and coordinated promotion of common prosperity.

For areas that reach the peak on time (weak) and may reach the peak beyond the schedule, the total industrial energy consumption is high. Therefore, it is necessary to strengthen the control of carbon emissions from energy consumption. On the one hand, it is necessary to improve the utilization rate of energy, and on the other hand, it is necessary to adjust the industrial structure and reduce the industrial enterprises with backward technology and serious pollution. Possible strategies include improving the energy structure, expanding the proportion of clean energy, and reducing the proportion of fossil energy such as coal. At the same time, it is necessary to promote a green and low-carbon lifestyle, expand the scope of urban public services such as public transportation, and improve the efficiency of urban public services. It is also necessary to initiate multi-party cooperation between the government, enterprises and the public to accelerate the development of low-carbon urbanization.

Table 9. Regional Management of carbon peak in Zhejiang Province.

Carbon peak division	The main body of management	Management measures	Management area
Take the lead in peak	Municipal ecological and environmental administrative agency	1. Successful treatment of black and odorous water bodies in built-up areas; 2. Comprehensive construction of eco-environmental protection industrial park; 3. New space restriction industry in territorial space planning to encourage the development of tertiary industry	Wenzhou city, Shaoxing city and Zhoushan cities
Peak reaching on schedule (strong)	Municipal ecological and environmental administrative agency	1. Basically, black and odorous water bodies will be eliminated in the built-up areas; 2.60% of industrial parks will be built into eco-friendly industrial parks; 3. Limit increment	Huzhou City, Jinhua City, Quzhou City, Lishui City
Peak reaching on schedule (weak)	Provincial ecological and environmental administrative institutions	1. Standard rate of black and odorous water treatment in 2023; 2,40% industrial parks to be built ecological and environmentally friendly industrial park; 3. New space inland space planning, encourage stock potential; 4. Reduce pollution and carbon, promote green development	Ningbo City, Hangzhou City, Jiaxing City

(continued)

Table 9. *(continued)*

Carbon peak division	The main body of management	Management measures	Management area
May exceed the peak	Provincial ecological and environmental administrative institutions	1. The treatment rate of black and odorous water bodies in built-up area reaches 30% in 2023; 2.30% industrial parks; 3. Limit new space and encourage stock potential exploration; 4. Build modern energy system; 5. Give priority to ecological compensation system	Taizhou

References

1. Liu, Z., Guan, D., Moore, S., et al.: Climate policy: Steps to China's carbon peak. Nature **7556**, 279–281 (2015)
2. Yaping, L.: Evaluation of low-carbon economy development level in Zhejiang Province based on DPSIR—TOPSIS. Times Finance **29**, 67–68 (2017)
3. Ma Zheng, M., Yuhong, C.: Research on decomposition of driving factors of China's carbon emission based on SDA model. Coal Econ. Res. **40**(07), 32–36 (2020)
4. Boqiang, L., Zhujun, J.: Environmental kuznets curve prediction of carbon dioxide in china and analysis of influencing factors. Manage. World **4**, 27–36 (2009)
5. Guoquan, X., Zeyuan, L., Zhaohua, J.: Decomposition model and empirical study of carbon emissions for China: 1995–2004. China Population, Res. Environ. **16**(6), 158–161 (2006)
6. York, R., Rosa, E., Dietz, T.: STIRPAT, IPAT and IMPACT: analytic tools for unpacking the driving forces of environmental impacts. Ecol. Econ. **46**(3), 351–356 (2003)
7. Wei, Y., Yaru, H., Xiaoying, Z., Mengfei, G.: Prediction of carbon emission in Shandong Peninsula blue economic zone based on STIRPAT Model. J. Univ. Jinan (Science and Technol.) **35**(02), 125–131 (2021)
8. Rui, H., Zheng, W.: Influencing factors of carbon emissions from energy consumptions in Chongqing based on STIRPAT model. Acta Sci. Circum. **33**(02), 602–608 (2013)
9. Yongfeng, S. : Adhering to the self-reliance and self-improvement of science and technology to shape the new advantages of innovation-driven development of Hangzhou during the "14th Five-Year Plan" - Science and Technology Chapter of "The 14th Five-Year Plan for Hangzhou's National Economic and Social Development and the Outline of Long-term Goals for 2035". Interpretation. J. Hangzhou Sci. Technol. **2021**(1), 4 (2021)
10. Enyan, Z.: Study on the spatial-temporal pattern of carbon emission and its response to urbanization in Zhejiang Province. ZheJiang University (2020)
11. Yan Xinjie, Y., Hui, S.: Research on prediction and realization path of "carbon peak" in Xinjiang based on STIRPAT model. J. Xinjiang Univ. (Nat. Sci. Ed. Chinese English) **38**(6), 1–8 (2021)
12. Heya, S., Xiang, L., Yichen, W.: Prediction of Carbon Emissions in Inner Mongolia Based on STIRPAT Model. J. Mongolia Normal Univ. (Natural Sci. Ed.) **50**(04), 347–352 (2021)

13. Tong, W., Xingyu, Z., Xingxing, C., et al.: Analysis and prediction of industrial carbon emission of Linyi City based on STIRPAT model. Huadian Technol. **43**(06), 47–54 (2021)
14. Guili, L., Yi, W.: Temporal and spatial prediction and zoning management of the realization of dual carbon targets in Chengdu-Chongqing area. Environ. Prot. **49**(16), 60–62 (2021)

Peak Projections and Solutions for Carbon Neutrality and Carbon Peaking in the Construction Industry

Baoxi Zhang, Zhiqiang Kang(✉), and Tongxin Jin

Shenyang Jianzhu University, Princeton, Liaoning 110168, Shenyang, China
kzq222@sju.edu.cn

Abstract. China's carbon emissions from buildings account for about 20% of the total emissions during operation. We analyze and point out the main factors that affect the carbon emission in the operation stage of construction industry. On this basis, the long-term prediction model of carbon emission from building operation based on LEAP model is constructed, and the contribution rate of different industries to dual-carbon enterprises is analyzed. The results show that under the current development model, China will not be able to peak its carbon emissions by 2030. Only a combination of control measures, such as electrification of heating systems and efficiency improvements, will ensure that China's carbon emissions peak at 2.7 billion tons in 2030. Between 600 and 800 million tons of carbon need to be neutralized by 2060.

Keywords: Construction department · Carbon up to peak · Carbon neutral · Carbon emission forecasting model

1 Background

The year 2020 has been an extraordinary year. Since industrialization, the global climate has continued to warm, COVID-19 has been raging, and mankind is facing huge ecological problems [1]. According to estimates by the UN's Intergovernmental Panel on Climate Change (IPCC), in order to reach the 2 °C or even 1.5 °C temperature target [2], the world would have to achieve net zero CO2 emissions, or carbon neutrality, by 2050. In September 2020, At the general debate of the 75th Session of the United Nations General Assembly [3], President Jin ping,X solemnly pledged that China would strive to peak carbon dioxide emissions by 2030 and become carbon neutral by 2060. However, the construction department does not have accurate specifications for the control objectives of carbon peak and carbon neutrality. According to the analysis of the whole life cycle of the building, since the carbon emission of the related infrastructure is about 100–150 million tons of CO2 [4], which is relatively small compared with the carbon emission in the construction and operation stage, it is assumed that there is no significant change in the carbon emission [5]. Therefore, The control objective of our study is mainly to establish a prediction model of building carbon emissions based on

W. Guo and K. Qian (Eds.): GBCESC 2022, LNCE 211, pp. 166–175, 2023.
https://doi.org/10.1007/978-981-19-5217-3_15

LEAP Framework analysis and Calculate the long-term development trend of carbon emissions in the construction industry Under the baseline scenario for carbon emissions caused by energy activities in The building operation stage.

2 Research Methods and Carbon Emission Prediction Model

The model proposed by us aims to more accurately and reasonably predict the development trend of carbon emissions in the construction sector in the future. The technical route of quantifying the target by combining carbon peak and carbon neutralization is shown in Fig. 1.

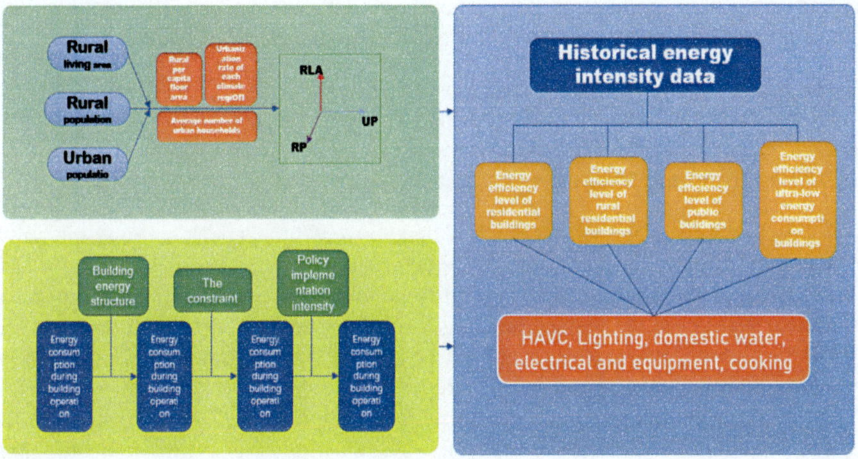

Fig. 1. A framework for long-term prediction of carbon emissions from building operations

Therefore, in our study, the range of energy consumption and carbon emissions is the energy consumption and carbon emissions caused by the use of building functions and cooking by the occupants or users. The calculation scope of energy consumption and carbon emission includes the sum of energy emissions from direct (coal, oil and natural gas) and indirect (electricity and heat) consumption in theprocess of building operation, which can be calculated according to the various energy and carbon emission coefficients consumed by the construction sector:

$$C_{building} = \sum EC_j F_j \tag{1}$$

where: $C_{building}$ is the carbon emission of the construction sector, kg, EC is the classification energy Source consumption (standard coal), kg, F is carbon emission factor, kg/kg.

Where, the energy consumption demand of the construction sector can be calculated according to energy activity intensity and energy activity level. Activity intensity refers to the building Energy per unit area or per household, and activity level refers to the total amount of building area or number of households.

$$EC_{building} = \sum_i \sum_j \sum_k \left(e_{i,j,k,m} + e_{i,j,k,d} + e_{i,j,k,q}\right) \cdot AL_{i,j,k} \quad (2)$$

where, e stands for building energy intensity, AL is activity level: (i = severe cold, cold, hot summer and cold winter, hot summer and warm winter, mild, j = urban residential buildings, rural residential buildings, public buildings, k = new, existing).

It can be seen from Eq. (2) that the energy used in building calculation needs to be obtained Different climate zones, different building types, different building classifications and different The energy intensity of formal energy consumption and its corresponding floor area, for The increase in energy consumption per household also requires analysis of population data. The above data can be based on the results of China's population census and the following statistical method of building area.

3 Construction Industry Control Objectives

3.1 Urban Buildings Run Carbon Neutral Paths

According to the development model assumed above, China's peak carbon time will be far behind the country's original strategic deployment of peak carbon by 2030. In order to reach the peak carbon in the original time, we must propose the implementation path under the constraint of the peak carbon target through a combination of policy regulation and technical solutions, starting from macro factors such as population urbanization, floor area, energy intensity, and energy structure.

The implementation path of the carbon peak target is proposed through a combination of policy regulation and technical solutions. China's total carbon emissions are mainly generated during the operation of existing and new buildings, with a large portion of consumption coming from winter heating in the north. Therefore, the main pathway for urban buildings to reach peak carbon emissions by 2030 is to reduce building energy consumption, followed by technological improvements. Therefore, by evaluating the

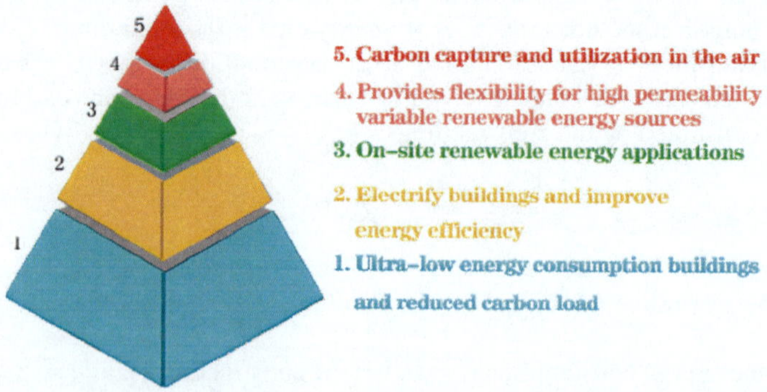

Fig. 2. Five measures to achieve carbon neutral urban building operation pyramid

existing methods for energy saving and emission reduction, we have sorted them into the following five approaches in a comprehensive manner.

Figure 2 shows the five basic measures for operating carbon neutralization. The pyramid shape shows the priorities of these initiatives, from bottom to top.

- The realization of ultra-low energy consumption buildings must be extended to urban areas. The use of passive technologies, renewable energy and load leveling all require coordination, synergy and resource sharing within the urban area.
- Performance-based energy consumption limit design. Load regression and energy system optimization, building integrated photovoltaic (BIPV) design, urban passive technical environment analysis (sunshine, solar energy availability, wind environment and pollutant diffusion).
- Low implicit carbon bearing negative load. BIM selection of low-carbon materials, material reduction treatment, the adoption of assembly technology, design of long-life buildings.
- A large proportion of renewable energy will be used in rural areas. The abundant renewable resources in rural areas, such as solar, wind and biomass energy, provide support for the low-carbon operation of rural houses. The establishment of reasonable low-cost renewable energy utilization mode is the key to realize low carbonization of rural residential energy.
- The clean heating in the north. China's northern cities and towns already have central heating area of about 15 billion m^2, under the carbon neutrality target, in the future.The development direction is to replace coal and natural gas with low-carbon or even zero-carbon energy. At the same time, flexible building power system is used.

3.2 Building Carbon Reduction Target Setting and Result Analysis

Taking into account the carbon emission status of the building sector, the national strategic goals of peak carbon emission and carbon neutrality, and the technology base and development trend of low-carbon and zero-carbon buildings, two comprehensive technology scenarios, namely the policy acceleration scenario and the peak control scenario, are set up in this study. Among them, the development and promotion of the above-mentioned low-carbon building technologies is the strongest under the policy acceleration scenario. Among them, the development and promotion of the above-mentioned low-carbon building technologies is the strongest under the policy acceleration scenario, while the promotion of various technologies is slowed down under the peak control scenario considering potential barriers such as economic level, difficulty of policy implementation, and technology diffusion. However, in both scenarios, peaking in 2030 at the latest, while depth will reach carbon in 2060 at the latest is the constraint. Table 1 describes in detail the main technical conditions for scenario setting.

Policy Acceleration Scenarios. (1) During the 14th Five-Year Plan period, the mandatory building energy efficiency standard shall be revised and upgraded once to reach the energy efficiency index level of ultra-low energy consumption buildings; before 2030, the mandatory building energy efficiency standard shall be revised and upgraded once to

Table 1. The main boundary conditions of carbon peak control scenario and policy acceleration scenario

	In 2025		**In 2030**		**In 2035**		**In 2060**		Reference
	Policy acceleration scenario	Peak control scenario	Policy acceleration scenario	Peak control scenario	Policy acceleration scenario	Peak control scenario	Policy acceleration scenario	Peak control scenario	
New buildings are more energy-efficient	100% ultra-low energy consumption	Ultra-low energy consumption in conditional areas	100% near zero energy consumption	Near zero energy consumption in conditional areas	100% of zero energy consumption	50% Near zero power consumption 50% Zero power consumption	100% of zero energy consumption	50% Near zero power consumption 50% Zero power consumption	[5]
Low-carbon transformation of existing buildings	North completed 50%	North completed 50%	North completed 100%	North completed 100%					[6]
			Depth: 1.5% / a	Depth: 1% / a	Depth: 1.5% / a	Depth: 1% / a	Depth: 1.5% / a	Depth: 1% / a	
Photovoltaic installations in rural areas	25GW	15GW	65GW	45GW	100GW	75GW	350GW	250GW	[7]
Improve energy electrification	55%	55%	65%	60%	70%	65%	95%	85%	[7]
Clean heating in northern China	70%	60%	85%	70%	100%	80%	100%	100%	[8]
Building a new power system	0.522	0.522	0.47	0.47	0.417	0.417	0.072	0.072	[9]

reach the energy efficiency index level of near-zero energy consumption buildings; after 2035 [10], the new building shall fully implement the zero-energy building standard.

(2) By 2030, 220 million m^2 of existing buildings in cold and cold areas will be renovated every year, with a total of 2.2 billion m^2 of renovation, and a total of 1.86 billion m^2 of old buildings with no renovation value will be demolished [11]. By 2030, no energy-saving buildings in north China will be completely removed; After 2030, the construction industry will transition from large-scale construction to renovation and renovation. Each year, 1.5% of the total amount of non-energy saving standards and 50% of the standard (with renovation value) in the whole climate zone will be deeply renovated to achieve ultra-low/near zero/zero energy building standards [12].

(3) Accelerate the electrification of building energy. In addition to northern heating, electric energy is used to replace the terminal energy of buildings, and the electrification of the construction sector will increase from 51% now to more than 90% by 2060.

(4) Large-scale promotion of renewable energy utilization in rural areas. By 2060, a total of 350 GW of p v will be installed (corresponding to 12 billion m^2 construction area and 4 billion m^2 roof area) [13].

(5) For central heating in the north, ultra-low/near zero/zero energy buildings have realized clean heating in the form of decentralized heat pumps. For the remaining common buildings, the thermal power unit combined heat and power generation and industrial waste heat retained by the power department are used to supplement them [14]. With the support of deep emission reduction in the power sector and CCUS technology, the construction sector will achieve net zero carbon emissions by 2060 [15].

Peak Control Scenarios. (1) The beijing-Tianjin-Hebei region, Yangtze River Delta and other relatively developed areas will implement ultra-low energy building standards in 2025, and near-zero energy standards in 2030. Other areas will achieve ultra-low energy standards in 2030, and near-zero energy standards in 2035. Half of all new buildings are built to zero energy building standards every year [16].

(2) After 2030, the annual proportion of deep renovation shall be 1%, and the total renovation shall be controlled under 10 billion m^2.

(3) In rural areas, solar photovoltaic is mainly promoted, and the trend of biomass energy being replaced by commodity energy is not strictly controlled [17].

Figure 3 shows the long-term development trend of carbon emissions in the construction sector under BAU and two emission reduction scenarios, policy acceleration scenario and peak control. Under the scenario, the peak time of carbon emission in the construction sector is 2027 and 2030 respectively, and the peak time of carbon emission is 2.374 billion tons of CO2 and 2.617 billion tons of CO2 respectively. Under the policy acceleration scenario, the carbon emission of the construction sector will enter the fluctuating plateau in 2024, and by 2060, the CO2 emission of the building operation phase will be 309 ~ 326 million tons, which is 45.9 billion tons less than that of the BAU scenario [19]. Under the peak control scenario, there will be 710–779 million tons of CO2 emission in the building operation stage by 2060, which will reduce 31 billion tons of CO2 emission compared with the BAU scenario [20]. With the increase of policy implementation and the depth of power grid emission reduction, the impact of area on carbon emissions will gradually decrease.

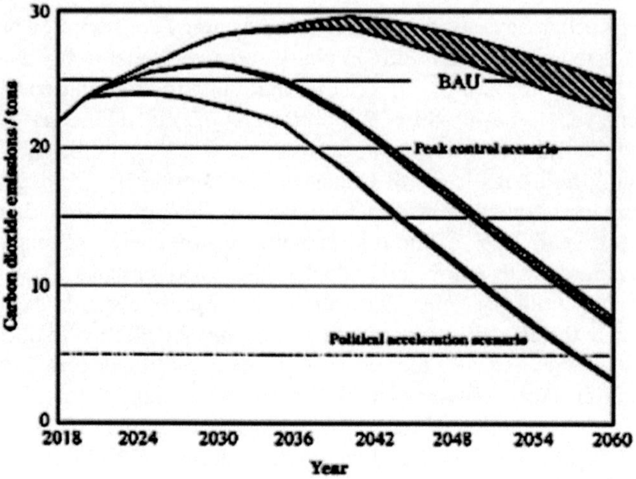

Fig. 3. Medium - and long-term development trends of carbon emissions in three scenarios [18]

Under the policy acceleration scenario, the peak can be advanced through the comprehensive improvement of building energy efficiency standards, but the construction sector should achieve "early peak and low The action plan with the guiding principles of peak value, dead-end pressure, short platform and deep balance needs to take measures such as vigorously developing and utilizing renewable energy and transforming existing buildings. By combining policy and technology, the policy acceleration scenario can control the platform period from 2024 to 2027. In the peak-control scenario, the intensity of various technical measures is relatively lower, and the plateau period is from 2028 to 2033. Compared with the policy acceleration scenario, the peak is higher and the plateau period is longer. By analyzing the long-term development trend of building carbon emissions and considering the carbon emissions of infrastructure, it is suggested that the corresponding control targets for policy acceleration scenario and peak control scenario are 2.5 billion to 2.6 billion tons of CO_2 and 2.7 billion to 2.8 billion tons of CO_2 respectively. From the perspective of the time and control target of carbon neutrality in 2060, the residual emission of building operation and infrastructure itself is 600 million to 1 billion tons of CO_2, which needs to be removed by CCUS technology. Therefore, it is suggested to control the residual emissions of 600 million to 800 million tons of CO_2 in 2060 [20].

In the policy acceleration scenario, from 2025, strong building energy conservation regulations will be comprehensively improved, and there will be 30.6 billion m2, 24.2 billion m^2 and 18.1 billion m^2 of new buildings under the three types of area stock respectively, designed and built according to GB /T 51350–2019 "Technical Standards for Buildings with Near Zero Energy Consumption". By 2060, the proportion of buildings meeting the energy efficiency level of GB /T 51350–2019 "Technical Standard for Buildings with Near Zero Energy Consumption" in the three area scales will be 60%, 56% and 52% of the total building volume respectively through new construction and

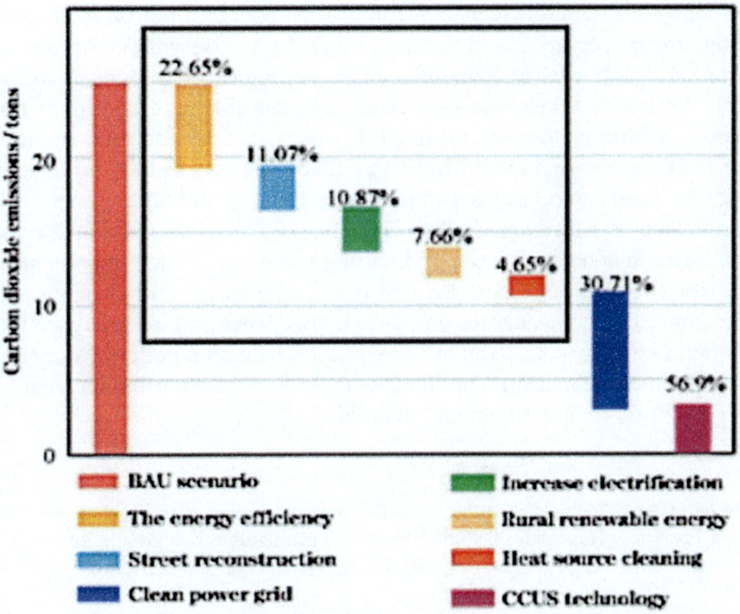

Fig. 4. Contribution rate of each technology under "30·60" carbon neutral target

in-depth renovation. The voluntary contribution rate of energy conservation and emission reduction in the construction sector reached 56.9% ~ 50.3%. In the peak control scenario, under three kinds of area stocks, 26.5 billion m^2, 20.1 billion m^2 and 14 billion m^2 of new buildings will be built in accordance with GB /T 51350–2019 "Technical Standard for Buildings with Near Zero Energy Consumption". By 2060, through new construction and in-depth transformation, The proportion of buildings conforming to GB /T 51350–2019 Technical Standard for Buildings with Near Zero Energy consumption accounted for 46%, 41% and 36% of the total building volume under the three sizes, and the promotion of various technologies was weakened, so the independent contribution rate of building energy conservation and emission reduction was about 40% [21].Taking the scenario of 85 billion m^2 of total construction volume and policy acceleration as an example, Fig. 4 shows the carbon neutrality of each sub-technology in the construction sector when carbon neutrality is realized Contribution rate.

4 Conclusion

(1) Under the baseline scenario, carbon emissions from China's construction sector and related infrastructure will peak around 2040, with a peak of 3.11 billion tons CO_2. The growth of carbon emissions from the construction sector is mainly due to the growth of indirect power emissions caused by the increase in electricity consumption.

(2) The policy acceleration scenario with priority given to the promotion of energy saving strong regulation and the peak control scenario with comprehensive consideration of the level of economic and technological development are proposed. Under the policy acceleration scenario, the peak time and control target are 2.5 billion to 2.6 billion tons of CO_2 in 2027. Under the peak control scenario, the peak time and control target are 2.7 billion to 2.8 billion tons of CO2 emission in 2030.
(3) Under the peak control scenario, the contribution rate of building energy efficiency improvement to carbon neutrality is 40%; Under the policy acceleration scenario, the maximum contribution rate of building energy efficiency improvement to carbon neutrality is 56.9%, and the rest is contributed by grid cleaning and CCUS technology. The two scenarios proposed in this paper provide reference for the formulation of energy conservation and emission reduction policies. They are not the only combination measures, nor do they mean that the contribution rate of building energy conservation work is limited to this.

Acknowledgement. This project comes from the project of Science and Technology Department of Liaoning Province. Research on multi-objective optimization design and key technology of zero carbon and zero energy building in cold region. Project Number: LJKZ0577.

References

1. Youwei, W.: "Carbon" – reflections on building energy efficiency in China under the vision of carbon peak and carbon neutrality. Build. Energy Conserv. **4901**, 1–9 (2021)
2. Hu'ai, G.: China's goal of achieving carbon peak by 2030 and main approaches. J. Beijing University of Technol. (Social Science edition), 2103, pp. 1–15 (2021)
3. Xudong, S., Leixin, Z., Bo, Z.: Research on the development and transformation of China's coal industry under the background of carbon neutrality. China Mining **3002**, 1–6 (2021)
4. Junfeng, L., Guang, L.: Carbon neutrality: opportunities and challenges in China's development transition. Environ. Sustainable Dev. **4601**, 50–57 (2021)
5. Hanzhi, G.: Analysis of opportunities, challenges and countermeasures faced by financial industry from the perspective of green development. Trade Show Economics **08**, 68–70 (2021)
6. Qingchen, C.: Scientific connotation of "carbon peak and carbon neutrality" and Policy measures in China. Environ. Sustainable Dev. **4602**, 14–19 (2021)
7. Hao, L., et al.: Carbon peak and carbon neutralization in Urban buildings in China. Hvac, Strategic position and development suggestion of energy storage under carbon neutral goal. China Power Enterprise Management, 10, pp. 36–39 (2021)
8. Li, H.: Development and transformation of China's coal industry in the context of carbon neutrality. Low Carbon World **1108**, 23–24 (2021)
9. Jinguang, G., Rui, W.: Network correlation characteristics and optimization path of Energy consumption in China: from the perspective of carbon peak and carbon neutra. Jianghai Journal **04**, 85–91 (2021)
10. Zhang, S., Wang, K., Yang, X., Xu, W.: Research on the target of carbon peak carbon neutralization emission control in building sector. Build. Sci. **3708**, 189–198 (2021)
11. Zhiyuan, O., et al.: Carbon peak carbon neutrality - Challenges and countermeasure. J. Hebei Univ. Econ. Bus. **4205**, 1–11 (2021)
12. Lin, Z., Zhen, L.: China Environmental Protection Industry **9**, 22–28 (2021)

13. Haisheng, C., Chang, L., Yujie, X., Fen, Y., Wei, L., Zhenhua, Y.: Energystorage Science and Technology **1005**, 1477–1485 (2021)
14. Jun, W., Yunpeng, C.: Thoughts on the Countermeasures of Transportation Transformation and Development under the Goal of "Peak Carbon Dioxide Emissions and Carbon Neutralization" Transportation energy conservation and environmental protection, 1705, pp. 33–36 (2021)
15. Jin, L., Tegen, C.: Study on green and low-carbon development of Liaoning under the background of carbon peak and carbon neutrality. Liaoning Economy **03**, 18–20 (2021)
16. Jianguo, Z.: "13th five-year" building energy saving and low-carbon development effect and "14th five-year" development path research prefabricated building technology and near-zero energy building integration become industry hot spot. China Energy **4306**, 31–38 (2021)
17. Yang, L., Heyang, W., Yongzhen, W., Jun, Z.: Huadian Technol. **4311**, 5–14 (2021, in Chinese)
18. Jianzhong, S.: Policies and Regulations: Outline of the 13th Five-Year Plan for Urban and Rural Construction of housing, pp. 578–600. China Property Management Industry Yearbook, China Market Press (2018)
19. Dewu, J.: Journal of Anhui University of Technology (Social Science Edition) **2306**, 12–17 (2021, in Chinese)
20. Liu, Y., et al.: China Building Energy Efficiency Annual Development Report. China Architecture and Architecture Press, Beijing (2013)
21. Liu, Y., et al.: China Building Energy Efficiency Annual Development. China Architecture and Architecture Press, Report. Beijing (2014)

Synthesis and Application of Tackifying Resin in Silane Modified Polyether Adhesive for Prefabricated Building

Yangqing Chen[1,2,3], Yuliang Qi[1,2(✉)], Hengchao Huang[3], Jianjun Chen[3], and Keke Huang[1,2]

[1] Guangzhou Research Institute of Construction Industry Co., Ltd., Guangzhou 510663, China
cupid.7@163.com
[2] Guangzhou Construction Engineering Co., Ltd., Guangzhou 510030, China
[3] Guangzhou Baiyun Chemical Industry Co., Ltd., Guangzhou 510540, China

Abstract. An branched polyether epoxy resin called FBEPO was synthesized by the ethoxylated pentaerythritol and epichlorohydrin. It can be added to the silane modified polyether sealant (SMP adhesive) as a tackifier. The effects of tackifier (FBEPO) which added to the SMP adhesive on the tack-free time, tensile strength, elongation at break, elastic recovery were studied. Research shown the effect of the bonding between SMP adhesive prepared by only adding FBEPO or aminosilane coupling agent and the precast concrete board (PC board) was general, but adding aminosilane coupling agent and FBEPO at the same time to the SMP adhesive could significantly improve the mechanical properties, elastic recovery rate and adhesion of SMP adhesive. The effect is the best when the amount of aminosilane is 0.5% (w/w) and the amount of FBEPO is 1–2% (w/w).

Keywords: Prefabricated building · Tackifier · Branched polyether epoxy resin · Silane modified polyether sealant

1 Introduction

With the rapid development of prefabricated buildings, sealant as the first line of defense for seam waterproofing of prefabricated buildings has been attracted more and more people's intense attention and discussion [1]. Silane modified polyether adhesive, referred to as SMP adhesive or MS adhesive in the industry (hereinafter collectively referred to as SMP for ease of description), has the characteristics of low VOC emissions, no pollution, excellent adhesion, aging resistance and excellent resistance to displacement [2]. Therefore, it is used widely in construction, home decoration and industrial fields. At present, SMP is also the preferred type of sealant for seam waterproofing in prefabricated building [3]. The precast concrete board (PC board) used in prefabricated building is a porous material. If it is used in humid environment or soaked in water, The moisture will slowly penetrates into the bonding surface of the sealant and destroys its interface layer, resulting in SMP adhesive bonding Failure, while the PC board surface is rich in residual

W. Guo and K. Qian (Eds.): GBCESC 2022, LNCE 211, pp. 176–184, 2023.
https://doi.org/10.1007/978-981-19-5217-3_16

treatment agent and release agent, which will also affected the adhesion of the silane-modified polyether adhesive. In this paper, We Prepared the one-component Silane Modified Polyether Adhesive for Prefabricated Building by using silane-terminated polyether polymer, calcium carbonate, thixotropic agent, water removing agent, plasticizer, coupling agent, catalyst and the branched polyether epoxy resin (FBEPO) as a tackifier. And also we discussed the effect of the tackifier (FBEPO) which added to the SMP adhesive on the tack-free time, tensile strength, elongation at break, elastic recovery.

2 Experimental

2.1 Materials and Instrument

Ethoxylated pentaerythritol (PP150) was supplied by Perstorp AB Co., Ltd.; Sweden. Epichlorohydrin was purchased form Guangdong Shantou Xilong chemical plant, Guangdong, China; Boron trifluoride-diethyl etherate was obtained from Zhejiang Huangyan chemical plant, Zhejiang, China; Sodium hydroxide was purchased from Shanghai Aladdin Biochemical Technology Co., Ltd, Shanghai, China; Silane-modified polyether resin (brand: 380S, 3623T) were obtained from Jiangsu Ruiyang Antai New Material Technology Co., Ltd., HuaiAn, China; PPG2000 was supplied by Jinan Jingcheng Chemical Co., Ltd., Jinan, China; Calcium carbonate (Brand CCS25) was purchased from Guangxi Warner New Material Technology Co., Ltd., Hezhou, China; Vinyltrimethoxysilane (trade number A171) was provided by Hubei xinlantian New Material Co., Ltd, Hubei, China; N-(β-aminoethyl)-γ-aminopropyltrimethyl(eth)oxysilane (trade number KH-792) was obtained from United Carbon Co., Ltd, U.S.A; Polyamide wax SLX was supplied by Arkema Co., Ltd., France; DBTDL (two Dibutyltin dilaurate) was purchased from Shanghai McLean Biochemical Technology Co., Ltd., Shanghai, China.

Mechanical tensile tests were performed on a universal machine (GMT 4304, Shenzhen Sansi Aspect Technology Co., Ltd.;) at 23°C and around 50% room humidity(RH). The shore hardness tester was obtained by LX-A Rubber Hardness Tester (Manufacturer: Shanghai Liuling Instrument Factory); Adhesive was prepared by Planetary mixer and planetary kneader which made by Foshan Golden Galaxy Intelligent Equipment Co., Ltd. Foshan, China; Dimension test by Vernier caliper(Asimeto, Co., Ltd., German); The obtained branched polyether epoxy resin was characterized by FTIR (Perkin-Elmer Corporation, U.S.A;) with KBr salt tablet liquid membrane method; Heating and drying was made by electric heating blast drying oven supplied by Shanghai Yiheng Scientific Instrument Co., Ltd., Shanghai, China.

2.2 Synthesis of Tackifier Branched Polyether Epoxy Resin FBEPO

The synthetic of branched polyether epoxy resin FBEPO was reported in reference [4]. The synthetic route was shown in Fig. 1. The specific synthetic process as follows:

(1) Weighed 1500 g pp150, heated it to 105 °C, vacuum dehydrated it for 0.5 h, and cooled it to room temperature with nitrogen; Added 10 g of $BF_3O(C_2H_5)_2$ as catalyst and stired evenly;

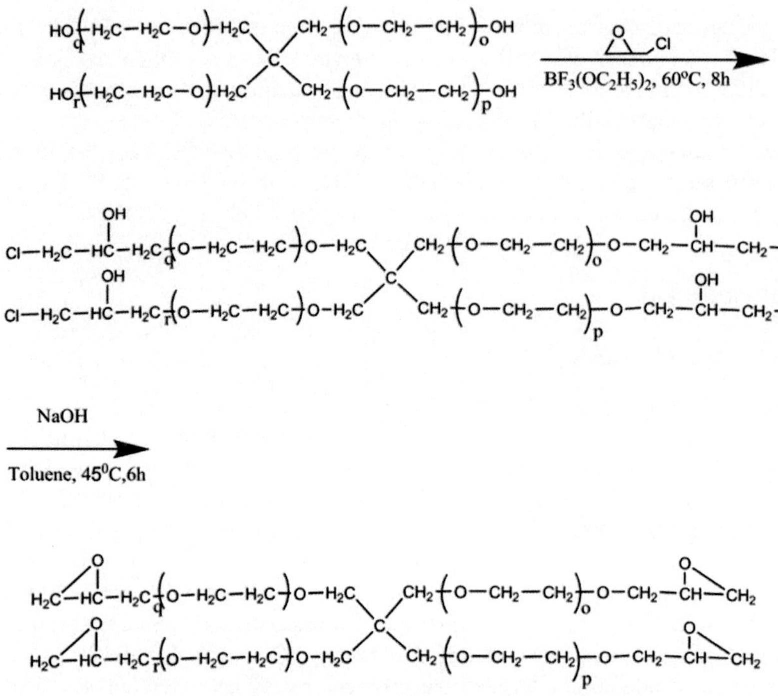

Fig. 1. Synthetic route of branched polyether epoxy resin FBEPO, where $1 \le o \le 15$, $1 \le p \le 15$, $1 \le q \le 15$, $1 \le r \le 15$, $4 \le o + p + q + r \le 40$.

(2) Raised the temperature to 60 °C and stired evenly. Added the 700 g epichlorohydrin dropwise by dropwise with the constant pressure funnel. The dropping time was required no more than 2 h and the temperature should be maintained for 8 h. Then the unreacted epichlorohydrin should be removed in vacuum;
(3) After cooling to 45 °C, The toluene should be added for dilution, then the 220 g solid NaOH should be added for several times, but the whole time should not exceed 2 h. and the temperature should be kept for 6 h;
(4) Filtered the product, collected the filtrate with a separating funnel, extracted the organic phase with toluene, and finally washed it with distilled water to neutral;
(5) Put the organic phase in a beaker, then put the beaker with organic phase into a blast drying oven which had been heated to 110 °C For removing the solvent about 8 h, and then put it into a vacuum drying oven which had been heated to 110 °C for removing the solvent about 0.5 h to obtain a low viscosity polymer with good fluidity. That is the tackifier branched polyether epoxy resin FBEPO described in this paper.

2.3 Preparation of Silane Modified Polyether Adhesive

The basic formula of the silane-modified polyether (SMP) adhesive was shown in Table 1, where the total mass of the silane coupling agent and the tackifier branched polyether epoxy resin accounts for 0 to 5%, and at least one of the two should be added. Mixed the base polymer, plasticizer, reinforcing agent, thixotropic agent, and tackifier in a planetary mixer according to the mass ratio in Table 1. Dispersed at high speed and heated the temperature to 100 °C, vacuum dehydration for 2 h, and then reduced the temperature to 50 °C. Gradually added water removing agent, silane coupling agent and catalyst, vacuumize and stiring evenly; Discharged and packed to obtain the silane modified polyether adhesive.

Table 1. Typical formula of SMP sealant

Component	Material	Mass percentage
Polymer	380S:3623T = 3:1	23%
Plasticizer	PPG2000	21%
Thixotropic Agent	SLX	1%
Reinforcing agent	Calcium carbonate	50–55%
Silane coupling agent	KH792	0–0.5%
Tackifier	FBEPO	0–5%
Dewatering agent	A171	1%
Catalyst	DBTDL	0.5%

2.4 Characterization and Performance Testing

Structural Characterization. The obtained branched polyether epoxy resin was characterized by FTIR with KBr salt tablet liquid membrane method and performed infrared characterization within the wave number 4000–400 cm^{-1};

Sealant Performance Testing. Tensile strength, 100% tensile modulus, surface dry time and elongation at break were tested according to standard JC/T 881-2017.

Elastic recovery rate: tested according to standard GB/T 13477.17-2017 method B.

Surface drying time: In the condition of (23±2) °C and (50±10)%RH, applied the sealant on a clean glass plate or film with a thickness about 2 mm. Then touched the glue surface every 1 min with finger. The time when the glue were not sticky was surface drying time.

Water immersion bonding performance test: According to the standard GB/T 13477.14—2017, the prepared sealant was made into H-type specimens and immersed by distilled water in standard environment for 4 days after curing for 28 days in a standard environment. Removed and placed in standard environment for 1 days, then stretched at 5–6 mm/s speed. The bonding base materials were concrete blocks which satisfied the standard of GB/T 13477.1-2017 (size: 75 mm × 25 mm × 12 mm).

3 Results and Discussion

3.1 FITR Characterization of Branched Polyether Epoxy Resin

Figure 2 is the infrared spectrum of the branched polyether. It can be seen that 1090 cm^{-1} and 940 cm^{-1} are the characteristic infrared absorption peaks of C-O-C bonds; 1456 cm^{-1} and 1371 cm^{-1} are the characteristic infrared absorption peaks of C-C bonds; Fig. 3 shows the infrared spectrum of branched polyether epoxy FBEPO. Compared with the figure 2940 cm^{-1}, the peak was folded by the peak at 900 cm^{-1}, which should belong to the infrared characteristic absorption peak of epoxy group, which indicated that most of the terminal hydroxyl groups of PP150 had been converted to epoxy

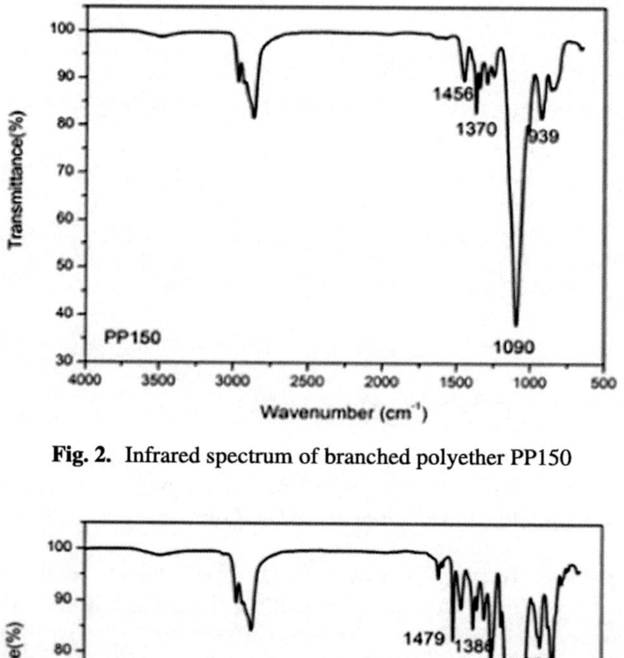

Fig. 2. Infrared spectrum of branched polyether PP150

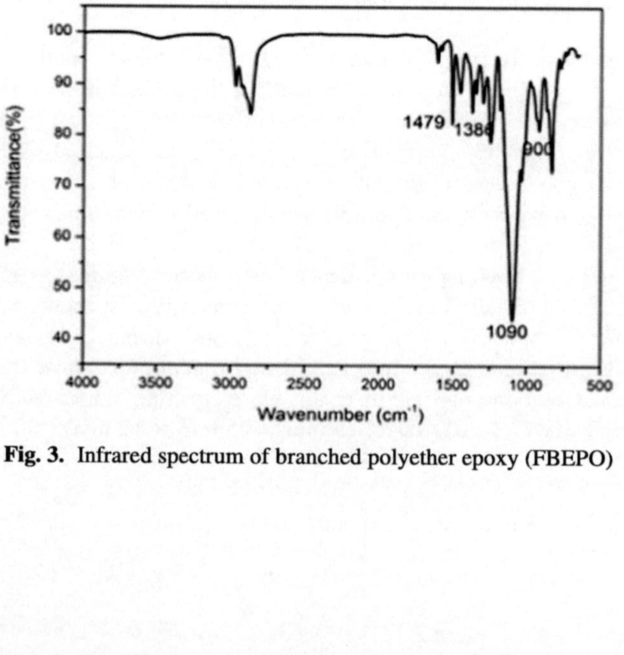

Fig. 3. Infrared spectrum of branched polyether epoxy (FBEPO)

groups. And also it confirmed that the tackifier FBEPO was successfully synthesized by the branched polyether PP150 and epichlorohydrin.

3.2 Effect of the Tackifier FBEPO on the Properties of SMP Adhesive

As adhesion promoter, the silane coupling agent contains both inorganic structure and organic group. This special molecular structure gives them both inorganic and organic functions which can increase the adhesion of those two materials [5]. The aminosilane coupling agent is a common adhesion promoter in SMP adhesive. In order to investigate the effect of the amount of tackifier FBEPO on the properties of SMP adhesive, SMP adhesive with different parts of tackifier branched polyether epoxy FBEPO (corresponding to reducing the amount of calcium carbonate in the same mass ratio) was prepared and the mass ratio of KH792 aminosilane coupling agent was 0.5%. For comparative analysis, The SMP adhesive with only 1.5% FBEPO but no KH792 as a comparison was also prepared. The results were shown in Table 2.

The results in Table 2 shown compared with the SMP adhesive added with the tackifier FBEPO and KH792 amino silane coupling agent at the same time, SMP adhesive added with FBEPO only had longer tack-free time and the lowest tensile strength, elongation at break and elastic recovery rate. The reason is that the amino silane coupling agent is alkaline. The acidic or alkaline conditions can accelerate the curing of SMP polymer and moisture in the air [6], which made the crosslinking density and mechanical properties get stronger. When the KH792 was 0.5%, the different amounts of FBEPO was added to SMP adhesive. With the amount of FBEPO increasing, the tensile strength of the sealant gradually increased from 0.73 Mpa to 0.83 Mpa, then slowly decreased and gradually stabilized. At the same time, its elastic recovery rate, shore A hardness and elongation at break also increased first, then decreased, and finally stabilized. The reason is that the aminosilane coupling agent hydrolyzed rapidly in a humid environment and produced a large amount of silanol groups [7]. On the one hand, the silanol groups formed a chemical bond with the substrate which led to a better adhesion; on the other hand, the silicon hydroxyl groups or amino groups formed by hydrolysis reacted with the epoxy group in the tackifier FBEPO to form a network structure, so that it improved the mechanical properties of silane modified polyether adhesive [8], However, with the addition of excessive tackifier FBEPO, the moisture transmission was relatively difficult after the epoxy group and amino group were crosslinked and cured into a film, which affected the moisture curing reaction of SMP polymer, resulting in the decrease of crosslinking density of the whole SMP adhesive system, causing the slow decline of mechanical properties and elastic recovery rate of sealant, and finally tended to be stable. In order to obtain silane modified polyether adhesive with excellent comprehensive properties, when the amount of aminosilane coupling agent is 0.5%, the optimal amount of tackifier FBEPO should be 1–2%.

Table 2. Effect of the tackifier FBEPO on the properties of SMP adhesive

Test items		Different FBEPO mass percentage + (0.5% KH792)							add 1.5% FBEPO
		0	0.5	1	1.5	2	3	4.5	
Tack-free time /min		65	60	55	53	55	58	66	86
Tensile Strength/MPa		0.73	0.77	0.80	0.83	0.82	0.79	0.74	0.66
Elongation at break /%		487	492	523	529	525	520	490	432
Elastic recovery rate /%		75.6	80.1	82.3	83.6	81.9	78.7	76.2	71.4
100%	23 °C	0.25	0.25	0.25	0.26	0.27	0.25	0.26	0.25
Tensile modulus	– 20 °C	0.26	0.27	0.27	0.27	0.28	0.27	0.27	0.25
Shore A hardness		24	25	26	27	26	25	25	23

Note: The test samples for tensile strength and elongation at break are all H-type test pieces

3.3 Effect of Tackifier FBEPO on Bonding Properties of SMP Adhesive

In order to investigate the effect of tackifier FBEPO on the adhesion between SMP adhesive and precast concrete board (PC board), SMP 1 with only 0.5% KH792, SMP 2 with only 1.5% FBEPO and SMP 3 with 1.5% FBEPO +0.5% KH792 were prepared respectively. The water immersion test of the three SMP adhesives was carried out according to GB / T 13477.9, and their properties are shown in Table 3.

Table 3. Effect of immersion on adhesion properties of SMP adhesive

		SMP 1	SMP 2	SMP 3
Before soaking	Tensile Strength/MPa	0.73	0.66	0.83
	Elongation at break/%	487	432	529
	100% tensile modulus /MPa(23 °C)	0.25	0.25	0.26
	Form of destruction	50%CF + 50%AF	60%CF + 40%AF	100%CF
After soaking	Tensile Strength/MPa	0.59	0.57	0.74
	Elongation at break/%	273	314	421
	100% tensile modulus/MPa(23 °C)	0.26	0.25	0.25
	Form of destruction	100%AF	80%AF + 20%CF	100%CF

Note: CF refers to cohesive failure; AF refers to Interface destruction

As the results from Table 3, we can see that SMP 1 with aminosilane coupling agentonly and SMP 2 with tackifier FBEPO only had cohesive failure and gradually became interface failure before and after soaking, and the mechanical properties of these two SMP adhesives had a large decline before and after soaking. The results of these two SMP adhesives had a large decline before and after soaking. The results of SMP 2 with tackifier resin were slightly better than SMP 1 with amino silane coupling agent. The bonding of SMP 3 with tackifier FBEPO and amino silane coupling agent to PC board after soaking in water can still maintain 100% cohesion failure after tensile test and the decline of mechanical properties before and after soaking in water was relatively slight. The reason may be: First, the inherent polar groups such as hydroxyl and ether bonds in branched polyether epoxy resin greatly enhanced the adhesion between the sealant and the precast concrete board. Second, the silicon hydroxyl group formed by hydrolysis of aminosilane coupling agent was easy to interact with the surface of the concrete prefabricated board (PC board), thereby improved the adhesion between sealant and substrate. Third, the amino group reacted with the epoxy group in the tackifier FBEPO to form an interpenetrating network structure which let the the damage of water molecules to the internal chemical bond of the sealant get weak and ensure the adhesion between the sealant and the concrete substrate before and after soaking which was also accord with the results of mechanical properties of SMP adhesive prepared by adding different amounts of tackifier.

4 Conclusion

The synthetic branched polyether epoxy resin FBEPO can be used as a tackifier for silane-modified polyether sealants. The bonding effect between SMP adhesive prepared by adding only FBEPO or aminosilane coupling agent and PC board is general, but adding aminosilane coupling agent and FBEPO at the same time can significantly improve the mechanical properties, elastic recovery and adhesion of SMP adhesive. The comprehensive properties of SMP adhesive are the best when the amount of aminosilane is 0.5%, and the amount of FBEPO is 1–2%.

Acknowledgement. This research was supported by special funding for Assembly Buildings of Guangzhou Municipal Construction Group Co., Ltd (2019-KJ033; 2019-KJ019; 2019-KJ030) and funded by the Science and Technology Project of Guangzhou (Project No. 201802010068)].

References

1. Yangqing, C., Jian-jun, C., Min-hua, G.: Preparation and elastic recovery rate of one-component silane-modified polyether sealant for prefabricated buildings. Silicone Material **35**(5), 16–20 (2021)
2. Ruihua, Z., Pei-yu, J., Shu-qin, F., et al.: Preparation of silane modified polymer and its sealant. Silicone Material **33**(4), 292–295 (2019)
3. Yangqing, C., Jian-jun, C., Min-hua, G.: Study on preparation and storage stability of silane modified polyether sealant for prefabricated building. China Adhesives **29**(12), 37–39 (2020)
4. Pinghui, L.: Synthesis and application of aliphatic glycidol ether. Thermosetting Resin **4**, 46–53 (1995)

5. Xiancheng, T., Yong, L.: Research on influencing factors of bonding strength of single component MS sealant after being soaked in water. China Build. Waterproofing **11**, 14 (2020)
6. Fei, L., Jian-jun, C., Jin-bo, J., et al.: Study on the influence of silane coupling agent on silane-modified polyether sealant. Guangdong Chem. Ind. **44**(5), 64–65 (2017)
7. Yuzhe, L.: Study on application of self-made high strength silane modified polyether adhesive in automotive glass. Adhesion **40**(05), 114–117 (2019)
8. Guoxian, X., Dajian, Q., Chaoyang, L., et al.: The influence of aminosilane coupling agent on the adhesion of epoxy coating. Material Protection **41**(3), 22–24 (2008)

Analysis on the Regeneration Strategy of Shikumen Residential Building

Ranran Sun(✉)

Tongji Architectural Design (Group) Co., Ltd., No. 1230 Siping road, Yangpu District, Shanghai, China
ranransun66@163.com

Abstract. Shikumen Lanes are the most memorable and unique residential building for the old citizens in Shanghai. With the changes of the times, which will gradually fade away, and a large number of Shikumen Lanes have been demolished due to the rebuilding, and thence the protection of old buildings have been paid more attention and explored by scholars. In this article, taking the reconstruction project of Shikumen residential building in Shanghai Huaanfang as an example, a meaningful discussion on the extraction and rebuilding of traditional buildings is presented.

Keywords: Shikumen · Residential building · Lanes

1 Introduction

The Shikumen Lanes in Shanghai originated in the 1960s. After many years, they gradually become a shelter for ordinary residents from the houses of the rich. The original scenery of Shikumen no longer appeared [1]. From 1920s to 1930s, foreign real estate agents also carried out a large number of constructions. The building features were mainly enclosed, the building carvings were gradually decreased, and the multi-entry has changed into single entry. Shikumen houses with Chinese and western characteristic can be seen on everywhere [2]. This kind of buildings are usually made of stone as the door frame and black-lacquered wood as the door, so they are called "Shikumen". Shikumen Lanes are often divided into older Shikumen, old Shikumen and new Shikumen [3], as shown in Fig. 1. In each period, the decoration of their windows has been relatively stable. At the earliest stage, it had a strong Chinese features, and later it was implanted with western-style in the decorative details. Shikumen Lanes gradually are developed into the residential buildings with convex balcony and then concave balconies have appeared one after another. The convex balcony usually has a slope top, and this kind of design can keep broad vision and is convenient for ventilation and lighting. Concave balcony is more concealed, rainproof and more applicable.

Older Shikumen is from 1870 to 1908, they have inherited the style of traditional houses in the south of the Yangtze River [4, 5]. The first entrance was the courtyard, which was rectangular or square, and the guest room through the courtyard. The back of the guest room was a horizontal staircase, followed by the back courtyard, with toilets

W. Guo and K. Qian (Eds.): GBCESC 2022, LNCE 211, pp. 185–189, 2023.
https://doi.org/10.1007/978-981-19-5217-3_17

and stoves on both sides. All wing rooms were arranged on both sides of the well and guest room. Shikumen has stone door frame, black paint and thick wood door, and the top of the door head is decorated with blue tile brick carving. Because there are the characteristics of cold winter, hot summer, humid and rainy in Shanghai, most of Shikumen Lane houses adopt sloping roofs with a slope of 30 degrees, which is beneficial to drainage and also plays a role of ventilation and lighting. Old Shikumen is from 1909 to 1919, this stage was greatly influenced by western architectural culture. Shikumen residence buildings have some changes. The courtyard area was reduced, the wing room was increased. With the reduction of family scale, a relatively economical single house was designed, and the spatial structure was courtyard-guest room-rear courtyard-annex room [6]. There is a strong western architectural decoration, and there is a western stone hairpin decoration on the lintel of the door head, which also reduces the traditional flower-and-bird patterns and brick carvings [7]. All decorations is exquisitely carved, including Greek triangle and Roman semi-circular flowers, as well as complex curved polygonal flowers [8]. New Shikumen is from 1919 to 1930, the residential buildings are changed from the second floor to the third floor, with several doors separating each household and increasing living rooms to meet the demand of social population growth at that times. In order to increase the space for lighting and ventilation, some protruding "tiger" windows are added to the roof, and there are street buildings and pavilions in the building lanes. The pavilion is placed between the second floor and the balcony, or between the second floor and the third floor. The space of the room can be increased. So the effective space are expanded by this kind of house structure. Cross-street building is a building across a street or alley, in which people can pass below and also obtain enough space.

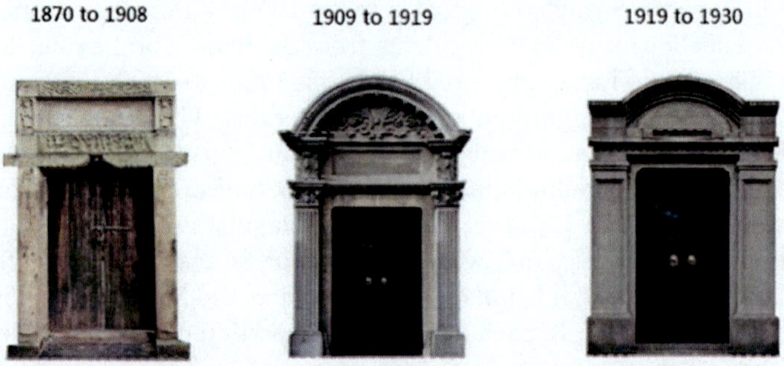

Fig. 1. Shikumen gates in different periods.

2 Regeneration Strategy of Residential Buildings in Shikumen Lane

Take Shanghai Huaanfang Shikumen Residential Rehabilitation Project as an example. The location is in Tianmu Community, Suzhou River, Jingan District. It can be connected between People's Square and Tibet North Road, they have same historical and cultural features. The function in the project is mainly residential buildings, accounting for more than 77% (including 59.6% of existing residential buildings, and most of the demolished buildings are also residential functions), and the commercial function along Zhejiang North Road accounts for more than 20%. The architectural style mainly references to that of Shikumen Lanes, and a small number of buildings are style of the Guangzhou Lane, other commercial buildings are along the street.

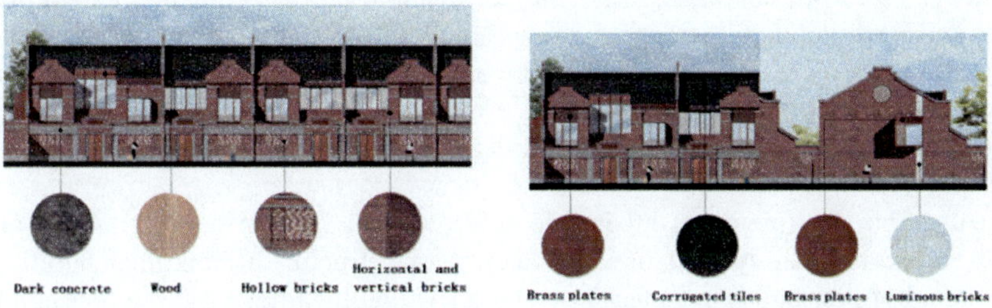

(a) old materials (b) New materials

Fig. 2. New and old materials of Shikumen building

In this project, the traditional horizontal and vertical layout of main lane and branch Lane is maintained. The original spatial layout of patio - Lobby - rear patio - annex room and central axisymmetric features are also retained, and some changes have been made on the original house types, which are more suitable for modern life. Keep the original form of the wall, the height of the courtyard wall is reduced for well lighting, and it is more fashionable and lively than the traditional form of the wall. The original red brick is retained, only the bricklaying method is changed and the material of concrete is used. Keep the original patio style and move the patio space down to ensure the lighting of the underground floor. The closed balcony, which is more suitable for modern people, replaces the original one-step platform, and it becomes more simple, sealed and private.

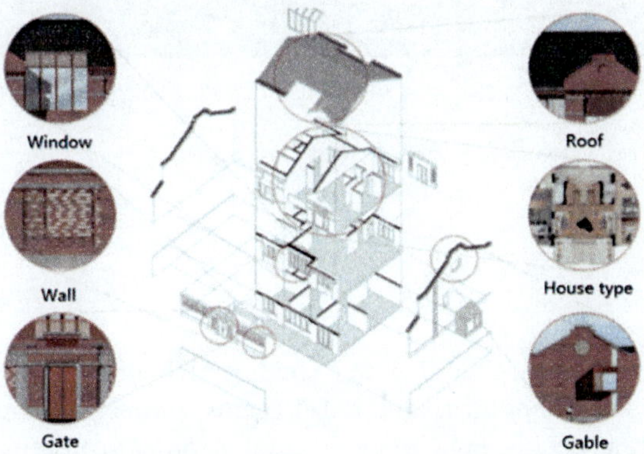

Fig. 3. Extraction of architectural elements in Shikumen

In the design of this project, not only the original sloping roof is retained, but also the sloping roof angle is adopted. The exterior wall is decorated with red brick and gray brick, and the building materials such as red brick, dark concrete and wooden doors are used. The entrance of the lane is equipped with a traditional archway. Many original building materials are also replaced, and the traditional tiles are replaced with corrugated tiles, which makes the building appearance more fashionable, as shown in Fig. 2. Both proportion and scale of the original traditional Shikumen are retained, and the original architectural structure is simplified, which makes the form more modern and simple. The traditional gables are also simplified, we are inspired by the original Guanyin pocket gables and herringbone gables, and the new gables of this project are optimized, as shown in Fig. 3. This project shows that traditional architectural elements have been presented in modern architectural structures, and the evolution of new gables is more modern. For the convenience of lighting, the modern skylight is used to replace the traditional 'Tiger' window.

To these building materials, all dark concrete, wood, hollow bricks and horizontal and vertical bricks are mainly used, in which keep spatial structure of Shikumen building. In every detail of new building materials, red bricks and blue bricks and red tiles and blue tiles are alternately used, it is shown the true colors of the materials. The entrance door is an important part of the courtyard house, and the door head with rich decoration is carved, as shown in Fig. 3. New materials, such as brass plates, corrugated tiles, luminous bricks, etc., have also been added, connecting the inner courtyard with the private courtyard, replacing the traditional closing inner courtyard, and creating a livable courtyard, a comfortable space between lanes, showing the characteristics of Shikumen and meeting the space needs of modern people, has been designed, and a new scene is seen with the combination of new and old materials, as shown in Fig. 4.

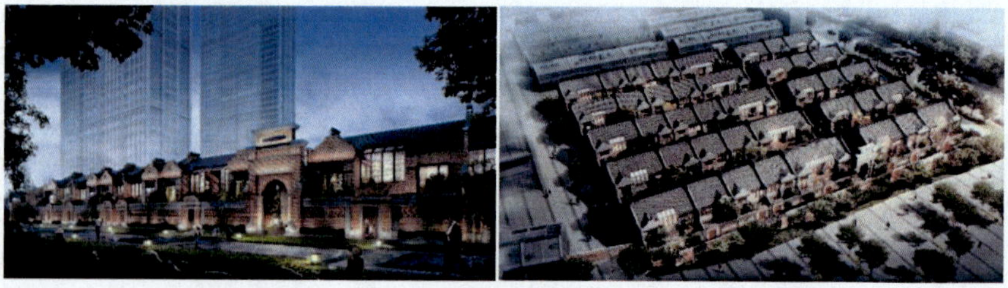

(a) Gates (b) Lanes

Fig. 4. New project

3 Conclusion

Whether reconstructing or rebuilding the residential buildings in Shikumen Lanes, Shikumen architecture retains the traditional elements, maintains the residents' memory, and prevents aging and decline in building material. In order to meet the needs of modern people, it is necessary to implant new elements with historical development and update.

So Shikumen architecture can last longer and better. Let this ancient building with the vicissitudes of Shanghai in the past century show a new attitude and the scenery and feelings of a modern city.

References

1. Zheqi, C.: Research of the place in Shanghai Li-long housing. Shanghai Jiao Tong University (2007)
2. Wenbing, F.: Protection and Renewal of Shanghai Lilong. Shanghai Science and Technology Press, Shanghai (2004)
3. Shaozhou, W., Zhimin, C.: Lilong Architecture. Shanghai Science and Technology. Literature Press, Shanghai (1987)
4. Simmonds, R.: Lessons from the Shikumen of Shanghai. In: Proceedings of 33rd PLEA International Conference, vol. 3, pp. 5116–5123. Edinburgh, United kingdom (2017)
5. Yansong, H., Characteristic and Vicissitude of Shanghai Residential Quadrangle with Stone-ringed Door, Anhui Architecture, pp. 41–43(2006)
6. Jiayan, Y., Shanghai Shikumen landscape: Weaving past, present, and future. In: Proceedings of the History of the Future: 52nd World Congress of the International Federation of Landscape Architects Congress, pp. 690–695. Saint-Petersburg, Russia (2015)
7. Zhenglan, M., Qunxing, L., Xi, Z., Wenying, M., Brief analysis on the house classification condition of Shanghai Shikumen lane. Housing Science, 46–49(2020)
8. Wu, Q., Zhang, Z., Xu, L.: Shanghai Shikumen cultural and creative product design based on design thinking. In: Fukuda, S. (ed.) AHFE 2018. AISC, vol. 774, pp. 33–40. Springer, Cham (2018). https://doi.org/10.1007/978-3-319-94944-4_4

Research on the Transformation and Development of Modern Western-Style Villas in Suzhou Under the Concept of Sustainable Development

Yonglan Wu, Yao Wu(✉), and Yishan Mao

Soochow University, 199 Renai Road, Suzhou, China
wuyaoseu@foxmail.com

Abstract. Architectural heritage carries history, accumulated culture and inherited spirit, and its protection has attracted more and more attention. Effective architectural heritage conservation and restoration measures are not only sustainable development of cultural relics, but also play an important role in the sustainable development of social civilization. This paper extracts the natural and social environment that affect the development of modern architecture in Suzhou, then selects the modern Western-style villas in Suzhou as the breakthrough point, summarizing the transformation process of its construction concept, Green construction technology and materials from the aspects of morphological layout, structural technology and residential culture, which could provide basic theoretical support for its repair and protection. Finally, based on the sustainable development strategy, outlooks on ecological methods of heritage protection are promoted, which may contribute to the inheritance and innovation of modern western-style villas in Suzhou, So as to promote the sustainable development of Suzhou, a famous Jiangnan city.

Keywords: Sustainable development · Heritage conservation · Characteristics of transformation

1 Introduction

1.1 Background

As one of the world cultural heritage sites, Suzhou has a long history of more than 2,000 years. Traditional craftsmen represented by "Xiangshanbang" have created a series of local buildings with typical styles of traditional dwellings in the South of the Yangtze River. After the Opium War in 1840, there emerged a large number of western-style buildings with excellent methods and unique shapes, especially western-style villas. On the one hand, they follow the traditional architectural style, on the other hand, they have distinct modern characteristics. It can be said that it is the materialized expression of the western and Chinese architectural and cultural exchanges, which contains rich historical and humanistic values, and is a living material heritage. There are many researches on

W. Guo and K. Qian (Eds.): GBCESC 2022, LNCE 211, pp. 190–201, 2023.
https://doi.org/10.1007/978-981-19-5217-3_18

Chinese early modern architecture from different angles. Liang Sicheng, Liu Xiandjue and Xu Jingzhi are the representatives of holistic architectural ontology research, such as "Architecture in the Late Qing Dynasty and After the Republic of China", "Chinese Architecture—Past and Contemporary", etc. Or pay attention to the development of modern architecture in a certain region, such as "Shanghai Modern Architecture Style", "The Early Modern Architecture in Jiangsu" [1], "Research on Modern Lingnan Architectural Decoration"; Or to reveal and create characteristics from a comparative perspective, such as "Comparison·Communication·Inspiration-Research on the History of Modern Architecture in China and Japan". In recent years, the theoretical research and practical research of modern architecture and related architectural heritage protection in China is gradually entering a benign development stage from exploration. Drawing on advanced research and protection experience and methods at home and abroad, the research perspectives are more diversified and more in-depth and specific, but for Suzhou early modern building type villa related research has failed to cause the attention of people. Most of them are scattered in local Chronicles, archaeological documents and other ancient books. Under the concept of sustainable development, it is an inevitable trend to promote the green and low-carbon development of construction heritage. This paper is based on the textual research of historical data, combined with the field investigation. Analyze the characteristics of the modern western-style villas and summarize how to achieve modern transformation. Thus, in the context of low-carbon new materials and new technologies, it can provide new ideas for the protection of modern architectural heritage in Suzhou, reduce energy consumption for the normal operation of old buildings, and promote the sustainable development of modern buildings and even the city of Suzhou.

1.2 Definition of Research Object Concept

The research object of this paper is the modern western-style villa, the time span is 1840–1949. The modern western-style villa refers to the residential function on the basis of basically maintaining the traditional residential space shape and system, with two main characteristics of external westernization and vertical space building. No matter in monomer, group, garden and other aspects, they reflect the ability to absorb and integrate the external influence of traditional architectural culture, and contain the local traditional aesthetic characteristics and cultural value orientation.

2 The Social and Economic Background of the Early Modern Suzhou

Since Shanghai opened its commercial port in 1843, Suzhou has begun a great modern urban transformation due to its geographical proximity to Shanghai and along the rivers (Fig. 1). The invasion of western church forces has also changed the cultural consciousness and values of Suzhou people. It not only brings direct and considerable changes in western buildings, roads, facilities and so on, but also brings along the lifestyle (such as car dealers), educational concepts (such as sports fields), hygiene habits (such as water purification treatment) to Suzhou. Especially during the First World War and in the years after the war, national capitalist industry and commerce developed rapidly, and the number of cement, stone and other building material factories continued to expand. The introduction of Western architectural techniques made local craftsmen begin to explore how to perfectly integrate Chinese and Western architectural cultures. Both in terms of design techniques and architectural techniques, it has distinct characteristics of times, culture and region.

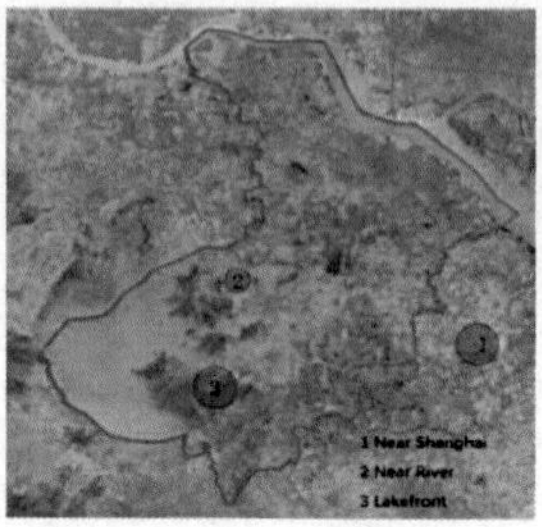

Fig. 1. Geographical environment in Suzhou **Source.** self-painted

3 Characteristics of the Transformation of Modern Western-Style Villas in Suzhou

3.1 Changes in Form and Layout

Shift in Floor Plan. Before the early modern times, traditional dwellings in Suzhou showed the characteristics of dwellings in the south of the Yangtze River, which were reflected in the combination of cubicles and depths in the plane, the courtyard as a transitional space for lighting and ventilation, and most of them were built near the water [2]. Suzhou traditional dwellings are divided into three types: small, medium and large. Traditional single dwellings generally have three to five rooms, and some large-scale residential buildings can reach dozens of rooms, which are accompanied by the combination of courtyard, compartment and patio. The courtyards are arranged along the axis, the primary and secondary are obvious, and the main axis is "Zhengluo", Zhengluo has a higher status and lives for the elders. The two sides of the "Zhengluo" are "Ciluo", the right side is equipped with auxiliary rooms such as study and storage, and the left side is equipped with the women's hall. For example, The Chen house in Fulangzhong Lane is a typical large-scale residential house with three falls and five entrances (Fig. 2).The main line is prominent, and the rooms are arranged in strict accordance with the primary and secondary order, which is the embodiment of the feudal society's way of life and consciousness. It is successively constructed as the entrance hall (gatehouse), sedan chair hall (teahouse), main hall (hall), inner hall (women's hall), etc. [3]. The central axis of the side fall is not connected, and the planes in the side fall are not exactly the same as the normal fall. The halls you enter must pass through the courtyard to enter. Some Suzhou dwellings also have gardens on the left and right sides, forming a private residence full of garden interest.

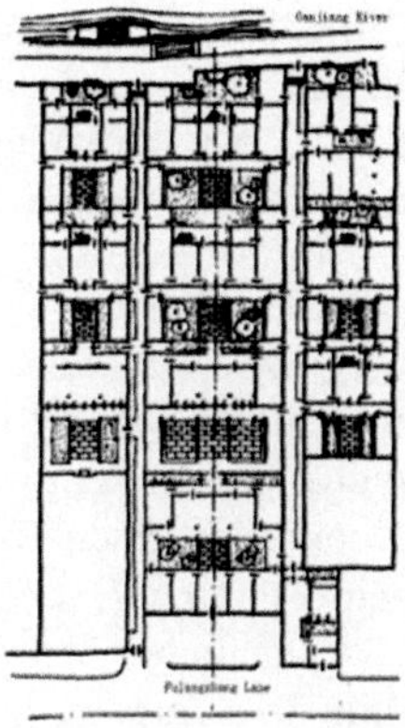

Fig. 2. The Chen house in Fulangzhong Lane **Source.** Mingsu, X, Suzhou Folk House

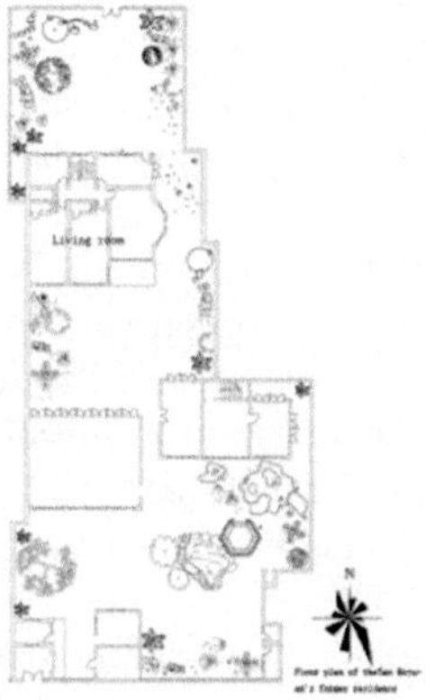

Fig. 3. Tao Boyuan's former residence **Source.** Xiaoxiang, T, Research on Suzhou Modern Architecture

In the early modern times, there were three main forms of residence in Suzhou: garden style villa, courtyard style villa and lane style villa. In addition to the passive absorption of the houses from Shanghai, most of them are adaptive changes on the basis of traditional dwellings. For example, Tao Boyuan's former residence (Fig. 3), located in No. 8 Of Shengjiabang, was built in the year of the Republic of China (1919). The floor plan is based on the traditional Suzhou courtyard layout, and at the same time, it integrates some characteristics of Western architecture, creating a kind of residential form in which you can enjoy Chinese gardens in a small foreign building. From the perspective of plane layout and streamline treatment, except for the main hall "Songying Hall" in Tao Boyuan's former residence, the plane layout of other single buildings is relatively free. In addition to the living function, the house also bears the functions of living, receiving guests, study and dining room. The exterior is provided with a corridor. Because of the corridor, the streamline of the master, guests and servants in the villa is more reasonable. From the perspective of the choice of axis and orientation, although the base site of the last bungalow in Tao Boyuan's former residence is consistent with the main axis, the main door of the bungalow opens on the east side, which breaks the sequence of the original axis to a certain extent, making the the whole space atmosphere of Tao Boyuan's former residence is relatively relaxed and free [4].

Modeling Externalization. During the Ming and Qing Dynasties, the whole ancient city of Suzhou was black, white and gray with white walls and black tiles. After entering the early modern times, People gradually recognize western culture and accept it to a certain extent. The aesthetic taste changes from introverted to open, and the people's psychological worship of foreign countries gradually emerges. Earliest western-style buildings to enter Suzhou were mainly churches, schools and hospitals. With the introduction of Western entertainment life, a number of new public buildings, such as grand theaters and cinemas, were also built. To a certain extent, it enriched the urban color of the ancient city of Suzhou at that time, which had a certain influence on the shape of Suzhou's western-style buildings [5].

Most of the western-style villas in Suzhou were built by industrialists who returned to Suzhou after their successful careers. They unabashedly show the wealth and strength of the owners. The villas of various Western architectural styles directly reflect the preferences of the owners. The complex and exquisite decoration details the new materials imported from abroad reflect the higher social status of the homeowners, and the villa construction greatly meets the needs of showing the Western style. Although the industrialists in Suzhou are open-minded and pragmatic in thinking, there is still a certain contradiction between their worship of western life and their deep-rooted traditional thoughts. On the one hand, the owners want to highlight their status and let western-style architecture lead the fashion. On the other hand, the owner wants to preserve the traditional architectural culture of Jiangnan, so that the fine traditions can be passed on.

In the early times, "patterns", "Western columns", "mosaics", "Western-style decorations" were combined with blue bricks and gray tiles to play the role of partial decoration. Taking Suzhou Wuzhai as an example, the main building of Wu's house is a Western-style fort building built by Wu Guting in Suzhou during the Guangxu period of the Qing Dynasty. It sits on the second floor facing north and south, with five semi-hexagonal rooms at the south, east and west ends. The exterior wall is built with clear water, lime jointed, inlaid with red and blue bricks, and the doors, Windows and floors are basically retained. Backyard building white wall black tile, facade simple and elegant. Bainan corner has bluestone octagonal and one, the existing two-color brick inlaid diamond pattern wall for its characteristics (Fig. 4).

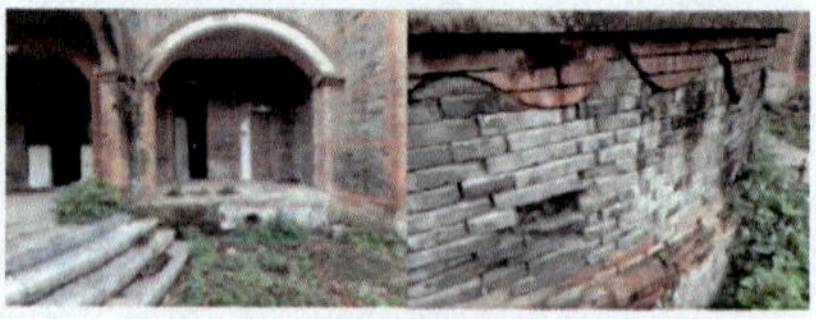

Fig. 4. Wu Zhai **Source.** Self-photographed

After 1920, as a result of the rapid and violent development of new building technology, outstanding performance such as cement, mechanism brick and tile output growth of new building materials, all kinds of decoration materials also emerge in endlessly. It is mainly manifested in the emergence of new roof tiles, the development of wall masonry materials and the renewal of other materials such as glass balustrades. The technology of making bricks and tiles in Suzhou kiln was relatively mature and widely used in the Han Dynasty. In traditional Suzhou buildings, the materials used for bricks and tiles are blue tiles, blue bricks, square bricks, glazed tiles, etc. The color of the glass and the pattern pasted on the glass began to be adopted. This facade aesthetic effect led to a new aesthetic taste (Fig. 5).

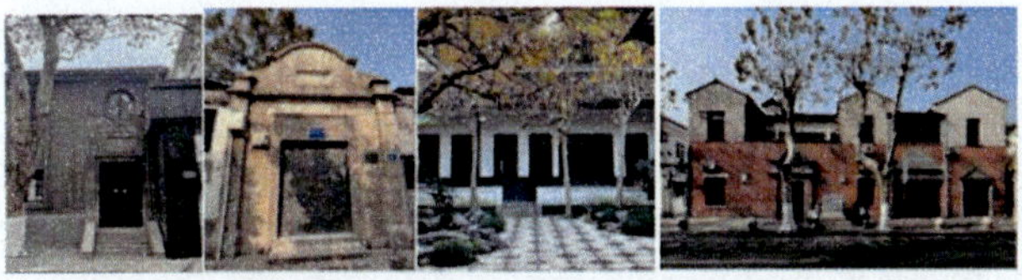

Fig. 5. Appearance of the form **Source.** Self-photographed

It can be seen that the changes in the morphological layout characteristics are mainly reflected in three aspects:

First, the change of living habits. In the early modern times, the urban environment of Suzhou was gradually deteriorated due to the war of various parties. The introduction of western sanitary concepts and urban management experience brought a new turning point to the urban living environment of Suzhou. With the establishment of hospitals and public health facilities, related changes are also reflected in the layout of western-style villas. For example, the placement of toilet, bathroom, kitchen and other sanitary facilities.

Second, the population boom is moving towards compact forms. Population change is one of the reasons for the rapid development of the early modern times. Under the influence of the real estate industry in Shanghai, get apartments emerged in Suzhou, such as Zhirenli Republic of China Building, Xinfuli, Tongyili, Tongdeli and so on. In terms of plane layout, the traditional dwellings have changed from one to multiple households, and the area of single-family residences has decreased accordingly, and the building density is higher than that of traditional residences. In order to meet the needs of urban development, the traditional design and construction methods have also changed.

Third, the advantages of structural technology. The introduction of advanced construction methods of foreign buildings is an important channel for the formation of modern mansions, which is mainly the acceptance and absorption of new structures, new materials, new technologies and new equipment. In the plane layout, a large number of arch structures are used to form the outer corridor, and western methods of wall masonry are introduced, such as plain brick walls.

3.2 Changes in Structural Technology

In the modern period, building structure technology generally has two forms. The first is a non-single-source building technology system that combines Western-style building technology with local Chinese building technology or other regional building technologies. It is mainly manifested in the use of Western-style building structural systems such as brick and stone walls for load-bearing, while the decoration or structural treatment of roofs and walls adopts non-Western-style methods. The second is purely Western architectural techniques. It mainly means that the main structure, roof structure and main decoration methods are all adopted by Westerners, such as brick (stone) wall bearing, wooden truss roof, etc.

Like most traditional buildings, the traditional building structure in Suzhou area is a wooden frame structure, that is, the frame of the whole house is built by wooden beams and columns. The roof frame and the column are a unified structural whole. The roof load is transmitted to the beam and column to the foundation. The wall does not bear load, but only plays the role of enclosure and space separation. The wood frame structure is divided into bucket type, lift beam type and hybrid type. In the modern period, the "load-bearing wall" structure appeared. Roofs are not limited to sloping roofs. Flat roofs, triangular roof trusses and truss structures are gradually accepted by modern architecture. The stimulation of foreign culture and the development of technology have promoted the change of structure and form, which has a great impact on the space, shape and interior shape of modern western-style villas in Suzhou.

The technology of making bricks and tiles in Suzhou kiln was relatively mature and widely used in the Han Dynasty. In traditional Suzhou buildings, the materials used for bricks and tiles are blue tiles, blue bricks, square bricks, glazed tiles, etc. After 1920, due to the rapid development of emerging building technologies, the output of new building materials such as cement, machine-made bricks and tiles has increased, and various decoration materials have emerged one after another. It is mainly reflected in the emergence of new roof tiles, the development of wall masonry materials and the update of other materials such as glass railings. Building roofing materials have broken through the limitations of traditional materials and become more abundant. In particular, cement-made blue (red) tiles have become the roofing materials of many modern western-style buildings in Suzhou; The flat type of grouping forms transitioned to the British style, fishbone, diced brick and other building methods; stained windows glass, mosaic exterior decoration, and iron railings became the mainstream facade decoration methods.

In addition, the development of technology has also made modern architecture develop to vertical space. The traditional buildings in Suzhou area use wood as the main material. The characteristics of wood's resistance to shear and pressure determine the upper limit of the vertical development of traditional buildings. However, in modern times, compared with the inner halls and bedrooms in traditional dwellings, Suzhou western-style villas use the site space more intensively and efficiently, and the plane functions are more abundant and freer. Vertically centralized western-style buildings became a major trend in the functional layout of residential buildings at that time. At the same time, the thickness of the walls began to decrease, and three to four-story buildings appeared. Yin Lu (Fig. 6), for example, is a garden villa that combines Western architecture with Chinese antique gardens. The building structure is a mixed brick and wood structure. Western structural technology makes the building height reach two floors outside and three floors inside.

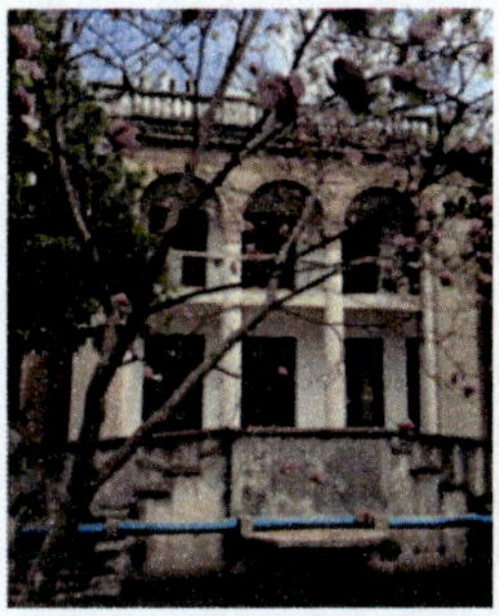

Fig. 6. Yin lu **Source.** Self-photographed

3.3 Change of Residential Culture

Weakening of Traditional Hierarchy. Although Suzhou traditional dwellings are divided into different architectural specifications and architectural forms, they are deeply influenced by the traditional feudal hierarchy, and there is basically an obvious axis between them. From the entrance to the main hall and even the back hall, the symmetrical layout and the series of steps are connected, showing the unique sense of sequence of traditional dwellings. In the modern period, under the influence of Western architectural culture, the residents' living concept changed, the hierarchy gradually weakened, and they began to pay more attention to the quality of life. The division of internal functional locations of modern western-style villas in Suzhou has gradually become flexible and free.

Weakening of Formal Metaphor. The architectural space and function of most of the western-style villas in Suzhou basically took into account the needs of Chinese traditional culture and western customs. As early as the Tang and Song dynasties, the reclusive culture of "big seclusion in the dynasty, middle seclusion in the city, and small seclusion in the wild" is prevalent, and it is also reflected in the traditional dwellings of Suzhou: one aspect is to devote yourself to nature, and the other aspect is to stay away from secular life. Expressed in the courtyard of Suzhou, it is to use the courtyard to create a landscape outside the world. You can feel the natural landscape without leaving the door of the house. In terms of material usage, the courtyard should try to use original ecological materials such as stone, water, trees, soil, etc. These original ecological materials in nature can shorten the distance between man and nature. Since the early modern times, influenced by the trend of Western modernism, most of the buildings have a fluid layout. Within this challenge, the past shows us possible ways to fill the gap between tradition and modernity [6].

The common auspicious patterns in the courtyards of traditional Suzhou dwellings include echo pattern, swastika character, tuanfu character and so on [7]. Most of them use images to express their meanings and metaphors to place people's yearning for a better life. Skillful craftsmen make full use of some waste materials, such as broken porcelain, broken cylinder, broken brick, residual tile, etc., through careful organization, waste into treasure, pile up these clever patterns for people to appreciate, for people to place the feelings of seeking good fortune, but also to add fun to the ordinary life.

Since the early modern times, modern western-style villa buildings in Suzhou have gradually gotten rid of the constraints of traditional dwellings and passively absorbed a large number of Western composition methods. With the further development of society and economy, the unification of architectural function and form has become the main theme of the development of western-style villa buildings. The facade form is no longer subject to traditional dwellings, and the metaphorical method is no longer used to express more freely and changeable. Correctly understand the local form of early modern western-style villas in Suzhou, and promote better urban renewal through relevant research on its adaptive reuse [8, 9].

4 Research on Conservation and Utilization Under the Concept of Sustainable Development

4.1 The Main Problem Facing the Current Western-Style Villas

Suzhou has a long history and rich architectural heritage. However, with the destruction of the war and the development and construction of modern cities, the historical buildings were seriously damaged, such as Zhirenli building complex and Yan Jiagan's former residence, and other buildings were piled up with debris and dilidated, which were in urgent need of repair. In recent years, many existing architectural heritage conservation and restoration projects in Suzhou have been underway. There are 2 state-level cultural relics protection units in Suzhou's current residential western-style buildings, including Tianxiang Xiaozhu and Jinxizhi's former residence, 5 provincial-level key cultural relics protection units.There are 28 municipal-level cultural relics protection units, such as modern residences in Waiwujing Lane, Lanshi Xiaozhu, Wu Zhensheng's former residence, etc., but some of the protected buildings have not been used sustainably.

4.2 The Concept of Sustainable Development

Sustainable development strategy refers to the general term for enabling countries to achieve sustainable development in many fields, and to achieve social, economic and ecological and environmental coordination in accordance with the principles of equity, commonness and sustainability.

Adaptive in use is the best sustainable way to use existing modern western-style villas. The protection of architectural heritage must be integrated into the strategy of sustainable urban Sustainable development involves three aspects: environment, economy and society. The sustainability of modern western-style villas in Suzhou requires coordination among the three to achieve effective protection and inheritance. From an environmental point of view, modern architecture needs to maintain a good living environment, adopt the most appropriate and economical repair methods for damaged buildings to ensure their existence, and awaken the public's memory for a long time. In terms of economy, Suzhou's modern western-style villas contain unique social resources such as history and culture, and enhance economic value through ecological transformation. Such as replacing historical buildings with exhibition halls and photo studios, commercial development is carried out on the basis of preservation, and economic benefits are generated.

4.3 Adaptive Reuse Strategies

Authenticity of Architectural Heritage. To study the authenticity of architectural heritage, it is not only necessary to study the preservation of the remains of the building, but also to study the environment on which the architectural life depends, but also to study the growth and change of the building, and to consider it from the perspective of the ecosystem and make it continue in the future time.

For modern western-style villas, "the layout, structure and decoration of the original building shall not be arbitrarily changed and destroyed, and the historical imprints retained on the building entity shall be continuously excavated through historical research and documentary records, such as architectural styles, construction techniques, building materials and the artistic characteristics displayed, practical functions in modern life, etc. In order to give full play to the practical value of the building, it can be partially changed according to the development of society and the needs of the modern life of residents, and the layout and interior decoration and shape can be made, which is also to protect its authenticity. To sum up, Suzhou's modern western-style villa buildings will be able to better continue the unique style, coordinate the space layout, external modeling, technical means, place the overall protection under the urban development process, highlight the superiority of architectural cultural resources, and make it permanent alive.

Continue the Green Mind. Ecological concept of harmony between man and nature is reflected in the courtyards of traditional Suzhou dwellings, which is to adjust the structure of the courtyard itself to conform to the laws of nature and create more favorable conditions [10] (Fig. 7).

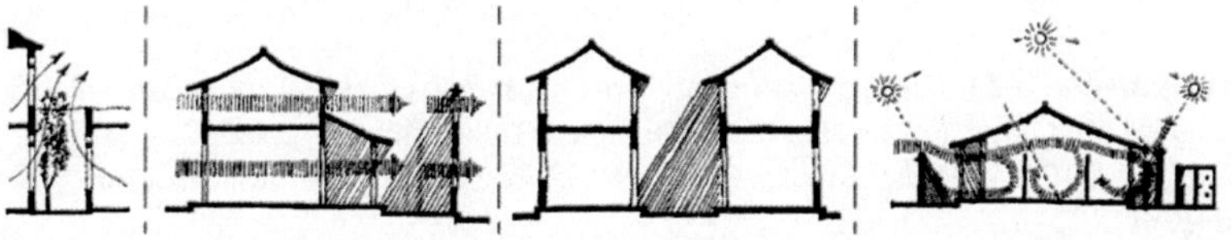

Fig. 7. Schematic diagram of the patio ventilation and shadow **Source.** Yufeng, Z, Research on Courtyard Space of Traditional Dwelling House in SuZhou

For example, the Suzhou area is hot in summer. In order to improve this situation, the courtyards of its residences are mostly small in depth to reduce the courtyard temperature.

From the construction that conforms to the natural conditions and local conditions, to the natural state of Suzhou residential buildings as the basic criterion for building modern western-style villas, the exploration of its green ideas can excavate deeper value significance in the collision of ancient and modern "green ideas".

The Intervention of New Technologies. At the social level, under the background of the rapid development of new technologies, technical means such as intelligence, BIM, GIS, and digitalization have been widely used in urban planning, renewal and other fields in recent years. Suzhou modern western-style villa buildings can also use digital technology to store real historical information for better inheritance of cultural connotation [11].

5 Conclusion and Inspiration

In the context of the rapid development of modern cities, Western culture is the incentive for the development of modern architecture in Suzhou. However, due to the stability and conservatism of traditional culture, the transformation of modern western-style villas is not passively following the West, but the collision and fusion of Chinese and Western cultures. It is the modern transformation of traditional consciousness culture and the expression and reflection in architectural form. As a representative residential building in early modern times, it was built by traditional craftsmen to adapt to the specific historical environment, technical conditions and construction materials at that time, affecting the traditional spatial pattern and traditional construction techniques in Suzhou, and has great research value and significance. Based on the modern urban development of Suzhou, this paper compares the changes of residential western-style buildings before and during the modern period, combines field research to study their type characteristics, and puts them into a relatively clear residential system framework to obtain the transformation characteristics. The construction, restoration and reconstruction of historical buildings provide a certain theoretical reference.

The development and decline of modern western-style villas in Suzhou are closely related to the process of modern cities. After the demolition and construction of the city, a serious "cultural fault" phenomenon has been caused. The construction of architectural heritage must be integrated into the strategy of sustainable urban development, and the ecological and green construction ideas in traditional dwellings can be absorbed in order to achieve better good construction.

Acknowledgments. The writing and research of this paper obtain support from National Key R&D Program of China (2021YFE0200100) and 2021 Policy Directed Program of Jiangsu Province (BZ2021015), Postgraduate Research & Practice Innovation Program of Jiangsu Province (KYCX22_3199).Thanks.

References

1. Xianjue, L., Xin, W.: The Early Modern Architectural in Jiangsu. Jiangsu Science and Technology Press, Nanjing (2008)

2. Xue, Z., BenJian, L.: Research on the application of traditional houses in contemporary architectural space design in Jiangnan. Chin. Foreign Archit. **24**(06), 52–54 (2018)
3. Mingsu, X.: Suzhou Folk House. China Building Industry Press, Beijing (1991)
4. Kai, X.: Study on the Western Style Reconstruction of Traditional Residences in Taihu Lake Basin from the Late Qing Dynasty to the Republic of China, Dissertation, Suzhou University of Science and Technology, Suzhou, Jiangsu, China (2020)
5. Guoqing, S.: Study on the Modernization Process of Urban Space in Suzhou, Hangzhou, Zhejiang, China (2020)
6. Moriset, S., Rakotomamonjy, B., Gandreau, D.: Can earthen architectural heritage save us? Built Heritage **5**(1), 58–60 (2021)
7. Okpalanozie, O.E., Adetunji, O.S.: Architectural heritage conservation in Nigeria: The need for innovative techniques. Heritage **4**(3), 2124–2139 (2021)
8. Yabo, Z., Peng, Z., Xuemei, Z.: The latest progress in urban renewal research at home and abroad. Urban Dev. Res. **26**(10), 42–48 (2019)
9. Adamopoulos, E., Volinia, M., Girotto, M., Rinaudo, F.: Three-dimensional thermal mapping from IRT images for rapid architectural heritage NDT. Buildings **10**(10), 187 (2020)
10. Masood, K., et al.: Small but well-established architecture? Sustainable development of small heritage sites in Iran. Chin. Cult. Heritage **06**, 109–110 (2021)
11. Hyunjung, L.: Sustainable cultural heritage conservation and utilization: Focusing on herculaneum case in Italy. Korean J. Folk Stud. Null **45**, 187–218 (2019)

A Multi-objective Optimization Approach of Green Building Performance Based on LGBM and AGE-MOEA

Yuxuan Shen[1] and Yue Pan[1,2](✉)

[1] Shanghai Key Laboratory for Digital Maintenance of Buildings and Infrastructure, Department of Civil Engineering, Shanghai Jiao Tong University, 800 Dongchuan Road, Shanghai, China
panyue001@sjtu.edu.cn

[2] Anhui Province Key Laboratory of Intelligent Building and Building Energy Saving, Anhui Jianzhu University, Hefei, China

Abstract. This paper proposes a novel multi-objective optimization approach for green building performance combining Building Information Model (BIM) and artificial intelligence (AI) algorithms. LGBM and AGE-MOEA algorithm are respectively utilized to construct the prediction and optimization model, and finally the multi-objective optimization process for total site energy consumption, CO2 emission and discomfort hours can be achieved by changing values of 12 corresponding level-2 indicators. The pareto front and the optimal solution are obtained. The applicability and effectiveness of the optimization method are verified by a practical case, and the results suggest that: (1) the accuracy of LGBM prediction model is as high as 99.975%; (2) under the multi-objective optimization of AGE-MOEA algorithm, the green performance metrics of building information model has improved by 14.32%, which is better than that of widely used NSGA-II and NSGA-III algorithms under the same conditions; (3) the setting of HAVC system is the biggest factor affecting the performance of green buildings.

Keywords: Multi-objective optimization · Green building performance · Artificial intelligence algorithms

1 Introduction

The energy consumption and carbon emission effects of buildings are receiving increasing attention from building practitioners and environmentalists all over the world [1]. According to the China Association of Building Energy Efficiency, the construction industry (upstream and downstream included) accounts for 51% of China's overall carbon emissions, 22% of which comes from the building operation phase [2].

Green building plays a role in saving resources, protecting the environment and reducing pollution, which greatly promotes sustainable development [3]. The performance optimization of green buildings requires reasonable analysis and rational use of space, building materials, equipment, climate and other resources without comfort loss,

W. Guo and K. Qian (Eds.): GBCESC 2022, LNCE 211, pp. 202–210, 2023.
https://doi.org/10.1007/978-981-19-5217-3_19

and finally reduces the waste and loss of resources caused by improper design or operation and management absence of buildings, thus the performance on building energy and carbon emission reduction can be greatly enhanced.

Nowadays, with the promotion of AI algorithms, one can build and train an algorithmic framework for automatic prediction of building energy consumption and carbon emissions with only limited model characteristics and simulation test results [4]. However, optimizing single desirable target values requires designers to analyze the extent and mechanism to approach the expected values through multiple attempts, which is tedious and uncreative [5]. Therefore, new optimization methods need to be proposed to take designers and researchers out of such tasks.

In this paper, attempts would mainly be focused on the performance optimization of green buildings, which generally includes: (1) the orthogonal test setting of feature sets for the model; (2) the implementation of building performance prediction algorithms; (3) the seeking of multi-objective optimization solutions. Finally, the reliability of the adopted LGBM prediction algorithm and AGE-MOEA optimization model is verified with an actual green building designed model as the case study.

The innovation of the research lies in: (1) proposing a whole-process building performance optimization framework to realize the synergy between BIM and DesignerBuilder, thus providing a database for the establishment of targeted prediction algorithm; (2) utilizing an efficient MOO algorithm based on adaptive geometry estimation to enhance the performance of most crucial indices for the green building and elevate the level of automation integration.

2 Preliminaries

2.1 LightGBM

LightGBM (Light Gradient Boosting Machine) is a framework of GBDT algorithm. It uses a histogram-based decision tree algorithm, which discretizes the continuous feature values into n bins and then constructs a feature histogram, traverses the data to accumulate certain statistics and then finds the optimal segmentation point [6]. Therefore, the LightGBM algorithm will be adopted for the prediction of green building performance.

2.2 AGE-MOEA Algorithm

AGE-MOEA solves the problems of degraded optimization performance and poor generality of the algorithm [7]. AGE-MOEA adopts the framework of the classical non-dominated ranking genetic algorithm named NSGA-II, but it replaces the crowding distance with a survival score that combines both diversity and proximity of the non-dominated fronts. Furthermore, AGE-MOEA differs from NSGA-III algorithm in the proximity. AGE-MOEA considers the use of the L_p norm instead of L_2 norm for an accurate measurement in the generalized M-dimensional space. The Euclidean norm (L_2 norm) represents a special case of the L_p norm where $p = 2$. Different values of p correspond to different measures of the distance in $\mathbb{R}^M$. In AGE-MOEA, the value of p can be calculated as:

$$p = -\frac{\log(M)}{\log\left(\frac{1}{M}\sum_{i=1}^{M} C_i\right)} = \frac{\log(M)}{\log(M) - \log\left(\sum_{i=1}^{M} C_i\right)} \quad (1)$$

where M is the number of objectives; C_i is the i-th coordinate of C in the objective space. Survival score strongly depends on the diversity and proximity:

$$score(S) = \frac{diversity(S, \mathbb{F}_1)}{proximity(S)} \tag{2}$$

In this study, AGE-MOEA will be used as the main multi-objective algorithm for optimization solution, and the results will be compared with NSGA-II and NSGA-III in a transverse manner to check their effectiveness and applicability.

3 Methodology

3.1 Determination of Evaluation Objectives and Featured Indicators

Firstly, the total building energy consumption, carbon emission and indoor thermal discomfort hours are set as the metrics of green building evaluation. An ideal green building should reduce the total building energy consumption and carbon emission without increasing the indoor thermal discomfort hours [8].

After the determination of performance evaluation metrics, indicators that affects the three evaluation metrics require further search and research. Determined parameters can be divided into three major categories (envelope, openings, and HVAC) based on their properties, and these three categories are referred to as level-1 indicators [9]. They can be detailed into 12 level-2 indicators. Users can freely change the numerical magnitude of 12 indicators in DesignBuilder (a comprehensive user-graphic interface simulation software developed for EnergyPlus [10]), thus ultimately changing the value of evaluation metrics [11] (three yellow boxes shown on the left of Fig. 1(a)).

3.2 Building Information Model Customization and Orthogonal Test

The model in Revit software is exported as a gbxml format file and then imported into DesignBuilder for custom setting of each level-2 indicator in the form of designing orthogonal tests. The selection of the level-2 indicators follows the design principle of orthogonal test, and the values of 12 parameters will become a point in the multi-dimensional space after selection [12].

3.3 Prediction for Evaluation Objectives

The collection and cleaning of hundreds of sets of orthogonal test data are completed by the written programs and other specialized software. Data processed by the standardized process can be used as the raw data for the LightGBM prediction algorithm. The prediction algorithm can directly find out the evaluation metrics of green building without going through the custom parameter setting and time-consuming simulation process in DesignBuilder when level-2 indicators are known, which lays a solid foundation for the subsequent implementation of automatic optimization of AI algorithms.

In addition, the LightGBM algorithm can also quantify the importance of the 12 features. The larger the value of its F-Score, the greater role this feature plays in predicting the metrics, and vice versa. It can be used as a reference and should be a proof of the rationality of the selection of level-2 indicators.

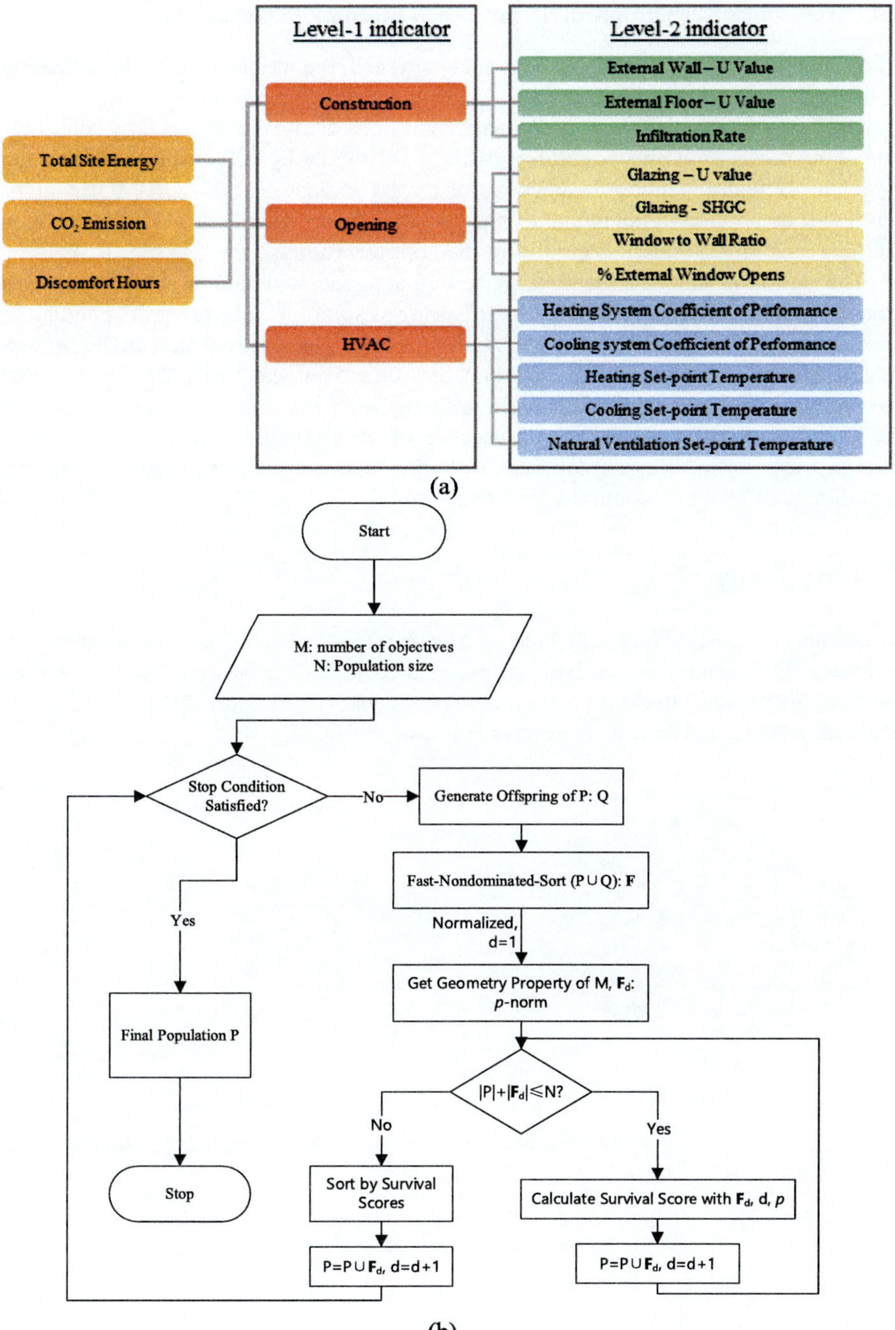

Fig. 1. (a) Indicators of green building performance; (b) Flowchart of AGE-MOEA algorithm.

3.4 Multi-objective Optimization for Green Building Performance

After the prediction algorithm of the model is trained, the multi-objective optimization of the building model can be started.

Optimization problem can be summarized as minimizing the total energy consumption, carbon emission and discomfort of the BIM model by reasonably adjusting the values of 12 features under 3 categories of level-1 indicators. the AGE-MOEA algorithm will be applied to the model to improve the performance indicators of the green building, and the flowchart of multi-objective optimization is shown in Fig. 1(b).

The algorithm starts the search from the initially generated random solution and uses crossover and mutation to generate new offspring as solutions. The offspring population Q is combined with the existing population P to form a new population. AGE-MOEA calculates the L_p-norm based on the geometry of the front, calculates the diversity and proximity to obtain the survival score, iterate to increase the solution set, and terminates the iteration when the front solution exceeds M. AGE-MOEA selects the remaining solutions after sorting the front solutions according to the survival score until the stopping condition is satisfied to output the final population P.

4 Case Study

An ordinary school building with three-story is selected to evaluate its green building performance. The 3D model of the building established in Revit is shown in Fig. 2(a). External walls are attached to the floors or roof. The model can be exported to DesignBuilder in gbxml format, and the export process is shown in Fig. 2(b).

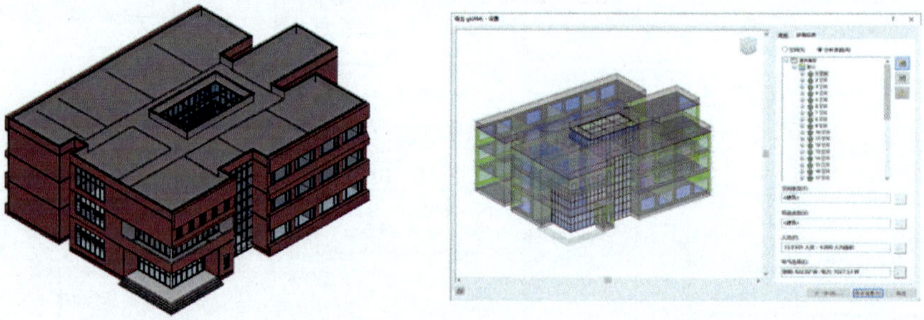

(a) (b)

Fig. 2. Building and exporting procedure: (a) Revit Model; (b) exporting gbxml file.

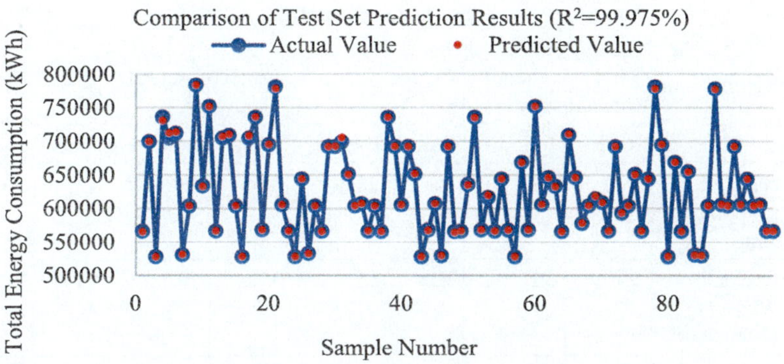

Fig. 3. Training results of energy consumption prediction on LGBM model.

Set the initial values of the cooling and heating temperatures and the energy density of equipment and other parameters according to the Design Standard for Energy Efficiency of Public Buildings (GB 50189-2015). Some detailed parameter settings and values after simulation can be found in the "Initial Design" column of Table 2.

Once the performance simulation data set of the building is obtained, the LGBM model can be used to train the building performance metrics. The prediction results of the LGBM model for the test set are shown in Fig. 3, which shows that the predicted values are consistent with the simulated values, and the R^2 fitness is 99.975%.

Table 1. Feature importance ranking of three optimization targets.

No	Features	Total site energy consumption		CO2		Discomfort (All Clothing)		Total	
		Rank	F-Score	Rank	F-Score	Rank	F-Score	Rank	F-Score
I	Heating set-point temperature	I	261	I	197	I	211	I	655
II	Cooling set-point temperature	II	192	II	162	II	167	II	516
III	Heating system seasonal CoP	III	134	III	142	III	163	III	418
IV	Infiltration rate (at 4 Pa)	IV	82	IV	113	IV	70	IV	308

(*continued*)

Table 1. (*continued*)

No	Features	Total site energy consumption		CO_2		Discomfort (All Clothing)		Total	
		Rank	F-Score	Rank	F-Score	Rank	F-Score	Rank	F-Score
V	Nat vent set-point temp	V	60	V	86	VI	66	V	232
VI	External Wall U-Value	VI	59	VI	85	V	59	VI	229
VII	Glazing U-value	VII	52	IX	48	XII	52	IX	136
VIII	External floor U-Value	VIII	43	X	48	IX	49	X	130
IX	Cooling system CoP	IX	40	XI	40	VIII	47	VIII	121
X	Window to Wall %	X	34	VIII	39	X	38	VII	114
XI	Glazing SHGC	XI	22	VII	31	XI	30	XI	102
XII	% External window opens	XII	22	XII	13	VII	27	XII	48

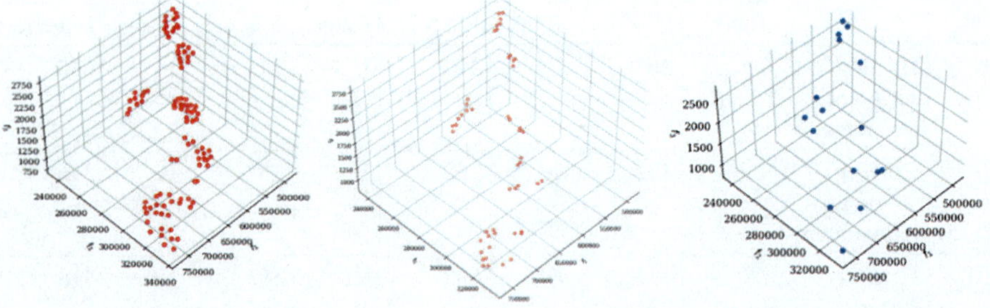

(a) (b) (c)

Fig. 4. Pareto front of optimization: (a) AGE-MOEA; (b) NSGA-II; (c) NSGA-III.

The importance of each feature in the LGBM prediction model is quantified and interpreted, and the data are organized as shown in Table 1. The results indicate that HVAC is an important control factor in evaluating the performance of green buildings, therefore its settings should be given priority in the multi-objective optimization.

Table 2. Summary of optimization results and corresponding feature sets.

Building parameters	Unit	Initial design	Optimization algorithm		
			NSGA-II	NSGA-III	AGE-MOEA
Window to wall ratio (WWR)	%	40	36	16	19
Heating system CoP	-	2	1	5	1
Cooling system CoP	-	2	5	5	5
% External window opens	%	20	14	6	22
Cooling set-point temperature	°C	24	26	27	28
Heating set-point temperature	°C	16	16	17	16
Nat vent set-point temp	°C	15	23	24	30
Infiltration rate (at 4 Pa)	$m^3/h\text{-}m^2$	9	20	21	14
Glazing U-value	-	1.99	4.53	4.60	4.62
Glazing SHGC	-	0.40	0.32	0.34	0.34
External wall U-value	-	0.62	0.13	0.89	0.15
External floor U-value	-	0.38	0.42	0.32	0.41
Total site energy consumption	kWh	567303	509106	507723	488588
CO2 emission	kg	286578	260170	240843	240431
Discomfort (All Clothing)	h	2774	2434	2661	2700
Distance	-	635584	587823	563764	544549
Ratio (Set Initial as Datum)	-	1.0000	0.9249	0.8870	0.8568
Optimization rate	-	-	7.51%	11.30%	14.32%

AGE-MOEA, NSGA-II and NSGA-III algorithms are used to solve this multi-objective optimization problem, and the pareto front solution obtained by the three algorithms are shown in Fig. 4. The final optimization results of each algorithm are compared by the size of the distance from the origin to the pareto front solution. The optimal solution target values and feature combinations of each algorithm are shown in Table 2, which states that AGE-MOEA has the best optimization effect, then NSGA-III and NSGA-II. Thus the superiority of AGE-MOEA is clearly presented.

5 Discussions

From the variation trend of the optimal solution value, f_1, f_2 and f_3 are contradictory optimization objectives. With the rapid reduction of building energy consumption and carbon emissions, the hour of building thermal discomfort gradually increases, thus the three objective functions cannot be realized at the same time. In addition, the target values of the three algorithms are improved compared with the original design scheme. Among them, the optimization amplitude of NSGA-III and AGE-MOEA is 11.30% and

14.32% respectively. The effectiveness of multi-objective optimization of green building performance indicators has been fully proved.

6 Conclusions

In this study, a multi-objective optimization approach combining BIM and LGBM-AGE-MOEA is proposed. By changing the values of 12 features subordinate to level-1 indicators, the metrics performance is improved. Then, a case model of a building is used to verify the applicability and effectiveness of AGE-MOEA. The accuracy of LGBM prediction model reaches 99.975%, and the building model achieves 14.32% improvement of green performance metrics under the MOO of AGE-MOEA. The optimization effect is better than NSGA-II and NSGA-III under the same conditions. The more important factors to affect the HAVC system performance can be revealed.

Acknowledgements. The current study is supported by grants from the Shanghai Sailing Program, China (No: 22YF1419100) and the open foundation of the key lab of Anhui Province Key Laboratory of Intelligent Building & Building Energy Saving (No: IBES2021KF06).

References

1. Liu, L.: Sustainability: Living within one's own ecological means. Sustainability **1**, 1412–1430 (2009)
2. China Building Energy Consumption Research Report. https://www.cabee.org/site/content/24021.html. Accessed 09 Feb 2022
3. Hossain, M.F.: Green science: Advanced building design technology to mitigate energy and environment. Renew. Sustain. Energy Rev. **81**, 3051–3060 (2018)
4. Huang, S., Zuo, W., Sohn, M.D.: A bayesian network model for predicting cooling load of commercial buildings. Build. Simul. **11**(1), 87–101 (2017). https://doi.org/10.1007/s12273-017-0382-z
5. Asl, M.R., Zarrinmehr, S., Bergin, M., Yan, W.: BPOpt: A framework for BIM-based performance optimization. Energy Build. **108**, 401–412 (2015)
6. Ke, G., Meng, Q., Finley, T., Wang, T., Chen, W., Ma, W., Ye, Q., Liu, T.Y.: Lightgbm: A highly efficient gradient boosting decision tree. Adv. Neural Inf. Process. Syst., 3146–3154 (2017)
7. Panichella, A.: An adaptive evolutionary algorithm based on non-euclidean geometry for many-objective optimization. In: Proceedings of the Genetic and Evolutionary Computation Conference (2019)
8. Li, Q., Zhang, L., Zhang, L., Wu, X.: Optimizing energy efficiency and thermal comfort in building green retrofit. Energy **237**, 121509 (2021)
9. Guo, K., Li, Q., Zhang, L.M., Wu, X.G.: BIM-based green building evaluation and optimization: A case study. J. Cleaner Prod. **320**, 128824 (2021)
10. DesignBuilder Software Ltd – Home. https://designbuilder.co.uk/. Accessed 09 Feb 2022
11. Pan, Y., Zhang, L.M.: Data-driven estimation of building energy consumption with multi-source heterogeneous data. Appl. Energy **268**, 114965 (2020)
12. Chen, B., et al.: Multiobjective optimization of building energy consumption based on BIM-DB and LSSVM-NSGA-II. J. Cleaner Prod. **294**, 126153 (2021)

Carbon Emissions of Passive Building During Production and Operation in Cold Region of China

Zhiqiang Kang, Bo He(✉), and Xiaodong Xia

Shenyang Jianzhu University, Princeton 110168, SY, China
hgf199524@163.com

Abstract. Passive buildings have long been regarded as energy-efficient and environmentally friendly buildings. This kind of buildings can reduce the heat loss of the building by improving the thermal resistance of building envelope and improving the air tightness of the buildings. In this paper, carbon emissions of passive buildings in cold regions and traditional building models based on the example of passive building in the same geographical conditions were calculated and analyzed in the production section and operation section of the building. This article analyzes the carbon emission characteristics of passive buildings in the production section and operation section of building materials, which account for the vast majority of carbon emissions in the whole life cycle of passive buildings. According to the case calculations, the carbon emission of the passive building is 0.77t CO_2/m^2 in the production section of building materials, and 3.07t CO_2/m^2 in the operation section. Compared with traditional buildings, the carbon emissions of passive buildings is 37.5% higher in the production section of building materials, 53.27% lower in the operation section. The analysis shows that passive buildings have low carbon characteristics compared with traditional buildings in the whole life cycle.

Keywords: Passive building · Carbon emissions · Software simulation · Carbon emission characteristics

1 Introduction

With the increasing concern of carbon emissions, there are more and more studies in the field of carbon emissions from the construction sector in recent years. Most of these studies were based on the whole life cycle theory of buildings (LCA) to study the carbon emissions in each section of the whole life cycle of buildings. More scholars have studied the production section and operation section of building materials. For example, through hotel construction, Su Xing [1] calculated and analyzed the differences between passive buildings and traditional buildings in the production section and operation section of the building. Chen Ying and Zhu Yan [2] calculated the carbon emissions of each stage of the life cycle of Chinese traditional buildings and constructed a mathematical model for carbon emissions calculation. Ma Caiyun [3] et al. made a comparative analysis of

W. Guo and K. Qian (Eds.): GBCESC 2022, LNCE 211, pp. 211–220, 2023.
https://doi.org/10.1007/978-981-19-5217-3_20

the calculation results of carbon emissions in production section between prefabricated buildings and traditional cast-in-place buildings in Fujian, China. Bribian [4] from Spain directly divided architecture into two sections of construction and use for research. Bonamente E [5] from Italy study showed that energy consumption and carbon emissions of buildings mainly come from the operation section. Malindu Sandanayake [6] from Canada established a model to estimate carbon emissions during building construction.

According to the data obtained from China Building Energy Consumption Report 2020 [7], in the whole life cycle of a building, the production section of building materials and the operation section of a building are two sections that consume more energy and emit more carbon emissions. In 2018, the total life-cycle energy consumption and carbon emissions of buildings in China were 2.147 billion tce and 4.93 billion t CO2. Among them, the energy consumption of building materials production section accounted for 51.3% of the building energy consumption, the energy consumption of building operation section accounted for 46.6% of the building energy consumption. Corresponding to the energy consumption, the carbon emission in the production section of building materials accounts for 55.2% of the carbon emission in the whole life cycle of the building, and the carbon emissions in the building operation section accounts for 42.8%. In the whole life cycle of a building, the carbon emissions in the production section and operation section of building materials accounts for 98% of the carbon emissions in the whole life cycle of a building. Therefore, in the process of studying the carbon emissions of buildings, the production section of building materials and the operation section of buildings are two phases that cannot be ignored and need to be studied emphatically.

In fact, from the perspective of the whole life cycle of buildings, passive buildings do not have the advantage of energy saving and emissions reduction compared with traditional buildings in the production section of building materials. In order to obtain low heat transfer coefficient and better air tightness, a large number of thermal insulation materials and components are applied in the construction process of the passive buildings. The production of these materials and constructions will bring a large amount of energy consumption and carbon emissions. In the use section of buildings, the energy consumption of passive buildings is much lower than that of traditional buildings due to the thermal insulation and energy saving characteristics of passive buildings [8]. The research focus of this paper is to analyze whether the carbon emission reduced by passive buildings in the operation section can compensate for the carbon emission increased in the production section of building materials.

2 Architectural Introduction

2.1 Introduction to Passive Building

The ultra-low energy consumption building studied in this paper is the multi-function hall project of Xiongan (Anxin) Relocation Headquarters Office located in Xiongan New Area (Fig. 1). It is a single-story building with a floor area of 54 m^2, mainly providing office, reception and conference space for relocation workers. The multi-function hall adopted brick strip foundation, steel structure modular frame system, prefabricated light steel and light slurry non-thermal bridge walls, opaque envelope heat transfer coefficient of roof, ground and external walls $\leq$0.13 W/(m^2·k), heat transfer coefficient of doors and windows $\leq$0.8 W/(m^2·k), and building air tightness <0.6 h^{-1}.

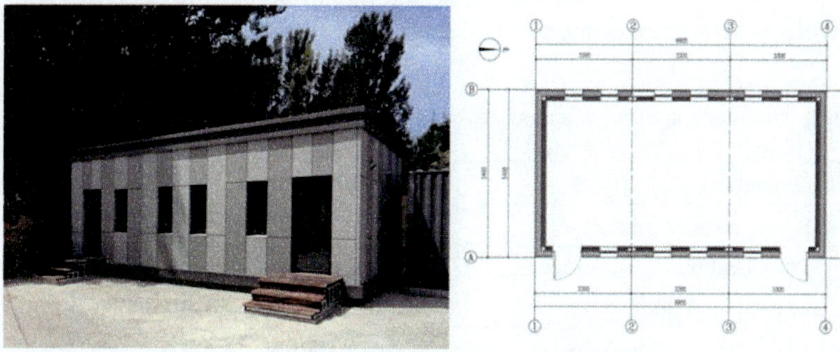

Fig. 1. A case of passive building

The building used two separate HVAC technology systems. The energy consumption of heating and refrigeration and indoor temperature and humidity environment are monitored and measured. The project was designed with the help and technical guidance of Professor Wolfgang Fester, director of the German Passive House Research Institute and founder of the German passive house theory. All aspects of the building have reached the standard of passive building [8], and the specific parameters are shown in Table 1.

Table 1. Some specific parameters of passive building

Building parameters	Data
Area	54 m^2
Building height	3.3 m
Service life	50 years
External wall heat transfer coefficient	0.13 W/(m^2·K)
External window heat transfer coefficient	0.8 W/(m^2·K)
Roof heat transfer coefficient	0.13 W/(m^2·K)
Surface heat transfer coefficient	0.13 W/(m^2·K)
Air tightness	0.43 h^{-1}

2.2 Traditional Architectural Model

In order to evaluate the energy use and carbon emission of the passive building and the traditional building in building production section and operation section. From the perspective of energy consumption and carbon emissions, a traditional building model was established. Changed only the insulation structure and windows of the building model. According to the design specification of the area, the heat transfer coefficient of the building maintenance structure in the traditional building model was set. The different thermal properties of the two buildings were shown in Table 2.

Table 2. Comparison of parameters between passive building and traditional building

Building parameters	Passive building	Traditional building
External wall heat transfer coefficient	0.13 W/(m^2·K)	0.35/(m^2·K)
External window heat transfer coefficient	0.8 W/(m^2·K)	1.5/(m^2·K)
Roof heat transfer coefficient	0.13 W/(m^2·K)	0.40/(m^2·K)
Ground floor heat transfer coefficient	0.13 W/(m^2·K)	0.60/(m^2·K)
Air tightness	0.43 h^{-1}	0.5 h^{-1}

3 Carbon Emissions Calculation of Building Materials Production Section

3.1 Collection of Building Materials

In the process of carbon emissions calculation of passive building production section, the biggest difficulty was how to obtain the building material list. In order to obtain accurate and comprehensive data, we contacted the design leader of the case passive building and counted the construction materials and building components for the site construction. As for the carbon emissions factors of related building materials, they were obtained from the Carbon Emission Calculation Standard and THE IPCC carbon emission factor database. However, due to the difficulty of statistics, it was difficult to accurately collect some data during construction, including the utilization of some energy and the production of some construction raw materials. Therefore, for some data that were difficult to be counted, representative data commonly used in the current industry collected from authoritative literature were used instead [9–11]. The specific project list was shown in Table 3.

Table 3. Engineering inventory of case building

Construction site	Building materials	Unit	Amount	Carbon emission factor ($kgCO_2$/unit)
Ground floor	100 mm thick C30 cushion	m^3	14.4	321.3
	Waterproofing material	m^2	54	2.38

(continued)

Table 3. (*continued*)

Construction site	Building materials	Unit	Amount	Carbon emission factor ($kgCO_2$/unit)
	Extruded polystyrene board	m^3	14.56	669
	60 mm thick C30 protective layer	m^3	3.36	321.3
Exterior wall	Steel structural module	t	0.358	2380
	Light steel keel	t	0.325	2310
	Light concrete (foamed cement)	m^3	6.18	297
	Fiber reinforced cement board	t	0.845	977
	Extruded polystyrene board	m^3	6.5	669
	10 mm gypsum board	m^2	65	2.465
	Polyurethane insulation board	t	0.702	5220
	OSB plate structure	m^3	0.65	12.34
	Plaster mortar	m^3	0.325	356.126
	Polystyrene granule slurry	m^3	1.3	482.77
The roof	Light steel structure	t	0.27	2310
	OSB plate structure	m^3	0.54	12.34
	Polyurethane insulation board	t	0.324	5220
	Extruded polystyrene board	m^3	10.8	669
	Waterproofing materials	m^2	54	2.38
	SLATE waterproof	m^2	54	1.838
	6 mm thick cement board	t	1.004	977
	10 mm thick gypsum board	m^2	54	2.465
Door and Window	Polyurethane window frames	t	0.015	5220
	Low-e insulating glass	t	1.163	1150
	Aluminium alloy fixed window	m^2	15.5	46.3
	Wood fire door	m^3	0.27	12.34

For traditional buildings, through the established model, we used calculation software to calculate the amount of the main building materials in the construction process. According to the carbon emission factor of building materials, the carbon emission of the production part of traditional building was calculated (Table 4).

Table 4. Engineering inventory of traditional building

Building materials	Unit	Amount	Carbon emission factor ($kgCO_2$/unit)
Steel	t	3.240	2380
Concrete	m^3	30.6	321.3

(*continued*)

Table 4. (*continued*)

Building materials	Unit	Amount	Carbon emission factor ($kgCO_2$/unit)
Block	block	6480	0.2945
Mortar	m^3	9.945	356.125
Glass	t	1.163	1150
Insulation board	m^3	5.95	669
Coating	m^3	0.436	11
Wood fire door	m^3	0.27	12.34
Aluminium alloy fixed window	m^2	15.5	46.3
Polyurethane window frames	t	0.015	5220
Waterproofing materials	m^2	108	2.38

3.2 Carbon Emission Calculation and Data Analysis

The building materials lists of passive buildings and traditional buildings were obtained, and the carbon emissions of the two buildings in the production of building materials were calculated by combining the carbon emission factors of building materials. The carbon emissions of passive buildings in the production of building materials was 41.79tCO_2, while that of traditional buildings was 30.09tCO_2. The proportion of different building materials in the production of building materials of the two buildings is shown in Fig. 2. The carbon emissions of passive buildings in the production of building materials is 38.9% higher than that of traditional buildings. It can be seen that passive buildings are not low-carbon in the production of building materials.

It could be seen that in the production section of building materials, passive buildings did not have low carbon performance compared with traditional buildings. On the contrary, due to the use of a large number of insulation materials and construction, the

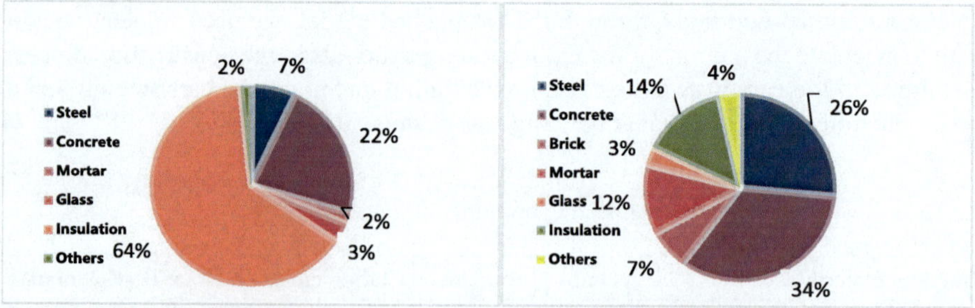

(a) Passive building (b) Traditional building

Fig. 2. Carbon emission of passive building and traditional building materials in production section carbon emission of passive buildings in the production section of building materials was much higher than that of traditional buildings. For passive buildings, carbon emissions in the production section of building materials mainly came from insulation materials (64%) and concrete (22%). For traditional buildings, the proportion of thermal insulation materials in carbon emissions was only 14%, while steel (26%) and concrete (34%) occupy a large proportion in the production section of building materials.

4 Calculation of Carbon Emissions During Operation

4.1 Calculation Method

In order to reflect the low-carbon performance of passive buildings, a traditional building model was simulated and compared with passive buildings in the research process, and the difference of carbon emissions between passive buildings and traditional buildings in each section of the whole life cycle was studied, so as to point out the direction for the future low-carbon analysis.

In order to reduce errors in the calculation process, when constructing traditional building models, the building coefficients of the model such as shape coefficient, window-wall ratio and door and window orientation should be consistent with passive building cases. In the process of building energy consumption simulation calculation, it is also necessary to set up traditional buildings in accordance with the building codes of the climate zone where passive buildings are located.

In practice, however, there are too many factors affecting the difference in a building's carbon emissions to control for everyone. Therefore, when comparing the carbon emissions of traditional and passive buildings, only the heat transfer coefficient and air change times of the building maintenance structure are studied.

4.2 Calculation Results and Data Analysis

In the calculation of carbon emissions in the building operation section, it is difficult for us to directly obtain the carbon emission data of the building, and only the energy consumption in the building operation section can be calculated first. After that, the energy consumption was uniformly converted into electricity, and the carbon emissions of the building during operation was calculated according to the carbon emission factor of electricity (1.128 $kgCO_2/kWh$). We calculated the total annual kilowatt-hours of

Table 5. Building operating energy consumption

Title	Passive building (kWh/m^2 yr)	Traditional building (kWh/m^2 yr)
HVAC	38.58	97.20
Lighting equipment	7.66	7.84
Ventilation	3.58	6.17
The hot water	4.65	5.32

electricity used for HVAC systems (heating, cooling, ventilation), hot water, and lighting by simulating. The results are shown in Table 5.

Through the simulation calculations, we get the total annual electricity consumption of the building, and then according to the carbon emission factor of unit electricity, we can get the carbon emission data of the building in the operation section. Passive buildings (54.47 kWh/m^2 yr) were 53.26% lower than conventional buildings (116.53 kWh/m^2 yr). In passive buildings, heating and cooling requirements accounted for the largest share (70.8%) of the total electricity demand in the operational phase, followed by lighting (14.06%), hot water (8.53%) and ventilation (6.57%). In traditional buildings, although the contribution of heating and cooling is still the largest (83.40%), the total energy consumption of heating and cooling is larger than that of passive buildings, so the energy consumed by heating and cooling is much larger than that of passive buildings.

There is a big difference in the power demand of HVAC systems for the two types of buildings, accounting for 70.8% of the total power demand of passive buildings and 83.40% of the total power demand of conventional buildings. In contrast, In the heating section of the building, the energy required for passive building heating is converted into electricity 13.28 kWh/m^2 per year which meets the standard of $\leq$15 kWh/m^2 per year for passive building heating terminal heat consumption [12]. However, for traditional buildings, the energy heating in the room is converted into electricity for 69.88 kWh/m^2 year. Converting electricity consumption into carbon emissions, during the 50-year service life of the building, the operating carbon emission of passive building is 3.07t CO_2/m^2, while the operating carbon emission of traditional building is 6.57t CO_2/m^2.

5 Results Analysis

In this paper, LCA method was used to study the carbon emissions in the production section of building materials for passive buildings in cold regions of China. The results were compared with traditional buildings in the same climate and structure. Specific comparison results are shown in Fig. 3.

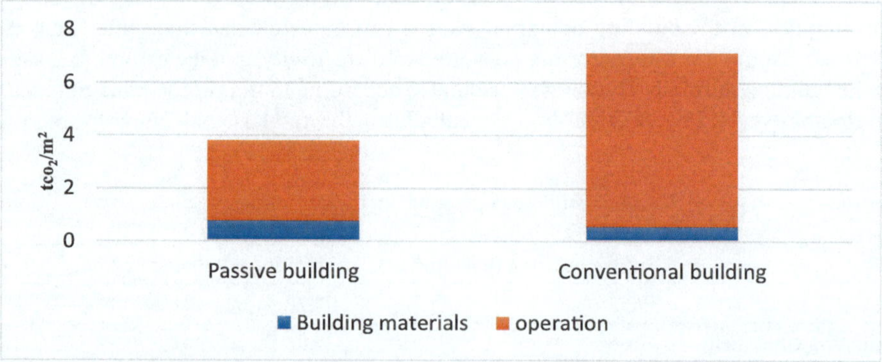

Fig. 3. Comparison of carbon emissions between passive building and traditional building

It can be seen that the carbon emissions of passive buildings in the production section of building materials (0.77t CO_2/m^2) is 37.5% higher than that of traditional buildings (0.56t CO_2/m^2). During the building's operational phase, carbon emissions from passive buildings (3.07t CO_2/m^2) were 53.27% lower than those from traditional buildings (6.57t CO_2/m^2). The main conclusion of this case study is that, compared with conventional buildings, the carbon emissions reduced by the large amount of energy saved during the operation phase of passive buildings can compensate for the carbon emissions during the production phase of building materials.

Although in the production section of the building, because the passive building needs the outer envelope with good thermal insulation performance, many thermal insulation materials and building components with high carbon emission factor are applied in the building materials. As a result, the production phase of passive buildings has higher carbon emissions than traditional buildings. However, in the later operation process, passive buildings can save a lot of energy compared with traditional buildings by virtue of their excellent energy saving characteristics. The carbon emissions from these energy production can more than offset the additional carbon emissions of passive buildings during the production phase of the building. As the service life of the building increases, the energy saving and environmental protection characteristics of passive buildings will gradually be reflected compared with traditional buildings.

The results also show that as passive building technology becomes more sophisticated, buildings will consume less energy during operation. With the reducing of operating energy consumption, the carbon emissions in the operation section will account for a lower proportion in the whole life cycle of the building, while the carbon emissions in the production section of building materials will account for an increasing proportion in the future. According to the calculation, for traditional buildings, the carbon emissions in the production section of building materials accounts for 7.85% of the total carbon emissions in the two sections, while the carbon emissions in the production section of building materials of passive buildings accounts for 20.05%. It can be seen from the analysis that while pursuing better energy-saving performance of buildings, we also need to pay attention to reducing carbon emissions in the production section of building materials.

6 Conclusions

Through the calculations of carbon emissions of passive building materials in the production section and operation section, we find that passive buildings not only have efficient energy saving characteristics compared with traditional buildings in the operation section. At the same time, passive buildings also have low carbon characteristics compared with traditional buildings in the life cycle of buildings.

In the calculation of carbon emissions during the operation of the building, the carbon emissions during the operation of the building could not be measured directly, but could only be used to simulate the energy consumption of HVAC, ventilation, hot water and lighting by computer software. It would be meaningful to have a complete set of operational carbon emission data to compare with the calculated results.

Project Source. This project comes from the project of Science and Technology Department of Liaoning Province. Research on multi-objective optimization design and key technology of zero carbon and zero energy building in cold region. Project Number: LJKZ0577.

References

1. Su, X., Lushao, X.: Embodied and operational energy and carbon emissions of passive building in HSCW zone in China: A case study. Energy & Buildings **222**, 110090 (2020)
2. Caiyun, M.: Research on carbon footprint evaluation and emission reduction strategy of prefabricated buildings in physical and chemical section. Fujian Agriculture and Forestry University (2019)
3. Yan, Z., Ying, C.: Life cycle energy consumption and environmental emissions of residential building. Sci. Technol. **50**(03), 330–334 (2010)
4. Bribian, Z., Alfonso, U., Sabina, S.: Life cycle assessment in buildings: State-of-the-art and simplified LCA methodology as a complement for building certification. Build. Environ. **44**(12), 2510–2520 (2009)
5. Bonamente, E., Merico, C., Rinaldi, S., et al.: Environmental impact of industrial prefabricated buildings: Carbon and energy footprint analysis based on an LCA approach. Energy Procedia **61**, 2841–2844 (2014)
6. Malindu, S., Guomin, Z., Sujeeva, S.: Environmental emissions at foundation construction section of buildings-two case studies. Energy and Buildings **95**, 189–198 (2016)
7. Chen, Y., Wang, Y., Wang, Y., et al.: Energy efficiency of buildings in China. Excerpt: Editorial Office of Building Energy Conservation **49**(360) (2021)
8. Feist, W., Schnieders, J., Dorer, V., Haas, A.: Re-inventing air heating: Convenient and comfortable within the frame of the passive house concept. Energy and Buildings **37**(11), 1186–1203 (2005)
9. National Bureau of Statistics: China Energy Statistical Yearbook 2009/2010. Science Press, Beijing (2011)
10. National Bureau of Statistics: Yearbook of China building material industry 2011. China Building Materials industry Press, Beijing (2012)
11. GB/T51366–2019: Building carbon emission calculation standard 2019. China City Press, Beijing (2020)
12. Mengyue, P.: European ultra-low energy building and passive housing standard system. Constr. Sci. Technol. **21**, 43–47 (2014)

Strength of an Improved Connection for Modular Concrete Structures Without Onsite Casting

Chengguang Xu[1](✉), Xudong Qian[2], Ran Tao[1], and Rong Wang[1]

[1] China Harbour Engineering Co., Ltd., Beijing 100027, China
cgxu@chec.bj.cn

[2] Department of Civil and Environmental Engineering, National University of Singapore, Singapore 117576, Singapore

Abstract. This paper presents an improved design for connections to be applied in modular concrete structures. The proposed the connection removes the onsite casting procedure, which improves the productivity in the modular construction. The installation of the connection requires only bolt connections between the two adjacent prefabricated units. The current study aims to investigate the structural performance of the connection under ultimate loading conditions through the calibrated finite element analyses. The validation of the finite element analyses employs a reported experimental study in the literature. Based on the parametric investigation, the proposed connection demonstrates satisfactory ultimate strength with sufficient ductility. The parametric study also examines a number of important parameters for the connection.

Keywords: Modular construction · Dry connection · Ultimate strength

1 Introduction

Prefabricated modular concrete structures entail numerous advantages, such as short production cycles, flexible building layouts, low construction costs, and enhanced environmental protection. The connection between the modules remains a critical factor in determining the integrity and robustness of the entire structure. However, the on-site casting inevitably slows down the construction to some extent, as the concrete material typically requires 28 days of curing to achieve its full strength. If the composite connection requires shear studs or connectors to develop the interfacial strength between the two materials, the construction work will be further delayed, should there be a need for onsite installation of the shear studs or connectors. Such delays compromise the advantages of adopting modular constructions.

To enhance the construction productivities, some researchers have proposed different connection designs to be incorporated in modular concrete structures. Park et al. [1] have proposed an embedded steel column-to-foundation connection for modular structural systems. The structural performance of the connection depends on the embedded length

W. Guo and K. Qian (Eds.): GBCESC 2022, LNCE 211, pp. 221–229, 2023.
https://doi.org/10.1007/978-981-19-5217-3_21

of the steel column, the shape of the end plate welded to the column and the shear studs. They have examined the resistance of the connection under reversed loading conditions to characterize the strength of the connection under seismic actions. Their experimental study reveals that extensive cracking of the concrete foundation is the primary failure mode observed for this type of connections. The embedded length of the column is the most critical parameter in the connection design. Insufficient embedded length leads to pull-out failure of the connection, as observed in the experiment. The shape of the end plate does not impose a significant effect on the seismic resistance of the connection. The presence of shear studs enhances the interaction between the steel and concrete material and constrains the separation of the steel and concrete interface. This, however, leads to a brittle failure as compared to the connections without shear studs. The connection design proposed by Park et al. [1] demonstrate good seismic resistance and sufficient ductility for the seismic design. Nevertheless, the connection design also entails a number of weaknesses, which can be further improved. The construction of the connection requires casting of the concrete on site. The steel column requires shear studs to develop the interfacial strength between the concrete and steel material. Such connection design may delay the construction progress. An alternative design is to remove the on-site casting requirement by using bolt connections. The precast foundation can also embed a steel section, which extends out of the foundation to be bolt connected to the steel column on top.

Ro et al. [2] have proposed another composite connection for precast concrete modular beam system. Their connection requires on-site casting using non-shrink mortar and the connection strength develops over the curing period of the concrete. Their experimental study covers two connections with different lengths under a four-point bend loading. Cracking of concrete in the connection region remains the main failure mechanism. Compared to a fully continuous monolithic beam without the connection, the specimens with the proposed connection are able to achieve around 80% of the static capacity. The ductility of the beam specimens connected with the proposed joint shows 20 to 30% in-crease compared to the monolithic beam. However, the experimental study focuses on the structural performance of the connection under monotonically increasing load only. The structural resistance of the connection under reversed loadings requires further validations.

The other type of connection used in modular concrete structures involves the design of strengthened joint through the use of reinforcing steels. This type of connections often entails good ductility and seismic resistance, at the sacrifice of prolonged construction time and reduced productivity due to on-site casting. Parastesh et al. [3] proposed a moment resisting beam-to-column connection with excellent ductility suitable for high seismic zones. Their connection design entails sophisticated reinforcement details to ensure the structural performance of the connection. The experimental study reveals that the proposed connection with closed stirrups demonstrates sufficient rotational capacity and sufficient residual strength. The ductility of the joint also exceeds that of the monolithic specimens by more than 45%. The hysteretic energy dissipation of the proposed connection exceeds the corresponding monolithic specimen by 30%.

With a similar objective, Kim et al. [4] have examined the effect of using steel sections in the connection of concrete modular structures. The advantage of embedding a steel section in the connection is to increase the connection strength and ductility. However, this type of connection requires on-site casting, similar to the previous connections proposed by Parastech et al. [3]. Other researchers have also proposed different types of steel-concrete composite connections for concrete modular structures. [5–7].

Different from the above connection designs, Hu et al. [8] have proposed a dry mechanical connection for fully-restrained moment connections. Their connection design includes two steel plates welded to a H section embedded in the prefabricated concrete column. This connection requires onsite bolt installation and removes completely the requirement for on-site casting and welding. Hu et al. [8] have performed an experimental investigation on the proposed connection and have confirmed the full moment capacity of the connection under both monotonic and reversed loading conditions. The plate thickness remains a critical parameter in the connection. The stiffness and strength of the connection depends significantly on the plate thickness. The full moment capacity development requires a significant large thickness of the plate. For the five specimens tested by Hu et al. [8], only the connection with the plate thickness of 45 mm satisfies the full moment requirement.

This study proposes an improved connection design based on the original design proposed by Hu et al. [8] and performs a parametric study to understand the performance of the proposed connection under ultimate loading conditions, through calibrated finite element analyses.

2 Proposed Connection Scheme

Inspired by the connection design proposed by Hu et al. [8], this study aims to improve this connection design to enhance its performance in two aspects: 1) the strength of the connection; 2) the repairability of the connection. To address these two problems, a new improved connection design is shown in Fig. 1. Compared to the connection design by Hu et al. [8], the design shown in Fig. 1 entails the following two changes: 1) the connection has an additional steel square hollow section (SHS) between the reinforced concrete column units; 2) the H-section inside the reinforced concrete (RC) column is replaced by a SHS steel section.

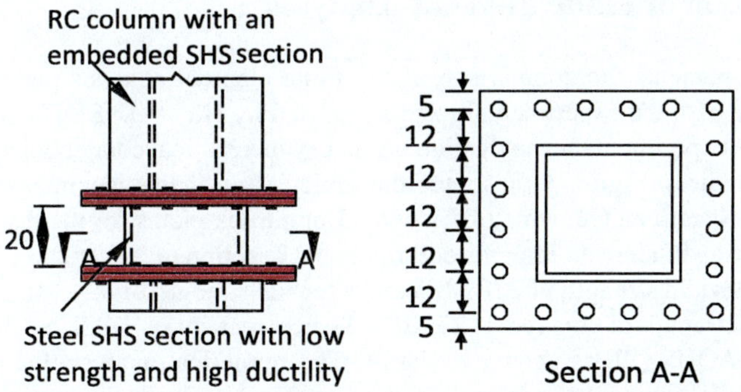

Fig. 1. A new dry connection design proposed for concrete modular structures.

The additional SHS steel section is fabricated from lower strength steel than the SHS inside the RC column. The use of lower strength SHS in the new connection helps to 1) facilitate the failure under extreme loading conditions in the steel SHS section so as to enhance the repairability of the connection; and 2) improve the energy absorption capacity of the connection since the steel materials with a lower yield strength typically has higher ductility. The length of the SHS section between the RC columns remains at a minimum value 200 mm to allow for bolt strengthening during the installation.

Figure 2 illustrates the installation procedure. Both units 1 and 2 are parts of the pre-fabricated modular unit manufactured in a factory. The onsite assembly procedure involves only the bolting pro-cess to connect unit 1 and unit 2 into an integrated unit. The bolt connection between the two steel plates utilizes the high strength bolt with a yield strength of 1000 MPa.

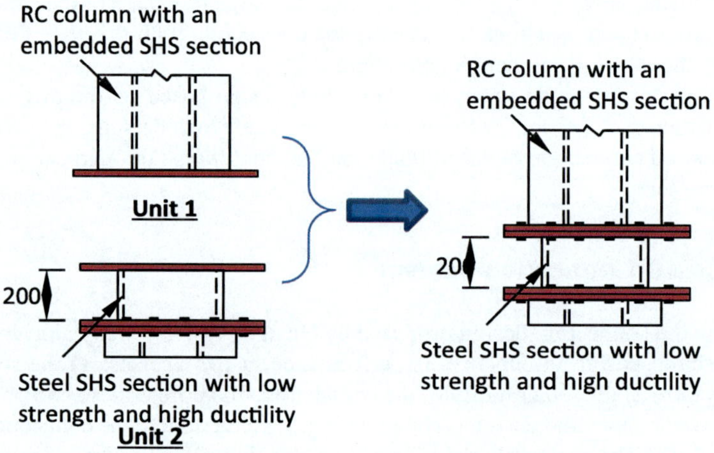

Fig. 2. Construction procedure for the proposed dry connection.

3 Validation of Finite Element Analyses

This section presents the comparison of the finite element analysis performed in the current study with the experimental results reported by Hu et al. [8]. Hu et al. [8] have presented the experimental investigation on six dry mechanical connections for concrete modular structures. Figure 3 illustrates the dimensions of the specimens tested. This study selects specimen C2, which uses two 45 mm thick plates for the dry connection. The steel section in-side the concrete column is an H-section of 200 × 200 × 8 × 12. The H section has yield strength pf 378 MPa and a tensile strength of 494 MPa. The 28-day compressive strength of concrete is 27 MPa. The current study builds the finite element mode in ABAQUS [9], as shown in Fig. 4. The numerical analysis incorporates the damaged plasticity model of concrete. In the numerical analysis, the material properties follow the properties reported in by Hu et al. [8]. The steel plate makes use of the SM490 grade with a yield strength of 315 MPa and a tensile strength of 490 MPa.

The steel plate has a dimension of 700 × 700 mm, with the bolts of 20 mm in diameter. The center-to-center spacing of the bolt equals 120 mm. The specimen is restrained against movement in the x-direction at the two ends of the base.

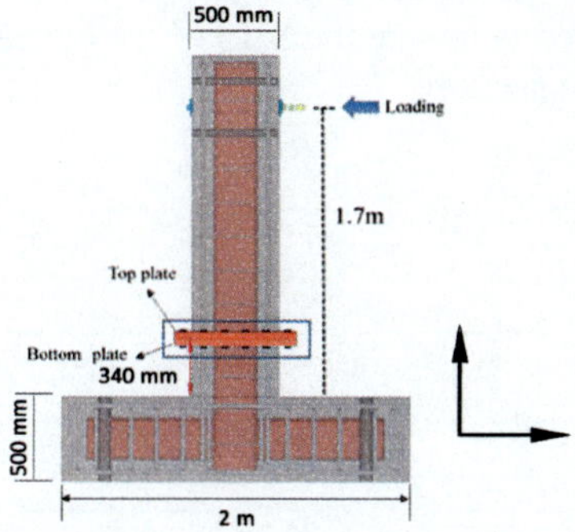

Fig. 3. The test specimen proposed by Hu et al. [8].

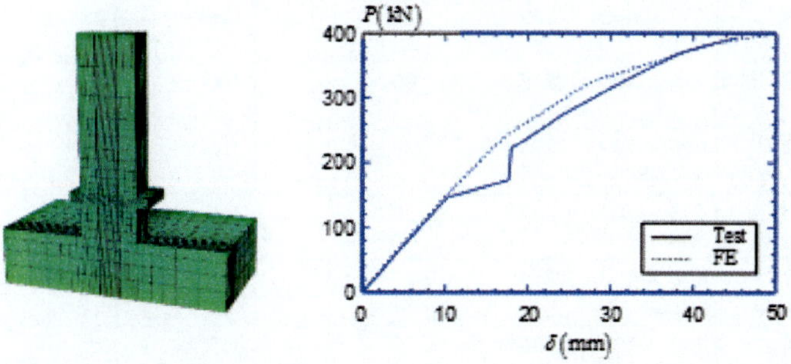

Fig. 4. The current FE model and comparison of the current finite element analyses with the experimental results.

Figure 4 compares the load-deformation curve computed from the current FE analysis and the experimental results reported by Hu et al. [8]. The current FE analysis predicts closely the load-deformation responses of the connection and thus confirms the accuracy of the finite element model used in this study. The current analysis does not include the modeling of fracture and fatigue failure in the connections [10, 11], which will be the focus of future studies.

4 Parametric Investigation on the Connection Performance

This study examines the performance of the proposed connection covering a range of parameters. Table 1 lists the parameters investigated in this study. The concrete material and the steel SHS inside the column follow the same material utilized by Hu et al. [8]. Table 1 includes a reference connection model, P0, and two groups of connection models. The connection models in Group 1 focuses on the properties of the steel plate including the thickness of the steel plate and the yield strength of the steel plate. Group 2 examines the effect the SHS section on the connection performance. The connection model P8 makes use of a low yield strength, which has become increasingly popular in the structural design for strong seismic regions. When using low-yield steels, the application of in-filled concrete in-side the SHS of the connection can help to enhance the axial capacity of the connection.

Table 1. Connection parameters investigated in the current study.

Connection model		SHS		Steel plate		Steel bolt	
		Section	σ_y (MPa)	Thickness t_p (mm)	σ_y (MPa)	D (mm)	σ_y (MPa)
Ref	P0	400 × 400 × 16	275	20	315	25	1000
Group 1	P1	400 × 400 × 16	275	**30**	315	25	1000
	P2	400 × 400 × 16	275	**15**	315	25	1000
	P3	400 × 400 × 16	275	20	**275**	25	1000
	P3	400 × 400 × 16	275	20	**460**	25	1000
Group 2	P5	**400 × 400 × 12.5**	275	20	315	25	1000
	P6	**400 × 400 × 8**	275	20	315	25	1000
	P7	400 × 400 × 16	**235**	20	315	25	1000
	P8	400 × 400 × 16	**100***	20	315	25	1000

*Use of low-yield strength steel for seismic applications [12]

Figure 5 shows a half-symmetric finite element model constructed in ABAQUS using C3D8 elements [9]. The loading and boundary conditions applied in the numerical analysis follow those illustrated in Fig. 3, which is the same as the experimental setup in Hu et al. [8]. The concrete material utilizes a concrete damaged plasticity model, which is based on the experimental measurement performed in the structural laboratory of National University of Singapore, as shown in Fig. 6.

Figure 7 compares the change in the steel plate thickness and material yield strength on the connection response. In contrast to the findings by Hu et al. [8], who have reported a significant effect of the plate thickness on the connection response and capacity. The current study indicates marginal variations in the load-deformation responses of the connection with changes in both the plate thickness and the yield strength of the steel end plate.

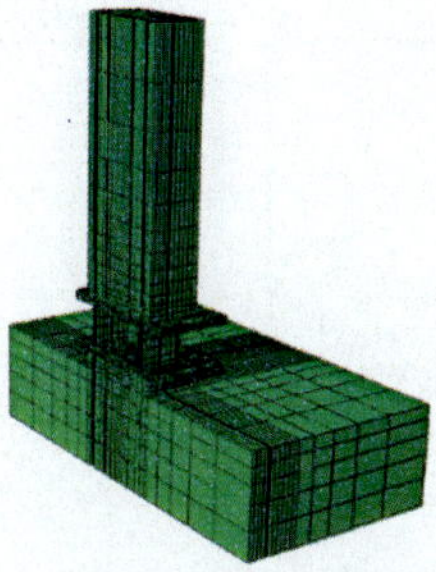

Fig. 5. A half-symmetric finite element model for the proposed connection (No. of nodes: 35217, No. of elements: 25684).

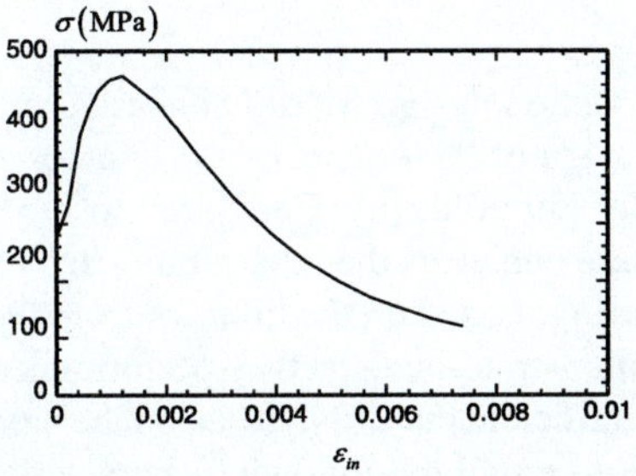

Fig. 6. Compression behavior of the concrete simulated in the finite element analysis.

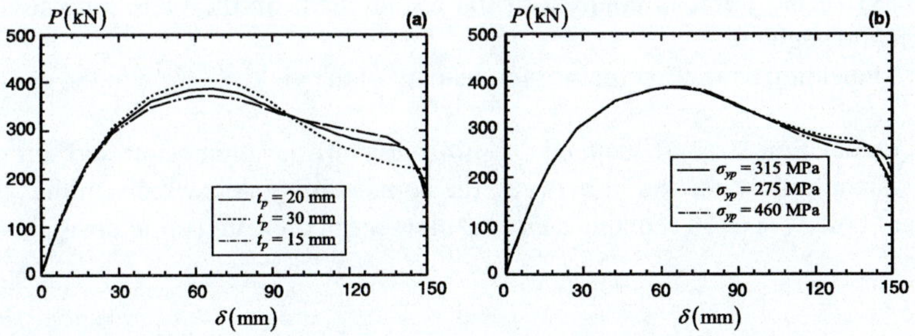

Fig. 7. Effect of (a) plate thickness and (b) the yield strength; of the end plates on the load-deformation response of the connection.

Figure 8a demonstrates the significant effect caused by the size of the SHS section in between the two steel plates on the connection strength. The peak resistance of the connection decreases as the wall thickness of the SHS section reduces. Figure 8b indicates that the yield strength of the SHS is insignificant on the resistance of the connection unless a low yield steel with the yield strength of 100 MPa is used for the SHS section.

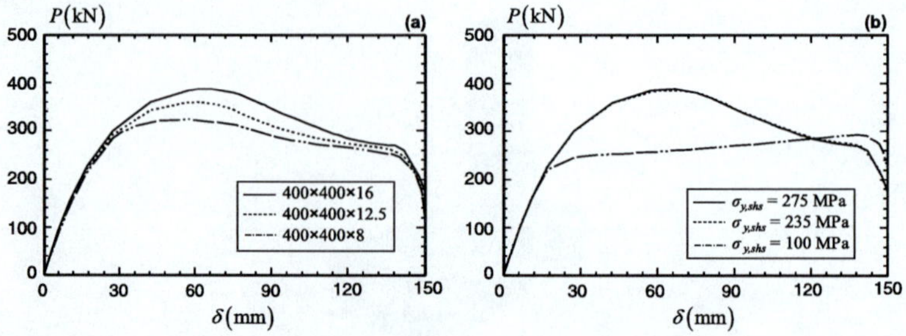

Fig. 8. Effect of (a) section size and (b) the yield strength; of the square hollow section in between the steel end plates on the load-deformation response of the connection.

5 Conclusions

This study proposes a connection design for dry construction in the modular reinforced concrete structures. The design of the connection originates and extends from the connection design proposed by Hu et al. [8]. Compared to the connection design by Hu et al. [5], the proposed design enhances the repairability and ease of replacement during the post-earthquake retrofitting. Based on the findings from the calibrated finite element analyses, the proposed demonstrates satisfactory performance on the ultimate capacity for a rigid connection with sufficient ductility. Based on the parametric study, the strength of the proposed modifications entail the following characteristics:

1) The connection does not require a very thick connection plate to achieve the full capacity of the joint, in contrast to the requirement on the plate thickness in the original design.
2) The connection strength remains independent of the yield strength of the connection plates.
3) The connection strength depends significantly on the dimension and strength of the square hollow section in between the connection plates, which implies that the square hollow section remains a critical element in the connection design.

References

1. Keum-Sung, P., Jiho, M., Sang-Sup, L., Kyu-Woong, B., Charles, R.: Embedded steel column-to-foundation connection for a modular structural system. Eng. Struct. **110**, 244–257 (2016)
2. Kyong-Min, R., Min-Sook, K., Chang-Geun, C., Young-Hak, L.: An experimental study on the flexural behavior of precast concrete modular beam systems using inserted steel plates. Appl. Sci. **11**, 3931 (2021)
3. Hossein, P., Iman, H., Reza, R.: A new ductile moment-resisting connection for precast concrete frames in seismic regions: an experimental investigation. Eng. Struct. **70**, 144–157 (2014)
4. Jisoon, K., Won-Kee, H., Ji-Hun, K.: Experimental investigation of the influence of steel joints upon the flexural capacity of precast concrete columns. Struct. Des. Tall Special. Build. **26**, e1340 (2017)

5. Olivia, M., Sakdirat, K.: Influence of shear bolt connections on modular precast steel-concrete composites for track support structures. Steel Compos. Struct. **27**(5), 647–659 (2018)
6. Cheng, L.W., Jian Cheng, K., Qi Hui, W., Ji Ming, L., Zun Qiang, L.: FEM analysis of the modular prefabricated steel-concrete composite beam-column internal joint under reciprocating action. Steel Compos. Struct. **41**(1), 45–64 (2021)
7. Cheng Long, W., Jian Cheng, K., Ji Ming, L., Mou, B.: Non-linear FEA of mechanical properties of modular prefabricated steel-concrete composite joints. Steel Compos. Struct. **40**(4), 533–554 (2021)
8. Ju-Yun, H., Won-Kee, H., Seon-Chee, P.: Experimental investigation of precast concrete based dry mechanical column–column joints for precast concrete frames. Struct. Des. Tall Special. Build. **26**, e1337 (2017)
9. Dassault Systemes: ABAQUS User's Guide (2020)
10. Liuyang, F., Xudong, Q.: An adaptive learning approach to determine and update crack sizes from strain relaxation data for welded plate joints. Eng. Fract. Mech. **259**, 108165 (2022)
11. Tianyao, L., Xudong, Q., Wei, W., Yiyi, C.: A ductile tearing assessment diagram to estimate load resistance versus crack extension for welded connections with surface cracks. Thin-Walled Struct. **169**, 108435 (2021)
12. Samans, S., Tetsuhiko, A., Takushi, K., Katsuhiro, Y.: Applicability of low-yield strength steel for ductility improvement of steel bridge piers. Eng. Struct. **27**(7), 1064–1073 (2015)

Research on Application Effect of Built-In Louver Insulating Glass Window

Aofeng Shi and Guoqing Yu(✉)

University of Shanghai for Science and Technology, Shanghai, China
yuguoqinghvac@126.com

Abstract. This article focuses on the built-in louvered insulating glass. This article focuses on the built-in louvered insulating glass, Analyzed heat transfer of the built-in louvered insulating glass window. Through the simulation calculation of the whole summer, when the louver of the built-in louver is pulled downward and the angle of the louver is 90°, in Shanghai, Beijing, from the perspective of the cumulative total heat gain of summer windows, single-layer LOW-E glass > double-layer LOW-E glass > insulating glass + indoor blinds > built-in louvered insulating glass. Through the simulation calculation of winter heating period, it is found that the single-layer Low-E glass is the smallest in different directions in the two areas, while the total heat gain of hollow shutter glass and double-layer Low-E glass is different.

Keywords: Energy · Louvered · Glass · Radiation · Build

1 Introduction

The thermal performance of windows is the worst among the three building envelope components, and is one of the most important factors affecting the indoor thermal environment and building energy consumption. Therefore, more and more buildings will consider using energy-saving exterior windows in the future. The built-in louver hollow glass window, as the most convenient type of adjustable sunshade device, can better meet the above requirements compared with ordinary windows.

The research on the built-in louver insulating glass mostly focuses on its own performance. At present, there is no research on the photothermal performance of the venetian blinds under two different seasonal operating conditions in summer and winter heating periods. Domestic and foreign research on building exterior window systems has done a lot of work. The research methods are mainly divided into theoretical calculation, software simulation and performance testing. The Lawrence Berkeley National Laboratory in the United States has obtained the detailed calculation formula of the solar radiation heat gain coefficient of the internal and external shading through rigorous theoretical analysis [1, 2]. Kim et al. developed a test setup that can measure U-value and solar heat gain coefficient under different fenestrations [3]. Although the domestic related research started late, it has also made great progress. Dong Zhizhong [4–6] believed that radiant heat gain is the primary cause of the deterioration of indoor thermal environment, and

W. Guo and K. Qian (Eds.): GBCESC 2022, LNCE 211, pp. 230–238, 2023.
https://doi.org/10.1007/978-981-19-5217-3_22

effective control of radiant heat gain and the use of shading devices are the keys to solving the problem. However, the research on the comprehensive influence of different types of heat gain and external working conditions for built-in hollow louver windows needs to be further studied.

Dana Luo et al. [7] Conducted the experimental study on the photothermal performance of the built-in louver insulating glass in winter and found that the shading capacity of the built-in louver insulating glass is quite different when the louver is placed at different angles. Chuan Zhao et al. [8] Proposed a new double-layer ventilation photovoltaic louver (PV blind), and simulated the comprehensive energy consumption and power generation performance of the double-layer ventilation photovoltaic louver by using energy plus, A new angle control optimization strategy is proposed.

2 Thermal Analysis of Built-In Louver Insulating Glass

2.1 Structure of Built-In Louver Insulating Glass

Adding a shutter that can be adjusted at will between two pieces of ordinary white glass, constitutes a simple built-in shutter insulating glass window, which can block a large number of shutters in summer by adjusting the shutter angle mechanically or manually. It can also transmit sunlight into the room through diffuse reflection, so that the room has better lighting and good comfort.

There are three heat transfer methods for sunlight entering the room through the built-in louver hollow glass window: heat conduction, convection and radiation [9]. In fact, the air entering and leaving the room through the window gap also transfers heat to the indoor air. If the indoor air conditioning system gives enough positive pressure or strictly seals the window frame, air infiltration can be effectively controlled [10]. In this article, ignoring the air infiltration factor, it can be considered that the heat transferred to the room through the built-in louver insulating glass mainly includes two aspects: the solar radiation heat gain and the temperature difference heat transfer of the window.

2.2 Built-In Louver Insulating Glass Calculation Model

The simulation calculation in the paper mainly uses WINDOW7.7 software. The software is capable of calculating the total U-value and solar heat gain coefficient of a window product in compliance with the National Window Rating Board. When the louver is in the pulled-up state, its thermal calculation process is consistent with that of the double-layer insulating glass; when the louver is partially pulled up or closed, its thermal calculation process can be based on the solar radiation model below.

Since the thickness of the air interlayer of the built-in louver insulating glass is about 14mm, the author chooses to use the double-layer built-in louver insulating glass with the specification of 3 + 14A + 3 (3 mm white glass + 14 mm air layer + 3 mm white glass) as the research object. The louver is a 90° straight louver blade, the transmittance of white glass is $\tau = 0.82$, the absorption ratio $\alpha = 0.1$, and the reflection ratio $\gamma = 0.08$. For the convenience of calculation, the solar radiation is regarded as 1000 W/m^2, and the reflected radiation of three times and above is not considered. It is assumed that the

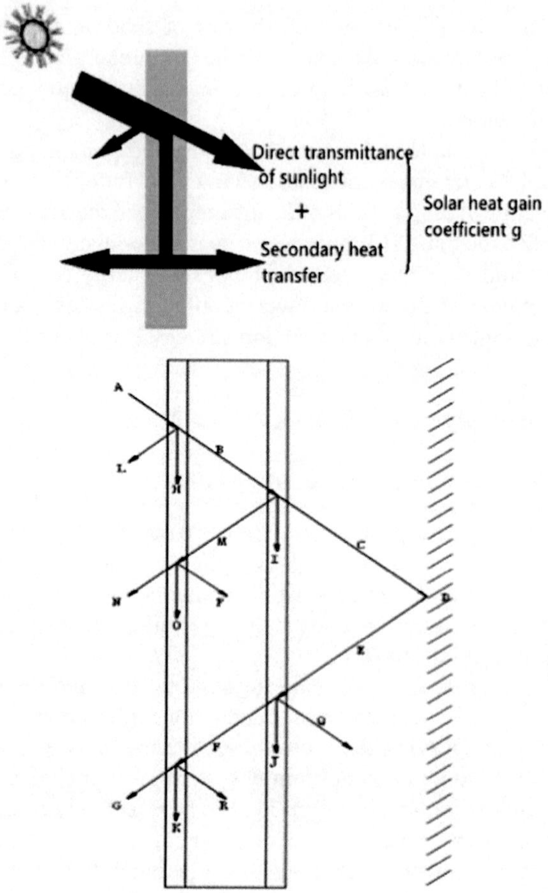

Fig. 1. Built-in louver insulating glass solar radiation model

main line of solar radiation is the radiation transmitted through the insulating glass and absorbed by the glass, the solar radiation in the A to E sections shown in Fig. 1.

It should be noted that in the calculation, it is assumed that the influence of a certain section of the main line of solar radiation is proportional to the radiation reflected back to the outside and the radiation absorbed by the glass and louvers. Taking the radiation in the B section of the main line as an example, the radiation with a ratio of K/B is absorbed by the outer glass, the radiation with a ratio of (I + J)/B is absorbed by the inner glass, and the radiation with a ratio of D/B is absorbed by the louver., the radiation with a ratio of G/B is reflected back to the outdoor; the influence caused by the secondary radiation P, Q, R is consistent with the influence caused by the radiation in the same layer of the main line, that is, the influence caused by the radiation in the P and R sections is the same as the radiation in the B section. The impact caused by the Q-segment radiation is the same as that caused by the C-segment radiation.

3 Overall Thermal Performance of Different Types of Glass Windows

The zoning method based on annual total radiation is one of the most widely used solar radiation zoning methods in China. It divides my country into four categories: resource-rich area, resource-rich area, general resource area, and resource-scarce area according to the annual total radiation amount of each region. In the region [12], the author chooses Shanghai with average annual total radiation resources and Beijing with abundant resources to analyze the shading performance of built-in louvered insulating glass windows, and explore and analyze the heat gain of windows in summer and winter heating periods. compare.

Different shading forms have very different shading effects. In the comparison of summer application effects, this chapter selects three types of shading forms: built-in louver insulating glass, insulating glass + indoor venetian blinds, and double-layer LOW-E glass. Analysis; in the comparison of application effects in winter, this chapter selects three types of shading forms: built-in louver insulating glass, single-layer LOW-E glass, and double-layer LOW-E glass to analyze the heat gain. The heat transfer coefficients of different types of glass windows and the solar heat gain coefficient values are calculated by the WINDOW7.7 software, and the calculation results are shown in Table 1.

Table 1. Influence of different glass thicknesses on thermal performance of windows

Glass window	Heat transfer coefficient	Solar heat gain coefficient
Built-in louver insulating glass	2.74	0.161
Insulating glass + indoor blinds	3.207	0.305
Double layer Low-e glass	2.702	0.598
Single layer Low-e glass	3.671	0.715

4 Research on Application Effect in Shanghai

Shanghai is located at the mouth of the Yangtze River, with an average altitude of about 2.19 m. It has a subtropical monsoon climate with four distinct seasons, abundant sunshine and abundant rainfall. In a typical year in Shanghai, the annual average temperature is between 15 °C and 22 °C, the summer temperature is between 19 °C and 26 °C (Table 2).

Table 2. Cumulative total solar radiation in different months in summer in Shanghai

Mouth	Horizontal total radiation
June	470
July	490
August	431
Summer	1391

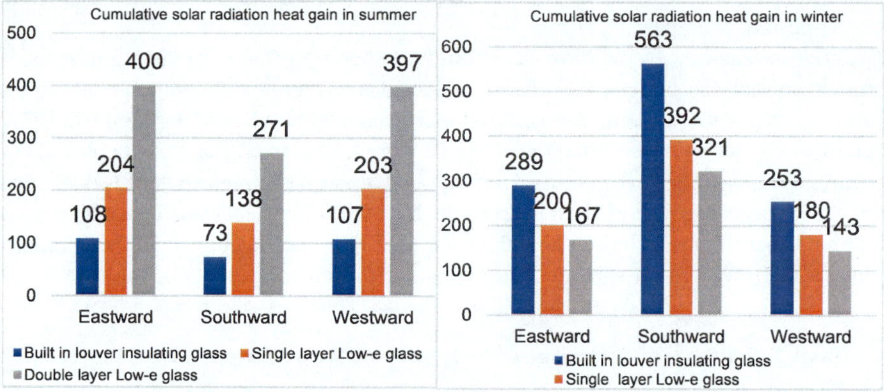

(a) Summer (b) Winter

Fig. 2. Cumulative solar heat gain for different shade types

Low-E Glass, single-layer Low-E glass has the worst shading ef in summer, when calculating the solar radiation heat gain entering the room through windows, multiply the solar heat gain coefficient of different types of windows by the cumulative total solar radiation received by the vertical planes in all directions, and then the cumulative solar radiation entering through different types of windows can be obtained. The heat gain from solar radiation is shown in Fig. 2(a). The winter heating time in Shanghai in winter is generally from November to February of the following year. After calculation, the direct solar energy transmission of the glass at different times of the day can be obtained, and the calculation results can be accumulated to finally obtain the cumulative solar energy of different types of glass during the heating period. direct transmission. The specific results are shown in Fig. 2(b). It can be seen from the summer map that when the glass windows are facing east, south or west, the solar radiation heat entering the room is the least when the built-in louver insulating glass is used, followed by insulating glass + indoor venetian blinds, double-layer LOW-E. From the solar radiation heat gain entering the room, the cumulative solar radiation heat gain obtained by the east-facing window is similar to that of the west-facing window, and the difference is very small, while it is much higher than that of the south-facing window. It can be seen from the winter map that under the same orientation, the cumulative direct solar energy transmission of hollow louver glass is the largest, followed by single-layer LOW-E glass, and the least is double-layer LOW-E glass.

Using simulation calculation, after the calculation, the heat transferred into the room by the temperature difference of the windows can be obtained throughout the summer. Through the same method, the daily temperature difference heat transfer during the heating period was calculated respectively, and after the accumulation, the temperature difference heat transfer of different types of glass during the heating period from November to February was obtained. The specific results are shown in Table 3. It can be seen that in the same month, the cumulative heat transfer amount of single-layer LOW-E glass is the largest, followed by hollow louver glass, and the least double-layer LOW-E glass.

Table 3. Cumulative temperature difference heat transfer of windows in Shanghai

Window type	Heat transfer in summer	Heat transfer in winter
Built-in louver insulating glass	115	383
Insulating glass + indoor blinds/Single layer Low-e glass	135	285
Double layer Low-e glass	114	213

The solar radiation heat gain and temperature difference heat transfer of different types of windows in Shanghai in summer have been calculated in the previous section, and the total heat gain of different types of windows in Shanghai in different orientations can be calculated. The heat transferred into the room by the temperature difference of the window, double-layer LOW-E glass > built-in louver insulating glass window > insulating glass + indoor venetian blinds. The total heat gain of the glass during the heating period in Shanghai can be obtained from the accumulated direct solar transmittance and the accumulated temperature difference heat transfer. The specific results are that in the case of east and south orientation, the total heat gain of hollow louver glass > double-layer LOW-E glass > single-layer LOW-E glass. The three types of glass will be in a state of heat gain or heat loss in different months.

5 Research on Application Effect in Beijing

When calculating the solar radiation heat gain entering the room through the window, multiply the solar heat gain coefficient of different types of windows by the total solar radiation received by the vertical plane in all directions, and then the solar radiation entering different types of windows can be obtained. heat, the calculation results are shown in Fig. 3(a).

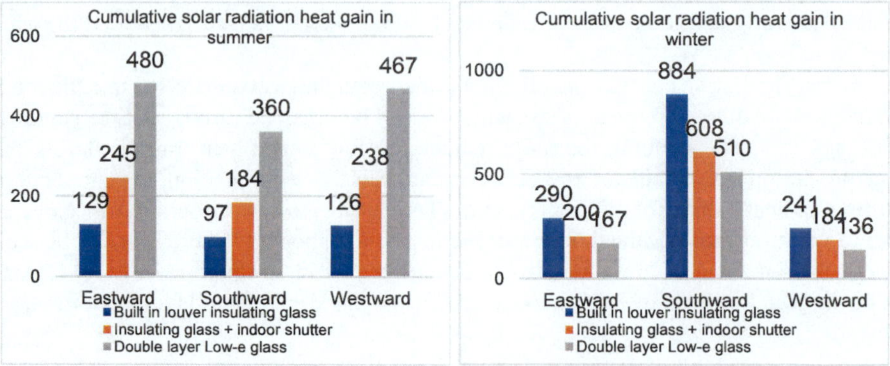

(a) Summer (b) Winter

Fig. 3. Solar heat gain of different shade types in summer

The heating time in winter in Beijing is generally between mid-November and mid-February of the following year. According to the calculation method of a typical year, the direct solar energy transmission through different types of glass in 4 months was cumulatively calculated. When the glass windows are facing east, south or west, the shading effect of the built-in louver insulating glass far exceeds that of other types of shading. When the same type of glass window is used in the east, south and west orientations, the cumulative solar radiation heat gain obtained by the east-facing window is equivalent to that of the west-facing window, but much higher than that of the south-facing window. It can be seen from the winter chart that the cumulative direct solar transmission of different types of glass with the same orientation has the same trend, and the cumulative direct solar transmission of hollow louver glass is the largest.

After the author calculates with the same method as in Shanghai, the total amount of heat transfer from the window temperature difference from June to August in summer and winter can be obtained. The results are shown in Table 4.

Table 4. Cumulative temperature difference heat transfer of windows in Beijing

Window type	Heat transfer in summer	Heat transfer in winter
Built-in louver insulating glass	67.5	690
Insulating glass + indoor blinds/Single layer Low-e glass	79	507
Double layer Low-e glass	66.6	379

The total heat gain of different types of windows in Beijing in summer can be obtained from the solar radiation heat gain and temperature difference heat transfer of different types of windows in Beijing in summer. When the same type of windows are used, the Beijing area gets total heat gain throughout the summer with east-facing windows > west-facing windows > south-facing windows. For different types of windows, according to the value of the heat transfer into the room by the temperature difference of the window, double-layer LOW-E glass > built-in louver insulating glass window > insulating glass + indoor venetian blinds. The total heat gain of a single layer of LOW-E glass in different orientations is the smallest, and it is basically negative throughout the heating period. In the east and west directions, the total heat gain of the hollow louver glass and the double-layer LOW-E glass is equivalent, and in the south direction, the total heat gain of the hollow louver glass is significantly higher than that of the double-layer LOW-E glass.

6 Conclusion

Focusing on four different types of windows, this paper starts from the two aspects of solar radiation heat gain and temperature difference heat transfer, analyzes the total heat gain of windows in summer and winter heating periods, and studies their application effects in different regions. get conclusion. In Shanghai and Beijing, the transmission heat gain and temperature difference heat transfer in the whole summer were analyzed, and it was found that the total heat gain when the shutters of the built-in shutter insulating glass window were closed was the smallest, followed by the insulating glass + indoor venetian blinds, after that is double layer Low-e glass. The sunshine time received by south-facing windows in different regions is the longest, but the total heat gain of summer windows is the smallest. From the cumulative total heat gain value of windows, east-facing windows > west-facing windows > south-facing windows, among which East-facing windows are slightly larger than west-facing windows. Analysis of the three types of windows in the winter heating period, the total heat gain of the single-layer LOW-E glass is the smallest in different orientations in the two regions, and the total heat gain of the hollow louver and double-layer Low-e glass is different. Regions vary.

References

1. Klems, J.H.: A new method for predicting the solar heat gain of complex fenestration systems: I. Overview and derivation of the matrix layer calculation. ASHRAE Trans. **100**(1), 1065–1072 (1993)
2. Klems, J.H.: A new method for predicting the solar heat gain of complex fenestration systems: II. Detailed description of the matrix layer calculation. ASHRAE Trans. **100**(1), 1073–1086 (1993)
3. Kim, T.J.: Development and evaluation of a measurement method and system for solar heat gain coefficient and thermal transmittance. Mater. Res. Innov. **S5**, 964–9S5, 967 (2015)
4. Zizhong, D.: Research on energy-saving measures for windows in hot areas in summer. Chongqing University (2002)
5. Zizhong, D., Yongguang, X., Yongling, W., Qigao, C.: Study on the thermal process of windows in hot areas in summer. HVAC **2003**(03), 93–96 (2003)
6. Zizhong, D.: Analysis of Shading Measures for Building Exterior Windows (Energy-Saving Window Technology). China Building Industry Press, Beijing (2003)

7. Daina, L., Jikang, Y., Guoqing, Y.: Experimental study on shading and heat transfer performance of hollow glass window with built-in louver Building science (2021)
8. Chuan, Z., Qing, J., Nian, P.: Study on angle control optimization and energy efficiency simulation of photovoltaic shutter Building science (2021)
9. Ruicheng, Z.: Civil Building Solar Water Heater System Engineering Technical Manual, 2nd edn. Chemical Industry Press, Beijing (2011)

Micro-scale Regional Spatial Low-Carbon Planning System Framework Based on Scenario Analysis

Zhexuan Chen and Junting Chen(✉)

School of Civil Engineering and Architecture, Guangxi University, Nanning 530004, China
chjt12@gmail.com

Abstract. As the development of national space continues to accelerate, the pressure of carbon emission is increasing day by day. It is urgent to consider how to guide the low-carbon development and construction of territorial space scientifically and directly from the planning department. Therefore, this article studies the relative lack of micro-scale of land and space as the research object, trying to build based on the scenario analysis method, index constraint method as the core, supporting system and standard as the safeguard of micro-scale space low carbon technology framework, finally realizes "Harmony between man and land for Natural harmony" the overall framework of the target. It is hoped that this study can provide some reference value and inspiration for low-carbon planning at the micro-scale.

Keywords: Scenario analysis · Micro-scale · Low-carbon · Territorial space

1 Introduction

With the rapid growth of China's economy, the process of industrialization and urbanization is also accelerating, and the increase in population makes the energy demand constantly increase. At present, China has become the second-largest carbon emission country in the world. Reducing emissions and optimizing carbon emission structure are not only effective measures to actively respond to changes in the interterritorial situation, but also effective ways to enhance territorial competitiveness. From the perspective of a functional relationship, planning is the originating subject of human influence, while the ecological environment and territorial space are the receiving objects. Starting from the planning can more directly reduce the disturbance degree and negative impact of human activities on the natural ecological environment and territorial space, formulate scientific carbon emission planning and provide guidance for green construction. It is in China's interest to reduce emissions and optimize the carbon emission structure. Controlling carbon emissions through land use management in territorial space planning will help implement eco-first policies and promote all-around sustainable development. However, most of the existing studies are confined to the macro and micro-scale territorial space, and most of the research content is the calculation and deconstruction

W. Guo and K. Qian (Eds.): GBCESC 2022, LNCE 211, pp. 239–247, 2023.
https://doi.org/10.1007/978-981-19-5217-3_23

of the existing established carbon emission structure, without sufficient research on the specific implementation level and application ideas.

Therefore, based on scenario simulation analysis of planning expectation construction, this paper determines carbon emission constraint indexes at the micro-scale from different perspectives, to guide the detailed level territorial spatial planning to achieve low-carbon planning objectives through carbon emission structure optimization. This will promote the implementation of low-carbon national space, help China fulfill its commitment to green development, and contribute to the global low-carbon cause.

2 Current Status and Analysis of Carbon Emission Related Study

In the existing carbon research, research object is diverse, and in different industries or industry as the research object, such as Wang Yaxian research on the carbon intensity of China's coal industry [1], Wang Xiulan took agricultural production as the research object to analyze the characteristics of carbon emission [2], and Ding Baogen's study on the measurement, decoupling characteristics and driving factors of carbon emission from Chinese seed and cropping industry [3]; Settlements in different scale for the object of study is has an important role in the study of the carbon, especially the macro and medium regional or urban agglomeration, city and county scales, For example, Li Tianwei took China's three urban agglomerations as examples to study the spatial correlation network structure and influencing factors of carbon emissions [4]; Zhu Enyan took Zhejiang Province as the research object to analyze the relationship between carbon emission and urbanization [5], Feng Tianwei studied the carbon emission effect of land-use change in Zhengzhou [6]. In research on the relationship between human-specific behavior and carbon emissions, have Yao Dan in Yangtze river delta tourism human behavior and the intensity of carbon space correlation research [7]; In terms of carbon emission accounting methods, Yubo calculated the spatial spillover effect of carbon emission in Harbin and Changchun cities based on NPP-VIIRS model [8], Xu Guoquan calculated the difference of carbon emission in Jiangsu Province with LMDI model [9], and Zhang Lifeng empirically-analyzed the influencing factors of carbon emission in Beijing-Tianjin-Hebei region with VAR model [10]. In addition, there are also studies on the impact of carbon emission from an economic perspective. For example, Yuan Yujuan analyzed the spatial differentiation of carbon emission in commuting based on housing prices [11].

In summary, there are several problems in the existing studies: (1) Scale problem: Few existing studies on carbon emissions have carried out quantitative comprehensive assessment and planning at the micro-spatial scale; (2) Research methods: Traditional regulatory plan methods and carbon emission method are difficult to adapt to the needs of current urban and regional development, so more scientific carbon emission prediction needs to be considered; (3) Application issues: how to combine carbon emission prediction methods with regulatory planning practices, and provide scientific suggestions for the determination of government planning schemes.

3 Lack of Study on Carbon Emission at Micro-scale in Territorial Space Planning

The core of territorial space planning is to provide basic protection for all kinds of development and construction activities, the role of not only confined to the space planning guidance but also combined with a background of carbon emissions has been paid more and more attention to the environment, seeking to build national spatial planning as the core of the index system of carbon emissions constraints, to guide solve problems encountered in the process of practice to carbon and carbon-neutral goals are met.

Territorial space is a regional life coexisting body in which natural elements and human activities blend and coexist. The layout and implementation of territorial space planning should reflect its organic coordination. According to the above content, existing studies on the correlation between territorial space planning and carbon emissions mostly analyze the status quo after the implementation of the planning, which belongs to the evaluation and analysis of the planning results. From the perspective of planning, it is necessary to guide and constrain regional low-carbon development and construction activities before implementation, so that the final planning results can effectively guide low-carbon and green development and construction activities in micro territorial space.

4 Applicability of Scenario Analysis in Micro-scale Territorial Spatial Planning

As a programmatic document for regional and spatial development, territorial space planning has the characteristics of guidance, long-term, complexity, comprehensiveness, and strategy, and can determine the level and quality of regional carbon emission construction in the future to a certain extent. The construction of a regional development carbon emission index system in territorial space planning can fundamentally reduce carbon emissions and optimize the structure of carbon emissions.

The current urban planning system focuses on projects that produce economic benefits, but ignores environmental benefits, especially those related to carbon emissions. The formulation of deterministic standards in planning is not inclusive of all kinds of problems that may exist in regional future development, and will even affect regional economic development because of the lagging standards. Therefore, it is necessary to seek a method that can face the uncertainty, and identify, analyze and deal with the uncertainty.

Scenario analysis focuses on the characteristics of long-term assumptions, allowing it to provide early warning, describe pathways to different scenarios, and conduct simulations and re-tests. Therefore, scenario analysis can predict carbon emissions more diverse and effectively solve the uncertainty in development.

5 Study on Micro-scale Ecological Low-Carbon Planning of Territorial Space Based on Scenario Analysis

5.1 The Overall Framework

The core of micro-scale territorial space ecological low-carbon planning lies in the realization of Harmony between man and land for Natural harmony, and through institutional innovation and technological renewal, the real ecological priority is achieved (Fig. 1).

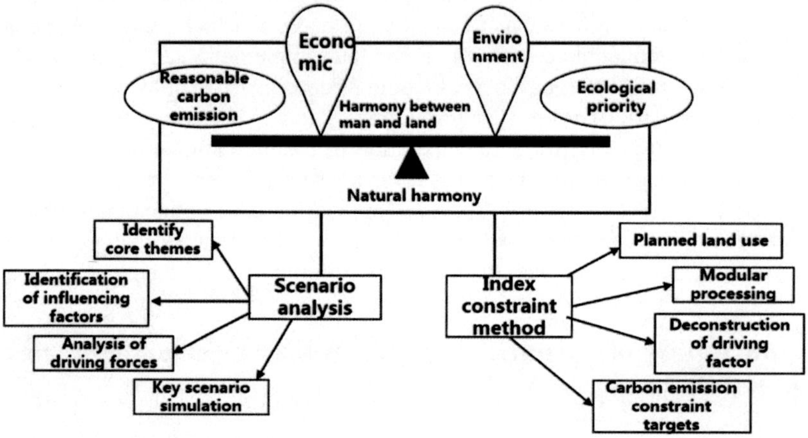

Fig. 1. The overall framework of ecological low-carbon planning for territorial space at the micro-scale.

In the overall framework, "Ecological priority" is the basis for low-carbon territorial space planning at the micro-scale. Based on the natural ecological environment of the region and the basis of reasonable assessment of the natural resources of the region, it is necessary to define the control bottom line and development boundary of the regional development and construction at the current scale, reasonably control the development intensity, determine the construction constraint target, and ensure that the ecological security is not damaged.

Second, "Reasonable carbon emission" is the core of low-carbon territorial space planning at the micro-scale. Ecology and economy should be considered in territorial space planning. First, according to the characteristics of the region to tap the advantages of economic development, so that the compilation of territorial space planning to meet the actual needs of regional economic development. Secondly, in the preparation process, carbon emission constraint targets should be defined according to the natural conditions of the region, and reasonable carbon emission constraint indicators should be formulated based on the actual industrial conditions of planned development to guide the low-carbon development direction of the region.

Finally, institutional innovation and technological renewal should be taken as an important support for low-carbon territorial space planning at the micro-scale. Institutional innovation is the guarantee for the implementation of national space low-carbon planning, which can improve the efficiency of planning and effectively supervise the implementation of low-carbon work. The updating of technology is an important driving force for the planning work, and the scientific assessment of regional natural resources and environmental conditions will promote the continuous updating of the technical process of territorial space planning.

5.2 Technical Framework

Based on the overall framework of micro-scale territorial space low-carbon planning mentioned above, a technical framework for micro-scale territorial space low-carbon planning can be constructed in combination with specific contents in planning practice, including three parts: scenario analysis, index constraint method, and System and standard support. Among them, scenario analysis is the basis, index constraint method is the core, and system and standard support is the guarantee of implementation (Fig. 2).

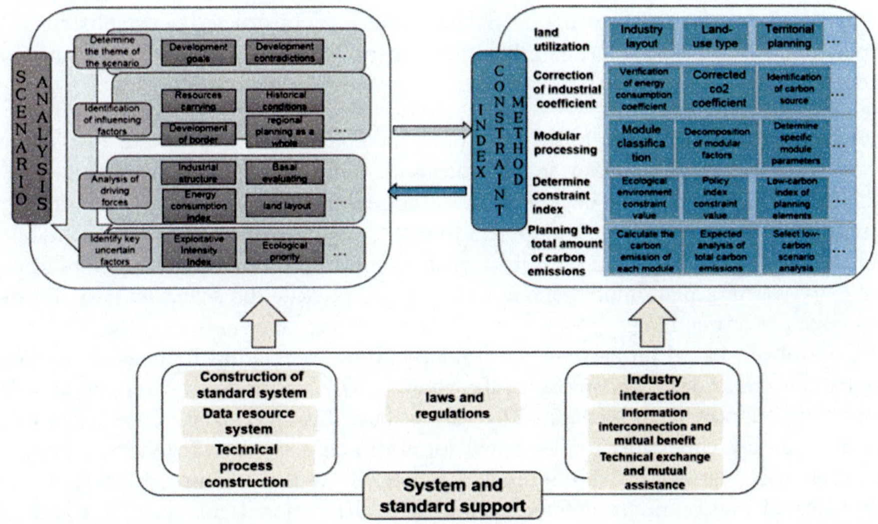

Fig. 2. The technical framework of ecological low-carbon planning for territorial space at the micro-scale

Scenario Analysis. Scenario analysis is the basis and premise of micro-scale territorial space low-carbon planning. According to its early warning and strategic characteristics, it can effectively deal with the uncertainty caused by a series of practical problems such as the selection of development goals and the positioning of leading industries. The optimal regional development scheme and low-carbon planning and construction scheme are selected by simulating the regional planning and construction situation in each mode.

In practice, the earlier scenario analysis is involved in the preparation of low-carbon planning, the smaller the cost will be, so that different preparation plans can coordinate or even merge and improve. In the process of low-carbon planning, the first core problem to be solved is to determine the theme and time range of regional planning, and predict whether the implementation of regional low-carbon planning can meet expectations, and what impact it will have on the environment and economy, to determine its effectiveness. Secondly, the factors affecting the theme should be identified from more angles. The influencing factors can include politics, economy, and other aspects, as well as the definition of development boundary and other specific factors. Many factors that seem insignificant at present will play a key role in the future, so it is important to identify them as much as possible. Thirdly, as the "external force" affecting low-carbon planning, the driving force will increase the uncertainty in the implementation process. Its content includes social and environmental aspects, and the driving force must be identified according to the actual situation of the region. Next, locate several key uncertainties that are most important and uncertain. The existence of key uncertainties can directly determine the development of scenarios, and the key uncertainties of different themes are also different. Finally, the initial scenario is constructed, If there are n key uncertainties, 2^n scenarios can be constructed. When there are only 2 or 3 key factors, a scenario matrix is often used for scenario construction. Scenario plans are constructed by establishing two opposite scenario dimensions, and then the scenario plans are analyzed and compared in detail.

Index Constraint Method. Index constraint method is the core of the low-carbon micro-scale territorial planning, in the process planning, through the quantitative analysis method, formulates the development of the regional industry layout scientifically, to determine the low carbon development forecast, planning of total carbon calculation, and the result was compared with the target value of constraint analysis, monitoring, and early warning plan implementation process, to provide the scientific basis for the improvement of planning.

First of all, based on each scenario plan and the correlation framework of "land use-carbon emissions", according to the characteristics of industrial sectors and the carbon emission behavior generated by various activities, the carbon emissions of the general plan are concretely implemented for each land-use type. Secondly, according to different scenarios, the carbon emission behavior, energy consumption type, and other related indicators are defined according to the regional characteristics and the different industrial types introduced in the scenarios, and the coefficients are modified according to the regional characteristics. Again, for the connections between land layout and industrial modular processing, the overall situation of land layout can be divided into several modules, such as residential, industrial and decompose the content of each module and driving factors, determine the different modules of carbon or carbon uptake calculation formula and the specific application, to implement the results of the factor to specific geographic spatial land-use level. Then, it is compared and evaluated by relevant constraint indicators. Policy constraints index, for example, the main strategic carbon rafting according to the regional government to release documents, to gain total energy consumption, total carbon dioxide emissions targets, and other related data, and calculate the regional carbon constraints, energy consumption per unit GDP total amount related to binding forces, as evaluation standard. Based on the content of the constraint index, the factors affecting carbon emission are decomposed, and the low carbon index of planning elements of specific modules is obtained based on the correlation framework of "land use-carbon emission", which is used as the evaluation basis of different scenarios in the end. Finally, the obtained low-carbon index of planning elements of each module can be compared with the total amount of planned carbon emission of each scenario plan. The carbon emissions of different modules in each scenario plan are calculated to obtain the total carbon emissions of different scenarios, and the best scenario low-carbon planning scheme is obtained after comparative analysis.

System and Standard Support. From setting the theme of the scenario to the simulation of different scenarios, the standard system must be strictly followed. The establishment of a sound technical process and the establishment of the scene theme must conform to the construction needs of regional development. The positioning should not be too high so that the planning has no basis for implementation, nor should the positioning be too low to ignore the needs of regional development to achieve the low-carbon goal. Therefore, in the construction of a data resource system, it is necessary to ensure that the planning data of different territorial spaces are standardized from format to reference standard, to make the assessment and positioning analysis of regional territorial space scientific, and the theme and scheme of scenario simulation effective.

Secondly, the timeliness of the accounting method and calculation basis should be ensured in the process of carbon emission calculation and determination of related constraint value in the index constraint method. Therefore, it is necessary to ensure industrial intercommunication, timely and accurately obtain the variation information of industrial energy consumption coefficient and carbon emission coefficient, and ensure the scientific calculation process by sharing information. And through technical exchanges and mutual assistance, the regional carbon emission structure is constantly optimized.

Finally, it is necessary to improve the legal system related to micro-scale territorial space low-carbon planning and promote its legislative process. The validity and necessity of low-carbon planning for territorial space should be clarified from the legal level, and the whole process of preparation, approval, implementation, and supervision involved in low-carbon planning should be planned, with clear requirements for transmission and control between upper and lower levels.

6 Conclusion and Discussion

Micro-scale territorial spatial low-carbon planning is based on the concept of green development ecological priority, in the face of the current carbon emissions disputes, in the concrete practice of the territorial spatial planning level gives an exploration train of thought, as a comprehensive framework needs to overall consideration of the regional natural resources, economy, society, history, humanities, policy guidance, and space layout, and other aspects. Based on the problems that may be encountered in the practice of micro-scale territorial space low-carbon planning, this paper discusses the overall thinking of low-carbon planning based on the uncertainty of planning and the actual needs of construction and development, which has great reference value for understanding the practical level of low-carbon planning and the implementation of specific work. At the same time, as the first exploration of micro-scale low-carbon planning, it also has reference significance for other countries to carry out similar planning projects. It will not only help China fulfill its commitment to green development, but also contribute to global low-carbon development.

The application of scenario analysis plays an important role in the preparation and implementation of micro-scale territorial space low-carbon planning. With the increasing complexity, refinement, and diversification of regional development issues, the cost of approval and implementation of low-carbon territorial space planning will continue to rise, so strict requirements must be set in the preparation and evaluation stage. Through the simulation of a multi-scenario scheme, the problems and contradictions faced by regional development are accurately captured, so that the regional low-carbon planning can truly consider ecology and economy. The index constraint method can accurately evaluate the scientificity and rationality of the scenario simulation scheme by quantitative analysis. Based on scientific evaluation of regional resource ecology, the low-carbon index system of planning elements in line with regional construction and development is established in combination with policy-oriented objectives and requirements, to correctly evaluate the applicability and authenticity of each scenario planning scheme.

Micro-scale territorial spatial low-carbon planning as a new concept, new methods, and new ideas of preparation method, needs the support of strong system and specification and to ensure that the work, ensure technical process specifications are fully followed, ensure that effect is the acceptance of science, these are all low carbon planning to implement security. In addition, looking to the future, the implementation of micro-scale territorial space low-carbon planning needs the joint participation of more disciplines and professionals, to make the technical information interconnection and planning more scientific and reasonable through different ways of collaboration.

References

1. Yaxian, W.: Research on carbon emission and total carbon intensity reduction of Thermal power industry in China (2020)
2. Xiulan, W., Yanxin, M., Yuhong, S., Yaotong, G.: Carbon emissions from farmers' agricultural production and its influencing factors: a case study of Hubei Province. Bull. Soil Water Conserv. **40**(06), 160–167+174 (2020)
3. Baogen, D., Yu, Z., Junhong, D.: Carbon emission measurement,decoupling characteristics and driving factors of planting industry in China. Agric. Resour. Region. Plann. China 1–14 (2021)
4. Ai, L., Yanan, W., Meng, L., Bowen, W., Wei, C.: Spatial correlation network structure and influencing factors of carbon emissions: a case study of three urban agglomerations in China. Environ. Sci. Technol. 1–8 (2021)
5. Enyan, Z.: Spatial-temporal pattern dynamics of carbon emissions and its response to urbanization in Zhejiang Province. Zhejiang University (2020)
6. Tianwei, F.:Study on carbon emission effect of land use change in Zhengzhou City. Henan University (2017)
7. Dan, Y., Liyan, R., Renfeng, M., Zekun, L., Caijuan, W.: Spatial pattern and influencing factors of tourism carbon emission intensity in Yangtze River Delta. Ecol. Sci. **40**(02), 89–98 (2021)
8. Bo, Y., Xu, Y., Xiangli, W.: Spatial spillover effects and influencing factors of carbon emissions at county level in Harbin and Changchun Urban agglomeration: an empirical study based on NPP-VIIRS nighttime light data. Chin. J. Environ. Sci. **40**(02), 697–706 (2020)

9. Guoquan, X., Zhu, C., Shiwei, F.: Spatial-temporal differences and influencing factors of carbon emissions based on two-stage LMDI model: a case study of Jiangsu Province. Soft Sci. 1–12 (2021)
10. Yujuan, Y., Qingchun, L., Hanqing, M.: Spatial differentiation of commuting carbon emissions based on housing price: a case study of Jinan city. J. Nat. Resour. **36**(08), 2081–2094 (2021)
11. Zheng, H., Jinyi, W.: Measurement and analysis of the impact of household housing structure on consumer carbon emissions: based on STIRPAT model. China Prices (07), 99+109 (2021)

The Optimization of High-Density Campus Transportation Based on Green Transport

Ming Liu(✉)

Shandong Jianzhu University, Jinan 250101, Shandong, China
liuming971010@163.com

Abstract. With the increasing importance of the concept of sustainable development and environmental protection, green transportation has become one of the core ideas of modern transportation planning, but it is mainly applied to urban transportation rather than campus transportation. The travel characteristics of campus transportation are different from those of urban transportation, and high-density campuses with higher population density and building density are more in need of green transportation methods. The article takes the Xinzhuang campus of Nanjing Forestry University as an example, and investigates the campus in three aspects: spatial form, dynamic traffic and static traffic; by drawing on the neighborhood community model, emphasizing the status of slow traffic and adopting green technology, it proposes optimization methods such as dividing campus clusters, creating campus public space, optimizing road classification and cross-sectional form, supplementing public transport and compounding the layout of parking facilities to alleviate the paradox of traditional campus transport. The aim is to provide a healthy and safe travel environment for students and teachers, and to provide a reference for other high-density campuses in terms of traffic optimization.

Keywords: Green transport · Campus transport · High density campus

1 Introduction

It was a period of concentrated construction in American colleges and universities in 1960s. R.P. Dober specified the design requirements of roads and car parks on campus, and analyzed the planning of campus transportation systems from the perspective of landscape design science [1]. Chinese scholars have also summarized a design system in terms of planning, landscape and architecture for local campus [2]. These leads to a study of common problems such as the pedestrian environment [3], traffic safety [4], the adaptation of traditional campuses to new technologies [5], and a case study of a campus under special conditions [6], where most high-density studies targeting congested cities. In recent years many scholars have focused on this feature of high density on campus planning and design [7]. Most of them have focused on the design of the spatial form, with relatively little research on the campus road system. Green transport has been widely used in urban transport planning [8]. However, the circumstance is quite different in

W. Guo and K. Qian (Eds.): GBCESC 2022, LNCE 211, pp. 248–254, 2023.
https://doi.org/10.1007/978-981-19-5217-3_24

the campus transport. This study investigates the campus transport system, and then proposes optimization strategies based on the transport characteristics of high-density campuses with green transport as the guide, especially for the medical problem like takeaway, enriching the study of the transport system of high-density campuses.

2 Basic Concepts and Characteristics

2.1 Green Transport

Green transport requires to satisfy the travel needs while protecting the environment [9]. To implement this concept on campus, firstly, the proportion of slow transport should be increased; secondly, gas and noise pollution should be reduced; the road network on campus need to improve the accessibility [4].

2.2 High-Density Campus

In urban planning, the floor area ratio is often used as the primary indicator [10]. When this indicator is too high, the problems of static space tension, intersection traffic conflicts and high-intensity traffic flow will become more acute [7]. This will result in greater demands on the planning of a efficient transport system.

3 Analysis of Traffic Problems on High-Density Campuses

3.1 Rationale for Case Selection

The Xinzhuang campus of Nanjing Forestry University covers a relatively low area among many universities in Nanjing, and students generally report that the campus is crowded, which is one of the typical representatives of a high-density campus.

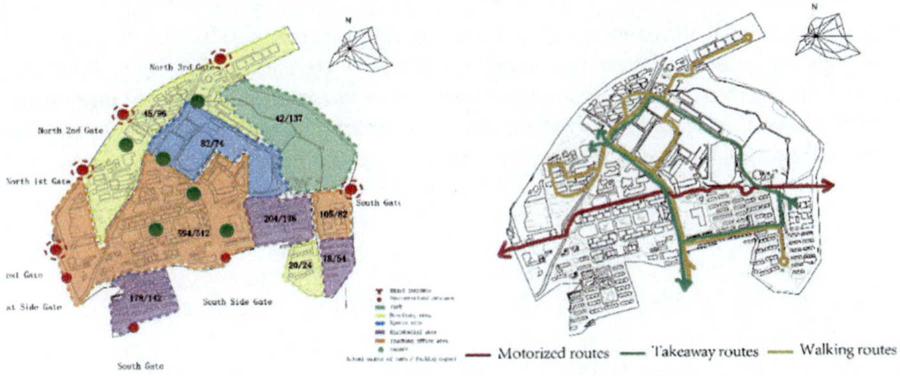

Fig. 1. Current space form, static traffic and traffic flow in peak hours.

3.2 Campus Space Form

Entrance. The main problems with the traffic at the entrances are: firstly, the width of the passage space at the entrances and exits is insufficient, many different traffic types compete with each other for the right of way; secondly, there is a lack of traffic space, cause it can only meet the needs to pass quickly, while being unable to stay. Thirdly, some entrances are unsupervised, causing traffic chaos (see Fig. 1).

Long Internal Commuting Distances. More added dormitory areas and teaching offices making what was once a clearer functional zoning more chaotic. The distance from the new buildings to the central living areas become larger, causing much unnecessary commuting traffic (see Fig. 1).

3.3 Campus Dynamic Traffic

Daily Travel Survey. Campus traffic is dominated by students, staff and social outsiders. Traffic activity on campus is strongly regular, with a surge in pedestrian and vehicular traffic during peak periods of the school day and in a single direction. Campus traffic consists of pedestrians, non-motorized vehicles and motorized vehicles, so it is necessary to overlay the traffic flow of both in order to better analyze the overall situation of campus traffic (see Fig. 1).

Takeaway Issues. Research shows that takeaways are delivered during peak meal times, and there will be a spike in delivery during snow and rain. Students' destinations after classes are similar to the deliveries. The massive flow of people and traffic to the same area simultaneously, coupled with the lack of road space, is a significant cause of traffic accidents caused by takeaway deliveries.

3.4 Campus Static Traffic

After field research, the existing motor vehicle parking spaces is 1319, and the number of legal parking spaces and the actual number of it were different (see Fig. 1). It shows that the parking pressure in the teaching office and residential areas are higher in comparison. Half of the staff reflected on the problems of on-street parking. There are lots of long-standing unused non-motorized vehicles that take up parking space.

4 Strategies for Optimizing High-Density Campus Traffic

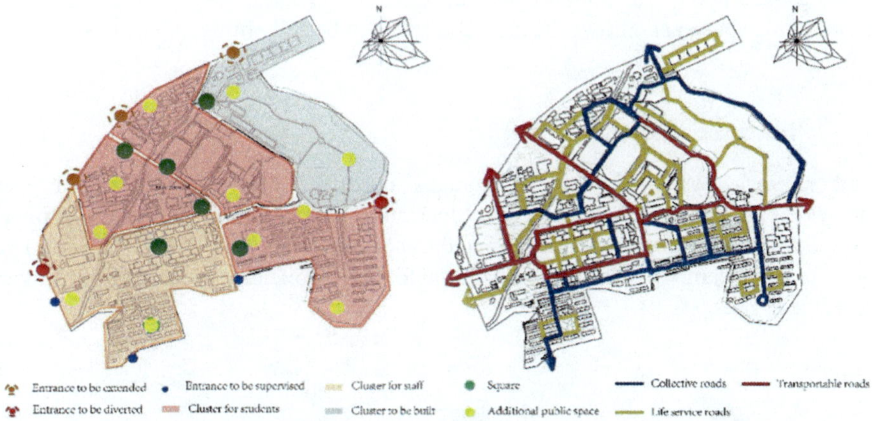

Fig. 2. Nanjing Forestry University Xinzhuang Campus traffic optimization plan.

4.1 Optimization of Space Form

Entrance. For the entrances for motor vehicles, the West Gate, which has a large traffic flow, should provide a special lane for pedestrian based on the separation of vehicles; in the East Gate, the buffer zone should be spared to form a small square cambinated with surrounding landscape, to improve its buffering capacity; for entrances focusing on non-motor vehicles, a more professional duty system is required (see Fig. 2).

Function. In 1920s New Urbanism hopes to reshape the atmosphere of urban planning, highlighting the humanity. TND abandoned mechanical functional zoning that ignores the diverse needs of the community, and calls for a compact and walkable neighborhood structure. Many educators believe that the university is a unique community, where students live on a specific size campus and form different campus cultures. That will be, therefore, suitable for student-centered high-density campuses.

Division of Campus Clusters. Using a composite functional cluster as an integrated functional cell of the campus, can avoid the inconvenience of dividing the campus into functional zonings, which leads to excessive distances between sub-zones. This method is more applicable to new campuses under construction. For old campuses the existing conditions can be fully utilized through the grouping pattern of some disciplines and the relative concentration of some campus spaces.

Creation of Dynamic Public Space. In addition to large squares, small squares, small green areas should be enlarged, and more building spaces should be opened up for various clusters. Then students will shorten the distance to have a meet, reducing the traffic load on the main roads. For outdoor spaces, the landscape of the campus can be fully utilized, or construct new public spaces; for indoor spaces, the connection between the ground floor space and the public space can be taken into account, which is called 'shared ground floor'. The method gives up some of the building space, but adds to the vitality of the space and allows the crowded space to be fully utilized (see Fig. 2).

4.2 Optimization of Dynamic Traffic

Road Network. Within the campus, students' needs for space to do activities is even more vital. Therefore, it is necessary to determine the traffic function as the primary function and ensure that the road is created to accommodate slow-moving traffic. This study classifies campus roads into transportable roads, collective roads and life service roads (see Fig. 2).

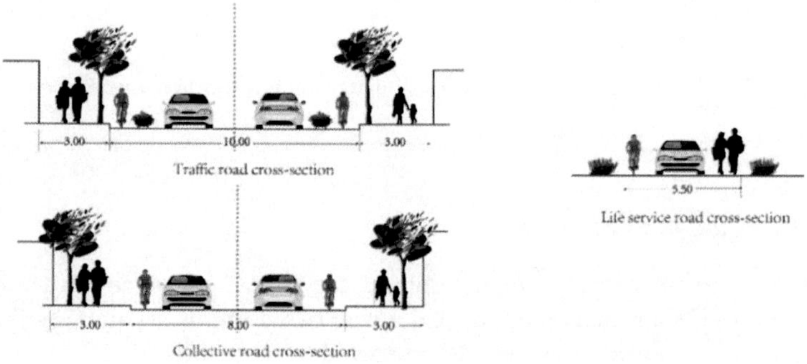

Fig. 3. Cross-section of three kinds of roads.

Transportable Roads. As the highest level of the entire campus road network, its function should be mainly motorized traffic, so that it should satisfy the two-way motor vehicle traffic, reduce the parking on both sides, weaken the slow traffic function. Green landscaping can be placed on both sides of the road, to reduce the negative impact of harmful gases and noise generated by cars (see Fig. 3).

Collective Roads. The collective road serves as a link between the campus center and the surrounding clusters. It needs to satisfy the demand of motor and slow traffic. Its road width still needs to meet the two-way motorized traffic, but also needs to increase the space for non-motorized vehicles and pedestrians(see Fig. 3).

Living Service Roads. It mainly serve short-distance traffic, penetrating into various groups and plots, and the emphasis should be on slow-moving traffic. The building complex will be the future of high-density campus architecture: buildings may run through the pedestrian corridor as a whole, allowing the pedestrian paths to leave the ground and integrate with the buildings. The pedestrianisation of high-density campuses should be combined with the 'ground floor sharing' of public spaces so that the pedestrian paths penetrate deeper into the building (see Fig. 3).

Road Refinement Design

Design for Traffic Stabilization. Currently, traditional campuses arrange speed bumps and installing turnouts in complex zones at intersections. It can be added humane and less costly stabilization designs, such as speed tables and the raising of intersections.

Sponging. The main way that use permeable paving to reduce rainwater infiltration into the ground, can control rainwater runoff at source. Permeable paving can be used on waterlogged sections, gardens and car parks, while the drainage pipes can be expanded. It also need to increase the number of drainage outlets and check blockages regularly.

Public Transport. Campus bus routes can be designed according to the daily communication routes, which is convenient for teachers and student. Campus buses can be purely electric, in order to reduce pollution. Public bicycle rental points can be centrally located in teaching and residential areas.

Campus Takeaway Management. Improve delivery routes. Campus management department can use the access control system to manage the path of the delivery vehicle after entering the school, to avoid the conflicts; or try to choose the school gate nearest to the delivery address to enter, thus avoiding penetration within the school.

In addition, the delivery will be carried out uniformly by the delivery stations after entering the campus, which can not only alleviate the intense traffic conflicts within the campus, but also become one of the ways for students to work and study [7].

4.3 Optimization of Static Traffic

Size Forecast. The demand for parking on campus can be divided into two parts: daily demand and flexible demand. Daily parking demand forecasts can use the generation rate method based on the nature and size of the building (Eq. 1), while elastic demand forecasts are based on daily demand forecasts multiplied by an elasticity factor [11].

$$y_i = \sum_{k=1}^{k} a_i^k X^k \tag{1}$$

In the equation:

y_i is the parking demand (in units) for vehicle type i (motor vehicle, bicycle).

a_i^k is the parking demand generation rate for the i-th vehicle type in the k-th building.

X^k is the architectural (business) scale of the building for building type k.

Composite Layout. It should be focused on making full use of existing resources. For example, a small parking strip can be formed at the side or back of a building The campus parking dynamic guidance system can be built based to provide real-time query service for parking information. It is necessary to mark of non-motorized parking areas to regulate parking, and regularly remove unused bicycles to avoid taking up space. The construction of elevated parking on the ground floor of new buildings and the construction of large centralized underground parking garages are reasonable ways to organize static traffic in a three-dimensional manner.

5 Conclusion

In response to the current problems of high-density campus traffic, this study, based on the concept of green transportation, draws on the neighborhood community model to achieve "safe travel and green environment". However, this study token the Xinzhuang Campus as an example, has certain limitations for the application in other high-density campuses. Therefore, there is still room for exploring and deepening in practice, and continuous optimization and innovation are needed.

References

1. Dober, R.P.: Campus Design. Wiley, Hoboken (1992)
2. Jianguo, W., Qing, Y.: The role of urban design in campus planning. Journal **5**, 29–32 (2002)
3. Yawei, D., Peng, Q., Qing, Y.: Evaluation of university campus walking environment based on principal component analysis. Journal **35**(04), 97–103 (2020)
4. Ting, L.: Research on the optimal design strategy of campus traffic based on the diversion of people and vehicles—a case of the west campus of Wuhan University of Technology. Journal **45**(20), 23–25 (2020)
5. Yishu, Y.: New campus slow traffic system optimization research based on the shared bikes. Journal **45**(20), 27–28 (2020)
6. Xiyuan, Z., Jingshi, X., Rixue, J.: Research on campus transportation planning based on green traffic in severe cold area. Journal **11**, 25–26 (2019)
7. Jinqiu, S., Ning, M.: Be obedient to situation and the mountain, benefit together of harmonious: the exploration of planning and design in intensive mountain campus. Journal **08**(28), 67–71 (2010)
8. Huapu, L.: Approaches towards realization of urban green transportation. Journal **7**(06), 23–27 (2009)
9. He, W., Chunying, G.: Retrospect & prospect of urban green transport planning in China. Journal **1**, 5–12 (2020)
10. Haixiu, L., Andong, H.: Analysis on the index and form of high-density university campus planning. Journal **17**(16), 76–79 (2020)
11. Chuanlong, G., Baorui, H., Yang, X., Gu, H.: Research on campus transportation planning from the perspective of traffic safety–a case of Nanjing Forest University. Journal **02**, 182–186 (2016)

Design of Wave Energy Capture Structure and Research on Mooring System

Mingjun Li[1], Xinhui Duan[1], Zheng Huang[2(✉)], and Yan Wei[1]

[1] Guangdong Power Grid Co., Ltd., Guangzhou 510080, China
[2] Electric Power Research Institute of Guangdong Power Grid Co., Ltd., Guangzhou 510080, China
zhhuang8831@163.com

Abstract. The wave energy in the ocean contains huge energy and how to convert its unstable energy into stable energy with sustainable output is a significant problem to be solved. In this study, a truss-type trilateral wave energy capture device is designed and modeled. More importantly, the influence of different mooring systems on the device is clarified. The results show that the six mooring line system can effectively improve the anti-wave performance and vibration reduction capability of the wave energy capture device, and reduce the surge response displacement of the device by about 37% under regular waves.

Keywords: Wave energy · Mooring system · Energy conversion

1 Introduction

With the continuous development of human society and the world economy, traditional energy sources are increasingly scarce, the ecological environment has been seriously polluted and destroyed, and the use of large-scale fossil fuels has caused serious ecological and environmental problems. According to the basic principles of ocean energy development and utilization and its own characteristics, ocean energy can be further subdivided into ocean tidal kinetic energy, tidal kinetic energy, wave energy [1–3], temperature difference change energy and seawater salt concentration difference energy [4], etc. Ocean energy mainly has the following Several advantages: first, huge reserves. According to statistics, the average energy reserve of the earth's wave energy is about 29, 500 TWH, which is about twice the global power consumption in 2006; second, the comprehensive utilization of marine energy equipment will hardly affect the ecological environment [5, 6]. It will promote the improvement of nature and realize the harmonious development of man and nature, and also alleviate the shortage of fossil fuels and the environmental damage caused by them; third, the social benefits are obvious. The development and utilization of marine clean energy can also effectively promote the development of other healthy industries such as shipbuilding, fishing and coastal defense in China. Furthermore, the rational use of ocean energy to generate electricity can effectively solve the major problems of our country's coastal defense forces garrisoned and some remote areas of coastal defense and living energy shortages.

W. Guo and K. Qian (Eds.): GBCESC 2022, LNCE 211, pp. 255–262, 2023.
https://doi.org/10.1007/978-981-19-5217-3_25

China has vast inland waters and long and winding coastline, which is about 180,000 km long, in addition, the coastline of the islands is about 140,000 km long. As a result, the reserves of marine resources are extremely rich. According to experts' estimates, the total energy density of the ocean is about 8×10^8 kw, which shows its huge comprehensive utilization potential of resources. Taking ocean tidal power generation technology as an example, it started and developed rapidly in southern my country, and gradually built a number of tidal power stations, including Guangxi Guozishan tidal power station, Zhejiang Jiangxia tidal power station and Fujian Pingtan tidal power plants, etc. The annual average wave height in the coastal waters of my country is about 2 m, and the total energy of the waves reaches 5×10^7 kw. Among them, the global wave energy density that can be freely developed and comprehensively utilized is 1×10^7 kw. Thus, it is of great significance to develop resource.

This study mainly discusses the influence of mooring types such as bifurcated catenary, mooring chain parameters, mooring line angle, number of mooring points and other factors on the stability of mooring system and mooring chain tension, clarifies the adaptability of the mooring system in different marine environments, and conducts optimization selection. Finally, the dynamic modeling research on the coupling of wind, wave, flow, device structure and mooring system is carried out. Through dynamic modeling, hydrodynamic calculation, frequency domain motion analysis, and time domain mooring analysis, the influence of each factor on the system and the influence of mutual coupling on the system are clarified. Meanwhile, the motion response of wave power generation device under wind and wave flow load, such as pitch, yaw, heave, roll, pitch and bow, is analyzed. The layout scheme of sensor measuring terminals such as acceleration, displacement, strain, pressure and inclination angle is studied to provide basic data for the safety monitoring of the device structure.

2 Mooring System Calculation Theory

This study mainly involves two catenary mooring systems, including the three-segment sub-fork mooring system and the six-segment mooring system. The system has great flexibility, and it is generally believed that the mooring lines of the floating body do not bear the action of shear force and bending moment. Under the action of external loads, the floating body device has a static equilibrium position at each moment. The static analysis of the mooring system mainly studies the linear shape and tension of the catenary in this equilibrium state, and calculates the restoring force provided by the mooring system. Generally, the lumped mass method is used to establish the catenary equation, and the quasi-static theory is used to calculate the line shape and tension of the catenary, so as to carry out the static analysis of the mooring system. The catenary can be divided into the catenary without undercover length and the catenary with undercover length according to the length of undercover, so there are two types of catenary control equations, the force of a single catenary [7, 8] is shown in Fig. 1.

$$X_F(H_F, V_F) = L_B + \frac{H_F}{w} \ln\left[\frac{V_F}{H_F} + \sqrt{1 + \left(\frac{V_F}{H_F}\right)^2}\right] + \frac{H_F L}{EA}$$

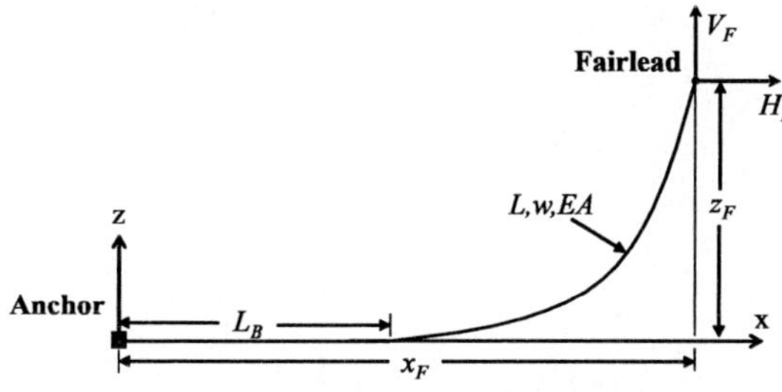

Fig. 1. Force analysis of a single catenary

$$+\frac{C_B w}{EA}\left[-L_B^2+\left(L_B-\frac{H_F}{C_B w}\right)\max\left(L_B-\frac{H_F}{C_B w},0\right)\right] \tag{1}$$

$$Z_F(H_F,V_F)=\frac{H_F}{W}\left[\sqrt{1+\left(\frac{V_F}{H_F}\right)^2}-\sqrt{1+\left(\frac{V_F-wL}{H_F}\right)^2}\right]$$

$$+\frac{1}{EA}\left(V_F L-\frac{wL^2}{2}\right) \tag{2}$$

where, x_F and z_F are the horizontal and vertical coordinates of the fairlead relative to the anchor point, respectively, H_F and V_F are the horizontal and vertical components of the effective tension in the catenary on the fairlead, and w is the unit length of the catenary, L is the total unstretched catenary length, EA is the tensile and compressive stiffness of the catenary, $L_B = L - V_F/w$ is length of catenary relaxation on the seabed, C_B is the static friction coefficient between the catenary and the seabed. Using Newton-Raphson method to iteratively solve Eqs. (1) and (2) in the local coordinate system, the effective tensioning force at the catenary fairlead can be obtained. Then the line type and tension of the catenary can be obtained. Given any point on the catenary, the length from the point to the anchor in the horizontal and vertical directions is x and z, the line tension is T_e, the tension and the mooring the coordinates of each point on the line can be obtained by the following formula:

$$x(s)=\begin{cases} s & 0\le s\le L_B-\frac{H_F}{C_B w} \\ s+\frac{C_B w}{2EA}\left[s^2-2s\left(L_B-\frac{H_F}{C_B w}\right)+\lambda\left(L_B-\frac{H_F}{C_B w}\right)\right] & L_B-\frac{H_F}{C_B w}<s\le L_B \\ L_B+\frac{H_F}{w}\ln\left[\frac{w(s-L_B)}{H_F}+\sqrt{1+\left(\frac{w(s-L_B)}{H_F}\right)^2}\right]+\frac{H_F s}{EA}+\frac{C_B w}{2EA}\left[\lambda\left(L_B-\frac{H_F}{C_B w}\right)-L_B^2\right] & L_B<s\le L \end{cases} \tag{3}$$

$$z(s)=\begin{cases} 0 & 0<s\le L_B \\ \frac{H_F}{w}\ln\left[\sqrt{1+\left(\frac{w(s-L_B)}{H_F}\right)^2}-1\right]+\frac{w(s-L_B)}{2EA} & L_B<s\le L \end{cases} \tag{4}$$

$$T_e(s)=\begin{cases} \max[H_F+C_B w(s-L_B),0] & 0\le s\le L_B \\ \left[\sqrt{H_F^2+[w(s-L_B)]^2}\right] & L_B<s\le L \end{cases} \tag{5}$$

$$\lambda = \begin{cases} L_B - \frac{H_F}{C_{BW}} \text{ when } \left(L_B - \frac{H_F}{C_{BW}}\right) \\ 0 \quad \text{otherwise} \end{cases} \tag{6}$$

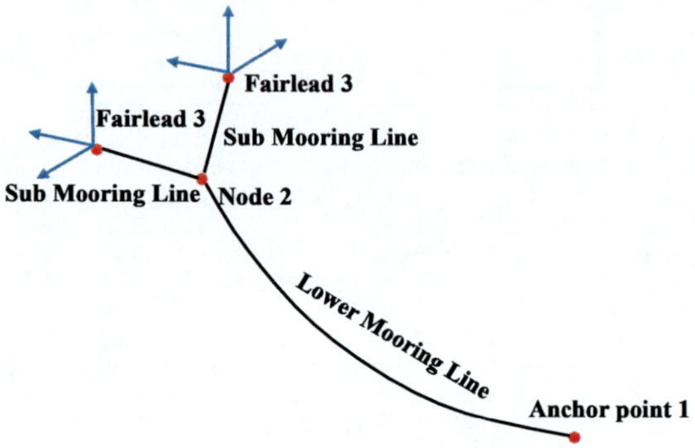

Fig. 2. Multi-segment catenary model

where s is the length from the catenary point to the anchor point, similarly, H_A and V_A are the horizontal and vertical components of the effective tension at the anchor point, respectively, $H_A = \max[H_F - C_B w L_B, 0]$ and $V_A = 0$.

By using the multi-segment mooring line theory, the mooring line can be divided into two segments as shown in Fig. 2. The first part is from the fairlead 3 to the position of node 2, and the second part is from node 2 to seabed anchorage point 1. Each segment is solved by quasi-steady theoretical analysis, and the system satisfies the equilibrium condition at node 2. The multi-segment mooring line system can be solved iteratively using the Newton–Raphson method.

3 Case Settings

For the reason that the environmental load of the wave energy capture device is quite complex while operating in the ocean, the wave energy capture device will shake violently when subjected to the environmental load, resulting in a large motion response. This leads to a significant reduction in power generation, and even the phenomenon of foundation overturning damage, so it is necessary to take measures to improve the stability of the wave energy capture device in water. This section combines the truss-type trilateral wave energy capture device with the original mooring system, as well as two three-segment sub-branch mooring systems with different parameters and a six-mooring system, as shown in Fig. 3, which are Model 1, Model 2, Model 3 and Model 4, to study its dynamic response in wind, wave and current environmental loads. The software is used for numerical simulation calculation, the environmental parameters in the calculation process are set, the motion response of the device under the combined action of wind and waves is calculated, and finally the calculation results are compared and analyzed.

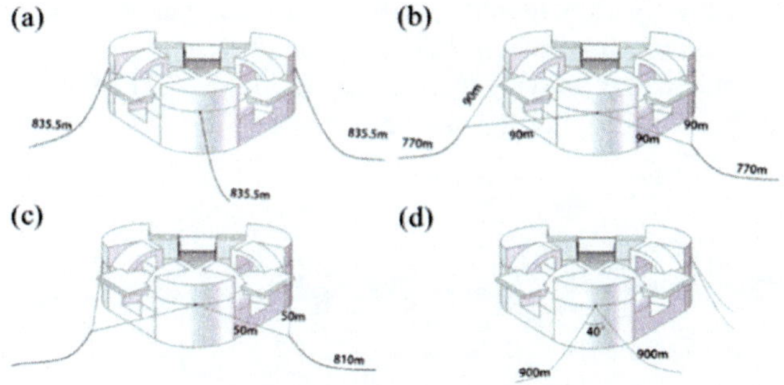

Fig. 3. Multi-segment catenary model

In order to compare the advantages and disadvantages between the optimized models, the motion responses of the above four models under different wave conditions were calculated. According to the relevant DNV specifications [9], the loads applied to the structure are determined by their respective key parameters. In the process of using the software for analysis, it is only necessary to set these key parameters according to the actual environmental requirements, and the loads can be simulated and applied to the structure. The load conditions are shown in Table 1.

Table 1. Environmental load case setup parameter

Case	Type	H or Hs(m)	T or Tp(s)	Wind environment
LC1	Regular wave	2.44	6.2	Stable and uniform wind, $V = 8$ m/s
LC2	Regular wave	5.32	8.4	Stable and uniform wind, $V = 8$ m/s
LC3	Regular wave	7.56	10.6	Stable and uniform wind, $V = 8$ m/s
LC4	Irregular waves	3.5	7.2	Stable and uniform wind, $V = 8$ m/s
LC5	Irregular waves	5.5	9.4	Stable and uniform wind, $V = 8$ m/s
LC6	Irregular waves	7.5	11.6	Stable and uniform wind, $V = 8$ m/s

4 Result and Discussion

Under the load case of regular waves, the time domain analysis method is used to obtain the time history of the motion response of each improved mooring system model under the selected marine environmental conditions, as shown in Fig. 4. Under the load case of irregular waves, for the time-history curve of the motion response of the device is random, it is difficult to see the influence of the new mooring system on the motion response of the device through the time-history curve. Therefore, it is necessary to convert the time domain signal into the frequency domain through the Fast Fourier Transform (FFT). In the frequency domain curve, the relative magnitude of the motion response of the device and the frequency corresponding to the amplitude can be clearly reflected, and the change of the motion response of the device can be analyzed, as shown in Fig. 5, where the horizontal axis represents the frequency, and the vertical axis represents the Power Spectral Density (PSD) of the motion response.

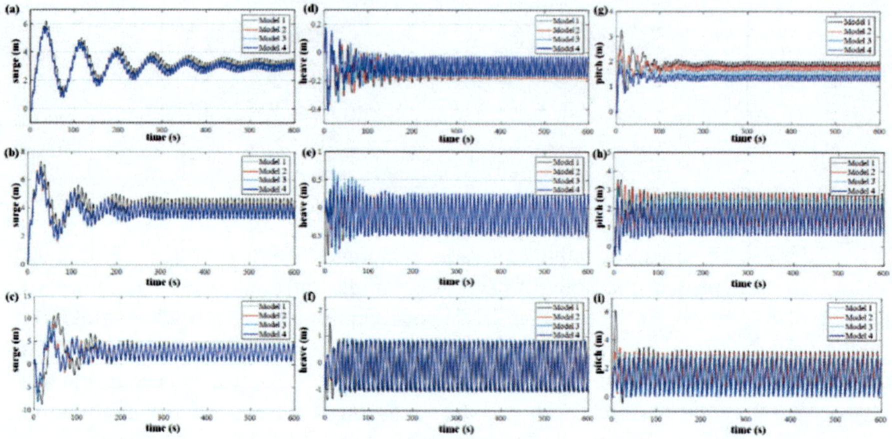

Fig. 4. Time-history curve of the model under regular wave loading, (a)–(c)surge, (d)–(f)heave, (g)–(i)pitch, (a)(d)(g)LC1, (b)(e)(h)LC2, (c)(f)(i)LC3

As shown in Fig. 4 (a)–(c), from Model 1 to Model 4, not only for LC1, but also for LC2 and LC3, the reduction of the device's surge displacement is obvious and gradually decreases. It illustrates that the mooring system model of the bifurcated catenary can reduce the surge displacement of the device, and the shorter the upper line length of the bifurcated catenary, the greater the reduction of the surge displacement of the device, of which Model 4 has the largest reduction, which is reduced by about 38.5%, indicating that Model 4 has the best effect on reducing the surge motion of the device. As shown in Fig. 4 (d)–(f), similar to the heave motion response, the heave motion response of the device also decreases gradually from Model 1 to Model 4, but the decrease of the heave motion is very small, and in the working condition in LC2, compared with Model 1, the average heave motion of Model 4 is reduced by about 10%, indicating that changing the length of the upper catenary does not have much effect on the heave motion of the device. For the pitch motion shown in Fig. 4 (g)–(i), from Model 1 to Model 4, the pitch motion of the device slightly increased by about 9.2%, indicating that the bifurcated catenary mooring system the pitch motion of the device cannot be reduced.

As shown in Fig. 5 (a)–(c), the first peak of the device's surge response appears at about 0.01 Hz, which is relatively close to the device's natural frequency of surge. Due to the improved mooring system model, the surge stiffness is increased, therefore, from Model 1 to Model 4, there is a slight shift in the frequency of the spectral peaks of the surge response. It can be seen from the figure that by adjusting the wave frequency close to the natural frequency of the structure, the additional motion in the wave frequency region can be enhanced. Similarly, in the frequency range of 0–0.03 Hz, the surge PSD

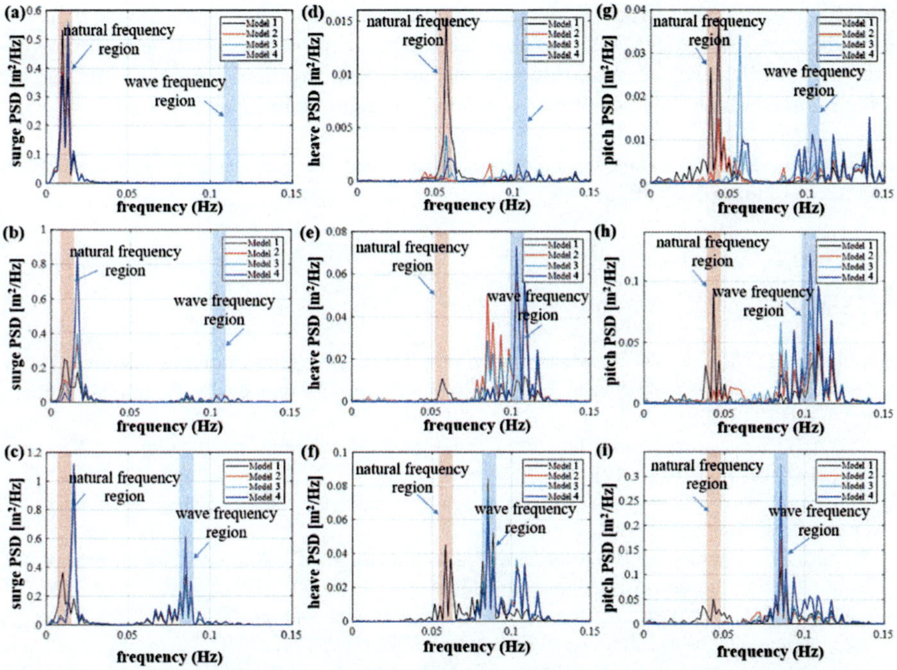

Fig. 5. Frequency domain curve under irregular wave loading, (a)–(c)surge, (d)–(f)heave, (g)–(i)pitch, (a)(d)(g)LC1, (b)(e)(h)LC2, (c)(f)(i)LC3

curve of Model 4 is the lowest under each working condition, which is consistent with the results under the action of regular waves in the previous section. As can be seen from Fig. 5 (d)–(i), from Model 1 to Model 4, the heave PSD curve of the device decreases slightly, while the pitch PSD curve almost overlaps, indicating that the new mooring system model changes the length of the upper catenary. The heave and pitch responses of the device have little effect, which is also similar to that found under regular waves. In addition, it is also found that as the wave frequency approaches the natural frequency of the structure, additional motions that enhance heave and pitch in the wave frequency region are enhanced.

5 Conclusion

To sum up, the hydrodynamic calculation is carried out after combining the mooring system and the wave energy device, it was found that under the action of regular waves, the surge response displacement of the improved mooring system models 2–4 to the device was reduced compared with the original Model 1. Among them, Model 4 has the largest reduction, that is, the six-mooring system, which is reduced by about 37%. For the heave displacement of the device, the reduction effect of the improved mooring system model is not large, and the Model 4 has the largest reduction in the heave displacement, about 10%. For the pitch response of the device, the improved mooring system model is slightly increased, and Model 4 increases the pitch displacement of the device by about 9.2%, but has little effect. It shows that the six-mooring system model can effectively reduce the surge displacement of the device, and the heave displacement of the device is not reduced much, but the pitch displacement cannot be reduced.

Acknowledgement. The work was supported in part by National Key Research and Development Program of China (2019YFB1504400).

References

1. Christie, D., Neill, S.: Measuring and observing the ocean renewable energy resource (2021)
2. Ward, S.L.: 8.07 - Tidal Stream Energy Letcher T M. Comprehensive Renewable Energy, pp. 104–122 (2022)
3. Czech, B.: Wave energy converter concepts: design challenges and classification. Ind. Electron. Mag. IEEE **6**(2), 4–16 (2012)
4. Chuankun, W., Wei, L.: Ocean Energy Resources Analysis Method and Reserve Evaluation. Ocean Press, Beijing (2009)
5. Weimin, L., Changlei, M.: Development and utilization of marine renewable energy and technological progress. Adv. Marine Sci. **36**(01), 1–18 (2018)
6. Changlei, M., Dengwen, X.: Research on development countermeasures of ocean energy development and utilization. Ocean Dev. Manag. **33**(03), 51–56 (2016)
7. Jonkman, J.M., Butterfield, S., Musial, W.: Definition of a 5MW reference wind turbine for offshore system development. Technical report, Office of Scientific & Technical Information (2009)
8. Masciola, M., Jonkman, J., Robertson, A.: Implementation of a multisegmented. Quasi-static cable model. In: Proceedings of the 23th International Ocean (Offshore) and Polar Engineering Conference, Anchorage, AK, USA (2013)
9. Specification for Classification and Construction of Offshore Single Point Moorings. China Communications Press, Beijing (1996)

The Concept and Strategy of Territorial Spatial Planning System for Construction Waste Reduction and Resource Utilization

Shuaibo Li[1] and Yingying Zhou[2,3](✉)

[1] Jinshui Bureau of Zhengzhou Natural Resources and Planning Bureau, Zhengzhou, China
[2] Suan Sunandha Rajabhat University of Thailand, Bangkok, Thailand
lilyzhou321@outlook.com
[3] Henan University of Chinese Medicine, Zhengzhou, China

Abstract. This study intends to combine the content and methods of current construction waste resourceful disposal technology, and connect the work of construction waste reduction and resourceful utilization with the territorial spatial planning system. In the process of management, preparation, implementation and supervision and inspection of territorial spatial planning, the objectives, processes and requirements of construction waste reduction and resource utilization are formulated in terms of laws and regulations, norms and standards, overall territorial spatial planning, special planning, zoning planning, control plan preparation, urban design, construction approval, planning verification and legal responsibility, and implemented in planning texts, control plans, planning design conditions, urban design guidelines and In the chart of construction waste reduction and resourcefulness, the construction waste reduction and resourcefulness is implemented into the spatial planning and source management. Through scientific and reasonable planning, the work of construction waste reduction and resource utilization is unified into the spatial planning system of the country, which is more conducive to making guidance from the top, controlling and prejudging construction waste from the source, reducing waste of resources, lowering transportation costs, protecting the construction environment, realizing the source reduction of construction waste and lowering disposal costs, changing passive end management into active source research and recycling, and fundamentally solving the problem of construction waste reduction and resource utilization.

Keywords: Construction waste · Resource reduction · Territorial spatial planning system · Concept and strategy

1 Background

1.1 High Volume of Construction Waste Generation

With the rapid development of urbanization in China, the pace of urban construction has accelerated, and Chinese cities have entered the era of stockpile renewal, and construction waste has increased. According to the Ministry of Housing and Construction, China

W. Guo and K. Qian (Eds.): GBCESC 2022, LNCE 211, pp. 263–275, 2023.
https://doi.org/10.1007/978-981-19-5217-3_26

Construction Industry Research Institute and Shenzhen University, China's construction waste generation is on the rise year by year, climbing from 1.17 billion tons in 2010 to 3.5 billion tons in 2019, with an annual growth rate of 6% to 20% in recent years, (shown in Fig. 1) and will increase year by year in the 14th Five-Year Plan period. According to the recognized data of the construction industry, every 1 square meter of building demolition is expected to produce 1.3 tons of construction waste; every new 1 square meter of building is expected to produce 0.3 tons of construction waste; in the renovation process, if it is residential renovation, 1 square meter of building will produce 0.1 tons of construction waste, and if it is public building renovation, more construction waste will be produced. By the end of 2020, the total amount of construction waste stockpiled in China reached about 20 billion tons. In the face of the massive amount of construction waste, many cities are ill-prepared, no special planning for construction waste, no disposal sites, no treatment facilities, management is not in place, construction waste is piled up at will, and the problem of urban construction waste surrounding the city is highlighted.

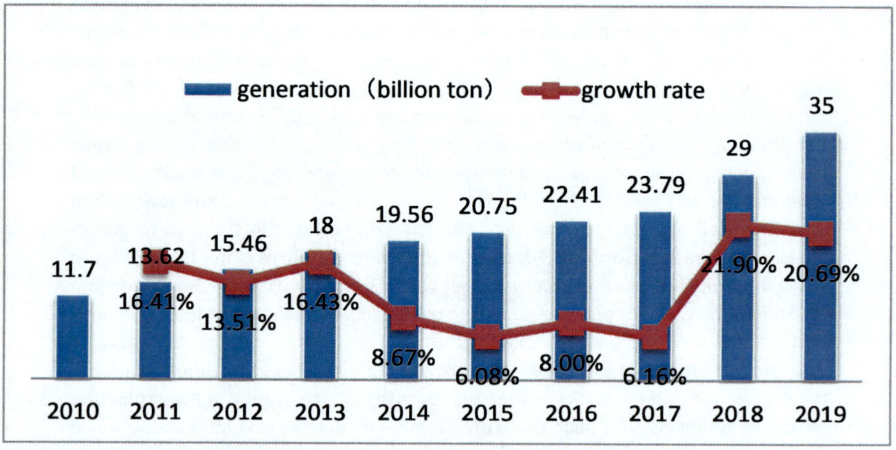

Fig. 1. Construction waste generation and growth rate, 2010–2019

1.2 The Great Harm of Construction Waste

First, the impact on the ecological environment. Construction waste in the pile, loading and unloading, transportation, landfill and other links, will have a negative effect on the surrounding air quality; construction waste in the open suburban site pile or landfill, after long-term rainwater erosion, groundwater infiltration and fermentation and other effects, heavy metals, paints, paints and other substances in the harmful elements will be through the ground surface leaching into the natural water and soil, pollution of soil and water; certain sulfate containing Some decorative materials containing sulfate will ferment and produce foul-smelling gas in the anoxic environment of the pile, which will have a negative impact on the surrounding ecological environment. Secondly, the impact on residents' life and social economy. The volume of construction waste generated is huge, and its dumping and disposal sites will encroach on urban construction and development land, and cause "garbage siege" in the suburban areas around the city for a long time.

1.3 China's Construction Waste Reduction and Resource Utilization Rate is Low

At present, there are 867 construction waste disposal plants in China's provinces and cities, of which 717 are landfills, 83 are landfills, and only 67 are resource-based disposal. China's construction waste resource disposal-related enterprises are basically private enterprises, the scale of enterprises is small, due to poor public awareness of environmental protection, construction waste resource production lines often face the dilemma of lack of raw materials, production efficiency is insufficient, most enterprises fail to make a profit, or even perennial losses. At present, the overall resourceization rate of China's construction waste is about 10%, compared with 95% in Europe, America, Japan and South Korea and other developed countries, a far cry.

1.4 Policy Support, Steady Development of the Industry

In March 2021, the National Development and Reform Commission, together with nine departments, issued the "Guidance on the Comprehensive Utilization of Bulk Solid Waste in the 14th Five-Year Plan". It is clearly stipulated that by 2025, the comprehensive utilization rate of new bulk solid waste will reach 60%, green construction will be promoted in the field of engineering construction, in situ recycling of waste road materials and demolition waste will be promoted, and construction waste classification management, source reduction and resource utilization will be implemented.

2 Research Progress

2.1 Foreign Research Related to Construction Waste Reduction and Resource Recovery

The UK's Department of Environmental Resources Assessment and associated businesses have developed a calculation methodology for calculating construction waste landfill recovery rates, and this methodology now forms the basis for government testing of the sector's implementation of construction waste minimisation. In 2008, 《The Sustainable Construction Strategy》 stated that the UK was committed to achieving the 《Waste Framework Directive》 target of recycling at least 70% of building construction and demolition waste by 2020. At the same time the UK government inspectorate and the construction waste industry have a joint target to reduce the amount of construction waste to landfill by 2020 to half of what it was in 2008. Currently, the UK has achieved a 93% construction waste reduction and resource recovery rate, which has exceeded the previously set 2020 target.

Japan is a small country with scarce resources, so it attaches great importance to construction waste reduction and resource recovery. In 1960, the Japanese government began to implement strict management of construction waste from the perspectives of laws and regulations and the construction waste resource recovery industry. In 1994, Japan formulated the 《21Century Recycling Plan》 in response to changes in the economy and society, proposing several countermeasures against construction waste: thorough prevention of construction waste in the design of buildings; maximum promotion of recycling by sharing information between projects, etc.; promotion of appropriate disposal of waste that is difficult to reuse; and active promotion of technology development, etc. Now, in many areas of Japan, the recycling efficiency of construction waste has reached more than 99% (He 2011).

The source reduction and resourceization of construction waste in the United States was carried out at the earliest, and now the U.S. construction waste reduction and resourceization industry is the world leader, and the resourceized products of construction waste are widely promoted and applied in both construction and transportation fields. The U.S. tried to recycle waste asphalt generated by highway renovation in 1915 and recycle it resourcefully. After a century of promotion, currently the U.S. waste concrete and asphalt can reach all resourceful recycling. 《The Superfund Act》 of the United States adopts the extended producer responsibility system, which extends the responsibility of waste disposal to the generator, and each industry should deal with the waste generated by itself, so each industry has sufficient enthusiasm to carry out the research related to waste reduction and resource recovery spontaneously.

Germany was the first country in the world to enact laws on waste recycling economy. In 1972, Germany promulgated 《The Waste Management Act》 and established waste treatment stations to implement incineration treatment. In 1986, the German government realized that end-of-pipe treatment of construction waste could not solve the fundamental problem, so it enacted a new law on construction waste management, trying to solve the problem of source reduction and resource utilization of construction waste. Germany's 《Circular Economy and Waste Removal Act》 provides for the establishment of a construction waste disposal program system and a construction waste disposal supervision system, especially supervising units and individuals who produce more than 2 tons of construction waste per year, and requiring them to formulate plans for source reduction, removal and recycling, and resource-based disposal for each category of construction waste.

2.2 Domestic Research Related to Construction Waste Reduction and Resource Recovery

Since the development of urbanization in China is slower than that of western developed countries, China faced the problem of construction waste siege late, so the relevant research on construction waste disposal in China is decades later than that of western developed countries. 1980 years later, experts and scholars in China started the relevant research on construction waste disposal.

Lyv Shuangru, Jin Yanjun and Ren Fumin studied the quantification methods of construction waste based on BIM, which classified some literature based on waste sources, quantification levels and quantification processes, and identified three types of construction waste quantification methods, including waste generation rate method, classification and accumulation method and variable model method. In addition, using the project budget as a bridge, the construction waste quantification was combined with BIM technology to predict the amount of construction waste on the basis of material consumption and to point out the shortcomings in the current progress (Lyu et al. 2021). Chen Rui, Yang Kai, Xiao Wei, Gao Dehong, and Ren Fumin proposed a resource-based treatment and disposal model for engineering residue to provide a systematic and feasible direction for its resource-based treatment, and also proposed a production process for recycled sand and gravel aggregates from construction waste; analyzed the content of each element in shield mud cake, and explored the feasibility and required conditions for burning residue into ceramic granules; proposed two methods for making bricks from mud cake, and reusing mud cake into planting soil The method of planting soil was proposed (Rui et al. 2020). Wang Ruofei, Zhang Guangzhi, Wang Ning, Ren Fumin, Chen Rui and Xi Chenggang listed the patents and standards related to engineering mud, engineering sludge and renovation waste at home and abroad, analyzed some of the standards and patents that have been implemented, further clarified the shortcomings of China's construction waste control, and put forward suggestions and bases for improvement and reasonable control in terms of patent innovation. It is pointed out that the development of construction waste standards and norms, theory and technological innovation are the main tasks for the future development of the industry (Ruo-fei et al. 2020). Wang Ning, Lou Dai, Chen Daqing, Yang Bing, Guo Hongzhi and Ren Fumin introduced "BIM + GIS" technology into the field of construction waste supervision and established a real-time monitoring and intelligent control information management platform based on this technology, which can realize the accurate control of construction waste (Ning et al. 2020). Liu Zheng and Ma Tangli introduced the way of construction waste reduction and resource treatment, and proposed the idea of construction waste recovery and recycling APP, which can realize the information communication among the sellers of construction waste, buyers of recycled resources, the disposal site and the fleet through various subsystems of APP, so as to reduce the waste of resources, reduce pollution, improve the utilization rate and realize the construction waste reduction and resource utilization led by circular economy. Utilization (Zheng et al. 2018). Li Hao and Zhai Baohui proposed to improve the technical system of construction waste reduction and resource utilization by sorting out the technical standards and production norms of China's construction waste inprocess products; the state established the Construction Waste Recycling Building Materials Promotion Committee to form a closed policy chain of construction waste reuse (Hao et al. 2015). Lu Guangyou and Ma Hesheng analyzed the current situation of construction waste production and management in Shenzhen, and proposed that reduction is the most effective measure to solve the current waste dilemma in Shenzhen, and efforts should be made in terms of policies and regulations, reclamation of sludge acceptance sites, and mud and water separation of sludge in preconstruction (Guanyou 2019). Dai Lou, Guangzhi Zhang, Fumin Ren, Chenggang Xi et al. took the construction of underground pipe corridor in Ningbo, Zhejiang Province as the research object, and established a quantitative prediction model for three types of construction waste: engineering residue, waste concrete, and engineering slurry through on-site research and construction progress tracking (Dai et al. 2020).

3 The Concept of National Spatial Planning System for Construction Waste Reduction and Resource Utilization

This paper connects the work of construction waste reduction and resourceful utilization with the national land spatial planning system, combines the content and methods of current construction waste resourceful disposal technology, implements construction waste reduction and resourceful utilization into spatial planning and source management in the process of management, preparation, implementation and supervision and inspection of national land spatial planning, controls the generation and emission of construction waste through scientific and reasonable planning, and anticipates the later resourceful disposal, so as to realize healthy and sustainable urban development.

- Research on construction waste reduction and resourcefulness in the preparation and implementation system of territorial spatial planning. The relevant content of the planning at all levels is clearly integrated into the concept of construction waste reduction and resourceization and related research and analysis, which breaks through the original thinking that urban planning does not consider construction waste and better connects construction waste reduction and resourceization with planning at all levels and better meets the development needs of modern cities. Starting from China's national conditions, we improve the classification of construction land, clarify the land for construction waste reduction and resourcefulness facilities, and ensure the legality of construction waste reduction and resourcefulness sites; incorporate the concept of construction waste reduction and resourcefulness into the preparation and implementation of the national spatial planning system, and construct a perfect planning system for construction waste reduction and resourcefulness: incorporate the goals and requirements of construction waste reduction and resourcefulness in the spatial master planning objectives, and in the In the control indexes of the detailed control plan, the relevant indexes of construction waste reduction and resourcefulness are included in the control indexes, the guidelines of construction waste reduction and resourcefulness are fully implemented in the urban design, the charts of construction waste reduction and resourcefulness are implemented in the approval of traffic roads, municipal pipelines and building construction plans, and the construction contents are reviewed and verified in the planning verification in accordance with the control plan, urban design guidelines and the approved charts of construction waste reduction and resourcefulness. Through the above-mentioned means, the planning content of construction waste reduction and resourcefulness will be integrated and implemented.
- Research on construction waste reduction and resourcefulness under urban planning management and supervision. Firstly, at the level of urban planning laws and regulations control, there is an urgent need to update the concept of legislation, implement construction waste reduction and resourcefulness, establish a circular society, and implement the economic concept of circular economy into the concept of construction waste related legislation. As soon as possible, promulgate the "Management Measures for Construction Waste Reduction and Resourcefulness" to ensure the smooth implementation of construction waste reduction and resourcefulness from all aspects of urban planning, construction, operation management and maintenance, and establish a perfect legal system for construction waste management. The target of construction

waste reduction and resource utilization will be included in the assessment index of energy saving and emission reduction. At the level of urban planning supervision and inspection, it is necessary to innovate the construction of institutional mechanisms, establish a digital management platform for construction waste, establish a market operation mechanism under government supervision, clarify the responsibilities of each participant in the process of construction waste reduction and resource utilization, improve the economic incentive system for construction waste reduction and resource utilization, and take multiple measures to realize the smooth implementation of construction waste reduction and resource utilization under the control of urban planning and implementation.

4 Strategies for the Implementation of the Territorial Spatial Planning System for the Reduction and Resourceization of Construction Waste

Combined with the existing construction waste resourceful disposal technology, the management, preparation, implementation, supervision and inspection of the normative standards and territorial spatial planning system at the level of implementation of construction waste reduction and resourcefulness to achieve the goals, requirements, processes, ways and so on, so that it becomes an important part of the spatial planning system and determine its planning status. At the same time, texts, diagrams, guidelines and diagrams for construction waste reduction and resourcefulness are prepared and planned at the stages of city general planning, special planning, zoning planning, control planning, urban design and construction plan (shown in Table 1).

First, in terms of planning and management, update the legislative concept, establish a recycling-oriented society, promulgate the 《Measures for the Management of Construction Waste Reduction and Resourcefulness》 as soon as possible, and establish a perfect legal system for construction waste management and urban management system. The management of construction waste reduction and resourcefulness is realized through innovative institutional mechanism construction, establishment of digital management platform for construction waste, establishment of market operation mechanism under government supervision, clarification of responsibilities of each participant in the process of construction waste reduction and resourcefulness, and improvement of economic incentive system for construction waste reduction and resourcefulness.

Secondly, in terms of standardization criteria, the *Guidelines for Land and Space Survey, Planning and Use Control Land and Sea Classification (for Trial Implementation)* and *Urban Land Classification and Planning Construction Land Standards* need to specify the land for construction waste reduction and resourceful disposal and related facilities to ensure the legality of construction waste reduction and resourceful sites.

Thirdly, in terms of planning preparation: at the level of overall territorial space planning, the overall target of construction waste reduction and resource utilization rate is formulated in combination with the overall development requirements of the city, and special planning for construction waste reduction and resource utilization is prepared, while relevant contents are implemented in the planning yellow line. In the zoning plan of construction waste reduction and resource utilization, the rate of construction waste reduction and resource utilization is decomposed and implemented according to the environmental attributes and target requirements of each area, and the zoning control principles are proposed, covering all aspects of construction waste collection, transportation, consumption and resource utilization. At the level of detailed control planning, the targets and indicators of construction waste reduction and resource utilization rate determined in zoning planning are implemented to each specific plot, and detailed control planning figures for construction waste reduction and resource utilization are formed to guide and control the implementation of construction waste reduction and resource utilization. At the level of urban design, the research and analysis of the basic conditions of the land parcels are strengthened, the types and quantities of construction waste generated within the land parcels are determined, the requirements of various indicators such as green building design, recycling of construction materials and green construction are formulated, the earth in the project area is reasonably planned, while guidance is given in the aspects of sponge city and green landscape design, and urban design guidelines for the reduction and resource utilization of construction waste are formed. As an important intermediate link in the process of planning and implementation, urban design carries on the top and bottom, and the content of the design guidelines directly affects the concrete implementation of construction waste reduction and resource utilization.

Table 1. Strategies for the implementation of the territorial spatial planning system for the reduction and resourceization of construction waste

Strategies for the implementation of territorial spatial planning system for construction waste reduction and resource utilization		
Level	Stage	Content
1 Normative standards		
	1.1 《Guidelines for land and space survey, planning and use control of land and sea use classification (for trial implementation)》	Specify the land for construction waste reduction and resource disposal and related facilities
	1.2 《Standard for urban land use classification and planning and construction land》	
	1.3 《Technical standards for urban construction waste control》	Clarify the norms and standards for construction waste control and disposal
2 Planning management		
	2.1 《Construction waste reduction and resourcefulness management measures》	Clarify the construction waste reduction and resource management measures
	2.2 《Construction waste detection and pollution control technical provisions》	Clarify the content of construction waste pollution detection and treatment means
	2.3 《Construction waste recycling products utilization tracking technical regulations》	Establishment of construction waste recycling products regulation

(continued)

Table 1. (*continued*)

Strategies for the implementation of territorial spatial planning system for construction waste reduction and resource utilization		
Level	Stage	Content
	2.4《Construction waste operation and disposal electronic joint management technical provisions》	The whole process of detecting the transportation of construction waste
	2.5 Economic incentive system for construction waste reduction and resourcefulness	Incentives for construction waste reduction and resource treatment
	2.6 Responsibilities of each participant in the process of construction waste reduction and resource utilization	Clarify the division of responsibilities among various departments of construction waste reduction and resource management, and enhance the efficiency of collaborative management
	2.7 Construction waste digital management platform	Establish a big data platform for construction waste to enhance government supervision and control
3 Planning preparation		
	3.1 Spatial planning of town system	Combined with the regional development level, set regional construction waste reduction and resourcefulness targets and macro measures
	3.2 Territorial spatial master plan	Based on the regional construction waste reduction and resourcefulness target, set the urban construction waste reduction and resourcefulness target and principle measures by combining with the city development level
	3.3 Yellow line planning	Combined with the city's construction waste reduction and resourcefulness goals, focusing on the development of various measures methods and content requirements
	3.4 Special planning	Formulate the objectives and requirements of each special plan based on the objectives of construction waste reduction and resource utilization in the master plan
	3.5 Zoning planning	According to the actual situation of each region, decompose and implement the target of reducing and resourcefulness of construction waste in the master plan, and formulate the target and method measures of reducing and resourcefulness of each sub-district

(*continued*)

Table 1. (*continued*)

Strategies for the implementation of territorial spatial planning system for construction waste reduction and resource utilization		
Level	Stage	Content
	3.6 Detailed control planning	Further decompose and implement the zoning plan's construction waste reduction and resourcefulness target, put forward various specific mandatory, prescriptive and guiding measure requirements, and implement them in the plan. It is an important link to implement the construction waste reduction and resourcefulness
	3.7 Urban design	According to the control plan on construction waste reduction and resourcefulness objectives and requirements, the specific implementation methods and processes and requirements are formulated and clarified in the design guidelines
4 Planning implementation		
	4.1 Notification of planning and design conditions	Incorporate the requirements of the control plan on construction waste reduction and resource utilization into the land transfer contract, with legal effect
	4.2 Architectural design scheme 4.3 Municipal engineering design scheme 4.4 Traffic engineering design scheme	Based on the urban design of construction waste reduction and resourcefulness objectives and measures and requirements, fully implemented in the building program design and construction drawing design, and in the construction project planning permit submission program to increase the construction waste reduction and resourcefulness chart, become an important planning and approval content

(*continued*)

Table 1. *(continued)*

Strategies for the implementation of territorial spatial planning system for construction waste reduction and resource utilization		
Level	Stage	Content
	4.5 Planning verification, completion acceptance	Based on the land transfer contract and construction project planning permit in the construction waste reduction and resourcefulness icon approval content for verification, to ensure that the resourcefulness rate target and the implementation of measures in place
5 Supervision and inspection		
	5.1 Supervisory mechanism	Establish a supervision mechanism for construction waste disposal, encourage the public to supervise and report construction waste violations, and strengthen the competent departments to inspect and curb construction waste disposal violations
	5.2 Spot check verification mechanism	The establishment of random selection of inspection objects, random selection of law enforcement inspectors "double random", sampling inspection results want to society "a public" regulatory mechanism
	5.3 Accountability penalty provisions	Develop accountability and punishment methods, establish a ledger of non-compliant design units and non-compliant disposal enterprises for construction waste reduction and resourceization, include them in the credit blacklist, and impose penalties and accountability on the enterprises and relevant departments involved to enhance the social effect of construction waste reduction and resourceization

Fourthly, in terms of planning and implementation: Issuance of planning and design conditions notice, based on the control plan to clarify the objectives and measures of construction waste reduction and resource utilization, so that it becomes a necessary clause in the land transfer contract and is strictly enforced. When the construction project planning permit for buildings, traffic roads and municipal pipelines is approved, the construction project design plan must be implemented in accordance with the superior planning (detailed control planning figures and urban design guidelines) and construction waste reduction and resourceful disposal technology to implement the planning requirements related to construction waste reduction and resourcefulness, and the construction plan needs to reflect the quantity of each construction material and the amount of construction waste generated at a later stage as well as the type and quantity of construction waste generated by construction (which can be combined with BIM and other technical means), and construction waste reduction and resource utilization projections, while developing various specific measures for green construction, from the source of construction waste reduction and end of construction waste resourceization planning prediction, the development of late construction waste resourceization content and process, and the formation of construction waste reduction and resourceization chart, as an important content of the planning program drawings common approval for the record. In the project completion for planning verification, strictly in accordance with the planning approval of the content of the chart of construction waste reduction and resource utilization verification of indicators, to ensure the implementation of the project construction in place.

Fifthly, in terms of planning supervision and inspection, clarify the relevant legal responsibilities, develop relevant supervision mechanisms and penalties, impose penalties for violations of the preparation and implementation of construction waste reduction and resourcefulness, and promote the widespread and effective implementation of construction waste reduction and resourcefulness.

The study of the national spatial planning system of construction waste reduction and resource utilization has important value, unifying the work of construction waste reduction and resource utilization to the national spatial planning system, which is more conducive to making guidance from the top, controlling construction waste from the source, reducing resource waste, lowering transportation cost, protecting the construction environment, realizing the source reduction of construction waste and lowering disposal cost, changing passive end treatment to active source It is important for the sustainable development of the city to study the recycling and fundamentally solve the problem of construction waste reduction and resource utilization.

5 Conclusion

From the perspective of construction waste reduction and resource utilization, the study extends the research on construction waste in the national spatial planning system. At present, most of the research on construction waste reduction and resource utilization is focused on end-of-pipe management, and there is a lack of forward-looking macro-level management research. The research on the management of construction waste at the level of territorial 14 spatial planning system is conducive to the guidance from the top, pollution control at the source and planning for resourceful disposal at the later stage. The concept and strategy will have some reference significance for the promotion and management of construction waste reduction and resourceization in the future.

From the perspective of the territorial spatial planning system, the study extends the preparation, implementation and urban planning management, supervision and inspection of spatial master plans, zoning plans, detailed control plans, urban design, construction approval and planning verification, and incorporates the preparation of special planning for construction waste reduction and resource utilization, control plans, urban design guidelines and construction plan charts, further improving the territorial spatial planning system. So that urban planning can also keep pace with the times to scientifically respond to various urban diseases that appear in the process of development.

References

Dai, L., Guang-zhi, Z., Fu-min, R., Cheng-gang, X.: Establishment of quantitative prediction model based on characteristics of waste generated in underground pipeline gallery construction. Environ. Eng. 9–14 (2020)

Guanyou, L.: Analysis on current situation and reduction measures of construction waste management in Shenzhen. Ready-Mixed Concrete 69–70 (2019)

He, Y.: Construction waste reducing management method and the game analysis. ChongQing University(2011)

Hao, L.,et al.: Policy research on promoting recycling construction waste in China. Urban Dev. Stud. 119–124 (2015)

Lyu, S., Jin, Y., Ren, F.: Research on quantitative method of construction waste based on BIM. Build. Technol. Dev. 55–57 (2021)

Ning, W., Dai, L., Da-qing, C., Bing, Y., Hong-zhi, G., Fu-min, R.: Preliminary study on the integration control platform of construction waste based on "BIM+GIS" technology. Environ. Eng. 46–50 (2020)

Rui, C., Kai, Y., Wei, X., De-hong, G., Fu-min, R.: Analysis on recycling treatment and disposal of engineering slag. Environ. Eng. 22–26 (2020)

Ruo-fei, W., Guang-zhi, Z., Ning, W., Fu-min, R., Rui, C., Cheng-gang, X.: Analysis on domestic and foreign construction and demolition waste related standards and patents. Environ. Eng. 27–32 (2020)

Zheng, L., Tang-li, M., Jing-jing, H., Bing-xue, L., Fang-ya, W., Yu-ya, S.: Constructing idea of APP for recycling construction waste. Build. Energy Efficiency 99–104 (2018)

Check for updates

The Study on the Development of Contemporary International Timber Heritage Conservation

Xi Chen and Mei Huang(✉)

School of Architecture, Soochow University, No. 199, Jen Ai Road, Suzhou Industrial Park, Suzhou 215123, China
meihuang_0822@163.com

Abstract. Wood built heritage has always been the focus of international heritage conservation because of its special material properties and rich geographical and cultural characteristics. Taking the 2017 revision of the International Council on Monuments and Sites (ICMONUMENTS) Code for The Protection of Wooden Built Heritage and the main content of IIWC conferences in past years as the research object, this paper analyzes the changes in the Code and explains the deep-seated reasons for the changes. It also explains the response and adjustment of the international conservation community to the conservation concept, method, technology and management of wooden heritage in the past 20 years. The paper then looks forward to the development trend of international wooden built heritage protection since 2018 and compares the programmatic documents in China to propose mutual reference areas.

Keywords: IIWC · Principles for the conservation of wooden built heritage (2017) · Principles for the preservation of historic timber structures (1999) · Culture heritage conservation

1 Wood Built Heritage Overview

1.1 Timber Built Heritage Criteria Development Process

Wood built heritage has always been the focus of international heritage conservation because of its special material properties and rich geographical and cultural characteristics. Although China has a long history of construction and a large wooden building complex, and new conservation technologies are constantly emerging, research on the theories and paradigms of the preservation of wooden built heritage is still lacking.

In 2017, the International Council on Monuments and Sites completed the revised Principles for the Conservation of Wooden Built Heritage. Based on the Principles for the Preservation of Historic Timber Structures (1999), the contents concerned by previous IIWC (The International Timber Council of the International Commission on Monuments and Sites) conferences are integrated, such as the concerns on structure

W. Guo and K. Qian (Eds.): GBCESC 2022, LNCE 211, pp. 276–287, 2023.
https://doi.org/10.1007/978-981-19-5217-3_27

mentioned in the 15th session and the re-discussion on authenticity and integrity in the 16th session (Fig. 1). The release of the revised edition reflects the development trend of international conservation concepts and technologies for wooden heritage conservation.

The new guidelines respond to the basic problems existing in the protection of wooden built heritage. By studying the reasons for the changes in this document, this paper hopes to understand the challenges faced by the new guidelines in the application process and their sustainable development characteristics. At the same time, we will make a horizontal comparison between China's wooden heritage protection work and that of other parts of the world, to provide a reference for the development of China's protection cause and promote the scientific development of wooden heritage protection.

1.2 A Comparison of the New and Old Protection Criteria

The 1999 edition of the guidelines is divided into eight parts, for a total of 15 articles, including "Introduction", "inspection and management", "monitoring and maintenance", "intervention", "repair and replacement", "historical forest reserves" , "contemporary materials and technology", and "education and training". Compared with the 1999 edition of standards, the integrity and logic of the 2017 edition of standards, it will content expanded to 36, and the old "inspection and management" partial dismantling became "inspection, investigation and research" and "record and archive", "repair and replacement parts in the "intervention" and added "analysis and assessment" section (Fig. 2). The new guidelines highlight the importance of cultural significance and value and are more consistent with the objective law of wooden heritage protection system work.

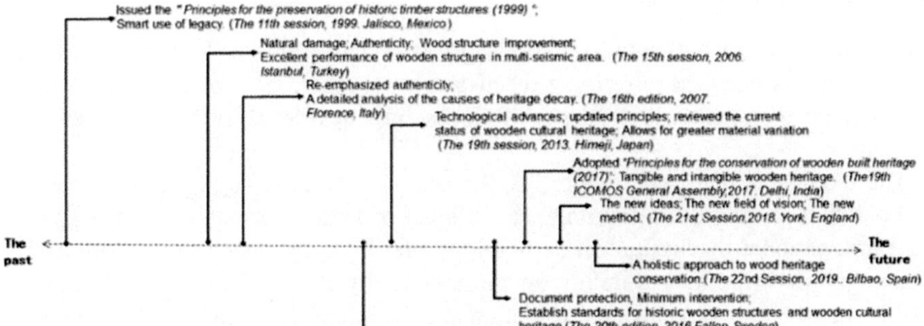

Fig. 1. IIWC Important meetings and themes.

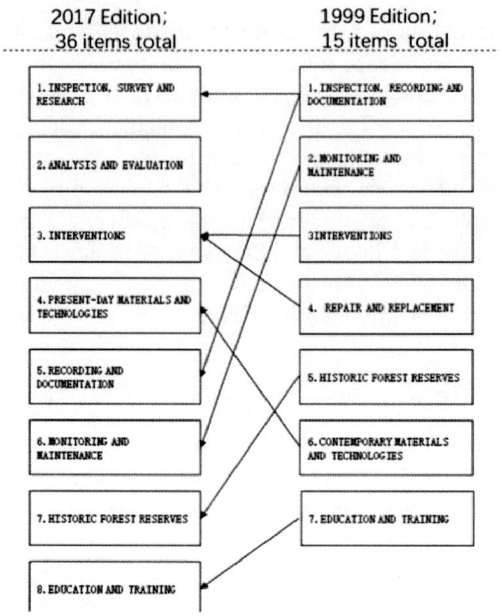

Fig. 2. Comparison table between the two editions of standard changes.

2 Analysis of Reasons for the Change of New and Old Protection Standards

2.1 New Value Paradigm

A new awareness of authenticity and integrity—What is true and what is complete? This has always been the core issue discussed by international conservation theorists, especially for wooden heritage, especially at the practical level, and is still a persistent problem.

The 1999 guidelines only emphasize the authenticity and integrity of the protection and continuation of cultural heritage at the level of structural materials. It triggered a fierce international discussion on authenticity and integrity and began to think about whether the description and definition of authenticity in the Nara Document are reasonable. Principles for Analysis, Protection and Structural Restoration of Architectural Heritage (ISCARSAH, 2003) held that: "The value and authenticity of architectural heritage cannot be established on fixed standards. In line with the principle of respect for all cultures, its material heritage needs to be considered in the context of the culture to which it belongs" [1].

The NARA Document was reflected in the 2005 revision of the Operational Guidelines for the Implementation of the World Heritage Convention, and people began to realize that without collective recognition, many wooden heritage sites would be abandoned by The Times. At the 20th IIWC in 2016, Mikel LANDA and Alazne OCHANDIANO proposed the limitations of the previous guidelines: To understand authenticity as a text conveying the history of a building, it is important to understand not only its actual condition but also its history, construction and relationship with other materials when preserving a wood-framed building [2]. The so-called minimum intervention is the best solution made after a comprehensive value judgment of the building. For example, Medieval Tower in Donamaria, Navarra, in the process of protection and restoration, experts resolutely decided to repair the damaged components and replace those that could not be repaired after a comprehensive and systematic investigation.

In the process of revision, the guidelines supplement the description of authenticity and integrity, emphasizing that the primary task of conservation work is to preserve the authenticity of historical sites, and the value of "character-defining features" must be clearly defined before the intervention, rather than blindly repairing for the sake of renovation.

Although the 1999 guidelines identified wooden heritage as an important part of the World's Cultural Heritage, they did not change their gradual disappearance from public view. In the face of the great pressure brought by economic globalization, people begin to rethink the historical value of traditional culture, which is embodied in the 2017 edition of the guidelines. It emphasizes that the object of protection (wooden heritage) is the witness of each historical era, and the value is to build a bridge between the public and the protection. Therefore, the protection behaviour should be considered by the local spiritual culture, to make it have regional and ethnic characteristics.

Re-examination of the Three Traditional Values. The Australian Charter of Barra once put forwards the concept of "cultural significance" and used this concept to expand traditional values such as historical value and artistic value. In the 1999 version of the guidelines, it was only pointed out that the basic principles of conservation should be practised based on attaching importance to heritage value.

In the IIWC conference held in Fallon, Sweden in 2016, Gennaro TAMPONE mentioned "value" again, and he believed that: the value of the ancient wooden structure is mainly historical, aesthetic, scientific, technological, anthropological, symbolic, etc., and the purpose of using the ancient structure for modern practical use is always problematic [3].

The 2017 code then explains the types of wooden heritage values and defines the scope of application. Cultural value is evolving with the change of The Times, it depends on people's cognition and social development level and other factors. In line with the American scholar Wd. LIPE's analysis of the value of cultural resources in the whole social system and conservation theory experts represented by David THROSBY, heritage value is divided into aesthetic, spiritual, social, historical, symbolic, and other development trends.

In addition to the above expansion of value types, the Venice Charter also discussed the historical value of historical sites, "they are the historical sites of people from generation to generation, full of information of the past years remained until now, and become the ancient living testimony of people" [4]. Emphasizes the protection of architectural relics and historical information. Genealogy Tempon suggests that old, load-bearing wooden heritages are specimens that have survived to the present, allowing us to investigate their history because they have a strong heritage and over time, the superior properties of wooden architecture will become more prominent [5].

2.2 Extension of the Protected Object Range

From Wood Structure to Wood Construction Technology. The protection of wood structures should carry forward the traditional culture. "Technologies also need to be preserved and brought back, especially those that are already lost," said Alejandro MARTINEZ at the 20th IIWC conference in 2016. The protection of wood construction technology is also a continuation of culture. Since the 20th century, the cause of heritage protection in Japan has developed rapidly and began to have systematic and scientific protection methods. In addition to revising traditional techniques into the cultural protection Law, educational institutions for traditional techniques, protection of traditional techniques, reproduction of historical artefacts, and subsidies for inheritors of traditional techniques have been organized (Fig. 3).

From Wooden Structure to Wooden Built Heritage. The 1999 guidelines offer protection for ancient wooden structures, which are of high cultural value or form part of a historic site. However other wooden structures in addition to this need not be protected?

At the 20th SESSION of IIWC, Maria de Guadalupe Zedepa MARTINEZ said, "What is more important in the current research work is the traditional characteristics of wooden buildings, while the value of handicrafts and the durability of temporary materials are not mentioned at all" [6]. Detachable and temporary timber structures also need attention, as they extend the traditional concept of wooden heritage. The revised code adds a new definition of "Timber Built Heritage" "It refers to all types of wooden buildings, as well as other wooden structures of cultural significance, including those that are temporary, mobile and continuously evolving."

The La Petatera bullring in Colima, for example, is restored or rebuilt every February and has a history of more than 150 years. Since the 17th century, the building has been the main venue for local intangible cultural heritage events such as bullfighting rituals, and it has grown from a palisade in the city centre to a building that can accommodate 5,000 people at a time. Therefore, in all respects, these detachable but valuable monuments need to be protected by clearer international principles.

2.3 Advances in Protection Methods and Tools

The Interaction between Technology and Theory. Blind adherence to contemporary theories and official protection documents may have an impact on practice. At the 15th IIWC conference in 2006, Gennaro Tampone mentioned that: due to progress in other fields, conservation technology precedes theoretical examples, and intervention measures that do not conform to conservation culture are usually the product of a lack of technical knowledge. Only by taking measures first can more advanced policies for active conservation be put forwards [7]. This paper discusses the relationship between conservation technology and theory by taking the seismic technology of historic buildings in Italy as an example.

A similar interaction occurred in the restoration process of the Monastery of SAN Francesco de Mazara in Italy, which was severely damaged by an earthquake in 1968. In 1988, experts commissioned by the provincial government of Trapani began to restore the monastery. The final maintenance plan adopted traditional metal rod technology and reinforced ring beams. Carbon fibre reinforced bandage (CFRP) technology was also used. Among them, a metal rod is a traditional Italian repair technology. Similar to it, the technical experience passed down after improvement is wall nails and buttresses. These methods of applying traditional techniques to historic buildings appeared in early Italian regulations, such as the Royal Decree No. 2600/1884. Later, a series of Italian documents issued in the 1990s further stated that "the preservation of historic buildings does not require strict compliance with seismic codes", For example, The Italian Ministry of Cultural and Environmental Heritage Circular No.1032/86 of 1986.

Traditional restoration codes, determined, selected and followed by experience and intuition, have been the basis of modern seismic building theory and legislation for decades. Only in this way can more advanced conservation policies be proposed, and the restoration work needs to conform to the local theoretical codes and practical cases that have been adopted for a long time.

New Protective Materials and Technologies. Due to the uncertainty of the factors affecting timber construction. Before a more concise method of repair is consolidated with spikes, wood, joinery, fasteners, or a combination of both are used to repair glue or reinforce the wooden components, (Fig. 4), and the type, quality, production with the original wood moisture and size of the same methods of the prosthesis are considered.

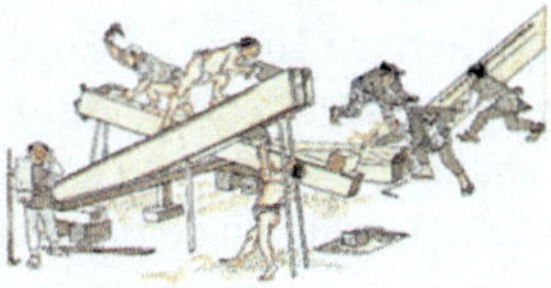

Fig. 3. Japanese traditional woodcraft scene

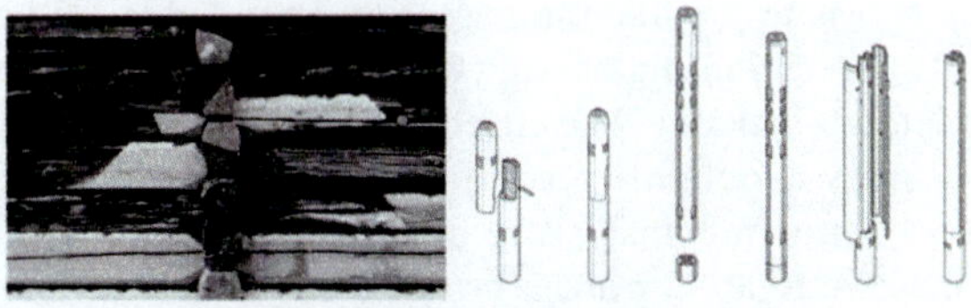

Fig. 4. Different sockets

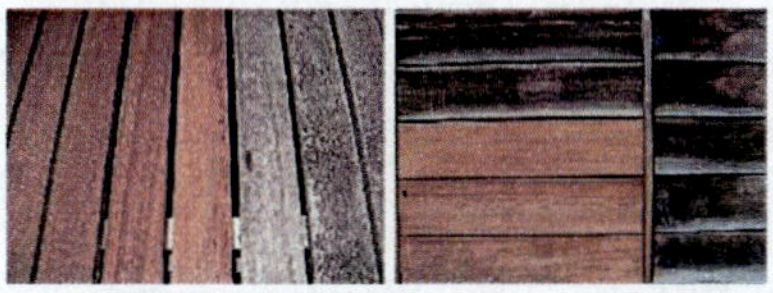

Fig. 5. Uv oxidation resistance of nano-silicate treated wood (left) versus untreated wood (right)

In addition to the traditional materials mentioned above, materials such as steel and carbon and glass fibre reinforced polymers (CFRP/GFRP) have also been popular in recent decades to improve the various properties of traditional wooden buildings. However, due to the poor air permeability of CFRP, the reinforcement effect of the CFRP material will gradually decrease in the later stage, so it needs to be replaced in time. Therefore, in addition to the use of FPR for the reinforcement of wood structures, some new stable chemical materials have also been applied to the repair of wood structures, such as polyethene glycol (PEG), organic montmorillonite (OMMT), hyper-branched polyacrylic acid and inorganic silicate nanomaterial fillers. In addition to the depth of materials, wood protection and detection technologies have emerged in an endless series, such as near-infrared spectroscopy, energy scattering X-ray fluorescence spectroscopy (EDXRF), scanning electron microscopy energy spectrum analysis (SEM-EDS), microscopic Raman spectroscopy (μ-Raman), microscopic Fourier transform infrared spectroscopy (μ-FTIR), etc. (Fig. 5).

2.4 Conservation Management and Social Participation

The Improvement of the Management and Recording Requirements of Timber Construction. As the types of protected objects become richer, the complexity of related collation work increases, which puts forwards higher requirements for management work. How to implement these interventions and management measures is not covered in the material compilation of the 1999 version of the guidelines, and some practices are not in line with the development of The Times.

Take the Jesuit neighbourhood and the House of Cordoba. It is an outstanding example of a religious, political, economic, legal and cultural system, all elements of which have retained their typological, spatial character and appearance of life. It was listed as a World Heritage Site in 1999. But even though its outstanding value has been recognized, a master plan management plan for the estate has yet to be approved, with only photos of damaged buildings showing the need for repairs.

Then, in 2014, ICOMOS issued Guidelines for On-site Assessment of Historic Timber Structures. The aim is to provide the necessary knowledge for historical analysis, structural safety and intervention engineering with minimal impact on the structure of the building. Specific methods include: 1) To clarify its historical heritage value through case analysis; 2) Visual survey to obtain a general structural form; 3) Survey and determine the main problems; 4) Structure analysis to determine the stress level; 5) Preliminary report, describing the investigation work to be done; 6) Detailed investigation, including biological and environmental research; 7) Diagnostic report on structural condition and cause of distress [8](Fig. 6).

The revised guidelines refine document protection, management and recording, and add factors of time, human and natural force majeure. It was also recognized that, with full documentation, valuable coverings on structures could be temporarily removed so that damaged structures could be easily detected.

Public Participation and Interdisciplinary Discussion. In 2014, the Florence Conference mentioned that the preservation of architectural heritage needs multidisciplinary participation to promote development. In 2016, the IIWC conference pointed out that we need to cooperate with other teams when conducting on-site evaluations of wooden heritage. Moreover, for a long time, official functions have been emphasized in the protection of cultural relics. Currently, all walks of life and social strata show great enthusiasm for the protection of cultural heritage, so the concept of heritage communities has emerged.

Giving full play to social power is the direction of the development of cultural heritage protection, and education is a core way to enhance awareness of wooden heritage protection. It is suggested in the 1999 edition that the value regeneration of historical timber heritage is the premise of sustainable heritage conservation, and relevant educational activities should be vigorously developed and publicized. The 2017 guidelines further emphasize the importance of education, while encouraging wider community participation and activities other than wooden heritage to promote sustainable development. Projects that contribute to an in-depth understanding of the salient features and social and anthropological aspects of wood built heritage should be encouraged. Recognize the relevance of community involvement in wooden heritage conservation and its role in sustainable development. Social participation can play a role in all aspects of conservation work, but the benefits should also be shared.

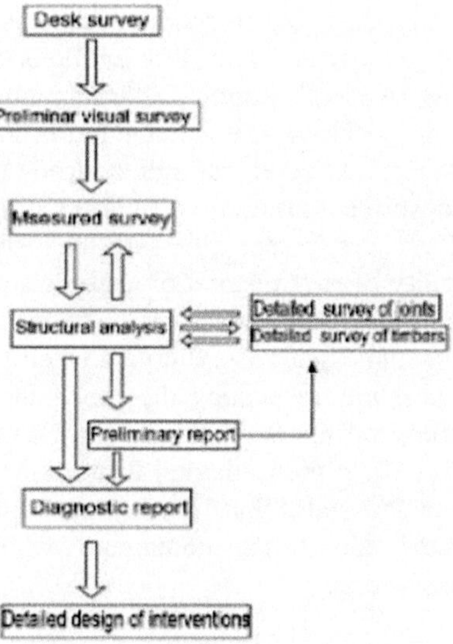

Fig. 6. Repair flow chart

3 The Development Trend of Wooden Heritage Protection and Its Influence on China

3.1 International Action on Wooden Heritage Protection After the Release of the 2017 Edition of the Code

At the 19th ICOMOS General Assembly held in Delhi, India on December 15, 2017, ICOMOS endorsed the new IIWC Wood Built Heritage Protection Guidelines, which replaced the 1999 guidelines.

On September 12, 2018, the 21st IIWC Conference was held in York, UK, aiming to promote the exchange of ideas and experience on the "new vision of protecting Wood Built Heritage" [9] The New Horizons workshop invited a diverse group of experts, professionals and practitioners to explore new disciplinary perspectives and potential related to conservation. The progress of laser and geospatial imaging and the application of new technologies such as digital microprobes are discussed. The recording, excavation, preservation and display of shipwrecks (marine heritage) and polar heritage, also consider whether traditional handicrafts such as woodworking can compete economically with modern building techniques. Whether it is the use of new technologies and materials or the practical application of new protection concepts, the problems that need to be faced have become the focus of this conference.

The 22nd IIWC conference held in Bilbao, Spain, in 2019 proposed "going beyond disciplines" and focused on a holistic approach to wooden heritage conservation. This approach is not limited to the structure but also focuses on the materiality of timber structures and their complex spiritual aspects, as well as the adaptability of local timber heritage, with multidisciplinary participation. It is hoped that professionals from different perspectives, disciplines and industries can conduct conservation actions with common goals, including exchange and discussion of different theories and processes, methods and intervention techniques for wooden heritage conservation.

On April 15, 2019, a fire at Notre Dame cathedral in Paris severely damaged the cathedral, causing the iconic spire to collapse and the lead-covered wooden roof above the vaulted ceiling to be destroyed, bringing to light the difficulties of preserving cultural heritage. The 22nd IIWC conference discussed the restoration of Notre Dame De Paris, including the durability of roof restoration materials and how to combine modern restoration methods with the maintenance methods of Notre Dame over the years. Association member Emmanuel. "Scientists should do their best to reconstruct as much of the old framework as possible, respecting the appearance of history and the need for existing wooden materials to form the present," suggests Emmanuel Maurin, one of the researchers. On July 16, 2019, the Cathedral Restoration Bill was approved by the French Parliament. The new law states that "restoration work must preserve the historical, artistic and architectural value of the monument", which embodies the values of international heritage conservation.

3.2 Comparison of Wooden Heritage Protection in China

In recent years, China's wooden heritage protection has also developed rapidly. The authenticity of heritage and the protection of the cultural heritage environment mentioned in the 2017 edition of the guidelines are highly consistent with the guidelines for the Protection of Cultural Relics and Historic Sites in China (revised Edition) and some ideas in the Beijing document. However, the combination of heritage protection and traditional culture in China has its characteristics, which are closely related to the existing cultural relics, architecture, tradition, cultural background and other elements in China, including the understanding of architectural form and architectural practices. Chinese people do not cling to the fragmented material level of authenticity but have their requirements on the major historical level, such as the form and system. The text contains the tao,

tao is more important than the implements, Tao resides in the implements, and Tao is integrated with the implements.

For example, the Beijing Document discusses authenticity as follows:

"The purpose of any repair and restoration shall be to preserve the authenticity of the source of such information intact" [10]. Historical information should be orderly and valuable, otherwise, there can be no information. It is a kind of intangible cultural heritage. In Chinese culture, intangible culture has broader significance and more lasting vitality.

According to the 2017 version of the guidelines, have such a request for authenticity: "keep the authenticity of historical heritage, including the layout structure, materials, assembly, integrity, architecture and cultural heritage value, respect history", [11] which for the value of the intangible cultural protection and authenticity of historical information, screening and transmission is consistent with China's protection theory. But the 2017 guidelines also include areas not covered by the Guidelines, such as the emphasis on the elements that define heritage characteristics and the protection of temporary, mobile and continuously evolving structures.

In terms of specific technical measures, the newly released Technical Standard for The Maintenance and Reinforcement of Wooden Structures in Ancient Buildings (GB/T50165-2020) added the monitoring of wooden structures in ancient buildings, checking the bearing capacity of wooden structures, safety appraisal of wooden structures in ancient buildings, and revised the seismic appraisal part of wooden structures in ancient buildings. The standard also covers interventions related to timber replacement and preservation of ancient buildings, which are generally in line with the 2017 guidelines.

In addition, the "standard" on the maintenance of the wooden structure, repair and reinforcement system from the material, frame type, column, beam fang, brackets and other different types made a distinction. The protection of wooden heritage is more localized in the Standard, which clearly divides the protection methods and levels, and different components also have their protection methods. It can be seen that China's wooden heritage protection not only follows the steps of the international community but also constantly improves in combination with China's unique cultural background, environment and other factors.

4 Conclusion

The 2017 edition of the guidelines enriches the contents of the 1999 edition of the guidelines based on the practical experience of international wooden heritage protection and relevant theoretical discussions. It has the following characteristics:

1) In the process of reshaping value cognition, it is aware of the lack of contemporary spiritual civilization and historical needs, respects historical changes, regards authenticity as a token to convey the historical context of architecture, and periodically reviews how authenticity is defined to adapt to the changing cognition and attitude of human beings.

2) Respecting different local traditions, construction practices and conservation methods, inheriting traditional crafts to establish the mark of The Times, while some other types of wood monuments and skills of different types and levels need diversified methods to protect them.
3) Can use modern technology to intervene in the appropriate place to minimize losses, ensure to keep as much as possible the authenticity and integrity, but should pay attention to the rigour of use, and the mechanical properties of interventions should not be changed, each about the history of the building components is meaningful, needs fully to understand the whole.

There are far more commonalities than differences in heritage protection between the East and the West. On the whole, the international cultural heritage protection system is inclusive, which has great overlap with China's conservation concept and conforms to China's cultural characteristics. However, it is not enough to guide China's wooden heritage protection only by relying on the theories of international literature. It should be expounded and dealt with based on China's characteristics. In connecting with the international conservation literature, our conservation theory should have the characteristics of traditional Chinese and realistic thoughts, establish the subject consciousness of self-value, emphasize the cultural characteristics of our own, and try to construct the heritage conservation path with Chinese characteristics.

The evolution of timber structure protection criteria is closely related to how people view them. The international heritage protection standards of western components are constantly absorbing practical experience from all over the world and adjusting their universality, while China has formed its own unique wooden structure culture due to its profound cultural background and complex influencing factors. Therefore, in the future, international norms need to carry out more topics such as the feelings of the subject, in reality, expand a broader space and add more multidimensional perspectives.

Acknowledgement. The work was supported in part by the National Key R&D Program of China (2021YFE0200100) and the 2021 Policy Directed Program of Jiangsu Province (BZ2021015).

References

1. ICOMOS: Principles for the Analysis, Conservation and Structural Restoration ofArchitectural Heritage. http://www.international.icomos.org/charters.htm.pdf. Accessed 15 Jan 2022
2. Landa, M., Ochandiano, A.: Towards a wooden heritage conservation theory in Spain. In:20th IIWC International Conference and Symposium, Falun, Sweden (2016)
3. Tampone, G.: Ancient timber structural systems: an attempt to define the main components of the paradigm. In: 20th IIWC International Conference and Symposium, Falun, Sweden, pp. 1–12 (2016)
4. Venice Charter (1964): Selected Documents of international Cultural Heritage Protection. Cultural Relics Publishing House, Beijing (2007)
5. Tampone, G.: Ancient timber structural systems: an attempt to define the main components of the paradigm. In: 20th IIWC International Conference and Symposium, Falun, Sweden (2016)

6. Martinez, M.G.Z.: The tradition and the updating of the ICOMOS principles for the preservation of historic timber structures: case: dismantling wooden monuments in Mexico. In: 20th IIWC International Conference and Symposium, Falun, Sweden, pp. 17–28 (2016)
7. Tampone, G., Messeri, B.: Compliance of the practice of strengthening ancient timber structure in seismic areas with official documents on conservation. In: Proceedings of the 15th International Symposium of IIWC-Why Save Historical Timber Structures, İstanbul (2006)
8. Cruz, H., Yeomans, D.: Guidelines for on-site assessment of historic timber structures. Int. J. Archit. Herit. **9**, 277–289 (2015)
9. New Vision of Protecting Wood Built Heritage. http://iiwc.icomos.org/papers.html. Accessed 01 Feb 2022
10. Beijing Document. Selected Documents of International Cultural Heritage Preservation. Cultural Relics Publishing House, Beijing (2007)
11. ICOMOS: Principles for the Analysis, Conservation and StructuralRestoration of Architectural Heritage (2003). http://www.international.icomos.org/charters.htm.pdf. Accessed 05 Mar 2022

Analytical Evaluation for Optimization of Landside Traffic System Planning at a Hub Airport

Hongjing Ge and Jiang He(✉)

College of Civil and Architectural Engineering, Guangxi University, Nanning 530004, China
20100081@gxu.edu.cn

Abstract. With the explosive growth of airport passenger and cargo traffic, the operational efficiency of airport ground transportation affects the overall development of the airport. It is important to correctly understand the operation of airport ground transportation networks, evaluate key road nodes, and identify bottlenecks in transportation development for the sustainable development of airport ground transportation in the future. This study takes the land traffic of a hub airport in China as the research object and conducts a simulation analysis of the land traffic of this airport. The operational performance of the landside traffic is discussed by analyzing the simulation results using VISSIM traffic simulation. Potential problems, such as traffic congestion at the entrance and exit of the parking lot, are identified. Suggestions for improvements are also given in this paper.

Keywords: Hub airport · Landside traffic · Simulation analysis · Evaluation

1 Introduction

Under the rapid development of air transportation, airport landside traffic closely related to the airport operation efficiency is also bearing new pressure. As the large-scale airport passenger transport hub connects various modes of transportation in the comprehensive transportation system, it shoulders important tasks such as passenger travel and smooth transfer [1]. Landside traffic at an airport involves multiple modes of transportation, and the streamlining of various means of transportation is relatively complex. Therefore, how correctly understanding the operation of the current airport landside traffic network and the evaluation of road critical nodes is a basis for sustainable development of airport landside traffic in the future. Since traffic simulation is non-reproducible and experimental, even if renewable and experimental, it needs to pay high costs and unknown risks. Thus, using traffic simulation software as a traffic planning and design tool for rationality verification has become a current mainstream. The traffic simulation technology is used to simulate and evaluate the current traffic operation, find traffic bottlenecks, and optimize the current traffic layout. The ultimate purpose is to realize efficient management of traffic rationalization, high efficiency, and intensification based on strengthening the comprehensive traffic hub at the airport.

W. Guo and K. Qian (Eds.): GBCESC 2022, LNCE 211, pp. 288–298, 2023.
https://doi.org/10.1007/978-981-19-5217-3_28

The landside of an airport is the working area relative to the airside. The landside transportation system of an airport provides transportation facilities for passengers or cargo arriving or departing from the airport, including the landside distribution system in front of the airport terminal, and parking facilities. Usually, road and rail transportation are essential components of landside transportation at hub airports in terms of transportation modes. Nowadays, most of the research on airport landside traffic has centered on the development of landside traffic systems, in terms of landside traffic simulation. However, for simulation, mostly the reports stay in simulating the operation of the current traffic conditions, lacking the study of post-simulation evaluation and improvement measures, and consequently cannot directly guide the application to the current airport traffic. This study focused on landside traffic operation performance and was discussed through the analysis of simulation results. Potential problems such as traffic congestion at the entrances and exits in the parking lot were found. Meanwhile, improvement suggestions are also given in this paper.

2 Description of the Target Airport

The target airport is located in southwest China and is one of the aviation hubs with passenger and cargo distribution functions in inland China. There are two terminals (T1, and T2) at this airport and landside traffic is mainly road traffic and rail traffic. The rail transit connects the airport with the urban south railway station and the urban east railway station as well as the surrounding cities. There are general roads and airport expressways to connect the airport and the urban areas. The airport expressways cover the outer ring roads around the city and connect the surrounding cities, to realize the road traffic route network radiating the whole city with the airport as the core. Its main transportation is in front of T2 Terminal, with various transportation modes such as taxi, online taxi, bus and so on. The two terminals are connected by ferrying electric vehicles.

3 Airport Landside Traffic Organization and Operation Scheme

Landside traffic at this airport refers to the traffic connecting the airport and the urban road transportation [2], mainly including the entrance and exit roads of the parking lot. Traffic organization on the land side discussed in this paper mainly refers to the traffic organization on the lane side around the entrance and exit roads of the parking lot in front of the terminals. The roadside in front of the terminal is responsible for passager transfer and the intersection of traffic flow so that the vehicles in and out of the airport can conveniently allow passengers on and off. The traffic organization in and out of the parking lot is the place where all kinds of vehicles enter the airport temporary park [3] and come together, resulting in a relatively complicated traffic situation at the entrance and exit of the parking lot. By setting a queuing detector at the entrance and exit, the traffic flow at the entrance can be monitored by detecting the queuing situation of vehicles in the parking lot. This can lay the foundation for the late reasonable planning of driving routes in the parking lot.

3.1 Raffic Organization Scheme for the Front Lanes of Terminal Buildings

The roadside in front of the terminal is the main traffic line in the terminal area, and it bears 60%–70% of the loading and unloading of passenger cars. The running streamlines of various vehicles are shown in Fig. 1.

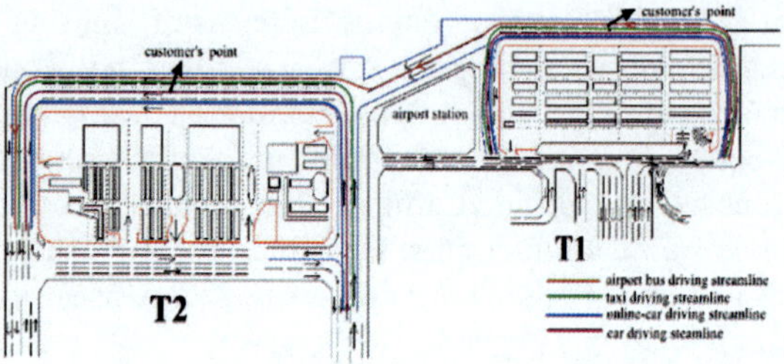

Fig. 1. Driving routes of various vehicles along the front lines of the terminals

3.2 Organization Plan of Traffic Roads Connected with the Parking Lot

According to the statistical data of the target airport over the years, many online vehicles are going to the airport parking lot. Therefore, the traffic organization plan of the parking lot designs different vehicle function zones in the parking lot according to the target airport and arranges specific online vehicle flow lines to reduce the traffic congestion caused by online vehicles, as illustrated in Figs. 2 and 3.

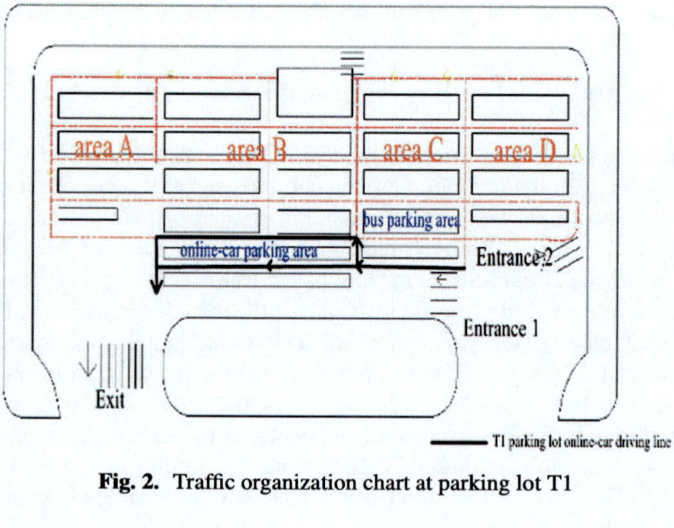

Fig. 2. Traffic organization chart at parking lot T1

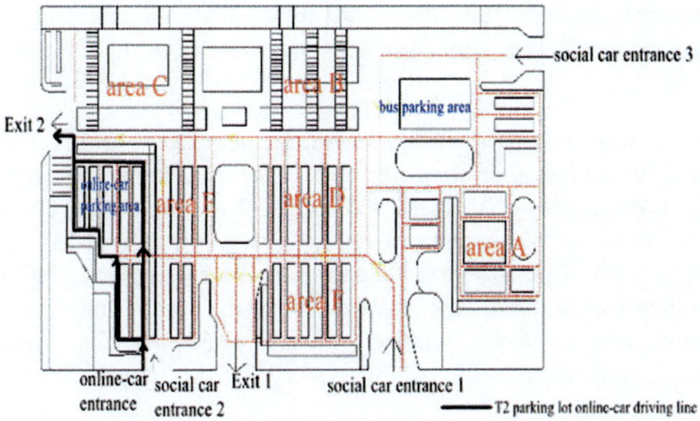

Fig. 3. Traffic organization chart at parking lot T2

4 Simulation Analysis of Landside Traffic Organization

4.1 VISSIM Traffic Simulation Software

VISSIM is a simulation program for studying traffic information flow. It is also a micro-simulation modeling tool based on the time interval and driving behavior [4], which is often used to analyze urban traffic and other traffic. It can evaluate current traffic situations under various conditions, being an effective tool for evaluating traffic design and urban planning schemes [5]. This study uses VISSIM to conduct simulation modeling of the lane edge and the entrance and exit of the parking lot in front of the terminal at the target airport and analyzes traffic conditions on the landside of the target airport.

4.2 Establishment and Analysis of Landside Traffic Simulation Model

Establishment of the Road Network. In the micro traffic simulation, the establishment of the road network is the basis of simulation and it is necessary for outputting

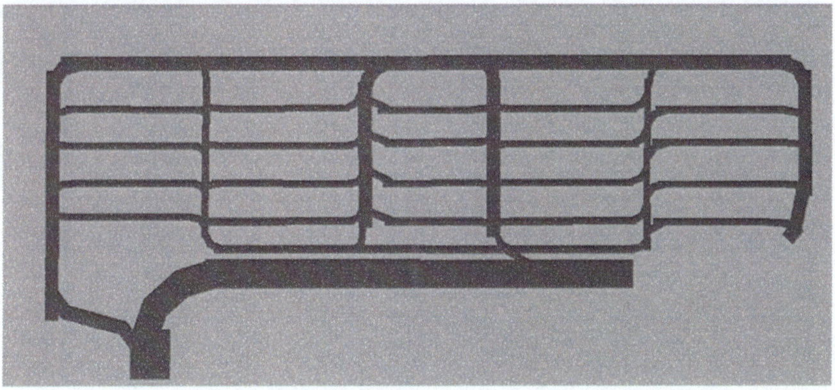

Fig. 4. Road network of the parking lot at T2

various evaluation indicators. The subsequent simulation modeling can be completed by establishing a road network, e.g. the road network of the T2 parking lot can be designed as shown in Fig. 4.

Traffic Flow Parameter Setting and Behavior Rules. Micro traffic flow parameter setting is mainly to set behavior characteristics of vehicles in the road network such as acceleration and deceleration. VISSIM adopts the physiological-psychological model established by Professor Wiedemann in 1974, i.e. when the car driver thinks that the distance from the car in front is shorter than the safe distance (psychological), the driver starts to slow down, until the distance between two cars reaches another safe distance; the driver then starts to accelerate, and this process is repeated [4]. The simulation in this study used the default micro traffic parameters in VISSIM, namely The Wiedemann74 model.

Macro traffic flow parameters include vehicle composition, route selection, etc. According to the statistics at the target airport, the vehicles can be classified as social vehicles, rental vehicles, online ride-hailing vehicles, shuttle buses, and airport buses, and their proportions are 25%, 30%, 5%, and 5% respectively. The number of input vehicles at the lane edge in front of each terminal during rush hours can be obtained from the revised calculation based on the passenger numbers for each vehicle type, i.e. 1.5 for social vehicles, 1.5 for taxis, 1.5 for online ride-hailing vehicles, and 30 for large vehicles, as indicated in Table 1. Similarly, according to the relevant guidance documents for the airport planning and design, the traffic input of vehicles in the parking lot is 1:1, as shown in Table 2.

Table 1. Number of vehicles in each mode at each terminal during rush hours

Terminal	Social car	Taxi	Shuttle bus	Airport bus	Online car
T1 (32.58%)	558	447	6	6	1009
T2 (64.2%)	1003	801	10	10	1809
Total	1561	1248	16	16	2818

Table 2. Number of vehicles entering and leaving the parking lot

The proportion of inbound and outbound	1:1	
	Enter the port	clearance
Social-car	1102	1102
Taxi	588	588
Bus	14	14
Online-car	1312	1312

Performance Effect. The number of vehicles is input into the simulation model for dynamic simulation, as shown in Fig. 5. Then evaluation indexes are determined and their simulation results are output for later evaluation and optimization.

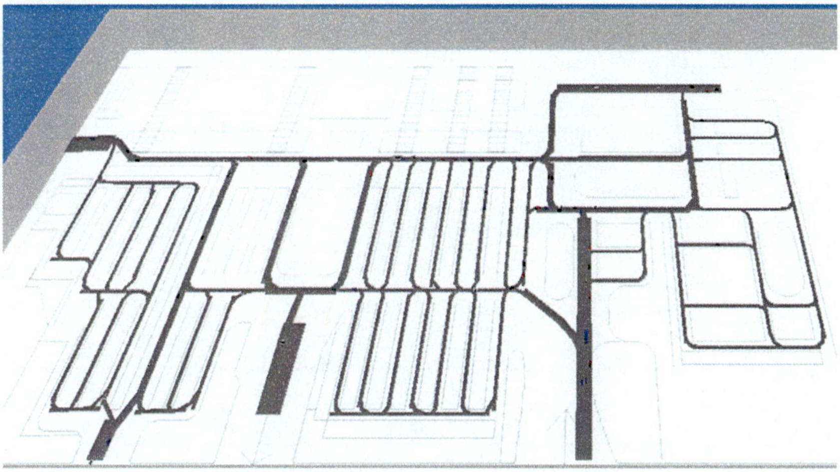

Fig. 5. Schematic diagram of the simulation model

Output and Analysis of Simulation Results

Simulation Results for Entrance and Exit Roads. Travel time (including parking and waiting time) is a simple and reliable indicator for evaluating network performance, while delay refers to the time lost by vehicles from the starting point to the endpoint related to free-running traffic [5]. The trip time and average delay time at the road section can be detected by setting the trip time detector. Therefore, the trip time and delay are considered together in this study, and the trip time and average delay are selected as the indicators to evaluate the current lane edge system. The output results are shown in Tables 3, 4, 5 and 6.

Table 3. Traffic operation on the departure floor of T1 during peak hours

Evaluation index	Social car	Taxi	Bus	Online car
Average delay (s)	6.8	20.3	5.4	7.0
Average queue length (m)	51	142	0	50

Table 4. Traffic operation on the departure floor in the T2 station building during peak hours

Evaluation index	Social-car	Taxi	Bus	Online car
Average delay (s)	7.9	24.2	6.7	8.1
Average queue length (m)	63	156	0	68

Table 5. Traffic operation on the arrival floor in the terminal during peak hours

Terminal		Taxi	Bus
T1	Average delay (s)	8.1	0
T2		19.3	0

Table 6. Output list of lane side travel time

Time	T1 lane side	T2 lane side
30–330	50.2	60.5
330–630	49.8	62.1
630–930	51.2	62.3
930–1230	50.0	62.9
1230–1530	51.0	61.5
1530–1830	49.5	71.8
1830–2130	51.8	68.2
2130–2430	49.7	65.3
2430–2730	51.3	66.4
2730–3030	50.0	66.3
3030–3330	50.3	66.4
3330–3600	50.4	66.5

It was found that the average delay time for taxis and online cars is longer than for other vehicles. This is because taxis always are required to drive to the taxi waiting area for the next passengers after dropping passengers at the airport, which causes the average delay time and queuing length longer. Online ride-hailing is a popular travel mode at present, with a large number and a lack of designated drop-off points on the departure floor, resulting in a small amount of queuing. Nevertheless, such queuing does not increase over time, and congestion hardly occurs. Combined with an analysis of the simulated travel time, it can be found that about one minute is needed for driving a vehicle to the lane side in front of the terminal to complete the loading and unloading.

In addition, using the saturation index to evaluate the service level of the road can better reveal the relationship between design speed and road traffic flow to a certain extent. Usually, saturation is obtained using the ratio of the traffic volume in peak hours to the design capacity of the road. In this paper, the design speed and traffic capacity are referred to in the "Capacity Manual" of the United States (see Table 7), and the road capacity is referred to in the "Code of Urban Road Design" (see Tables 8 and 9).

Table 7. Theoretical capacity of single lane

Design speed (km/h)	20	30	40	50	60
Traffic capacity	1380	1550	1640	1690	1730

Table 8. Classification index of road service level

Level	Level 1	Level 2	Level 3	Level 4	Level 5	Level 6
V/C	≤0.33	0.34–0.5	0.51–0.65	0.66–0.8	0.8–1.0	>1.0
Running status	Smooth and orderly	Slight delay	Acceptable delay	Bearable delay	Slightly crowded	Crowded

Table 9. Analysis of lane service levels

Road	Traffic (pcu/h)	Design speed (km/h)	Number of lanes	Saturation	Service level
T1 lane side	2026	30	6	0.22	Level 1
T2 lane side	3633	30	6	0.39	Level 2

According to the analysis and evaluation results, the service level of lane side roads in front of the T2 terminal is Level 2, and the operation is slightly delayed. Combined with the average delay for various vehicles, the airport should strengthen vehicle management in front of the T2 terminal. Although there are no major problems in the simulated results and the number of lanes can also bear the traffic flow at the airport. Future measures are still needed for the sustainable development of airport traffic.

Simulation Results for the Roads at the Parking Lot. The simulation for the roads at the parking lot focused on the maximum queue length of each entrance and exit at the parking lot. The maximum queue length was used to reflect the traffic situation at the entrance and exit of the parking lot. The simulated results are shown in Tables 10 and 11.

Table 10. Maximum parking line queue length (m) at T1 parking lot

Time	Entrance 1	Entrance 2	Exit
30–330	0	0	0
330–630	0	0	0
630–930	3.1	4.0	0
930–1230	3.2	0	0
1230–1530	0	0	0
1530–1830	0	3.5	0
1830–2130	3.5	0	0
2130–2430	0	0	0
2430–2730	0	0	0
2730–3030	2.9	0	0
3030–3330	0	0	0
3330–3600	0	0	0

Table 11. Maximum parking line queue length (m) at T2 Parking lot

Time	Entrance 1	Entrance 2	Entrance 3	Exit 1	Exit 2
30–330	0	0	8.2	4.9	4.2
330–630	0	0	7.3	5	4.3
630–930	0	0	6.9	6.2	4.6
930–1230	0	0	7.6	5.5	4.2
1230–1530	1	0	8.0	4.8	4.7
1530–1830	0	1	7.5	5	4.6
1830–2130	1	1	6.7	6.1	5.1
2130–2430	0	0	8.1	5.7	5.3
2430–2730	0	1	7.8	5.1	4.5
2730–3030	1	0	7.5	4.9	4.9
3030–3330	0	0	8.0	4.8	4.6
3330–3600	0	0	7.8	4.5	4.2

From comparison analysis of two parking queue lengths, it can be found that there are a few times in the queue at the T1 parking lot, and the maximum queue length does not increase with time, which suggests that such queue line with a short time would not cause a great influence on the vehicles into the parking lot. The reason partly results from only one exit at the T1 parking lot. For this reason, the vehicles can only leave the parking lot through one exit, which to some extent causes vehicles to queue at the exit. At the T2 parking lot, compared with Entry 1 and 2, the queue length for Entrance 3 is much longer. A major reason is that there are parking areas for online cars near Entrance 3, where the specific flow lines are designed for online cars. This temporary queuing does not cause a great influence on the subsequent traffic, within a reasonable range.

4.3 Suggestions for Optimization of Future Transportation Organization

According to the simulation results and road traffic service level evaluation described above, a major traffic flow in the airport terminal area should mostly concentrate on the terminal front driveway. Since the elevated feeder system has only two lanes, with the development of future aviation business, this would affect the landside traffic efficiency to a certain extent. Therefore, reasonable traffic organization planning for different vehicle types in front of the terminal and strengthening the channelization function of roads are conducive to efficiency improvement of landside traffic operation. For this reason, planning suggestions can be summarized below:

Reasonable planning should be prepared for different vehicles in front of the terminal. Because online cars are a cause of traffic congestion at the driveway edge and entrance/exit, the driving lines for online cars should be designed independently from social vehicles. With the growth of future aviation business, online cars will share more airport landside traffic. In landside traffic reconstruction at the airport, online-car driving lanes and bus parking zones can be designed in front of the terminal building, where traffic management should be further strengthened and each car must abide by the mandatory provision of "stop and go" to avoid disorderly vehicle driving and congestion.
Functional zoning should be strengthened in the parking lots. As found in the simulation results, traffic congestion is serious at the entrances and exits of the T2 parking lot. To address this issue, parking spaces for online cars should be increased in future renewal planning. In addition, parking areas for shared cars and battery-charging stations for electric vehicles should be constructed in the subsequent airport landside traffic planning to strengthen the function of the airport parking lot. Furthermore, automatic charging systems should be popularized at the entrances and exits of the parking lot to speed up vehicle circulation.

5 Conclusion and Shortcomings

After the airport is put into operation, land transportation must keep pace with the development of the airport. Therefore, to improve the airport infrastructure and give full play to the airport's function as an integrated transportation hub, it is necessary to evaluate whether landside traffic can still undertake the airport's operation. This paper discusses the landside traffic of the airport using a traffic simulation tool (VISSIM). By analyzing the travel time of roadside vehicles in front of the terminal building and the queue length of each parking lot, it is found that the landside traffic design scheme of the target airport can still meet the basic demand of traffic volume. With the development of the aviation business and the enhancement of airport hub functions, the landside traffic at the target airport may become overburdened, especially in the case of high parking demand. The current parking lots, such as T2, are no longer able to meet the future parking demand for all types of vehicles. To clarify this point, further investigation is necessary.

In the article, online cars are considered a popular travel tool in the future and are used as a specific object for the simulation analysis. By analyzing the average delay time and the performance of internet-hailed vehicles on the curb of the terminal limb, future demand can be predicted to guide the vehicle management and the functional zoning of the parking lot, thus strengthening the integrated transportation function of the airport. At the same time, the airport is a large integrated transportation hub, it is not enough to consider vehicles as a factor affecting the future development of airport ground transportation. Passengers also play an important role in the development of airport ground transportation. For example, pedestrians can influence the switching area per capita or the departure frequency of public transport. Therefore, as our future research topic, we will use Anylogic or Legion to combine pedestrian flow with traffic flow to fully show the airport landside traffic operation. In addition, traffic engineers can use econometric methods, trend extrapolation methods, and tourism development index methods to effectively and reasonably forecast airport traffic in the future, as well as to improve the public transportation sharing rate and optimize the frequency and grouping of public transportation in future planning, so as to reduce the congestion of ground roads.

References

1. Shang, L.: Planning and design of Xiaoshan airport traffic center based on simulation technology. Traffic Transp. **36**(02), 6–10 (2020)
2. Yaofeng, X.: Characteristics and problems analysis of landside traffic organization of hub airport in China. Huazhong Archit. **30**(6), 69–73 (2012)
3. Yuan, L., Chao, W., Han, J.:Beijing Daxing airport landside traffic planning and design of simulation assessment. In: The 14th Chinese Wisdom Can Transport Conference Proceedings, pp. 707–716 (2019). https://doi.org/10.26914/Arthurc.nkihy.p.013402(2019)
4. Zhibiao, W.: Research on Applicability of VISSIM Traffic Simulation. Jiaotong University, Beijing (2007)
5. Dong, L., Xiaokuan, Y., Chao, G.: VISSIM-based simulation analysis on road network of CBD in Beijing, China. Procedia. Soc. Behav. Sci. **96**, 461–472 (2013)

Study on the Protection and Development of Farming-Type Traditional Village Landscape Under the Concept of Sustainability – A Case Study of Western Henan

Sai Yin(✉)

Henan University, Kaifeng, China
bigyinsai@outlook.com

Abstract. For traditional village landscapes, sustainability can be seen as a complex system consisting of four dimensions: environmental, social, economic, and cultural. Based on the concept of sustainability, this paper takes farming-type traditional villages in western Henan as an example. By evaluating resource characteristics and distribution, conservation, and utilization of 11 typical traditional village samples, we summarize the resource characteristics of the villages and the sustainability and shortcomings of the existing development model. At the same time, through integrating the survey results, we propose strategies for the renewal of western Henan farming-type village landscapes in three aspects: environmental sustainability, economic and social sustainability, and cultural sustainability.

Keywords: Concept of sustainability · Farming-type traditional villages · Update strategy

1 Research Background

The traditional village is a special category of the rural landscape. From the perspective of heritage conservation, it belongs to the subcategory of "cultural landscape", "rural landscape" in the subcategory of "continuous landscape". It has the characteristics of "settlement landscape" at the same time: On the one hand, the traditional village, like other rural landscapes, consists of human, agricultural and natural landscapes [1], is closely linked to the vernacular culture and traditional way of life [2], and is formed through a long process of evolution of the rural geographical natural and socio-cultural environment [3]. On the other hand, compared with other rural landscapes, traditional village landscape has higher regional characteristics and historical and cultural connotations and has more obvious advantages of the natural landscape and humanistic landscape [4].

The sustainability of traditional villages has become a hot topic of discussion in the field of heritage conservation in recent years. This is because, for the traditional village, it is a "living heritage". In the conservation of a village, the tangible traditional architecture and intangible traditional intangible culture with Outstanding Universal

W. Guo and K. Qian (Eds.): GBCESC 2022, LNCE 211, pp. 299–310, 2023.
https://doi.org/10.1007/978-981-19-5217-3_29

Value are important. However, it is more important to integrate cultural and natural conservation into the economic and social development of the village and to achieve sustainable development of the village for as long as possible. In its Guidelines on the Heritage of Rural Landscapes, adopted in 2017, the International Council on Monuments and Sites - International Union of Landscape Architects Scientific Committee on Cultural Landscapes (ICOMOS - ISSCL) has highlighted the multiple values of rural landscapes in the conservation of biodiversity, the maintenance of land and food security, and the transmission of cultural traditions [5]. The sustainability of traditional village landscapes can be seen as a complex system consisting of four dimensions: environmental, social, economic, and cultural (Fig. 1). It encompasses the sustainability of material systems such as natural resources, land and agriculture, industry and economy, architecture and public space, as well as the sustainability of non-material systems such as regional culture, customs, and morals, etc., which should be fully considered. In addition, the decisive role of community groups in the conservation of cultural heritage and in the manifestation of heritage values must be fully recognized [6].

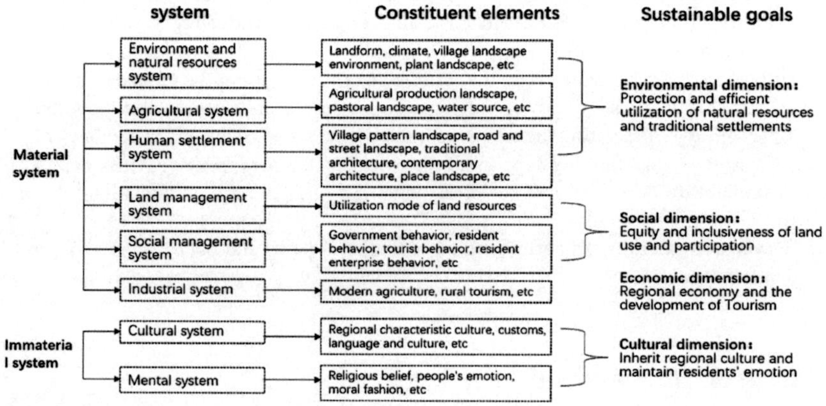

Fig. 1. System of sustainable development of traditional village landscape and its components (self-drawn)

Based on the concept of sustainability of village landscape heritage, this paper takes farming-type traditional villages in western Henan as the research object, hoping to sort out the landscape characteristics of traditional villages and the shortcomings of the existing renewal model on the basis of in-depth research. Based on in-depth research, we hope to propose feasible strategies for the conservation and development of traditional villages on this basis.

2 Farming-Type Traditional Villages in Western Henan Landscape Features and Existing Renewal Evaluation

2.1 Overview of Farming-Type Traditional Villages in Western Henan

Western Henan contains the western areas of Sanmenxia, Luoyang, and Zhengzhou in Henan Province, including the administrative area of 25 counties and districts in 3 cities. Most of the landforms are mountainous and loess hilly areas. The Yi and Luo rivers cross the area and merge into the Yellow River, forming a vastly different landscape of loess ravines, mountains, and plains. Since ancient times, there has been a saying that "five mountains, four hills, and one river." Since the Neolithic period, western Henan has been the main habitat of human beings in the Yellow River basin and has been the core of the Central Plains cultural circle. The unique natural environment of the region and the influence of human factors have formed the unique regional habitat of western Henan. At present, there are 47 national-level traditional villages and 133 provincial-level traditional villages in the region, among which the number of national-level traditional villages accounts for 1/4 of the total number of selected villages in Henan Province, and it is also one of the most densely populated areas of traditional villages in Henan Province. The current classification of western Henan traditional village types, can be divided into farming-type, commercial-type, and defensive-type based on functions. The term "farming-type traditional villages" in this study is based on the function of villages and refers to those villages where the primary function is the human settlement and agricultural production.

2.2 Research Methods and Sample Selection

This study is based on a qualitative research method, with the aim of sorting out the resource characteristics of farming-type traditional villages in the western Henan landscape and the inheritance and utilization of the traditional style in the village renovation. The specific research process is as follows: (1) Through literature review, fieldwork, and browsing relevant website materials, a number of cases of farming-type villages were selected as samples for specific research and analysis. (2) The characteristics and advantageous resources of each village were summarized, and the richness of their influence factors was graded and marked. (3) The villages were categorized into different types according to the conservation and utilization patterns of the sample villages. The performance of the 11 villages in the four dimensions of environment, society, economy, and culture in heritage renewal was evaluated using the evaluation criteria of sustainable development. The main characteristics of each renewal pattern and the problems existed were summarized.

The following principles were followed in the selection of the case sample: 1) Selected from national and provincial traditional village lists to ensure the typicality of the villages. 2) Selected traditional villages that have farming as their initial function in their own resource type and have a certain degree of originality and integrity. (3) The traditional villages in western Henan that are better developed in terms of conservation and utilization, higher in the number of tourists, search popularity, and government support, and can represent the existing western Henan cultural characteristics and development direction was selected. Finally, from the matched cases, Yangjiamen Village, Renmazhai Village, Shiba Village, Youdian Village, Houshangzhuang Village, Beiying Village, Dawangmiao Village, Haiqiao Village, Miaosang Village, Weipo Village, and Siposhan Village, a total of 11 villages, were selected as the scope of the study (Fig. 2).

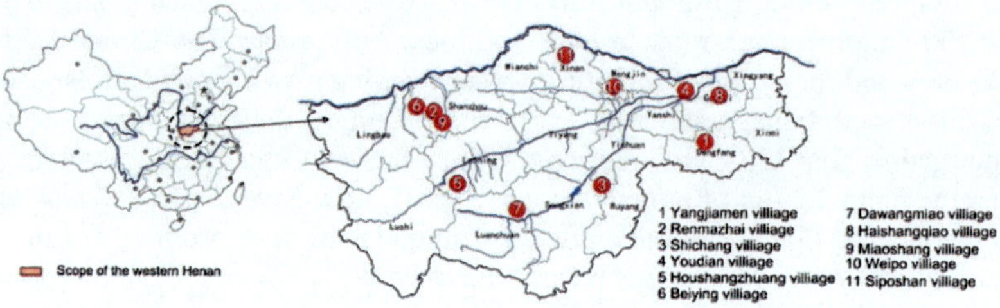

Fig. 2. Distribution of traditional village samples (self-drawn)

2.3 Analysis of Landscape Features and Advantageous Resources

The landscape of farming-type traditional villages includes both material heritage and intangible heritage, and the material heritage can be divided into natural resources, built heritage resources, and agricultural resources. The farming-type traditional villages in western Henan landscape material heritage was subdivided into five factors expressing resource richness. The richness of the intangible heritage of the village landscape is divided into three factors. On this basis, each factor in the factor layer of the sample resources of the 11 villages was graded in terms of richness, with "I", "II" and "III" indicating three grades of poor, average and good richness, and factors without relevant resources were not marked, as shown in Table 1.

Table 1. Resource characteristics and distribution of typical farming-type traditional villages in western Henan landscape (self-made)

Village name	Material heritage					Intangible heritage		
	Natural resources	Built heritage resources			Agricultural resources			
		Age	Quantity	Integrity		Folk culture	Agricultural culture	Historical culture
Yangjiamen village	II	I	I	II		I		I
Renmazhai village	I	III	I	II	II	III		III
Shichang village	III	II	II	III	II			I

(continued)

Table 1. (*continued*)

Village name	Material heritage					Intangible heritage		
	Natural resources	Built heritage resources			Agricultural resources			
		Age	Quantity	Integrity		Folk culture	Agricultural culture	Historical culture
Youdian village	III	II	III	III	III		I	III
Houshangzhuang village	II	III	II	III	III		II	II
Beiying village	III	II	I	II		I		I
Dawangmiao village	III	III	II	II	III	I	II	III
Haishangqiao village	III	II	III	III	II	I	I	II
Miaoshang village	III	III	II	II	II	II		II
Weipo village	II	III	III	III	I		III	III
Siposhan villiage	III	II	II	II	II			I

As can be seen from Table 1: 1) In terms of natural resources, the landscape of western Henan is rich and diverse, especially the yellow landform area is very rich in landforms. 2) In terms of built heritage resources, the remaining buildings in the villages are very rich in terms of age, quantity, and integrity. 3) In terms of agricultural resources, the western Henan traditional village farmland resources are mainly reflected in the various farmland forms shaped by man in interaction with the natural landscape. 4)In terms of intangible heritage, almost every village has a rich history and culture, including historical legends, historical events, and famous people, etc. Some villages also retain folk culture, and agricultural culture. In general, traditional agricultural villages in western Henan are characterized mainly by their built heritage, followed by topography, farmland resources, and historical, folk cultures.

2.4 Conservation and Utilization Model of Existing Landscape and Sustainability Evaluation

In recent years, as the protection and revitalization of traditional national villages and traditional provincial villages have become national policies, almost every traditional village has undergone a few rounds of protection and renovation and village renewal attempts. Some villages have also become "online celebrity" attractions through the transformation of tourism. As the research samples, all of these 11 traditions villages have also undergone conservation and renewal efforts. According to the different resource conditions and different starting points of utilization, the conservation and utilization patterns of village landscapes can be summarized into five different types, namely, heritage exhibition type, natural tourism type, folklore experience type, integrated scenic area type, and cultural interpretation type. In the following, the sustainability of each village style is evaluated by introducing the four dimensions of environment, society, economy, and culture in the theory of sustainable development and eight evaluation factors. At the same time, the conservation and utilization methods of each renewal mode are summarized (Table 2). The evaluation table used "√" and "◊" to indicate whether sustainability was achieved.

It can be seen from Table 2 that heritage exhibition type, nature tourism type, and folklore experience type are sustainable in development, but they are also easy to cause the problems of insufficient highlights and attraction. For example, Haishangqiao Village, Houshangzhuang Village and Youdian Village retain a complete architectural community of the Ming and Qing Dynasties. Shichang Village and Siposhan Village, as mountain villages, have unique residential forms. They are mainly updated with architectural repair, display and street style, but they can only attract some tourists interested in traditional architecture; As the village with the richest intangible cultural heritage in Sanmenxia area, Renmazhai Village has Chengni inkstone, printing techniques, paper cutting, clay sculpture, and the art of building a Di Keng Yuan, etc. at present, there are many folk custom exhibition halls, but there is a lack of in-depth display and product development of intangible culture.

Through comprehensive and large-scale development, the integrated scenic area type has driven the local employment and industrial transformation. For example, the Di Keng Yuan in Miao Shang Village is a national cultural preservation unit. Part of the core village is concentrated and protected as a scenic spot and combined with folk culture for display. Yangjiamen Village, with the core of a boutique lodging, converted from traditional buildings, integrates field tourism, folklore experience, and Buddhist Zen martial culture. Beiying Village, based on the prototype of Di Keng Yuan, integrates accommodation, sightseeing, experience, and outdoor adventure. Weipo Village has the highest standard government-style courtyard complex in western Henan, around which several commercial clusters have been developed, forming a landscape similar to a tourist complex. However, scenic development brings damage to the village landscape pattern, street pattern, and even the integrity and originality of traditional architecture.

The cultural interpretation type is to take the most valuable part of the village resources for tourism development as the leading resource and to reconstruct the village spatially by means of spatial narrative. For example, Dawangmiao Village is a typical mountainous farming village. Still, in the renewal, the event that Henan University ran a school in exile here during the Republican period is taken as the center of the village narrative. The disadvantage of this model is that it highlights a certain culture too much and inadvertently weakens the characteristics of other natural and human resources.

Table 2. Conservation and utilization patterns and evaluation of typical farming-type traditional villages in western Henan landscape (self-made)

Update mode	Village name	Sustainability evaluation							
		Environment			Social		Economy	Culture	
		Nature	Agriculture	Habitat for humanity	Land	Social		Culture	Spirituality
Heritage exhibition type	Haishangqiao Village	√	√	√	√	√	√	√	√
	Houshangzhuang Village	√	√	√	√	√	√	√	√
	Youdian Village	√	√	√	√	√	√	√	√
	Protection mode: Repairing key ancient buildings, protecting the original texture of the village, and improving the image along the street **Utilization method:** Part of the courtyard as an exhibition space for farming culture								
Nature tourism type	Shichang Village	√	√	√	√	√	√	√	√
	Siposhan Village	√	√	√	√	√	√	√	√
	Protection mode: Repairing key ancient buildings, setting up visitor service centers, and upgrading the infrastructure along the tour route **Utilization method:** Renovate part of the courtyard as a cultural exhibition hall and B&B, and establish an overall tour of the village								
Folklore experience type	Renmazhai Village	√	√	√	√	√	√	√	√
	Protection mode: Repairing key ancient buildings, improving the image along the street, establishing a folk culture museum and folk culture exhibition hall **Utilization method:** Using the folklore museum and cultural exhibition hall to display a variety of non-traditional items in the village								
Integrated scenic area type	Miaoshang Village	√	◊	◊	◊	√	√	√	√

(*continued*)

Table 2. (*continued*)

Update mode	Village name	Sustainability evaluation							
		Environment			Social		Economy	Culture	
		Nature	Agriculture	Habitat for humanity	Land	Social		Culture	Spirituality
	Yangjiamen Village	√	◊	√	◊	√	√	√	◊
	Beiying Village	◊	◊	◊	◊	√	√	◊	◊
	Weipo Village	√	√	◊	◊	√	√	√	◊
	Protection mode: Comprehensive restoration of ancient buildings, partial reconstruction, and new construction setting up core protection areas and coordination areas **Utilization method:** According to the needs of tourism, the core protection is used for cultural display, other areas are used for commercial development and tourism support, and the participation and experience of the village's characteristic culture and folk culture is the focus								
Cultural interpretation type	Dawangmiao Village	√	√	√	√	√	√	◊	√
	Protection mode: Repairing key ancient buildings and upgrading the environment along streets and waterways **Utilization method:** Highlighting the "theme culture" in the spatial image and architectural display of the village								

3 Update Strategy of Farming-Type Traditional Villages in Western Henan Under the Concept of Sustainability

From the above survey of typical village samples, it can be seen that farming-type traditional villages in western Henan have rich natural resources, architectural heritage resources, agricultural resources, and cultural resources, which have shaped the village with distinctive regional characteristics but distinctive landscape. There have been many attempts to preserve and utilize the village landscape, and different models have been developed. Based on the development concept of landscape sustainability, the following is a strategy for regenerating the landscape of western Henan farming-type villages in terms of environmental sustainability, economic and social sustainability, and cultural sustainability, taking into account the useful experiences and shortcomings in the regeneration of existing villages.

3.1 Environmental Sustainability

Environmental sustainability includes the sustainability of natural ecology in the material system, the sustainability of agriculture, and the sustainability of human settlement system, and the following three strategies are proposed in terms of ecological protection and utilization principles, development of agriculture, and control of village renewal.

1) The construction of ecological protection circles and landscape security pattern priority renewal approach. Western Henan is mostly in the Yellow River Basin yellow landform area. The ecological environment is very fragile. The construction should be based on the village landscape area characteristics to determine the regional sensitivity, divided into different regional protection circles. In addition, the relationship between natural and artificial elements in the long-term development of traditional villages constitutes the landscape safety pattern of the village itself. In the renewal, it should take the landscape safety pattern as the skeleton of the village, avoiding the way of overall demolition or piecemeal protection. Weipo Village and Miaoshang Village are two opposite cases, the former retaining only the core of the village buildings and rebuilding the rest, while the latter enclosing part of the overall village courtyard as a scenic display, both of which destroy the village landscape security pattern.
2) Encourage agricultural development and increase the development and utilization of agricultural resources. For farming-type villages, agriculture is the root of village development, encouraging agricultural development is an important basis for preserving traditional culture and maintaining a dynamic balance of human-land relationship in the long term. Agriculture is still an important resource. Western Henan is rich in agricultural resources. Still, in the actual utilization, there are few other ways to use it except for the display of farming tools in the form of a farming museum. To enrich the local agricultural culture, it should also try different forms such as sightseeing agriculture, fruit tree picking, and ecological technology agriculture.

3) Strategy of stratification and classification in village protection and regeneration. At the level of village pattern protection, the protected area is divided into core protection zone, construction control zone, and environmental coordination zone, and a categorized control is implemented for the renewal of buildings and public space in the area. When dealing with important village historical elements (architectural elements and environmental elements), it is not enough to refer only to the method of regional control but to systematically evaluate their value and apply the evaluation results in the judgment of the degree and direction of the renewal of traditional village historical elements. The British "traffic light model" overlay analysis method can be used to propose preservation, repair, strengthening, innovation, and other renewal methods for different historical elements.

3.2 Economic and Social Sustainability

Sustainable economic development is the premise on which the development of traditional village conservation depends, while social sustainability involves fairness and inclusiveness of all parties in land use. The following strategies are proposed from three aspects: the choice of conservation and development model, the development of special industries, and the multi-party construction of the village.

1) The selection of the conservation and development model of traditional village development is a comprehensive weighing process: Firstly, it is necessary to consider the resource situation.Secondly, we should consider the location conditions. The integrated scenic spot type can be chosen for traditional villages with multiple resources, close to the city, and surrounded by scenic spots, which can be guaranteed in terms of the number of visitors. In addition to this, it is important to consider whether the chosen model can complement other villages within the county or town to build a rural tourism cluster. From the research on the renewal of existing villages, the heritage exhibition type, nature tourism type, and folklore experience type are still in the stage of gradual improvement, but they are doing better in terms of sustainability. In contrast, several villages in the integrated scenic area type of development generally suffer from problems such as destruction of heritage integrity, over-commercialization, and fragmentation of the relationship between local residents and the village. The model should be continuously refined and improved in the selection process.
2) Expand new characteristic industries and optimize industrial structure. In addition to tourism, the traditional village should also create special development industries to complement each other with tourism development: Establish an industrial base for processing agricultural products and use the Internet to strengthen cross-system, cross-industry, and cross-region ties and cooperation. Strengthen the development of the tertiary industry and strive to realize the joint development of the service industry and tourism to achieve the transformation of the modern characteristics of the village while providing jobs for local residents and improving the living standards of local residents.
3) The villagers are the center of the village, and the multiple parties are promoted to build the village together. "People" are the core of village conservation and development, and their perceptions and behaviors are derivatives of emotions [7]. on

the one hand, we should try to promote the positive emotions of local residents and encourage tourism activities that promote public awareness and support for the preservation of cultural heritage [8]. On the other hand, we should fully understand the needs and wishes of local residents and improve the living standards and the rights of indigenous residents to participate in village conservation and development and management decisions. While retaining local residents, it is important to actively introduce social capital to multiple talents and handle the relationship between the government, social capital, and villagers. For example, in the case of Shichang Village, through the village enterprise cooperation between the village and China Energy Conservation and Environmental Protection Group, the capital was introduced, tourism and photovoltaic industry were developed, and a sustainable development path was developed taken.

3.3 Cultural Sustainability

Cultural sustainability is the goal of traditional village renewal, which has the role of transmitting regional culture and maintaining residents' emotions. The following strategies are proposed in three aspects: cultural transmission methods, cultural commercialization, and maintenance of spiritual culture.

1) Establishing a cultural museum to organize and display the regional culture. In the protection of traditional villages, the registration of village architecture and cultural relics, the organization of resource characteristics, and the combing of village history changes are the basic work of traditional village culture display and dissemination and should be displayed in one village museum. The research found that only tourist-oriented villages currently have cultural museums, and the government has to achieve universal access in every traditional village.
2) Strengthening the development of cultural and creative products. Under the current back, traditional village cultural values should be based on the attributes of commercialization, i.e., the relationship between buying and selling of village culture. The unique cultural value of goods can be explored, and secondary development can be carried out according to the attractive historical and cultural traditions of the traditional village to create a brand with its own characteristics, such as special food, special fabrics, special weaving, special handicrafts, etc. In the case of Renmazhai Village, for example, both the Chengni inkstone, printing skills and paper-cutting are the continuation of traditional crafts while lacking sufficient attraction for young people.
3) Strengthen the construction of village spiritual civilization. In addition to passing on history, a positive village culture can also play a role in shaping moral morality, and the construction of village spirit should be actively promoted to ensure that a good village culture can effectively contribute to the good and orderly operation of social governance order. For example, the village of Jiangli in Luoning County has maintained the tradition of settling disputes by the patriarch, and a new "council hall" has been built in the village to settle disputes among villagers, giving this spiritual culture a vehicle to be passed on.

4 Concluding Remarks

From the concept of sustainable development, this paper explains the relationship between the composition system of traditional villages in western Henan and the four sustainable dimensions of environment, society, economy and culture. Based on the field investigation of the current situation and existing renewal of typical samples of agricultural traditional villages in western Henan, this paper analyzes the sustainability, advantages and disadvantages of several existing renewal models of traditional villages under the measurement of these four sustainability dimensions. On this basis, this paper attempts to put forward the renewal strategies of traditional villages in western Henan in three aspects: environmental sustainability, economic, social sustainability and cultural sustainability. It is hoped that this study will give some enlightenment to the protection and development of traditional villages in western Henan.

References

1. Yangang, F., Jisheng, L.: Exploration of urban texture organization based on complex system theory. **33**(10), 32–37 (2008)
2. Chanchan, D., Binzhang, Y., Shuangjiao, S.: Study on the ways of rural landscape improvement combined with the improvement of villagers' quality – taking Zhangyang village in Yixing City as an example. **34**(05), 19–22 (2018)
3. Lijie, W., Rui, N., Shuyang, W.: Research on rural landscape protection and development strategy based on regionality. **32**(10), 65–67 (2016)
4. Junjie, L., Deyin, L.: Traditional villages need "weak landscape" – exploration on the practice of traditional village landscape construction. **25**(05), 26–31 (2018)
5. Scazzosi, L.: Rural landscape as heritage: reasons for and implications of principles concerning rural landscapes as heritage ICOMOS-IFLA 2017. **2**(3), 39–52 (2018)
6. The Florence Declaration on Heritage and Landscape as Human Values (2014). http://www.icomos.org. Accessed Nov 2017
7. Fazio, R.I.H., Michae, A.: Attitudes: Foundations, Functions, and Consequences. Sage, London (2003)
8. Siem Reap Declaration on Tourism and Culture-Building a New Partnership Model (2014). http://cf.cdn.unwto.org

Surrogate Based Multi-objective Optimization for Energy-Saving Building Design

Yuchen Chen[1](✉) and Xuhua Shi[2]

[1] Yisheng Architectural Design Consulting Co., Ltd., Hangzhou, China
chenyuchen0301@163.com

[2] Faculty of Electrical Engineering and Computer Science, NingBo University, Ningbo, China

Abstract. In order to coordinate building energy consumption and indoor comfort for architectural design, a surrogate based multi-objective optimization approach is proposed in this paper. Taking a rectangular planar layout of civil housing as an example, several design parameters closely related to building energy consumption and indoor comfort are selected. With the help of EnergyPlus energy consumption simulation analysis software, the sample data can be generated by simulating the software. Thus, a three-layer radial basis function (RBF) neural network is used to train and learn the sample data, and the models of building energy consumption and indoor comfort can be established. The established energy consumption and comfort models are verified by the experiments. On this basis, NSGA-II multi-objective optimization algorithm is used to balance the building energy consumption and indoor comfort to obtain the value range of design parameters and the Pareto front. So, the optimization results can be used to guide energy saving architectural design.

Keywords: Building energy consumption · Indoor comfort · Neural networks · Multi-objective optimization

1 Introduction

With the improvement of living standards, healthy living become important topics in the future architectural design [1]. Because the average temperature in northeast China is about 20 °C lower than that in the same latitude area of the world from January to March, and the average temperature is about 15 °C lower in the middle of the Yellow River, about 10 °C lower in the south of the Yangtze River, while the average temperature from July to September is about 5 °C higher than that in the same latitude region of the world [2], to maintain comfortable building environment, air conditioning which will produce a lot of energy consumption is strongly needed. So, vigorously promoting green building, and providing a series design standards in the early stage of building scheme willgive help for energy-saving building design.

Building energy-saving design is a typical multi-objective optimization problem. There exist some conflicting performance metrics such as building energy consumption and user discomfort [3, 4]. So, evolutionary optimization approaches can be used

W. Guo and K. Qian (Eds.): GBCESC 2022, LNCE 211, pp. 311–318, 2023.
https://doi.org/10.1007/978-981-19-5217-3_30

in building energy saving design. For example, Facundo et al. used EnergyPlus software to evaluate the individual objective function value, and applied NSGA-II [5] for the optimization design of typical residential buildings [6]. Ascione et al.proposed a multi-objective optimization approach for selecting the optimal transformation configuration, considering the construction cost, incentive mechanism, heating and requirements [7]. Furthermore, Ascione et al. studied the cost optimization of building renovation and proposed a multi-stage cost optimization analysis framework integrating multi-objective evolutionary optimization and artificial neural network [8]. Hamdy et al. compared the advantages and disadvantages of seven multi-objective evolutionary optimization algorithms from the aspects of execution time, search performance, convergence and diversity [9]. In recent years, domestic researchers used multi-objective evolutionary optimization technology for building energy conservation. Reference [10] took economy and comfort as optimization objectives, and gave the optimal optimization scheme for HVAC system. Reference [11] to the lowest energy consumption and the lowest cost as the goal, to optimize the residential buildings in Wuhan. Reference [12] conducted a multi-objective optimization design for the window size of rural houses in hot summer and cold winter regions. Reference [13] deeply studied the digital energy-saving design buildings in severe cold regions. Therefore, based on NAGA-II multi-objective algorithm, this paper uses RBF neural network to establish the surrogate models of energy consumption and comfort respectively, and uses multi-objective optimization for building energy consumption and comfort. References [14–16] are other applications for building energy consumption. The contributions of this article: (1) The models of the building energy consumption and indoor comfort have been established with the help of EnergyPlus energy consumption simulation software; (2) We can obtain the balance parameters for the building energy consumption and indoor comfort using NSGA-II multi-objective optimization algorithm. The optimization results may be applied to the energy saving architectural design.

The rest of this paper is organized as follows: Sect. 2 presents multi-objective optimization of building energy consumption and comfort problem. Section 3 represents the main steps of the proposed approach, and Sect. 4 gives the application of the energy-saving building design. Concluding remarks are given in Sect. 5.

2 Multi-objective Optimization for Building Energy Saving Design

Various factors may influence building energy consumption, such as external wall heat transfer coefficients, external window and roof of the buildings, the ratios of window to wall, the shape coefficients, orientation and lighting of the buildings. Generally, the calculation of the building energy consumption adopts three methods, namely, the law of thermodynamics, artificial neural network and integrated calculation platform [13]. In recent years, some integrated computing platforms will provide effective approaches for energy consumption calculation, such as Energyplus, DeST and DOE-2. As an open energy consumption simulation analysis software, Energyplus software is powerful and has been fully recognized in the fields of building energy conservation design.

Taking the type of rectangular plane building layout as an example, the dual-objective optimization design of building energy consumption and indoor comfort can be described as:

$$\begin{cases} y_1 = min(En(x)) \\ y_2 = max(S(x)) \end{cases}$$
$$s.t.\ \ S(x) > \alpha$$
$$M(x) > \beta \quad (1)$$

where y1, y2 represent building energy consumption and comfort respectively, x is decision variable, *En*, *S*, *M* represent building energy consumption, comfort and cost function respectively, α, β represent comfort and operation cost threshold respectively.

3 Surrogate Aided Multi-objective Optimization Approach

3.1 Principle and Description

In this paper, we proposed a novel multi-objective evolutionary framework as shown in Fig. 1. The framework consists of three parts, *i.e.* optimizer, modeling system and model manager. In the process of optimization, the Optimizer using NSGA-II in this paper completes evolutionary optimization operations, such as population generation, replication, crossover, mutation, preselection, fitness evaluation, etc. Modeling process mainly completes the training of the surrogate models, and Models management completes the call of the surrogate models. Optimizer, Modeling system and Models management are collaborated to complete evolutionary optimization tasks.

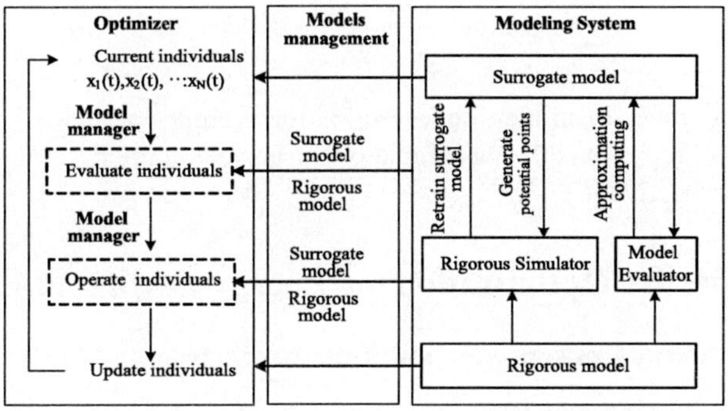

Fig. 1. Multi-objective evolutionary framework for surrogate model

The basic steps of NSGA-II multi-objective optimization algorithm are as follows:

Step 1: Initialize decision variables and intermediate storage variables required by NSGA-II to generate multi-objective evolutionary initial population;

Step 2: Randomly generate a sampling point according to Latin square sampling, establish the initial proxy model;

Step 3: Perform surrogate model management program, call the surrogate models, evaluate the fitness value of each individuals;

Step 4: Select parents, perform gene evolutionary operations, namely crossover and mutation operations;

Step 5: Perform surrogate model management program, evaluate the fitness value of each individual;

Step 6: Perform fast non-dominated sorting operation to search the non-dominated solution set;

Step 7: Is the termination condition satisfied ? If not, execute Step8, otherwise execute Step 10;

Step 8: Update the proxy model and evaluate the accuracy of the model;

Step 9: Return Step 3;

Step 10: End the program, output the non-dominated solution and the target value, and obtain the optimization results.

3.2 Establishment of Surrogate Models

Surrogate models are approximate models for complex actual objects, which are used to solve time-consuming calculation problem of evolutionary algorithm. Common surrogate models include polynomial regression model, radial basis function (RBF) model and Gaussian process model. The polynomial regression model is fast and convenient, but the model accuracy is low, while RBF neural network model needs a large number of sample data, which has the characteristics of good robustness, and can meet the requirements of model accuracy to a certain extent, and Gaussian model has strong nonlinear ability, and unbiased estimation at the training sample points. This paper we choose RBF agent model, as shown in Eq. (2):

$$y = \sum_{i=1}^{h} w_{ij} \exp(-\frac{\left\| x_p - c_i \right\|^2}{2\delta^2}), j = 1, 2, \cdots, n \tag{2}$$

where h is the number of hidden nodes, w_{ij} is the weight coefficient, x_p is the input samples, and δ, c_i represents the base function variance and the base function center of the first node, respectively.

4 Multi-objective Optimization for Energy-Saving Building Design

4.1 Building Energy Consumption and Comfort Surrogate Models

The training sample data of the RBF neural network model can be obtained by repeatedly simulating using Energyplus software. We construct a three-layer RBF neural network. The number of input layer is 21, hidden layer is 25, and the output layer is 2, namely, building energy consumption and comfort. The upper and lower bound range of input parameters are shown in Table 1. By selecting 21 parameters within the specified input range and running Energyplus software. Thus, 500 sets of input-output sample data can be generated. Among them, 450 sets of data are used to train RBF network, and another 50 sets of data are used to test the established RBF model.

Table 1. Inputs for neural networks

NO.	Design parameters	Basel values	Upper bound	Lower bound
1	Building shape coefficient	0.15	0.4	0.1
2	Space area	120	300	60
3	Building storey	8	12	1
4	Window-wall ratio east of building	0.3	0.8	0.2
5	Building west window-wall ratio	0.3	0.8	0.2
6	Building south window-wall ratio	0.3	0.8	0.2
7	Building north window-wall ratio	0.3	0.8	0.2
8	Outer wall heat transfer C. (East) (W/m^2k)	1.2	4	1
9	Outer wall heat transfer C. (West) (W/m^2k)	1.2	4	1
10	Outer wall heat transfer C. (South) (W/m^2k)	1.2	4	1
11	Outer wall heat transfer C. (North) (W/m^2k)	1.2	4	1
12	Outer window heat transfer C. (East) (W/m^2k)	2	3	1
13	Outer window heat transfer C. (West) (W/m^2k)	2	3	1
14	Outer window heat transfer C. (South) (W/m^2k)	2	3	1
15	Outer window heat transfer C. (North) (W/m^2k)	2	3	1
16	Roof heat transfer C. (W/m^2k)		4	1
17	Outer window heat gain solar radiation C. (East)	0.5	0.8	0.3
18	Outer window heat gain solar radiation C. (West)	0.5	0.8	0.3
19	Outer window heat gain solar radiation C. (south)	0.5	0.8	0.3
20	Outer window heat gain solar radiation C. (north)	0.5	0.8	0.3
21	Depth of peripheral zone(m)	8	14	4

4.2 Models' Training and Verification

Here, the learning rate of RBF is set at 0.5, the training error accuracy is 0.01, and the maximum number of iterations is set to 1000. In order to improve the prediction accuracy and convergence performance, all sample data need to be normalized before network training, which is shown as Eq. (3), and denormalized as Eq. (4):

$$\bar{x}_i = \frac{x_i - min(x)}{max(x) - min(x)}, i = 1, 2, \cdots, m \tag{3}$$

$$y_i = \bar{y}_i * (\max(y) - min(y)) + min(y), i = 1, 2 \tag{4}$$

The results show that the simulated and predicted values of building energy consumption are in good agreement. The average training absolute accuracy is 0.73 kWh/m^2, and the average relative accuracy is 0.52%, while the average absolute accuracy of indoor comfort is 0.16, and the average relative accuracy is 0.36%. Then, the trained RBF is used to predict another 50 set of samples. The average absolute error of building energy consumption prediction is 1.15, and the relative error is 0.81%, while the average absolute error of comfort is 0.17, and the relative error is 0.42%, which are shown as Figs. 2 and 3. From the above results we can see, the RBF has high precision and can meet the optimization requirements.

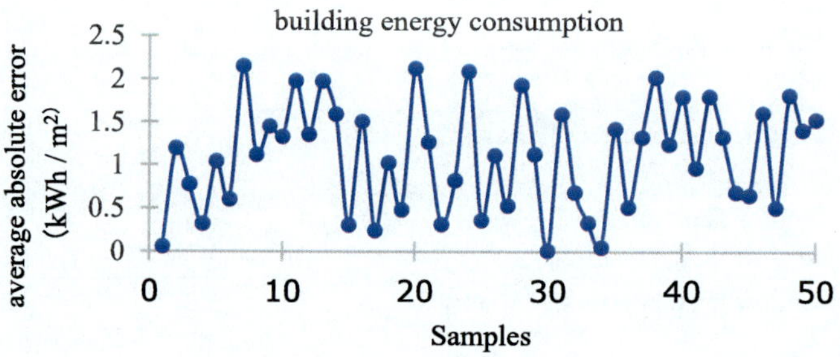

Fig. 2. The average absolute error of building energy consumption prediction

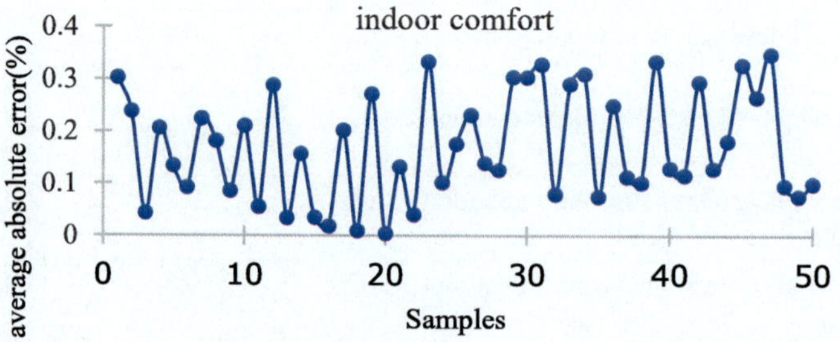

Fig. 3. The average absolute error of comfort

The design parameters of NSGA-II include 21 inputs such as the number of building layers and the ratio of east window to wall, and the output is building energy consumption and comfort. The NSGA-II parameters are set as follows: the solved decision variables are encoded by real numbers, the population size P = 100, the maximum number of iterations is 1000, the crossover probability is 0.8, the mutation probability is 0.05, and the number of non-dominated solutions is 100, $\alpha = 40\%$, $\beta = 500$. After 20 consecutive operations, the optimization frontier of building energy consumption and comfort in the evolution of NSGA-II is obtained. Following are the average statistics for 20 optimized runs. Table 2 and Fig. 4 list the compromise results obtained by the proposed approach.

Table 2. Compromise solutions obtained by the proposed approach

Approach	Compromised solutions	Objective Values
Surrogate based NSGA-II	(0.25, 135, 10, 0.5, 0.5, 0.5, 0.5, 2.5, 2.5, 2.5, 2.5, 2, 2, 2, 2, 3, 0.6, 0.6, 0.6, 0.6, 10)	142.2 (kWh/m^2), 42.6(%)

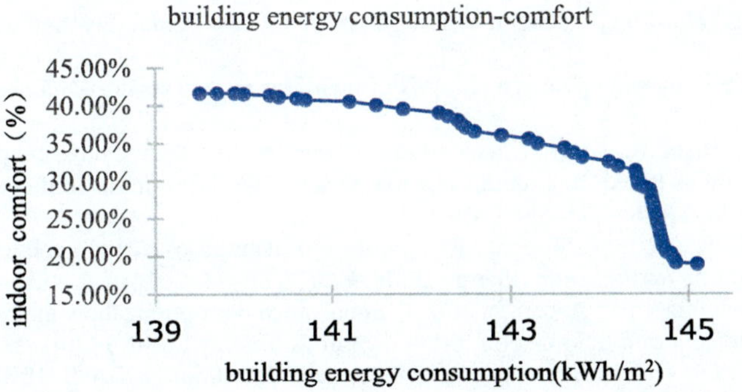

Fig. 4. Compromise results obtained by the proposed approach

We take a civil residence with rectangular layout as an example, and fix some parameters to verify the rationality of the optimization results. The building orientation is set to be south, the building area is 120 m^2, the number of layer is set to 8, and the building height is 3.0 m. The wind speed in the inflow direction is 2.0 m/s, and is uniformly distributed, which direction is parallel to the normal direction of the building. It can be seen that when the window-wall ratio in the east, west and north directions of the building increases from about 0.38 to about 0.62, the energy consumption of the building increases from about 142.2 kwh/m^2 to about 142.7 kwh/m^2, and the indoor comfort decreases from about 39.2% to 37.6%. When the outer window heat transfer coefficient changes, the impact on building energy conservation and indoor comfort is not obvious, with the continuous increase of the outer wall heat transfer coefficient, from 3.02 w/m^{2}k to 3.52 w/m^{2}k, the change of building energy consumption are not obvious, and the comfort decreases. The influence of roof heat transfer coefficient on building energy consumption is more sensitive.

5 Conclusions

Low energy consumption and high comfort is the goal of energy-saving buildings. In this paper, by selecting several design parameters closely related to the building energy consumption and comfort, taking a typical south rectangular plane civil building as an example, with the help of EnergyPlus full energy consumption simulation analysis software, 500 sets of sample data with wide coverage are generated, and the RBF neural network is used to learn and train the sample data so as to establish the proxy model of building energy consumption and indoor comfort, it is used to test the building energy consumption and indoor comfort. NSGA-II multi-objective optimization algorithm is used to optimize the energy consumption and comfort of the building, and the design parameters are obtained. The parameters can be used to verify the rationality of the results. In the future research, if we can further explore the factors that affect building energy consumption and comfort, establish more perfect sample data and improve the neural network model.

References

1. Minghao, L., Huadong, L.: Green building strategy under sustainability theme. Archit. J. **3**(2), 16–17 (2003)
2. Junyi, W.: Energy-saving design analysis of buildings in cold regions. Sci. Rep. **5**(7), 28–31 (2015)
3. Zhang, A., Bokel, R., Andy, V., et al.: Optimization of thermal and daylight performance of school buildings based on a multi-objective genetic algorithm in the cold climate of China. Energy Build. **139**(6), 371–384 (2017)
4. Madias, E., Kontaxis, P., Topalis, F.: Application of multi-objective genetic algorithms to interior lighting optimization. Energy Build. **125**(7), 66–74 (2016)
5. Bre, F., Fachinotti, V.: A computational multi-objective optimization method to improve energy efficiency and thermal comfort in dwellings. Energy Build. **154**(4), 283–294 (2017)
6. Deb, K.: A fast and elitist multi-objective genetic algorithm: NSGA-II. IEEE Trans. Evol. Comput. **120**(6), 182–197 (2002)
7. Ascione, F., Bianco, N., Masi, R., et al.: Energy retrofit of educational buildings: transient energy simulations, model calibration and multi-objective optimization towards nearly zero-energy performance. Energy Build. **144**, 303–319 (2017)
8. Ascione, F., Bianco, N., Stasio, C., et al.: CASA, cost-optimal analysis by multi-objective optimization and artificial neural networks: a new framework for the robust assessment of cost-optimal energy retrofit, feasible for any building. Energy Build. **146**, 200–219 (2017)
9. Hamdy, M., Nguyen, A., Hensen, J.: A performance comparison of multi-objective optimization algorithms for solving nearly-zero-energy-building design problems. Energy Build. **121**, 57–71 (2016)
10. Kaiyun, Z.: Multi - objective optimization model of HVAC system. Enterp. Rep. **12**, 70 (2016)
11. Tao, L., Yaolin, L., Wei, Y.: Green building optimization design based on genetic algorithm. Build. Energy Efficiency **44**(6), 53–57 (2016)
12. Xuhong, F., Zhaochang, Z., Jun, W, et al.: Multi-objective optimization design of rural housing window size in hot summer and cold winter areas. Build. Energy Efficiency **45**(12), 80–84 (2017)
13. Yunsong, H.: Research on digital energy-saving design of office buildings in cold regions. Harbin Institute of Technology (2016)
14. Caglayan, S., Yigit, S., Ozorhon, B., Ozcan-Deniz, G.: A genetic algorithm based envelope design optimization for residential buildings. Eng. Sustain. **173**(6), 1–10 (2019)
15. Frey, G.: Multi-objective techno-economic optimization of design parameters for residential buildings in different climate zones. Sustainability **14**(1), 1–30 (2021)
16. Zhang, K.: Energy-saving parameterized design of buildings based on genetic algorithm. Int. J. Build. Pathol. Adapt. **38**(5), 785–795 (2020)

Research on the Development of Modern Mosaic Wall Decoration Materials Based on Nostalgic Memory—Taking Wumi Congee (Congee Shop) in Cantonese Area of China as an Example

Qiang Tang[1], Yile Chen[2](✉), Junzhang Chen[3], and Liang Zheng[2]

[1] Shunde Polytechnic, Shunde District, Foshan City 528300, Guangdong Province, China
[2] Faculty of Humanities and Arts, Macau University of Science and Technology, Avenida Wai Long, Taipa, Macau 999078, China
2009853gat30001@student.must.edu.mo
[3] Faculty of Innovation and Design, City University of Macau, Avenida Padre Tomás Pereira, Taipa, Macau 999078, China

Abstract. Mosaic is an ancient method of building wall decoration. Based on the memory of nostalgia, this paper develops and designs the decorative materials for the walls of the Wumi Congee (Congee shop), which was designed from the renovation of the village house in the Cantonese Area of Lingnan, China. By introducing in detail the design and development process of the red series ceramic mosaics in the restaurant, the modular combination method of ceramic mosaics, the material properties and the experience of innovative design practices, it provides a reference for the subsequent building wall material decoration.

Keywords: Mosaic · Wall decoration materials · Nostalgic memory · Congee shop · Cantonese area

1 The Background of Rural Development

With the passage of time, the process of urbanization has gradually accelerated in China. As the so-called "one side of water and soil, one side of people", the city is gradually surrounded by reinforced concrete, many architectural designs are "the same", and the appearance of the city is also "a thousand cities". People have been unable to find some familiar shadows in their memory [1–4], which will make people feel special attachments to certain feelings in their hometown. Therefore, in addition to meeting material needs, architectural design also needs to be based on people's spiritual pursuits. Based on the memory of nostalgia, this paper develops and designs the decorative materials for the walls of the Wumi Congee (Congee shop), which was designed from the renovation of the village house in the Cantonese Area of Lingnan, China [5–7].

W. Guo and K. Qian (Eds.): GBCESC 2022, LNCE 211, pp. 319–328, 2023.
https://doi.org/10.1007/978-981-19-5217-3_31

2 The Development History of Red Brick Houses in Cantonese Area

Red bricks are very common in rural areas of China, and they have been used for thousands of years. It can be said that it is an element of rural memory. However, China banned red bricks as early as 2012. The main reasons are as follows:

① The raw material for red bricks in the firing process is clay, which can only be excavated underground. After kneading and stirring, this soil will be made into a red brick model, and then left to dry naturally. It then needs to be fired at 900° to form the solid red brick we see.
② Pollution to the environment will lead to a large amount of gas emissions during the firing process, especially the great demand for coal. Under such circumstances, the content of PM2.5 in the air is very high, so the content of metal ions and toxic gases in the vicinity is very high, which will cause harm to the growth of crops, and breathing into the human body will also lead to various respiratory diseases.
③A lot of waste of land resources It should be known that the production process of red bricks requires highly viscous soil, which is generally deep underground. Some places even use hills for mining, and such a production environment will lead to the disappearance of millions of acres of fertile fields and a lot of waste of land resources. Farmers' arable land is also threatened and harmed.

Because of its high cost performance and long history of use, it has become an indispensable material for building houses in the countryside (Fig. 1).

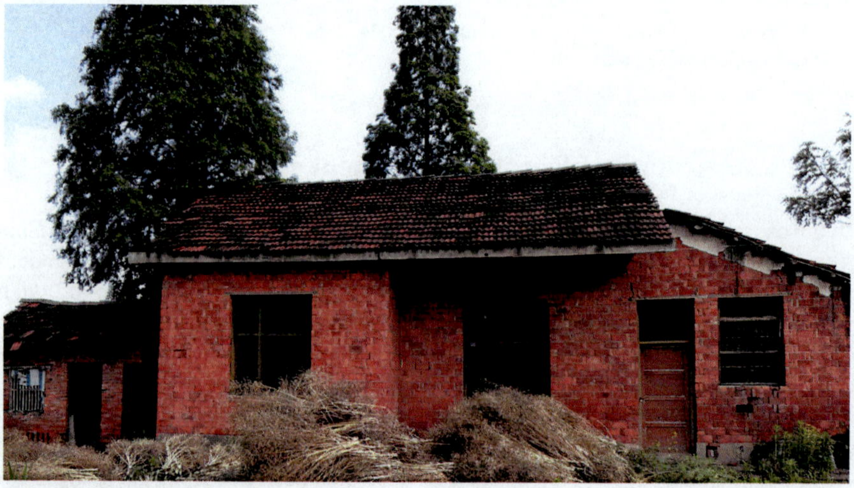

Fig. 1. The exterior wall of the red brick house in the village ancestral home (Image source: Photographed by the author).

3 Properties of Mosaic Materials

Mosaic is an ancient method of building wall decoration. The original meaning of mosaic is mosaic pattern and mosaic technology. It originated in ancient Greece and experienced extraordinary development in the Roman period. As early as in the Babylonian kingdom, "mosaic" was used as an architectural decoration on the Ishtar Gate, the main gate of the capital. It is dedicated to the goddess Ishtar. The walls are covered with colored glazed bricks. The shape and size of the tiles are almost the same as the daily red bricks. On the blue tiles, bulls, dragons and lions are inlaid with glaze. Its shape is a side image, with a strong decorative effect.It has been widely used in the walls and floors of various spaces and decorative art mosaics. Although the unit area of a single piece of mosaic is small, it has a wide variety of colors, and infinite combinations create an unpredictable visual space, which can fully express the designer's inspiration. For thousands of years, human beings have perfectly interpreted its practicality and artistry, and finally evolved a technology into an enduring art. Among them, ceramic mosaic is an important member of the mosaic family, and its formation principle is relatively special. "Ceramic mosaic is a kind of ceramic surface formed by firing and then spraying glaze. Bricks, which have the advantage of never fading, are also known as "ceramic mosaic tiles".

Therefore, ceramic mosaic has many advantages: ① non-fading, flame retardant, fireproof, no radiation; ② wear-resistant, hard texture, no falling off, no deformation, long-lasting effect; ③ good moisture resistance, no water seepage, and thermal insulation to the wall The effect of heat; ④ high temperature resistance, anti freezing, can adapt to extreme cold and hot climate sudden changes; ⑤ easy to make personalized space accessories. The advantages of ceramic mosaic in its artistic effect are also very obvious. "Ceramic mosaic: If it is divided by style, ceramic mosaic itself is considered to be a transition of style because it retains the simplicity and thickness of pottery and the delicate and moisturizing of porcelain. It has bright colors and a wide variety of varieties." On the whole, ceramic mosaic has many advantages, just like ceramic products that have been indispensable to human beings since ancient times, it is a high-quality decorative material that is energy-saving and environmentally friendly.

4 Mosaic Material Innovative Combination Process

Wumi Congeeis a Congee shop with regional flavors in Cantonese area. In order to create a unique personality space and highlight the regional nostalgic memory, the author designed and developed the red series of ceramic mosaics. At the beginning of the design, a detailed development plan was drawn up in consideration of the standardized and systematic unit combination characteristics of the newly developed mosaic. First draw the mosaic construction drawing (see Fig. 2), and design a total of 100 * 100 * 30 mm, 50 * 50 * 10 mm, 50 * 100 * 30 mm, 50 * 50 * 20 mm, 100 * 50 * 30 mm, 100 * 50 * 20 mm, 100 * 50 * 10 mm and 25 * 25 * 10 mm and other eight specifications, the above data show that the mosaic has a modulus relationship on the length, width and thickness, forming a certain proportion of various modules. Then, use the wooden plywood to release the complete model at a scale of 1:1, and try a variety of patchwork methods to simulate a variety of mosaic combinations (see Fig. 2). After demonstration, the simulated mosaic achieved the expected effect and began to proof. For the sake of safety, let the factory produce a small batch of samples, and continue to use the samples to conduct various simulation combinations on the construction site, and finally install the selected combination modules on the theme wall in a large area. The above process is like returning to the ancient history of mosaic and mosaic, but using modern technology to continue to explore the application and innovation of ceramic mosaic.

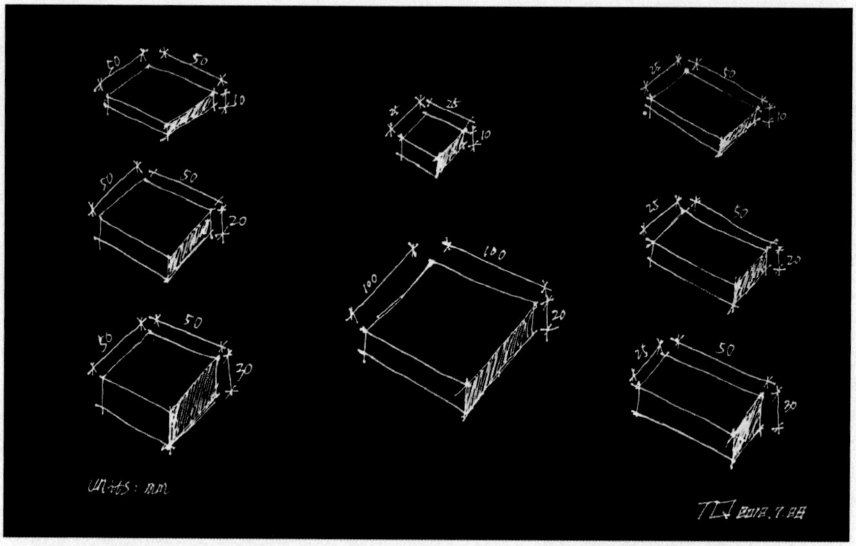

Fig. 2. The exterior wall of the red brick house in the village ancestral home (Image source: Photographed by the author).

Through careful design, the end products of the same ceramic mosaic can produce diverse patterns, and form numerous modular combinations through orderly and disorderly combinations. Orderly arrangement can produce a neat visual effect, forming a three-dimensional small space and pattern with a variety of changes; random arrangement looks disordered, but the thickness varies, resulting in a concave and convex layering. The combination of many disordered small modules also produces an orderly aesthetic sense, forming a natural and harmonious situation. The combination of ceramic mosaics pays attention to the natural construction method. Its special texture and elegant texture can be combined infinitely in the shape, and its unique artistic charm and personality temperament can be fully displayed in the space. Wumi Congee (Congee shop) "Congee" interprets the world, conveys culture in the way of "food", connects love, gives people the comfort of nature, the transcendence of emotions and the conclusion of spirit. The use of ceramic mosaics echoes the theme of the original food culture of the Congee, and is close to the needs of modern people's emotional return to nature (Fig. 3).

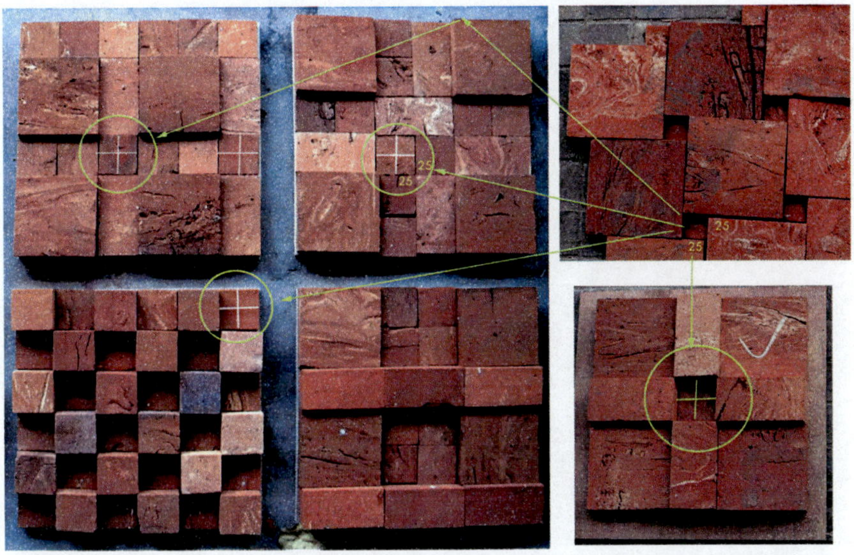

Fig. 3. 25 × 25 stitching method (Image source: Photographed by the author).

5 Design Experience of Mosaic Decorative Materials

5.1 Similar Mosaics Pay Attention to Clever Combination

Mosaic design should break the shortcomings of traditional single form, dull color and rigid collocation, and should pursue rich and changeable visual effects of mosaic. Mosaic has been widely used in various indoor and outdoor decorations. For example, gradient mosaic pays attention to natural and ingenious combination. The so-called gradient is a gradual and natural transition in color tone, presenting a harmonious picture. Especially those mosaics with smaller particles have the advantages of flexibility, controllability and convenient combination.

5.2 Mosaic Should Pay Attention to Mix and Match Design

Mosaic mashup is not a chaotic patchwork, even if it is chaotically stacked, it often refers to a decorative body with certain visual rules that mosaics are carefully combined through multiple materials. Mix and match can be divided into two categories: the first type is the combination of different varieties of mosaics. The market is commonly used to mix and match different varieties of mosaics such as stone, glass, ceramics, aluminum, wood mosaics. Mixing and matching needs to determine the main mosaic of the mixing and matching according to the main color of the space and the surrounding environment, and then match two to three other varieties of mosaics according to a certain proportion. If it is determined that a certain color is the main color, it can take up 50% of the proportion, and then configure about 30% of other types of mosaics with similar colors that are slightly darker or lighter, plus 20% of other mosaics are only used as decorations.

Visually, the overall color tends to be clear and easy to unify naturally. The second type is the combination of similar mosaics. Mosaics of the same color of different sizes or heights can also be mixed and matched. Through the gradient and contrast of small areas of color blocks, visual differences that are impossible for decorative materials of equal area are formed (Fig. 4).

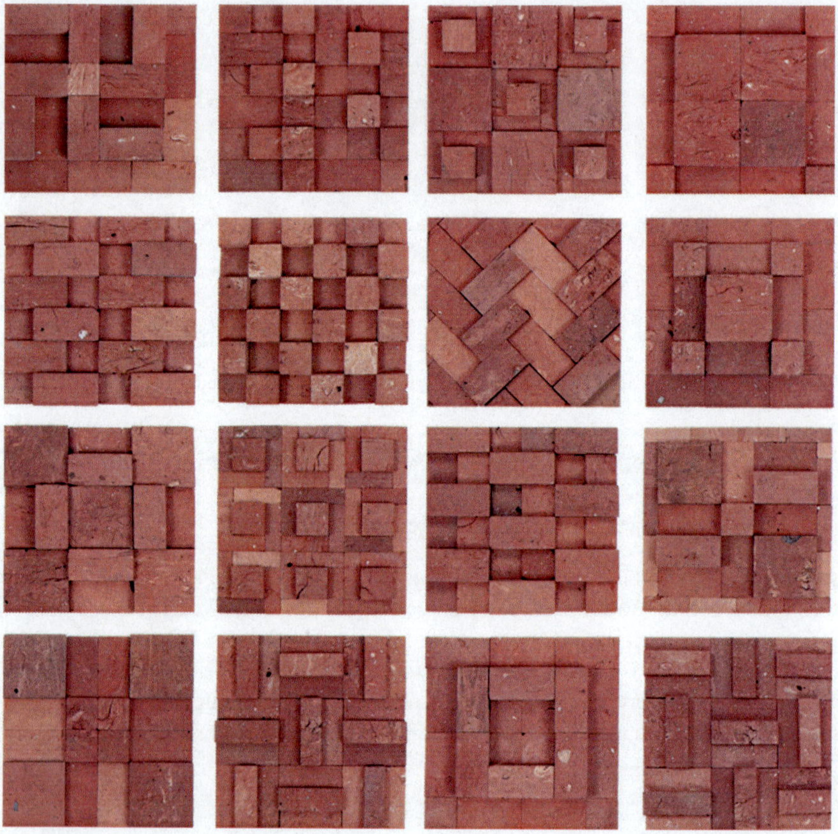

Fig. 4. Clever combination design (Image source: Photographed by the author).

5.3 Unleash the Artistic Effect of Mosaics

Mosaic mosaic is to arrange each mosaic color piece in a regular and orderly manner, and combine them to form various forms of artistic patterns. Exquisite mosaic has super plasticity, and designers can produce patterns of various art forms through regular and orderly arrangement and combination, which can produce both figurative patterns and abstract patterns. This technique is similar to the oil painting art of pointillism, and it is only a color piece when viewed up close, which is not only ingenious but also very interesting. The unique personality and temperament of mosaic patterns give play to the unparalleled advantages of mosaics as decorative materials and make up for the deficiencies of general decorative materials. Designers freely sway their unique creativity through mosaics, and pave a spiritual space with complete self-individuality. It can be either a delicate and elegant space scene, or a rough and unrestrained decorative picture, so that the mosaic can interpret thousands of styles.

5.4 Mosaic Decoration Should Pay Attention to the Excavation of Chinese Culture

"The design of mosaics pays more attention to the overall application and carrying cultural elements. The value of mosaics lies in the puzzles. Because the creativity of the puzzles is infinite, more cultural elements can be integrated into the mosaics. This will greatly deepen the cultural connotation of mosaics." In today's pursuit of individual characteristics of design, the form of mosaic decorative art should develop in a diversified direction. Modern mosaic decorative art can draw the essence of Chinese decorative art patterns and deeply explore the real "Chinese ancient pattern culture". The ancient Chinese decorative art has a long history, creating a broad and profound decorative art pattern. There was once an art master who brought the art of Chinese murals into full play through mosaic collage. Designers should have a deep understanding of the cultural and artistic connotations to be interpreted by ancient Chinese patterns, and then recreate them by absorbing beneficial elements according to the needs of the design, rather than blindly imitating them. This is the real excavation of Chinese culture, not superficial imitate.

5.5 Combination of Light and Shadow Effects

The interior walls of Wumi Congee (Congee shop) use these red brick mosaic materials developed by the author, which are more effective under the combination of light and shadow (see Figs. 5, 6, 7 and 8). When matching, it is necessary to consider the light and shadow effects presented by the sunlight received during the day. At night, it is even more important to consider the coordination effect with each different type of lighting in the architectural space.

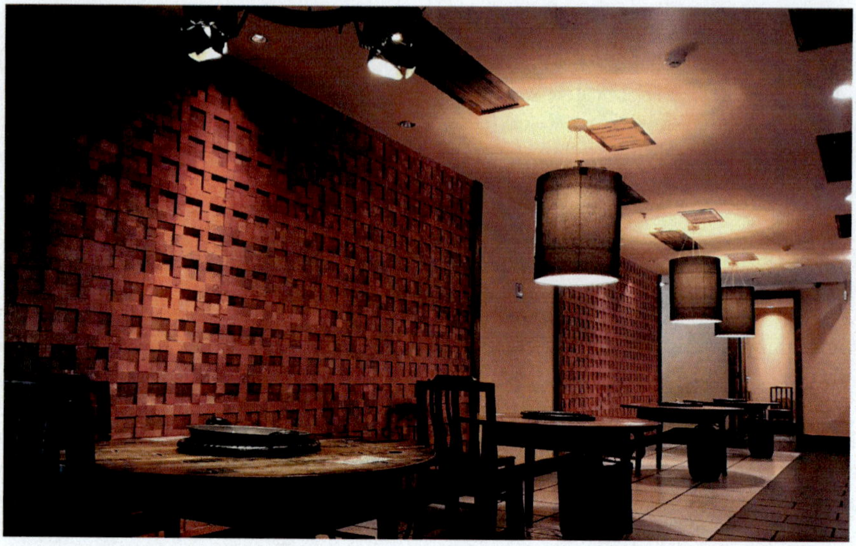

Fig. 5. The combination of natural sunlight (Image source: Photographed by the author).

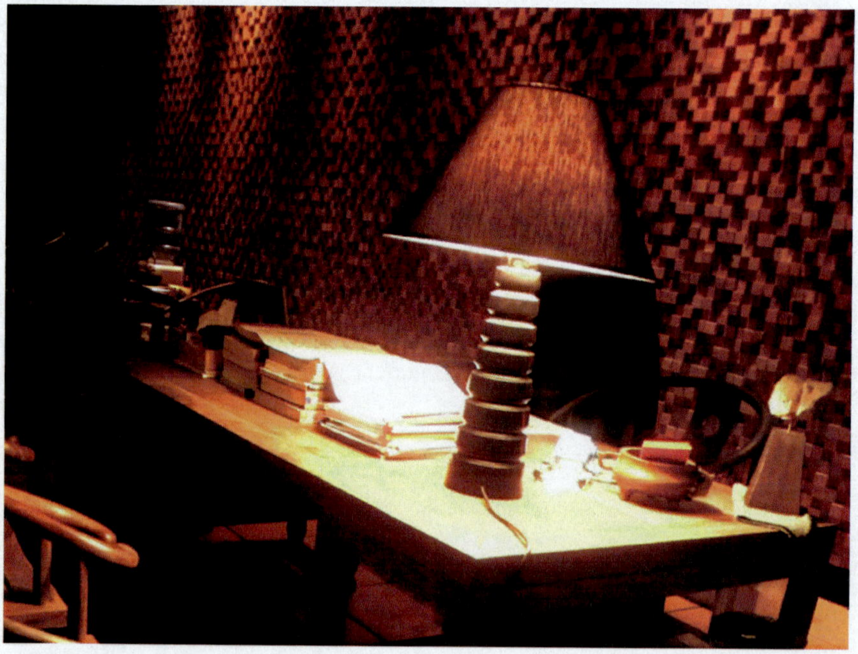

Fig. 6. Matching with table lamp lighting (Image source: Photographed by the author).

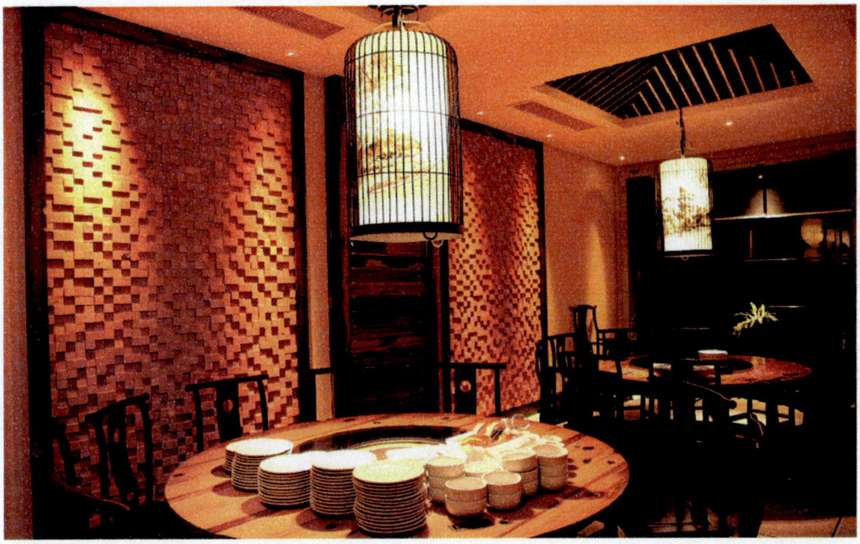

Fig. 7. Matching with the ceiling lighting in the lobby (Image source: Photographed by the author).

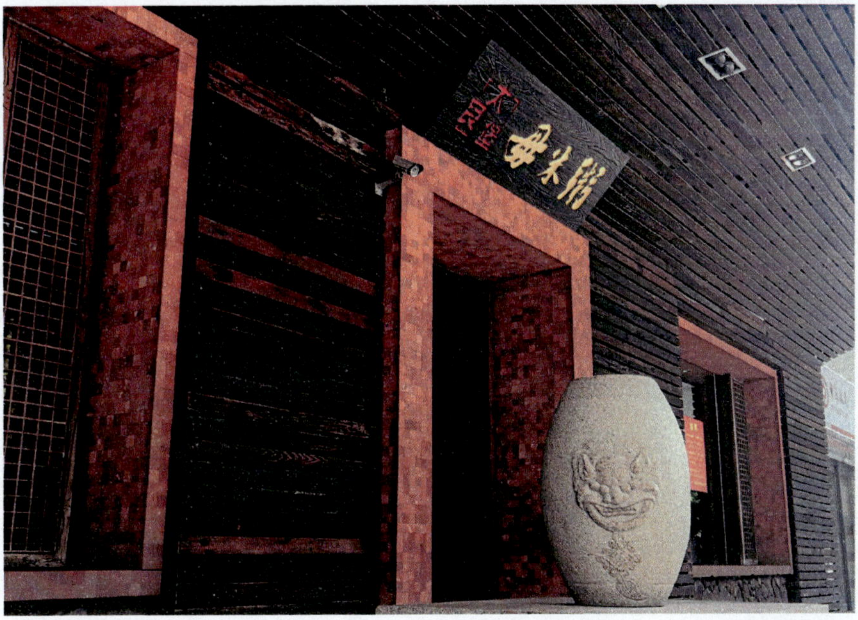

Fig. 8. Front door entrance effect (Image source: Photographed by the author).

6 Conclusion

This research shows a new development form of mosaic design innovation in today's decoration industry through in-depth research on the modular combination of red series ceramic mosaics. However, this is only one direction of the innovative development of mosaics. We hope to inspire more designers to participate in the research of mosaic materials, to develop broader innovative ideas of mosaic art, to discover newer forms of expression, and to perfectly interpret its practicality and artistry, make mosaic play more charming charm.

Acknowledgement. This article is based on the research staff from 2018 to 2020 from the 2020 Guangdong Province Ordinary Universities Characteristic Innovation Project (No. 2020WTSCX286); 2019 Guangdong Province Ordinary University Philosophy and Social Science Project (No. 2019GXJK131); 2018 Guangdong Province Philosophy and Social Science Planning Discipline Co-construction with the result of the paper funded by the project (No. GD18XYS40).

References

1. Allen, J.R.L., Fulford, M.G.: Early Roman mosaic materials in southern Britain, with particular reference to Silchester (Calleva Atrebatum): a regional geological perspective. Britannia **35**, 9–38 (2004)
2. Arletti, R., Vezzalini, G., Fiori, C., Vandini, M.: Mosaic glass from St Peter's, Rome: manufacturing techniques and raw materials employed in late 16th-century Italian opaque glass. Archaeometry **53**(2), 364–386 (2011)
3. King, S.: Mosaic Techniques & Traditions: Projects & Designs from Around the World. Sterling Publishing Company, Inc. (2003)
4. Flügel, E.: Microfacies-based provenance analysis of Roman imperial mosaic and sculpture materials from Bavaria (Southern Germany). Facies **41**(1), 197–208 (1971). https://doi.org/10.1007/BF02537466
5. Vandini, M., Fiori, C., Cametti, R.: Classification and technology of Byzantine mosaic glass. Anal. Chim. **96**(9–10), 587–599 (2006)
6. Moore, R.E. Some factors governing ancient mosaic design. Bull. Inst. Class. Stud. 87–91 (1964)
7. Sarhangi, R.: Interlocking star polygons in Persian architecture: the special case of the decagram in mosaic designs. Nexus Netw. J. **14**(2), 345–372 (2012)

Exploration and Practice of the Application of Natural Lighting and Shading Technology in Building Energy Conservation

Jinghua Shen[1,2], Jie Zhang[1,2], Zhaoyang Xue[1,2], and Jie Wu[1,2](✉)

[1] School of Architecture, Soochow University, Suzhou, China
jwu@suda.edu.cn

[2] International Research Center for Green Building of Soochow University, Suzhou, China

Abstract. Natural lighting and shading technology, as an important factor in building energy conservation, has attracted much attention in building design and renovation in recent years. Taking the renovation design of a college building in Taizhou as an example, this paper focuses on analyzing the design methods and influencing factors of building natural lighting and shading technology, and analyzes the renovation objectives, design methods and performance comparison before and after the renovation, Emphasize the importance of natural lighting and shading technology, and the appropriate renovation strategy and design concept are given. so as to provide relevant reference for the application of building lighting and shading technology. It also provides experience for building sustainable development in the future.

Keywords: Building renovation · Natural lighting · Sunshade technology · Building energy efficiency

1 Natural Lighting Design

1.1 Factors Affecting Natural Lighting

The main factors affecting natural lighting are natural light characteristics and architectural design. Factors such as weather conditions and solar altitude Angle will affect natural illumination. At the same time for the building itself, the building orientation, form and plane planning will affect the natural lighting effect of the building. Therefore, in order to improve the energy-saving performance of building lighting, the shape design of building engineering should be optimized to make full use of natural lighting conditions and minimize the energy consumption of artificial lighting [1].

1.2 Natural Lighting Method

Natural lighting means is divided into two kinds, one kind is to open natural lighting window, one kind is to set lighting auxiliary system. Natural lighting window can be divided into side window lighting, skylight lighting and mixed lighting. Increasing the

W. Guo and K. Qian (Eds.): GBCESC 2022, LNCE 211, pp. 329–336, 2023.
https://doi.org/10.1007/978-981-19-5217-3_32

size of the window will benefit daylight availability, so choosing an appropriate window design is a challenging task [2]. The lighting auxiliary system is to use some advanced technical means to optimize the lighting effect, which are commonly used light guide pipe lighting system and reflector. The working principle of the light pipe lighting system is to collect outdoor natural light, and use the reflection principle of light to transmit natural light to the interior through a series of specially made pipes during the day, which is supplemented by diffuser. Which is a non-electric lighting system. The main equipment in the lighting system is the lighting hood, which is installed outdoors to collect sunlight. The inner wall of the light guide tube is made of highly reflective material, and the reflectivity can reach 92%–95% [3]. The reflector is a mirror reflector mounted directly above the window. Sunlight is reflected through the reflective panels to the ceiling of the interior, and the diffuse reflection of the ceiling leads light into deeper areas of the interior.

2 Building Shading Technology

2.1 Shading Form of Architecture

There are various forms of shading for buildings, which can be divided into internal shading, external shading, self-shading and mutual shading. Exterior shading is the most direct and effective shading method among all shading forms. Therefore, in recent years, most of the research and development and promotion of new sunshade products are carried out on the basis of outer sunshade. Exterior shading of a building can be defined as a complex of components tightly integrated with the building envelope, located outside the building to protect the building from excessive solar radiation [4]. According to the relationship between building structure itself and exterior shading, exterior shading can be further divided into exterior shading and embedded shading. The difference between the two lies in whether they can exist independently from buildings.

2.2 Architectural Shading Design Strategy

In architectural shading design, the choice of materials is particularly important. The shading component is exposed to the sun in a large area and is exposed to wind erosion and sun exposure for months and years, which requires the material to have excellent corrosion resistance and durability. Secondly, it is necessary to consider whether the shading component is easy to clean or replace; Finally, the shading equipment is required to be easy to adjust. In different areas or different types of buildings, the shading requirements are different. For example, Office buildings in hot summer and cold winter areas often have large areas of windows, even if low-e glass is used, shading conditions are different in different orientations and time periods. Especially in the north, active shading is advocated, which can be adjusted according to the season or the needs of users.

From the perspective of materials, metal shading and fabric shading are the most commonly used exterior shading methods in the market, both of which are exterior shading. Metal sunshade takes metal as the sunshade main body, often in the form of metal louvers, namely metal louvers sunshade (see Fig. 1). Excellent metal shutter shade

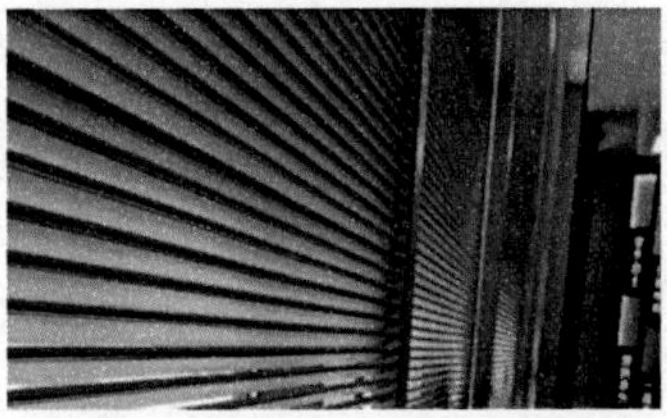

Fig. 1. Metal shutter shading

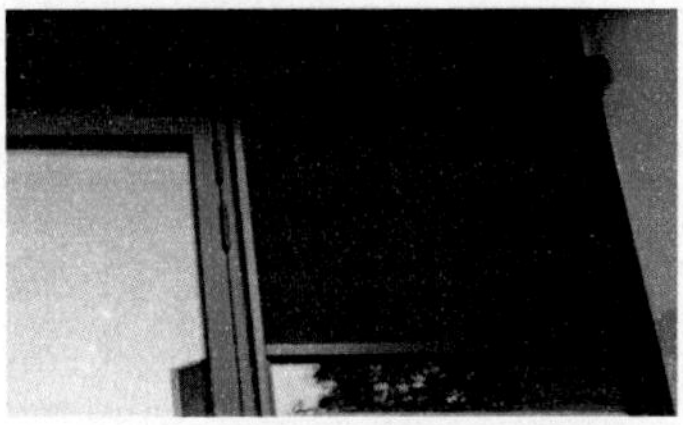

Fig. 2. Fabric shutter shading

have higher flexibility, can be adjusted under the condition of different Angle of the sun louver Angle, to achieve better shading effect (see Table 1), not only could lead to lower energy consumption of air conditioning, but also make indoor natural lighting increasing by using the light reflection refraction principle of the indoor natural light through the metal blade into the deeper area, effectively reduce the energy consumption of lighting. Is the mainstream sunshade market at present. Compared with metal shading, fabric shading takes fabric fabric as the main body and often uses the form of fabric rolling shade (see Fig. 2). At present, the sunshade fabric is used through the combination of various advanced materials, changing the process mode, so as to maximize its sunshade performance. Although the most used on the market is still metal shutter shading, but fabric shutter shading has a huge space for development. Table 2 shows the fabric parameters of advanced fabrics on the market at present. It is obvious that the tensile resistance, fire resistance, uv linear resistance and other properties of fabrics have been greatly improved. Compared with metal shutter shading, the dead weight of fabrics is lighter and the market price is closer to the people. In recent years, with the rapid development of textiles, fabric shading has gradually been covered in the market and become a potential stock in the shading market.

Table 1. Total solar transmittance of a metal sunshade louver product

Total solar transmittance		Shutter angle						
		Composite sync	Composite asynchronous	0°	30°	45°	60°	75°
Solar altitude angle	0°	0.591	0.554	0.522	0.409	0.319	0.213	0.19
	15°	0.58	0.532	0.469	0.352	0.272	0.193	0.19
	30°	0.526	0.453	0.377	0.245	0.211	0.22	0.19
	45°	0.428	0.352	0.267	0.193	0.238	0.207	0.186
	60°	0.29	0.252	0.209	0.251	0.302	0.235	0.19
	75°	0.161	0.175	0.188	0.296	0.163	0.15	0.149

Table 2. Technical parameters of a light blocking fabric

Technical indicators	Parameter values
Porosity	4%
Weight per square meter (g/m^2)	520 ± 5%
Thickness (mm)	0.75 ± 5%
The tensile strength is meridional and latitudinal (10 N/5 cm)	220, 200
Folding strength meridional and latitudinal (10 N/5 cm)	100, 100
Fracture elongation	≤5%
Antimicrobial treatment	More than 99% of the bacteria were killed
Fire rating	B1
Color fastness	7/8
Uv radiation	达96%
Transmittance (related to porosity)	4–21%
Reflectivity (porosity related)	4–66%
Absorption rate (relative to porosity)	92–13%

3 Case Analysis

3.1 Project Overview

The renovation project is a teaching building located in Taizhou, Zhejiang Province. The building is 84 m long and 30 m wide, covering an area of 2460 m^2, with a total area of about 11508 m^2. The building has 5 floors and the total building height is 21 m. Floors 1 to 3 are all for classrooms, while floors 4 and 5 are mostly for offices and other spaces.

3.2 Climatic Conditions

Taizhou is located in the central coastal area of Zhejiang Province, located in hot summer and cold winter areas. According to "Code for Thermal Design of Civil Buildings", It is clearly stated in the code that the buildings in this area must meet the requirements of heat prevention in summer and winter [5]. According to "Standard for daylighting design of buildings", Zhejiang province belongs to class IV light climate zone [6], and its daylight climate coefficient is 1.10 and the standard value of indoor natural illuminance should not be less than 300 Lx.

4 Transformation Design

The scheme is to use Swire software for simulation and calculation, based on the analysis results of the software for transformation, and get the relevant content of the analysis diagram (Figs. 3 and 4).

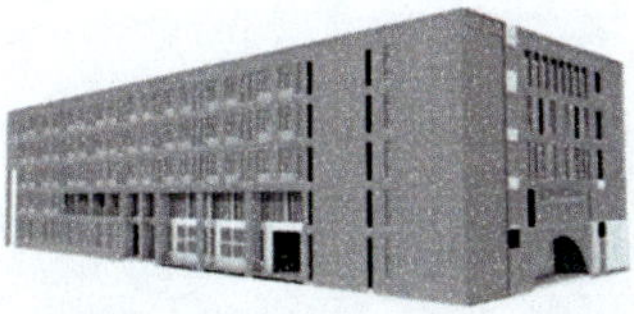

Fig. 3. Building models (before design)

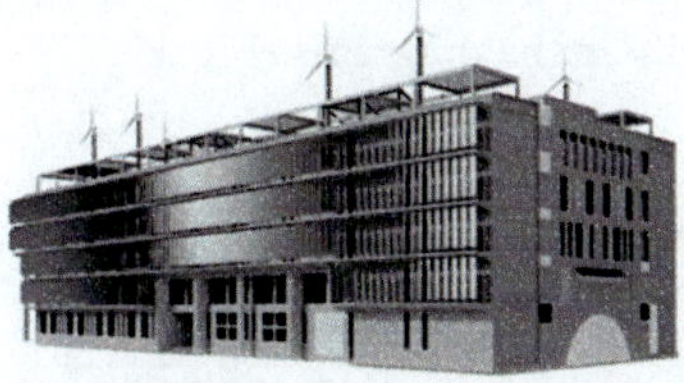

Fig. 4. Building models (after design)

5 Natural Lighting Design

Indoor Layout. The interior layout of the building is orderly (see Fig. 5), and the interior traffic is organized in the way of corridors. The inner corridor is long, and the east and west lighting openings do not bring light into the inner space. In the process of renovation design, in terms of plane layout, a classroom in the north is transformed into an open space (see Fig. 6), which serves as a rest platform or an open teaching and communication area. After the renovation design, there are more light enter the corridor space inside of the building, while increasing the activity space of students and enhance connection between indoor and outdoor.

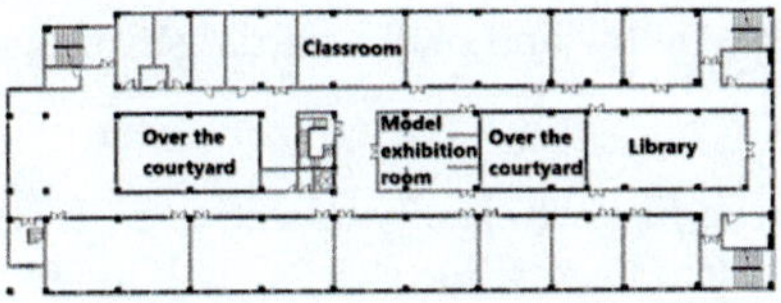

Fig. 5. 2–3 floor plan before design.

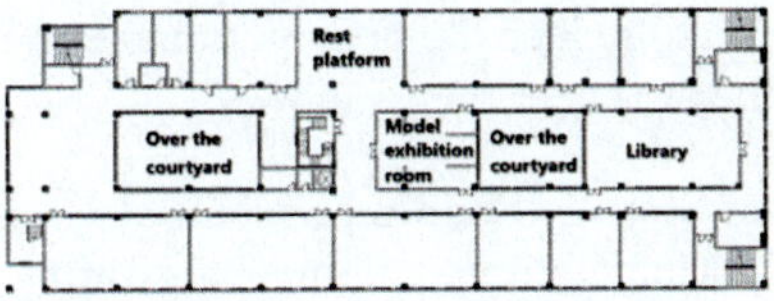

Fig. 6. 2–3 floor plan after design.

The lighting effect of the interior plane of the building is not only affected by the outdoor environment, but also affected by internal factors [7]. For example, ceiling, wall, ground, outer surface, etc., the difference between various materials and colors respectively corresponding to different reflection ratio. In the renovation design, to improve insufficient lighting room, replace the finishing materials or coatings with materials with relatively higher reflection, using the principle of light action such as reflection refraction, so that the natural light as much as possible to reflect into the room, improve the value of indoor illumination.

Tubular Day Lighting System Setup. On the west side of the courtyard is the Model exhibition room and on the east side is the library (see Fig. 5). Due to this layout, these two rooms rely on lamps most of the time. Software simulation of lighting shows the model exhibition room and the library is IV. Compared to the standard lighting coefficient of the room of 1.1%, the lighting coefficient of the library of 0.24%, and the lighting coefficient of the model exhibition room is 0.18%. lighting guide pipe systems are proposed the lighting coefficient of the showroom then turns to 1.69%, and that of the exhibition hall turns to 2.35%. It also shows the rooms with light guide pipe meets the lighting requirements and the illumination is more uniform than before, as shown visually in the diagrams (Fig. 8).

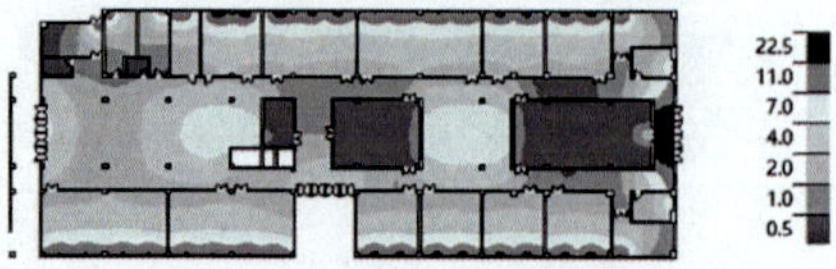

Fig. 7. Lighting diagram before design.

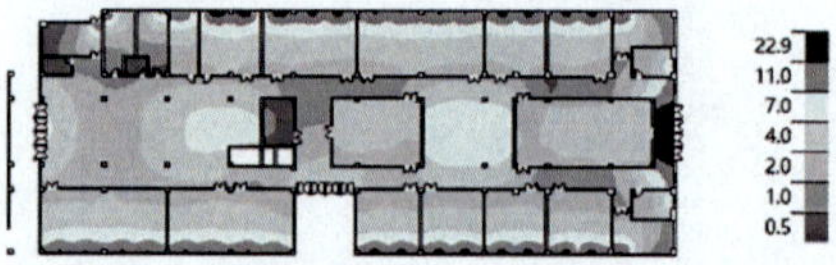

Fig. 8. Lighting diagram after design.

5.1 Shading System Design

Reflector Settings. Foreign studies have shown that exposure to the right amount of natural light improves occupants' mood, alertness and overall health, and enhances the synchronicity of our circadian rhythms with day and night [8]. In the renovation design scheme, the author used software to simulate the lighting of the project before the renovation design (see Fig. 7). The lighting illumination in the south of the teaching building is uneven, and the illumination value in the area near the outer window is 22.9, much higher than that near the inner wall is 4.0. Due to the advantageous orientation to south, the illumination is sufficient. In the the orientation to south, the reflective board is added to reduce the heat accumulation near the window and play a partial shading role in the area near the window.

Heat Storage Shading System. Lighting and shading are closely related to each other. Increasing lighting will inevitably bring solar radiant heat, while shading will inevitably affect indoor lighting. Therefore, in the design of shading measures in the reconstruction, shading and lighting should be considered as a whole to the greatest extent. As shown in Fig. 3, the original building faces south without any shading measures. According to the requirements of "Code for Thermal Design of Civil Buildings", the climate region in which it is located is hot in summer and cold in winter, so shading equipment needs to be added. In the renovation design, an adjustable heat storage shading system combining sunshade, solar energy storage and natural ventilation is set up on the south facade of the building. The main body of the system is metal shutter shading, which combines the advantages of reflector and external sunshade (see Fig. 9). However, it has not been implemented at the design stage for the time being. The principle is that the transverse perforated reflector introduces solar rays into the room to increase illumination while avoiding glare at the window. The louvers of the external sunshade use phase change materials for energy storage, and in the transition season, the Angle of the single louvers can be adjusted to guide the wind direction independently to improve indoor ventilation. The concept of the whole system lies in the combination of light guiding and sun shading. The components of the system cooperate with each other to ensure the overall operation and energy efficiency of the building by utilizing the characteristics of renewable energy.

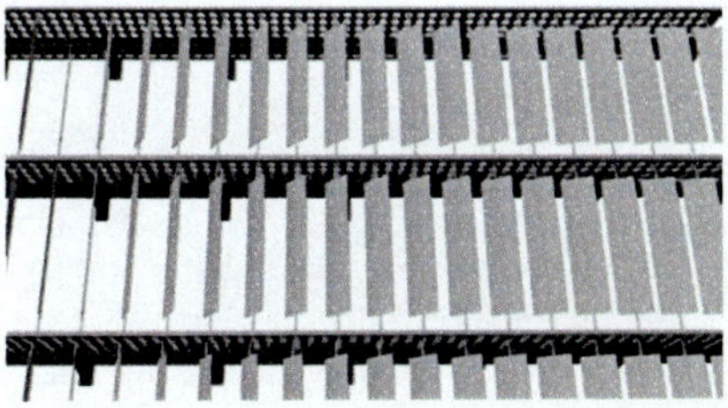

Fig. 9. Adjustable heat storage shading system

6 Results

Natural lighting and shading technology are two complementary means in architectural design. Especially in architectural design in areas with hot summer and cold winter. In the study by Zixuan Chen et al., it was proved that Windows account for a significant proportion of the total energy lost in buildings [2]. In addition, according to the survey results, most university buildings' shading measures are mostly internal shading such as curtains [9]. For the renovation design of lighting part, windows and tubular day lighting systems are added, which greatly improved the value illumination of the interior corridors and rooms. Through simulation, all rooms after the renovation design meet the requirement of lighting and the illuminance value of a few rooms can be evenly floated between 4–7. The adjustable vertical shading louvers are mainly used for shading and guiding ventilation, according to relevant research, in areas with hot summer and cold winter, compared with buildings without shading measures, buildings with shading louvers can save more than 26% of energy consumption in summer.

7 Conclusion

With the continuous development of society and economy, more and more attention has been paid to the sustainability of development, energy conservation has become an important factor that cannot be ignored in architecture. Globally, the construction industry consumes more than 30% of the final energy, and lighting accounts for 30% of the total power consumed in office buildings [10]. In the outer skin of buildings, the glass will increase the indoor solar radiation heat while obtaining sufficient lighting. This article embarks from the theory, expounds related concepts of natural lighting and shading technology, analyzes the reconstruction scheme of Taizhou project, and puts forward some suggestions, conceptual design emphasizes the role of natural lighting and shading technology in building energy efficiency, and provides reference for application in this regard.

Acknowledgments. This research was funded by China-Portugal Joint Laboratory of Cultural Heritage Conservation Science, National Key R&D Program of China (2021YFE0200100); The 2021 Jiangsu Construction System Science and Technology Project, Research on energy efficiency design method of passive ultra-low energy office building based on health and infection prevention (2021ZD03).

References

1. Bing, L.: Analysis on key points of building lighting energy saving Design. Housing Real Estate (15), 55+81 (2019)
2. Zixuan, C., Hammad Ahmed, W.A., Imriyas, K., Assed, H.: Optimising window design on residential building facades by considering heat transfer and natural lighting in nontropical regions of Australia. Buildings **10**(11) (2020)
3. Guoxin, S., Yuan, Z.: Discussion on building natural lighting methods. Energy Conserv. **29**(06), 32–35+2 (2010)
4. Gongshu, Q.: Research on building exterior shading and building integration design in hot summer and cold winter area. Zhejiang University (2008)
5. GB50176: Code for thermal design of civil buildings. Ministry of Housing and Urban-Rural Development, PRC, Beijing (2016)
6. GB50033: Building lighting design standards. Ministry of Housing and Urban-Rural Development of the People's Republic of China, Beijing (2013)
7. Lifei, D.: Research on natural lighting design of small and medium-sized buildings. Tianjin University (2006)
8. Zahra, H., Ebrahim, S., Trevor, H., Henry, S., Gillian, I., Ruwan, F.: Lighting for work: a study of the relationships among discomfort glare, physiological responses and visual performance. Build. Environ. **167** (2020)
9. Qibo, L., Juan, R.: Research on the building energy efficiency design strategy of Chinese universities based on green performance analysis. Energy Build. **224** (2020)
10. Overen, O.K., Meyer, E.L., Makaka, G.: Indoor daylighting and thermal response of a passive solar building to selective components of solar radiation, **11**(1), 34–34 (2021)

Analysis of the Effect of Stress Intensity Factors on Temperature-Shrinkage Reflective Cracks

Qinshou Huang(✉)

Guangxi Communications Design Group Co. Ltd., Guangxi 530029, China
kz101216@163.com

Abstract. A 6-parameter generalized Maxwell model is established considering the viscoelasticity of the asphalt mixture, using the second phase of the reconstruction of the East Ring Road of the existing motorway in Nanning, Guangxi as the engineering background. ABAQUS software is used to analyse the changes in stress intensity factor at the tip of temperature-shrinkage reflective cracks under the changes in initial temperature, temperature drop range, temperature drop rate, pavement structure layer, interlayer contact state and crack length. The results show that the stress intensity factor for the same relaxation time gradually decreases as the initial temperature increases, and the relaxation curve gradually slows down; the stress intensity factor for the same relaxation time gradually increases with the increase of the cooling rate of the road surface, and the relaxation curve gradually slows down; there is a positive correlation between the length of the reflective cracks and the stress intensity factor under the same relaxation time condition; there is a positive correlation between the stress intensity factor, the cooling rate of the road surface and the length of the reflective cracks under the same relaxation time condition.

Keywords: Asphalt overlay · Temperature load · Stress intensity factor

1 Introduction

A large number of highways or primary roads around cities are continuously upgraded and expanded into urban expressways with rapid urban development. At present, the most widely used in cement concrete pavement reconstruction and expansion project is the addition of asphalt surface layer, revealing the expansion mechanism of reflection crack is one of the main problems solved. Li Zilin et al. [1] carried out an analysis based on the finite element method during their study to determine the correlation between pavement structure and reflective cracking under temperature stress. Deng Hongliang et al. [2] used singular unit and fracture mechanics theory to calculate and analyse the extension mechanism of temperature-based reflective cracking stresses. Wang Xuelian et al. [3, 4] established a discrete element model of the pavement structure to study the influence mechanism and analysis of the expansion rate of reflection cracks in asphalt mixtures. Huang Likui et al. [5] applied the extended finite element method to establish a fatigue expansion model to analyse the influence of pavement structure parameters

W. Guo and K. Qian (Eds.): GBCESC 2022, LNCE 211, pp. 337–346, 2023.
https://doi.org/10.1007/978-981-19-5217-3_33

on the expansion of temperature-shrinkage reflective cracks. In this paper, based on the existing research results, the typical structure of the asphalt overlay layer in the second phase of the reconstruction of the existing motorway in Nanning, Guangxi is used as the research background, and the generalized Maxwell model is established on the basis of simplified processing by considering the characteristics of the asphalt mixture, followed by the correlation between the crack tip stress intensity factor and each parameter based on the finite element analysis software to determine the correlation between the crack tip stress intensity factor and each parameter under certain temperature stress, and the analysis parameters including initial temperature, temperature drop range, temperature drop rate, pavement structural layer, interlayer contact state and crack length.

2 Basic Theory

2.1 Determination of Type I Stress Intensity Factor $K_I(T, t)$

To investigate the correlation between the road surface variation and the structure of the paving layer, a symmetric model is chosen, with the stress intensity factor uniquely determined by interpolation of the displacement of the crack tip singular unit, see Eq. (1) [6].

$$K_I(T,t) = \frac{\sqrt{2\pi}E(T,t)}{4\left(1-\mu^2\right)}\frac{v(T,t)}{\sqrt{r}} \tag{1}$$

In Eq. (1): $E(T, t)$ is the modulus of stiffness; μ is the Poisson's ratio; r is the radial coordinate.

2.2 Mathematical Simulation of Relaxation Modulus

Different methods can be chosen in the process of viscoelastic analysis, mainly including the creep flexibility configuration method and the mechanical model method [7]. The relaxation modulus of the mixture under isothermal conditions can be described based on the generalized Maxwell model, and at the reference temperature T0, the stress relaxation modulus can be determined by the following expression as [8]:

$$E(T_0, t) = \sum_{i=1}^{N} E_i e^{(-t/\tau_i)} \tag{2}$$

In Eq. (2): E_i and τ_i correspond to the modulus of elasticity and relaxation time respectively.

Asphalt mixes can be analyzed in certain intervals based on the WLF principle of time-temperature equivalence, which corresponds to the following expression:

$$E(T,t) = E\left(T_0, t/A_\theta\right) \tag{3}$$

In Eq. (3): A_θ is the shift factor, which takes the value 20110 (0K) and is a dimensionless constant. In this paper, a 6-parameter generalized Maxwell model is used, see Fig. 1.

3 Determination of Analytical Model and Temperature Field

3.1 Determination of Analytical Model

The calculation process is considered to be viscoelastic for the paving layer material for simplicity of treatment and wired elastic for the rest of the material. In the literature [9], the modulus of stiffness of AC-16I asphalt mixes at a certain reference temperature was determined and the data is shown in Table 1. The data in the table is analyzed in Eq. (2) to determine how the modulus of stiffness changes over time, and then the modulus of stiffness for each temperature condition is determined based on Eq. (3). Figure 2 shows the finite element model correlation and Fig. 3 shows the division diagram of crack tip mesh with the relevant parameters as shown in Table 2 below.

Table 1. Generalized Maxwell model parameter [10]

Modulus of stiffness *E*/MPa		Relaxation time *t*/s	
E_1	659.95	τ_1	4215497.30
E_2	10406.02	τ_2	32.12
E_3	2534.92	τ_3	986.37
E_4	431.99	τ_4	145124.72
E_5	4148.22	τ_5	191.58
E_6	1612.87	τ_6	7569.68

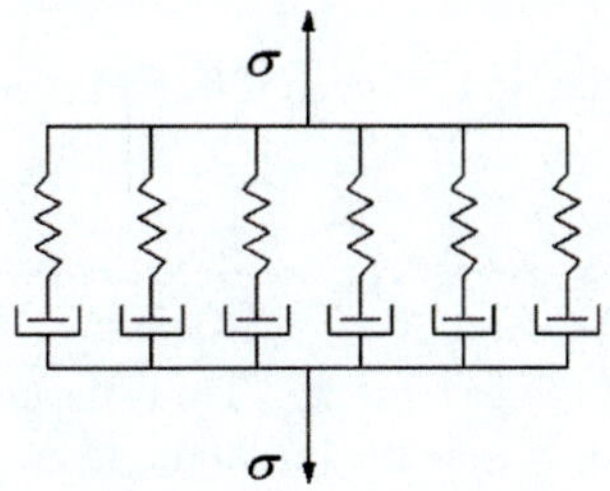

Fig. 1. Generalized Maxwell model with six parameters

3.2 Determination of Temperature Field

The road surface temperature in Naning, Guangxi, is around 10 °C at cold nights. The temperature loading parameters used in this paper assume that the initial temperature of the road surface is 10 °C and 15 °C cooling occurs, and the distribution of the pavement temperature field obeys an exponential decay mathematical model [8, 11]. Temperature decay along the depth is shown in Eq. (4), when the initial temperature of the road surface

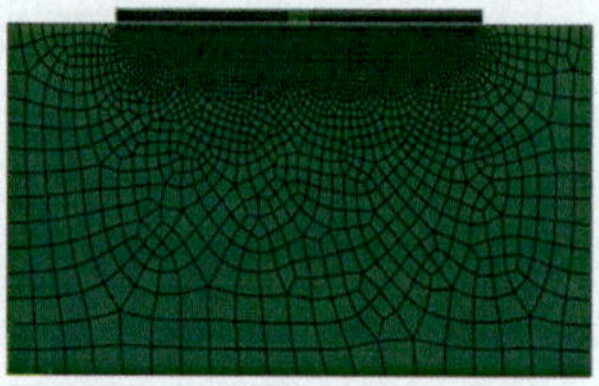

Fig. 2. Finite element model

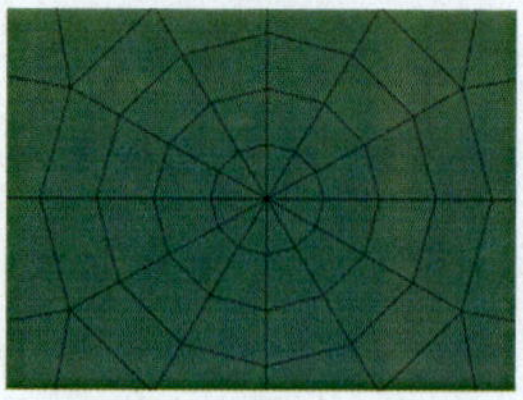

Fig. 3. Division diagram of crack tip mesh

Table 2. Major calculation parameters

Structural layer	Thickness *h*/cm	Modulus of elasticity *E*/MPa	Poisson's ratio μ	Thermal conductivity/w/m°C	Temperature shrinkage factor/1/°C
Asphalt paving	9 (9–20)	—	0.25	1.2	2.1×10^{-5}
Stress absorbing layer	2.5 (1–4)	800 (400–1600)	0.25	1.2	2.1×10^{-5}
Old cement concrete slab	24	30000	0.15	1.5	1.0×10^{-5}
Base	—	100	0.3	1.0	0.5×10^{-5}

Note: ① Parameters in brackets are the magnitude of change

is 10 °C, the temperature distribution curve at different road surface cooling is shown in Fig. 4.

$$T(t, z) = T_0 + \Delta T_m \times \frac{t}{t_m} \times \exp\left(-z/z_0\right) \tag{4}$$

In Eq. (4): z is the depth of the pavement, ΔTm is the continuous cooling amplitude, tm is the cooling experience time, z_0 is the initial thickness and takes the value of 0.2 to 0.3 m.

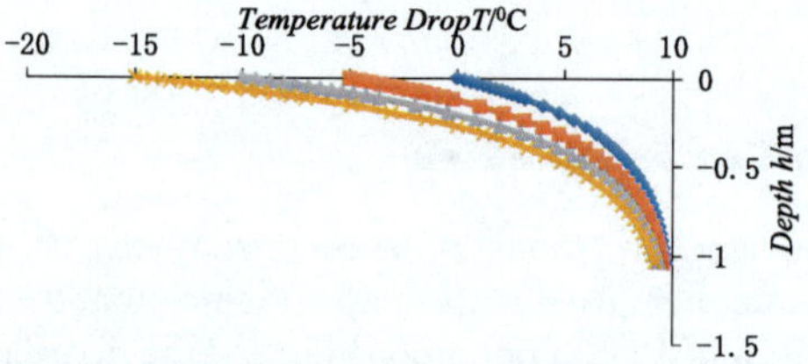

Fig. 4. Temperature distribution curve of different surface cooling

4 Effect of Parameter Changes on the Stress Intensity Factor of Temperature-Shrinkage Type Reflective Cracks

Considering the viscoelasticity of the asphalt mixture, the law of change of the stress intensity factor at the tip of temperature-shrinkage reflective cracks is analyzed under changes in parameters such as initial temperature, temperature drop range, temperature drop rate, pavement structural layer, interlayer contact state and crack length.

4.1 Effect of Initial Temperature Change on Stress Intensity Factor

Assuming that the initial temperatures of the road surface are 0 °C, 10 °C, 20 °C and 30 °C, respectively, and a fixed cooling rate of 15 °C constant, the calculation results are shown in Fig. 5. According to the results of this figure, it can be seen that there is a negative correlation between the tip stress intensity factor and the initial temperature. Taking a crack length of 1 cm as an example, the stress intensity factor at an initial temperature of 0 °C increases by 92.3% and 150.1% compared to the stress intensity factor at an initial temperature of 20 °C and 30 °C respectively. The analysis shows that the lower the initial temperature, the more detrimental it is to the structure and therefore this should be controlled appropriately to meet the structural performance requirements. Figure 5 shows that the stress intensity factor increases non-linearly as the initial temperature decreases, mainly because the decrease in initial temperature corresponds to an increase in cooling; the modulus of elasticity of the asphalt concrete increases non-linearly as the temperature decreases.

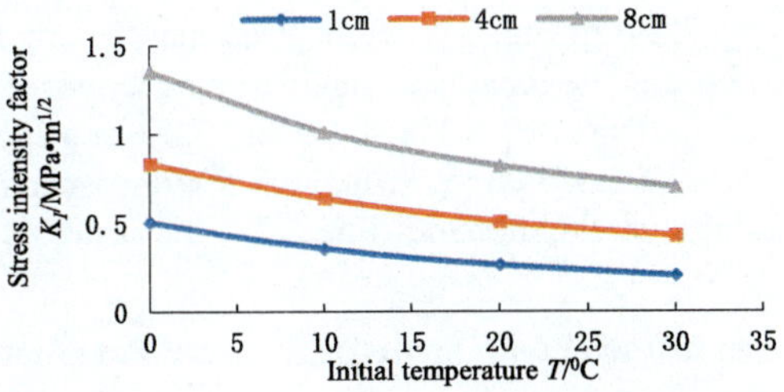

Fig. 5. Initial temperature impact on temperature shrinkage stress intensity factor

4.2 Effect of Changes in the Magnitude of Cooling on the Stress Intensity Factor

Taking the initial temperature of the road surface as 10 °C, it is assumed that cooling of the road surface occurs at 10 °C, 15 °C, 20 °C and 25 °C respectively. The results of the calculations are shown in Fig. 6, which shows that KI increases non-linearly with the increase in surface temperature difference of the paving layer. The stress intensity factor KI increases from0.232 MPa·m1/2 to 0.589 MPa·m1/2 when the temperature drop increases from 10 °C to 25 °C for a 1cm crack length, for example, the latter being 2.54 times greater than the former. The stress intensity factors for crack lengths of 4 cm increased by 49.0%, 97.7% and 146.4% when the surface temperature was increased from 10 °C to 15 °C, 20 °C and 25 °C respectively. The analysis shows that the change in temperature has a significant effect on the stress intensity factor of the pavement, which also proves that cold winters are one of the main causes of the expansion of reflective cracks in asphalt pavements.

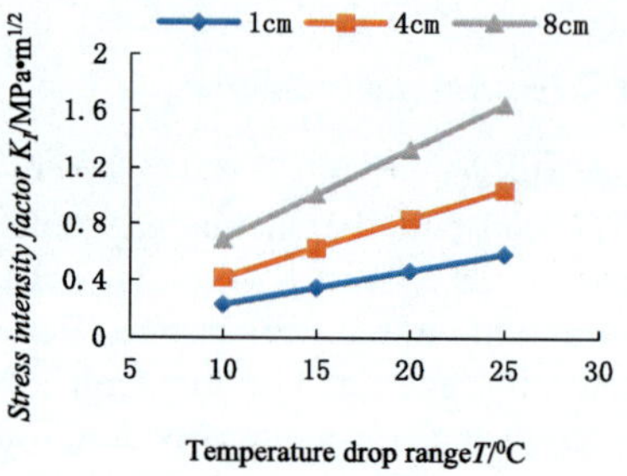

Fig. 6. Temperature drop range impact on temperature shrinkage stress intensity factor

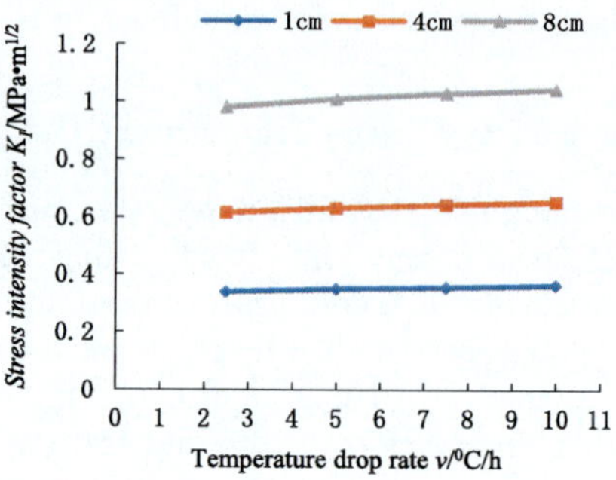

Fig. 7. Temperature drop rate impact on temperature shrinkage stress intensity factor

4.3 Effect of Changes in Temperature Drop Rate on Stress Intensity Factor

Assuming an initial road surface temperature of 10 °C, a 15 °C cooling occurs and the cooling rate increases in steps from 2.5 °C/h by 2.5 °C/h to 10 °C/h. The calculation results are shown in Fig. 7. From Fig. 7 it can be seen that the crack tip stress intensity factor increases as the cooling rate increases. Using a crack length of 1cm as an example, the stress intensity factor increases by 6.5%, 3.4% and 2.0% at a cooling rate of 10 °C/h compared to cooling rates of 2.5 °C/h, 5 °C/h and 7.5°C/h respectively. This indicates that the higher the cooling rate, the more detrimental to the structure of the paving layer.

4.4 Effect of Changes in Pavement Structural Layer Parameters on Stress Intensity Factors

Effect of Variation in Thickness of Asphalt Paving Layers on Stress Intensity Factor. The results are shown in Fig. 8, which shows that there is a negative relationship between this factor and the thickness of the surface layer. When the thickness was increased from 9 cm to 20 cm, the reduction in stress intensity factor was 31.0%, 31.9% and 48.8% for 1 cm, 4 cm and 8 cm crack lengths respectively. From the results in Fig. 8, it was found that the stress intensity factor levelled off after the thickness of the additional pavement reached a high level. It can be seen that the increase in thickness can reduce the stress intensity factor to a certain extent.

Effect of Change in Temperature Shrinkage Coefficient of Asphalt Paving Layers on Stress Intensity Factor. Changing the asphalt paving layer temperature shrinkage coefficient from 1 × 10−5 °C−1 in steps of 1 × 10−5 °C−1 to 5 × 10−5 °C−1, the

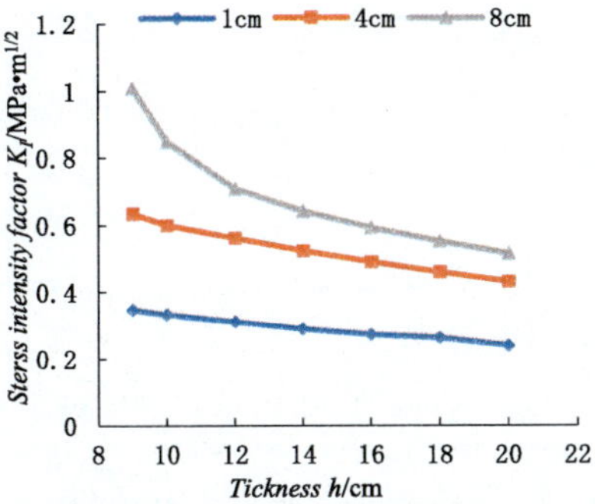

Fig. 8. Surface thickness impact on temperature shrinkage stress intensity factor

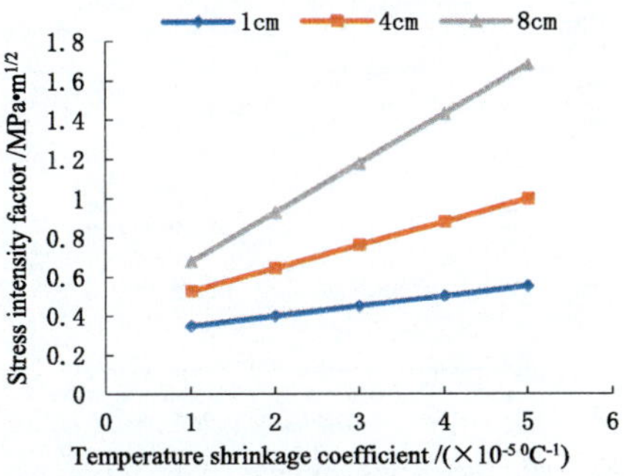

Fig. 9. Surface temperature shrinkage coefficient impact on stress intensity factor

calculation results are shown in Fig. 9. From the results related to Fig. 9, it can be found that the corresponding strength factor continues to increase during the increase of the paving layer temperature shrinkage coefficient. Taking the crack length of 1cm as an example, the stress intensity factor increases from 0.351 MPa·m1/2 to 0.558 MPa·m1/2 when the temperature shrinkage factor of the paving layer increases from 1 × 10−5 °C−1 to 5 × 10−5 °C−1, an increase of 59.0%. The factors influencing the temperature shrinkage coefficient of this material include the grade composition, the nature of the bitumen, the temperature, etc. The grade composition should therefore be optimized according to the regional pavement temperature parameters, on the basis of which the expansion of layer reflection cracks can be effectively suppressed.

Effect of Stress Absorbing Layer Thickness Variation on Stress Intensity Factor. The correlation was analyzed by taking the thickness of the absorbing layer and increasing it from 1 cm to 4 cm. The results of the calculations are shown in Fig. 10, where it can be seen that increasing the thickness of the stress absorbing layer reduces the temperature-based stress intensity factor at the crack tip, with the trend of reduction slowing down as the thickness increases. When the thickness of the absorbing layer was varied from 1–2.5 cm under a crack length of 1cm, the KI was reduced by 0.189 MPa·m1/2, a reduction of 35.1%, and from 2.5–4 cm, the KI was reduced by 14.9%. The analysis shows that under temperature loading, increasing the thickness of the stress absorbing layer can reduce the stress intensity factor to a certain extent, but the effect is no longer obvious after increasing the thickness to a certain level, so the thickness should be determined by a comprehensive trade-off.

Effect of Changes in Elastic Modulus of the Stress Absorbing Layer on the Stress Intensity Factor. In the course of this study, the modulus of elasticity was set to increase from 400 Mpa to 1600 Mpa, and the results of the analysis under these conditions are shown in Fig. 11 below, which shows that there is a negative correlation between this strength factor and the modulus of the absorption layer, and that the strength factor basically stabilizes after the modulus of the absorption layer reaches a certain threshold. When the crack length is 4 cm, the modulus of the stress absorbing layer increases

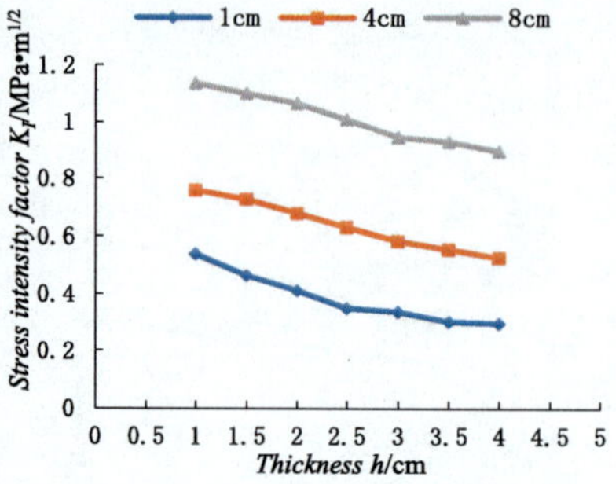

Fig. 10. Stress absorbing layer thickness impact on temperature shrinkage stress intensity factor

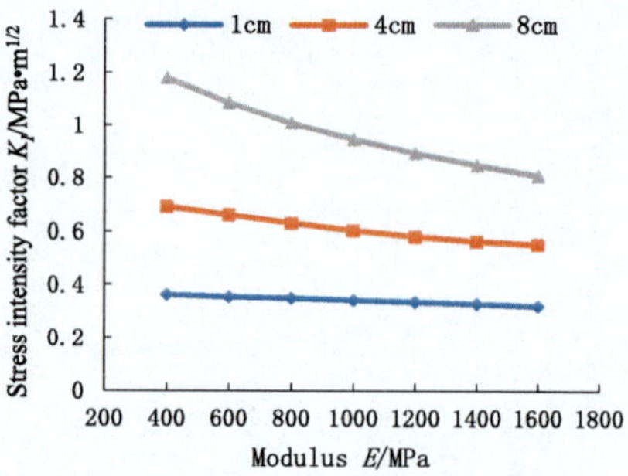

Fig. 11. Stress absorbing layer modulus impact on temperature shrinkage stress intensity factor

from 400Mpa to 1000 MPa, the KI decreases by 0.090 MPa·m1/2, a reduction of 12.9%; when the modulus increases from 1000 Mpa to 1600 MPa, the KI decreases by 0.052 MPa·m1/2, a reduction of only 8.6%. The analysis shows that under temperature loading, increasing the modulus of the stress absorbing layer can reduce the stress intensity factor to a certain extent, but the rate of reduction slows down after increasing to a certain modulus.

Effect of Changes in Interlaminar Contact State on Stress Intensity Factor. It is assumed that the bituminous paving layer and the stress absorbing layer are completely continuous, and the contact state between the old cement concrete slab and the foundation is a bonding factor f = 0.5, changing the contact state f between the stress absorbing layer and the old concrete slab between 0 and 1. The results of the calculation are shown in Fig. 12, which shows that the stress intensity factor increases with the transition from smooth to continuous interlayer contact and slows down when the interlayer bonding factor reaches a certain level. When the interlayer bond factor increases from 0 to 0.7, the stress intensity factor increases by 12.9%; when f increases from 0.7 to 1, the increase is only 8.6%. This is because when the structural layer is completely continuous, the greater the tensile stress on the crack tip under temperature loading, the greater the stress intensity factor and the faster the crack expansion. Therefore, from the point of view of temperature loading, the smoother the layers are, the better it is for the structure of the paving layer.

Effect of Crack Length Variation on Stress Intensity Factor. Vary the length of the reflective cracks in increments of 1 cm to 8 cm from 0 cm. The results of the calculations are shown in Fig. 13, which shows that the stress intensity factor at the tip of the crack increases with the length of the reflective crack in three stages: a rapid increase, a slow increase and a rapid increase. The 8cm crack length corresponds to the KI of 1.010 MPa·m1/2, which is 10.3 times greater than the 0.098 MPa·m1/2 corresponding to a crack length of 0 cm. The analysis shows that as the length of the reflective crack grows, the stress intensity factor increases and that it is most effective to take measures

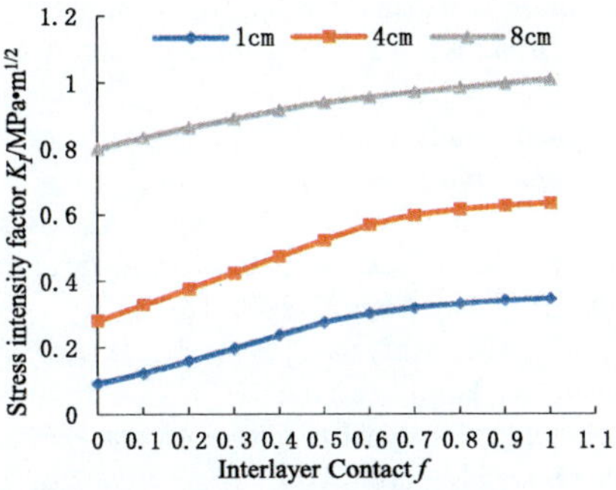

Fig. 12. Interlayer contact impact on temperature shrinkage stress intensity factor

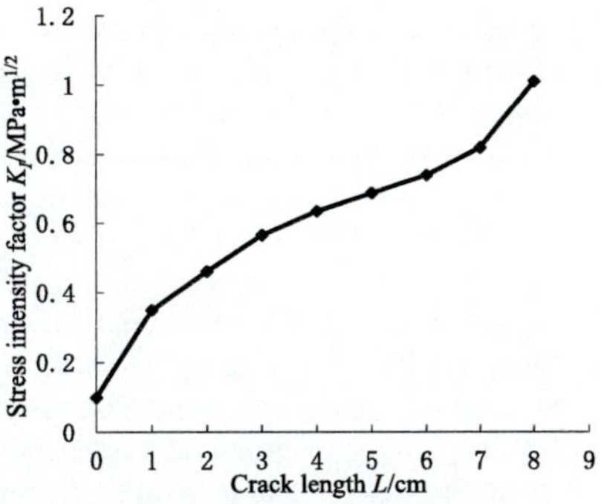

Fig. 13. Crack length impact on temperature shrinkage stress intensity factor

at the initial stage of crack expansion in terms of slowing down the expansion of the reflective crack.

5 Conclusion

In this paper, based on the second phase of the reconstruction of the East Ring of the existing motorway in Nanning, Guangxi, considering the viscoelasticity of the asphalt mixture, the 6-parameter generalized Maxwell model is established in combination with the finite element method to analyze the change law of the stress intensity factor of temperature-shrinkage type reflective cracks under different initial temperatures, temperature drop ranges, temperature drop rates, pavement structural layers, inter-layer contact states and crack lengths, and the following conclusions were drawn through numerical simulations: the calculation parameters that can reduce the stress intensity factor under temperature loading, taking into account the viscoelasticity of asphalt concrete, are: increasing the thickness of the asphalt bedding layer, reducing the modulus of the asphalt bedding layer and increasing its thickness and modulus. In addition a higher initial temperature of the road surface, a decrease in the temperature drop rate and a slower temperature drop rate can also effectively reduce the stress intensity factor of the crack tip temperature shrinkage. The higher the contact state between the layers, the higher the value of the stress intensity factor.

Acknowledgement. The work was supported in part by NSFC project 51968006.

References

1. Zilin, L., Nengfei, G., Xiaobing, L.: Analysis of the expansion mechanism of temperature-shrinkage reflective cracks in semi-rigid base asphalt pavements. J. Highway Transp. Res. Dev. **25**(1), 43–47 (2008)

2. Hongliang, D., Dan, L., Zhengnian, W.: Analysis of the expansion mechanism of temperature-based reflective cracks in semi-rigid base asphalt pavements. J. Water Resour. Archit. Eng. **7**(4), 4–7 (2009)
3. Xuelian, W., Xiaoming, H., Guojian, B.: Mechanistic analysis of LSPM for prevention and control of reflection cracks in semi-rigid base asphalt pavements. J. Highway Transp. Res. Dev. **33**(7), 8–15 (2016)
4. Xuelian, W., Xiaoming, H.: Analysis of reflection crack expansion path of semi-rigid base asphalt pavement based on numerical simulation. Highway **63**(5), 1–6 (2018)
5. Likui, H., Pan, Z., Zhiqiang, H., et al.: Fatigue extension analysis of temperature shrinkage reflection cracks in semi-rigid base asphalt pavements. Highway Eng. **45**(5), 79–83 (2020)
6. Jinchang, W., Xiangrong, Z., Junneng, Y., et al.: Stress intensity factors of cracked road structures under dynamic loading. J. Vibr. Eng. **16**(1), 114–118 (2003)
7. By Yangxian, H.: Road Surface Analysis and Design, pp. 89–103. China Communication Press, Beijing (1998)
8. Jianlong, Z., Qisen, Z.: The nolinear analysis of low temperature shrinkage cracking in ashpalt pavements. In: Proceedings of the Asian Pacific Conference on Comput. Mech, Hong Kong (1991)
9. Guoping, Q., Zhongyin, G., Jianlong, Z., et al.: Calculation of thermal viscoelastic temperature stresses in asphalt pavements under ambient conditions. J. Tongji Univ. **31**(2), 150–155 (2003)
10. Guangxi Zhuang Autonomous Region Expressway Administration, School of Civil Engineering, uangzhou University: Study on the Typical Structure of Asphalt Pavement of Old Cement Concrete Pavement in Guangxi. General report of the study, vol. 2, pp. 61–63 (2006)
11. Jianlong, Z., Zhigang, Z., Qisen, Z.: Principles and Methods of Crack-resistant Design of Asphalt Pavements, pp. 15–64. China Communication Press, Beijing (2002)

Exploration of the Emergency Renovation of Sports Venues Resilience Space Based on Green Building Technology Under the Normal Epidemic Situation

Jie Liu and Tong Yang(✉)

School of Management, Xi'an Polytechnic University, Xi'an 710048, Shaanxi, China
yangtong1268@163.com

Abstract. The sudden outbreak of COVID-19 has caused a surge in medical demand. It has inspired people to continuously explore how to transform public buildings such as gymnasiums in a fast, low-cost and green way during emergencies. The article studies the feasibility of applying gymnasium to sudden public events, discusses the design methods for the renovation of gymnasium space, water supply and drainage system, ventilation system and intelligent system in emergency situations. The focus is on preventing cross-contamination, preventing backflow contamination, pressure shaving, airflow organization, and system control. Through these design methods, the gymnasium has the characteristics of efficient, adaptable and inclusive epidemic prevention. The application prospect of green building technology in emergency reconstruction was explored, and a reference is put forward for the design and reconstruction of gymnasiums in the post-epidemic era with "combination of epidemic control" and improving the resilience space of gymnasiums.

Keywords: Green building technology · Passive energy saving technology · Backflow prevention · Ventilation system · Emergency transformation

1 Preface

At the beginning of the prevention and control of the COVID-19, Wuhan, which was on the front line of anti-epidemic, went from the temporary site selection of the "Liang shen shan Hospital" to the transformation of large public buildings such as exhibition halls and stadiums into temporary hospitals. This indirectly expresses the elastic service ability of urban public venues [1]. But the construction industry cannot ignore its professional responsibilities just because the "Wuhan-style Fangcang shelter hospitals" seems to have solved the problem. Under the normal situation of the epidemic, what strategy should the gymnasium adopt at the beginning of the design to be suitable for emergency reconstruction and construction is a question worth thinking about in architecture at the moment [2]. In the post-epidemic era, it is of great practical significance to use green building technology to carry out resilient space design and emergency renovation of stadiums.

W. Guo and K. Qian (Eds.): GBCESC 2022, LNCE 211, pp. 347–355, 2023.
https://doi.org/10.1007/978-981-19-5217-3_34

2 The Feasibility of Applying Sports Venues to Emergency Public Events

As an important urban infrastructure, the gymnasium is a position for mass sports, as well as an important place for sports training, sports competition and mass sports. According to the building level, it can be divided into four types of venues: premium, A, B, and C [3]. The safety performance of each level of gymnasium is shown in Table 1.

Table 1. Safety performance table of gymnasium classification.

Building grade	Requirements	Main structure service life	Fire-resistant level	Audience
Premium	World-class competition	>100 years	≥1st level	
Grade A	National event			6000–10000 people
Grade B	Regional events	500–100 years	≥Level 2	3000–6000 people
Grade C	Local and mass events	25–50 years		≤3000 people

It can be concluded from Table 1 that according to the architectural requirements of the gymnasium, the gymnasium has considered good use effects and relatively safe features in the early stage of design and construction. The gymnasium is basically a large-span space with independent entry and exit channels, which is convenient for people to enter and exit and flow and has strong accessibility. The well-equipped facilities are easy to use and the lighting is better for rehabilitation which can meet the needs of a variety of unexpected public events.

3 Resilient Space Design and Emergency Renovation Strategy of Gymnasium Based on Green Building Technology

3.1 Renovation of Space

In order to make full use of the gymnasium in peacetime and during the epidemic, the inclusiveness of the gymnasium should be increased at the beginning of the design, and the needs of epidemic prevention should be added to it. Two copies of the design drawings are drawn, one for the normal operation and the other for the renovation during the epidemic. The division of "three zones and two passages" and "pressure gradient zone" in the drawings should be clear and clear, and the integral plane ceiling should be used as far as possible in the area where the pressure gradient is prepared [4]. The key to the prevention and control of temporary hospitals is to prevent cross-infection. Therefore, the transformation of the indoor space of the stadium is mainly the division and utilization of the large space. The use of tenon-and-mortise technology allows the gymnasium to be transformed into the "three zones and two passages" required by the transformation of the various independent functional divisions of the gymnasium and the entrance passage into the infectious disease treatment hospital to avoid crossing each other [5]. The doors and windows that need to be blocked during the epidemic are designed to be airtight doors and windows in each room, and secret doors are added, which are normally closed and connected to the outdoors during the epidemic.

In the whole process of space renovation, passive energy-saving technology is used to rationally arrange the space in the gymnasium to reduce energy consumption. At the same time, green, sustainable and reversible materials should be used as much as possible. Such as using the Woven Image invented by the British Institute of Materials. This material uses sustainable materials and can be assembled and infinitely changed according to actual needs [6].

3.2 Water Supply and Drainage System Transformation

Water Supply System. According to the requirements of the *Guidelines for the Design of Emergency Treatment Facilities for Novel Coronavirus Pneumonia (Trial)*, independent water supply mains can be set up for large gymnasiums "combined with epidemic prevention" at the beginning of construction, and the location of the cut-off water tank can be reserved for epidemic situations [7]. During the epidemic, using green building technology, the water supply transformation of the cut-off water tank is carried out for the lower area of the first floor, and the frequency conversion pressurized water supply is used for the second floor and above, that is, the water supply method of the cut-off water tank is adopted. Add a decompression type backflow preventer during the supply process to prevent contamination caused by water backflow.

Hot Water System. Combined with the idea of green energy saving and saving transformation cost, there are two options for heating mode: commercial gas water heater and commercial electric water heater [8]. The advantages and disadvantages are as follows (Table 2):

Table 2. Comparison of heating equipment schemes for hot water system.

Heating method	Advantage	Shortcoming
Commercial gas water heater	Basically no electricity consumption, relatively high energy consumption, low operating costs	Exhaust smoke and equipment must be placed outdoors; municipal gas supply conditions are required
Commercial electric water heater	Keep clean and safe equipment indoors	The energy consumption is relatively low, the operating cost is high, the electricity load is large, and the requirements for the power supplier are high

When the conditions of the venue meet the requirements of gas water heaters, commercial gas water heaters can be given priority for heating. When the municipal gas supply conditions are not met, but the electric water heaters are used to prepare hot water, the decentralized electric water heaters should be used to supply the patients' toilets and showers. The water heaters on the market are generally instant type and volumetric type [9]. We check the two types of electric water heaters. Take the case of a toilet and a shower as an example (there are 12 sensor taps and 4 showers).

Calculation formula of power consumption N (kW) of instant electric heating equipment [10]:

$$N = (1.10 \sim 1.20)3600q(t_r - t_l)C\rho_r/(3617\eta) \quad (1)$$

Calculation formula of power consumption N (kW) of volumetric electric heating equipment (except heating before use and continuing to heat during use):

$$N = (1.10 \sim 1.20)(3600qT_1 - V)(t_r - t_l)C\rho_r/(3617\eta T_1) \quad (2)$$

The parameter values are based on the *Building Water Supply and Drainage Design Standards.* Winter is the time when the new coronavirus is at its highest, so the cold water temperature is set at 4 °C.

Table 3. Calculation of power consumption in toilets and showers in wards under instant electric heating equipment.

Water appliance	Use water temperature /°C	Cold water temperature /°C	Power consumption /kw/piece	Number of appliances /piece	Total power /kw
Induction faucet	30	4	13.0	12	156
Shower	40	4	18.0	4	72

Table 4. Calculation of power consumption in toilets and showers in wards volumetric electric heating equipment.

Water appliance	Water storage volume /L Power/kw	Use water temperature /°C	Cold water temperature /°C	Power consumption /kw/piece	Water time	Number of appliances /piece	Total power /kw
Induction faucet	400/9	30	4	0.6	12	12	7.2
Shower	800/12	40	4	2.5	5	4	10

According to the rated power of the instant electric heating equipment, the heat supply obtained from Table 3 is far from enough to cover the designed hourly heat consumption. It can be seen from Tables 3 and 4 that the volumetric equipment not only consumes less power, but also can meet the continuous water consumption of medical staff in the gymnasium during the epidemic period.

Drainage System. The most difficult part of the renovation during the epidemic is the reasonable discharge of sewage. The main sources of sewage are domestic sewage in medical and nursing areas, domestic sewage in wards, and medical wastewater [11]. Direct discharge will cause environmental pollution, so it must be treated up to standard. Using green building technology, sports venues can be transformed as follows. Because the medical care area has not been infected, the medical care area on the ground floor can use the original toilet of the gymnasium. A single row of pipes is added to the shower room on the ground floor to go directly out of the house, and a separate drain riser is added to the shower room on the high-level for drainage on the same floor and the riser is set outdoors against the outer wall.

The toilets in the wards are replaced by mobile toilets, which can not only be installed quickly but also facilitate recovery after the epidemic. In this renovation, the mobile toilet will be equipped with pipeline drainage, and UPVC drainage pipes (bonded) will be used to lay along the ground to reduce the amount of pipeline excavation at the construction site, and a drainage riser will be installed outdoors against the outer wall. The drainage pipes are determined by DN200 drainage pipes for every 20 toilets with a slope of 0.01. All drainage risers in the ward are extended to the top for ventilation, and in order to protect the surrounding environment, ultraviolet lamps need to be installed at the top for disinfection [12]. The layout of the local drainage of the toilets and toilets in the ward is as follows (Fig. 1).

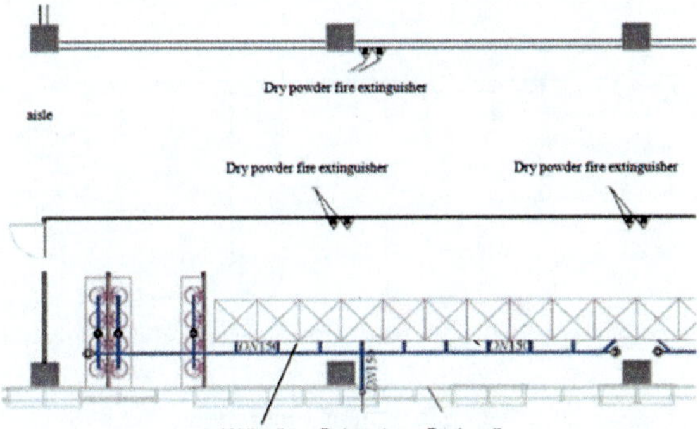

Fig. 1. Layout plan of local drainage renovation of toilet in ward.

3.3 Ventilation System Transformation

Many of the original building ventilation and air-conditioning systems in the gymnasium are positive pressure systems, which cannot meet the medical requirements of infectious diseases such as new coronavirus pneumonia. Respiratory infectious disease wards are divided into clean areas, semi-contaminated areas and polluted areas according to whether they are contaminated by pathogenic microorganisms [13]. In the process of reconstruction, the original ventilation and air-conditioning equipment and facilities are fully utilized. Using the power distributed ventilation system, part of the power is distributed at the end of each branch, and the adjustment is flexible and stable [14]. A certain pressure gradient is formed between the clean area, the semi-polluted area and the polluted area, as shown in Table 5 below. The clean area only supplies air and does not exhaust air, forming a unique airflow from the clean area to the polluted area, thereby avoiding air pollution.Adding coarse, medium and high-efficiency filters to the air supply system to further protect the health of medical staff. The flow chart of the ventilation system is shown in Fig. 2.

The airflow organization in the gymnasium can be transformed according to the following Fig. 3: U-shaped airflow direction is adopted. Use the air inlet of the high side window on the far left to naturally enter the air, collect it into the middle aisle, and flow from left to right. The indoor exhaust system is distributed on both sides of the ward area and the public health area, and the air outlet is arranged in the lower part of the room, 500 mm from the ground, and the air intake of the aisle is supplied to the ward area on both sides. In the air inlet system, the safety of the air intake should be ensured, and the air inlet should be kept away from the pollution source and the waste flow line. The exhaust system adopts the method of dispersing the indoor exhaust outlets to reduce the air flow between the intake air and the exhaust air and the airflow interference between the cabins, so that the polluted air can be discharged as soon as possible.

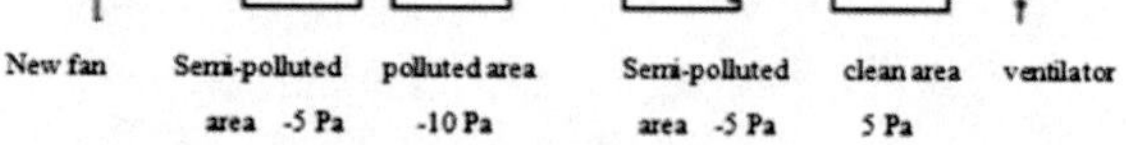

Fig. 2. Schematic diagram of ventilation system flow.

3.4 Intelligent System Transformation

Environmental Monitoring Transformation. Environmental monitoring can realize real-time monitoring and emergency response of the environmental conditions in the stadium during the epidemic. On air-conditioning equipment, automatic control is adopted

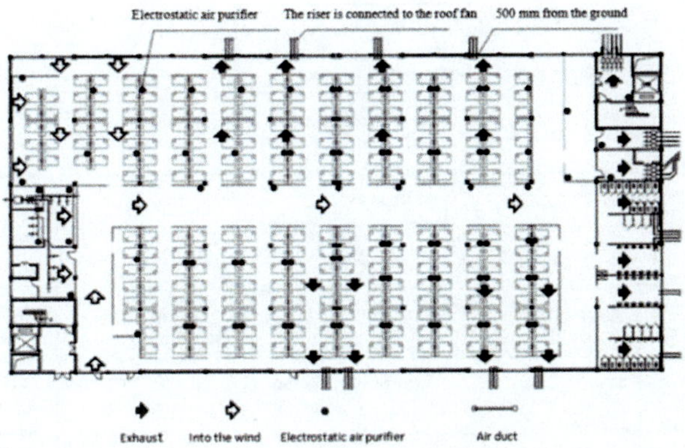

Fig. 3. Indoor airflow organization diagram after renovation.

Table 5. Hospital pressure difference for respiratory infectious diseases.

Ribbon	Pressure
Ward, Ward restroom	−15 Pa
Contaminated areas such as buffer rooms and patient walkways	−10 Pa
Semi-contaminated areas such as medical corridors	−5 Pa

for polluted and semi-polluted areas. The ventilation system is controlled by a digital control system, the system is easy to adjust, the air volume is easy to balance, the system runs stably and monitors the differential pressure state of each negative pressure unit in the polluted area and semi-polluted area. Each pressure unit is equipped with a temperature and humidity detection device to supply and exhaust air. The system's filter sets the differential pressure sensor [15]. According to the monitoring of the pressure difference between the ward and the buffer room or between the ward and the medical corridor by the differential pressure sensor, the automatic adjustment operation of the branch exhaust fan is linked and controlled, and then the variable air volume operation of the main exhaust fan is linked. The control logic is shown in Fig. 4 below.

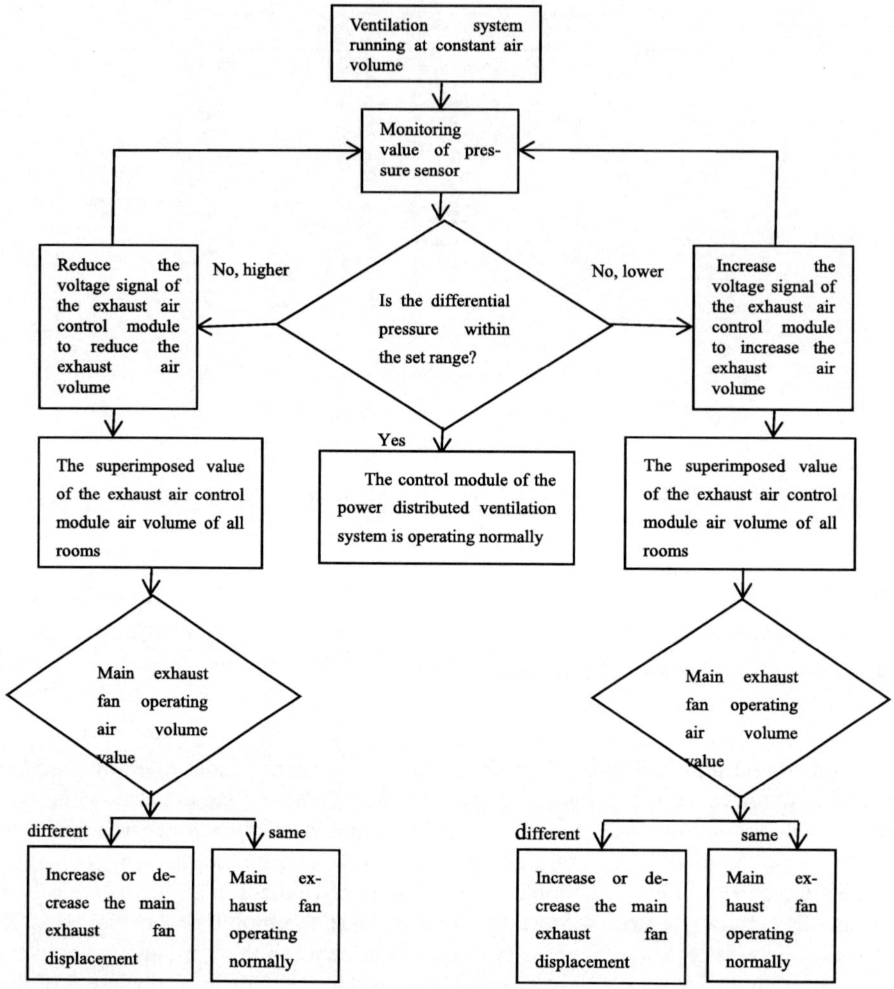

Fig. 4. Control logic diagram of ventilation system after renovation.

4 Concluding Remarks

Compared with the temporary construction of hospitals, it is more convenient and efficient to make full use of the larger-scale public buildings in the existing cities. It is the inevitable result of the future global "dual-track peace and war" thinking through renovation and construction to make it a temporary treatment place. To implement this idea, all government departments have to pay attention to the important role played by the resilience of public buildings in the event of major natural disasters [8]. In the initial design of related projects, the use requirements for future disaster response should be added to it and considered together. And in the process of device research and development and construction, make reasonable planning and relevant strategic preparations in response to the epidemic, firmly establish the awareness of "full cycle management", and improve the city's comprehensive ability to deal with sudden disasters.

Acknowledgement. The work was supported by Shaanxi Science and Technology Department Soft Science Project (2021KRM127).

References

1. Yaoxing, L., Yi Min, S.: Research on the value connotation and renewal mode of existing gymnasiums in my country. Archit. J. (s1), 165–168 (2021)
2. Hao, L., Ke, X.: Discussion on the anti-epidemic plan of architectural design of large-space public buildings under the background of healthy cities–taking large-scale gymnasium buildings as an example. Shanghai Urban Plann. (02), 31–37 (2020)
3. Hao, W.: Thoughts on the application of gymnasium to emergent public event. Contemp. Sports Technol. **11**(24), 145–147 (2021)
4. Yanying, L.: Research on the flexible design of buildings with combination of ping and epidemic taking the renovation design of the Fangcang hospital of Lianjiang county as an example. Fujian Archit. (05), 28–31 (2021)
5. Sihong, L., Long, Z.; Research on the reconstruction design of urban resilient space makeshift hospital under the new crown epidemic. Beij. Plann. Constr. (04), 39–41 (2020)
6. Xiaoru, Z., Yang, M., et al.: Study on the design of the field hospital under the epidemic situation. Creativity Des. (02), 33–39 (2021)
7. Jianbin, W., Feng, X.: Discussion on the main points of water supply and drainage design of "combination of epidemic control' in general hospitals. Water Supply Sewerage **57**(10), 85–90 (2021)
8. Qiyuan, Z., Ying, H., et al.: Main points of water supply and drainage design in shelter hospitals. Huazhong Constr. **38**(04), 123–125 (2020)
9. Bin, L., Ming, W., et al.: Discussion and summary of hot water system design in Fangcang cabin hospital. Water Supply Sewerage **56**(05), 18–21 (2020)
10. Zengzhang, W., Yu Fei, G.: Building Water Supply and Drainage Engineering. China Construction Industry Press, Beijing (2016)
11. Xiaomei, Q., Ying Hui, H., et al.: Design of water supply, drainage and fire protection system in Fangcang shelter hospital–taking the renovation project of table tennis gymnasium in sportscenter as an example. Water Supply Sewerage **56**(05), 32–351 (2020)
12. Notice on Doing a Good Job in the Supervision of Medical Sewage and Urban Sewage in the Pneumonia Epidemic of Novel Coronavirus Infection. https://www.mee.gov.cn/xxgk2018/xxgk/xxgk06/202002/t20200201_761163.html. Accessed 01 Feb 2020
13. Zaipeng, Z., Yanhua, C.: Design of ventilation and air-conditioning system in a shelter hospital. HVAC **50**(11), 62–65+87 (2020)
14. Hao, F., Jingping, W.: Discussion on reform measures and operation management strategies of hospital air-conditioning ventilation system. Chin. Hosp. Build. Equip. **22**(04), 101–103 (2021)
15. Jing, L., Sen, L.: Research on standardization of intelligent system construction in epidemic prevention hospital. Stand. Sci. (11), 57–59+89 (2021)

Research on Boundary Space Design of Urban Residential Community in the Pearl River Delta

Junrui Cao[1,2](✉) and Yichenfei Guo[3]

[1] Faculty of Humanities and Arts, Macau University of Science and Technology, Avenida Wai Long Taipa, Macau 999078, China
415392373@qq.com
[2] Industrial Design Institute, Zhao Qing University, Zhaoqing 526061, China
[3] CIECC, 25 Cheongzhuang West Road, Beijing 100048, China

Abstract. The essence of boundary space is the transition and fusion zone of two different substances, space or regions. As a form of living closely related to people's life, the way of presentation of public space affects people's every move. The author hopes that through the research of urban residential community boundary space types in Pearl River Delta region, taking different types of common problems as the starting point, from the perspective of urban protection and renewal, through "architectural design", "art design", "landscape aesthetics" etc. explore the new model of harmonious coexistence and greening between people and nature. This is not only the transformation and continuation of the ecological environment, spatial environment, cultural environment, visual environment, and recreation environment, but also the continuation and renewal of the social network structure, psychological setting, and emotional attachment of the neighbors.

Keywords: Guangdong-Hong Kong-Macao Greater Bay Area · Flexible boundary · Urban renewal · Neighborhood social network

1 Research Background and Significance

Research background and significance: after more than 40 years of reform and opening up, China's economy has developed rapidly, and urban development will change from extensive extension to connotation promotion. Due to the gradual natural aging of urban buildings and facilities, the original architectural design standards and methods have fallen behind, and the original functional layout, architectural facilities and spatial environment of some cities no longer meet the higher living needs of residents. They need to be transformed and updated to make them look new and improve the people's sense of acquisition, happiness and security, However, the boundary space of residential areas is often the most easily neglected zone. With more and more refinement and development, the boundary space will release great potential, which is not only in line with the requirements of the new era, but also the growing needs of the people for a better life.

W. Guo and K. Qian (Eds.): GBCESC 2022, LNCE 211, pp. 356–364, 2023.
https://doi.org/10.1007/978-981-19-5217-3_35

2 Existing Problems in the Boundary Space of Residential Areas

The boundary space can be distinguished according to the physical level and the psychological level. From the physical point, the boundary space is an objective physical space, which is affected by adjacent and different spaces or regions. At the same time, its use function and spatial attribute are also influenced by the contradiction and mutual integration between different subject spaces. It has the properties of transition and connection, and it also has the characteristics of the space between adjacent subjects. From the psychological point, the boundary space is psychological boundary space generated by people's subjective feelings. For example, the boundary space produced by vision, touch, smell and hearing caused by color, material, smell and sound, and even the psychological boundary space between people [1].

In order to have a comprehensive and in-depth understanding of the current situation of the boundary space of urban residential areas in the Pearl River Delta region, the author conducted random visits to the residential community in the Pearl River Delta region, including Guangzhou, Shenzhen, Zhuhai, Zhongshan, Jiangmen, Huizhou, Zhaoqing, Foshan and Dongguan. Research to residential boundary space as the main research object. The main elements include architecture, landscape, roads, etc. different boundary spaces are formed according to the relative positions of different subjects, through the author's data sorting and research found that there are many common problems, after summary mainly have the following:

2.1 The Boundary Space is Separated in a Rigid Way

The boundary space between hard media lacks flexible transition, integration and interaction, which has caused it to become a negative space, making the "harder" interface more stiff, giving people a "cold" space atmosphere and greatly reducing people's emotional demand for space.

For example, simple walls are often used in the front space of the house. In order to ensure the privacy of their lives, low-rise residents have taken measures such as making closed balconies, hard maintenance structures and heightening windowsills. The boundary space formed between the landscape and the road in the residential area is usually separated by stone steps (Fig. 1). Roads and buildings in residential areas are usually "naked" and directly connected to roads.

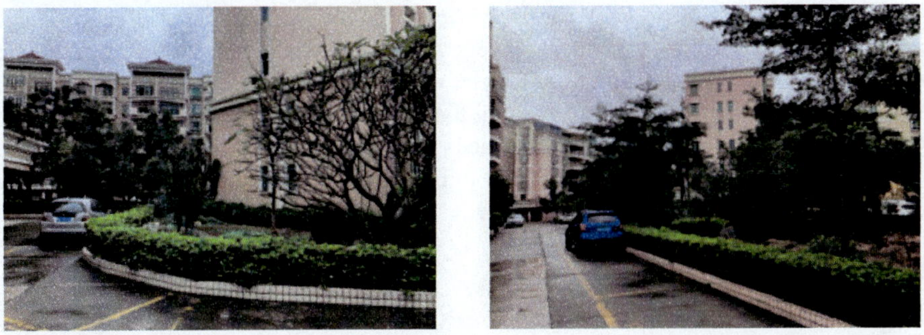

Fig. 1. World partition of landscape area of residential area (from the author)

2.2 The Boundary Space Has a Single Functional Form and a Low Utilization Rate

Most of the boundary spaces in residential areas only play the role of dividing space, including only separated hard media, only ornamental landscape, and space that can only pass quickly, but it has nothing to do with human activities.

For example, residents' activity space from buildings, roads to landscape is not direct enough, and too clear boundary of landscape area is easy to lead users to choose "enter or not enter", and even bypass to find the entrance in order to enter the landscape. The space in front of the house is also occupied by garage and warehouse, but most of them are stacked with old things, and the space utilization rate is very low (Fig. 2). It also gives the passing residents a "cold" feeling, which not only affects the overall beauty of the environment, but also indirectly reduces the humanity of the neighbors. The form and function of the unit door are too single and lack of necessary residence space and facilities, which greatly reduces people's communication and activities.

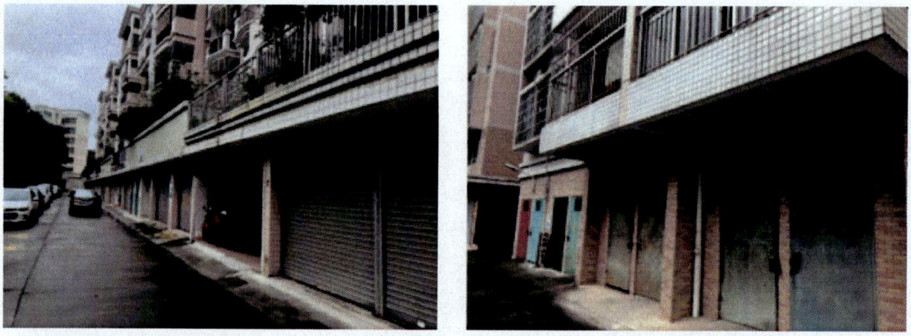

Fig. 2. Low utilization rate of front space in residential areas (from the author)

2.3 The Attributes of Boundary Space Are Unclear and Lacking Overall Design

Due to the long construction time and the lack of unity in the overall spatial planning of the old residential area, it affects the overall beauty.

For example, under the influence of different living habits and use preferences of different residents, the front space forms a variety of space forms, some are directly abandoned, some are used to plant fruits and vegetables and accumulate sundries, and some park vehicles, with chaotic function distribution and unclear space attributes (Fig. 3). The road design does not take into account the connection between the traffic flow line of motor vehicles and the parking area, does not achieve the diversion of people and vehicles, and lacks road guidance signs or indicative carriers, so it is difficult to form a sense of domain and identity of road space.

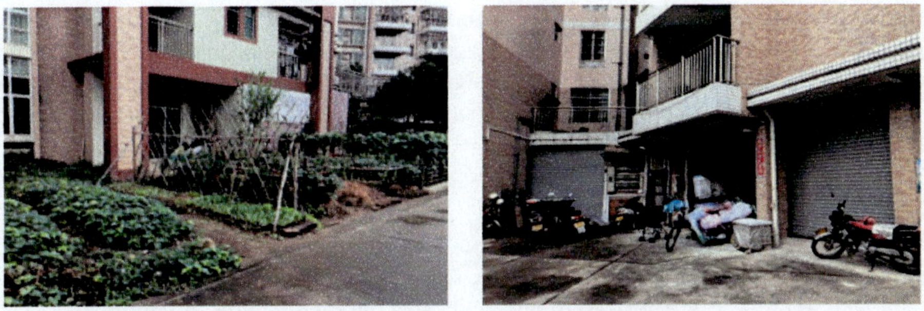

Fig. 3. Overall status of the front space in residential areas (from the author)

2.4 The Boundary Space Lacks Interactive Public Areas

Due to neglect of management, public space has also been gradually abandoned, resulting in a sharp decline in the popularity of the space, forming a vicious circle.

For example, the level of landscape vegetation is single, the collocation of tree species is not rich, and some areas will be overgrown with weeds or exposed ground cover; The accessibility of paths and corridors between landscapes is limited. The landscape boundary space lacks residents' daily activity sites, and the places with public space attributes such as semi enclosed space and enclosed space are scarce, which can not meet the needs of residents for stay, rest, entertainment and communication. The activity function area set in the central landscape area of the residential area has no ground laying, the land is muddy, and the environment after the rain is particularly unbearable, which is inconvenient for residents to use. Fitness equipment, landscape kiosks and other structures have become the "furnishings" of the central area, and most people stay away (Fig. 4).

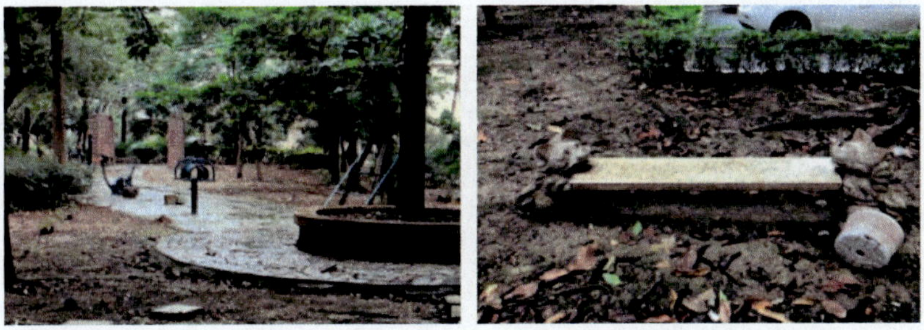

Fig. 4. Current status of boundary space use between landscapes in residential areas (from the author)

3 Design Strategy of Residential Boundary Space

It brings more available space for residents, promotes the communication between human and nature, improves residential environment, life and the quality of city which using flexible boundary space design gimmick to weaken tough dividing line and forming a good space transition. The specific optimization strategy of residential boundary space mainly starts from architectural space design, optimization of landscape facilities and strengthening of residential area management [2]. The specific performance is as follows:

3.1 Architectural Design Level

From the aspect of architectural space: define spatial attributes, enrich spatial functions, strive to provide residents with more public space for communication, leisure and entertainment, improve space utilization rate, and carry out rectification and optimization of space.

The space in front of the house belongs to the activity space of semi-public property in the residential district. According to the survey, residents have different uses of the space in front of the house, which makes its spatial attributes fuzzy. The optimization needs to define the boundary space property of the semi-public space as the premise, as a good transition medium of indoor and outdoor space, to complete the natural transition of space. Semi-public Spaces are uniformly set in front of the house to ensure the integrity of the domain level. For example, platforms or yards are added outside the ground floor, and fences, hedges, shrubs and other media with low and semi-permeable media properties are set as the transition of the house front space [2].

Unit entrance, as the boundary space between interior and exterior of residential district, is one of the Spaces where residents communicate most frequently. In view of the small scale of the canopy and the cramped space at the entrance of the unit, the abandoned warehouse and garage space are utilized to widen the size of the entrance of the unit and provide a larger space buffer area for people to enter and exit the unit. Conditional village still can expand the scale of vestibular, complementary with interior design, wait like laid of lamplight, furniture, ground, vestibular design which can offer resident to stay, the public space of association (Fig. 5).

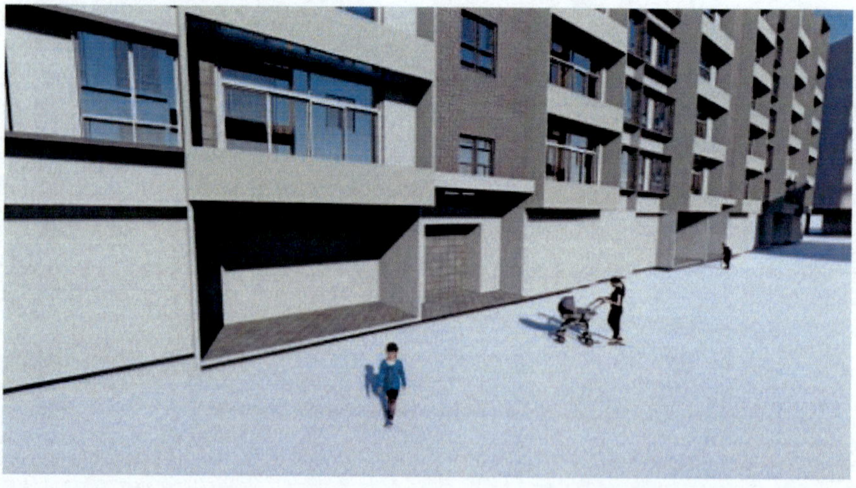

Fig. 5. Design of buffer space inside unit entrance of Xinlinli residential district, Zhaoqing City (from the author)

For the large-scaled space at the entrance of the unit, we can adopt the spatial circuitousness processing technique, draw on the function of the screen wall in the courtyard of traditional folk houses to partially block the interior, so that the hallway of the unit has a turning point in space, so as to slow down the rhythm of people entering and leaving inside and outside, and provide enough time for people's psychological transition between inside and outside. In addition to the foyer, the boundary space at the entrance of the unit also contains the entry buffer, or canopy and other elements. According to the form, material and scale of the canopy, a corridor is set up between the canopy at the entrance of the unit to form an integrated space between the entrance of the unit and the house.

The integrated design of architecture, landscape and road is a transition from a system composed of landscape, road, landscape and architecture to a public space – semi-public space – private space. Open the limited and narrow entrance of the landscape area to make an open landscape area, directly connected with the road, so that the landscape space on both sides of the boundary can transition naturally (Fig. 6). The entrance and exit are set in the place where the flow of people is most concentrated in the community. The linear landscape boundary is curvilinear, the monotonous linear boundary is weakened, and the concave and convex areas are used to create more possibilities and diversity of space.

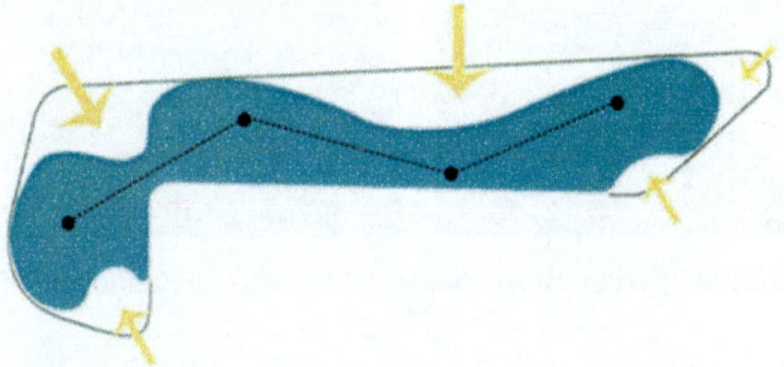

Fig. 6. Schematic diagram of open landscape area of Jinshengyuan Residential District, Zhaoqing City (from the author)

The boundary space between the landscape of residential areas, starting from the function of the landscape area, because the current situation of the landscape area of the residential area is basically abandoned, the accessibility of the road in the landscape area is poor, it is necessary to replan the functional distribution of the landscape area, the landscape area of the residential area into an integrated consideration. Taking Jinshengyuan residential area as an example, starting from the main entrance and the secondary entrance, the landscape area of the entire residential area is divided into functional activity area and landscape area, and the residents are guided to reach the public activity area through the emphasis and transition of landscape area. Water features are set up in two landscape areas near the entrance, gradually transitioning to landscape structures, and functional activity areas are set up in the middle, so as to guide the route and transition of sight.

3.2 Landscape Facilities Units

From the landscape facilities: rich vegetation, increase the sense of vertical visual hierarchy; By means of landscape design, flexible boundary is used to transition boundary space, for "soften" the hard interface in residential area and improve the quality of residential space. Targeted optimization and renovation will be carried out from the aspects of greening, water body, ground pavement and sketch facilities to meet the needs of daily activities and communication of residents and improve the overall atmosphere of residential space [3].

Taking "rigid boundary flexibility" as the main design strategy and landscape as the flexible medium, it can play a transitional and flexible role in the boundary space. When dealing with the boundary space adjacent to the internal roads in the community, we should mainly weaken the sense of boundary, choose a variety of plants for reasonable collocation, and plant in the way of landscape multi-layer, so as to increase the sense of visual hierarchy [2] (Fig. 7).

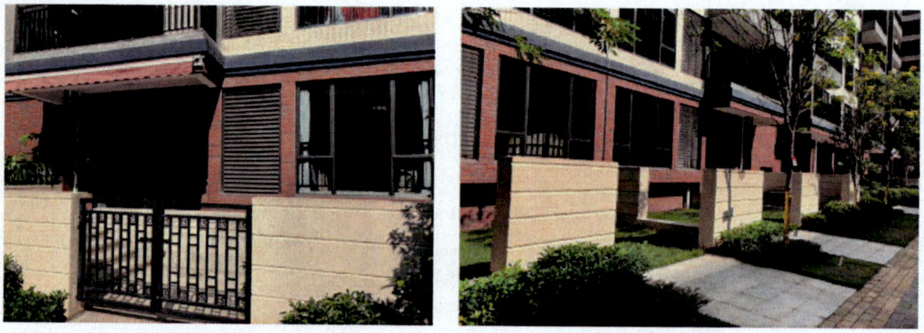

Fig. 7. Front space design of xingfuli residential community in Zhaoqing City(from the author)

Through greening, water body, ground pavement, sketch facilities and other elements, the public space can set up sketches or shape micro terrain in the way of "point" distribution, so as to create a richer landscape level and provide different activity spaces for different people (Fig. 8).

Flexible media are skillfully used to weaken the boundary space. Compared with the softer material of the main object, such as colonnade, corridor, canopy and concave convex surface of the building, the media that can penetrate and have a sense of hierarchy in the line of sight can be used to weaken and transition the front space of the house, so as to achieve the effect of shielding the bottom space [4].

Arrange public facilities in various spaces in the residential area. A good rest area can be set in the area close to the front space of the house, or steps, seats, pavilions, etc. can be formed along the boundary to provide people with a space to stop and rest [5]. In the large open space, it can be considered to set up corresponding facilities for residents' activities, such as children's entertainment facilities, fitness equipment, etc., which can not only meet the activities needs of residents, but also facilitate parents' care of their children and promote communication between parents.

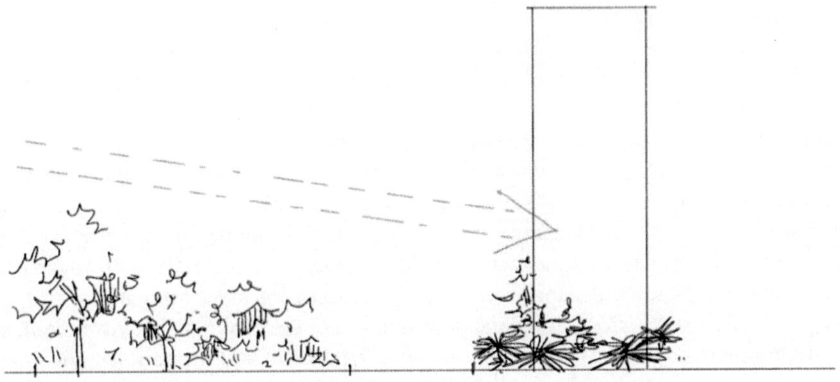

Fig. 8. Schematic diagram of landscape height change of integrated design of architecture, landscape and road in Jinshengyuan residential area (from the author)

3.3 Residential Management Level

From the residential area management: unified planning management, reasonable planning, overall and integrated design under the premise of meeting people's daily use functions, to keep the style of residential area unified, or neat and uniform without losing the original characteristics.

Re-plan the existing roads in the residential area, implement the diversion of people and vehicles, and strengthen the boundary space between roads to ensure travel safety. The boundary space formed between carriageway and sidewalk needs to be emphasized and separated to ensure the safety of pedestrians. Double layer planting is carried out on both sides of the motorway to reduce noise and filter tail gas pollution. At the same time, the height of street trees planted on both sides of the road should be at the same interface as much as possible, which can not only meet the role of shading and cooling, but also create a beautiful road boundary space. In the boundary space between carriageway and sidewalk, adding landscape isolation belt can play a role of separation. The boundary space can also be distinguished and strengthened through different ground pavement methods to strengthen the boundary space between roads and attract the attention of residents (Fig. 9).

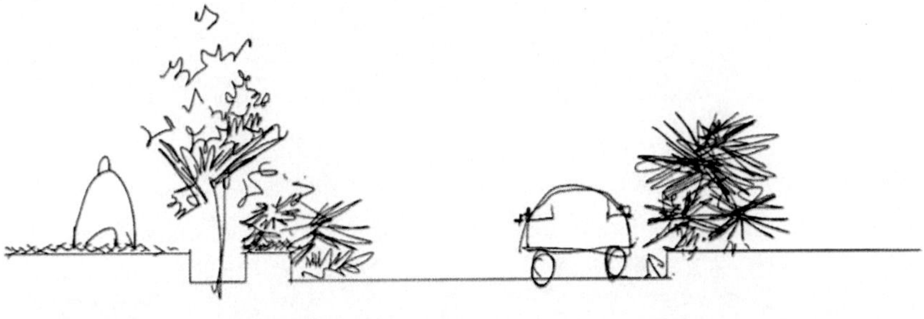

Fig. 9. Boundary space design of carriageway and sidewalk in Jinshengyuan residential community, Zhaoqing City (from the author)

4 Conclusion

Centering on the theme of urban renewal, the author studies the boundary space which is easily neglected by people in urban residential area. Taking residential areas in the Pearl River Delta region as the research object, four common problems are sorted out. It connects hard medium boundary space to flexible boundary space, stimulates the potential energy boundary space, makes its real "live" rise, attracts more people are willing to stop in the border area, the exchange and sharing, which bring people more sense of gain and happiness by art design, landscape aesthetics, landscape design" etc. from the architectural space design, landscape facilities, residential area management three aspects. This is not only the transformation and continuation of the ecological environment, spatial environment, cultural environment, visual environment and recreation environment, but also the continuation and renewal of the neighborhood's social network structure, psychological set and emotional attachments.

Acknowledgement. This article is based on the In 2020, Zhao Qing University's school-level scientific research fund project "Research on the Transition Space Optimization of Urban Old Campus in Guangdong-Hong Kong-Macao Greater Bay Area". The author would like to thank her co-authors for their lasting moral support to make the article a success.

References

1. Cao, J.: Preliminary study on transitional space types of old residential areas in Guangdong-Hong Kong-Macao Greater Bay Area. Beauty Age **03**, 40–42 (2021)
2. Zheng, J., Ling, Y.: Study on flexible boundary design of urban street space. Mod. Decoration (Theory) **12**, 155 (2016)
3. Han, X.: Analysis of "boundary effect" among the neighborhood space in high-rise residence-in the modern sprit of humanities. HuaZhong Archit. **03**, 78–80 (2007)
4. Zhang, J., Qing, Y.: "Flexible boundary" in residential space. Cryog. Constr. Technol. **05**, 30–31 (2006)
5. Pan, H., Fu, Z.: Research on function of urban transition space. Archit. Cult. **11**, 71–72 (2017)

Distributed Ledger for Construction Material Providers to Control Carbon Emissions, a Case Study of Glulam Material

Wenhao Liu and Zhan Shu(✉)

Shanghai University, Shanghai 200444, China
shuz@shu.edu.cn

Abstract. To combat global warming, the European Union pioneered an emissions trading system (ETS) that regulates emissions in government-selected industries and has proven to be successful. In many countries, the construction industry is enforced to join the ETS because of its high energy consumption and emissions. Nonetheless, most existing ETSs only cover the operation phase of a typical project while the materialization phase is rarely considered. The materialization phase includes emission from both construction work and the construction material production. Currently, there are few instruments to evaluate carbon emission within the manufacturing process of construction materials. As the accurate data is not available, the benefits of low-carbon construction materials such as the glulam are veiled to the public. To address such issues, this paper proposes a carbon emission assessment tool for wood products such as glulam during the production stage. The new tool is also supported by the distributed ledger which is highly compatible with future ETSs under Industry 4.0 framework. The system enables construction material manufacturers to evaluate, upload, and share information in a distributed manner. Carbon emissions and materials properties simultaneously updated on the chain when producing constructional materials. Such a system serves as a foundation of subsequent construction material and carbon allowance trading.

Keywords: Wood product · Glulam · Sustainable material · Construction material · Emission assessment · Distributed ledger

1 Introduction

The European Union first established the Emissions Trading System (ETS) in 2005 [1] to reduce rising greenhouse gas emissions. The ETS uses a market mechanism to regulate each entity's carbon emissions, effectively turning carbon allowances into a commodity [2]. From 2005 to 2019, the European ETS (EU ETS) resulted in a 35% reduction in emissions within the regulated electrical system [3].

W. Guo and K. Qian (Eds.): GBCESC 2022, LNCE 211, pp. 365–372, 2023.
https://doi.org/10.1007/978-981-19-5217-3_36

The construction industry is notorious for its high energy use and emissions, accounting for 40% of global energy use and 36% of global carbon emissions [4]. In China, the construction industry accounted for about 51% of total life cycle carbon emission in 2018 and 2019 [5]. The carbon emissions of the materialization phase are complex and are not fully evaluated compared with the operation phase. Therefore, the materialization phase has the largest potential for emission reduction. Furthermore, the production of construction materials is responsible for the majority of carbon emissions during the materialization phase [6]. ETS covering the materialization phase become an ongoing trend.

Meanwhile, renewable, recyclable, and biodegradable materials such as wood have not received much attention in terms of carbon emission. In fact, when compared with other construction materials, the production of wood requires up to 75% less energy consumption. Timber structure construction requires no heavy equipment and could be quickly assembled on-site. This corresponds to less dust, less pollution, and lower energy consumptions. Besides, after timber structures are demolished at the end of their life cycle, 90% of the materials can be recycled and reused. Figure 1 shows the relationship between wood material and carbon reductions.

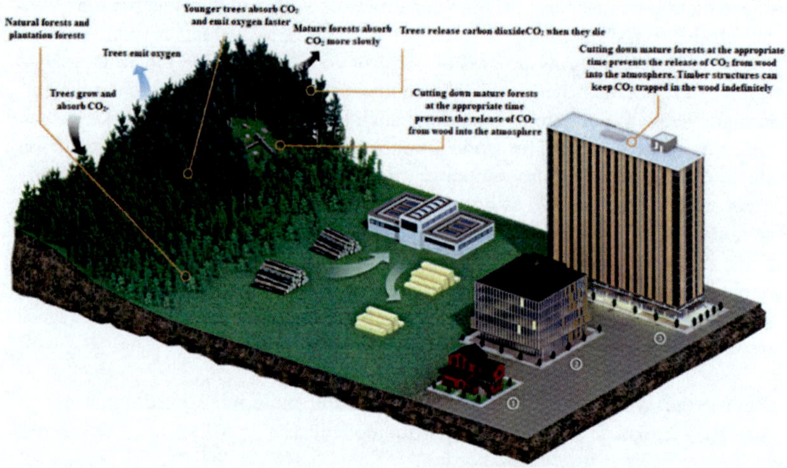

Fig. 1. The relationship between wood material and carbon reductions.

Nonetheless, the carbon emission of timber structures during construction is not included in most of the current ETSs. Most of the ETSs do not accurately record the carbon emissions of wood construction materials. One of the reasons is the lack of a standard tool to assess carbon emissions from the manufacturing of materials. Besides, the current database of carbon emission is neither accurate nor easily available. Finally, current data in the existing ETS that were collected from various sources might be easily tampered by a third party.

To address these aforementioned issues, this paper focuses on a novel ETS for material providers that is enhanced by distributed ledgers (or Blockchain), namely the Blockchain-enhanced ETS for Material Providers (BEMP) [7]. Such system combines the distributed ledger technology with existing ETS. In this paper, low-carbon construction materials represented by wood are extensively studied. The proposed system can automatically collect and distribute data on carbon emissions from all construction materials, making it available to all participants in real time. It also makes it easy to track and audit the carbon emission of manufacturers.

2 Reduce Carbon Emissions for Material Providers in the Construction Industry

2.1 ETS and Distributed Ledger Technology

ETS is a complete platform that establishes a quantitative limit and a price on emissions. Its "currency" is carbon allowances, which are normally granted and regulated by the government. Carbon allowances allow holders to emit greenhouse gases [8]. However, the current ETSs have management and fraud issues. The blockchain, which is a type of distributed ledger, is all the rage these days. The information in the blockchain is transparent and traceable, and it can be used to secure the exchange of goods like cash, property, and contracts without the use of third-party intermediaries like banks or governments [9]. To address the management and fraud issues that plague traditional ETS, Khaqqi *et al.* [10] developed a new ETS model that combines blockchain technology with reputation systems. Mandaroux *et al.* [11] created a framework that combines EU ETS and blockchain technology. In addition, more types of blockchain-enhanced ETS are also being explored, including the ETS for fashion industry [12] and the ETS with smart contracts (SC) [13].

2.2 Production Process of Timber and Emission Sources

Classification of Wood. Greenpeace, a global environmental organization, divides wood into "green wood" and "illegal wood". The proportion of "green wood" cut is reasonable, causing no forest degradation, animal and plant extinctions, or other negative consequences, and the use of "green wood" to construct buildings can also help with carbon sequestration, which is beneficial to the environment. Furthermore, the use of "illegal building materials" is harmful to the environment. The Forest Stewardship Council recommends 11 different types of "green wood" to consumers, as shown in Fig. 2.

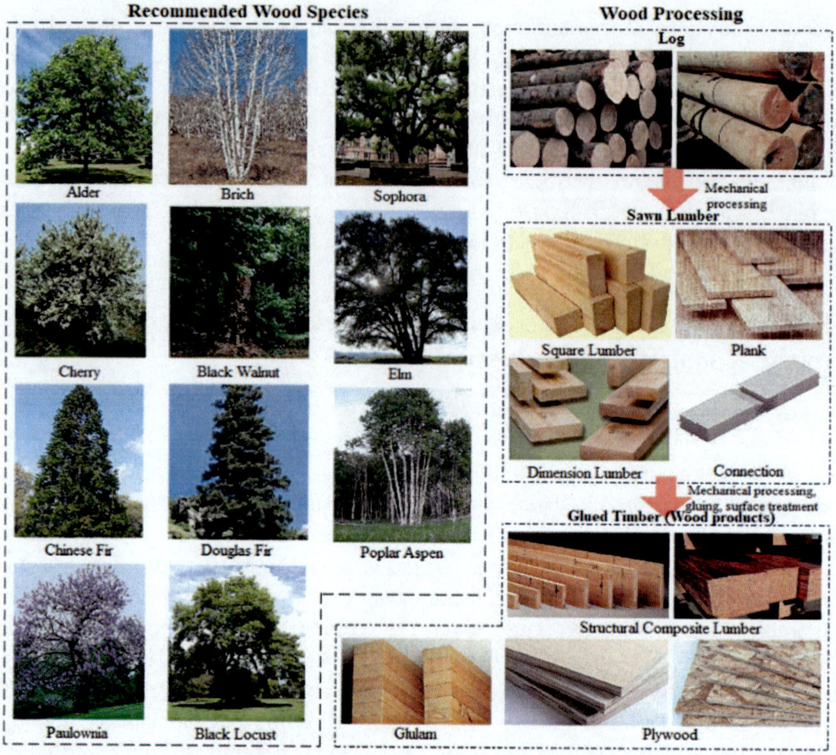

Fig. 2. Recommended wood species and their processing.

Wood Processing and Emission Sources. The wood used in design is currently divided into two categories: wood-based panels and natural wood. Cutting, drying, gluing, and surface processing are the four main steps in the wood process.

Wood Cutting. Wood cutting equipment is used to remove the chips from the workpiece, and the chips or remaining parts can become products, such as planning, veneer cutting, and milling. Therefore, the main source of carbon emissions at this stage of production is the operation of machinery.

Wood Drying. In addition to the direct use of logs in the design, wood should be dried when processed into boards and other products. Natural or artificial drying is usually used. Natural drying uses the sun's radiation to directly evaporate the water in the wood, thus producing no carbon emissions. Artificial drying requires the use of a drying kiln and an air energy heat pump dryer. Operation steps include preheating, dehumidification, and drying. Carbon emissions here are mainly generated by the operation of the air energy heat pump dryer.

Wood Gluing. Wood gluing is the technology of gluing wood to wood into an organic whole. The manufacture of wood-based panels such as plywood is based on gluing technology. Gluing technology can compensate for material flaws such as making short materials longer and thin materials thicker, it has a significant impact on wood processing. The main steps of wood gluing are gluing, embryo formation, drying, closure, heating, and post-treatment. The source of carbon emissions here is the use of glue.

Wood Surface Processing. The wood and its products must be surface treated at the end of the materials manufacturing process, according to the use and visual requirements. Wood surface treatment is mainly by mechanical processing (grinding, polishing). The carbon emission source for this step is the operation of machinery.

Wood Special Treatment and Its Emission Sources

Fire Prevention of Wood. Wood fire treatment must be carried out for all kinds of wood. Commonly used treatment methods are flame retardant impregnation treatment and surface coating treatment. Wood fire treatment is usually completed in the wood surface treatment stage, and the carbon emission source is the use of fire retardants.

Wood Preservation, Mildew and Insect-proof Treatment. 1. Drying: This method is capable of resolving the majority of issues. This method is commonly used in domestic wood processing, and the carbon emissions produced by this method come from the operation of heating machinery. 2. Fumigation is a moth-killing technique. The consumption of fumigant and the energy consumption of the fumigating chamber account for the method's carbon emissions. 3. Injection of drugs: Insect prevention and anticorrosion are achieved by injecting preservatives into the wood. The use of preservatives is a source of carbon emissions for this method. 4. Surface coating: To prevent wood mildew, an antimildew agent is used in the wood surface coating, and this method of carbon emission source is the use of mildew inhibitor.

3 Case Study and Uncertainties

3.1 Case Study

As a case study, the carbon emission assessment and automatic information uploading of the glulam are presented. Figure 3 depicts the manufacturing process, inputs, and outputs. In Fig. 3, the manufacturing of this product is divided into eight processes. Commonly used machines in the production of glulam include the grading machine, the spray glue machine, the finger-jointers, etc. The direct carbon emission sources for each step are mining and transportation of raw materials, equipment operation, human labor, and equipment maintenance. Finally, the carbon emissions from each step as well as information about the glulam will be automatically uploaded to the proposed system.

The consumption of two raw materials in the manufacturing process, Douglas fir and gum, is represented by $M_{1,1}$ and $M_{3,1}$. Assuming that the mechanical equipment consumes only electricity and the personnel consume both electricity and water, the consumption of each step is $E_{1,1}$,..., $E_{8,1}$ and W_1,...,W_8. Since the carbon emissions of equipment maintenance are difficult to calculate accurately, this paper uses the maintenance cost to evaluate the carbon emissions, i.e., C_1,..., C_8. Each step yields a temporary result (P1,... P7). The semi-finished product input in the next step is the temporary product output in the previous step. Douglas fir glulam is the outcome of the final step.

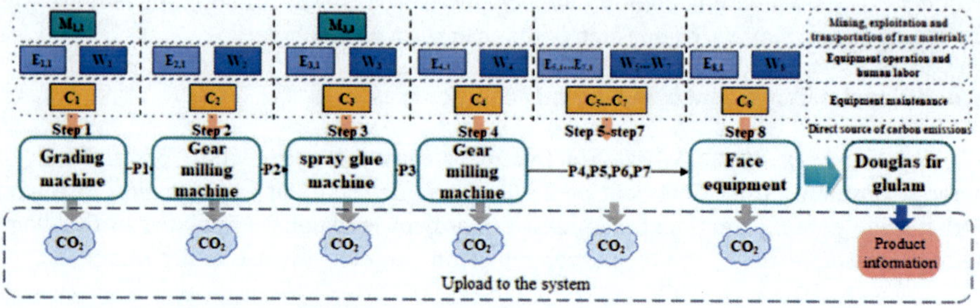

Fig. 3. The manufacturing process and carbon emission source of the glulam.

The electricity and water consumption of the personnel (not the managers) involved in the production of glulam are used to calculate their carbon emissions. The following equation can be used to calculate the carbon emission of the glulam:

$$CE = \sum_{i=1}^{p}[\sum_{jm=1}^{m} M_{i,jm}mf_{jm} + \sum_{jk=1}^{e} E_{i,jk}ef_{jk} + W_i f_w + L_i(\eta f_e + \zeta f_{lw}) + C_i f_c] \quad (1)$$

where CE is the total carbon emission of product A, $M_{i,jM}$ represents the amount of material jm in step 1; mf_{jm} is the process-based carbon emission factor of material j_m. $E_{i,jk}$ is the amount of energy jk in step 1; $E_{f,jk}$ is the process-based carbon emission factor of energy jk; W_i represents the water resource consumption at step 1; f_w is the process-based carbon emission factor of water resources. Assuming that participants' carbon emissions come from the use of electricity and water resources, L_i is the number of participants in step 1; η and ζ are, respectively, the amounts of electricity and water resources consumed by a worker in one step. These two parameters should be determined by the construction materials manufacturer through measured data. f_e and f_{lw} are process-based carbon emission factors of electricity and water resources, respectively. Although equipment maintenance costs may not be generated in each step, this part of the formula is included to ensure the formula's integrity. C_i represents the maintenance cost of equipment in each step; f_c is the input-output carbon emission factor of the machinery industry.

Equation (1) can be used to calculate the carbon emissions of the glulam:

$$\begin{aligned} CE &= mf_1 M_{1,1} + mf_3 M_{3,3} \\ &+ ef_1 (E_{1,1} + E_{2,1} + E_{3,1} + E_{4,1} + E_{5,1} + E_{6,1} + E_{7,1} + E_{8,1}) \\ &+ f_w (W_1 + W_2 + W_3 + W_4 + W_5 + W_6 + W_7 + W_8) \\ &+ (L_1 + L_2 + L_3 + L_4 + L_5 + L_6 + L_7 + L_8)(\eta f_e + \zeta f_{lw}) \\ &+ f_c (C_1 + C_2 + C_3 + C_4 + C_5 + C_6 + C_7 + C_8) \end{aligned} \tag{2}$$

3.2 Calculation of the Uncertainties

When assessing carbon emissions, the influence of uncertainties on the calculation results should be determined [14]. Uncertain factors can have a parallel or serial relationship. The total uncertainty for N parallel uncertainties and the total uncertainty for N series uncertainties can be calculated by the following formula:

$$U_c = \frac{\sqrt{\sum_{n=1}^{N} (U_{sn} \cdot e_{sn})^2}}{\left| \sum_{n=1}^{N} e_{sn} \right|} \qquad U_c = \sqrt{\sum_{n=1}^{N} (U_{sn})^2} \tag{3}$$

where U_c is the total uncertainty (%), U_{sn} is the uncertainty range (%) of the nth calculation result, and e_{sn} is the nth carbon emission calculation result.

Take step 1 in the production process of the glulam as an example. Assuming that the carbon emission of consumed electricity in step 1 is $20 \pm 5\%$ tCO_2e and that of water is $10 \pm 2\%$ tCO_2e, the total uncertainty is:

$$U_c = \frac{\sqrt{(20 \times 0.05)^2 + (10 \times 0.02)^2}}{|20 + 10|} = 3.40\% \tag{4}$$

4 Conclusion

Most of existing ETSs in the construction industry only assess carbon emissions during the operation phase of a typical project and rarely address carbon emissions during the materialization phase whose boundary covers the construction material manufacturing stages. Low-carbon materials such as timber are also not valued to the level as they deserve. Inaccurate data recording, inconvenient data sharing, and a lack of standardized calculation tools to evaluate carbon emissions in the manufacturing stage of construction materials characterize the current ETSs. Distributed ledger's security, transparency, disintermediation, low threshold, and other attributes can effectively address the inadequacies of existing ETSs.

In order to assess carbon emissions during the manufacturing stage, this paper analyzes the main manufacturing processes and emission sources of wood products. Various wood treatments were considered. It is shown that the carbon emission sources for wood products are relatively straightforward, i.e. the lumbering of raw materials and the energy consumed by mechanical equipment. Then, this paper provides a case study of the glulam products shown how the carbon emissions were evaluated and recorded in the distributed ledgers. The carbon emissions of each step in the production stage are calculated using a hybrid method, which combines the processed-based method and the input-output method. In addition, this carbon evaluation procedures are compatible BEMP. This ensures the accuracy of the data input into BEMP and provides a basis for the next stage of transactions in the BEMP.

References

1. Jun-Seok, P.: Analysis of the positions of domestic and international countries' emissions trading schemes through principal component and cluster analysis. KSCE J. Civ. Eng. **7**, 1–9 (2017)
2. Yue-Jun, Z.: Research on carbon emission trading mechanisms: current status and future possibilities. Int. J. Global Energy Issues **39**(1/2), 89–107 (2016)
3. EU Emissions Trading System (EU ETS): https://ec.europa.eu/clima/policies/ets_en:~:text=The%20EU%20Emissions%20Trading%20System%20has%20proven%20to%20be%20an,35%25%20between%202005%20and%202019 Accessed 26 Feb 2022
4. Bo, L.: Feasibility assessment of the carbon emissions peak in China's construction industry: factor decomposition and peak forecast. The Science of the Total Environ. **706**(Mar.1), 135716.1–135716.13 (2020)
5. China Building Energy and Emissions Database http://www.cbeed.cn/#/calculate Accessed 26 Feb 2022
6. Xiaocun, Z.: Reducing building embodied emissions in the design phase: a comparative study on structural alternatives. J. Clean. Prod. **243**, 118656 (2020)
7. Zhan, S.: Blockchain-enhanced Trading Systems for Construction Industry to Control Carbon Emission. Clean Technologies and Environmental Policy (2022)
8. Climate explained: How emissions trading schemes work and can they help us shift to a zero carbon future https://www.stuff.co.nz/environment/climate-news/115310076/climate-explained-how-emissions-trading-schemes-work-and-can-they-help-us-shift-to-a-zero-carbon-future/ Accessed 26 Feb 2022
9. Mayank, R.: Sok of used cryptography in blockchain. IEEE Access, 1–1 (2019)
10. Khamila, K.: Incorporating seller/buyer reputation-based system in blockchain-enabled emission trading application. Applied Energy **209**(JAN.1), 8–19 (2018)
11. Rahel, M.: A european emissions trading system powered by distributed ledger technology: an evaluation framework. Sustainability **13**(4), 2106 (2021)
12. Bailu, F.: Blockchain enhanced emission trading framework in fashion apparel manufacturing industry. Sustainability **10**(4), 1105 (2018)
13. Alia Al, S.: A Hierarchical Blockchain of Things Network For Unified Carbon Emission Trading (HBUETS): A Conceptual Framework. 2020 IEEE International Conference on Technology Management, Operations and Decisions (ICTMOD). IEEE (2020)
14. Data Analysis Toolkit #5: Uncertainty Analysis and Error Propagation http://seismo.berkeley.edu/~kirchner/eps_120/Toolkits/Toolkit_05.pdf Accessed 26 Feb 2022

Dynamic Response Analysis of Stress Intensity Factor of Reflective Cracking in Asphalt Pavement Addition

Qinshou Huang(✉)

Guangxi Communications Design Group CO. LTD, Guangxi 530029, China
kz101216@163.com

Abstract. Taking the second phase of the reconstruction of the East Ring Road of the existing motorway in Nanning, Guangxi as an example, this paper employs ABAQUS software based on kinetic theory and the plane strain finite unit method to analyse the dynamic stress intensity factor variation law with time under dynamic load for reflection cracks. The results show that: the time course curve of dynamic stress intensity factor slows down as the thickness or modulus of asphalt paving layer increases; the time course curve slows down as the thickness or modulus of stress absorbing layer increases; the amplitude gradually becomes smaller as the vehicle speed increases; the amplitude gradually becomes larger as the axle load increases; the vibration curve gradually tends to level off as the action time continues, and the time course curve does not completely decay to zero when the load action ends.

Keywords: Asphalt overlay · Reflection crack · Dynamic stress intensity factor · Dynamic response

1 Introduction

With the rapid development of the city, a large number of urban peripheral highways or primary roads are constantly upgraded and expanded into urban expressways, and currently the most used in the old cement concrete pavement reconstruction and expansion projects at home and abroad is the addition of asphalt surface layer. The existence of joints in the old pavement leads to reflection cracks in the asphalt surface layer, shortening the life of the pavement, so road workers at home and abroad conduct a lot of research on the mechanism of reflection crack expansion. Explored the change law of dynamic stress intensity factor and the effect on the working characteristics of road structure under the change of road structure parameters and foundation parameters under the action of dynamic vehicle load based on dynamic theory and plane strain finite unit method [1–4]. Studied the steady state response of infinite slabs on viscoelastic foundations under moving loads [5, 6]. Dug up the dynamic response of multi-layered asphalt pavements under moving loads based on structural dynamic theory [7, 8]. Analyzed the dynamic characteristics of asphalt pavement structures containing cracks under vertical

W. Guo and K. Qian (Eds.): GBCESC 2022, LNCE 211, pp. 373–384, 2023.
https://doi.org/10.1007/978-981-19-5217-3_37

loads by combining the dynamics theory [9, 10]. Rooted in the theoretical basis on fracture mechanics and dynamics, the dynamic response of reflective cracks under dynamic loading is analyzed in conjunction with the finite element method, and the law of change in dynamic stress intensity factor over time under changes in travel speed, thickness of asphalt paving layer, modulus of elasticity of asphalt paving layer, thickness of stress absorbing layer, modulus of elasticity of stress absorbing layer and vehicle axle load is investigated.

2 Basic Theory, Simplification of Analytical Models and Determination of Load Field

2.1 Basic Theory of Kinetics [11]

The controlling equations for the analysis of vehicle loads acting on the pavement body using the finite unit method [11] are Eqs. (1) and (2):

$$M\ddot{u}(t) + C\dot{u}(t) + Ku(t) = F(t) \tag{1}$$

$$C = \alpha M + \beta K \tag{2}$$

In Eqs. (1) and (2): M, C and K are the system mass matrix, damping matrix and stiffness matrix respectively; $\ddot{u}$, $\dot{u}$ and u are the node acceleration, velocity and displacement vectors respectively; $F(t)$ is the dynamic load vector acting on the node; α and β are the Rayleigh damping coefficients.

2.2 Determination of Dynamic Stress Intensity Factor $K_I(t, \theta)$ [2]

The expression for the displacement field near the crack tip [2] is Eq. (3):

$$\begin{aligned} u(t,\theta) &= \frac{(1+\mu)K_I(t,\theta)}{2E}\sqrt{\frac{r}{2\pi}}\left[(2s-1)\cos\left(\frac{\theta}{2}\right) - \cos\left(\frac{3\theta}{2}\right)\right], \\ v(t,\theta) &= \frac{(1+\mu)K_I(t,\theta)}{2E}\sqrt{\frac{r}{2\pi}}\left[(2s+1)\sin\left(\frac{\theta}{2}\right) - \sin\left(\frac{3\theta}{2}\right)\right], \end{aligned} \tag{3}$$

In Eq. (3): $u(t, \theta)$ and $v(t, \theta)$ are the displacements along the x-axis and along the y-axis, respectively; $s = 3 - 4\mu$; μ is the Poisson's ratio. Letting $\alpha = 180^0$ yields an expression for the stress intensity factor expressed by the displacement perpendicular to the reflected crack face [2], see Eq. (4). With the singular element set up at the crack tip, $v(t, \theta)/\sqrt{r}$ can be approximated as a linear function of r [2], see Eq. (5).

$$K_I(t,\theta) = \frac{\sqrt{2\pi}E}{4(1-\mu^2)}\lim_{r\to 0}\left(\frac{v(t,\theta)}{\sqrt{r}}\right) \tag{4}$$

$$v(t,\theta)/\sqrt{r} = A(t,\theta) + B(t,\theta)r \tag{5}$$

Taking the limit [2] on Eq. (5), Eq. (6) is obtained as:

$$\lim_{r\to 0}\left[v(t,\theta)/\sqrt{r}\right] = A(t,\theta) \tag{6}$$

The vertical displacements at nodes J and N on the singular element near the crack tip are determined by Eq. (6) for $A(t, \theta)$, and then by Eqs. (5) and (6) for $K_I(t, \theta)$.

3 Analytical Model and Load Field

3.1 Simplification of the Analytical Model

The plane strain calculation model is established along the direction of travel, the finite element model is shown in Fig. 1, the mesh division near the crack tip is shown in Fig. 2 and the calculation parameters are shown in Table 1, where the values of the damping matrix coefficients are given in the literature [12].

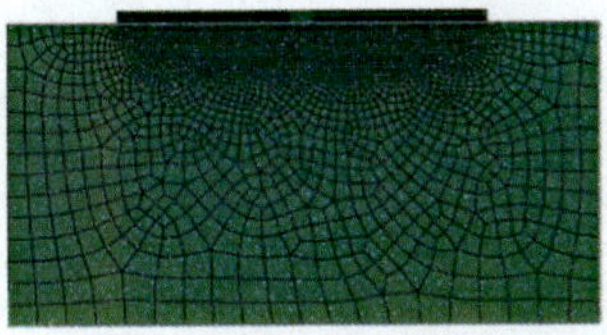

Fig. 1. Finite element model

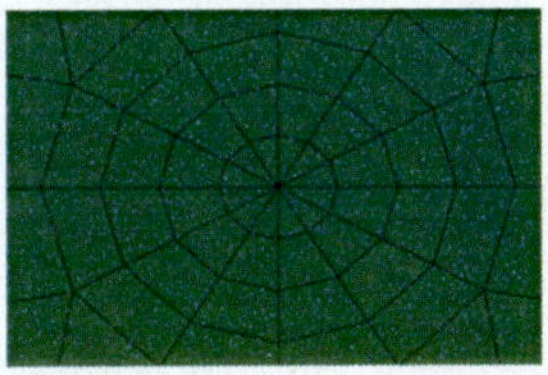

Fig. 2. Division diagram of crack tip mesh

Table 1. Major calculation parameters

Structural layer	Thickness h/cm	Modulus of elasticity E/MPa	Poisson's ratio μ	Density $\rho/kg/m^3$	Damping matrix coefficient α [8]	Damping matrix coefficient β
Asphalt paving	9 (9 –20)	–	0.25	2400	0.4	0.005
Stress absorbing layer	2.5 (1–4)	800	0.25	2300	0.4	0.005
Old cement concrete slab	24	30000	0.15	2400	0.4	0.005
Base	–	100	0.3	1900	0.4	0.005

Note: ① Parameters in parentheses are magnitudes of change; ② The dynamic modulus is used for the modulus of elasticity of the asphalt pavement.

The factors affecting the dynamic properties of asphalt concrete are mainly vehicle loading, temperature changes and time of action, etc. The dynamic modulus values of asphalt mixture at 20 °C are selected [13] for calculation. The dynamic modulus of asphalt concrete at different loading frequencies is shown in Table 2 [14].

Only the dynamic modulus of asphalt mixture corresponding to the four loading frequencies are listed in Table 2, and the regression analysis shows that there is a good power multiplication relationship between the dynamic modulus and the loading frequency, see Fig. 3. Using the reciprocal relationship between the loading period and the loading frequency, the relationship between the asphalt concrete and the loading period

Table 2. Dynamic modulus of asphalt mixture of different frequency [14] (0.7MPa, 20°C)

	Loading frequency μ/Hz			
	1	4	10	16
Dynamic modulus of asphalt mixture *E*/MPa	998	1149	1237	1453

is obtained after the formula conversion, see Eq. (7):

$$M_A = 982.36T^{-0.1229} \quad r^2 = 0.9215 \tag{7}$$

In Eq. (7): M_A and T are dynamic modulus of asphalt mixture, loading period respectively.

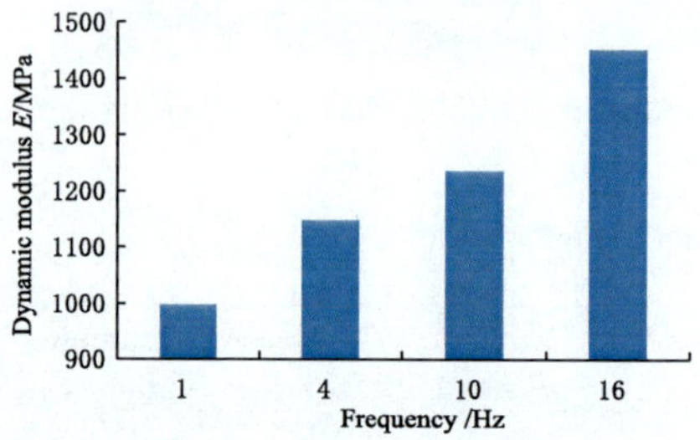

Fig. 3. Relationship between material dynamic modulus and loading frequency

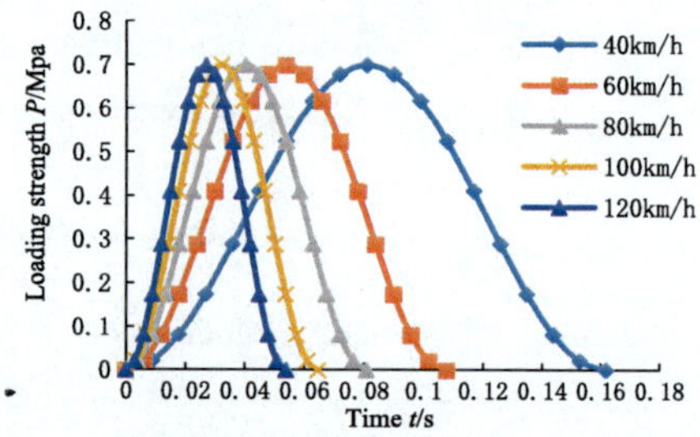

Fig. 4. Temporal variation of different velocities loading strength

3.2 Determination of the Load Field

The standard axle load BZZ-100 is used for the load and 0.7MPa is used for the Wheel grounding pressure [13]. The law of variation of the load was in the simplified form of the Kenlayer procedure [15] for a sinusoidally distributed load acting symmetrically directly above the reflective cracks in the asphalt paving layer, see Eq. (8).

$$q(t) = q_{\max} \sin^2(\frac{\pi}{2} + \frac{\pi t}{T}), T = \frac{12L}{V} \tag{8}$$

In Eq. (8): T is the period of load action; L is the radius of the wheel contact area (15cm); V is the vehicle travel speed. Converting the load action time to the interval [0, T] yields Eq. (10):

$$q(t) = q_{\max} \sin^2(\frac{\pi t}{T}), T = \frac{12L}{V} \tag{9}$$

The dynamic modulus corresponding to different vehicle speeds is obtained from Eq. (9) and then from Eq. (7), see Table 3. The dynamic load half-period variation values are shown in Table 4, and the graph of the variation of load intensity with time for different vehicle speeds is shown in Fig. 4.

Table 3. Load period and asphalt mixture of different vehicle speed

Vehicle speed *V*/km/h	40	60	80	100	120
Load period *t*/s	0.162	0.108	0.081	0.0648	0.054
Dynamic modulus *E*/MPa	1229	1291	1338	1375	1406

Table 4. Dynamic load half-period change value (T/2)

Time *t*/s	0	*T*/18	*T*/9	*T*/6	2*T*/9	5*T*/18	*T*/3	7*T*/18	4*T*/9	*T*/2
Wheel pressure *P*/MPa	0	0.0211	0.0819	0.1750	0.2892	0.4108	0.5250	0.6181	0.6789	0.7000

3.3 Determination of Load Strength

Take the vehicle driving speed as 40 km/h, 80 km/h and 120 km/h respectively, and take the power response analysis time period T as 1s, the form of load intensity distribution is shown in Fig. 4, the load intensity distribution curve under different vehicle speed is shown in Fig. 5.

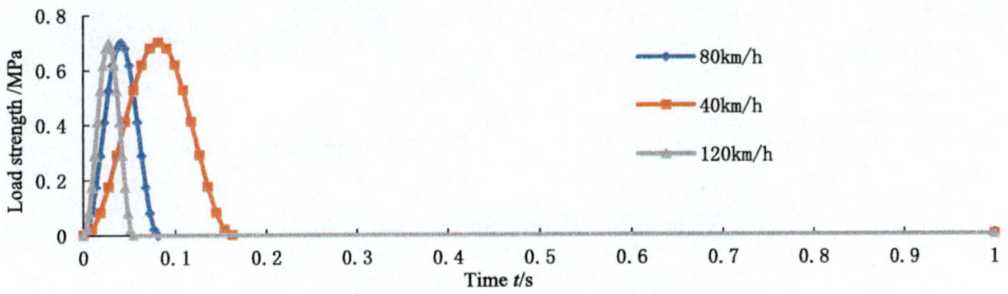

Fig. 5. Load distribution curve of vehicle speed

4 Dynamic Response Analysis of Stress Intensity Factors Under Dynamic Loading

Calculate and analyze of the variation of the dynamic stress intensity factor with time under single cycle dynamic loading with parameters such as travel speed, asphalt pavement thickness, asphalt pavement modulus of elasticity, stress absorbing layer thickness, stress absorbing layer modulus of elasticity and vehicle axle load.

4.1 Effect of Travel Speed Variation on the Dynamic Response of Asphalt Pavement Structure

The effect of the change in driving speed on the dynamic response of the asphalt pavement structure was analyzed at speeds of 40 km/h, 80 km/h and 120 km/h. The time course of the dynamic stress intensity factor at different speeds is given in Fig. 6.

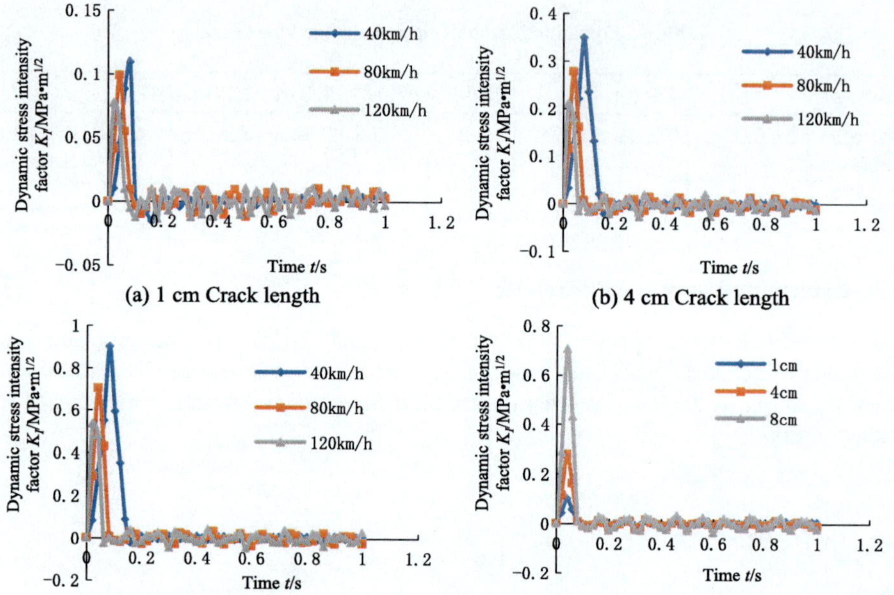

(a) 1 cm Crack length

(b) 4 cm Crack length

(c) 8 cm Crack length

(d) SIF time course curves for different crack lengths

Fig. 6. SIF with time for different vehicle

It can be seen from Fig. 6 that during the three periods of crack expansion, the dynamic stress intensity factor change curve gradually progresses downwards with increasing vehicle speed, and the maximum dynamic stress intensity factor gradually decreases with increasing vehicle speed. With the continuation of the analysis time, the maximum amplitude gradually decreases, the greater the vehicle speed, the faster its decay rate, and the vibration soon tends to level off in the second half of the period. The stress intensity factor does not return to zero as the pavement structure stabilizes from vibration, indicating the existence of a residual stress intensity factor. Taking the crack length of 4cm as an example, the maximum dynamic stress intensity factor decreases by 20.2% as the vehicle speed increases from 40 km/h to 80 km/h to 39.6% at 120 km/h. The time course of the dynamic stress intensity factor for different crack lengths at 80 km/h is also shown in Fig. 6(d), indicating that the amplitude of the dynamic stress intensity factor time course increases as the crack length increases.

4.2 Influence of Changes in Parameters of Asphalt Paving Layers and Stress Absorbing Layers on the Structural Dynamic Response of Asphalt Paving Layers

The pavement structural layer parameters were varied on the working condition of the pavement layers by changing one material parameter at a time, as shown in Table 1, where the thickness of the asphalt pavement layer was taken as 9 cm, 12 cm and 15 cm; the modulus of elasticity of the asphalt pavement layer was taken as 800 MPa, 1200 MPa and 1600 MPa; the thickness of the stress absorbing layer was taken as 1cm, 2.5 cm and 4 cm; and the modulus of elasticity of the stress absorbing layer was taken as 600 MPa, 900 MPa and 1200 MPa. The calculation results are shown in Figs. 7, 8, 9 and 10.

Figures 7, 8, 9 and 10 show the changes in the time course of the dynamic stress intensity factor with increasing thickness and modulus of the asphalt paving layer and stress absorbing layer respectively. The maximum amplitude gradually decreases with time, and the larger the structural parameters, the faster the decay rate. The stress intensity factor tends to level off quickly in the second half of the vibration, and the value does not return to zero at the end of the vibration when it stabilizes, indicating the existence of a residual stress intensity factor. Figure 7 gives the time course curve of the dynamic stress intensity factor as the thickness of the asphalt paving layer varies. KI decreases with increasing thickness. The maximum dynamic stress intensity factor decreases by 43.4% and 62.2% as the thickness of the asphalt paving layer increases from 9 cm to 12 cm and 15 cm respectively, taking the mid-crack expansion period as an example. Figure 10 shows the time course of the dynamic stress intensity factor as the modulus of the asphalt paving layer changes. The time course of the dynamic stress intensity factor

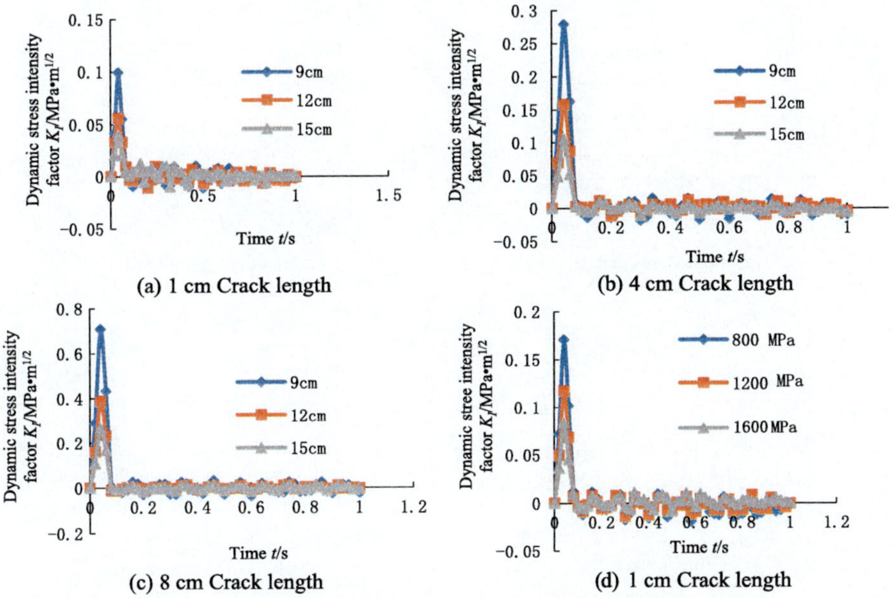

(a) 1 cm Crack length (b) 4 cm Crack length (c) 8 cm Crack length (d) 1 cm Crack length

Fig. 7. SIF with time for different asphalt overlay thickness as the modulus of the asphalt paving layer changes is given in Fig. 8. As can be seen from Fig. 8, KI decreases as the modulus of the asphalt paving layer increases. For example, the maximum dynamic stress intensity factor decreases by 22.6% and 35.8% when the modulus of the asphalt paving layer increases from 800 MPa to 1200 MPa and 1600 MPa respectively, in the middle of crack expansion. Figure 9 indicates the time course curve of the dynamic stress intensity factor as the thickness of the stress absorbing layer is varied. From Fig. 9, it is known that the KI decreases as the thickness of the stress absorbing layer increases. For example, the maximum dynamic stress intensity factor decreases by 13.8% and 33.4% when the thickness of the stress absorbing layer increases from 1 cm to 2.5 cm and 4 cm respectively in the middle stage of crack expansion. Figure 10

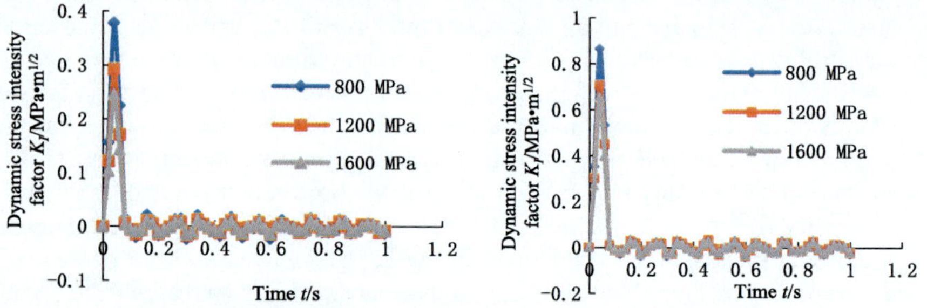

(a) 4 cm Crack length (b) 8 cm Crack length

Fig. 8. SIF with time for different asphalt overlay modulus

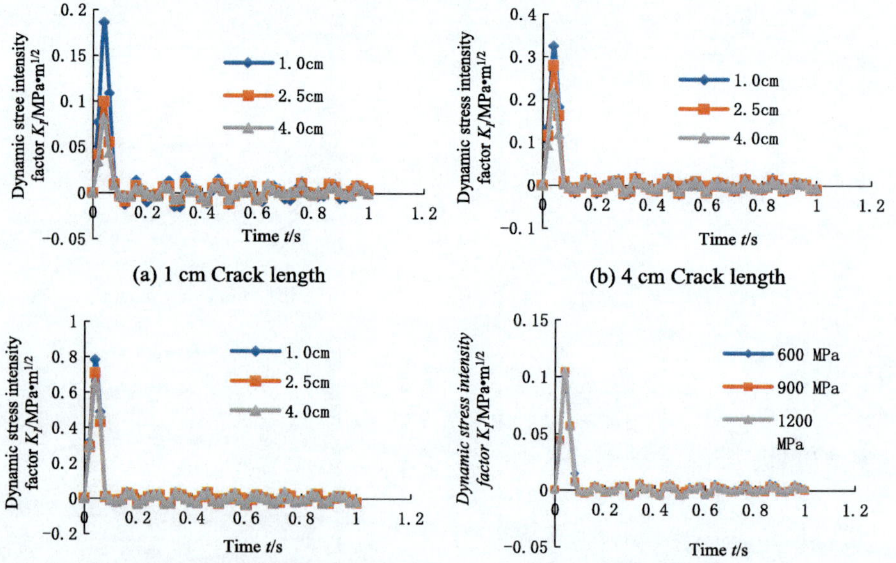

(a) 1 cm Crack length (b) 4 cm Crack length

(c) 8 cm Crack length (d) 1 cm Crack length

Fig. 9. SIF with time for different stress absorbing layer thickness expresses the time course curves of the dynamic stress intensity factor with the modulus of the stress absorbing layer. From Fig. 10, it is known that KI decreases as the modulus of the stress absorbing layer increases, but the trend is not obvious, for example, when the modulus of the stress absorbing layer increases from 600 MPa to 900 MPa and 1200 MPa respectively, the maximum dynamic stress intensity factor only decreases by 1.3% and 2.8% respectively in the middle of crack expansion.

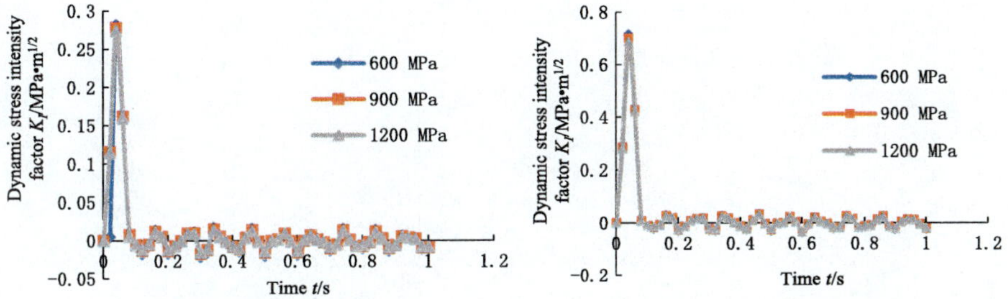

(a) 4 cm Crack length (b) 8 cm Crack length

Fig. 10. SIF with time for different stress absorbing layer modulus

4.3 Effect of Axle Load Variation on the Dynamic Response of Asphalt Pavement Structures

The dynamic response analysis is carried out by taking different axle loads of 100 kN, 160 kN and 220 kN respectively. The peak loads and their distribution are shown in Table 5. The time course curves of the stress intensity factors at different axle loads are shown in Fig. 11.

Table 5. Relationship between axle load and ground contact area, wheel pressure

Axle load *P*/kN	100	120	140	160	180	200	220	240
Single wheel ground area *A*/cm^2	356	392	432	472	512	552	592	632
Single circle load radius *r*/cm	15	15.8	16.6	17.3	18	18.7	19.4	20
Wheel pressure *P*/MPa	0.7	0.765	0.81	0.847	0.879	0.906	0.929	0.949

Note: The standard value of 100 kN axle load is adopted from the Specifications for Highway Design of Asphalt Pavements (JTG D50–2017) [2]

As can be seen from Fig. 11, the dynamic stress intensity factor time curve rises upwards with increasing vehicle axle load in all three periods of crack expansion. KI decreases rapidly with the end of the loading time and stabilizes at the end of the vibration period and does not return to zero, indicating the existence of a residual stress intensity factor. The maximum dynamic stress intensity factor increases by 25.3% and 34.2%

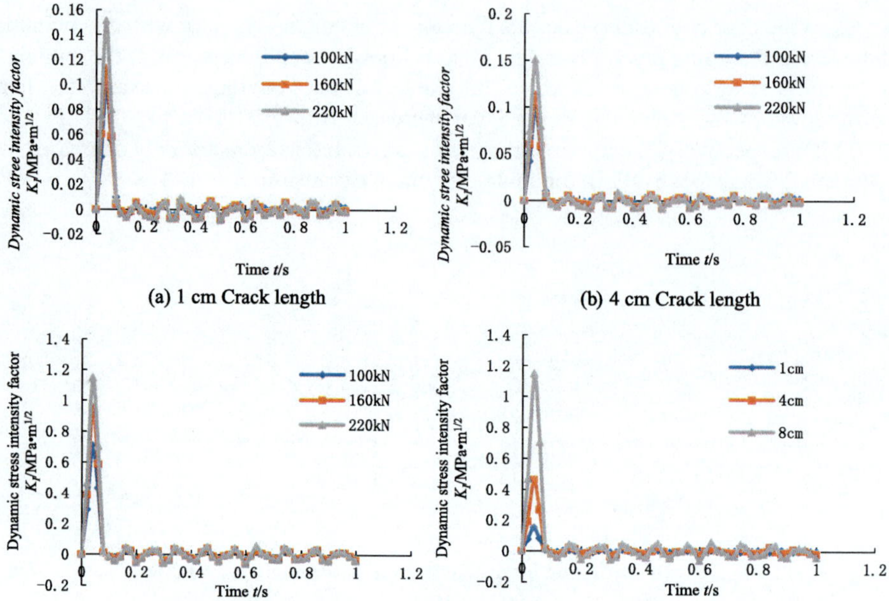

(a) 1 cm Crack length

(b) 4 cm Crack length

(c) 8 cm Crack length

(d) SIF time course curves for different crack lengths

Fig. 11. SIF with time for different axle load

when the axial load is increased from 100 kN to 160 kN and 220 kN respectively, taking the crack extension length of 1cm as an example. The analysis shows that the increase in axle load will greatly accelerate the crack extension. The graph also shows the time course of the dynamic stress intensity factor at the three stages of crack expansion when the axle load is 220 kN. It can be seen from the graph that the time course of the dynamic stress intensity factor increases with the crack length and decreases rapidly with the end of the load, the shorter the crack length the faster it decreases.

5 Conclusion

This paper is based on the second phase of the reconstruction of the East Ring Road of the existing motorway in Nanning, Guangxi, and employs the plane strain finite unit method to analyse the dynamic response of the reflective cracks in the asphalt paving layer under the action of single period dynamic load based on the kinetic theory, the analysis parameters are the driving speed, the thickness of the asphalt paving layer, the elastic modulus of the asphalt paving layer, the thickness of the stress absorbing layer and the elastic modulus of the stress absorbing layer, the following conclusions are obtained through the calculation and analysis:

(1) The dynamic stress intensity factor change curve gradually progresses downwards with the increase in vehicle speed, with the value gradually decreasing with the

increase in vehicle speed. As the analysis continues, the maximum amplitude gradually decreases, and the greater the vehicle speed, the faster it decays, and the vibration soon plateaus in the second half of the period. The stress intensity factor does not return to zero when the vibration of the pavement structure stabilizes, indicating the presence of a residual stress intensity factor.

(2) As the thickness of the asphalt paving layer, the modulus of the asphalt paving layer, the thickness of the stress absorbing layer and the modulus of the stress absorbing layer increase the time course curve of the dynamic stress intensity factor slows down. With the continuation of the analysis time, the maximum amplitude gradually decreases, the larger the structural parameters, the faster the decay rate, to the second half of the vibration soon tends to level off, there is a residual stress intensity factor exists.

(3) The dynamic stress intensity factor time curve rises upwards with increasing vehicle axle load in all three periods of crack expansion. The KI decreases rapidly at the end of the loading time and stabilizes at the end of the vibration period and the stress intensity factor does not return to zero. As the action time continues, the vibration curve tends to level off, but at the end of the load action, the time course curve does not completely decay to zero.

Acknowledgement. The work was supported in part by NSFC project 51968006.

References

1. Jinchang, W., Xiangrong, Z., Junneng, Y., et al.: Stress intensity factors of cracked road structures under dynamic loading. J. Vibration Eng. **16**(1), 114–118 (2003)
2. Jinchang, W., Xiangrong, Z.: Dynamic response analysis of asphalt roads with reflection cracks on soft ground. China J. Highway Trans. **1**(17), 1–6 (2004)
3. Zhiyi, H., Jinchang, W., Xiangrong, Z.: Analysis of viscoelasticity of asphalt pavement structures containing reflective cracks under dynamic loading. J. Zhejiang Univ. (Eng. Sci.) **41**(1), 114–119 (2007)
4. Jianqun, J., Huafei, Z., Tuqiao, Z.: Steady-state response of infinite slabs on viscoelastic foundations under moving loads. China J. Highway Trans. **19**(1), 26–29 (2006)
5. Hui, L., Hongping, Z., Xingzhou, H.: Numerical analysis of dynamic response of asphalt pavement. J. Traffic Trans. Eng. **7**(4), 44–47+53 (2007)
6. Yu, M., Yundong, W., Shaomin, Z.: Dynamic response analysis of asphalt pavements containing reflective cracks. Rock and Soil Mech. **30**(8), 2511–2516 (2009)
7. Xianyong, M., Weiwen, Q., Zejiao, D.: Analytical solution for dynamic response of transverse isotropic viscoelastic asphalt pavement. China J. Highway Transport **33**(10), 135–145 (2020)
8. Yong, Y., Lu, C., Huiqin, Y.: Dynamic response analysis of asphalt pavement under different layer contact state. Highway **5**(5), 43–48 (2021)
9. Yao, Z., Chen, K., Huimei, Z.: Dynamic response of uneven asphalt pavement under vehicle vibration model. Highway **6**(6), 71–76 (2021)
10. Xianyong, M., Weiwen, Q., Zejiao, D., et al.: Dynamic response analysis of vehicle-asphalt pavement under random irregularity excitation. J. Mechanical Eng. **57**(12), 40–50 (2021)
11. Yanzhu, L., (ed): Higher Dynamics. Higher Education Press, Beijing 10–16 (2003)

12. Jingsong, S., Xiaoming, H., Gongyun, L.: Stress response analysis of pavement structures under moving loads. J. Highway Trans. Res. Dev. **24**(1), 10–13 (2007)
13. Industry Standard of the People's Republic of China, Specifications for the Design of Highway Asphalt Pavements (JTG D50–2017). China Communication Press Co.Ltd, Beijing
14. Xudong, W., (ed): Dynamic characteristics and Dynamic Parameters of Asphalt Pavement Materials. China Communication Press, Beijing 88–96 (2002)
15. By Yangxian, H.: Road Surface Analysis and Design. China Communication Press, Beijing 89–103 (1998)

Study on the Evaluation Index System of Reinforcement and Renovation of Existing Campus Buildings: Take Yueyi Building as an Example

Xiaoxue Tang[1], Jie Wu[1,2(✉)], Youqing Zhu[1], Ruijie Xu[1], Shunuo Li[1], and Qianqian Yan[1]

[1] School of Architecture, Soochow University, Suzhou, China
jwu@suda.edu.cn
[2] International Research Center for Green Building of Soochow University, Suzhou, China

Abstract. In the context of the constant perfection of the green building system,a substantial proportion of old campus buildings need to be reinforced and renovated to meet the requirements of safety, use, energy saving and so on. However, due to the unique characteristics of the existing campus buildings in terms of cultural inheritance, functional improvement, and usage and so on, the existing index system of reinforcement and renovation cannot fully and purposefully evaluate the reinforcement and renovation of campus buildings. Taking Yueyi building of Soochow University as the object of practical measurement, studying the green building concept, analyzing the relevant evaluation index system of buildings with a comprehensive subjective and objective evaluation method and searching for the correlation between the renovation of ultra-low energy consumption buildings and the building reinforcement project, the study aims to build a scientific and reasonable evaluation index system for existing campus buildings. Lastly, an overall evaluation index system with two levels of reinforcement and renovation is built to provide a reference for the selection of reinforcement and renovation plans and to guarantee the sustainable development of existing campus buildings.

Keywords: Renovation · Evaluation index · Weight

1 Introduction

Campus buildings are of high cultural, historical and functional values. However, with the rapid progress of socialization and urbanization, a large number of campus buildings are facing functional renewal and regeneration.Nowadays China's green building evaluation index system is increasingly perfect, but the combination of green building concept and reinforcement and renovation of campus buildings still lacks a lot of theoretical support and practical basis [1]. In this context, how to carry out reasonable reinforcement and renovation of campus buildings under the concept of green building is of great significance for realizing green transformation of campuses.In accordance

W. Guo and K. Qian (Eds.): GBCESC 2022, LNCE 211, pp. 385–393, 2023.
https://doi.org/10.1007/978-981-19-5217-3_38

with GB/T50378–2019 "Evaluation standard for green building" and the existing building evaluation system, also with field research and expert interviews, the following conclusions can be drawn. 1) The reinforcement and renovation projects of campus buildings are not consistent with those of other types since there is much more complexity, including the protection of campus culture, teaching needs, functional positioning, etc. Different from the current evaluation index system, stress of the reinforcement and renovation programs of campus buildings is placed on the historical values, cultural values and the comfort (e.g., lighting) after reinforcement and renovation instead of the traditional museum-style preservation. 2) While the reinforcement and renovation are usually designed and evaluated respectively, the reinforcement and renovation should be considered comprehensively. For instance, ultra-low energy renovation techniques (e.g., design of insulation layer) are significantly correlated with the strengthening of structure (e.g., selection of materials for beams, columns and walls) [2].

2 Analysis and Establishment of the Evaluation Index System

2.1 The Principle of Constructing the Evaluation Index System

The reinforcement and renovation of campus buildings have been considered a complex systematic project, which consists of sociality, economic efficiency, humanity, building performance and other aspects. Thus, the evaluation system should be comprehensive and scientific and show certain hierarchical structure characteristics [3].

2.2 Determination of Evaluation Indicators

Literature Survey Method. Using the keywords of "evaluation", "index system", "reinforcement and renovation" and others, the websites of CQVIP, CNKI and Wanfang Data Knowledge Service Platform are searched. As a result, more than 30 representative articles are selected, in combination with nearly 20 correlation codes, to determine first-level indicators (i.e., economic rationality, technical construction feasibility, reliability of reinforcement effect and excellent overall effect).

Field Survey. Investigation and research show that the Yueyi building of Soochow University have received sporadic restoration, functional transformation, decoration and other renovation activities several times over the past hundred years. Since the previous programs lacking systematic structure and performance improvement primarily have focused on building structure reinforcement, the original spatial pattern, structural practices, appearance and performance have changed, which is not consistent with the trend of sustainable development of China's buildings [4].

Taking Yueyi building as a typical case, the performance are measured and analyzed. The evaluation indicators of the reinforcement and renovation of campus buildings are explored by subjective and objective discussion and analysis simultaneously.

Performance Satisfaction. The research team distributed 100 questionnaires targeting the users of Yueyi building totally, 82 of which are valid and well collected. The overall environmental satisfaction of these buildings is far from satisfaction, since only 44%

of the respondents are satisfied with the indoor physical environment in summer (see Fig. 1). To be specific, the number of people dissatisfied with the indoor natural lighting and the indoor sound insulation accounts for a higher proportion (see Fig. 2).

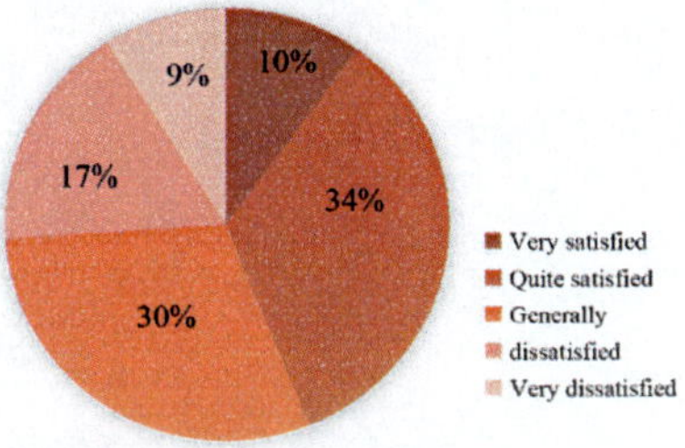

Fig. 1. Overall environmental use evaluation

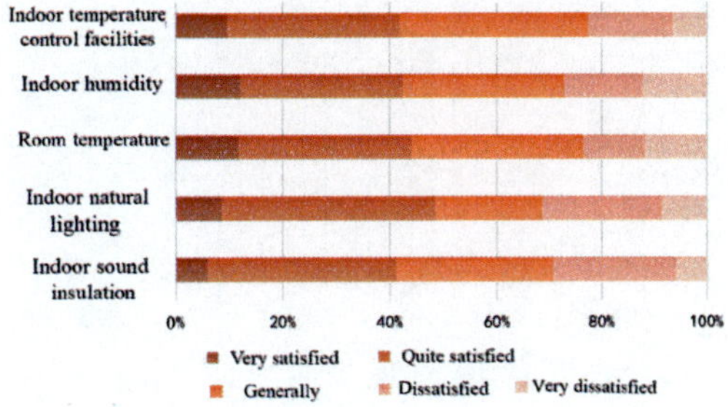

Fig. 2. Satisfaction with each use performance

Data Analysis. Fixed-point time-by-time measurements were performed in accordance with GB/T 5699 "Method of daylighting measurements", GB/T 18883 "Indoor air quality standard", JGJ/T 132–2009 "Standard for energy efficiency test of residential buildings", as well as GB/T 50785–2012 "Evaluation standard for the indoor thermal environment in civil buildings". The period was concentrated on sunny days in August 2021.

a. Lighting measurement analysis
The natural lighting area of the conference room in Yueyi Building largely lies beside the room's eastern and northern windows. The light goes weakener with the room going deeper. Although the window opening area is not remarkably small, the indoor lighting is significantly insufficient due to the shade of the surrounding plants. The natural illumination value of the office core area is lower than 450 lx [5], so the use requirement cannot be met.
b. Air quality measurement and analysis
The indoor air quality significantly changes with the change of the surrounding atmospheric air quality under the indoor natural ventilation: the concentrations of CO_2, the PM10 and the PM2.5 are lower than the 800 ppm, 100 ug /m^3, and 50 ug /m^3 proposed by ASHRAE [6], respectively. The indoor PM2.5 and PM10 contents are significantly lower than those in the atmosphere due to the adsorption and purification of green plants around the building, which suggests an excellent overall air quality (see Fig. 3). The above result reveals that the green environment should be designed reasonably for the subsequent reinforcement and renovation projects to improve the indoor air quality.

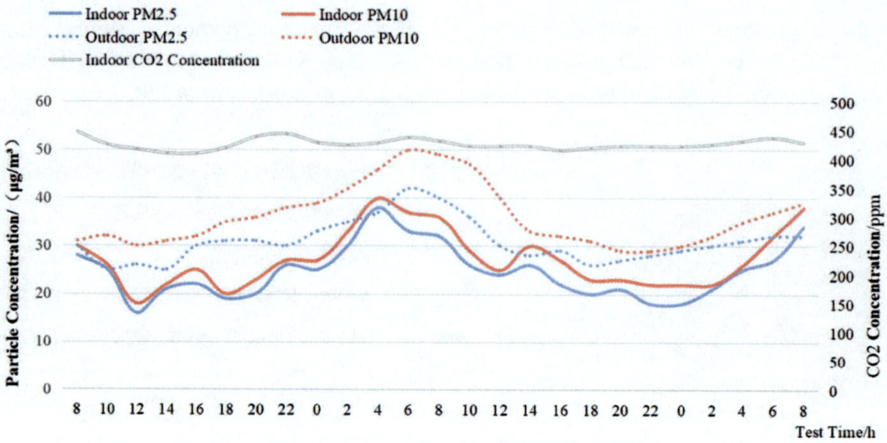

Fig. 3. Indoor air quality under natural ventilation

c. Measurement and analysis of thermal environment
The indoor environment under the operation of air-conditioning chillers in summer (9:0015:00) is affected by the atmospheric temperature and humidity. The indoor temperature fluctuates from 23°C to 27 °C while the relative humidity fluctuates from 70% to 80%. The average PMV of this office space is 1.2 with the PPD around 36%. The result shows that the expected subjective indoor feeling is hot following a high rate of dissatisfaction with the expected cold and heat comfort, which, though meeting the overall evaluation level of indoor hot and humid environment of III, (see Table 1) should be improved.

In general, the indoor relative humidity is 60%80% throughout the day, which has a certain gap in comparison with the "relative humidity of office space of 40%60%" proposed by ASHRAE [6] (see Fig. 4).

Table 1. Overall evaluation indicators for non-heated public buildings

Level	Overall Evaluation Metrics	
Grade I	$PPD \leq 10\%$	$-0.5 \leq PMV \leq +0.5$
Grade II	$10\% < PPD \leq 25\%$	$1 \leq PMV < -0.5$or $+ 0.5 < PMV \leq +1$
Grade III	$PPD > 25\%$	$PMV < -1$ or $PMV > +1$

d. Analysis of the thermal insulation performance of the wall.
By using heat flow testers, the envelope structure which meets the measurement requirements was selected for a continuous 48-h test [7]. The heat transfer coefficient of most of the envelope structure of the Yueyi is significantly higher than the limit

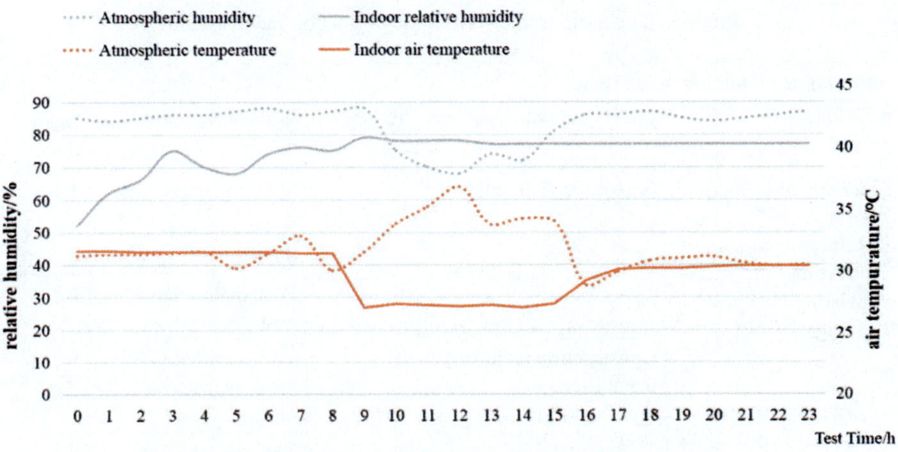

Fig. 4. Normal office state indoor temperature and humidity

value of 0.8W/(m^2·K) of heat transfer coefficient of the wall in hot summer and cold winter areas, as stipulated in China GB 50189–2015 "Energy-saving Design Standard for Public", indicating a poor heat insulation.The energy-saving requirements of green buildings and the comfort requirements for users should be valued in the following reinforcement and renovation(see Table 2) [8].

Table 2. Heat transfer coefficient of maintenance structure

Testing object	Thermal Resistance (m^2·K)	Thermalc onductivity [W/(m·K)]	Heat transfer coefficient [W/(m^2·K)]
Conference exterior wall	0.39	2.56	1.56
Conference interior wall	0.15	6.6	2.56
office exterior	0.78	1.28	0.98

2.3 Framework of Evaluation Index System

The basic framework of the evaluation index system is set on the foundation of a comprehensive analysis combined with specific construction plan analysis, typical case data measurement, expert consultation and literature review (see Table 3).

Questionnaire Collection and Data Collation. With the framework of the initial evaluation index system, the questionnaire was designed referring to the classical five-level Likert scale scoring method. 62 questionnaires were distributed, of which 50 were valid,

Table 3. The basic framework of the evaluation index system

Selection of evaluation index items.				
Target layer	First-level indicator	Type no	Secondary indicators	Index Code
The Evaluation Index System of Reinforcement and Renovation of Existing Campus Buildings	Economic rationality	A	Reinforcement cost	A1
			Reinforcement duration	A2
			Post-maintenance cost	A3
	Construction technical feasibility	B	The complexity of the construction process	B1
			The impact of construction on the surrounding environment	B2
			Influence of construction on synchronous use of structures	B3
			construction risk	B4
	Effectiveness and reliability	C	Meet safety requirements	C1
			Meet suitability requirements	C2
			Meet durability requirements	C3
	Excellent overall effect	D	Comfort requirements	D1
			Energy-saving requirements	D2
			The impact of the project on the overall appearance	D3

including 10 from project investment managers, 20 from university researchers engaged in relevant projects, 9 from construction units engaged in related projects, and 11 from design and research units engaged in relevant projects.

Reliability Analysis of Questionnaire Data. The reliability of the questionnaire data was investigated using α reliability coefficient method. The Cronbach a coefficient value is 0.728, thus indicating a high data reliability and capability for subsequent research.

Indicators Based on Relative Importance Index and Coefficient of Variation. By the analysis of questionnaires data from experts, university researchers, design and construction parties, the coefficient of variation of this indicator of reinforcement duration(A2) is higher than 0.25 whose relative importance index is lower than 75 [9], leading to the confirmation that the indicator is less important and can be eliminated (see Fig. 5).

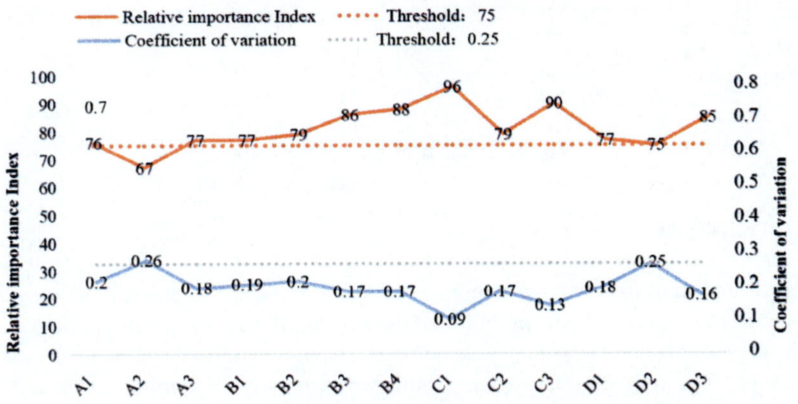

Fig. 5. Coefficient of variation and relative importance index of indicators

3 Research on the Weight of the Evaluation Index System

Based on the completion of the evaluation index system, the weights are determined using the fuzzy comprehensive evaluation method (FCE) [10]. The problem that the AHP is extremely subjective when the weight is being determined is solved in accordance with the fuzzy concept, which ensures the accuracy of the calculation result.

The weight coefficients reveal the concern of the respective index item in the design of the reinforcement and renovation plan of the existing campus buildings. In accordance with the measured environmental assessment report and consideration of experts' and users' concern regarding the comfort index after reinforcement and renovation(D1) (the weight value is 7.80), we suggest that the weight value of the comfort index after reinforcement and renovation should increase (the increment is indicated within 10% obtained on the basis of the weight from questionnaires targeting users) during the selection of the scheme [11]. In the practical evaluation model, the obtained index weights (see Fig. 6) are adopted to quantitatively assign the importance of the indexes, which is recognized as a vital part of conducting the evaluation and preference of the reinforcement and renovation programs for existing campus buildings [12].

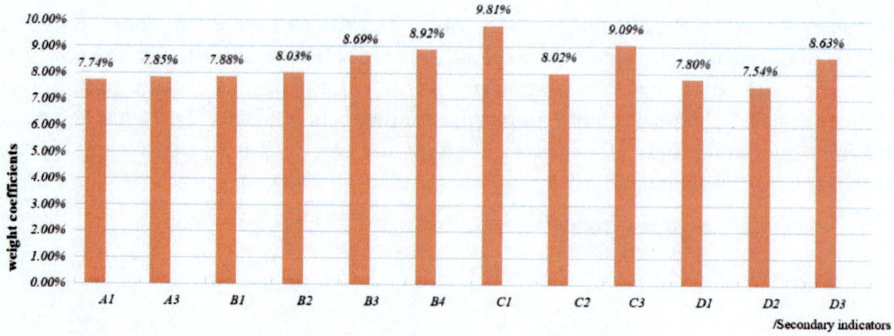

Fig. 6. Weights of various indicators

4 Conclusion

The construction of the evaluation index system and the determination of its weight values provide the basis for the formulation of the conservation plan of campus buildings, so that these precious and non-renewable cultural heritage materials can be used rationally in terms of functional continuity, social utility optimization,economic value and green campus construction. The results in the study indicate: (1) The performance of indoor thermal comfort, air quality, natural lighting, heat insulation and heat preservation in the use of Yueyi buildings is significantly deficient. The moisture-proof layer, the reduction the shading of outdoor plants from sunlight and other measures ought to be employed to improve the performance of the building. (2) Relevant measures for building comfort and green performance are generally missing in most of the existing reinforcement and renovation plans for campus buildings, thus revealing that the existing reinforcement and renovation plans have rarely considered performance and comfort, which should have not been ignored (3) Ecology and humanity are the development trend of reinforcement and renovation project, which implies that this study on these buildings should not only attend to the fundamental needs, but also a higher pursuit of spirit, emotion and quality of life.

Acknowledgement. This research was funded by China-Portugal Joint Laboratory of Cultural Heritage Conservation Science, National Key R&D Program of China(2021YFE0200100); The 2021 Jiangsu Construction System Science and Technology Project, Research on energy efficiency design method of passive ultra-low energy office building based on health and infection prevention (2021ZD03); "Undergraduate Training Program for Innovation and Entrepreneurship, Soochow University" (202010285082Y).

References

1. Cho, H.M., Yun, B.Y., Yang, S., Wi, S., Chang, S.J., Kim, S.: Optimal energy retrofit plan for conservation and sustainable use of historic campus building: Case of cultural property building. Applied Energy **275**(10), 115313 (2020)

2. Chowdhury, M., Sabrina, H., Zzaman, R.U., et al.: Green building aspects in Bangladesh: A study based on experts opinion regarding climate change. Environ. Dev. Sustain. (2021)
3. He, B.-J., Zhao, D.-X., Gou, Z.. Integration of low-carbon eco-city, green campus and green building in China. Green Building Dev. Countries, pp. 49–78. Springer (2020) https://doi.org/10.1007/978-3-030-24650-1_4
4. Chodnekar, H., Yadav, P., Chaturvedi, H.: Review and assessment of factors associated with green building rating systems. IOP Conference Series: Earth and Environmental Sience. **795**(1), 012033 (2021)
5. Li, D.H., Lam, J.C.: Evaluation of lighting performance in office buildings with daylighting controls. Energy and Buildings **33**(8), 793–803 (2001)
6. McGregor, G.: Biometeorology for Adaptation to Climate Variability and Change. Springer-Verlag.Germany (2009) https://doi.org/10.1007/978-1-4020-8921-3
7. Budimir, S., Bratislav, I., Vladimir, M., Aleksandar, S.A.: Heat flux transmission assessment of a vegetation wall influence on the building envelope thermal conductivity in Belgrade climate. J. Cleaner Prod. **223**(6), 907–916 (2019)
8. Yuan, P.: Application of heat flow measuring sensor in building energy saving detection **30**(08), 68–72 (2015)
9. Xiaolong, Z.: Evaluation and optimal research on reinforcement schemes of historic building structures. Eng. Science Technol. II 40–46(2017)
10. Wenming, W., Ming, M.: Evaluation system of green building and its application in engineering practice. In: Civil Engineering and Architecture.2019 4th International Conference on Landscaping, pp. 131–135 (2019)
11. Author, S.L., Author, P.C.: Research on fuzzy comprehensive evaluation in practice teaching assessment of computer majors. Int. J. Modern Educ. Computer Sci. **7**(11), 11–19 (2015)
12. Yang, L., Hongyu, C., Xianjia, W.: Research on green renovations of existing public buildings based on a cloud model –TOPSIS method. J. Building Eng. **34**, 101930 (2021)

In-Process Reinforcement Method for 3D Concrete Printing: Status, Potentials and Challenges

Billy Kazadi[1], Liang Yao[2], and Li Wang[1](✉)

[1] Hebei University of Technology, Tianjin 300401, China
wangl1@hebut.edu.cn
[2] Yaobai Special Cement Technology R&D Co. Ltd., Xi'an 710100, China

Abstract. 3D concrete printing (3DCP) as a revolutionary technology has reached the construction industry with numerous achievements. While the key to promote the structural application of 3D printing lie in the effective reinforcement approaches that ensure the mechanical capacities of printed structures. Many researchers have tried to find solutions to reinforce 3D concrete printing elements. Hence, the need to review the different types of in-process reinforcement solutions based on their mechanical performances, their application process challenges, and their performance on different structures. Our primary focus is to examine ten different reinforcement solutions used in 3D concrete printing over the past ten years. This paper shows the improvement of each reinforcement strategy over time and propose a reinforcement solution that can be considered for future research.

Keywords: 3D concrete printing · Reinforcement method · Mechanical property

1 Introduction

In recent years, 3D concrete printing (3DCP) has made great progress in cementitious materials, automation equipment, software control system and construction methods, etc. The 3D printing intelligent construction technology system is in a fast development stage, with a rapid increase in market share and technical maturity, showing a diversified and promising development trend [1, 2].

3DCP has the potential to revolutionize the construction industry [3]. The development of increasing numbers of 3DP projects in the construction industry was presented in Fig. 1. The past decade has witnessed the rapid development of 3D printing in the field of civil architecture. Before the year of 2014, 3DP was still under development and few large-scale demonstration projects were constructed through the 3D printing technology. 3DCP was mostly applied to landscape architectures, street furniture, and non-load-bearing maintenance structures etc. While in the last five years from 2016, many excellent engineering project are reported, and revolutionary development of printing projects is as expected as the unique characteristics of 3DCP. 3DCP practices are changing to load-bearing structures and have witnessed remarkable achievements in structural engineering.

W. Guo and K. Qian (Eds.): GBCESC 2022, LNCE 211, pp. 394–402, 2023.
https://doi.org/10.1007/978-981-19-5217-3_39

3DCP construction compares with the convention construction method exhibits orthopedic the absence of an effective integration method for reinforcement. Moreover the brittleness behavior of concrete and its low mechanic performance in terms of tension and flexural strength, there is a great need to find an effective reinforcement solution that will encounter the automation process of printing and meet the rheological properties for easy concrete extrusion, buildability, and binding with the reinforcement. Therefore, research and testing of reinforcement methods suitable for 3D printing process automation and flexibility have tremendous significance for expanding the engineering application of 3D printing.

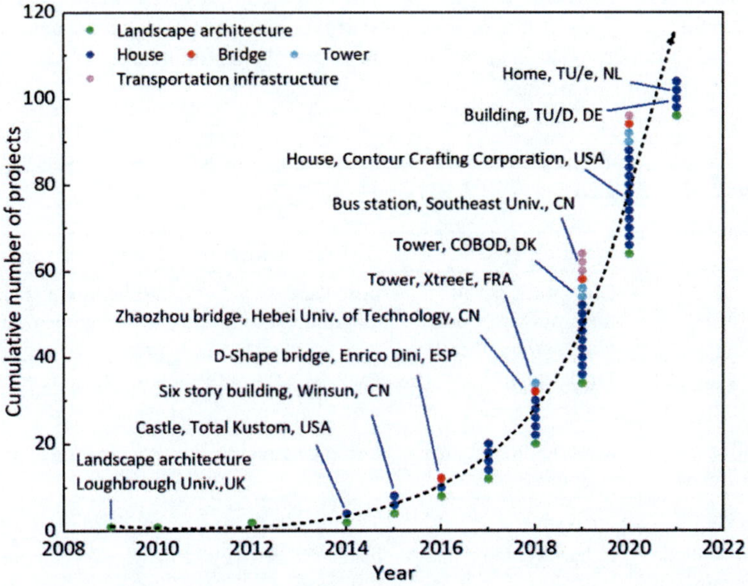

Fig. 1. Development of large-scale 3DP for application in the construction sector

2 In-Process Reinforcement Solution for 3D Printing Concrete

Many researchers have tried to find solutions to reinforce 3D concrete printing elements such as the incorporation of micro cable in the printer layers, the addition of fibers in the concrete mixture, printing the contour crafting that allows manual installation, nailing printed layer, and using the printed concrete as permanent formwork, etc. According to the different stages of printing, different reinforcement methods are proposed before printing, during printing and after printing. The main reinforcements are shown in Table 1, including all construction phases from before printing to after printing.

In-process reinforcement method makes the reinforcement measure and printing can be carried out synchronously, which greatly simplifies the process of the reinforcement measure. So this review mainly analyzes the in-process reinforcement methods. Figure 2 shows the proportional distribution of the various reinforcement methods studied during the printing process in the current published papers, it can be seen that more than half of relevant studies are focused on the influence of fiber incorporation on mechanical properties of materials. In the papers we searched for, most of the reinforcement methods were based on adding fibers. Because fibers are directly incorporated into the 3D-printed cement-based material mix, the difficulty of the printing process is minimized and the mechanical properties of the material can be significantly reinforced. The continuous cable, recently proposed reinforcements. During the printing process, the placement of steel cables and concrete printing are carried out synchronously, it also simulates reinforcement of steel rebar in concrete to a great extent. There's a lot of interest in it now. Other reinforcements like UHPC reinforcement, nail reinforcement and mesh reinforcement, all of them are new reinforcement methods that have been proposed recently, the number of studies on them is relatively small, but they will gradually become a focus for researchers.

Table 1. Illustration of reinforcement method based on the material used, embedding location and moment of application.

Method	Materials used	Embedding location	Moment of application
Fiber reinforcement	Glass fiber, basalt fiber, polypropylene fiber, carbon fiber	Inside and longitudinal to layers	While mixing concrete or during the printing process
Entrained micro cable	Micro steel cable, steel chain	Inside layers	While printing
Preinstalled Reinforcement	Mesh mold, polymer, steel mesh	Inside de structure	Prior printing
Pre-stressed cable	Steel cable	Outside layers	After concrete hardened
Manual installed reinforcement	Rebar, welded rebar	Inside the permanent formwork	After layers hardened
3D printed rebar reinforcement	Steel, iron, polymer	Outside layers	Before printing or while printing
Polymer reinforcement	Polymer, carbon fiber	Inside or outside	While printing concrete
Nail reinforcement	Iron, steel, and carbon	Transversal to the layers	While printing or after hardening for screwing
Bar inserting reinforcement	Iron or steel	Transversal and longitudinal to the layers	While printing or screwing after printing

From publications about in-process reinforcement method available on Science Direct, most scholars' research ideas are to reinforce the mechanical properties of 3D-printed concrete by adding various fibers. Its good construction workability, low cost and significant enhancement effect make it widely used in engineering.

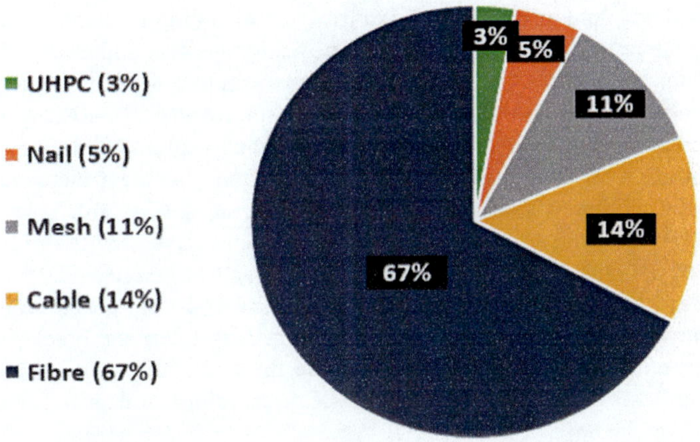

Fig. 2. Proportion of papers related to each in-process reinforcement method

2.1 Aligned Short Fiber for Reinforcement

Fiber strengthening is a reinforcing method of adding fibers during the mixing stage or while printing the layers trough the print head to strengthen concrete materials. Many research works have been accomplished up to now in this area with the use of different types of fiber materials. For example, steel fiber, basalt fiber, PVA fiver and glass fibers etc. as reinforcement for concrete composition to meet the requirement of 3DCP [4]. It is necessary to mention that, in this method the good design of the print head and a specific extruded nozzle is essential to allow the free flow of the raw materials and avoid material stacking during the printing process.

However, most of the reinforcement solution are still limited to micro-crack control, and it is difficult to exert the toughening effect of steel-like steel. The fibers are short and oriented in the same direction of printing layers deposition. Thus they do not provide continuity of reinforcement in the printing element. Also, they do not give a rigid bridge between the consecutive layers, which is very important to prevent the weak interface's cracking. The limit of fibers' actions across the layers is related to the weak fiber bridge in the interlayers interface due to their orientation in the printing direction. The strength continuity is also affected by the fiber size. Based on these limitations of this method, it is necessary to find a new approach that will provide continuous reinforcement in the printed layer. Fiber concrete can be combined with steel structure to achieve toughness from two levels of material and structure.

2.2 Vertical Penetration of nail/bar for Reinforcement

Nail insertion is a reinforcement method in which nails can be automatically inserted in subsequent layers that serve as the reinforcement for printing layers. The steel nails are generally inserted orthogonally to the printing direction to serve as a bridge between layers and strengthen the weak interface interlayers. This technology has several advantages for 3D concrete structures because to its automated procedure, readily available materials, and low cost. It enables the user to specify the location of reinforcement.

Wang et al. [5] demonstrated the feasibility of embedding U-shaped nails into concrete during the printing process (Fig. 3a). The bridging effect and dowel action of the applied U-nails were studied to have a better understanding of the toughness increase. The 3D printed concrete's ultimate tensile strength improved by 145.0%, while its shear strength increased by 220.0%. Geneidy et al. [6] have explored a strategy for in-situ printing that attempts to embed discrete U-shaped reinforcement elements "staples" vertically interlocking layers simultaneously while printing (Fig. 3b). Freund et al. [7] have noticed tremendous progress in this reinforcement method with the invention of integration of short reinforcement bars perpendicular to Shotcrete 3D printed layers (Fig. 3c). Marchment et al. [8] explored the feasibility of the bar penetration technique on the interlayer reinforcement, as presented in Fig. 3(d). Lauri and Freek [9] proposed a screw-type reinforcement method to decrease the voids around reinforcement elements and furtherly increase the bonding between the reinforcement with the concrete, as shown in Fig. 3(e).

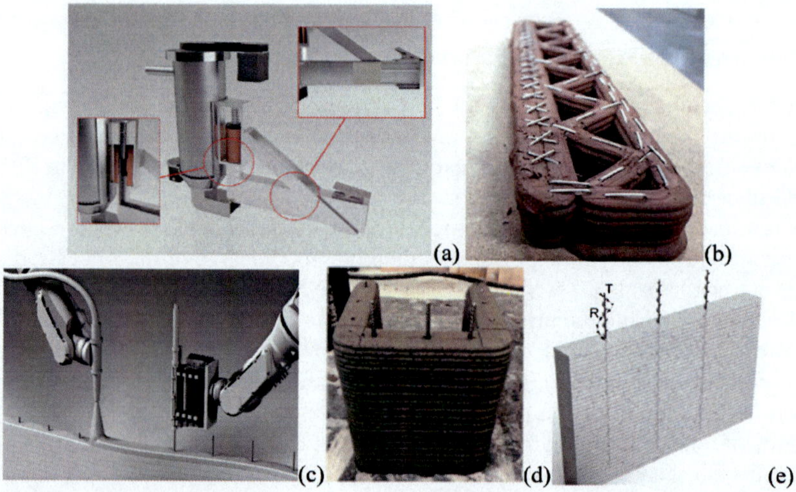

Fig. 3. (a) Automatic device for U-nail insertion, (b) U nail reinforcement, (c)robot-guided integration of short reinforcement bars, (c) bar penetration technique, (d) screw reinforcement.

2.3 Entraining of Micro-cable for Reinforcement

A simultaneous reinforcement solution entails the incorporation of a continuous steel micro cable in the middle of the concrete layer from a specific device attached to the print head while printing concrete. Recently, this technic emerges with a high tensile steel cable in conjunction with post-tensioned reinforcement applied for the pedestrian bridge by TU/e. Also, the novel design of the printed head allows the incorporation of single and multi-cable. While the concrete is being extruded, the steel wire rope is fed into the print head and extruded from the nozzle. A wire rope or micro-cable extruder is mounted on the material barrel for the wire rope conveyance. Because the steel wire rope is used as reinforcing material, its rigidity is large, and the size of the wire rope which can be put into concrete is small. Li et al. [10] experimentally explored the feasibility of directly entraining a continuous micro steel cable (1.2 mm) during filaments (12 mm) deposition process, forming a reinforced geopolymer composite material. Five different reinforcements for 3D printing were optimized and the strength and ductility enhancement was validated by printing complex structures (Fig. 4).

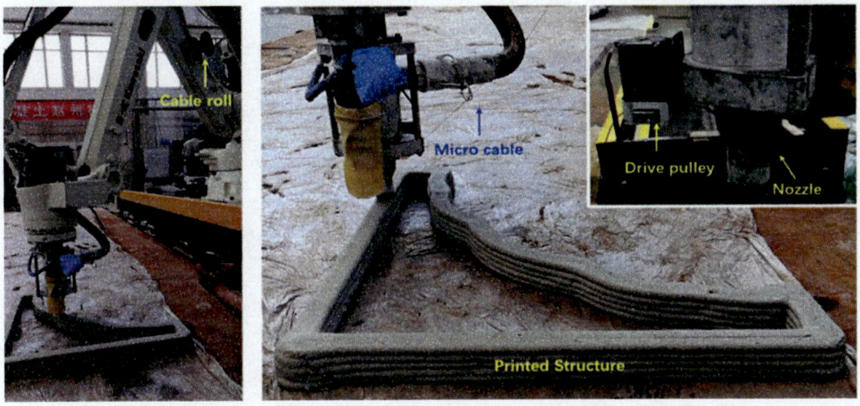

Fig. 4. Entrained micro cable reinforcement method for 3DCP.

2.4 Incorporation of Mesh for Reinforcement

Steel mesh toughening is a method of toughening 3D printed concrete using horizontal, vertical, and spatial grids. To adapt to the characteristics of 3D printing flexibility, the researchers explored the synchronous reinforcement method of 3D printing concrete [11, 12]. This method can effectively improve the mechanical properties of 3D printed concrete, but its degree of automation is not high, and it relies more on manual labor, and further research is needed for the interlayer bonding performance.

2.5 UHPC Reinforcement Method

An in-process reinforcing method that involves printing both ultra-high performance concrete (UHPC) and standard 3DP-C simultaneously from a unique print head. The UHPC performs as reinforcement for the 3D concrete composite. Bai et al. [13] presented a dual extrusion print head capable of printing both UHPC and conventional concrete concurrently (shown in Fig. 5). The experimental result obtained using the four points bending test indicates that this approach may improve the ultimate bending strength by 160.5% and convert brittle bending failure to ductile bending failure.

However, this idea of twofold extrusion, synchronous and simultaneous printing has received little attention; moreover, there is an obvious limit to strengthen in the transverse direction. Additionally, this reinforcement is not continuous in vertical direction and does not span the interlayer's weak contact.

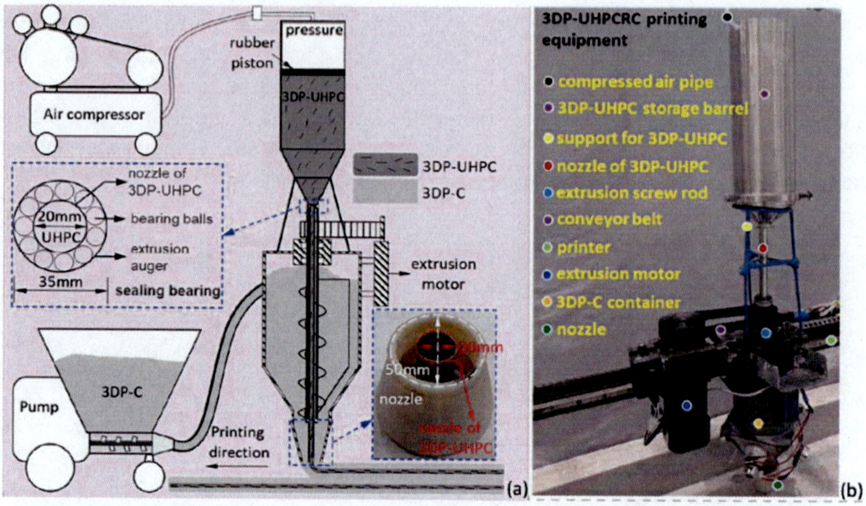

Fig. 5. Extrusion system for dual 3D concrete printing: (a) schematic; (b) set-up.

3 Application Potentials and Challenges of 3DCP

3.1 Application Potentials

3D printing of concrete definitely shows high potential, but the best way to benefit from it still needs to be explored. Based on the evaluation of adaptability, coordination and economy among 3D printing materials, equipment, methods and structures, the potential development and application scenarios of 3DCP can be divided into four modes: (1) Intelligent plant for prefabricated construction, (2) On site 3D printing and assembly construction, (3) In site 3D printing of entire building, (4) Swarm robot for the in situ construction.

3.2 Technology Fusion, 3D Printing +

3D printing belongs to a construction process that automatically distributes building materials to the designed and optimized structures. From structural fabrication to intelligent construction, 3D printing is only one of the important links, which still needs other automation, intelligent and digital technologies as auxiliary and complementary, such as topology optimization, artificial intelligence technology, UAV and 3D laser scanning technology, Internet of things (IoT) system, etc. The integration of multiple technologies and the intersection of various disciplines are the general trend and necessity of the development of civil architecture.

3.3 Challenges to Promoting the Sustainable Development of 3DCP

According to the aforementioned engineering projects, 3D printing has its unique applicability because of its flexible construction processes. It shall to find the accurate orientation and application areas of 3DCP, rather than taking 3DCP just as one fabrication method. The integration of architects and structural engineers of 3D printing can achieve the multiplier effect, which is very important to promote the sustainable development of 3D printing. As aforementioned, researchers have been investigating and refining techniques for 3D printing in construction in the past decade and various projects are put into practices. The final application goal of 3DP technology is to combine with and eventually adapt to the conventional construction processes. It is expected that the 3D printed elements and structures are designed and fabricated with standard features to promote the engineering application. Currently, the questions regarding the implications of 3D printing in construction are still unanswered, and only several specifications are proposed worldwide.

4 Conclusion

Among all the strengthening methods, steel micro cable can improve the tensile performance of components to the greatest extent by 290%, but the interlayer shear performance is still very weak. The interlayer shear strength can be increased by 220% by adding U-nail in the interlaminar. If the two methods can be used together and appropriate fibers are added, they do not conflict with each other. This should be the most comprehensive way to improve the mechanical properties of 3D-printed concrete components.

References

1. Xiao, J., et al.: Large-scale 3D printing concrete technology: Current status and future opportunities. Cement Concr. Compos. **122**, 104115 (2021)
2. Buswell, R., Leal, W.R., Jones, S.Z., Dirrenberger, J.: 3D printing using concrete extrusion: A roadmap for research. Cem. Concr. Res. **112**, 37–49 (2018)
3. Wang, L., Liu, Y., Yang, Y., Li, Y., Bai, M.: Bonding performance of 3D printing concrete with self-locking interfaces exposed to compression–shear and compression–splitting stresses. Addit. Manuf. **42**, 101992 (2021)
4. Ma, G., Li, Z., Wang, L., Wang, F., Sanjayan, J.: Mechanical anisotropy of aligned fiber reinforced composite for extrusion-based 3D printing. Constr. Build. Mater. **202**, 770–783 (2019)
5. Wang, L., Ma, G., Liu, T., Buswell, R., Li, Z.: Interlayer reinforcement of 3D printed concrete by the in-process deposition of U-nails. Cem. Concr. Res. **148**, 106535 (2021)
6. Geneidy, O., Kumarji, S., Dubor, A., Sollazzo, A.: Simultaneous reinforcement of concrete while 3D printing. In: Bos, F.P., Lucas, S.S., Wolfs, R.J.M., Salet, T.A.M. (eds.) DC 2020. RB, vol. 28, pp. 895–905. Springer, Cham (2020). https://doi.org/10.1007/978-3-030-49916-7_87
7. Freund, N., Dressler, I., Lowke, D.: Studying the Bond Properties of Vertical Integrated Short Reinforcement in the Shotcrete 3D Printing Process, pp. 612–621. Springer International Publishing, Cham (2020)
8. Marchment, T., Sanjayan, J.: Bond properties of reinforcing bar penetrations in 3D concrete printing. Autom. Constr. **120**, 103394 (2020)
9. Hass, L., Bos, F.: Bending and Pull-Out Tests on a Novel Screw Type Reinforcement for Extrusion-Based 3D Printed Concrete, pp. 632–645. Springer International Publishing, Cham (2020) https://doi.org/10.1007/978-3-030-49916-7_64

10. Li, Z., Wang, L., Ma, G.: Mechanical improvement of continuous steel microcable reinforced geopolymer composites for 3D printing subjected to different loading conditions. Compos. B Eng. **187**, 107796 (2020)
11. Marchment, T., Sanjayan, J.: Mesh reinforcing method for 3D concrete printing. Autom. Constr. **109**, 102992 (2020)
12. Liu, M., Zhang, Q., Tan, Z., Wang, L., Li, Z., Ma, G.: Investigation of steel wire mesh reinforcement method for 3D concrete printing. Archives Civil Mech. Eng. **21**(1), 1–18 (2021)
13. Bai, G., Wang, L., Wang, F., Ma, G.: In-process reinforcing method: dual 3D printing procedure for ultra-high performance concrete reinforced cementitious composites. Mater. Lett. **304**, 130594 (2021)

Systematic Framework of Economic KPIs for Prefabricated Building Construction

Bingyan Wang(✉)

Department of Economics and Management, North China Electric Power University, Baoding, China
13032687982@163.com

Abstract. Prefabricated construction has been increasingly adopted in building projects all over the world because prefabrication technology has tremendous advantages over the cast-in-situ construction. However, the development of prefabrication technology in China has encountered many challenges such as cost overrun. Economic performance determines the success of a project, which has raised the attention of construction project stakeholders. Therefore, this study aims to develop a systematic framework of economic key performance indicators (KPIs) for measuring prefabricated buildings. First, this paper establishes a systematic framework of eight KPIs through a comprehensive literature review. The KPIs cover unit cost, cost variance, etc. Second, to demonstrate the use of the framework, a cross-case study is conducted to measure and compare the economic performance of a prefabricated building and a building built using cast-in-situ construction. Throughout the case analysis, the economic performance of a building project using prefabricated construction is much better than that using cast-in-situ construction. For example, the overall delivery of prefabricated buildings is 30%–50% faster than that of traditional buildings. The paper should assist construction enterprises in measuring the economic performance of prefabricated building projects and could also facilitate the development of prefabricated construction in China.

Keywords: Prefabricated construction · Economic performance · KPIs · Case study

1 Introduction

Compared with traditional construction technology, prefabricated construction has a number of benefits such as reduced construction cost and shortened construction time [1–3]. Therefore, it has attached the construction industry to widely adopt the prefabricated construction in building projects worldwide [4]. Nevertheless, the adoption of prefabricated construction has been facing severe challenges such as a lack of systematic understanding on its economic performance and evaluation of the cost barrier to the adoption of the precast construction.

To address the challenges of using prefabricated construction, many researchers have conducted academic studies. For example, Hong has investigated the major impact

W. Guo and K. Qian (Eds.): GBCESC 2022, LNCE 211, pp. 403–411, 2023.
https://doi.org/10.1007/978-981-19-5217-3_40

factors on the financial performance of prefabricated building [5]; and Lou has studied the key cost drivers and cost control paths of prefabricated buildings [7]. However, the previous studies have not systematically addressed the measurement of economic performance of prefabricated construction.

Therefore, this paper aims to develop a systematic framework of key performance indicators (KPIs) for measuring the economic performance of prefabricated buildings. To achieve the research aim, this study has adopted a mixed-research strategy, including a comprehensive literature review to develop the KPIs framework and a cross-case study to analyze the economic performance of prefabricated construction in building projects. The findings should clearly indicate the differences of economic performance between prefabricated and cast-in-situ construction.

Following this introduction, Sect. 2 presents the results of the literature review. Section 3 illustrates the research methodology for developing the KPIs framework and conducting the case study. Section 4 elaborates on the results of the KPIs framework. Section 5 applies the framework to carry out a cross-case study to measure and compare the economic performance of a prefabricated building and a building built using cast-in-situ construction. Section 6 discusses the implications and limitation of the study. Finally, Sect. 7 concludes the study with major findings and contributions of the study.

2 Literature Review

2.1 Features of Prefabricated Construction in Buildings

Prefabrication construction is a method that manufactures construction components in the factories, transports the components to the construction site, and finally connects the components together to construct the whole building [1]. Compared with cast-in-situ construction, prefabricated construction has many advantages such as lower cost, energy consumption and better environmental protection [8, 9] (Table 1). Therefore, prefabricated construction has attracted the attention worldwide in building projects because of their important role in the process of future urbanization [5].

Nevertheless, as a new construction method, prefabrication construction is different from traditional methods in production, transportation and on-site construction [10]; Prefabrication construction use unique standardized design, new technology and new material, which mostly adopt on-site installation process and reduce the wet operation on the construction site [11]. Therefore, the construction industry has facing great challenges to the adoption of prefabricated construction in building projects. In China, the high construction cost has become the main factor restricting the development of prefabricated buildings [7]. In addition, the evaluation method of economic performance of prefabricated construction in buildings is fragmented and not systematic, e.g., Hwang measured the impact of rework on construction cost performance [12]. Therefore, it is necessary to measure and understand the economic performance of prefabricated construction in building projects.

Table 1. Differences between conventional and prefabricated constructions

Differences	Conventional construction	Prefabricated construction
Cost	Low initial construction cost	Cost saving due to repetitive and standard modular production
Time	Longer	Shorter
Quality	Difficult to control the quality as the site condition always varies	Better quality achieved at the factory production
Productivity	Labor intensive	Most of the construction elements are prefabricated in the factory/precast yard

2.2 Economic Performance Measurement of Construction Project

Three major indicators, consisting of cost, time and quality, can measure construction project performance [13]. Therefore, based on the current research, scholars in the field of construction most often divided economic performance into three parts: cost, schedule, and quality performance. The research findings of the literature review on performance measurement are elaborated below.

First, to explore the basic cost composition of prefabricated construction and test the impact of prefabricated construction on the total cost of actual construction projects, Hong has established a cost-benefit analysis framework [5]. Based on the previous research results, Lou defined and screened the influencing factors of prefabricated construction cost from dynamic formation process and determined the key cost performance indicators of prefabricated building through sensitivity analysis [7].

Second, the progress of construction projects which directly affects time performance is often used as a KPI for project management [13]. Dissanayaka has studied the methods of time performance measurement [14].

Although there are a lot of studies on prefabricated building, these studies are insufficient in addressing the economic performance of prefabricated construction in building projects. Limitations of the existing research can be summarized as follows. First, most of the existing studies focus on a single aspect, for example, cost only [15]. Therefore, there is a lack of systematic understanding on economic performance which includes cost, schedule, and quality performance. Second, a part of studies analyzed economic KPIs without applying the research results to actual cases such as Lou and Guo [6]. Third, some case studies were limited to prefabricated buildings, without a conventional case for benchmarking, e.g. Hong et al. [5] and Ji et al. [6].

3 Methodology

3.1 Overview of the Research Design

This study has adopted a mixed research strategy, including a literature review and a cross-case study (see Fig. 1). Through a comprehensive literature review, the characteristics of prefabricated construction were identified, and the KPIs of the economic performance of prefabricated were investigated. Based on the results of the literature review, a systematic framework of KPIs was established. Finally, a real-life project demonstrated and verified the effectiveness and efficiency of the proposed framework. The methods of data collection and analysis are elaborated in the following section.

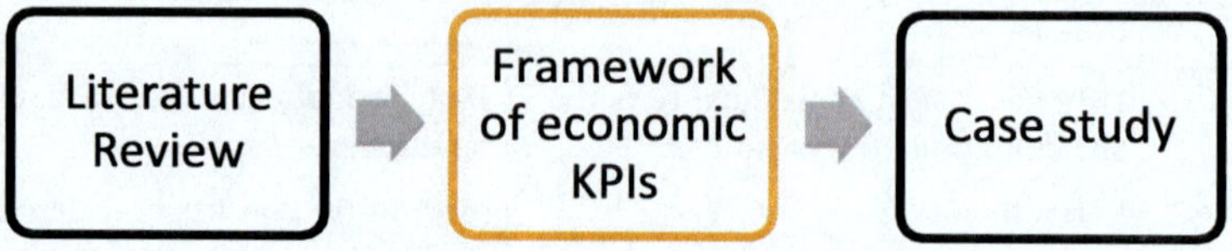

Fig. 1. Flowchart of the research design.

3.2 Methods of Data Collection and Analysis

Literature Review. The literature review covered more than 30 journal articles related to prefabricated construction, economic performance, and the differences of economic performance between prefabricated and conventional construction. First, the characteristics of prefabricated construction were identified and compared with conventional construction. Second, the existing economic KPIs of prefabricated construction were investigated and their limitations were summarized. Third, a preliminary framework of economic KPIs was established.

Case Study. The case study was conducted to demonstrate the use of the KPIs, and then to verify the effectiveness and efficiency of the proposed framework. Considering location, type and scale, construction methods and complexity, a prefabricated construction project was selected. The project was selected mainly because 1) it is a typical prefabricated building project in China with a common-used precast rate; 2) all aspects of the building are in line with typical residential buildings; and 3) data of the case project can be accessed. In consideration of these characteristics, the project was pondered a suitable case for this research. To analyze the economic performance of the project, the established KPIs were adopted and calculated using an Excel table.

4 Framework of Economic KPIs

Based on the comprehensive literature review, the framework of economic KPIs for measuring prefabricated building projects was developed (see Fig. 2).

4.1 Cost-Related KPIs

The 'cost' for construction consists of project expenditures from construction planning to handover, e.g., the original tender sum and variation cost. Construction cost efficiency has great impact on the project success [16]. Two KPIs were selected to represent the cost efficiency, i.e., 'unit cost' and 'cost variance'.

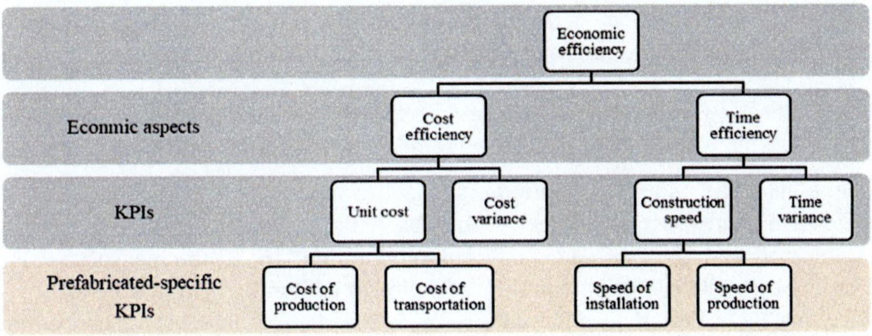

Fig. 2. Framework of KPIs for measuring economic performance of prefabricated construction.

"Unit cost" is a measure of cost efficiency to indicate the relative resources (i.e., labour, machine, and material) input. It can be measured using Eq. 1, i.e., the total construction cost and the construction quantities (e.g., construction floor area, number of units, tons of units). For example, when comparing the cost of production of various precast elements, 'tons of units' can be used to represent the construction quantities, to indicate the production efficiency. In addition, 'cost of transportation' was perceived to be great important for prefabricated building projects, as it greatly affects the overall construction cost, and it should be considered during the decision-making process of factory selection.

$$Unit\ cost = \frac{Total\ construction\ cost}{Construction\ quantities} \tag{1}$$

"Cost variance" is another measure of the cost efficiency of a construction project, to reveal the percentage of variation of actual cost compared with the estimated cost [16]. Cost variance can be further divided into labour cost variance, material cost variance and equipment cost variance. The estimated and actual cost of the various cost components were used to calculate the overall cost variance (Eq. 2). The indicator can help the project team to check cost overruns and judge the cause of cost overruns. If it is negative, there are cost overruns. Whilst, if it is positive, there are cost savings.

$$Cost\ variance = \frac{The\ estimated\ cost - actual\ cost\ of\ work\ performed}{The\ actual\ cost\ of\ work\ performed} \tag{2}$$

4.2 Time-Related KPIs

"Time" of construction can be defined as the duration of completing a construction project [17], which affects the economic efficiency of project delivery to a great extent. Time-related KPIs includes "speed of construction" and "time variance". 'Speed of construction' is a relative time consumption of construction activities, which can be measured using Eq. 3. To measure the speed of installation of precast components, "total of construction durations" and "number of units" can be used. Furthermore, the

KPI of "speed of production" is of great importance to prefabricated building projects, as it may great affect the logistics of transportation and site installation.

$$Speed\ of\ construction = \frac{Construction\ quantities}{Total\ construction\ duaration} \quad (3)$$

"Time variance" is another important aspect to measure the time efficiency of the project. It can be measured using Eq. 4, to indicate the percentage of variation of actual time compared with the estimated construction duration [16]. If the measured value of time variance is negative, it indicates that the project is running behind schedule. "Time variance" can be further divided into different projects such as engineering, procurement, and construction. The "time variance" of the various projects can be used to calculate the overall time variance. The indicator should help the project team to check project schedule and find the possible reasons of delay.

$$Time\ variance = \frac{The\ estimated\ time - the\ actual\ time\ of\ work\ performed}{The\ actual\ time\ of\ work\ performed} \quad (4)$$

5 Case Study and Analyses

The case project was a 34-storey building in Shenzhen. It adopted various concrete precast elements such as precast staircase, precast façade, and balcony (Fig. 3). There were three flat units in a typical floor and the total construction floor area (CFA) of a typical floor is around 480 m^2. The precast rate was around 23%. The overall construction cost of a typical floor was around 0.5 million; and the construction floor cycle was around 4 days. The production cost of the precast façade for a typical floor was around 44 thousand.

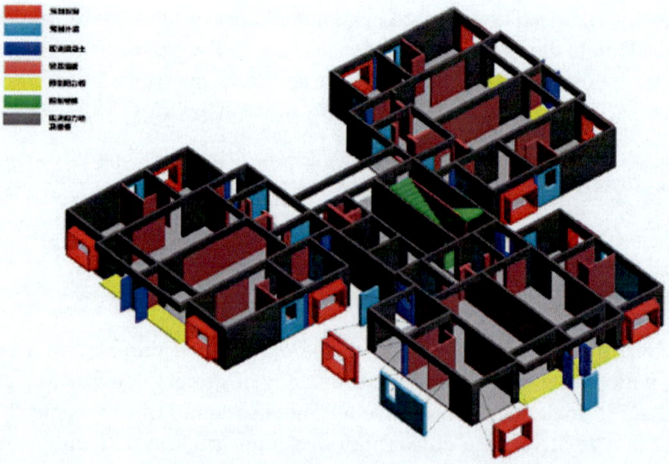

Fig. 3. Typical floor of the case building.

According to the data collected, 'unit cost' of a typical floor and 'unit cost' for precast façade were calculated. They were \$900/m^2 and \$27/ton, respectively. Comparatively, the cost using conventional construction is a little bit lower and was established as around \$900/m^2 (Table 2). Based on the data collected from the case project, the higher unit cost was derived mainly from factory production, transportation, and site connection. Nevertheless, the cost may also be reduced by fast construction, enhanced quality, and also other costs such as reduced site labor.

Table 2. Measured cost performance.

Project performance comparison	Cost \$	CFA m^2	Unit cost \$/m^2
A typical floor using prefabricated construction	523200	480	1090
A typical floor using conventional construction	432000	480	900
Difference of a standard layer	91200	480	190

In addition, the 'speed of construction' for a typical floor was calculated using Eq. 3, and can reach 119 m^2/day (4-day cycle). Comparatively, it may take over 5 days to complete a floor using conventional construction. The case project further indicated that the building can be delivered faster using prefabricated construction method.

The established framework of KPIs was presented to the industry experts. All of them agreed that the KPIs can effectively and comprehensively illustrate their economic concerns on prefabricated buildings. They commented that the developed framework could help major stakeholders to identify the benefits and barriers of adopting prefabricated construction in buildings.

6 Discussion

The developed KPIs in this study were found to enable comprehensively measure the economic performance of prefabricated construction in high-rise building projects, which were compared with other KPI frameworks introduced in the previous studies (e.g., Chan and Chan [16]). Specifically, (1) the economic performance framework established in this study includes two aspects: time and cost; (2) under time and cost, 2nd tier indicators lists KPIs in the two aspects, such as unit cost; and (3) under time and cost, the study establishes 3rd tier indicators specifically for prefabricated construction, e.g., cost of transportation.

Using the 3-tier framework, users can compare prefabricated construction with conventional construction practices, and can also benchmark various prefabricated building projects. For example, 'cost of transportation' varies among different prefabrication projects and has an important impact on cost performance, which fills the knowledge gaps of the previous studies that did not consider the off-site construction (e.g., Li et al. [18]).

Results of the case study indicated that there exist trade-offs in terms of economic performance between conventional and prefabricated construction, which echoes the previous findings by Mao et al.[19]. Compared with traditional buildings, prefabricated buildings have higher prices but shorter construction time. Therefore, time and cost performance should be comprehensively considered when selecting the construction method.

7 Conclusions

This study has proposed a systematic framework of economic KPIs for measuring prefabricated buildings. The framework was developed using a combined method, including a literature review and a cross-case study. The key findings of the study are concluded as follows.

First, the economic KPIs framework established in this study includes two aspects (i.e., cost and time) and eight KPIs (e.g., unit cost under the cost aspect, and construction speed under time aspect). Using the equations provided in this study can quantitatively measure all these KPIs. Four KPIs were identified to apply only to prefabricated buildings, e.g., cost of production.

Second, results of the case study indicated the differences of economic performance between prefabricated and conventional construction. Compared with conventional construction for building projects, prefabricated construction may have higher cost but shorter construction time. Therefore, time and cost performance should be comprehensively considered when selecting the construction method.

By using the framework developed in this study, the major project stakeholders can systematically evaluate the performance of prefabricated construction projects, thus can identify opportunities for further improvement. The paper should therefore facilitate better uptake of prefabricated construction in high-rises.

References

1. Vivian, W.Y., Chi-Ming, T., et al: Towards adoption of prefabrication in construction. Building and Environ. **42**, 3642–3654 (2007)
2. Yinghui, S., Junwu, W., Feng, G., Jiequn, L., Sen, L.: Research on Supplier Selection of Prefabricated Building Elements from the Perspective of Sustainable, **13**(11) (2021)
3. Navaratnam, S., Tuan, N., Tharaka, G., David, H.: Performance review of prefabricated building systems and future research in Australia. Buildings **9**(2) (2019)
4. Zhengdao, L., Geoffrey Qiping, S., Xiaolong, X.: Critical review of the research on the management of prefabricated construction. Habitat International **43**, 240–249 (2014)
5. Jingke, H., Geoffrey Qiping, S., Zhengdao, L., Boyu, Z., Wanqiu, Z.: Barriers to promoting prefabricated construction in China: A cost–benefit analysis. J. Cleaner Production **172**, 649–660 (2018)
6. Yingbo, J., et al.: BIM-based life-cycle environmental assessment of prefabricated buildings. Engineering, Construction and Architectural Manage. **27**(8), 1703–1725 (2020)
7. Na, L., Jingjuan, G.: Study on key cost drivers of prefabricated buildings based on system dynamics. Advances in Civil Engineering (2020)

8. Yuhong, Z.: Features and construction of prefabricated buildings. Applied Mechanics and Mat. 580–583, 2262–2265 (2014)
9. Chao, M., Qiping, S., Liyin, S., Liyaning, T.: Comparative study of greenhouse gas emissions between off-site prefabrication and conventional construction methods: two case studies of residential projects. Energy and Buildings **66**, 165–176 (2013)
10. Ting, L., Xiaolong, X., Yuna, W., Weirui, X., Yongtao, T.: A systematic overview of prefabricated construction policies in China. J. Cleaner Prod. **280** (2021)
11. Weishu, Z., Beibei, Z., Yang, Y.: Empirical study of comprehensive benefits for prefabricated buildings: a case study of Hefei city. Int. J. Electrical Eng. Educ. pp. 1–17 (2020)
12. Bon-Gang, H., Stephen, R.T., Carl, T.H., Carlos, H.C.: Measuring the impact of rework on construction cost performance. J. Construction Eng. Manage. **135**, 187–198 (2009)
13. Youngsoo, J., Seunghee, K.: Knowledge-based standard progress measurement for integrated cost and schedule performance control. J. Construction Eng. Manage. **133**, 10–21 (2007)
14. Sunil, D., Mohan, K.: Comparing contributors to time and cost performance in building projects. Building and Environ. **34**, 31–42 (1998)
15. Ephremg, S., Tabish, S., Jha, K.: Critical factors affecting cost performance: a case of Ethiopian public construction projects. Int. J. Construction Manage. **18**, 108–119 (2018)
16. Albert, C., Ada, C.: Key performance indicators for measuring construction success. Benchmarking: An International J. **11**, 203–211 (2004)
17. Meeampol, S., Ogunlan, O.: Factors affecting cost and time performance on highway construction projects: evidence from Thailand. J. Financ. Manag. Prop. Constr. **11**, 3–20 (2006)
18. Yuan Yuan, L., Po-Han, C., David Ah Seng, C., Chee Chong, T., Rong Gui, D.: Critical project management factors of AEC firms for delivering green building projects in Singapore. J. Construction Eng. Manage. **137**, 1153–1163 (2011)
19. Mao, C., Fangyun, X., Lei, H., Peng, W., Jun, W., Xiangyu, W.: Cost analysis for sustainable off-site construction based on a multiple-case study in China. Habitat Int. **57**, 215–222 (2016)

Civil Engineering

Numerical Simulation of Resists Progressive Collapse of the Elevated Station

Xiaolan Yuan[2], Weizhuo Huang[2], Guangtao Li[2], Zhi Li[1,2], and Xiaofang Deng[1,2](✉)

[1] GuangXi Key Laboratory of New Energy and Building Energy Saving, Xuzhou 541004, Jiangsu, China
XiaofangDeng@126.com

[2] College of Civil Engineering and Architecture, Guilin University of Technology, Guilin 541004, Guangxi, China

Abstract. To study the progressive collapse resistance of the elevated station, LS-DYNA software was used to establish the pure frame numerical simulation model for the elevated station in Xi'an Station, and 1–3 columns in the corner area and in the middle area were removed for numerical simulation analysis. The results show that the elevated station has good robustness. Under the conditions of single column removal and double columns removal in the middle area, the beam mechanism is the main resistance mechanism, and the catenary mechanism is developed under the three columns removal condition. Finally, the influence of the floor on the collapse resistance of the high-speed railway station was analyzed, and the results show that the deformation of the structure can be effectively reduced, and the collapse resistance of the structure can be improved by the tensile membrane mechanism of the floor under the large deformation of the two columns and three columns removal.

Keywords: Progressive collapse · Railway station · LS-DYNA · Numerical simulation

1 Introduction

China has built large railway stations in various hub cities. The new large railway station with the characteristics of large volume, multiple functions and dense personnel which integrate tourism. In recent years, railway station accidents are common. On November 14, 2015, a high-speed train derailed during the test in the northeast of France (see Fig. 1(a)), resulting in 11 deaths and 42 injuries. On February 9, 2016, a train collision occurred in Bavaria, southern Germany (see Fig. 1(b)), resulting in 12 deaths and 85 injuries.

Different from foreign countries, modern railway stations usually use online type in China, the high elevated waiting hall is above the railway track. If the large railway station house lacks sufficient resistance to progressive collapse, once the train running in the building derails and hits the station house, it may lead to large area collapse

W. Guo and K. Qian (Eds.): GBCESC 2022, LNCE 211, pp. 415–426, 2023.
https://doi.org/10.1007/978-981-19-5217-3_41

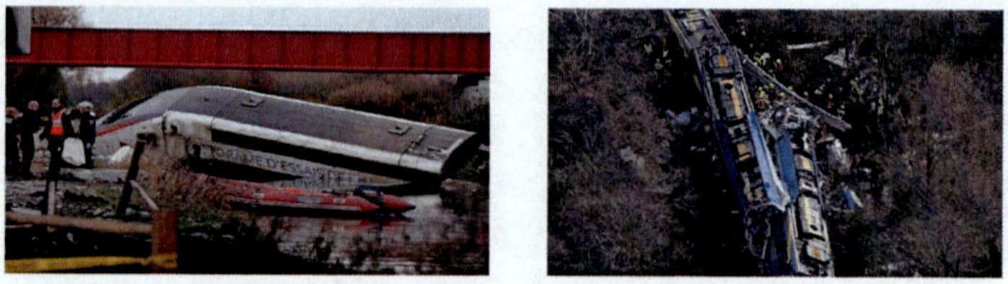

(a) (b)

Fig. 1. Railway station accident: (a) Train accident in northeastern France; (b) Train accident in Bavaria, Germany

of the station building. Causing serious casualties and property losses, and leading to significant negative social impact.

Many scholar have done a lot of experimental and theoretical research on the progressive collapse resistance of building structures [1–16]. At the same time, domestic scholars have paid attention to the collapse of railway stations and conducted research on it. Yuepeng [17] applied LS-DYNA software to study the failure mode of the platform column and the dynamic response of the structure when a short-distance explosion occurred in a subway station. The results show that the progressive collapse of the station structure will not occur when a single column failure. Huaijin et al. [18] analyzed the progressive collapse resistance of the overall spatial steel structure. The results show that the structure is safe and reliable even after the failure of a single key member. Rui [19] conducted a linear static analysis of a high-speed railway passenger station house by using the method of dismantling components, and the research results showed that the tree-shaped column of the side-type station house was a general structural member, and its failure would not cause structural damage. Qian [20] used nonlinear dynamic component removal method to analyze the progressive collapse resistance performance of Y shaped reinforced concrete typical units of single column awning on railway platform with 12 single beam and column members failing as collapse condition. The research results show that edge columns are the weak parts of single column awning structure. In addition, some scholars [21–25] conducted collapse resistance analysis for large-span structures.

To summarize, since a large number of modern large-scale railway station buildings have been built in recent years in combined with the rapid development of the high-speed railway, a large number of station buildings will be reconstructed in the future with this goal in mind. However, the research on the progressive collapse resistance of the viaduct waiting station is still very scarce, and most of the research is focused on the numerical study of a single column failure condition, while the research on the failure condition of different column removal positions and multiple columns is less. Considering this, this work employs the recently built Elevated waiting area of Xi'an Station as the research object, simplifies the structural system into a fibre model, and applies the deconstruction approach to evaluate its anti-progressive collapse performance using ANSYS/LS-DYNA explicit dynamic analysis software.

2 Finite Element Modeling

This paper chooses the recently built elevated waiting area of Xi'an Station as the engineering object to build a model. The elevated waiting station span the whole yard. The plane column network size is 123 m × 198 m, the distance of the longitudinal column is 12 m × 4 + 13.5 m × 2 + 12 m × 4, and the vertical column distance is 21.0 m–26.4 m. The main structure of the elevated waiting station is a reinforced concrete frame structure. To improve the bearing capacity and seismic ductility of the frame, the beams and columns (Columns that support the grid, beams and columns with large partial loads) at crucial parts are SRC beams and columns. The floor structure adopts cast-in-place reinforced concrete slab and the roof structure adopts steel structure grid frame. Table 1 shows some important beam-column sections and simplified reinforcement ratios. The locations are shown in Fig. 4.

Table 1. Critical section and simplified reinforcement ratio

Position	Section size/mm	Simplified reinforcement ratio
13 Shaft column	1600 × 1600	2.2%
14, 16 Shaft column	1600 × 2200	10.2%
15 Shaft column	1600 × 1600	10.2%
13 Axle beam (A–F)	800 × 2000	1.2%
13 Axle beam (F–G)	1100 × 2200	5.2%
14 Axle beam (A–B, E–H)	1100 × 2200	5.2%
14 Axle beam (B–E)	1100 × 2200	1.2%
15 Axle beam (A–B, E–F)	1000 × 2200	5.2%
15 Axle beam (B–E, F–H)	1000 × 2200	1.2%
16 Axle beam	1200 × 2200	5.2%

Using the ANSYS/LS-DYNA explicit dynamic finite element software to simulate the collapse of the station. Simulation calculation of dynamic analysis centre sill, column, rack adopts Hughes-liu beam element. Among them, the beams and columns using elastic-plastic fibre model definition, the fibre model as shown in Fig. 2, according to the actual beam and column section steel, concrete and the position of the core concrete custom integration points. Concrete integral point of material use *MAT_RC_BEAM, and the parameters in the model FRACR is set to 0 said plain concrete. Bilinear models *MAT_PLASTIC_KINEMATIC are used for the material of reinforcing bar integral points and steel frame materials. Imposed constraints on column bottom of six degrees of freedom as the boundary conditions of the model. Figure 3 shows the finite-element model finally established.

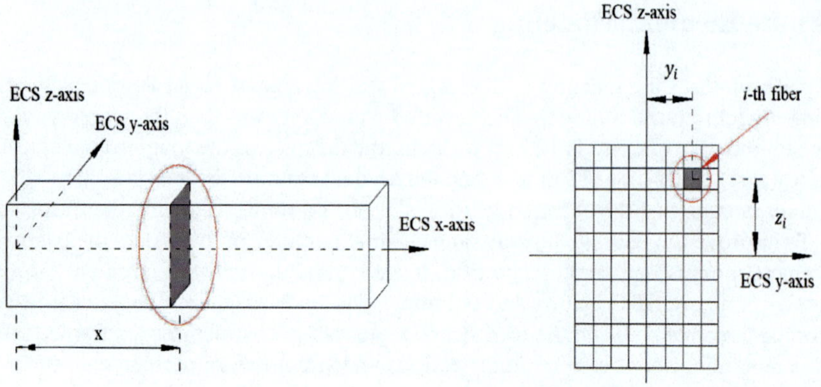

Fig. 2. Sketch map of fibre model

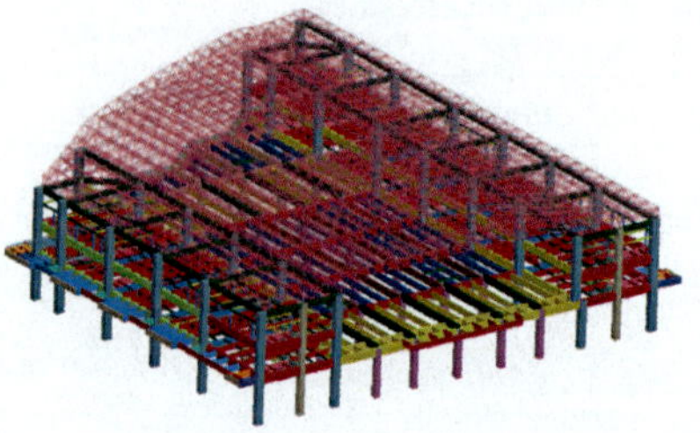

Fig. 3. Finite element model

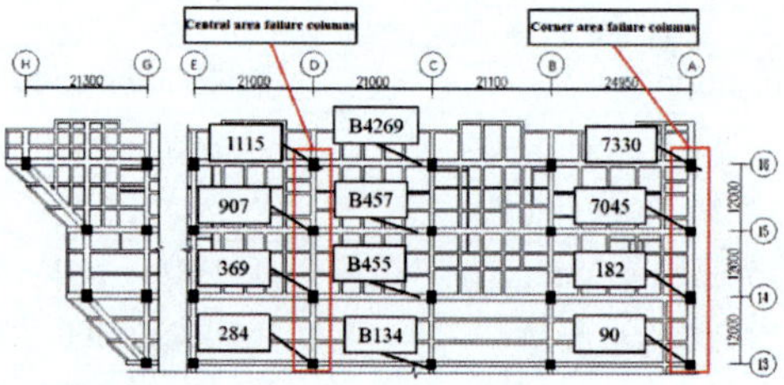

Fig. 4. Removed column position

Equivalent load is determined according to CSA2003 [26] and UFC4-023-03 [27] in the progressive collapse analysis. The long-span structure is mainly subjected to constant load and live load, and the equivalent load combination is 1.2 dead load + 0.5 live load. According to the relevant provisions of DoD2010 [28] and Chinese code [29], the following definitions are used to determine whether the frame member is finally invalid: when the vertical displacement of the top of the removed column exceeds 1/5 of the shortest span frame beam in the internal force redistribution area, the structure is considered to have progressive collapse.

3 Simulation Analysis of Pure Frame Model

In this paper, the anti-progressive collapse condition of 1 to 3 columns in the middle and corner areas failure conditions were analyzed. After the load was applied and stabilized, the keyword *MAT_ADD_EROSION was used to simulate the instant failure of the column. The location of the failure column and corresponding displacement record joint is shown in Fig. 4, and the operating conditions of each column removal are shown in Table 2.

Table 2. Details of column removal conditions

Failure condition	Failure columns	Vertical displacement recording node
Single column failure in corner area	A16	7330(A16)
Double columns failure in corner area	A15, A16	7045(A15)
Three columns failure in corner area	A14, A15, A16	182(A14)
Single column failure in the middle area	D15	1115(D16)
Double columns failure in the middle area	D15, D16	907(D15)
Three columns failure in the middle area	D14, D15, D16	369(D14)

3.1 Failure of Corner Column

Figure 5 shows the displacement-time travel curve of column failure in the corner area. The node displacement curves of each failure condition tend to converge, that is, it remains stable after the displacement decreases. Under 1 to 3 columns failure conditions, the maximum vertical displacement of the failed node position is 40.8 mm, 349.9 mm and 1,553.4 mm, respectively, and the ratio to the shortest span frame beam (10.1 m) is 0.004%, 3.46% and 15.38%, respectively, all less than 20%. Its deformation does not reach the critical collapse condition, Therefore, it is considered that the structure has not collapsed.

3.2 Failure of Middle Column

Figure 6 shows the displacement-time travel curve after column failure in the middle area. The structure is similar to the column failure in the corner area under the failure of single and double columns in the middle area. The displacement of the joint remains stable after a section of displacement is reduced. 1 to 2 columns failure condition, the maximum vertical displacement of the failed joint position is 168.5 mm and 690.6 mm, respectively. The maximum vertical displacement of the failed joint is 1,396.3 mm, although the joint displacement is not stable for 6 s under the three columns failure condition, the curve tends to converge. The ratios of the 1 to 3 column failure conditions to the shortest span frame beam (10.1 m) are 1.67%, 6.84%, 13.82%, respectively, which are less than 20%, and their deformation does not reach the critical collapse condition, so it is considered that the structure has not collapsed, too. Figure 7 shows the displacement nephogram of structure at 6 s. It can be seen that the displacement is the most obvious under the failure condition of three columns, but the structure does not collapse under the failure condition of 1 to 3 columns. Indicating that the high-speed railway waiting station room has good robustness.

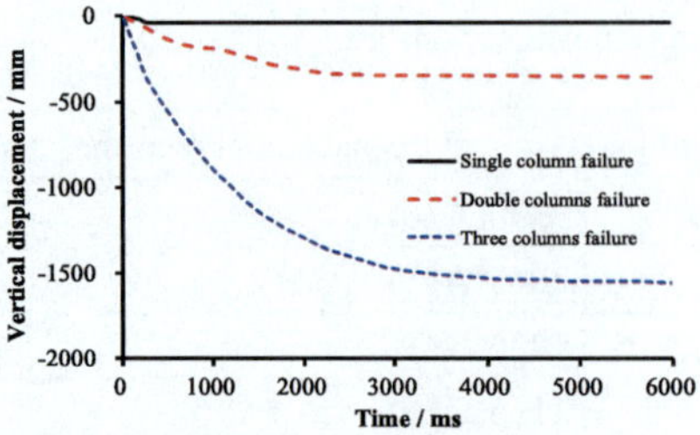

Fig. 5. Displacement-time travel curves of column failure in corner area

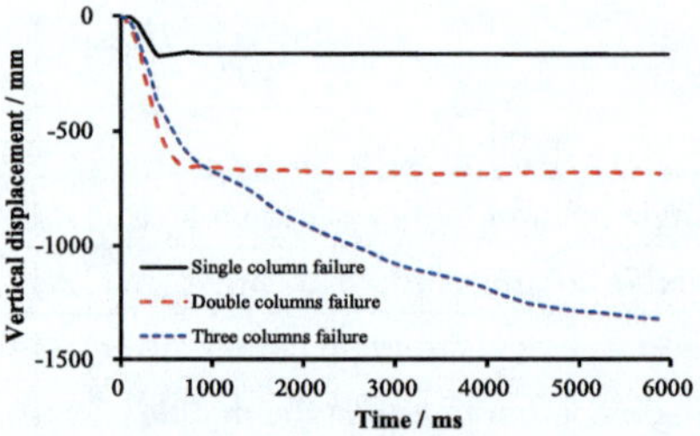

Fig. 6. Displacement-time travel curves of column failure in middle area

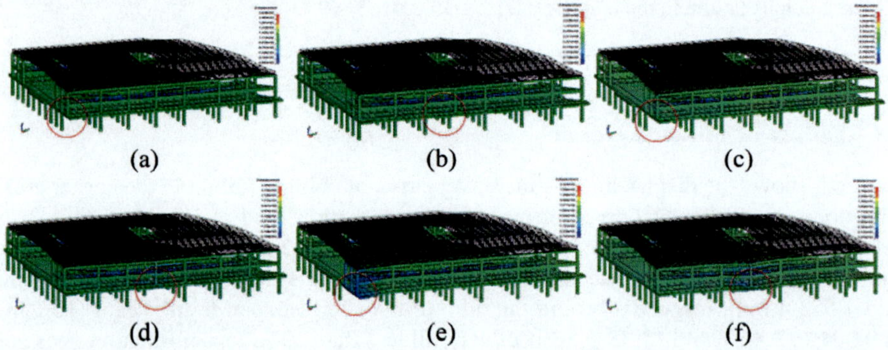

Fig. 7. Displacement program of simulation analysis results: (a) Single column failure in corner area; (b) Single column failure in middle area; (c) Double columns failure in corner area; (d) Double columns failure in middle area; (e) Three columns failure in corner area; (f) Three columns failure in middle area

3.3 Result Analysis

Figure 8 shows the extracted axial force and bending moment of some beam elements, and the position of each beam element is shown in Fig. 4. Under the conditions of single column failure and double columns failure, the maximum vertical displacement of the failed node in the middle area is larger than that in the corner area, while it is just the opposite under the condition of three columns failure. The resistance analysis is as follows:

In the single column failure condition, for column failure in the corner area, there is no horizontal constraint after removing the column, and the failure area can only provide collapse resistance by beam mechanism. For column failure in the middle area, it can be seen from Fig. 8(a), that the axial force and bending moment of beam element 4,269 stabilized at about 32,000 kN and 22,000 kN·m, respectively. However, due to the small displacement of 168.5 mm, the resistance contribution provided by the catenary mechanism was much smaller than the beam mechanism. It can be considered that the resistance mechanism of column failure in the middle region is the same as that in the corner area, and the load in the middle area is larger. Therefore, the displacement of column failure in the middle area is large.

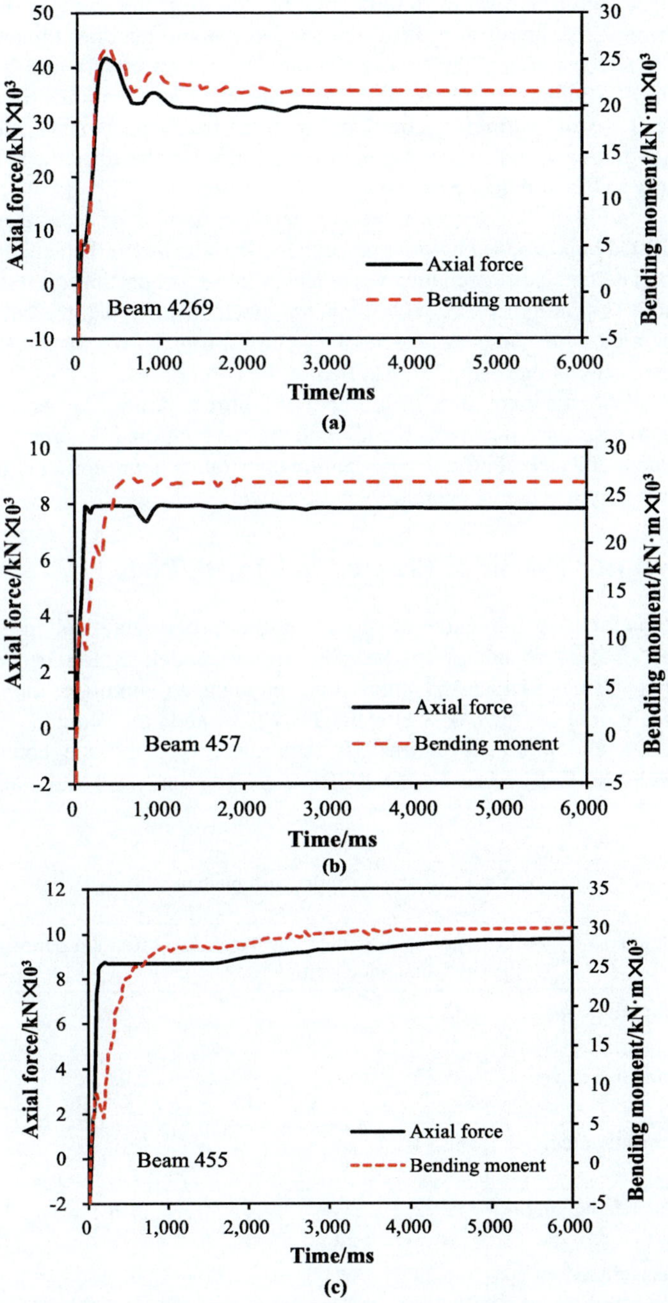

Fig. 8. Diagram of beam element axial force and bending moment: (a) Single column failure in middle area; (b) Double columns failure in middle area (c) Three columns failure in middle area

The resistance mechanism of double columns failure is the same as that of single column failure. As shown in Fig. 8(b), the axial force and bending moment of beam element 457 is stable at about 7,800 kN and 21,000 kN·m, respectively, and the maximum displacement of structure is 690.6 mm. The resistance contribution provided by the catenary effect is still insufficient compared with the beam mechanism, and the beam mechanism is mainly used to resist collapse, resulting in the displacement larger of column failure in the middle area.

As shown in Fig. 8(c), the axial force and bending moment of beam element 455 is stable at 9,700 kN and 28,000 kN·m, respectively, the maximum displacement reaches 1,396.3 mm. The catenary mechanism provides a larger proportion of resistance contribution and acts simultaneously with the beam mechanism, When the column failure in the corner area, only the beam mechanism plays a role, so the displacement of the column in the corner area is smaller than that in the corner area.

Therefore, With the increase of the number of failure columns, the beam mechanism is still the main resistance mechanism, although the contribution of catenary mechanism increases. Due to the lack of effective horizontal constraints in the corner column failure condition, catenary mechanism cannot play effectively.

4 Simulation Analysis of Frame Model with Plate

In order to study the influence of floor on the progressive collapse resistance of structure, slabs were added to the pure frame model, which become a floor frame model. Using SHELL163 thin shell element to simulate the floor slab, the material model is *MAT_CONCRETE_EC2, and the contact relationship between beams and plates is defined as Automatic face to face contact *CONTACT_AUTOMATIC_SURFACE_TO_SURFACE. The analysis results are shown in Table 3.

Table 3. Analysis results with plate model

Failure condition	Maximum vertical displacement of failed node/mm	Ratio to pure frame model
Single column failure in corner area	43.12	1.06
Single column failure in middle area	165.32	0.98
Double columns failure in corner area	162.34	0.46
Double columns failure in the middle area	319.72	0.46
Three columns failure in corner area	896.29	0.58
Three columns failure in the middle area	490.99	0.35

Figure 9 shows the comparison of failed node displacements between the pure frame model and the floor frame model, analysis results can be seen in Table 3. Under single column failure in corner area and single column failure in middle area, floor frame displacement is 1.06 times and 0.98 times that of pure frame, respectively. Mainly because column failure in the corner area, floor model cannot effectively play a role of tensile membrane mechanism, and column failure in the middle area, floor model can play a role of tensile membrane mechanism. Due to the small deformation of the structure, the floor did not play the role of tensile membrane mechanism or small role. At this time, the resistance caused by the beam mechanism accounts for a large proportion of the structural resistance.

The displacement of the floor frame is 1.06 times that of the pure frame when the double columns failure in corner area and double columns failure in middle area. Mainly because the floor can play the role of tensile membrane mechanism and reduce the maximum vertical displacement. At this time, the resistance mechanism is the combined action mechanism of beam catenary and tensile membrane mechanism, and the tensile membrane mechanism accounts for a large proportion in the structure resistance.

Under three columns failure in corner area and three columns failure in middle area, floor frame displacement is 0.58 times and 0.35 times that of pure frame, respectively. Due to the large deformation of the structure, the floor plate plays a role of tensile membrane mechanism and reduces the maximum vertical displacement. At this time, the resistance mechanism is the combined action mechanism of beam catenary and plate tension film, the resistance generated by the action of tension film accounts for a large proportion in the structure resistance.

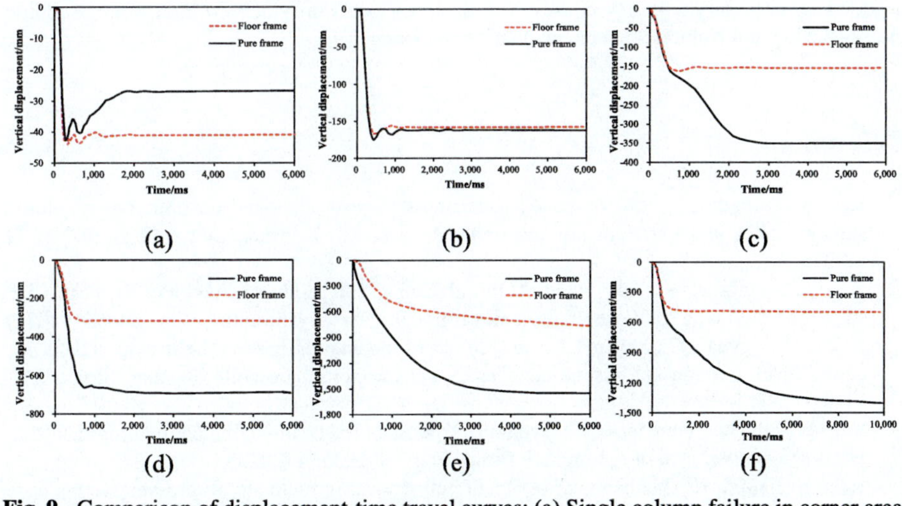

Fig. 9. Comparison of displacement-time travel curves: (a) Single column failure in corner area; (b) Single column failure in middle area; (c) Double columns in corner area; (d) Double columns in middle area; (e) Three columns in corner area; (f) Three columns in middle area

5 Conclusions

(1) The structure did not progressive collapse under the failure of single column, double columns and three columns, so the high-speed railway waiting station room model has good robustness.
(2) In the condition of single column failure in the middle area and double columns failure in the middle area, the beam mechanism is the main mechanism of the structure, and the displacement of the failed node is larger than that of the column failure in the corner area. In the condition of three columns failure in the middle area, the resistance proportion provided by catenary mechanism increases, resulting in a smaller displacement of the failed node than that of the columns failure in the corner area.
(3) Under small deformation, the resistance provided by the tensile membrane mechanism of the floor frame is very small; Under large deformation, the floor frame can effectively reduce the vertical displacement of failed joints by providing tensile membrane mechanism, which can reduce the maximum vertical displacement by 42%–65% compared with the pure frame.

Acknowledgements. The authors gratefully acknowledge the financial support provided by Guangxi Key Laboratory Project of New Energy and Building Energy Saving (Gui Ke Neng22-J-21-4), the Natural Science Foundation of Guangxi (No. 2021GXNSFFA196001), the Guangxi Science and Technology Program Project (Gui Ke AD21220071), the China Postdoctoral Science Foundation (2021M693814) and the Guangxi Young and Middle-aged Teacher Basic Ability Improvement Project (2021KY0254). Any opinions, findings and conclusions expressed in this paper do not necessary reflect the view of Guangxi Key Laboratory of New Energy and Building Energy Saving and Natural Science Foundation of Guangxi.

References

1. Kai, Q., Dongqiu, L., Lu, Z., et al.: Robustness of post-tensioned concrete beam-column subassemblies under various column removal scenarios. J. Struct. Eng. **148**(5), 04022032 (2022)
2. Kai, Q., Dongfang, W., Ting, H., et al.: Initial damage and residual behavior of RC beam-slab structures following sudden column removal-numerical study. Structures **36**, 650–664 (2022)
3. Kai, Q., Songyuan, G., Shilin, L., et al.: Effects of loading regimes on the structural behavior of RC beam-column sub-assemblages against disproportionate collapse. Eng. Struct. **251**, 113470 (2022)
4. Kai, Q., Xinyi, C., Ting, H., et al.: Dynamic response of RC beam-slab substructures following instantaneous removal of columns. J. Build. Eng. **45**, 103554 (2022)
5. Kai, Q., Jianfei, C., Yunhao, W., et al.: Effect of loading methods on progressive collapse behavior of RC beam-slab substructures under corner column removal scenario. J. Build. Eng. **44**, 103258 (2021)
6. Kai, Q., Xi, L., Zhi, L., et al.: Effects of steel braces on robustness of steel frames against progressive collapse. J. Struct. Eng. **147**(11), 04021180 (2021)

7. Kai, Q., Xi, L., Zhi, L., et al.: Behavior of steel moment frames using top-and-seat angle connections under various column-removal scenarios. J. Struct. Eng. **147**(10), 04021144 (2021)
8. Kai, Q., Dongqiu, L., Shunkai, L., et al.: Effects of infill walls on load resistance of multi-story RC frames to mitigate progressive collapse. Structures **33**, 2534–2545 (2021)
9. Kai, Q., Shilin, L., Fu, F., et al.: Progressive collapse resistance of emulative precast concrete frames with various reinforcing details. J. Struct. Eng. **147**(8), 04021107 (2021)
10. Kai, Q., Pan, H., Honghao, L., et al.: Impact of concrete strength on seismic behavior of T-shaped double-skin composite walls. Eng. Struct. **236**, 112039 (2021)
11. Kai, Q., Yunhao, W., Fu, F., et al.: Numerical evaluation of the reliability of using single-story substructures to study progressive collapse behaviour of multi-story RC frames. J. Build. Eng. **33**, 101636 (2021)
12. Decheng, F., Sicong, X., Jun, X., et al.: Robustness quantification of reinforced concrete structures subjected to progressive collapse via the probability density evolution method. Eng. Struct. **202**, 109877 (2020)
13. Kai, Q., Shilin, L., Fu, F., et al.: Progressive collapse resistance of precast concrete beam-column sub-assemblages with high-performance dry connections. Eng. Struct. **198**, 109552 (2019)
14. Kai, Q., Xide, Z., Fu, F., et al.: Progressive collapse-resisting mechanisms of planar prestressed concrete frame. ACI Struct. J. **116**(4), 77–90 (2019)
15. Kai, Q., Yi, L.: Strengthening and retrofitting precast concrete buildings to mitigate progressive collapse using externally bonded GFRP strips. J. Compos. Constr. **23**(3), 04019018 (2019)
16. Feng, F., Qin, F.: Numerical investigation on load redistribution capacity of flat slab substructures to resist progressive collapse. J. Build. Eng. **29**, 101109 (2020)
17. Yuepeng, J.: Study on simplified load model and dynamic response of subway station structures under explosion. Tianjin University (2018). (in Chinese)
18. Huaijin, S., Qiao, L.: Stability analysis and anti-continuous collapse analysis on complex supporting roof structure of Jiangmen railway station. Build. Struct. **52**(1), 17–23 (2022). (in Chinese)
19. Rui, J.: Anti-progressive collapse analysis of steel structure in a railway station. J. Railway Eng. Soc. **38**(7), 60–65 (2021). (in Chinese)
20. Qian, C.: Study on performance and design method for single-column canopy structures to resist progressive collapse. Chongqing University (2019). (in Chinese)
21. Chenghao, T., Cheng, D., Ming, L.: Dynamic analysis for anti-progressive collapse of large span steel structure in station building. J. Railway En. Soc. **33**(4), 22–25+35 (2016). (in Chinese)
22. Chih-Chieh, L., Jin-Hung, H.: Nonlinear collapse simulation of Daikai Subway in the 1995 Kobe earthquake: necessity of dynamic analysis for a shallow tunnel. Tunn. Undergr. Space Technol. **87**, 78–90 (2019)
23. Guibo, N., Chenxiao, Z., Dongfang, L., et al.: Collapse of the spatial double-layer cylinder shell by experimental study. Eng. Struct. **245**, 112862 (2021)
24. Carpinteri, A., Bazzucchi, F., Manuello, A.: Nonlinear instability analysis of long-span roofing structures: the case-study of Porta Susa railway-station. Eng. Struct. **110**, 48–58 (2016)
25. Limin, T., Jianpeng, W., Jiping, H.: Optimisation of long-span single-layer spatial grid structures to resist progressive collapse. Eng. Struct. **188**, 394–405 (2019)
26. GSA 2003: Progressive collapse analysis design guidelines for new federal office buildings and major modernization projects. United States General Services Administration, Washington, D.C. (2003)
27. UFC4-023-03, Design of buildings to resist progressive collapse. US Army Corps of Engineering Washington, D.C. (2010)

28. DoD 2010: Design of structures to resist progressive collapse. Department of Defense, Washington, D.C. (2010)
29. CECS 392-2014: Code for anti-collapse design of building structures. China Planning Press, Beijing (2015). (in Chinese)

Numerical Simulation Method of Dynamic Coupling of Complicated Mountains and the Wake of Multiple Wind Turbines

Chi Ma[1], Shuanglong Fan[2](✉), and Zhenqing Liu[2]

[1] CGN New Energy Holdings Co. Ltd., Beijing 100070, China
[2] School of Civil and Hydraulic Engineering, Huazhong University of Science and Technology, Wuhan 430074, Hubei, China
fshuanglong@hust.edu.cn

Abstract. The computational fluid dynamics (CFD) method is widely employed in the numerical simulation of wind turbine wake with the advancement and improvement of computer technology. In the simulation of wind turbine wake, the actuator disk model (ADM) has been verified. This method, however, can only isolate the wake flow field of a single wind turbine and does not take into account the mutual coupling effect between numerous wind turbines' wakes. As a result, this paper proposes a calculation method that takes into account the dynamic coupling of multiple wind turbines' wakes, and explores the coupling effect between the wakes of multiple wind turbines and the terrain by dynamically updating the resistance source term in real time according to the incoming wind speed during numerical simulation. The revised dynamic updating source term approach can detect the incoming wind speed on the basis of ensuring the quality of numerical simulation, making it more suited for linking numerical modeling of complex terrain flow field and the wake of many wind turbines.

Keywords: Wind turbine wake · ADM · Complex terrain · Numerical simulation

1 Introduction

Wind energy has garnered a lot of attention in the realm of energy development in recent years. The design, siting, development, and construction of wind farms has spread over the globe. Increased wind energy consumption necessitates the use of wind turbine clusters and grid-connected power supplies. Wake zones, or wind speed decay zones, are created by wind passing over wind turbines, a phenomenon known as the wake effect [1]. This impact can significantly reduce the power generated by downstream wind turbines in clustered wind farms. As a result, research into wind turbine wake flow is critical for improving wind farm power generation.

Yan et al. [2] and Gao et al. [3] performed full-model modeling of wind turbines and calculated the velocity distribution, pressure distribution, and turbulence distribution results for the entire flow field. However, when full-model modeling is performed, the wind turbine blade advective meshing is difficult and the mesh requirement is high,

W. Guo and K. Qian (Eds.): GBCESC 2022, LNCE 211, pp. 427–435, 2023.
https://doi.org/10.1007/978-981-19-5217-3_42

and the numerical simulation computational efficiency will be significantly reduced [4]. Therefore, simplified wake models have been widely used in recent years. Porté-Agel et al. [5], based on the actuated disk model with rotation (ADM-R) and Blade Element Momentum Method (BEM), achieved wind turbine disturbance of wind fields in flat terrain by introducing a volume force source term in CFD numerical simulations. The effect of wind turbine perturbation on the wind field in flat terrain is achieved by introducing a volumetric force source term in CFD numerical simulation. The model accounts for the fluctuation of axial and tangential generated forces at various places on the blade and can enhance computational efficiency greatly [6]. Shamsoddin et al. [7] applied the model to typical hilly terrain and analyzed the evolution of wind turbine wake in typical complex terrain. Later, the model was also applied to real complex terrain and its reliability was further verified [8].

During the numerical simulation, the volume force source term added to the grid cells contained in the actuator disk is a constant value, which is a limitation of this technology. The magnitude of this source term is related to the incoming wind speed in front of the actuator disk, and the incoming wind speed downstream of the wind turbine will be influenced by the upstream wind turbine wake, which is in a constant state of change during the calculation process, so the corresponding volume force source term of each circle will change as a result [9]. To address the above ADM-R model's shortcoming, this paper proposes a coupled method for calculating the topographic flow field and multiple wind turbine wake flows, which establishes a link between the incoming wind speed and the wind turbine wake flow, resulting in a more realistic distribution characteristic of the wind turbine wake flow.

2 Fundamental Theory

2.1 Actuator Disk Theory

The sweep range of the wind wheel forms a cylindrical thin plate (wind wheel disc) during the rotation of the blades of a conventional horizontal axis wind turbine, which is the actuating disc. The actuator disk can be thought of as having an endless number of blades. It has two effects on entering wind: (1) transforming wind energy into mechanical energy, and (2) obstructing the incoming wind. As a result, the wind speed before and after the actuator disk will abruptly shift, resulting in a pressure step change.

The method of adding volume force source term is used in CFD numerical simulation to represent the blockage of the actuator disk to the incoming wind, and the size of the source term is computed using Eq. (1).

$$f = \frac{C_T A_f \cdot \frac{1}{2}\rho u^2}{A_f \cdot \Delta x} = \frac{1}{2\Delta x} C_T \rho u^2 \quad (1)$$

where, C_T represents the thrust coefficient of the wind turbine, A_f is the area of the actuating disc, ρ is the air density, Δx represents the size of the grid in the X direction, and u represents the incoming wind speed. In this theory, the average wind speed at the hub center of the wind turbine is taken.

2.2 Blade Element Momentum Theory

As shown in Fig. 1, the blade element momentum theory discretizes the blade radially into N micro segments with a width of dr, the blade element. The leaf element revolves around the axis to generate a number of annular elements, and the annular element is based on the following assumptions: (1) Its radial properties are independent of each other, that is, what happens in one element does not affect other elements; (2) The blade's force on the air flow is constant in each annular unit.

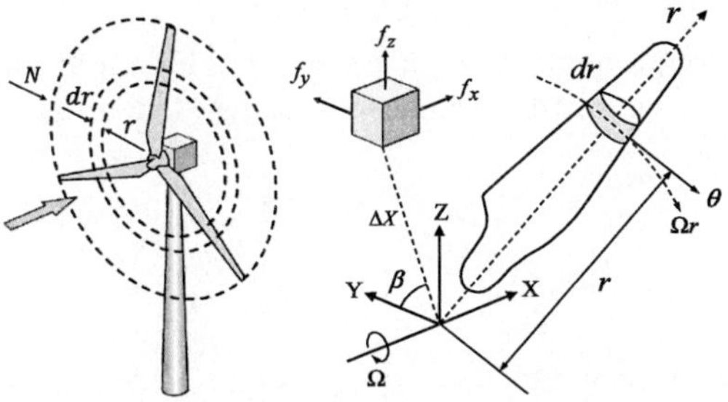

Fig. 1. Schematic diagram of blade element on wind wheel disc

Chord length c and pitch angle of wind turbine β will change along the blade axis. It is assumed that the rotation angle speed of the wind turbine is Ω and the incoming wind speed is U_0. Then the sum of the tangential velocity Ωr at the blade element and the tangential velocity $a'\Omega r$ of the air flow in the middle of the thickness of the wind wheel disc is the relative tangential flow velocity of the two, and its value is $\Omega r(1 + a')$. The velocity and force on the blade element at radius r are shown in Fig. 2.

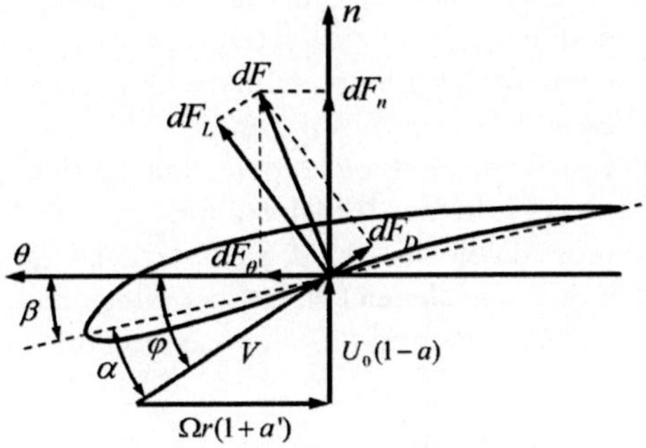

Fig. 2. Schematic diagram of blade element distribution and stress

The combined velocity of air flow on the blade is expressed as the following formula:

$$V = \sqrt{U_0^2(1-a)^2 + \Omega^2 r^2 (1+a')^2} \tag{2}$$

Inflow angle φ is the angle between the relative air velocity and the rotating plane, which is calculated by the following formula:

$$\varphi = \arctan\frac{U_0(1-a)}{\Omega r(1+a')} \tag{3}$$

Angle of attack α expressed as inflow angle φ and pitch angle β the difference of is expressed by the following formula:

$$\alpha = \varphi - \beta \tag{4}$$

Considering the inflow angle φ, calculate the axial lift coefficient C_n and tangential resistance coefficient C_θ under the plane coordinate system of the wind turbine:

$$C_n = C_l \cos\varphi + C_d \sin\varphi \tag{5}$$

$$C_\theta = C_l \sin\varphi - C_d \cos\varphi \tag{6}$$

The axial force dF_n and normal force dF_θ under the blade element coordinate system are expressed as formulas (7) and (8):

$$dF_n = \frac{1}{2}\rho V^2 N c C_n dr \tag{7}$$

$$dF_\theta = \frac{1}{2}\rho V^2 N c C_\theta dr \tag{8}$$

2.3 Actuator Disc Model with Rotation

The rotating actuator disk model considers not only each annulus element's axial resistance to the incoming wind, but also the tangential resistance of each blade element to the incoming wind due to the blade's movement in the disk plane. In the numerical computation, the total wind speed V is determined from Eq. (2), and the associated axial and tangential forces at each blade element are computed using Eqs. (7) and (8), assuming that the actuator disk is evenly divided. Finally, along the radial micro ring, the axial and tangential induced forces on the three blades are averaged. Similarly, to evaluate the two resistance impacts of each leaf element on the incoming wind, the method of adding source term is utilized, and Eqs. (9) and (10) are used to compute the value of source term.

$$f_n = -\frac{B}{2\pi r}\frac{dF_n}{\Delta x dr} \tag{9}$$

$$f_\theta = -\frac{B}{2\pi r}\frac{dF_\theta}{\Delta x dr} \tag{10}$$

2.4 Dynamically Update the Source Item Actuation Disk Method

On the basis of the actuator disk model with rotation, the method of dynamically updating the source term actuator disk suggested in this study is enhanced. Based on FLUENT software's user-defined function (UDF) module, the linkage of wind turbine wake dynamic calculation process in numerical simulation domain is done by writing UDF program.

To begin, each actuator disk is partitioned into N equidistant rings after the hub center of each wind turbine is located in the calculation domain. Then, using the fixed source term approach, go through all of the model's grid cells, discover the cells that are contained in each ring, and set the volume force source term value for each cell. Following the start of the numerical simulation, the average axial wind speed v_n of the elements contained in each ring and the average tangential wind speed v_θ relative to the actuator disk are retrieved in each iteration step, and the resultant blade speed V is determined using formula (2). Then the inflow angle φ is calculated by formula (3). Then, the lift coefficient C_l and drag coefficient C_d are calculated by interpolation according to the aerodynamic parameters of the blade, and then the axial lift coefficient C_n and tangential drag coefficient C_θ of each ring are calculated by formulas (5) and (6). Axial force dF_n and tangential force dF_θ of each ring are calculated by (7) and (8). Finally, it is assigned to each unit and decomposed to obtain the size of the resistance term in the three directions of XYZ of each unit.

3 Numerical Simulation

3.1 Wind Turbine Model

In this study, a wind turbine model based on an actual Mitsubishi Heavy Industries MWT-1000 scaled down to 1/1000 is used, and the key parameters of this wind turbine are shown in Table 1. In addition, the chord lengths and pitch angles of the blades at different positions along the wingspan direction required for this study are consistent with the study by Ishihara et al. [10], and are not repeated in this section.

Table 1. MWT-1000 wind turbine parameter table

Wind turbine type	Tip speed ratio λ	Thrust coefficient C_T	Blade diameter D (m)	Hub height H (m)
MWT-1000	5.52	0.81	57	70

3.2 Computational Domain Model

In this study, the wind turbine wake flow is simulated numerically using a flat terrain and a complex mountainous terrain computational domain model, respectively. Two wind turbines are arranged in series in a flat terrain, and a wind farm consisting of 10 wind turbines is constructed in a complex terrain. The effect of wake flow of two wind turbines in tandem in flat terrain and the coupling effect between the wake flow of multiple wind turbines and the topographic wind field in complex terrain are investigated using the traditional fixed source term actuator disc method and the dynamic updated source term actuator disc method proposed in this paper, respectively. The computational domain models for flat and complex terrain are shown in Fig. 3.

Among them, the length of the calculation domain of flat terrain is 1500 m, the width and height are 200 m, the wind turbine WT1 is arranged at the origin, and the wind turbine WT2 is arranged at 7D position directly behind WT1; the length, width and height of the grid model in complex terrain are set to 5000 m, 5000 m and 2000 m, respectively, and the range of 500 m radius with the origin as the center is the grid encryption area, and the minimum grid size in this area is 4 m. The minimum grid size in this area is 4 m, and 10 wind turbines are arranged in staggered rows in the grid encryption area. In addition, 1/1000 scale down was performed for both terrain models. In addition, the inlet wind speed is set to $U_{ref} = 10$ m/s for both terrain models.

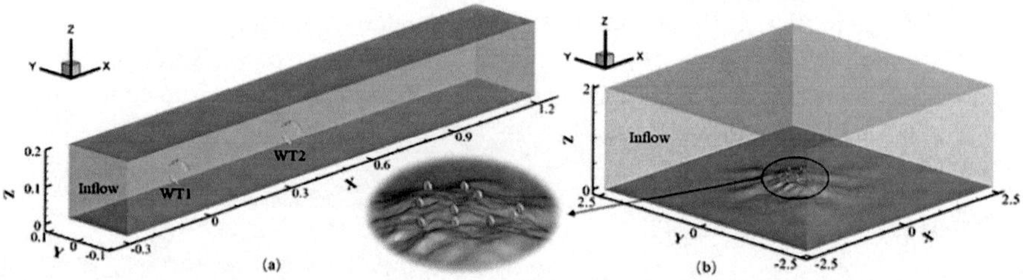

Fig. 3. Computational domain model, (a) flat terrain, (b) complex terrain

4 Result Analysis

4.1 Flat Terrain

Figure 4 shows the wake distribution after velocity normalization for the $y = 0$ profile. Comparing (a) and (b) of Fig. 4, it can be seen that the wind turbine wake distributions calculated using the two methods are similar overall. For the upstream wind turbine WT1, the wind speed loss of the improved dynamic source term method is slightly smaller than that of the traditional fixed source term method in the near wake region, while for the downstream wind turbine WT2, the wind loss of the wake obtained by the traditional method is significantly higher than that of the improved method, probably because the wind speed of the incoming wind is overestimated in the calculation of the fixed source term, resulting in a large value of the added source term and thus a high axial drag on the incoming wind. However, after $x = 6D$, the WT2 wake profile shape of the downstream wind turbine calculated by both methods transitions from a "double hump" state to a "single hump" shape, and the velocity loss gradually decreases and converges.

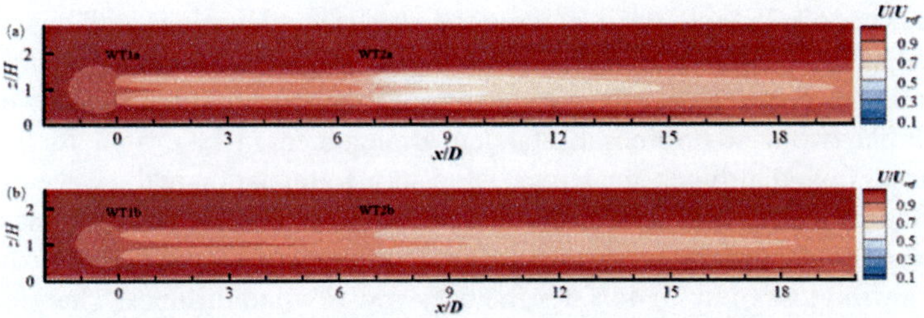

Fig. 4. Cloud map of wake velocity distribution in $y = 0$ profile of flat terrain, (a) traditional fixed source term method, (b) improved dynamic source term method

In order to investigate the effect of using two different methods of adding source terms on the distribution pattern of the vortex structure of the front and rear wind turbine wake, the vortex distribution clouds in the Y-direction of the wind turbine wake for the two cases are presented in Fig. 5. It can be clearly observed that the vortices at the tip and root of the front wind turbine can be transmitted to the downstream very steadily in both cases, and it is also obvious that the vortex values at the tip and root of the vortex obtained by using the fixed source term method are larger than those obtained by using the dynamic source term simulation, especially in the rear wind turbine location. Comparing the wake of WT2a and WT2b, it can be seen that the vortex structure of WT2a is more abruptly changed than that of WT2b, and the vortex structure expands in this profile, and at $x = 11D$, the vortex structure at the root of the leaf is almost completely broken and disappears, while the vortex structure at the root of WT2b extends to a longer distance. It is not until $x = 14D$ that the vortex values at the tip and the root of the blade gradually return to zero. From the above comparison, it can be judged that the wind turbine wake flow obtained by using the dynamic source term shows a smoother transition in the generation-sudden change-disappearance process.

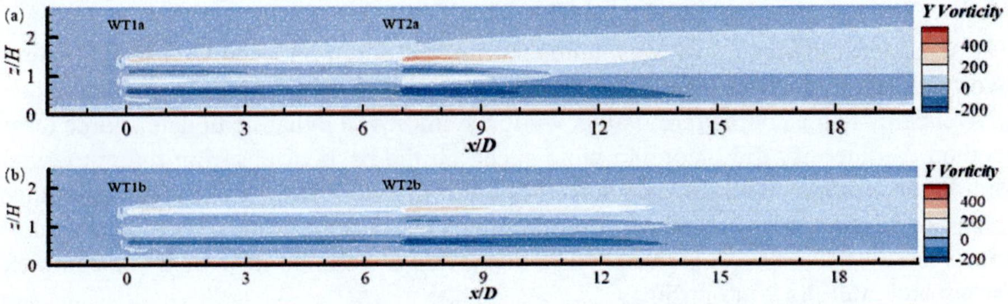

Fig. 5. Cloud map of wake vorticity distribution in $y = 0$ profile of flat terrain, (a) traditional fixed source term method, (b) improved dynamic source term method

4.2 Complex Terrain

Figure 6 shows the velocity distribution on the longitudinal height $z = H$ contour, which shows that higher velocity distributions occur at higher altitudes in the mountains, with wind speeds up to 1.25 times the inlet wind speed. The wind turbines with high incoming wind speed have a longer tail flow, and similarly, the higher the incoming wind speed is, the higher the wind speed loss after passing through the wind turbine, which can reach up to 20%. Unlike two wind turbines in a tandem arrangement in flat terrain, for a staggered arrangement of wind turbines, the largest wind speed attenuation in the wake of the back row of wind turbines is located at the intersection of the wake of the front and back wind turbines, which, together with the effect of the leeward slope caused by the terrain, results in a wind loss of up to 44% for the back row of wind turbines. Therefore, in the actual wind farm layout, special attention should be paid to the staggered arrangement distance of the front and rear rows of wind turbines. In addition, the results obtained from numerical simulations using the two methods are consistent overall, but based on the improved method will cause the wind turbines to identify the magnitude of the incoming wind speed, as shown by the attenuation of the wake wind speed of the front row wind turbines will increase by about 8% compared to the results calculated using the fixed source term method.

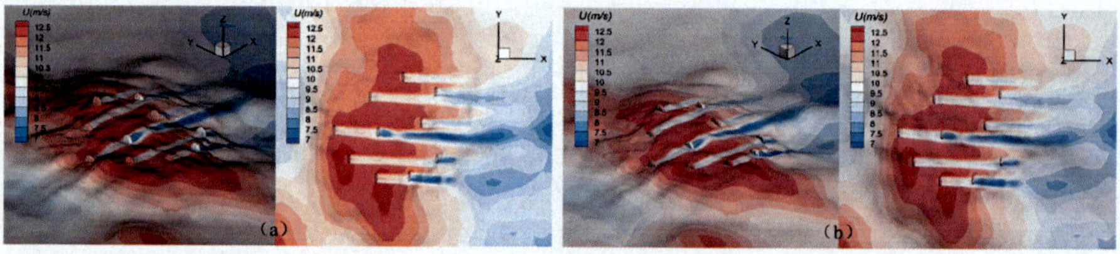

Fig. 6. Visualization of wind turbine wake in complex terrain, (a) traditional fixed source term method, (b) improved dynamic source term method

5 Conclusion

The wind turbine wake flow was simulated in flat terrain and complex terrain using two methods, and it was found that the traditional fixed source term method is prone to overestimate the source term value, while the improved dynamic update source term method can identify the incoming wind speed on the basis of ensuring the numerical simulation accuracy, and the calculated wind turbine wake flow will show a smoother transition in the process of generation-abrupt change-disappearance. This tail flow distribution law will be more suitable for the analysis of coupled flow fields of complex terrain and multiple wind turbines.

References

1. Liu, Z., Lu, S., Ishihara, T.: Large eddy simulations of wind turbine wakes in typical complex topographies. Wind Energy **24**, 857–886 (2021)
2. Yan, S., Shi, S., Chen, X., Wang, X., Mao, L., Liu, X.: Numerical simulations of flow interactions between steep hill terrain and large scale wind turbine. Energy **151**, 740–747 (2018)

3. Gao, Z., et al.: Modelling the nacelle wake of a horizontal-axis wind turbine under different yaw conditions. Renew. Energy **172**, 263–275 (2021)
4. Hornshøj-Møller, S.D., Nielsen, P.D., Forooghi, P., Abkar, M.: Quantifying structural uncertainties in Reynolds-averaged Navier-Stokes simulations of wind turbine wakes. Renew. Energy **164**, 1550–1558 (2021)
5. Porté-Agel, F., Wu, Y.T., Lu, H., Conzemius, R.J.: Large-eddy simulation of atmospheric boundary layer flow through wind turbines and wind farms. J. Wind Eng. Ind. Aerodyn. **99**, 154–168 (2011)
6. Nedjari, H.D., Guerri, O., Saighi, M.: CFD wind turbines wake assessment in complex topography. Energy Convers. Manage. **138**, 224–236 (2017)
7. Shamsoddin, S., Porté-Agel, F.: Large-eddy simulation of atmospheric boundary-layer flow through a wind farm sited on topography. Bound. Layer Meteorol. **163**(1), 1–17 (2016). https://doi.org/10.1007/s10546-016-0216-z
8. Murali, A., Rajagopalan, R.G.: Numerical simulation of multiple interacting wind turbines on a complex terrain. J. Wind Eng. Ind. Aerodyn. **162**, 57–72 (2017)
9. Qian, G.W., Song, Y.P., Ishihara, T.: A control-oriented large eddy simulation of wind turbine wake considering effects of Coriolis force and time-varying wind conditions. Energy **239**, 121876 (2022)
10. Ishihara, T., Yamaguchi, A., Fujino, Y.: Development of a new wake model based on a wind tunnel experiment. Global Wind Power (2004). http://windeng.t.u-tokyo.ac.jp/ishihara/proceedings/2004-5_poster.pdf

Numerical Simulation of Substructure Shake Table Testing for Base-Isolated Structure Using Model-Based Integration Algorithm

Bo Fu, Futai Zhang, and Jin Chen(✉)

Chang'an University, Xi'an 710061, Shaanxi, China
jinchen2019@chd.edu.cn

Abstract. Base isolation is an effective method of reducing the seismic responses of structures. As a promising experimental measure, substructure shake table testing (SSTT) can be used to investigate the seismic performance of base-isolated structures. Model-based integration algorithms have the advantages of explicit formulation and unconditional stability, and thus they are very suitable for SSTT. In this study, the SSTT using a family of generalized CR (GCR) algorithms for the base-isolated structures is numerically studied. Firstly, time history analyses of a six-story base-isolated structure are conducted and compared with the corresponding non-isolated structure. Then, the SSTT approach using the GCR algorithms is proposed for the base-isolated structures. The numerical simulation of a series of SSTTs with different time steps, integration coefficients and time lags are finally conducted. The results indicate that the GCR algorithms with relative large time step still have excellent accuracy. The SSTT results of using the subfamily of $\kappa_1 = 1$ are less accurate than the subfamily of $\kappa_1 = 1/2$. In addition, the time lag has obvious adverse effect on the SSTT results. There is an instable tendency of the SSTT results as the time lag increases.

Keywords: Substructure shake table testing · Base isolation · Integration algorithm · Model-based

1 Introduction

Base isolation [1] is an effective method of reducing the seismic responses of structures. There are a lot of relevant studies [2–4] regarding the seismic performances of base-isolated structures. Considering the vibration characteristics of base-isolated structures, especially for velocity-dependent property, it is an excellent measure to conduct substructure shake table testing (SSTT) [5], which is burgeoning in the field of structural engineering, for the base-isolated structures. Computationally efficient integration algorithms are required in implementing a SSTT because it is a real-time test. The model-based integration algorithms [6–8] have the advantages of explicit formulation and unconditional stability, and thus they are very suitable for SSTT. There are some applications of SSTT using model-based integration algorithms for the structures with dampers [9–12] and soil-structure-interaction [13, 14], but the relevant applications for

W. Guo and K. Qian (Eds.): GBCESC 2022, LNCE 211, pp. 436–444, 2023.
https://doi.org/10.1007/978-981-19-5217-3_43

the isolated structures are limited [15]. In this study, the SSTT using the GCR algorithms [16] is applied for the base-isolated structures. In Sect. 2, time history analyses of a six-story base-isolated structure are conducted and compared with the corresponding non-isolated structure subjected to the ground motion. The SSTT approach using the GCR algorithms is proposed for the base-isolated structures in Sect. 3. Section 4 provides the numerical simulation of a series of SSTTs with different time steps, integration coefficients and time lags. Some conclusions are summarized in Sect. 5.

2 Time History Analysis of Base-Isolated Structure

A six-story non-isolated RC frame is firstly designed according to the Chinese seismic code. The frame is regarded as a shear-type structure with a mass of 4.06×10^5 kg and a lateral stiffness of 9.11×10^8 N/m for all six stories. The fundamental period is determined as 0.55 s. The Rayleigh damping model with the damping ratio of 5% for the first two modes is adopted to construct the damping matrix. In addition, an elastic perfectly-plastic shear force-drift relationship with a yielding drift of 60 mm is used to consider the structural nonlinearity.

The structure is then base-isolated by fitting a group of lead rubber bearings (LRBs). The isolation layer is simplified as a SDOF system with $m = 2.24 \times 10^5$ kg and $k = 5.12 \times 10^7$ N/m. The fundamental period of the isolated structure becomes 1.52 s. Suppose the equivalent damping ratio of the isolation layer is 15%. As a result, a non-proportional damping model is used to simulate the damping property of the base-isolated structure.

Both the two structures are subjected to the El Centro ground motion. The corresponding time history analyses are conducted using the Newmark explicit (NE) algorithm with $\Delta t = 0.001$ s. The numerical results indicate that the structural responses especially the drifts of the isolated structure are significantly reduced.

Figure 1 shows the time history curves of the first story drift for the two structures. It can be drawn from Fig. 1 that throughout the entire process of earthquake, the drifts of the isolated frame are always smaller than those of the non-isolated frame. The maximum

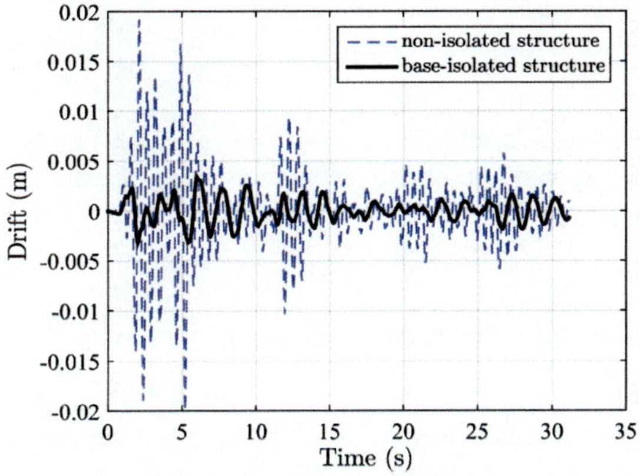

Fig. 1. First story drift time history curves.

drifts of all six stories for the two structures are compared in Fig. 2. Figure 2 indicates that the descent degrees of the maximum story drifts due to base isolation range from 82.9% to 85.9%.

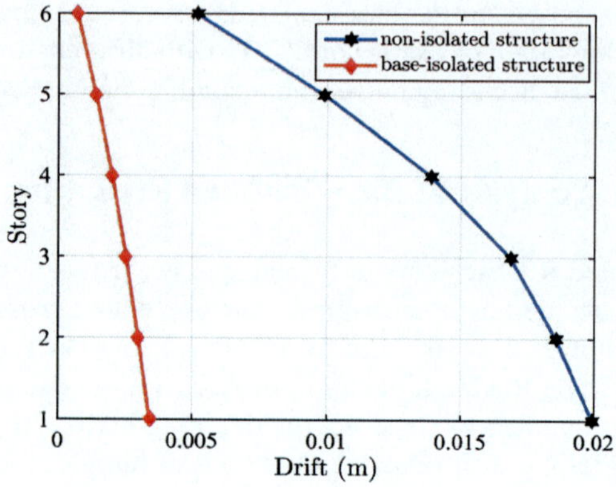

Fig. 2. Maximum drifts of all stories.

3 Substructure Shake Table Testing Approach for Base-Isolated Structure

3.1 Model-Based Integration Algorithm

To achieve the SSTT for the base-isolated structures, the GCR algorithms [16] is selected. As for a SDOF system with the mass m, the stiffness k and damping coefficient c, the expressions of the GCR algorithms are shown in Eqs. (1–2).

$$\dot{x}_{i+1} = \dot{x}_i + \Delta t \alpha_1 \ddot{x}_i; \; x_{i+1} = x_i + \Delta t \dot{x}_i + \Delta t^2 \alpha_2 \ddot{x}_i \quad (1)$$

$$\alpha_1 = m / \left(m + \kappa_1 \Delta t c + \kappa_2 \Delta t^2 k \right); \; \alpha_2 = (0.5 + \kappa_1)\alpha_1 \quad (2)$$

where Δt is the time step; $\dot{x}$, x, and $\ddot{x}$ are velocity, displacement and acceleration, respectively; κ_1 and κ_2 are integration coefficients determining the numerical properties of the integration algorithms, the CR algorithm is a representative member of the GCR algorithms when $\kappa_1 = 1/2$, $\kappa_2 = 1/4$.

3.2 Substructure Shake Table Testing Approach

As for the base-isolated structure, there are two substructuring schemes. For one, the main structure and the isolation layer are taken as the experimental/physical and analytical/numerical substructures, respectively. The other scheme is just the opposite. In this study, the first substructuring scheme is adopted as shown in Fig. 3. In Fig. 3, the entire base-isolated structure is split into the upper experimental substructure (main structure) and lower analytical substructure (isolation layer). The two substructures share the interface force f_{i+1}^{I} with reverse directions. The isolation layer is numerically simulated. The relative displacement of the analytical substructure x_{i+1}^{A} is predicted by using the integration algorithms and then converted to total displacement at the interface u_{i+1}^{I}. The

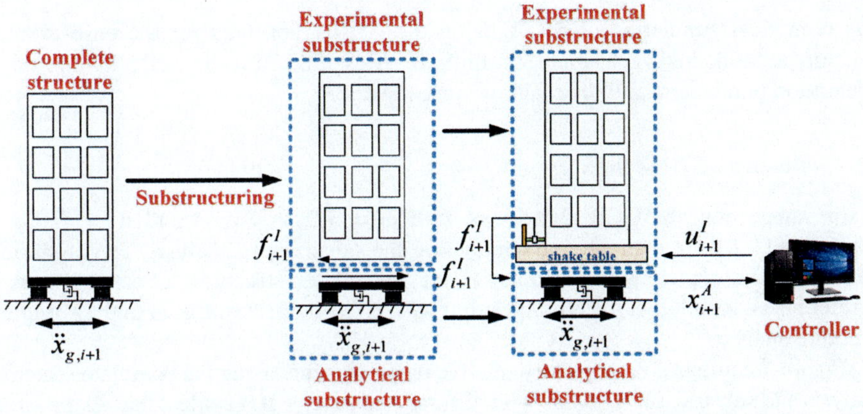

Fig. 3. Substructure shake table testing scheme for base-isolated structure.

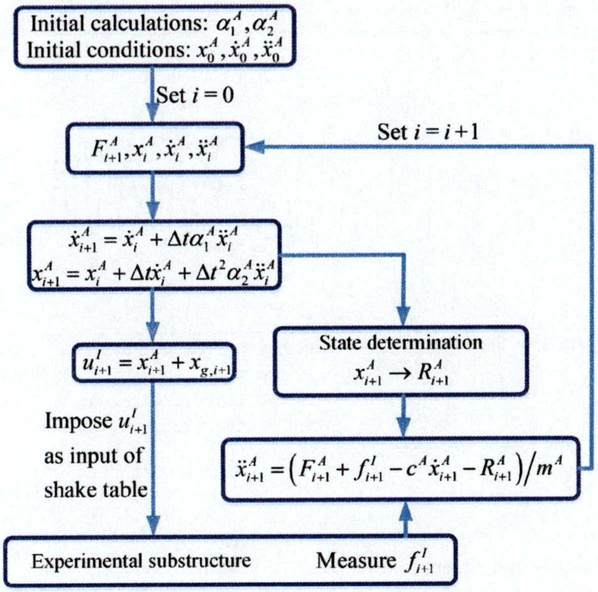

Fig. 4. Entire SSTT process using GCR algorithms.

experimental substructure (main structure) is mounted on the shake table, which is actuated by u_{i+1}^I. Then, the interface force f_{i+1}^I is measured and the signal is transmitted to the analytical substructure. The procedure of SSTT using the GCR algorithms is depicted in Fig. 4. In Fig. 4, all the superscripts A represent the analytical substructure.

4 Numerical Simulation

The numerical simulation of SSTT using the GCR algorithms for the base-isolated structure is conducted via Matlab/Simulink. A series of SSTTs with different time steps, integration parameters and time lags are investigated.

4.1 Influence of Time Step

In this subsection, the GCR algorithms with $\kappa_1 = 1/2$, $\kappa_2 = 1/4$ and $\Delta t = 0.001$ s, 0.005 s and 0.01 s, are adopted in conducting the substructure analysis. The numerical results of the complete structure analysis for the isolated structure in Sect. 2 are taken as references. Recall that the NE algorithm with $\Delta t = 0.001$ s was used in the complete structure analysis.

Figure 5 compares the top story relative displacements using the complete structure analysis and substructure analysis with different time steps. It is evident that all the structural responses from the substructure analysis agree well with those from the complete analysis.

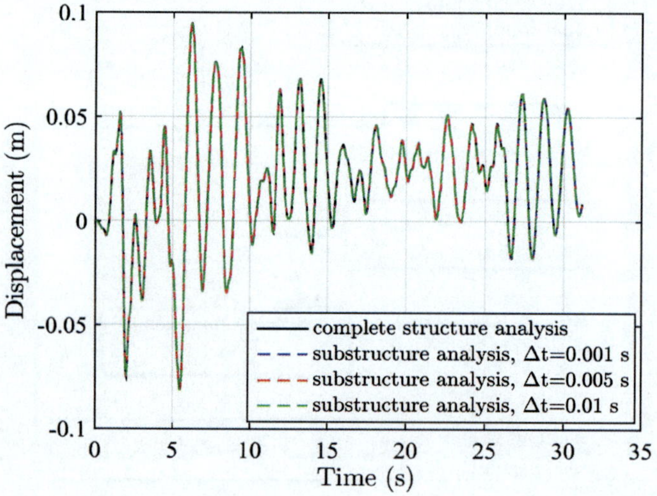

Fig. 5. Top story relative displacement responses based on complete structure analysis and substructure analysis with different time steps.

The normalized energy error (NEE), as defined in Eq. 3, is introduced to quantify the matching degree. Table 1 gives the NEE values of displacement responses of six stories with different time steps.

$$\mathrm{NEE} = \left| \frac{\sum_{1}^{N}\left[(C_i)^2 - (R_i)^2\right]}{\sum_{1}^{N}(C_i)^2} \right| \tag{3}$$

where C_i and R_i are computed and referenced values, respectively; N is the total amount of samples. It can be concluded from Table 1 that although the errors increase with the increasing of the time step, the errors are very small and negligible even when the time step is 0.01 s. A larger time step means a higher computational efficiency and is more competitive in implementing the SSTT. As a result, the GCR algorithms with $\Delta t = 0.01$ s will be adopted for the following analysis.

Table 1. NEE of displacement responses of six stories with different time steps.

Story	Time step (s)		
	0.001	0.005	0.01
1	4.00×10^{-4}	2.48×10^{-3}	6.70×10^{-3}
2	3.82×10^{-4}	2.38×10^{-3}	6.41×10^{-3}
3	3.69×10^{-4}	2.30×10^{-3}	6.19×10^{-3}
4	3.59×10^{-4}	2.24×10^{-3}	6.04×10^{-3}
5	3.53×10^{-4}	2.20×10^{-3}	5.93×10^{-3}
6	3.50×10^{-4}	2.18×10^{-3}	5.88×10^{-3}
Average	3.69×10^{-4}	2.29×10^{-3}	6.19×10^{-3}

4.2 Influence of Integration Coefficients

There are two integration coefficients, i.e., κ_1 and κ_2, of the GCR algorithms. Different combinations of them result in diverse numerical characteristics of integration algorithms. The subfamily of $\kappa_1 = 1/2$ are second-order accurate, while the subfamily of $\kappa_1 \neq 1/2$ only possess first-order accuracy. The GCR algorithms with $\kappa_1 = 1/2$, $\kappa_2 = 1/4$ and $\Delta t = 0.01$ s is taken as the reference model. The influence of the integration coefficients are investigated by comparing the reference model with three other combinations of integration coefficients, i.e., $[\kappa_1, \kappa_2] = [1/2, 1/2]$, $[1, 1/2]$ and $[1, 1]$.

Figure 6 shows the top story relative displacements based on substructure analysis with different integration coefficients. It can be seen from Fig. 6 that the structural responses of $\kappa_1 = 1/2$, $\kappa_2 = 1/2$ coincide with the reference model. While there exists some observable amplitude differences when using $[\kappa_1, \kappa_2] = [1, 1/2]$ and $[1, 1]$. The errors are caused by the first-order accuracy of the subfamily of $\kappa_1 = 1$. Table 2 summarizes the NEE of displacement responses of six stories with different integration coefficients. It can be found from Table 2 that the NEE of the subfamily of $\kappa_1 = 1$ can reach 4.8%. Therefore, the subfamily of $\kappa_1 = 1$ should be carefully selected if high accuracy is required.

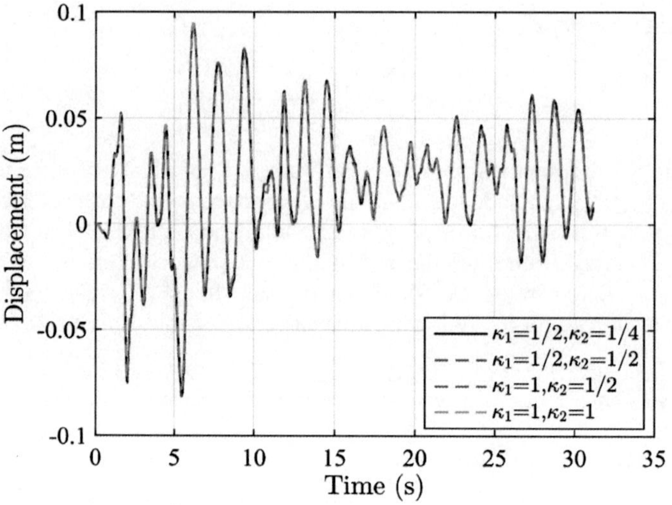

Fig. 6. Top story relative displacement responses based on substructure analysis with different integration coefficients.

Table 2. NEE of displacement responses of six stories with different integration coefficients.

Story	Integration coefficients $[\kappa_1, \kappa_2]$		
	[1/2, 1/2]	[1, 1/2]	[1, 1]
1	8.84×10^{-4}	4.45×10^{-2}	4.42×10^{-2}
2	8.39×10^{-4}	4.57×10^{-2}	4.55×10^{-2}
3	8.03×10^{-4}	4.67×10^{-2}	4.65×10^{-2}
4	7.75×10^{-4}	4.74×10^{-2}	4.72×10^{-2}
5	7.57×10^{-4}	4.78×10^{-2}	4.76×107^{-2}
6	7.47×10^{-4}	4.81×10^{-2}	4.79×10^{-2}
Average	8.01×10^{-4}	4.67×10^{-2}	4.65×10^{-2}

4.3 Influence of Time Lag

The time lag, or time delay, of a SSTT is caused by the shake table dynamics, computation time and communication time. It is well known that the time lag has similar function as a negative damping, which adds energy into the SSTT system. Therefore, excessive time lag may cause severe errors and even instability of the testing. In this study, the influence of the time lag is investigated by introducing different amplitudes of time lags. The GCR algorithms with $\kappa_1 = 1/2$, $\kappa_2 = 1/4$ and $\Delta t = 0.01$ s is employed. The case without time lag is adopted as a contrast for the cases with the time lags $\tau = 0.01$ s, 0.05 s and 0.1 s.

Figure 7 gives the top story relative displacement time history curves with different time lags. It can be drawn from Fig. 7 that when the time lag $\tau = 0.01$ s, the errors become obvious after 3 s. As for larger time lag of 0.05 s and 0.1 s, the results are completely inaccurate and have a tendency of losing stability as the time lag increases. Therefore, it is necessary to adopt lag compensation to eliminate the adverse effects of time lag.

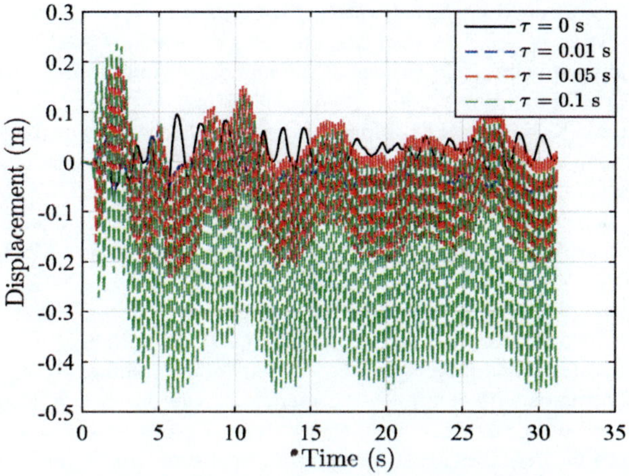

Fig. 7. Top story relative displacement responses based on substructure analysis with different time lags.

5 Conclusions

A series of numerical simulation of SSTT using the GCR algorithms for base-isolated structure are conducted in this study. Three conclusion are summarized as follows.

1) The GCR algorithms with relative large time step still have excellent accuracy in implementing the SSTT for base-isolated structure. The average NEE of relative displacements of six stories with $\Delta t = 0.01$ s is 6.19×10^{-3}.
2) The subfamily of $\kappa_1 = 1$ should be carefully selected if there is a high requirement for accuracy. The average NEE of relative displacements of six stories with $\kappa_1 = 1$ can reach 4.8%.
3) The adverse effects of time lag are significant. As for larger time lag of 0.05 s and 0.1 s, the SSTT results are completely inaccurate and have a tendency of losing stability as the time lag increases.

Acknowledgements. Financial support from the National Natural Science Foundation of China (Grant Nos. 51908048, 52108432) and the Young Talent Fund of University Association for Science and Technology in Shaanxi, China (Grant No. 20200412) are gratefully acknowledged.

References

1. James, M.K.: Aseismic base isolation: review and bibliography. Soil Dyn. Earthq. Eng. **5**(4), 202–216 (1986)
2. Ruijun, Z., Aiqun, L.: Experimental study on the performance of damaged precast shear wall structures after base isolation. Eng. Struct. **228**, 111553 (2021)
3. Chunfeng, Z., et al.: Preliminary study on the periodic base isolation effectiveness and experimental validation. Eng. Struct. **226**, 111364 (2021)
4. Chunwei, Z., Amir, A., Li, S.: Investigation on low-cost friction-based isolation systems for masonry building structures: Experimental and numerical studies. Eng. Struct. **243**, 112645 (2021)
5. Bo, F., Chinmoy, K., James, R.M., Huanjun, J., Tao, W.: Stability analysis of substructure shake table testing using two families of model-based integration algorithms. Soil Dyn. Earthq. Eng. **126**, 105777 (2019)
6. Chinmoy, K., James, R.M.: Assessment of explicit and semi-explicit classes of model-based algorithms for direct integration in structural dynamics. Int. J. Numer. Methods Eng. **107**(1), 49–73 (2016)
7. Cheng, C., James, R.M.: Development of direct integration algorithms for structural dynamics using discrete control theory. J. Eng. Mech. **134**(8), 676–683 (2008)
8. Chinmoy, K., James, R.M.: Development of a family of unconditionally stable explicit direct integration algorithms with controllable numerical energy dissipation. Earthq. Eng. Struct. Dyn. **43**(9), 1360–1380 (2014)
9. Jinting, W., Yao, G., Fei, Z., Feng, J., Mengxia, Z.: Real-time hybrid simulation of multi-story structures installed with tuned liquid damper. Struct. Control Health Monit. **23**(7), 1015–1031 (2016)
10. Bo, F., Huanjun, J., Tao, W.: Experimental study of seismic response reduction effects of particle damper using substructure shake table testing method. Struct. Control Health Monit. **26**(2), e2295 (2019)
11. Bo, F., Huanjun, J., Tao, W.: Comparative studies of vibration control effects between structures with particle dampers and tuned liquid dampers using substructure shake table testing methods. Soil Dyn. Earthq. Eng. **121**, 421–435 (2019)
12. Bo, F., Huanjun, J., Jin, C.: Substructure shake table testing of frame structure-damper system using model-based integration algorithms and finite element method: numerical study. Symmetry **13**(9), 1739 (2021)
13. Qiang, W., Jinting, W., Feng, J., Fudong, C., Chuhan, Z.: Real-time dynamic hybrid testing for soil-structure interaction analysis. Soil Dyn. Earthq. Eng. **31**(12), 1690–1702 (2011)
14. Zhenyun, T., Hua, M., Jun, G., Zhenbao, L.: Effect of soil-structure interaction on seismic performance of long-span bridge tested by dynamic substructuring method. Shock Vib. **2017**, 1–12 (2017)
15. Alireza, S., Andreas, H.S., Matthew, J.S., Gilberto, M., Stephen, A.M.: Real-time hybrid simulation of seismically isolated structures with full-scale bearings and large computational models. Comput. Model. Eng. Sci. **120**(3), 693–717 (2019)
16. Bo, F., De-Cheng, F., Huanjun, J.: A new family of explicit model-based integration algorithms for structural dynamic analysis. Int. J. Struct. Stab. Dynamics **19**(6), 1950053 (2019)

Effect of Ceramic Waste Powder on Mechanical Properties of Recycled Coarse Aggregate Concrete

Ronghui Yang[1,2,3], Xin Min[1,2,3], Feixiang Chen[1,2,3](✉), and Gang Li[1,2,3]

[1] CCCC Second Harbor Engineering Company LTD., Wuhan 430040, China
chenfeixiang@cccltd.cn

[2] Key Laboratory of Large-Span Bridge Construction Technology, Wuhan 430040, China

[3] CCCC Highway Bridge National Engineering Research Centre Co. Ltd., Wuhan 430040, China

Abstract. In this paper, the recycled coarse aggregate concrete (RCAC) was prepared by replacing cement with ceramic waste powder (CWP) in mass fractions of 0, 10%, 20%, 30% and 40%. Its mechanical properties, hydration products, microstructure, and morphology were investigated by means of isothermal calorimetry, thermogravimetric analysis (TGA), scanning electron microscopy (SEM), and universal testing machine. The results show that the early compressive strength of RCAC is reduced by the addition of CWP, but the compressive strength of RCAC with 10% and 20% CWP increased by 12.53% and 4.78% respectively. In addition, after 56d of curing, the splitting tensile strength and flexural strength of RCAC with 10% CWP are higher than that with none CWP, but lower than natural aggregate concrete (NAC). The incorporation of CWP would accelerate the hydration of blended paste slightly, which leads to earlier exothermic peak. On the other hand, CWP could markedly reduce the total hydration heat and the hydration exothermic rate. The TG results showed that the content of Ca(OH)2 in CWP-cement paste decreased with the increase of CWP content after 56d of curing. It can be seen from SEM that the incorporation of CWP makes the microstructure of hardened paste more compact.

Keywords: Ceramic waste powder · Recycled coarse aggregate concrete · Mechanical properties · Microstructure

1 Introduction

Recycled coarse aggregate (RCA) is spheroid particle generated by crushing old concrete from C&D wastes, after sieved to meet certain gradation, it can be used to substitute the coarse aggregate concrete. Due to secondary crushing and attached with old mortar, the surface of RCA has numerous micro-crack which leads to a series of imperfection such as large water absorption [1], high porosity [2] and complicated interfacial transition zone [3], comparing to natural aggregate. So, the mechanical properties of RCAC made with RCA are usually not good enough for large-scale application. To promote the

W. Guo and K. Qian (Eds.): GBCESC 2022, LNCE 211, pp. 445–454, 2023.
https://doi.org/10.1007/978-981-19-5217-3_44

application of RCAC in engineering has become the focus issue of many researchers. Two main methods have already been proven the validity to enhance the mechanical properties of RCAC, one is to pretreat RCA, and the other is adding supplementary cement materials (SCM). Using physical, chemical, or microbial approach [4–6] can compensate for microcracks on the surface of RCA, decreasing its water absorption. Thus, the mechanical properties of RCAC can be improved. The SCM rich in silicon dioxide or aluminum oxide such as glass powder [7], plant shell ash [8, 9], brick dust [10] etc. can react with the hydration products such as Ca(OH)2 to generate more CSH gel, and increase the strength of RCAC.

Due to its excellent performance, ceramics has been widely used in our daily life, but its waste brings great environmental pressure, especially under the social background of green development pursuit. According to statistics, 25% of the total domestic product of ceramics, about 13 million tons, is discarded due to shipping damage and some other reasons. Most of these wastes are usually piled up or landfilled as solid waste, which not only takes up a lot of land but also causes environmental pollution, so finding a solution for resource utilization of ceramic waste is an urgent need. It has been suggested that ceramic waste powder (CWP) is a potential SCM [11, 12]. It is feasible to make ceramic waste into CWP and apply in concrete. Pitarch [10] et al. used different proportions of CWP which made by ceramic waste or ceramic sanitary ware in cement mortar and found that the compressive strength of mortar with 15% ceramic waste CWP or 25% ceramic sanitary ware CWP is close to that with none CWP. Li [13] studied the effect of CWP on strength of mortar, and indicated that as the incremental replacement of CWP ranging from 10% to 40%, the early compressive strength of mortar decreased, but after curing for 90d, the strength of that with 10%–30% replacement ratio of CWP increased. Kannan [14] studied the mechanical performance of concrete at different ages with 10%–40% cement replaced by CWP. Results showed that with the increase of CWP content, the compressive strength of concrete decreased before 28d, but increased after 90d. Likewise, other researchers had come up with similar results [15, 16]. However, the research about using CWP as SCM in RCAC is quite limited.

Using CWP of 0, 10%, 20%, 30% and 40% by mass to replace cement to prepare concrete, this paper studied the influence of CWP on the mechanical performance such as compressive strength, split strength and flexural strength of RCAC. Additionally, hydration exothermic process, hydration products and microstructure of hardened blended CWP cementitious paste were analyzed by using thermogravimetric analyses (TGA), X-ray diffraction (XRD) and scanning electron microscopy (SEM). This was beneficial for the use of CWP in RCAC, promoting the recovery of ceramic industry waste to produce sustainable green concrete.

2 Experimental

2.1 Materials

Natural coarse aggerate (NCA) and river sand (NFA) were provided by a project of CCCC (Guangzhou) Engineering Co., LTD. The particle sizes of NCA and NFA were 5–25 mm and 0–4.75 mm, respectively. The particle size of RCA was 5–25 mm, continuously graded provided by Wuhan Hanyang Municipal Construction Group Co., LTD.

The physical properties and gradations of aggerates were given in Table 1 and Fig. 1, respectively. CWP was obtained by grinding waste ceramic tiles for 1 h and sieving. X-ray diffraction pattern of CWP was present in Fig. 2, which indicated that the main phase of CWP were quartz and mullite. The physical characteristics of ordinary Portland cement (P·O 42.5) were listed in Table 2 and its particle size distribution (PSD) was presented in Fig. 3, as well as the CWP's.

Table 1. Physical properties of aggregate

Aggregate type	Apparent density/kg·m^{-3}	Bulk density/kg·m^{-3}	Water absorption/%	Crushing value/%
NCA	2679	1623	1.15	11.6
RCA	2581	1230	5.8	16.8
NFA	2559	1480	1.1	—

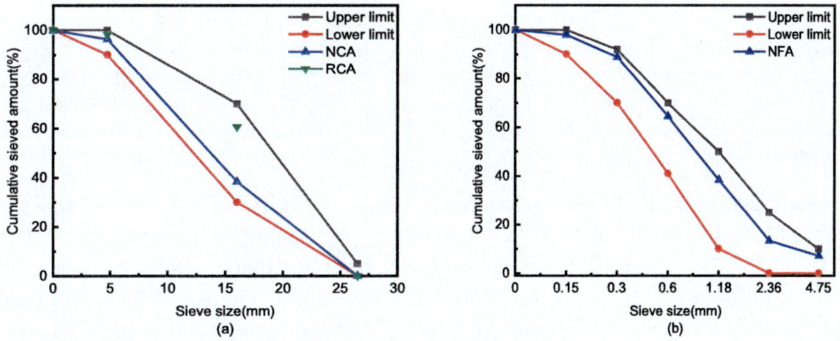

Fig. 1. PSD of aggregate

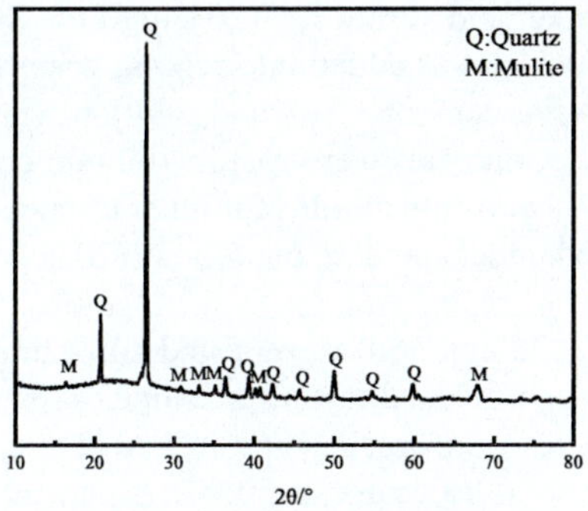

Fig. 2. XRD pattern of CWP

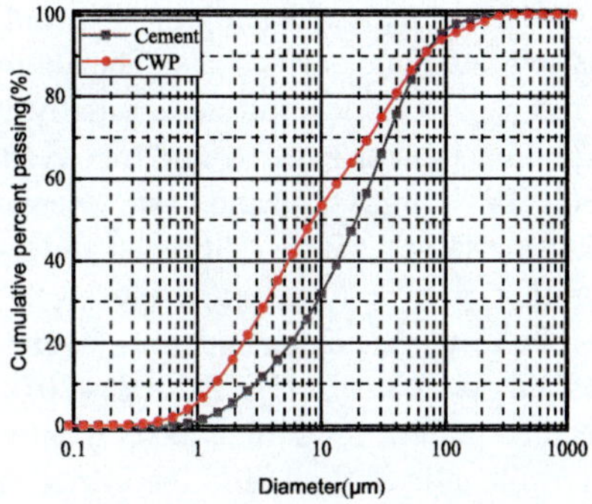

Fig. 3. PSD of cement and CWP

Table 2. Physical characteristics of cement

Cement type	Water quantity needed for standard consistence/%	Setting time/min		Compressive strength/MPa		Flexural strength/MPa	
P·O 42.5	28.1	Initial	Final	7d	28d	7d	28d
		224	296	37.1	44.3	5.3	7.7

Table 3. Mix proportion of concrete (kg/m^3)

NO	OPC	CWP	NCA	RCA	NFA	W	excess water	SP
NAC	400	0	1115	0	655	180	0	4.0
RCAC-0	400	0	446	669	655	180	28.6	4.0
RCAC-10	360	40	446	669	655	180	28.6	4.0
RCAC-20	320	80	446	669	655	180	28.6	4.0
RCAC-30	280	120	446	669	655	180	28.6	4.0
RCAC-40	240	160	446	669	655	180	28.6	4.0

2.2 Methods

NCA sample represented the concrete made with none RCA or CWP. RCAC-0, RCAC-10, RCAC-20, RCAC-30 and RCAC-40 samples represented the concrete made with 0, 10, 20, 30 and 40% dosage of CWP in the blend cementing material system, respectively, and with 60% amount of RCA replacing NCA by weight. The water-to-binder ratio in concrete was fixed to 0.45. Besides, due to the high-water absorption of RCA, excess water was added when preparing RCAC. The proportions of concrete were displayed in Table 3. The corresponding cement pastes labeled as OPC, CWP10, CWP20, CWP30 and CWP40 were cast for isothermal calorimetry, TGA, XRD, SEM.

RCA and excess water were first put into mixer and stirred for 120 s until the RCA was saturated and surface-dry. Then, superplasticizer was added into mixing water and stirred well to form solution A. Finally, NCA, NFA, cement, CWP and solution A were poured into mixer in order and mixed for 3 min. Concretes were cast in 100 mm cubic molds for strength tests and pastes were cast in 40 mm cubic molds for other tests. After 1 day, the samples were demolded and then cured under standard conditions (20 ± 2 °C, humidity $\geq 95\%$) to assigned ages.

The mechanical strengths of concretes (at 7, 28 and 56d) were tested according to the Chinese standard GB/T 50081–2019 by using universal testing machine. After the specified period, CWP blended cementitious pastes were crushed and soaked in ethanol to terminate hydration, and then dried in a vacuum desiccator at 50 °C. A part of the sample was fine ground for TGA and XRD. The rest part was broken for scanning SEM.

XRD was taken on an RIGAKU D/max-RB X-ray diffractometer, with Cu Ka radiation at a scanning speed of 0.02°/min. The TGA was performed via a simultaneous thermal analyzer (STA449F3) with a heating rate of 10 °C/min from 40 to 800 °C under an N_2 atmosphere, at a flow rate of 30 mg/min. A JSW-5510LV scanning electron microscope (SEM) was used to exam the sample microstructure at the accelerating voltage set as 20 kV, current set as 0.5 nA. Isothermal calorimetry was conducted by using the calorimeter (TAM Air 8CH, Waters, USA) for 72 h at 20 °C.

3 Results and Discussion

3.1 Mechanical Properties

Figure 4 presents the compressive strength of samples for various replacement levels of CWP at different curing times (7, 28 and 56d). Comparing with RCAC-0, the 7d compressive strength of RCAC-10–40 decreased by 0.72%, 13.56%, 16.98% and 46.81%, respectively. While the 28d compressive strength of RCAC-10 was 5.13% higher than that of RCAC-0, and the compressive strength of the other samples with higher replacement levels of CWP were still lower than that of RCAC-0. After 56d curing, the compressive strength of RCAC-10 and RCAC-20 were 12.53% and 4.78% higher than that of RCAC-0, achieved 36.01 MPa and 33.53 MPa, respectively. Otherwise, the compressive strength of RCAC-10 was still 11.37% lower than that of NAC after curing for 56d. This is because in the early curing stage, CWP is mainly acted as filler to fill the microcracks [14]. And cement content reduction caused by CWP replacement resulted in decrement of concrete strength. Along with the increment of hydration time, the hydration reactivity of CWP was enhanced. It can react with the hydration products of cement to form more CSH gel [17], making the matrix denser, and increased the compressive strength of concrete.

Figure 5 presents the spilt tensile strength of samples NAC and RCAC after curing for 56d. Similar to the influence on compressive strength, the split tensile strength of RCAC with 10% and 20% CWP dosage were 3.29 MPa and 3.06 MPa respectively, which were 10.03% and 2.34% higher than that with none CWP. With the increase of CWP dosage, the split strength of RCAC occurred decline trend. Comparing with RCAC-0, the 56d spilt strength of RCAC-30 and RCAC-40 declined 10.1% and 13.38%, respectively. However, although the spilt strength of RCAC-10 was higher that of all other RCAC samples, it's still 15.86% lower than that of NAC. This is due to the high pozzolanic activity of CWP, more CSH gel was generated, which improving the microstructure of RCAC and enhancing the bonding force of ITZ. Thus, the split tensile strength of RCAC was improved. When the CWP content was too high, hydration products reduced due to the cement content decreased sharply, would result in the decrement of bonding force of ITZ. Thus, with the increase of CWP dosage, the splitting tensile strength of RCAC decreased obviously.

Figure 6 presents the 56d flexural strength of samples NAC and RCAC. Generally speaking, the flexural strength of samples RCAC firstly increased and then decreased with the increasing of CWP dosage. The flexural strengths of RCAC-10 and RCAC-20 were 7.12 MPa and 6.93 MPa, 4.4% and 1.62% higher than that of RCAC-0 which with none CWP, respectively. When the CWP dosage continued going up to 30% and 40%, the flexural strengths were 7.48% and 13.2% lower than that of RCAC-0. Although the flexural strength of RCAC-10 was highest in all RCAC samples, it's still decreased by

3% compared to NAC sample. This is because the fracture when the specimen withstood bending force not only occurred in the ITZ, but also occurred in the coarse aggregate itself. As microcracks spread all over the surface of RCA, the crushing value of RCA is much higher than NCA, so the flexural strengths of samples RCAC were lower than that of NAC sample. CWP and the secondary hydration reaction products generated by CWP could fill the pores of RCAC matrix to make the microstructure denser. But when CWP was overdosed, there was too less hydration products to fill up the pores inside RCAC matrix. This is the reason why the right amounts of CWP dosage could improve flexural strength of RCAC samples.

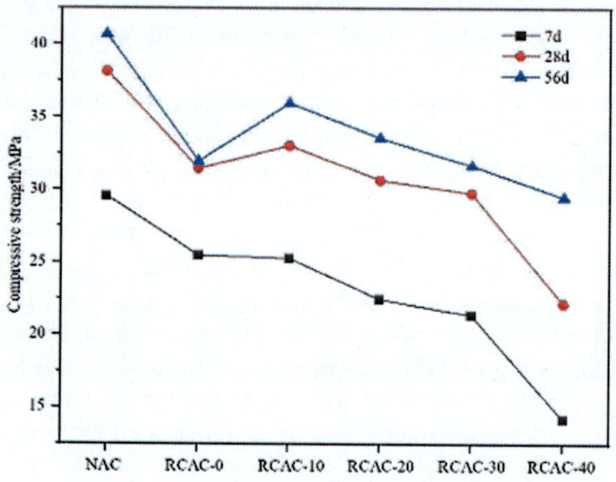

Fig. 4. Compressive strength of NAC and RCAC

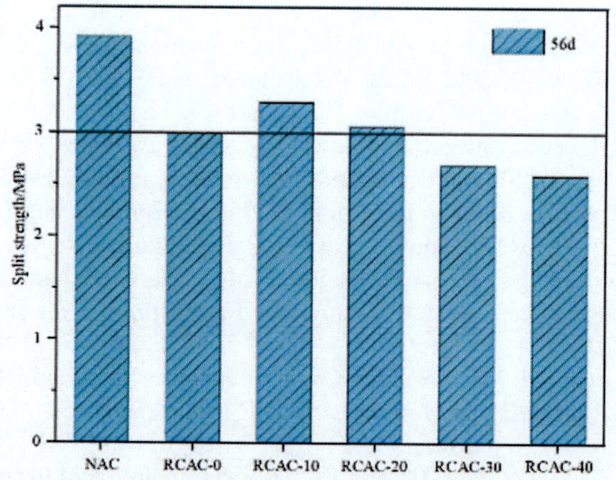

Fig. 5. Split tensile strength of NAC and RCAC

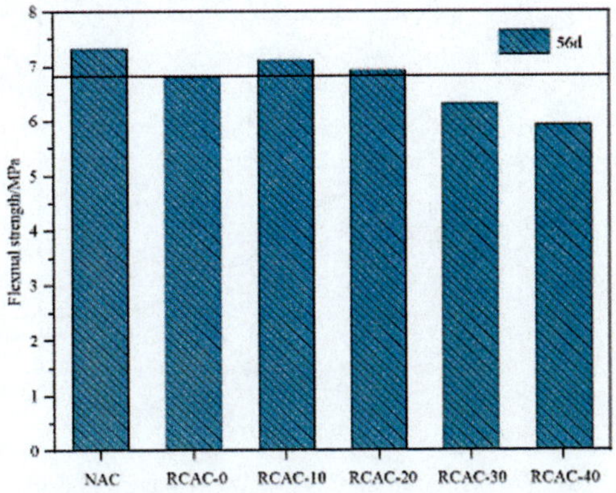

Fig. 6. Flexural strength of NAC and RCAC

3.2 TGA

Figure 7 shows the TG curves of the CWP-cement pastes. The weight loss below 150 °C implies the decomposition of the ettringite and the beginning of CSH dihydroxylation. The region between 400 °C and 550 °C represents the dehydration of portlandite. The weight loss around 650–720 °C corresponds to the decomposition of carbonates. TG can be used to quantify the portlandite content in paste and the degree of secondary hydration [18, 19]. The normalized portlandite content of samples containing various amounts of CWP after curing for 56d was given in Fig. 8. In general, the amount of portlandite decreased with increasing dosage of CWP. In detail, the portlandite content of samples OPC, CWP10, CWP20, CWP30 and CWP40 were 10.62%, 9.28%, 7.01%, 5.34% and 4.29%, respectively. The main reason was that the replacement of cement led to its decrement, which means less portlandite generated from cement hydration. Other than that, the secondary hydration of CWP consumed an amount of portlandite. This was in good accordance with the XRD and mechanical strength results.

3.3 Isothermal Calorimetry

The heat flow curves within 48 h of different compositions were illustrated in Fig. 9. The blended cements hydration would be accelerated slightly due to the incorporation of CWP, since the two exothermic peaks were advanced in CWP-cement blended sample compared to those of the control samples. Meanwhile, increasing CWP content led to the effect of promoting hydration even more pronounced. This was attributed to CWP providing more nucleation sites for CSH gel formation. The cumulative heat release within 48 h of samples CWP10 and CWP20 were 6.22% and 14.66% lower than that of neat OPC sample without CWP, respectively, calculated by integrating. The decrement of hydration heat was obviously attributed to the dilution effect of CWP, which decreased the cement content and hydration products formed.

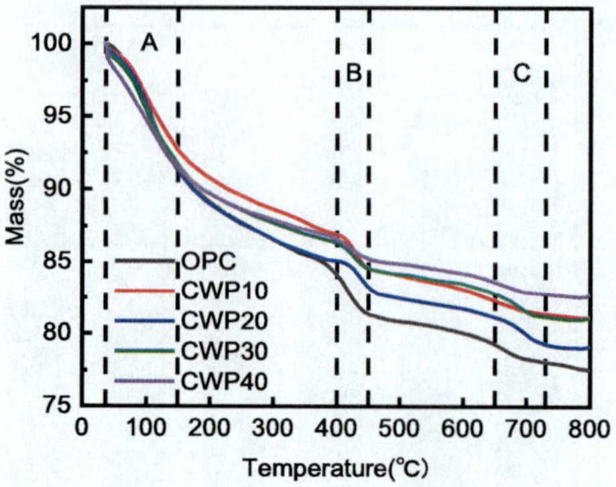

Fig. 7. TG curve of CWP-cement pastes at 56d

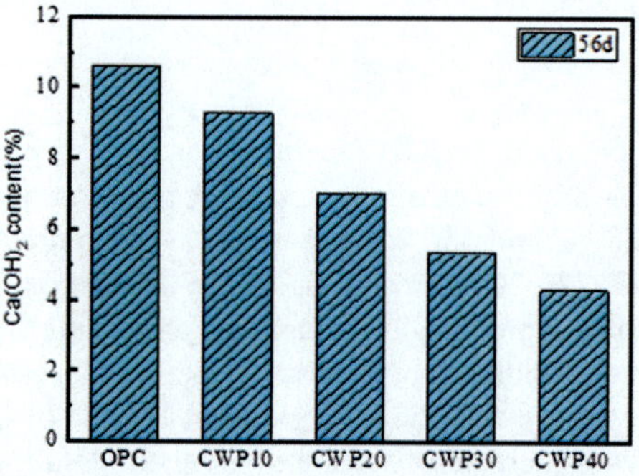

Fig. 8. Portlandite content in the mixed pastes at 56d

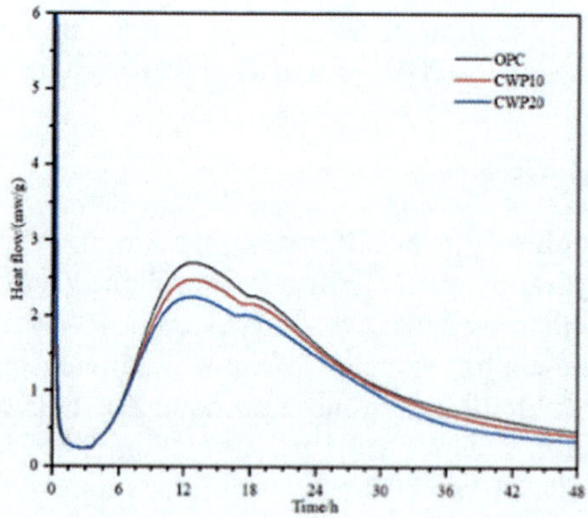

Fig. 9. Heat flow of hydration for CWP- cement paste within 48 h

3.4 SEM

Figure 10 shows SEM images of OPC and CWP10 pastes cured for 56d. Obviously, after hydrated for 56d, the microstructures of the two samples were rather dense with plenty of CSH gel, portlandite and ettringite in the hydrated pastes. In Fig. 10(a), CSH gel distributed evenly in the paste with plate-like portlandite and a bit of needle-like ettringite inserted in the OPC sample without CWP incorporation. There still were some pores not filled up by hydration products. Thanks to the dosage of CWP, some portlandite was consumed as it can react with CWP to produce more CSH gel and fill up the pores in the matrix. That's why the microstructure was denser and plate-like portlandite was more less in Fig. 10(b).

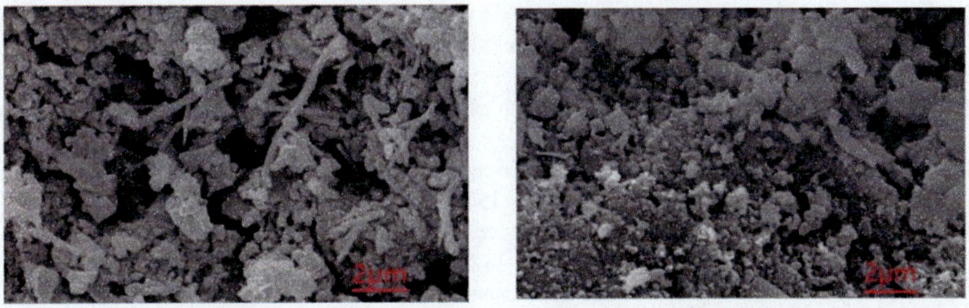

(a) OPC (b) CWP10

Fig. 10. SEM images of CWP- cement paste at 56d

4 Conclusions

This research addressed the effects of ceramic waste powder on the mechanical properties of recycle coarse aggregate concrete. Isothermal calorimetry, TGA and SEM were used to investigated the influence of CWP on the hydration process of blended system. The following conclusions can be drawn:

1. The replacement of cement by CWP decreased the early strength of RCAC. After curing for 56 days, although the compressive strength of samples RCAC-10 and RCAC-20 are still lower than that of sample NAC, but 12.53% and 4.78% higher than that of RCAC without any CWP incorporation. Alike, with an increase of CWP dosage, the splitting tensile and flexure strength of RCAC first increased and then decreased.
2. CWP could accelerate the hydration of blended paste slightly and markedly reduce the total hydration heat and the hydration exothermic rate. With the increase of CWP replacement rate, the content of $Ca(OH)_2$ in CWP-cement paste after 56d cured decreased. As CWP reacted with portlandite and generated more CSH gel, the incorporation of CWP makes the microstructure of hardened paste more compact as showed in SEM images.

3. In some projects that the requirement for early strength of concrete is not strict, CWP can be used to replace part of cement to produce recycle aggregate concrete, which is positive and meaningful in the utilization of C&D waste and the sustainable development of ceramic industry.

References

1. Thomas, C.: Macro- and micro-properties of multi-recycled aggregate concrete. J. Clean. Prod. **245**, 118843 (2020)
2. Layachi, B.: Mechanical and durability properties of concrete based on recycled coarse and fine aggregates produced from demolished concrete. Constr. Build. Mater. **246**, 118421 (2020)
3. Chunran, W.: Pore size distribution and ITZ performance of mortars prepared with different bio-deposition approaches for the treatment of recycled concrete aggregate. Cem. Concr. Compos. **111**, 103631 (2020)
4. Katz, A.: Treatments for the improvement of recycled aggregate. J. Mater. Civ. Eng. **16**(6), 597–603 (2004)
5. Grabiec, A.M.: Modification of recycled concrete aggregate by calcium carbonate bio-deposition. Constr. Build. Mater. **34**, 145–150 (2012)
6. Mengdi, L.: Improving the properties of recycled aggregate pervious pavement blocks through bio-mineralization. Constr. Build. Mater. **262**, 120065 (2020)
7. Pitarch, A.M.: Reutilization of hazardous spent fluorescent lamps glass waste as supplementary cementitious material. Constr. Build. Mater. **292**, 123424 (2021)
8. Shakouri, M.: Pretreatment of corn stover ash to improve its effectiveness as a supplementary cementitious material in concrete. Cem. Concr. Compos. **112**, 103658 (2020)
9. Tekin, İ: A regional supplementary cementitious material for the cement industry: pistachio shell ash. J. Clean. Prod. **285**, 124810 (2020)
10. Pitarch, A.M.: Pozzolanic activity of tiles, bricks and ceramic sanitary-ware in eco-friendly Portland blended cements. J. Clean. Prod. **279**(1), 123713 (2020)
11. Yining, D.: Study on pozzolanic activity of ceramic waste powder. J. Build. Mater. **18**(5), 867–872 (2015). (in Chinese)
12. Yaxian, C.: Application and research progress of solid waste ceramics in cement based materials. China Ceram. **56**(12), 53–60 (2020). (in Chinese)
13. Le, L.: Waste ceramic powder as a pozzolanic supplementary filler of cement for developing sustainable building materials. J. Clean. Prod. **259**, 120853 (2020)
14. Kannan, D.M.: High performance concrete incorporating ceramic waste powder as large partial replacement of Portland cement. Constr. Build. Mater. **144**, 35–41 (2017)
15. Li, L.G.: Cementing efficiency factors of ceramic polishing residue in compressive strength and chloride resistance of mortar. Powder Technol. **367**, 163–171 (2020)
16. Li, L.G.: Reutilizing ceramic polishing waste as powder filler in mortar to reduce cement content by 33% and increase strength by 85%. Powder Technol. **355**, 119–126 (2019)
17. Lasseuguette, E.: Chemical, microstructural and mechanical properties of ceramic waste blended cementitious systems. J. Clean. Prod. **211**, 1228–1238 (2019)
18. Weerdt, K.D.: Fly ash-limestone ternary cements: effect of component fineness. Adv. Cem. Res. **23**, 203–214 (2011)
19. Tajra, F.: Properties of lightweight concrete made with core-shell structured lightweight aggregate. Constr. Build. Mater. **205**, 39–51 (2019)

Construction Technology Innovation of Light-Weight Steel-Concrete Composite Cable-Stayed Bridge

Xueyun Kang[1,2,3](✉), Qi You[1], Gang Li[1,2,3], and Huijun Shen[1,2]

[1] CCCC Second Harbour Engineering Co., Ltd., Wuhan 430040, China
220171028@seu.edu.cn
[2] Key Laboratory of Large-Span Bridge Construction Technology, Wuhan 430040, China
[3] CCCC Highway Bridge National Engineering Research Centre Co. Ltd., Wuhan 430040, China

Abstract. Jiangxinzhou Yangtze River Bridge in Nanjing, built at the end of 2020, is one of the important river channels in Nanjing. The bridge is composed of a central double-cable plane three-pylon Light-weight composite beam cable-stayed bridge with two main spans of 600 m. In view of the new structures such as steel shell concrete pylons and UHPC-steel Light-weight composite beam with coarse aggregate, the innovative construction technologies and equipment such as segmental manufacturing and installing of steel shell concrete pylon, industrial precast UHPC bridge deck with coarse aggregate and integral manufacturing and installing of Light-weight steel-concrete composite beam are systematically developed, which significantly improve the engineering quality and construction efficiency, and promote the great transformation for bridges from field construction to industrialized lean construction.

Keywords: Light-weight steel-concrete composite cable-stayed bridge · Steel shell concrete pylon · UHPC bridge deck with coarse aggregate · Construction technology

1 Introduction

Jiangxinzhou Yangtze River Bridge in Nanjing (Fig. 1) is a central double-cable plane three-pylon steel-concrete composite beam cable-stayed bridge with a span of (80 + 218 + 600 + 600 + 218 + 80) m, and it is the first light-weight steel-concrete cable-stayed bridge in the world. The main beam is a flat streamlined integral box-shaped steel-concrete composite beam, and the pylons is a longitudinal diamond-shaped.

In order to respond to the national high-quality development strategy and improve the level of industrial construction of bridges, a number of world or domestic pioneering technologies is adopted in this bridge: (1) a new structure of the steel shell concrete composite pylon (Dang and He 2013) and its assembly construction technology are innovatively adopted, which not only meets the stiffness requirements of the mid-pylon of the multi-span cable-stayed bridges, but also gives full play to the advantages of steel

W. Guo and K. Qian (Eds.): GBCESC 2022, LNCE 211, pp. 455–464, 2023.
https://doi.org/10.1007/978-981-19-5217-3_45

pylons for manufacturing and installation; (2) UHPC with coarse aggregate is developed and applied to long-span composite beam of cable-stayed bridge, and its weight was reduced by at least 25% compared with the traditional steel-concrete composite beam.

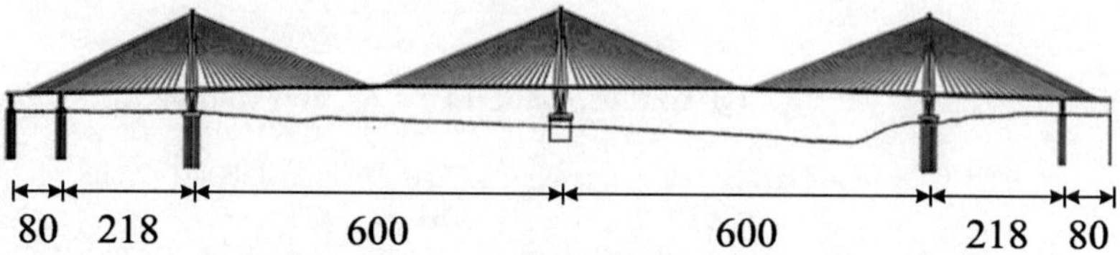

Fig. 1. Jiangxinzhou Yangtze River Bridge in Nanjing (unit: m).

2 Construction of Steel Shell Concrete Pylon

2.1 Structural Design

The pylon is a longitudinal diamond-shaped steel shell concrete composite structure. The middle pylon with a height of 175.4 m is divided into 37 segments and the side pylons with a height of 167.7 m is divided into 36 segments. The height of standard pylon segment is 4.8 m and the highest segment is 5.2 m. The steel shell of the pylon consists of inner and outer wall plates, vertical stiffening rib, horizontal stiffening rib, connecting angle steel, shear stud, etc. The inner and outer wall plates and stiffening rib are made of Q345C, the others are made of Q235B. The C50 shrinkage compensating concrete is used in pylons and HRB400 reinforcement is used in the steel shell. In order to realize the cooperative working between steel structure and concrete, the vertical reinforcement passes through the reinforcement hole of the horizontal stiffening rib in turn, and the horizontal reinforcement passes through the reinforcement hole of the vertical stiffening rib in turn to form the reinforced concrete tenon. Figure 2 shows the structure of the middle pylon.

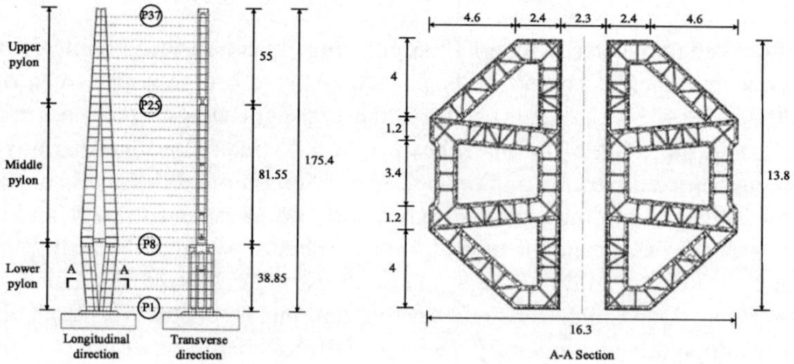

Fig. 2. Structure of the middle pylon (unit: m).

2.2 Overall Construction Technology

The steel shell in the pylon is not only involved in force, but also the concrete casting form. The reinforcement of the pylon is assembled with the steel shell in the yard, which eliminates the field reinforcement binding process, so as to realize the rapid construction. According to the height and weight of the pylon segment, the steel shell of the lower pylon segment is hoisted by floating crane, and the steel shell of the middle and upper pylon segment is hoisted by tower crane. The steel structure processing and reinforcement installation of segments are completed in the yard. The connection between steel structure and reinforcement and concrete casting are completed on site. The specific construction process is showed in Fig. 3.

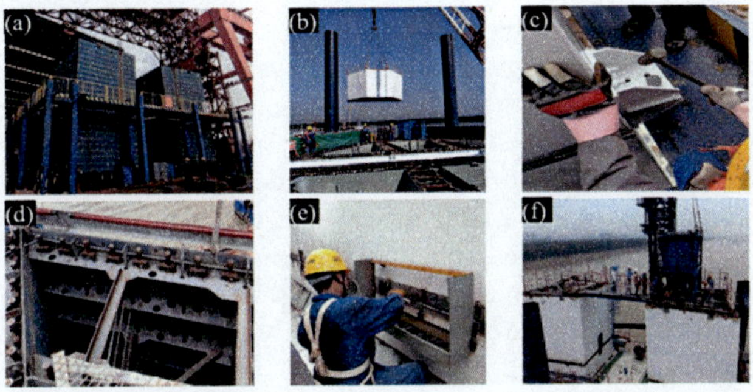

Fig. 3. Pylon construction process: (a) manufacturing and pre assembling steel shell in the yard; (b) hoisting standard segment of steel shell; (c) matching and positioning segment of steel shell; (d) reinforcements connection in steel shell; (e) welding between steel shell segments; (f) concrete casting in steel shell.

2.3 Installing Technology of the First Steel Shell Pylon Segment

The precise positioning and installing of the steel shell pylon segment are the foundation to ensure the construction alignment of the pylon. Since the steel shell segment has been processed in the yard, the field installation is only a reproduction of the assembly in the yard. The spatial position of the bottom plate and top surface of the steel-concrete joint section must be very accurate because the spatial position of the steel shell pylon cannot be adjusted step by step like the concrete pylon segment. Therefore, the BIM technology is used to establish the reinforcement model for collision analysis and optimize position of the reinforcement. Then, two layers of reinforcement positioning plates are set in the cap to improve the accuracy of embedded reinforcement. At the same time, vertical and horizontal positioning frames are set at the foot of the pylon to ensure the installation accuracy of the first steel shell pylon segment. In view of the special-shaped steel shell of the first segment and the complicated internal structure, a taper lock bush is used to connect the vertical reinforcement. Figure 4 shows the hoisting of the first steel shell pylon segment.

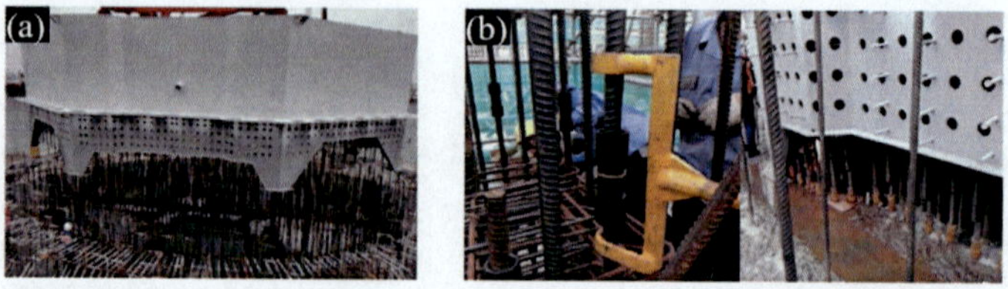

Fig. 4. Installing of the first steel shell pylon segment: (a) hoisting of the first shell pylon segment; (b) connecting of the vertical reinforcements.

2.4 Hoisting and Positioning Technology of Special-Shaped Steel Shell Segment

Since the steel shell segment is a special-shaped structure, the center of pylon segments is different. A precision adjustable strut-type hanger (Fig. 5) is developed. By setting the adjustment length on the longitudinal and transverse main beams of the hanger and setting the adjustment shackle at the end of the hanger, the hoisting problem of steel shell pylon segment with special-shaped structure, a certain degree of inclination, continuous change of the section size, and different gravity centers of the segment is solved.

At the meantime, in order to enable the segment to be installed easier to locate and dock with the installed segment, the matching parts are installed in place during the pre-assembly in the yard, and removed after the assembly between the pylon segments is completed. When adjusting the position accurately on site, priority to ensure that the groove surface measuring point deviation within the design allowable range.

2.5 Development and Application of Pylon Construction Platform

According to the structural characteristics and functional requirements of steel shell pylon, a light-weight automatic climbing operating platform (Fig. 6) is developed for steel shell hoisting, inter-segment welding and appearance painting. The platform realizes the synchronous climbing of the track and simplifies the operation steps and reduces the overall weight. At the same time, the platform is scalable to avoid manual repeated disassembly.

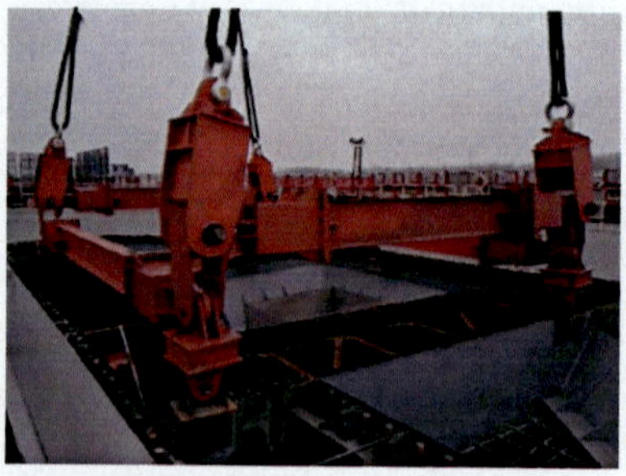

Fig. 5. Strut-type hanger.

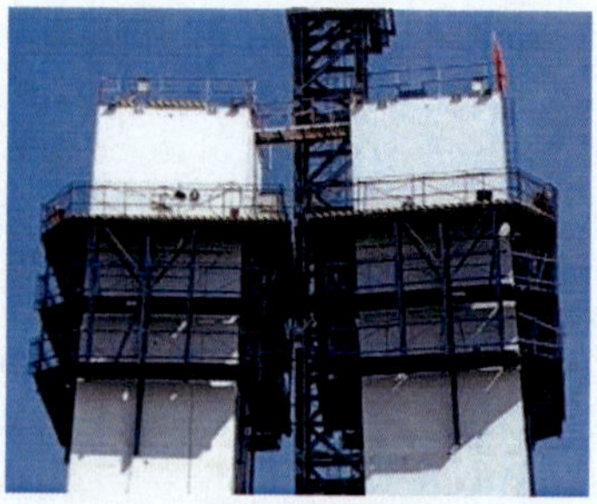

Fig. 6. Automatic climbing operating platform.

2.6 Concrete Casting and Curing in Steel Shell

The hydration heat of concrete in steel shell pylon is large. According to the preliminary test, it is necessary to strictly control the core temperature of concrete not more than 65 °C (Peng 2019). During summer construction, the molding temperature is reduced by setting cold storage to cool sand and stone and ice machine to cool the mixing water. In order to ensure the density of concrete casting, concrete vibration holes and vent holes are set on the horizontal and circumferential stiffening plates of the pylon wall. In order to prevent the cracks between concrete and steel shell caused by the rapid loss of water on the surface of concrete, the top concrete is curing by water storage with a depth of 10 cm in time after initial setting of concrete. Figure 7 shows the concrete casting and curing in steel shell.

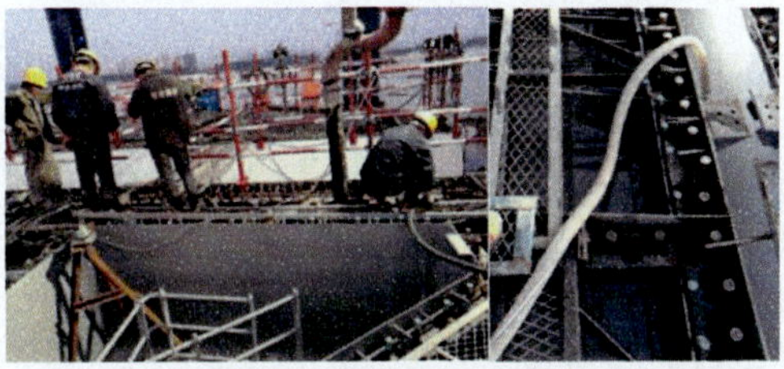

Fig. 7. Concrete casting and curing in steel shell.

3 Integral Segment Manufacturing and Installation of Light-Weight UHPC Composite Beam

3.1 Structural Design

The main beam is a steel-UHPC streamlined flat integral box composite beam (Lu et al. 2021) (Fig. 8). The self-weight of standard segment with a size of 35.6 m × 14.6 m × 3.6 m is about 407 t, 25% lower than the traditional steel concrete composite beam. UHPC bridge deck with coarse aggregate is divided into four precast slabs, three longitudinal wet joints casted in yard, one transverse wet joint casted in yard and two transverse wet joints casted in field. The dovetails and connection steel plates are set around precast slab with a size of 6.92 m × 11.3 m × 0.17 m.

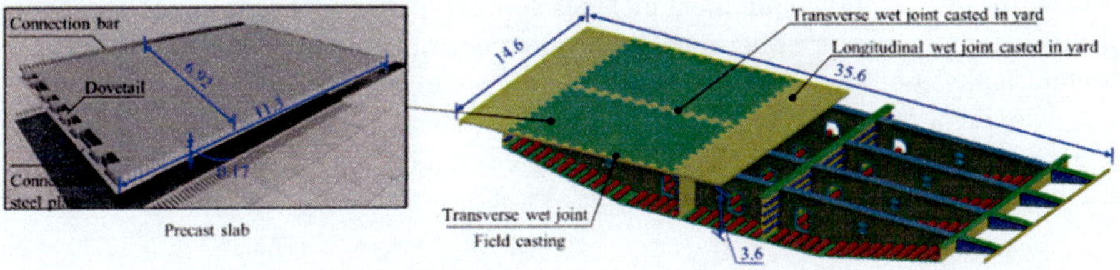

Fig. 8. Standard segment of main beam.

3.2 Overall Construction Technology

The light-weight composite beam is constructed by precast assembly process. The steel box beam segment and precast bridge deck are transported to the assembly jig after production. An integral segment can be formed by successively completing the installation of precast bridge deck, the butt welding of the embedded T-shaped parts at the bottom and the steel beam diaphragm, and the casting of longitudinal and transverse wet joints in the segment. All the beam segments are transported to the bridge site by ship, and installed by the floating crane or the erection crane. The bolt welding connection of the steel beam, the casting of transverse wet joint, and the tensioning of stayed cable are completed successively.

3.3 Industrial Production Technology of UHPC Bridge Deck

Materials of UHPC with Coarse Aggregate and Its Construction Technology. In order to reduce the cost of raw materials, UHPC in the project added gravel with particle size less than 8 mm on the basis of conventional mix proportion, and replaced quartz sand with ordinary sand (Table 1) (Brühwiler 2019; Graybeal et al. 2020). For UHPC with coarse aggregate, its fluidity is lower than that of conventional UHPC, so it is necessary to study reasonable vibration process to ensure the density of UHPC and the uniformity of steel fiber distribution (Shao et al. 2019; Zhu et al. 2018).

Table 1. Initial mix proportion of UHPC with coarse aggregate

Material	Coarse aggregate	Fine aggregate	Reactive powder	Steel fiber	Admixture	Water
Weight/kg	480	598	1,174	195	23.63	155.29

Industrial Production of UHPC Bridge Deck with Coarse Aggregate. According to the requirements of casting process for UHPC with coarse aggregate and the precast accuracy of bridge deck, an industrial production line (Fig. 9) based on standardized production process and automation equipment was developed, including high precision mold system and automation equipment. The high precision mold system with the functions of precise positioning and tensioning of steel mesh and automatic releasing from mold is composed of plates system, support system, hydraulic jacking system and locking mechanism. Automation equipment includes vertical shaft planet type mixer, automatic spreader, digital array type plug-in vibrator, automatic flat vibrator, automatic laminating machine and so on. The centralized control system cooperates with the running time of each equipment to improve the casting efficiency.

Practice shows that it only takes 10 h to precast a deck using this production line, and the precast accuracy reaches millimeter level.

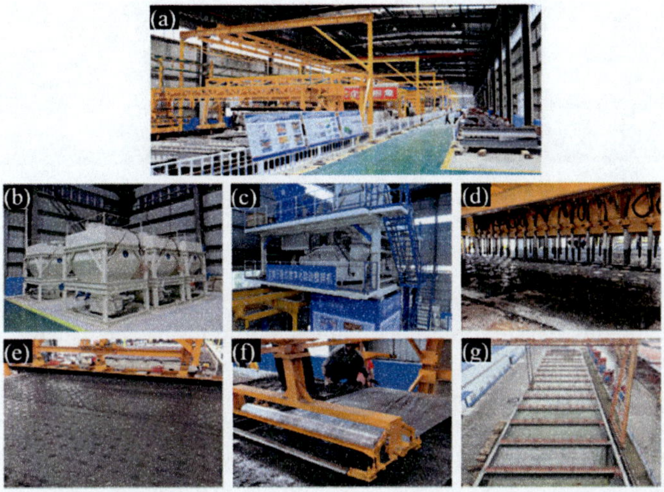

Fig. 9. Bridge deck production line and automatic equipment: (a) bridge deck production line; (b) raw material hopper; (c) automatic mixer; (d) row vibrator; (e) surface leveling; (f) membrane for curing; (g) water curing.

3.4 High Precision Assembly Technology of Steel-Concrete Composite Beam

Bridge Deck Installation. The jacks and supporting connectors are used to ensure the precise positioning of the reinforcement joint and the precise matching of the embedded steel plate with the diaphragm of steel beam and the flange plate of the web. In order to reduce the influence of welding on the shrinkage of precast plate, the embedded T-shaped part and the diaphragm of steel beam are welded first, then the embedded steel plate and the flange plate of the web are welded, and finally the splicing plate and reinforcement at the transverse wet joints are connected.

Horizontal Wet Joint Casted in Yard. The small vertical shaft planet type mixer is used to stir and distribute concrete in the transverse wet joints, and then the small row type vibrator is used to carry out the plug-in vibration. The small digital automatic flat plate vibration leveler is used to re-vibration and leveling the concrete, and the concrete surface after leveling is immediately covered with the film for water curing.

Longitudinal Wet Joint Casted in Yard (Fig. 11). A beam segment is casted in a special production line, and the posture is adjusted after being transported to the casting station by a flat car. The form of longitudinal wet joint is installed and the reinforcement is bound. The concrete is stirred by two vertical shaft planet type mixers with a capacity of 3 cubic. The gantry crane hoists the concrete hopper for transportation. The concrete placing boom, the row vibrator and the plate vibration leveler are integrated to continuously cast along the longitudinal strip of the composite beam.

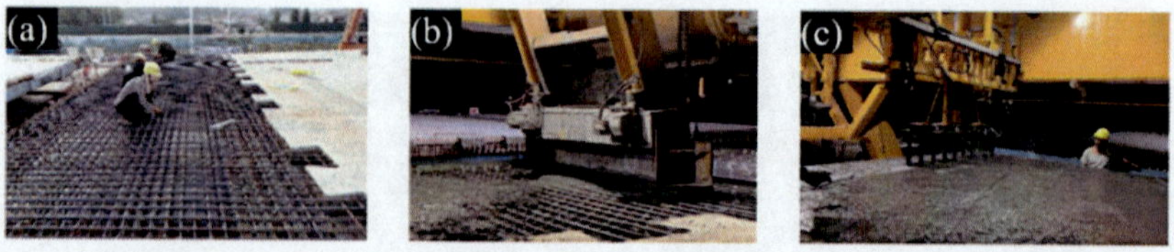

Fig. 11. Longitudinal wet joint casted in yard: (a) reinforcement binding; (b) concrete casting; (c) concrete vibrating.

Accuracy Control of Inter-segmental Matching. Complex processes with assembling on the tig, removing the tig, reset, longitudinal wet joints casting is easy to introduce errors, which makes it difficult for adjacent segments to match connections. To this end, a rigid support device of beam segment is developed. As shown on the Fig. 12, the rigid support of concrete is set at the position of each diaphragm and the end of the beam. After the beam is placed, the top surface of the beam is adjusted by steel gasket, so that the overall size of the segment is consistent with the size in assembly.

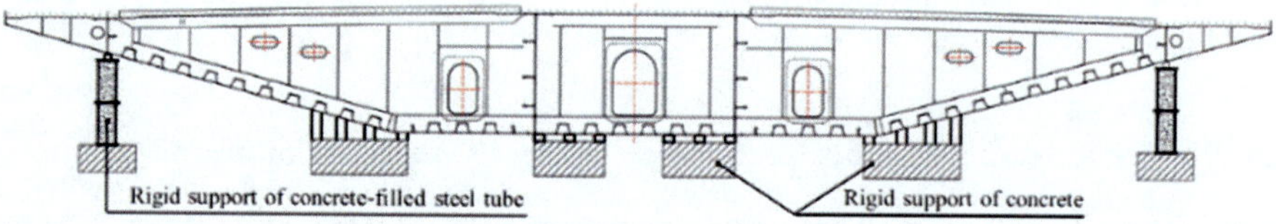

Fig. 12. Rigid support device of beam segment.

3.5 Integral Installation Technology of Steel-Concrete Composite Beam

Standard Segment Installation by Deck Erection Crane. In addition to the area of pylon and support, the standard segment is cantilevered installed by 450 t deck erection crane (Fig. 13). The bearing frame of the new deck erection crane with a working range of 9.5 m and a lifting height of 55 m is designed as a single main truss, which is especially suitable for the central cable plane cable-stayed bridge. The deck erection crane equipment with walking type pushing device has the functions of lifting point vertical and horizontal shift, slope adjustment, whole machine forward and track self-reversal.

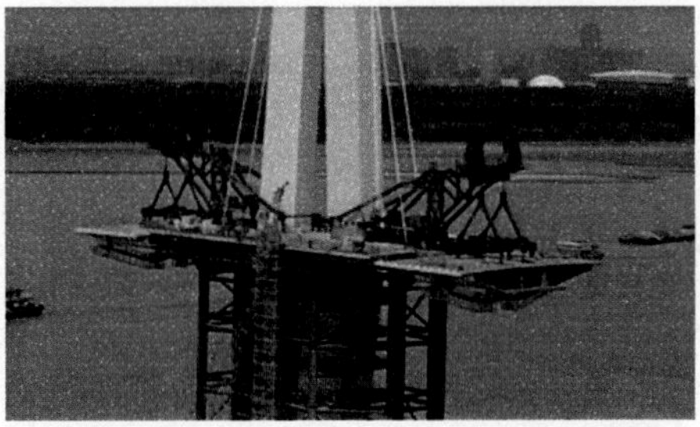

Fig. 13. Deck erection crane with single main truss.

UHPC Wet Joints Casted-in-Place. The lower edge of transverse wet joint with a standard width of 30 cm and thickness of 17 cm is laid with 12 mm thick steel plate. Before casting the transverse wet joint, the embedded steel plate is butt welded first. The casting equipment of transverse wet joint used in the yard is used for distribution and vibration. After the completion of the heat curing of the upper and lower edges of the joint, the curing temperature is not less than 90 °C, and the heating curing time is not less than 48 h. In order to avoid the tensile stress of concrete at the section of beam segment, temporary rods are set on the top of the transverse wet joint of beam segment.

4 Conclusions

Jiangxinzhou Yangtze River Bridge in Nanjing has novel structural design and huge engineering scale. In accordance with the structural characteristics of the bridge and the requirements of the construction environment, with the guidance of factorization, assemblage, and intelligence, a large number of new technologies, new techniques and new equipment have been developed and applied. The precision adjustable strut-type hanger and light-weight hydraulic self-climbing operation platform were invented to ensure the rapid installation and connection of special-shaped steel shell pylon. An automatic production line of UHPC bridge deck with coarse aggregate was developed, and the deck manufacturing accuracy can reach millimeter level. A 450 t deck erection crane with single main truss was developed to realize the rapid installation of the integral segment of the central cable-stayed bridge. A series of new technologies significantly improved the level of industrial construction of the bridge, providing a beneficial reference for subsequent engineering construction.

References

Dang, D., He, S.H.: Spatial mechanical behavior research of cable-pylon anchorage zone of steel box concrete tower. Appl. Mech. Mater. **256**(1), 1466–1473 (2013)

Peng, Q.: Full-scale model test of construction process for steel shell-concrete composite pylon of Fifth Changjiang River Bridge in Nanjing. Bridg. Constr. **49**(3), 46–50 (2019). (in Chinese)

Brühwiler, E.: UHPFRC technology to enhance the performance of existing concrete bridges. Struct. Infrastruct. Eng. **16**(1), 94–105 (2019)

Graybeal, B., Brühwiler, E., Kim, B., et al.: International perspective on UHPC in bridge engineering. J. Bridg. Eng. **25**(11), 04020049 (2020)

Shao, X., Deng, L., Cao, J.: Innovative steel-UHPC composite bridge girders for long-span bridges. Front. Struct. Civ. Eng. **13**(4), 981–989 (2019). https://doi.org/10.1007/s11709-019-0531-9

Zhu, Z., Yuan, T., Xiang, Z., et al.: Behavior and fatigue performance of details in an orthotropic steel bridge with UHPC-deck plate composite system under in-service traffic flows. J. Bridg. Eng. **23**(3), 04017142 (2018)

Lu, K., Xu, Q., Li, W., et al.: Fatigue performance of UHPC bridge deck system with field-cast dovetail joint. Eng. Struct. **237**, 112108 (2021)

Numerical Simulations of the Effects of the Radiant Floor Combined with the Displacement Ventilation of the Spread of Exhaled Contaminants in the Confined Space

Jie Wu[1,2], Liang Xu[1,2], Jing Hua Shen[1,2](✉), António Candeias[3], and Wei Zhang[1,2]

[1] School of Architecture, Soochow University, Suzhou, China
shenjh@suda.edu.cn
[2] International Research Center for Green Building of Soochow University, Suzhou, China
[3] Institute for Advanced Studies and Research, Evora University, Évora, Portugal

Abstract. The COVID-19 pandemic has seen the importance of confined space ventilation to reduce the risks of cross infection. To evaluate and compare the relative impacts of different mitigation strategies is important in order to reduce the risk of infection in a given situation. Using CFD methods, this study aimed to modulate the spread of exhaled contaminants in a floor-heated and ventilated space. Three different inlet velocities and four floor temperatures were used to assess the effect of the radiant floor combined with the displacement ventilation (RFDV) on room airflow and pollutant spread. Results show that RFDV reduced exposure to infection from 87% to 50% compared to the reference case. The inlet velocity is required to increase when the floor temperature is higher to decrease the contaminant exposure risk to in the room. This research provides a timely and necessary study of the ventilation and heating systems. These findings are expected to be useful for designing future of RFDV.

Keywords: Displacement ventilation · Radiant flooring · CFD · Tracer-gas

1 Introduction

Human-to-human transmission of COVID-19 is the primary form occurs through respiratory droplets, particularly in poorly ventilated rooms, as outlined in the World Health Organization (WHO) report [1]. Various studies emphasize the importance of airborne transmission of COVID-19 [2]. A case In Switzerland, the proximity to work and the celebration of birthdays in the conference room caused 11 of the 13 members of the team to become infected [3]. A research has been investigated extensive infected cases in China various cities found that all the infections occurred indoor [4]. There are considerable evidences to show that the effective strategies applied in poorly ventilated rooms to reduce infection risk is imperative.

W. Guo and K. Qian (Eds.): GBCESC 2022, LNCE 211, pp. 465–473, 2023.
https://doi.org/10.1007/978-981-19-5217-3_46

The adequate indoor ventilation is of great importance to reduce the risk of infectious disease [5], especially in confined rooms. Displacement ventilation (DV) is considered an interesting alternative in comparison with other ventilation strategies, as it is efficient and energy efficient [6]. Quite a few researchers have revealed that DV has obvious advantages in removing the exhaled contaminant [7]. However, the effect of only the DV is not ideal, which is mainly due to the supply air short-circuiting at high temperature [8].

The auxiliary heating system is recommended for use of the DV in winter. In comparison with traditional auxiliary heating system, the radiant floor (RF) has its energy-efficient and comfortable advantages [9]. The temperature requirement for indoor thermal comfort because of radiant heat exchange can be lower, while reducing the temperature gradient vertically [10]. These features of RF offer great advantage and the potential to combine with DV in winter. Thus, the Spread of exhaled contaminants with the radiant floor system combined with DV was investigated in this study. More specifically, we sought to assess the effect of different floor temperature and inlet velocities on the Spread of exhaled contaminants in the confined space by this system. The Spread of exhaled contaminants was simulated using tracer gas in this study. To quantitatively assess the risk of infection, the concentrations in different conditions will be normalized relative to the contaminant concentration with only DV.

2 Methods

2.1 CFD Model Description

All numerical simulations in this study were calculated by using Fluent (version 2021). A steady-state solver, the RNG k-ε model with pseudo-transient was used to the airflow. The radiation flux was calculated using the Fluent P1 radiation model and the energy equation that calculated the convective heat transfer on each surface.

In addition, we selected the tracer-gases to simulate Aerosols carrying viruses. This model has three benefits as described below: (1) airborne transmission of the disease, mainly by the tiny droplet, (2) the tracer-gases can be used well for these particles (<3–5 μm), (3) this way can also save more computational resources than other methods [11].

2.2 Case Description and Numerical Parameters

Our conference room geometry was an indoor environmental chamber facility at International Research Center for Green Building of Soochow University. Assuming that precaution during the COVID-19 period would be taken into consideration when setting up a conference room; we place six people to face each other in two rows, with the people shown red represent potential infection. The conference room is 6 m long, 4 m wide, and 3.6 m tall, as illustrated in Fig. 1. This room has an inlet diffuser on the south wall and a single pressure outlet in the ceiling. The temperature of the wall was set 12 °C to represent the heat losses to the cold environment, and the ceiling was assumed to adiabatic. The humans were set a heat transfer rate of 39 W/m^2.

To track the airborne disease transmission by the potential infection, Carbon monoxide (CO) was selected in this study as a tracer gas, as CO had similar density and viscosity.

In this study, In this study, each nose was modeled as a steady velocity inlet of 1.3 m/s and the temperature of 32 °C, resulting in a constant mass flow rate was applied to correspond to the average human breathing rates. The potential infection's exhaled air was 100% CO and the inlet supplied the fresh air without CO. Notably, healthy individual was considered to have no respiratory activity because the effect of breathing from a healthy occupant was found to be negligible in the literature.

We considered four different floor temperature (22 °C, 26 °C, 30 °C and 34 °C) and three different inlet velocities (0.3 m/s, 0.4 m/s and 0.5 m/s). The supplied air temperature of 18 °C in test cases was chosen based on the investigation of Yu Zhou et al. [12]. In the meantime, the reference case was supplied 35 °C warmer air of 0.3 m/s without radiant floor to compare the infection probabilities of the case which used RF. The details of temperature and velocity were shown in Table 1.

Table 1. Initial setup of temperature and velocity

Ventilation mode	Floor temperature	Inlet velocity and temperature
Reference case	Adiabatic	v = 0.3 m/s
		T = 35 °C
Test case	T = 22 °C, 26 °C, 30 °C,	v = 0.3 m/s, 0.4 m/s, 0.5 m/s
	34 °C	T = 18 °C

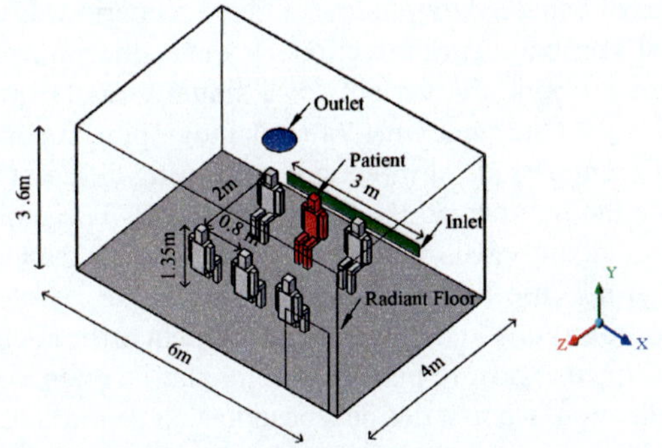

Fig. 1. Conference room geometry for simulations. The people in red represent potentially infected.

2.3 Meshing and Grid Independence Test

The conference room was meshed in ICEM (version 2021) with hexahedral un-structured meshes. For the purpose of a smaller detail to be captured in important points, the mesh along the inlet, outlet, human surfaces and nose is refined as is visualized Fig. 2. To determine the appropriate mesh element size, a mesh independence study was conducted in Reference case. We run this case with different meshes: a fine, medium, and coarse mesh with a total number of 1.2, 2.5 and 3.9 million cells.

By carrying out precise convergence tests, with a difference in speed between the average and fine mesh remaining less than 4% error limit.

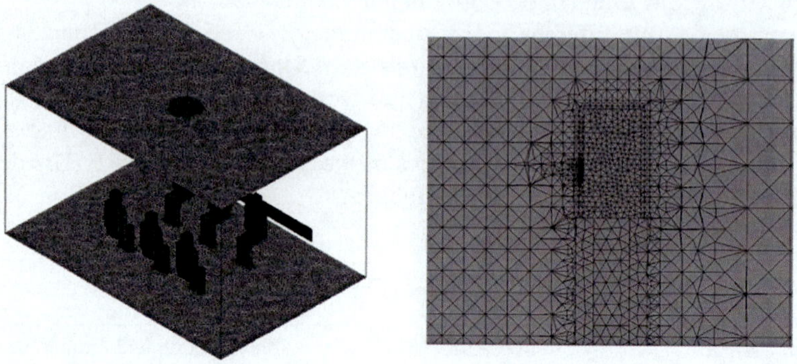

Fig. 2. Visualization of the grid division of all models. (a) View of conference room from outside. (b) Sections to view the mesh surrounding the manikin.

2.4 Evaluation Metrics

Typically, the number of expected to inhale in a given scenario was used to estimate the susceptible individual's probability of infection. However, the number of viruses exhaled may differ significantly among the variants of a single virus [13], not to mention the different pathogens [14]. At the same time, other demographic factors such as variability between persons, different periods of infection, and during different breathing activities may greatly influence the number of exhaled viruses [15]. Therefore, we focus on the proportion of exposure at any given time in this study, rather than restricting infection to specific diseases at specific times, at specific times. We equated the contaminant concentration dimensionless or standardized measurement to the equivalent standardized value of exposure for fixed exposure periods and inhalation rates.

For data standardization, we first define a baseline concentration, $C_{R0.}$ In the present study, C_{R0} stands for the room-averaged concentration. When other cases are normalized by this value, it can be considered exposure in relation to the average exposure of the room in our reference case, and it will be referred to as "relative exposure"

3 Results and Discussion

3.1 Reference Case

The result average temperature in the reference room is 18.09 °C and satisfies comfortable zone defined by the specification [16]. The average CO mass fraction in the reference room is 0.00187993, as will be the baseline concentration in the further study.

From the Fig. 3 (a) and Fig. 3 (b), we can find down-supply warm air ventilation is potential to cause "short-circuiting". Here it is clear that this temperature gradient is not typical for DV, where Fresh air straight to outlet without forming a cold air lake. This phenomenon is similar to flow field pattern of mixed ventilation, which also leads to Fresh air is unable straight to the body area and unable to create an upward flow by the human thermal plume to carry away exhaled contaminant gases.

Meanwhile, we also observed that a large number of contaminants permeate the room, and a significant concentration in the human activity zone. Usually speaking, exhaled contaminant gas migrates upwards due to be at a higher temperature and affected by the human thermal plume, however, the unreasonable ventilation systems possibly carry infectious contaminant gas to around the room [17]. The high temperature fresh air rises the top and forming a zone of high temperature and velocity at the top, which subsequently hits a cold wall and flows downward, forming the indoor air circulation. This means the possibility that contaminants into this circulation may be diffused into the inner room. Moreover, there are many vortices formed in the human breathe zone, where exhaled contaminants are mixed horizontally with the vortices, to cause a large aggregation of contaminants in the breathe zone.

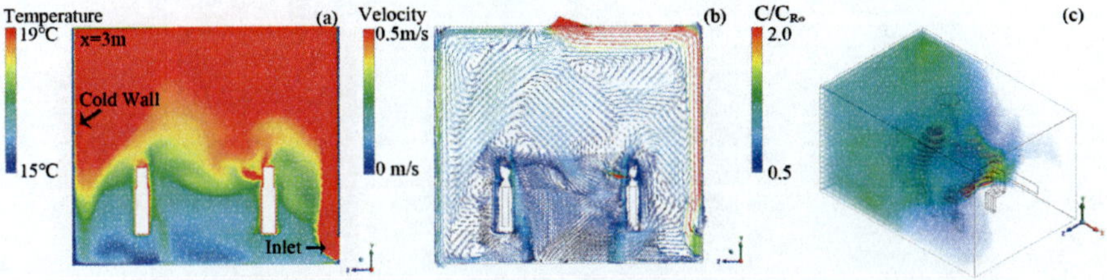

Fig. 3. (a) Contours of temperature in the reference case (x = 3 m). (b) Velocity field reference case (x = 3 m). (c) Volume rendering of "relative exposure". The transparent limit is set at 0.5xC_{R0} to 2xC_{R0}.

3.2 Airflow Pattern and Temperature Distribution

In all cases with floor heating, these average temperature of 16.02–16.95 °C, respectively, with each case can satisfy the requirements for the comfort zone defined by GB 50736-2012. From the Fig. 4, it can be seen that clean air is transferred to the human activity zone, rather than straight to the outlet. These are interesting outcomes.

Further analysis of Fig. 4 reveals inlet velocity and floor temperature have a large effect on the thermal stratification in the vertical direction. This can further affect the transmission path of the air jet. There are two primary airflow structure in the vertical plane: one is the air jet comes into contact with the heat source (potentially infected) and moves up by the thermal plume, which carries away contaminants as well, and the other is the air rises up between the two people, forming a protective barrier of clean air. It can be found that, in the same floor temperature, higher inlet velocities result in a farther air jet. At the inlet of 0.3 m/s, significant differences on the flow field emerged with different floor temperature. When the floor temperature of 34 °C, the clean air migrates upward between the two. This is room temperature heightened by increased floor temperature, air jet is weakened by natural convection (buoyancy by temperature difference) effect in the vertical direction and travels further upward horizontally at the floor temperature of 34 °C and the inlet velocity of 0.3 m/s. A similar situation is happening in the supplying air with an initial velocity at 0.4 m/s. The airflow is much influenced by the forced convection (i.e. Initial supply velocity) when the inlet velocity of 0.5 m/s, and it travels further in the horizontal direction. With velocity decay, Natural convection plays a dominant role in airflow transmission, causing the airflow to curl upward. Meanwhile, we also notice that the region where clean air mixed well with surrounding air is different with different floor temperatures. This may well affect the effectiveness of contaminant removal, and we will discuss the detail in Sects. 3.3 and 3.4.

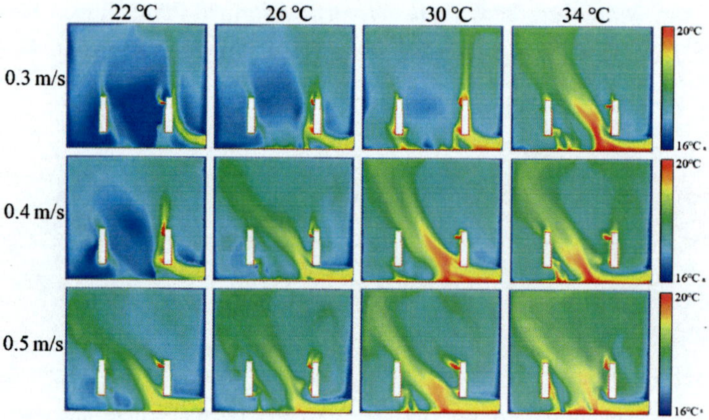

Fig. 4. Contours of temperature at different floor temperatures and inlet velocities (x = 3 m)

3.3 Exhaled Contaminant Transmission Routes

Figure 5 presents the spread of contaminant concentration on the breath level with different inlet velocity and different floor temperatures. When the inlet velocity set at 0.3 m/s, the spread is clearly increasing with the floor temperature increasing. However, there does not appear to be a correlation with increasing the floor temperature, when increasing the velocity of the inlet. Notably, the contaminant concentration is high level compared to other cases at the temperature of 22 °C and the inlet velocity of 0.4 m/s. This may be due to a couple of reasons: (1) Natural convection plays a leading role in airflow pattern when the RFDV with high floor temperature and low ventilation rate. This may causes turbulence in the room, the room temperature tends to be uniform, and the contaminant concentration diffuses evenly. (2) From the undesirable results, it was found that low velocity air jet, easily affected by natural convection (including the thermal plume of the patient), floats directly the top and diffuses horizontally along the ceiling. The infected air downward flow along the wall due to the influence of the cold wall, and diffuses following the air jet. From the satisfactory results, with the increasing of the velocity of the inlet, the air jet is mainly influenced by the forced convection, and maintains the jet shape in the breathing zone. The air jet and the infected air meet in the breathing zone and curve upward, allowing the infected air almost directly to the room outlet.

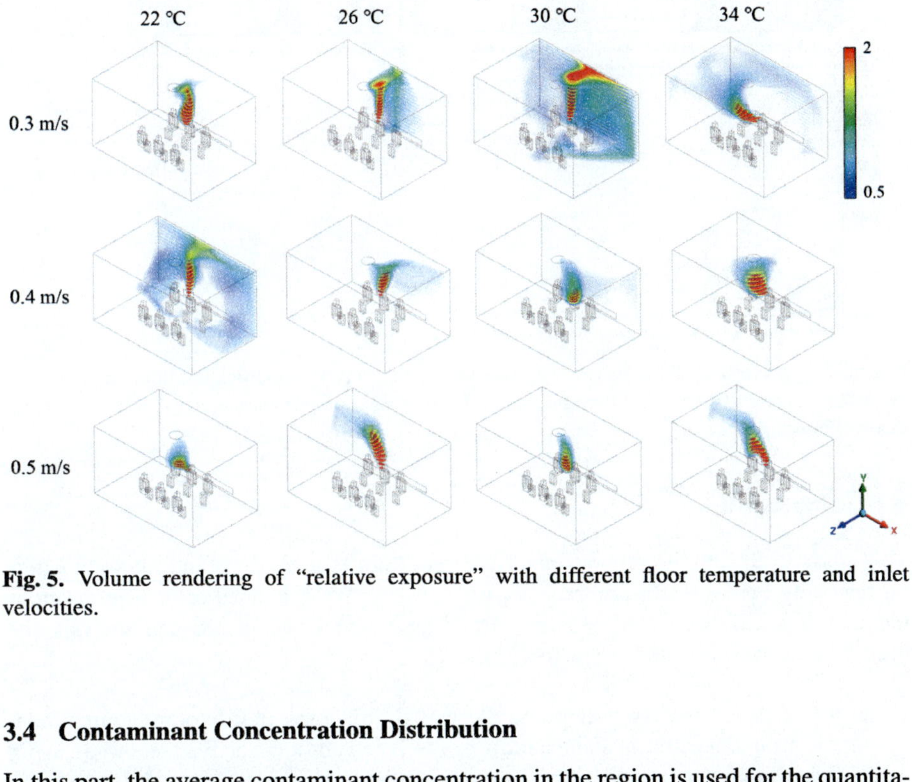

Fig. 5. Volume rendering of "relative exposure" with different floor temperature and inlet velocities.

3.4 Contaminant Concentration Distribution

In this part, the average contaminant concentration in the region is used for the quantitative analysis of "relative exposure". Figure 6 represents the comparison of area -averaged "relative exposure" in three zones with different conditions. The most intuitive outcome is that the average exposure of the room, in all of 12 cases, is below the reference case. This effect can still reduce exposure by approximately 50% in the worst result. While the inlet velocity is 0.4 or 0.5, the differences of room relative exposure gradually decreasing with increasing floor temperature, where relative exposure approaches 0.3.

Meanwhile, the results of room area-averaged "relative exposure" were further confirmed at the contaminant concentration distribution in correlation with floor temperatures and inlet velocities in Sect. 3.3. The same trends are obtained of activity zone and breathing zone. The average exposure of breathing zone is below than room and activity zone in all conditions. It is believed that the breathing zone was not formed as a contaminant containment zone, where contaminants are transported by buoyancy at breathing height, resulting in long-range transport of contaminants.

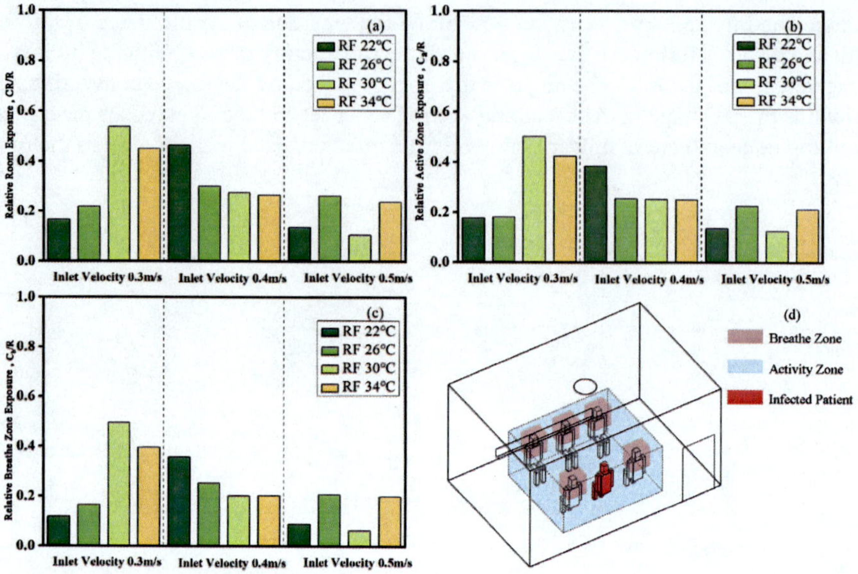

Fig. 6. Comparison of relative zone exposure with different inlet velocities and floor temperatures (a) room; (b) activity zone; (c) breathe zone. (d) Different colored rectangles represent different analysis zones.

4 Conclusion

This study evaluated the influences of different floor temperatures and inlet velocities on relatively exposures to arbitrary aerosol sources. We evaluated four different floor temperatures and three inlet velocities. From these simulation results and analysis, the main conclusions obtained are as follows:

1. In the winter, it is easy to lead the thermal stratification in the room, contaminant spread around the house, and a build-up of infected air of the active zone when using DV only in winter
2. The RFDV helps to significantly reduce both the thermal stratification in vertical direction and the exposure of exhaled contaminants in this room.
3. Where Different floor temperatures can affect the indoor airflow distribution. For the 0.3 m/s inlet velocity, Natural convection plays an increasing role due to the floor temperatures increasing, which also leads to the homogenization of the room temperature and the contaminant spread. When the inlet speed was 0.4 m/s or 0.5 m/s, there is no significant exhaled contaminant spread in concentration with increasing floor temperature.
4. Increasing inlet velocity typically decreases the average exposure of the room, but increasing the inlet velocity rate when forced convection intensity is insufficient may increase exposure risk. With the increase of the floor temperature, this space is close to being well-mixed, and the effect of inlet velocity on contaminant removal may be less influential.

Acknowledgments. This research was funded by National Key R&D Program of China (2021YFE0200100); 2021 Construction System Science and Technology Project of Jiangsu Province (Guidance) (2021ZD03); 2021 Policy Directed Program of Jiangsu Province (BZ2021015).

References

1. Sotomayor-Castillo, C., et al.: Australian dentists' knowledge, preparedness, and experiences during the COVID-19 pandemic. Infect. Dis. Health **27**, 49 (2022)
2. Vernez, D., Schwarz, S., Sauvain, J., Petignat, C., Suarez, G.: Probable aerosol transmission of SARS-CoV-2 in a poorly ventilated courtroom. Indoor Air **31**, 1776 (2021)
3. Weissberg, D., et al.: Does respiratory co-infection facilitate dispersal of SARS-CoV-2? Investigation of a super-spreading event in an open-space office. Antimicrob. Resist. Infect. Control **9**, 1–8 (2020)
4. Qian, H., Miao, T., Liu, L., Zheng, X., Luo, D., Li, Y.: Indoor transmission of SARS-CoV-2. Indoor Air **31**, 639 (2021)
5. Yang, J., Sekhar, S.C., Cheong, K.W.D., Raphael, B.: Performance evaluation of a novel personalized ventilation-personalized exhaust system for airborne infection control. Indoor Air **25**, 176 (2015)
6. Skistad, H.: Displacement Ventilation. Research Studies Press, Italy (1994)
7. Zhuang, C., Yang, G., Long, T., Hu, D.: Numerical comparison of removal and deposition for fully-distributed particles in central- and split-type air-conditioning rooms. Build. Environ. **112**, 17 (2017)
8. Javed, S., Ornes, I.R., Dokka, T.H., Myrup, M., Holos, S.B.: Evaluating the use of displacement ventilation for providing space heating in unoccupied periods using laboratory experiments, field tests and numerical simulations. Energies **14**, 952 (2021)
9. Werner-Juszczuk, A.J.: Experimental and numerical investigation of lightweight floor heating with metallised polyethylene radiant sheet. Energy Build. **177**, 23 (2018)
10. Lin, B., Wang, Z., Sun, H., Zhu, Y., Ouyang, Q.: Evaluation and comparison of thermal comfort of convective and radiant heating terminals in office buildings. Build. Environ. **106**, 91 (2016)
11. Ai, Z., Mak, C.M., Gao, N., Niu, J.: Tracer gas is a suitable surrogate of exhaled droplet nuclei for studying airborne transmission in the built environment. Build. Simul. **13**(3), 489–496 (2020). https://doi.org/10.1007/s12273-020-0614-5
12. Zhou, Y., Deng, Y., Wu, P., et al.: The effects of ventilation and floor heating systems on the dispersion and deposition of fine particles in an enclosed environment. Build. Environ. **125**, 192–205 (2017)
13. How the Delta variant achieves its ultrafast spread. Nature. https://www.icpcovid.com/sites/default/files/2021-07/Ep%20156-5%20a%20How%20the%20Delta%20variant%20achieves%20its%20ultrafast%20spread.pdf. Accessed 21 July 2021
14. Zemouri, C., Awad, S.F., Volgenant, C.M.C., Crielaard, W., Laheij, A.M.G.A., de Soet J.J.: Modeling of the transmission of coronaviruses, measles virus, influenza virus, Mycobacterium tuberculosis, and Legionella pneumophila in dental clinics. J. Dent. Res. **99**, 1192 (2020)
15. Santurtun A., Colom M.L., Fdez-Arroyabe P., Real A.D., Fernandez-Olmo I., Zarrabeitia M.T.: Exposure to particulate matter: direct and indirect role in the COVID-19 pandemic. Environ. Res. **206**, 112261 (2022)
16. China, M.: GB 50736-2012, Design Code for Heating Ventilation and Air Conditioning of Civil Buildings. China Construction Industry Press, Beijing (2012)
17. Lu, J., Yang, Z.: COVID-19 Outbreak associated with air conditioning in restaurant, Guangzhou, China, 2020 response. Emerg. Infect. Dis. **26**, 2790 (2020)

Research on Impact Collapse Resistance of RC Beams After Fire

Zhi Li[1,2](✉), Kaiqian Huang[2], and Xiaolan Yuan[2]

[1] GuangXi Key Laboratory of New Energy and Building Energy Saving, Guilin University of Technology, Guilin 541004, China
Lizhi@glut.edu.cn

[2] College of Civil Engineering and Architecture, Guilin University of Technology, Guilin 541004, Guangxi, China

Abstract. In order to explore the mechanical properties of improve concrete (RC) beams of the combined action of impact load and fire, numerical simulation of the high temperature test of RC slabs conducted by ZHANG et al. was carried out based on ABAQUS finite element software to verify the effectiveness of sequential thermal-mechanical coupling modeling. Then, numerical simulation of impact tests such as FUJIKAKE was carried out and the influence of temperature deterioration effect was considered, and the influence of different fire duration, drop height of drop weight and beam reinforcement ratio on impact resistance of RC beam after high temperature was studied. Finite element analysis shows that with the fire time, the RC beam changes from bending failure to shear failure, which significantly reduces the impact resistance of RC beam. When the drop height of the drop weight and the beam reinforcement ratio are the same, the maximum impact force of the beam and the maximum reaction force of the support decreased by 21.03% and 24.62%, respectively, compared with the case without fire in the cases of 90 min of fire exposure. Under fire conditions, increase the reinforcement ratio can effectively reduce mid-span displacement, but has limited impact on maximum impact force and the maximum reaction forces of support.

Keywords: RC · Beam · Hammer down · Fire time · Deformation capacity · Finite element analysis

1 Introduction

In this year, accidents such as fires and explosions caused by natural disasters or terrorist attacks have occurred frequently of high-rise. In most cases, the loss of resistance of building due to the destruction of its key components after fire and high temperature can cause more serious damage to the engineering structure [1, 2]. Compared with the static load, the secondary impact time is shorter and the impact strength is higher, resulting in the failure of the RC structure different from the static response.

In previous studies, scholars have in-depth study fruitful studies on effects of fire and impact on structures [3–7]. Among them, Zhang et al. [8] carried out the static loading test of the RC plate after high temperature, and verified the change of the performance of

W. Guo and K. Qian (Eds.): GBCESC 2022, LNCE 211, pp. 474–483, 2023.
https://doi.org/10.1007/978-981-19-5217-3_47

the RC plate at high temperature and the extended analysis of the refractory performance. Fujikake et al. [9] carry out the impact resistance tests of RC beam at room temp, the results shown the reinforcement ratio and the height of the drop weight are importance factors affecting the impact response of the beam. Qian et al. [10–24] conducted relevant research on continuous resistance of RC beam-slab structures. The study showed that the failure of side columns of RC beam-slab structures can form an effective beam mechanism, compression arch mechanism, catenary mechanism and Pull-membrane mechanism resists collapse. Qin et al. [25] study of the effect of the impact rate the falling hammer, the curvature of the hammer head and the shape of the hammer head on the impact damage of the RC beam. Liu et al. [26] conducted related research on dynamic response and the mechanism of the RC beams of the impact. The research showed that the damages of RC beams mainly occurred in local response stage, and the deformation of the beam body occurred in over response stage. Zhang et al. [27] proposed the failure mechanism of SFRC beams under the combined action of drop weight impact with fire. Results show the impact damage is more concentrated under the action of high temperature and constant load, and brittle failure occurs. Xiao et al. [28] conducted tests on RC slabs under the low impact loads to study the effects of impact energy, impact area diameter and impact head shape on the damage of RC specimens. As the impact energy increases, the damage degree of the plate under low-speed impact increases. Jin et al. [29] studied the interfacial bonding performance of recycled steel tube concrete after high temperature, and proposed the calculation method of the interface bonding strength and the bond-slip constitutive of recycled steel tube recycled concrete after high temperature.

To sum up, many studies about the mechanical properties and the direction of continuous collapse resistance of concrete members without structural damage at home and abroad, but there are few studies about the coupling effect of RC beams at high temperature and impact. Therefore, the paper is based on research results of research group on continuous collapse resistance of concrete structures and steel frame structures in the early stage [10–24]; the high temperature test of Zhang et al. [8] and the impact test of Fujikake et al. [9] were verified by the finite element software ABAQUS, and then Fujikake et al. [9] Based on the impact test, considering the high temperature deterioration reaction, dynamic response trend of the RC beam under different fire conditions was predicted, and the impact of the drop height and the beam reinforcement ratio of low impact resistance of RC beam after high temperature was studied.

2 Finite Element Modeling and Verification of Heating Test

In this section, the experiments of Zhang et al. [8] are simulated to verify the effectiveness of the sequential thermo-mechanical coupling modeling method adopted in this paper.

2.1 Experimental Details

As shown in Fig. 1, the test piece verified in this paper is a rectangular plate with a side length of 1880 mm, and the bearing column in the middle of the plate is the static load area. The No. 3 steel bar with a diameter of 9.5 mm and a yield strength of 520.0 MPa is used for the reinforcement of the bottom plate. The No. 4 steel bar with a diameter of 12.7 mm and a yield strength of 497.0 MPa is used.

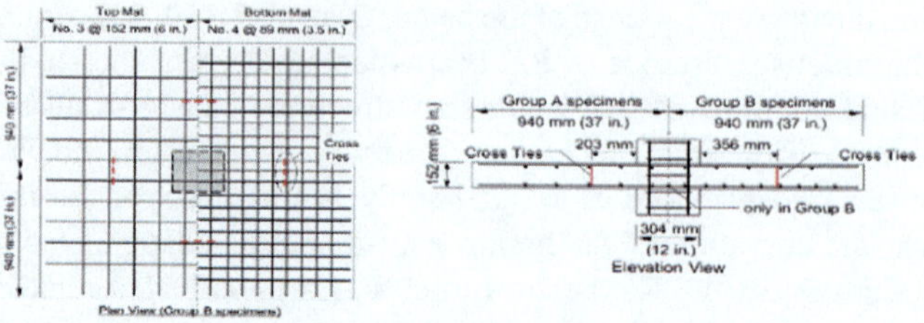

Fig. 1. Dimensions and reinforces details of specimen [8]

2.2 Finite Element Modeling and Verification

In paper, the ABAQUS is used to simulate and the impact resistance of the RC plate at high temperature. In the FEM, the finite element model of the RC plate without considering the hightemperature and impact is first established. On this basis, the temperature field under the action of high temperature and he initial velocity of the drop weight impact are applied sequentially.

Finite Element Constitutive Model. In this paper, the calculation of concrete material parameters in ABAQUS software refers to "Concrete Structure Design" (GB 50010-2010) [30].

Thermomechanical Coupling Parameters. The heat conductivity and heat capacity of concrete are given as suggested in the Eurocode [31], including the elastic modulus and strength model of the steel bar after high temperature. Thermal parameters of steel bars are defined by (CECS200:2006) [32–34]: the density of steel bars is 7850 kg/m^3, the thermal conductivity is 45 W/(m°C).

High Temperature Degradation Effect. Since the strain rate effect of concrete materials at high temperature has not been deeply studied, according to Liu et al. [27], the strain rate effect of concrete at high temperature is consistent with that at room temperature, so this paper only considers the effect of the impact load. For the amplification behavior of concrete strength, the dynamic amplification factor of concrete compressive strength (CDIF) and the dynamic amplification factor of tensile strength (TDIF) given in CEB-FIP specification [35] are used. According to the recommendations in Eurocode [36], the three-sided heat convection exchange coefficient of the RC beam exposed to fire was set to 25 W/(m^2 °C).

Finite Element Verification of RC Plate Heating Test. As shown in Fig. 2, the simulation results of the top, 30 mm from the top, the midpoint, and the bottom of the RC board comparative test results (Fig. 2a) and the center column deflection curve of the RC board at high temperature (Fig. 2b). The difference between the peak force and maximum displacement at midspan and the numerical simulation is within 5%.

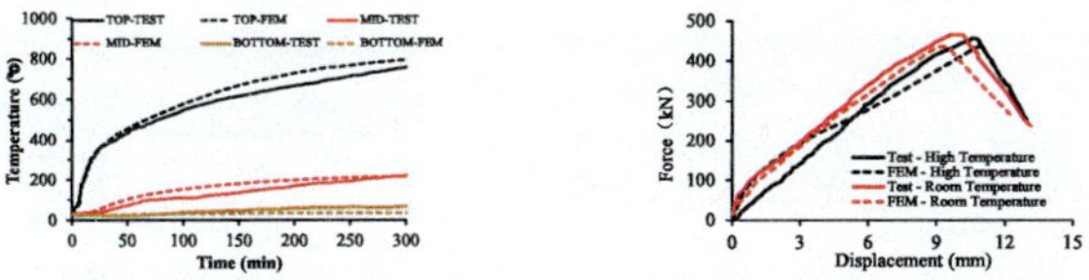

(a) Temperature curve (b) Displacement curve

Fig. 2. Comparison between test results and modeling results

3 Impast Test Finite Element Modeling and Verification

3.1 Introduction to the Experiment of FUJIKAKE [9]

The reinforcement and cross-sectional dimensions of test specimen S1322 are show on Fig. 3.

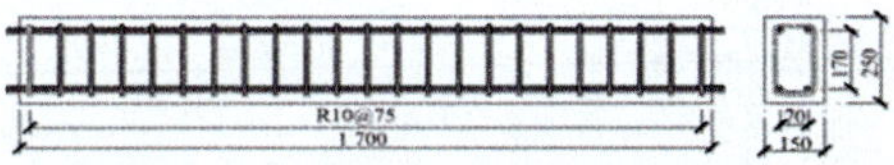

Fig. 3. Cross-section size and rebar arrangement [25]

The material constitutive tested in this paper has been introduced in Sect. 2.2. Test data from reference [9]. The model steel frame is composed of three different types of longitudinal bars, including two kinds of steel bars with diameters of 13 mm and diameters of 22 mm. The strength of the 13 mm bar is 397 MPa and the yield strength of the 22 mm bar is 418 MPa. Beam stirrups are steel bars with diameter of 10 mm, a yield strength of 295 MPa, a uniform distribution of 75 mm spacing (Table 1).

Table 1. Reinforcement ratio of longitudinal reinforcement [9]

Specimen number	Compressed steel bars	Cross-sectional area	Tension steel	Cross-sectional area
	(mm)	(mm^2)	(mm)	(mm^2)
S1322	2-D13	265.3	2-D22	759.9

3.2 Finite Element Modeling and Verification

Figure 4 shows the comparison of crack failure mode between the test and FEM. It can be see of the maximum displacement of the beams of 0.3 m and 0.6 m is small under the drop weight impact, while the concrete of the top of the RC beam is compressed under the conditions of 1.2 m and 2.4 m. The failure mode of the test can be well simulated by the finite element model.

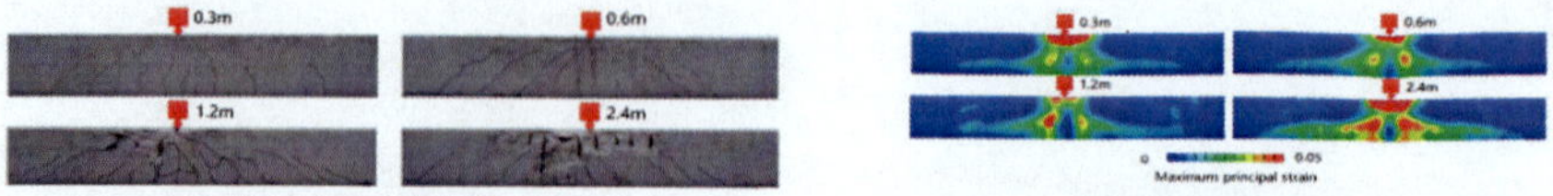

(a) Test results[9] (b) FEM results

Fig. 4. Finite element simulation results

Figure 5 is a comparison diagram of impact time-history curve between the test and the FEM. The change trend of the impact force time-history curve in elastic stage, the platform stage and descending stages is basically consistent, and the error between the maximum value of impact force simulated by FEM and the test is: 5%–10%.

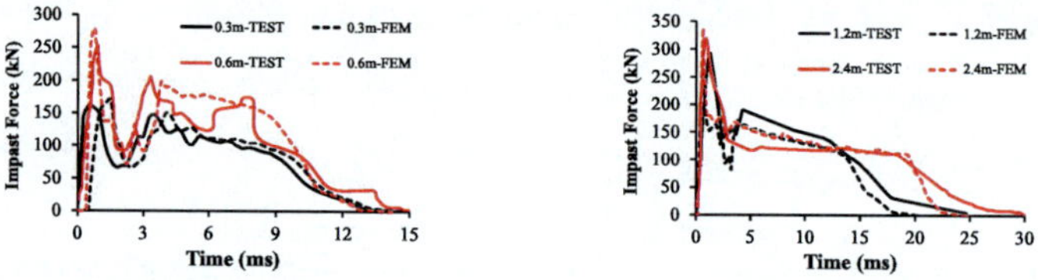

(a) 0.3 m/0.6 m (b) 1.2 m/2.4 m

Fig. 5. Comparison of impact force time history curve

4 High Temperature and Shock Coupling Finite Element Expansion

Based on the verified thermal analysis method and high temperature degradation model in Sect. 2, this section extends the analysis of the drop weight impact test of RC beams in Fujikake et al. [9] to learn the impact resistance of RC beams after high temperature. This section expands the analysis into the original test specimen S2222.

4.1 S1322 Expansion Analysis After High Temperature

Figure 6 is the impact deflection diagram of S1322 after high temperature. When the drop height of the model S1322 is 0.6 m, the fire exposure time increases from 0 min to 30 min, 60 min and 90 min, and the mid-span peak displacement increases by 17.69%, 27.98% and 72.27% respectively compared with 0min.

Figure 7 is the time and history diagram of the impact force of S1322 after high temperature. It shown the figure that at the same drop height, the maximum value of the impact force decreases with increase of the high temperature times. When the drop height of drop weights is 0.6 m, the fire exposure time increases from 0 min to 90 min, the peak impact force decreases by 33.97%, 35.57% and 37.51% respectively compared with 0 min.

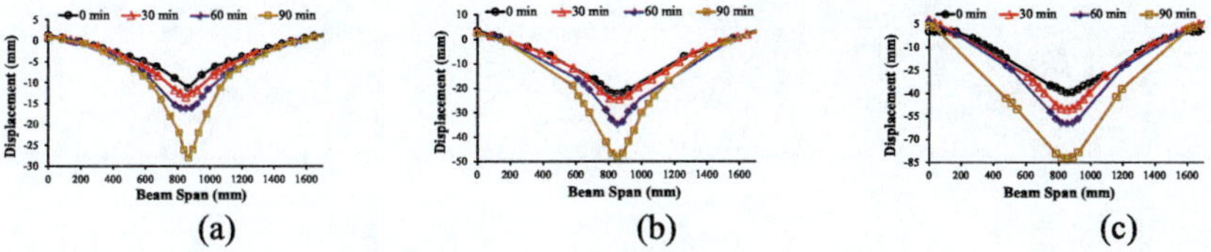

Fig. 6. S1322 impact deflection diagram after high temperature: (a) 0.6 m height (b) 1.2 m height (c) 2.4 m height

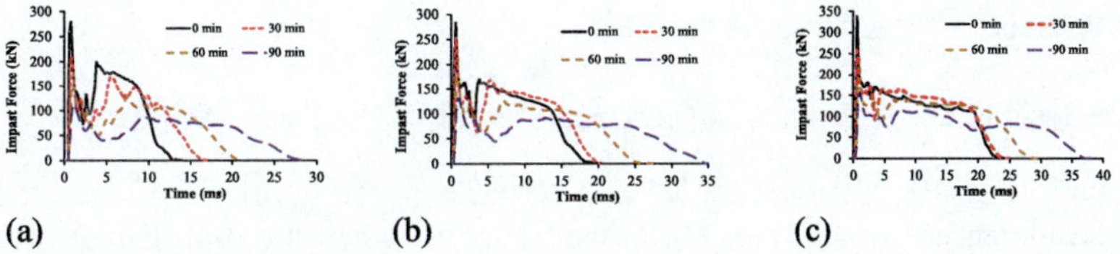

Fig. 7. S1322 impact force time history after high temperature: (a) 0.6 m height (b) 1.2 m height (c) 2.4 m height

4.2 S2222 Expansion Analysis After High Temperature

Figure 8 is the impact deflection diagram of S2222 after high temperature. Under the same drop height of the drop, the fire time of the specimen S2222 increased from 0 min to 30 min, 60 min and 90 min, and the peak displacement at mid-span increased by 17.72% and 25.07% respectively compared with 0 min.

Figure 9 is a time history diagram of the impact force of S2222 after high temperature. For example, when the drop height of the drop weight is 1.2 m, the fire exposure time increases from 0 min to 30 min, 60 min and 90 min, and the peak impact force decreases by 17.37%, 45.32% and 37.14% respectively compared with 0min.

Fig. 8. S2222 impact deflection diagram after high temperature: (a) 0.6 m height (b) 1.2 m height (c) 2.4 m height

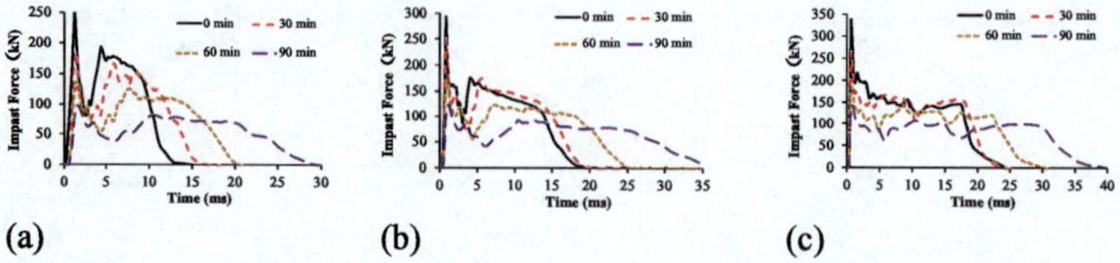

Fig. 9. Time history diagram of s2222 impact force after high temperature: (a) 0.6 m height (b) 1.2 m height (c) 2.4 m height

5 Extended Parameter Analysis

5.1 Analysis of the Influence of the Drop Height of the Drop Weight

As shown in Fig. 10, when the model S2222 was exposed to fire for 30 min, the peak deflection increased by 35.57%, 80.7% and 69.17% when the drop height increased from 0.3 m to 0.6 m, 1.2 m and 2.4 m. When the exposure time was 60 min, the peak deflection increased by 53.84%, 67.97% and 65.04% when the drop height increased from 0.3 m to 0.6 m, 1.2 m and 2.4 m. When the exposure time is 90 min, when the drop height increases from 0.3 m to 0.6 m, 1.2 m and 2.4 m, the peak deflection increases by 20.06%, 71.45% and 68.56%. The best reason of this phenomenon is internal material of the RC beam has deteriorated after 90 min of high temperature, and there has been a large displacement in the impact of 0.3 m drop weight, so the impact of 0.6 m has been affected. During subsequent impacts, the displacement continues to increase due to the increased impact energy.

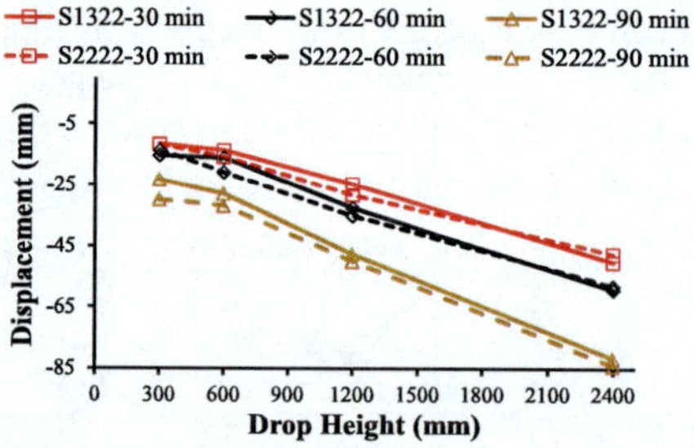

Fig. 10. Influence of falling height of drop hammer on peak displacement in midspan

5.2 Influence Analysis of Reinforcement Ratio

According to the literature [9], this section considers the effect of the reinforcement ratios of three specimens on impact resistance of RC beam after fire, that the reinforcement ratio of S1322 is 0.34%, the reinforcement ratio of S1616 is 1.06%, and the reinforcement ratio of S2222 is 2.06%.

Figure 11 shown the increase of reinforcement ratio can effectively reduce the mid-span deflection, and this beneficial effect becomes more obvious with the increase of fire time and the increase of impact energy.

Figure 12 shown the reinforcement ratio is increased from 0.34% to 2.06%, the maximum impact force is increased by 8.84% on average, and the maximum reaction force of the bearing is increased by 12.88% on average. The impact forces of the maximun reaction forces of the support is limited.

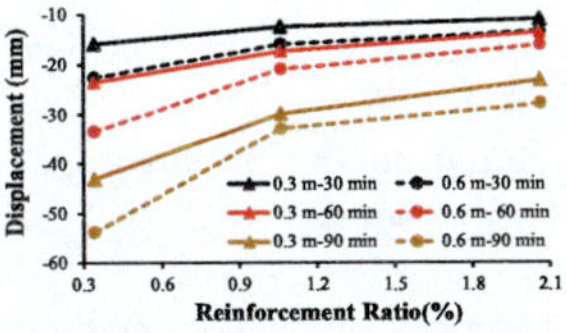

Fig. 11. Influence of different reinforcement ratio on peak displacement in mid span

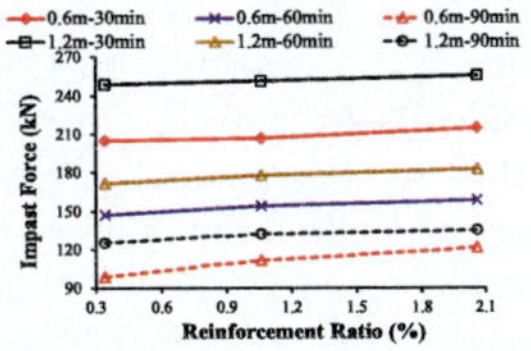

Fig. 12. Influence of reinforcement ratio on peak impact force

6 Conclusion

In article, the software of ABAQUS is used to verify the high temperature loading test of ZHANG et al. [8] and the drop weight impact test of FUJIKAKE et al. [9]. The effects of different drop heights or reinforcement ratios on the impact resistance of RC beams were studied. The conclusion of this paper are the following three points:

(1) The temperature model and impact model established in this paper can better simulate the temperature field distribution in the RC beam after fire and the rate effects of concrete and steel materials of the impact load, and are suitable for the impact resistance of RC beam after fire.
(2) Under the same drop weight drop height condition, the longer the fire time, the more serious the punching of the RC beam, and the continuous increase of the mid-span peak displacement.
(3) Under the same impact load, when the fire time is 90 min, compared with the no fire condition, maximum impact of the beam is reduced by 21.03% on average.

Acknowledgements. The authors gratefully acknowledge the support provided by Guangxi Key Laboratory Project of New Energy and Building Energy Saving (Gui Ke Neng22-J-21-4), the Natural Science Foundation of Guangxi (No. 2021GXNSFFA196001), the Guangxi Science and Technology Program Project (Gui Ke AD21220071), the China Postdoctoral Science Foundation (2021M693814) and the Guangxi Young and Middle-aged Teacher Basic Ability Improvement Project (2021KY0254).Any findings and conclusions expressed in this paper do not necessary reflect the view of Guangxi Key Laboratory of New Energy and Building Energy Saving and Natural Science Foundation of Guangxi.

References

1. Qin, F., Jiankui, Z., Li, C.: Numerical analysis of fire resistance of steel beams under combined action of explosion and fire load. J. Civ. Eng. **43**(S2), 62–68 (2010)
2. Fuxing, W.: Analysis on the causes of frequent fires in high-rise buildings and their thinking urban architecture (12), 219 (2013)
3. Bo, W., Guihe, W.: State of the art of fire resistance study on concrete structures in recent years. J. Build. Struct. **31**(6), 110–121 (2010)
4. Wei, G., Kexu, H.: Progress of research on fire resistance of self-compacting concrete structures. J. Build. Struct. **37**(3), 1–9 (2016)
5. Daudevile, L., Malecot, Y.: Concrete structures under impact. Eur. J. Environ. Civ. Eng. **15**(S1), 101–140 (2011)
6. Jianguo, N., Fenghua, Z., Zhihua, Z., et al.: Constitutive model, failure mechanism and numerical method for reinforced concrete under intensive impact loading. Scientia Sinica Technologica **46**(4), 323–331 (2016)
7. Hong, H., Yifei, H., Jun, L.: Review of the current practices in blast-resistant analysis and design of concrete structures. Adv. Struct. Eng. **19**(8), 1193–1223 (2016)
8. Zhang, C., Ma, W., Liu, X., et al.: Effects of high temperature on residual punching strength of slab-column connections after cooling and enhanced post-punching load resistance. Eng. Struct. **199**(15), 1–12 (2019)
9. Fujikake, K., Li, B., Soeun, S.: Impact response of reinforced concrete beam and its analytical evaluation. J. Struct. Eng. **135**(8), 938–950 (2009)
10. Kai, Q., Bin, L.: Testing and simulation of 3D effects on progressive collapse resistance of RC buildings. Mag. Concr. Res. **67**(4), 163–178 (2015)
11. Kai, Q., Zhi, L., Yunhao, W., et al.: Study on continuous collapse resistance of reinforced concrete beam slab substructure. Eng. Mech. **36**(6), 239–247 (2019)
12. Kai, Q., Ding, L., Sheng, H., et al.: Progressive collapse performance of reinforced concrete frame structures subjected to ground two-column removal scenario. J. Build. Struct. **39**(1), 61–68 (2018). (in Chinese)
13. Kai, Q., Zhi, L., Yunhao, W., et al.: Behavior of RC beam-slab substructures to resist progressive collapse. Eng. Struct. **36**(6), 239–247 (2019). (in Chinese)
14. Kai, Q., Zhi, L., Pan, H., et al.: Progressive collapse mechanism of PC beam-slab substructure with bolted connections. J. Build. Struct. **41**(1), 173–180 (2020). (in Chinese)
15. Kai, Q., Songyuan, G., Shilin, L., et al.: Effects of loading regimes on the structural behavior of RC beam-column sub-assemblages against disproportionate collapse. Eng. Struct. **251**, 113470 (2022)
16. Yunhao, W., Kai, Q., Feng, F., et al.: Numerical investigation on load redistribution capacity of flat slab substructures to resist progressive collapse. J. Build. Eng. **29**, 101109 (2020)
17. Xiaohui, Y., Kai, Q., Dagang, L.: Analysis on the ability of reinforced concrete frame structures to resist continuous collapse considering catenary effect. J. Build. Struct. **38**(04), 28–34 (2017)
18. Xiaohui, Y., Kai, Q., Dagang, L.: Probabilistic assessment of the ability of reinforced concrete frame structures to resist continuous collapse based on stochastic pushdown method. J. Build. Struct. **38**(02), 83–89 (2017)
19. Kai, Q., Da, L., Hesheng, Y., et al.: Research on the failure resistance of two adjacent columns at the bottom of reinforced concrete frame structure against continuous collapse. J. Build. Struct. **39**(01), 61–68 (2018)
20. Kai, Q., Pan, H., Xiaofang, D., et al.: Research on continuous collapse resistance of self-resetting prefabricated structures under failure of central column. J. Build. Struct. 1–14 (2022)
21. Kai, Q., Xinyu, T., Zhi, L., et al.: Impact performance and the influence factors of reinforced concrete slabs under high temperatures. Eng. Mech. **39**, 1–13 (2022)

22. Kai, Q., Yi, D., Yonghang, T., et al.: Clinical characteristics of 2019 novel infected coronavirus pneumonia: a systemic review and meta-analysis. MedRxiv (2020)
23. Kai, Q., Yueyu, W., Zhi, L., et al.: Seismic performance of shear wall with openings with flanges. J. Beijing Univ. Technol. **47**(07), 736–747 (2021)
24. Kai, Q., et al.: Numerical investigation on load transfer mechanism of bonded post-tensioned concrete beam-column substructures against progressive collapse. Adv. Struct. Eng. **24**(8), 1569–1582 (2021)
25. Yuan, Q., Kang, Q., Youzhe, T., et al.: Effect of hammer curvature radius on impact resistance of RC beams. J. Guangxi Univ. (Nat. Sci. Ed.) **46**(3), 545–556 (2021). (in Chinese)
26. Fei, L., Qizhi, L., Zhigang, J.: Study on dynamic response and failure mechanism of RC beams under low velocity impact. Eng. Mech. **32**(5), 155–161 (2015)
27. Renbo, Z., JinLiu, J., Xiuli, D., et al.: Mechanical behavior of SFRC beam under combined action of impact load and fire. Explos. Impact **39**(9), 63–75 (2019)
28. Yuan, X., Bin, L., Fujikake, K.: Behavior of reinforced concrete slabs under low-velocity impact. ACI Struct. J. **114**(3), 643–658 (2017)
29. Hengrui, J., Zongping, C., Junrui, C.: Study on interfacial bond slip behavior and constitutive equation of recycled concrete filled circular steel tubular after high temperature. Eng. Mech. **38**(10), 119–133 (2021)
30. Ministry of housing and urban-rural development of the people's republic of China, code for design of concrete structures, GB 50010-2010. Architecture & Building Press, Beijing (2010)
31. European Committee for Standardization. Eurocode 2: Design of concrete structures-part 1-2: General rules-structural fire design (EN1992-1-2:2004). The Committee, Brussels (2004)
32. China association for engineering construction standardization. Technical code for fire safety of steel structure in building, CEC200: 2006. China Planning Press, Beijing (2006)
33. Zhenghai, G., Xudong, S.: Behaviour of Reinforced Concrete at Elevated Temperature and Its Calculation. Tsinghua University Press, Beijing (2010)
34. Özkal, F., Polat M., Yağan, M., et al.: Mechanical properties and bond strength degradation of GFRP and steel rebars at elevated temperatures. Constr. Build. Mater. **184**, 45–57 (2018)
35. Comite Euro-International Du Béton: CEB-FIP Model Code 1990. Redwood Books, Trowbridge, Wiltshire, UK (1993)
36. European Committee for Standardization, Eurocode 1: Actions on structures-part 1-2: General actions-actions on structures exposed to fire, EN 1991-1-2:2002, The Committee, Brussels (2002)

Method Research on Traffic Volume Forecasting Based on Bio-LSTM

Xiao Li, Chaoyang Li, Tao Wang(✉), Yi Zhang, and Changqin Xi

Shanghai Jiao Tong University, Shanghai 200240, People's Republic of China
wangtao127@sjtu.edu.cn

Abstract. Nowadays, urban traffic management plays an important role to improve traffic system efficiency and ensure safety, and decision-making about that is based on the adequate understanding for traffic condition. Traffic volume is the representation of current traffic operation, so it is of great significance to master traffic volume in advance and accurately. The developments of computer and Deep-Learning provided support for traffic forecasting. Firstly, the principle of Long Short-Term Memory neural network (LSTM) to overcome "gradients vanishment/explosion" is analyzed. And then Bio-LSTM is used to forecast the traffic volume of a certain road in the next 24 h. Finally, the Wavelet Neural Network is used to predict the same data of this road. The two series of prediction results are compared, it is showed that Bio-LSTM for prediction has the advantages of simple operation, convenient adjustment of key parameters and higher prediction accuracy.

Keywords: Traffic volume forecasting · Gradients vanishment/explosion · Bio-LSTM · WNN

1 Introduction

The key to solve traffic congestion is to improve the traffic resources supply and reduce the demand dynamically through efficient managements. Because the traffic control strategies usually take a period to work, its effect lags traffic conditions of this moment, so it is necessary to predict the traffic condition of target road in advance to modify the future traffic demand. With the substantial improvement of computer processing capacity, Deep Learning has developed rapidly and been widely used in various engineering fields, it includes Recurrent Neural Network (RNN) which mainly dealing with temporal objects and Convolutional Neural Network (CNN) which mainly dealing with images. Both methods are widely used in traffic big data processing and driving-behavior analysis [1, 2]. Long short-Term Memory neural network (LSTM) is a special RNN, which can prevent gradients from vanishing or exploding effectively when dealing with longer sequences [3]. In traffic prediction, a kind of LSTM network which considers temporal–spatial correlation in traffic system via a two-dimensional network which is composed of many memory units is proposed [4]. In case of missing traffic data, the multiscale

This research was supported by Shanghai Commission of Science and Technology, China (19DZ1208800, 20DZ1202900, 21DZ1200800).

W. Guo and K. Qian (Eds.): GBCESC 2022, LNCE 211, pp. 484–495, 2023.
https://doi.org/10.1007/978-981-19-5217-3_48

temporal smoothing is employed to infer lost data or using LSTM to correct missing temporal and spatial values [5, 6]. And a deep stacked bidirectional and unidirectional LSTM (SBU- LSTM) neural network architecture is proposed to predict network-wide traffic speed [7]. Meanwhile model "Transformer" is developing, "attention" is also being widely introduced into LSTM [8]. Similarly the Autoencoder is used to obtain the internal relationship of traffic flow by extracting the characteristics of upstream and downstream traffic flow data, so a novel traffic flow prediction method, called Autoencoder Long Short-Term Memory (AE-LSTM) prediction method is proposed [9]. It is not only used for prediction, but also for more complex lane changing behavior [10]. Wavelet Neural Network (WNN) based on the topology of traditional BP neural network [11], which sets the activation function in neurons of middle layer as "wavelet function", which integrates the advantages of neural network and wavelet transform, it has the characteristics of converging rapidly and not easy to fall into local optimization. These two methods all have strong adaptive learning ability for different data and are widely used in engineering for a long time.

1.1 RNN

RNN is composed of input layer, middle layer, and output layer, which is different from the input-output mechanism of traditional neural network. RNN takes time as the baseline, takes the output of the middle layer at the previous time O_{t-1} as another input this moment [12], and inputs it to middle layer together with the current input X_t. Then, the output of middle layer now O_t is regarded as another input at the next time, input to middle layer at the next time together with the input X_{t+1}, moving in circles to get the final output. The final output in the ideal state is a composite function of each historical intermediate state and historical input.

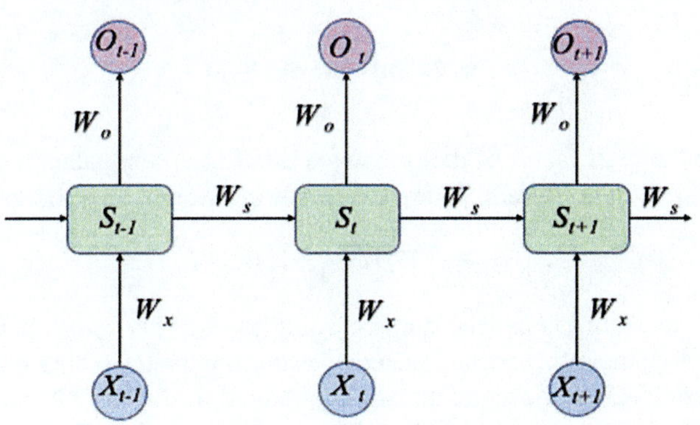

Fig. 1. Structure of RNN

The structure of RNN is shown in Fig. 1, where S_t represents the current state of the middle layer, X_t is the current input, O_t is the current output, W_s is the transform weight from the state before to this moment, W_x is the input weight of the middle layer now, W_o is the output weight of the current middle layer. The output of this moment will be obtained after current S_t being given weight W_o. Parameters W_i in each epoch are sharing in RNN all the same.

1.2 LSTM

LSTM was proposed by Sepp Hoch Reiter and Juergen Schmid Huber in 1997 [13]. Because the "Gate" adopted, the information in historical epochs can be introduced into the current calculation step to inherit the historical information in the process of predicting. So, compared with typical RNN, LSTM has better performance in processing long sequence information. The structure of LSTM is shown in Fig. 2, it is expanded by time axis, the intermediate unit is the main structure of it. Compared with RNN, LSTM introduces the time duration "cell", the purpose of the training process is to get more accurate C_t at the final moment. At the same time, "forgetting gate" is introduced to the process of state transfer to comprehensively filter some information of C_{t-1}, "input gate" is introduced to filter its input, "output gate" is aimed to select output at this moment, where h_{t-1} and x_t is the previous output and input now, like S_{t-1} and X_t in RNN.

The combination of previous information and the current input $[h_{t-1}, x_t]$ is the input of these "gate".

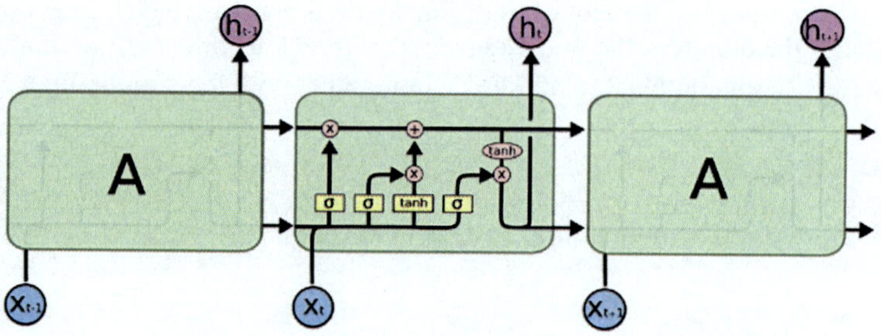

Fig. 2. Structure of LSTM

Let f_t to be the filter factor of the previous output C_{t-1}, and taken $f_t \times C_{t-1}$ pooled to the current C_t. Where, W_f and b_f are the forgetting weight and offset.

$$f_t = \sigma(W_f[h_{t-1}, x_t] + b_f) \tag{1}$$

Then input screening is carried out according to $[h_{t-1}, x_t]$, *tanh* is selected as the activation function, the information is selected according to (2) to obtain the preselected input information $\approx C_t$, at the same time, i_t is obtained for further filtering of $\approx C_t$, take $i_t \times \approx C_t$ to current C_t. And then combine the two parts to update C_{t-1} to get the current C_t. Where, W_c and b_c are the input preselected weight and offset, W_i, b_i are the input filter weight and offset.

$$\widetilde{C}_t = tanh(W_c[h_{t-1}, x_t] + b_c) \tag{2}$$

$$i_t = \sigma(W_i[h_{t-1}, x_t] + b_i) \tag{3}$$

$$C_t = (f_t \times C_{t-1}) + (i_t \times \widetilde{C}_t) \tag{4}$$

At this point, C_t contains the preference of historical status and input information, and it will be selectively output. In the process of output, *tanh* is used as the activation function of C_t, then calculate o_t according to formula (5), o_t is used for filtering the output information, finally take $o_t \times C_t$ as the output this epoch, as shown in formula (6). Where W_o and b_o are output weight and offset.

$$o_t = \sigma(W_o[h_{t-1}, x_t] + b_o) \tag{5}$$

$$h_t = o_t \times tanh(C_t) \tag{6}$$

1.3 Bi-LSTM

Bi-LSTM is based on the structure of LSTM. It is widely used in the field of natural language processing. Structurally, Bi LSTM is composed of two LSTM structures, but the propagation directions of the two structures are different, one is forward propagation, and the other is back propagation. For the prediction results of one target, the two results of forward prediction and reverse prediction are combined.

As we know, LSTM can consider more distant historical information when forecasting. Bi-LSTM is used to predict from two directions, which has been proved to achieve better prediction effect [14].

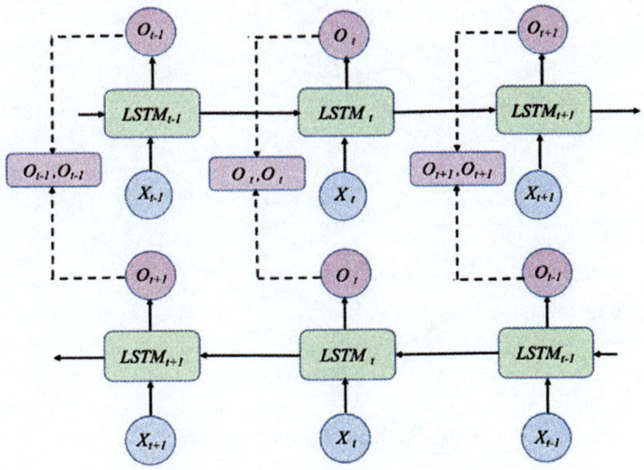

Fig. 3. Structure of Bi-LSTM

1.4 WNN

WNN adopt "Morlet wavelet function" as the activation function of neurons in middle layer, as shown in (7), its structure is shown in Fig. 3. It has center b_j and width a_j.

When the input is [−1, 1], there is obvious output, and the maximum output up to 1. The WNN structure is shown in Fig. 5, the full connection structure is adopted from the input layer to the middle layer and to the output layer. The transfer weights are W_{ij} and W_{jk} respectively (Fig. 4).

$$y = \cos(1.75x)e^{-\frac{x^2}{2}} \tag{7}$$

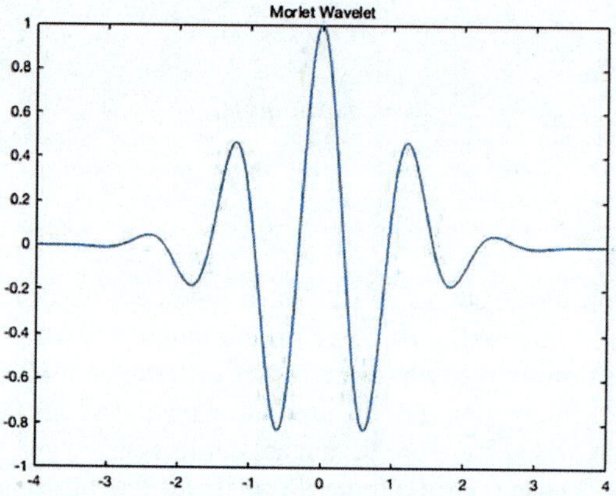

Fig. 4. Morlet wavelet

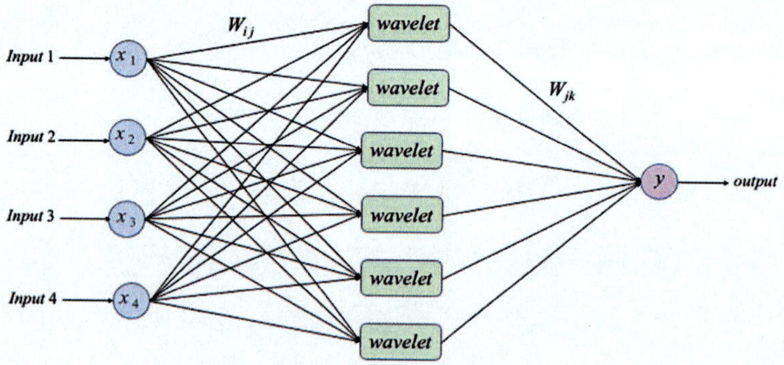

Fig. 5. Structure of WNN

Give weight W_{ij} to X_i and input it into the middle layer, the values of all input nodes are collected into each neuron of middle layer, and the input value of each neuron is $\sum_{i=1}^{k} X_i W_{ij}$. For each neuron j, the distance of this input from b_j determines the output of this neuron. Finally, the output of all neurons is given weight W_{jk} respectively and output to each node of the output layer. Because the wavelet function can better fit the nonlinear data and characterize the temporal data, WNN has been widely used in the field of prediction.

2 Related Analysis

The singleness of RNN structure and the "sharing of parameters" lead to "gradients vanishment or explosion" with the extension of time, which indicates that the ability of RNN to retain information from a more distant era is limited.

2.1 Gradients Vanishment or Explosion

The "gradients vanishment or explosion" of RNN are caused by the derivation of the loss function in the process of using "gradient descent" strategy to update the parameters W_i. This process is to constantly update W_i depending on the partial derivative of current loss function over W_i. Since loss function of RNN is determined by the output at each epoch, and the output is a composite function composed of historical state S_t and historical input X_t, it is necessary to calculate the composite derivative over every S_i that is far away from now. According to the structure shown in Fig. 1, assume that the activation function of the intermediate state is $f = 1$, and the state and output of each epoch are shown below. Where, S_{t-1} is the state of the previous time, b_1 and b_2 is the offset.

$$S_t = W_s S_{t-1} + W_x X_t + b_1 \tag{8}$$

$$O_t = W_o S_t + b_2 \tag{9}$$

Assume that the actual output is y_t, the loss function is obtained easily as shown in formula (10).

$$L_t = Y_t - O_t \tag{10}$$

When "gradient descent" strategy is used to update W_i in RNN, calculate the partial derivatives of current loss function over W_s, W_x and W_o, as formula (11), (12) and (13). Here L_t is a composite function, W_s, W_x and W_o are regarded as independent variable, W_o is only related to the output, but W_s and W_x are included in the subfunctions of every historical state. Therefore, it is necessary to calculate each partial derivative of S_i over W_s and W_x. Firstly, calculate the partial derivative from the outer function to the inner function, and then the partial derivatives of internal and external functions are assembled. With the increase of the partial derivative of the current loss to the previous parameters, the length of the formula increases, and the intermediate polynomial increases exponentially. Finally, the phenomenon of "gradients vanishment or explosion" will appear.

For LSTM, the partial derivative formula of loss to the previous parameters is more complex, because there are more parameters in LSTM. The "adjustability of parameters" enables neural networks to adjust f_t, i_t, o_t, $\approx C_t$ by adjusting W_i and b_i autonomously in the training process, these parameters can help avoid "gradients vanishment/explosion" eventually.

Furthermore, these parameters can even independently decide when the gradient should "disappear" to weaken its influence, or when the gradient should be "exploded" to enhance its influence at a certain moment. In this way, information with long time distance achieves the optimal output.

2.2 Bio-LSTM

Bio-LSTM has been used for pedestrian three-dimensional walking posture and gait prediction [15]. In fact, a two-layer stacked LSTM recurrent neural network followed by a densely connected neural network (NN) layer are implemented as the basic network architecture of Bio-LSTM, and the second layer of the stack is Bi-LSTM.

3 Experiment

3.1 Characteristics of Traffic Flow

Traffic flow refers to the amount or equivalent of traffic entities passing through a certain place or section of the road in a selected period. Like the analysis to evolution characteristics of people flow shown in [16, 17], predicting traffic volume of the target region accurately is the basis of formulating reasonable management strategies, which is of significance to reduce congestion and improve traffic efficiency and safety. Traffic flow on each road has certain characteristics affected by the environment and function of the road. At the same time, traffic flow also contains strong randomness and fluctuation due to the disturbance of weather, road grade, pedestrian behavior, and so on. It can be inferred that the traffic flow has characteristics of randomness, periodicity, space-time, and network.

3.2 Experiment Data Pre-preprocessing

The selected data takes the traffic volume of a tunnel section of highway in Shanghai urban area every 5 min as a group. And each group of data is divided by lane, includes the volume of different types of passenger and vehicles. In the experiment, three groups of data on the same section are selected to increase the repeatability of the experimental results. Use MATLAB processing the data, considering that the main vehicle types in the road are minivan and minibus, so other types of vehicles can be ignored, according to the standard vehicle conversion coefficient, the minivan is taken as 1.5 and the minibus is taken as 1, the number of standard vehicles in a single lane is calculated, and then the volumes of each single lanes are added together, forming 2015 groups of data as the input. Before inputting the neural network, normalize the data to make the data range between $[-1, 1]$ for neural network training.

3.3 Bio-LSTM and WNN Rraining

Divide the data into training set and test set. Take the first 90% of the data as the training set and the rest as the test set. For Bio-LSTM, when the input is x_1, the predicted output is h_1, corresponds to actual data x_2, so the training set is divided into two groups, the first group of data is from the first training data to the penultimate training data, the second group of data is from the second training data to the last training data, each output of the first group of data corresponds to the second group of data, the test set is similarly. The dimension of traffic volume is 1, so the output dimension of Bio-LSTM is set to 1. Set the parameters of hidden units in Bio-LSTM, the number of hidden units of LSTM and Bi-LSTM is set to 100, which determines the prediction effect. Considering that this number is the dimension of x_t and h_t is multiplied by the number of layers of the four "gates", set it as 540. The training process is shown in Fig. 6, through 150 rounds of iteration, RMSE and the loss each round tend to converge.

WNN is used to predict with the same data. The data division method is same as Bio-LSTM, but different from it, WMM only takes historical data within a certain time step from this moment for input, the more historical data taken each time, the higher the accuracy of prediction results, but the calculation will be more complex. Here, 4 historical data are taken for each prediction, that is, when the input is x_1 to x_4, the prediction output y_1 is the prediction of the future time data x_5. Therefore, the data of the training set is divided into two groups, the first group is the first data to the penultimate data, the second group is the fifth data to the last training data, and so is the test set. Set the number of WNN neurons as 6, or cluster the input data by clustering method

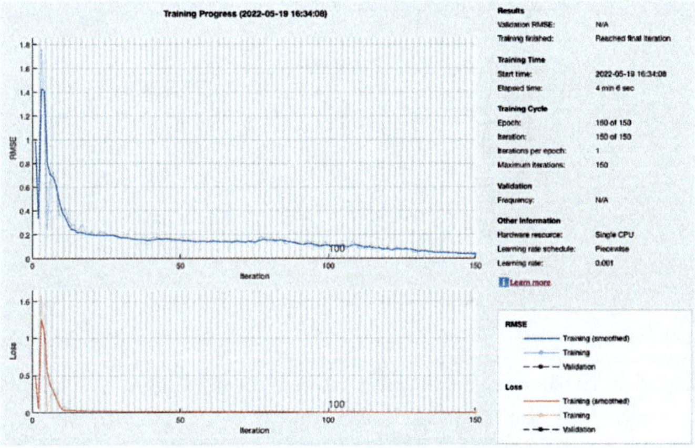

Fig. 6. Bio-LSTM training process

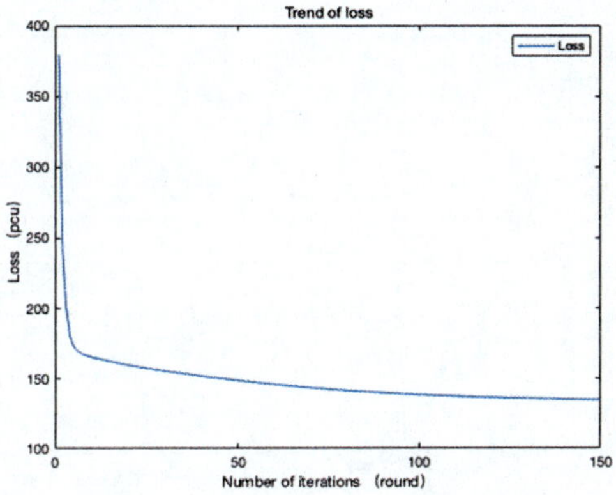

Fig. 7. Convergence process of Loss

and take the number of cluster centers as the number of neurons. The training process is shown in Fig. 7, through 150 rounds of iteration, the loss of the neural network tends to converge.

3.4 Prediction Results

The prediction result using Bio-LSTM is shown in Fig. 8. The blue line of Fig. 8 is the volume of training set, and the yellow part is test set.

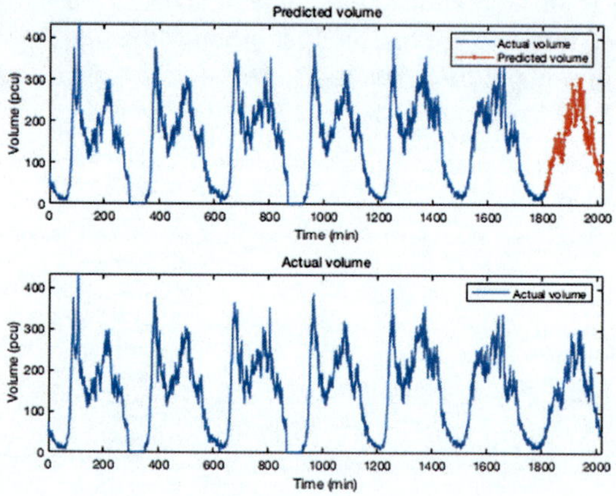

Fig. 8. Bio-LSTM prediction result

WNN is used to predict with the same set of data, as shown in Fig. 8, which is the same as Fig. 8. In Fig. 9, the blue line is the volume of the training set, and the yellow part is the test set.

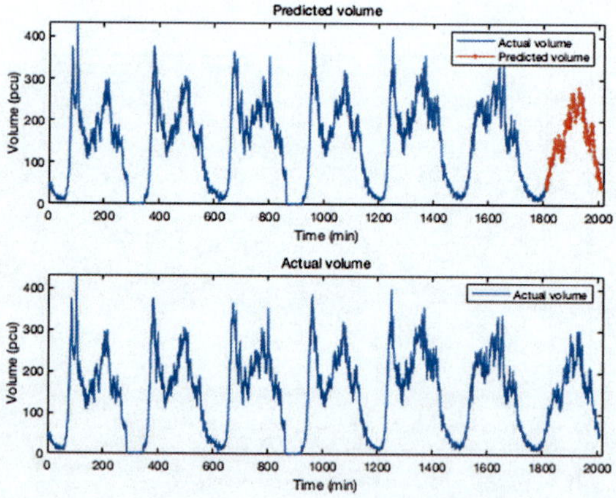

Fig. 9. WNN prediction result

3.5 Evaluation

There are mainly three metrics for evaluation, they are MAE, MAPE, and RMSE respectively. Figure 10 and Fig. 11 compared the predicted results with their actual data, RMSE of the two methods are: Bio-LSTM 7.3476 and WNN 22.65, but the sum of the difference between predicted results with actual data is obvious, Bio-LSTM is −6.1746 and WNN is −51.1278.

The two indexes of Bio-LSTM are smaller than those of WNN. In addition, it can also be seen from the figures below that Bio-LSTM has higher data fitting than WNN.

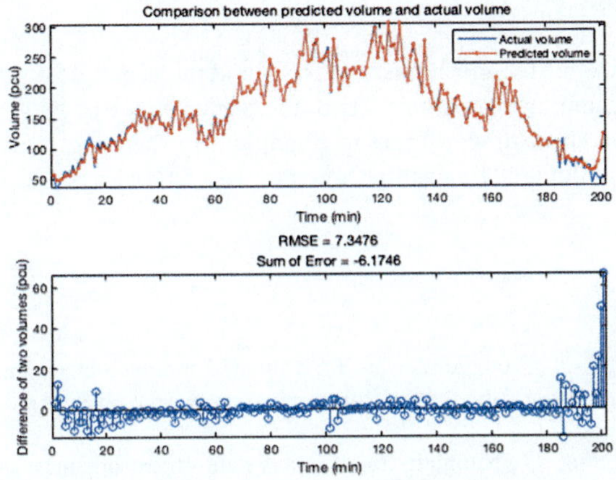

Fig. 10. Comparison of Bio-LSTM prediction result

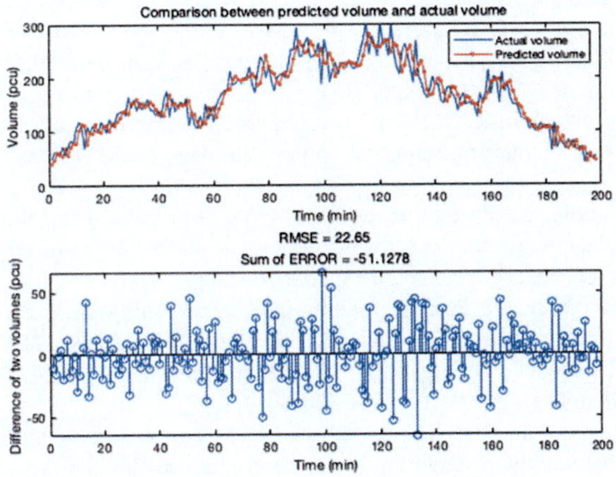

Fig. 11. Comparison of WNN prediction result

The other three groups of volume are used to predict with these two methods, and the RMSE and the difference between predicted results with actual data of every group of data are compared. The results are shown in Table 1, Bio-LSTM has better performance.

Table 1. Comparison of prediction effect of two methods.

No	Bio-LSTM		WNN	
	RMSE	Difference	RMSE	Difference
1	6.4264	−43.0215	23.777	−101.1888
2	7.7584	−34.1309	23.6602	−115.0928
3	7.2853	43.7925	23.3560	−165.2145

4 Conclusions

In this paper, we deduce the mechanism of RNN and the principle of LSTM overcoming "gradients vanishment or explosion". Then, the performance of Bio-LSTM is tested by using a road traffic short-time volume in Shanghai. In the experimental, WNN is also used for the same volume. The results show that Bio-LSTM has excellent performance in prediction.

References

1. Zilin, H., Lunhui, X., Yongjie, Li., et al.: Citywide metro-to-bus transfer behavior identification based on combined data from smart cards and GPS. Appl. Sci. **9**(17), 3597 (2019)
2. Ling, H., Hengcong, G., Ronghui, Z., et al.: A data-driven operational integrated driving behavioral model on highways. Neural Comput. Appl. **32**(16), 13017–13033 (2020)
3. Navin Kumar, M.: Deep Learning with Applications Using Python. Springer, Berkeley (2018). https://doi.org/10.1007/978-1-4842-3516-4
4. Zheng, Z., Weihai, C., Xingming, W., et al.: LSTM network: a deep learning approach for short-term traffic forecast. IET Intell. Transp. Syst. **11**(2), 68–75 (2017)
5. Yan, T., Kaili, Z., Jianyuan, L., et al.: LSTM-based traffic flow prediction with missing data. Neurocomputing **318**(27), 297–305 (2018)
6. Dong-Hoom, S., Kyungyong, C., Roy, C.: Park prediction of traffic congestion based on LSTM through correction of missing temporal and spatial data. IEEE Access **8**, 150784–150796 (2020)
7. Zhiyong, C., Ruimin, K., Ziyuan, P., et al.: Stacked bidirectional and unidirectional LSTM recurrent neural network for forecasting network-wide traffic state with missing values. Transp. Res. Part C Emerg. Technol. **118**, 102674 (2020)
8. Pan, W., Zilin, H., Yuzhuang, P., et al.: A Combined deep learning method with attention-based LSTM model for short-term traffic speed forecasting. J. Adv. Transp. **2020**, 0–15 (2020)
9. Wangyang, W., Honghai, W., Huadong, M.: An autoencoder and LSTM-based traffic flow prediction method. Sensors **19**(13), 2946 (2019)
10. Ling, H., Hengcong, G., Ronghui, Z., et al.: LSTM-based lane-changing behavior model for unmanned vehicle under environment of heterogeneous human-driven and autonomous vehicles. China J. Highw. Transp. **33**(7), 156–166 (2020)
11. Yuehui, C., Bo, Y., Jiwen, D.: Time-series prediction using a local linear wavelet neural network. Neurocomputing **69**(4–6), 449–465 (2006)
12. Song, X., Zhou, H., Liu, G.: Human behavior recognition based on multi-feature fusion of image. Clust. Comput. **22**(4), 9113–9121 (2018). https://doi.org/10.1007/s10586-018-2073-7

13. Sepp, H., Jürgen, S.: Long short-term memory. Neural Comput. **9**(8), 1735–1780 (1997)
14. Huiying, W., Dongran, Z.: Highway traffic volume prediction based on Bi-LSTM model. Highw. Eng. **44**(06), 51–56 (2019)
15. Xiaoxiao, D., Ram, V., Matthew, J.R.: Bio-LSTM: a biomechanically inspired recurrent neural network for 3-D pedestrian pose and gait prediction. IEEE Robot. Autom. Lett. **4**(2), 1501–1508 (2019)
16. Zilin, H., Lunhui, X., Yongjie, L.: Multi-stage pedestrian positioning using filtered Wi-Fi scanner data in an urban road environment. Sensors **20**(11), 3259 (2020)
17. Zilin, H., Xianyuan, Z., Yongjie, L., et al.: A novel WIFI-oriented RSSI signal processing method for tracking low-speed pedestrians. In: 5th International Conference on Transportation Information and Safety (ICTIS), Liverpool, UK, pp. 1018–1023. IEEE (2019)

Application of BIM in Automatic Typesetting of Universal Segments in Shield Tunnel

Liujie Zhu[1], Yuxian Zhang[2(✉)], Jixun Zhang[2], and Zichang Ma[2]

[1] Water Conservancy and Irrigation District Engineering Construction Administration of Xixiayuan Project, Zhengzhou, Henan, China
[2] Hohai University, Nanjing 210098, Jiangsu, China
hhuzhangyuxian@163.com

Abstract. Shield method is an advanced construction method of underground engineering, and the typesetting of segments is an important procedure in the construction process. The traditional segment design method is not easy to modify, low efficiency. This study proposes a universal segment automatic typesetting method based on BIM technology. Firstly, a parametric lining ring template is established in the BIM platform according to the segment geometry. A python script was created to realize the secondary development of the BIM platform and then select the appropriate point of the lining ring based on the distance between the center point of the lining ring and the tunnel axis. A shield tunnel project is used to evaluate the automatic segment assembly algorithm. The results show that the proposed algorithm can quickly establish a shield tunnel model, which improves the design efficiency of segment to a certain extent. The method proposed in this paper has certain significance for guiding the design of the segment and the assembly in the construction process.

Keywords: Shield tunnel · BIM · Segment typesetting

1 Introduction

BIM is to provide a digital management method for the whole life cycle of project design, construction, operation, and maintenance [1]. Over the past few years, the characteristics of BIM technology such as synergies, parameterization, and visualization have shown great potential in the AEC (Architecture, Engineering & Construction) industry [2]. For example, during the design phase of the new HQ project of KLH in South Korea, BIM was used to establish a three-dimensional model and shop drawings, and to simulate possible collisions during the construction process [3]. State Grid Research Center of Jiangsu province of China had developed a BIM-based collaborative platform for different disciplines during the construction phase [4]. With the digital transformation of the AEC industry, the application of BIM technology is only a starting point [5].

Different from the construction industry, the construction of tunnel projects often has to traverse different strata, and the information obtained during the design and construction phases is usually limited [6]. Tunnel construction can be divided into conventional

W. Guo and K. Qian (Eds.): GBCESC 2022, LNCE 211, pp. 496–503, 2023.
https://doi.org/10.1007/978-981-19-5217-3_49

construction and shield method according to whether TBM is adopted. Shield method is an advanced construction method developed with the level of modern transportation, which has the characteristics of fast construction speed and strong adaptability to the ground. During the construction of the shield method, the prefabricated segments are assembled to form the tunnel axis (FTA) according to the alignment of design tunnel axis (DTA). Due to the characteristics of the prefabricated segment ring, there is bound to be a deviation between FTA and DTA. Normally, a lining ring is composed of 6–10 segments. By adjusting the attitude of the lining ring, the axis with the smallest deviation from the DTA can be fitted. In the current design phase, the lining ring is still mainly two-dimensional, which is inconvenient to express the typesetting information of the lining segment [7]. During the construction phase, due to the uncertainty of the stratum and the possible float of the lining of the as-built tunnel, the DTA needs to be adjusted, and the re-modification of the two-dimensional drawing will consume a lot of time and cost, resulting in a decrease in construction efficiency. In addition, the attitude of the shield machine is closely related to the operator's technology. The assembly point can only be adjusted depending on the operator's experience when axis deviation occurs.

BIM technology provides a new way for the design and typesetting of shield tunnel segments. The traditional design method of shield assembly point requires the DTA to be divided into different alignments, and the minimum distance to the design assembly point is calculated according to the different attitudes of the lining segment. This usually requires a lot of matrix operations, especially in long-distance diversion tunnels, which are difficult to modify when the axis changes. The purposes of this study are to use BIM technology to establish the parameterized geometric model of the segment ring and the method of automatic typesetting.

2 Literature Review

Shield segments can be divided into ordinary segments and universal segments. The commonly used universal segment is a double-sided wedge-shaped isosceles trapezoidal ring (see Fig. 1). During the construction phase, FTA is formed by assembling segments under the condition of specific assembly points. Therefore, the assembling of shield segments is actually the determination of the assembly point of each ring. Currently, the calculation process of assembly point is to divide the tunnel axis into a vertical alignment and a horizontal alignment. It can be further divided into a straight line, a circular curve, and an eased curve according to the type of curve. The initial point is used as a known condition to calculate the assembly point of the segment according to the geometric relationship of the curve [8–10], and this design method cannot be expressed in an intuitive way. In the construction process, the floating of the segment will cause the deviation of axis, and the assembly points need to be adjusted in time. Although there are empirical formulas to estimate the deviation of the shield segments, they are rarely used in practice [11]. Normally, the adjustment of the attitude of shield machine mainly relies on the experience of the operator, which makes that time-consuming and error-prone.

There have been studies on applying BIM technology to shield construction. Li et al. [12] used CATIA to establish a parametric model of shield tunnel segments and modify the assembly information in a data-driven manner. Zhong et al. [13] extended the description of the shield segment model in the IFC standard, and realized the digital expression of geometric information, which provided support for data exchange between different platforms. Ninić et al. [14] used Revit to realize a parametric model of the shield lining, and established a set of design procedures connected with numerical calculations. Borrmann et al. [15] established shield tunnel models with different levels of detail to realize data transfer between BIM and GIS. Getuli et al. [16] mainly focused on generating as-built models of shield tunnels based on construction information. In short, the current application of BIM technology in the shield tunnel design phase requires the assembly information to be used as a parameter to form a three-dimensional model, which cannot realize the automatic assembly of segments. In this paper, the BIM platform is used to establish the parameterized model of the universal wedge-shaped segment, and the tunnel model is established through secondary development.

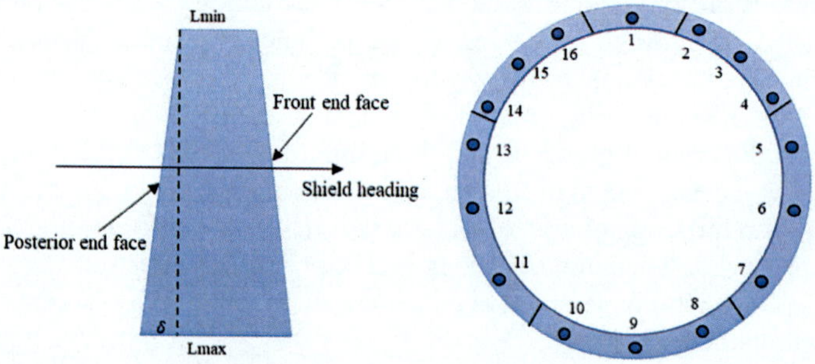

Fig. 1. Universal segment structure.

3 Typesetting Technology of Tunnel Segments

3.1 Universal Segment Geometry Model

BIM technology has the characteristics of parameterization and visualization and provides designers with convenient and flexible design methods. Currently, the commonly used BIM design platforms include AutoDesk, Bentley, and Catia. The tunnel axis is a curve in space. Therefore, we choose the Catia platform, which is more adaptable to curve and surface modeling. The universal wedge ring is composed of a K-block, two B-block, and three to five A-block (see Fig. 2). The number of lining ring longitudinal joint connecting bolts determine the number of assembly points that could be chosen for the lining ring. For example, there are 16 longitudinal joints connecting bolts of the lining ring in Fig. 1, and 16 points can be selected when assembling. Shield tunnel is formed by assembling multiple lining rings, therefore, a lining ring template must be established first. In this paper, the geometric model of lining ring is established in the generative surface design module of Catia. The input conditions of the model include the width of the lining ring (w), the thickness of the segment (t), the wedge angle (β), the number of A-blocks, and the chord length of the segment. The lining ring could be created by the multi-section entity command after the sketches of the front end face and the posterior end face are established. The specific establishment flow of the lining ring template is shown in Fig. 3.

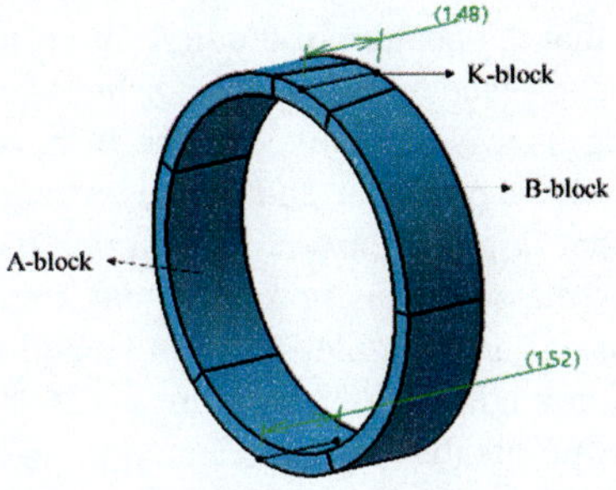

Fig. 2. Three-dimensional geometric model of lining ring.

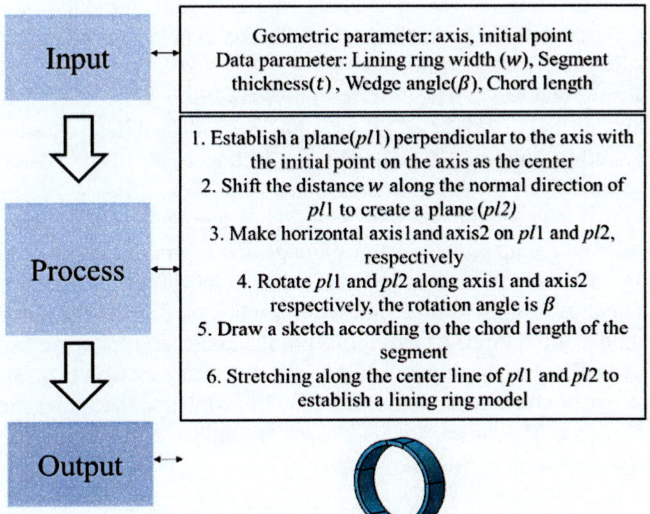

Fig. 3. Establishing flow of lining ring geometric model.

3.2 Automatic Typesetting Technology of Shield Tunnel Segments

As shown in Fig. 1, during the construction of the shield tunnel, the appropriate assembly points of each lining ring are selected to fit the DTA. Therefore, the current segment typesetting is based on the previous ring. In the traditional assembly point design process, according to the alignment of the tunnel axis, coordinate rotation is used to determine the assembly point with the closest distance between the center point of the front end face of the lining ring and the design assembly point. The calculation procedure often involves a large number of matrix operations. It is necessary to divide the three-dimensional axis into a horizontal alignment and a vertical alignment when calculating the design assembly points, and the typesetting process cannot be displayed intuitively.

This study uses BIM technology to realize the automatic typesetting of shield segments under the condition that the posture of the first ring is known. When the previous ring is assembled, the front end face of previous ring should coincide with the posterior end face of the current ring, that is, the center of the front end of the previous ring is consistent with the center of the posterior end of the current ring. When selecting the assembly point of the current ring, the current ring around the normal line of front end face of the previous ring is rotated by an appropriate angle (see Fig. 4). Taking a lining ring with 16 assembly points as an example, assuming that the rotation angle at the first point is 0°, then the rotation angle at the n th point is $(360° \cdot n/16)$. Therefore, there are 16 front end face center points that can be selected for each ring. All the alternative front-end face centers constitute a cone with the front-end face normal of the ring as the axis and the front-end face center of the previous ring as the apex. When the rotation angle of the current ring is determined, the polyline connecting the center of the posterior end face and the front end face of the current ring is used as the polyline of the fitting DTA. In order to decrease deviation from the DTA, each polyline of the FTA must meet the minimum distance from the tunnel axis.

The typesetting process is based on the previous lining ring. Therefore, the center of the posterior end face of the current ring has been determined. Assuming that the current ring assembly point is i, the coordinates of the center of the posterior end face and the point coordinates on the tunnel axis are (x, y, z) and (x_0, y_0, z_0), respectively. The distance (d_{imin}) between the two points could be obtained by Eq. 1. In order to ensure the rigidity of the lining and prevent water seepage, the segment can not produce straight joint. Each ring segment has 6 alternative assembly points, after excluding the assembly points that may cause straight joints. The minimum distances from the tunnel axis could be calculated under the 6 kinds of assembly points, and the point corresponding to the minimum value will serve as the assembly point of the current ring (Eq. 2).

A Python script has been created to realize the insertion and the automatic typesetting of the shield segment. The script flow is shown in Table 1. Catia could be operated by out-of-process access with python script.

$$d = \sqrt{(x - x_0)^2 + (y - y_0)^2 + (z - z_0)^2} \tag{1}$$

$$d_{imin} = min\{d_i\}, i = 1, 2 \ldots, 6 \tag{2}$$

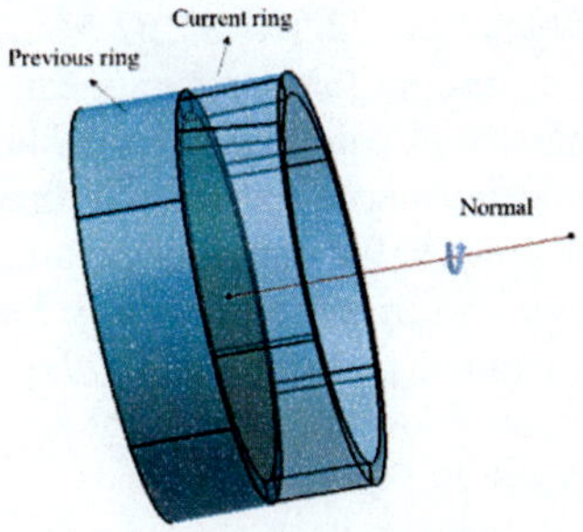

Fig. 4. Schematic diagram of segment typesetting.

Table 1. Algorithm for typesetting of lining segments.

Simulation of lining segment typesetting	
Instantiate lining ring template	
Set the initial assembly point i	
Calculate the shortest distance between the center of the front face and the axis of the tunnel d_{imin}	
For k in { Alternative assembly points } Calculate the shortest distance between the center of the front face and the axis of the tunnel d_{kmin}	
$d_{kmin} < d_{imin}$	
Yes	No
$i = k$	$i = i$
next k	

4 Case Study

The automatic typesetting algorithm of segments proposed in this paper was used in a construction project. The project is located in Henan Province, China. The total length of the tunnel is 2,850 m, and it is constructed shield-driven. Rein-forced concrete was used in this project with an outer diameter of 6600 mm and an inner diameter of 5900 mm, which are assembled in staggered joints. The wedge shape of the tube segment is 39.6 mm, and the thickness and width are 350 mm and 150 mm, respectively.

The sketch of the tunnel axis in the horizontal and vertical planes was drawn in the generative surface design module of Catia, and the tunnel space axis was created by the function of folding the curve. According to the description in Sect. 3.1, a three-dimensional geometric model of the tunnel lining ring is shown in Fig. 2. The lining ring is composed of 1 K-block, 2 B-blocks, and 3 A-blocks. Engineers can directly modify the parameters to drive the model to update automatically when the design size changes. Each lining ring has 16 optional assembly points, so there are 15 optional rotation angles between two adjacent rings, which were 22.5°, 45°, 67.5°…337.5°. During the modeling process, avoid the points where straight joints may appear.

When assembling segments, first determine the posture of the initial lining ring, and select the next ring point according to the normal direction of the front end of the initial ring. Taking the starting point of the tunnel axis as the insertion point of the initial lining ring (see Fig. 5), the python script established in Sect. 3.2 is used to realize the automatic typesetting of the segments, which avoids the calculation of design points and the complicated matrix conversion. The universal segment assembly effect of the established shield tunnel is shown in Fig. 6.

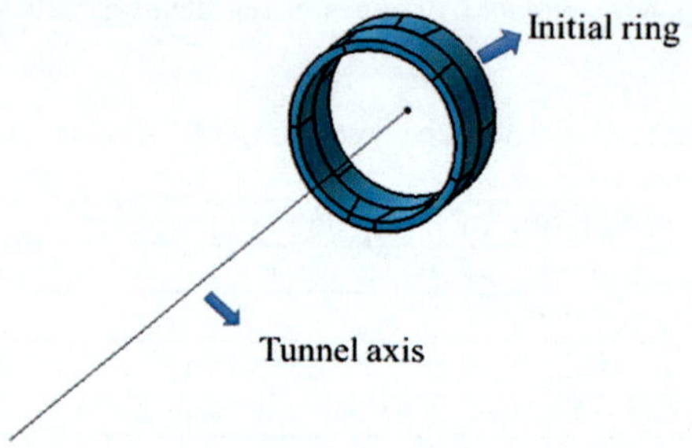

Fig. 5. Initial ring of Tunnel axis.

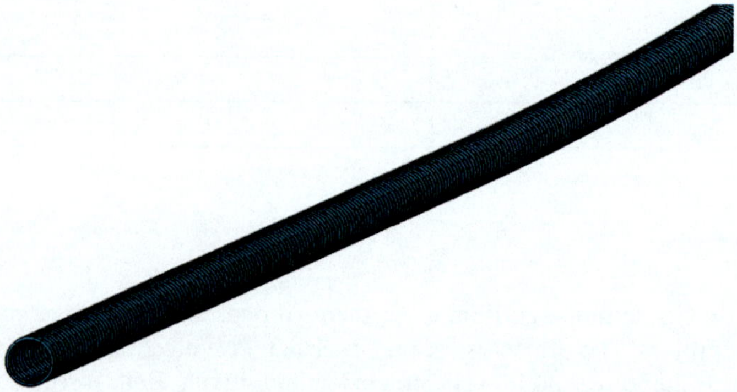

Fig. 6. Universal segments assembly effect.

5 Conclusion

This study focused on the automatic typesetting of universal segments of shield tunnel. Aiming at difficult modification and large calculation in the geometric model and typesetting of traditional shield tunnel segment design, BIM-based technology to realize segment modeling and automatic typesetting was proposed. Firstly, a double-sided wedge-shaped universal segment model was established on the Catia platform to realize automatic update of the parameter-driven model. According to the geometric relationship between the annulus of lining ring and the tunnel axis, a python script was proposed to select appropriate assembly points. The proposed segment assembly algorithm is verified through a shield tunnel project. The result shows that the tunnel model established by the algorithm is easy to modify and can be displayed on the BIM platform in an intuitive manner, which improves the design of shield segments to a certain extent efficient.

References

1. Bryde, D., Broquetas, M., Volm, J.M.: The project benefits of building information modelling (BIM). Int. J. Proj. Manag **31**, 971–980 (2013)
2. Azhar, S.: Building information modeling (BIM): trends, benefits, risks, and challenges for the AEC industry. Leadersh. Manag. Eng. **11**, 241–252 (2011)
3. Kim, S., Chin, S., Han, J., Choi, C.: Measurement of construction BIM value based on a case study of a large-scale building project. J. Manag. Eng. **33**, 5017005 (2017)
4. Li, L., Yuan, J., Tang, M., Xu, Z., et al.: Developing a BIM-enabled building lifecycle management system for owners: architecture and case scenario. Autom. Constr. **129**, 103814 (2021)
5. Zabin, A., González, V.A., Zou, Y., Amor, R.: Applications of machine learning to BIM: a systematic literature review. Adv. Eng. Inform **51**, 101474 (2022)
6. Raymond, L.S.: The development of underground engineering improves the quality of modern life. Eng. PRC **3**, 9–12 (2017)
7. Luo, H., Li, L., Chen, K., Parametric modeling for detailed typesetting and deviation correction in shield tunneling construction. Autom. Constr., 104052 (2021)
8. Zhang, Z.: Research on optimized typesetting method of shield segments. Master thesis. Huazhong University of Science and Technology, Wuhan, Hubei, China (2012)
9. Lin, T.W.: Research on key technology of shield tunneling attitude control. Doctor thesis. Zhejiang University, Hangzhou, Zhejiang, China (2014)
10. Rui, H.S., Xing, C.H., Han, Z.L., Ji Wen, Z.: Research and development of general segment typesetting and deviation correction software for shield tunnels. J. Undergr. Sp. Eng. **5**(03), 557–562 (2009). (in Chinese)
11. Wang, P., Kong, X., Guo, Z., Hu, L.: Prediction of axis attitude deviation and deviation correction method based on data driven during shield tunneling. IEEE Access **7**, 163487–163501 (2019)
12. Yong, M.L., Kai, T.Z., Zhe, L.Z., et al.: Parametric modeling and segment layout of wedge shield tunnel segment based on software CATIA. Tunn. Constr. **39**(3), 391–397 (2019). (in Chinese)
13. Yu, Z., Shao, D.Z., Jian, C., Gou, L.Z., Jia, M.W.: Research on data model of shield tunnel structure based on IFC standard. J. Undergr. Sp. Eng. **13**(S2), 613–622 (2017). (in Chinese)
14. Ninić, J., Bui, H.G., Meschke, G.: BIM-to-IGA: a fully automatic design-through-analysis workflow for segmented tunnel linings. Adv. Eng. Inform. **46**, 101137 (2020)
15. Borrmann, A., Kolbe, T.H., Donaubauer, A., Steuer, H., et al.: Multi-scale geometric-semantic modeling of shield tunnels for GIS and BIM applications. Comput. Aided Civ. Inf. **30**, 263–281 (2015)
16. Getuli, V., Capone, P., Bruttini, A., Pour Rahimian, F.: On-demand generation of as-built infrastructure information models for mechanised tunnelling from TBM data: a computational design approach. Autom. Constr. **121**, 103434 (2021)

Rating Techniques Research on Operation Performance of Continuous Box Beam with Common Span in China High Speed Railway

Wei Wang, Penghui Liu, Xin Meng(✉), Zhensheng Dong, Jing Yin, and Yigan Wang

Railway Engineering Research Institute, China Academy of Railway Sciences Corporation Limited, Beijing 10081, China
sdjxmx@sina.com

Abstract. Rating techniques for operation performance of continuous box beam with common span are discussed based on measured data and theoretical research on dynamic performance of bridge in China high speed railway in recent years. Rating tasks are analyzed and the main technical parameters to evaluate the operation performance of continuous box beam are proposed. According to the statistical analysis on measured data of the bridge dynamic performance by the testing and commissioning of high speed railway with the reliability of 97.5%, the suggested ordinary value of main parameters of prestressed concrete continuous box beam under 250 and 350 km/h high speed train operation speed are put forward: the vertical natural frequency is determined by the power function of span, the vertical damping ratio is 0.5% to 2.0%, the ratio of deflection to span is 1/5500 to 1/3500 and 1/6000 to 1/4000 respectively, the vertical rotation angle at girder end is 0.60‰ and 0.50‰ rad respectively, the dynamic factor is 1.17 to 1.26 and 1.20 to 1.30 respectively, the vertical and lateral amplitude at midspan of beam is 0.25 to 0.80 and 0.10 to 0.15 mm respectively, the vertical vibration acceleration at midspan of beam is 0.25 to 0.40 m/s^2, the lateral amplitude at pier top is chosen as the ratio of the height of pier and the lateral average width of the pier in the range of 0.5 to 4.2, the relative lateral displacement of rail support on both sides of adjacent girder end of ballastless track is 0.5 mm.

Keywords: Continuous box beam · Operation performance · High speed railway

1 Introduction

High speed railway is a new passenger dedicated railway designed to multiple unit trains of 250 km/h (including reserved) and above, and the initial operation speed is not less than 200 km/h. The design grades of China high speed railways are mainly 250 km/h and 350 km/h.

In order to ensure the high reliability of the line and the comfort of passengers, and achieve the purposes of saving land and protecting the environment, China high speed

W. Guo and K. Qian (Eds.): GBCESC 2022, LNCE 211, pp. 504–518, 2023.
https://doi.org/10.1007/978-981-19-5217-3_50

railways largely adopt the form of "replacing roads with bridges", and the specific weight of bridges on some lines is more than 80% [1]. China high speed railway bridges with the design speed 250 km/h can be divided into two types: one is the high speed railway that takes into account freight cars with ZK live load and CS-live load adopted as the design live load; the other is passenger dedicated line railway with ZK live load adopted as the design live load; and ballasted or ballastless track is adopted. The bridge design live load for 350 km/h is ZK live load, which is generally ballastless track.

From the perspective of bridge structure, prestressed concrete box girder accounts for the majority. In addition to specially designed structures, large-span continuous box beam is often used when crossing valleys, rivers, railways, roads and other special sections. The general drawing of cast-in-situ prestressed concrete continuous box girder, including spans (32 + 48 + 32)m, (40 + 56 + 40)m, (40 + 64 + 40)m, (48 + 80 + 48)m, (60 + 100 + 60)m prestressed concrete double-line continuous box beam with different section heights, was prepared by CHINA RAILWAY (former Ministry of Railways).

The rating of technically complex and important bridge shall be conducted no less than once every ten years specified by *Regulations on Railway Technology Management (high speed railway)* [2]. It has been 13 years since the opening of Beijing Tianjin Intercity Railway in August 2008. It is urgent to carry out the rating of high speed railway bridges.

The existing *code for rating existing railway bridges* (TYH [2004] No. 120) is applicable to the *rating* of existing bridges on mixed passenger and freight railways with a maximum running speed of 160 km/h; When the maximum speed of passenger train is 200 km/h, it can be implemented by reference [3]. However, there are no relevant regulations on the test and evaluation of bridge operation performance according to the characteristics of high speed railway train operation and bridge structure.

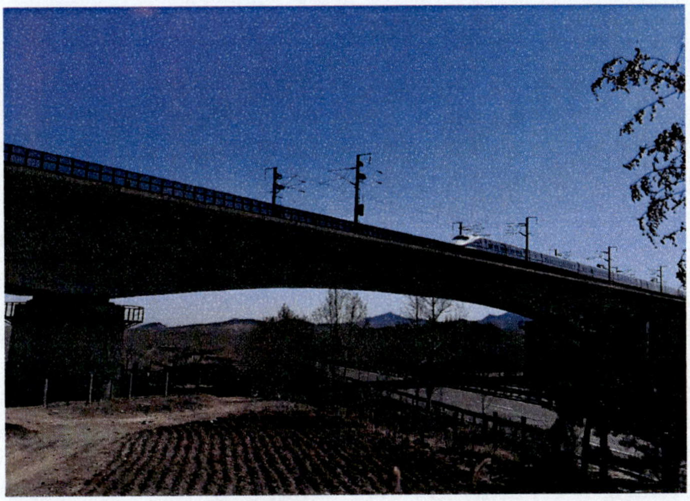

Fig. 1. Field test of (60 + 100 + 60)m prestressed concrete double-line continuous box beam

In this paper, the object of high speed railway bridge operation performance rating is proposed combining with the characteristics of high speed railway bridges and rating practice of general railway bridge in China. Based on related theoretical research and measured data on dynamic performance of high speed railway bridge carried out in China in recent years, the main technical parameters that should be paid attention to in the evaluation of the operation performance of continuous box beam (Fig. 1) with common span are analyzed. The ordinary value of technical parameters are proposed considering the design speed grade and so on, which may be used as evaluation basis for the operation and maintenance of the common span continuous box beam of China high speed railway.

2 Task of Operation Performance Rating

The Chinese Railway Department issued the *code for rating existing railway bridges* in 1978, which was revised in 2004 based on the summary of rating practice and the analysis of test data. The key objects that should be paid attention to in the rating of existing railway bridges were given in the 2004 specification, including bridges whose operation performance needs to be determined due to train speed increase or overload, bridges with damage, etc. [3]. The operation performance rating of existing railway bridges provides an important means for comprehensively judging the operation state of bridge structures and determining the operating conditions of bridges. The operation performance rating object of high speed railway bridge shall be determined for the purpose of comprehensively judging the operation state of bridge structure and in combination with specific operation requirements, which can be roughly divided into the following categories:

1. Periodically rating bridges. A certain number of prestressed concrete double track continuous box girders are selected for periodic rating to determine the actual working state of the bridge, which can not only ensure the normal operation, but also accumulate the basic data of the bridge operation state.
2. Bridges with increased operation speed of MUs (the maximum operation speed shall not exceed the design speed). At present, the maximum operation speeds of some high speed railways with design speeds of 250 km/h and 350 km/h in China at the initial stage of operation are 210 km/h and 310 km/h respectively. If the MU operation speed increases but does not exceed the design speed, representative bridges shall be selected for operation performance rating according to defects and diseases of bridge structure, combined with daily operation conditions, to determine whether the bridges meet the operation requirements.
3. Bridges with abnormal operation status, major defects, damage, etc. The abnormal operation status of bridges or major defects or damages directly affect the operation of MUs. The operation performance rating of these bridges can provide a basis for the formulation of subsequent operation measures.

3 Parameters of Operation Performance Rating

When the train passes through the bridge at high speed, the dynamic response of structure is intensified, and the stability of train operation and the comfort of passengers are more important. The selection of operation performance evaluation parameters of high speed railway bridges can refer to the design specifications. The corresponding vertical and lateral evaluation parameters should ensure the stability of track structure on the bridge, so as to ensure the stability and ride comfort of trains crossing the bridge [4].

3.1 Vertical Rating Parameters of Bridge

Studies have shown that the low natural vibration frequency of the beam will lead to excessive vibration or resonance when the MU passes through at high speed, which will not only affect the stability and ride comfort of the train on the bridge, but also cause damage to the bridge structure itself. Vertical stiffness of the beam, including the deflection and vertical rotation angle at girder end, is the main static index affecting the track structure and ride comfort of train on the bridge [4, 5]. For the continuous box beam with ballasted track, the excessive vertical vibration acceleration of the beam deck will lead to the pulverization of the ballast on the bridge and affect the stability of the track roadbed. For the continuous box beam with ballastless track, the excessive vertical vibration acceleration of the beam deck will affect the stability of the ballastless track structure. Therefore, the vertical evaluation parameters of bridge should include the vertical natural vibration characteristics of the beam, the vertical stiffness of beam and the vertical dynamic response [4–11]. The main purpose are listed in Table 1.

3.2 Lateral Rating Parameters of Bridge

The lateral stiffness and torsional stiffness of continuous box beam used in China high speed railway are significantly greater than that of simple supported T girder in general railway. Previous studies have shown that the lateral and torsional limits of the beam do not control the design of the bridge structure, however, excessive lateral displacement of the beam will cause significant track irregularities, which will affect the operation stability and ride comfort of vehicle. Generally, it is difficult to measure the lateral displacement of the beam, and the method of measuring the latearl vibration of the beam and pier is usually used to evaluate the lateral state of the bridge. For the bridge with ballastless track, the lateral relative displacement between two rail supports separately at adjacent beam ends shall be evaluated [4–11]. The main purpose are listed in Table 1.

Table 1. Technical parameters and purpose to evaluate the operation performance of continuous box beam with common span

	Parameters		Purpose
Vertical	Vertical natural vibration characteristics of beam	Vertical natural frequency	To avoid resonance of vehicle and bridge and excessive vibration
		Damping ratio	
	Vertical stiffness of beam	Vertical deflection	To ensure passenger comfort and stability of trains at high speed
		Vertical rotation angle at girder end	To the static strength of fastening system of ballastless track at girder end and ensure the stability of ballasted track
	Vertical dynamic response	Dynamic factor of operating trains	To avoid excessive vibration of bridge
		Vertical amplitude of beam	
		Vertical acceleration	To ensure the stability of ballastless track on bridge and avoid ballast pulverization of ballasted track
Lateral	Lateral amplitude of beam		To avoid excessively large track alignment irregularity that affect the passenger comfort and stability of train
	Lateral amplitude of pier		
	Lateral relative displacement between two rail supports at adjacent beam ends of ballastless track		

4 Ordinary Value of Operation Performance Evaluation

Ordinary value and limit value for traffic safety are used to evaluate the operation performance of bridge in *Code for rating existing railway bridges* [2]. The ordinary value is the upper limit of deflection, vertical rotation angle at girder end, amplitude, acceleration, dynamic factor, the lower limit of frequency and the range of damping ratio [5], which were measured during operation performance of bridge.

Based on the test samples of dynamic performance of high speed railway bridge, the maximum inferred value is calculated according to the credibility of 97.5%, and the recommended ordinary value for the evaluation of operation performance parameters of double line continuous box beam with common span are proposed.

4.1 Vertical Natural Frequency and Damping Ratio of Beam

The track structure type has a great influence on the vertical natural frequency of continuous box beam. Therefore, the relationship between the measured vertical natural frequency and span are analyzed according to linear, exponential, logarithmic, and power regression respectively considering the design speed of the bridge and the type of track structure. The results show that the vertical natural frequency of the beam and the span are a power function. The relationship between vertical natural frequency and span of prestressed concrete double line continuous box beam with common span used in 250 km/h high speed railway with ballasted track and 350 km/h high speed railway with ballastless track is shown in Fig. 2 and Fig. 3 respectively. The vertical natural frequency of the beam is inferred according to the confidence level of 97.5%, and the inferred lower limit is taken as the proposal ordinary value. The results are shown in Table 2.

Table 2. Proposal ordinary value of vertical natural frequency of continuous box beam with common span

Bridge design speed (km/h)	Type of track structure	First-order frequency (Hz)	Second-order frequency (Hz)	Available span (m)
250	Ballasted track	$340/L^{1.2}$		$32 \leq L \leq 100$
350	Ballastless track	$400/L^{1.2}$		

Note: 1 The unit of span L is m.
2. The ordinary value in the table are applicable to the continuous box gbeam with three spans and one connection using the general reference drawing. The ordinary value of the first-order vertical natural frequency is calculated by the middle span, and the ordinary value of the second-order vertical natural frequency is calculated by the side span.

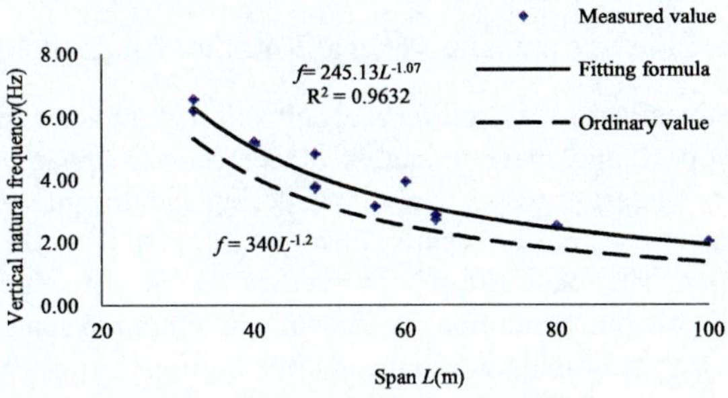

Fig. 2. Relation between vertical natural frequency and span of continuous box beam with ballasted track used in 250 km/h high speed railway

In the test, the vertical damping ratio of beam is usually analyzed according to the assumption of viscous damping. Statistical analysis is made based on the measured first-order vertical damping ratio of more than 30 prestressed concrete double line continuous box beam with common span of multiple high speed railways. The inferred lower limit

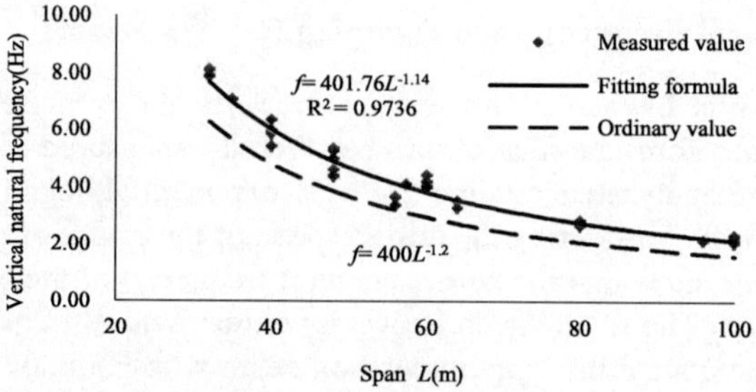

Fig. 3. Relation between vertical natural frequency and span of continuous box beam with ballastless track used in 350 km/h high speed railway

value and upper limit value range are taken as the proposal ordinary value. The results are shown in Table 3.

Table 3. Statistical table of first vertical damping ratio of continuous box beam with common span

Bridge design speed (km/h)	Type of track structure	First vertical damping ratio				
		Average value	Variance	Inferred lower limit	Inferred upper limit	Proposal ordinary value
250 and 350	Ballasted/ballastless track	1.25%	0.00168%	0.45%	2.05%	0.5%–2.0%

4.2 Ratio of Deflection to Span and Vertical Rotation Angle at Girder End

The relationship between the measured vertical deflection of midspan (deducting vertical displacement of support) and span are analyzed according to linear, exponential, logarithmic, and power regression respectively considering the design speed of the bridge and the type of track structure. The results show that the vertical deflection of mid span is linear with the span. The relationship between the vertical deflection and span of pre-stressed concrete double line continuous box beam with common span used in 250 km/h high speed railway with ballasted track and 350 km/h high speed railway with ballastless track is shown in Fig. 4 and Fig. 5 respectively. The vertical deflection of midspan is inferred according to the confidence level of 97.5%, and the ratio of deflection to span converted form inferred upper limit is taken as the proposal ordinary value. The results are shown in Table 4.

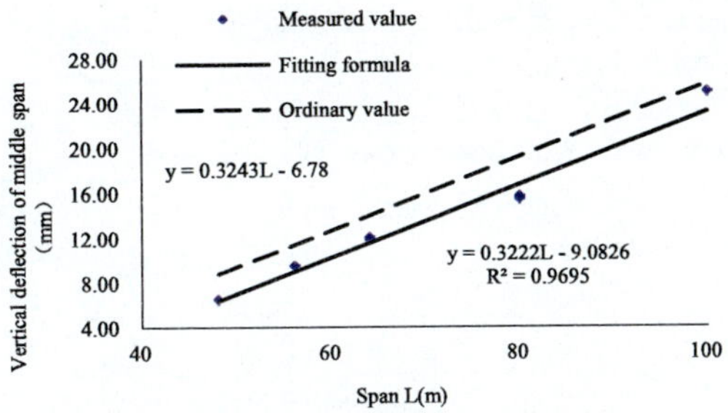

Fig. 4. Relation between vertical deflection of middle span and span of continuous box beam with ballasted track used in 250 km/h high speed railway

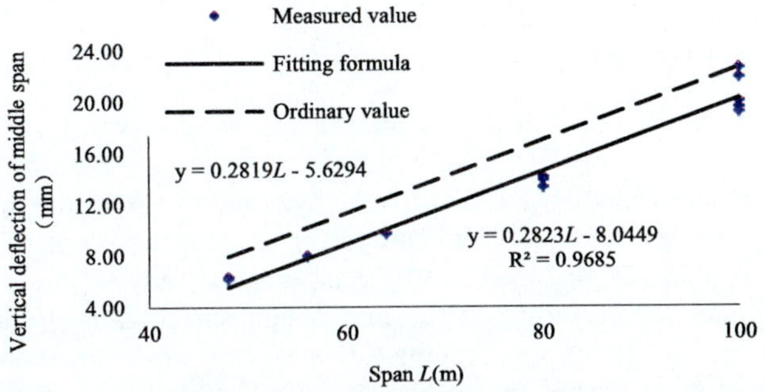

Fig. 5. Relation between vertical deflection of middle span and span of continuous box beam with ballastless track used in 350 km/h high speed railway

Table 4. Proposal ordinary value of ratio of deflection to span of continuous box beam with common span (conversation to ZK Live Load)

Bridge design speed (km/h)	Type of track structure	Span (m)				
		48	56	64	80	100
250	Ballasted track	1/5500	1/5000	1/4500	1/4000	1/3500
350	Ballastless track	1/6000	1/5500	1/5000	1/4500	1/4000

Statistical analysis is made based on the measured vertical rotation angle at girder end of more than 30 prestressed concrete double line continuous box beam with common span of 25 high speed railways. The inferred upper limit value and the proposal ordinary value are listed in Table 5.

Table 5. Proposal ordinary value of vertical rotation angle at girder end of continuous box beam with common span (conversation to ZK Live Load)

Bridge design speed (km/h)	Type of track structure	Vertical rotation angle at girder end				Available side span (m)
		Average value	Variance	Inferred upper limit	Proposal ordinary value	
250	Ballasted track	0.35‰	1.4e−8	0.58‰	0.60‰	$32 \leq L \leq 60$
350	Ballastless track	0.31‰	9.5e−9	0.50‰	0.50‰	

4.3 Dynamic Response

Vertical Dynamic Response

Dynamic Factor. In the *code for design of high speed railway*, the dynamic factor is not considered for the bridge structure with loading length greater than 68 m, but the measured results show that there is an obvious dynamic effect for the bridge structure with loading length greater than 68 m.Due to the difference between the line design speed and the bridge design speed in China high speed railway, the dynamic factor is statistically analyzed according to the line design speed. Considering the type of track structure and line design speed, statistical analysis is made based on the measured dynamic factor of more than 30 prestressed concrete double line continuous box beam with common span of 25 high speed railways. The inferred upper limit value and the proposal ordinary value during the single line operation of the MU are listed in Table 6.

Vertical Amplitude of Beam. Studies have shown that the vertical amplitude at midspan of beam is affected by train speed, axle load, line condition, etc. [6]. Considering the type of track structure and line design speed, statistical analysis is made based on the measured maximum test results of continuous box beam with common span of 250 km/h and 350 km/h high speed railways under the action of trains with different axle loads. The inferred upper limit value and the proposal ordinary value during the single line operation of the MU are shown in Table 7.

Vertical Vibration Acceleration of Beam. Similar to the vertical amplitude, statistical analysis of vertical vibration acceleration is made based on the test maximum value. The inferred upper limit value and the proposal ordinary value during the single line operation of the MU are shown in Table 8.

Lateral Dynamic Response. The lateral dynamic response of the bridge under the action of MU includes both the lateral vibration of the pier and the lateral bending vibration contribution of the beam. The lateral vibration characteristics of the bridge are directly related to the lateral stiffness or natural frequency of the pier and beam. The box beam, which has a large lateral stiffness, is used in China high speed railway. However, when the beam and piers form a beam-pier coupling system, the lateral natural frequency

Table 6. Statistical table of dynamic factor of continuous box beam with common span

Line design speed (km/h)	Type of track structure	Loading length L_φ (m)	Dynamic factor			
			Average value	Variance	Inferred upper limit	Proposal ordinary value
250	Ballasted track	48	1.14	0.0039	1.26	1.26
		59	1.17	0.0013	1.24	1.24
		64	1.18	0.0005	1.22	1.22
		80	1.12	0.0018	1.20	1.20
		100	1.11	0.0011	1.17	1.17
350	Ballastless track	48	1.20	0.0026	1.30	1.30
		59	1.14	0.0039	1.26	1.26
		64	1.16	0.0024	1.26	1.26
		80	1.14	0.0020	1.23	1.23
		100	1.13	0.0016	1.20	1.20

Note: L_φ is influence line loading length (m). For continuous box beam with three spans and one connection, L_φ is the average span multiplied by 1.3, while the calculating L_φ is less than the maximum span, the maximum span shall be taken as L_φ.

Table 7. Statistical table of vertical amplitude at midspan of continuous box beam with common span

Line design speed (km/h)	Type of track structure	Span (m)	Vertical amplitude at midspan of beam			
			Average value (mm)	Variance (mm^2)	Inferred upper limit (mm)	Proposal ordinary value (mm)
250	Ballasted track	32	0.17	0.001	0.23	0.25
		40	0.27	0.003	0.38	0.40
		48	0.21	0.001	0.27	0.30
		56	0.29	0.002	0.38	0.40
		60	0.22	0.003	0.33	0.35
		64	0.72	0.003	0.83	0.85
		80	0.58	0.002	0.67	0.70
		100	0.49	0.002	0.58	0.60

(continued)

Table 7. (*continued*)

Line design speed (km/h)	Type of track structure	Span (m)	Vertical amplitude at midspan of beam			
			Average value (mm)	Variance (mm^2)	Inferred upper limit (mm)	Proposal ordinary value (mm)
350	Ballastless track	32	0.14	0.002	0.23	0.25
		40	0.17	0.003	0.28	0.30
		48	0.19	0.008	0.37	0.40
		56	0.37	0.020	0.65	0.65
		60	0.20	0.003	0.31	0.35
		64	0.28	0.007	0.44	0.45
		80	0.30	0.006	0.45	0.45
		100	0.28	0.006	0.43	0.45

Table 8. Statistical table of vertical vibration acceleration at midspan of continuous box beam with common span

Line design speed (km/h)	Type of track structure	Span (m)	Vertical vibration acceleration at midspan of beam			
			Average value (m/s^2)	Variance (m^2/s^4)	Inferred upper limit (m/s^2)	Proposal ordinary value (m/s^2)
250	Ballasted track	32	0.15	0.003	0.26	0.30
		40	0.21	0.005	0.35	0.35
		48	0.15	0.006	0.30	0.30
		56	0.23	0.004	0.35	0.35
		60	0.13	0.004	0.25	0.25
		64	0.24	0.005	0.38	0.40
		80	0.16	0.003	0.27	0.30
		100	0.13	0.001	0.19	0.20
350	Ballastless track	32	0.13	0.001	0.19	0.20
		40	0.10	0.001	0.16	0.20
		48	0.11	0.001	0.17	0.20
		56	0.18	0.002	0.27	0.30
		60	0.10	0.002	0.19	0.20

(*continued*)

Table 8. *(continued)*

Line design speed (km/h)	Type of track structure	Span (m)	Vertical vibration acceleration at midspan of beam			
			Average value (m/s^2)	Variance (m^2/s^4)	Inferred upper limit (m/s^2)	Proposal ordinary value (m/s^2)
		64	0.11	0.001	0.17	0.20
		80	0.11	0.002	0.20	0.20
		100	0.11	0.001	0.17	0.20

of the pier and beam is smaller than the lateral natural frequency of the beam, and the stiffness of the pier affects the vibration characteristics of the whole bridge system. The measured lateral excellent forced frequency when the MU passes at a speed of 200–360 km/h is about 1.5–5.3 Hz, which may be close to the lateral natural frequency of some high pier bridges, resulting in the peak value of lateral amplitude of beam and pier. In this case, the ordinary value is generally appropriately enlarged.

Lateral Amplitude of Beam. The measured lateral amplitude of beam are small due to the large lateral stiffness and torsional stiffness of the bridges in China high speed railway, and the measured lateral amplitudes are discrete to some extent.statistical analysis is made based on the measured lateral amplitude of beam. The inferred upper limit value and the proposal ordinary value during the single line operation of the MU are listed in Table 9.

Table 9. Statistical table of lateral amplitude at midspan of continuous box beam with common span

Line design speed (km/h)	Type of track structure	Span (m)	Lateral amplitude at midspan of beam			
			Average value (mm)	Variance (mm^2)	Inferred upper limit (mm)	Proposal ordinary value (mm)
250	Ballasted track	32	0.06	0.0003	0.09	0.10
		40	0.04	0.0002	0.07	0.10
		48	0.04	0.001	0.10	0.10
		56	0.05	0.002	0.14	0.15
		60	0.06	0.002	0.15	0.15
		64	0.04	0.001	0.10	0.10
		80	0.05	0.0003	0.08	0.10
		100	0.08	0.001	0.14	0.15

(continued)

Table 9. (*continued*)

Line design speed (km/h)	Type of track structure	Span (m)	Lateral amplitude at midspan of beam			
			Average value (mm)	Variance (mm^2)	Inferred upper limit (mm)	Proposal ordinary value (mm)
350	Ballastless track	32	0.04	0.0002	0.07	0.10
		40	0.05	0.0003	0.08	0.10
		48	0.06	0.0002	0.09	0.10
		56	0.07	0.001	0.13	0.15
		60	0.06	0.002	0.15	0.15
		64	0.04	0.0002	0.07	0.10
		80	0.05	0.0005	0.09	0.10
		100	0.09	0.001	0.15	0.15

Note: When the lateral forced frequency is close to the lateral natural frequency of the pier and beam, the maximum lateral amplitude at midspan of beam should not be greater than 2 times the ordinary value

Lateral Amplitude of Pier. The design of China high speed railway piers mostly adopts round-end solid piers, and some adopt double-column, rectangular and circular piers. The factors affecting the lateral amplitude at the top of pier mainly include the height of the pier and the lateral width of the pier body. The higher the pier height, the greater the amplitude, and the larger the lateral width of the pier, the smaller the amplitude. Considering the influence of the full height of the pier and the lateral width of the pier body, statistical analysis is made based on the measured lateral amplitude at the top of pier. The inferred upper limit value and the proposal ordinary value during the single line operation of the MU are shown in Table 10.

Table 10. Proposal ordinary value of lateral amplitude at the top of pier

Line design speed (km/h)	Proposal ordinary value (mm)	Scope of application
250	$\frac{H_p}{55B} + 0.02$	$0.5 \le \frac{H_p}{B} \le 4.2$
350	$\frac{H_p}{60B} + 0.03$	

Note: 1. H_p is the full height of the pier (m), which is from the bottom of enlarged foundation or the pile foundation cap to the pier top, B is horizontal average width of pier (m);
2. When the lateral forced frequency is close to the lateral natural frequency of the pier and beam, the maximum lateral amplitude at the top of pier should not be greater than 2 times the ordinary value.

Lateral Relative Displacement Between Two Rail Supports at Adjacent Beam Ends of Ballastless Track. Statistical analysis is made based on the measured lateral relative displacement between two rail supports at adjacent beam ends of ballastless track. The inferred upper limit value and the proposal ordinary value during the single line operation of the MU are shown in Table 11.

Table 11. Statistical table of lateral relative displacement between two rail supports at adjacent beam ends of ballastless track

Line design speed (km/h)	Type of track structure	Lateral relative displacement between two rail supports at adjacent beam ends of ballastless track			
		Average value (mm)	Variance (mm^2)	Inferred upper limit (mm)	Proposal ordinary value (mm)
350	Ballastless track	0.23	0.013	0.45	0.50

5 Concluding Remarks

Rating techniques research on operation performance of continuous box beam with common span are made based on measured data and theoretical research on dynamic behaviors of bridge in China high speed railway in recent years.

1. Rating tasks and the main technical parameters to evaluate the operation performance of continuous box beam with common span used in China high speed railway are clarified.
2. Considering the bridge design speed, the type of track structure and other factors, the proposal ordinary value of vertical natural frequency and damping ratio of beam, ratio of deflection to span and vertical rotation angle at girder end of continuous box beam with common span used in China high speed railways are given.
3. Combined with the design speed of the line, the dynamic response of the continuous box beam with common span under the action of the MU is statistically analyzed, and the proposal ordinary value for the evaluation of the dynamic response of the continuous box beam with common span is given.

The research results provide technical support for professionals to carry out the operation performance rating of continuous box beam with common spans of China high speed railways. However, the statistically measured samples are limited, these recommendations ordinary values usually require more actual operation verification and timely revised, so as to guide the maintenance of bridge in high speed railway much better.

Acknowledgement. The work was supported in part by China Academy of Railway Sciences, Co., Ltd Fund project (No. 2020YJ087).

References

1. Shuli, S.: Design and Practice of High Speed Railway Bridge. China Railway Publishing House, Beijing (2011)
2. China Railway Corporation: Regulations on Railway Technology Management (High Speed Railway). China Railway Publishing House, Beijing (2014)
3. The Ministry of Railways of the People's Republic of China: TYH [2004] No. 120 code for rating existing railway bridges. China Railway Publishing House, Beijing (2004)
4. The Ministry of Railways of the People's Republic of China: TB 10621-2014 Code for Design of High Speed Railway. China Railway Publishing House, Beijing (2014)
5. China Railway Corporation: TG/GW 209-2014 Regulations on Rating of Operation Performance for High-Speed Railway bridges. China Railway Publishing House, Beijing (2014)
6. Wang, W., Yin, J., Yang, Y., et al.: Rating techniques research on operation performance of bridge with common span in China high speed railway . In: International Association for Bridge Maintenance and Safety 2014, pp. 1604–1608. Taylor & Francis Group, London (2014)
7. Wei, W.: Discussion on rating technology for operation performance of high-speed railway long span cable-stayed bridge. J. China Railw. Soc. **40**(1), 103–110 (2018)
8. Jingchuan, Y.: Rating of operation performance for simply supported box beam with common span for high speed railway. China Railw. Sci. **38**(2), 34–41 (2017)
9. Comité Européen de Normalisation. EN 1991-2-2003: Eurocode 1: Actions on structures, Part 2: Traffic loads on bridges (2003)
10. Md Riasat, A., Mustafa, G.: Damage detection of steel-truss railway bridges using operational vibration data. Eng. Struct. **146**(3), 04020008 (2020)
11. Xin, M.: Vertical resonance condition analysis for simply supported box girder with common span used in high speed railway. J. Vib. Shock **38**(23), 19–24, 32 (2019)

Static Simple Shear Tests on Mechanical Behavior of Interface Between Coral Sand and Concrete

Chunyong Jiang[1,2], Xuanming Ding[1,2](✉), and Gang Yang[1,2]

[1] College of Civil Engineering, Chongqing University, Chongqing 400045, China
dxmhhu@163.com

[2] Key Laboratory of New Technology for Construction of Cities in Mountain Area, Chongqing University, Chongqing 400045, China

Abstract. There are various concrete foundations and retaining structures on the islands and reefs, which are subjected to various static and dynamic loads. The interaction characteristics of the contact surface between coral sand and concrete is the key to affect the stress and deformation of rock and soil mass and structure. The static simple shear test of the contact surface between coral sand and concrete was carried out using the simple shear tester. The influence of normal stress, different roughness, particle size and relative density on the static shear characteristics of the surface of contact was studied. The relation between axial displacement, shear stress and tangential displacement of interface. The experimental results indicate that the magnitude of the normal stress has a certain effect on the normal displacement and shear stress of the contact surface, and there is a direct proportional relationship between shear strength and applied normal stress.; the roughness of the contact surface is relatively dense with the coral sand The degree also has a positive effect on the shear stress of the contact face; The particle size has a certain influence on the shear stress and normal displacement of the contact surface.

Keywords: Coral sand · Concrete interface · Static simple shear · Roughness

1 Introduction

After the death of coral group, the rock and soil formed by its remains through long geological process is coral reef. The general structure of coral reef is that the surface layer is mainly composed of Holocene loose reef gravel layer, and the following is consolidated reef limestone. The loose sand gravel layer is mainly coral sand, which is a sediment formed by physical, chemical and biological processes [1]. The mineral composition of coral sand is mainly aragonite, dolomite and calcite, and the chemical composition is mainly calcium carbonate. It is widely distributed in the Red Sea, the western continental shelf of Australia and the islands in the South China Sea [2]. Scholars at home and abroad have done a lot of research work on the properties of coral sand. Including physical and mechanical properties [3], internal pore characteristics of coral

W. Guo and K. Qian (Eds.): GBCESC 2022, LNCE 211, pp. 519–526, 2023.
https://doi.org/10.1007/978-981-19-5217-3_51

sand structure [4], particle crushing characteristics [5], shear characteristics [6]. The above research shows that coral sand is a special kind of rock and soil with irregular shape, porous and easy to break, as a result, coral sand foundation presents the characteristics of high compressibility and low bearing capacity. Pile foundation has great advantages in enhance the bearing capacity of soft soil, so pile foundation in coral sand is widely used. Because the soil is granular, its stiffness is much less than that of the structure, and the soil adjacent to the contact surface will be more or less affected by the contact surface, so the shear characteristics of the contact face between soil and structure deserve our attention. Many scholars have also done corresponding research on the contact face between soil and structure [7, 8].

In conclusion, there is fewer investigate on the shear characteristics of the interface between coral sand and structures with special properties. Therefore, in this paper, the static single shear characteristics of the contact face between coral sand and concrete are studied by single shear instrument, and the influence of normal stress, different roughness, granular size and relative compactness on the static shear properties of the contact surface are systematically studied.

2 Test Instruments and Materials

The instrument used in this study is KTL -EMDSSA cyclic single shear instrument. It can be used for static and dynamic single shear tests of soil, strong earthquake test, cyclic shear strength and other soil dynamic indexes. KTL is equipped with horizontal (shear direction) and axial (normal direction) dynamic actuators (servo motor), which are controlled by high-precision encoder. Each actuator can realize stress control or strain control mode. The load range is ±5 kN and the loading frequency is 0–5 Hz.

The material used in this study is coral sand, which is retrieved from one island reef in the South China Sea. Its mineral composition is mainly aragonite and magnesium containing calcite, and its chemical composition is mainly calcium carbonate. Three kinds of coral sands with single granular sizes of 0.25–0.5 mm, 0.5–1 mm and 1–2 mm and two kinds of compactness (D_r = 50%, 70%) were used in the test. Its physical properties are shown in Table 1.

Table 1. Physical properties of experimental coral sand

Size of particles (mm)	G_s	e_{max}	e_{min}
0.25–0.5	2.77	1.22	0.89
0.5–1.0	2.77	1.53	1.06
1.0–2.0	2.77	1.42	1.01

3 Static Single Shear Test of Contact Face

For investigate the shear characteristics of pile-soil interface, the self poured concrete base and coral sand were used to simulate the concrete pile-soil interface. The manually prepared concrete base has the surface tooth shape as shown in Fig. 1. The face roughness of concrete can be changed only by changing the valley peak distance of the tooth shape on the concrete surface [9]. The valley peak distance of the surface tooth profile is the roughness of the structural surface, which is recorded as R, as shown in Fig. 1. In this study, the concrete base with two roughness of 0 mm and 4 mm is used.

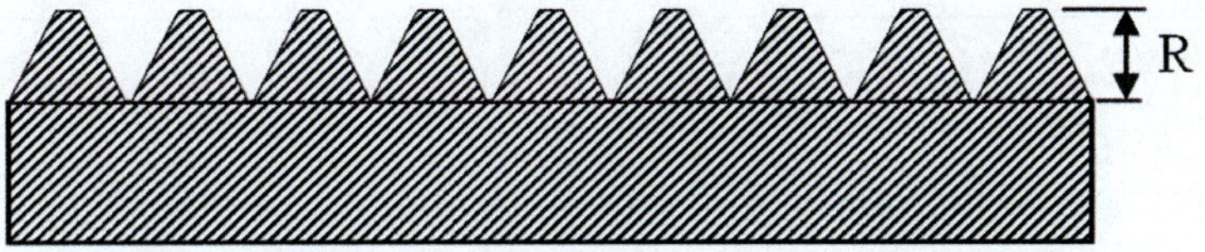

Fig. 1. Sketch map of shape and roughness of concrete base.

The self-made concrete base with two roughness (R = 0, 4 mm) forms the contact surface between coral sand and the structure. The static single shear test of each contact surface is carried out under the normal stress of 100, 200, 300, 400 kPa. In this study, the samples were made by layered static pressure method. In the process of sample preparation, more dense samples can be prepared by gradually knocking the side of the mold and pressing and compacting. Before monotonic shear, the specimen is consolidated by applying the target normal stress. Strain-controlled was used in this test, the shear rate is 0.1 mm/min, and the test is stopped when the shear strain reaches 20%.

4 Test Results and Discussion

4.1 Static Single Shear Characteristics of Coral Sand and Contact Surface Under Different Normal Stresses

To analyze the influence of normal stress on the static single shear properties of coral sand and contact surface, the results are grouped and analyzed by taking R = 0, $D_r = 50\%$ and R = 4, $D_r = 70\%$ as examples. Figure 2 shows the static simple shear test results of the contact surface between coral sand and concrete base under different normal stress conditions. The test results include the relationship curves between shear stress and tangential displacement, and between normal displacement and tangential displacement of contact surface. The normal displacement is positive in compression and negative in expansion. Because the area of the contact face remains unchanged during the test, the change of normal displacement is consistent with the change of volume near the contact surface.

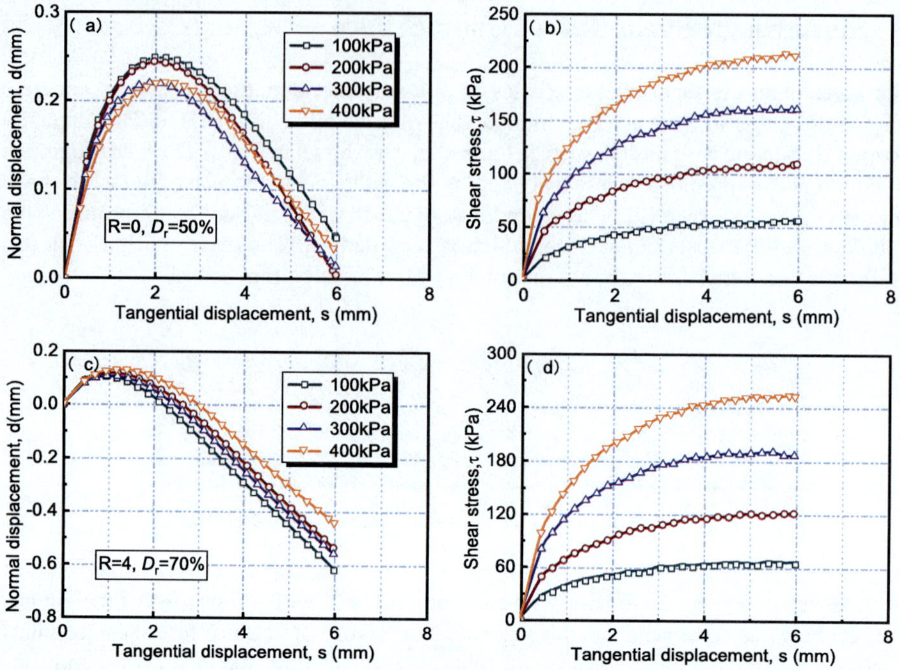

Fig. 2. Soil-structure interface static simple shear test curves under different normal stresses

The results reveal that under the same conditions, the shear stress of the contact surface increases with the increase of tangential displacement and finally tends to be stable, and it is obviously that the shear stress has no obvious strain softening phenomenon. The relationship between normal displacement and tangential displacement of contact surface has different trends with the change of normal stress, that is, the normal displacement first increases with the tangential displacement, reaches the peak point at about 2 mm of shear displacement (corresponding to shear strain of 6.67%), and then decreases with the continuous increase of tangential displacement until the end of the experiment. This means that the soil near the contact surface is shear shrinkage before reaching the peak point, and then shear expansion. It can be seen from the figure that when the roughness R = 0 and $D_r = 50\%$, the amount of shear shrinkage and shear expansion are almost the same. Under other test conditions, the shear shrinkage is obviously less than the shear expansion, that is, the shear expansion is the main factor.

Figure 3 reveals the relationship between static simple shear stress and normal stress of the contact surface between coral sand and concrete in different roughness and the linear fitting results. The shear strength value corresponds to the peak shear stress of the contact face under monotonic shear. The test results show that there is a positive proportional relationship between shear strength and normal stress. Therefore, the shear strength of the contact face can be calculated by the friction angle of the Mohr Coulomb strength criterion ϕ_m to describe. As shown in the graph that the roughness of the contact face and the relative compactness of coral sand have a certain impact on the friction angle.

With the same roughness, the friction angle with large compactness is relatively large. When the compactness is the same, the friction angle of the corresponding contact face with large roughness is relatively large.

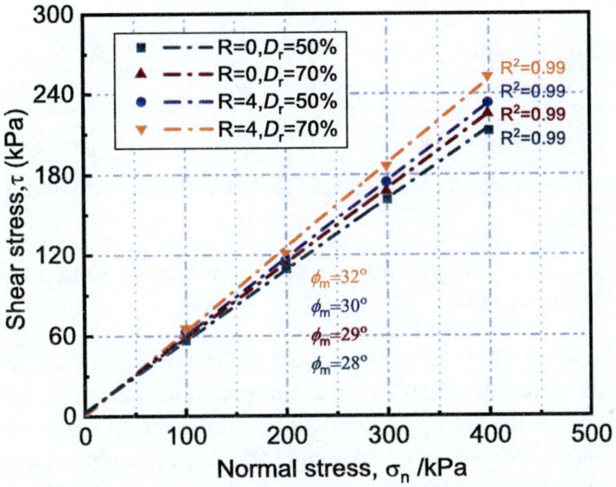

Fig. 3. Relationship between shear strength and normal stress of coral sand concrete interface

4.2 Static Simple Shear Properties of Coral Sand and Contact Surface Under Different Roughness

In this test, two static single shear tests with different roughness are carried out, and the normal stress is 100 kPa, $D_r = 50\%$, as revealed in Fig. 4. The test results indicate that under the same roughness, the shear stress of the contact surface increases with the tangential displacement, and finally it hardly changed. It can be clearly seen from different test results that roughness can enhance the shear stress. The shear stress when roughness R = 4 is greater than that when roughness R = 0. The normal displacement of the sample begins to increase with the tangential displacement of the contact face, and then reaches the peak point. Then the normal displacement begins to decrease until the end of the test, that is, the process from initial shear shrinkage to shear expansion. It can be seen from the figure that the change of normal displacement when roughness R = 4 is significantly greater than that when roughness R = 0. When the roughness R = 0 and the relative compactness $D_r = 50\%$, the normal displacement first increases and then decreases, and finally almost returns to the initial state.

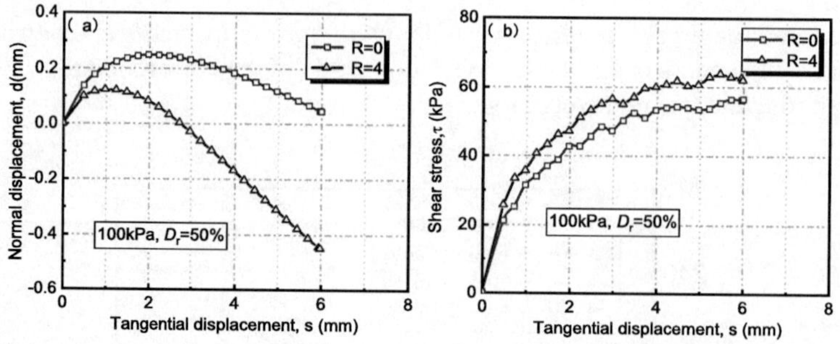

Fig. 4. Soil-structure interface static simple shear test curves under different roughness

4.3 Static Simple Shear Properties of Coral Sand and Contact Surface Under Different Compactness

Figure 5 shows the static single shear test results of the contact surface between coral sand and concrete under different relative compactness conditions when the normal stress is 300 kPa and R = 4. From the figure that for the same relative compactness, the shear stress of contact face increases with the tangential displacement of the contact surface. When a certain tangential displacement is reached, the static simple shear stress of the contact face does not change and is a relatively constant value. It can also be seen that the relative compactness has a certain impact on the shear stress of the contact face, and the greater the relative compactness, the greater the corresponding shear stress. When the normal stress is 300 kPa and the roughness R = 4, the shear stress when the relative compactness $D_r = 70\%$ is about 8.5% higher than that when $D_r = 50\%$.

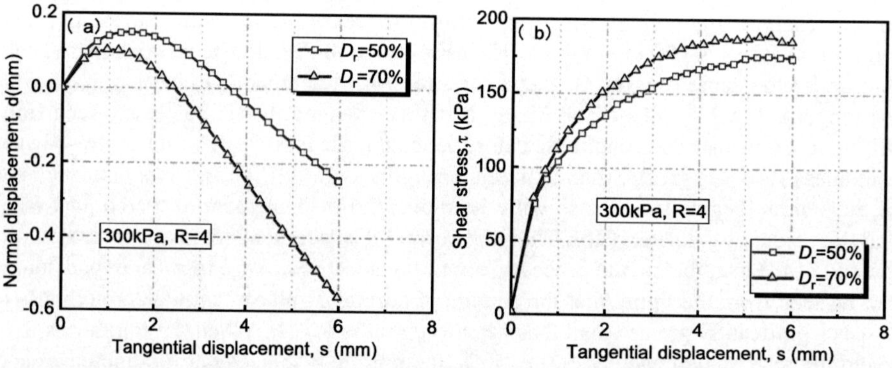

Fig. 5. Soil-structure interface static simple shear test results under different relative density

4.4 Static Simple Shear Properties of Coral Sand and Contact Surface Under Different Particle Sizes

For analyze the effect of different particle sizes on the monotonic shear properties of the contact surface between coral sand and structure, some test results are sorted and revealed in Fig. 6.

From the figure that under each granular size, the curves of the shear stress with the tangential displacement is roughly the same, and its variation law is similar to the above test results. The variation curves of normal displacement and shear displacement of contact surface under different particle sizes are also resemble to the above results. From the test results, the particle size has a certain effect on the shear stress and normal displacement of the contact surface. If the particle size is large, the normal displacement curve is located above, that is, the shear shrinkage before reaching the peak point is large. For the stress-strain relationship, the larger the particle size, the smaller the shear stress.

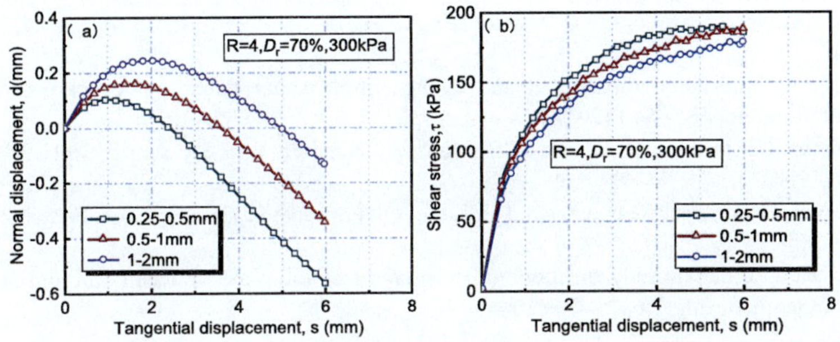

Fig. 6. Soil-structure interface static simple shear test curves under different particle sizes

5 Conclusion

In this study, the static simple shear experiments of the contact surface between coral sand and concrete structure are carried out, and the variation curves of normal displacement, shear stress and tangential displacement are obtained. The effects of roughness, normal stress, relative density and granular size on the shear characteristics of the contact face were studied. According the results, the following conclusions are drawn:

(1) The normal stress has a noteworthy effect on the normal displacement and shear stress of the contact surface. The greater the normal stress, the greater the shear stress of the corresponding contact face. The shear strength of contact surface is directly proportional to the normal stress.

(2) Roughness and compactness have a certain impact on the shear stress of the contact face. The shear stress when R = 4 is greater than that when roughness R = 0. When the normal stress is 400 kPa and the R = 4, the shear stress of the contact face when the $D_r = 70\%$ is about 8.5% higher than that when $D_r = 50\%$.

(3) The particle size has a certain effect on the shear stress and normal displacement of the contact face.
(4) When the roughness R = 0 and the relative compactness $D_r = 50\%$, the normal displacement first increases and then decreases, and finally almost all return to the initial state.

References

1. Manal, S.: Static and cyclic behavior of North Coast calcareous sand in Egypt. Soil Dyn. Earthq. Eng. **55**, 83–91 (2013)
2. Celestino, V.: Comparison of the Dynamic Properties and Undrained Shear Strengths of Offshore Calcareous Sand and Artificially Cemented Sand, pp. 133–141 (2013)
3. Matthew, C.: The mechanics of uncemented carbonate sands. Géotechnique **40**, 607–626 (1990)
4. Changqi, Z.: Experimental microscopic study of inner pores of calcareous sand. Mater. Res. Innov. **18**, S2-207 (2014)
5. Matthew, C.: Particle breakage during shearing of a carbonate sand. Géotechnique **54**, 157–163 (2004)
6. Daniela, P.: Undrained monotonic and cyclic simple shear behaviour of carbonate sand. Géotechnique **58**, 635–644 (2008)
7. Morimichi, U.: Influential factors of friction between steel and dry sands. Soils Found. **26**, 33–46 (1986)
8. Morimichi, U.: Friction between sand and steel under repeated loading. Soils Found. **29**, 127–137 (1989)
9. Ga, Z.: Experimental study on monotonic behavior of interface between soil and structure. Chin. J. Geotech. Eng. **26**, 21–25 (2004)

Concrete and Steel Structure of High-Rise Modular Buildings: A Systematic Review

Zhen Yu(✉)

Suzhou University of Science and Technology, No. 298 Binhe Road, Huqiu District, Suzhou, Jiangsu, China
Zhenyuxy9903@163.com

Abstract. Modular construction has been increasingly adopted in high-rise building projects recently and various studies have been carried out on high-rise modular buildings. However, the majority of the previous research focused either on steel-framed modular structures or on concrete modular building. There is still a lack of systematic review on both steel and concrete modular systems for high-rise building development. Therefore, this paper aims to systematically review the research and development on high-rise modular buildings by investigating the state-of the-art practices and academic research to point out the future research directions. A comprehensive and critical literature review is conducted, including over 30 journal papers and 10 case projects all over the world. Content-based analyses is utilized to analyze the identified journal papers and the case projects. Results of the review reveal that the previous research mainly focused on steel modular system; and most of the case projects are low- to mid-rise, while few of them are high-rise buildings. In future, the paper suggests focusing more on innovative structural connection system for both concrete and steel modular buildings. Throughout the systematic review, this paper contributes to both the knowledge and the practices of high-rise modular construction.

Keywords: High-rise modular building · Concrete and steel · Systematic review

1 Introduction

Modular construction is an innovative construction method, which represents the highest level of off-site construction [1]. Compared with traditional construction methods, it has significant advantages such as faster project delivery, safer construction, and lower costs [2]. However, the structure design of high-rise modular building is complicated in terms of its overall stability and the connection systems [3].

To address the complex structural design of high-rise modular buildings, many studies have been conducted recently. For example, Wang et al. (2020) conducted a computer simulation to develop a new concrete modular system [4]; Chen et al. carried out lab experiment on an innovative type of steel modular connection system [5]. Nevertheless, few studies have been done to systematically examine and summarize the features of various modular systems. Therefore, this paper aims to systematically

W. Guo and K. Qian (Eds.): GBCESC 2022, LNCE 211, pp. 527–535, 2023.
https://doi.org/10.1007/978-981-19-5217-3_52

review the research and development on high-rise modular buildings by investigating the international projects and academic studies to explore the future research.

For achieving the research aim, the study has adopted content-based and statistical analyses. First, a comprehensive literature was searched including both case projects and the academic journal papers. Second, the literature was sorted out from the databased followed by contents analysis. Third, the major findings from the literature were summarized. Fourth, the research gaps were identified and the research directions were proposed.

Following this introduction, the paper presents the research methodology in detail. Then, the paper elaborates on the review results in terms of the case projects and the academic journal papers. The paper then discusses the implications from the systematic review, and finally concludes the study with the major findings.

2 Methodology

This research has adopted a content-based analysis for the systematic review. The review followed a systematic process from literature search to content analysis (see Fig. 1).

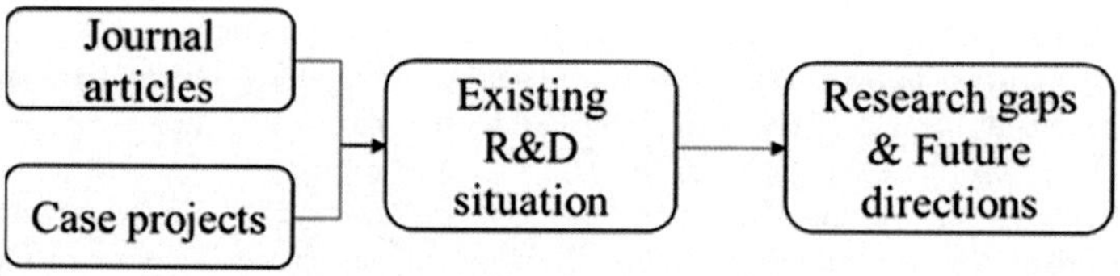

Fig. 1. Overall process of the review

To collect the data used for this research, journal articles were searched and identified from various database, e.g. Scopus. Specifically, Scopus and Web of Science databases were used for searching journal papers. The search criteria are summarized as follows:

- Search scope: title, abstract, keywords
- Search keywords: Modular construction, modular building, structure design, concrete modular building, steel modular building
- Restriction: 1/2000–12/2021 (20 years, to provide comprehensive coverage)
- Subject area: Engineering
- Books, industry reports, and conference papers were deleted in the searching process.

A total of 57 papers were identified. These papers were published in journals including but not limited to Journal of Constructional Steel Research, Journal of Building Engineering, Engineering Structures, Structures, Structural Design of Tall and Special Buildings, Thin-Walled Structures, Steel and Composite Structures, Engineering, Construction and Architectural Management. In order to ensure a high-quality review, 37 papers were selected in the second stage. The criteria of screening output included: (1) the title and abstract do not contain the above search keywords; and or 2) the article content after a simple visual inspection appears to be less relevant or irrelevant to the topic studied. Last, 30 articles that are most relevant to the topic were selected from well-known and recognized scientific journals for the final content analysis.

Over 20 case projects were searched, and 10 case projects were used for the analysis. Data of these case projects were collected mainly from the public domain such as the BCA websites of Singapore, and some other relevant information were collected from the interviews with modular practitioners. The case projects covered both concrete and steel modular buildings, and are all perceived as high-rise building that is over 10-storeys.

To analyze the data collected, content-based analysis was adopted [6], following the process of theme identification, categorization, and detailed analysis. To summarize the findings, statistical analysis was then conducted using an Excel table. For example, the publications in the field of modular construction have been quantitatively sorted out based on their research methods (e.g. lab test and simulation); and the distribution (%) of the case projects in different regions was derived.

3 Results and Analyzes of the Review

3.1 Review on the International Case Projects

The 10 case projects reviewed include 4 concrete modular building and 6 steel modular building. An overview of the case projects is provided and summarized in Table 1. Detailed analyses and findings are presented as below.

Table 1. Basic information of the case projects reviewed

Cases	Region/Country	Type	Story	No. of modules	Floor cycle	Completion
1	Singapore	Concrete	56	3069	7	2023
2	Singapore	Concrete	40	1899	7	2019
3	Singapore	Concrete	14	3711	7	2019
4	Hong Kong	Concrete	16	3800	5	2021
5	Hong Kong	Steel	17	418	3	2020
6	Hong Kong	Steel	17	884	4	2023
7	UK	Steel	44	1500	6	2020
8	UK	Steel	29	679	7	2017
9	UK	Steel	23	653	7	2019
10	China	Steel	18	1285	5	2018

Overall, most of the high-rise modular buildings worldwide used steel-framed modules. For concrete modular building, they were constructed with a central core and modules surrounded. Most of these concrete case projects were designed as shear-wall system. Conventionally, the shear-wall system was built as a single wall to sperate two rooms. For the modular buildings, the shear walls were designed either within the modules (see Fig. 2 left) or built separately as in-situ structures. In some cases, two module walls were combined into a single shear wall (see Fig. 2 right).

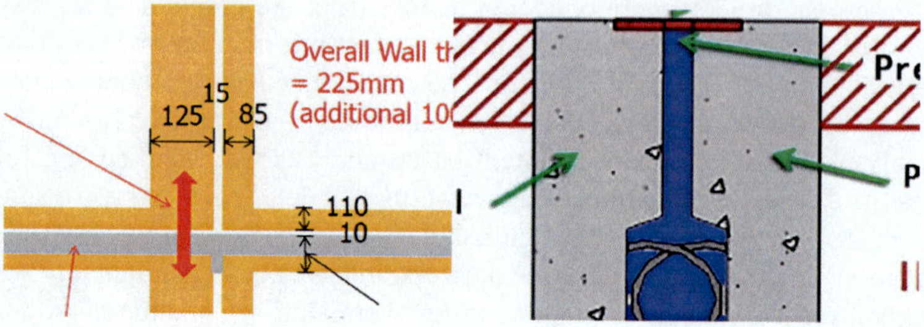

Fig. 2. Diagram of the wall-wall connection system of the concrete modular buildings

In terms of the connections, there are various vertical and horizontal connection systems. Different systems have different features. For example, the connection system in Fig. 2 left was easy to construct, while the system in Fig. 2 right was difficult to connect as it has an in-situ grouting. Nevertheless, the system in Fig. 2 right was better in terms of thickness control, i.e. to reduce the wall thickness by combing two module walls.

For high-rise steel modular buildings, most of them were constructed with a concrete central core and steel-framed modules surrounded (see Fig. 3 left). Different from concrete modular building, most of the steel modular case projects were designed as a system with 1) steel-framed modularized area and 2) concrete core wall as the lateral support system. In terms of the connections, most of the vertical connections used bolt and nuts (see an example in Fig. 3 right).

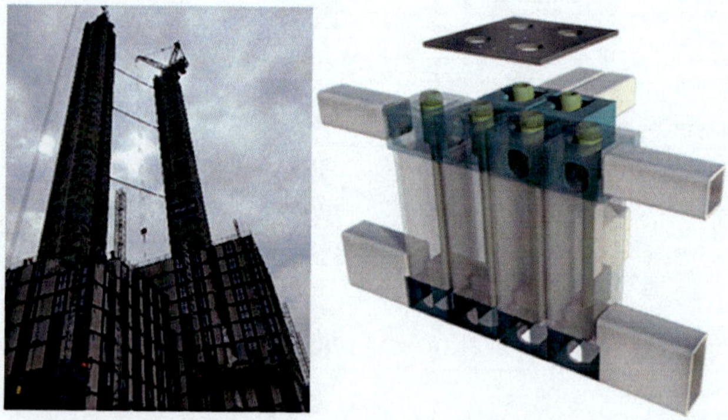

Fig. 3. Diagram of the structural system of the steel modular buildings

As analyzed above, different systems have different features. These features were summarized in terms of their structural and construction performance as follows. In terms of concrete structures, some designs were the same as conventional design, while some were different in terms of the connections, e.g. the sandwich wall system (e.g. Fig. 2 right). During construction, some were easier for module installation with innovative design (e.g. Fig. 2 left). Similarly, for steel modular structures, most connections systems were similar to that of the conventional building, using bolt-and-nut system for module-to-module connection. Compared with concrete modular structure, most of the steel modular buildings can be built faster. For example, the construction floor cycle of case project 5 (using steel modules) achieved 3 days, which was 2 days faster than project 4. Reasons behind are mainly reside in that the late case of concrete modules took a long time in the critical path of construction.

3.2 Review on the Academic Journal Papers

Over 30 journal articles were selection and analyzed. An overview of the journal papers was summarized and provided in Table 2. Detailed analyses and findings are presented as below.

Table 2. Overview of the articles reviewed

	Lab test	Simulation	Review	Case study
Steel	(Dai et al. 2020) [4] (Dai et al. 2021) [17]	(Wang and Pan 2020) [1] (Zhang et al. 2021) [10]	(Thai et al. 2020) [2] (Nadeem et al. 2021) [8]	(Hough and Lawson 2019) [23]
Concrete	N/A	(Wang et al. 2020) [22] (Wang et al. 2021) [35]	(Ferdous et al. 2019) [28] (Lacey et al. 2018) [37]	(Jaillon and Poon 2009) [38]

Various steel modular systems were studied by the previous researchers. Most of them have proposed innovative structural designs, such as an innovative modular steel structure connection with bolted cover plates [7] (see Fig. 4 left) and a shear-keyed grouted sleeve connection (SK-GSC) [8] (see Fig. 4 right). The innovations are reflected by 1) the primary purpose of the localized haunching retrofit solution is to mitigate damage in the column and protect the connecting weld from premature cracking, in order to improve the seismic performance of the connection and achieve ductile failure; and 2) compared with existing bolted connections, SK-GSC connections have sufficient axial resistance, stiffness and ductility to vertically integrate modular columns. They were robust enough to withstand potential strains from unintended effects in modular building design.

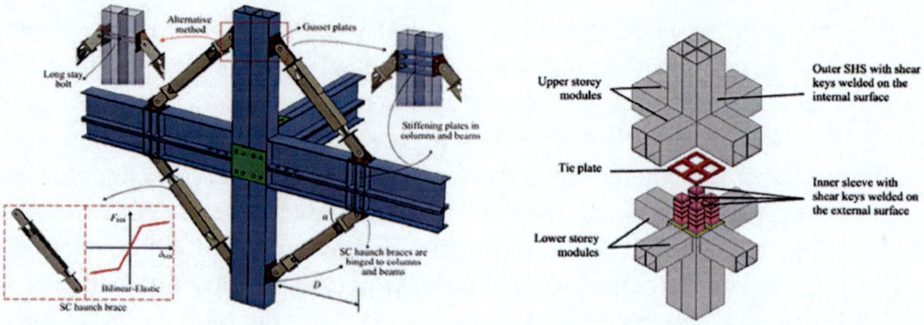

Fig. 4. Innovative steel modular structures (Extracted from Zhang et al. 2021 and Dai et al. 2020)

For concrete modular systems, the previous researchers have proposed some innovative structural designs, such as Modular Integrating System (MIS) [9] (see Fig. 4 left) and A new lateral force resisting system using precast shear walls as part of the modules for high-rises.[4] (see Fig. 5 right). The innovations are mainly reflected by 1) the innovative easy-to-install, self-assembly and self-locking mechanisms of the modular system can facilitate the automated assembly of modular buildings and provide an effective solution for controlling construction tolerances; and 2) this new anti-lateral force system has sufficient strength to resist wind and seismic loads.

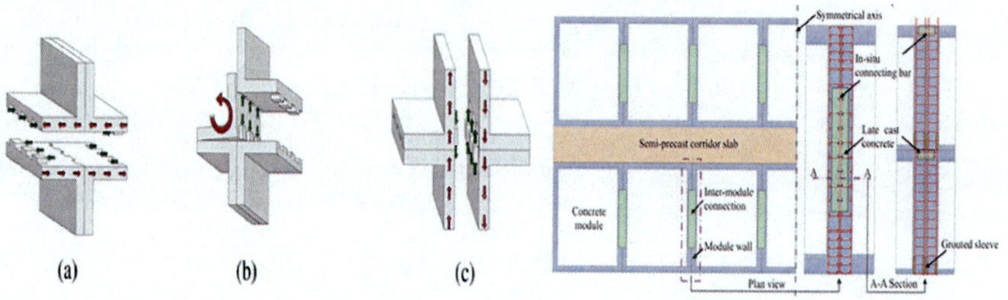

Fig. 5. Innovative concrete modular structures (Extracted from Sharafi et al. 2018 and Wang et al. 2020)

Comparatively, most of the previous research focused on steel modular systems, while only a few on concrete structural modules. One of the primary reasons is that most of the modular high-rise projects worldwide were using steel structures. In addition, most of the research on concrete modular structures were using simulation modelling instead of lab test, which was mainly due to that it is difficult to conduct the lab test for concrete structures, especially for a full-size test.

4 Discussion

4.1 Practical Implications from the Review on Case Projects

Most of the exiting modular building projects in the European countries such as UK used steel-framed modules; while most of them in Singapore were concrete, which may be due to the market demand and the maturity of supply chain. Most of the residential projects for sale were built using concrete modules, e.g. the 40-storey residential building in Singapore and the 16-storey staff quarter in Hong Kong, which was mainly because concrete structure is more durable and can also provide better sound insulation than steel-framed structures.

From the literature review on the multiple case projects worldwide, it can be summarized that (1) different types of modules are suitable for different types of projects, e.g. concrete for residential and steel-framed for hotel and student hostel; and (2) buildings using steel-framed modules can be delivered faster than concrete structure. Nevertheless, the speed of construction depends on the regions (see Table 1).

4.2 Academic Implications from the Review on Journal Articles

Many researchers have conducted studies on high-rise modular buildings in terms of innovation structural design and connections, e.g. Sendanayake et al. (2021) [10] and Peng et al. (2020) [11]. Nevertheless, most of the research used simulation only while few conducted lab tests, especially for concrete modules [2]. The reasons of lacking lab test for concrete modules are mainly due to that it is challenging to build a full-size concrete module in the lab and is difficult to apply load. In addition, most of the innovations have not been adopted in real-life projects to validate the simulation and lab experiments.

Researchers are therefore suggested to conduct lab tests using scale-model instead of 1:1 models for concrete modular structures. Also, they are expected to conduct more research on hybrid system, e.g. using concrete wall in the steel frame, to benefit from both concrete and steel structures.

5 Conclusions

This study has systematically reviewed the research and development on high-rise modular buildings by investigating the international projects and academic studies to explore the future research. The major findings are concluded as below.

First, results from the practical applications revealed that different systems have different features in terms of their structural and construction performance. For example, in terms of concrete structures, some designs were the same as conventional design, while some were different. For steel modular structures, most connections systems were similar to that of the conventional building, using bolt-and-nut system for module-to-module connection. Compared with concrete modular structure, most of the steel modular buildings can be built faster.

Second, most of the previous research focused on steel modular systems, while only a few on concrete structural modules. One of the primary reasons is that most of the modular high-rise projects worldwide were using steel structures. In addition, most of the research on concrete modular structures were using simulation modelling instead of lab test, which was mainly due to that it is difficult to conduct the lab test for concrete structures, especially for a full-size test.

Third but most importantly, researchers are suggested to conduct more lab tests to validate the simulation results, and to conduct more research on hybrid system to benefit from both concrete and steel structures.

Throughout the systematic review, this paper contributes significantly to the body of knowledge of high-rise modular buildings. The paper should assist both practitioners and academic researchers to better understand the state-of-the-art R&D on high-rise modular buildings.

References

1. Huu-Tai, T., Tuan, N., Brian U.: A review on modular construction for high-rise buildings - ScienceDirect. Structures **28**, 1265–1290 (2020)
2. Wahid, F., Yu, B., Tuan-Duc, N., et al.: New advancements, challenges and opportunities of multi-storey modular buildings – a state-of-the-art review. Eng. Struct. **183**, 883–893 (2019)
3. Zhen, W., Wei, P., Yang, Z.: Parametric study on module wall-core system of concrete modular high-rises considering the influence of vertical inter-module connections. Eng. Struct. **241**, 112436 (2021)
4. Hough, M., Lawson, M.: Design and construction of high-rise modular buildings based on recent projects. In: Civil Engineering, pp. 1–38 (2019)
5. Zhihua, C., Jiadi, L., Yujie, Y., et al.: Experimental study of an innovative modular steel building connection. J. Constr. Steel Res. **139**, 69–82 (2017)
6. Yuanshan, L., Di, W., et al.: Lift path planning for a nonholonomic crawler crane. Autom. Const. **44**, 12–24 (2014)
7. Ge, Z., Long-He, X., Zhong-Xian, Li.: Development and seismic retrofit of an innovative modular steel structure connection using symmetrical self-centering haunch braces. Eng. Struct. **229**, 111671 (2021)
8. Ziquan, D., Cheong, T.Y., Dai Pang, S., et al.: Experimental study of grouted sleeve connections under bending for steel modular buildings. Eng. Struct. **243**, 112614 (2021)
9. Rafaela, S., Junjie, T., Amirahmad, F., et al.: Investigation of the seismic performance of braced low-, mid- and high-rise modular steel building prototypes. Eng. Struct. **234**, 111986 (2021)
10. Sendanayake, S.V., Thambiratnam, D.P., Perera, N.J., et al.: Enhancing the lateral performance of modular buildings through innovative inter-modular connections. Structures **29**, 167–184 (2021)
11. Jiahao, P., Chao, H., Luming, S.: Lateral resistance of multi-story modular buildings using tenon-connected inter-module connections. J. Constr. Steel Res. **177**, 106453 (2020)
12. Wang, Z., Pan, W.: A hybrid coupled wall system with replaceable steel coupling beams for high-rise modular buildings. J. Build. Eng. **31**, 101355 (2020)
13. Heshachanaa, R., Keerthan, P., Perampalam, G., Pg, A., et al.: A-State-Of-The-Art review on modular building connections. Structures **34**, 1903–1922 (2021)
14. Ziquan, D., Sze Dai, P., Liew, J.R.: Axial load resistance of grouted sleeve connection for modular construction. Thin-walled Struct. **154**, 106883 (2020)
15. Sung-Gul, H., Bong-Ho, C., Kyung-Soo, C., et al.: Behavior of framed modular building system with double skin steel panels. J. Constr. Steel Res. **67**, 936–946 (2011)
16. Yi-Fan, L., Guo-Qiang, L., Ke, C., et al.: Behavior of splice connection during transfer of vertical load in full-scale corner-supported modular building. Eng. Struct. **230**, 111698 (2021)
17. Alembagheria, M., Sharafia, P., Hajirezaeib, R., Samali, B., et al.: Collapse capacity of modular steel buildings subject to module loss scenarios: the role of inter-module connections. Eng. Struct. **210**, 110373 (2020)

18. Gohar, N., Nor Azizi, S., Nabilah Abu, B., et al.: Connection design in modular steel construction: a review. Structures **33**, 3239–3256 (2021)
19. Yang, D., EnFeng, D., Liang, Z., et al.: Cyclic tests on corrugated steel plate shear walls with openings in modularized-constructions. J. Constr. Steel Res. **138**, 675–691 (2017)
20. Perampalam, G., Keerthan, P., Shanmuganathan, G., et al.: Development of affordable steel-framed modular buildings for emergency situations (Covid-19). Structures **31**, 862–875 (2021)
21. Andrew William, L., Wensu, C., Hong, H., et al.: Effect of inter-module connection stiffness on structural response of a modular steel building subjected to wind and earthquake load. Eng. Struct. **213**, 110628 (2020)
22. Xiao-Huang-Can, H., TakMing, C., Kwok-Fai, C.: Effect of inter-module connections on progressive collapse behaviour of MiC structures. J. Constr. Steel Res. **185**, 106823 (2021)
23. Renwei, M., Junwu, X., Hongfei, C., et al.: Experimental and numerical investigation of mechanical properties on novel modular connections with superimposed beams. Eng. Struct. **232**, 11858 (2021)
24. Zhihua, C., Jiadi, L., Yujie, Y.: Experimental study on interior connections in modular steel buildings. Eng. Struct. **147**, 625–638 (2017)
25. Xiao-Meng, D., Liang, Z., Yang, D., et al.: Experimental study on seismic behavior of a novel plug-in self-lock joint for modular steel construction. Eng. Struct. **181**, 143–164 (2019)
26. Zhihua, C., Kashan, K., Arsalan, K., et al.: Exploration of the multidirectional stability and response of prefabricated volumetric modular steel structures. J. Constr. Steel Res. **184**, 106826 (2021)
27. Zhen, W., Wei, P., Zhiqian, Z.: High-rise modular buildings with innovative precast concrete shear walls as a lateral force resisting system. Structures **26**, 39–53 (2020)
28. Pezhman, S., Mina, M., Bijan, S., et al.: Interlocking system for enhancing the integrity of multi-storey modular buildings. Autom. Constr. **85**, 263–272 (2018)
29. Andrew, L., Wensu, C., Hong, H., et al.: Lateral behaviour of modular steel building with simplified models of new inter-module connections. Eng. Struct. **236**, 112103 (2021)
30. Chua, Y.S., Liew, J., Pang, S.D.: Modelling of connections and lateral behavior of high-rise modular steel buildings. J. Constr. Steel Res. **166**, 105901 (2019)
31. Andrew, L., Wensu, C., Hong, H., et al.: New interlocking inter-module connection for modular steel buildings: Simplified structural behaviours. Eng. Struct. **227**, 111409 (2021)
32. Andrew, L., Wensu, C., Hong, H., et al.: New interlocking inter-module connection for modular steel buildings: Experimental and numerical studies. Eng. Struct. **198**, 109465 (2019)
33. Jiahao, P., Chao, H., Luming, S.: Numerical analysis of corner-supported composite modular buildings under wind actions. J. Constr. Steel Res. **187**, 106942 (2021)
34. Jun-Yi, L., EnFeng, D., JinMing, H., et al.: Numerical analysis on seismic performance of corner fitting connection in modular steel building. Structures **33**, 1659–1676 (2021)
35. Deng, E.F., Zong, L., Ding, Y.: Numerical and analytical study on initial stiffness of corrugated steel plate shear walls in modular construction. Steel Compos. Struct. **32**, 347–359 (2019)
36. Srisangeerthanan, S., Hashemi, M.J., Rajeev, P., et al.: Numerical study on the effects of diaphragm stiffness and strength on the seismic response of multi-story modular buildings. Eng. Struct. **163**, 35–37 (2018)
37. Lacey, A.W., Chen, W., Hao, H., Bi, K.: Structural response of modular buildings – an overview. J. Build. Eng. **16**, 45–56 (2018)
38. Jaillon, L., Poon, C.S.: The evolution of prefabricated residential building systems in Hong Kong: a review of the public and the private sector. Autom. Constr. **18**, 239–248 (2009)

Dynamic Properties of Hydrophobic Sand as a Filling Material in Road Subgrade

Gege Huang and Hongjie Lin(✉)

Sun Yat-Sen University, Zhuhai 519082, China
linhj33@mail.sysu.edu.cn

Abstract. In order to ensure the safety and comfort of vehicle transportation, a great concern has been given on the subgrade of road about its deformation and stability. Water is one of the most important contributing factors to the diseases of subgrade such as frost heave in cold areas and collapse in saline soil subgrades. Addressing water-induced subgrade damage is usually costly and time-consuming. This study proposed a novel solution by using hydrophobic sand as a subgrade filling material. Hydrophobic sand can impede water infiltration with low cost. Although such sand has been proofed to hinder the water movement effectively, its dynamic mechanical properties are still unclear which are important for the assessment of subgrade seismic response. This study tests four hydrophobic sands by resonant column test to investigate the effect of hydrophobized treatment and outdoor weathering on their dynamic properties. Hardin-Drnevich model is used to fit relationship curves of dynamic shear modulus with shear strain. The result showed that hydrophobized treatment increased both dynamic shear modulus and damping ratio, depending on the type of water repellent agent and outdoor weathering duration. Hardin-Drnevich model is applicable for all tested subjects.

Keywords: Dynamic property · Hydrophobic sand · Outdoor-weathering · Resonant column test

1 Introduction

Water in subgrade engineering usually causes soil soaking, saturation and erosion which reduce the strength of soil especially in saline soil area and frozen soil area, therefore subgrade defects come out, such as mud pumping, settlement and freezing damage of foundation, slide and collapse of subgrade slope and so on [1]. In consequence, to ensure subgrade stability, keeping drying is needed for subgrade.

To mitigate the disasters caused by water in subgrade engineering, three types of solutions are proposed:

(a) Raising roadbed elevation: This method separates the subgrade fillings and groundwater by increasing the heigh of subgrade [2];
(b) Decreasing groundwater level [3];
(c) Isolating groundwater [4].

W. Guo and K. Qian (Eds.): GBCESC 2022, LNCE 211, pp. 536–541, 2023.
https://doi.org/10.1007/978-981-19-5217-3_53

These methods are costly and prolong the construction. This study proposes a novel method to solve the water infiltration by adding hydrophobic sand layer. Hydrophobic sand has been applied in mining engineering, ground engineering [5] and cover system of engineered landfills [6]. The capillary rise of extremely hydrophobic soil is negligible. When applied in subgrade, such soil can prevent the movement of groundwater into subgrade.

Adding hydrophobic soil layer has some advantages. It has a lower cost. Synthetic water repellency generates a sealing effect albeit with increased surface erosion [7]. Its stability and durability have been studied to guarantee a 70-year lifetime in application environment [8].

In the assessment of construction seismic stability and soil seismic response, dynamic parameters, such as dynamic shear modulus (G) and, damping ratio (D) are indispensable [9]. G represents the ability of a material to resist deformation while D represents the attenuation form after excitation force.

Resonant column test is an applicable method to measure dynamic shear modulus and damping ratio. Resonant column test has clear end restriction and simple instrument structure. The test has little disturbance to soil sample. Domestic and foreign scholars did many resonant column tests. They think resonant column test is a dependable way to measure soil dynamic characteristic at small strain [10]. Thus, resonant column test is a reliable method to measure dynamic shear modulus and damping ratio at small strain (0.0001% to 0.01%).

Dynamic constitutive models can be classified into two types, viscos-elasticity theory and elastic-plasticity theory. In viscos-elasticity theory, equivalent linear model and Masing type nonlinear model are wildly-used. At present, equivalent linear model is more popular. Hardin-Drnevich model, Ramberg Osgood model, bilinear model and some combined curve model belong to equivalent linear model [11]. This article studies sand dynamic properties at small strain. Hardin-Drnevich model can simulate soil dynamic properties especially cohesionless soil [12]. Hardin-Drnevich model is hyperbola model, and has wide application. By using H-D model, maximum dynamic shear modulus in any conditions can be concluded.

$$G = \frac{G_{max}}{1 + \gamma/\gamma_{ref}} \tag{1}$$

where G is dynamic shear modulus in γ, G_{max} is maximum dynamic shear modulus, γ is shear strain in G, and γ_{ref} is shear strain in $G = 0.5G_{max}$. γ is directly proportional to 1/G. The intercept of the λ and 1/G curve is $1/G_{max}$. G_{max} manifests the pure elastic deformation characteristics. This study tests five sands dynamic shear modulus and damping ratio by using resonant column test. Hardin-Drnevich model simulates relationship of dynamic shear modulus with shear strain. This article analyzes hydrophobic agent and degradation effect on soil dynamic properties.

2 Materials and Methods

2.1 Materials

Five materials, Fujian sand, DMDCS-sand, TO-sand, degraded DMDCS sand and degraded TO-sand are tested. Fujian sand (Xiamen ISO Standard Sand Co. Ltd, China), with 96% of SiO_2 was used. Hydrophobicity can be influenced by particle size, so, this study used uniformly graded sands (63 to 425 μm) by wet sieving. For DMDCS-sands, DMDCS adds to 100.0 g of air-dried Fujian sand by a single channel pipette (Pipetman P100 from Gilson®, France). Then mixed it and put into a fume hood to get rid of the generated HCl. Tung oil was added to 100.0 g of Fujian sand by percentage weight followed by mixing. Hydrophobized sands were placed in outdoor for one year, then obtained the degraded samples [9].

2.2 Methods

Outdoor Weathering of Hydrophobized Sand. The one-year outdoor weathering of hydrophobized sand was conducted in Hong Kong from November 1st, 2019 to November 30th, 2020. DCDMS-treated and Tung oil-treated hydrophobized sand was firstly placed in cylindrical containers (diameter of 15 cm and height of 10 cm) with permeable bottoms for drainage. At the height of 5 cm, a moisture sensor (EC-5) connecting to a data logger (Emb50) was buried in the soil column in order to detect water infiltration. In addition, to prevent the loss of surface soil caused by rainfall and wind during outdoor weathering, the surface of the sample was covered with a 2 mm diameter wire mesh which allow water to flow in and avoid erosion. For each type of hydrophobized sands, 8 columns were set up and 2 replicate columns were collected for analyzing every quarter.

Resonant Column Test. Resonant column test measures dynamic shear modulus and damping ratio. The resonant column used in this study is GDS-RCA test system. Basic test parameters are listed in Table 1.

Table 1. Basic parameters of five tested sands.

	Weight m(g)	Volume v(cm^3)	Density ρ(10^3 kg/m^3)	Moment of inertia I(kg·m^2)
Fujian sand	320.0	196.35	1.633	1.0×10^{-4}
DMDCS-sand	320.0	196.35	1.633	1.0×10^{-4}
TO-sand	310.0	196.35	1.582	9.7×10^{-5}
Degraded DMDCS-sand	320.0	196.35	1.632	1.0×10^{-4}
Degraded TO-sand	310.0	196.35	1.582	9.7×10^{-5}

In this study, a fixed-free resonant column is used with 100 kPa of confining pressure. Each condition tested triple samples and error was controlled within 10% for each batch. The machine exerts torsional vibration on samples, and changes excitation voltage gradually. Damping ratio was calculated by Eq. 2 [13].

$$\lambda = \frac{1}{2\pi} \times \frac{1}{N} ln \frac{A_1}{A_{N+1}} \tag{2}$$

where λ is damping ratio, N is vibration times of calculation. A_1 is the first amplitude after stop vibration. A_{N+1} is the $(N + 1)$th amplitude after stop vibration.

3 Results

Figure 1(a) demonstrates the fitting curves of dynamic shear modulus with shear strain. Figure 1(b) is curves of damping ratio with shear strain. These fitting curves are contained by using B spline interpolation.

In Fig. 1(a), shear strain increases, dynamic shear modulus decreases. The rate of dynamic shear modulus decrease is faster and faster. The dynamic shear modulus of DMDCS-sand is the highest. Dynamic shear modulus of DMDCS-sand is higher than degraded DMDCS-sand, however, dynamic shear modulus of TO-sand is less than degraded TO-sand.

As shown in Fig. 1(b), shear strain exponentially increases with the damping ratio. The outdoor weathering resulted in a negligible change of damping ratio in DMDCS-sand. The damping ratio of fresh TO-sand is higher than that of degraded TO-sand. The damping ratio of Fujian sand is the lowest.

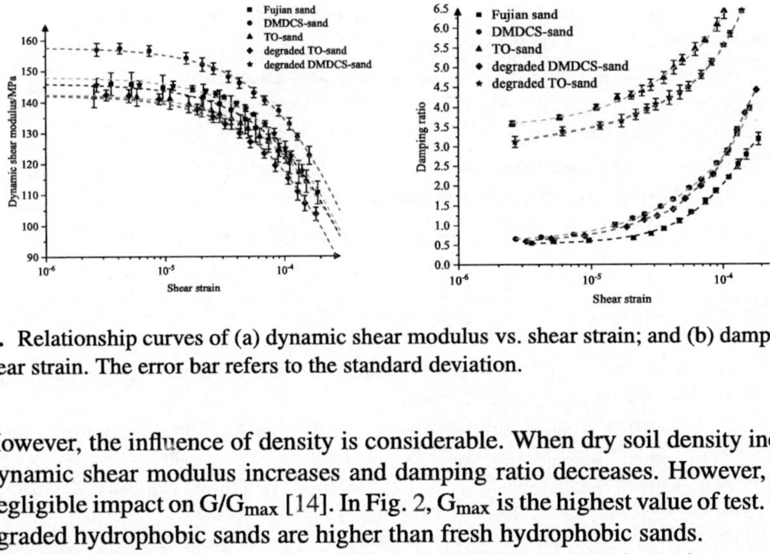

Fig. 1. Relationship curves of (a) dynamic shear modulus vs. shear strain; and (b) damping ratio vs. shear strain. The error bar refers to the standard deviation.

However, the influence of density is considerable. When dry soil density increases, the dynamic shear modulus increases and damping ratio decreases. However, density has negligible impact on G/G_{max} [14]. In Fig. 2, G_{max} is the highest value of test. G/G_{max} of degraded hydrophobic sands are higher than fresh hydrophobic sands.

TO-sand coating is thicker than DMDCS-sand. Thus, the dynamic properties of TO-sand are more influenced by coating. Because of the coatings, damping ratio of hydrophobic sand is higher than Fujian sand. After degradation, hydrophobic coatings appear different degrees of breaking [15]. Soil dynamic shear modulus and damping ratio have different changes.

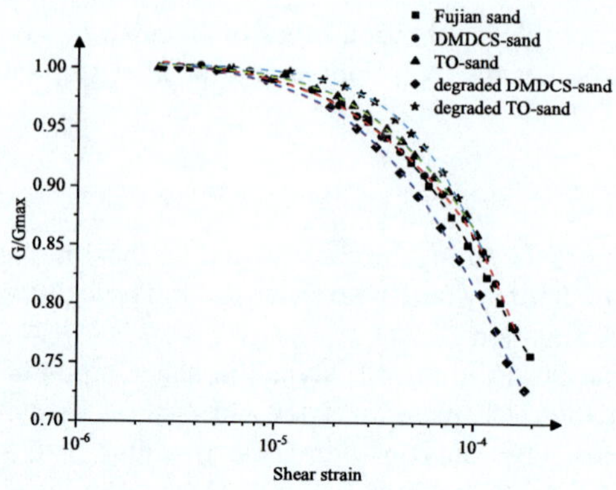

Fig. 2. Relationship curves of G/G_{max} vs. shear strain.

4 Primary Finding and Future Work

The primary findings of this work are as follows:

a) Hydrophobized treatment increases the damping ratio. Tung oil-induced sand has a higher damping ratio than DMDCS-sand.
b) Hardin-Drnevich model is applicable for the dynamic characteristics of hydrophobized sands.

The effect of hydrophobization and outdoor weathering on the dynamic properties is highly dependent on the hydrophobic agent type, i.e., the thickness and viscosity of the hydrophobic coatings. A further study is required to clarify the mechanism.

References

1. Jiankun, L., Qiaoling, Z., Yongfeng, H., Jun, Y.: Subgrade Engineering. China Building Industry Press, Beijing (2014).(in Chinese)
2. Yong, Q.: Construction technology of roadbed settlement and mud-pumping in highway engineering. Sichuan Cem. **07**, 246 (2019). (in Chinese)
3. Hairui, L.: Prevention and treatment of roadbed mud-pumping. Highways Transp. Inner Mongolia **03**, 16–17 (1994). (in Chinese)
4. Lourenço, S.D.N., et al.: Soil wettability in ground engineering: fundamentals, methods, and applications. Acta Geotech. **13**(1), 1–14 (2017). https://doi.org/10.1007/s11440-017-0570-0
5. Shuang, Z., Xin, X., Lourenço, S.D.N., Peter, J.: Cover systems with synthetic water-repellent soils. Vadose Zone J. **20**(1), 20093 (2021)
6. Lourenço, S.D.N., Gong-Hui, W., Toshitaka, K.: Processes in model slopes made of mixtures of wettable and water repellent sand: implications for the initiation of debris flows in dry slopes. Eng. Geol. **196**, 47–58 (2015)
7. Hongjie, L., Lourenço, S.D.N.: Accelerated weathering of hydrophobized sands. Acta Geotechnica **17**(2), 377–390 (2021)

8. Weimin, H., Deqing, L., Jie, Y., Yiqing, L.: Recent progress in research on dynamic shear modulus, damping ratio, and Poisson ratio of soils, China Earth Quake. Eng. J. **38**(02), 309–317 (2016)
9. Chengwei, L.: Resonant Column Experimental Study on Dynamic Properties of Cemented Clay, Hubei University of Technology (2016)
10. Wuping, P., Yi, Y., Chunlin, L., Yuyan, C.: A review of soil dynamic properties. In: The 4th National Hydraulic Engineering Aseismic Disaster Prevention Academic Exchange Conference, Xi'an, Shaanxi, China, pp. 469–474 (2013). (in Chinese)
11. Shichun, Chi., Xiaoxia, Guo., Jun, Y., Gao, L.: Small strain characteristics and threshold strain of dynamic Hardin-Drnevich model for soils. Chin. J. Geotech. Eng. (02), 243–249 (2008)
12. Dingyi, X.: Soil Dynamics, Xi 'an Jiaotong University Press, Xi'an, Shaanxi, China (1988)
13. Zhengyin, C., Xiangsong, L.: Effects of material state on the small strain behavior of dry sand. Rock Soil Mech. **01**, 10–14 (2004)
14. Hongjie, L., Lourenço, S.D.N.: Physical degradation of hydrophobized sands. Powder Technol. **367**, 740–750 (2020)

Effects of Redispersible Latex Powder on the Performance of 3D Printing Cementitious Materials

Huan Wang, Xinchen Tian, Huixian Wei, Xueyan Dai, Yujun Che, and Huashan Yang(✉)

School of Materials and Architecture Engineering, Guizhou Normal University, Guiyang 550025, China
201510003@gznu.edu.cn

Abstract. To investigate the effects of redispersible latex powder (RLP) on the workability and mechanical properties of 3D printing cementitious materials (3DPC), the fluidity, extrudability, open time, buildability, green strength, flexural strength, and compressive strength of cement mortar with 0%–4% of RLP were investigated. Results show that RLP can improve the fluidity, buildability, and open time of 3DPC, but it negatively influences the strength of 3DPC. When the dosage of RLP is about 2%–3%, the strength loss of 3DPC is not significant.

Keywords: 3D printing cementitious materials · Redispersible latex powder · Workability · Mechanical properties

1 Introduction

3D printing is a rapid prototyping technology based on many advanced technologies such as information technology, modeling technology, control technology, and material science, and run by the cooperation of computers and 3D printers. It can significantly reduce production and manufacturing time and save labor. In recent years, 3D printing technology has been rapidly developed and applied in the field of construction engineering [1, 2].

Concrete 3D printing technology, a new construction technology that organically combines 3D printing and building construction, has the advantages of reducing costs, speeding up construction, mold-free construction, reducing construction waste, etc. [3]. The principle is to segment the designed building model by computer, convert the 3D graphic information into a printing path, then send the concrete or mortar mix to the printing head through the conveying system, and print it sequentially through the concrete 3D printer according to the pre-processed path, so that the materials are stacked layer by layer, and finally get the actual building components. As the core of architectural 3D printing technology, 3D printing concrete has significant differences in the workability requirements from ordinary concrete materials. Easy-extrusive, easy-flowing, well-buildable, and proper open time are essential for 3DPC, and they allow the material

W. Guo and K. Qian (Eds.): GBCESC 2022, LNCE 211, pp. 542–551, 2023.
https://doi.org/10.1007/978-981-19-5217-3_54

to extrude smoothly from the printer nozzle until the end of the entire printing process without collapsing [4]. The admixtures used for 3DPC are usually viscosity modifiers, nanomaterials, set promoters, retarders, etc. [5]. In order to formulate 3DPC with excellent properties, many scholars have studied the effects of some industrial wastes and admixtures on the performance of 3DPC. Chen et al. [6] obtained 3D printing mortars with good printability by replacing cement with 30% of ceramic sand. Yu et al. [7] investigated the effects of basalt fibers on the properties of 3DPC. Zhang et al. [8] used bentonite for admixture and concluded that about 1%–3% of bentonite could improve the buildability of mortar.

Redispersible latex powder (RLP) is a water-soluble redispersible latex powder, which can be divided into different types such as polyvinyl acetate and vinyl acetate/ethylene (EVA) copolymer powder, polyacrylic acid system, etc. [9]. It is usually obtained from polymer emulsions after special treatment and spray drying. It is mainly used for repairing mortars, self-leveling materials [10], and tile adhesives. Ordinary Portland cement mortar has the problem of easy water secretion, which does not meet the printability of 3DPC. Van [11] found that RLP has a thickening effect. Early literature states that RLP can improve the fluidity of cement mortar [12]. Over the past decades, RLP has been widely used to bind substrates, improve the workability and permeability of mortars, etc. [13]. Therefore, the appropriate dosage of RLP can improve the workability of 3DPC and alleviate the problem of interlayer "weak interface," thus enhancing the interlayer tensile strength and interlayer shear strength. The effects of RLP on the workability and mechanical strength of 3DPC were investigated in this paper.

2 Materials and Methods

2.1 Materials and Mixtures Proportions

In this study, the raw materials were ordinary Portland cement (PC, P. O 42.5), fly ash (FA), silica sand (S), RLP, nano SiO_2 (NS), high-efficiency water reducing agent (SP), hydroxypropyl methylcellulose (HPMC), and water. The average particle sizes of PC and FA were 12.6 μm and 8.5 μm, respectively (Table 1).

Table 1. Mixture proportions

Sample	PC/g	FA/g	S/g	Water/g	SP/g	HPMC/g	NS/g	RLP/g
M-0	378	162	1350	216	2.16	0.54	5.4	0
M-1	378	162	1350	216	2.16	0.54	5.4	5.4
M-2	378	162	1350	216	2.16	0.54	5.4	10.8
M-3	378	162	1350	216	2.16	0.54	5.4	16.2
M-4	378	162	1350	216	2.16	0.54	5.4	21.6

2.2 Methods

In this work, 30% of cement was replaced by fly ash, and the effects of 0–4% of RLP on the fluidity, extrudability, open time, buildability, green strength, flexural strength, and compressive strength of 3DPC were investigated.

Printability. Fluidity is a critical parameter in assessing the printability of material for 3D printing. It refers to the ability of the printing mortar to extrude continuously and without interruption from the nozzle. Controlling and improving the fluidity of cementitious materials can be considered from three aspects: raw materials, admixtures, and mix ratios. Grzeszczyk and Lipowski [14] pointed out that replacing a certain amount of Portland cement with fly ash can significantly improve the fluidity of cementitious materials, and finely ground fly ash has a better effect than that of coarse fly ash. However, an excessive amount of finely ground powder will instead increase the consistency of the material and thus reduce the fluidity. Li [15] found that incorporating an appropriate amount of retarder can give good fluidity to cement mortar. Kwan and Wong [16] concluded that the fluidity of mortar increase with the increase of the water to cement ratio. In this paper, mini-slump and jump table flowability were used to evaluate the fluidity of 3DPC.

Extrudability refers to the ability of 3DPC to be continuously extruded from the nozzle without interruption of blockage. The yield stress of concrete and the mechanical extrusion pressure are two essential factors that determine the extrudability of 3DPC, so the extrudability can be tested quantitatively using the experimental method reported in the literature [17]. The extrudability of the 3DPC was evaluated by plotting the relationship between the extrusion force and displacement by applying pressure at a rate of 60 mm/min.

The open time is the period during which the material can be continuously pumped and extruded. The test method is shown in Fig. 1. Print strips of 30 cm in length at 20 min intervals until the width of the strip is significantly reduced or interrupted, then record the corresponding time as the open time.

Buildability refers to the ability of cementitious materials to have sufficient stability to withstand the load of the upper layer of printed material without deformation. Le et al. [3] evaluated the buildability of 3DPC in terms of the number of layers of printable long strips. As shown in Fig. 2, the buildability of cementitious materials was tested in this paper using the method of Sonebi, M [18] et al. Fresh cement mortar was filled into a conical mold with an upper diameter of 70 mm and a lower diameter of 100 mm, and after removing the mold, a 600 g weight was placed on the upper surface of the cement mortar, and the deformation of the upper and lower diameters was measured after 1 min. The smaller variation in diameter, the better buildability of the mortar.

Green strength is an essential factor in ensuring the good buildability of cementitious materials. The green strength was tested using the method reported in the literature [19] and measured every 30 min, and the average of four measurements for each sample was taken as the test result.

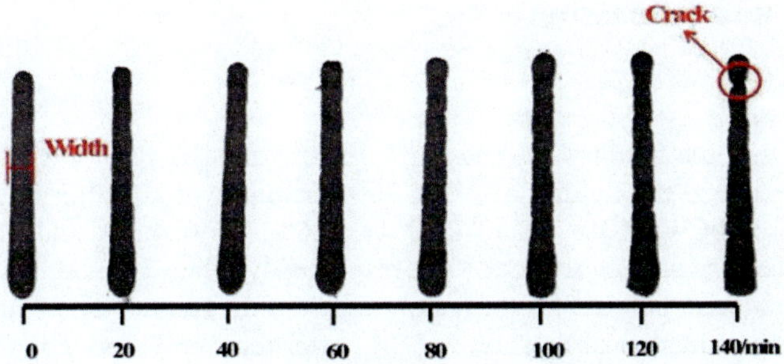

Fig. 1. Open time

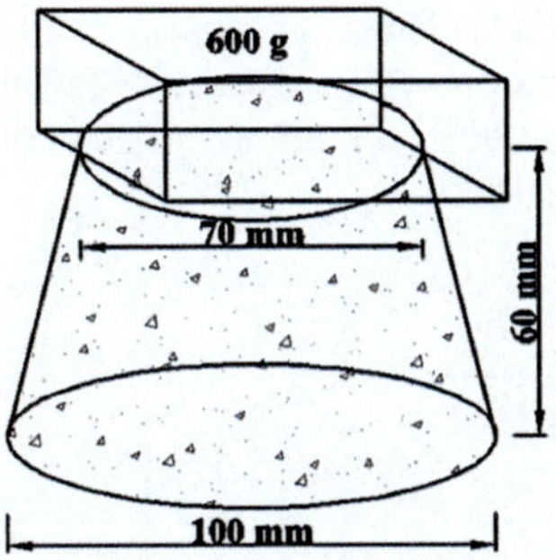

Fig. 2. Buildability

Mechanical Properties. Three consecutive layers of specimens were printed by a 3D printer and cut into standard specimens of 160 mm × 40 mm × 40 mm after 24 h. The specimens were placed in the cement concrete standard curing box and cured to the specified age. According to GB/T17671–1999, the strength was tested for 7 d and 28 d, and the loading direction is shown in Fig. 3 (the x-axis is the printing direction).

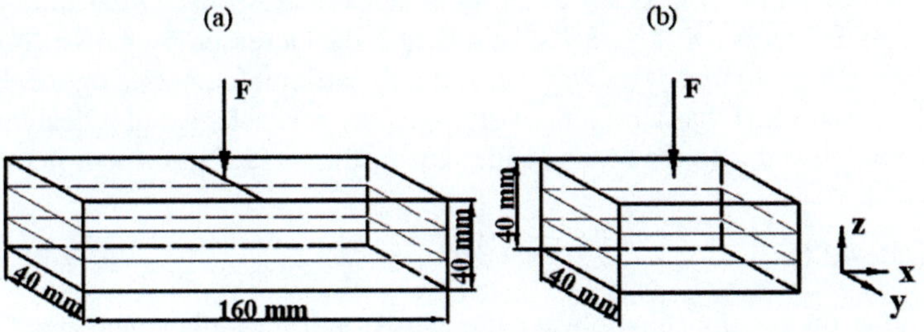

Fig. 3. Load direction. (a) flexural strength, (b) compressive strength.

3 Results and Discussion

3.1 Fluidity

The test results of the fluidity are shown in Fig. 4 (a) and (b). From Fig. 4 (a), the RLP significantly reduced the fluidity of 3DPC. For example, at a resting time of 15 min, the fluidity of 3DPC with 1%–4% of RLP was 3.6%, 3.6%, 4.3%, and 2% lower than that of the reference sample without RLP, respectively. The reason is that RLP affects the particle gradation before film formation, resulting in gradual thickening of cement mortar and thus reduction of flowability [20]. After the film is formed, a "water film" is wrapped around the aggregate and cement particles, reducing the friction between the particles and improving fluidity [21]. The fluidity of the samples with 1% RLP was slightly higher than those of the reference sample at 0 min and 75 min, with an increase of 1.36% and 1.35%, respectively. This is because a small dosage of RLP facilitates the improvement of particle gradation and increases the flow of cement mortar. Moreover, under the control of both RLP film formation and cement hydration at the later stage, the fluidity showed an increasing trend.

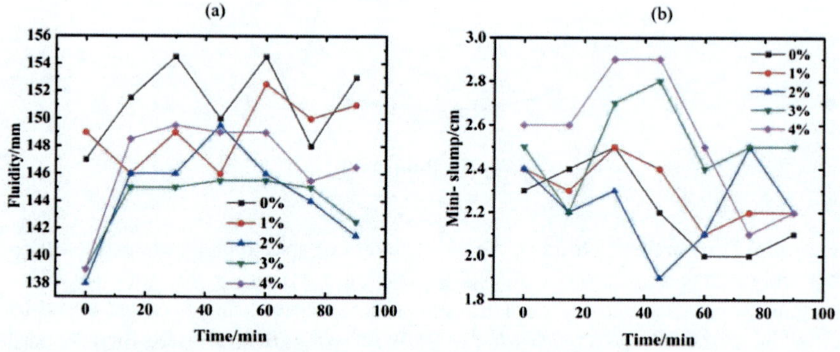

Fig. 4. Test results of fluidity

As shown in Fig. 4 (b), RLP has a significant effect on the initial mini-slump of 3DPC. The 3DPC with 1%, 2%, 3%, and 4% of RLP increased by 4.4%, 4.4%, 8.7%, and 13.0% compared to that of the reference sample without RLP. This figure also shows that the 3DPC with RLP has a high mini-slump at the early stage, but it decreases at the later stage, which is due to the redispersion, adsorption, and film formation of RLP.

3.2 Extrudability

Figure 5 shows the relationship between the extrusion force and displacement of 3DPC with RLP. The extrudability of 3DPC was evaluated by the extrusion force. The larger the extrusion force, the more difficult the extrusion; on the contrary, the smaller the extrusion force, the better the extrudability. From the figure, the extrusion force increases sharply when the extrusion displacement is within 10 mm, which is caused by the uneven local force of the 3DPC at the beginning of the pressure. When the extrusion displacement was greater than 10 mm, the 3DPC with 3% and 4% of RLP presented a smaller extrusion force and varied around 0.03 kN with the increase of the extrusion displacement. This is attributed to the lubrication effect between RLP particles, which improves the workability of the fresh mortar [22]. Therefore, the addition of a suitable amount of RLP is beneficial to the extrudability of 3DPC. Moreover, when the dosage of RLP is about 3%–4%, 3DPC has suitable extrudability.

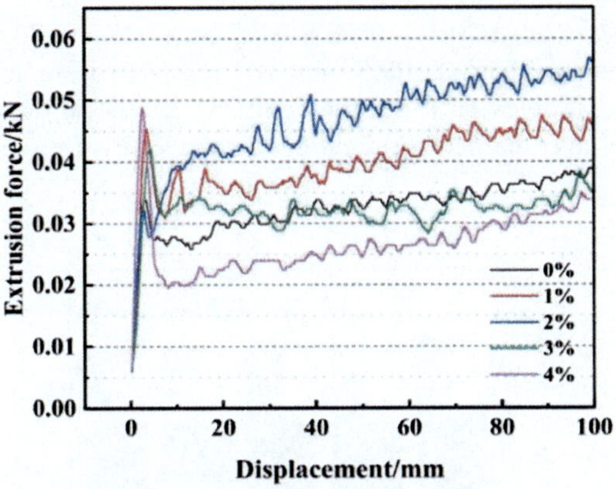

Fig. 5. Test results of extrudability

3.3 Open Time

Figure 6 shows the test results of the open time of 3DPC with 1%-4% of RLP. With the increase of RLP, the open time of 3DPC shows a parabolic trend. The open time of 3DPC with RLP was higher than that of the reference sample without RLP. The open

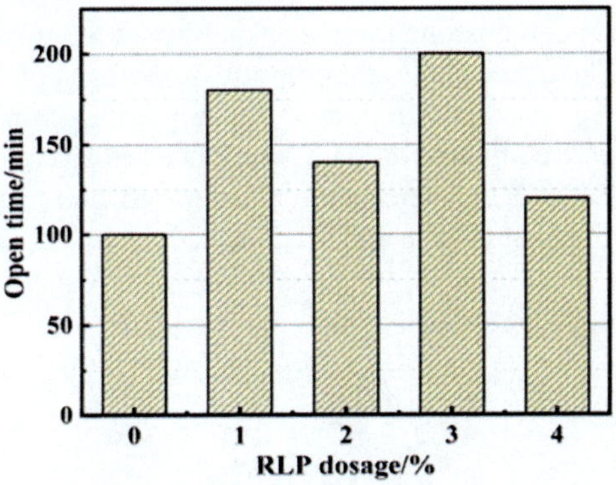

Fig. 6. Test results of open time

time of 3DPC with 1%, 2%, 3%, and 4% was higher 80%, 40%, 100% and 20% than the reference sample, respectively. This can indicate that RLP enhances the water retention of the mortar and prevents the loss of water. This property enables 3DPC containing RLP to maintain good printing performance for a long period of time.

3.4 Buildability

Figure 7 shows the buildability test results for the samples. The upper diameter and lower diameter of the reference sample without RLP are 76 mm and 111 mm. The upper diameter of the 3DPC with 1%–4% of RLP is 76 mm, 76 mm, 76.5 mm, and 76 mm, and that of the lower diameter is 108.5 mm, 105 mm, 107 mm, and 105 mm. As can be seen, the incorporation of RLP has no significant effect on the upper diameter but slightly reduces the lower diameter, indicating that the RLP improves the buildability of the 3DPC.

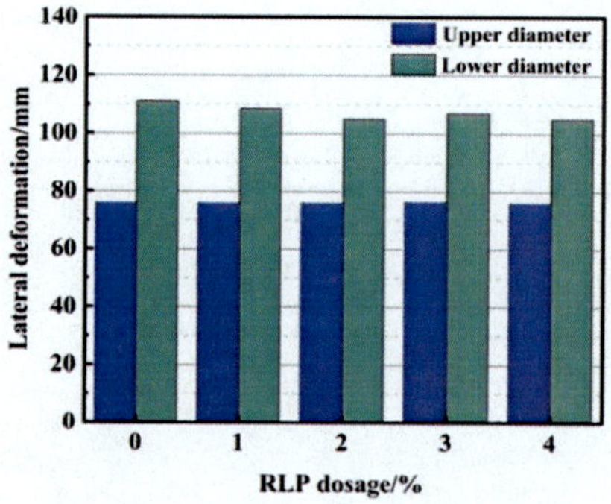

Fig. 7. Test results of buildability

3.5 Green Strength

Figure 8 (a), (b), and (c) demonstrate the green strength of 3DPC with 0%, 1%, and 3% of RLP, respectively. The figure shows that the stress of all samples gradually increased with time, and stress increased approximately linearly with strain. The stress of 3DPC with 1% of RLP decreases compared to 3DPC without RLP. The slope of the stress-strain curve for 3DPC with 3% of RLP increases at a faster rate, and the stress is greater than that of the reference sample at 60 min and 90 min, with increases of 15.51% and 1.34%.

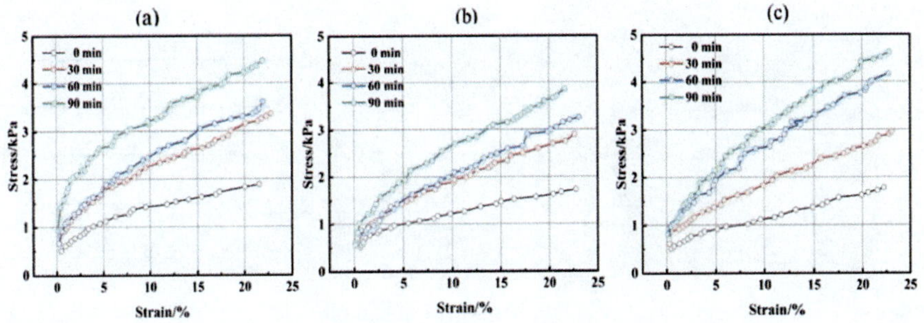

Fig. 8. Test results of green strength. (a) 3DPC without RLP, (b) 3DPC with 1% of RLP, and (c) 3DPC with 3% of RLP.

3.6 Strength

Figure 9 (a) shows the results of the flexural strength. As can be seen from the figure, the addition of RLP decreases the flexural strength of 3DPC. The flexural strengths of the cement mortar in the reference sample at 7 d and 28 d were 2.88 MPa and 3.60 MPa, respectively. The lowest flexural strength of 3DPC was reached at 1% RLP admixture, and its 7 d and 28 d flexural strengths were 2.31 MPa and 2.75 MPa, which were reduced by 19.79% and 23.61%. At the RLP admixture of 2%, the strength loss rate was lower compared to the cement mortar without RLP admixture, which was 3.47% and 2.22%.

As illustrated in Fig. 9 (b) for the test results of compressive strength, the variation pattern is basically the same as that of flexural strength. However, it is clear that the compressive strength at 2% RLP admixture loses 17.39% and 13.35% at 7 d and 28 d, compared with the reference sample. It can be seen that the admixture of RLP will reduce the strength of cement mortar. The decrease in strength of 3DPC after RLP admixture was correlated with their pore structure. As shown in Fig. 10, with the increase of RLP admixture, its air-entraining effect [11] increased both the pore space and pore size of the specimens, thus decreasing the compactness of the material.

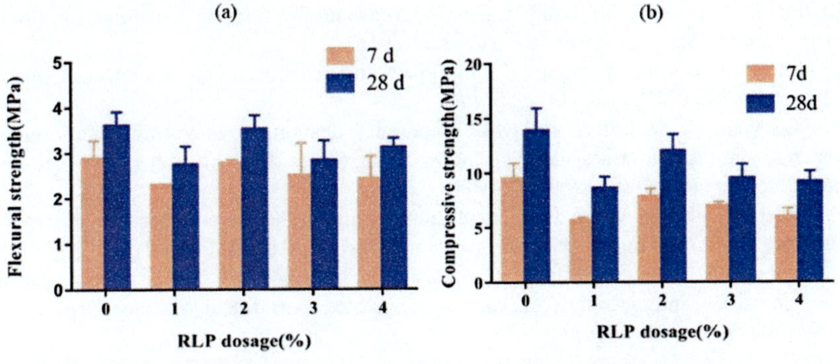

Fig. 9. Results of flexural strength and compressive strength

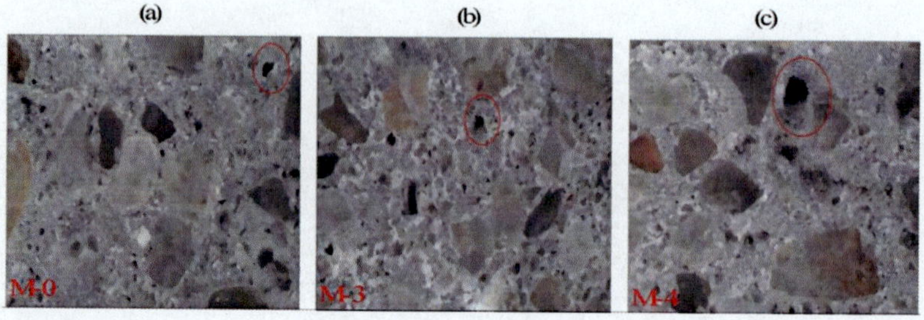

Fig. 10. Air voids of sample

4 Conclusions

(1) The admixture of RLP can significantly improve the printability of 3DPC, but it has a negative effect on the strength, and the reduction rate of the strength is low when the dosage is 2%.
(2) The admixture of RLP can improve the buildability and open time of 3DPC. At 2–3% of RLP admixture, 3DPC has suitable buildability and open time.
(3) The admixture of RLP reduces the green strength of 3DPC, and the green strength loss rate is small when the dosage is 3%.

Acknowledgments. This study was funded by the Guizhou Provincial Science and Technology Foundation (grant no. [2020] 1Y246).

References

1. Ovsianikov, A., Mironov, V., Stampfl, J., Liska, R.: Engineering 3D cell-culture matrices: multiphoton processing technologies for biological and tissue engineering applications. Expert Rev. Med. Devices **9**(6), 613–633 (2012)
2. Kading, B., Straub, J.: Utilizing in-situ resources and 3D printing structures for a manned mars mission. Acta Astronaut **107**, 317–326 (2015)
3. Le, T.T., et al.: Hardened properties of high-performance printing concrete. Cem. Concr. Res. **42**(3), 558–566 (2012)
4. Ma, G., Wang, L.: A critical review of preparation design and workability measurement of concrete material for largescale 3D printing. Front. Struct. Civ. Eng. **12**(3), 382–400 (2017). https://doi.org/10.1007/s11709-017-0430-x
5. Zareiyan, B., Khoshnevis, B.: Effects of interlocking on interlayer adhesion and strength of structures in 3D printing of concrete. Autom. Constr. **83**, 212–221 (2017)
6. Chuanfeng, C., Chenxi, L., Siqi, L., Chuanzhen, S., Jingjing, F.: Influence of ceramic sand on mechanical and microscopic properties of 3D printed mortar. China Concr. Cem. Prod. **01**, 7–11 (2022). (in Chinese)
7. Ying, Y., Ben, L., Liangjun, X., Yuxuan, D., Yaona, M., Yongfei, X.: Research on the influence of basalt fiber on the properties of 3D printed recycled cement-based materials. China Concr. Cem. Prod. **12**, 45–49 (2021). (in Chinese)

8. Yi, Z., Yanmei, Z., Qiang, R., Zhengwu, J.: Effects of bentonite on printability of 3D printing mortar. J. Chin. Ceram. Soc. **50**(2), 420–428 (2022). (in Chinese)
9. Jinghui, L., Wenjie, Z.: Research development of dispersible latex powder. Bull. Chin. Ceram. Soc. **35**(12), 4038–4043 (2016). (in Chinese)
10. Jie, Z.: Application of redispersible emulsion powder in self-leveling ground materials. New Build. Mater. **6**, 28–30 (2003). (in Chinese)
11. Phan, V.T.: Relationship between the adhesive properties and the rheological behavior of fresh mortars. Cachan Ecole Normale Supérieure (2012)
12. Walters, D.G.: VAE redispersible-powder hydraulic-cement admixtures. ACI Compilation **23**, 8–12 (1992)
13. Zhang, Y., Zhengwu, J., Yanmei, Z., Jie, Z., Tao, H.: Effects of redispersible polymer powders on the structural build-up of 3D printing cement paste with and without hydroxypropyl methylcellulose. Constr. Build. Mater. **267**, 1–17 (2020)
14. Grzeszczyk, S., Lipowski, G.: Effect of content and particle size distribution of high-calcium fly ash on the rheological properties of cement pastes. Cem. Concr. Res. **27**(6), 907–916 (1997)
15. Li, G., He, T., Hu, D., Chen, S.: Effects of two retarders on the fluidity of pastes plasticized with aminosulfonic acid-based superplasticizers. Constr. Build. Mater. **26**(1), 72–78 (2012)
16. Kwan, A., Wong, H.: Effects of packing density, excess water and solid surface area on flowability of cement paste. Adv. Cem. Res. **20**(1), 1–11 (2008)
17. Huashan, Y., Yujun, C.: Recycling of aggregate micro fines as a partial replacement for fly ash in 3D printing cementitious materials. Constr. Build. Mater. **321**, 1–10 (2022)
18. Sonebi, M., Abdalqader, A., Fayyad, T., Perrot, A., Bai, Y.: Optimisation of rheological parameters, induced bleeding, permeability and mechanical properties of supersulfated cement grouts. Constr. Build. Mater. **262**, 120078 (2020)
19. Huashan, Y., Weiwei, L., Yujun, C.: 3D printing cementitious materials containing Nano-$CaCO_3$: workability, strength, and microstructure. Front. Mater. **7**, 1–15 (2020)
20. Ohama, Y.: Polymer-based admixtures. Cement Concr. Compos. **20**(2–3), 189–212 (1998)
21. Miyake, J.I., Matsushita, H.: Evaluation method for consistencies of mortars with various mixture proportions. J. Adv. Concr. Technol. **5**(1), 87–97 (2007)
22. Liguang, X., Jiancheng, Z.: Influence of redispersible polymer powder on properties of construction mortar. J. Jilin Inst. Archit. Eng. **19**(4), 19–24+42(2002). (in Chinese)

Study on the Deterioration Law of Sleeve Grouting Defects on the Lateral Bearing Capacity of Assembled Shear Walls

Qizhou Liu, Yuhan Lin, Jinzhou Li, and Min Zhang(✉)

College of Civil Engineering and Architecture, Guilin University of Technology, Guilin 541004, Guangxi, China
emailzm@126.com

Abstract. Sleeve grouting defects in assembled buildings leave a major safety hazard for building safety. Assembled shear walls are the most widely used key lateral force resisting element in assembled buildings and their lateral bearing capacity is critical to the safety of the building. However, the deterioration pattern of the defects on the lateral bearing capacity of assembled shear walls is still unclear, which makes it impossible to make an accurate assessment of the performance of defective assembled shear walls and thus to develop reasonable reinforcement solutions. In order to be able to accurately predict the lateral bearing capacity of defective assembled shear walls. A non-linear finite element model of the defective assembled shear wall structure was developed using the finite element software ABAQUS and compared with tests to verify the correctness of the model. Subsequently, the influence of the defects distribution and the defect rate on the ultimate lateral bearing capacity of shear walls was investigated. The results show that defects concentrated on one side of the wall are more harmful than uniformly dispersed defects, and the degree of deterioration effect of individual defects decreases as the number of defects in the same part increases. Finally, the deterioration coefficient, which reflects the degree of deterioration of defects on the lateral load bearing capacity of shear walls, is obtained, and can be used to calculate the lateral bearing capacity of assembled shear walls with defects, providing a reference basis for the performance assessment of assembled buildings.

Keywords: Assembled shear wall · Grouting defect · Finite element model · Lateral bearing capacity · Deterioration coefficient

1 Introduction

A large number of scholars have carried out research work on the seismic performance of assembled shear walls, as they are essential for the seismic safety of high-rise assembled buildings. Soudki et al. [1, 2] compared the seismic performance of assembled shear walls with sleeve grouting connections and bolted connections and made design recommendations for both connection methods. Qian et al. [3] investigated the force transfer

W. Guo and K. Qian (Eds.): GBCESC 2022, LNCE 211, pp. 552–561, 2023.
https://doi.org/10.1007/978-981-19-5217-3_55

performance of through-length reinforcement in grouted sleeve-connected assembled shear walls and demonstrated that the assembled shear walls could have similar seismic performance to that of cast-in-place shear walls through low-cycle repeated loading tests and quasi-dynamic test. Xue et al. [4] investigated the seismic performance of a bolt-sleeve hybrid connection prefabricated shear wall through experiments. Li et al. [5] carried out experimental research work on a T-part precast shear wall and showed that the T-part precast concrete shear wall had good ductility, stiffness and energy dissipation capacity. Min et al. [6] investigated the seismic performance of an assembled shear wall with grouted sleeve grout connections through low-cycle repeated loading tests.

All the above studies assume that the grouting sleeve is completely dense, however, sleeve grouting requires on-site construction and due to the complex site conditions, it is easy to have various grouting defects due to various reasons resulting in incompatible grouting. At present, sleeve grouting defects have become one of the research hotspots in academia. Xu et al. [7] investigated the bond strength between reinforcement and grout in grouting sleeves with defects under uniaxial tensile loading. Guo et al. [8] conducted a uniaxial tensile test on grouted sleeve connection joints and the results showed that the damage mode of grouted sleeve connections shifted from reinforcement fracture to reinforcement pull-out when the defect rate reached 20%. Zhang et al. [9] investigated the mechanical properties of defective grouting sleeve joints after fire. Zheng et al. [10] demonstrated through cyclic loading tests that cyclic loading causes a change in the damage pattern of grouted sleeve joints compared to monotonic loading.

There has been some progress in research on defective grouting sleeves, but there is still very little research into defective shear walls with sleeve grouting. Li et al. [11] investigated the seismic performance of assembled shear walls with defects through low cyclic loading tests and showed that there is a significant deterioration in the seismic performance of shear walls when the defects reduce the flexural load capacity of the wall by 85%. Apart from this, no other research has been reported on the seismic performance of assembled shear walls with defects. As the distribution of defects in the wall is so varied that it is difficult to derive general rules from the few test results available, there is an urgent need to carry out a large number of finite element simulations and theoretical analyses of assembled shear walls with defects, and to obtain a computational model that can characterise the effect of defects on the performance of assembled shear walls from a large amount of analytical data.

Based on the above reasons, this paper firstly establishes a non-linear finite element model of the defective assembled shear wall and verifies the correctness of the finite element model by comparing it with the test results. Afterwards, a large number of finite element simulations are carried out to investigate the effect of defect rate and defect distribution on the lateral bearing capacity of shear walls with defects. Finally, a theoretical calculation model for the lateral bearing capacity of assembled shear wall with defects is developed. This provides a theoretical reference for the performance assessment of assembled shear walls with defects.

2 Finite Element Modelling

2.1 Experimental Details

In order to investigate the effect of defects on the seismic performance of assembled shear walls, Li et al. [11] conducted a low cyclic loading test study on assembled shear walls with defects. In this paper, specimens PW1 and PW2 were selected as the calculation models. Specimens PW1 and PW2 had identical dimensions, reinforcement and number of defects, and defect distribution sites, as shown in Fig. 1, with the difference that the grouting defect rates of PW1 and PW2 were 50% and 100%, respectively.

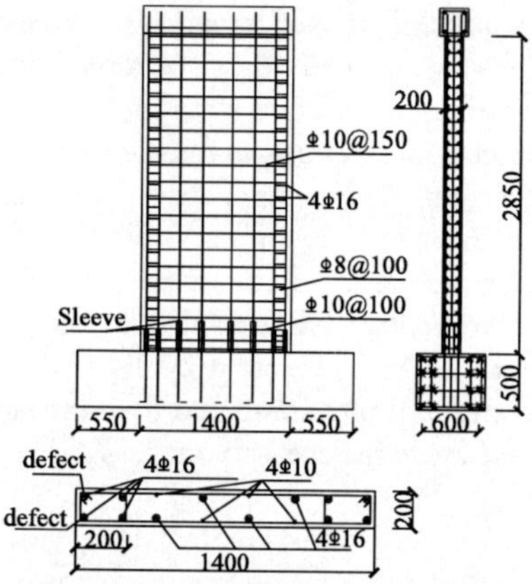

Fig. 1. Assembled shear wall details

The measured cubic compressive strength of the concrete was 55.4 MPa. The yield strengths of 16 mm and 10 mm bars are 439 MPa and 459 MPa, respectively. The tensile strengths of 16 mm and 10 mm bars are 616 MPa and 641 MPa, respectively.

2.2 Element and Boundary Condition

The finite element model was built using the finite element software ABAQUS. The linear reduced integration three-dimensional solid element C3D8R were used to simulate the concrete and grouting sleeve. The reinforcement were simulated by the space truss element T3D2. Fixed constraints were applied at the bottom of the finite element model, vertical forces were applied at the top of the model, and the side edges of the top beam were coupled to a reference point where horizontal displacements were applied. The finite element models of the concrete shear wall and internal reinforcement are shown in Fig. 2.

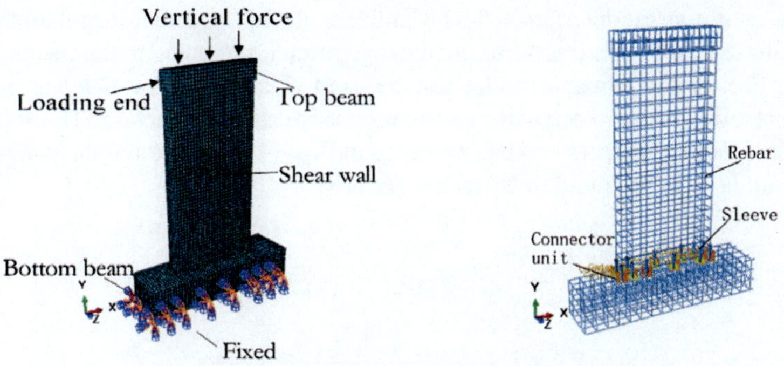

Fig. 2. Finite element models

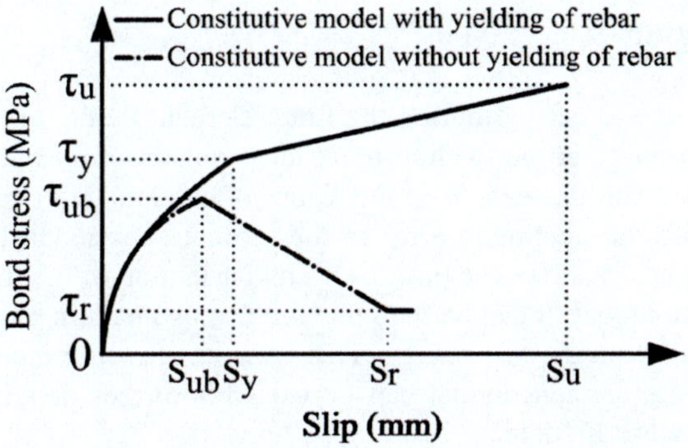

Fig. 3. Constitutive model of bond stress-slip relationship.

Connection units were used to simulate the connection between the reinforcement and the grouting sleeve. The bond-slip constitutive model between reinforcement and grout can be modelled by defining the stiffness and stress-strain relationships of the connection units. For defective grouting sleeves, the key data points of the bond-slip constitutive model in Fig. 3 can be determined in terms of the experiments reported in literature (Xu et al. [7]).

2.3 Material Constitutive Model

An ideal elastic-plastic model is used to simulate the stress-strain relationship of the reinforcement. After the reinforcement has yielded, the stress-strain relationship can be simplified to a horizontal straight line.

The constitutive model proposed by Mander et al. [12] is used to simulate the stress-strain relationship of concrete. A plastic damage model is adopted to define the concrete material. The plastic damage model can be used to simulate the damage behaviour of concrete by defining the tensile and compressive damage factors. The tensile and compressive damage factors can be expressed in Eqs. (1) and (2), and the parameters in Eqs. (1) and (2) can be found in Birtel's work [13].

$$d_t = 1 - \frac{\sigma_t E_c^{-1}}{\varepsilon_t^{pl}(1/b_t - 1) + \sigma_t E_c^{-1}} \tag{1}$$

$$d_c = 1 - \frac{\sigma_c E_c^{-1}}{\varepsilon_c^{pl}(1/b_c - 1) + \sigma_c E_c^{-1}} \tag{2}$$

3 Test Validation

A pushover analysis was carried out on the finite element model to extract the lateral force-displacement curve at the loading point and compare it with the test results, as shown in Fig. 4. As can be seen from the figure, the calculated curves fit well with the test curves, with the maximum error in the ultimate horizontal load capacity not exceeding 10%. Figure 5 shows the plastic strain distribution of the final damage stage of the finite element model. It can be seen that the damage pattern of the finite element model matches well with the test results. This indicates that the modelling process is correct and the finite element model can be used for further parametric analysis of assembled shear walls with defects.

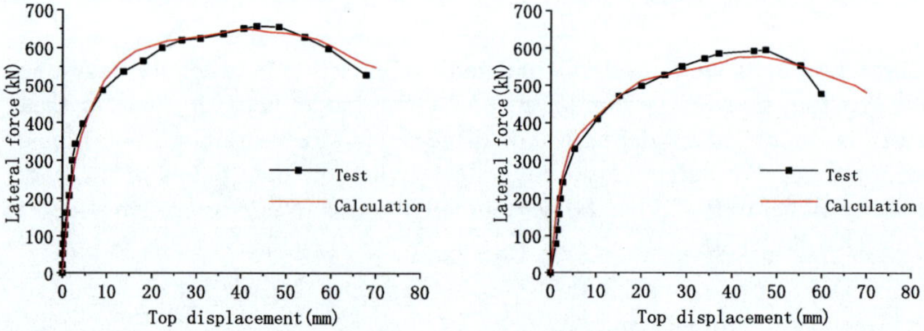

(a) PW1 (b) PW2

Fig. 4. Comparison of lateral force-top displacement skeleton curves

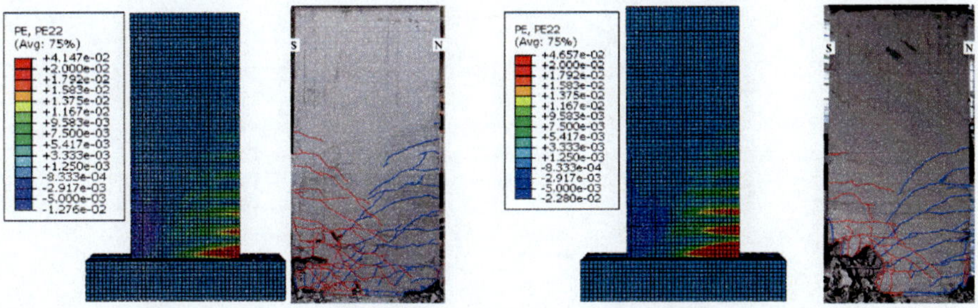

(a) PW1 (b) PW2

Fig. 5. Comparison of damage at the final stage

4 Defective Deterioration Law

In order to further investigate the influence of the defect distribution on the lateral bearing capacity of the assembled shear wall, the finite element models PW7-PW12 were established by further changing the defect distribution site on the basis of the original test specimens PW0-PW6, as shown in Table 1 and Fig. 6. Pushover analysis of the finite element model was carried out to obtain the lateral force-top displacement skeleton curves of the finite element model, as shown in Fig. 7. The ultimate lateral bearing capacity of the assembled shear walls obtained from the test and finite element simulation is shown in Table 2, where the data of PW0-PW6 can be referred to the research results of Li [11].

Table 1. Key parameters for assembled shear walls.

No.	PW 0	PW 1	PW 2	PW 3	PW 4	PW 5	PW 6	PW 7	PW 8	PW 9	PW 10	PW 11	PW 12
Locations of defects	/	① ②	① ②	① ② ③ ④	① ② ③ ④	①② ③④	⑤ ⑥ ⑦ ⑧	①	③	⑤	⑥	①⑪	①② ③④ ⑨⑩ ⑪⑫
Number of defects	0	2	2	4	4	4	4	1	1	1	1	2	8
Defect rate R(%)	0	50	100	25	50	100	50	50	50	50	50	50	50

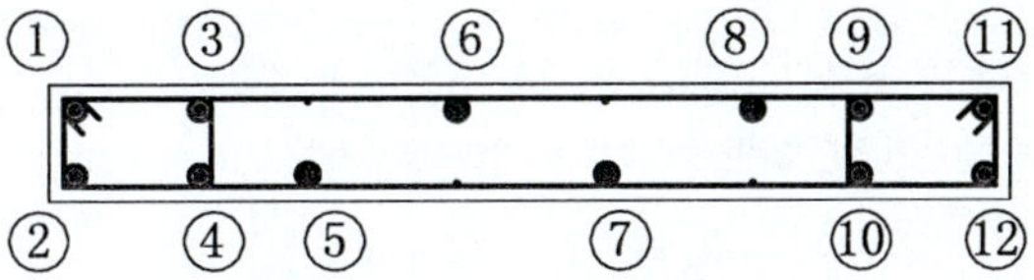

Fig. 6. Location of defects

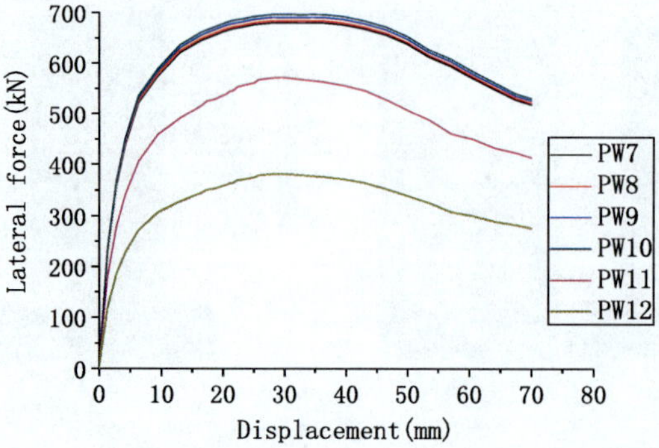

Fig. 7. Lateral force-top displacement skeleton curves

In Table 2, $F_{\mathrm{PWi\text{-}max}}$ represents the ultimate lateral load capacity of shear walls numbered PWi (i = 0–12). α_i, β_i, γ_i and k_i can be expressed by the following equations:

$$\alpha_{\mathrm{i}} = \frac{\sum_{i=1}^{n_1} f_{\mathrm{yi}} A_{\mathrm{si}}}{\sum_{i=1}^{n} f_{\mathrm{yi}} A_{\mathrm{si}}} \tag{3}$$

$$\beta_{\mathrm{i}} = \alpha_{\mathrm{i}} \cdot R \tag{4}$$

$$\gamma_{\mathrm{i}} = (F_{\mathrm{PWi\text{ - }max}} - F_{\mathrm{PW0\text{ - }max}})/F_{\mathrm{PW0\text{ - }max}} \tag{5}$$

$$k_{\mathrm{i}} = \gamma_{\mathrm{i}}/\beta_{\mathrm{i}} \tag{6}$$

where n and n_1 are the number of all sleeve connections and the number of defective connections in the assembled shear walls, respectively. $\sum_{i=1}^{n_1} f_{\mathrm{yi}} A_{\mathrm{si}}$ represents the sum of the tensile bearing capacity of the vertical connecting reinforcement at the location of the defects. $\sum_{i=1}^{n} f_{\mathrm{yi}} A_{\mathrm{si}}$ represents the sum of the tensile bearing capacity of all the vertical connecting steel bars in the assembled shear wall. α_{i} represents the percentage of the vertical tensile bearing capacity of the reinforcement where the defect is located. β_{i} can be considered as the equivalent average defect rate. γ_{i} is the deterioration factor of the ultimate lateral bearing capacity of the shear wall with defects. k_{i} represents the deterioration factor per unit equivalent defect rate. The larger the value of k_{i}, the more significant is the average effect of defects for a given defect rate and defect distribution conditions. If the value of k_{i} is known, the ultimate lateral bearing capacity of the assembled shear wall with defects can be expressed as:

$$F_{\mathrm{PWi}} = F_{\mathrm{D}} \times (1 - k_{\mathrm{i}} \beta_{\mathrm{i}}) \tag{7}$$

where F_{D} is the design lateral bearing capacity of the defect-free shear wall.

Table 2. Analysis of key data.

Number	Number of defects	α_i (%)	Defect rate R (%)	β_i (%)	$F_{\text{PWi-max}}$ (kN)	γ_i (%)	k_i
PW0	0	/	/	/	702	/	/
PW1	2	16.6	50	8.3	654	0.068	0.0082
PW2	2	16.6	100	16.6	600	0.145	0.0088
PW3	4	33.3	25	8.3	672	0.043	0.0051
PW4	4	33.3	50	16.7	556	0.208	0.0125
PW5	4	33.3	100	33.3	480	0.316	0.0095
PW6	4	33.3	50	16.7	568	0.191	0.0115
PW7	1	33.3	50	4.2	679	0.033	0.0079
PW8	1	8.3	50	4.2	683	0.027	0.0065
PW9	1	8.3	50	4.2	689	0.019	0.0045
PW10	1	8.3	50	4.2	695	0.010	0.0024
PW11	4	33.3	50	16.7	572	0.185	0.0111
PW12	8	66.6	50	33.3	381	0.457	0.0137

As can be seen from Table 2, the *k*-value of PW7 is much greater than that of PW10 under the condition that the number of defects is only one, indicating that the deterioration of the shear wall is greatest when the defects are distributed in the edge area. *k*-values of PW7 and PW8 show a certain increase relative to PW1 and PW2 respectively, indicating that the average deterioration impact of defects increases as the number of defects in the same area increases. Comparing the *k* values between PW3, PW4 and PW5 shows that there is a significant increase in the average deterioration impact of defects when the defect rate is increased from 25% to 50%, while the average deterioration impact of defects even decreases slightly when the defect rate is increased from 50% to 100%. Compared to PW11, there is a certain increase in the k-value for PW4, indicating that defects are more harmful when they are concentrated on one side than when they are evenly dispersed on both sides. When the equivalent average defect rate β_i exceeds 16%, the ultimate lateral bearing capacity deterioration factor γ_i of the shear wall is all close to or greater than 15%, therefore, when the equivalent average defect rate β_i exceeds 16%, necessary strengthening measures should be taken for the shear wall with defects.

5 Conclusion

(1) The comparison with the experimental results shows that the finite element models developed in this paper agrees well with the experimental results and can be used for further parametric analysis.

(2) The deterioration effect is greater when the defects are distributed in the edge of the shear wall than those distributed in the central area of the shear wall.

(3) When the number of defects in the same area increases, the average deterioration effect of the defects increases.
(4) Defects concentrated on one side of the shear wall have a more significant deterioration of the lateral carrying capacity of the shear wall than those evenly dispersed on both sides of the wall.
(5) The equivalent average defect rate is defined. When the equivalent average defect rate is higher than 16%, the reduction in the lateral load capacity of the shear wall approaches or even exceeds 15%, and in such cases the necessary strengthening measures should be given to the shear wall.

Acknowledgement. The work was supported in part by the Guangxi Science and Technology Base and Special Fund for Talents Program (Grant No. Guike AD19110073), the Guangxi Natural Science Foundation Program (Grant No. 2018GXNSFAA050145), the National Natural Science Foundation of China (Grant No. 52068013) and the Innovation Project of Guangxi Graduate Education (Grant No. YCSW2021212).

References

1. Soudki, K.A., West, J.S., Rizkalla, S.H., et al.: Horizontal connections for precast concrete shear wall panels under cyclic shear loading. PCI Journal **41**(3), 64–80 (1996)
2. Soudki, K.A., West, J.S., Rizkalla, S.H., et al.: Horizontal connections for precast reinforced and prestressed concrete walls. J. Struct. Eng. **129**(3), 286–296 (2003)
3. Peng, Y.-Y., Qian, J.-R., Wang, Y.-H.: Cyclic performance of precast concrete shear walls with a mortar–sleeve connection for longitudinal steel bars. Mater. Struct. **49**(6), 2455–2469 (2015). https://doi.org/10.1617/s11527-015-0660-0
4. Weichen, X., Mingxiao, C., Yanan, L., et al.: Seismic performance of new precast concrete shear wall under high axial compression ratio. J. Harbin Eng. Univ. **39**(257), 58–66 (2018). (in Chinese)
5. Jianbao, L., Qiaoqiao, F., Zheng, L., et al.: Experimental study on seismic performance of T-shaped partly precast reinforced concrete shear wall with grouting sleeves. Struct. Des. Tall Spec. Build. **28**(13), e1632.1-e632.13 (2019)
6. Min, W., Xiang, L., Hongtao, L., et al.: Seismic performance of precast short-leg shear wall using a grouting sleeve connection. Eng. Struct. **208**, 110338 (2020)
7. Feng, X., Kai, W., Shuguang, W., et al.: Experimental bond behavior of deformed rebars in half-grouted sleeve connections with insufficient grouting defect. Constr. Build. Mater. **185**, 264–274 (2018)
8. Hong, G., Jiaxin, Z., Changwei, W.: Experimental study on influence of connection defects on joint strength of half-grouted sleeve splicing of rebar. Adv. Civ. Eng. **2020**(3), 1–15 (2020)
9. Wangxi, Z., Jia, W., Jinyi, Z., et al.: Experimental study on post-fire performance of half grouted sleeve connection with construction defect. Constr. Build. Mater. **244**(1), 118165 (2020)
10. Guanyu, Z., Zhiping, K., Jianzhuang, X., et al.: Mechanical performance for defective and repaired grouted sleeve connections under uniaxial and cyclic loadings. Constr. Build. Mater. **233**, 117233 (2020)
11. Shun, X., Zhuolin, W., Xiangmin, L., et al.: Study of effects of sleeve grouting defects on the seismic performance of precast concrete shear walls. Eng. Struct. **236**, 111833 (2021)

12. John, B.M., Mjn, P., Robert, P.: Theoretical stress-strain model for confined concrete. J. Struct. Eng. **114**(8), 1804–1826 (1988)
13. Veit, B., Peter, M.: Parameterized finite element modeling of RC beam shear failure. In: 2006 ABAQUS User's Conference, pp. 95–108. Taiwan (2006)

Monitoring Technology of Runway Pavement with Holes Based on Vertical Shear Strain

Jing Cai[1], Hanxue Wang[1], Yue Li[1(✉)], and Zhiqiang Du[2]

[1] School of Transportation Science and Engineering, Civil Aviation University of China, 300300 Tianjin, China
Leoliyue@163.com

[2] Shaanxi Railway Engineering Vocational and Technical College, Weinan 714000, China

Abstract. Analytical solution of flexural-tensile strain and vertical shear strain in slab on elastic foundation were obtained based on theory of moderate thickness slab and the relation between vertical shear strain and flexural-tensile strain on slab bottom was obtained by the solution. The finite element model of slab with holes was built by ABAQUS to analyze the distributing characteristics and influence factors of vertical shear strain in holes under aircraft load. The model test of slab with holes was designed to obtain the vertical shear strain and its distribution characteristics in the holes. The research results show that there is a linear correlation between vertical shear strain and flexural-tensile strain with the same location of aircraft load. The distributing characteristics of vertical shear strains are different in different holes. In addition the results obtained from the model experiment show that distributing characteristics of vertical shear strain in holes are in accord with theory resolution and finite element analysis results. The shear strain in the holes increases with the load's increasing. Flexural-tensile strains on the slab bottom calculated by vertical shear strain in holes differ from the measured values with maximum value of 16% and this value is conform to the engineering test require, which indicates that the monitoring technology of flexural-tensile strain calculated by shear strain is feasibility and the state of the existing pavement slab can be monitored with this technique.

Keywords: Slab of moderate thickness · Runway pavement with holes · Vertical shear strain · Model experiment · Flexural-tensile strain

1 Introduction

As an important infrastructure of airport operation, the damage of airport pavement structure is a major threat to the safe and efficient operation of aircraft. At present, in the aspect of airport pavement structure state detection, the International Civil Aviation Organization (ICAO) and China's civil airport operation safety management regulations [1, 2] require a comprehensive evaluation of airport pavement security every five years. However, due to the rapid increase in the number of flights taking off and landing at domestic airports, the above-mentioned test period is obviously too long. At the same time, in the daily maintenance of the runway, it is impossible to perceive the stress state

W. Guo and K. Qian (Eds.): GBCESC 2022, LNCE 211, pp. 562–570, 2023.
https://doi.org/10.1007/978-981-19-5217-3_56

of the pavement structure simply by visual inspection, and it is impossible to judge and evaluate the stress state of the pavement structure in time. According to China's "Technical Specification for the Evaluation of Civil Airport Road Surface Management [3], the Falling Weight Deflectometer testing (FWD) is adopted to test the existing pavement structure performance [4], but the test interval of this method is 1–5 years, obviously, this method cannot be used to evaluate the pavement structure performance in time. Therefore, it is especially important to monitor the stress and strain in the runway panel in real time and evaluate the state of the pavement structure in real time. The health monitoring technology of engineering structure has been widely studied, and the health monitoring technology of airport runway has also received more and more attention. In 1992, the United States Federal Aviation Administration (FAA) established the cement concrete pavement test area during the construction of Denver international airport, and buried 123 strain gauges to measure the pavement strain and 50 displacement sensors to measure the vertical displacement of the structure of each layer of the pavement, and collected real-time performance parameters of the pavement layer when different types of aircraft passed [5, 6]; National Airport Pavement Test Facility (NAPTF) of FAA in the United States buried multiple dynamic sensors in 3 kinds of rigid pavement and 6 kinds of flexible pavement, conducted long-term load cycling test of aircraft, and monitored and analyzed the mechanical response of rigid pavement [7, 8]; During the construction of the fourth runway of pudong airport, more than 300 sensors have been embedded in each section and depth of the runway and the contact track to collect five kinds of parameters such as temperature, humidity, strain, displacement and load in real time, providing support for the safety evaluation, prediction and early warning of airfield facilities in the flight area [9]. Yang Laihua, based on the Jilin Airport Pavement Project, used temperature and strain sensors and an intelligent acquisition platform to carry out research on the distribution law of pavement temperature and strain [10]. Based on the above analysis, it is found that all the research objects are pavement structures under construction or made in test, and all use embedded sensors, so the survival rate is difficult to be guaranteed and cannot be maintained or replaced. However, more than 90% of the runways in China are existing runways in operation, so it is impossible to embed sensors for stress monitoring. However, more than 90% of runways in China are in operation and maintenance, and pressure monitoring sensors cannot be embedded. In order not to damage the surface structure of the runway, a stress monitoring sensor shall be installed in the hole where the navigation light is installed. The sensor shall be installed vertically along the hole wall to pick up the vertical shear strain in the hole.Therefore, based on Finite element model of panel with holes, the vertical shear strain distribution law and its influencing factors under the plane load are analyzed. The model test with orifice plate proposed a non-destructive real-time monitoring technique based on the vertical shear strain in the hole.

2 Analysis of Shear Strain in the Hole of Runway Panels

2.1 Finite Element Model of Panel with Holes

According to the provisions on navigation visual aids in Annex 14 of the Convention on International Civil Aviation [2], its specific layout is shown in Fig. 1. Based on the relevant data, the finite element model is established by using ABAQUS software, Fig. 2 shows the distribution and load position of light holes (A, B, C).

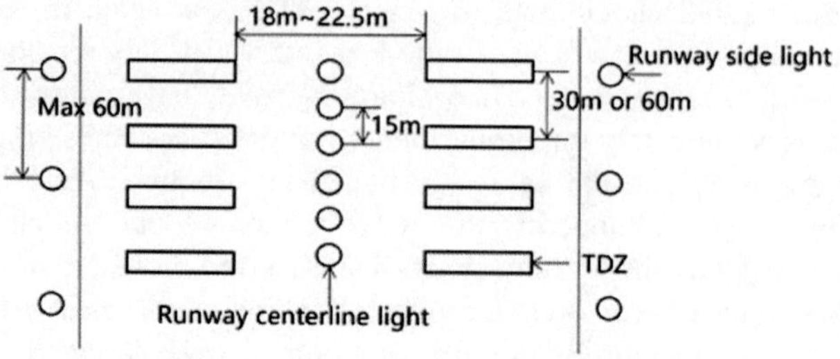

Fig. 1. Runway lighting arrangement

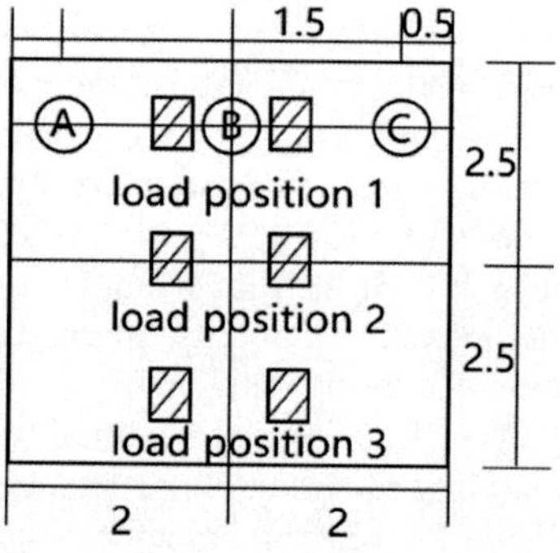

Fig. 2. Model grid distribution map

2.2 Distribution Rule of Vertical Shear Strain in Holes

Influence of Opening on the Vertical Shear Strain of the Runway Panel. The open-hole plate has stress concentration under the external load due to the existence of the hole [11]. The distribution rule of shear strain in the plate after the opening has changed, leading to the change in the direction of shear strain at some points. The ratio between the vertical shear strain in the hole of the perforated plate and the vertical shear strain at the corresponding position of the unperforated plate is defined as the correction factor k of shear strain.

Influence of Aperture on Vertical Shear Strain of Runway Panel. The Sect. 2.1 model is used to analyze the variation rule of the vertical shear strain in holes A and B with aperture d of 10, 20 and 25 cm at load position 1, as shown in Fig. 3. The vertical shear strain distribution of holes at different positions is different. The side hole a is oval, while the center hole B is double circular. The strain of the center hole B is greater than that of the side hole a. In hole B, the shear strain on the side away from the edge of the plate is about twice the strain on the side close to the edge of the plate. It can be seen that the hole diameter D has little effect on the shear strain in the hole. Therefore, the effect of pore size on shear strain can be ignored.

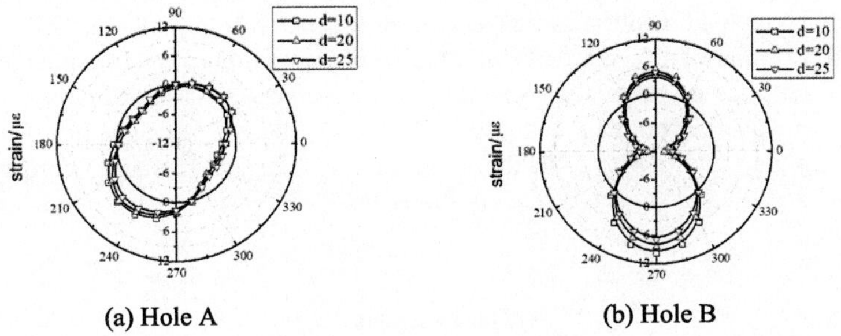

Fig. 3. Vertical shear strain distribution curves around different apertures

Influence of Loading Position. The variation law of plate hole shear strain with loading position is analyzed. Three different loading positions of loading positions 1, 2 and 3 are taken, as shown in Fig. 2. The distribution of shear strain in hole a and hole B is shown in Fig. 4 the hole is small; When the wheel load is at load position 1, the shear strain in the hole can reach 10 micro strains. The research results provide a basis for the layout of test points.

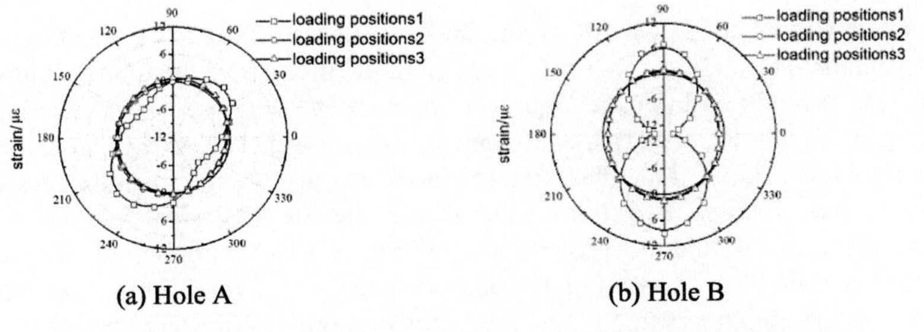

Fig. 4. Vertical shear strain distribution curves at different load locations

3 Pavement Stress Detection Test Based on Vertical Shear Strain

3.1 The Design of Test Board and Arrangement of Measuring Points

Taking the single runway panel at the grouting belt as the research object, the designed perforated plate is 4.0 m long, 0.1 M thick and 2.0 m wide along the driving direction of the wheel. Three load positions are selected, three holes with a diameter of 0.1 M and a spacing of 1.5 m are arranged, and FBG strain sensors are placed in the hole and at the bottom of the plate respectively. Considering the symmetry of position and strain, the shear strain measuring points are located in hole a and hole C respectively. The specific arrangement is shown in Fig. 5. At the same time, two bending strain measuring points G1 and G2 are set 5 cm and 10 cm from the bottom of the plate to the edge of the plate respectively.

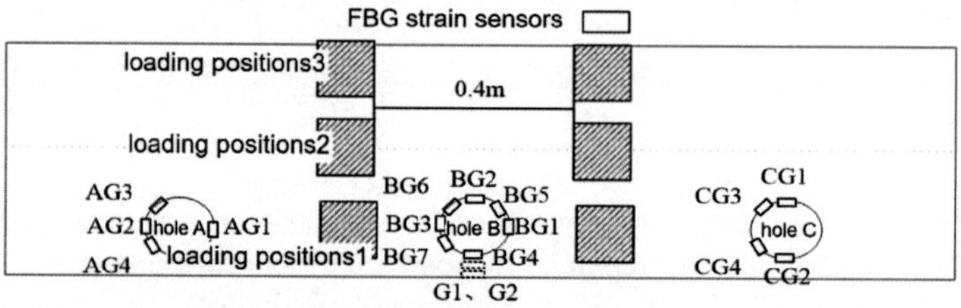

Fig. 5. Load and monitoring points arrangement in model experiment

3.2 Analysis of Strain Test Results in the Plate

Analysis of the Bending Strain of the Bottom of the Plate. The bending tensile strain is obtained through the theoretical solution. the theoretical flexural-tensile strains obtained were modified in consideration of the concentrated effect of the hole on the runway panel, and the correction coefficient k was set as 1.1. The comparison results between the test value and the theoretical value of the plate bottom bending are shown in Table 1. It can be seen from Table 1 that the test results are basically consistent with the theoretical results, and the maximum difference is 13%. With the increase of load, the bending strain at the bottom of the plate increases. Due to different positions, the strain values are slightly different, indicating the reliability of the test sensor.

Table 1. Comparison of test and theoretical values of flexural-tensile strain on slab bottom

Load/kN	Measuring point	Load position 1			Load position 2			Load position 3		
		Test /με	Theory /με	Error /%	Test /με	Theory /με	Error /%	Test /με	Theory /με	Error /%
24	G1	11.8	13.4	11.9	9.8	10.0	2.0	7.7	8.1	4.9
	G2	12.8	13.8	7.2	9.3	10.0	7.0	8.0	8.2	2.4
72	G1	41.4	40.3	2.7	25.2	29.1	13.4	19.8	22.4	11.6
	G2	45.2	41.5	8.9	27.6	30.1	8.3	22.0	24.8	11.3
96	G1	92.9	107.4	13.5	64.7	80.4	19.5	54.5	63.1	13.6
	G2	111.6	110.7	0.8	82.6	80.4	2.7	72.7	66.3	9.6

Vertical Shear Strainin the Hole. Vertical shear strain in the hole. The test data of shear strain under load conditions of 72 kn and 96 kn are shown in Table 2. The distribution law of vertical shear strain along the hole wall (expressed in angle) of hole a and hole B under different load conditions and different load positions is analyzed, and compared with the theoretical solution and finite element simulation analysis results. Figure 6 shows the results when the load is 96 kn.

Table 2. Vertical shear strains in holes A,B and C

Load position	Load/kN	A/με				B/με						C/με			
		1	2	3	4	1	2	3	5	6	7	1	2	3	4
1	72	−1.32	2.42	3.36	−1.75	1.89	−4.14	4.27	−4.20	7.83	4.36	2.30	4.82	−3.66	3.78
	96	−2.31	3.04	4.05	−2.03	2.68	−6.44	5.37	−4.90	9.83	5.44	4.25	5.42	−5.99	4.56
2	72	1.17	2.09	2.25	2.01	1.26	−1.89	3.61	2.25	3.73	2.85	1.31	2.41	2.41	1.38
	96	3.01	4.23	4.77	4.79	2.31	1.48	6.60	5.19	10.07	5.39	2.78	4.99	4.99	3.00
3	72	1.66	1.06	2.69	−1.07	0.03	−0.33	2.24	2.02	5.74	2.47	−1.07	−0.13	−0.13	0.46
	96	2.21	3.99	7.09	6.37	1.69	−0.55	6.84	5.97	7.98	5.17	−7.20	−2.07	−2.07	3.18

It can be seen from Table 2 and Fig. 6 that the shear strain in the pores increases with the increase of load. Although there are some errors, the shear strain data are basically consistent with the theoretical solution and finite element analysis results. It can be considered that the data collected by FBG sensor is effective and real, and can be used to calculate the bending and tensile strain at the bottom of the plate.

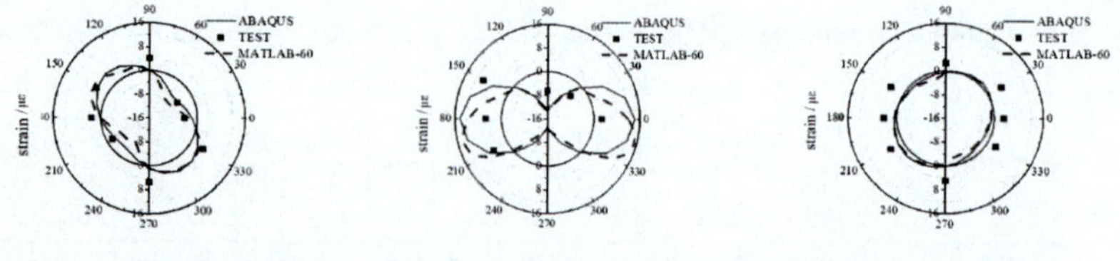

(a) 96kN Load Position1 A (b) 96kN Load Position1 B (c) 96kN Load Position2 A

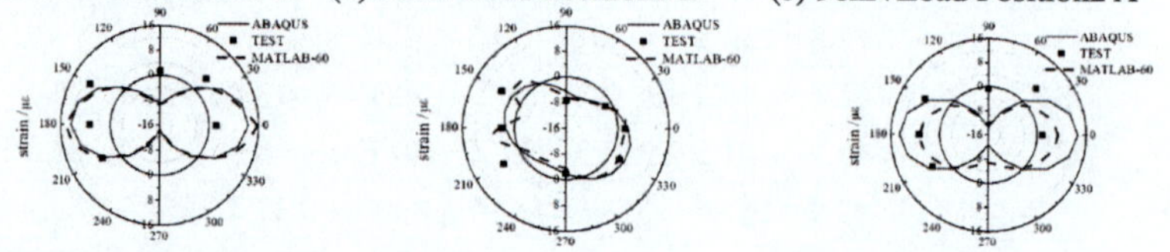

(d) 96kN Load Position2 B (e) 96kN Load Position3 A (f) 96kN Load Position3 B

Fig. 6. Vertical shear strain distribution curves with 96kN load

3.3 Calculation of Flexural-Tensile Strain at Bottom of Plate by Vertical Shear Strain in Hole

The correlation between the vertical shear strain of the plate bottom and the flexural tensile strain is:

$$\varepsilon_x(x, y, z) = \delta(a1, a2, b1, b2) \times \gamma_{xy}(x, y, z) \quad (1)$$

According to the parameters of the test model, the flexural-tensile strain at the G1 measuring point at the bottom of the plate and the shear strain at the positions of AG1–4, CG1–4 and BG1–7 in the hole, δ1, δ2, δ3 are shown in Table 3 and Table 4, respectively corresponding to three different loading position.

Table 3. The ratio between flexural-tensile strain and shear strain in holes A and C

Coefficient	AG1	AG2	AG3	AG4	CG1	CG2	CG3	CG4
δ1	−36.5	207.7	18.4	−18.1	206.1	60.9	−13.8	24.9
δ2	−81.7	90.4	−107.4	46.9	80.37	−20.61	121.8	−36.2
δ3	26.1	7.5	18.1	−22.5	−10.4	−15.4	−27.3	19.9

Table 4. The ratio between flexural-tensile strain and shear strain in hole B

Coefficient	BG1	BG2	BG3	BG5	BG6	BG7
δ1	8.9	−6.3	9.1	−189.1	338.8	8.4
δ2	5.1	−8.1	6.5	56.7	12.2	12.5
δ3	9.4	−4.8	9.4	−64.3	37.9	22.4

The vertical shear strain measured in different holes was substituted into Eq. (1) to obtain the bending tensile strain. Table 5 and Table 6 show that the maximum error between the calculated bending strain and the measured value at the bottom of the plate is 16%, which is consistent with the engineering test. It is pointed out that the shear strain can be used to calculate the bending tensile strain at the bottom of the plate, so as to obtain the bending tensile stress at the bottom of the plate.

Table 5. Comparison between the calculated and test flexural-tensile strain with 72 kN

Project	Position1		Position2		Position3	
	AG1 (με)	AG3 (με)	BG7 (με)	BG3 (με)	CG3 (με)	CG4 (με)
Measured shear strain	−1.32	4.27	5.39	6.84	−1.27	2.24
Measured flexural-tensile strain	46.26	46.26	27.61	27.61	21.99	21.99
Calculated flexural-tensile strain	48.18	38.88	23.47	32.76	24.08	21.06
Error (%)	14.07	15.9	15.00	15.72	8.68	4.23

Table 6. Comparison between the calculated and test flexural-tensile strain with 96kN

Project	Position1		Position2		Position3	
	AG1 (με)	AG3 (με)	BG7 (με)	BG3 (με)	CG3 (με)	CG4 (με)
Measured shear strain	−2.31	−5.99	5.39	6.84	−2.07	3.18
Measured flexural-tensile strain	92.94	92.94	64.68	54.49	54.49	54.49
Calculated flexural-tensile strain	84.32	82.62	67.36	64.26	56.48	63.36
Error(%)	9.27	11.10	3.98	15.20	3.52	14.00

4 Conclusion

(1) Based on relevant theories and MATLAB software, the relationship between shear strain and flexural tensile strain at the bottom of the plate is obtained.
(2) The distribution of vertical shear strain of holes in different positions is different. The lateral hole is elliptic, and the central hole is double circular, and the strain of the central hole is greater than that of the lateral hole.
(3) The influence of pore diameter on the shear strain in the hole is small, with a maximum difference of 2 microstrains, so the influence of pore diameter on shear strain can be ignored.

(4) With the increases of load, the shear strain in the hole increases, and the maximum can reach 10 $\mu\varepsilon$. The strain distribution regularity in the hole are basically consistent with the theoretical solution and finite element analysis results. The flexural-tensile strain and the measured value of the bottom of the plate are up to 16%, calculated by the vertical shear strain in the hole, which is in line with the engineering test requirements. It is indicated that the shear strain in the hole can be used to estimate the flexural-tensile strain of the bottom of the plate, thereby monitoring the internal stress state of the existing runway panel.

Acknowledgement. This work was supported by the key project of Tianjin Science and Technology Support Program (14ZCZDGX00001).

References

1. Civil Aviation Administration of China. CCAR-140 Regulations on the operation of civil airports. Civil Aviation Administration of China, Beijing (2007). (in Chinese)
2. International Civil Aviation Organization. Convention on international Civil Aviation attachment 14 Airport. International Civil Aviation, Chicago, USA (2013). (in Chinese)
3. Civil Aviation Administration of China. MH/T5024–2009 Technical specifications of aerodrome pavement evaluation and management. Civil Aviation Administration of China, Beijing (2009). (in Chinese)
4. Yuan, J.: Study on inversion of rigid pavement structure parameters of airport. Tongji University, Shanghai (2007). (in Chinese)
5. Dong, M., Gordon, G.F. and Fang, Y.: Runway Instrumentation at Denver International Airport: Dynamic Sensor Data Processing. Aircraft/Pavement Technology in the Midst of Change, ASCE (1997)
6. Xingtao, W., Jianfeng, C., Guanbao, Y., et al.: Mechanical response of Boeing 747 running on runways. J. Traffic Transp. Eng. **16**(2), 1–9 (2016). (in Chinese)
7. U.S.A Department of Transportation Federal Aviation Administration, Airport pavement design and evaluation, Chicago (2009)
8. Gordon, F., Hayhoe, R.C., Garg, N.: Slow-rolling response tests on the test pavements at the national airport pavement test facility (NAPTF). In: Advancing Airfield Pavements, pp. 30–58 (2001)
9. Hongduo, Z., Can, W., Xiaohong, W., Yuefeng, Z.: Pavement Condition Monitoring System at Shanghai Pudong International Airport Pavement Materials, Structures, and Performance GSP 239. pp, 281–295. ASCE (2014)
10. Laihua Y.: Temperature warpage test and deformation control of concrete pavement in Jilin Songyuan Airport **65**(09), 65–69 (2020). (in Chinese)
11. Shulun, L., Jianghong, X.: Deformation in rectangular plates with a stress-concentrating circular hole **20**(3), 48–57 (1988). (in Chinese)

Research on Vibration Control for Transformers of Off-Shore Wind Turbine

Xianqing Guo[1], Chengyuan Wang[2(✉)], and Hongju Wu[1]

[1] Guangdong Mingyang Electric Co, Ltd, Zhongshan 528400, Guangdong, China
[2] School of Civil and Hydraulic Engineering, Huazhong University of Science and Technology, Wuhan 430074, Hubei, China
m202171426@hust.edu.cn

Abstract. Offshore wind power has gradually become the mainstream of wind power generation, which has the characteristics of more resources, less turbulence and far away from human living areas. However, offshore wind turbines bear great wind and wave loadings in complex working environment, and various components are also prone to damage in work. Among them, transformer is an important working component in offshore wind turbine, with complex internal structure and large size. Therefore, it is necessary to study the motion response of transformer in operation. In order to design a larger offshore transformer that can adapt to a more complex environment to configure today's larger offshore wind turbine and ensure its safety and stability, a damping device can be added to the usual transformer to enhance its anti-vibration ability and reduce its motion response. This paper introduces the modal analysis theory, uses SolidWorks to model the transformer, and uses ANSYS software to calculate and analyze the modal of the transformer before and after setting the vibration damping device and the vibration life at the main frequency. It can be found that the vibration damping device can greatly improve the vibration resistance of offshore wind turbines.

Keywords: Wind Turbine · Transformer · Modal analysis · Vibration · Damping

1 Introductions

With the increasing demand of human society for energy consumption and the deepening awareness of environmental protection, wind power has gradually become a popular direction of energy development, especially offshore wind power [1, 2]. By the end of April 2021, China's offshore wind power grid connected capacity has reached 10.42 million KW. According to industry statistics, the average annual utilization hours of offshore wind power in China are about 2500 h, which is about 500 h higher than the average annual utilization hours of onshore wind power. Thus, offshore wind power has gradually become the mainstream of wind power industry.

Compared to onshore wind power, offshore wind power has the advantages of richer wind resources, stronger predictability, less offshore turbulence, less topographic changes and obstacles. At the same time, the majority of offshore wind power turbines are far away from human society and will not produce potential negative effects such

W. Guo and K. Qian (Eds.): GBCESC 2022, LNCE 211, pp. 571–578, 2023.
https://doi.org/10.1007/978-981-19-5217-3_57

as noise, leading to the opposition of local residents [3]. However, the offshore wind turbines especially the floating offshore wind turbines are far from the land, the technology is relatively immature, the offshore wind and waves are large, and the working environment is complex, so it is very easy to damage the internal components under the wind and wave load [4].

At home and abroad, there have been some studies on the structural strength and anti-vibration ability of wind turbine blades [5], gearbox [6] and tower [7]. However, the research on transformer is very few. Transformer is an indispensable working component of wind turbine, and its internal structure is complex. In the case of external vibration, the internal electrical components are easy to be damaged. In order to ensure the safe operation of large offshore floating wind turbine, it is very meaningful to study the strength and response of transformer working at sea. Previously, Marzieh, Reza et al. have studied the contribution of dampers to the fatigue life of offshore wind turbine [8]. According to this idea, a damping device can also be set on the transformer model to analyze its contribution to the anti-vibration capacity of transformers of wind turbine. This paper studies the changes of mechanical properties of offshore floating wind turbine transformer by using ANSYS simulation analysis under the condition of installing vibration damping device.

2 Governing Equations

For general multi degree of freedom structural system, any motion can be synthesized by its free vibration mode. Finite element modal analysis is the process of establishing modal model and numerical analysis. The essence of modal analysis is to solve the modal vector of the motion equation without damping and without external load with limited degrees of freedom (because the damping of the structure has little influence on its modal frequency and vibration mode, it can be ignored). The matrix expression of the undamped free vibration equation of the system is as follows:

$$[M]\{u\} + [K]\{u\} = \{0\} \tag{1}$$

For linear structural systems, [M] and [k] in Eq. (1) are real symmetric matrices, and the equation has the solution of the following simple harmonic motion form, which is:

$$\{u(x, y, z, t)\} = \{\Phi(x, y, z)\}e^{i\omega_n t} \tag{2}$$

$[K - \lambda_i M]\{\Phi\} = \{0\}$ $\{\Phi\ (x, y, z)\}$ is the amplitude of displacement vector, which defines the spatial distribution of displacement vector $\{u\}$; ω_n is the angular frequency of simple harmonic motion. After substituting Eq. (2) into Eq. (1), the following and $\{\Phi\}$ And ω_n relevant equations:

$$[K - \omega_n^2 M]\{\Phi\}\exp(i\omega t) = \{0\} \tag{3}$$

Equation (3) holds at any time t, so the term containing t is removed to obtain:

$$[K - \omega_n^2 M]\{\Phi\} = \{0\} \tag{4}$$

Equation (4) becomes a typical real eigenvalue problem. The condition that $\{\Phi\}$ has a non-zero solution is that the value of its coefficient determinant is zero, that is:

$$\left|K - \omega_n^2\right|\{\Phi\} = \{0\} \tag{5}$$

or

$$\left|K - \lambda^2\right|\{\Phi\} = \{0\} \tag{6}$$

where, $\lambda = \omega_n^2$. The left side of Eq. (6) is λ Polynomial, a set of discrete roots can be solved λ_i ($i = 1, 2, \ldots, n$), replace Eq. (6) back to Eq. (4) to obtain the corresponding vector $\{\Phi_i\}$ ($i = 1, 2, \ldots, n$), so that the following equation holds:

$$[K - \lambda_i]\{\Phi\} = \{0\}, i = 1, 2, \ldots, n \tag{7}$$

where, λ_i is the ith eigenvalue of the structural system, $\{\Phi\}$ is the corresponding i^{th} eigenvector.

3 Simulation Settings

3.1 Numerical Models

The structure of transformer is complex, so the model should be simplified in modeling. The transformer is mainly composed of steel frame, support, coil, pull rod, base, shell, iron core, etc. the shell is not considered here for the time being. Therefore, the shell is simplified into a counterweight of equal quality, and a three-dimensional model is established through SolidWorks. The simplified structure of the transformer in this calculation is shown in Fig. 1. Then import the 3D model into ANSYS Workbench software. The properties of the main materials of the transformer are shown in Table 1.

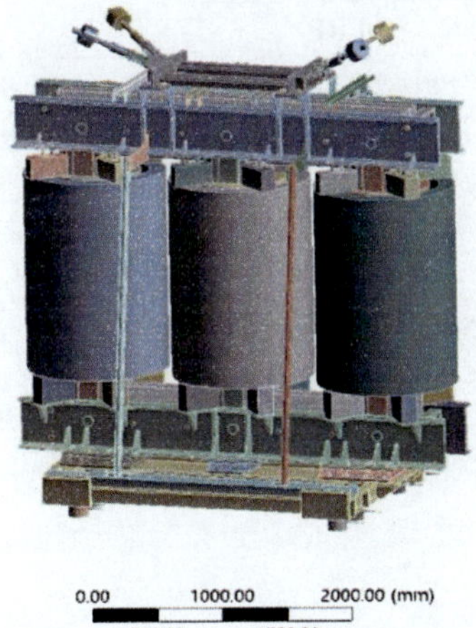

Fig. 1. Schematic diagram of transformer structure

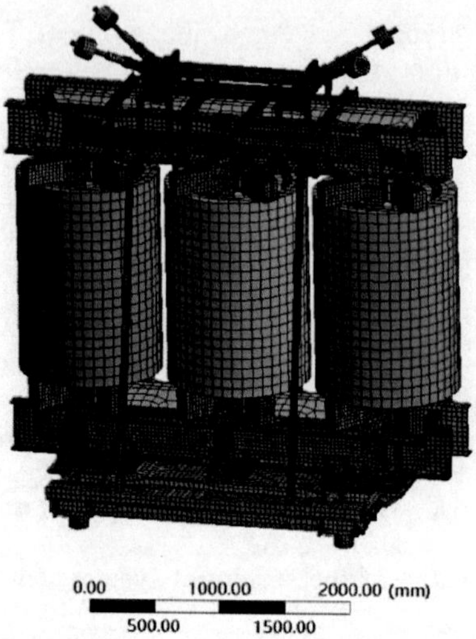

Fig. 2. Diagram of structural grid distribution

Table 1. Material list

Serial number	Part name	Material	Remarks
1	Steel frame	Structural steel	
2	Base	Structural steel	
3	Support	Structural steel	
4	Pull rod	Structural steel	
5	Iron core	Silicon steel sheets	
6	Cushion block	FRP	
7	Insulating plate	FRP	
8	High voltage coil	Self-built materials	Quality is the same as the actual
9	Low voltage coil	Self-built materials	Quality is the same as the actual
10	Shell	self-built materials	quality is the same as the actual

After the transformer entity is transformed into the finite element model, it is divided by hexagonal grid, including 935368 nodes and 240944 elements, as shown in Fig. 2.

The x-axis is the transverse direction of the transformer, the z-axis is the longitudinal direction of the transformer, and the y-axis is the vertical direction of the transformer.

3.2 Damping Device Setting

The vibration damping device is added on the pull rod and support, which on the support is set in the x direction and z direction, and the y direction remains constrained and on the pull rod is arranged along the axial direction of the pull rod. The parameters of vibration damping device are shown in Table 2.

Table 2. The parameters of vibration damping device

Position	Direction	Rigidity (N/mm)	Damping (Ns/mm)
Support	X	7698.6	31.5
Support	Z	5293.8	36.9
Pull rod	Axial	5991	2.49

3.3 Working Cases

When the offshore floating wind turbine is working, the vibration of the wind turbine is mainly caused by the impact of waves and wind on it, so it is necessary to carry out modal analysis. In this analysis, the modal analysis of the transformer model is carried out, its natural frequency is calculated, and the transformer structure is checked. The results of modal analysis are shown in Table 3.

Table 3. The results of modal analysis

Mode	Frequence (Hz)	
	Whether damping device is set?	
	Y	N
1	5.5404	22.003
2	6.9188	24.675
3	8.2418	29.443
4	22.021	29.657
5	22.264	30.149
6	25.506	31.02

Next, we conduct frequency sweep analysis of the model in x, y and z directions with acceleration of 0.5 g, find the frequency with the largest response of the model structure in the range of 2–60 Hz (generally on the main frequency), conduct fatigue analysis at this frequency, and output the life of the structure. The output results are shown in Table 4. and Fig. 3. The analysis results show that the life of structural steel material is much lower than that of other materials. At the same time, due to the limited space, Table 4 only shows the life of transformer structural steel.

Table 4. The life of structural steel of the transformer

Direction	Mode	Life(s)	
		Whether damping device is set?	
		Y	N
X	1	16559	29758
	2	13260	6246.5
	3	14422	1.0466
	4	45411	0.92732
	5	44916	0.71619
	6	39206	0.76451
Y	1	1.7968e5	45448
	2	1.4236e5	40527
	3	1.4444e5	1432.1
	4	46878	1186.1
	5	30435	793.58
	6	30313	878.36
Z	1	56942	14.856
	2	45598	2.6640
	3	52976	2.6640
	4	45411	1.4805
	5	44916	1.4805
	6	39206	1.4805

4 Results and Discussions

Because structural steel is the material with the lowest life, life of structural steel can be regarded as the life of the whole transformer. It can be seen that under the condition of the same structural model, the vibration life of the transformer model equipped with vibration damping device at the main frequency is basically much longer than that of the rigidly connected transformer.

It can be found from the results that the effect of the vibration damping device in the x and z directions is much more obvious than that in the y direction. This is because the vibration damping support in the vibration damping device is only set with stiffness and damping in the x and z directions, and it is still fixed in the y direction. This is considering that in the operation of the transformer, due to the impact of wind and waves, the vibration and displacement in the x and z directions will be very large. Moreover, the size of the transformer in the y direction is very large, and the coil with great mass is wound on the iron core, which is easy to overturn. Therefore, its vibration response in the x and z directions must be controlled.

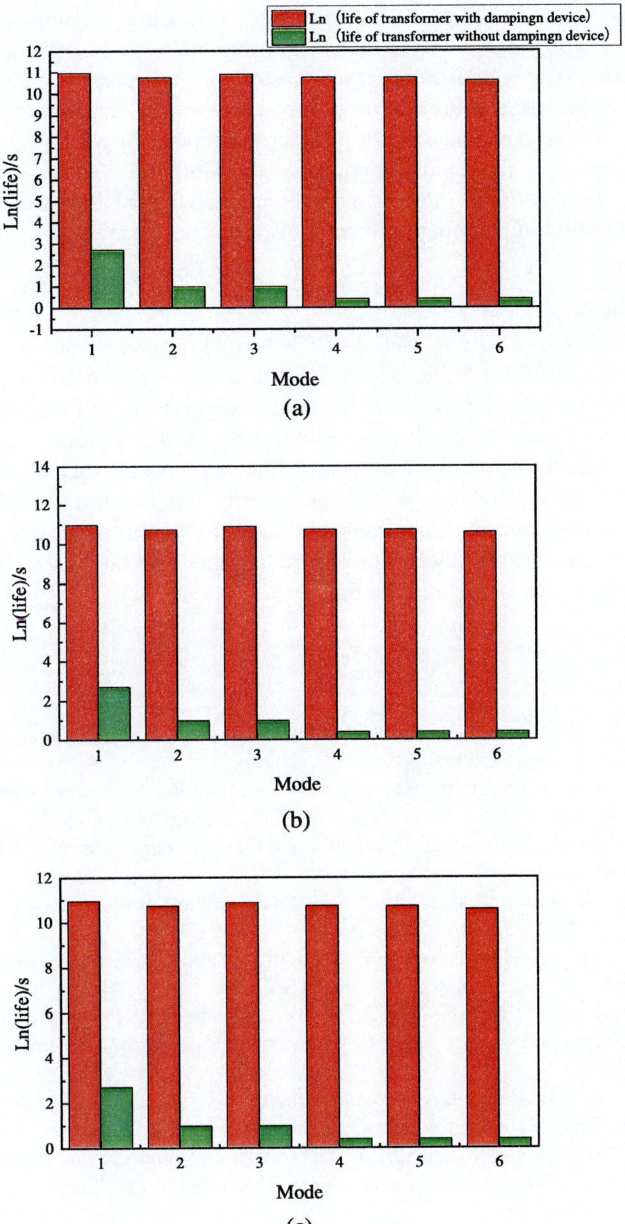

Fig. 3. Transformer life under frequency sweeping condition (a) X direction; (b) Y direction; (c) Z direction.

The damping effect in y direction is mainly due to the action of damping pull rod. The setting of damping tie rod is mainly to reduce the vibration response of the superstructure, help the vibration energy consumption of the structure and prevent the structure from overturning. The direction of the damping tie rod is along the axial direction, which can reduce the vibration and enhance the structural performance in the x, y and z directions. Under the condition of fully fixed connection, the vibration life of the structure in y direction is much longer than that in x and z direction, but it is still lower than expected. Therefore, the setting of damping pull rod is also very necessary.

5 Conclusions

This paper introduces the significance of structural analysis of offshore floating wind turbine transformer, and compares the modal analysis results and vibration life at the main frequency of transformer under the condition of whether vibration damping device is configured or not through SolidWorks modeling and ANSYS analysis. It can be seen that the isolation bearing and damping rod can greatly improve the overall anti vibration performance of the transformer. The analysis conclusion of this paper can provide some ideas for the design of offshore wind turbine transformer with larger size, higher quality and worse working environment in the future.

References

1. Liu, Y., Hajj, M., Bao, Y.: Review of robot-based damage assessment for offshore wind turbines. Renew. Sustain. Energy Rev. **158**, 1364–0321 (2022)
2. Buljac, A., Kozmar, H., Yang, W., Kareem, A.: Concurrent wind, wave and current loads on a monopile-supported offshore wind turbine. Eng. Struct. **255**, 0141–0296 (2022)
3. Sovacool, B.K., Enevoldsen, P.: One style to build them all: corporate culture and innovation in the offshore wind industry. Energy Policy **86**, 402 (2015)
4. Diamond, K.E.: Extreme weather impacts on offshore wind turbines: lessons learned. Nat. Resour. Environ. **27**(2), 37–41 (2012)
5. Hamed, B., Vahid, F.: Edgewise vibration reduction of small size wind turbine blades using shunt damping. J. Vib. Control **26**(3–4), 186–199 (2020)
6. Vamsi, I., Sabareesh, G.R., Supradeepan, K., Penumakala, P.K.: Integrated condition monitoring scheme for bearing fault diagnosis of a wind turbine gearbox. J. Vib. Control **25**(12), 1852–1865 (2019)
7. Martynowicz, P.: Vibration control of wind turbine tower-nacelle model with magnetorheological tuned vibration absorber. J. Vib. Control **23**(20), 3468–3489 (2017)
8. Emami, M., Dezvareh, R., Mousavi, S.A.: Contribution of fluid viscous dampers on fatigue life of lattice-type offshore wind turbines. Ocean Eng. **245**(110506), 0029–8018 (2022)

Hybrid Simulation Platform for Onshore Wind Turbine Under Coupled Wind-Earthquake Load

Hao Hu[1], Zhu Mei[2](✉), Yang Liu[2], and Kao-shan Dai[1]

[1] Department of Civil Engineering, Sichuan University, Chengdu 610065, Sichuan, China
[2] School of Architectural Engineering, Wenzhou University, Wenzhou 325035, Zhejiang, China
zhu.mei@wzu.edu.cn

Abstract. Onshore wind turbines are usually faced the challenges of multiple disasters such as wind and earthquake. With the continuous growing demands on the generating capacity of wind turbine, the length of the tower and blades is gradually increasing. Due to the significant superiority of conducting physical tests on large scaled specimen, such as tower and blades, hybrid simulation is widely used in structural seismic analysis. Combining the numerical simulation and physical testing, it has the ability to solve the problem of multiple loads facing by wind turbines. A software platform suitable for standard hybrid simulation of onshore wind turbine is developed based on Simulink. In order to consider the interaction between the blade deformation and the wind load in the process of hybrid simulation, the software OpenFAST for dynamic analysis of wind turbine under wind load has been developed to realize the transmission of data with Simulink step by step, which provides a new way to implement the hybrid simulation of wind turbines under operating conditions. The accuracy of the software platform is numerically verified, and the cost of time on computation of numerical substructure and data transmission between Simulink and OpenFAST is also tested.

Keywords: Wind turbine · Hybrid simulation platform · Wind-earthquake coupling · Blade deformation · OpenFAST development

1 Introduction

Wind power generation has the advantages of low cost, short cycle, renewable. The construction of wind farm is gradually developing to mountain and ocean, and the tower height and blades size are increasing [1]. The cost of wind tunnel and shaking table test of high-rise structure is too high, and it is difficult to simulate the influence of these two loads. The application of hybrid simulation method to the field of wind turbine provides a way to solve these problems [2].

For the seismic analysis of civil engineering structures, numerical simulation is easy to conduct and repeat, the accuracy however needs to improve when nonlinearity occurs. Due to the limitation of site and equipment, it is difficult to carry out full-scale or large-scale physical tests on large-size and complex structures in laboratory. The complex mechanical behavior of structures is usually limited to a small part of

W. Guo and K. Qian (Eds.): GBCESC 2022, LNCE 211, pp. 579–590, 2023.
https://doi.org/10.1007/978-981-19-5217-3_58

the components, which makes it possible to study the dynamic behavior of the rest by numerical simulation; this is the basis of substructures hybrid simulation. As a relatively economical and accurate research method [3], hybrid simulation reduces the cost of seismic research of building structures [4, 5].

In order to carry out research by hybrid simulation, scholars have developed a variety of hybrid simulation software platforms. Takahashi [6] developed an open source test platform OpenFresco which is able to interact with finite element software OpenSEES and compatible with a variety of loading systems; Kwon [7] developed UI-SimCor based on Matlab language to support a variety of finite element software such as ABAQUS, OpenSEES, etc., but the computing speed is restricted; Xiao [8] developed a networked structure laboratory NetsLab to support remote collaborative experiments in multiple locations. Yang [9] developed a HyTest software platform, which can work with finite element software OpenSEES and loading system MTS to carry out standard hybrid simulation.

These hybrid simulation platforms above promote the development of the hybrid simulation, but it is difficult to analyze the dynamic characteristics of the wind turbine and the load other than earthquake for the time being. Different from seismic load, which is related to the mass of the structure and acceleration time history, the wind load on the wind turbine tower will change after the large deformation of the blades. Most of the existing research regard the blade as a rigid body, generate wind load in advance and then load it [10, 11]. There is a lack of research on wind turbine considering both complex coupling load and large deformation of blades, and this method can simulate the performance and response of the structure more accurately.

In order to carry out the hybrid simulation research of wind turbine under the action of multiple disasters, it is necessary to establish a hybrid simulation software platform. In this paper, a hybrid simulation platform of wind turbine is developed based on Simulink, which ensures the coordination of physical substructure and numerical substructure. The coupling analysis software of wind turbine OpenFAST is secondary developed to realize the data transmission between OpenFAST and Simulink, which ensures that the wind load calculated in each step is accurately sent to Simulink to calculate the structural response, and the displacement of tower is returned to OpenFAST software to calculate the next wind load. The platform provides a software basis for the subsequent hybrid simulation research under wind-earthquake coupling.

2 Architecture of Standard Hybrid Simulation for Wind Turbine

The hybrid simulation was originally developed for seismic engineering without considering other forms of load. Compared with the general building structure, the onshore wind turbine mainly needs to consider the influence of wind load; compared with the offshore wind turbine, it should consider the accidental seismic action without considering the wave and other loads. Therefore, for the hybrid simulation of onshore wind turbine, the multi-degree-of-freedom motion equation of the structure can be written as follows:

$$\mathbf{M}\ddot{\boldsymbol{x}}(t) + \mathbf{C}\dot{\boldsymbol{x}}(t) + \boldsymbol{R}_{\mathrm{n}}(t) + \boldsymbol{R}_{\mathrm{e}}(\boldsymbol{x}, \dot{\boldsymbol{x}}, \ddot{\boldsymbol{x}}, t) = \boldsymbol{F}_{\mathrm{wind}}(t) + \boldsymbol{F}_{\mathrm{eq}}(t) \tag{1}$$

Among them, **M** and **C** are the mass and damping matrices of wind turbine structure. $\boldsymbol{F}_{\text{wind}}$ for the wind load vector of each joint, including static wind load and fluctuating wind load. $\boldsymbol{F}_{\text{eq}}$ represents seismic load, related to structure mass matrix **M** and input seismic acceleration $\ddot{x}_g$. $\ddot{\boldsymbol{x}}$, $\dot{\boldsymbol{x}}$ and $\boldsymbol{x}$ are the acceleration vector, velocity vector and displacement vector of the structure. $\boldsymbol{R}_{\text{n}}$ represents a vector composed of restoring forces at each degree of freedom (DOF) of numerical substructure (NS). $\boldsymbol{R}_{\text{e}}$ for the restoring force vector of experimental substructure (PS), it is generally measured by physical test. The displacement of the structure is obtained by solving the equation of motion, which is sent to PS and NS at the same time, and the restoring force $\boldsymbol{R}$ of the corresponding substructure is calculated or loaded respectively. The response of the whole structure can be obtained by returning to the coordinator for solution, and the next step of the cycle can be obtained, as shown in Fig. 1.

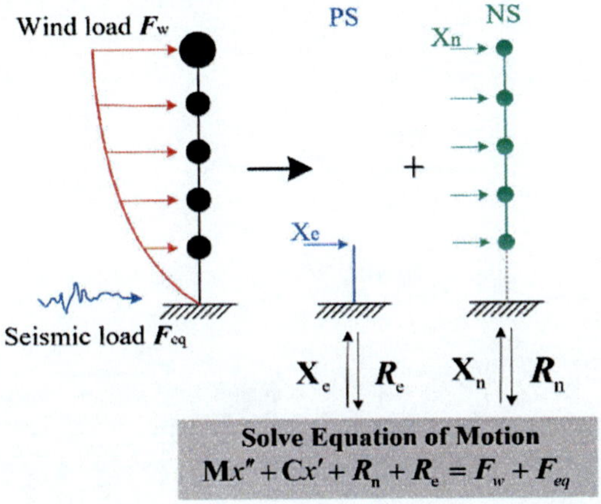

Fig. 1. Schematic diagram of standard hybrid simulation

2.1 Simulink Based Software Platform for Wind Turbine Hybrid Simulation

In order to carry out the hybrid simulation of onshore wind turbine structure, it is necessary to establish a general software platform. Simulink is a MATLAB simulation toolbox developed by MathWorks. Its main functions are modeling, simulation and dynamic system simulation. In this section, under the environment of Matlab, the software platform HySIM-WT which can carry out standard hybrid simulation is established by using Simulink. The platform consists of four main modules for hybrid simulation: load input module, coordinator module, PS module and NS module, as shown in Fig. 2.

2.2 Function of Hybrid Simulation Software Platform

As a fully coupled analysis software in the field of wind turbine, OpenFAST is a program software based on Fortran language, so it has a strong advantage of operation speed. OpenFAST has nine modules. For onshore wind turbine, in the modeling process, the main input files, ElastoDyn, InflowWind, ServoDyn, AeroDyn are mainly used, and the related module relationship and analysis flow are shown in Fig. 3 [12].

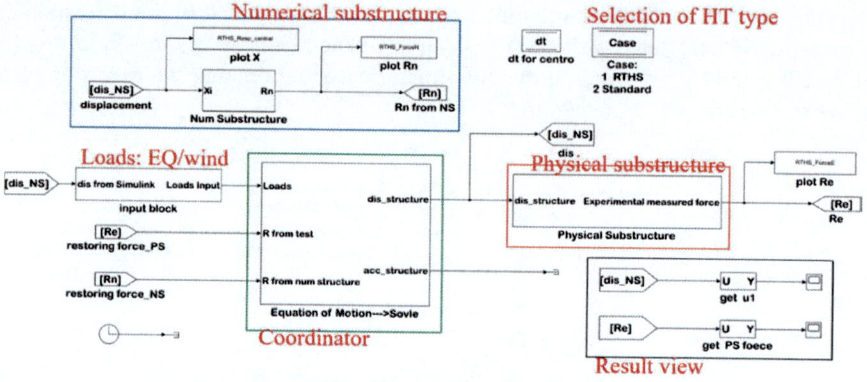

Fig. 2. HySIM-WT software platform

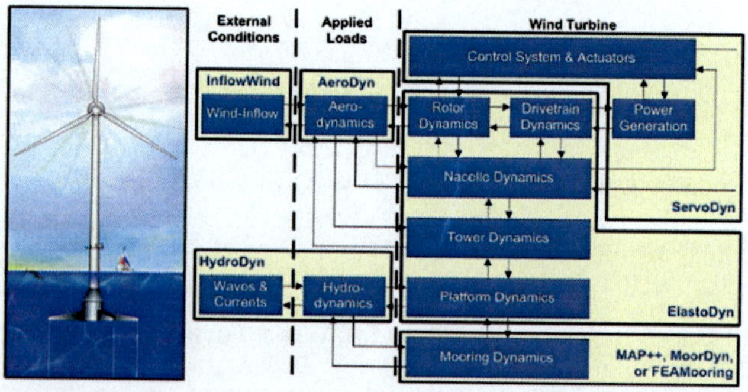

Fig. 3. Calculation flow of each module of wind turbine

As shown in Fig. 4, the BeamDyn module [13] can more accurately consider the change of the load on the tower after the blade is deformed.

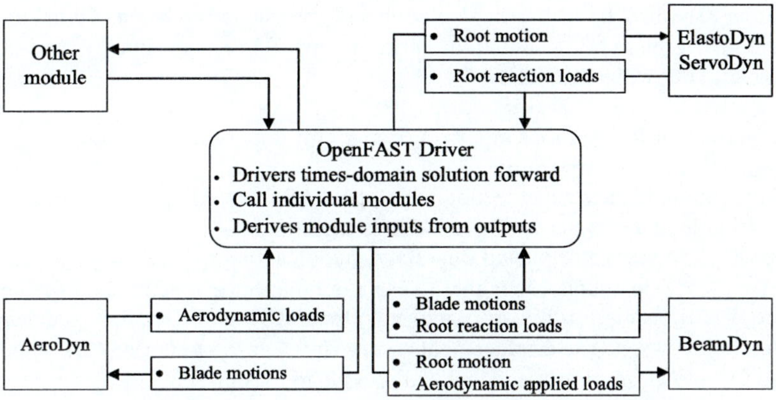

Fig. 4. OpenFAST flow chart considering blades deformation

OpenFAST is powerful in wind field generation, wind load calculation and so on. It can consider a variety of loads and calculate fast. However, for the analysis of tower structure, OpenFAST is calculated by mode superposition, and can't consider the fatigue under cyclic load and the nonlinear characteristics of structure subjected to high speed wind or extreme earthquake. Because its code is completely open source, it provides the conditions for the subsequent secondary development to implement other functions.

In order to analyze the character of the structure under wind load more accurately, the platform adds wind load input module to the standard hybrid simulation and obtains it step by step from OpenFAST, and the specific flow is shown in Fig. 5.

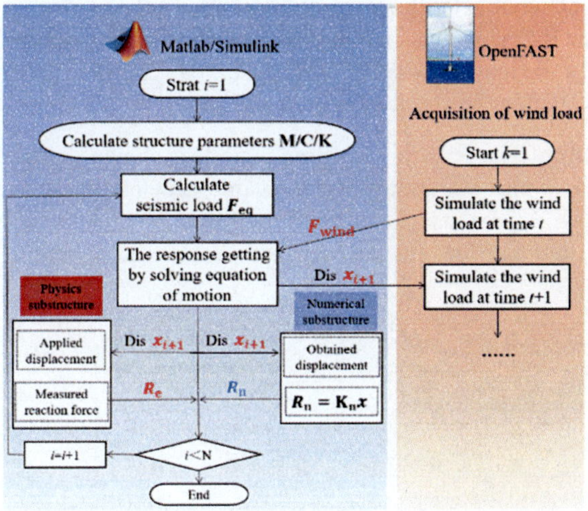

Fig. 5. Hybrid simulation flow of HySIM-WT

In order to realize the accurate calculation and gradual transmission of wind load and the operation of the hybrid simulation platform, OpenFAST and Simulink are required to realize the following functions:

(1) Simulink has the functions of solving equations of motion, controlling calculation, suspending and terminating;
(2) Simulink establishes a connection with OpenFAST according to the user's demand for wind load and waits for OpenFAST to transfer data;
(3) OpenFAST uses AeroDyn and BeamDyn modules (blade model of geometric exact beam theory) to couple each other to carry out numerical calculation of wind load;
(4) The data transmission function between OpenFAST and Simulink to ensure that Simulink can send the displacement results of tower top to OpenFAST accurately, and OpenFAST returns the calculated wind load to Simulink;

TCP/IP protocol is a connection-oriented and reliable communication protocol. On the basis of TCP/IP protocol, this paper uses C++ language and Fortran language, and combines Matlab function, S function and other functions in Simulink to write related programs, and realized the above functions.

3 Submodule of Hybrid Simulation Platform

3.1 Load Input Module

In dynamic analysis, seismic is generally related to mass distribution and input time history of ground acceleration, so seismic waves can be selected in advance. The wind load of the tower is generally related to the wind field and the shape of the tower, while the wind load on the top tower is mainly related to the blade airfoil and size. In addition, the wind load will change after the large deformation of the blades.

In the Simulink hybrid simulation platform, the load input module includes the input of earthquake and wind load. Before operation, users can choose whether to input seismic load or wind load, define the simulation time and so on.

Seismic Load Input. The seismic recording module is shown as Fig. 6. The developed platform prepares several commonly used seismic acceleration selections (amplitude modulation to PGA = 1.0g, users can input their own coefficients to adjust amplitude again as needed), and provides the user-defined function of uploading seismic records for seismic analysis.

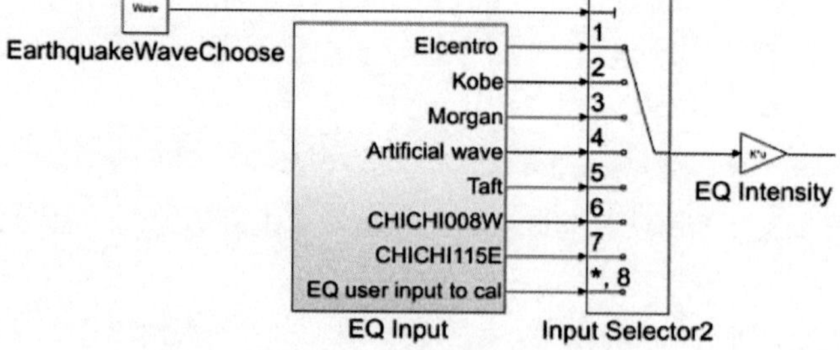

Fig. 6. Earthquake records choose module

Wind Load Input. In this section, the wind load on the top tower is inputted step by step, the structure response is calculated, and then the displacement of the top tower is returned to OpenFAST to calculate the next wind load and go through all the time, so that more accurate results can be obtained. Therefore, this paper modifies the FAST_prog, FAST_Solver and other files of OpenFAST. Based on the Socket communication with TCP/IP protocol, the communication function is added in the original OpenFAST cycle. When the Simulink platform is started, if the wind load is considered, the communication module in the internal program will arouse the OpenFAST and listen and wait for the connection to be established; after the OpenFAST is started, the input fully coupled numerical model of the wind turbine will run at the same time, and the data transmission between the Simulink platform and OpenFAST will begin step by step; finally, connection will be closed after the simulation.

It is worth noting that although the data type transmitted is double, including the current steps, current simulation time, displacement, load and other data, in order to ensure the accuracy and efficiency of transmission, "memset", "memcpy" and other related functions in C++ language are used to convert the data into strings first, and then to convert strings into double type data when received. The relevant data sent by OpenFAST (shown in Fig. 7 red box) are processed on the Simulink platform and combined with seismic loads to be sent as loads to the rest of the modules for subsequent operations.

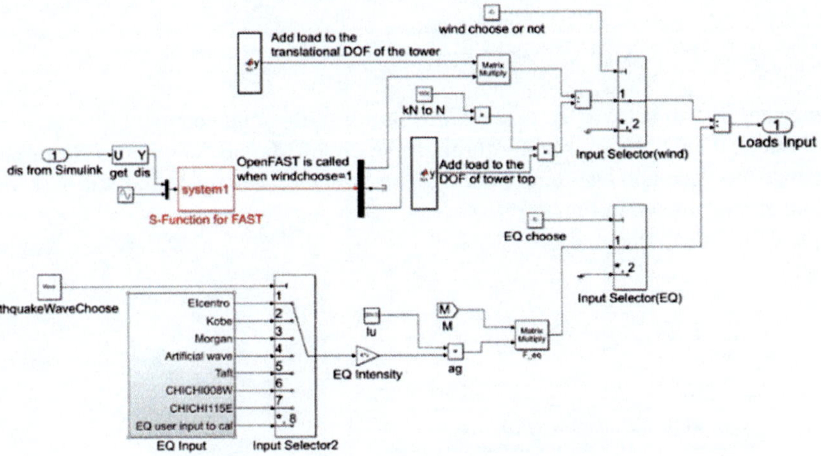

Fig. 7. Simulink-OpenFAST flow chart

3.2 Other Modules

Coordinator Module. The coordinator is responsible for the important process of hybrid simulation, including the solution to differential equation of motion and data transmission. According to the restoring force R and the input load F transmitted back by PS and NS (the rest of the structure), the equation of motion is solved and the next cycle is entered.

Figure 8 shows the concrete content of the calculation flow of the module. The solver module program of this platform is mainly the central difference method. The formula (2) (3) can be brought in (1) to solve the next displacement and send it to two substructures. This method has the advantage of unconditional convergence, but reduces the time step and increases the number of cycles.

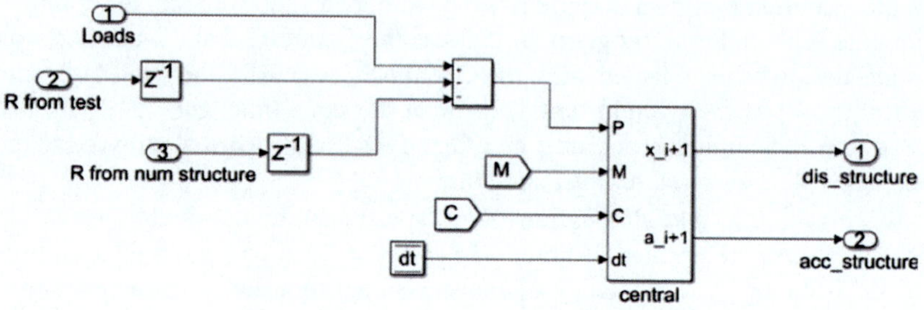

Fig. 8. Numerical integration algorithm

$$\dot{x}_i = \frac{x_{i+1} - x_{i-1}}{2\Delta t} \tag{2}$$

$$\ddot{x}_i = \frac{x_{i+1} - 2x_i + x_{i-1}}{\Delta t^2} \tag{3}$$

Numerical Substructure. At present, the platform uses finite element method to carry out numerical calculation, mainly through the code (Matlab function in Simulink) to Calculate the Mass and Damping Matrix of the Structure, and the Initial Stiffness Matrix $\mathbf{K_e}$ and $\mathbf{K_n}$, as Shown in Fig. 9.

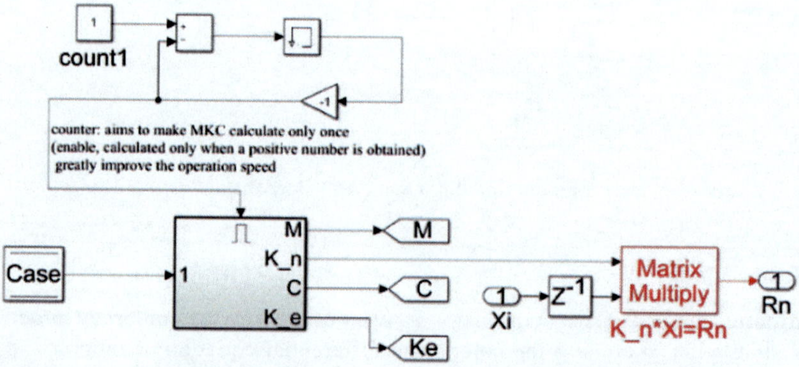

Fig. 9. NS module used to obtain restoring force

In order to ensure the operation rate, the mass stiffness and damping matrix are calculated only at the first time by judging the cycle. The module calculates the $\boldsymbol{R}_{\mathrm{n}}$ through the displacement sent from the coordinator (shown in the red box in Fig. 9) and sends it to the coordinator for the next step of the loop.

Virtual Physical Substructure. The purpose of this paper is to develop a hybrid simulation software platform suitable for wind turbine under wind-earthquake coupling load. At present, the physical substructure is mainly replaced by numerical method, the displacement obtained based on numerical integration algorithm $\mathrm{K}_{\mathrm{e}}x_i$ will be regarded as the restoring force R_{e} of the current step, and the code and physical experiment of control system will be completed in the following research.

4 Platform Test and Verification

In order to verify the stability and effectiveness of the platform, a virtual hybrid simulation is carried out. Based on a common example of 5 MW wind turbine given by OpenFAST, the corresponding finite element model is established in the software platform of Simulink to carry out virtual hybrid simulation. The wind turbine is 87.6 m high, and acted on by El-Centro wave (0.1 g) and wind load (normal operation state, wind speed at hub is 12 m/s) in the horizontal direction. The 1:2 Scale of the 20 m part at the bottom tower is selected as the virtual physical substructure to carry out the simulation and verification of the hybrid simulation.

4.1 Data Transmission Test

After the Simulink simulation begins, listen and wait to establish a connection to the OpenFAST. After the connection is established, the simulation starts automatically. A case study of 10-s wind load was carried out and the execution results of the OpenFAST running are shown in Fig. 10(a).

It can be seen that the wind load calculation data of OpenFAST based on Fortran language can be sent to Simulink accurately, and the next step of tower top displacement data can be received accurately. The running time of the whole program, the wind load generation module and the data transmission module are collected and shown in Fig. 10(b). Though these two modules increase the executing time of hybrid simulation, they are not large compared to the time of actuator loading time.

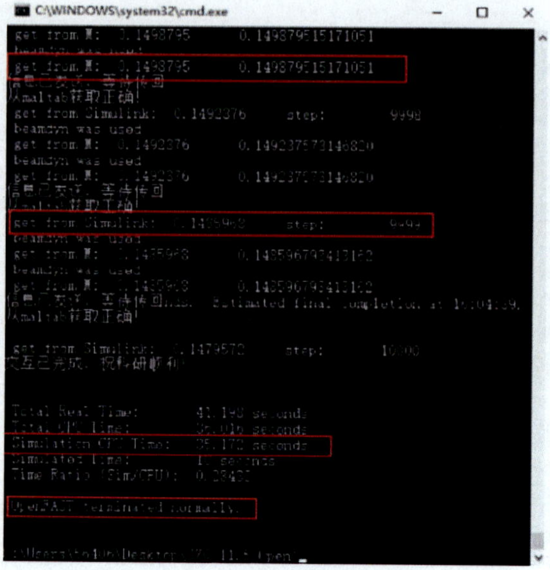

(a) OpenFAST test

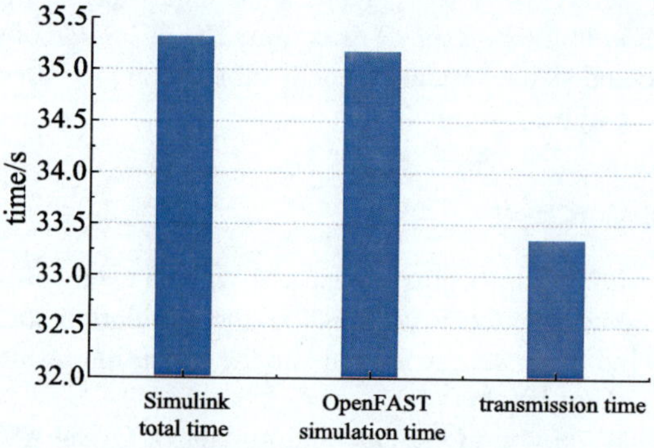

(b) Simulated time statistics

Fig. 10. Data transmission testing

4.2 Comparison of Numerical Results

In order to test the accuracy of the hybrid simulation, the numerical simulation (linear elasticity) of the whole structure is completed by Matlab. Figure 11 compares the wind load time history at the top hub of the wind turbine and the horizontal displacement of the degree of freedom at the top tower.

It can be seen that in this case, the hybrid simulation results are in good agreement with the pure numerical simulation results. Considering that the load of hybrid simulation varies with blade deformation, this two can't be completely consistent.

It can be seen that after the OpenFAST and Simulink transmit step by step through the secondary development, the hybrid simulation platform developed at present can accurately realize the data exchange, and the results of numerical simulation are in good agreement with the hybrid simulation results, which verifies the accuracy and feasibility of the hybrid simulation software platform and provides sufficient software support for the subsequent development of the real hybrid simulation.

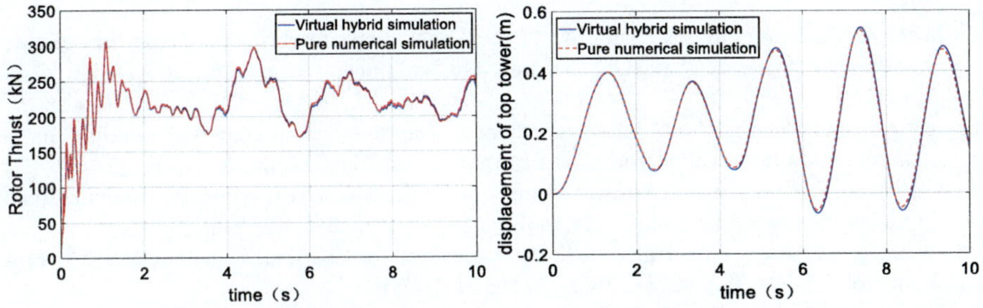

(a) Comparison of wind load at hub (b) Comparison of displacement of tower top

Fig. 11. Comparison of load and displacement time history at the top tower

5 Conclusion

In this paper, a software platform suitable for the standard hybrid simulation of onshore wind turbine under wind-earthquake coupling is developed, which is verified by virtual hybrid simulation. The specific conclusions are as follows:

1) The platform mainly includes load input module, communication module, control module and substructure module, which achieves the basic functions required for the standard hybrid simulation of onshore wind turbine under wind-earthquake coupling;
2) After the secondary development of OpenFAST, the functions of waiting, calculation and data transmission with Simulink step by step required by the hybrid simulation are realized, the interaction of blade deformation and wind load is considered in the testing process, and wind load inputs in hybrid simulation is more exact;
3) The wind-earthquake coupling hybrid simulation of wind power tower was carried out and compared to the pure numerical results. The consistent results confirmed the accuracy and stability of the platform.

Acknowledgement. The work was supported by the National Natural Science Foundation of China (51908384), the Scientific Research Fund of Institute of Engineering Mechanics, China Earthquake Administration (2019D18), and Science and Technology Project of Sichuan Province (2020YJ0079).

References

1. Kao-shan, D., Anthony, B., Chao, L., et al.: Environmental issues associated with wind energy–a review. Renew. Energy **75**, 911–921 (2015)
2. Eric, R.L., Kao-shan, D., Wensheng, L., et al.: Wind turbine testing methods and application of hybrid testing: a review. Wind Struct. **29**(3), 195–207 (2019)
3. McCrum, D.P., Broderick, B.M.: Evaluation of a substructured soft-real time hybrid test for performing seismic analysis of complex structural systems. Comput. Struct. **129**, 111–119 (2013)
4. Javad, H.M., Yassamin, A., Riadh, A., et al.: Application of hybrid simulation for collapse assessment of post-earthquake CFRP-repaired RC columns. J. Struct. Eng. **143**(1), 04016149 (2017)
5. Justin, A.M., Mehrdad, S.: Collapse resistance of a seven-story structure with multiple shear-axial column failures using hybrid simulation. J. Struct. Eng. **143**(5), 04017012 (2017)
6. Yoshikazu, T., Gregory, L.F.: Software framework for distributed experimental–computational simulation of structural systems. Earthq. Eng. Struct. Dyn. **35**(3), 267–291 (2006)
7. Oh-Sung, K., Amr, S.E., Billie, F.S.: A framework for distributed analytical and hybrid simulations. Struct. Eng. Mech. **30**(3), 331–350 (2008)
8. Yan, X., Qing, H., Yurong, G., et al.: A network platform for remote pseudo-dynamic testing. J. Build. Struct. **26**(3), 122–129 (2005)
9. Ge, Y., Bin, W., Ge, O., et al.: HyTest: platform for structural hybrid simulations with finite element model updating. Adv. Eng. Softw. **112**, 200–210 (2017)
10. Maxime, T., Valentin, C., Thomas, S., et al.: Real-time hybrid model testing of a semi-submersible 10MW floating wind turbine and advances in the test method. In: International Conference on Offshore Mechanics and Arctic Engineering, vol. 51975, pp. V001T01A013. San Francisco (2018)
11. Zhu, M., Hao, H., Kao-shan, D., et al.: Dynamic response analysis and preliminary verification of hybrid test of wind power tower under the coupling of long period ground motion and fluctuating wind. Eng. Mech. **38**(S), 58–65 (2021)
12. OpenFAST Documentation. https://www.nrel.gov/wind/nwtc/fast.html. Accessed 5 Feb 2022
13. BeamDyn User's Guide and Theory Manual. https://www.nrel.gov/wind/nwtc/beamdyn.html. Accessed 5 Feb 2022

Risk-Based Inspection Planning for Support Structure of Offshore Wind Turbine

Weiheng Zhang[1,2], Dagang Lu[1(✉)], Jianjun Qin[3,4], Min Liu[1,2], and Michael Havbro Faber[2]

[1] School of Civil Engineering, Harbin Institute of Technology, Harbin, China
ludagang@hit.edu.cn
[2] Department of the Built Environment, Aalborg University, Aalborg, Denmark
[3] State Key Laboratory of Ocean Engineering, School of Naval Architecture, Ocean and Civil Engineering, Shanghai Jiao Tong University, Shanghai, China
[4] Shanghai Key Laboratory for Digital Maintenance of Buildings and Infrastructure, School of Naval Architecture, Ocean and Civil Engineering, Shanghai Jiao Tong University, Shanghai, China

Abstract. The optimal allocation of deterioration control in risk-based inspection (RBI) planning is based on the pre-posterior analysis using Bayesian decision theory and expected utility theory. This paper applies the RBI planning in the area of offshore wind turbine foundation structure, aiming to minimize the life-cycle cost which takes the direct cost and the indirect consequence into account. The direct cost refers to the direct loss and the indirect consequence refers to the power loss due to the downtime of wind turbine. A monopile offshore wind turbine structure is analyzed in OpenFAST simulation package to demonstrate how to utilize RBI planning in the field of support structure of offshore wind turbine, in which the environmental condition is taken from the statistical data in North Sea. The case study show that the RBI planning can effectively reduce the expected life-cycle cost of offshore wind turbine.

Keywords: Risk-based inspection · Offshore wind turbine · Life-cycle cost

1 Introduction

Utilization and development within the field of renewable energy sources is crucial to maintain an environmentally sustainable society, especially in the context of increased climate change in recent years. As an important renewable energy source, offshore wind turbines (OWTs) have been developed significantly in recent years, with the significant advantages of less spatial restrictions, less turbulence, and higher wind speeds, while causing fewer casualties and social losses. the safety of the OWT foundation directly affects the overall safety of the wind turbine, which will be subjected to complex environmental loads (wind and waves) during its service life, so the corresponding fatigue reliability analysis is a matter of concern.

W. Guo and K. Qian (Eds.): GBCESC 2022, LNCE 211, pp. 591–598, 2023.
https://doi.org/10.1007/978-981-19-5217-3_59

Risk-based Inspection planning (RBI) can be utilized to optimize the integrity management through utilizing Bayesian pre-posterior decision analysis. RBI planning has been appreciated in various field of civil engineering including steel structures [1], offshore structures [3], oil and gas pipeline [7], etc. Since the operation and maintenance costs of offshore wind turbines account for a large portion of the total investment, it is necessary to optimize the operation and maintenance strategy to reduce the expected value of costs. Furthermore, since SHM information could also be employed to further improve the operation and maintenance strategies for offshore wind turbine. reduce the uncertainties in the probabilistic model associated with the performance of offshore wind turbine. Quantifying the potential economic value of SHM information can be achieved by Value of Information (VoI) analysis [8].

However, in the area of offshore wind turbine, there are only a few investigations have been conducted, see e.g. [6, 7]. In these contributions, RBI planning has been successfully applied in this area with an illustrative case study. In this paper, in order to apply the RBI planning for a specific benchmark model, OpenFAST simulation package is used to get the fatigue stress of the hot spot of the 5MW NREL monopile offshore wind turbine.

This paper starts with a brief illustration of RBI planning with two heuristic analysis approaches, then introduce the methodology of fatigue reliability analysis of offshore wind turbine. Then, an illustrative example is demonstrated how to apply the RBI planning for a specific offshore wind turbine support structure.

2 An Outline of Risk-Based Inspection Planning

The optimum allocation of deterioration control is the focus of RBI planning., it is based on the structural reliability analysis and preposterior analysis. Planning of inspections is to identify what, how, where and how often to inspect [1]. The purpose of RBI inspection planning is to make optimal maintenance planning to achieve the lowest engineering life-cycle risk.

2.1 Assessment of Life-Cycle Cost

The structural life-cycle cost consists three costs with a constraint related to annual failure probability $\Delta P_{\mathrm{F}}^{\max}$. The formulation for identifying the total expected life-cycle cost, $\mathrm{E}(C_T)$, represented by:

$$\mathrm{E}(C_T) = \mathrm{E}(C_{\mathrm{F}}) + \mathrm{E}(C_{\mathrm{CR}}) + \mathrm{E}(C_{\mathrm{PR}}) + \mathrm{E}(C_{\mathrm{I}}) \quad (1)$$

where C_{F}, C_{CR}, C_{PR} and C_{I} are the costs of failure, corrective maintenance, preventive maintenance and inspection, respectively. Based on probabilistic models of each event and corresponding cost, the life-cycle risk can be estimated by the expected utility theory. However, the computation cost of full decision scenario is huge due to large numbers of possible combination, the corresponding simplified decision tree is proposed to solve the problem as shown in Fig. 1.

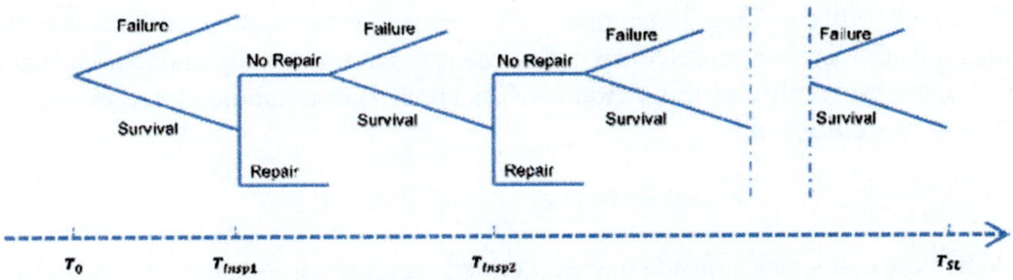

Fig. 1. RBI planning in the form of decision tree.

The following formulations can be used to calculate the costs associated with a specific inspection and repair strategy for a given service life period:

$$E(C_{\text{I}}(T_{\text{SL}})) = \sum_{i=1}^{n} \begin{pmatrix} p_{\text{S}}(t_{\text{insp,i}}) \times c_{\text{I}} + p_{\text{R}}(t_{\text{insp,i}}) \\ \times E(C_{\text{I}}(T_{\text{SL}} - t_{\text{insp,i}})) \end{pmatrix} \times \frac{1}{(1+r)^{t_{\text{insp,i}}}} \tag{2}$$

$$E(C_{\text{PR}}(T_{\text{SL}})) = \sum_{i=1}^{n} p_{\text{R}}(t_{\text{insp,i}}) \times \left(c_{\text{PR}} + E(C_{\text{R}}(T_{\text{SL}} - t_{\text{insp,i}}))\right) \times \frac{1}{(1+r)^{t_{\text{insp,i}}}} \tag{3}$$

$$E(C_{\text{CR}}(T_{\text{SL}})) = \sum_{t=1}^{T_{\text{SL}}} \begin{pmatrix} \Delta p_f(t) \times (1 - p_{\text{col|fat}}) \times c_{\text{CR}} \\ +p_{\text{R}}(t) \times E(C_{\text{CR}}(T_{\text{SL}} - t)) \end{pmatrix} \times \frac{1}{(1+r)^t} \tag{4}$$

$$E(C_{\text{F}}(T_{\text{SL}})) = \sum_{t=1}^{T_{\text{SL}}} \begin{pmatrix} \Delta p_f(t) \times p_{\text{col|fat}} \times c_{\text{F}} \\ +p_{\text{R}}(t) \times E(C_{\text{CR}}(T_{\text{SL}} - t)) \end{pmatrix} \times \frac{1}{(1+r)^t} \tag{5}$$

where $\Delta p_{\text{F}}(t)$ is the probability of survival till the $t-1$ year and failure at t year, $p_{\text{R}}(t)$ is the probability of no repair till the $t-1$ year and repair at t year and $p_{\text{S}}(t)$ denotes the survival probability till the t year, $[t_{insp,1}, t_{insp,2}, \ldots, t_{insp,\text{n}}]$ stands for the scheduled inspection year, $p_{\text{col|fat}}$ stands for conditional failure probability given the crack size has exceeded the critical crack size, r is the interest rate.

With reference to the above illustration, the purpose of RBI planning is minimizing the overall expected cost given by:

$$\begin{aligned} &\min_{\boldsymbol{I},\boldsymbol{d}} \text{E}(C_T(\boldsymbol{I}, \boldsymbol{d}, T_{\text{SL}})) \\ &\text{s.t.} \Delta p_{\text{F}}(\boldsymbol{I}, \boldsymbol{d}, t) \le \Delta p_{\text{F}}^{\max}, t = 0, \ldots, T_{\text{SL}} \end{aligned} \tag{6}$$

2.2 Modelling of Inspection and Repair

The ability of a particular inspection method is illustrated by the probability of detection (PoD). Herein, a log-logistics model of PoD model is used, which can be described by:

$$\text{PoD}(a) = \frac{\exp(\alpha_{\text{D}} + \beta_{\text{D}} \ln(s))}{1 + \exp(\alpha_{\text{D}} + \beta_{\text{D}} \ln(s))} \tag{7}$$

The probability of false indication (PFI) represents the probability of observing a defect indication in the non-defect spot, also has influence on the repair decision-making. Thereby, the probability of indication (PoI) is adopted to combine above two types of probability together, with:

$$\mathrm{PoI}(a) = \mathrm{PoD}(a) + (1 - \mathrm{PoD}(a))\mathrm{PFI} \tag{8}$$

When the crack size exceeds the pre-defined repair requirement s_c the maintenance action will be taken. The uncertainties in the indicated crack sizes is also considered here, which is denoted by ε_i. Correspondingly, the event of repair action can be defined as:

$$g_M(a) = a_c - a - \varepsilon_i \tag{9}$$

3 Fatigue Reliability Analysis for Offshore Wind Turbine

The 3-bladed National Renewable Energy Laboratory (NREL) 5MW [9] is investigated, which consists of a wind turbine, tower, transition piece, and monopile structure. All dimensions were implemented in a finite-element model according to the original OC3 specifications. The loads on the OWT structures is calculated in the time domain.

A S-N curve is a plot of the amount of an fatigue stress range vs the failure cycles for a given detailed component in fatigue analysis:

$$N_F = C \cdot \Delta S^{-m} \tag{10}$$

where N_F stands for stress cycles to failure for a given ΔS, C and m define the inverse slope and intercept of a fatigue critical detail, which are determined by experiments.

Design criteria against fatigue failure are generally based on S-N curve, which is obtained from experiments or tests on typical structural details. However, it is not possible to directly measure damage on existing structures with available sensors. As such, the fracture mechanics (FM) approach, which is mainly based on the linear elastic fracture mechanics (LEFM), is preferred to develop the an inspection planning. a formulation of one-dimension crack size growth, which is based on the Paris' Law, is utilized to describe the relationship between crack size a and the stress cycles n:

$$a(n) = \left[(1 - \frac{m}{2})C\pi^{m/2}\Delta S_e^m Y^m n + a_0^{1-m/2}\right]^{(1-m/2)^{-1}} \tag{11}$$

where a_0 is the initial crack size; ΔS_e is the equivalent stress range; Y is the geometric correction factor; C and m represent the crack growth parameters which depend on the material.

To develop an inspection and maintenance planning, the semi-empirical FM model is generally calibrated to S-N curve because of all the information for S-N model can be obtained from the experiment. The calibration is performed by least squares fitting to ensure that the difference in failure probabilities under the two models is within acceptable limits, expressed as:

$$\min_{x_1,x_2\ldots,x_n} \sum_{t=1}^{T_{\mathrm{SL}}} \left(\left(p_{f,\mathrm{SN}}(t) - p_{f,\mathrm{FM}}(t; x_1, x_2 \ldots, x_n)\right)\right)^2 \tag{12}$$

where $x_1, x_2, \ldots, x_n$ are the fitting parameters in FM model. Herein, the initial crack size a_0 together with the crack growth parameter C are used as fitting parameters.

4 Case Study

4.1 Environmental Condition

The representative environmental conditions are adapted from the Upwind Design Basis (K13 Shallow Water Site) [10]. The mean wind speed of 10 min V_w using a Weibull distribution:

$$F(V_w) = 1 - \exp\left(-\left(\frac{V_w}{A_w}\right)^{B_w}\right) \tag{13}$$

where A_w and B_w are the scale parameter and the shape parameter, respectively. The turbulence intensity is regarded as a function of mean wind speed V_w, which is given as:

$$TI = \frac{\sigma_w}{V_w} \tag{14}$$

where σ_w is the S.D value of the time varying wind speed, V_w is the mean value of wind speed. Generally, for fatigue analysis, the associated model of turbulence intensities for all wind speeds can be derived based on the IEC 61400-1 standard [11]. While, in the present work, a turbulence intensity model proposed in [10] is used, expressed as:

$$TI(V_w) = \frac{(15 + aV_w)}{(1 + a)V_w} \cdot I_{15};\ a = 5 \tag{15}$$

where $I_{15} = 0.14$ is the reference turbulence intensity corresponding to the mean wind speed of 15 m/s.

For the wave parameters, a scatter diagram, containing the 10-min wind speed V_w, the significant wave height H_s and peak spectral period T_p, can be utilized to express the wave climate data. For the sake of simplicity, the full scatter diagrams are not given here, more details can be found in [10].

To address this wind speed range, the IEC standard requires minimum intervals of 2 m/s. Since this work mainly considers the fatigue damage of wind turbines in operational conditions, the range of mean wind speeds considered in the next fatigue reliability analysis is between the cut-in and cut-out wind speeds. As such, 11 lumped environmental conditions, on the basis of the lumped environmental conditions summarized in [10], are regarded as the representative condition of the fatigue loading.

4.2 Modelling Details

The details of the parameters for the fatigue reliability analysis are summarised in Table 1. In this case study, a_0 and $\ln C_m$ in Eq. (10) are calibration parameters, and their mean and standard deviation are determined by Eq. (11).

Table 1. Details of parameters in fatigue reliability analysis and inspection & maintenance.

Variable	Distribution	Mean value	S.D value	Description
a_0	Exponential	0.343*	0.343*	Initial Crack Size (mm)
a_{th}	Normal	60	3	Critical Crack Size (mm)
ΔS_e	Normal	10.83	1.083	Equivalent Stress Range (MPa)
m	Deterministic	3	–	FM Model Parameter
$\ln C_{\mathrm{m}}$	Normal	−29.373*	0.296*	FM Model Parameter
v	Deterministic	5×10^7	–	Annual Load Cycles
Y_{G}	Deterministic	1	–	Geometric Correction Factor
α_{D}	Deterministic	0.63	–	Parameters of PoD
β_{D}	Deterministic	1.16	–	Parameters of PoD
PFI	Deterministic	0.138	–	Probability of False Indication
ε_{i}	Normal	0	0.1	Random Error of Inspection (mm)

The costs associated with the maintenance actions involved are assessed in two terms in the present work, i.e. direct cost and indirect consequence. For the LCC assessment, the indirect consequence (i.e. the monetary loss due to the wind turbine shutdown caused by failure, maintenance and inspection) is taken into account in the cost models for these events. The cost model are summarized in Table 2.

Table 2. Parameters value in cost assessment analysis.

Variable	Description	Value
$c_{\mathrm{F,d}}$	Direct Failure Cost	1×10^7
$c_{\mathrm{CR,d}}$	Direct Corrective Maintenance Cost	1×10^6
$c_{\mathrm{PR,d}}$	Direct Corrective Maintenance Cost	2×10^4
$c_{\mathrm{I,d}}$	Direct Inspection Cost	1×10^4
τ_{F}	Downtime of Failure	2160 h
τ_{CR}	Downtime of Corrective Maintenance	34 h

(continued)

Table 2. (*continued*)

Variable	Description	Value
τ_{PR}	Downtime of Preventive Maintenance	20 h
τ_I	Downtime of Inspection	4 h
E_{per}	Power Production Per Hour	2.76×10^3 KWh
P_e	Power Price	0.05 /KWh
$p_{col\|fat}$	Conditional Probability of Failure	0.3
r	Discount Interest	0.02

4.3 Optimum Inspection Planning Results

Two heuristic approaches stated in [2] are utilized for the optimization of RBI planning in this case study. For the RBI planning considering the LCC metric, the trends of LCC under different decisions are shown in Fig. 2. The optimal decision for both different strategies seems to similar. It can be seen that with the optimization of inspection planning, the expected value of LCC can be effectively reduced.

5 Conclusion

RBI planning is applied to the monopile structure of offshore wind turbine to minimize the expected value of life cycle costs considering both direct costs and indirect consequences. An example considering the fatigue cracks in NREL 5MW monopile offshore wind turbine structure is illustrated to investigate the details. It can be seen from the corresponding analysis results, RBI planning can provide substantially potential benefits for the operation & maintenance strategy of offshore wind turbine.

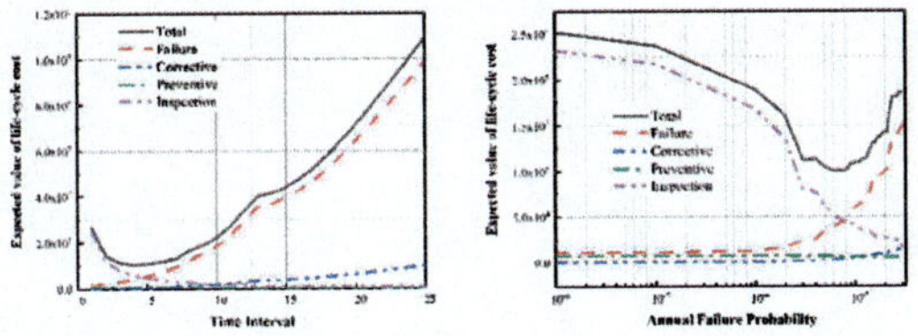

Fig. 2. Life-cycle cost for the two heuristic approaches.

Acknowledgement. China's National Key R&D Plan is funding this project (Grant No. 2021YFB2600500). The China Scholarship Council (CSC) (Grant No. 201906120281) supports the first author's research at Aalborg University.

References

1. Faber, M.H.: Risk-based inspection: the framework. Struct. Eng. Int. **12**(3), 186–195 (2002)
2. Straub, D., Faber, M.H.: Computational aspects of risk-based inspection planning. Comput.-Aided Civ. Infrast. Eng. **21**(3), 179–192 (2006)
3. Straub, D., Goyet, J., So/rensen, J.D., Faber, M.H.: Benefits of risk based inspection planning for offshore structures. In: International Conference on Offshore Mechanics and Arctic Engineering, vol. 47489, pp. 59–68, January 2006
4. Zhang, W.H., Qin, J., Lu, D.G., Thöns, S., Faber, M.H.: VoI-informed decision-making for SHM system arrangement. Structural Health Monitoring, 1475921720962736 (2020)
5. Dey, P.K., Ogunlana, S.O., Naksuksakul, S.: Risk-based maintenance model for offshore oil and gas pipelines: a case study. J. Qual. Maintenance Eng. **10**(3), 169–183 (2004)
6. Rangel-Ramírez, J.G., Sørensen, J.D.: Risk-based inspection planning optimisation of offshore wind turbines. Struct. Infrastruct. Eng. **8**(5), 473–481 (2012)
7. Nielsen, J.J., Sørensen, J.D.: On risk-based operation and maintenance of offshore wind turbine components. Reliab. Eng. Syst. Saf. **96**(1), 218–229 (2011)
8. Zhang, W.-H., Lu, D.-G., Qin, J., Thöns, S., Faber, M.H.: Value of information analysis in civil and infrastructure engineering: a review. J. Infrastruct. Preserv. Resilience **2**(1), 1–21 (2021). https://doi.org/10.1186/s43065-021-00027-0
9. Jonkman, J., et al.: Definition of a 5-MW reference wind turbine for offshore system development. No. NREL/TP-500-38060. National Renewable Energy Lab. (NREL), Golden, CO (United States) (2009)
10. De, Vries, WE., et al.: Final report WP 4.2: support structure concepts for deep water sites: deliverable D4. 2.8 (WP4: offshore foundations and support structures) (2011)
11. International Electrotechnical Commission. Wind energy generation systems-Part 3–1: Design requirements for fixed offshore wind turbines. International Standard IEC 61400-3-1 (2019)

Upper Bound Limit Analysis for the Stability of Elliptical Tunnels Based on the Node-Based Smoothed Finite Element Method and Second Order Cone Programming

Feng-Tao Liu[1], Xi-Wen Zhou[2], and Yan Zhang[1(✉)]

[1] College of Civil Engineering and Architecture, Guilin University of Technology, Guilin 541004, China
zhangyan@glut.edu.cn

[2] School of Earth Sciences and Engineering, Sun Yat-Sen University, Guangzhou 510275, China

Abstract. The stability of 2D elliptical tunnels in cohesive-frictional soils are studied by using upper bound approach of limit analysis based on smoothed finite element method (SFEM) and node-based strain smoothing techniques (NS-FEM). According to the upper bound theorem of limit analysis, the stability of tunnels will be expressed as a second order cone programming (SOCP) problem following the Mohr-Coulomb (for 2D) yield criterion and the associated flow rule. This convex programming can be solved in fast and robust way by primal-dual interior point algorithm. In order to analyze the effectiveness of NS-FEM-based upper bound approach for the elliptical tunnel stability, especially to overcome the performance of volumetric locking under incompressible condition, several typical numerical tunnel models are established and analyzed. From the numerical results, failure mechanism of the elliptical tunnels can be revealed without the volumetric locking problem.

Keywords: Tunnel stability · Upper bound limit analysis · Node-based smoothed finite element · SOCP

1 Introduction

Tunnel stability is one of the most important issues in geotechnical engineering and has been probed by using upper bound approaches of limit analysis [1, 2]. Compared with the incremental elastic-plastic, limit analysis predicts directly the limit load and failure mechanism of engineering structures using upper bound (UB) theorem of plasticity, and this method has been proved to be more effective. Especially, the upper bound approach of limit analysis (UBLA) based on finite element method (UBFEMLA) [3, 4] is more powerful to solve complicated problems including the stability of tunnel. The UBFELA methods developed by Sloan [3] were applied to estimate the stability of tunnel [5, 6].

W. Guo and K. Qian (Eds.): GBCESC 2022, LNCE 211, pp. 599–606, 2023.
https://doi.org/10.1007/978-981-19-5217-3_60

However, some key details of FELA should be further improved, especially in upper bound approach of FELA, to ensure the existence of limit load for a purely cohesive material, the finite element must be volumetric "locking-free" [7–9]. To overcome the volumetric locking phenomena, various smoothed strain techniques over different smoothing subdomains generated from three-node triangular element (T3-FEM). For example, node-based (NS-FEM) [10–12], especially stable NS-FEM (SNS-FEM), has the advantages of high accuracy and efficiency, and has the property of locking free [13, 14]. These methods have been applied into the complex problems, such as geotechnical large deformation problems.

Compared with the classic finite elements, the smoothed finite elements can give more accurate, fast and robust solutions under the same mesh, and are generally insensitive to the distortion of element [11]. Recently, Mohapatra and Kumar [8] improve the SFEM-based UBLA of cohesive-frictional materials by using T3-FEM which is more suitable for adaptive analysis, and compare the property of various strain smoothing techniques in UBLA. According to the results of comparative analysis, node-based finite element is proved to be highly efficient and accurate in solving large scale problems and may be the best choice for UBFEMLA [8]. Coincidentally, Vo-Minh et al. [15] also used stable NS-FEM-based UBLA to analyze the stability of dual circular tunnels in cohesive-frictional soils.

Following the works of Vo-Minh et al. [15], we will try to apply the adaptive NS-FEM-based UBLA (ANS-UBFEMLA) to the stability analysis of non-circular tunnel sections. Firstly, the NS-FEM without stabilized technique [14] can also perform the stability analysis within the framework of UBLA in high accurate and efficient way. Secondly, the plastic dissipation adaptive mesh method is applied to simulate the failure mechanism of the elliptical tunnel, and the different features between cross section and face stability of elliptical tunnel has also been found. These results improve the technical support for the stability evaluation or design of non-circular tunnels in practical engineering.

2 Methodology of ANS-UBFEMLA

2.1 UB Limit Theorem of Ideal Plastic Materials

The UB limit theorem claims that an ideal plastic solid or structure V with surface S will reach the limit state and failure if there is a kinematically admissible velocity field (KAV) $\mathbf{u}$, which the rate of external work equals to plastic dissipation rate. From a mathematical viewpoint, the loading multiplier of limit state μ is the minimum of the *nonlinear* programming problem [3], i.e.,

$$\begin{aligned} &\mu = \min_{\mathbf{u}\in U} \int_V k\lambda \mathrm{d}V \\ &\text{s.t.} \quad \int_V \mathbf{g}^T \mathbf{u} \mathrm{d}V + \int_{S_t} \mathbf{t}^T \mathbf{u} \mathrm{d}V = 1 \\ &\quad \lambda \geq 2\sqrt{J_2(\mathbf{e})}, \quad \operatorname{div} \mathbf{u} = a\lambda \\ &\quad \mathbf{u} = \bar{\mathbf{u}} \quad \text{on } S_u \end{aligned} \tag{1}$$

where $S = S_u \cup S_t$, $S_u \cap S_t = \emptyset$, S_u denotes the velocity boundary, $e_{ij} = \frac{1}{2}(u_{i,j} + u_{j,i}) - \frac{1}{N}\text{div}\,\mathbf{u}\delta_{ij}$, the remaining part of boundary surface is S_t which subjected to tractions **t**, **g** is body forces, $\bar{\mathbf{u}}$ is the given velocities at the boundary S_u, and U is defined as the set of KVAs, i.e.,

$$U = \{\mathbf{u} = \bar{\mathbf{u}} \text{ on } S_u, \quad W_{\text{ext}}(\mathbf{u}) > 0\} \tag{2}$$

The nonlinear programming problem Eq. (1) for a practical structure with finite domain is firstly discretized by finite element, and then the approximate problem was further solved by a numerical optimization algorithm. In the next subsection, a combination of SOCP and NS-FEM will be used for stability analysis of non-circle tunnels.

2.2 Second Order Cone Programming of ANS-UBFEMLA

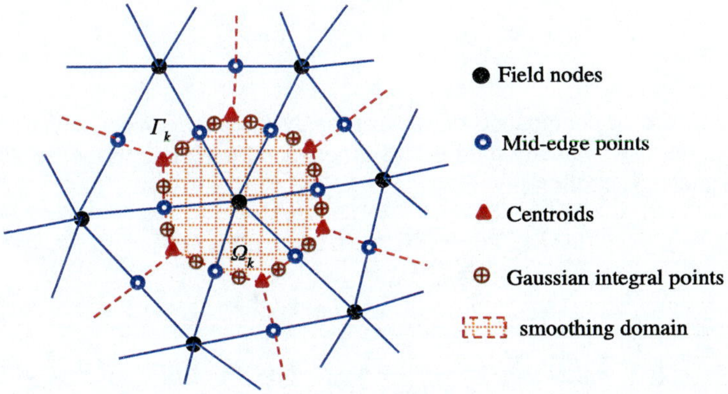

Fig. 1. Triangular elements and strain smoothing domains based on nodes (after Nguyen-Xuan et al. [18])

In ANS-UBFEMLA, the ideal plastic structure is firstly discretized into N_e finite elements. The KVA $\tilde{\mathbf{u}}$ is approximated in terms of the following nodal velocities $\tilde{\mathbf{d}}$,

$$\tilde{\mathbf{u}} = \mathbf{\Phi}(\mathbf{x})\tilde{\mathbf{d}} \tag{3}$$

where $\mathbf{\Phi}(\mathbf{x})$ is the shape function. In order to ensure that the strain rate is compatible over elements, each finite element is further subdivided into several node-based smoothing subdomains Ω_k^s ($k = 1, 2, \ldots, N_s$) with the boundary Γ_k^s such that $\Omega = \sum_{k=1}^{N_s} \Omega_k^s$ and $\Omega_i^s \cap \Omega_j^s = \emptyset$, $i \neq j$, where N_s denotes the number of all the subdomains (See Fig. 1 for 2D problems). Based on these smoothed domains, the smoothed rate is defined as follows

$$\begin{aligned} \bar{\boldsymbol{\varepsilon}}(\mathbf{x}_k) &= \int_{\Omega_k^s} \tilde{\boldsymbol{\varepsilon}}(\mathbf{x}) w(\mathbf{x} - \mathbf{x}_k) \mathrm{d}\Omega \\ &= \int_{\Omega_k^s} \mathbf{L}_d \tilde{\mathbf{u}}(\mathbf{x}) w(\mathbf{x} - \mathbf{x}_k) \mathrm{d}\Omega \end{aligned} \tag{4}$$

where $\bar{\boldsymbol{\varepsilon}}(\mathbf{x}_k)$ denotes the smoothed strain rate at the location $\mathbf{x}_k$, $\mathbf{L}_d$ denotes the differential operator matrix, and the weight function $w(\mathbf{x}-\mathbf{x}_k)$ over the smoothed subdomain Ω_k^s is usually chosen as [8]:

$$w(\mathbf{x}-\mathbf{x}_k) = \begin{cases} \frac{1}{A_k^s}, & \mathbf{x} \in \Omega_k^s \\ 0, & \mathbf{x} \notin \Omega_k^s \end{cases} \tag{5}$$

where A_k^s is the volume (for 3D) or aera of the subdomain Ω_k^s. Based on the Eq. (5) and Eq. (4), according to the divergence theorem, we can obtain the following formulation for calculating the smoothing strain rate, i.e.,

$$\bar{\boldsymbol{\varepsilon}}(\mathbf{x}_k) = \frac{1}{A_k^s\left(\text{or } V_k^s\right)} \int_{\Gamma_k^s} \mathbf{n}_k^s(\mathbf{x})\tilde{\mathbf{u}}(\mathbf{x})\mathrm{d}\Gamma \tag{6}$$

where $\mathbf{n}_k^s(\mathbf{x})$ is the outward normal matrix on Γ_k^s, i.e.,

$$\mathbf{n}_k^s(\mathbf{x}) = \begin{bmatrix} n_{kx}^s & 0 \\ 0 & n_{ky}^s \\ n_{ky}^s & n_{kx}^s \end{bmatrix} \quad \text{for 2D} \tag{7}$$

where n_{kx}^s, n_{ky}^s are the components of $\mathbf{n}_k^s(\mathbf{x})$ along the x, y axis, respectively. Considering the approximation of velocity field Eq. (3), the smoothing strain in the Eq. (6) can be further formulated as follows:

$$\bar{\varepsilon}(\mathbf{x}) = \overline{\mathbf{B}}\tilde{d} = \left[\overline{\mathbf{b}}_1(\mathbf{x})\ \overline{\mathbf{b}}_2(\mathbf{x}) \cdots \overline{\mathbf{b}}_N(\mathbf{x})\right]\tilde{\mathbf{d}} \tag{8}$$

where

$$\overline{\mathbf{b}}_i = \begin{bmatrix} \bar{b}_{ix} & 0 \\ 0 & \bar{b}_{iy} \\ \bar{b}_{iy} & \bar{b}_{ix} \end{bmatrix}, \quad \bar{b}_{ij} = \frac{1}{A_k^s\left(\text{or } V_k^s\right)} \int_{\Gamma_k^s} \mathbf{n}_j(\mathbf{x})\boldsymbol{\Phi}_i(\mathbf{x})\mathrm{d}\Gamma, \quad (i = x, y) \tag{9}$$

The integral of Eq. (9) can be calculated by using Gaussian quadrature, for example, if one Gaussian point is set on each boundary edge Γ_k^s of smoothed subdomain, the Eq. (9) can be expressed as follows

$$\bar{b}_{ij} = \frac{1}{A_k^s} \sum_{m=1}^{N_\Gamma^s} n_{j,m} \boldsymbol{\Phi}_i\left(\mathbf{x}_m^G\right) l_m \tag{10}$$

where N_Γ^s is the number of boundary edges of smoothed subdomain, $n_{J,p}$ denotes the component of outward normal along the j direction on the m-th boundary edge, l_m is the length of the m-th boundary edge. In addition, the strain rates over elements are actually constant due to using strain smoothing technology, therefore, according to the associated flow rule $a\lambda = \theta$, where $\theta = \varepsilon_x + \varepsilon_y + \varepsilon_z$ is volumetric strain rate, the plastic multiplier λ is also constant over a finite element if the shear strength parameter a is constant. Finally, the objective function of Eq. (1) is simplified as

$$\int_{\Omega_k^s} k\lambda \mathrm{d}V = \bar{k}_k A_k^s \lambda_k \tag{11}$$

where $\overline{k}_k = \sin \varphi_k$ is the shear strength parameter.

In summary, based on the smoothed strain rate, the nonlinear programming problem (1) is formulated as follows:

$$\begin{aligned} \mu &= \min\begin{bmatrix} \mathbf{K}_1 & -\mathbf{F}_0 \end{bmatrix}\begin{bmatrix} \mathbf{Z} \\ \mathbf{U} \end{bmatrix} \\ \text{s.t.} &\begin{bmatrix} -\mathbf{K}_2 & \mathbf{K}_3 \\ 0 & \mathbf{F} \end{bmatrix}\begin{bmatrix} \mathbf{Z} \\ \mathbf{U} \end{bmatrix} = \begin{bmatrix} \mathbf{0} \\ 1 \end{bmatrix} \end{aligned} \tag{12}$$

In Eq. (12) the variable **Z** is constrained by a series of quadratic cones, and the rest of constraints and objective function are linear equations, it can be obviously seen that the nonlinear programming problem (12) is a classical SOCP. This SOCP can be efficiently optimized by the primal-dual interior point algorithm (PDIP) in MOSEK toolbox [10].

3 Tunnel Stability Analysis Using ANS-UBFEMLA

In the following section, the stability analysis of single non-circular tunnels is performed using the proposed method. Particularly, the plastic dissipation-based adaptive meshing method [6] is used to refine the mesh near the zones involving high shear strain rates.

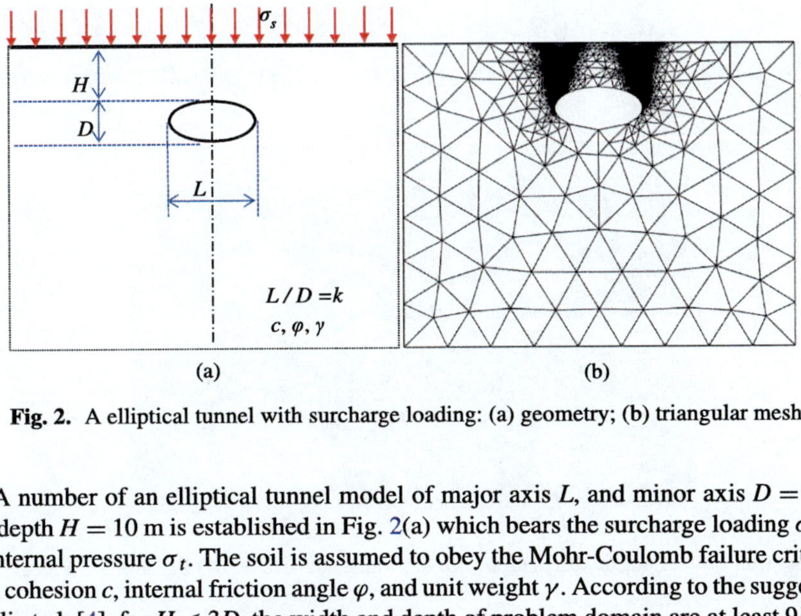

Fig. 2. A elliptical tunnel with surcharge loading: (a) geometry; (b) triangular mesh

A number of an elliptical tunnel model of major axis L, and minor axis $D = 10$ m and depth $H = 10$ m is established in Fig. 2(a) which bears the surcharge loading σ_s but no internal pressure σ_t. The soil is assumed to obey the Mohr-Coulomb failure criterion with cohesion c, internal friction angle φ, and unit weight γ. According to the suggestion of Ali et al. [4], for $H \leq 3D$, the width and depth of problem domain are at least $9D$ and $7D$, were gradually enlarged to minimize the boundary effect. Consequently, the soil domain is also discrete by T3-FEM as illustrated in Fig. 2 (b). Based on these elements, the velocity field is approximate, and the smoothed strain rates are constructed with the node-based strain smoothing domain. Finally, a SOCP programming problem for the analysis of stability of elliptical tunnel is obtained and solved using PDIP algorithm, and the dimensionless load parameter σ_s/c with the different frictional angle and ratio $k = L/D$, can be obtained as shown in Fig. 3.

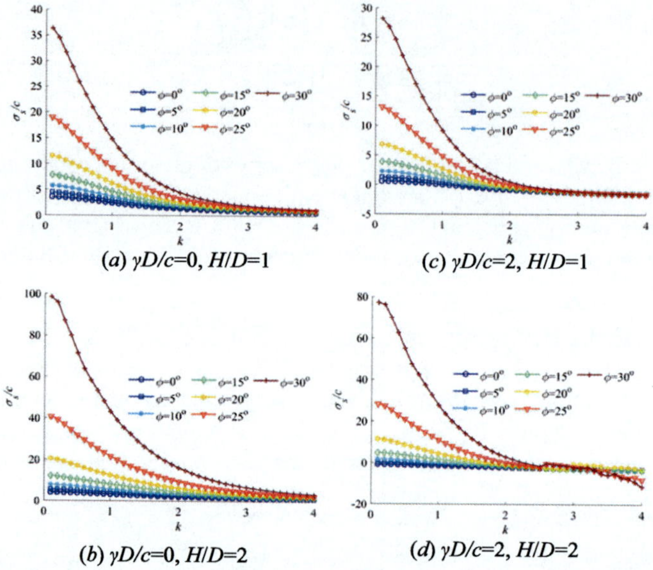

Fig. 3. Dimensionless load parameter σ_s/c versus the ratio k with different parameters

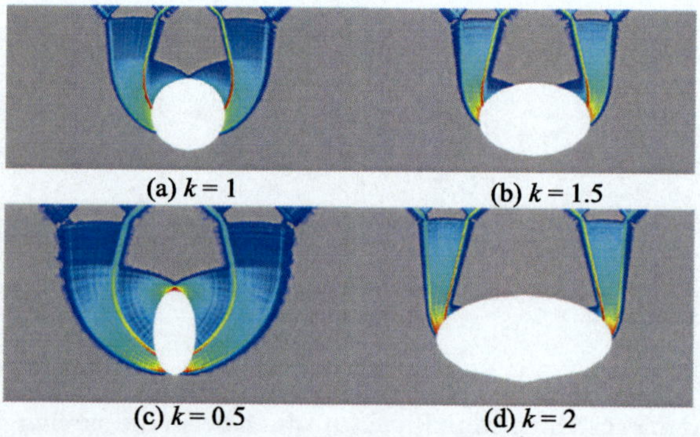

Fig. 4. Contours of plastic dissipation with different ratio k ($c = 10$ kPa, $\varphi = 5°$, $H = 10$ m)

With increasing the ratio k, the dimensionless load parameter gradually decreases. This is different from the conclusion drawn by Pan and Das [1] based on the stability analysis of the tunnel face. There exists on monotonic relationship between the stability of the tunnel face and the section shape, but the most unfavorable section of the tunnel is the circular shape which gives the smallest dimensionless load parameter. When k is equal to 4, the σ_s/c approximately reaches to zero in Fig. 3 (a)–(c), that is, the elliptical tunnel tends to collapse due to only gravity. On the other hand, in Fig. 3 (e) the σ_s/c even appears negative, it illustrates that σ_t is needed to keep the stability of tunnel. This also shows that the long axis of the elliptical tunnel is beneficial to the stability of the tunnel section when it is along the section height direction, which is different to the face stability of the tunnel. Furthermore, Fig. 4 demonstrates the contours of plastic dissipation with different ratios of k. When $k < 1$, the failure mechanism reflected by the contour is similar, and it is different from the one when $k > 1$.

4 Conclusions

The stability of a 2D elliptical tunnel is investigated by using ANS-UBFEMLA. Trough the stability analysis of elliptical tunnel, we can infer that the proposed method can not only overcome the volume locking of incompressible materials, but also it is very easy to realize the adaptive analysis, and it is easier to find the limit loading and failure mechanism of elliptical tunnel. Furthermore, the failure mechanism of the elliptical tunnel is obviously different from that of the circular tunnel, demonstrating that the geometry of the tunnel section has an obvious influence on stability.

Acknowledgement. The financial contribution of the National Natural Science Foundation of China (No. 52068016), Guangdong provincial Natural Science Foundation of China (No. 2018A030313897), the Science and Technology Program of Guangzhou, China (201707010082) and Research Startup Project of Guilin University of Technology (No. RD2100002541) are gratefully acknowledged.

References

1. Pan, Q., Dias, D.: Upper-bound analysis on the face stability of a non-circular tunnel. Tunn. Undergr. Space Technol. **62**, 96–102 (2017)
2. Li, T., Gong, W., Yang, X.: Stability analysis of a non-circular tunnel face in soils characterized by modified Mohr-Coulomb yield criterion. Tunn. Undergr. Space Technol. **109**, 103785 (2021)
3. Sloan, S.W.: Upper bound limit analysis using finite elements and linear programming. Int. J. Numer. Anal. Meth. Geomech. **13**, 263–282 (1989)
4. Makrodimopoulos, A., Martin, C.M.: Upper bound limit analysis using simplex strain elements and second-order cone programming. Int. J. Numer. Anal. Meth. Geomech. **31**, 835–865 (2007)
5. Ali, A., Lyamin, A.V., Huang, J., Sloan, S.W., Cassidy, M.J.: Undrained stability of a single circular tunnel in spatially variable soil subjected to surcharge loading. Comput. Geotech. **84**, 16–27 (2017)
6. Xiao, Y., Zhao, M.H., Zhang, R., Zhao, H., Peng, W.: Stability of two circular tunnels at different depths in cohesive-frictional soils subjected to surcharge loading. Comput. Geotech. **112**, 23–34 (2019)
7. Nagtegaal, J.C., Parks, D.M., Rice, J.R.: On numerically accurate finite element solutions in the fully plastic range. Comput. Methods Appl. Mech. Eng. **4**, 153–177 (1974)

8. Mohapatra, D., Kumar, J.: Smoothed finite element approach for kinematic limit analysis of cohesive frictional materials. Eur. J. Mech. A. Solids **76**, 328–345 (2019)
9. Zhou, X.W., Liu, F.T., Yin, Z.Y., Jin, Y.F., Zhang, C.B.: A mixed constant-stress smoothed-strain element with a cubic bubble function for elastoplastic analysis using second-order cone programming. Comput. Geotech. **145**, 104701 (2022)
10. Nguyen-Xuan, H., Rabczuk, T., Nguyen-Thoi, T., Tran, T.N., Nguyen-Thanh, N.: Computation of limit and shakedown loads using a node-based smoothed finite element method. Int. J. Numer. Meth. Engng. **90**, 287–310 (2012)
11. Zeng, W., Liu, G.R.: Smoothed finite element methods (S-FEM): an overview and recent developments. Arch. Comput. Meth. Eng. **25**(2), 397–435 (2016). https://doi.org/10.1007/s11831-016-9202-3
12. Yuan, W.H., Zhang, W., Dai, B., Wang, Y.: Application of the particle finite element method for large deformation consolidation analysis. Eng. Comput. **36**, 3138–3163 (2019)
13. Jin, Y.F., Yin, Z.Y.: Two-phase PFEM with stable nodal integration for large deformation hydromechanical coupled geotechnical problems. Comput. Methods Appl. Mech. Engrg. **392**, 114660 (2022)
14. Jin, Y.F., Yin, Z.Y., Zhou, X.W., Liu, F.T.: A stable node-based smoothed PFEM for solving geotechnical large deformation 2D problems. Comput. Methods Appl. Mech. Engrg. **387**, 114179 (2021)
15. Vo-Minh, T., Nguyen-Son, L.: A stable node-based smoothed finite element method for stability analysis of two circular tunnels at different depths in cohesive-frictional soils. Comput. Geotech. **129**, 103865 (2021)
16. MOSEK. https://www.mosek.coms

Response of Soil-Structure with Respect to Different Shapes of Pile Cross-section Under Cyclic Loads in the Context of CO_2 Emission

Linfeng Li and Jianjun Qin(✉)

State Key Laboratory of Ocean Engineering, Shanghai Key Laboratory for Digital Maintenance of Buildings and Infrastructure, School of Naval Architecture, Ocean and Civil Engineering, Shanghai Jiao Tong University, Shanghai 200240, China
jianjunqin@sjtu.edu.cn

Abstract. In the process of prosperous development of the construction industry, the problem of carbon emissions caused by material consumption is severe and cannot be ignored. The pile cross-section is an essential factor influencing the lateral resistance, while a reasonable pile cross-section can reduce material consumption and CO_2 emissions. Thus, the behavior of piles with different cross-sections embedded in Toyoura sand under cyclic lateral load was studied numerically in this paper. This paper described how to use cyclic loads to simulate earthquakes, how to find the most unfavorable load, and the numerical analysis results, such as the lateral load-displacement curves and the cyclic accumulation of horizontal displacement of the head of the pile. This paper presented a new indicator (ρ) to represent the trade-off between the CO_2 emission (represented by the consumption of the materials) and the seismic behavior, which helps in making engineering decisions. By comparing the results, it can be concluded that the lateral resistance of the solid square pile is higher than that of the solid circular pile by 39.98% under the most unfavorable cyclic load and ρ_s is 7.573, ρ_c is 16.066.

Keywords: Soil-structure interaction · CO_2 emissions · Pile cross-sections · Numerical analysis

1 Introduction

Currently, the construction sector accounts for more than 20% of annual global CO_2 equivalent emissions. Between a quarter and half of these emissions are due to cement production, while the rest comes from the manufacture, transport and construction processes of other construction materials. Therefore, selecting the appropriate pile cross-sections can reduce construction material consumption and thus reduce CO_2 emissions [1]. Many studies have been conducted to reduce CO_2 emissions in terms of construction materials. Using 3D-printed formwork elements made from recycled mineral foam, researchers at ETH Zurich had created a precast concrete slab that is lighter and better insulated while using 70% less material [2]. However, this paper will introduce a new indicator to help engineers reduce CO_2 emissions by reducing the consumption of

W. Guo and K. Qian (Eds.): GBCESC 2022, LNCE 211, pp. 607–615, 2023.
https://doi.org/10.1007/978-981-19-5217-3_61

construction materials in the design process. Besides, the seismic design is emphasized to ensure the safety of buildings in natural disasters. One crucial step is to analyze the response of the structure under earthquakes [3]. Therefore, this paper used cyclic loads to simulate earthquakes according to [4, 5].

This paper first introduced a new indicator (ρ) to describe the trade-off between the CO_2 emission (represented by the consumption of the materials) and the seismic behavior. Then, this paper explained how to use cyclic loads to simulate earthquakes. After collecting the relevant information on soil and piles, a numerical analysis was carried out. Finally, it can be found from investigations on the proposed indicator that the performance of the solid square pile is much better than that of the solid circular pile due to smaller displacement under the same cyclic load.

2 Proposal of an Indicator to Represent the Trade-off Between the CO_2 Emission and the Seismic Behavior

A new indicator (ρ) was proposed here to represent the trade-off between the CO_2 emission (represented by the consumption of the materials) and the seismic behavior. Its physical interpretation is the performance (represented by the maximum induced displacement under the lateral load) of different engineering decisions, assuming that the CO_2 emission (represented by the consumption of the materials) is the same. Since ρ reflects the maximum induced displacement, the smaller the value of ρ, the better lateral resistance. The following formula can express ρ:

$$\rho = (\textit{the maximum displacement})/(\textit{the consumption of the construction material}) \tag{1}$$

As the piles of different cross-sections chosen in this paper are made of concrete and of the same length, the formula can be simplified as:

$$\rho = [\textit{the maximum displacement}(\Delta)]/[\textit{the cross} - \textit{section area of pile}(S)] \tag{2}$$

3 Modeling of Cyclic Loads

3.1 Simulation of Earthquakes

According to [4], the cyclic load can be used to simulate the earthquake, and it can be assumed as sinusoidal:

$$F(t) = F_0 * \sin(\omega t) \tag{3}$$

where F_0 is the equivalent static load and ω is a frequency corresponding to the predominant frequency of the earthquake.

Besides, the earthquake motion can be regarded as the force applied in the form of a prescribed horizontal acceleration. Because the earthquake induces horizontal and vertical accelerations, the latter's amplitude usually is considerably less.

3.2 The Most Unfavorable Scenario

The following equation can define the cyclic load used in this paper:

$$F(t) = H_{\max} * \sin[2\pi * (t/T)] \quad (4)$$

where T is the period of the cyclic load and H_{max} is the maximum magnitude of the cyclic load. According to [5], H_{max}, which characterized the power production load, was 13.5% of the lateral capacity at serviceability (at which rotation angle is limited to 0.25°). H_{max} in this paper was taken as 32 kN. And N is the number of loading cycles. The upper limit of the number of loading cycles N is 12 due to the computational cost [5, 6]. Considering the computer solution time and computer storage space limitations, $N = 5$ was chosen for this paper, and [7] confirms that $N = 5$ is also an acceptable number of loading cycles.

During an earthquake, resonance will occur when the frequency is close to the natural frequency of the building, and it may cause severe damage. Thus, to find the most unfavorable load, it is necessary first to find the structure's natural period.

The structure's natural period can be obtained by applying an instantaneous lateral load to the model and then recording the change in acceleration at the top of the pile after the load is removed. The time interval between two adjacent maximum negative accelerations is taken as the natural period of this structure. (See Fig. 1 for details).

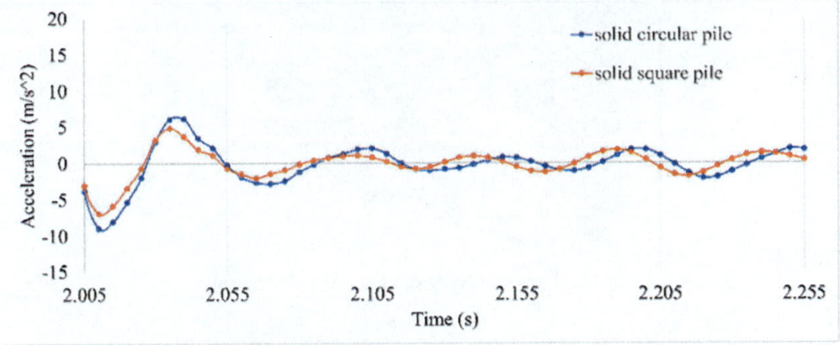

Fig. 1. Acceleration at the head of the pile under free vibration

β is the ratio of the natural period of the structure (κ) and the period of the cyclic load (T). By applying cyclic loads with different periods, it can be found that the maximum displacement of the head of the pile (Δ) tends to reach peak value when β close to 1. Therefore, it can be assumed that Δ is a function of κ and T.

4 Procedures for Numerical Analysis

4.1 The Geometry of the Soil-Pile Model

This paper focused on the soil-pile interaction with two different pile cross-sections. The first is a solid square section with a side length of 1 m and a cross-sectional area of $Ss =$

1 m^2; the second is a solid circular section with a diameter of 1 m and a cross-sectional area of $Sc = \pi/4(\mathrm{m}^2)$.

Because of the symmetry of the problem, this paper only modeled half of the pile and soil systems to save computing time. According to [8], the soil model can be designed as shown in Fig. 2. The horizontal boundaries cannot produce displacement in both x and y directions, except for the profile of the whole system, which can produce displacement in the x-direction. The bottom of the boundary is fixed in all directions. After sensitivity testing, the final determined boundary of the soil model has a diameter of 15*D* and a depth of 2*D* beneath the pile tip (*D* is the diameter of the pile), which means the soil model has a diameter of 15 m and a depth of 17 m.

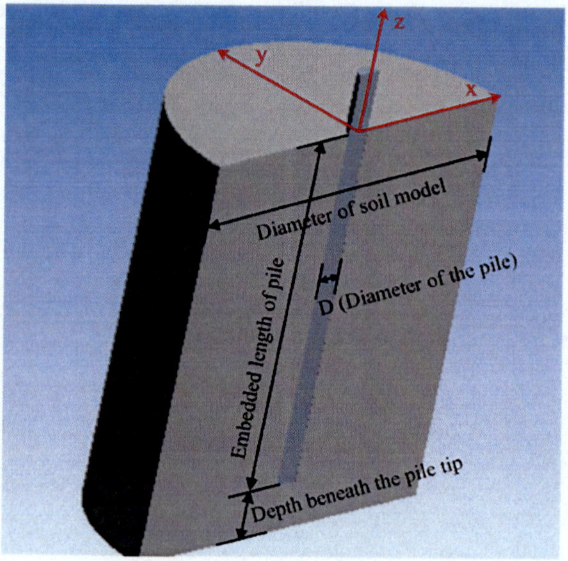

Fig. 2. The geometry of the soil-pile model

According to [8], the equivalent loading eccentricity of horizontal force (*e*) can be varied from 0.167–3.33*L* (*L* is the embedded length of the pile). Thus, the cyclic load is applied on the top of the pile, which is at the height of 2.5 m ($e = 0.167L$) above ground level, with the horizontal to the right specified as the positive direction. Only half of the lateral force is applied to the 3D model because of the symmetry of the problem.

4.2 Material Properties

When modeling the soil, it is necessary to define its density and isotropic elasticity. Besides, this paper selected the Mohr-Coulomb as the constitutive model, which requires a set of parameters: inner friction angle, cohesion and dilatancy angle. This project is performed in medium dense Toyoura sand, and the solid piles are made of concrete. The material properties used in this paper can be found in Table 1. The residual inner friction angle and cohesion are assumed to be equal to the initial values.

Table 1. Material properties of soil & pile

Material properties	Toyoura sand	Concrete
Density (kg/m^3) [9]	1450	2400
Young's Modulus (MPa) [10]	157.9	45000
Poisson's Ratio [10]	0.15	0.23
Initial Inner Friction Angle (°) [9]	31	
Initial Cohesion (Pa) [8]	4000	
Dilatancy Angle [11]	5	
Residual Inner Friction Angle (°)	31	
Residual Cohesion (Pa)	4000	

Besides, the soil-pile interface was modeled as frictional in this paper, and in the model, the tangential frictional stress is linearly proportional to normal stress by a frictional coefficient. The contact friction angle (δ) is taken as two-thirds of the soil friction angle (φ) [11]. Besides, the contact friction angle (δ) can be taken as 0.5 to 0.7 times the soil friction angle (φ) according to [12].

Thus, this paper takes the contact friction angle as $\delta = 31° \times 2/3 = 20.67°$, and the friction coefficient is $\tan(\delta) = \tan(20.67°) = 0.377$.

5 Numerical Investigation Results

5.1 Investigation of the Most Unfavorable Scenario

Table 2 shows the natural period of the different soil-pile systems and the most unfavorable cyclic load period. The values of β are less than 1 for the worst load cases due to damping in the system.

Table 2. Natural period and frequency of different piles

	Solid circular pile	Solid square pile
The natural period of the structure (s)	0.06	0.055
The natural frequency of the structure (Hz)	16.667	18.182
The period of the most unfavorable dynamic loads (s)	0.08	0.08
β	0.75	0.688

In Fig. 3, by applying cyclic loads with different periods, the pile produced the most significant displacements when the period of the applied load was close to the natural period of the soil-pile system.

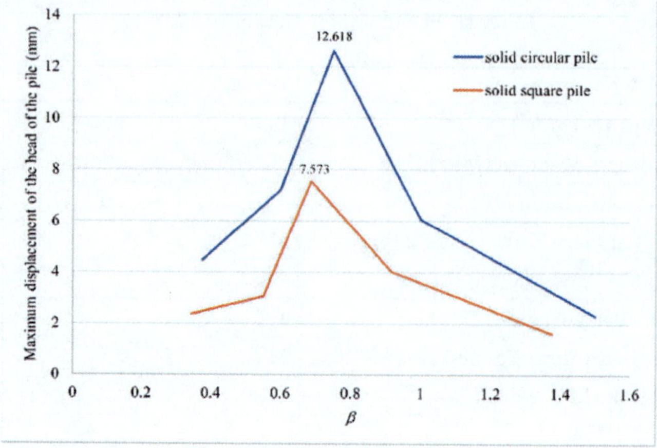

Fig. 3. The maximum displacement of the head of the solid piles with different β

It can also be concluded from Fig. 3 that the solid square pile has better lateral resistance than the solid circular pile. And when solid piles resonate with the lateral input load, the lateral resistance of the solid square pile is higher than the solid circular pile by about (12.618–7.573)/12.618 $\approx$ 39.98%.

Since the most dangerous load conditions are considered in practical engineering, the subsequent analysis will be based on the most unfavorable cyclic load.

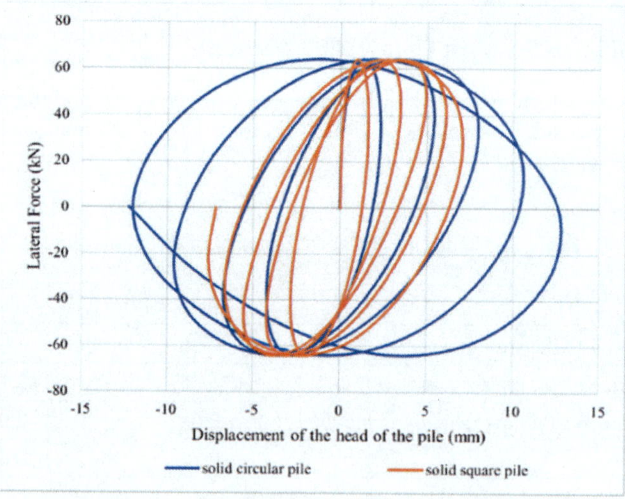

Fig. 4. Cyclic variation of y_h with F for solid piles

Figure 4 records the changes in displacement at the head of the pile as the lateral force changes for solid piles. This figure shows that the solid circular pile has more significant lateral deflection at the head of the pile after five cycles than the solid square pile. Besides, the curve corresponding to the solid circular pile is more dispersed than that corresponding to the solid square pile. These phenomena illustrate that the solid square pile has better lateral resistance than the solid circular pile.

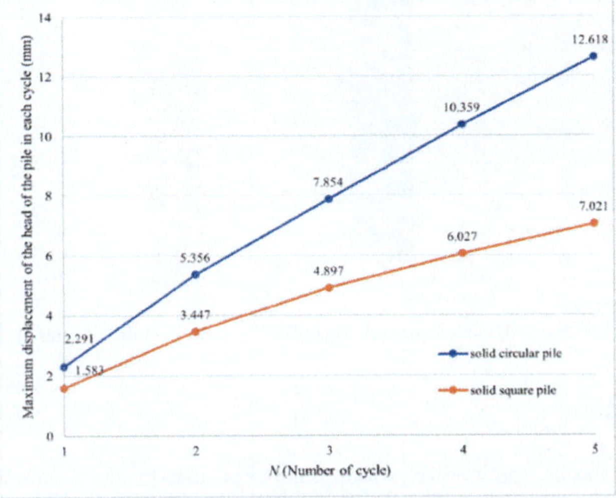

Fig. 5. Cyclic accumulation of y_h for solid piles

As shown in Fig. 5, during the first five cycles, the accumulation of horizontal displacement of the head of the pile (y_h) in the positive direction for the solid circular pile is higher than that for the solid square pile. Furthermore, the slope of the orange curve is lower than the slope of the blue curve, which agrees with the experimental finding of [13, 14] that solid square piles have a larger contact area with the soil than solid circular piles. Another speculation for the difference in slope is that the solid square pile has a greater moment of inertia, resulting in better bending resistance than the solid circular pile.

5.2 Values of the Proposed Indicator Under the Most Unfavorable Scenario

The maximum displacements produced by piles of different cross-sections under the most unfavorable cyclic loads can be obtained following the above steps. Then, substituting the maximum displacements into the formula given in Sect. 2 of this paper, the ρ for solid square piles and solid circular piles can be found.

The calculation result can be found in Fig. 6. The new indicator proposed in the previous section corresponding to the solid square pile ρ_s is $7.573/1 = 7.573$, and that corresponding to the solid circular pile ρ_c is $12.618/(\pi/4) \approx 16.066$.

The calculated result demonstrates that although more material is used to construct a solid square pile than a circular pile, the resulting lateral resistance compensates for using more material.

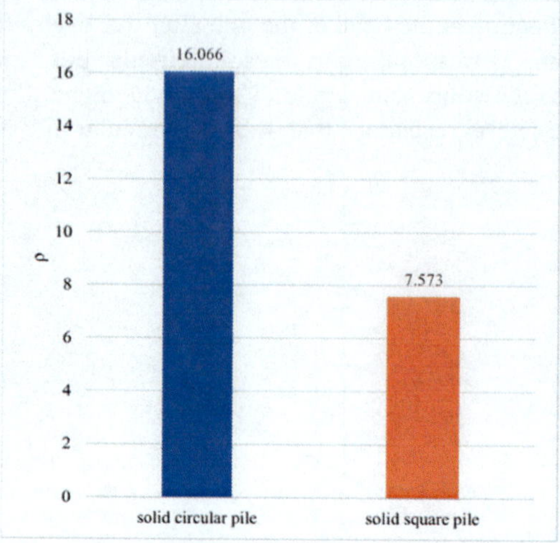

Fig. 6. Values of the proposed indicator for the soil-structure with circular and square piles

6 Conclusion

In response to the internationally proposed concept of reducing carbon emissions, the consumption of construction materials should be reduced while ensuring safety. Therefore, this paper proposed a new indicator to represent the trade-off between the CO_2 emission (represented by the consumption of the materials) and the seismic behavior. Then, this paper discussed how to use cyclic loads to simulate earthquakes and how to find the most unfavorable load. Finally, this paper used a specific example to do an analysis.

Based on cyclic loading tests of two piles with different cross-sections, it is necessary to consider the cross-section shape when designing the laterally loaded piled foundation. For this paper, because ρ_s is 7.573, which is smaller than ρ_c is 16.066, it is more economical and environmentally friendly to use the solid square pile than the solid circular pile.

The research undertaken in this paper can be extended to other cross-sections, as the methodology developed and discussed is of application to all piles. Besides, as the data obtained in this paper are generated by numerical simulations and not real engineering data, the results obtained from numerical simulations can be compared with those obtained from actual engineering in future studies to judge whether the method proposed in this paper is in line with reality and to further optimize the method proposed in this paper.

References

1. GLOBE - Global Consensus on Sustainability in the Built Environment. https://www.rilem.net/globe. Accessed 10 Feb 2022

2. How can we cut the emissions from concrete use? | World Economic Forum. https://www.weforum.org/agenda/2022/01/eth-zurich-3d-printer-concrete-carbon-emissions?utm_source=linkedin&utm_medium=social_video&utm_term=1_1&utm_content=24830_scientists_geometry_concrete_emissions&utm_campaign=social_video_2022. Accessed 21 Feb 2022
3. Menglin, L., Huaifeng, W., Xi, C., Yongmei, Z.: Structure-soil-structure interaction: literature review. Soil Dyn. Earthq. Eng. **31**(12), 1724–1731 (2011)
4. Keshavan, N.: Dynamic and Earthquake Forces on Deep Foundations. Performance of Deep Foundations. ASTM International (1969)
5. Aliasger, H., Masanari, N., Gopal, M.: Cyclic lateral loading of monopiles for offshore wind turbines. In: 3rd International Symposium on Frontiers in Offshore Geotechnics, pp. 711–716 (2015)
6. Spyros, G., Nikos, G., George, G.: Cyclic lateral response of piles in dry sand: finite element modeling and validation. Comput. Geotech. **44**, 116–131 (2012)
7. H.Ercan, T., Stavros, S.: Behaviour of cyclic laterally loaded large diameter monopiles in saturated sand. In: 7th European Conference on Numerical Methods in Geotechnical Engineering, pp. 905–910. Taylor & Francis Group (2000)
8. Huan, W., Lizhong, W., Yi, H., Amin, A., Ben, H., Hualin, P.: Influence of pile diameter and aspect ratio on the lateral response of monopiles in sand with different relative densties. J. Marine Sci. Eng. **9**(6), 618–639 (2021)
9. Mehmet, I.O., Seyfettin Umut, U., Volkan, O., Mustafa, T.: An approach to determine the initial shear modulus of clean sandy soils. Global J. Res. Eng.: E Civ. Struct. Eng. **17**(2), 31–39 (2017)
10. Laxmi, S., Reiko, K.: Statically and dynamically measured Poisson's ratio of granular soils on triaxial laboratory specimens. Geotech. Test. J. **36**(4), 493–505 (2013)
11. Martin, A., Yu-Shu, K., Khalid, A.-R.: Behavior of monopile foundations under cyclic lateral load. Comput. Geotech. **36**(5), 725–735 (2009)
12. Haider, A.-J., Sarmad Abdullah, A.: Finite element analyses of laterally loaded pile groups in um Qaser port. Int. J. Sci. Eng. Res. **5**(11), 662–668 (2014)
13. Jasim, M.A., Qasim Ibrahim, H.: The effect of pile cross section on the lateral behavior of piles under combined loading. J. Eng. Sci. Technol. Rev. **11**(3), 174–179 (2018)
14. Lymon, C.R., William F. Van, I.: Single Piles and Pile Groups Under Lateral Loading, 2nd edn. CRC Press, Netherlands (2011)

Simplified Calculation of Waist Hole of Straddle Monorail Pin Bearing

Hui Wang[1], Shuyi Duan[2], and Yanqiang Ji[2](✉)

[1] Hunan Boyu Construction Engineering Consulting Co. Ltd, Changsha 410017, China
[2] School of Civil Engineering, Central South University, Changsha 410075, China
a19911588371@163.com

Abstract. To realize the horizontal displacement function of the bearing, a kind of pin bearing with a waist hole is applied to the straddle monorail viaduct system for the first time. Because there is no stipulation on the calculation of the pin bearing with waist hole in the domestic and foreign codes, based on the pin design formula of the steel structure design standard, this paper analyzes the key parameters that affect the force of the waist hole and establishes a simplified calculation formula of the waist hole, which can be used as a supplement to the relevant provisions of the code and can be used as a reference for the corresponding design.

Keywords: Pin bearing · Waist hole · Simplified formula

1 Introduction

With the rapid development of domestic tourism, as a tool of scenic spot transportation and sightseeing, the straddle monorail elevated tour bus is increasingly widely used. Its viaduct structure usually adopts a simply supported single-chamber steel box girder with a span of 12–20 m [1–4]. To solve the problem of temperature deformation of the beam, a pin bearing with a waist hole is designed, which is composed of the upper pendulum, lower pendulum, and pin, and adopts a double-ear plate single shear structure, as shown in Fig. 1. The pin shaft can achieve limited horizontal displacement in the waist hole. The bearing has the advantages of simple manufacture, convenient installation, and deformation, will not affect the tracking smoothness, and can better meet the use of monorail steel box girder.

(a) Side view (b) Front view

Fig. 1. Pin bearing with waist ear plate

W. Guo and K. Qian (Eds.): GBCESC 2022, LNCE 211, pp. 616–628, 2023.
https://doi.org/10.1007/978-981-19-5217-3_62

The calculation formulas of the ear plate pinhole for pin connection in the Chinese steel structure code are all aimed at the round hole, and there is no basis for the calculation of the pinhole as a waist hole. To solve the above problems, based on the calculation formula of round hole ear plate in China's steel structure code, this paper deduces the simplified calculation formula of waist hole ear plate by analyzing the key parameters affecting the stress of waste hole.

2 Simplified Calculation of Pin Shaft

The stress state of the pin shaft is very complex. In the actual design, the approximate allowable stress method is often used to calculate the bending stress, shear stress, and bearing stress of the pin shaft, and the safety factor or empirical coefficient is used to enlarge or reduce it. At the same time, the corresponding structural requirements are put forward for the ear plate and gap. Literature [5, 6] studied the failure modes of pin joint, ear plate, and pin shaft. According to the different loading conditions, there are five main failure forms of pin connection: pin bending failure, pin shear failure, pin bearing pressure failure, ear plate bearing pressure failure, and ear plate tear failure. The tear damage of the ear plate can be divided into the following three cases [8]: pulling at the net section of the hole, splitting at the end, and shearing at the end. The failure form of the ear plate is shown in Fig. 2. *China's Steel structure Design Standard (GB50017-2017)* [9] has detailed provisions on the calculation of conventional pin joints. In addition to the structural requirements, the ear plate mainly checks the bearing capacity of the net section tensile strength, end section tensile (splitting) strength, and shear strength, and the pin shaft also has the corresponding checking formula.

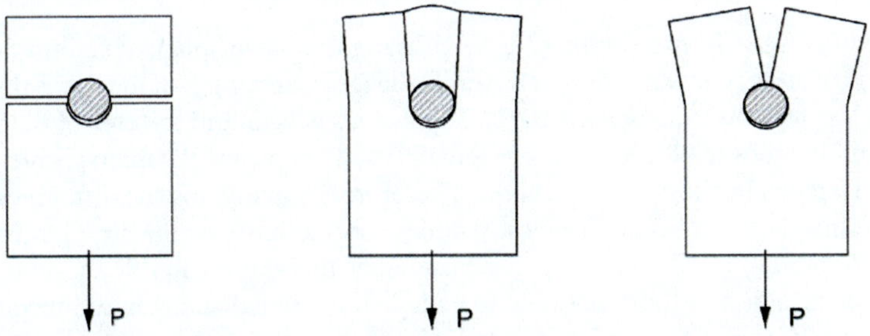

(a)Break at the net section (b) Shear at the end (c) Split at the end

Fig. 2. Destruction form of ear plate

The pin and round hole ear plate of the pin bearing used in this paper have no different from the conventional pin bearing, and these can be checked and calculated directly according to the code. For the case that the ear plate pinhole is a waist hole, because there is no standard calculation method, the finite element method can be used for the corresponding analysis and calculation in the actual design [10]. However, the workload is heavy and there is no unified standard. It is not conducive to the standard design

[11, 12]. Based on the calculation formula of round hole ear plate of pin connection in steel structure design standard and combined with finite element analysis, the proposed calculation formula of waist hole is established in this paper for design reference.

3 Calculation and Analysis of Waist Hole

3.1 Tensile Strength at the Net Section of the Ear Plate

The calculation diagram of the tensile strength of the net section of the ear plate in the code is shown in Fig. 3, which shows the standard structure of the pin ear plate. In the figure, *d0* is the diameter of the pinhole, *b* is the gabarit between the edges of both sides of the ear plate and the edge of the pinhole, and *a* is the minimum distance between the edge of the pinhole and the edge of the ear plate in the direction of the force. Its structure requires that the width-to-thickness ratio on both sides of the ear plate should meet $b/t \leq 4$, and the geometric dimensions should be in accordance with the provisions of formulas (1) and (2):

$$a \geq \frac{4}{3} b_{\varepsilon} \tag{1}$$

$$\mathrm{b}_e = 2t + 16 \leq b \tag{2}$$

Tensile strength of pin hole net section:

$$\sigma = \frac{N}{2\mathrm{t}b_1} \leq f \tag{3}$$

$$\mathrm{b}_1 = \min\left(2t + 16, b - \frac{d_0}{3}\right) \tag{4}$$

As can be seen from formula (3), considering the out-of-plane instability of the ear plate and the stress concentration of the hole, the tensile strength of the net section adopts *b1*, which is less than the actual size *b*. There is no substantial difference in the out-of-plane stability between the lumbar hole and the round hole, and the main problem focuses on the difference in stress concentration. Therefore, by analyzing the difference of stress concentration between the two types of holes, on the basis of the strength calculation formula of the circular hole, this paper introduces the stress amplification factor of the stress concentration of the waist hole to establish the tensile calculation formula of the waist hole. The method is well connected with the existing steel structure specifications and can reflect the influence of the structural parameters of the waist hole on the shear capacity.

Referring to *the Handbook of stress concentration Factors* [13], consider a lath with a width of *B* and length of *L*, and thickness of *t*. A waist hole is set on its center line, and its two ends are connected by an arc of diameter *d0*. The length of the waist hole is *c*, and the long axis of the hole is perpendicular to the edge of the board. When the slab bears the tensile stress of σ, the maximum stress occurs at point *A* at the edge of the hole, as shown in Fig. 4. The stress concentration factor and the reference stress are:

$$\alpha_\sigma = \frac{\sigma_{\max}}{\sigma_n} = \frac{\sigma_A}{\sigma_{\mathrm{n}}} \tag{5}$$

$$\sigma_n = \frac{\sigma}{1 - \frac{c}{B}} \tag{6}$$

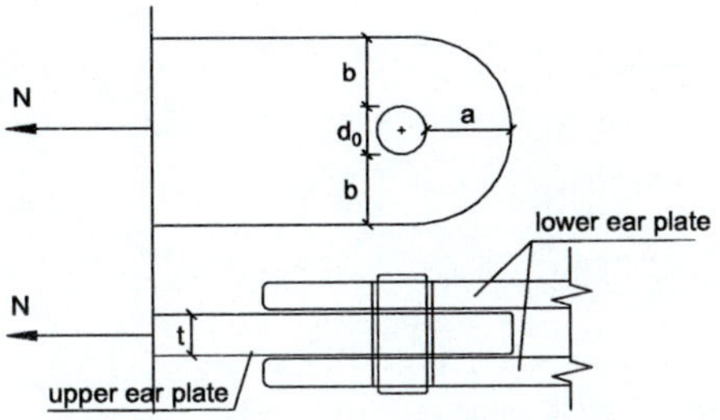

Fig. 3. Diagram of calculation of net section tensile strength of the lug plate of the pin bearing

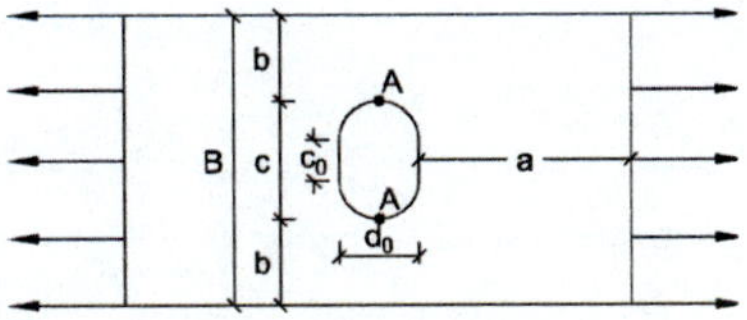

Fig. 4. Stretching of slats with a waist hole

According to the photoelastic experimental results of Nishida Masahiro [14, 15], the relationship between α_σ and *c/r* is:

$$\alpha_\sigma = \left[2 + \left(1 - \frac{c}{B}\right)^3\right]\left\{1 + \frac{2}{3}\left[1 - \left(\frac{c}{B}\right)^{1.6}\right]\left[\sqrt{\frac{c}{2r}} - 1\right]\right\} \tag{7}$$

According to the experimental results of Neugebauer, the relationship between α_σ^1 and *d0 /B* is as follows:

$$\alpha_\sigma^1 = 2 + \left(1 - \frac{d_0}{B}\right)^3 \tag{8}$$

From formulas (7) and (8), it can be concluded that the stress concentration factor of the lumbar hole is related to *c*/*B* and *c*/2*r*. Although the above boundary conditions are different from the force of the pin bearing, they can reflect the degree of stress concentration of the two kinds of the hole. Formula (4) shows the parameters that affect the net section of a circular hole in tension are *b* and *d0*. When the net section is constant, the stress depends on *d0*. For waist holes, hole length *c* needs to be considered. Therefore, this paper uses *c* and *d0* as parameters to construct several pin bearing's ear plate models. Through the finite element analysis of the stress value at the net section of the ear plate with a round hole and waist hole, when the ear plate is under tension, the ratio of the stress value of the waist hole to the round hole is obtained, which is marked as *K1* as the stress magnification factor of the waist hole load effect.

According to the basic assumption of Hertz's theory [16, 17], the size of the contact area is much smaller than the size of the object. When using ABAQUS finite element software for analysis, it can be approximately loaded with concentrated force [18, 19]. The calculation diagram of net section tension is shown in Fig. 5 and 6.

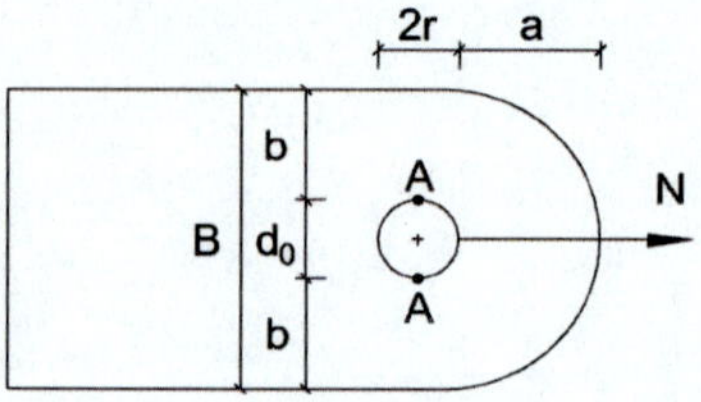

Fig. 5. Diagram of calculation of net section tension of lumbar plate with a circular hole

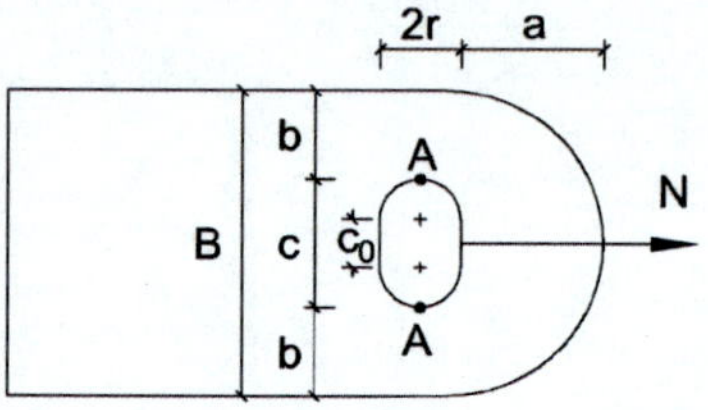

Fig. 6. Diagram of calculation of net section tension of lumbar plate with a waist hole

The detailed structure parameters of the ear plate and the ABAQUS stress calculation results are shown in Table 1, and part of the stress cloud diagram is shown in Fig. 7. Statistics of the original *K1* data are shown in Fig. 8 and the fitting plane is shown in Fig. 9. The formula for the adjustment coefficient *K1* is as follows:

$$K1 = 0.9191 + 0.0039c - 0.003d0 \tag{9}$$

After the regression equation of *K1* is adjusted, its determinable coefficient *R2* is 0.93343, indicating that the percentage of the variability of the dependent variable that it can explain is greater than 93%, and the fitting degree of the sample value is very well.

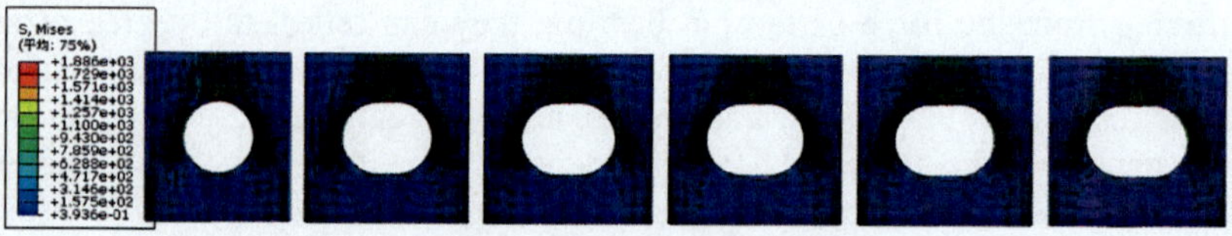

Fig. 7. Partial stress cloud diagram

After considering *K1*, the formula for calculating the net section tensile strength of the ear plate with lumbar hole established in this paper is shown in formula (10):

$$\sigma = K_1 \frac{N}{2tb_1} \leq f \tag{10}$$

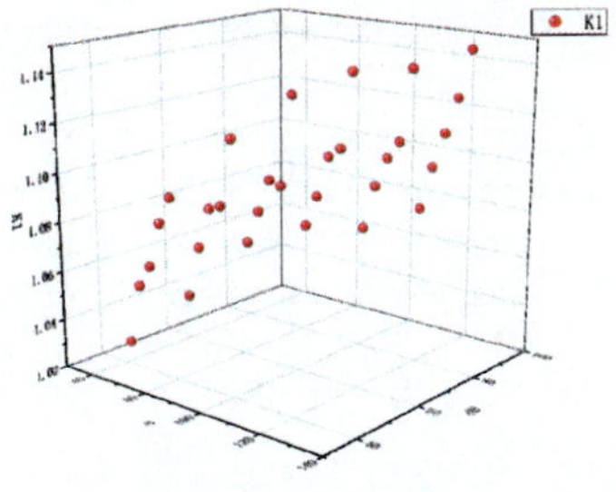

Fig. 8. Original data of *K1*

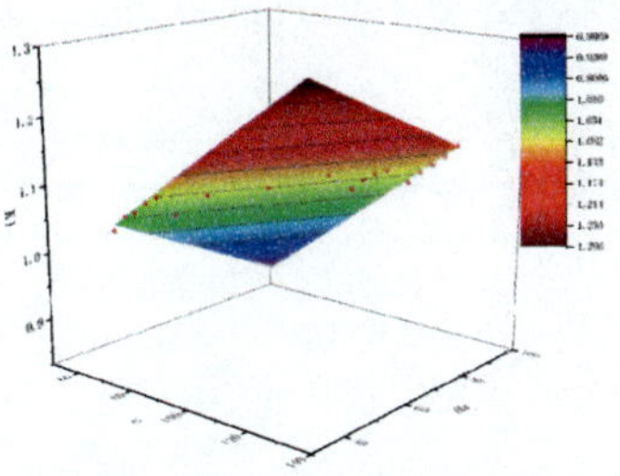

Fig. 9. The fitting plane of *K1*

The ABAQUS finite element verification results of formula (10) are shown in Table 2. The theoretical bearing capacity value calculated by the simplified formula is basically consistent with the value of the finite element simulation., and the relative error is less than 8%, indicating that the formula is reasonable.

Table 1. The structural parameters of the lug plate and the calculation results of ABAQUS stress

$a = 60, b = 45, t = 12$	c	σ/Mpa	*K1*	$a = 60, b = 45, t = 12$	c	σ/Mpa	*K1*
d0 = 40	0	29.06452		*d0* = 50	0	29.06428	
	62	29.91421	1.029234		72	30.42639	1.046865
	66	30.63588	1.054065		76	31.04697	1.068217
	70	30.90258	1.063241		80	31.54331	1.085295
	74	31.4443	1.081879		84	31.60109	1.087283
	78	31.77478	1.09325		88	32.40746	1.115027
d0 = 60	0	29.33371		*d0* = 70	0	29.91455	
	82	31.3207	1.067737		92	32.10654	1.073275
	86	31.72764	1.08161		96	32.50179	1.086487
	90	32.13289	1.095425		100	33.01238	1.103556
	94	32.09495	1.094132		104	33.14202	1.107889
	98	33.18626	1.131335		108	34.08444	1.139393
d0 = 80	0	30.59139		*d0* = 90	0	31.04729	
	102	32.75732	1.070802		112	33.45328	1.077495
	106	33.32575	1.089383		116	34.02321	1.095851
	110	33.70581	1.101807		120	34.48809	1.110825
	114	33.93636	1.109344		124	34.9652	1.126192
	118	34.86118	1.139575		128	35.58252	1.146075

[unit: mm].

Table 2. The ABAQUS finite element verification results of the tensile strength at the net section of the ear plate

c0	*d0*	The bearing capacity value of ABAQUS/*N*	The bearing capacity value of the formula /*N*	The relative error/%	*d0*	The bearing capacity value of ABAQUS/*N*	The bearing capacity value of the formula /*N*	The relative error/%
22	40	285906.427	308600	7.93741	50	283455.567	303300	7.00090
26		281684.808	301200	6.92802		279305.49	297150	6.38889
30		277586.046	298500	7.53422		275275.183	292200	6.14833
34		273604.854	293200	7.16184		271359.533	292000	7.60632
22	60	281046.369	294570	4.81189	70	278677.779	287416	3.13560
26		276966.031	291372	5.20135		274665.436	283976	3.38978
30		273002.477	287121	5.17157		270766.991	279676	3.29029
34		269150.764	287511	6.82154		266977.662	278580	4.34581
22	80	276348.779	287415	4.00444	90	274058.385	275767	0.62345
26		272402.746	283976	4.24858		270177.031	271168	0.36678
30		268567.819	279676	4.13608		266404.082	267469	0.39974
34		264839.37	278580	5.18829		262735.058	263988	0.47688

Note: The relative error = the value of the finite element the value of the formula.1

3.2 Tensile (Splitting) Strength of the End Section of the Ear Plate

The formula for calculating the tensile strength (splitting) at the end of the ear plate in the steel Structure design standard is as follows:

$$\sigma = \frac{N}{2t\left(a - \frac{2d_0}{3}\right)} \leq f \tag{11}$$

According to the above formula, for the round hole, the tensile strength of the end of the ear plate is related to the minimum distance a from the edge of the pinhole to the edge of the plate and the hole diameter *d0*, while *d0* mainly reflects the non-uniformity of the stress distribution. For the waist hole, the force in the direction of the hole length is different from that of the round hole under the pin load. With the increase of the whole length, the stress of the ear plate tends to be more inclined to the bending member. The length of the waist hole *c* can be regarded as the span of the beam, *a* is the height of the beam, and the calculation diagram of the tension at the end of the ear plate is shown in Fig. 10. It can be seen that the size of *a* and *c* is an important parameter that affects the tensile strength of the end of the ear plate with the lumbar hole.

According to the above content, this paper takes the length of the straight section of the waist hole *c0* (*c0* = *c* − *d0*) and the pinhole diameter *d0* as parameters and constructs several pin bearing's ear plate models according to engineering practice and data analysis. Through the finite element analysis of the stress values at the end of the ear plate with a round hole and lumbar hole, the stress adjustment coefficient *K2* is calculated. The detailed structural parameters of the ear plate and the calculated results of ABAQUS stress are shown in Table 3 and part of the stress cloud diagram is shown in Fig. 11. The original data of *K2* is shown in Fig. 12 and the fitting plane is shown in Fig. 13. The formula for the adjustment coefficient *K2* is as follows:

$$\mathrm{K2} = 1.398 - 7.649\mathrm{e} - 4c0 - 4.8\mathrm{e} - 4d0 \tag{12}$$

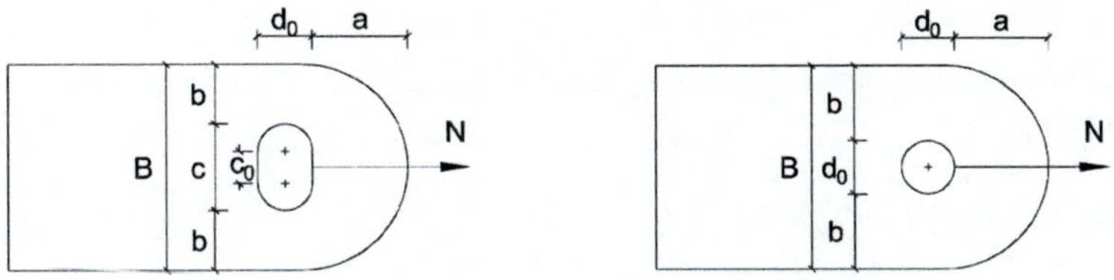

(a) Ear plate with a round hole (b) Ear plate with a waist hole

Fig. 10. Diagram of calculation of tension at the end of ear plate

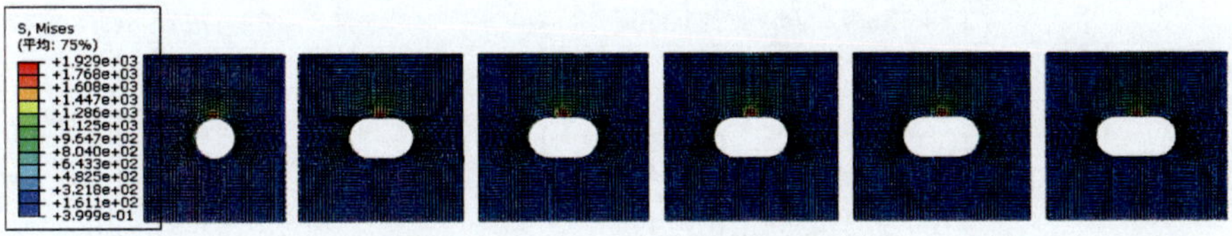

Fig. 11. Part of the stress cloud diagram

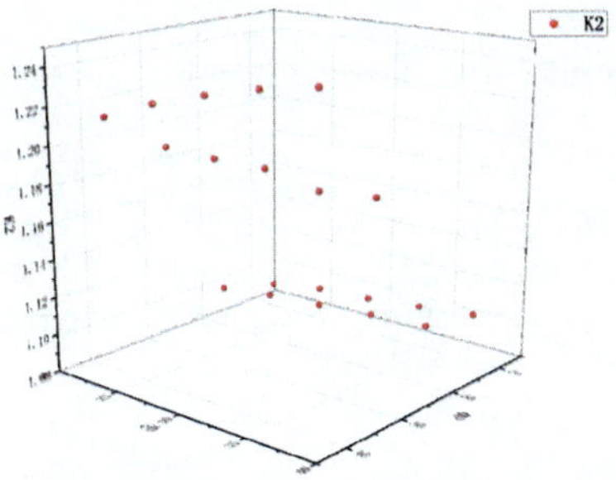

Fig. 12. Original data of *K2*

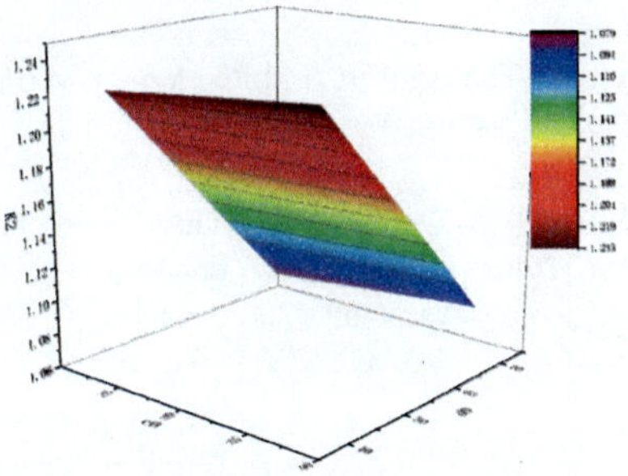

Fig. 13. Nonlinear surface fitting of *K2*

The adjusted *R2* is 0.90689, indicating that the percentage of the variability of the dependent variable explained by the formula is more than 90%, and the regression equation fits the sample observations well.

After considering *K2*, a computational formula of the tensile (split) strength at the end section of the lumbar ear plate established in this paper is shown in formula (13):

$$\sigma = \frac{K_2 N}{2t\left(a - \frac{2d_0}{3}\right)} \le f \tag{13}$$

The ABAQUS finite element verification results of formula (13) are shown in Table 4. The theoretical bearing capacity value calculated by the simplified formula is basically consistent with the value of the finite element simulation, and the relative error is less than 5.1%, indicating that the formula is reasonable.

Table 3. The structural parameters and stress calculation results of the tensile model at the end of the ear plate

$a = 60, b = 45, t = 12$	c	σ/Mpa	K2	$a = 60, b = 45, t = 12$	c	σ/Mpa	K2
$d0 = 40$		31.00012		$d0 = 50$	0	31.00003	
	22	37.63389	1.213996		22	37.63389	1.190907
	26	37.98437	1.225302		26	37.98437	1.190072
	30	38.25357	1.233986		30	38.25357	1.190439
	34	38.47274	1.241056		34	38.47274	1.184717
$d0 = 60$		31.00001		$d0 = 70$	0	31.00001	
	22	34.14978	1.101606		22	33.87681	1.0928
	26	34.25624	1.10504		26	34.02557	1.097598
	30	34.32437	1.107238		30	34.07921	1.099329
	34	34.39416	1.109489		34	34.17386	1.102382

[unit: mm]

Table 4. The ABAQUS finite element verification results of the tensile strength of the ear plate at the end section

c0 /mm	d0 /mm	The bearing capacity value of ABAQUS/N	The bearing capacity value of the formula/N	The relative error/%	d0 /mm	The bearing capacity value of ABAQUS/N	The bearing capacity value of the formula/N	The relative error/%
22	40	198014	202360.1	2.14769	50	161472	168926	4.41260
26		196100	201856.1	2.85159		160747	168487.1	4.59388

(continued)

Table 4. (*continued*)

c0 /mm	d0 /mm	The bearing capacity value of ABAQUS/*N*	The bearing capacity value of the formula/*N*	The relative error/%	d0 /mm	The bearing capacity value of ABAQUS/*N*	The bearing capacity value of the formula/*N*	The relative error/%
30		194613	201354.6	3.34815		163747	168050.4	2.56080
34		193606	201354.6	3.84826		163525	167616	2.44071
22	60	133030	132093.1	0.70930	70	90806	91981.43	1.27790
26		132568	131735.2	0.63214		90372	91721.21	1.47100
30		132273	131379.4	0.68020		90216	91462.46	1.36281
34		131971	131025.4	0.72170		89940	91205.17	1.38717

Note: The relative error = the value of the finite element/ the value of the formula -1

3.3 Shear Strength of the End Section of Ear Plate

The calculation diagram of the shear strength of the ear plate in the steel structure code is shown in Fig. 14, where *Z* is the width of the shear section at the end of the ear plate, and the formula for calculating the shear strength of the ear plate with round hole is as follows:

$$\tau = \frac{N}{2tZ} \leq f_v \tag{14}$$

$$Z = \sqrt{(a + d_0/2)^2 - (d_0/2)^2} \tag{15}$$

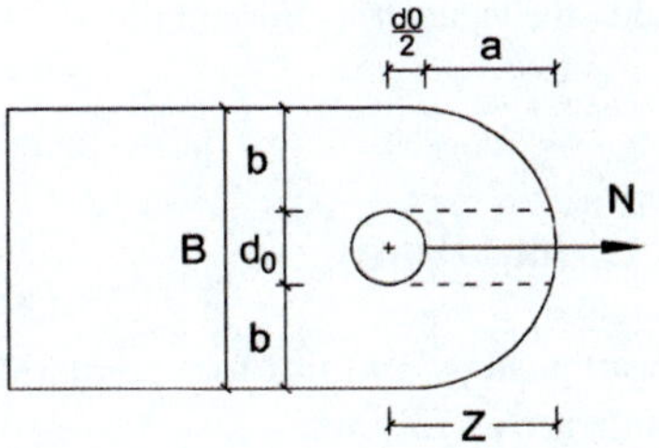

Fig. 14. Illustration of ear plate shearing with a round hole

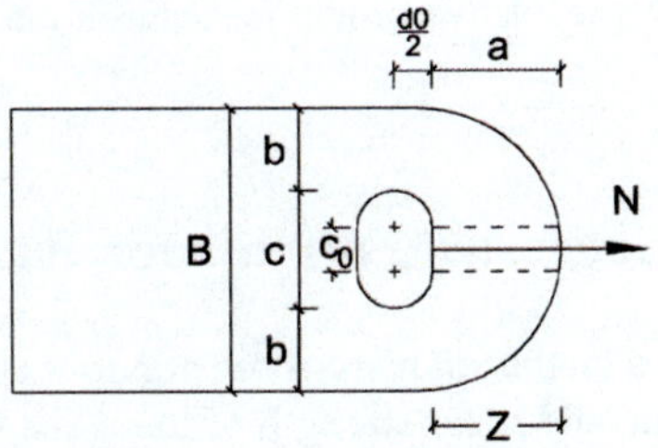

Fig. 15. Illustration of ear plate shearing with a waist hole

It can be seen from the above formula that the shear section at the end of the ear plate is the main factor affecting the shear strength of the ear plate. For the waist hole, because of its opening length, its shear failure surface is more likely to occur in the straight section. Therefore, we will no longer analyze the stress amplification factor, but directly take the Z value as the minimum distance between the edge of the hole and the edge of the plate, namely, $Z = a$, as shown in Fig. 15. The calculation formula for the shear strength of the ear plate with waist hole is:

$$\tau = \frac{N}{2ta} \leq f_v \tag{16}$$

The ABAQUS finite element verification results of formula (16) are shown in Table 5. The theoretical bearing capacity value calculated by the simplified formula is basically consistent with the value of the finite element simulation, and the relative error is less than 6.5%, indicating that the formula is reasonable.

Table 5. The ABAQUS finite element verification results of the shear strength of the ear plate at the end section

c0 /mm	*d0*/mm	The bearing capacity value of ABAQUS/*N*	The bearing capacity value of the formula/*N*	The relative error/%	*d0*/mm	The bearing capacity value of ABAQUS/*N*	The bearing capacity value of the formula/*N*	The relative error/%
22	40	269055	259200	3.80208	50	269387	259200	3.93017
26		270189	259200	4.23958		270292	259200	4.27932
30		270286	259200	4.27701		270282	259200	4.27546
34		270555	259200	4.38079		275840	259200	6.41975
22	60	265075	259200	2.26659	70	269630	259200	4.02392
26		270539	259200	4.37461		271000	259200	4.55247
30		270799	259200	4.47492		270160	259200	4.22840
34		270329	259200	4.29360		270357	259200	4.30440

Note: The relative error = the value of the finite element/ the value of the formula 1

4 Suggested Formula for Waist Hole Calculation

Based on the analysis of the previous section, the suggested formulas for calculating the tensile and shear strength of the waist hole ear plate are as follows:

The tensile strength of the net section of the ear plate with a waist hole:

$$Kl = 0.9191 + 0.0039c - 0.003d0$$

$$\sigma = K_1 \frac{N}{2tb_1} \leq f$$

Tensile (splitting) strength of the end section of the ear plate with a waist hole:

$$K2 = 1.398 - 7.649e - 4c0 - 4.8e - 4d0$$

$$\sigma = \frac{K_2 N}{2t\left(a - \frac{2d_0}{3}\right)} \leq f$$

Shear strength of the ear plate with a waist hole:

$$\tau = \frac{N}{2ta} \leq f_v$$

In the formula, N—design value of axial tension of member (N);
$b1$—calculated width (mm);
f—design value of tensile strength of ear plate(N/mm^2);
fv—design value of shear strength of steel(N/mm^2).

5 Conclusion

The pin bearing of the straddle monorail tour car adopts the form of waste hole, and its force is different from that of the ear plate with a round hole, so we can not simply use the formula for calculating the bearing capacity of the ear plate with a round hole in the code. Based on the calculation formula of the ear plate with the round hole in the steel structure code, this paper analyzes the key parameters affecting the force of the waist hole and establishes a simplified calculation formula when the pinhole of the ear plate is a waist hole, taking into account the particularity of the force of the waist hole, and organically combined with the code formula. The formula is clear in concept, convenient and practical, and can be used as a reference for design.

References

1. Qingzhong, T., Weifeng, X.: Research on manufacturing technology of rail steel beam of monorail tourist car. Mech. Eng. **05**, 154–156 (2018)
2. Fengqi, G., Keyuan, C., Fagen, G., et al.: Reviews on current situation and development of straddle-type monorail tour transit system in China. Changsha: J. Central South Univ. (Nat. Sci. Ed.) (2021)
3. Weifeng, Y., Huachun, P., Jiashan, H., Mingyang, L., Wengkai, C.: Research on Design of straddle track beam support. Railway Eng. **58**(06):53–56+96 (2018)
4. Meiling, Z., Yongkun, L., Xiyao, F.: Study on the contact stress of support of straddle type PC rail beam. Railway Stand. Des. **62**(09), 94–99 (2018)
5. Ping, Z.: Experimental study and FEM analysis on spherical plain bearing joint in a grand theatre. Const. Technique **47**(11), 66–69+87 (2018)
6. Dayi, D., Bin, P., Yuanqing, W., Yong, S., Wei, L.: Application and research of pin connection in steel structure engineering. Tianjin University. In: Collection of papers on the celebration of the 60th anniversary of Professor Liu Xiliang's teaching and the 11th National Symposium on Modern Structural Engineering. Tianjin University: Academic Committee of the National Symposium on Modern Structural Engineering, vol. 801–806 (2011)

7. Tianyi, Y.: The design method of domestic and foreign bridge pin joints. World Bridge (02), 22–25 (2011)
8. Shuai, W., Xianzhong, Z., Yiyi, C.: Analysis and design of pin bearing performance. Build. Struct. **39**(06), 77–81 (2009)
9. GB50017–2017, Steel structure Design Standard. Higher Education Publishing House, Beijing (2017)
10. Yuanqing, W., Junzhu, L., Dayi, D., Gang, S., Rihong, A., Qiuzhuan, F.: Finite element analysis of beam-column pin joints of large-span heavy-duty steel structure. J. Shenyang Jianzhu Univ. (Nat. Sci. Ed.) **32**(05), 769–777 (2016)
11. Yongchi, L.: Plastic analysis and design suggestions of ear plate force. Building Technol. Dev. **32**(09), 13–21 (2005)
12. Jiasen, L.: A practical method of pin shaft design. Build. Struct. **47**(06), 24–28 (2017)
13. Science and Technology Committee of the Ministry of Aviation Industry. Stress concentration factor manual. Higher Education Press (1990)
14. Kefeng, W., Chunlei, C., Feng, L.: Research progress of stress concentration at the hole edge of connectors. Jiangsu Airlines **S2**, 72–74 (2010)
15. Anan, L., et al.: Nishida Masahiro. Stress concentration. translator. Machinery Industry Press, Beijing (1983)
16. Xisheng, Z., Zhen, W., Jingchun, Z.: Contact stress calculation of mechanical parts. Machinery **01**, 26–28 (2000)
17. Marvin: Numerical calculation and finite element analysis of rolling bearing contact problems. Mech. Eng. Autom. **12**(6), 51–53 (2011)
18. Xiao, G., Xitian, T., Jun, Z., Zhongwei, Z.: Stress analysis of pin shaft structure based on elastic contact. Mech. Sci. Technol. **33**(03), 322–325 (2014)
19. Bin, Z., Wenbo, S.: Research on the modified design method of pin shaft connection based on Hertz theory. J. Build. Struct. **39**(S1), 410–417 (2018)

Experimental and Numerical Studies on the Seismic Performance of Crawler-Type Connection Node

Yixin Zeng, Wei Lin, Jing Li(✉), Yixi Hu, Zhilong Jiang, and Peixing Xie

School of Civil Engineering and Transportation, South China University of Technology, Guangzhou 510640, Guangdong, China
cvjingli@scut.edu.cn

Abstract. A kind of prefabricated external wall panel connection node is proposed in this paper. The node is simplified as a crawler damper and analyzed by quasi-static tests to investigate the influence of different materials on the mechanical properties of the node. The test results indicate that specimens made of cold-rolled Q235 and hot-rolled Q235 have excellent energy dissipation effects, but the specimens made of hot-rolled materials have better fatigue resistance and are more in line with the needs of practical use. In addition, based on the finite element analysis software ABAQUS, a specimen model was established for finite element analysis and compared with the mechanical test results to verify the reliability of the finite element simulation method. The influence of the height, thickness, and arc segments' diameter geometric parameters of the crawler damper on its bearing capacity and energy dissipation capacity was studied. The variation of the load capacity and energy dissipation capacity of the damper with these geometric parameters is determined.

Keywords: Crawler damper · Energy dissipation capacity · Quasi-static test · ABAQUS finite element analysis

1 Introduction

Prefabricated structure building has become one of the green building systems in China because of its advantages of standardized production, green environmental protection, and short construction time [1]. The outer protective structure of prefabricated steel structure mostly adopts prefabricated external wall panels, which have more flexibility in installation and better adaptability [2], but their structural integrity is poor, and the prefabricated external wall panels are easy to fall off. Therefore, assembling steel frame and external wall panel connection nodes is significant in assembling steel structures.

The connection nodes connecting external wall panels and steel frames include rigid, semi-rigid, and flexible connection nodes. Rigid connection node is a mature connection form that can keep the connection between the wall panel and the steel frame well in the elastic stage. Chen modeled composite wallboard, conducted finite element analysis,

W. Guo and K. Qian (Eds.): GBCESC 2022, LNCE 211, pp. 629–648, 2023.
https://doi.org/10.1007/978-981-19-5217-3_63

and concluded that although the rigid connection node could have the greater bearing capacity, the destruction of the load on the wall panel would be greater under the earthquake action [3]. In non-earthquake-prone areas, rigid connection nodes can provide good endurance. However, the overall performance of rigid connection nodes would be weaker than that of flexible connections in earthquake-prone areas. Zhao thought that the steel structure itself had good deformation performance, which led to a good seismic performance [4]. This requires that the connection between the wall and the steel structure should be a flexible connection member. Cui studied the connection forms of the connecting nodes and the hanging points of the wall panels and considered that the wall panels should not affect the stiffness of the main structure, the connection design should meet the requirement that the wall panel can adapt to the coordinated deformation of the main structure under earthquake action [5]. Therefore, when a connection node is designed, unnecessary freedom should be released properly. Wu carried out a quasi-static test on a single-span frame structure, analyzed the result, and concluded that the use of rigid connection nodes could improve the lateral stiffness of the structure but also increase the stress value on the wall panel, which accelerated the destruction of the wall panel. Wu considered that the relatively flexible connection between the wall panel and steel frame can improve the lateral stiffness of the structure and ensure that the wall panel still has effective bearing capacity when the structure is in large horizontal displacement [6].

In the semi-rigid and flexible connection nodes related research, Fu carried out a full-scale seismic test on the multi-story fabricated steel structure building frame and verified that the flexible connection nodes could effectively reduce the in-plane deformation of the wall panels [7]. Li conducted a dynamic elastoplastic time history analysis on a 12-story frame structure and concluded that friction-slip connectors could effectively reduce the seismic response of the superstructure [2]. Wang used Ansys finite element analysis software to simulate the low-cycle displacement loading of the whole frame system with a semi-rigid sliding friction connection [8]. Moreover, he analyzed that although a semi-rigid sliding friction connection can dissipate part of the earthquake energy, its friction and slip are limited to microscopic deformation, and it is still insufficient to protect the wall panel under the earthquake action.

As can be seen from the above content, compared with the rigid connection nodes, although the semi-rigid and flexible connection nodes reduce the destruction of interlayer displacement on wall panels by releasing a degree of freedom, the protection of wall panels is still inadequate under earthquake action.

However, compared with the traditional connection nodes, the energy dissipation capacity of the damper is much more excellent. Since 1972 Kelly et al. proposed the concept of energy dissipation shock absorption [9], more energy dissipation damping components have been designed and used. As a connecting tool to connect the prefabricated external wall panels and structural steel frames, it is very important to have the ability of energy dissipation and shock absorption. Crawler dampers can exert a good energy dissipation effect among all kinds of steel dampers used for energy dissipation. Hiroshi Tagawa et al. explored the mechanical properties of a vibration control system with a crawler-type plate based on a quasi-linear motion mechanism [10]. They combined the energy dissipation system of a crawler-type plate with a steel strand support system to avoid the problem of buckling under the pressure of traditional support. Hou carried out quasi-static tests on crawler dampers, and the hysteresis curves obtained were relatively saturated, which showed that crawler dampers have a good energy dissipation capacity [11]. Chong used a crawler damper to connect the bottom frame beam to the precast wall panels [12]. After conducting quasi-static tests on a single frame, he concluded that the crawler damper could yield before the main structure was destroyed and continue to dissipate seismic energy through the expected crawler rolling deformation, thus serving as the first seismic defense line to protect the wall panels.

Based on the concept of protecting wall panels by releasing degrees of freedom and the good energy dissipation capacity of crawler dampers under reciprocating loads, an independent energy dissipation connection node is designed to connect the external wall panel and frame beams. This connection node can strengthen the integrity of the structure, and the design of crawler makes the node's energy dissipation capacity better, which could protect the wall panels from damage during earthquakes. Additionally, aiming at the bearing capacity and energy dissipation capacity, the connection node was studied by the method of combining mechanical test and finite element analysis in this paper.

2 Crawler-Type Connection Node and Mechanical Characteristic

As shown in Fig. 1, the crawler-type connection node comprises a crawler-type steel plate and an angle steel plate. The angle steel connection is welded in the crawler-type steel plates and bolted to the upper beam. The crawler-type steel plates are bolted to the prefabricated external wall panels (refer to Fig. 2). When a minor earthquake occurs, the crawler-type steel plate is not going to dissipate the energy, so the prefabricated external wall panels move with the steel frame, it makes full use of the lateral stiffness of the wall panels. However, as the main energy dissipation, the crawler-type steel plates would roll likes a crawler and produce plastic deformation to consume the energy when the structure interlayer displacement occurs under frequent earthquake action, to protect the prefabricated external wall panels from destruction.

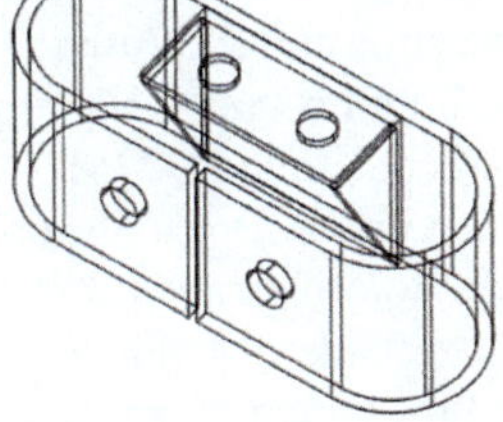

Fig. 1. Crawler-type connection node

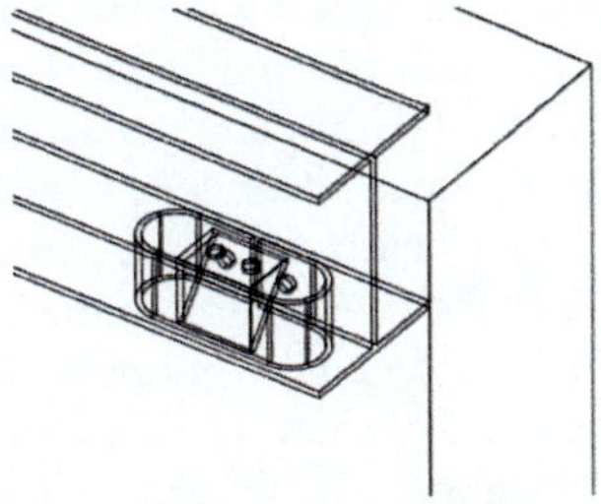

Fig. 2. Actual use of the Crawler-type connection

Compared to the connection nodes widely used in prefabricated buildings currently, the yield segment of the Crawler-type connection node will move along the crawler-type steel plate when rolling like a crawler, which avoids the concentration of plasticity at one cross-segment. The mechanism of the crawler-type connection node is beneficial to improve its fatigue resistance to consume more energy under seismic load.

In order to analyze the energy dissipation effect of the Crawler-type connection node when it suffers a load such as an earthquake, the Crawler-type connection node is simplified to a crawler damper, as shown in Fig. 3, and tested by applying reciprocating loads which simulate seismic action. The crawler damper is made of carbon structural steel. Considering the influence of material properties on mechanical properties, crawler dampers are made of hot-rolled Q235B and cold-rolled Q235B, respectively. This paper tests the crawler dampers made of these two materials and compares their energy dissipation effects and fatigue resistance. In addition, the finite element software ABAQUS is used to analyze the crawler dampers made of hot-rolled Q235B with different sizes to study the geometrical parameters of a crawler damper.

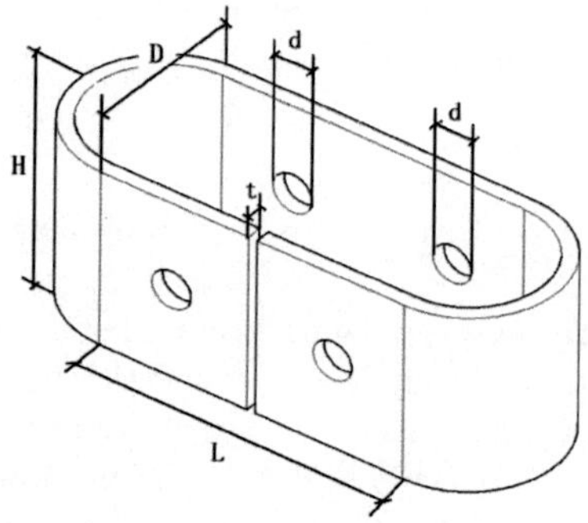

Fig. 3. Dimensional sketch of crawler damper

3 Test

This segment analyzes the influence of different materials on crawler dampers by carrying out quasi-static test and fatigue load test on the specimens made of hot-rolled Q235B and cold-rolled Q235B, and the material which is suitable for practical use is selected.

Before the quasi-static test, a tensile test was carried out on the tensile steel plates (refer to Fig. 4), which are made of hot-rolled Q235B and cold-rolled Q235B. As shown in Fig. 5, the tensile test obtained the material's stress-strain curve. Due to the cold-rolled Q235B plate being cold working hardened, its curve has no obvious yield stage. From a stiffness point of view, the peak stress of the steel plate made of cold-rolled Q235B is greater than that of steel plate made of hot-rolled Q235B, which means that the steel plate made of cold-rolled Q235B has a better bearing capacity under a single load.

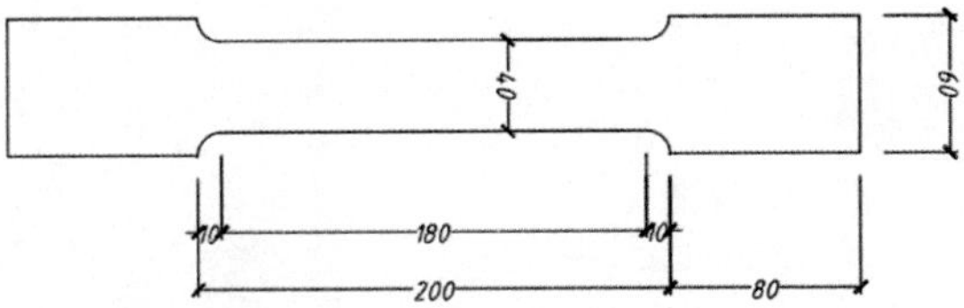

Fig. 4. Tensile steel plates

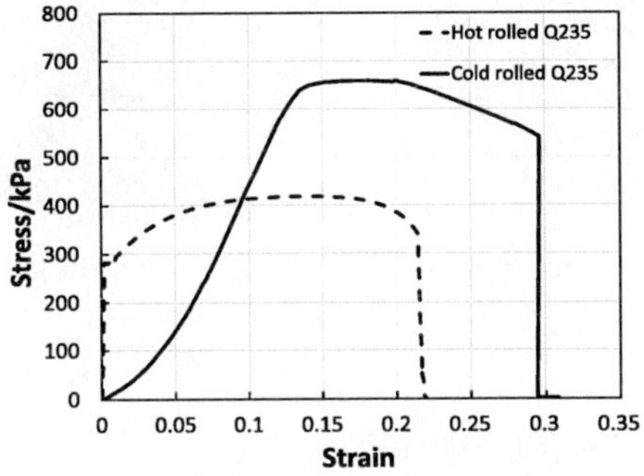

Fig. 5. Stress-strain curve of the specimen

3.1 Quasi-static Test Apparatus

MTS servo testing machine is used in the test. As shown in Fig. 6, The test apparatus includes two lateral restraint baffles, a steel plate mounted in the middle, and a base plate. All the components of the test apparatus are made of 45# steel. The base plate is connected to the bottom of the testing machine with a pin, and the steel plate is connected to the top of the test machine. The lateral restraint baffles are bolted to the baseplate to provide lateral stiffness for the crawler dampers during the test to prevent the specimen's deformation perpendicular to the direction of motion. The crawler dampers are bolted between the steel plate and the lateral restraint baffles. When the testing machine starts to work, an opposing force would be applied to both sides of each crawler damper through the steel plate mounted in the middle and the lateral restraint baffles. The working mechanism imitates the working mechanism when the crawler-type connection node is bolted to the prefabricated external wall panel and frame beam under the action of an earthquake and can reflect the actual result of the crawler-type connection node under the load. The above bolts are all M22 high-strength friction grip bolts. The actual test apparatus is shown in Fig. 7.

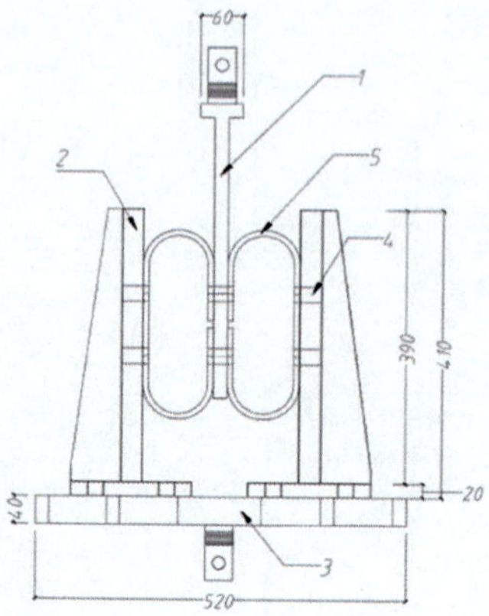

Fig. 6. Design of the test apparatus

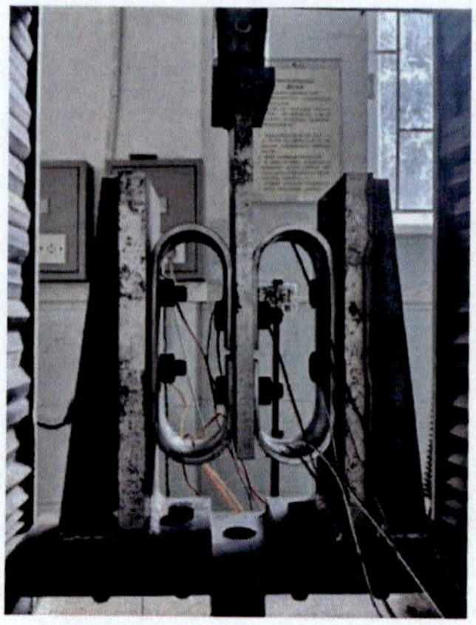

Fig. 7. Actual installation of the test apparatus

3.2 Test Group and Loading System

The test mainly studied the different mechanical properties of crawler dampers made of hot-rolled Q235B and cold-rolled Q235B. The crawler dampers used in the test are specimens with sizes L = 270 mm, H = 100 mm, D = 100 mm, d = 22 mm, and t = 8 mm. The quasi-static test and fatigue load test were carried out on the specimens made of above two different materials. The hysteresis curve of the crawler dampers could be obtained through the test to reflect the energy dissipation effect and compare the mechanical properties of two kinds of crawler dampers.

As shown in Fig. 8, the loading system of the quasi-static test follows the requirements of AISC [13]. Two turns were performed at each amplitude: 6, 12, 24, 36, and 48 mm. In addition, four turns were performed at amplitude 36 mm after 48 mm.

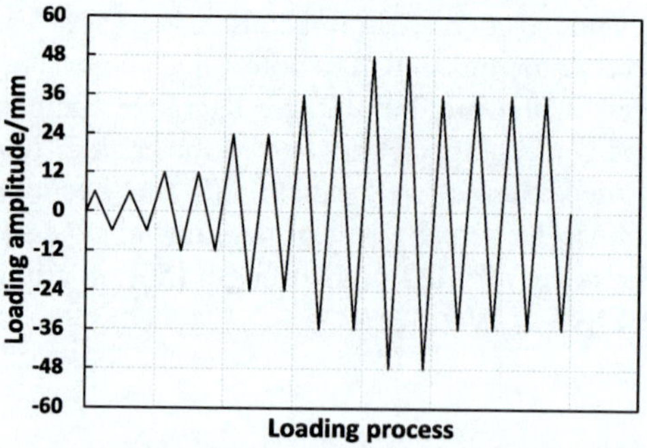

Fig. 8. Loading system of quasi-static test

After the quasi-static test, the fatigue load test was carried out on the crawler dampers to study the fatigue resistance of specimens. The loading system of the fatigue refers to JGJ 297-2013 [14]. For displacement-dependent energy dissipation devices, the peak stress attenuation of the energy dissipation device should not exceed 15% after 30 turns at the designed displacement amplitude of the energy dissipation device. In order to test the fatigue load-bearing capacity of specimens under rare earthquakes, the maximum allowable inter-story displacement of the building structure under seismic load, amplitude 60 mm (1/50 of the building's single-story height) is taken as the loading amplitude to load the specimens.

3.3 Test Result and Analysis

Quasi-Static Test. When the force was close to zero in the reciprocating loading, the specimen rattled could be observed, along with small amplitude vibration on the hysteresis curve. This is due to the change of force position on the connecting bolt, which transmitted the force. After the test, no obvious splitting was observed in the appearance of each specimen. The specimen's shape did not change significantly, and the surface of the specimen was worn at the connection of the bolt (refer to Fig. 9 and Fig. 10).

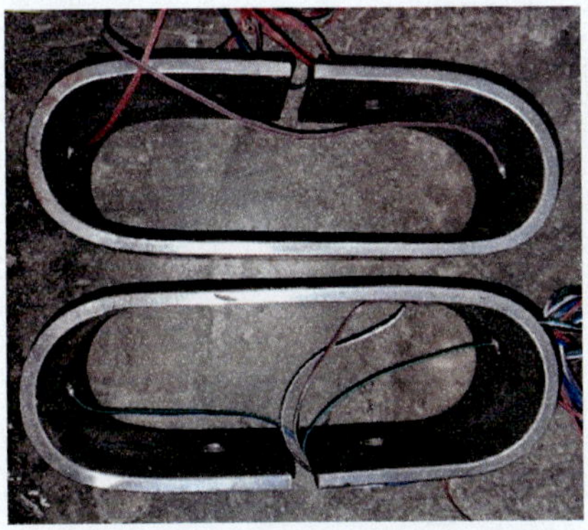

Fig. 9. Specimens' appearance after the test

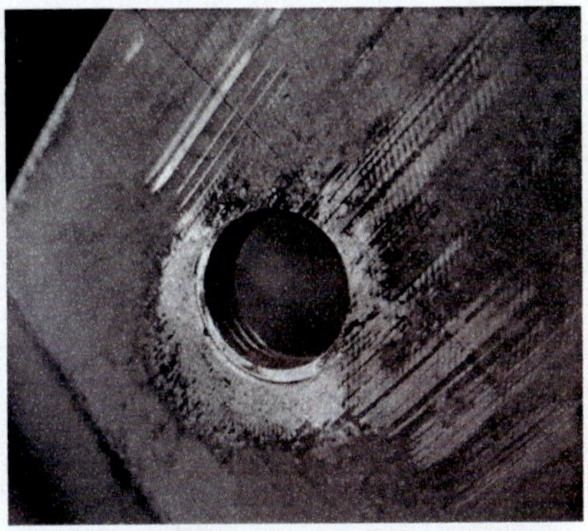

Fig. 10. Partially worn of the specimen

The hysteresis curve of a single specimen obtained by the quasi-static test is shown in Fig. 11 and Fig. 12. It can be seen from the hysteresis curves that the hysteresis curves of the two kinds of specimens are full.

In order to reflect the hysteresis curve more directly, the energy dissipation ratio and the equivalent viscous damping ratio were selected to show the energy dissipation effect of the specimen [15]. Both can well reflect the energy dissipation of structures under earthquake action. It can be considered that the energy dissipation effect of components is better when the energy dissipation ratio is closer to 1 and the equivalent viscous damping ratio is higher. Their calculation formula and diagram are as follows (Fig. 13).

$$\psi = \frac{A_1}{A_1 + A_2} \tag{1}$$

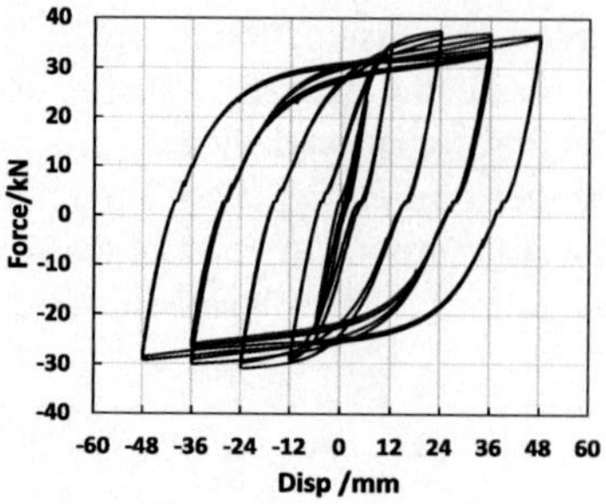

Fig. 11. Hysteresis curves of cold-rolled specimens under quasi-static test

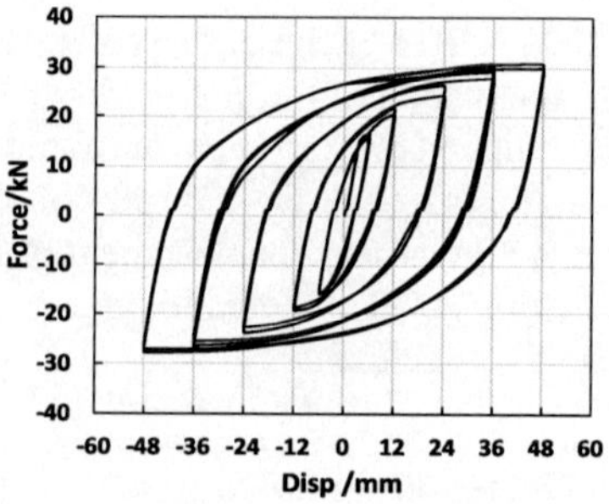

Fig. 12. Hysteresis curves of hot-rolled specimens under quasi-static test

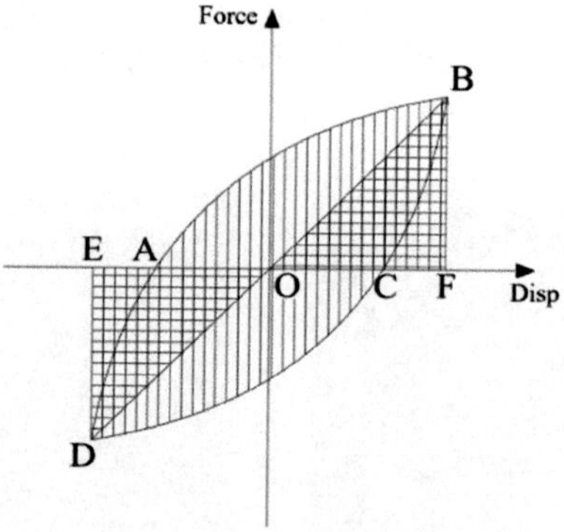

Fig. 13. Calculation diagram of hysteretic curve

$$\zeta = \frac{A_1}{2\pi A_3} \tag{2}$$

In the calculation formula, A_1 represents the energy absorbed by the specimen in one turn of loading (refer to the area of the ABCD), A_2 represents the energy released of the specimen when it is unloaded (refer to the area of the BCF and ADE), A_3 represents half of the product of loading amplitude and peak force (refer to the area of the OBF and ODE).

As shown in Table 1, it can be obtained by calculating the parameters of a hysteresis curve that the hot-rolled specimen's energy dissipation ratio and Equivalent viscous damping ratio are 0.962 and 0.447, and those of the cold-rolled specimen are 0.951 and 0.436, respectively. Moreover, due to the slow plastic development of the specimens in the later loading stage, the energy dissipation ratio of the two specimens increases and tends to be stable with the increase of the loading amplitude (refer to Fig. 14).

Table 1. Hysteretic curve parameters

Material	Energy dissipation ratio (ψ)	Equivalent viscous damping ratio (ζ)
Hot-rolled Q235B	0.962	0.447
Cold-rolled Q235B	0.951	0.436

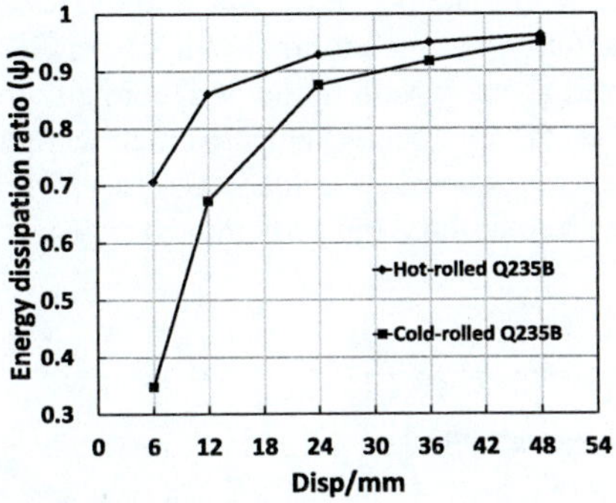

Fig. 14. Energy dissipation ratio

According to the hysteresis curve and the analysis results from the curve, it can be indicated that no matter what kind of materials the specimens are made of have an excellent hysteretic energy dissipation effect. However, the difference between the two types of specimens is that the hysteresis curves of the hot-rolled specimens show cyclic strengthening at the same displacement amplitude while the hysteresis curves of the cold-rolled specimens show cyclic degradation.

The first peak load of each loading amplitude on the hysteresis curve was selected to draw the skeleton curve, as shown in Fig. 15. It can be observed in the comparison of the skeleton curves of the two specimens that the specimen made of cold-rolled Q235B showed better energy dissipation capacity in the quasi-static test, which is consistent with the result of the tensile test, while the restoring force of the specimen made of hot-rolled Q235B is expected. The specimens are designed to resist seismic loads, so it is very important to conduct fatigue load tests on these two types of specimens.

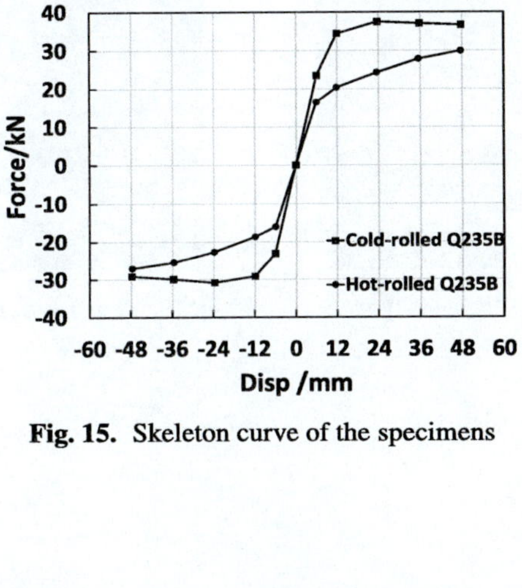

Fig. 15. Skeleton curve of the specimens

Fatigue Load Test. The hysteresis curve of the specimen after fatigue loading is shown in Fig. 16 and Fig. 17. In this test, to facilitate the experimental device's assembly, M20 bolts were chosen to connect the specimen with the steel plate and lateral restraint baffles instead of M22 bolts. Compared with the quasi-static test, it can be observed that the specimen rattled more frequently, and the amplitude of vibration in the hysteresis curve increased in the fatigue load test. At the initial stage of the fatigue load test, the crawler deformation of crawler dampers still occurs. With the increase of loading turns, the plastic hinge of the crawler-type steel plate began to appear at the junction of arc segment and straight segment and gradually lost the motion mode of crawler deformation. Finally, a complete fracture occurred at the plastic hinge of specimens. As shown in Fig. 18 and Fig. 19, the cold-rolled Q235B specimens began to crack when it was loaded to the 14th turn, fracture failure occurred when it was loaded to the 17th turn, and the hot-rolled Q235B specimen was cracked at the 22nd turn and damaged at the 23rd turn.

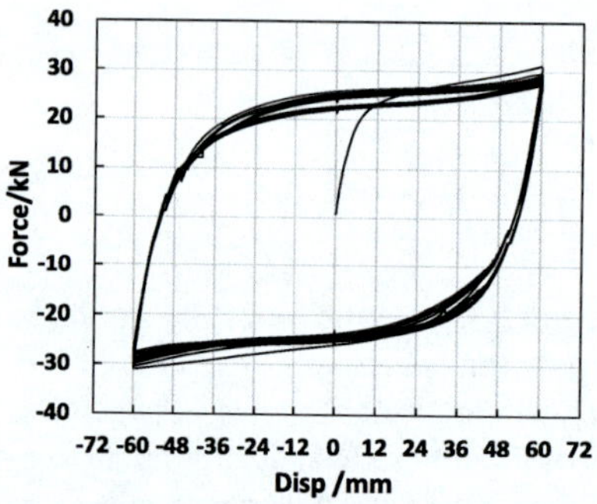

Fig. 16. Hysteresis curve of cold-rolled specimen under fatigue load

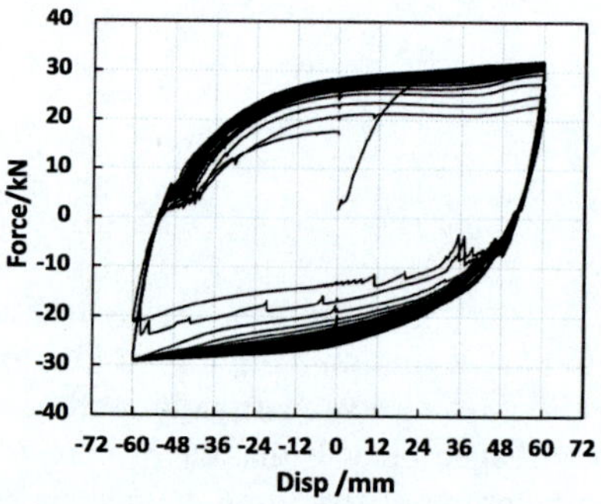

Fig. 17. Hysteresis curve of hot-rolled specimen under fatigue load

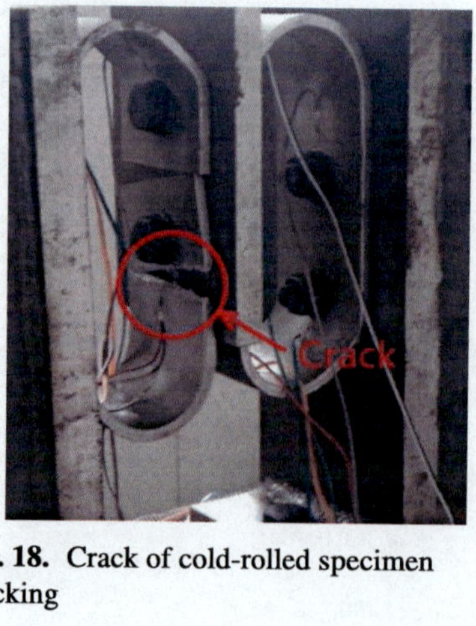

Fig. 18. Crack of cold-rolled specimen cracking

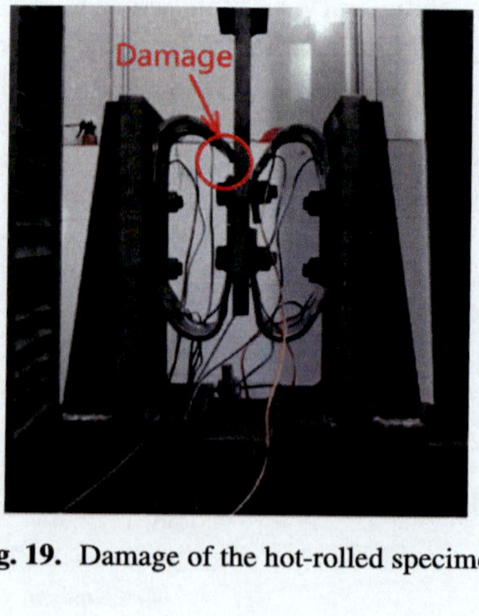

Fig. 19. Damage of the hot-rolled specimen

As shown in Fig. 20, the energy dissipation of the cold-rolled specimen is slightly better than that of the hot-rolled specimen in the first turn of fatigue load, but it is exceeded after the strengthening of the hot-rolled specimen in the first turn of loading. The energy dissipation capacity of the subsequent two specimens decreases simultaneously, and the hot-rolled specimen is loaded with more turns than the cold-rolled specimen. It can also be obtained from the degradation stiffness curves of two kinds of specimens that the stiffness of cold-rolled specimens degrades faster (refer to Fig. 21), which indicates that the fatigue load resistance of cold-rolled specimens is weaker.

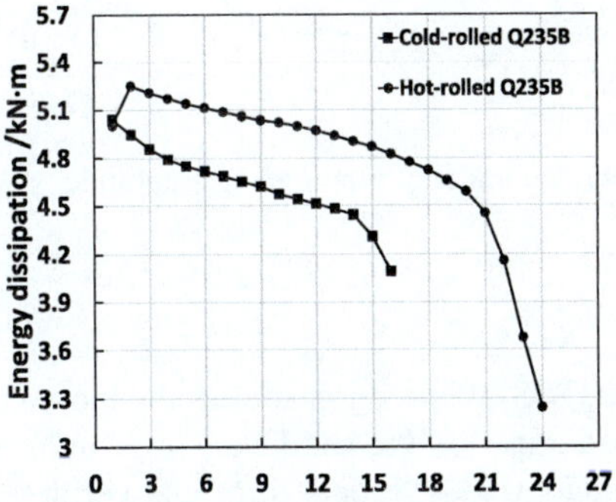

Fig. 20. Energy dissipation of the specimens

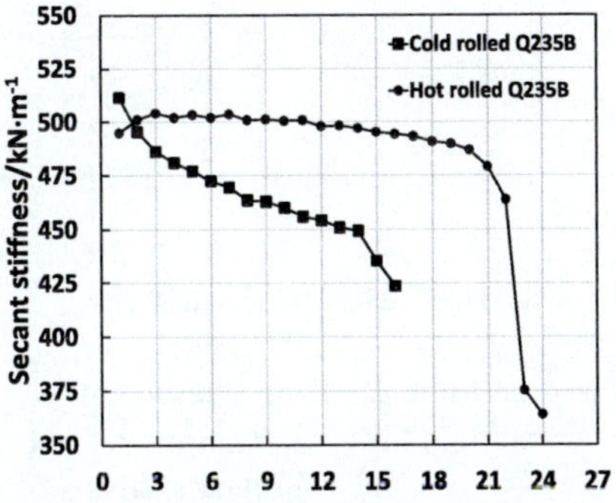

Fig. 21. Secant stiffness of the specimens

Conclusion. Above all, it can be concluded that although the cold-rolled specimen exhibits better energy dissipation capacity in the quasi-static test, it exhibits worse fatigue load resistance capacity in the fatigue load test due to its cold work-hardening characteristics. The connection nodes designed in this paper are designed for prefabricated structures in earthquake disaster-prone areas. Therefore, from the perspective of design and practical use, the nodes made of hot-rolled Q235B have more practical use-value.

To verify whether the damper can achieve desired energy dissipation under the action of frequent earthquakes, a fatigue load test with a loading amplitude of 30 mm was carried out on a hot-rolled Q235B specimen. The loading amplitude of 30 mm met the requirements of GB/T51232-2016 [16]; that is, the outer wall panels should not fall off when the inter-story displacement of the structure reached 1/100 under the action of a frequent earthquake. The hysteresis curve of the fatigue loading is shown in Fig. 22. After 30 turns of fatigue loading, the peak force of the specimen decreases by 1.16%, which is less than 15% required by the specification, and the specimen is intact. It shows that the specimen can exert a favorable energy dissipation effect and protect the wall panel under frequent earthquakes. Although the specimen was damaged when the load reached 23 turns under loads that simulate the action of rare earthquakes, the specimen had a good energy dissipation effect and delayed the failure time of the wall panel.

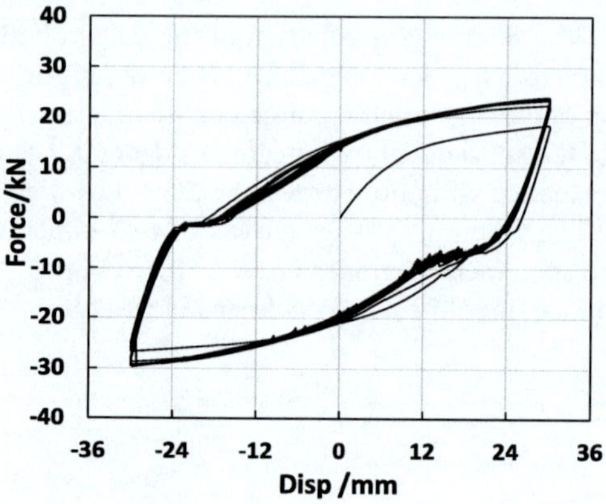

Fig. 22. Hysteresis curve of hot rolled specimen under 30mm loading amplitude

4 Finite Element Analysis

Based on finite element method software ABAQUS, this section studies the method of establishing a finite element model of crawler damper by the solid element, conducts a three-dimensional nonlinear analysis of the model under vertical force and verifies the rationality of the modeling method by comparing it with mechanical test results.

4.1 Finite Element Model

To consider the deformation of the lateral restraint baffles on both sides of the crawler damper due to insufficient stiffness in the mechanical test, a finite element model was established by ABAQUS, which was identical to the physical experiment device. As shown in Fig. 23, the whole model was established by 8-node linear brick, reduced integration with hourglass control (C3D8R) elements, including crawler damper, the lateral restraint baffles, force transmission unit and base plate. Hard contact is assumed between the crawler damper, the lateral restraint baffles, and the force transmission unit in the normal direction. The penalty contact with a 0.2 coefficient of friction is assumed in the tangential direction [17].

The lower part of the base is fixed, and the upper part of the transmission part is free. The loading process of the model is consistent with the mechanical test, and the displacement loading control is applied to the upper surface of the force transmission unit. According to the result of the mechanical test, the parameters of hot-rolled Q235 steel were input as the model's material. The specifications of the materials are given in Table 2, and Table 3 shows the stress-strain relationship of the material [2].

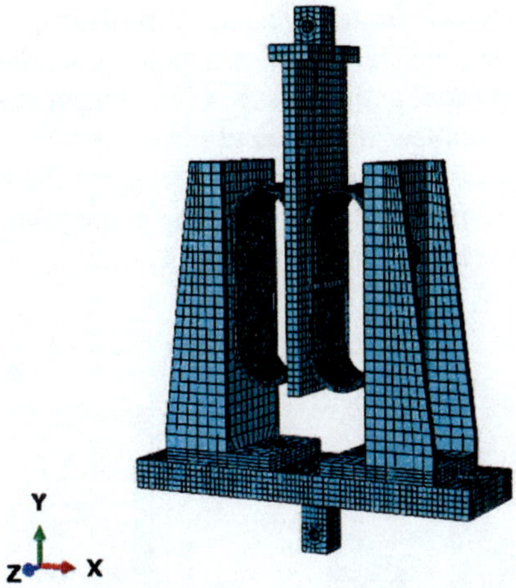

Fig. 23. Finite element model.

Table 2. Material Properties of the crawler Damper

Properties	Yield stress (σ_y/N · mm^{-2})	Ultimate stress (σ_u/N · mm^{-2})	Modulus of elasticity (E/N · mm^{-2})	Mass property (G/N · mm^{-3})	Passion ration (v)
Q235	240	410	2.1×10^5	0.078	0.3

Table 3. Mechanical Properties of Stress-strain of the crawler Damper

Mechanical characteristics	Plastic strain (ε_y/mm · mm^{-1})	Yield stress (σ_y/N · mm^{-2})
Initial Value	0	0
Yield Value	0.002	240
Ultimate Value	0.12	410

4.2 Comparison of Finite Element Analysis with Physical Experiment

Through the unidirectional tension-compression hysteresis simulation of the finite element model, its stress nephogram and plastic strain diagram were obtained (refer to Fig. 24 and Fig. 25). The steel plate has yielded at the junction of the arc and straight segments, and the plastic strain is also concentrated in this area, which is consistent with the mechanical test results. Figure 26 shows the comparison of hysteresis curves between finite element analysis results and test results, which shows that the two curves are in good agreement on the whole. There are two reasons for the discrepancy in a particular curve part: one is caused by the test error. In the physical test, the standard bolt load cannot be applied accurately, resulting in insufficient stiffness of the lateral restraint baffles during the physical test. Consequently, the lateral restraint baffles had a certain extent of deformation, and the shape of the hysteresis curve obtained is not as full as the finite element analysis. The second reason is that the sliding deformation of bolts occurs when the load is large in the test process, but the sliding deformation is not considered in the finite element simulation, so the post-yield stiffness of the specimen obtained by analysis is larger than the test value [18].

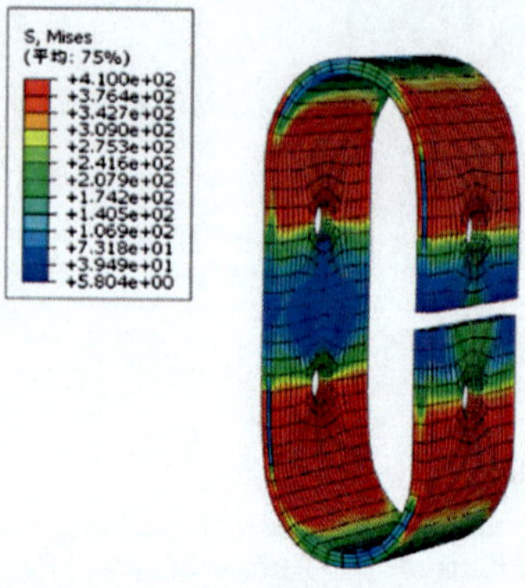

Fig. 24. Stress nephogram

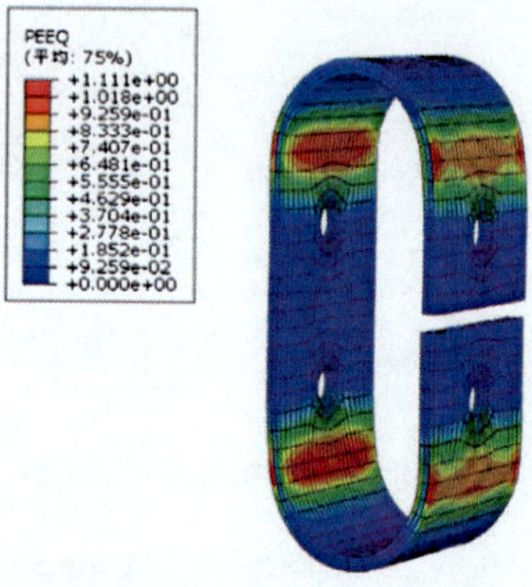

Fig. 25. Plastic strain diagram

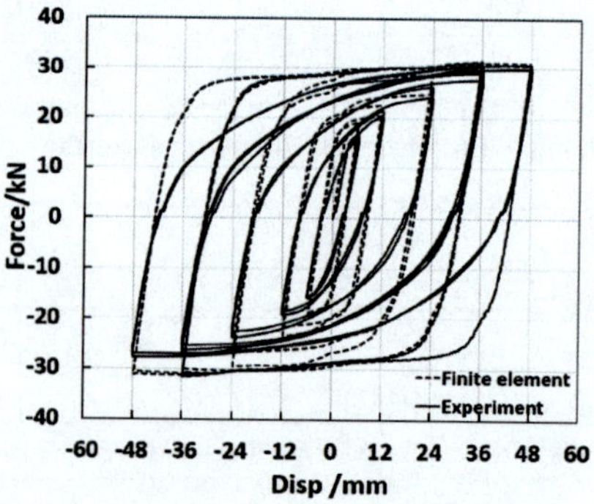

Fig. 26. Comparison of hysteresis curves between finite element analysis and test

4.3 Geometric Parameter Analysis

The geometrical parameters of a crawler damper have an important influence on its mechanical properties. Height (H), thickness (t), and the diameter of the arc segment (D) are three important geometric parameters of the damper which will largely affect the energy consumption capacity of the damper. These crawler damper parameters were analyzed using ABAQUS finite element software. As shown in Table 4, 9 samples were designed to study the effects of t, H, and D on the performance of the specimen under reciprocating loading. MX0 is the standard specimen in the test, which adopts energy dissipation plates with a length of 100 mm, a width of 100 mm, a thickness of 8 mm, and a diameter of 100 mm. All samples were defined with hot-rolled Q235 material.

Table 4. Geometric parameters of the model

Specimen	Thickness (t/mm)	Diameter of the curve steel plate(D/mm)	Height (H/mm)
MX-0	8	100	100
MX-1	6	100	100
MX-2	10	100	100
MX-3	8	80	100
MX-4	8	120	100
MX-5	8	150	100
MX-6	8	100	80
MX-7	8	100	120
MX-8	8	100	150

According to the finite element simulation results, 9 specimen's hysteretic curve is shown in Fig. 27. Further analysis of hysteretic curve results can obtain the skeleton curve and initial stiffness of each specimen (refer to Fig. 28 and Table 5, 6, 7), representing the specimen's bearing capacity and plastic deformation performance [19].

Figure 27 (a)–(c) shows the hysteresis curves of dampers of different thicknesses. When only the thickness is changed, as the steel plate thickness increase, the area of the hysteresis loop of the damper increases, and more energy can be dissipated. The damper with a large steel plate thickness can dissipate more energy [20]; however, too large thickness will lead to large stiffness, so it cannot produce crawler deformation well to dissipate energy. In the animation of the finite element simulation, it is seen that the MX-2 is less likely to undergo the expected crawler deformation than the MX-0 and MX-1 during loading.

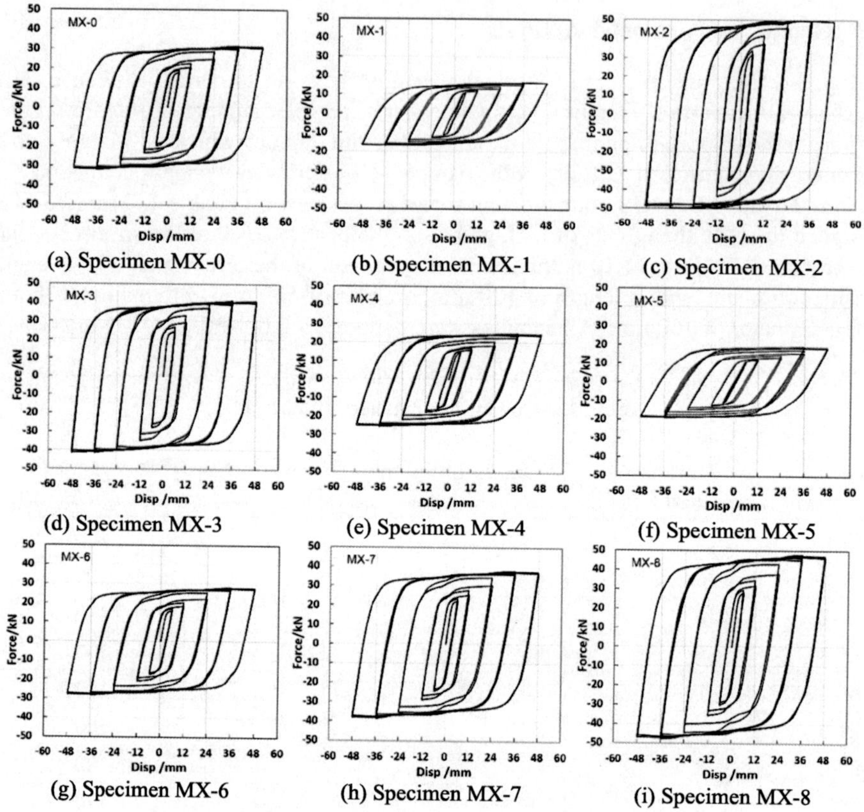

Fig. 27. Hysteresis curves of 9 specimens

Figure 27 (a), (d)–(f) show the hysteresis curves of dampers of different diameters of the arc segment of a damper. By comparing the data in Table 6 and the slope of the skeleton curve in Fig. 28 (b), it can be seen that the larger the diameter of the arc segment of the damper, the smaller the initial stiffness, the smaller the bearing capacity, and the weaker the energy dissipation effect.

Figure 27 (a), (g)–(i) show the hysteresis curves of dampers of different heights. As expected, with the increase of the height of the steel plate, the bearing capacity of the damper can be improved to obtain better energy dissipation performance. By comparing the data in Table 7, it can be found that the influence of the steel plate height on the initial stiffness is slighter than that of the control group with the thickness and arc segment diameter. Therefore, changing the steel plate height is not the best choice to obtain high initial stiffness through geometric parameters.

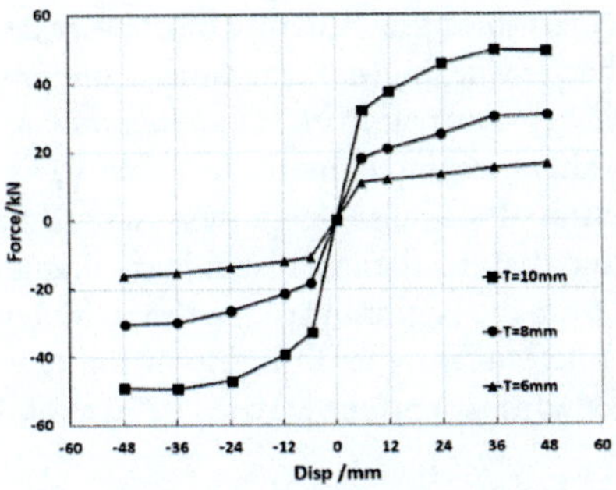

(a) Influence of thickness on skeleton curve

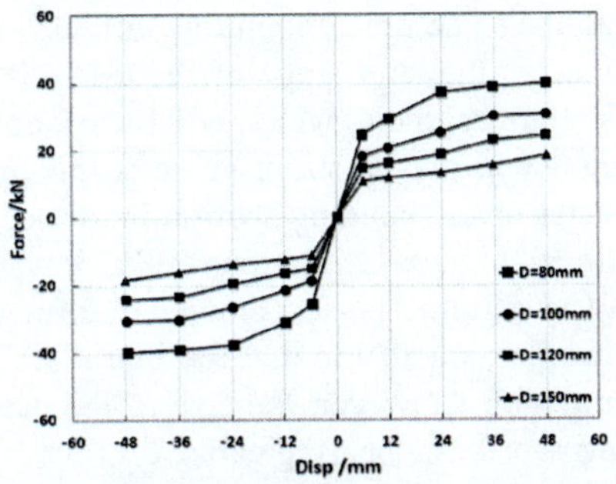

(b) Influence of arc segments' diameters on skeleton curve

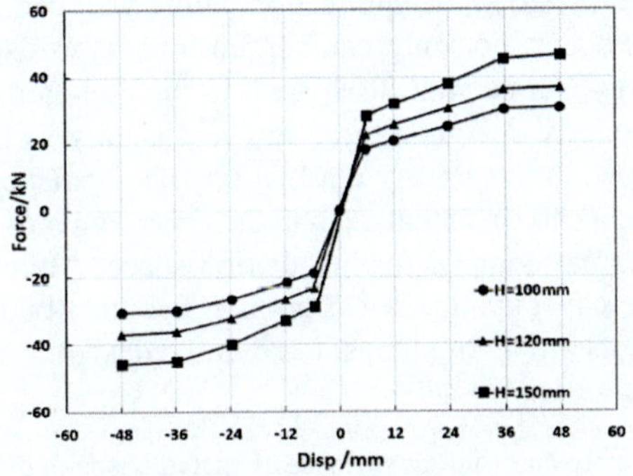

(c) Influence of height on skeleton curve

Fig. 28. Comparison of skeleton curves

In Table 5, the initial stiffness of the 10 mm specimen is 12.447 kN/mm, and the initial stiffness of the 6 mm specimen is 2.422 kN/mm. it is corresponding to the skeleton curve in Fig. 28 (a). The bearing capacity of the 10 mm specimen in the first turn is about three times that of the 6 mm model. It is because, before the specimen yielded, it mainly absorbed energy through elastic deformation. If the initial stiffness of a damper were large, it would have a higher bearing capacity than that of a specimen with small initial stiffness; that was, thus it can dissipate more energy under the same displacement, making the steel plate exhibit ideal energy dissipation effect during reciprocating loading after yielding.

At the same time, the initial stiffness also determines the difficulty of crawler deformation. At the later stage of the test, MX-2 had a warping deformation at the end of the straight segment, which will have a poor energy dissipation effect and aggravate the accumulated plastic strain value at the end of the straight segment. In Table 6, as the diameter of the bending section increases, the initial stiffness decreases. According to the finite element simulation results, it was found that the dampers with large diameters could have almost perfect crawler deformation from the beginning of the test, which well verifies the conclusion that the initial stiffness determines the difficulty of the crawler deformation. However, the cost of the dampers' almost perfect crawler deformation is losing part of the bearing capacity.

Therefore, to the pros and cons of the two, designers must adapt to local conditions. In areas with frequent earthquakes, the thickness of the damper could reduce, and the diameter of the damper could be increased to make it flexible. As the initial stiffness decreases, dampers can yield in the application before wall panels are destroyed. When frequent earthquakes occur, not only can certain interlayer displacement be released, but also seismic energy can be well dissipated. In the non-earthquake-prone areas, the steel plate can be properly designed to be thicker, usually as a rigid node to connect the wall panel. Because of the energy dissipation characteristics of the damper, it can protect the building to a certain extent even in the event of a rare earthquake. The height of the damper has a great demand for installation space. Although it can also change the energy dissipation capacity, it is not as good as the other two geometric parameters. Therefore, space factors are mainly considered in the design.

Table 5. Comparison of initial stiffness of specimens with different thickness

t (mm)	6	8	10
Initial stiffness(kN/mm)	2.422	6.680	12.447

Table 6. Comparison of initial stiffness of specimens with different arc segments' diameters

D (mm)	80	100	120	150
Initial stiffness(kN/mm)	11.681	6.680	3.959	2.098

Table 7. Comparison of initial stiffness of specimens with different height

H (mm)	80	100	120	150
Initial stiffness(kN/mm)	5.868	6.680	7.299	7.963

5 Conclusion

In this paper, a new type of steel damper, a crawler damper, is proposed. Both finite element analyses and physical tests were conducted to investigate the performance of the damper. According to the test and analysis results, the main conclusions of this study are as follows:

1. According to the quasi-static test results, energy dissipation ratios of crawler dampers, which were made from hot roll Q235B and cold roll Q235B are 0.962 and 0.951, respectively. The main way of energy consumption is the transition between the straight segment and arc segment, and the energy consumption effect is satisfactory.
2. The analysis of the fatigue load test shows that the cold-rolled specimens had weaker fatigue resistance in the fatigue test because of the cold work-hardening characteristics. The hot-rolled specimens show excellent fatigue resistance and energy dissipation capacity under the simulated loads of various earthquake levels, which can play a good role in protecting wall panels in application scenarios.
3. The finite element model established by solid elements can simulate the load-displacement hysteretic relationship of dampers, and the results have certain reliability, which can be used to analyze the structure of steel dampers.
4. The mechanical properties of the damper are affected by the key geometric parameters. These geometric parameters will first affect the initial stiffness of the damper, thus affecting its bearing capacity and energy dissipation capacity. As the radius of the arc segment of the component decreases, the width and the thickness increase; the larger the bearing capacity is, the stronger the energy dissipation capacity will be, and vice versa. In the design, the geometric parameters of the damper should be designed according to the frequency of seismic recurrence.

Acknowledgement. The project was supported by the National Innovation and Entrepreneurship Training Program for College Students (202010561001) and New Modular Building of Guangdong Foundation.

References

1. Guangxin, W.: Analysis and application of prefabricated structure. Sichuan Build. Mat. **44**(12), 80–81+83 (2018). (in Chinese)
2. Jiupeng, L.: Study on Vibration Damping Properties of External Wall Panels of Industrialized Housing. Tianjin University, Tianjin (2009).(in Chinese)
3. Peng, C.: Research on the Application of Composite Wallboard in Prefabricated Steel Structure Building in Highly Earthquake Sensitive Areas. Southwest University of Science and Technology, Sichuan (2019).(in Chinese)
4. Daifeng, Z.: Application of lightweight composite wallboard in steel structure residence. Build. Sci. **06**, 78–81 (2006). (in Chinese)
5. Hailong, C.: Study on the Performance of Precast Concrete External Wall Panel and Its Application in The Assembly Type Steel Structure Housing. Anhui Jianzhu University, Anhui (2018).(in Chinese)

6. Minglei, W.: Hysteretic Property Research on Prefabricated Structure of Steel Frame with Energy-saving Composite Panel. Shandong University, Shandong (2013).(in Chinese)
7. Weiqi, F.: Seismic demand on nonstructural components of floor-by-floor assembled steel braced frame. Shandong University, Shandong (2018).(in Chinese)
8. Haowen, W.: Finite Element Analysis and Study on Sliding Energy Dissipation of Double-rib Angle Semi-rigid Joint for PC Exterior Wall. Xihua University, Sichuan (2017).(in Chinese)
9. Kelly, J.M., Skinner, R.I., Heine, A.J.: Mechanisms of energy absorption in special devices for use in earthquake-resistant structures. Natl. Soc. Earthq. Eng. **5**(3), 63–88 (1972)
10. Tagawa, H., Gao, J.: Evaluation of vibration control system with U-dampers based on quasi-linear motion mechanism. J. Constr. Steel Res. **70**, 213–225 (2012)
11. Hetao, H., Jin, C., Xiaofang, L., Jianliang, L.: Quasi-static test and finite element analysis of the damper with u-shaped steel plates. Ind. Constr. **7**(10), 175–179+134 (2017). (in Chinese)
12. Xun, C., Huiling, S., Linlin, X.: Experimental and numerical studies on the seismic performance of precast concrete shear wall structures with an energy dissipation cladding panel. Journal of earthquake engineering **26**, 3264–3279 (2020)
13. AISC (American Institute of Steel Construction). Seismic Provisions for Structural Steel Buildings, Chicago (2005)
14. JGJ 297–2013 (Technical specification for seismic energy dissipation of buildings). Ministry of Housing and Urban-Rural Development of the People's Republic of China, China (2013). (in Chinese)
15. Shansuo, Z.: Experimental study on seismic behaviors of steel reinforced high strength and high performance concrete frame. Eng. Mech. **26**(05), 88–93+101 (2009). (in Chinese)
16. GB/T 51232–2016(Technical standard for assembled buildings with steel-structure). Ministry of Housing and Urban-Rural Development of the People's Republic of China, China (2016). (in Chinese)
17. Najari Varzaneh, M., Hosseini, M.: Cyclic performance and mechanical characteristics of the oval-shaped damper. KSCE J. Civ. Eng. **23**(11), 4747–4757 (2019). https://doi.org/10.1007/s12205-019-1382-6
18. Xun, Z., Linbin, H., Xi, C.: Study on mechanical performance and numerical simulation analysis method of U-shaped metal damper. Build. Struct. **51**(01), 114–120 (2021). (in Chinese)
19. Maoshe, L., Junlong. L.: Analysis of shear stiffness of the thin u-shape metal damper. J. Xi'an Univ. Technol. **29**(02), 207–210 (2013). (in Chinese)
20. Jinhe, G., Jiahuan, X., Yuwen, X.: Analysis of the mechanical properties and parameter sensitivity of a u-shaped steel damper. Front. Mat. **8**, 13 (2021)

Experimental Study on the Lateral Performance of Traditional Wooden Structure with Bamboo-Woven Mud Infill

Yunyang Qi[1](✉) and Hao Huang[1,2]

[1] School of Civil Engineering, Chongqing University, Chongqing 400045, China
201916131083@cqu.edu.cn

[2] Key Laboratory of New Technology for Construction of Cities in Mountain Area (Chongqing University), Ministry of Education, Chongqing 400045, China

Abstract. It is a typical practice to embed the bamboo-woven mud wall (also called wattle and daub wall, W&D wall) in the traditional Chuan-Dou wooden structure widely distributed in southwest China, such as Chongqing or Guizhou. Based on a large-scale field investigation in Chongqing, this paper had carried out detailed research on two specimens of bamboo-woven mud infilled wooden frames with different wall thicknesses for cyclic loading test, in order to study the influence of wall thickness on the lateral resistance of this kind of wall. The test results showed that bottom slip of all the bamboo-woven mud infilled wooden frames were obvious during the whole cyclic loading process. Increasing the wall thickness can obviously improve the bearing capacity and stiffness and energy dissipation capacity of the traditional Chuan-Dou timber frames, on the other hand, serious mud falling off may occur if the wall thickness is too large and this may become a safety hazard.

Keywords: Chuan-Dou wooden frame · Mortise and tenon joint · Wattle and Daub · Seismic performance

1 Introduction

Traditional wooden structure is an important symbol of Chinese civilization, in some southern areas of China, such as Chongqing, the Chuan-Dou wooden structure is widely seen in traditional village dwellings and ancient towns [1]. Despite excellent energy dissipation ability, the construction of Chuan-Dou frames mostly relies on experienced local craftsmen and lacks systematic design and analysis. Various earthquake damage investigations in the past had shown that there will be some regular form of damage to the Chuan-Dou structure, such as damage of M&T joints, wall collapse, column foot slip and overall incline, etc. [2]. Therefore, it is vital to conduct experimental research

W. Guo and K. Qian (Eds.): GBCESC 2022, LNCE 211, pp. 649–659, 2023.
https://doi.org/10.1007/978-981-19-5217-3_64

and theoretical analysis on such structures, to study their seismic performance, and lay a foundation for further structural treatment, operation and maintenance measures. The main achievements of this research are as follows: (1) extensive field investigation on 30 traditional ancient dwellings and villages in Chongqing was carried out on which the design of the whole experiment were based; (2) several traditional wooden frames with W&D infill were designed for the cyclic test based on the actual practice, the overall structural performance subjected to lateral force were evaluated, intending to provide corresponding theoretical reference and technical basis for similar traditional timber engineering.

2 Experimental Program

2.1 Specimen Design

The serial numbers and parameter information of all specimens are listed in Table 1. All the specimens were made by spruce from the southwest of China, the mortise and tenon joints were all set as full penetrate M&T joints. The column feet were constructed according to the actual practice and the dimensions of standard specimen and members are shown in Fig. 1.

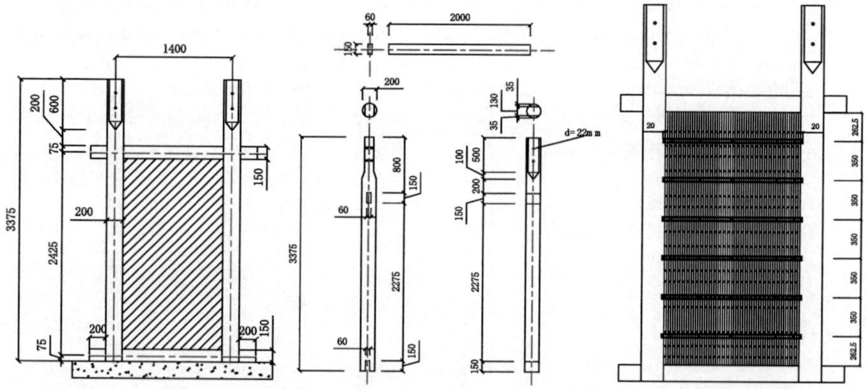

Fig. 1. Dimensions of infilled wall, beam-column members and inner bamboo net

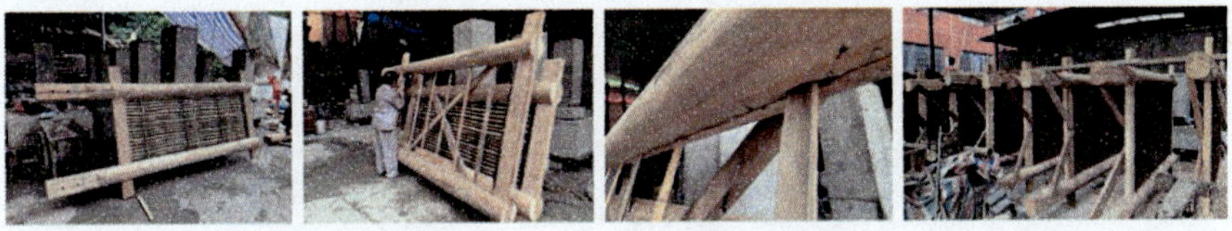

Fig. 2. The on-site construction of bamboo net and specimens after mud application

The inside bamboo net was composed of bamboo horizontal braces and vertical bamboo strips. The thickness of bamboo brace was 8–10 mm and the width was 40 mm. The two ends of the brace were stuck into the grooves on the column side with an embedded depth of 20 mm. The thickness of vertical bamboo strips was 4–5 mm and the width was 20 mm. The strips were interlaced and fixed between bamboo braces, covering the whole wall space (Fig. 2). The clay mix is made by mixing local clay and straw scraps at a weight ratio of 130:1. After the specimens are all dried, a layer of white lime should be applied on the mud surface to draw and observe the crack.

Table 1. Details of specimen numbers and parameters

Specimen numbers	Parameters of specimens					
	Wall span (mm)	Brace spacing (mm)	Wall thickness (mm)	Admixture and weight ratio		
				Clay	Straw scraps	Admixture
QH-50	1400	350	50	130	1	/
QH-60	1400	350	60	130	1	/

2.2 Test Setup and Loading Protocol

In accordance with the traditional practice of the Chuan-Dou structure, the load on each column was set as 10kN, imposed by four concrete blocks. The schematic diagram of the whole specimen after installation was shown in Fig. 3. Three square steel tubes with end plates were set in the middle of the two wooden columns for force transmission. In addition, in order to avoid large bending moment caused by extrusion of steel plates, the hinge end construction as shown in Fig. 3 is set. Two H-beams were also used as an supporting device to restrain the out-of-plane deformation of the specimen.

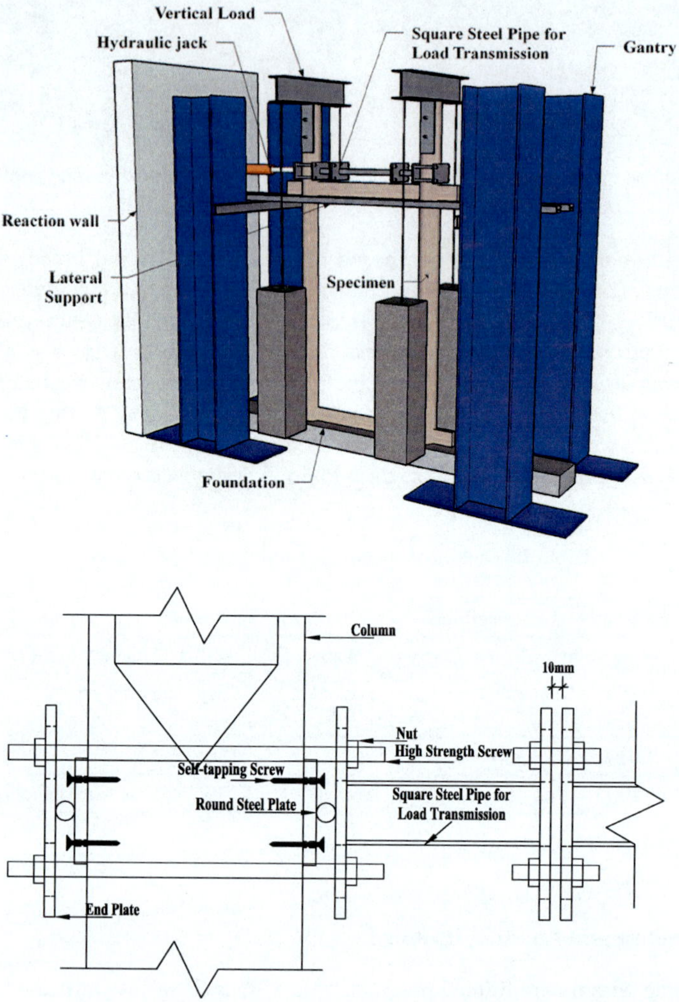

Fig. 3. Test model and the hinge end construction

According to ISO 21581-2010 [3], the loading protocol were set and shown in Fig. 4. The specimen was preloaded to 30mm before the experiment, and the initial state was restored after 5 min of load holding. The displacement was controlled at a rate of 25 mm/min during the formal loading, and the wall cracks were drawn after the completion of single cycle loading and each triple cycle loading. Here are the definitions of several terms: taking one cycle of the loading protocol as an example, 'ZT' is defined as a process when the specimen is pushed toward the target displacement from the equilibrium position, after which the 'ZL' happens when the model returns to the initial position from the 'ZT end'. Likewise, 'FL' is defined as the process when the specimen moves from 'ZL end' to the opposite target displacement and 'FT' happens when the model finishes the 'FL' and being pushed back to the equilibrium position (Fig. 4).

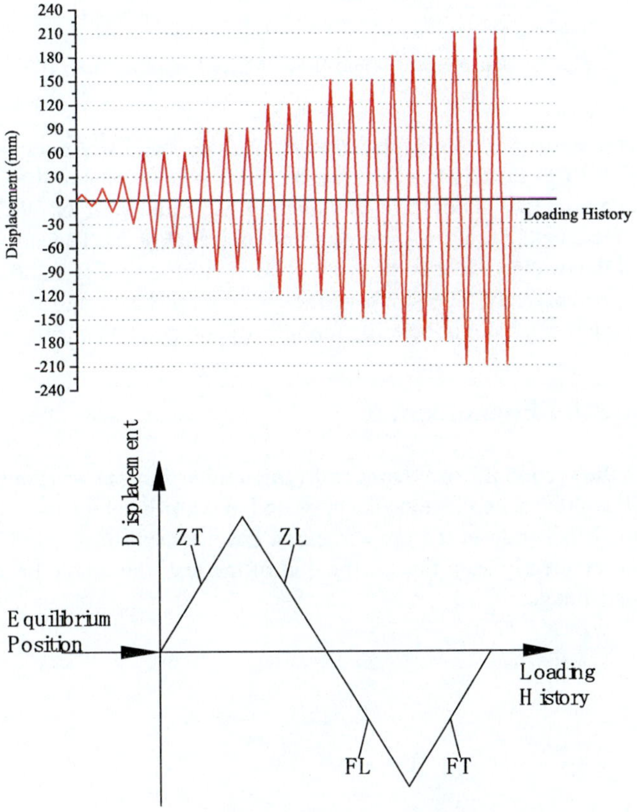

Fig. 4. Loading protocol and definition of terms

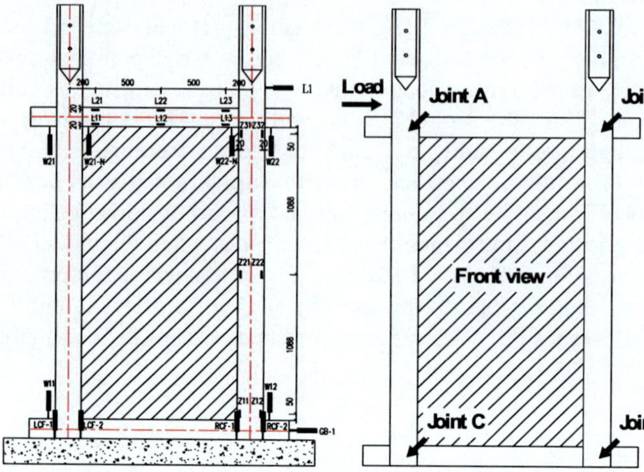

Fig. 5. Gauge arrangement (left) and joint numbers (right)

The arrangements of all the displacement transducers and strain gauges are shown in Fig. 5. LCF-1, LCF-2, RCF-1, RCF-2 were used to measure the uplift of left column foot and right column foot under the cyclic loading, W21 and W21-N, W22 and W22-N were used to calculate the rotation of joint A and joint B (Fig. 5). The slip at the bottom of the wall and the displacement at the top will be measured by GB-1 and displacement transducer L1, respectively. The strain gauges on the top beam and right column are aimed at studying the transfer of bending moment of components during cyclic loading.

3 Experimental Phenomenon

For QH-50, as the second ZT of 90 mm ends, the mud at Joint D was slightly extruded out of the wall, similar phenomenon also appeared in the front at the third ZT of 90 mm (Fig. 6). Figure 7 had showed the mud cracking development during l = 150 mm and 210 mm, it can be clearly seen that, at the end of the test, the whole back of the wall showed serious damage.

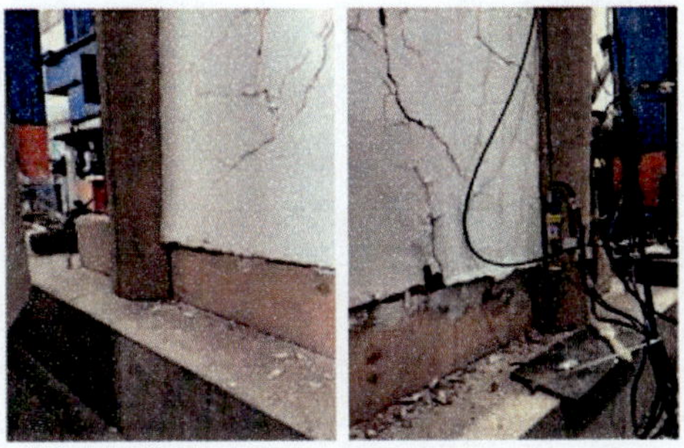

Fig. 6. Clay extruded out of the back (left) and front (right) of the wall ($l = 90$ mm)

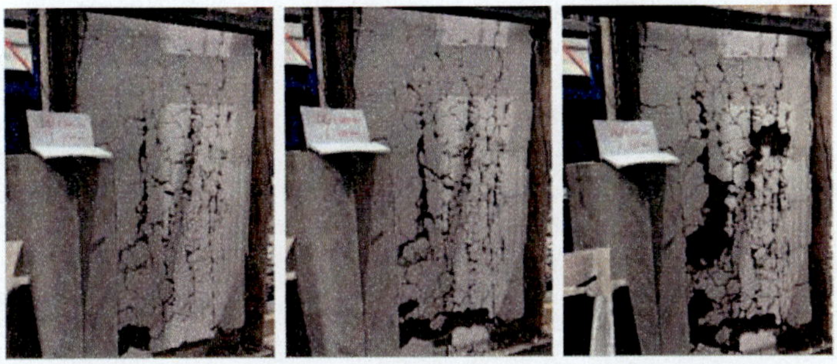

Fig. 7. The development of the crack of the clay mix of QH-50 ($l = 150$ mm, $l = 180$ mm, $l = 210$ mm)

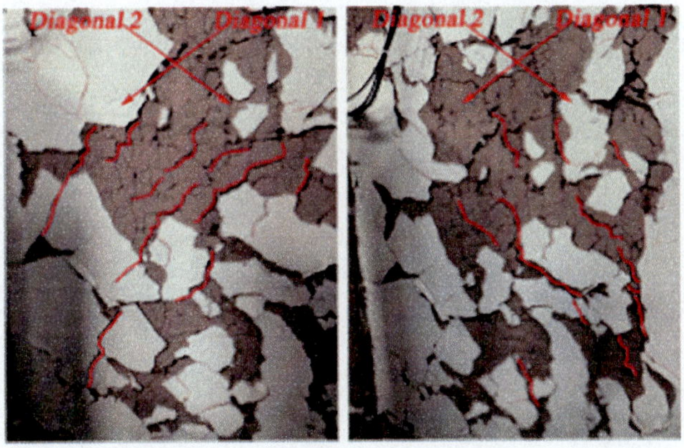

Fig. 8. Crack development of ZT end (left) and FL end (right) ($l = 120$ mm, first cycle)

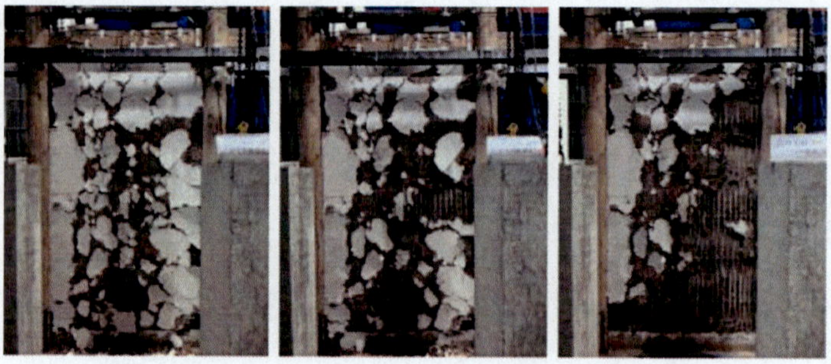

Fig. 9. The development of the crack of the clay mix of QH-60 (l = 150 mm, l = 180 mm, l = 210 mm)

For QH-60, the cracking tendency of the same area on the back wall (l = 120 mm) were shown in Fig. 8. It can be clearly seen that before the end of ZT, the gap of mud along the diagonal 1 direction gradually increases, after which gaps along the diagonal 1 began to close while the crack between the mud along the diagonal 2 direction continued to expand. During l = 150 mm–210 mm, large chunks of mud had undergone serious damage. In the end, almost half of the mud of the front wall had fallen off (Fig. 9).

The readings of GB-1 also showed that the final bottom slip of QH-50 and QH-60 was +3.01 mm and +4.56 mm, respectively.

4 Analysis of Results

4.1 Hysteretic Loops and Envelope Curve

The hysteretic loops of QH-50 were spindle shaped at l = 7.5 mm (Fig. 10), with the increase of displacement magnitude, the hysteresis loop gradually became Z shaped, with obvious pinching effect. The development of hysteretic loops of QH-60 (Fig. 10) is similar to that of QH-50.

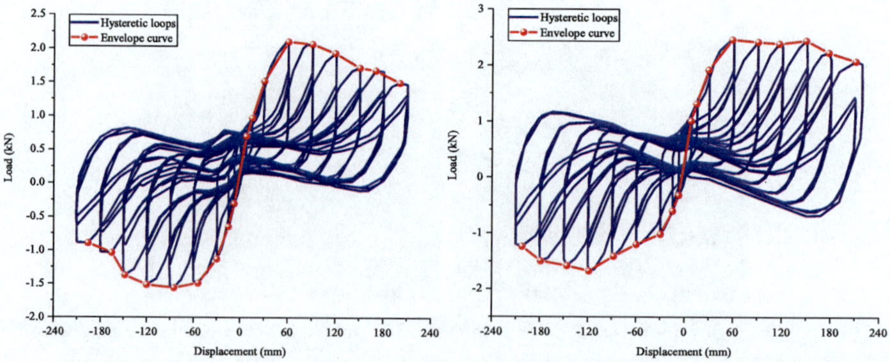

Fig. 10. The hysteretic loops and envelope curve of QH-50 (left) and QH-60 (right)

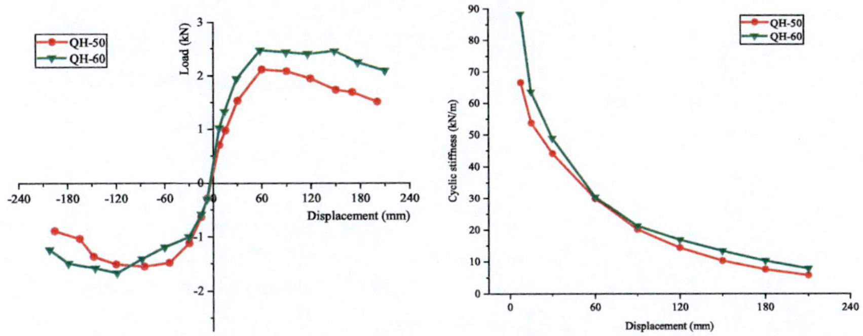

Fig. 11. Envelop curves (left) and stiffness degradation curves of the two specimens (right)

The envelop curves of two specimens are shown in Fig. 11, which showed that the larger the wall thickness, the higher the bearing capacity of the specimen, besides, the time when the load showed obvious drop of QH-60 was also later than QH-50. The influence of wall thickness on improving the bearing capacity of specimen is obvious, however, it should be noted that, compared with similar studies conducted by previous scholars [4–7], the overall bearing capacity of the W&D infilled frames made by referring to actual construction of traditional timber engineering is relatively not high.

4.2 Stiffness Degradation

The secant stiffness *Ki* under each displacement magnitude can be calculated by Eq. (1)[8], the degradation of stiffness of each specimen was illustrated in Fig. 11, which showed that QH-60 had higher initial stiffness (88.13 kN/m) than QH-50 (66.33 kN/m) and the stiffness of QH-60 was always higher than that of QH-50 during the whole test.

$$K_i = \frac{|+P_i| + |-P_i|}{|+\Delta_i| + |-\Delta_i|} \tag{1}$$

where Pi = peak load of each cycle and Δi = displacement magnitude of each cycle.

4.3 Strength Degradation

The strength degradation of QH-50 and QH-60 were shown in Fig. 12, where R2/R1 means the ratio of the peak load of the second loop to that of the first loop under the same displacement, R3/R2 means the ratio of the peak load of the third loop to that of the second loop under the same displacement. Figure 12 indicated that the main strength degradation happened on the second loop, the value of R2/R1 mainly fluctuated between 0.67 and 0.87 while that of R3/R2 were all above 0.79 and under 0.96.

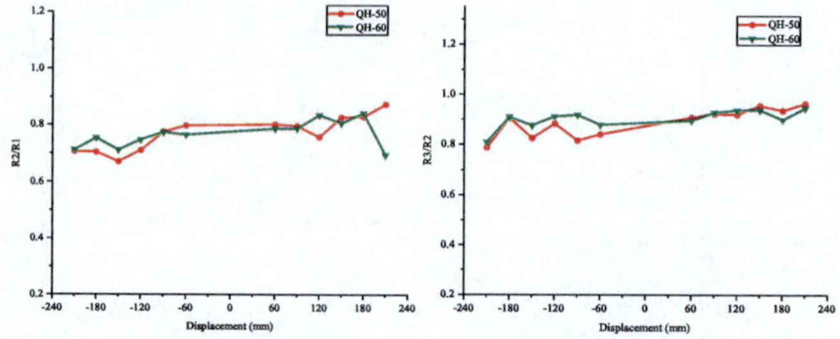

Fig. 12. Strength degradation of QH-50 and QH-60

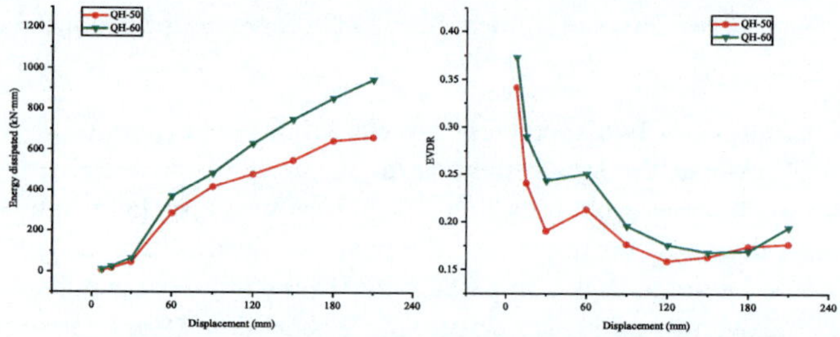

Fig. 13. Energy dissipation and EVDR of QH-50 and QH-60

4.4 Energy Dissipation Capacity

Figure 13 showed the energy dissipation growth and the variation of EVDR (Equivalent Viscous Damping Ratio) of the two specimens, it can be clearly seen that despite minor difference of energy consumption between QH-50 and QH-60 during the single cycle loading, the energy dissipated by QH-60 was always higher than QH-50 during the subsequent triple cycle loading. The total energy dissipated by QH-50 and QH-60 were 3118J and 4132J, respectively, which had proven that the larger the wall thickness, the more the dissipated energy of specimens.

The EVDR curves of QH-50 and QH-60 showed similar variation trend: constant descending during the single cycle loading while continued to decline until the gradual increase in the end, despite the obvious rise on $l = 60$ mm. This had to do with the energy dissipation mechanism of the specimen, the pinching effect during the early loading stage had caused the decline of energy efficiency due to the joints taking the main role of energy dissipation. As the squeezing between clay mix and timber frame grew, the infill wall had made obvious contribution to energy consumption on $l = 60$ mm while obvious decline of EVDR occurred due to the accumulated damage of the infills. The gradual increase of EVDR in the end may be explained by the energy dissipated by friction which helped to improve the energy efficiency. On the whole, the increase of wall thickness had obvious effect on the improvement of the energy dissipation capacity of specimens.

5 Conclusions

According to the above discussions, the following conclusions can be drawn.

(1) The bottom construction of Chuan-Dou timber frames adopted in practical engineering will cause obvious slippage away from the initial position under cyclic loading. Rough surface between the ground wooden beam and the foundation is recommended for slip restriction and may also contribute to energy consumption.
(2) Variation of wall thickness has obvious influence on the lateral performance of the specimens. The thicker the wall is, the higher the bearing capacity and stiffness and energy dissipation capacity of the specimens while also the more serious the mud damaging during loading, which may be a hidden danger for personal safety. In addition, the general strength of Bamboo-Woven Mud infilled wall is not high.

References

1. Yu, Z.: The Study on Folk Architecture in Sichuan and Chongqing Area. Chongqing University, Chongqing (2006).(in Chinese)
2. Zhe, Q., Andreea, D., Jiangrong, Z., Jingjiang, S.: Seismic damage to masonry-infilled timber houses in the 2013 M7.0 Lushan, China, earthquake. Earthquake Spectra **31**(3), 1859–1874 (2015)
3. ISO 21581 (2010): Timber structures—Static and cyclic lateral load test method for shear walls. International Organization for Standardization (ISO), Geneva, Switzerland (2010)
4. Hao, H.: Lateral Performance of Chuan-Dou Type Timber Frame with Wood Panel Infill. Chongqing University, Chongqing (2019).(in Chinese)
5. Hao, H., Yuntian, W., Zhao, L., Zewei, S., Zhixiong, C.: Seismic behavior of Chuan-Dou type timber frames. Eng. Struct. **167**, 725–739 (2018)
6. Qifang, X., Long, W., Shengying, L., Lipeng, Z., Weibing, H.: Infuence of wood infll walls on the seismic performance of Chinese traditional timber structure by shaking table tests. Bull. Earthq. Eng. **18**, 5009–5029 (2020)
7. Qifang, X., Yanyang, T., Lipeng, Z., Shengying, L., Long, W.: Seismic behavior of Chinese traditional timber frames with masonry infill wall: experimental tests and hysteretic model. Int. J. Arch. Heritage **15**(8), 1130–1144 (2021)
8. (JGJ101–2015): Specification of Testing Methods for Earthquake Resistant Building. China Architecture & Building Press, Beijing (2015). (in Chinese)

Seismic Response Analysis of Shallow Tunnel Under Simultaneous Oblique Incidence of P and SV Wave

Dong-Dong Zhang[1], Yang Liu[2(✉)], Feng Xiong[1], and Zhu Mei[2]

[1] Sichuan University, Chengdu 610065, China
[2] Wenzhou University, Wenzhou 325035, Zhejiang, China
yang.liu@wzu.edu.cn

Abstract. The incident angle of seismic waves can significantly affect the seismic response of shallow tunnels. Based on wave field separation technology and equivalent seismic load method, a general 2D numerical model whose accuracy of the model is verified considering oblique incidence characteristics is established. A series of time-history analysis are conducted to study the influence law of P wave oblique incidence alone, SV wave oblique incidence alone and simultaneous incidence of P and SV wave (P-SV wave) on the dynamic response of shallow tunnels. The results show that (1) The seismic response of tunnels increases with the increasing of incident angle in small range. (2) For the same intensity level, the seismic response of structures under the obliquely incident SV wave is higher than that under P wave, while the vault and arch foot are vulnerable parts. (3) The seismic response resulted in the oblique incidence of SV wave alone is less than that under P-SV wave, which indicates the necessity of considering the simultaneous incidence of P and SV waves.

Keywords: Seismic response · Oblique incidence of seismic waves · Shallow buried tunnel

1 Introduction

Due to the constraint of surrounding rock medium, the seismic response of tunnels is lower than that of the above-ground structure, and it is generally considered that there will be no serious earthquake damage. However, in a series of strong earthquakes [1–4], such as the Northridge earthquake in the United States, Kobe earthquake in Japan, Chi-Chi earthquake in Taiwan Province, China and Wenchuan earthquake in Sichuan Province, China. Many tunnels suffered serious earthquake damage, and the shallow buried section of the tunnel was one of the places where the seismic damage was concentrated, such as lining crack, arch and cave collapse, etc., and the seismic research of tunnel structures gradually became active. However, the research on seismic resistance of tunnel structures mostly assumes that seismic waves are incident vertically, and the investigation of earthquake damage shows that when the buried depth is shallow,

W. Guo and K. Qian (Eds.): GBCESC 2022, LNCE 211, pp. 660–671, 2023.
https://doi.org/10.1007/978-981-19-5217-3_65

the seismic waves of near-field strong earthquakes often incident obliquely at a certain angle, which increases the risk of tunnel damage.

In recent years, some progress has been made in the research of seismic analysis of underground structures such as tunnels under obliquely incident seismic waves. Liu et al. [5] transformed the problem of ground motion input into the problem of vibrating source, and converted the ground motion into equivalent load to deal with the oblique incidence of wave. Du et al. [6] preliminarily explored the seismic response the subway under seismic waves at different incident angles, and found that the law of the seismic response under obliquely incident waves was quite different from that under vertical waves. Huang et al. [7, 8] studied the effect of oblique incidence of SV wave and P wave at various angles respectively on the seismic response of a subway station, and proposed that incident angle could amplify the seismic response of subway station, Dunyu Lyu et al. [9] used a 3D numerical model to evaluate the nonlinear seismic response of arched tunnel, and suggested that the damage of tunnels increases with the increase of incident angles. Zhao et al. [10] studied the influence of obliquely incident P wave on the safety of cavern group of hydropower station, and the results showed that the stability of cavern group was highly affected by the oblique incident angle. Moreover, because P wave decays rapidly in the foundation, to improve the calculation efficiency, the incidence of P wave was often ignored in the past, and only SV wave was considered. But the tensile and compressive load caused by P wave on structures in near field area can not be ignored, which indicates that it is necessary to consider the coupling effect of simultaneous incidence of P and SV waves.

In summary, the research on the seismic response of shallow buried tunnels under obliquely incident seismic waves is not sufficient, this paper simulates obliquely incoming SV and P waves using the input method of waves together with viscos-spring boundary, and consider the oblique incidence of only P wave, only SV wave, P wave and SV wave in the meantime. two intensities of P wave are considered to discuss the influence of the intensity of coupled P wave.

2 Numerical Model

2.1 Problem Description

Taking the entrance section of a mountain tunnel in Tibet Autonomous Region of China as the research object, the main component of the shallow buried section of the tunnel site is granite, and the surrounding rock is classified as grade III. The cross-section of the tunnel is horseshoe-shaped and made of cast concrete, with a horizontal span of 11.3 m and a thickness of 0.5 m. According to the *Zoning Map of Peak Acceleration of Ground Motion in China* (GB18306-2015), the seismic intensity of the proposed line site is degree VIII, and the peak acceleration of ground motion is 0.3 g.

2.2 Establishment of the FEM Model

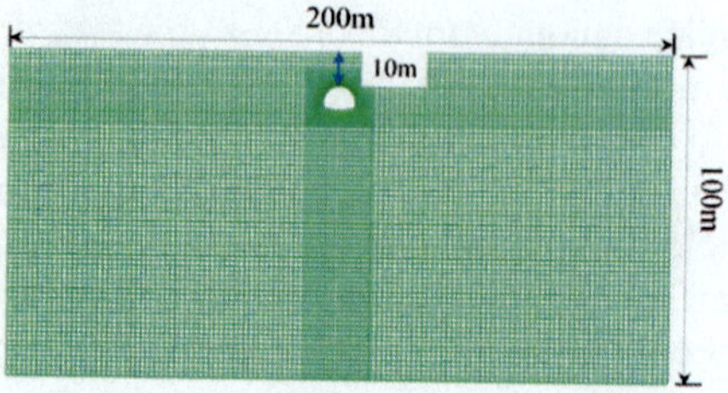

Fig. 1. Finite element model

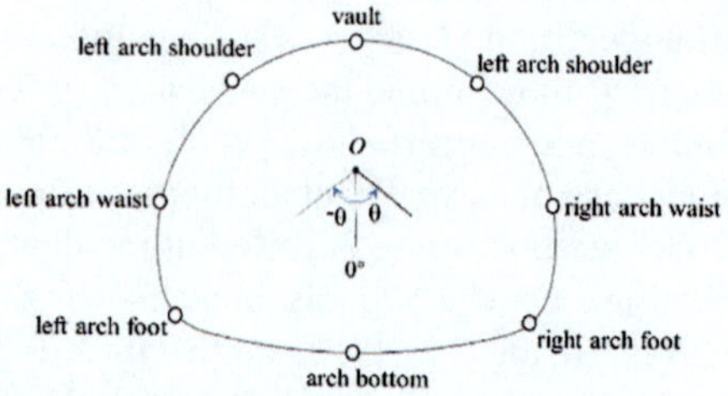

Fig. 2. Schematic diagram of monitoring points

Numerical simulation technology is one of the research methods widely applied to evaluate the seismic behavior of underground tunnels, which is a feasible and effective means. A two-dimensional calculation model whose width and height are 200 m and 100 m respectively is established for the portal section in Fig. 1, and it was seen that the buried depth of the tunnel is 10 m. The surrounding rock adopts four-node plane strain element discretization (CPE4), and the lining is discretized with beam element (B21). The separation and relative slip between the tunnel and surrounding rock are ignored, and bind their contact interface. The maximum size of the mesh in the model is 1.5 m to ensure the accuracy of calculation, and 11704 elements are meshed in total. Figure 2 shows eight monitoring points along the tunnel cross-section.

2.3 Material and Seismic Wave

Assuming that the surrounding rock is hard and complete, it is a continuous and uniform linear elastic medium. The plastic damage model of concrete (CDP) embedded in ABAQUS is used to simulate the plastic mechanical behavior of tunnel. The key material parameters in the numercial model are given in Table 1.

Table 1. Material parameters in tunnel model

Model material	Elastic modulus(GPa)	Density(kg/m^3)	Poisson's ratio	Tensile yield stress(MPa)	Compressive yield stress(MPa)
Tunnel lining	30	2500	0.2	2.03	14.07
Rock	10	2500	0.25	–	–

In this paper, Rayleigh damping is adopted, and the calculation method is as follow:

$$\boldsymbol{C} = \alpha \boldsymbol{M} + \beta \boldsymbol{K} \tag{1}$$

where $\boldsymbol{C}$ is damping matrix, $\boldsymbol{M}$ is mass matrix, $\boldsymbol{K}$ is stiffness matrix; α and β, respectively, are the mass coefficient and stiffness coefficient, and the calculation formula is as follows:

$$\alpha = \frac{2\xi\omega_i\omega_j}{\omega_i + \omega_j}, \quad \beta = \frac{2\xi}{\omega_i + \omega_j} \tag{2}$$

where ω_i and ω_j, respectively, are the frequencies of vibration modes (i and j-order) of the calculation model; ξ is the damping ratio, which is 0.05 for both structure and surrounding rock; The extracted 1st and 3rd natural frequencies of the model are substituted into the above formula for calculation. When performing modal analysis, the lateral boundary of the model is vertically constrained, and horizontally unconstrained, and the bottom is fixed.

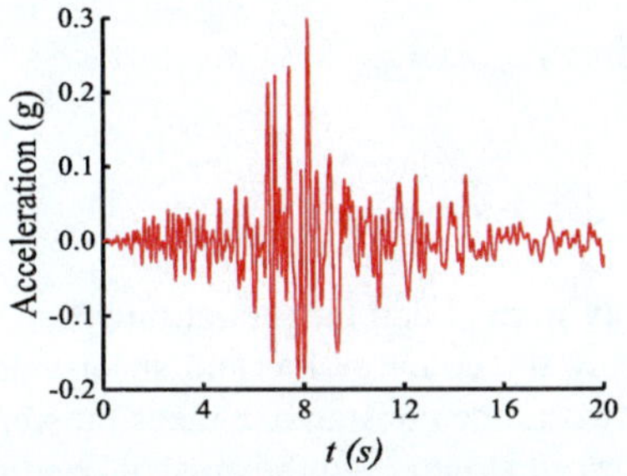

Fig. 3. Acceleration time history of Northridge wave

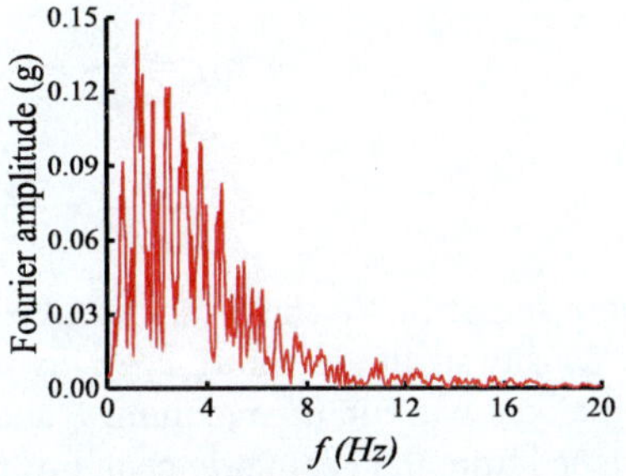

Fig. 4. Fourier spectrum of Northridge wave

The input ground motion is the Northridge wave recorded in California, America in 1994, with the acceleration peak amplitude modulated to 0.3 g for 20 s. After filtering and baseline calibration, the time history of acceleration and Fourier amplitude spectra can be seen in Fig. 3 and Fig. 4 respectively.

2.4 The Operating Mode of Calculation

In this paper, three cases, the incidence of only P wave, SV wave, both P wave and SV wave, are considered. To study the contribution of P wave intensity to the structural response, two intensities of P waves are set, and a total of 12 calculation conditions (0.3 g P, 0.3 g SV, 0.3 g P + 0.3 g SV, 0.1 g P + 0.3 g SV) are arranged, with the incident angles of 0°, 15° and 30°.

3 Input Method of Oblique Incident Seismic Waves

Liu Jingbo, et al. [5] transformed the problem of seismic wave input into the problem of wave source, and the equivalent seismic load acting on artificial boundary could be obtained from input ground motion to realize the incoming process of waves.

3.1 Viscous-Spring Boundary Conditions

Considering the cost of numerical calculation, it is impossible to model and analyze the whole foundation-structure system, so the finite calculation area must be cut from the infinite foundation. Because the seismic wave will produce meaningless wave reflection when it propagates to the boundary of the truncated calculation area, resulting in distortion of the calculation result, thus it is essential to set dynamic artificial boundary at the cutted line to simulate approximately the radiation damping of the semi-infinite foundation. Viscous-spring artificial boundary [6, 11] by setting the spring-damper parallel system with far-end grounding fixed at the boundary of the model performs well in simulating the radiation damping and elastic restoring force of the foundation, which has the advantages of clear physical concept, stable calculation and simple realization. The parameter of spring K and damping C can be got through the following.

$$K_N = \frac{1}{1+A} \cdot \frac{\lambda + 2G}{2r}, \quad C_N = B\rho c_p \tag{3}$$

$$K_T = \frac{1}{1+A} \cdot \frac{G}{2r}, \quad C_T = B\rho c_s \tag{4}$$

where subscripts N and T represent respectively normal and tangential direction; ρ is the density of the medium; c_p, c_s is respectively the compression and shear velocity of incident wave in the medium; λ and G are two Lame constants; r takes the vertical distance from the geometric center of the concerned structure to the artificial boundary where the node is located; A and B are dimensionless parameters, and the value is $A = 0.8, B = 1.1$, according to the better suggestion.

3.2 Equivalent Nodal Forces

The input method of seismic waves through converting the free wave field into equivalent loads applied on the nodes of artificial boundaries has a good precision. The equivalent node force can be obtained by the following:

$$F_{ij} = (K_{ij}u_{ij} + C_{ij}v_{ij} + \sigma_{ij})A_i \tag{5}$$

where subscript j represents the direction of node i; A_i is the equivalent area of the node i; K_{ij} and C_{ij} are the spring and damping coefficients respectively in direction j of node i; u_{ij} and v_{ij} are the displacement and velocity vector respectively in j direction of the node i; σ_{ij} is the stress vector. According to elastic wave mechanics, u_{ij},v_{ij} and σ_{ij} can be obtained by analytical or numerical methods.

The waveforms of the half-space free field under oblique incident waves are converted as shown in Fig. 5. The free wave field can be considered as the linear superposition of three parts: the incident wave and two reflected waves. Because the number of nodes at artificial boundary is usually large and the calculation of equivalent force is complicated, the ABAQUS software was further developed and equivalent concentrated force was rapidly obtained and added for calculation in time domain.

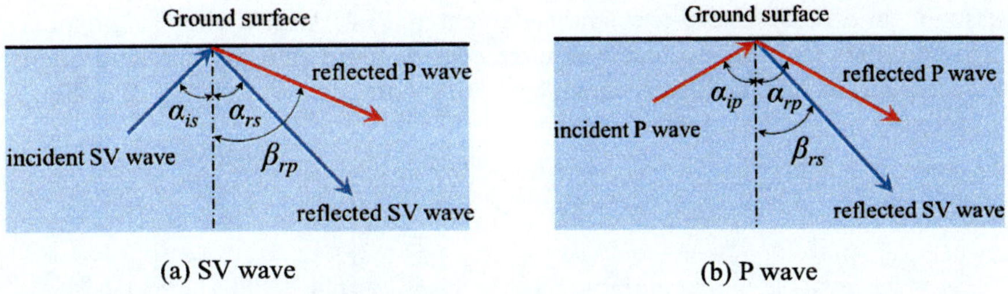

Fig. 5. Diagram of P wave and SV wave waveform conversion on the ground surface

4 Static-Dynamic Coupling Analysis

Because the underground structure already has initial stress under the static loads such as the gravity and earth pressure. To calculate the real earthquake response and simulate the static-dynamic coupling effect of the structure under complex loads, such as seismic load. It is usually necessary to conduct the static-dynamic coupling analysis of the underground structure. For the analysis problems involving nonlinear effects, the superposition principle should not be used, and a certain processing method is needed for nonlinear analysis. In this study, the artificial boundary transformation [12] is used to realize the static-dynamic coupling analysis. A complete time history analysis process with the help of ABAQUS is provided as follows:

(1) Establish a model soil-structure system model, and carry out static analysis according to the construction excavation process, with normal constraint of the lateral boundary of the model and fixed constraint of the bottom boundary. Then obtain information such as constraint reaction force and coordinates of boundary nodes.
(2) Import the information obtained in the previous step and displacement and velocity of seismic wave into the automatic modeling system. Then get the equivalent nodal forces and Viscous-spring artificial boundary.
(3) Add a dynamic analysis step after static analysis and set the time increment and output items. In added dynamic analysis step, the static boundary is deleted, and the viscoelastic boundary and equivalent nodal forces are added. Then, A complete static-dynamic coupling analysis model is built.

Notice: (1) Kill the spring and damping element in the static step, or there will be a problem of computational convergence. (2) Add a static step without load after static analysis to reduce the local unbalance force in the model, which could cause initial oscillation of calculation results.

5 Verification

A two-dimensional free-field model as shown in Fig. 6 is established, with point A and point B as monitoring points. Assuming that the half-space is an elastic medium, the material parameters are as follows: density 2200 kg/m^3, elastic modulus 1.2 GPa,

Poisson's ratio 0.35. The plane strain element (CPE4) with a side length of 3 m which complies with the accuracy requirements of finite element simulation is used for discretization, and the displacement curve of the impulse wave is drawn in Fig. 7.

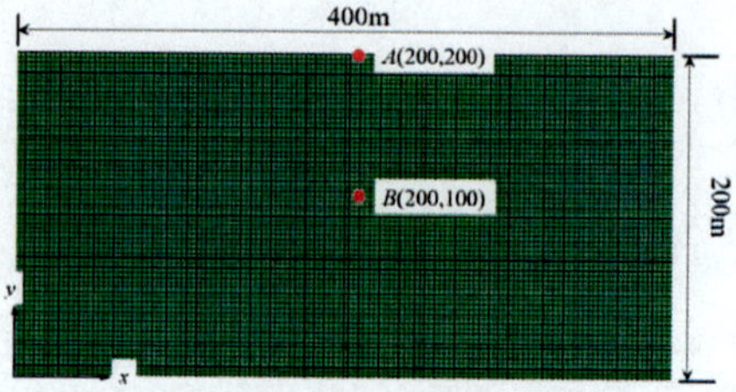

Fig. 6. Free-field model

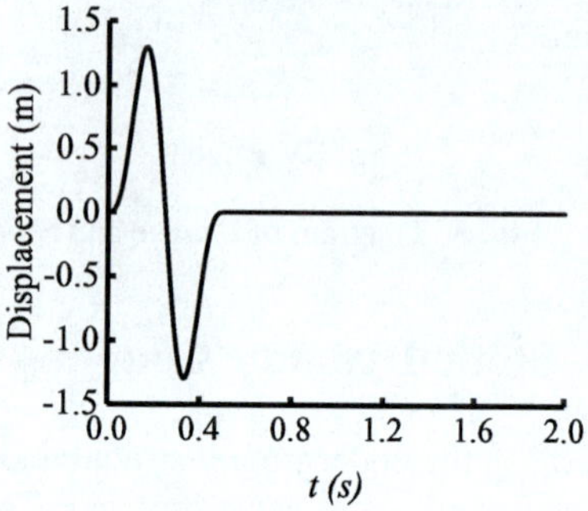

Fig. 7. Displacement curve of incident wave

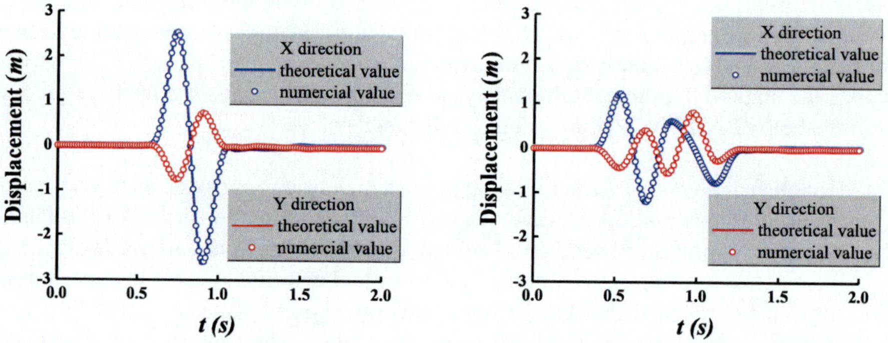

(a) point A

(b) point B

Fig. 8. Theoretical and numerical value of displacement under 20° incident SV wave

Calculate the incident situation of SV wave with a tilt of 20°. Figure 8 shows the recorded displacement curves of two monitoring points in x and y directions, respectively, compared with the analytical value. It can be clearly seen that the displacement amplitudes of two points under SV wave are in good agreement with the analytical solution calculated based on wave theory, which indicates that the applied viscous-spring boundary effectively absorbs the external scattered waves and accurately reflects the propagation law of oblique incidence of seismic waves.

6 Result Analysis

6.1 Relative Deformation

The deformation of tunnel lining is mainly affected by the deformation of surrounding rock. Previous work has paid much attention to the shear deformation of the lining.

In fact, for the waveform conversion of seismic waves at the ground surface, the free wave field tends to be complicated under the oblique incidence of seismic wave, thus deformation characteristics of lining structure changes a lot.

The formulas for calculating the relative deformation of different deformation characteristics are as follows:

$$\lambda_{hc} = \frac{U_{x\_D} - U_{x\_B}}{\overline{BD}}, \quad \lambda_{vs} = \frac{U_{y\_D} - U_{y\_B}}{\overline{BD}} \tag{7}$$

$$\lambda_{hs} = \frac{U_{x\_C} - U_{x\_A}}{\overline{AC}}, \quad \lambda_{vc} = \frac{U_{y\_C} - U_{y\_A}}{\overline{AC}} \tag{8}$$

where take λ_{hc} as an example, U_{x_D} and U_{x_B} are the displacements of points D and B respectively, and $\overline{BD}$ is the distance between points D and B without deformation.

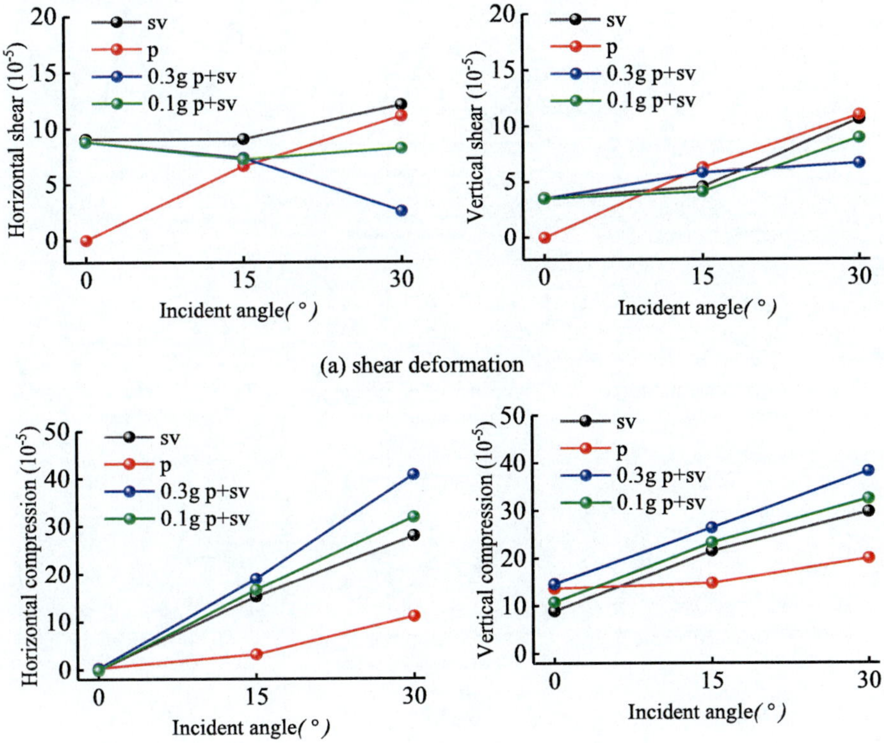

(a) shear deformation

(b) compression deformation

Fig. 9. Shear and compressive deformation of tunnel under Oblique incident seismic wave

The peak values of lining deformation in different calculation conditions are compared as shown in Fig. 9. It reveals a trend that when the angle of oblique incident P wave or SV wave increases, both the shear and compression deformation of tunnel lining show a nonlinear increasing trend, and the deformation of lining under SV wave is often higher than that under P wave at the same incident angle; Compared with SV wave incident alone, when P and SV waves simultaneously incident, the compressive deformation of tunnel lining increases, while the shear deformation decreases, and the smaller the intensity of P wave is, the less the relative displacement will increase or decrease.

6.2 Internal Force and Stress

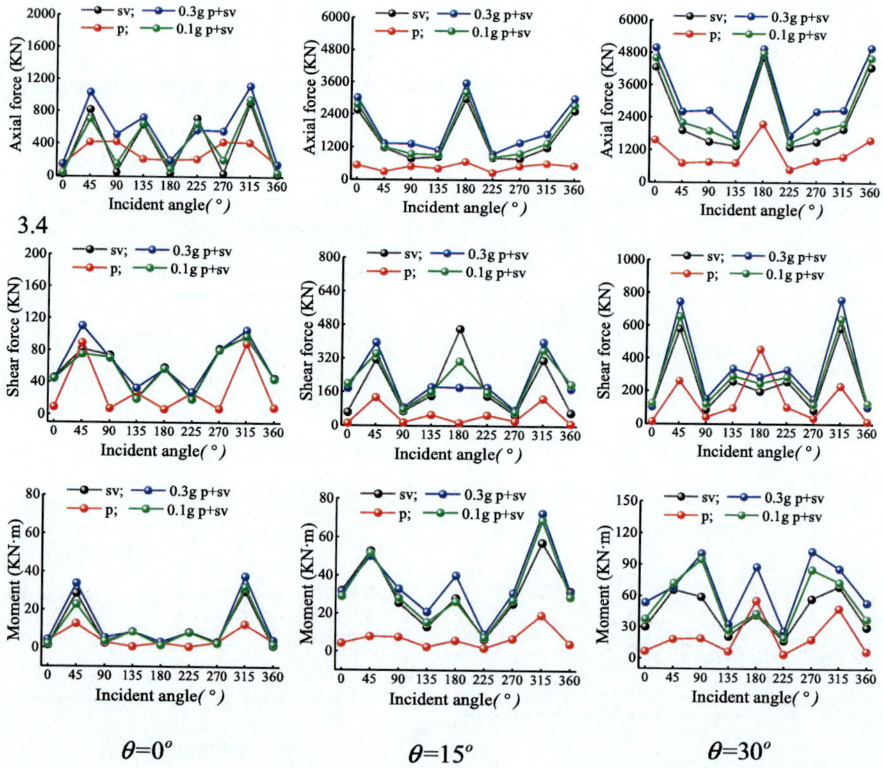

Fig. 10. Internal force distribution of tunnel lining under oblique incident seismic waves

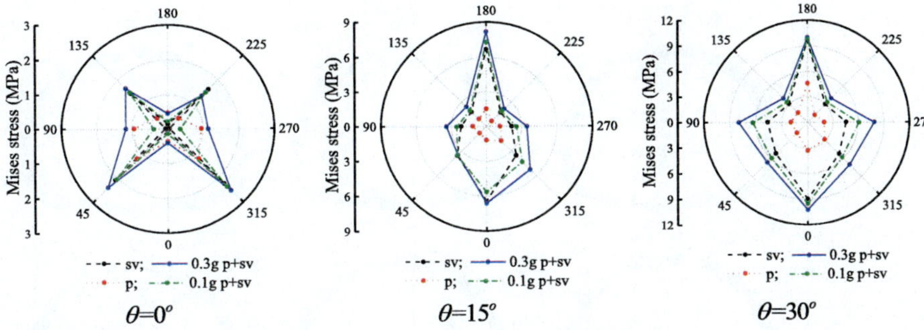

Fig. 11. Mises stress distribution diagram of tunnel lining under oblique incident seismic waves

Figure 10 shows the influence law of oblique incident angle and different wave types on the internal force of tunnel lining. Apparently, the amplitudes of internal force grow with the increasing of incoming angle, and the risk of vault and arch foot increases. It can be also seen clearly that the wave type has a great contribution on the internal force of the tunnel. The internal force of lining under SV wave is mostly larger than that under P wave at the same incident angle, which is consistent with the law of relative displacement. Moreover, the amplitude of internal force under oblique incidence of coupled wave is slightly larger than that under olique incident SV wave alone, which supports the rationality of assuming the incoming wave is shear wave in seismic analysis, and the distribution of internal forces in these two cases is very similar. The Mises stress variation law of the structure can be seen in Fig. 11. Likely the internal force, the distribution of stress changes with the increasing of incoming angle, and the vault becomes the most vulnerable part.

6.3 Damage Characteristics

Figure 12 shows clearly the tensile damage of the lining under different input angles and wave type at 20 s. When the incident angle is 0°, there is no damage to the lining under four working conditions. When the incident angle is 15°, except for the situation of incident P wave alone, the vault and bottom of the lining under the other three working conditions begin to be damaged, and the damage values are different. When the incident angle is 30°, the sections under four working conditions are all damaged, and the magnitude of maximum damage value is further increased, the damage range is expanded. The damage in the structure caused by respective oblique incidence of P and SV wave or simultaneous oblique incidence of P wave and SV wave increases with the increasing of incoming angle, and when P wave and SV wave appear at the same time, the damage of the structure is more serious than that under obliquely incident SV wave alone. In addition, Damage in section of tunnel decreases with the decreasing of P wave intensity.

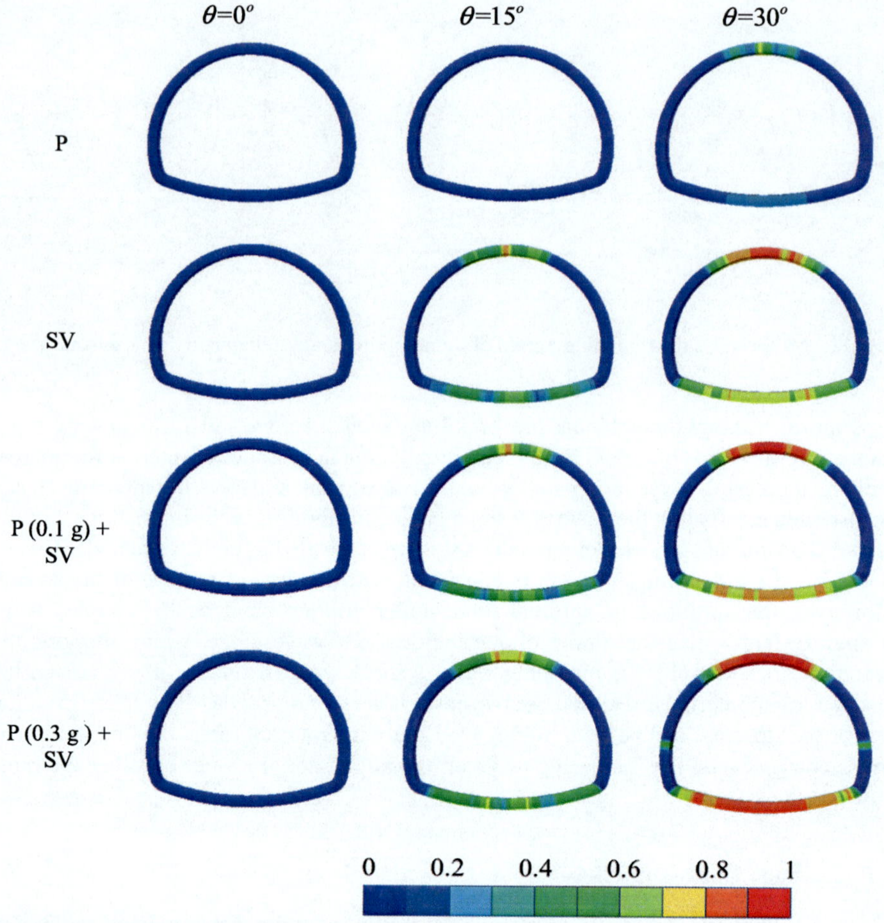

Fig. 12. Tensile damage mode of the lining subjected to obliquely incident seismic waves

7 Summary and Conclusions

This study concerns the effect of incident angle and type of seismic wave on the dynamic response of shallow tunnels. Taking the entrance of a tunnel in Tibet Autonomous Region of China as the research object, a series of dynamic analysis was conducted, and several conclusions are got as follows:

(1) If the incident angle is consistent, the compressive deformation of tunnels is larger under simultaneous oblique incidence of P and SV waves than that under obliquely incident SV wave alone, and the shear deformation is smaller.
(2) Considering the results of relative displacement, internal force and stress, when the inclination angle is less than 30°, the seismic response of tunnels under P wave, SV wave or P and SV waves increases with the increasing of incident angle.

(3) Under the oblique incidence of seismic waves, the damage of lining structure is mainly in vault and arch bottom, which indicates that this is the weak part of seismic fortification. Moreover, if the P wave is considered, the damage value and range of the structure are mostly higher and more extensive than that under SV wave alone, which indicates that the simultaneous incidence of P and SV waves is nonnegligible.

Acknowledgement. The financial supports received from National Natural Science Foundation of China (Grant No. 51908384), the Key R&D and Transformation Program of Tibet Autonomous Region (Grant No. XZ201801-GB-07), and China Postdoctoral Science Foundation Grant (Grant Nos. 2020M673223, 2019M663500).

References

1. Wang, W.L.: Assessment of damage in mountain tunnels due to the Taiwan Chi-Chi Earthquake. Tunn. Undergr. Space Technol. **16**(3), 133–150 (2001)
2. Kazuhide, Y.: Historical earthquake damage to tunnels in Japan and case studies of railway tunnels in the 2004 Niigataken-Chuetsu Earthquake. Q. Rep. RTRI **48**(3), 136–141 (2007)
3. Guangyao, C.: Study on different seismic intensities and earthquake damage to highway tunnels in Wenchuan area affected by earthquakes. Mod. Tunnel Technol. **51**(6), 1–6 (2014)
4. Tianbin, L.: Failure characteristics and influence factor analysis of mountain tunnels at epicenter zones of Great Wenchuan Earthquake. J. Eng. Geol. **16**(6), 742–750 (2008)
5. Jingbo, L.: A direct method for analysis of dynamic soil-structure interaction. J. Civ. Eng. **31**(3), 55–64 (1998)
6. Xiuli, D.: Preliminary study of time-domain seismic response for underground structures to obliquely incident seismic waves. Earthq. Disaster Prev. Technol. **2**(3), 290–296 (2007)
7. Jingqi, H.: Effect of the oblique incidence of seismic SV waves on the seismic response of subway station structure. Eng. Mech. **31**(9), 81–88 (2014)
8. Jinqi, H.: Impact of incident angles of P waves on the dynamic responses of long lined tunnels. Earthquake Eng. Struct. Dynam. **45**(15), 2435–2454 (2016)
9. Dunyu, L.: Effects of oblique incidence of SV waves on nonlinear seismic response of a lined arched tunnel. Shock. Vib. **2020**(8093804), 1–12 (2020)
10. Baoyou, Z.: Seismic response of a large underground rock cavern groups considering different incident angles of earthquake waves. J. Rock Mech. Eng. **29**(S1), 3396–3402 (2010)
11. Xiuli, D.: A stress artificial boundary in FEA for near-field wave problem. Chin. J. Theor. Appl. Mech. **38**(1), 49–56 (2006)
12. Feng, G.: Study on transformation method for artificial boundaries in static-dynamic analysis of underground structure. J. Vib. Shock **30**(11), 165–170 (2011)

Experimental and Numerical Investigations on the Structural Behaviour of the Joints in Aluminium Alloy Frames Supported Glass Curtain Wall

Mengyu Li[1], Chunyin Ma[1], Zhongxing Wang[1,2](✉), Guoxing Li[3], and Lulu Fan[3]

[1] Tianjin University, Tianjin 300350, China
zhongxing_wang@tju.edu.cn

[2] Key Laboratory of Earthquake Engineering Simulation and Seismic Resilience of China, Earthquake Administration (Tianjin University), Tianjin 300350, China

[3] Dadi Facade Technology, Shenzhen 518172, China

Abstract. In-plane and out-of-plane structural behaviour of aluminium alloy joints connected by stainless steel bolts were investigated through experiments and numerical analyses. Coupon tests were conducted prior to the tests of joints to obtain the material properties of aluminium alloys. Then, six full-scale tests on the joints were conducted, considering different loading conditions and different locations of joints in aluminium alloy frames. Key structural responses of joints, including failure modes, moment-rotation curves, initial stiffness, ultimate moment resistances and corresponding rotation at ultimate moment resistances, were obtained from the tests and the joints were classified by the specification in EN 1993-1-8. The test results indicated that the flange of the vertical member and the bolts in the horizontal member need to be strengthened for joints subjected to in-plane and out-of-plane loads, respectively. Following the tests, numerical models verified by the test results were established to mimic the structural behaviour of joints. Through discussion of test and numerical results, design suggestions were proposed.

Keywords: Aluminium alloys · Joints · Moment-rotation responses · Stainless steel bolts

1 Introduction

Aluminium alloys, with advantage of high strength-to-weight ratio, excellent corrosion resistances and recyclability, are increasingly utilized in constructional industry. Joints in frames, playing an important part in transferring internal forces, are also a research focus. Fasteners are recommended to be adopted in aluminium alloy joints since strength of aluminium alloys in heat affected zone, attributed to welding, degrades severely [1]. However, current studies on joints connected by fasteners focused on steel and stainless steel joints. Calado et al. [2] studied the structural response of steel bolted

W. Guo and K. Qian (Eds.): GBCESC 2022, LNCE 211, pp. 672–678, 2023.
https://doi.org/10.1007/978-981-19-5217-3_66

angle connections under monotonic and cyclic loads. Elflah et al. [3, 4] studied structural behaviour of stainless steel beam-to-column joints with different cross-section shapes of columns and different layouts of angle cleats. With regards to aluminium alloy beam-to-column joints, scarce investigations were performed. Wang et al. [5] conducted eight tests on aluminium alloy beam-to-column joints connected by swage-locking pins. The test results indicated that the existence of stainless steel web angle cleats improved the structural performance of joints obviously and the application of swage-locking pins in aluminium alloy joints was reasonable.

The brief literature review highlights the need for studies on structural behaviour of aluminium alloy joints, which may provide theoretical basis for the design of aluminium alloy frame structures. Hence, this study proposed a new type of aluminium alloy joints connected by stainless steel bolts. Six full-scale tests, considering different loading conditions and different locations of joints in aluminium alloy frames, were conducted. Key mechanical properties of joints, including moment-rotation curves, initial stiffness and ultimate moment resistances, were obtained and the tested joints were all classified by the specification in EN 1993-1-8 [6]. Numerical models replicating structural performance of joints were established and validated by test results. Through analyses of test and numerical results, some suggestions were proposed for further design of aluminium alloy joints.

2 Experimental Study

2.1 Details of Tested Joints

Two types of joints, employed in the mid and end span of aluminium alloy frames (MSJ and ESJ), were tested to investigate the structural behaviour of aluminium alloy joints connected by stainless steel bolts, as shown in Fig. 1 (a). The components of joints include vertical, horizontal members and connecting plates extruded from 6061-T6 aluminium alloy. The cross sections of the components are shown in Fig. 1 (b)–(e).

Vertical and horizontal members were connected by countersunk stainless steel bolts M8 with full thread, which avoiding the possible electrochemical corrosion between aluminium alloys and carbon steels [1]. The bolts were in Grade A4-70 and the diameter of clearance holes is 8.6 mm.

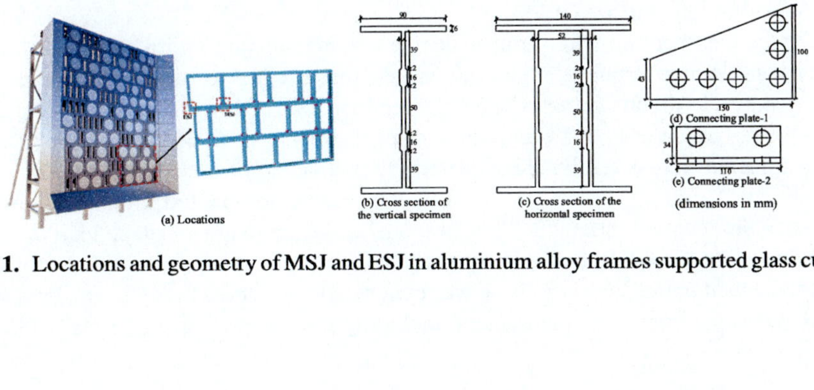

Fig. 1. Locations and geometry of MSJ and ESJ in aluminium alloy frames supported glass curtain wall

2.2 Material Tests

Material properties of the aluminium alloys were obtained through comprehensive coupon tests. Tensile tests were performed using the universal testing machines SHIMADZU according to the requirements specified in GB/T 228.1 [7]. The flat coupons were extracted from aluminium alloy connecting plate-2, the web and the flange of horizontal and vertical members. Coupon tests of each group repeated three times and a total of 15 tensile coupons were tested. Mean values of key mechanical properties were obtained from repeated tests including elastic modulus (E) equal to 67.5 GPa, yield (0.2% proof) strength ($f_{0.2}$) equal to 249.2 MPa and ultimate strength (f_u) equal to 269.1 MPa.

2.3 Test Setup and Testing Procedure

The test setup for joints subjected to in- and out-of-plane loads are shown in Fig. 2, respectively. MSJ and ESJ adopted the same experimental setup. A horizontal force, generated by a hydraulic actuator 1 m away from the edge of upper flange of the horizontal member, was applied to vertical member. Fixed beams were employed in the two sides of horizontal member to generate fixed end conditions. For MSJ, stiffeners, preventing flange crippling attributing to concentrated load, were installed at loading point and fixed beams, the locations are marked with numbers (1–6) in Fig. 2. With regards to ESJ, arrangement of stiffeners is similar to that of MSJ, while only one stiffener was installed at loading point.

Forces exerted by the hydraulic actuator was recorded by a load cell and linear variable displacement transducers (LVDTs) were utilized to measure the displacements. For joints subjected to in-plane load, 9 LVDTs were employed, as shown in Fig. 2 (a). The displacements of vertical members were recorded by L1–L2 and L5–L6. L3–L4 and L8–L9 were employed to measure the displacements of horizontal members. L7 was utilized to record the rigid displacement of joints. The layout of LVDTs for ESJ was similar to MSJ, while L3 was installed near the web of vertical members. With regards to joints subjected to out-of-plane load, 10 LVDTs were utilized, as reported in Fig. 2 (b). L1–L4 were utilized to record the displacements of vertical members and the displacements of horizontal members were measured by L5–L10. The layout of LVDTs for ESJ was the same as that of MSJ. Strain gauges were also employed at key locations to capture the strain development. A data acquisition system was applied and record readings of instruments every 1 s.

Vertical members were first connected to horizontal members with bolts, which were tightened utilizing a torque wrench to control the torque equal to 24.1 N·m. Loading was generated by a hydraulic actuator and recorded by a load cell with a maximum capacity of 300 kN. Then, the LVDTs and strain gauges were installed and the reliability of the instruments were carefully checked through preloading processes. Loading rate was approximate 1 mm/min during tests and the tests were terminated immediately when the maximum load was attained. Tests of joints subjected to out-of-plane load repeated twice due to the importance of this case in practical engineering, test of joint subjected to in-plane load under each condition was conducted only once. In total, six joints were tested and named with key parameters, including the location of the joints (MSJ and

ESJ), loading conditions (IP and OP) and the serial number. For example, MSJ-OP-1 represents joint in the mid span of aluminium alloy frames, subjected to out-of-plane load and this is the first test between the repeated tests.

3 Test Results

Key structural performance obtained from tests, including failure modes, the moment-rotation ($M - \Phi$) relationships, initial rotational stiffness and ultimate moment resistances, were reported and discussed in this section.

Failure modes of all the tested joints are shown in Fig. 3 (a)–(d).

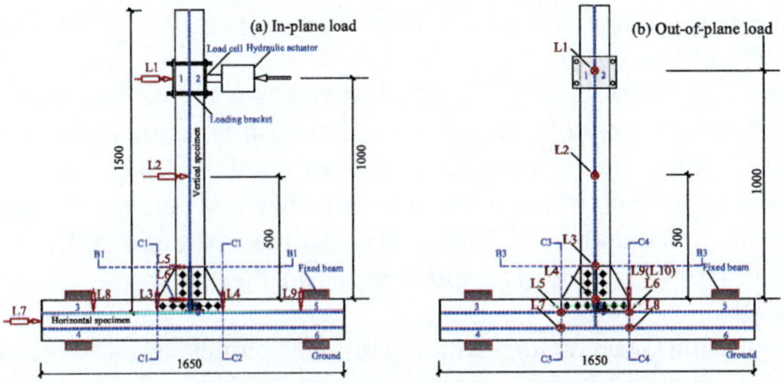

Fig. 2. Test setup for joints (dimensions in mm)

(a) MSJ-IP-1: Fracture occurred near the top bolts of the flange of vertical member due to the tensile force, as shown in Fig. 3 (a). The fracture happened suddenly with a great noise and the moment resistance of the joint decreased rapidly. The rotation of the vertical member was obvious while that of the horizontal member was neglectable.

(b) ESJ-IP-1: The failure of ESJ-IP was also attributed to the fracture of flange of vertical member, which is similar to that of MSJ-IP, as shown in Fig. 3 (b).

(c) MSJ-OP-1: Flange of horizontal member near the end bolts was teared and the junction between flange and upper web of horizontal member also fractured, as shown in Fig. 3 (c). The fracture happened suddenly with rapid decreasing of the moment resistance. Obvious rotation occurred in vertical member and localized deformation was observed near the connections in the horizontal member. The repeated test (MSJ-OP-2) showed the similar failure modes.

(d) ESJ-OP-1: The bolts located in the upper flange of horizontal member first fractured following a slight decreasing of the moment resistance. The test continued and then the fracture of the flange of vertical member was occurred near the top bolts, as shown in Fig. 3 (d). The repeated test (ESJ-OP-2) showed the similar failure modes.

The moment-rotation relationship ($M - \Phi$) is an important structural response of a joint. M is defined as the product of applied loading and the distance between the loading point and the edge of flange of horizontal member. Φ is equal to $\Phi_{\mathrm{v}} - \Phi_{\mathrm{H}}$, where Φ_{v} and Φ_{H} represent rotation of vertical and horizontal members, respectively, and can be derived from the readings of LVDTs. For joints subjected to in-plane load, Φ_{v} was calculated from the readings of L5 and L6 and Φ_{H} was calculated from the readings of L3 and L4. With regards to joints subjected to out-of-plane load, Φ_{v} was calculated from the readings of L1 and L2 and Φ_{H} was calculated from the readings of L9 and L10.

Figure 3 (i)–(j) showed the $M - \Phi$ curves for joints subjected to in-plane load and stiffness-reduction phenomenon is observed, which is attributed to the slippages of bolts when the friction is less than the external force and is common in bolted connections due to the gap between clearance holes and bolts. With regards to joints subjected to out-of-plane load, the whole moment-rotation curves were not obtained due to the obviously localized deformation occurred near L9 and L10, which resulted in inaccurate rotation of horizontal members.

The slope of the initial linear stage of M-Φ curves is defined as initial stiffness of the joints [6]. Linear regression analysis was adopted in the current study to obtain the initial stiffness of joints and the obtained results are summarized in Table 1.

Joints are commonly classified into rigid, semi-rigid and nominally pinned in accordance with their initial rotational stiffness and the stiffness of connected members (EI/L) according to EN 1993-1-8 [6]. The joints tested in this study are all categorised as semi-rigid joints.

Other key structural behaviour, including ultimate moment resistances (M_{u}) and the corresponding rotation (Φ_{u}) at M_{u}, are summarized in Table 1. As can be seen from Table 1, the initial stiffness of joints subjected to out-of-plane load is superior to that of joints subjected to in-plane load, and MSJ shows higher ultimate moment resistances and initial stiffness than those of ESJ. The repeated tests obtained similar results, the small difference may be attributed to the inevitable eccentricity in tests and different material properties of bolts.

4 Numerical Study

FE models considering the nonlinearity of material properties and geometry were established. The geometry was the same as the tested joints and the material properties employed the results of coupon tests. The reduced integration elements C3D8R were adopted in the simulation and the mesh near the connections was refined due to the stress concentrations. The size of refined mesh was approximate 2 mm. Normal and tangential interactions were considered in this model. The normal interaction was defined as 'Hard' contact and the 'Plenty' formulation with the friction coefficient equal to 0.2 was defined for the tangential interaction. For models of joints subjected to in-plane load, all degrees of freedom of the fixed beams and the translations of horizontal member parallel to the loading direction were restrained. With regards to joints subjected to out-of-plane load, all degrees of freedom of the fixed beams and the translations of vertical member perpendicular to the loading direction were restrained.

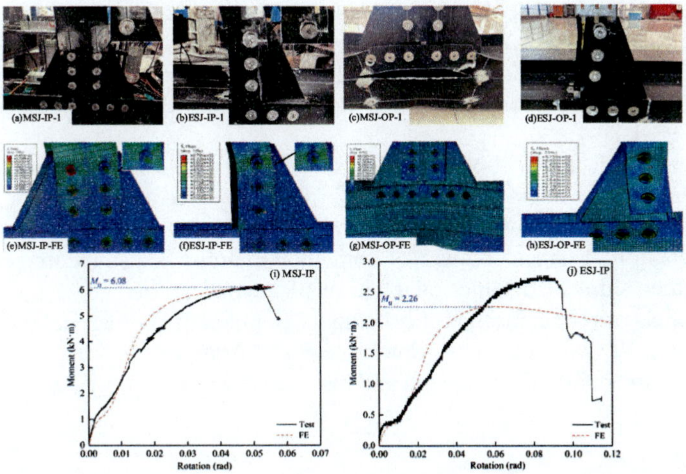

Fig. 3. Failure modes and $M - \Phi$ curves of joints from tests and numerical analyses

The established FE models were validated by test results, in terms of failure modes, moment-rotation curves, ultimate moment resistances and initial stiffness. The failure modes obtained from numerical analyses are shown in Fig. 3 (e)–(h) and the $M - \Phi$ curves are compared in Fig. 3 (i)–(j). Comparisons of ultimate moment resistances and initial stiffness from FE analyses and tests are summarized in Table 1. It may be concluded that the FE model can mimic the structural behaviour of tested joints accurately.

Table 1. Comparisons of key structural behaviour obtained from tests and FE analyses

Joint labelling	$S_{j,ini,test}$ (kN·m/rad)	$S_{j,ini,FE}$ (kN·m/rad)	$S_{j,ini,test}/S_{j,ini,FE}$	$M_{u,test}$ (kN·m)	$M_{u,FE}$ (kN·m)	$M_{u,test}/M_{u,FE}$	$\Phi_{u,test}$ (rad)
MSJ-IP-1	902.7	1004.50	0.90	6.15	6.08	1.01	0.051
ESJ-IP-1	126.4	112.90	1.12	2.73	2.26	1.21	0.087
MSJ-OP-1	2809.1	2832.50	0.99	18.46	18.28	1.01	–
MSJ-OP-2	3270.9	2832.50	1.15	18.24	18.28	1.00	–
ESJ-OP-1	250.0	258.40	0.97	9.06	9.63	0.94	–
ESJ-OP-2	220.7	258.40	0.85	8.24	9.63	0.86	–
Mean			1.00			1.00	
COV			0.12			0.12	

5 Conclusion

Structural behaviour of joints in aluminium alloy frames supported glass curtain subjected to in-plane and out-of-plane load were investigated in this study through experiments and numerical analyses. The following conclusions can be drawn:

(1) Mechanical properties of six joints employed in aluminium alloy frames supported glass curtain wall were comprehensively studied, key structural responses, including $M - \Phi$ curves, initial stiffness and ultimate resistances, were obtained. All the tested joints were classified as semi-rigid joints according to the specification in EN 1993-1-8.
(2) For joints subjected to in-plane load, it is recommended to employ vertical members with thick flange in practical design. For joints subjected to out-of-plane load, bolts located in the flange of horizontal member are recommended to be strengthened. The junction of flange and web of horizontal member is also recommended to be strengthened for MSJ subjected to out-of-plane load.
(3) FE models were established considering the nonlinearity of material properties and geometry. Key structural performance obtained from tests and numerical analyses were comprehensively compared and the accuracy of the FE models was verified.

References

1. GB 50429-2007, Code for Design of Aluminium Structures, MOHURD, Beijing (2007). (in Chinese)
2. Luis, C.: Experimental response of top and seat angle semi-rigid steel frame connections. Mater. Struct. **33**(8), 499–510 (2000)
3. Mohamed, E.: Behaviour of stainless steel beam-to-column joints-part 1: experimental investigation. J. Constr. Steel Res. **152**, 183–193 (2018)
4. Mohamed, E.: Structural behaviour of stainless steel beam-to-tubular column joints. Eng. Struct. **184**, 158–175 (2019)
5. Zhongxing, W.: Experimental investigation on the behaviour of aluminium alloy beam-to-column joints connected by swage-locking pins. Eng. Struct. **213**, 110578 (2020)
6. EN 1993-1-8, Eurocode 3: Design of Steel Structures-Part 1.8: Design of Joints, European Committee for Standardization (CEN), Brussels (2005)
7. GB/T 228.1, Metallic Materials – Tensile Testing-Part 1: Method of Test at Room Temperature, Standards Press of China, Beijing (2011). (in Chinese)

Discussion on Key Problems of 3D Printing Concrete in Coastal Environment

Xinyue Zhang[1], Weijiu Cui[1,2(✉)], Chengping Qu[1], Sheng Wang[3], and Qi Yu[3]

[1] School of Civil Engineering, Qingdao University of Technology, No. 777, Jialingjiang East Road, Qingdao 266520, China
cuiweijiu@163.com

[2] Innovation Institute for Sustainable Maritime Architecture Research and Technology, Qingdao University of Technology, Qingdao 266033, China

[3] Qingjian Group Co., Ltd., Qingdao 266011, China

Abstract. 3D printing concrete (3DPC) has the advantages of flexible design, high construction efficiency and less consumables, which meets the requirements of construction industrialization. The current research, forming the preliminary results of 3DPC, mainly focuses on the mix design, performance evaluation method and printing control technology. however, there are few studies considering the service environment condition. At the same time, the cement material consumption of 3DPC is too high currently, which is inconsistent with the dual-carbon target. Therefore, this study investigates and analysis the research literature of 3DPC at home and abroad, and summarizes its research progress, technical characteristics and problems, analyze the current problems of 3DPC, then a research framework and technical route is proposed for the green and low-carbon development of 3DPC in the coastal environment. and the key problems of them are point out, which is expected to provide exploration and reference for the expansion of 3DPC applications.

Keywords: 3DPC · Mix design · Performance evaluation · Coastal environment

1 Introduction

3D printing is a kind of "additive manufacturing" technology based on digital model through 3D printing equipment to complete stacking and forming layer by layer [1]. It has the advantages of no formwork, high construction flexibility, low raw material loss, which conforms to the requirements of industrial construction [2]. In 2021, Chairman Xi puts forward the dual carbon goal at the seventy-fifth United Nations General Assembly. Low carbon construction has become the future theme of construction industry. Therefore, it is urgent to further develop the 3D printing technology of concrete based on the requirement of low carbon construction.

Printing process control and printing material preparation are the core research points of 3D printing concrete. D-Shape, Contour Crafting and 3D concrete printing technology [3] are the main printing processes for this technology. Compared with printing

W. Guo and K. Qian (Eds.): GBCESC 2022, LNCE 211, pp. 679–688, 2023.
https://doi.org/10.1007/978-981-19-5217-3_67

process control, the preparation of printing materials has attracted more attention in the civil engineering field. Like the preparation of traditional concrete, 3D printing concrete is also guided by printing performance. Suitable printing performance is obtained by adjusting the dosage of additives or the proportion of cementitious materials. The development of preparation technology has also gradually formed the performance evaluation methods of 3D printing concrete at present. The extrusion performance and construction performance are the most important points, and these properties are decided by the rheological properties of concrete. In addition, how to adjust the printing process to achieve the excellent printing performance is also the focus of academic research. Based on the above researches, domestic and foreign research teams achieve many successful engineering practices. However, it is regrettable that the current 3D printing concrete is still through the "trial test method" to achieve printing control, and there is no uniform and applicable preparation technology, which is not conducive to large-scale and industrial application. In addition, the application scenarios of 3D printing concrete in current research are not discussed in detail. The corresponding 3D printing technology considering the service environment of concrete is rarely studied. With the continuous progress of industrialization and double carbon, such problems will inevitably become a hot research topic in the future.

Therefore, this review comprehensively selects some representative literatures at home and abroad, systematically discusses the research progress, the key problems of material selection, performance optimization and evaluation index of 3DPC technology, and extends the key technical framework of 3D printing concrete in coastal environment. It is proposed to use micro silica fume, fly ash, slag powder and other cementitious materials instead of some cement to realize the low carbonization development of 3DPC, in order to broaden the application of 3D printing concrete.

2 Research Hotspots of 3DPC in Domestic and Foreign Universities

The change of construction method puts forward new performance requirements for 3DPC. The material should have good fluidity during transmission and extrusion, and good shape stability after extrusion. From the rheological point of view, concrete extrusion stage should have lower dynamic yield stress and plastic viscosity, after extrusion should have higher static yield stress. Therefore, a large number of domestic and foreign academic researchers are improved the performance of concrete by improving the proportion of raw materials or adding new components, and successfully realize 3D printing of concrete.

This review summarizes the research hotspots, challenges and methods of representative universities at home and abroad in the 3DPC field. Biranchi P et al. [4], of Nanyang University of Technology, Singapore, are dedicated to the rheological study of 3D printing materials and successfully manufactured fly ash-based geopolymer mixtures. Jianzhuang X et al. [5] of Tongji University research the influence of recycled sand content on early performance of 3D printing mortar. Xinying Z et al. [6] by testing the water evaporation rate, setting time and rheological properties of 3DPC, the instability criterion of 3D printing construction based on static yield stress of rheological parameters were clarified. In addition. Li W et al. [7] propose a method of inserting U-shaped nails as reinforcement perpendicular to the printing layer in the process of 3D printing to improve the interlayer bonding performance. In the field of engineering practice, the team led by Professor Guowei M successfully printed a fabricated 3D printing Zhaozhou Bridge in Hebei University of Technology on October 13, 2019, realizing the promotion of 3D printing technology from laboratory research to practical engineering application [8].

The above research hotspots represent the current advanced research results. With the development of 3DPC technology, it still needs to solve the problems of preparing cement-based materials with low carbon and environmental protection, realizing the effective bonding between cement-based materials and steel bars in 3D printing, and using mechanical arm for synchronous and accurate reinforcement in the printing process, so as to realize the practical construction application of 3DPC.

3 Technical Characteristics of 3DPC

3.1 Mix Design Method of 3DPC

The mix design method of 3DPC is different from the traditional concrete, which needs to meet the printability and take into account the coordination of printing process and performance of materials [9]. Ma et al. [10] proposed a method of 3DPC mix design matching with the printing process. Firstly, according to the extrudability requirements, the cementitious material and aggregate are determined in combination with the nozzle size, then the additives, SCM and fibers were added to meet the buildability, setting time, strength and shrinkage requirements of 3DPC. A.V. Rahul et al. [11] proposed a design method based on yield stress. Preliminary design of cementitious materials composed of Portland cement and fly ash, and the aggregate content was determined by particle filling method. Then the amount of superplasticizer was determined by the test of workability and extrusion. Zhixin L et al. [13] proposed a mix design method based on rheological properties, and evaluated the influence of sand, cement and fly ash on yield stress.

Representative mix designs in each study are shown in Table 1. It can be seen from the table that the 3DPC mix designs mostly using ordinary Portland cement (OPC) as the main cementitious material, adding only 20–30% of other supplementary cementitious materials (SCM). Since OPC production is related to carbon dioxide emissions, Yu Chen et al. [14] find that the mixture of fly ash and silica fume can replace 45% of cement. Therefore, in order to avoid the increase of carbon dioxide emissions caused by the increase of cement content, more SCM mixtures should be developed when configuring concrete.

Table 1. Representative mix designs of 3DPC

Research team	Binder	Water-binder ratio	Sand-binder ratio	Sand size	Admixture (0.10% of binder)	Fibers (kg/m^3)
Guowei M et al. [10]	70%Portland cement, 20%fly ash, 10%silica fume	0.26	1.2	–	SP 0.29	PP fiber1.2
A.V. Rahul et al. [11]	80%Portland cement, 20%fly ash	0.32	1.5	0–1.3 mm	SP 0.83	PP fiber 1.8
M. Rubio et al. [12]	68%Portland cement, 24%fly ash, 8%silica fume	0.5	2	–	SP 0.55	PP fiber 1.2
Zhixin L et al. [13]	70%Portland cement, 20%fly ash, 10%silica fume	0.45	1.4	0–4.75 mm	SP 1.48 VMA 0.13	–

3.2 Performance Evaluation and Optimization of 3DPC

In this review, the performance requirements of 3DPC from mixing to hardening were sorted out, as shown in Fig. 1. In principle, materials used for printing should have

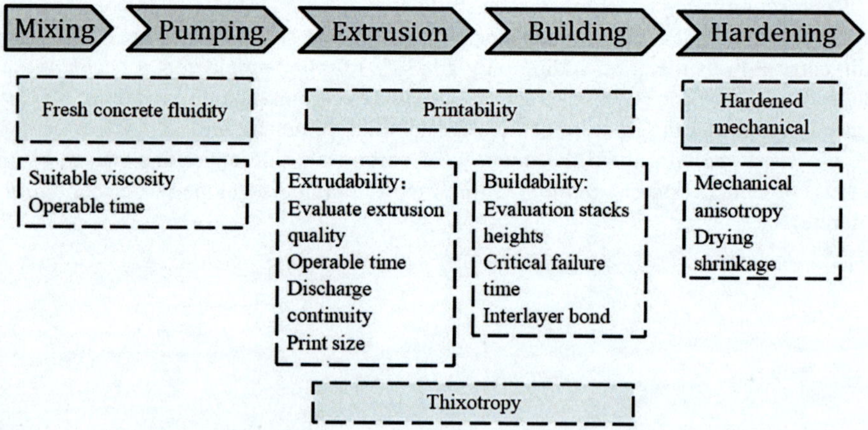

Fig. 1. Performance requirements of 3DPC in each process

the characteristics of easy extrusion, flow and construction, and must also have good mechanical strength and bonding properties. At the same time, the above physical and printing properties of 3DPC is decided by the composition of the material and its fresh mixing and hardening properties. For the pumping process of 3DPC, appropriate viscosity and operation time are crucial. During printing and construction, extrudability and buildability are regulated by rheology and yield stress. Therefore, the apparent viscosity, yield stress and thixotropy of printing mortar should be controlled within a reasonable range. Mechanical anisotropy and drying shrinkage are the focus of hardening properties of 3DPC.

Stirring, Pumping Performance. The setting time of 3DPC is an important parameter to characterize the condensation property, which needs to be controlled in two ways, sufficient operational time is often required in the pumping process, so that the nozzle can be extruded continuously without blockage, and therefore the setting time needs to be extended [15]. After extrusion from the print nozzle, the printing layer also requires the setting time of the printing material to be as short as possible to ensure that the printed layer obtains sufficient early strength to support the subsequent printing layer and improve the buildability. Yang et al. [24] studied the relationship between the setting time and printing height of 3D printing mortar, from which it can be concluded that the setting time of mortar suitable for printing is mainly 30–80 min.

In an effort to control the setting time of 3DPC within a reasonable range, it is necessary to use appropriate amount of accelerator, retarder and mineral admixture. Lin et al. [16] studied the effects of six common retarders on the setting time of extrusion printing materials. Among the six retarders, sodium tetraborate has the best effect. Weihong L et al. [17] showed that with the increase of silica fume content, the fluidity of 3D printing concrete decreased. Under the most suitable silica fume content, the initial setting and final setting time were prolonged by 35.7% and 16.7%. Chen et al. [18] used retarder and diatomite to improve sulphoaluminate cement, the printing time was extended to 35–39 min.

In order to meet the fluidity, pumping capacity and dimensional stability of concrete mortar, water - binder ratio should be controlled besides setting time. For the smooth conveyance of cementitious materials in the pipeline, a higher water demand is required, however to have good constructability after extrusion, a lower water binder ratio is required.

Printing and Building Performance. Printability of 3DPC includes extrudability and buildability. T.T. Le et al. [19] consider that extrudability and buildability are the two most important performance indicators of 3D printing concrete. Extrudability is defined as the ability of a material to be continuously and uniformly extruded from a 3D printer nozzle without blockage or breakage [20]. Buildability refers to the support and shape retention between printing layers in the printing process, which refers to the ability of the previous printing layer to support the subsequent printing layer and have good adhesion with the subsequent printing layer [21].

Printability of 3D printing cement-based materials is related to rheological properties. It is affected by both apparent viscosity and yield stress. The yield stress is too high, the material is difficult to flow out and the yield stress is too small to accumulate. Qianrong

Y et al. [24] found that when 3D printing construction mortar the apparent viscosity of 4.0–7.0 Pa · s, yield stress between 50–80 Pa has good extrudability and high printing height corresponding. In addition, the printability of materials is also related to printing parameters such as nozzle size, printing head walking rate and printing rate, and printing head lifting rate.

Based on the above analysis of 3DPC printability, it is revealed that the two most important properties of 3DPC printability are fluidity and buildability, and the optimization of these two key factors can be attributed to the regulation of its rheological properties. At present, the researches mainly use the rheometer or viscometer to test the rheological parameters of cement-based materials such as yield stress, viscosity and thixotropy, so as to realize the expression of the working performance of cement-based materials in the form of rheological parameters.

Post-hardening Properties. The properties of 3DPC after hardening mainly involve mechanical anisotropy and drying shrinkage. Bos et al. [22] test shows that the failure mode of 3DPC structure can be changed from brittle failure to ductile failure by increasing fiber. But the mechanical anisotropy and the existence of weak surfaces of 3D printing concrete structures are the inherent characteristics of the layer-by-layer 3D printing process. The addition of fibers cannot be completely eliminated, and when the number of inter-strip interfaces in the load direction increases, the material integrity on the fracture surface is weakened, and the mechanical properties are worse [23].

Yanfeng L et al. [8] found that the pore caused by the construction process and the weak interface between layers were the reasons for the difference of mechanical properties of 3D printing concrete. In addition, anisotropy is greatly affected by material parameters and printing process parameters, Paul et. al. [25] research shows that cement-based materials in the printing process, reduce the height of the print head, make concrete extrusion molding, can reduce porosity, is expected to achieve higher strength.

In addition to mechanical anisotropy, drying shrinkage is also an important evaluation parameter for 3DPC printing performance. The optimal control methods include internal curing, increasing the dosage of mineral admixtures, and adding additives containing expansion agents or shrinkage compensation [26]. Khatib et al. [27] found that adding fly ash in concrete mortar can significantly reduce the heat release of hydration process, thereby reducing the probability of external cracks. Sha et al. [28] using shrinkage inhibition additives in concrete can significantly reduce the surface tension caused by water evaporation of concrete external components, thereby reducing the shrinkage deformation of concrete.

3DPC is prone to brittle failure because it cannot be printed with steel bars at the same time. Adding fiber in the current researches are the widely recognized treatment method. However, excessive fiber content will lead to blockage and interruption in the printing process. Therefore, in the design of fiber reinforced 3DPC mixture ratio, it is necessary to take into account the improvement of mechanical bearing capacity and printability. In addition, the construction method without reinforcement is beneficial to the application of 3D printing concrete in coastal environment, so it is necessary to further explore its application scenarios and service performance.

4 Extended Application of 3DPC

4.1 Application Prospect of 3DPC

Qingdao has nearly 800 km of winding coastline. Among the coastline, from the trestle scenic spot to the Olympic Sailing Center, as one of the "gold coastline" in Qingdao, many landscape buildings have been built on the plank road, as shown in Fig. 2. 3DPC can realize the construction of these flexible structures due to the formless construction methods. In addition, the 3DPC structures are not equipped with steel bars, and the use of fiber reinforcement reduces the harm of chloride ion erosion in Qingdao coastal environment, making the 3D printing application of landscape architectures have broad prospects.

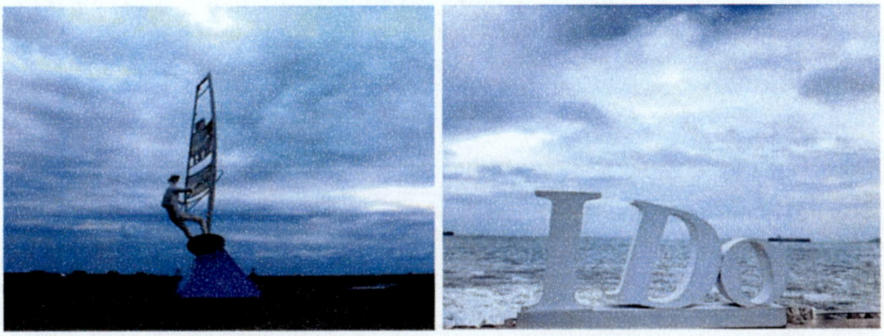

a. Canvas sculpture of the Olympic Sailing Center

b. Sculpture of Lover Dam

Fig. 2. Coastal plank road and landscape architecture in Qingdao [29]

4.2 Challenges of 3DPC Application in Coastal Environment

The factors that are against to the durability of concrete structures in coastal environments include freeze-thaw cycles, carbonation, chloride ion penetration, alkali-aggregate reaction, etc. Most of them are related to the water transpiration process in concrete pores [30]. The study find that the uniform and dense microstructure have stronger impermeability, and the printing interlayer time interval and the thixotropy of the material have different degrees of influence on the interface microstructure [31]. The research of Qian and Xu [32] shows that the impermeability of the interface decreased gradually with the increase of time interval, Wael and Kama [33] shows that the interface impermeability of high thixotropic materials is much lower than that of low thixotropic materials. The microstructure of 3DPC is loose, and there are many large pores, resulting in weak interfacial bonding [9], the influence of 3DPC on durability needs to be considered in future research.

In the current research, in order to meet the performance requirements of 3DPC such as printability, sufficient strength, a large amount of cement is generally used [10].

Replacing more than 45% of OPC with SCM is not a new way to reduce carbon dioxide emissions in traditional concrete, but rarely in 3D printing concrete [14].

At present, our team is committed to exploring the mix design and performance evaluation methods based on printing and service performance in coastal environment condition. Firstly, raw materials are selected according to the types and characteristics, Therefore, we utilize raw materials with regional characteristics and low carbon environmental protection such as fly ash, slag powder and micro-silica powder. Design the orthogonal experiments with the main control indexes of printing performance and rheological properties, so as to minimize the amount of cement, reduce costs and reduce carbon emissions. Then the mix design was optimized by impermeability and corrosion resistance. The team consists of six people. The research technical route is shown in Fig. 3. We mainly solve the problems in two directions. First, the control of material performance requirements is realized by numerical simulation method. In addition, the mix design method towards the goal of double carbon is continuously explored through experiments. Finally, we take the landscape architectures from Qingdao trestle to the Olympic Sailing Center as the goal to explore the practical application scenarios of 3DPC.

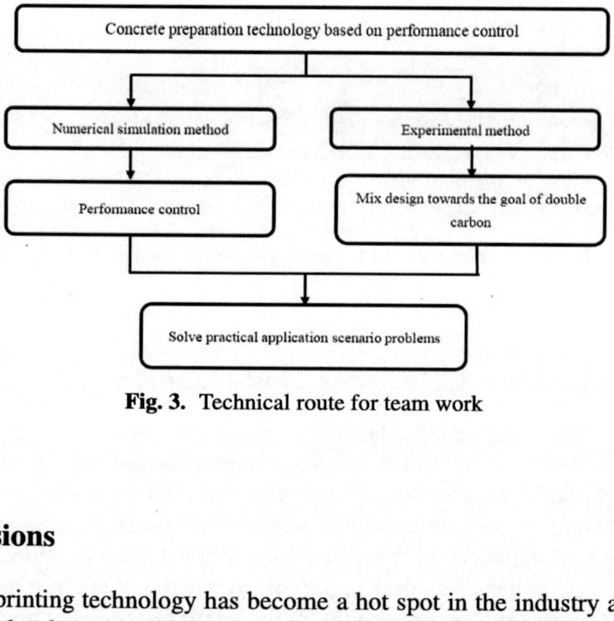

Fig. 3. Technical route for team work

5 Conclusions

Concrete 3D printing technology has become a hot spot in the industry and has broad prospects for development. Although many achievements have been made in recent years, in this paper, some representative literatures at home and abroad are comprehensively selected, and the research progress is systematically discussed. The key problems of 3DPC technology such as mix design methods, performance optimization and evaluation index are discussed. Finally, suggestions for 3D printing in coastal environment are given, and summarizes the following conclusions:

1. Suggestions for the future development direction of 3D printing concrete are given: 3D printing cement-based materials with good performance and meeting the requirements of green environmental protection should be further studied. At the same time, combined with the special coastal environment of Qingdao, it is of great significance to explore the application scenarios and service environment of 3D printing concrete.
2. The research methods of concrete preparation technology and performance evaluation towards dual carbon targets in coastal environment are proposed.
3. This paper briefly introduces the research team of 3D printing concrete in Qingdao University of Technology and the key technical framework for realizing 3D printing of concrete in coastal environment.

Acknowledgments. The authors would like to acknowledge the financial support from the National Natural Science Foundation of China (Nos. 52008224), and the Open Fund of innovation institute for Sustainable Maritime Architecture Research and Technology (iSMART), Qingdao University of Technology (Nu. 2020-031).

References

1. Gibson, I., Rosen, D.,Stucker, B.: Additive Manufacturing Technologies: 3D Printing, Rapid Prototyping, and Direct Digital Manufacturing, Springer, New York (2015). https://doi.org/10.1007/978-1-4939-2113-3
2. Weinstein, D.,Nawara, P.: Determining the applicability of 3D concrete construction (contour crafting) of low income houses in select countries, Cornell Real Estate Rev. **13**, 94–111 (2015)
3. Jianjun, W., Hongmei, L., et. al.: Application status and prospect of 3D printing technology in construction field. Archit. Technol. **50**(06), 729–732 (2019)
4. Panda, B., Unluer, C., Tan, M.J.: Investigation of the rheology and strength of geopolymer mixtures for extrusion-based 3D printing. Cem. Concr. Compos. **94**, 307–314 (2018)
5. Jianzhuang, X., Fei, Q., Tao, D., et. al.: Experimental study on early performance of 3D printing recycled mortar. J. Build. Mater. https://kns.cnki.net/kcms/detail/31.1764.TU.2021/09/16
6. Xinying, Z., Li, W., et. al.: Buildability prediction solution for 3D concrete printing. Concrete **381**(7), 38–42 (2021)
7. Li, W., Gma, B., Tl, A., et al.: Interlayer reinforcement of 3D printed concrete by the in-process deposition of U-nails. Cem. Concr. Res. **148** (2021)
8. Yanfeng, L., Ling, y., et al.: Study on mechanical anisotropy and meso damage mechanism of 3D printing concrete. J. Highway Transp. Res. Dev. **38**(5), 81–89 (2021)
9. Shaodan, H., Zhenhua, D., et al.: A review of 3D printed concrete: Performance requirements, testing measurements and mix design. Constr. Build. Mater. https://doi.org/10.1016/j.conbuildmat.2020.121745
10. Guowei, M., Li, W., et al.: Printable properties of cementitious material containing copper tailings for extrusion based 3D printing. Constr. Build. Mater. **162**, 613–627 (2018)
11. Rahul, A.V., Santhanam, M., Meena, H., et al.: 3D printable concrete: Mixture design and test methods, Cem. Concr. Compos. **97**, 13–23 (2019)
12. Rubio, M., Sonebi, M., Amziane, S.:3D printing of fibre cement-based materials: fresh and rheological performances. In: ICBBM (2017)

13. Liu, Z., Li, M., Weng, Y., et al.: Mixture design approach to optimize the rheological properties of the material used in 3D cementitious material printing. Constr. Build. Mater. **198**(20), 245–255 (2018)
14. Chen, Y., Veer, F., Copuroglu, O., et al.: Feasibility of using low CO2 concrete alternatives in extrusion-based 3D concrete printing. In: Wangler, T., Flatt, R. (eds.) DC 2018. RILEM Bookseries, vol. 19, pp. 269–276. Springer, Cham (2019). https://doi.org/10.1007/978-3-319-99519-9_25
15. Lu, B., Weng, Y., Li, M., et al.: A systematical review of 3D printable cementitious materials. Constr. Build. Mater. **207**(20), 477–490 (2019)
16. Xiqiang, L., Tao, Z.: Preparation, properties and application of cement-based building 3D printing materials. Coner. Australia **42**, 59–67 (2016)
17. Weihong, L., Xidong, C. et al.: Effect of mineral admixtures on properties of 3D printing cement-based materials. Bull. Chin. Silic. Soc. **39**(10), 3101–3107 (2020)
18. Chen, M.X., Guo, X.Y., Zhang, Y., et al.: Effect of tarie acid on the printable, rheological and mechanical properties of 3D printing sulphoaluminate cement paste. Materials **11** (2018)
19. Le, T.T., Austin, S.A., Lim, S., et al.: Mix design and fresh properties for high-performance printing concrete. Mater. Struct. **45**(8), 1221–1232 (2012)
20. Zhao, Y., Liu, W.S., Wang, H., et al.: Influence of limestone powder on performances of 3D printing cementitious materials D. Mater. Rep. **34**(2), 1217–1220 (2020)
21. Su, Q., Qingyuan, L., Xiaogpeng, C., et al.: Research status and prospect of 3D printed concrete materials. Concrete **375**(1), 36–39 (2019)
22. Bos, F.P., Bosco, E., Salet, T.: Ductility of 3D printed concrete reinforced with short straight steel fibers. Virtual Phys. Prototyping **14**(2), 160–174 (2019)
23. Jianguang, Y., Huaikui, Z., Fei, T.: Fracture mechanism experimental study of hardened 3D printed concrete. J. Build. Struct. https://doi.org/10.14006/j.jzjgxb.2021.0094
24. Gouwei, M., Li, W., et al.: Key technologies of 3D Printing of Cement-Based Materials. China Building Materials Industry Press, Beijing (2020)
25. Paul, S.C., Tay, Y.W.D., Panda, B., Tan, M.J.: Fresh and hardened properties of 3D printable cementitious materials for building and construction. Arch. Civil Mech. Eng. **18**(1), 311–319 (2017). https://doi.org/10.1016/j.acme.2017.02.008
26. Wu, L., Farzadnia, N., Shi, C., Zhang, Z., Wang, H.: Autogenous shrinkage of high performance concrete: a review. Constr. Build Mater. **149**, 62–75 (2017)
27. Khatib, J.M.: Performance of self-compacting concrete containing fly ash. Constr. Build. Mater. **22**(9), 1963–1971 (2008)
28. Shah, S.P., Karaguler, M.E., Sarigaphuti, M.: Effects of shrinkage-reducing admixtures on restrained shrinkage cracking of concrete. ACI Mater. J. **899**(3), 289–295 (1992)
29. Charming Qingdao coastline. https://www.meipian.cn/jvipca. Accessed 30 Jan 2016
30. Espinosa, R.M., Franke, L.: Influence of the age and drying process on pore structure and sorption isotherms of hardened cement paste. Cem. Concr. Res. **36**, 1969–1984 (2006)
31. Monteiro, P., Mehta, P.K.: Concrete: microstructure, properties and materials (2006)
32. Qian, P., Xu, Q., et al.: Experimental investigation on properties of interface between concrete layers. Constr. Build. Mater. **174**, 120–129 (2018)
33. Wael, A., Kamal, H., et al.: Effect of structural buildup at rest of self-consolidating concrete on mechanical and transport properties of multilayer casting. Constr. Build. Mater. **196**, 626–636 (2019)

Experimental and Numerical Investigations on the Compression Behavior of Calcareous Sand-Rubber Mixture

Ziye Zhou, Hongjie Lin, and Jiankun Liu(✉)

Sun Yat-sen University, Zhuhai, China
iujiank@mail.sysu.edu.cn

Abstract. Improving calcareous sand is of great engineering significance to the construction of man-made islands. At present, waste tire chips (*i.e.,* rubber) have been tested to improve geotechnical materials for infrastructures such as road subgrades and foundations. However, the influence of rubber on the compression behavior of calcareous sand is still unclear. In order to observe this, the one-dimensional compression tests of the mixture with rubber content of 0%,10%, 20%, 30% and 50% and the numerical simulation of MatDEM were carried out. The experimental results showed that the ductility and high compressibility of rubber resulted in a higher compressibility and lower strength of the mixture, which may mitigate the degree of particle breakage. As the rubber content is greater than 20%, the intensity of the calcareous sand-rubber mixture is too low to be recommended for engineering. Further study will focus on the particle-particle, particle-rubber and rubber-rubber stress to clarify the mechanism of particle breakage mitigation.

Keywords: Calcareous sand · Compression behavior · Rubber

1 Introduction

Calcareous sand (CS) is a sediment generated by the remains of shells and other creatures on the seabed through a series of physical and chemical actions, it has the properties of high porosity, irregular figure and easily to be crushed [1]. CS, which composed islands and reefs together with foundation rock [2], is mainly used as load bearing layer of a shallow foundation [3]. In recent years, the unique mechanical properties of CS have attracted the attention of scholars [1]. The confining pressure is related with the liquefaction resistance of CS [1]. Lv, Li *et al.* [4] observed the creepage law of CS under low stress function and found that creepage of CS mainly related to its irregular shape and pore. It is found that the loading rate had a certain influence on the crushing of CS particles [5], which will also affect its strength [8]. When the particles are subjected to impact load, the particle breakage increases as impact energy improves, and decreases first and then increases as the increase of moisture content [6]. Moisture content also has certain influence on compressibility of CS [7]. When artificial land is formed, fine grain convergence layer will form in CS foundation [9]. Studies have shown that the

W. Guo and K. Qian (Eds.): GBCESC 2022, LNCE 211, pp. 689–695, 2023.
https://doi.org/10.1007/978-981-19-5217-3_68

addition of fine grains in CS can reduce the permeability of CS [10], and also weakens its original dilatancy and shear strength [9]. In one-dimensional compression test, the pure CS particles are seriously broken [11], while adding fine-grained soil into CS can fill pores and reduce the breakage of CS particles [12].

At present, more waste tire chips are generated and most of the waste tire chips are simply piled up and landfilled in garbage dumps, which has caused great harm to the environment, due to its light weight, high permeability, high shear strength and great economic efficiency, waste tire chips can serve as lightweight filling materials for civil construction like road construction [13]. It is found that adding waste tire chips into CS under certain conditions can effectively improve the tensility and reduce shear strength of CS [13].

The purpose of this paper is to observe the compression behavior of CS-rubber mixture and find out the mechanism of cushion effect. In this study, one-dimensional compression experiments were conducted on the CS-rubber mixture with 5 diverse rubber contents, and the results were in comparison with the results of numerical simulation software MatDEM. This paper analyzed the effect of rubber on strength and particle breakage of the mixture.

2 Materials

2.1 CS

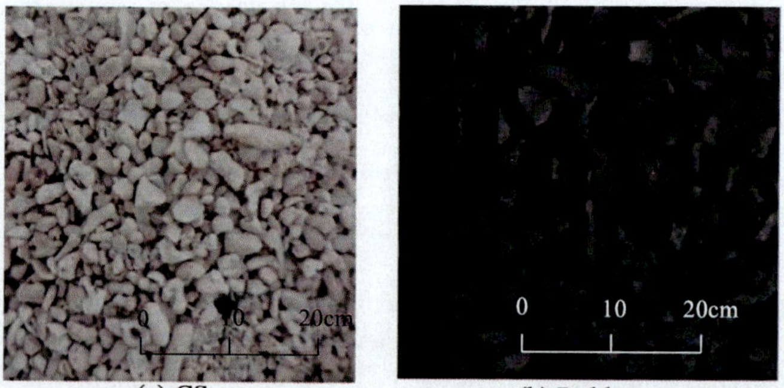

(a) CS (b) Rubber

Fig. 1. CS and rubber samples (scale = 1:10)

The CS in the research comes from Xisha Islands, as shown in Fig. 1(a). It can be found that they are off white, with many edge angles and pores, and the particle size is uneven, and they have various shapes such as long strips, ovals, flakes and etc. According to the test, Its specific gravity was 2.48. In this study, only CS particles smaller than 5 mm were used for the test. This is because CS as a granular material, the CS with a grain size below 5 mm can represent the properties of CS, and the maximum grain size allowed by triaxial tester used is 5 mm. The particle grading curve is shown in Fig. 2.

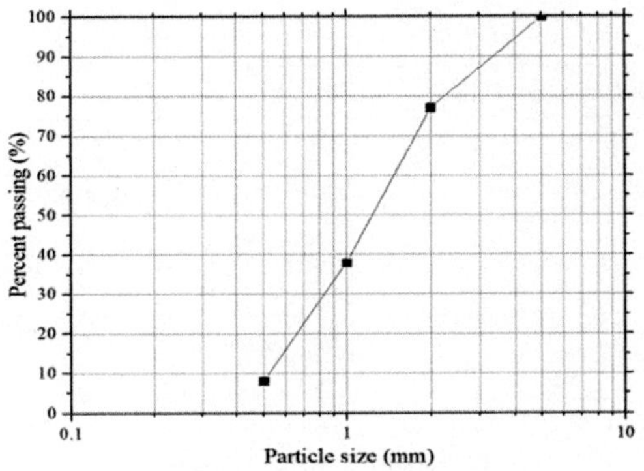

Fig. 2. Particle grading curve of CS

Table 1. Parameters of CS-rubber mixtures of diverse rubber contents

Rubber content (%)	Maximum dry density (g/cm^3)	Minimum dry density (g/cm^3)	Relative compactedness	Dry density (g/cm^3)
0	1.45	1.32	0.6	1.39
10	1.25	1.13	0.6	1.20
20	1.13	0.99	0.6	1.07
30	0.98	0.90	0.6	0.95
50	0.88	0.73	0.6	0.81

2.2 Rubber

The grain size of rubber of the test is between 1–3 mm, as shown in Fig. 1(b). In the test, the rubber contents were respectively 0, 10%, 20%, 30% and 50%.

3 Test Methods

3.1 One-Dimensional Compression Test

In the study, one-dimensional compression tests were carried out by changing the maximum axial load and rubber content respectively. The instrument used for the one-dimensional compression test was a triaxial tester, and the maximum axial force can be applied by the triaxial tester is 3.0 kN. Before the start of the test, the material was dried by dehydrator at 105 °C for 8 h to ensure that the material was in dry condition. The relative compactedness used by the test was determined to be 0.6. The dry density of the material with different rubber contents is shown in Table 1. The instrument used for the test sample preparation was a three-valve, diameter 39.1 mm and height 80.0 mm. The CS-rubber mixture was prepared by layered compaction method. In a compression test machine the stress control was used by setting the loading rate of the instrument to

Table 2. Test scheme of the loading stress

Rubber content (%)	Loading stress (kPa)
0	200, 400, 1000, 2000
10	200, 400, 600, 800, 1000
20	200, 400, 800
30	200, 400, 800
50	50, 200, 600

5.0 kPa/min and the data of stress-strain curve was saved. The loading stress is shown in Table 2.

3.2 MatDEM

MatDEM is a discrete element software. Software modeling code process is as follows: Firstly, set the basic material parameters (as Table 3). Then establish the model of one-dimensional compression test, and draw the stress-strain curve.

Table 3. Parameters of CS and rubber used for numerical simulation

Material types	Yang's modulus (Pa)	Poisson ratio	Tensile strength (Pa)	Compressive strength (Pa)	Internal friction coefficient	Density (kg/m^3)
CS	1.30E+08	0.20	1.00E+05	2.00E+06	0.82	2790
Rubber	7.84E+06	0.47	2.00E+03	5.00E+07	2.00	930

4 Preliminary Results

4.1 Results of One-Dimensional Compression Test

In the experiment, the stress-strain curve of the CS-rubber mixture changes greatly as the rubber content changed. In the overall proneness, it is found that the stress-strain curve is basically linear growth (Fig. 3). When the tests apply to the same strain, the greater the rubber content is, the smaller the axial stress can be withstood the CS-rubber mixture.

In the early stage of deformation, the deformation is caused by the compression of rubber, which is due to the high compressibility of rubber. In the later stage, the broken CS particles lead to compression. As is shown in the stress-strain loading-unloading curve (Fig. 3), the area of envelope was smaller with higher rubber content, indicating a lower energy dissipation and this may result in an amelioration of particle breakage of CS.

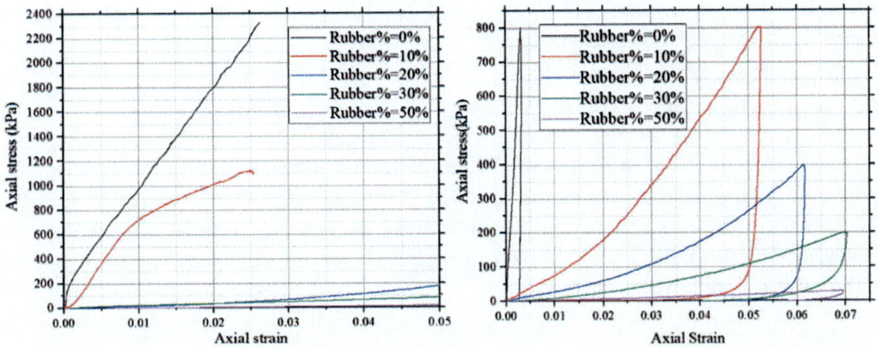

Fig. 3. Stress-strain curves under different rubber contents and stress-strain loading-unloading curves of materials with different rubber contents

This rule is similar to that of rubber-sand mixture. The addition of rubber reduces the shear strength and unit weight of the sample, and improves the compressibility and ductility [14]. Adding rubber to CS can effectively isolate the vibration [13]. The particle breakage of sand occurs mainly in the dilatancy period [15], and shear dilation effect can improve the shearing strength of CS [9].

In this test, the added rubber may wrap CS particles, which reduces the interlocking action between CS and particle breakage, weakening the shear dilation effect of CS. Therefore, the compressibility of the CS-rubber mixture is improved, and the shear strength is decreased.

4.2 Results of Numerical Simulation

Figure 4 shows the numerical simulation results. It can be found from the numerical simulation results of CS that the stress-strain curve roughly shows a linear growth, with low compressibility and high strength. Different from CS, rubber has had displacement

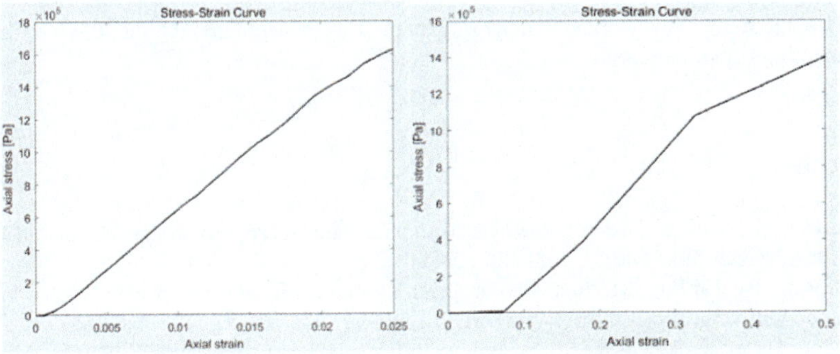

(a) CS (b) Rubber

Fig. 4. One-dimensional compression numerical simulation results before the stress is applied. This may be because rubber has many pores internally and strong ductility, high compressibility.

4.3 Comparison of Results Between Numerical Simulation and Tests

The numerical simulation results of CS are generally consistent with the experimental trend and results, which verifies the feasibility of using MatDEM software to simulate. The simulation results shows that the addition of rubber may indeed enhance the compressibility of the sample and weaken its strength. As rubber content gains, properties of the mixture are close to rubber.

5 Primary Findings and Following Works

In the research, the effect of rubber content on one-dimensional compression characteristics of CS was learned by one-dimensional compression test and numerical simulation software MatDEM. According to the results, the following preliminary summaries were drawn:

(1) When the tests apply to the same strain, the greater the rubber content is, the compressibility of the mixture increases, the axial stress it can withstand decreases, and the characteristics of the mixture are gradually close to rubber. The strength of the material is too low to be recommended as the rubber content is over 10%. The numerical results show that the rubber has strong ductility and compressibility, which verifies the test results: Adding rubber may indeed enhance the compressibility of the sample and weaken the strength.
(2) The increase of rubber content reduces the energy dissipation of the material during compression, which may result in an amelioration of particle breakage.
    This study will further investigate the relationship between rubber and CS particle breakage degree and strength through sand particle morphological analysis and MatDEM simulation.

Acknowledgement. This work was supported by the Key Project Fund (Grant 76140-42030018) of Guangdong Province, China.

References

1. Du-Min, K., et al.: A review of the physical and mechanical properties of coral reef.J. Xiangtan Univ. (Nat. Sci. Ed.) **40**(05), 108–126 (2018)
2. JianPing, W., LinJian, M.: Research progress of long-term safety theory and technology for reef engineering. Protect. Eng. **41**(3), 70–78 (2019)
3. JinJun, H., et al.: Key issues in seismic stability analysis of reef sites in South China Sea. China Earthq. Eng. J. **40**(2), 279–287 (2018)
4. Yaru, L., et al.: Comparative study of coral sand and silica sand in creep under general stress states. Can. Geotech. J. **54**(11), 1601–1611 (2017)

5. Du-Min, K., et al.: Experimental and numerical study on the fragmentation mechanism of a single calcareous sand particle under normal compression. Bull. Eng. Geol. Env. **80**(4), 2875–2888 (2021)
6. Yaru, L., et al.: Particle breakage of calcareous sand at high strain rates. Powder Technol. **366**, 776–787 (2020)
7. Yaru, L., et al.: Moisture effects on the undrained dynamic behavior of calcareous sand at high strain rates. Geotech. Test. J. **42**(3), 725–746 (2019)
8. JiaMing, Z., et al.: Research on influences of particle breakage and dilatancy on shear strength of calcareous sands. Rock Soil Mech. **30**(7), 2043–2048 (2009)
9. ChenYang, Z., et al.: Effect of fine particles content on shear strength of calcareous sand. Rock Soil Mech. **40**(S1), 195–202 (2019)
10. MingJian, H., et al.: Experimental study of the effect of fine particles on permeability of the calcareous sand. Rock Soil Mech. **40**(8), 2925–2930 (2019)
11. JiaMing, Z.: Study on the fundamental mechanical characteristics of calcareous sand and the influence of particle breakage. Inst. Rock Soil Mech. (Chin. Acad. Sci. P.R. China) (2004)
12. Xue, L., JianKun, L.: One-dimensional compression feature and particle crushability behavior of dry calcareous sand considering fine-grained soil content and relative compaction. Bull. Eng. Geol. Env. **80**(5), 4049–4065 (2021)
13. Nader, S., et al.: Geotechnical behaviour of the carbonate sand-granulated tire mixture. Indian J. Geo-Marine Sci. **49**(1), 149–155 (2020)
14. Ahmed, A., et al.: Mechanical properties and time-dependent behaviour of sand-granulated rubber mixtures. Geomech. Geoeng. Int. J. **13**(4), 288–300 (2018)
15. Shahnazari, H., Rezvani, R., Tutunchian, M.A.: The effects of dissipated energy on mechanical behavior of carbonate sands using monotonic triaxial tests. Jpn. Geotech. Soc. Spec. Publ. **2**(9), 397–400 (2016)

Stress Analysis of Narrow Steel Box Continuous Composite Beam Based on ABAQUS

Jing Huang[1], Shuai Liu[1], Yan Zheng[1,2](✉), and Xiaobei Liu[1]

[1] GuangXi Key Laboratory of New Energy and Building Energy Saving, Guilin University of Technology, Guilin 541004, China
2008064@glut.edu.cn

[2] College of Civil Engineering and Architecture, Guilin University of Technology, Guilin 541004, Guangxi, China

Abstract. The ABAQUS was used to analyze the stress of two narrow-width continuous composite steel box beams with different reinforcement ratios, and the rationality of the model was verified by comparing with the experimental data. Based on the model, by changing the strength grade of the composite beam wing plate and introducing the UHPC material, the stress condition of the beam negative moment area was improved. The buckling problem of the bridge deck was solved by filling the whole span with concrete. The simulation results showed that when the reinforcement ratio increased from 1% to 2%, the cracking load and yield load were significantly increased, and when the reinforcement ratio continued to increase, the enhancement effect was not significant. Only increasing the strength of the composite beam wing plate had no obvious optimization effect on the stiffness and yield strength of the composite beam. However, using UHPC material, the cracking load of the beam increased by 57.6%, and the deflection value decreased by 43.8% when reaching the cracking load. Compared with C40 composite beam, the yield load increased by 29.5%. The overall stress of composite beams was significantly improved.

Keywords: Composite beam · Partial filling · UHPC · Reinforced · Steel box buckling

1 Introduction

The steel-concrete composite beams relate to steel beams by bolts and other shear connectors, so that the concrete and the steel beams are subjected to joint force and coordinated deformation. The system makes full use of the compressive properties of concrete and the tensile properties of steel beams, and significantly improves the stiffness and stability of beams. Due to its reasonable structure, the composite beams have broad development prospects, and this kind of structure is widely used in bridges, houses and other structures [1]. Traditional composite beam structure of the concrete in the negative moment region of tensile cracks easily, thus affecting the durability of composite structures. At present, in order to alleviate the problem of crack development in the negative bending moment

W. Guo and K. Qian (Eds.): GBCESC 2022, LNCE 211, pp. 696–707, 2023.
https://doi.org/10.1007/978-981-19-5217-3_69

area, more than using the prestressed method, the constant loading method, optimizing the construction process, and can't pull resistant methods of shear stud connectors [2, 3], although these methods to improve the crack problem of the negative moment area, But for steel box girder of buckling problem is still not solved, therefore, when the Mo Shi Xu et al. [4] put forward such as infill narrow steel box - concrete continuous composite beams, which are at the edge of the shear force of the larger bearings, bearing in and through the partial filling concrete in the negative moment region to assist in steel box joint force, so as to improve the section flexural and shear bearing capacity of composite beams.

With the continuous improvement of finite element theory and the development of computer technology, the nonlinear finite element analysis of structures by numerical simulation can accurately simulate the combined forces of structures. As a kind of commonly used nonlinear finite element analysis software, ABAQUS realizes from simple linear elastic static problems to complex highly nonlinear problems, from the mechanics analysis of individual part to more complex physical field coupling analysis, which can effectively, quasi static analysis, static analysis of elastic-plastic analysis, geometric nonlinear analysis [5], because of its powerful functions. It has been applied in many fields. Naveen-Revanna et al. [6] used ABAQUS to model concrete beams and predict the overall stress process of reinforced concrete beams by deducing the damage plastic parameter (CDP) of ordinary concrete during compression and stretching. It is found that the simulation results of yield load, ultimate load, yield displacement and ultimate displacement are very close to the experimental results, which proves the applicability of the finite element software, and puts forward the guess whether the finite element software is suitable for other types of reinforced concrete beam stress analysis. In order to verify the above conjecture, Swoo-Heonlee et al. [7] used ABAQUS to analyze the overall stress process of post-tensioned prestressed concrete beams. The concrete damage plasticity (CDP) model in ABAQUS is also used to examine the damage behavior of concrete and the structural changes of reinforced concrete beams after external post-tensification (EPT). The results show that, based on the expansion angle, flow potential eccentricity, the ratio of compressive strength to uniaxial compressive strength under biaxial load, the analysis results of external bar strength, deflection and stress are similar to the experimental results, which further verifies the universality of ABAQUS in the stress analysis of beam structure. In the study of composite beams, Kong Tingting [8] used ABAQUS to analyze the internal force of the composite bridge during construction and operation, and obtained the most unfavorable result of the control section of the composite bridge. Ding Faxing et al. [9], through ABAQUS, developed a full-scale finite element model considering the force of the bolt to study its shear performance, and determined the limit shear-span ratio of flexural and shear failure of steel-concrete composite beams. Through the parameter analysis of the full-scale finite element model, Based on the superposition principle, a formula for calculating the shear capacity of composite I-steel-concrete beams considering the shear contribution of concrete slab, steel web and flange plate is proposed. Jinsong-zhu et al. [10] used ABAQUS to analyze the bending force process of steel-UHPC composite beams with waffle plates, and the analysis results were highly consistent with the experimental results. Through parameter study, a calculation model of ultimate bending capacity is proposed. The calculation model of the ultimate bending capacity is very close to the final result simulated by the finite element software, which verifies the accuracy of the finite element software for predicting the mechanical performance of the structure.

At present, a large number of simulation and experimental analysis of composite structures have been carried out, but the analysis of composite beams mostly focuses on I-steel beams, and the simulation of composite box beams, especially the parameter analysis of continuous composite beams with narrow steel boxes is less researched. In this paper, a continuous composite beam model of narrow steel box is established by ABAQUS, on the basis of verifying the validity of the model, the finite element parameter expansion analysis has been carried out to study the influence of the wing strength, reinforcement ratio and filled concrete span on the narrow-width steel box concrete composite beam.

2 Establishment and Verification of Finite Element Model

2.1 Specimen Processing

Two continuous composite beams PFSCB1 and PFSCB2 are designed. The reinforcement ratio of the two beams in the middle span was 1%, the reinforcement ratio in the middle support area was 1% and 1.5% respectively, and the shear connection degree was 1.0.

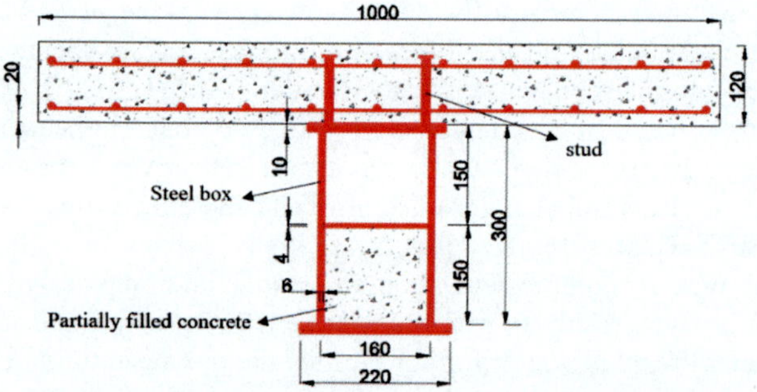

Fig. 1. Cross section of the concrete-filled narrow-width steel box composite continuous beam

As shown in Fig. 1, the test beams had the same sectional sizes, with a total length of 6400 mm and a beam height of 420 mm. Among them, the thickness of the wing plate was 120 mm and the width was 1000 mm, the width of the steel box was 220 mm and the height was 300 mm, and the height of the steel box was 150 mm. C40 concrete was used for the wing plate and the filling part of the steel box. The reinforcement ratio of the trans-central zone was set at 1%, and the reinforcement ratio of the central branch zone was set at 1% and 1.5% respectively. The longitudinal reinforcement was HPB400 hot-rolled ribbed reinforcement with a diameter of 12 mm, and the transverse reinforcement was hot-rolled smooth round reinforcement with a diameter of 8mm. PFSCB1 beam span and side support seat area arranged a layer of reinforcement, the longitudinal bar spacing was 96 mm, the distribution of reinforcement spacing was 200 mm, PFSCB2 in PFSCB1 bearing negative moment area is configured with the second layer length of 500 mm encryption reinforcement, encryption reinforcement and the first layer of steel mesh spacing was 80 mm, the longitudinal bar spacing was 192 mm. The steel box was made of Q235 steel plate welded together.

Vertical stiffeners with a thickness of 10 mm and a width of 30 mm are arranged at the support. The composite beam specimen with a shear connection degree of 1.0 was designed by the plastic method, and the concrete wing plate and the steel box were connected by studs. The studs were connected by 4.8 grade ordinary studs with a diameter of 13 mm. The relevant information of the test beam can be seen in Table 1.

Table 1. Specimen properties

Specimens	Reinforcement ratio		Degree of shear connection	Positive bending moment area stud		Negative moment area stud	
	Cross central region	Middle branch region		Quantity	Spacing	Quantity	Spacing
PFSCB1	1%	1%	1.0	28	116	38	81
PFSCB2	1%	1.5%	1.0	28	116	38	81

2.2 Mechanical Test Setup and Instrumentation

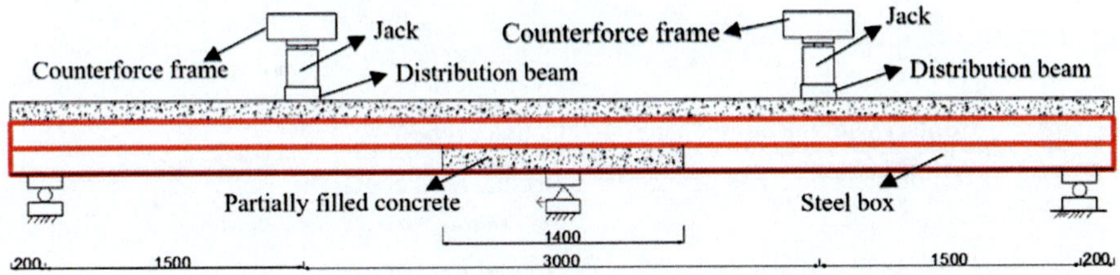

Fig. 2. Test setup

The loading device is shown in Fig. 2. Two 100T hydraulic jacks are set at the mid-span for centralized loading. The distance between the two loading points was 3000 mm, and the load data was recorded by a pressure sensor placed horizontally on the jack. Strain gauges and digital explicit dial gauges are arranged at 1/4 span, 1/2 span and 3/4 span of the beam to measure the deformation, deflection and bearing settlement of the specimen.

2.3 The Establishment of Finite Element Model

Due to the narrow-width steel-box concrete composite beam structure partially filled and the symmetry of loading, the 1/4 composite beam model was established by using

ABAQUS, and the efficiency of calculation analysis was improved by analyzing the 1/4 beam model. The finite element model is shown in Fig. 3

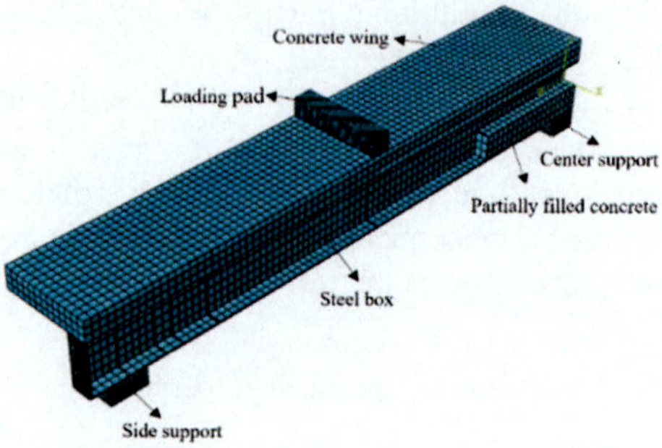

Fig. 3. Finite element model

In the model, steel beams, concrete, support pads and studs are all simulated by linear three-dimensional eight-node reduced integral solid element (C3D8R). This element type was not prone to shear self-locking phenomenon under bending load, and the solution of displacement was more accurate. Three dimensional linear truss element (T3D2) was adopted for reinforcing bar, and the model was meshed by structural meshing technology. In order to better simulate the stress of the bolt and its surrounding concrete, the mesh was encrypted near the bolt. The reinforcement and bolts are embedded into the concrete slab to make the reinforcement joints and bolts work cooperatively with the concrete slab. The main contact Settings used in the model are shown in Table 2.

Table 2. Model contact relation

Contact with the component	Contact type
Top of Steel beam and concrete slab	Normal hard contact Tangential friction coefficient 0.4
Filled concrete and steel box	Normal hard contact Tangential friction coefficient 0.6
Nail bottom and steel beam	Tie
The stud and concrete slab	Embed
Steel and concrete slabs	Embed
Load pad and concrete slab	Tie
Bearing pad and steel box	Tie

In the model, the displacement in the X and Y directions of the model was constrained by the side support, and the displacement in the X, Y and Z directions was constrained by the middle support. Due to symmetry, the 1/4 model was used to simulate, so the normal constraint was applied to the section at its truncation. In order to prevent the local failure of the beam caused by stress concentration, the vertical loading of the jack was simulated by applying concentrated load on the pad.

2.4 Material Parameters

C40 concrete constitutive model adopts the stress-strain curve of concrete structure design code (GB50010-2010) [11]. The values of other parameters are recommended in the literature [12], as shown in Table 3.

Table 3. Parameter selection

Dilation angle	Eccentricity	$\frac{f_{b0}}{f_{c0}}$	k	Viscosity parameter
30	0.1	1.16	0.6667	0.0005

The constitutive relation of steel bar is based on double broken line reinforced plastic model, The stress - strain relationship of reinforcement is as follows:

$$\sigma_s = \begin{cases} E_S\varepsilon & \varepsilon \le \varepsilon_y \\ f_y & \varepsilon_y < \varepsilon \le \varepsilon_t \\ f_y + 0.01E_s(\varepsilon - \varepsilon_t) & \varepsilon_t < \varepsilon \le \varepsilon_u \end{cases} \tag{1}$$

The bolt adopts the triple line constitutive model [13]:

$$\sigma_s = \begin{cases} E_{ss}\varepsilon, & \varepsilon \le \varepsilon_{sy} \\ f_{sy} + 0.01E_{ss}(\varepsilon - \varepsilon_{sy}), & \varepsilon_{sy} < \varepsilon \le \varepsilon_{su} \\ 1.2f_{sy}, \varepsilon_t < \varepsilon \le \varepsilon_u & \varepsilon > \varepsilon_{su} \end{cases} \tag{2}$$

2.5 Finite Element Model Validation

Figure 4 shows the mid-span load-deflection curves of PFSCB1 and PFSCB2. It can be seen from (a) and (b) that the test curve has a good fitting degree with the simulation curve. As shown in Fig. 4(a), the simulated stiffness of PFSCB1 beam at the elastic stage was greater than the experimental value. The reason for this phenomenon was that Tie constraint was used between the bottom surface of the stud and the steel plate during the simulation, which simplifies the stress on the stud. For PFSCB2, the fitting was better in the elastic stage, but there was a certain error between the simulated value and the experimental value in the later stage of beam crack development and the ultimate bearing capacity state. Compared with the experimental value, the slip amount of the simulated value increases by 4.13 mm, mainly because the damage factor after concrete cracking was different from the actual structure.

Taking PFSCB2 as an example, the mid-span load deflection curve of composite beams can be divided into three stages, namely elastic stage, crack development stage and yield stage. When P $< 0.28\,p_u$, the deflection increases linearly with the increase of the load. At this point, the reinforcement and concrete components cooperate with the load, and the structure does not produce obvious damage, and the stiffness remains unchanged. A few cracks appear in the negative moment area of the mid-span support, but none of them are connected, which was an elastic stage. When $0.28\,p_u \le P \le 0.87\,p_u$, it is the crack development stage. In this stage, the cracks are connected in the negative bending moment region of the support, and cracks begin to appear in the positive bending moment region and the number of cracks increases rapidly. The nonlinear characteristics of this stage are gradually strengthened. When P $> 0.87\,p_u$, it enters the ultimate state of bearing capacity, and the load does not change obviously with the change of deflection. Figure 5 for PFSCB2 bridge panel stress distribution, the figure shows that across in the stress of the concrete slab in the negative moment region was the largest, and for the tensile stress, and normal concrete tensile bearing capacity was poor, so central to support the optimization in the negative moment region of the section, to further improve the narrow steel box composite bearing capacity and durability of the bridge.

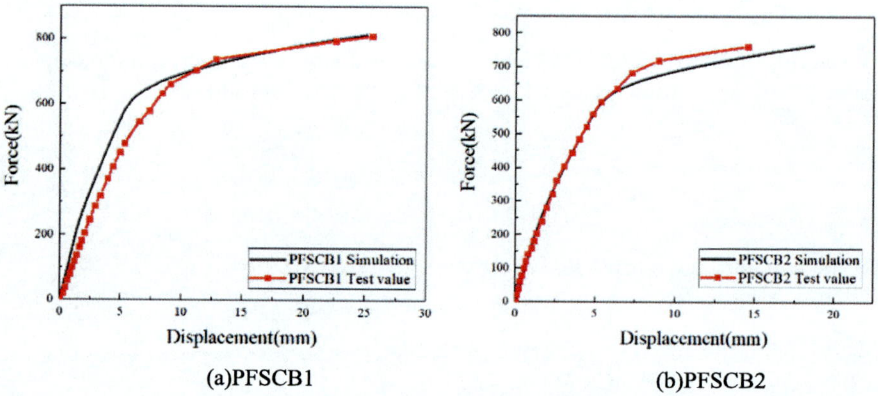

Fig. 4. Comparison between experimental and simulated values of mid-span load-deflection curves

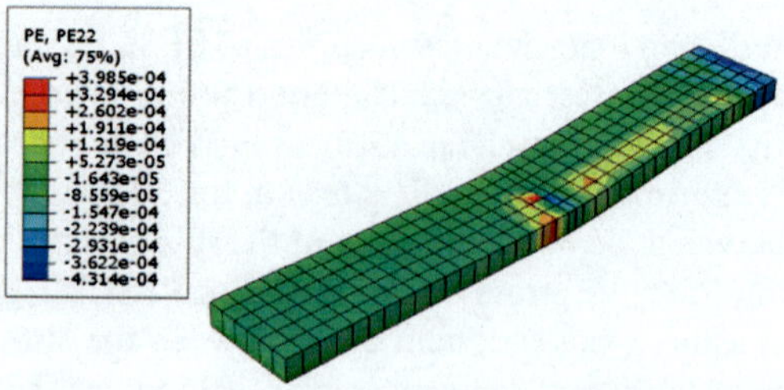

Fig. 5. Vertical stress distribution diagram of PFSCB2 bridge panel

In order to further analyze the tensile damage state of the composite beam, the tensile plastic damage model of the composite beam wing plate and the part partially filled with concrete was derived by using ABAQUS, as shown in Fig. 6. The tensile damage in the negative bending moment region of the concrete wing was much larger than that of the mid-span section, and the filled concrete part also has certain tensile damage. Therefore, it can be concluded that the filled concrete can not only limit the inbending of the steel box, but also bear part of the tensile stress. It can be seen from the tensile damage model, in the part of the upper wing also has certain tensile damage, analysis the reason, is because in the process of loading of specimens, the load was not directly applied on the specimens, but on the first block, after through the block to load of the specimen, block along with load will produce certain tensile damage of bridge panel. In order to solve the problems of wing cracking and steel yielding, two solutions were proposed, namely, increasing the reinforcement ratio and strengthening the strength of concrete wing. The optimal optimization effect of the composite beam structure was obtained through simulation by ABAQUS.

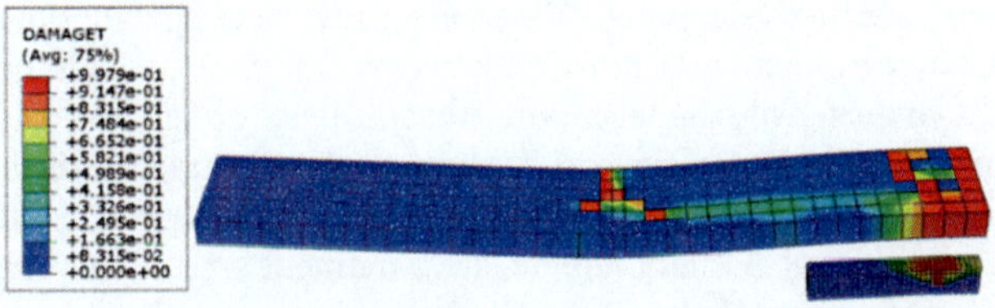

Fig. 6. Tensile damage model of concrete wing and filling part

3 Effect of Panel Reinforcement Ratio

Since the concrete at the lower part of the wing plate in the positive bending moment area was under tension, the cracks developed rapidly during the loading process, and there was a phenomenon of steel yielding. Considering the bottom reinforcement by the extended to the whole across, in the negative moment region of wing look adopted double reinforcement in the negative moment region, and the wing plate reinforcement ratio increased to 2%, 4%, used ABAQUS to simulate, composite load deflection curves, See Fig. 7. When the reinforcement ratio increased from 1% to 2%, the cracking load of the beam increased by 11%, the yield strength increased by 12.1%, and the ultimate bearing capacity increased by 21.2%. The optimization effect was significant, but the deflection of the composite beam at each stage was not significantly improved. When the reinforcement ratio increased from 2% to 4%, the cracking strength is almost unchanged and the yield strength increased by 1.5%, but the strengthening effect was not significant. Therefore, when the reinforcement ratio increases to 2%, the mechanical performance of the beam has a great improvement. When the reinforcement ratio was increased on the was baswas, the performance of the composite beam was poor. Therefore, in practical engineering, the reinforcement ratio of the composite beam partially filled with narrow steel box increases to 2% to improve the bearing capacity and durability of the beam.

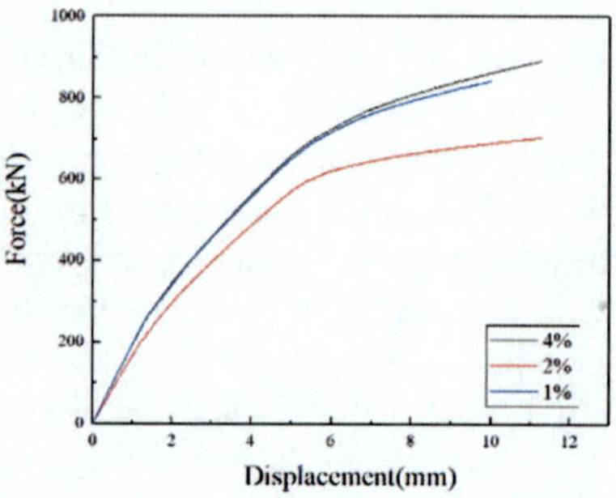

Fig. 7. Load-deflection curves of composite beams with different reinforcement ratios

4 Influence of Concrete Strength of Wing Slab

Obtained by the above analysis, partly filled the narrow steel box continuous composite beams in the negative moment area tensile damage was serious, It could be seen from the dynamic diagram of the tensile damage model that the tensile damage first appeared in the negative bending of the mid-span support, Later, with the increased of the load, the cracks in the positive moment area of the mid-span position began to develop, and the number of cracks increased sharply. Finally, the negative moment area of the mid-span support was pulled and cracked, and the steel box in the positive moment area was buckled. In order to improve the phenomenon that the concrete wing of composite beam structure was easy to crack under tension in the negative moment area of mid-span support, UHPC (Ultra high performance concrete) material and C60 concrete were considered to replace the composite beam wing of C40. UHPC material was characterized by high strength and durability, with compressive strength up to 150–200 MPa and the tensile strength can reach more than 7–10 MPa [14]. Li Wenguang et al. [15] studied the transverse flexural performance of steel-UHPC lightweight composite bridge deck system and found that it can withstand a tensile stress over 18 MPa under the action of negative bending moment, which was much larger than that of ordinary concrete. The steel fiber in UHPC material could still bear certain tensile stress even after the concrete wing was cracked, so as to improve the negative moment area of the wing when it was broken. The UHPC material was used in the negative moment area of the structure, which could improve the anti-cracking performance of the negative moment area at the same time Enhance the durability and bearing capacity of the structure. In order to study the overall optimization effect of UHPC material on the structure, on the baswas of the model of partially filled narrow-width steel box composite beam, the wing plate was replaced with UHPC material to replace the C40 concrete wing plate. The model is shown in Fig. 8. At the same time, in order to study the influence of the concrete strength improvement on composite beam wing slab, C60 concrete was considered to replace C40 concrete for analysis.

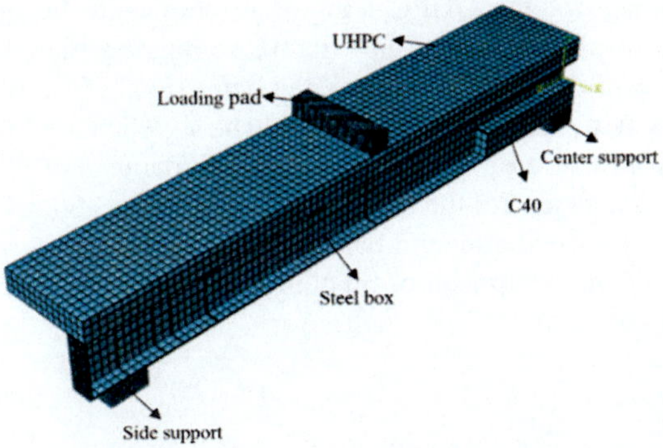

Fig. 8. UHPC composite beam model

The simulation results were compared with the composite beam of C40 concrete wing plate, and the results are shown in Fig. 9. It can be seen from Fig. 9 and simulation data, although the stiffness and yield strength of C60 concrete wing plate were also improved, the improvement effect was not obvious. The improvement of strength has almost no influence on the elastic stage of composite beams, and has no significant influence on the crack development stage and the yield stage of composite beams. The growth rate of deflection after cracking has not been effectively suppressed. Compared with C40 concrete wing, the yield load was increased by 3.4%, and the deflection decreases by only 3% at yield, indicating that the overall optimization effect was not significant.

Compared with normal concrete composite beams, the cracking load of UHPC wing plate composite beams increased by 57.6%, and the deflection of UHPC composite beams decreases by 43.8%. After cracking, the existence of steel fibers in the UHPC wing plate limits the development of cracks, so the post-crack deflection value increased relatively slowly. When the load reached 880 kN, the specimen enters the yield stage. Compared with C40 composite beams, the yield load increased by 29.5%. Therefore, the application of UHPC material to the narrow steel box composite beam structure has obvious optimization effect.

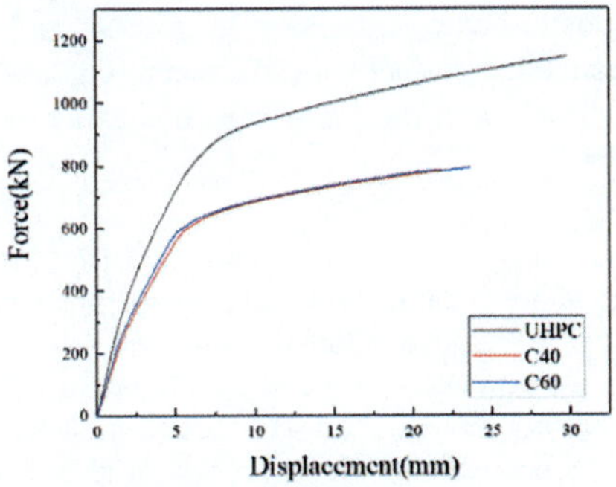

Fig. 9. The influence of the strength of the wing on the composite beam

The tensile damage of the UHPC wing-plate composite beam was analyzed by ABAQUS, and the results are shown in Fig. 10. comparing Figs. 6 and 10, it can be seen that under the same loading conditions, the deflection of beams with UHPC wings under the action of positive bending moment could be significantly improved. Compared with the composite structure of C40 wing, the tensile damage of the wing plate in the negative bending moment area of the bearing was reduced by about 43%, and the tensile damage of the partially filled concrete has also been greatly improved. Therefore, the overall stress state of the composite beam could be greatly improved by using UHPC instead of ordinary concrete.

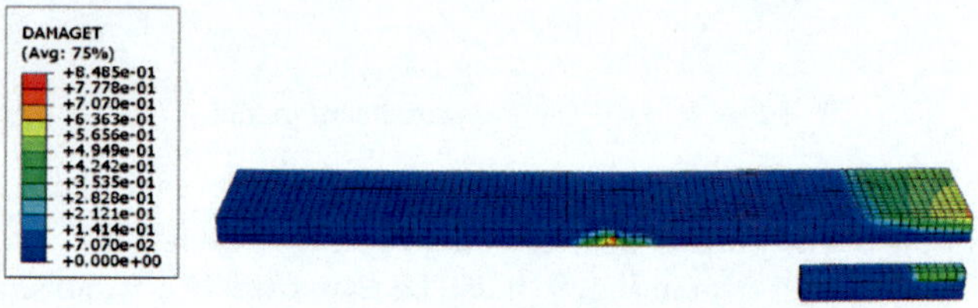

Fig. 10. Tensile damage model of UHPC wing and filling part

5 Conclusion

1) The reinforcement ratio can effectively increase the cracking strength and yield strength of the composite beams, when reinforcement ratio increased from 1% to 2%, the beam crack load increased by 11%, the yield strength increases by 12.1%, and the ultimate bearing capacity It increased by 21.2%, but after reinforcement ratio increased to 2%, the increase of the reinforcement ratio did not significantly improve the mechanical performance of the beam, so in the actual project, the reinforcement ratio of 2% can be used for design.
2) Improving the strength of concrete can improve the stiffness and other properties of composite beams to a certain extent, but the improvement was small.
3) Using UHPC material as the wing plate can greatly improve the mechanical properties of Bridges. Compared with normal concrete composite beams, the cracking load of UHPC composite beams increases by 57.6%, and the deflection value of UHPC composite beams decreases by 43.8% when reaching the cracking load. The deflection increases slowly and the yield load increases by 29.5% compared with C40 composite beams.

Acknowledgements. The authors gratefully acknowledge the financial support provided by the Natural Science Foundation of China (No. 52068012) and the Science and Technology Agency in Guangxi Province (2021GXNSFAA220101) and the project of GuangXi Key Laboratory of New Energy and Building Energy Saving (22-J-21-1). Any opinions, findings and conclusions expressed in this paper do not necessarily reflect the view of Guangxi Key Laboratory of New Energy and Natural Science Foundation of China and Science and the Science Technology Agency in Guangxi Province.

References

1. Lianguang, W.: Theory and Calculation of Steel and Concrete Composite Structures. Science Press, Beijing (2015). (in chinese)
2. Nie, J.G., Li, Y.X., Tao, M.X., et al.: Uplift-restricted and slip-permitted T-shape connectors. J. Bridge Eng. **20**(4), 30–42 (2015)
3. Lin, W., Yoda, T., Taniguchi, N., et al.: Mechanical performance of steel-concrete composite beams subjected to a hogging moment. J. Struct. Eng. **140**(1), 04013031 (2014). (in Chinese)
4. Shixu, M., Zitao, Z., Junyi, L.: Influence of different parameters and conditions on cracks of filling narrow-width steel-box-concrete composite continuous beams. Build. Struct. (2), 7 (2020). (in Chinese)
5. Zhan, L., Jingpin, Z., Yingli, Q., et al.: ABAQUS 6.6 Basic Tutorial and Detailed Examples. China Water Power Press (2008). (in Chinese)
6. Revanna, N., Moy, C., Krevaikas, T.: Verifying a finite element analysis methodology with reinforced concrete beam experiments. J. Appl. Math. Phys. **08**(11), 2549–2556 (2020)
7. Lee, S.H., Abolmaali, A., Shin, K.J., et al.: ABAQUS modeling for post-tensioned concrete beams. J. Build. Eng. **30**, 101273 (2020)
8. Tingting, K.: Theoretical analysis and simulation study on statics performance of narrow steel box girder-concrete composite bridge Shanghai university (2019). (in Chinese)
9. Faxing, D., En, W., Fei, L., et al.: Shear capacity of steel-concrete composite beams considering combined action. Eng. Mech. **38**(07), 86–98 (2021)
10. Jza, B., Xg, A., Jka, B., et al.: Numerical and theoretical research on flexural behavior of steel-UHPC composite beam with waffle-slab system Science-Direct. J. Constr. Steel Res. **171** (2020)
11. China Academy of Building Research: Code for Design of Concrete Structures GB50010-2010, 2015 edn. China Architecture and Architecture Press (2016)
12. Zinan, L., Xudong, S., Qiuheng, Q., et al.: Static test and finite element simulation of transverse bending of steel-ultra high performance concrete composite slab. J. Zhejiang Univ. (Eng.) **52**(10), 1954–1963 (2018). (in Chinese)
13. Faxing, D., Ming, N., Yongzhi, G., et al.: Experimental study on slip performance of stud shear connector and calculation of shear capacity. J. Build. Struct. **35**(09), 98–106 (2014). (in chinese)
14. Junyan, W., Li, P., Jun, G., Yuan, G., et al.: Study on axial tension performance and crack width control ability of UHPC. J. Harbin Inst. Technol. **49**(12), 165–169 (2017)
15. Wenguang, L., Xu, Dong, S., Heng, F., et al.: Experimental study on flexural performance of steel-UHPC composite plate. China Civil Eng. J. **48**(11), 93–102 (2015). (in Chinese)

Optimization Design and Simulation of TMD for a Wind Turbine Based on SIMPACK

Yang Li[1] and Chao Wang[2(✉)]

[1] CSSC Haizhuang Wind Power Co., Ltd., Chongqing 401123, China
[2] School of Civil and Hydraulic Engineering, Huazhong University of Science and Technology, Wuhan 430074, Hubei, China
wangchao_wc@hust.edu.cn

Abstract. TMD plays a great role in the wind turbine vibration control under various working conditions, which will reduce the fatigue damage and prolong the service life. However, previous research and design generally adopted the finite element simulation, which is not consistent with actual project in model and load calculation, causing difficulty when applying technology transformation. Meanwhile, the damping performance of TMD in all directions and limitation of installation space in nacelle need to be studied further. Consequently, this study proposes a new TMD design, and then completes the combined model of turbine and TMD in the commercial software SIMPACK. Innovatively, typical working conditions are analyzed to establish the radial basis function neural network, which contribute to global optimization of TMD by using genetic algorithm. Besides, comparison of simulation results with and without dampers is given to verify the performance of designed TMD, taking the reduction rate of time history standard deviation as the index to evaluate damping effect. The final results indicate that the installation of optimized TMD can reduce the average force at tower bottom by 23.84% and the average bending moment by 18.29%. The results have been applied to TMD design in real engineering.

Keywords: Wind turbine · TMD · Vibration control

1 Introduction

With the upgrading of the global energy industry, wind energy has been widely used as an great clean energy in recent years. Among them, wind turbine plays a great role in harvest wind energy. Owing to the working conditions, vibration control of wind turbine matters in correlation technology. Tuned mass damper (TMD), which is developed earlier, can be simply incorporated into wind turbine structure than other energy dissipation devices. Further, it is of great significance to improve vibration-resistance performance and prolong the service life of wind turbine by investigating TMD.

In the development of wind turbine generation, with the increase of turbine size, its vibration and fatigue load control has always been the main question. Consequently, as an effective vibration-resistance measure, lots of study on TMD has been carried out. Rezaei et al. [1] found that the fatigue load of wind turbine is directly related to the

W. Guo and K. Qian (Eds.): GBCESC 2022, LNCE 211, pp. 708–715, 2023.
https://doi.org/10.1007/978-981-19-5217-3_70

vibration bending stress which decreased with the increase of damping. Chen et al. [2] investigated the application of fore–aft tuned mass damper in nacelle for passive control. The results indicated that TMD vibration control obtained the best performance when the natural frequency of TMD was close to the first-order natural frequency of tower. Moreover, A large number of studies on different types of TMD were also carried out. Suvash et al. [3] proposed a pendulum impact tuned mass damper (PTMD) based on Hertz contact law. Subsequently, Colherinhas et al. [4] completed the TMD modeling of passive pendulum based on ANSYS, with the parameters optimized by genetic algorithm. Pirner [5] proposed a damping theory and model of horizontal motion that was compared with the pendulum model to verify its damping-efficiency. Both Zhang et al. [6] and Zhu et al. [7] have studied tuned mass-damper-inerter (TMDI), the former studied for the mitigation of edgewise blade vibrations in wind turbines, while the latter researched on wind-induced vibration control of linked high-rise buildings. They both verified better performance of TMDI compared with TMD. Etedali et al. [8] proposed the optimization of friction tuned mass damper (FTMD) for tall buildings under near-field earthquakes including SSI effects. Mensah et al. [9] applied TLD to the vibration control of wind turbine and verified its efficiency.

However, in previous research and design of TMD, the finite element simulation was generally adopted, of which the whole design process was complicated. It is difficult to be consistent with actual project in model and load calculation when applied technology transformation. Meanwhile, with the limitation of space in turbine nacelle, previous research seldom considered influence of the complex finite element simulation to TMD system installation. Previous research had focused on the vibration control of fore–aft direction, while the vibration-damping performance of TMD in all directions needs to be further studied.

Therefore, this study proposes a new TMD structure based on the commercial software SIMPACK and obtains the performance of the TMD through completing the modeling and standard working condition calculation according to technical standard. First, according to the field measured data of wind turbine vibration, the relevant parameters of the first-order mode are extracted, and model building of H136-2.2 MW wind turbine is completed in SIMPACK 2020. Based on the TMD entity model, TMD conceptual design is completed according to the derivation of theoretical formula, as shown in Fig. 1. Next, according to the IEC-1.2 working condition, a large number of damper parameter combinations are simulated and analyzed, and the radial basis function neural network is established according to the results. The damper parameters are globally optimized by genetic algorithm along with the deeper TMD conceptual design parameter combination. Then, on the basis of the TMD conceptual design parameter combination, the TMD system is designed in detail, including its hanging point position in tower, spring system and limit system, which are combined with turbine model. Last, this study compares the tower top displacement, tower bottom force and tower bottom bending moment in simulation results under various working conditions with and without damper, taking the reduction rate of the standard deviation of time history data as the index to evaluate the damping performance of TMD. The calculation can be used as the parameter reference in real engineering.

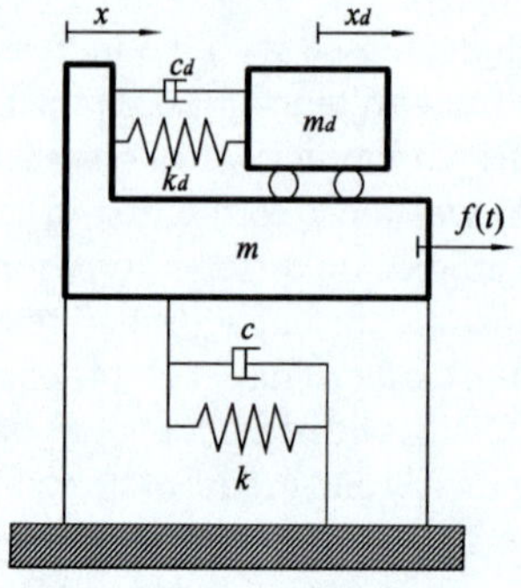

Fig. 1. TMD in SDOF system.

2 Model

2.1 Wind Turbine Model

This study imports the parameters of H136-2.2 MW wind turbine model in the commercial software GH bladed, and completes modeling based on the construction logic of body in SIMPACK. The rotor diameter of wind turbine model is 136.6 m, and the center hub height is 89.391 m, with a tilt angle of 5°. The model writes RBL file for SIMPACK blade modeling according to the blade attribute setting in GH-bladed, and defines the added mass on blades. Besides, taking the influence of unbalanced mass into account, the model in Simpack can ensure that the moment of inertia applied to hub center is equal. Finally, the whole turbine mass in SIMPACK is 350463 kg and that of GH bladed is 350535 kg, with the error of only 0.021% between them, which proves model in SIMPACK is reliable, as can be seen in Fig. 2.

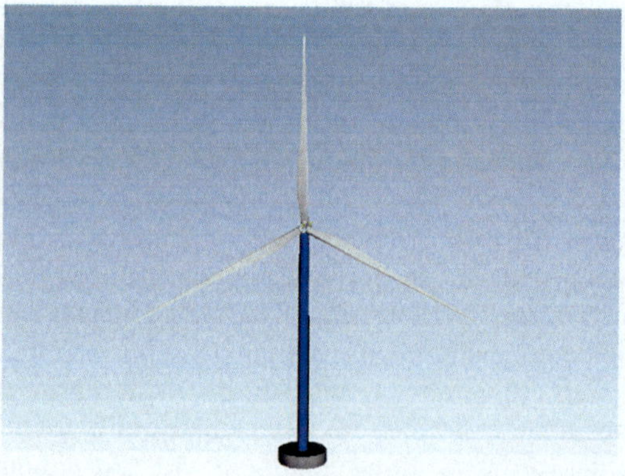

Fig. 2. Combined wind turbine model in Simpack

2.2 TMD Model

To get the results quickly in TMD model design, this study adopts radial basis function neural network to build the agent model of SIMPACK simulation model. On the basis, genetic algorithm is used to optimize the parameters. Finally, the damper parameters are optimized as vibrator mass 2.2 t, stiffness 5.42714 N/mm and damping ratio 0.05. Based on this parameter optimization, the detailed design is given in SIMPACK model: TMD is composed of vibrator spring damping system in two directions. Taking the damper vibrator closer to the tower top as an example, the total weight of all additional masses is 2040 kg, and the corresponding spring stiffness is about 1.3571 N/mm, which is prestressed to ensure that the vibrator is always under the tension of springs on both sides in reciprocating motion. In addition to vibrator stiffness damping system, the limiting device of damper should be designed to prevent the damage caused by rigid collisions. Therefore, the vibrator mass is 2.2 t, with stiffness of 5.428 N/mm and damping ratio of 0.05, which are consistent in both directions. The 3D entity and the corresponding SIMPACK model parameters of TMD is shown in Table 1.

Table 1. Working condition settings of IEC-1.2.

Parameter	Value	Parameter	Value
Vibrator number	2	Vibrator damping coefficient	345 Ns/m
Vibrator mass	2.2 t	Vibrator damping ratio	0.05
Spring number	4	Vibrator sliding distance	450 m
Spring stiffness	1357.1 N/m	Spring height	5.839 m

3 Load Cases

This study completes the calculation of relevant working conditions according to the design standard "IEC61400-1-2005" to verify service performance of the optimized design TMD. Different working condition calculations correspond to different requirements: In the fatigue analysis working condition, the cumulative fatigue values of tower bottom force and bending moment for 20 years are counted; In the extreme value analysis condition, the maximum values of tower bottom force and bending moment are counted; In all analysis conditions, the standard deviation reduction ratio of tower top displacement and the vibration amplitude reduction ratio are counted. Simulation conditions include normal power generation, both power generation and fault, normal shutdown, emergency shutdown and idling conditions, corresponding to standard conditions IEC-1, IEC-2, IEC-4, IEC-5 and IEC-6 respectively. Due to paper limitation, this study only gives the simulation results of each working condition near the turbine rated wind speed (about 8.8 m/s) as a typical working condition. As can be seen in Table 2, taking IEC-1.2 as an example, the specific working condition settings are given and the rest are omitted.

Table 2. Working condition settings of IEC-1.2.

Parameter	Value	Parameter	Value
Name	012_019	Simulation time	650 s
Wind type	NTM	Random seed	1
hub Wind speed	9 m/s	Inflow angle	8°
Turbulence intensity	13.72%	Vertical shear coefficient	0.23

4 Simulation Results

Before dynamic simulation analysis, the modal analysis of turbine tower with TMD is carried out to check the change of wind turbine natural vibration mode damping ratio after TMD is installed. Along with installation of TMD on turbine tower top, the turbine tower modal analysis is conducted. The modal comparison is shown in Table 3. As can be seen in Table 3, the tower first-order modes of downwind and lateral vibration are similar to the turbine first-order lateral frequency of natural vibration without TMD, while the corresponding damping ratio is increased by 11 times. However, the whole wind turbine damping mainly consists of aerodynamic damping provided by blades when wind turbine is in the normal power generation condition, limiting the damping effect provided by downwind damper. On the whole, with the installation of TMD, the whole turbine tower damping ratio can be increased by about 11 times, which plays an obvious role in the timely dissipation of turbine vibration energy and the prevention of turbine vibration divergence under resonance conditions.

Table 3. Tower first-order modal parameters with or without damper.

Case	Without TMD	With TMD
Lateral frequency	0.251842	0.257693
Lateral damping	0.005000	0.055200
Downwind frequency	0.254522	0.258886
Downwind damping	0.005000	0.055200

Taking the typical working condition IEC-1.2 as an example, the time history comparison of tower top displacement, tower bottom force and bending moment of wind turbine with or without damper is given, as shown in Fig. 3. It can be seen that after installing TMD, the tower top displacement, tower top acceleration, tower bottom force and tower bottom bending moment are reduced to varying degrees under different working conditions. Generally speaking, the designed TMD has an obvious inhibitory effect on the wind turbine tower vibration.

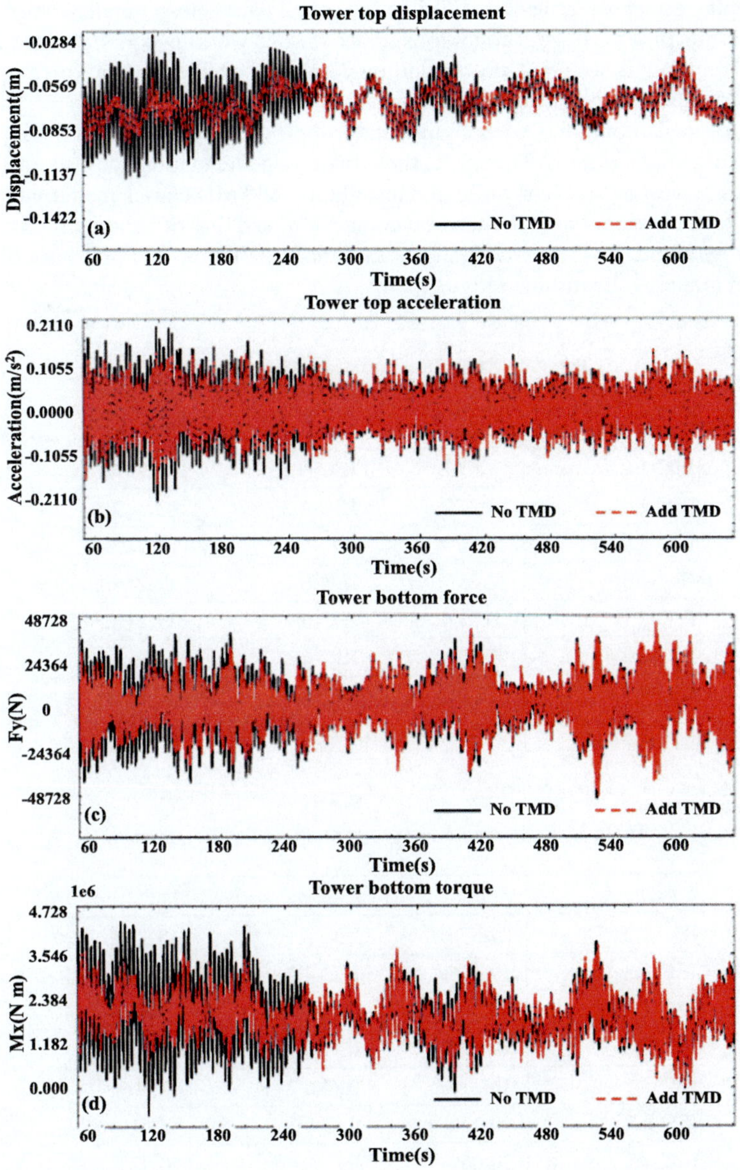

Fig. 3. Time history comparison of IEC-1.2 (a) tower top displacement; (b) tower top acceleration; (c) tower bottom force; (d) tower bottom torque.

Aiming at the working conditions above, through the simulation analysis of 124 working conditions including normal power generation, both power generation and fault, normal shutdown, emergency shutdown and idling, this study quantitatively evaluates the damping effect performance of TMD with actual parameters. Finally statistical evaluation results of typical wind conditions under various working conditions, as shown in Tables 4 and 5. It is seen that after installing damper, the tower bottom force is reduced by 23.84% and the tower bottom bending moment is reduced by 18.29%. The whole turbine tower damping has been significantly improved, from 0.005 to 0.055, with an increase of about 11 times. Therefore, the rate of load attenuation has also been greatly improved at least 57%. Meanwhile, the installation of TMD causes the fatigue value of force at tower bottom being reduced by about 53%, and that of bending moment being reduced by about 28.43%, which shows that the installation of TMD has an obvious effect on reducing the turbine fatigue load.

Table 4. Reduction rate of tower bottom force.

Case name	Value	Case name	Value
IEC-1.2	1.67%	IEC-4.1	8.90%
IEC-1.3	1.75%	IEC-4.2	43.38%
IEC-1.4	−0.96%	IEC-5.1	1.91%
IEC-1.5	48.25%	IEC-6.1	11.24%
IEC-2.1	6.21%	IEC-6.2	11.24%
IEC-2.2	−0.33%	IEC-6.3	5.75%
IEC-2.3	57.69%	IEC-6.4	52.76%
IEC-2.4	7.89%	IEC-1.2 F	53.23%

Table 5. Reduction rate of tower bottom bending moment.

Case name	Value	Case name	Value
IEC-1.2	27.55%	IEC-4.1	13.32%
IEC-1.3	9.43%	IEC-4.2	27.89%
IEC-1.4	−0.56%	IEC-5.1	3.50%
IEC-1.5	19.64%	IEC-6.1	11.50%
IEC-2.1	16.71%	IEC-6.2	11.50%
IEC-2.2	−0.22%	IEC-6.3	5.61%
IEC-2.3	20.66%	IEC-6.4	50.92%
IEC-2.4	9.36%	IEC-1.2 F	28.43%

5 Conclusion

This study completes the modeling of H136-2.2 MW wind turbine and TMD in the whole combined model based on SIMPACK. Taking the nonlinear and time-varying characteristics of wind turbine into account, TMD damping effect performance is evaluated through establishing the radial basis function neural network agent model, to optimize TMD parameters by genetic algorithm lastly. Subsequently, checking calculation of each working condition is given according to the design standard "IEC614001-2005". The final results indicate that the installation of optimized TMD can reduce the average force at tower bottom by 23.84% and the average bending moment by 18.29%.

References

1. Ramtin, R., Paul, F., Philippe, D.: Fatigue life sensitivity of monopile-supported offshore wind turbines to damping. Renew. Energy **123**, 450–459 (2018)
2. Yiming, C., Xin, J., Mengjie, L., Peng, C., Shuang, W.: Vibration reduction methods of large-scale wind turbines based on system-TMD coupled algorithm. Ocean Eng. **226**, 108832 (2021)
3. Suvash, C., Aly Mousaad, A.: Vibration attenuation in wind turbines: a proposed robust pendulum pounding TMD. Eng. Struct. **233**, 111891 (2021)
4. Colherinhas, G.B., Shzu, M.A., Avila, S.M., de Morais, M.V.: Wind tower vibration controlled by a pendulum TMD using genetic optimization: beam modelling. Procedia Eng. **199**, 1623–1628 (2017)
5. Miroš, P.: Actual behaviour of a ball vibration absorber. J. Wind Eng. Ind. Aerodyn. **90**, 987–1005 (2002)
6. Zili, Z., Breiffni, F.: Tuned mass-damper-inerter (TMDI) for suppressing edgewise vibrations of wind turbine blades. Eng. Struct. **221**, 110928 (2020)
7. Zhiwen, Z., Wei, L., Qinhua, W., Nayandeep, T., Budhaditya, H.: Study on wind-induced vibration control of linked high-rise buildings by using TMDI. J. Wind Eng. Ind. Aerodyn. **205**, 104306 (2020)
8. Sadegh, E., Morteza, A., Mohammad, S.: MOCS-based optimum design of TMD and FTMD for tall buildings under near-field earthquakes including SSI effects. Soil Dyn. Earthq. Eng. **119**, 36–50 (2019)
9. Mensah, A.F., Dueñas-Osorio, L.: Improved reliability of wind turbine towers with tuned liquid column dampers (TLCDs). Struct. Saf. **47**, 78–86 (2014)

Measurement and Analysis of Flow Velocity in a Large Air Duct Elbow System

Yi Tang[1], Sen Yu[2], Xiaocheng Cao[1], and Haidong Wang[2](✉)

[1] Shanghai Installation Engineering Group Co., LTD., Shanghai 200080, China
[2] University of Shanghai for Science and Technology, Shanghai 200093, China
whd@usst.edu.cn

Abstract. Increasing large air ducts (hydraulic diameter over 2 m) have been applied in various projects. In the air conditioning system composed of large air ducts, the air distribution of air ducts should be thoroughly analyzed by to analyze the influence of diversion plate setting of duct elbow on the pressure loss and energy consumption of the system, so as to realize energy saving of ventilation and air conditioning system of the building. The background project of this paper is the mechanical and electrical installation project of Shanghai XinZhuang Metro Station Project. In this paper, a scaled model test was carried out to measure the pressure loss of duct bend of a large air duct system. By measuring the air velocity and pressure at different positions along the duct under different working conditions, parametric study was conducted to provide guidance to reduce pressure loss and noise of large air duct system with different elbow flow diversion system.

Keywords: Large air Duct · Elbow airflow division · Scaled model · Numerical Simulation

1 Introduction

Large size air duct is extensively used in air distribution system of large building air conditioning systems, in which duct elbow accounts for a high proportion of the pressure loss as well as fan power consumption. Proper optimization of the right angle elbow can achieve the effect of drag reduction [1], such as shape optimization [2] and airflow deflection. The size, position, shape and number of deflectors will have different effects on the resistance and flow field in the duct system [3, 4]. Appropriate deflectors can improve the effect of flow field [5, 6]. In term of the shape of deflectors, the single-chip serpentine deflectors can achieve good drag reduction effect [7]. The optimization of air duct elbow deflectors can be performed through CFD numerical simulation [8] and necessary experimental validation is required. In addition, noise is another issue of duct elbow that can be tackled through resistance reduction through internal airflow deflector [9]. In this paper, the airflow distribution in a scaled experiment is conducted to provide some insights into resistance reduction.

W. Guo and K. Qian (Eds.): GBCESC 2022, LNCE 211, pp. 716–723, 2023.
https://doi.org/10.1007/978-981-19-5217-3_71

2 Experiment Study of Duct Elbow Pressure Loss

2.1 Specification of the Scaled Model Experiment

In this study, a scaled model air duct was firstly used for experiments, and under the guidance of similarity theory, the experimental conditions of pipeline with different wind speeds were carried out in the scaled model. The inlet section is 7.4 m long and the outlet section is 3.5 m long in the actual project. In order to simulate the flow field of the actual large air duct through the scale model, according to the similarity theory, according to the geometric ratio of 1:6, a set of 630 mm size is built ×200 mm air supply pipe test bench, see Fig. 1.

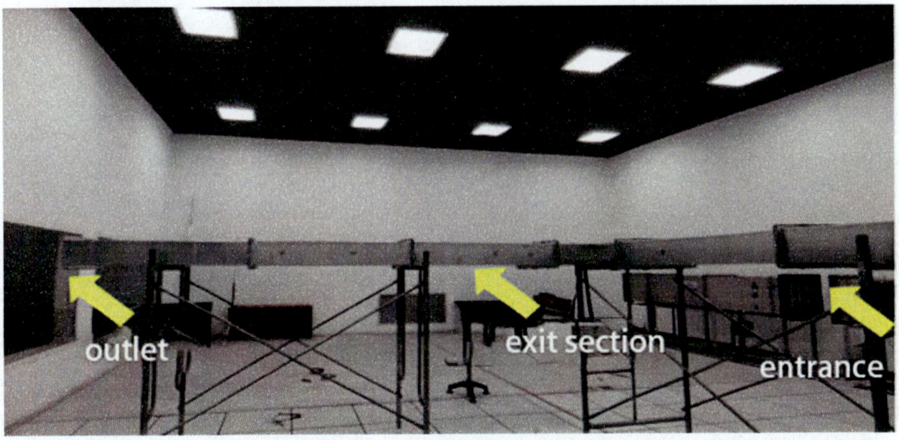

Fig. 1. Scene diagram of the experimental system

2.2 Measurement Positions and Instruments

Local components such as elbow will have a great impact on the air distribution state of the downstream pipe section, especially on the downstream flow field and has little relationship with the upstream wind speed. In the inlet pipe, the wind speed at the inlet of the system is relatively uniform. After the full development of 7.4 m, the cross-section wind speed is relatively consistent with the theoretical distribution. In terms of the arrangement of measuring points, there are few measuring points in the air duct section from the fan outlet to the inlet section of the local component, and a total of 3 groups of measuring points are arranged. These points are mainly to check the velocity distribution at the inlet end and verify whether the inlet section of the elbow is a fully developed flow to ensure the stability of the airflow before the subsequent elbow area. There are more measuring points from the outlet section of the local component to the end of the air duct, with a total of 6 groups of measuring points distributed in the straight section of the air duct to test the distribution of wind speed after the elbow. In order to

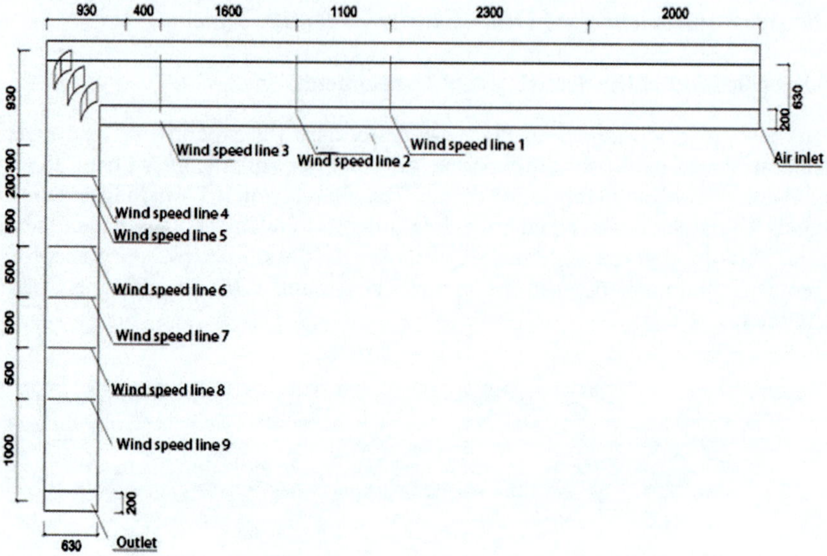

Fig. 2. Position diagram of pipeline measuring points

avoid interference of the measuring instrument probe with the resistance loss effect of local components, no measuring points are set in the elbow, as shown in Fig. 2.

The measuring points of all air duct sections are set at the center of the long edge of the top of the air duct. The cross sections at the end of each measuring point are tested with 10 readings in different directions according to the way of equal distance (depth position) to draw the wind speed measuring line of the cross section, so as to analyze the airflow uniformity at different sections of the duct. The air velocity distribution at the outlet of the pipe and pressure difference before and after the elbow were measured. The air velocity was tested by TSI5815 micro pressure meter. The data measured at 10 points in the inlet section and outlet section are weighted and averaged, and then multiplied by

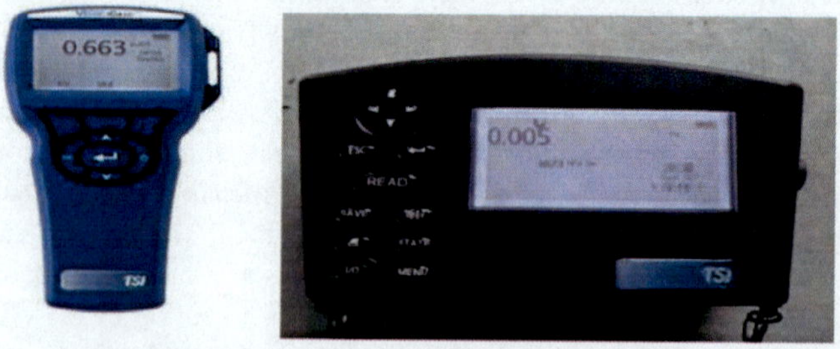

(a) TSI5815 (b) TSIDP-CALC

Fig. 3. Instrumentation of the experiment

the cross-sectional area of the air duct, the total air volume can be calculated. TSIDP-CALC high precision micro pressure gauge is used to measure pressure difference before and after elbow. The experimental instruments and performance parameters are shown in Table 1, and the corresponding instruments are shown in Fig. 3.

Table 1. Experimental instrument and its performance parameters

Instrument	Parameters	Range	Resolution	Precision
TSI5815	Air velocity	0.27–78.7 m/s	0.1 m/s	±1.5% below 10.16 m/s
TSI DP-CALC	Differential pressure	−3735–3735 Pa	0.001 Pa	±2% of reading

2.3 Selection of Experimental Conditions

The scale model test bench is used to effectively control the experimental variables, and the air supply volume can be controlled manually by adjusting the fan frequency. Experimental design conditions are shown in Table 2.

Table 2. Experimental design conditions

Parameter		CASE 1	CASE 2	CASE 3
Deflector	Yes/no	Yes	Yes	Yes
Supply air velocity	m/s	5	8	10
Delivery	m^3/h	2268	3623	4536

Air flow balance is judged by volume flow balance and mass flow balance. In the pre-experiment, the air system is opened first, and the wind speed in the tube is measured by anemometer after the air flow is stabilized, and the fan frequency is adjusted step by step. According to literature [10], the velocity on the outlet section is considered to be uniformly distributed with u0 value, and the flow is turbulent. Therefore, U0 is the air supply velocity vs of the nozzle, and the air supply volume can be obtained by using the formula below.

$$V_S = \mathrm{v}_s * \mathrm{A}_S \tag{1}$$

where:

vS -- the central wind speed measured by the air supply, m/s;

As -- Effective cross-sectional area of air supply main, m^2.

Through wind speed measurement, the air volume in CASE 1 in Table 2 above is the benchmark. Results the nozzle measurement method was used, and the results are shown in Table 3 below.

Table 3. Air volume measurement results of Scheme CASE1

Nozzle velocity	Measuring point	1	2	3	4	5
	m/s	4.4	5.15	5.21	5.25	5.2
	Measuring point	6	7	8	9	10
	m/s	5.24	5.18	5.22	5.16	5
The average velocity	m/s	5.101				
Calculate the air supply	m^3/h	2314				

The unbalance rate of volume air volume measured by the main measurement method is −2.03%, and the error is small. It can be seen that the air volume measured by the main measurement method is more accurate, so the main measurement method is used in subsequent experiments.

3 Analysis of Experimental Data

This chapter collects and analyzes the test data of the velocity field in the air duct under different wind speed conditions. According to result measured at each measuring point under different test conditions, the velocity field distribution diagram in the air duct is drawn. It provides qualitative theoretical basis for analyzing the exploration direction of drag reduction in the pipe in the future. The measurement of the elbow in the experiment is aimed at analyzing the air distribution and resistance loss in the pipe through the experiment.

It can be seen from Fig. 4 that the trend diagram of wind speed at different depths (the depth positions are numbered inward from the outside of the air duct) and Fig. 5 that the fluid has reached uniform flow state when it reaches the measuring line 1 of wind speed, which is a fully developed flow. At the wind speed measurement line 4, the airflow velocity near the outer arc wall is large, indicating that the airflow velocity axis has repeatedly shifted in the elbow. In the downstream straight pipe section, the flow state tends to be homogenized gradually, and the velocity measurement value of the wind velocity measurement line 9 has little change.

Through the velocity variation trend in the experimental model under different working conditions, the flow field distribution of air passing through the elbow is similar within the experimental working condition range of 5–10 m/s. Taking the wind speed flow field diagram of 8 m/s as an example, as shown in Fig. 5, the air flow axis deviates repeatedly at the elbow, and generally deviates to the outer arc wall at the straight pipe section at the outlet.

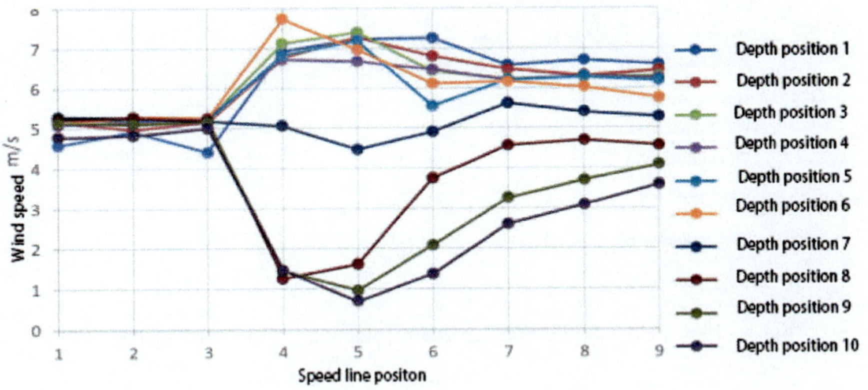

（a）5m/s

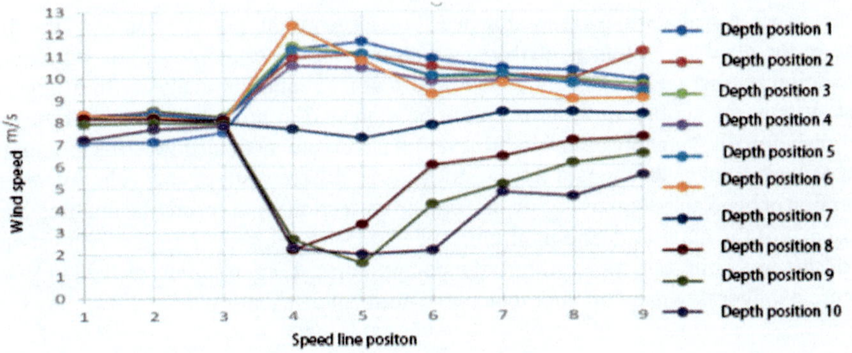

（b）8m/s

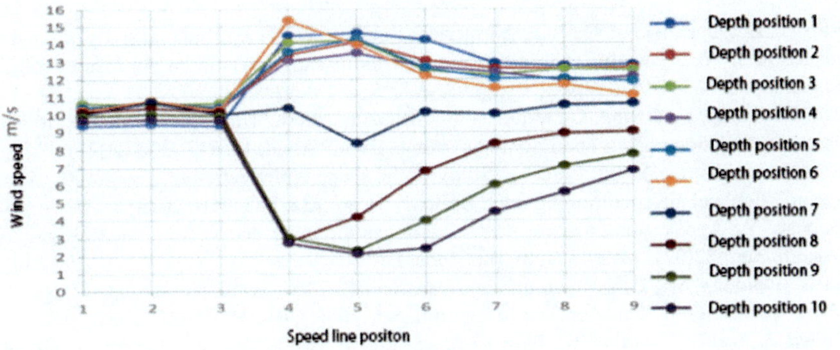

(c)10m/s

Fig. 4. 5 m/s velocity trend diagram of measuring points at different depths under different air supply conditions

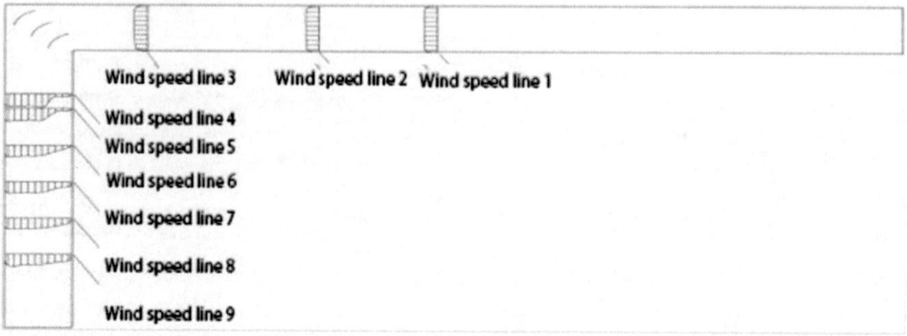

Fig. 5. Flow field diagram of plane wind speed under 8 m/s working conditions

4 Conclusion

In this paper, a scaled model is established for a large air duct in a project, and the flow state in the air duct system is analyze, which provides data reference for subsequent simulation prediction and energy consumption noise reduction analysis. In a large air duct system, the supply air system should be managed to provide a long straight pipe section before flowing into elbow to achieve full development of airflow, so as to stabilize flow velocity and reduce noise. In this study, three different supply air velocity, after a diversion of bend air becomes highly uniform cross section wind speed (the manifestation in point 4 of the maximum and minimum speed difference), proper air diversion should be designed to improve to avoid turbulence and noise. Larger supply air velocity will require a longer downstream duct to achieve uniform distribution.

References

1. Ran, G., Zhiyu, F., Angui, L., et al.: Numerical simulation and experimental study of the drag reduction of 90 elbows for ventilation and air conditioning tubes in an arc form. Procedia Eng. **205**, 3978–3984 (2017)
2. Chi, Z., Angui, L., Jigang, C., et al.: A low-resistance elbow with a bionic sawtooth guide vane in ventilation and air conditioning systems. Build. Simul. **15**, 117–128 (2021)
3. Ran, G., Kaikai, L., Angui, L., et al.: Study of the shape optimization of a tee guide vane in a ventilation and air-conditioning duct. Build. Environ. **132**, 345–356 (2018)
4. Haonan, Y.: Application of arc deflector in air conditioning and ventilation. Refriger. Air Conditioning **22**(02), 43–51 (2022). (in Chinese)
5. Yi, T., Haidong, W., Jian, P.: Simulation research of airflow in the prefabricated large duct elbow with different deflectors. Earth Environ. Sci. **760**(1), 012015 (2021)
6. Shuya, Z., Ke, Z., Yanming, K.: Numerical analysis of particle deposition in 90° rectangular elbow by deflector. J. Donghua Univ. (Nat. Sci. Ed.) **46**(01), 116–122 (2020)
7. Yifan, L., Ran, G., Zhiheng, Z., et al.: Study on resistance reduction in a jugular profiled bend based on entropy increase analysis and the field synergy principle. Build. Environ. **203**, 108102 (2021)

8. Manning, A., Wilson, J., Hanlon, N., et al.: Prediction of duct fitting losses using computational fluid dynamics. HVAC&R Res. **19**(4), 400–411 (2013)
9. Chi, Z., Angui, L., Jiaxin, L., et al.: Radiation noise control of a 90° rectangular elbow in ventilation and air conditioning systems. J. Build. Eng. **37**, 102157 (2021)
10. Zengji, C., Tianyu, L.: Hydrodynamic Pumps and Fans, 4th edn., pp. 270–280. China Construction Industry Press, Beijing (1999). (in Chinese)

Modal Identification of Underwater Precast Self-centering Segmental CFST Bridge Under Horizontal or Vertical Ground Motion Based on Covariance Driven Random Subspace Identification

Mo Zhang[1], Yunfei Wang[1], Dan Zhang[1](✉), Junfei Zhang[1], and Ning Li[2]

[1] School of Civil and Transportation Engineering, Hebei University of Technology, Tianjin 300401, China
zd951767323@tju.edu.cn
[2] School of Civil Engineering, Tianjin University, Tianjin 300072, China

Abstract. Precast segmental self-centering concrete-filled steel tubular (PSCFST) bridges have good seismic performance and resilience. In recent years, researchers have carried out certain research on the seismic performance of this kind of bridge, while there is still a lack of research on modal parameters under earthquakes action. In order to further understand the vibration characteristics of PSCFST bridge under earthquake action, based on the modal parameter identification process of covariance-driven random subspace identification (COV-SSI), this paper analyze the modal parameters of PSCFST bridge under different water depth conditions and different ground motion directions using the shaking table test of PSCFST single-span bridge. In three different water depth environments, namely: water depth of 1.5 m, water depth of 1.8 m, and no water, seismic ground motion in vertical and horizontal (along the direction of the bridge) directions was added through shaking table test to obtain the acceleration response of the bridge. The signal is used as the input data of the covariance random subspace method to obtain the natural frequencies and damping ratios of the first four modes under 6 different working conditions. The analysis results show that the algorithm has a good identification effect on the modal parameters of the PSCFST bridge, in which, water has a small effect on the natural frequency but a large effect on the damping ratio. The COV-SSI method provides theoretical support for the application of structural health monitoring system in PSCFST bridges.

Keywords: Self-centering bridge · Stochastic subspace method · Modal identification

1 Introduction

The segmental prefabricated assembly technology is a prefabricated application in bridge structures, which integrates standardized design, factory production and assembly construction, reduces construction period and labor costs, and improves construction speed

W. Guo and K. Qian (Eds.): GBCESC 2022, LNCE 211, pp. 724–733, 2023.
https://doi.org/10.1007/978-981-19-5217-3_72

[1–3]. Because the prefabricated assembly technology is factory-made and assembled on site, it can reduce the requirements for the construction site and reduce the impact of construction on the surrounding traffic and natural environment [4, 5]. Combined with the characteristics of self-centering and segmental prefabricated features, segmental self-centering bridge came into being.

Under the action of cyclic loads, reinforced concrete precast segmental columns are prone to cracking and failure of joints between segments, concrete spalling and buckling of steel bars at plastic hinges [6]. Researchers found that the combination of concrete filled steel tubular (CFST) and post-tensioned precast can reduce structural damage and resist earthquake-induced damage [7]. A well-designed CFST column has the advantages of high bearing capacity and good ductility, and the outer steel tube can be used as a formwork [8]. Hewes [9] studied the hysteretic behavior of four unbonded precast segmental concrete columns with CFST bottom segment, considering different aspect ratios, initial post-tensioning forces, and steel tube thicknesses. The results show that the overall damage of the column is small. Zhang et al. [8] conducted a quasi-static test study on the seismic performance of PSCFST bridge columns. The PSCFST columns with internal and external energy-dissipation steel bars are proposed and compared with the monolithic reinforced concrete (MRC) columns. The results show that the PSCFST specimen has no obvious damage after the test, and the average residual displacement of the PSCFST specimen is only 1/19.1 that of the MRC reference specimen. Based on the quasi-static test, Zhang et al. [10] established a fiber section analysis model, and then conducted a detailed parametric analysis of the energy dissipation performance and recovery performance of PSCFST columns. Furthermore, by introducing the energy dissipation coefficient λ_{ED} and the elastic coefficient ζ_{SC}, a design method considering both energy dissipation and self-centering is proposed. And it is proposed that when the ratio of energy-dissipation bars to prestressed tendon is 2:1, the self-centering column has an ideal flag-lag relationship and better energy dissipation.

There are many methods of modal parameter identification, which are mainly divided into three categories: time domain method, frequency domain method, and time-frequency method, such as: frequency domain method enhanced frequency domain decomposition method, time domain method characteristic system realization algorithm and random Space method, time-frequency method Hilbert-Huang transform method, etc. [11]. The enhanced frequency domain decomposition method uses the fast Fourier transform to convert the time domain to the frequency domain, while there is an error in the conversion process [12]. The eigensystem implementation algorithm has a rigorous derivation process and a small amount of calculation, but the modal parameter identification accuracy is poor [13]. In recent years, researchers have optimized various algorithms and proposed optimization algorithms such as the deterministic-random subspace method, the multivariate empirical mode decomposition-independent component analysis method, and the structural mode identification method based on machine learning [14–16]. The random subspace algorithm is one of the most widely used modal parameter identification algorithms in the field of bridge structural health monitoring. It does not need to convert time-domain signals and has certain anti-noise properties. It includes the data-driven stochastic subspace method (DATA-SSI) and the covariance-driven stochastic subspace method (COV-SSI) [17]. In the case of the same amount of data, the accuracy of the two algorithms is not much different, but the computational efficiency of the COV-SSI method is significantly higher than that of the DATA-SSI [18]. There is uncertainty in identifying modal parameters in random subspaces, and the stable map method can be used to eliminate spurious modes [19]. In recent years, big data analysis has developed rapidly. Continuous monitoring is carried out through intelligent equipment such as computers, and fully automatic operation and data information processing are carried out [20–22]. The modal parameters of bridge structures, namely frequency and damping ratio, are obtained through real-time data collection and analysis.

This paper analyzes the modal identification of the shaking table test of PSCFST bridge under the action of vertical and horizontal ground motion, respectively, using the process algorithm of covariance-driven random subspace modal identification based on MATLAB platform programming. The seismic responses under the three working conditions of water depth 1.5 m, 1.8 m and no water were analyzed, and the modal parameter analysis was carried out. The first 4 order natural frequencies and damping ratios of each working condition were compared and analyzed, and the effectiveness and robustness of the random subspace algorithm was verified.

2 PSCFST Bridge Shaking Table Test

2.1 Test Specimen

The prototype of the PSCFST bridge structure is one span of the Maobao Expressway Bridge in China. Actually built its 1/4 scale model for Tianjin University in the Key Laboratory of Binhai Civil Engineering Structure and Safety Ministry of Education. The PSCFST bridge model consists of three parts: the superstructure, the columns, and the support. For detail information of the test specimen, please refer to Zhang [19], Fig. 1 shows the test photos of the three water depth conditions.

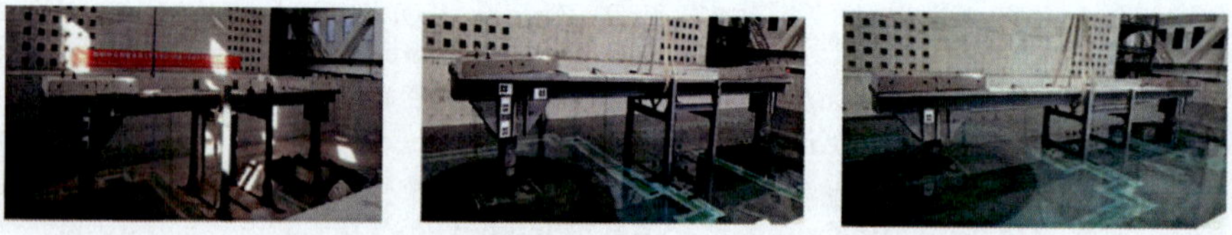

(a) No water (b) 1.5m water (c) 1.8m water

Fig. 1. Test conditions

2.2 Test Setup

Eleven three-direction acceleration sensors (PCB 351BC1) are installed in the PSCFST bridge model, all sensors are connected to the National Instruments (NI) video acquisition device, and synchronously trigger the whole shaking table controller and NI system acquisition. The sampling frequency is 256 Hz, the distribution of measuring points of the instrument is shown in Fig. 2.

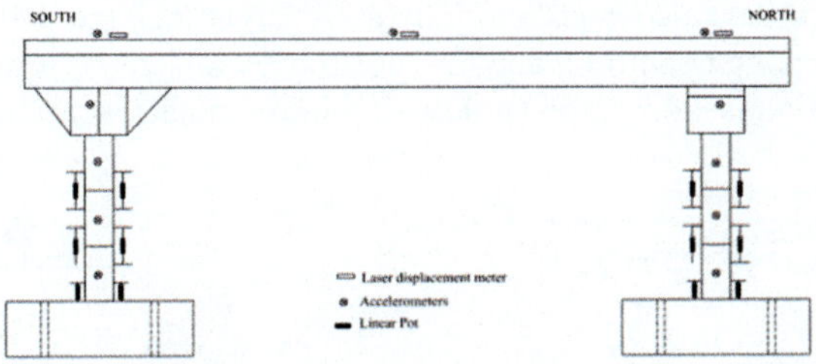

Fig. 2. Acceleration sensor setup

3 Analysis of Test Data

3.1 The Ground Motion

In this study, earthquake ground motion (GM) records were obtained from the Pacific Earthquake Engineering Research Center (PEER) strong motion database (http://peer.berkeley.edu/ngawest2). It is assumed that the bridge is located in a high earthquake-prone area in China, and the site is classified as Class II (according to the soil classification in "Code for Seismic Design of Buildings") [24]. Figure 3 shows the 30% amplitude acceleration time-history curve of the selected GM RSN953.

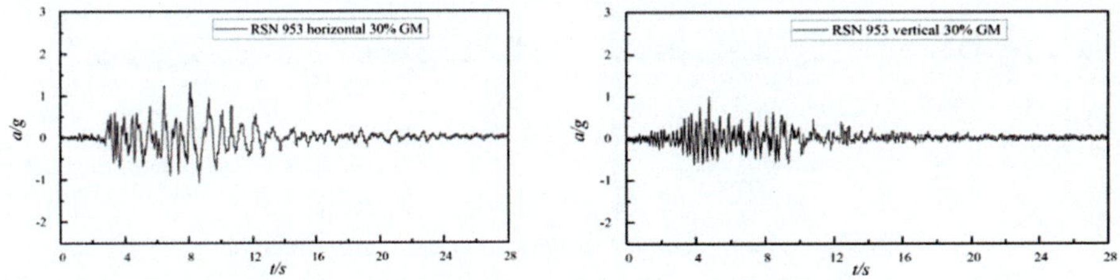

Fig. 3. Acceleration time-history curve of RSN 953

3.2 Acceleration Response of Test Specimen

Figure 4 shows the input ground motion and acceleration response in the horizontal direction under three conditions on the south and north side. The acceleration response at the water depth of 1.5 m fluctuates significantly more than other conditions during the entire sampling period. The largest peak acceleration is 3.09 g, and the smallest peak acceleration is 1.77 g. The acceleration response at the deep water 1.5 m working condition also fluctuates significantly more than other working conditions. Peak acceleration is 3.09 g, and the smallest peak acceleration is 1.92 g. Figure 5 shows the vertical input ground motion and acceleration response in three working conditions of superstructure. The acceleration response at water depth of 1.5 m fluctuates significantly more than other working conditions after 15 s. The maximum peak acceleration is 4.43 g, and the minimum peak acceleration is 1.09 g. In conclusion, the amplitude of the acceleration response of vertical seismic waves is significantly larger than that of horizontal seismic waves in each working condition, and the most obvious is at the water depth of 1.5 m, the peak acceleration is 4.43 g, four times the minimum peak acceleration.

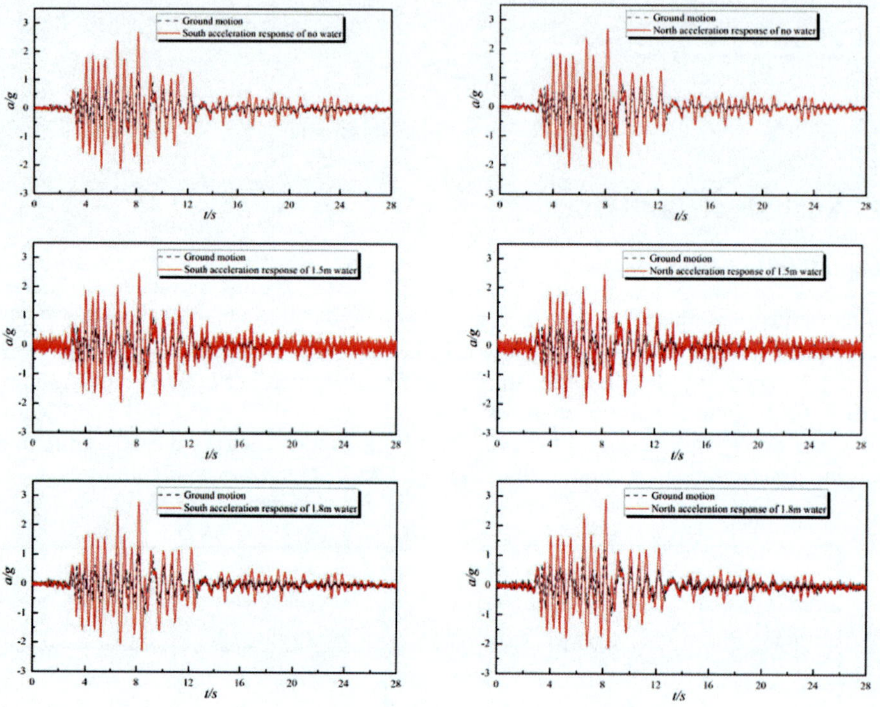

Fig. 4. Horizontal acceleration response of superstructure

3.3 Modal Parameter Identification

The basic idea of covariance-driven stochastic subspace method (COV-SSI) is to construct Toeplitz matrix by using the random response covariance matrix of the system state space equation, and then perform singular value decomposition. Finally, the eigenvalue decomposition of the obtained matrix is performed to obtain the modal parameters of the structure [25, 26].

The fundamental frequency of the test specimen is about 1.14 Hz by preliminary analysis of the acceleration response data using COV-SSI. In order to better obtain the modal parameters of the PSCFST bridge structure, the number of Toeplitz matrix blocks $M = 500$ ($M = 2\beta - 4\beta$, $\beta=$ sampling frequency/structure fundamental frequency), the maximum degree of freedom in the system is 3. In order to identify the modal components of the structure as much as possible, the order of the system is $N = 30$ ($N \geq 2n$, $n =$ the modal order of the study) [27]. Due to the existence of false modes, the

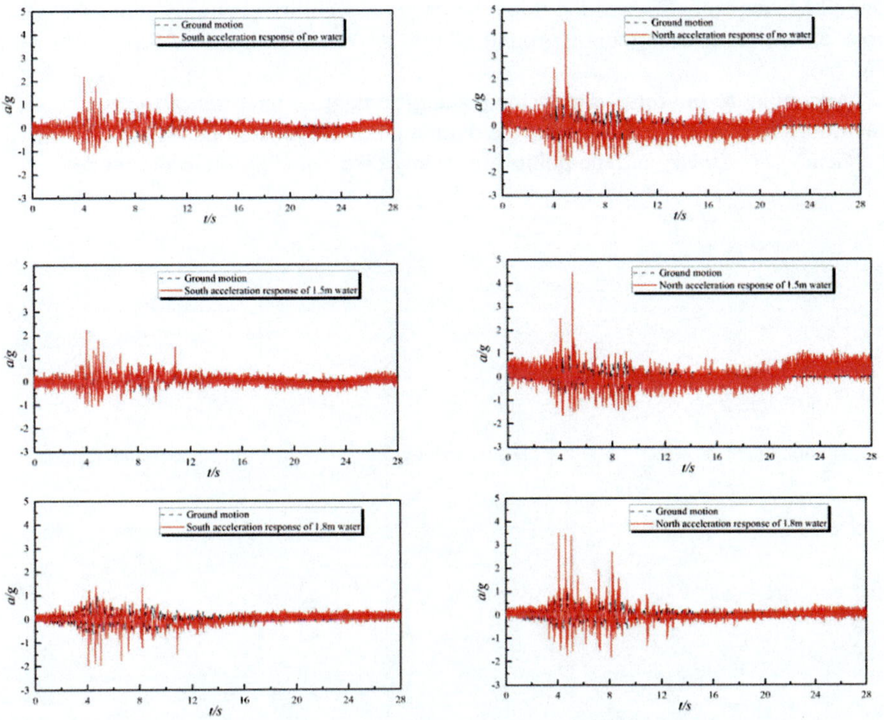

Fig. 5. Vertical acceleration response of superstructure

maximum true damping ratio in the algorithm needs to be determined in advance. The damping ratio of reinforced concrete structures is generally 3%–8%. Considering that the experiment is carried out under the water and no water condition, the damping ratio may be exceeding the range, in order to avoid missing structural modes, the maximum damping ratio was temporarily taken as 40%, and the frequency was taken as 128 Hz

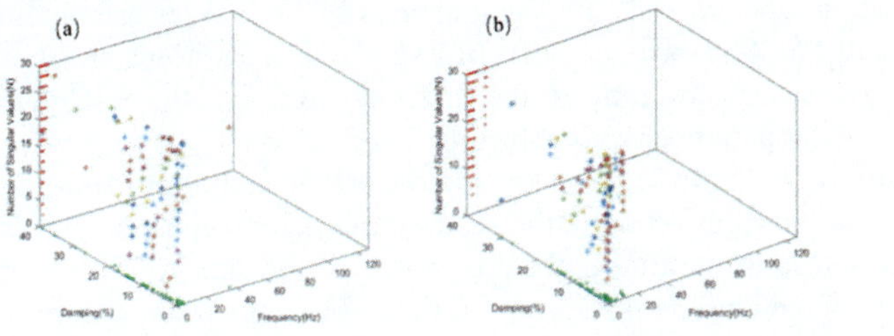

(a) Horizontal ground motion (b)Vertical ground motion

Fig. 6. 3D stability diagram assuming a maximum damping ratio of 40%

(sampling frequency/2), and the three-dimensional stability diagrams under the conditions of vertical and horizontal ground motion are obtained by COV-SSI, as shown in Fig. 6.

According to the three-dimensional stability diagram analysis of the two working conditions in Fig. 7, it is found that the damping ratio of most points is within 20%, the frequency is 0–20 Hz, and the points are mainly distributed in low-order modes.

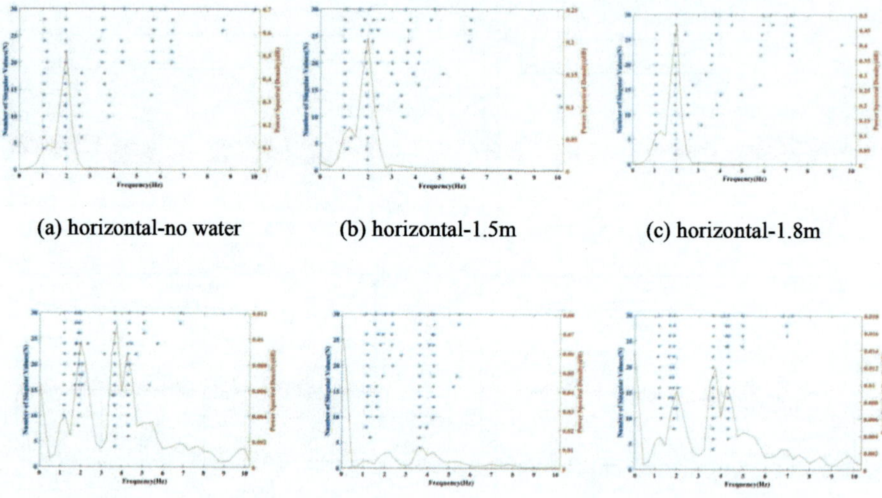

(a) horizontal-no water (b) horizontal-1.5m (c) horizontal-1.8m

(d) vertical-no water (e) vertical-1.5m (f) vertical-1.8m

Fig. 7. Stability diagram of the PSCFST bridge

In the structural vibration analysis, the low-order mode has a greater impact on the structure, and the effect on the structure is smaller as the modal order increases. There are many low-order modal points in the three-dimensional stability diagram, which has a greater impact on the structure, so the maximum damping ratio of the test can be taken as 20%. Using COV-SSI to analyze the data of each working condition in MATLAB, the modal parameters and stability diagrams of each order are obtained. The stability diagrams of PSCFST bridge under different water conditions are shown in Fig. 7. The frequency and damping ratio of the first four modes of the PSCFST bridge are taken from the modal parameters, as shown in Table 1.

According to Table 1, under the three working conditions of no water, water depth 1.5m, and water depth 1.8m, for the south side structure, under horizontal ground motion, the maximum and minimum absolute value of the first four-order natural frequency difference percentage is 4.73% and 0.63%. The maximum damping ratio difference percentage is 519% and the minimum is 80.18%.While, for the north side structure under horizontal ground motion, the maximum and minimum absolute value of the first four-order natural frequency difference percentage is 5.04% and 0.31%, respectively. The maximum percentage difference in damping ratio is 1292% and the minimum is 94.68%. In conclusion, the natural frequency of each order mode of PSCFST bridge has a small change, the maximum difference is 5.04%, and the rest are less than 5%. Within the allowable error range, water has little effect on the natural frequency of the structure, COV- SSI has good robustness and effectiveness for PSCFST bridge.The variation error of damping ratio of PSCFST bridge in each mode is large, the maximum is 1292%, the minimum is 94.68%, and most of the damping ratios are obviously less than 1.5 m water depth under waterless conditions. Under the water condition, the presence of water may increase the damping ratio of the PSCFST bridge.Therefore, it can be obtained that the natural frequency of each order mode of the PSCFST bridge has a small change, the maximum difference is 5.04%, and the rest are less than 5%, within the allowable error range. Water has little effect on the natural frequency of the structure. COV-SSI is proved to have good robustness and effectiveness for PSCFST bridge. The variation of the damping ratio of each mode of PSCFST bridge is large, the maximum is 1292%, and the minimum is 94.68%. Furthermore, the damping ratios in no water conditions are significantly smaller than those in water conditions.

Table 1. Comparative analysis of the first 4th order modes of PSCFST bridges

Direction	Parameter	Water conditions/orders	1st order	2st order	3st order	4st order
x	ω/Hz	No	1.0921	1.966	2.4613	3.5222
		1.5 m	1.0957	1.9464	2.5176	3.6827
		Difference 1/%	0.33%	−1.00%	2.29%	4.56%
		1.8 m	1.0527	1.9567	2.5122	3.3735
		Difference 1/%	−3.61%	−0.47%	2.07%	−4.22%
	ξ/%	No	7.29	5.02	0.59	10.69
		1.5 m	13.16	8.84	2.54	1.68
		Difference 1/%	180.52%	176.10%	430.51%	15.72%
		1.8 m	12.21	5.79	5.9	19.19
		Difference 1/%	67.49%	15.34%	900.00%	79.51%
z	ω/Hz	No	1.1864	1.9757	3.6413	4.358
		1.5 m	1.1658	2.0051	3.6265	4.2642
		Difference 1/%	−1.74%	1.49%	−0.41%	-2.15%
		1.8 m	1.187	1.9822	3.641	4.4033
		Difference 1/%	0.05%	0.33%	−0.01%	1.04%
	ξ/%	No water	5.44	1.17	1.32	2.36
		1.5 m	4.34	2.69	1.38	0.9077
		Difference 1/%	−20.22%	129.91%	4.55%	−61.54%
		1.8 m	7.24	6.47	1.23	2.56
		Difference 1/%	33.09%	452.99%	−6.82%	8.47%

4 Conclusion

Based on the covariance driven random subspace, this paper analyzes the modal parameter identification of PSCFST bridge under different ground motion action directions and different water depths, and the following conclusions can be drawn:

The covariance random subspace method is effective and robust for the identification of modal parameters of PSCFST single-span bridges under different water working conditions.

By analyzing the natural frequency and damping ratio of low-order modes of PSCFST bridge under different working conditions, it is found that the natural frequency changes within 5%.

The existence of water has little effect on the natural frequency of the structure, while has large infuence on the damping ratio. It may be that the presence of water increases the resistance around the bridge and increases the damping ratio of the overall structure.

Acknowledgement. The authors gratefully acknowledge the support of National Key R & D Program of China (grant Nos. 2019YFE0112500 and 2018YFC1504306), National Nature Science Foundation (grant Nos. 52178496, U2139207 and 51678407), and Natural Science Foundation of Hebei Province (Grant Nos. E2019202484 and 20373803D) towards this research.

References

1. Bu, Z.Y.: Investigation of the seismic performance of precast segmental tall bridge columns. Struct. Eng. Mech. **43**(3), 287–309 (2012)
2. Bojorquez, E.: Residual drift demands in moment-resisting steel frames subjected to narrow-band earthquake ground motions. Earthquake Eng. Struct. Dynam. **42**(11), 1583–1598 (2013)
3. Zhang, X., Hao, H., Li, C., Van Do, T.: Experimental study on the behavior of precast segmental column with domed shear key and unbonded Post-Tensioning tendon under impact loading. Eng. Struct. **173**(15), 589–605 (2018)
4. Tong, G.: Cyclic load tests on self-centering concrete pier with external dissipators and enhanced durability. J. Struct. Eng. **142**, 04015088 (2016)
5. Wang, J., Wang, Z., Tang, Y., Liu, T., Zhang, J.: Cyclic loading test of self-centering precast segmental unbonded posttensioned UHPFRC bridge columns. Bull. Earthq. Eng. **16**, 5227–5255 (2018)
6. Kian, M.J.T.: Reinforced concrete shear walls detailed with innovative materials: seismic performance. J. Compos. Const. **22**(6), 04018052 (2018)
7. Guerrini, G.: Seismic behavior of posttensioned self-centering precast concrete dual-shell steel columns. J. Struct. Eng. **141**, 04014115 (2015)
8. Dan, Z.: Experimental investigation and confinement model of composite confined concrete using steel jacket and prestressed steel hoop. Constr. Build. Mater. **256**, 119399 (2020)
9. Hewes, J.T., Priestley, M.J.N.: Seismic design and performance of precast concrete segmental bridge columns. Report No. PEER-SSRP-2001-25, Report, University of California at San Diego, La Jolla, CA (2001)
10. Dan, Z.: Energy dissipation and resilience of precast segmented concrete-filled steel tube self-centering column Parameter study and design methodology. Eng. Struct. **244**, 112747 (2021)

11. Yu, F.L.: Overview of modal parameter identification methods under ambient excitation. Eng. Mech. **31**(04), 46–53 (2014)
12. Jun, M.: Modal identifícation of a building structure by frequency domain decomposition. In: IOP Conference Series: Materials Science and Engineering, vol. 794, no. 1 (2020)
13. Rui, H.: Error analysis of improved NExT-ERA time domain modal identification method. J. Tsinghua Univ. (Nat. Sci. Ed.) **49**(06), 803–810 (2009)
14. Mohamed, B.: Multisensor-based hybrid empirical mode decomposition method towards system identification of structures. Struct. Control. Health Monit. **25**(5), e2147 (2018)
15. Yong Gao, C.: Modal parameter identification of bridge structures based on an improved deterministie-stochastic subspace identification method. J. Vibr. Shock **40**(2), 220–227 (2021)
16. Da Wei, L.: Machine-learning-based methods for output-only structural modal identification. Struct. Control Health Monit. **28**(12), e2843 (2021)
17. Jun Fei, X.: Data-driven and covariance-driven stochastic subspace method differentiation analysis. Vibr. Impact **32**(09), 1–4+20 (2013)
18. Jun, L., Wei Meng, L.: An improved random subspace method for identifying structural modal parameters. J. Xi ' an Univ. Electron. Sci. Technol. **44**(06), 26–30+58 (2017)
19. Reynders, E.: Uncertainty bounds on modal parameters obtained from stochastic subspace identification. Mech. Syst. Signal Process. **22**(4), 948–969 (2007)
20. Shao Fei, J.: Structural health monitoring-intelligent information processing and application. Eng. Mech. **26**(S2), 184–212 (2009)
21. Li Min, S.: Development and prospect of bridge structural health monitoring in the context of big data. China J. Highw. Transport **32**(11), 1–20 (2019)
22. Peng, L.: Real-time online automatic bridge modal identification. J. Vibr. Measur. Diagn. **41**(1), 76–84 (2021)
23. Dan, Z.: Seismic performance of bridge with unbonded posttensioned self-centering segmented concrete-filled steel-tube columns: an underwater shaking table test. Soil Dyn. Earthq. Eng. **138**, 106350 (2020)
24. Ministry of Housing and Urban–Rural Development of the People's Republic of China. GB 50011–2010 code for seismic design of buildings. Architecture and Building Press (2011)
25. Peter, V.O., Bart, D.M.: Subspace Identification for Linear Systems: Theory-Implementation-Applications. Kluwer Academic Publishers, Dordrecht, The Netherlands (1996)
26. Ting, Hua, Y.: Structural Health Monitoring Course. Higher Education Press, (2021)
27. Wei, L.: Parameter optimization of covariance stochastic subspace method based on sensitivity analysis. Eng. Mech. **38**(02), 157–67+78 (2021)

Earthquake Damages to Substation Buildings

Weixian Che(✉), Yanfeng Wang, Baichong Pan, Xinghua Wang, Wenwei Zhu, Jingen Guo, and Hantuo Dong

Grid Planning and Researd Center, Guangdong Power Grid Corporation, CSG, No.8, Shuijungang, Dongfengdong Road, Guangzhou, China
15915791023@139.com

Abstract. The present study highlights the significance of conducting the appropriate seismic design of substation buildings. First, the seismic damage of substation buildings during past several earthquakes is summarized to illustrate the damage features of common substation buildings. Then, the current seismic practice of substation buildings is presented, focusing on the seismic performance of substation buildings in some numerical and experimental studies. At last, the research trend from performance-based design to resilience-based design of substation buildings is discussed to shed a light the development of seismic design of substation buildings.

Keywords: Substation · Seismic design · Resilience · Earthquake damage

1 Introduction

Power system performs as a lifeline engineering that plays an important role in industrial production and daily life of citizens. Electrical system contains a large number of pieces of equipment that are seismically vulnerable. In recent decades, earthquake has been increasingly recognized as a major threat to the electrical system. In 1976, a 7.8 magnitude earthquake hit Tangshan, China, which lead to eight 110 kV substations destroyed and thirty-four transformer room collapsed. In 1994, Northridge, America was hit by a 6.7 magnitude earthquake, several large high-voltage substations suffered severe damage, which cause widespread and isolated power outages in seven districts of the Western United States and British Columbia province [1]. In 1995, the south of Hyogo Prefecture, Japan, forty-eight substations and thirty-eight High-Voltage lines has been destroyed and one million households without power after earthquake due to the 7.2 magnitude earthquake occurred [2]. In 2008, Wenchuan, China, There were 258 substations and 204 lines of 110 kV and above in the System of State Grid Corporation of China, which suffered damage to varying degrees. Maritime space near Puerto Rico was hit by a 6.5 magnitude earthquake in 2020, this earthquake causes a major power generation facility near the quake's epicenter to trigger a protection mechanism, causing power cuts to most of the island [3].

The electrical system contains two basic components: substations (nodes) and power transmission lines (links of the network). Substation plays the role of receiving, transforming, distributing, controlling the power flow and adjusting the voltage power facilities, and is the key link to determine whether the power grid can operate normally. During

W. Guo and K. Qian (Eds.): GBCESC 2022, LNCE 211, pp. 734–740, 2023.
https://doi.org/10.1007/978-981-19-5217-3_73

an earthquake, the seismic damage of substation buildings would cause malfunctioning of the substation, which in turn would significantly cause the failure of the electrical system and large area black-out. Meanwhile, it also greatly increases the difficulty of earthquake relief work as the post-repair and recovery of the buildings would be time-consuming. This study focuses on the seismic performance of substation buildings in past earthquakes. The current research and the future trends are further discussed to shed a light on the seismic investigation of substation buildings.

2 Damage of Substation Building in Past Earthquakes

This section examines the typical vulnerabilities that substation buildings have experienced in past earthquakes. There earthquakes have a moment magnitude larger than 6.1, covering a wide countries/regions from 1995 to 2020.

2.1 Columbia Earthquake (January 25, 1999, $M_{\omega} = 6.2$, the United State)

Two substations were found with significant damage [4]. In Substation Revisit, one of the transformer fall down and break completely and some others transformer were damaged at base. In Substation Sur, the walls and roof were damaged and large movement was detected.

2.2 Gujarat Earthquake (January26, 2001, $M_{\omega} = 7.9$, India)

Substation suffered varying degrees of damage in the high earthquake-intensity area including [5]:

- Substation building were severely damaged or even collapsed.
- Control equipment was seriously damaged or destroyed.
- Fragile equipment in high-voltage substation was badly damaged.

2.3 Bam Earthquake (December 26, 2003, $M_{\omega} = 6.5$, Iran)

Blackout occurred in three important cities (Kano, Raver and Gobat), Nonstructural elements, especially the batteries, of Abbasside Substation were slightly damaged. Some non-structural damage to the power company building [6].

2.4 Wenchuan Earthquake (May12, 2008, $M_{\omega} = 8.0$, China)

Substation buildings suffered some of the following damage in the earthquake through field visits and surveys:

- Many brick concrete substation buildings, which were built pre-1990s and not or fully seismic designed, collapsed during the earthquake. The structural walls were found with many "X" crack or horizontal cracking. The structure brackets were also damaged frequently.

- The reinforced concrete substation buildings, built after 1990s, suffered from slight or moderate damage in V-VII seismic intensity region while were severely damaged or even collapsed in regions of seismic intensity IX and above. The indoor equipment were broken.
- Frame column were broken in lantern shape and the concrete cracked and peeled off; infill wall was easy to crack and even collapse; shear failure of support, shear off of coupling beam of foundation and slip at the joint occurred in joint between the structure and the equipments [7, 8].

2.5 Haiti Earthquake (January 12, 2010, $M_\omega = 7.0$, Haiti)

The damage was found at Ancient Demas and Nouveau Demas, including shifted transformers, oil leaks, etc. The dispatch building was damaged, requiring slight repair [9].

2.6 Chile Earthquake (February 27, 2010, $M_\omega = 8.8$, Chile)

The support members of steel substation were deformed significantly and the joints were damaged. Structural and non-structural failure occurred in concrete structures. Owning to limit space between the substation partition and steel members of equipment, some pounding occurred the partition was damaged [10].

2.7 Christchurch Earthquake (September 4, 2010, $M_\omega = 7.2$, New Zealand)

Post-earthquake field conditions and found that several substations were damaged differently, as follows:

- Papanui Substation: Cracks in the wall distribution around the transformer.
- Springtown substation: Two transformer tripped and a cabinet door broke off.
- Hororata substation: Two transformers tripped and the windows was broken; current transformer toppled and foundation loosened.
- Bromley substation: cracks at the wall of control building was detected [11, 12].

2.8 Discussion

The damage of substation buildings in past earthquake reveals that the damage pattern depends highly on whether it is a masonry structure or a frame structure. For masonry structures, which were mostly constructed pre-2000s, suffered from severe seismic damage of the infilled walls. Typical damage includes the horizontal cracks, diagonal cracks, vertical cracks and "X" fissure damage [7]. For frame structures, to facilitate the placement of the equipment, it usually has a higher floor height, smaller lateral stiffness and less redundancy than masonry structures [13]. Typical damage of frame structures includes the horizontal cracks, diagonal cracks and the cracks in lantern shape with possible concrete spalling of the frame columns; the horizontal cracks and diagonal cracks at the frame beams or beam ends; the damage of the joint of the frame columns and beams [7, 14].

Electrical equipment are mounted on the floor of the substation buildings. Under earthquakes, the floor accelerations would amplify the seismic effects on the equipment. Thus, the substation buildings would also suffer from the damage to the transformer and the joint between building and transformer. The damage to the transformer includes oil leakage due to the strong shaking, short circuit, strong collision between transformers and even the collapse of the transformer. Meanwhile, it is not uncommon to detect the damage at the joint between building and transformer in terms of the shear failure of the bearings, the movement of the equipment bearing and the pull-out bolt at the joint.

3 Development of Seismic Design of Substation Buildings and Research Trends

3.1 The State-Of-The-Art Review

The structural damage of the substation buildings observed from past earthquakes has usually been the impetus for many improvements in seismic engineering codes and practice. Since 1971 when damaging earthquake, i.e. the M6.6 San Fernando Earthquake in USA, the seismic performance of electric power system has studied in the United States. The "IEEE Recommended Practice for Seismic Design of Substations" (IEEE 693-1971) which expounds the seismic design of substation as an independent chapter was compiled by Institute of Electrical and Electronic Engineers (IEEE) involves the seismic design of various type of electrical equipment. After that, the Standard 693 has been updated for several times in 1994, 2005 and the most recently 2018. In China, the "Code for Design of Seismic of Electrical Installations" was published and an independent chapter is used to guide the seismic design of substation building in 1996 in China. The new version, updated in 2013, made some major improvement and complement. The earliest specification in Japan is the "Guide for Seismic Design of Electrical Equipment" (JEAG 5003-1980) in 1980. The specification has also undergone several revisions and the latest version is JEAG 5003 published in 2010. HAZUS, a risk assessment software exploited by the United State, have four damage definitions for substation, which was used to project the damage situation of substation by Indranil Kongar in Christchurch earthquake.

In addition to the necessary attention to design philosophy and design criteria, researchers have made a great attempt to examine the seismic performance of substations. Early in the mid-1970s, Y Chen et al. [15] summarized the research status of building as the main structure and equipment as a second-order system and evaluated the research direction in the future. J Li [16], in 1994, also summarized the research status from several aspects, mechanical model, research methods, application and experimental research and made suggestions for the possible research directions. For the distribution building system composed by structure and equipment, Bo W [7] established the three-dimensional model and analyzed dynamic characteristics, studying the seismic performance of near-fault and concluding the greater damage appeared in near-fault, using base isolation technology to distribution building system achieved remarkable results. Shaking table test method for equipment-structure dynamic interaction was investigated by Xinliang J [17], who concluded that considering the equipment-structure interaction can reflect accurate the real seismic performance. Above studies indicate that the structure-equipment interaction has a pronounced effect on the seismic response of substation system and thus should be taken into consideration in seismic practice.

3.2 Research Trends

In view of the earthquake occurred frequently in recent year and result in irreversible damage to substation, which hindered the rescue work extremely. However, in the last decade, earthquake engineers have given more attention to the safety of substation buildings, often represented by the deformations of structures, rather than the post-earthquake recovery [18], which would be represented by the resistance of the structures. The concept of resilience was originated from ecology proposed by Holling [19] and was introduced in the seismic filed of engineering structure by Bruneau and Cimellaro [20, 21] et al. Since then, special focus have been placed on the strategies of resistance-based design of substation structures. Based on the post-earthquake recovery model of power system built by Cagnan [22], the skeleton curves of recovery was formed and the recovery situation was simulated for power system in Los Angeles. Rushan L [23] divided the resilience of the substation into five levels, which provides a certain standard for the evaluation of resilience level. They also proposed a formula for computing the resilience index of the buildings, taking into account the seismic intensity, the weighting coefficients of buildings and high-voltage electrical equipment. In addition to the resilience study of substation, the economic benefit of seismic strategy is another issue for consideration. Hinow M, Politano D [24, 25]stressed that in addition to the initial construction costs, post-operation and maintenance costs would not be neglected. Karl Telleen, Meijing Z et al. [26, 27] proposed to conduct post-earthquake quick assessment of substations to facilitate emergency rescue and decision making. Meanwhile, the post-earthquake quick assessment would also reduce the economy lose and inconvenience in rescue and life due to the substation damaged.

4 Conclusion

Substation is an important node of the power system and the damage of substation would lead to the failure of the whole power system. This paper summaries the seismic damage to substations under several major earthquakes. It reveals that the building structures, the electrical equipment like the transformers and the connection joint are all vulnerable components under a large earthquake. Thus, the seismic design of substations should be conducted on the buildings but also on the equipment, taking into account the interaction between them.

At present, the seismic design of substation is moving from performance-based design to resilience-based design. This involves the rapid post-earthquake assessment and recovery strategies, providing guidance for post-earthquake decision-making and deriving the formula of the economic benefit. More work are needed with respect to this issue.

References

1. Lund, L.V.: Lifeline utilities performance in the 17 January 1994 Northridge, California, earthquake. Bull. Seismol. Soc. Am. **86**, 11 (1996)
2. Luo, Q.: Damage to life-line systems caused by Hyogoken Nanbu, Japan, earthquake and their recovery. J. Catastrophol. **12**(1), 43–48 (1997)
3. Huayue, L.X.W.: Review of worldwide earthquake disasters and the impacts in 2020. Technol. Earthq. Disaster Prevention **16**, 6 (2021)
4. Asfura, A.P., Flores, P.J.: Quindio, Colombia Earthquake of January 25, 1999: Reconnaissance Report. terremotos (1999)
5. Edwards, C.: Lifeline Damage from the January 26, 2001 M7.7 Gujarat, India Earthquake. Sixth Us Conference & Workshop on Lifeline Earthquake Engineering (2003)
6. Takada, S., Bastami, M., Kuwata, Y., Javanbarg, M.B.: Performance of Electric Power Systems During the Bam Earthquake and Its Fragility Analyses. Memoirs of Construction Engineering Research Institute (2004)
7. Wen, B.: Study on Earthquake Response Analysis and Seismic Control of Switch Building-Electrical Installations System, Xi'an University of Architecture and Technology (2008)
8. Zhang, Z.Y., Zhao, B., Cao, W.W., Yong, L.I., Zheng-Xiang, G.E.: Investigation and Preliminary Analysis of Damages on the Power Grid in the Wenchuan Earthquake of M8.0. Electric Power Technologic Economics (2008)
9. Ba Nk, T.W.: Haiti earthquake PDNA (Post-Disaster Needs Assessment): assessment of damage, losses, general and sectoral needs (2012)
10. Cruz, E.F., Valdivia, D.: Performance of industrial facilities in the Chilean earthquake of 27 February 2010. Struct. Des. Tall Special Build. **20**, 83–101 (2011)
11. Fujisaki, E., Takhirov, S., Xie, Q., Mosalam, K.M.: Seismic vulnerability of power supply: lessons learned from recent earthquakes and future horizons of research. In: 9th International Conference on Structural Dynamics (EURODYN), pp. 345–350 (2014)
12. Watson, N.R.: Impact of the Darfield earthquake on the electrical power system infrastructure. Bull. New Zealand Soc. Earthq. Eng. **43**, 421–424 (2010)
13. Wen, B., Niu, D.T., Zhao, P.: Review of substation seismic performance research. Earthq. Resistant Eng. Retrofitting (2007)
14. Wu, P.C.K.: Investigation and Analysis of the Impact of Earthquake Disaster on Substation. Electrical Applications **30**, 38–41 (2011)
15. Chen, Y., Soong, T.T.: Seismic response of secondary systems. Eng. Struct. **10**, 218–228 (1988)
16. Jie, L.: Review of research on dynamic interaction between structure and equipment. World Inf. Earthq. Eng. **10**, 5–12 (1994)
17. Jiang, X., Zhang, C., Jiang, N., Tang, Z.: Shaking table test method for equipment-structure dynamic interaction. Zhendong yu Chongji/J. Vibr. Shock **38**, 108–115 (2019)
18. Cimellaro, G.P.: Urban Resilience for Emergency Response and Recovery - Fundamental Concepts and Applications. Geotechnical Geological & Earthquake Engineering. Springer, Cham (2016). https://doi.org/10.1007/978-3-319-30656-8
19. Holling, C.S.: Resilience and stability of ecological systems. Annual Rev. Ecol. Syst. **4**, 1–23 (1973)
20. Bruneau, M., Reinhorn, A.: Exploring the concept of seismic resilience for acute care facilities. Earthq. Spectra **23**, 41–62 (2007)
21. Cimellaro, G.P., Reinhorn, A.M., Bruneau, M.: Framework for analytical quantification of disaster resilience. Eng. Struct. **32**, 3639–3649 (2010)
22. Cagnan, Z., Davidson, R.A., Guikema, S.D.: Post-earthquake restoration planning for los angeles electric power. Earthq. Spectra **22**, 589–608 (2006)

23. Liu, R., Shu, R., Zhenxiu, H.U.: Study on the seismic resilience of substations. China Earthq. Eng. J. **41**, 827–833 (2019)
24. Hinow, M., Mevissen, M.: Substation maintenance strategy adaptation for life-cycle cost reduction using genetic algorithm. IEEE Trans. Power Deliv. **26**, 197–204 (2011)
25. Politano, D., Frohlich, K.: Calculation of stress-dependent life cycle costs of a substation Subsystem-demonstrated for controlled energization of unloaded power transformers. IEEE Trans. Power Deliv. **21**, 2032–2038 (2006)
26. Telleen, K., Yin, J.: Seismic Assessment of Buildings, Considering Post-Earthquake Safety. Improving the Seismic Performance of Existing Buildings & Other Structures (2014)
27. Zhang, M.: Study on Rapid Evaluation Method of Electric Power Facility Earthquake Damage and Its Hazard, Institute of Engineering Mechanics, China Earthquake Administration (2009)

Automatic Object Detection of Construction Workers and Machinery Based on Improved YOLOv5

Yuanzhi Xiang, Jiayue Zhao, Wenjing Wu, Caifeng Wen, and Yunzhong Cao(✉)

College of Architecture and Urban-Rural Planning, Sichuan Agricultural University, Chengdu 610000, China
caoyz@sicau.edu.cn

Abstract. Automatic detection and localization of workers and machinery on construction sites through surveillance video is important to supervise on-site safety and construction process, which could develop civil construction management and services. However, it is difficult to detect all instances due to the extremely complex construction environment and numerous multi-scale objects. This paper proposes an improved YOLOv5 model to automatically detect and localize construction workers and 11 common types of construction machinery. Firstly, use the bidirectional feature pyramid network (BiFPN) layer for better multi-scale feature fusion ability; Secondly, 3×3 convolution layer is replaced by RepVGG block, which shows favorable accuracy-speed trade-off. The experimental results indicate that the mAP (mean Average Precision) of our proposed method is 87.32%, which is 2.12% higher, and inference time reduce to 5.7 ms per frame.

Keywords: Construction workers · Machinery · Computer vision · YOLOv5

1 Introduction

On-site construction surveillance video may record a wide range of information about personnel and machines, which is useful for analyzing productivity and monitoring operations from a safety standpoint. However, due to the large variety and number of construction machines on the construction site and the small size displayed in the surveillance video, it is quite hard and inefficient for site managers to keep a constant eye on the surveillance cameras to recognize all the construction machinery, which is not helpful to monitoring on-site safety and the construction process [1–3]. It is demonstrated that deep learning object detection algorithms have ability to recognize a broad variety of instances in images and videos in near real-time. Therefore, construction industry may be able to gain from machine learning by integrating algorithms with on-site surveillance videos [4].

W. Guo and K. Qian (Eds.): GBCESC 2022, LNCE 211, pp. 741–749, 2023.
https://doi.org/10.1007/978-981-19-5217-3_74

In recent years, computer vision and image processing have developed and progressed. Object detection is one of the most crucial and difficult disciplines of computer vision. In last several years, most of the latest object detection algorithms have used convolutional neural networks (CNNs), and these have yielded a large number of promising results. Object detection algorithms can be divided into two types based on the flow of the algorithm: one-stage detectors and two-stage detectors. Faster R-CNN [5] is the most typical two-stage detector. The other type of detector is a one-stage detector, such as the YOLO series [6–8]. The two-stage detector is more accurate at localization and object recognition, but the one-stage detector is quicker at inference. However, there are numerous concerns that remain difficult. One of these is multi-scale and real-time detection, which is especially prominent when the image contains a high number of small-scale objects [9].

To address the above problems, this paper aims to improve the existing YOLOv5 model. On the one hand, add BiFPN [10] layer in order to learn the distribution of weights, which could strengthen the feature fusion ability of different scales. On the other hand, improves the detection speed via structural re-parameterization. The improved YOLOv5 algorithm is used for multi-scale object detection and localization of construction workers and machinery on-site. The classes of object detection include construction workers and 11 common types of construction machinery: excavators, loaders, tower cranes, vehicle cranes, rollers, bulldozers, dump trucks, concrete mixer trucks, pile drivers, pump trucks and hanging baskets.

2 Methods

2.1 Overview of YOLOv5

In accordance with the models depth multiple and layers channel multiple, YOLOv5 can be divided into four different models, including s, m, l and x [11]. In the meantime, YOLO5 is constantly being updated, and this paper uses the 5.0 version of YOLOv5. We use it as our baseline because it is the most notable and convenient one-stage detector. Generally, the basic framework YOLOv5 can be divided into four parts: Input, Backbone, Neck, and Prediction.

2.2 The Improved YOLOv5 Model

This paper uses the smallest model YOLOv5s since it is the fastest and easiest to deploy one. The depth multiple of YOLOv5s is 0.33, while the width multiple is 0.50. The network structure of improved YOLOv5 can be seen in Fig. 1. The red box is the section of network improvements. The origin Conv is replaced by RepVGG block, while the origin Concat is replaced by Concat_BiFPN block.

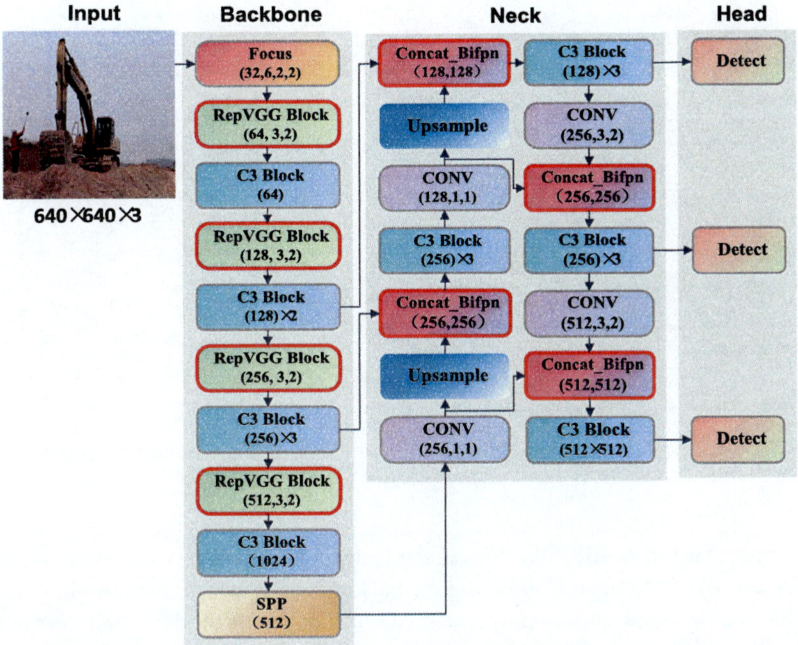

Fig. 1. Network architecture of improved YOLOv5(The **red box** is the section of network improvements)

RepVGG Block. We replaced convolution (3 × 3) with the RepVGG [12] block in the optimized YOLOv5 model. The aim of RepVGG block is to use a multi-branch model during model training (see Fig. 2(a)) and a multi-branch transformed into a single-way model during deployment and inference (see Fig. 2(B)). It uses different network architectures for the training and inference phases of the network, with the training phase being more concerned with accuracy and the inference phase being more concerned with speed. RepVGG heaps 20 layers of 3 × 3 convolution and divides it into 5 stages. The first layer of each stage is the down-sampling, and ReLU is used as the activation function for each convolution layer. Though the optimized model will take more time to train, it would take less time to infer than the original model. Therefore, the small increase in training time is worthwhile.

The network structure is associated with a set of parameters, for example, a convolution layer is represented by a 4th-order kernel tensor. If the parameters of one structure can be converted into another set of parameters coupled by another structure, the former can be replaced with the latter, changing the overall network architecture, in order to realize the aim of structural re-parameterization.

It needs to be noted that RepVGG is an efficient model designed for GPUs and dedicated hardware, seeking high speed and memory saving, with less attention to the number of parameters and theoretical computation. However, the performance on low-calculus devices is not impressive enough.

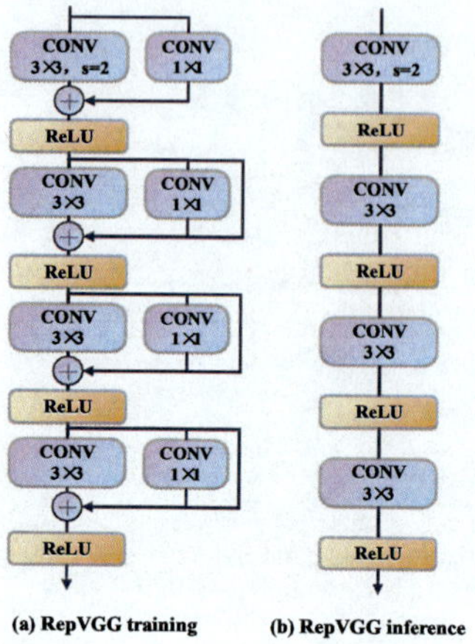

Fig. 2. The structure of RepVGG Block

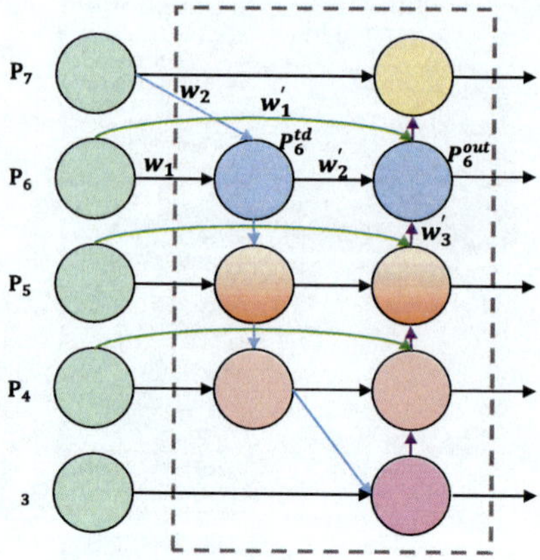

Fig. 3. The structure of BiFPN

Add feature fusion of BiFPN. We use the Concat_BiFPN layers to replace the initial Concat layers in this paper. Combining the backbone network's downsampling features with the Head network's upsampling features, and then fusing the downsampling features in the Head network, so that the network has stronger semantic information, therefore the model could solve the missed detection problem, which caused by small objects and repeated feature extraction in complex scenes. BiFPN is a weighted bidirectional feature pyramid network, which allows simple and fast multi-scale feature fusion. The aim of this structure is to pursue a more efficient multi-scale fusion ability [13]. Figure 3 illustrates the structure of BiFPN. Equation (1) and Eq. (2) show the formulas for fast normalized fusion and BiFPN network structure. w_i (Eq. (1)) is a weight that is learnable, and the Relu activation function, which would be applied every time after w_i. Equation (2) illustrates the BiFPN calculation by the example of the sixth level. On the top to down path, P_6^{td} is the intermediate feature of the sixth level. P_5^{out} and P_6^{out} are the output feature of the fifth and sixth levels on the bottom to up path. P_6^{in}, P_7^{in} are the input of the sixth and seventh levels feature. If the input and output nodes are of the same level, an extra edge is added to fuse more features without increasing cost. Resize is typically downsampling or up-sampling, and Conv is typically a feature processing convolution operation.

In general, BiFPN is equivalent to giving different weights to each layer for fusion, making the network pay more attention to important layers, and reducing some unnecessary layer node connections.

$$P_6^{td} = Conv\left(\frac{w_1 \cdot P_6^{in} + w_2 \cdot Resize\left(P_7^{in}\right)}{w_1 + w_2 + \varepsilon}\right) \quad (1)$$

$$P_6^{out} = Conv\left(\frac{w_1' \cdot P_6^{in} + w_2' \cdot P_6^{td} + w_3' \cdot Resize\left(P_5^{out}\right)}{w_1' + w_2' + w_3' + \varepsilon}\right) \quad (2)$$

3 Experimental Results and Comparative Analysis

3.1 Implementation Details

The improved YOLOv5 model was implemented on Pytorch 1.7.1, an open-source environment for a variety of tasks in the machine learning specialty. We use a desktop computer for all the experiments, and the computer configuration is[Intel(R) Xeon(R) CPU E5-2650 0 @ 2.2 GHz, 32GB RAM, GeForce RTX 2060 SUPER, Windows 7, 64 bit]. CUDA 10.1 and cuDNN 7.6.5 were used to accelerate the GPU. The specific parameters for the experiment were set as: learning rate with the value of 0.001, weight decay with the value of 0.0005, momentum with the value of 0.937, image size is 640 × 640; 100 epochs; other parameters are default.

3.2 Datasets

The dataset in this paper consists of three parts, the first of which is the MOCS dataset [14] of 23,404 images published by Tsinghua University. The second part is the ACID dataset [3] developed by AIRCon-Lab at the University of Alberta with a total of 2850 images. The third part is the addition type of construction machinery for on-site object detection: the construction hanging basket. The images in this section are from our online crawler collection of 704. Due to the small amount of data for the construction hanging basket, data enhancements (e.g. flip, gaussian noise, salt and pepper noise, rotate, darker, brighter, etc.) were done to avoid overfitting, increasing the final number to 5,608. In this case, the construction hanging basket dataset was manually labelled with labelImg. Detailed information about the dataset is shown in Fig. 4. The validation dataset is to verify whether the model is over-fitted during the training process, and the test dataset is to assess the performance of the model. The detailed division of the dataset is shown in Table 1. To guarantee that the results of these models are comparable, the training, test, and validation datasets utilized in every model are the same.

Table 1. The dataset division

Train and validation		Train		Validation		Test		Total
Number	Percent (%)	Number	Percent (%)	Number	Percent (%)	Number	Percent (%)	
28668	90	25483	80	3186	10	3186	10	31854

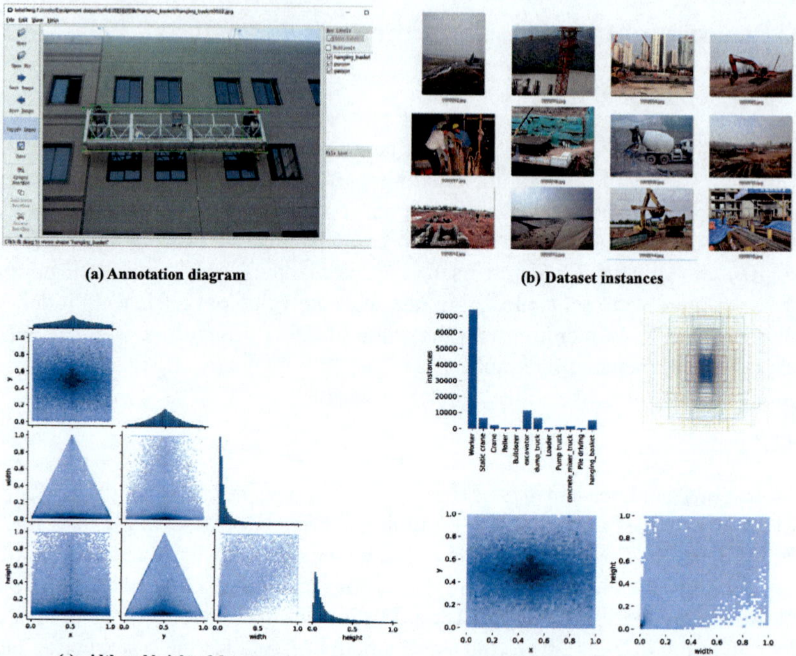

(a) Annotation diagram (b) Dataset instances

(c) width and height of data sample (d) Distribution of data sample

Fig. 4. The information about dataset

3.3 Evaluation Metrics

In this paper, the confidence threshold for the defect category is set as 0.5. Only if the box is legitimate and the box's category matches the ground truth is a detection considered true positive (TP). False positive (FP) means the misprediction of a positive example. False negative (FN) is the misprediction of a negative example. The mean average precision mAP (Mean Average Precision) is calculated using the accuracy P (Precision) and the recall R (Recall). The evaluation metrics can be calculated through TP, FP, and FN, as shown in Eqs. (3)–(5). Floating point operations (FLOPs) are a measure of model complexity; the larger the FLOP, the more parameters the model has, and the more complex the model.

$$P(Precision) = \frac{TP}{TP+FP} \tag{3}$$

$$R(Recall) = \frac{TP}{TP+FN} \tag{4}$$

$$mAP(MeanAveragePrecision) = \frac{\sum_{i=0}^{n} AP(i)}{n} \tag{5}$$

3.4 Comparative Analysis

Models were evaluated on the test datasets after training. The test results can be found in Table 2. In this paper, a total of 5 different object detection algorithms based on the YOLO v3, YOLOv5s, YOLOv5s-BiFPN, YOLOv5s-RepVGG and the proposed method were compared to analyze the construction workers and machinery detection performance. Generally, YOLOv5 performs much better than YOLOV3. The YOLOv5s-BiFPN model has the highest recall of 82.19%, owing to the BiFPN structure, which improves the feature fusion ability between different scales and allows for better usage of shallow features. However, because of the increased computation, the FLOPs are larger than the original algorithm, resulting in slower inference speed. The YOLOv5s-RepVGG model has the smallest Flops of 14.7G, which means this is the simplest inference model therefore the shortest inference time of 5.2 ms per frame. Our model (YOLOv5s-BiFPN-RepVGG) has the highest precision and mAP, which is 2.78 and 2.12 percentage points higher compared with YOLOv5.

Table 2. Comparison results

Model name	Precision (%)	Recall (%)	mAP@0.5 (%)	FLOPs (G)	Inference time(ms)
YOLOv3	78.62	79.13	79.27	116.9	20.9
YOLOv5s	88.74	80.15	85.95	16.4	5.8
YOLOv5s-BiFPN	91.48	**82.19**	87.26	17.8	7.8
YOLOv5s-RepVGG	88.52	80.14	85.75	**14.7**	**5.2**
Ours	**91.52**	81.72	**87.32**	16.1	5.7

Inputting images to the model for detection, which the model has never seen before, the experimental results of the improved model and the original model can be seen in Fig. 5. The picture in the first row shows that the confidence score in the improved network output images of construction workers and machinery is almost higher than the original YOLOv5. In addition, for complex scenes and images in camera's long perspective, the original YOLOv5 algorithm has missed detections for objects that are at small scale. As the red box shown on the bottom left of Fig. 7, a construction worker is missed. The recognition of small targets is more accurate after the addition of BiFPN, which are multi-scale feature layers, and the improved network detects all workers and machinery.

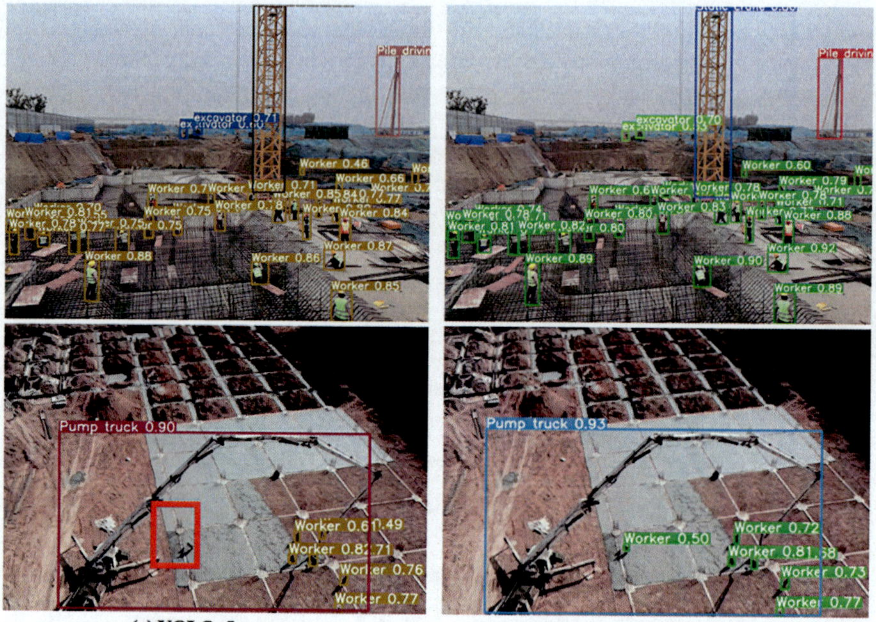

(a) YOLOv5 **(b) Improved YOLOv5**

Fig. 5. Detection results before and after improvements

4 Conclusion

YOLOv5 model has insufficient feature extraction ability in small-scale targets detection and can't make use of multi-scale feature. This paper proposes an improved YOLOv5 model for the automatic detection and localization of multi-scale construction workers and machinery, which aims to protect thses targets from hazardous situations such as collisions. Optimized models increase multi-scale feature fusion capabilities through BiFPN and reduce model inference time through structural re-parameterization of RepVGG Block. Experimental results show that mAP improves by 2.12% compared with the initial YOLOv5, and inference time reduce to 5.7 ms per frame. The improved YOLOv5 shows robustness in the detection of images in all construction environments of the dataset, indicating that the model is accurate and effective.

Acknowledgement. This paper is supported by the Scientific and Technological Innovation Research in the Field of Housing and Urban-rural Construction of Sichuan Province (No. CSJSKJ2021-21) and Sichuan Provincial Social Science Key Research Base, Systems Science and Enterprise Development Research Center (No. Xq20B09).

References

1. Zeng, T., Wang, J., Cui, B., Wang, X., Wang, D., Zhang, Y.: The equipment detection and localization of large-scale construction jobsite by far-field construction surveillance video based on improving YOLOv3 and grey wolf optimizer improving extreme learning machine. Constr. Build. Mater. **291**, 123268 (2021)
2. Kim, J., Chi, S., Seo, J.: Interaction analysis for vision-based activity identification of earthmoving excavators and dump trucks. Autom. Constr. **87**, 297–308 (2018)
3. Xiao, B., Kang, S.-C.: Development of an Image Data Set of Construction Machines for Deep Learning Object Detection. J. Comput. Civ. Eng. **35**, 05020005 (2021)
4. Chen, C., Zhu, Z., Hammad, A.: Automated excavators activity recognition and productivity analysis from construction site surveillance videos. Autom. Constr. **110**, 103045 (2020)
5. Ren, S., He, K., Girshick, R., Sun, J.: Faster R-CNN: towards real-time object detection with region proposal networks. ArXiv150601497 Cs. (2016)
6. Redmon, J., Divvala, S., Girshick, R., Farhadi, A.: You only look once: unified, real-time object detection. ArXiv150602640 Cs. (2016)
7. Redmon, J., Farhadi, A.: YOLOv3: an incremental improvement. ArXiv180402767 Cs. (2018)
8. Redmon, J., Farhadi, A.: YOLO9000: Better, Faster, Stronger. ArXiv161208242 Cs. (2016)
9. Zhao, Z.-Q., Zheng, P., Xu, S., Wu, X.: Object detection with deep learning: a review. ArXiv180705511 Cs. (2019)
10. Tan, M., Pang, R., Le, Q.V.: EfficientDet: scalable and efficient object detection. ArXiv191109070 Cs Eess. (2020)
11. Zhu, X., Lyu, S., Wang, X., Zhao, Q.: TPH-YOLOv5: improved YOLOv5 based on transformer prediction head for object detection on drone-captured scenarios. ArXiv210811539 Cs. (2021)
12. Ding, X., Zhang, X., Ma, N., Han, J., Ding, G., Sun, J.: RepVGG: making VGG-style ConvNets Great Again. ArXiv210103697 Cs. (2021)
13. Jing, Y., Ren, Y., Liu, Y., Wang, D., Yu, L.: Automatic extraction of damaged houses by earthquake based on improved YOLOv5: a case study in Yangbi. Remote Sens. **14**, 382 (2022)
14. Xuehui, A., Li, Z., Zuguang, L., Chengzhi, W., Pengfei, L., Zhiwei, L.: Dataset and benchmark for detecting moving objects in construction sites. Autom. Constr. **122**, 103482 (2021)

Check for updates

Study on Comparison and Selection of Grouting Pre Reinforcement Schemes for Underground Excavation Land Section of River Crossing Tunnel – Taking Rongjiang Fourth Road River Crossing Tunnel as an Example

Hongwei Zhang[1](✉), Helin Fu[2], Hongyu Guo[1], Guijun Luo[1], and Kaiyuan Zhang[2]

[1] China Construction Fifth Engineering Division Corp., Ltd., Changsha, Hunan, China
kaikxue@163.com
[2] Central South University, Changsha, Hunan, China

Abstract. Jiangxi Ganzhou Rongjiang 4th Road River crossing tunnel is the first river crossing tunnel constructed by mining method in Jiangxi Province. In order to ensure the construction safety of the underground excavation land section of the tunnel, based on the fuzzy analytic hierarchy process theory, this paper carries out the comparison and selection of grouting pre reinforcement schemes for the surrounding rock of the underground excavation land section of the river crossing tunnel. The research results show that: 1) according to the engineering properties of the surrounding rock in the concealed excavation land section of the Rongjiang fourth road river crossing tunnel, through the research on the grouting pre reinforcement scheme, the pre reinforcement engineering plans of high-pressure rotary jet grouting and sleeve valve pipe grouting are selected; 2) Based on the theory of fuzzy analytic hierarchy process, a comparison model of grouting pre reinforcement schemes for surrounding rock of underground excavation land section of river crossing tunnel is established. The comparison result of the model is that the high-pressure rotary jet grouting scheme is better than the sleeve valve tube grouting scheme.

Keywords: Underground excavation of land section of river crossing tunnel · Grouting prereinforcement · Fuzzy analytic hierarchy process theory · Jet grouting

1 Introduction

Generally speaking, the stability of surrounding rock depends on the tunnel construction method and its engineering properties, and it is also the decisive factor to ensure the safety of tunnel construction. Therefore, in the process of tunnel construction, how to improve the stability of surrounding rock around the tunnel is the key to the smooth progress of tunnel engineering. Compared with other tunnels, the buried depth of tunnel

W. Guo and K. Qian (Eds.): GBCESC 2022, LNCE 211, pp. 750–761, 2023.
https://doi.org/10.1007/978-981-19-5217-3_75

in the land section of river crossing tunnel is shallow and the engineering properties of surrounding rock are poor; Moreover, it is located near the river, with high groundwater level and abundant groundwater. The control of surrounding rock stability of underground excavation land section of river crossing tunnel is the top priority in engineering construction. Therefore, in the construction of river crossing tunnel, the theory and technology of tunnel surrounding rock pre reinforcement must be studied. In recent years, due to the rapid development of rock and soil treatment theory and technology, there are also many theories and technologies suitable for tunnel surrounding rock pre reinforcement, such as grouting method, air pressure method, freezing method, bolt support method and so on. However, in the face of so many grouting pre reinforcement technologies, it has become a difficult problem to select the grouting pre reinforcement technology in line with the specific engineering practice. The comparison and selection method of engineering schemes refers to solving the optimal solution of engineering problems by relying on scientific evaluation methods.

Hu Yong and Zhu Yuliang [1] selected the evaluation indexes such as experience, cost and construction difficulty, established the comprehensive evaluation model of slope rock and soil treatment scheme by using the fuzzy evaluation method, and applied the model to specific projects. The evaluation results show that the model is practical and effective. Based on the comprehensive analysis of the advantages, disadvantages and influencing factors of the collapsible loess treatment scheme, Li Linfang [2] established the influencing factor hierarchy diagram of the collapsible loess treatment scheme by using the method of "analytic hierarchy process + value theory". Shao Guoxia and Cao Zhengguo [3] established the optimization model of foundation reinforcement based on the triangular fuzzy number TOPSIS theory and applied it to the optimization of foundation reinforcement scheme of Qufu Linyi line close to Beijing Shanghai high speed railway. Li yanlei and sun Xiaohong [4] based on the foundation project of a large thermal power plant, adopted the fuzzy comprehensive evaluation method to optimize the treatment scheme of the foundation project, and put forward reliable measures for the settlement of the backfill of the plant. Li Qing and Ye Chaoliang [5] evaluated and optimized these four schemes from the aspects of technical applicability, settlement control, cost and foundation bearing capacity by using fuzzy analytic hierarchy process, and finally considered that the optimal scheme is the foundation treatment scheme of high-pressure jet grouting pile. Based on intuitionistic fuzzy theory, Chen Hongquan and Liming Yue [6] established a foundation treatment model based on Sugeno integral operator from four aspects: technology, economy, environmental impact and construction conditions. Based on the theory of Fuzzy AHP, Wang Feng [7] compared and optimized the grouting pre reinforcement schemes of surrounding rock in the land section of river crossing tunnel. From the above research, it can be found that the research on the optimization theory of soft soil reinforcement scheme is mostly concentrated in the reinforcement engineering of building foundation, while there is less research on the optimization of tunnel surrounding rock reinforcement engineering scheme. Based on the theory of fuzzy analytic hierarchy process, this paper will establish the comparison model of grouting pre reinforcement scheme of surrounding rock in the underground excavation land section of river crossing tunnel, and apply the model to the comparison and selection of grouting pre reinforcement scheme of surrounding rock in the underground excavation section of Rongjiang Fourth Road River Crossing tunnel.

2 Project Overview

Ganzhou Rongjiang 4th Road River crossing tunnel is the first river crossing tunnel constructed by mining method in Jiangxi Province. The tunnel project starts from the south of the West extension of Hakka Avenue and ends at the north of Huangjin road. The chainage range is K0 + 705 ~ K3 + 117, with a total length of about 2.412 km; It mainly includes the river crossing tunnel (tunnel 2) and Dongdong tunnel, with a length of 1745 m and a buried section of 949.5 m, of which the mining method length of the buried section is 795.5 m (EK1 + 224.5 ~ Ek2 + 020); The Xidong tunnel is 1739 m long and the buried section is 953 m, of which the mining method length of the buried section is 786 m (WK1 + 226 ~ WK1 + 012). The overall trend and location of the project are shown in Fig. 1.

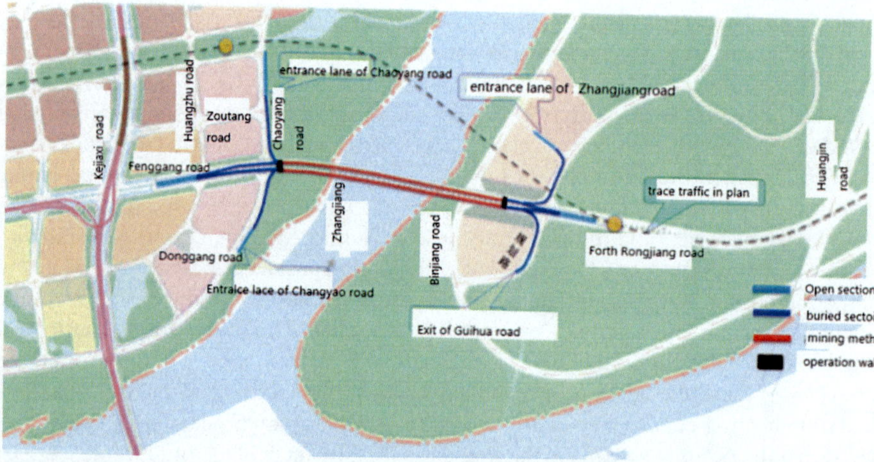

Fig. 1. A figure caption is always placed below the illustration. Short captions are centered, while long ones are justified. The macro button chooses the correct format automatically.

Among them, the construction of underground excavation land section of river crossing tunnel faces the following engineering problems:

(1) Due to the influence of river water, the engineering properties of surrounding rock in the underground excavation land section of the tunnel are poor, the groundwater is rich and connected with the river water;
(2) The underground excavation land section of the tunnel mainly crosses the moderately weathered argillaceous siltstone with poor engineering properties. The uniaxial saturated compressive strength is about 10 MPa and the local strength is

20–30 MPa. Moreover, the rock overburden in the land section of concealed excavation is relatively thin, and the surrounding rock grade is poor (grade V–VI). It belongs to soft rock, which is easy to soften in case of water, easy to dry crack and disintegrate after water loss, and the weathering speed is fast after excavation exposure;

(3) The self stability ability of tunnel surrounding rock is poor, and the thickness of overlying moderately weathered rock mass is less than one time of tunnel diameter, which is prone to roof collapse and water inrush.

At this stage, there is no construction experience of similar projects across the country. In addition, the distance between the left and right tunnels of the river crossing tunnel is small, the interaction between construction and excavation is large, and the stress redistribution is more complex. Therefore, it is necessary to study the theory and technology of pre reinforcement of surrounding rock in the land section of river crossing tunnel.

3 Selection of Grouting Pre Reinforcement Scheme

The underground excavation land section of the river crossing tunnel can also be called the portal section. Because it is located in the special geographical location near the river, generally speaking, the tunnel is buried shallowly, the engineering properties of the surrounding rock are poor, and the groundwater is rich. Groundwater is one of the most unfavorable engineering conditions in the process of tunnel construction. Its treatment principle is the combination of drainage and blocking. Drainage refers to reducing groundwater level and groundwater pressure; Plugging refers to reducing the permeability coefficient of surrounding rock and reducing water inflow, while strengthening surrounding rock and improving the engineering properties of surrounding rock. The theory and technology of tunnel surrounding rock pre reinforcement is to adopt a series of reinforcement technologies before tunnel excavation, such as grouting, anchor bolt and other means to reduce the permeability coefficient of surrounding rock and increase the strength of surrounding rock. The common tunnel surrounding rock pre reinforcement and waterproof technology is surrounding rock grouting water plugging technology.

Surrounding rock grouting is the most important means to strengthen the stratum and reduce groundwater discharge. It can be divided into surface grouting, advanced deep hole grouting before excavation, radial grouting or local grouting after excavation, etc. Generally, when groundwater discharge affects the ecological environment, high water pressure affects the stability of surrounding rock or construction safety, surface grouting technology and advanced deep hole grouting before excavation can be used for treatment. Otherwise, radial grouting or local grouting after excavation can be used. In view of the fact that the underground water in the surrounding rock of the land section of the river crossing tunnel is connected with the river water, the underground water is abundant in the excavation process, which is very easy to cause accidents such as water gushing, mud gushing and collapse in the tunnel excavation. Therefore, pre reinforcement measures should be taken for the surrounding rock before the tunnel is excavated into the tunnel, so as to reduce the harm of underground water and strengthen the surrounding rock.

According to the engineering hydrogeological conditions of tunnel surrounding rock, tunnel design documents and local engineering experience, there are two kinds of grouting pre reinforcement schemes for surrounding rock to be adopted.

Scheme 1: surface high-pressure jet grouting pre reinforcement treatment. Surface high-pressure rotary jet grouting has many functions, such as high quality and reliability of soil reinforcement, strengthening soil strength, water and seepage prevention, reducing soil water content and so on. The tunnel top overburden with a thickness of less than 8 m in the tunnel concealed excavation section shall be reinforced with a reinforcement depth of no less than 8 m to the moderately weathered rock layer. The section with a thickness of less than 8 m in the vault overburden shall be reinforced to the surface, and the reinforcement range is 8 m outside the excavation sideline.

Scheme 2: pre reinforcement treatment of surface sleeve valve pipe grouting. The grouting pipe of sleeve valve pipe grouting technology can carry out repeated, jumping and continuous grouting in the grouting reinforcement area because of its blocker, so as to improve the grouting quality.

The tunnel top overburden with a thickness of less than 8 m in the tunnel concealed excavation section shall be reinforced with a reinforcement depth of no less than 8 m to the moderately weathered rock layer. The section with a thickness of less than 8 m in the vault overburden shall be reinforced to the surface, and the reinforcement range is 8 m outside the excavation sideline.

4 Comparison and Selection of Grouting Pre Reinforcement Schemes

This paper combines the advantages of analytic hierarchy process and fuzzy mathematical evaluation method to form the fuzzy analytic hierarchy process theory, and uses this theory to construct the comparison model of grouting pre reinforcement schemes for the surrounding rock of the underground excavation land section of the river crossing tunnel, and then uses this model to study the comparison and selection of grouting pre reinforcement schemes for the surrounding rock of the underground excavation land section of the river crossing tunnel of the fourth Rongjiang road in Ganzhou, Jiangxi Province.

Concept of Fuzzy Mathematics Comprehensive Evaluation Method. Fuzzy mathematics comprehensive evaluation method, also known as FAHP method, is based on the theory of fuzzy mathematics and the principle of maximum membership. Through the comprehensive consideration of various factors affecting the research object, the fuzzy judgment matrix is established, and then the research object is comprehensively evaluated.

Comparison and evaluation of grouting pre reinforcement schemes for surrounding rock of underground excavation land section of river crossing tunnel.

The evaluation index system of grouting pre reinforcement scheme for surrounding rock in the concealed excavation land section of Rongjiang fourth road river crossing tunnel is constructed. According to the actual engineering geological conditions of the river crossing tunnel of Rongjiang 4th Road, there are no structures within 500m around the surrounding rock reinforcement area; Moreover, in the grouting pre reinforcement scheme of surrounding rock of Rongjiang fourth road river crossing tunnel, high-pressure rotary jet grouting and sleeve valve pipe grouting are applicable to weak surrounding rock, such as silt and clay. Therefore, the evaluation system of grouting pre reinforcement scheme for surrounding rock in the concealed excavation land section of Rongjiang fourth road river crossing tunnel is composed of four secondary evaluation indexes of technical feasibility, economic rationality, environmental impact and construction conditions, as well as 10 tertiary evaluation indexes.

Carry out expert evaluation of grouting pre reinforcement engineering scheme. Several experts were invited from the fields of rock and soil treatment and tunnel to form a 10 person expert review team to evaluate the evaluation indexes of grouting pre reinforcement scheme for the concealed excavation land section of Rongjiang fourth road river crossing tunnel. According to Table 3, 4 and 5, the expert group evaluates the four secondary evaluation indexes and their 10 tertiary evaluation indexes respectively, evaluates the ability to complete the evaluation indexes according to the reinforcement scheme, and gives the corresponding qualitative standards. The evaluation results of each expert on the indicators are recorded as one vote. Seen Table 1 and Table 2 for the evaluation results.

Table 1. Expert evaluation results of grouting pre reinforcement scheme I for surrounding rock of underground excavation land section of Rongjiang Fourth Road River Crossing Tunnel.

Evaluating indicator		Statistics			
		Excellent	Good	Optional	Not optional
Technical feasibility U1	Theoretical calculation and design methodU11	3	4	2	1
	Reinforcement quality levelU12	2	5	2	1
Economic rationalityU2	Engineering costU21	3	3	2	2
	Project durationU22	2	3	4	1
	Material supplyU23	2	4	3	1
Environmental effectU3	Vibration or noise pollutionU31	1	2	4	3
	Impact on air and waterU32	2	3	3	2

(*continued*)

Table 1. (*continued*)

Evaluating indicator		Statistics			
		Excellent	Good	Optional	Not optional
Construction conditionsU4	Construction difficultyU41	2	3	4	1
	Mechanical equipmentU42	2	2	4	2
	Professional quality of construction personnelU43	3	4	2	1

Table 2. Expert evaluation results of grouting pre reinforcement scheme II for surrounding rock of underground excavation land section of Rongjiang No. 4 River Crossing Tunnel

Evaluating indicator		Statistics			
		Excellent	Good	Optional	Not optional
Technical feasibility U1	Theoretical calculation and design methodU11	3	4	3	0
	Reinforcement quality levelU12	2	4	3	1
Economic rationalityU2	Engineering costU21	2	2	3	3
	Project durationU22	2	2	4	2
	material supplyU23	2	4	3	1
Environmental effectU3	Vibration or noise pollutionU31	2	4	3	1
	Impact on air and waterU32	2	3	3	2
Construction conditionsU4	Construction difficultyU41	2	2	4	2
	Mechanical equipmentU42	1	2	5	2
	Professional quality of construction personnelU43	1	3	4	2

The weight of each index is calculated by FAHP method. FAHP method is applied to calculate the weight of each index in scheme I: constructing the first mock exam matrix.

According to the corresponding evaluation results of each index by 10 experts, a single index evaluation matrix is established, and the U1 fuzzy consistent judgment matrix is obtained by adjusting according to the nature of the fuzzy consistent judgment matrix:

$$U_1 = \begin{bmatrix} 0.5 & 0.61 \\ 0.39 & 0.5 \end{bmatrix} \tag{1}$$

Similarly, the fuzzy consistent judgment matrix of other three-level evaluation indexes can be obtained:

$$U_2 = \begin{bmatrix} 0.5 & 0.47 & 0.83 \\ 0.53 & 0.5 & 0.86 \\ 0.17 & 0.14 & 0.5 \end{bmatrix} \tag{2}$$

$$U_3 = \begin{bmatrix} 0.5 & 0.71 \\ 0.29 & 0.5 \end{bmatrix} \tag{3}$$

$$U_4 = \begin{bmatrix} 0.5 & 0.41 & 0.36 \\ 0.59 & 0.5 & 0.45 \\ 0.64 & 0.55 & 0.5 \end{bmatrix} \tag{4}$$

The weights of evaluation indexes at all levels are calculated by formula (1), and the calculation results are normalized to obtain the weight set of each three-level index as follows:

$$a_{ki} = \frac{\sum_{j=1}^{n} r_{ij} + \frac{n}{2} - 1}{n(n-1)} \tag{5}$$

$$A_1 = (0.555, 0.445) \tag{6}$$

$$A_2 = (0.3716, 0.3983, 0.2301) \tag{7}$$

$$A_3 = (0.605, 0.395) \tag{8}$$

$$A_4 = (0.295, 0.34, 0.365) \tag{9}$$

Construct the fuzzy consistency matrix of each secondary evaluation index:

$$U = \begin{bmatrix} 0.5 & 0.61 & 0.81 & 0.91 \\ 0.39 & 0.5 & 0.7 & 0.8 \\ 0.19 & 0.3 & 0.5 & 0.6 \\ 0.09 & 0.2 & 0.4 & 0.5 \end{bmatrix} \tag{10}$$

The weight set of secondary evaluation index can be obtained by calculating the weight and normalizing:

$$A = (0.3192, 0.2825, 0.2158, 0.1825) \tag{11}$$

The weight and membership degree of each index in scheme I are shown in Table 3.

Table 3. Weight of indicators at all levels and subordinate degree of indicators at three levels in scheme I.

Secondary index (ratio Ak)	Third index (ratio Aki)	Membership degree of each index			
		Excellent	Good	Optional	Not optional
Technical feasibilityU1 (0.3192)	Theoretical calculation and design methodU11 (0.555)	0.3	0.4	0.2	0.1
	Reinforcement quality levelU12 (0.445)	0.2	0.5	0.2	0.1
Economic rationalityU2 (0.2825)	engineering costU21 (0.3716)	0.3	0.3	0.2	0.2
	Project durationU22 (0.3983)	0.2	0.3	0.4	0.1
	Material supplyU23 (0.2301)	0.2	0.4	0.3	0.1
Environmental effectU3 (0.2158)	Vibration or noise pollutionU31 (0.605)	0.1	0.2	0.4	0.3
	Impact on air and waterU32 (0.395)	0.2	0.3	0.3	0.2
Construction conditionsU4 (0.1825)	Construction difficultyU41 (0.295)	0.2	0.3	0.4	0.1
	Mechanical equipment U42 (0.34)	0.2	0.2	0.4	0.2
	Professional quality of construction personnelU43 (0.365)	0.3	0.4	0.2	0.1

Similarly, the weight and membership degree of indicators at all levels in scheme 2 are shown in Table 4.

Table 4. Weight of indicators at all levels of scheme II and membership degree of indicators at all three levels.

Secondary index (ratio Ak)	Third index (ratio Aki)	Membership degree of each index			
		Excellent	Good	Optional	Not optional
Technical feasibilityU1 (0.2283)	Theoretical calculation and design methodU11 (0.455)	0.3	0.4	0.3	0

(*continued*)

Table 4. *(continued)*

Secondary index (ratio Ak)	Third index (ratio Aki)	Membership degree of each index			
		Excellent	Good	Optional	Not optional
	Reinforcement quality levelU12 (0.545)	0.2	0.4	0.3	0.1
Economic rationalityU2 (0.2717)	Engineering cost U21 (0.3233)	0.2	0.2	0.3	0.3
	Project durationU22 (0.4133)	0.2	0.2	0.4	0.2
	Material supplyU23 (0.2634)	0.2	0.4	0.3	0.1
Environmental effectU3 (0.2116)	Vibration or noise pollutionU31 (0.585)	0.2	0.4	0.3	0.1
	Impact on air and waterU32 (0.415)	0.2	0.3	0.3	0.2
Construction conditionsU4 (0.2884)	Construction difficultyU41 (0.3483)	0.2	0.2	0.4	0.2
	Mechanical equipmentU42 (0.2433)	0.1	0.2	0.5	0.2
	Professional quality of construction personnelU43 (0.4084)	0.1	0.3	0.4	0.2

Establish single index evaluation matrix: According to Table 3, the three-level evaluation index membership matrix of scheme I can be obtained:

$$R_1 = \begin{bmatrix} 0.3\ 0.4\ 0.2\ 0.1 \\ 0.2\ 0.5\ 0.2\ 0.1 \end{bmatrix} \quad (12)$$

$$R_2 = \begin{bmatrix} 0.3\ 0.3\ 0.2\ 0.2 \\ 0.2\ 0.3\ 0.4\ 0.1 \\ 0.2\ 0.4\ 0.3\ 0.1 \end{bmatrix} \quad (13)$$

$$R_3 = \begin{bmatrix} 0.1\ 0.2\ 0.4\ 0.3 \\ 0.2\ 0.3\ 0.3\ 0.2 \end{bmatrix} \quad (14)$$

$$R_4 = \begin{bmatrix} 0.2\ 0.3\ 0.4\ 0.1 \\ 0.2\ 0.2\ 0.4\ 0.2 \\ 0.3\ 0.4\ 0.2\ 0.1 \end{bmatrix} \quad (15)$$

This paper makes a fuzzy comprehensive evaluation on the grouting pre reinforcement scheme of surrounding rock in the land section of river crossing tunnel. The fuzzy comprehensive evaluation of scheme I grouting pre reinforcement scheme is carried out. The evaluation U_1 results B_1 are calculated and normalized as follows:

$$B_1 = A_1R_1 = (0.555, 0.445)\begin{bmatrix} 0.3\ 0.4\ 0.2\ 0.1 \\ 0.2\ 0.5\ 0.2\ 0.1 \end{bmatrix} \\ = (0.2555,\ 0.4445, 0.2000, 0.1000) \tag{16}$$

Similarly, it can be calculated B_2,B_3,B_4:

$$B_2 = A_2R_2 = (0.3716, 0.3983, 0.2301)\begin{bmatrix} 0.3\ 0.3\ 0.2\ 0.2 \\ 0.2\ 0.3\ 0.4\ 0.1 \\ 0.2\ 0.4\ 0.3\ 0.1 \end{bmatrix} \\ = (0.2372, 0.3231, 0.3026, 0.1371) \tag{17}$$

$$B_3 = A_3R_3 = (0.605, 0.395)\begin{bmatrix} 0.1\ 0.2\ 0.4\ 0.3 \\ 0.2\ 0.3\ 0.3\ 0.3 \end{bmatrix} \\ = (0.1395{,}0.2466, 0.3139{,}0.3000) \tag{18}$$

$$B_4 = A_4R_4 = (0.295, 0.34, 0.365)\begin{bmatrix} 0.2\ 0.3\ 0.4\ 0.1 \\ 0.2\ 0.2\ 0.4\ 0.2 \\ 0.3\ 0.4\ 0.2\ 0.1 \end{bmatrix} \\ = (0.2365{,}0.3025, 0.3270{,}0.1340) \tag{19}$$

According the upper reslts, it will be gained:

$$R = \begin{bmatrix} B_1 \\ B_2 \\ B_3 \\ B_4 \end{bmatrix} = \begin{bmatrix} 0.2555\ 0.4445\ 0.2000\ 0.1000 \\ 0.2372\ 0.3231\ 0.3026\ 0.1371 \\ 0.1395\ 0.2466\ 0.3139\ 0.3000 \\ 0.2365\ 0.3025\ 0.3270\ 0.1340 \end{bmatrix} \tag{20}$$

The comprehensive evaluation result is:

$$B = AR = (0.3192, 0.2825, 0.2158, 0.1825) \\ \begin{bmatrix} 0.2555\ 0.4445\ 0.2000\ 0.1000 \\ 0.2372\ 0.3231\ 0.3026\ 0.1371 \\ 0.1395\ 0.2466\ 0.3139\ 0.3000 \\ 0.2365\ 0.3025\ 0.3270\ 0.1340 \end{bmatrix} \\ = (0.2218, 0.4583, 0.2153, 0.1046) \tag{21}$$

To sum up, according to the principle of maximum subordination, the evaluation result of grouting pre reinforcement scheme I for the surrounding rock of the concealed excavation land section of Rongjiang No. 4 river crossing tunnel is 0.4583.

Similarly, it can be concluded that the evaluation result of grouting pre reinforcement scheme II for the surrounding rock of the concealed excavation land section of Rongjiang No. 4 river crossing tunnel is 0.3595.

To sum up, the evaluation result of scheme I is 0.4583, which is greater than the evaluation result of scheme II is 0.3595. Therefore, the comparison and selection results of surrounding rock grouting pre reinforcement engineering schemes of river crossing concealed excavation land section of Rongjiang fourth road are: scheme I is better than scheme II.

5 Conclusions and Suggestions

In this paper, the fuzzy analytic hierarchy process (FAHP) is applied to the comparison and selection of surrounding rock grouting pre reinforcement schemes in the land section of river crossing tunnel, the evaluation index system of surrounding rock grouting pre reinforcement scheme is constructed, and the evaluation model of surrounding rock grouting pre reinforcement scheme in the land section of river crossing tunnel is established. Finally, the model is applied to the engineering practice of the comparison and selection of grouting pre reinforcement schemes for the surrounding rock of the concealed excavation land section of Rongjiang fourth road river crossing tunnel. The model evaluation result is that the high-pressure rotary jet grouting pre reinforcement scheme is better than the sleeve valve pipe grouting pre reinforcement scheme, which provides a basis for the actual engineering reinforcement.

References

1. Yong, H.: Stability analysis of slope embankment and comparison of foundation treatment schemes. Water Transp. Eng. (11), 151–155+169 (2019)
2. Linfang, L.: Study on Optimization of foundation treatment scheme in collapsible loess area. Transp. World (13), 86–87+103 (2019)
3. Guoxia, S.: Optimization of reinforcement scheme for adjacent foundation of existing high-speed railway based on triangular fuzzy number TOPSIS. Railway Stand. Des. **63**(02), 40–44 (2019)
4. Yanlei, L.: Optimization of treatment scheme for collapsible loess foundation. J. Wuhan Univ. (Eng. Ed.) **51**(S1), 205–208 (2018)
5. Qing, L.: Optimal design of composite foundation with temporary support of kailant bridge. Railway Const. **05**, 47–49 (2015)
6. Hongquan, C.: Optimization of foundation treatment scheme based on intuitionistic fuzzy set. J. Eng. Manage. **25**(05), 553–558 (2011)
7. Feng, W.: Study on grouting pre reinforcement and safety step of underground excavation land section of small clear distance tunnel across the river. Central South University (2021)

Study on Impact Resistance of Corroded Concrete Beams

Xiaofang Deng[1,2], Longmei Jin[2], and Zhi Li[1,2](✉)

[1] GuangXi Key Laboratory of New Energy and Building Energy Saving, Guilin 541004, China
Lizhi@glut.edu.cn

[2] College of Civil Engineering and Architecture, Guilin University of Technology, Guilin 541004, Guangxi, China

Abstract. To study the mechanical properties and failure modes of corroded RC beams under impact, this paper verifies the accuracy of finite element model established by finite element software ANSYS/LS-DYNA through impact test of concrete beams conducted by Fujikake et al. Based on the verified finite element model, the influence of corrosion rate of reinforcement on the impact resistance of concrete beams was researched. The finite element analysis shows that the concrete spalling degree of concrete member under impact effect increases with the increase of steel corrosion rate. The higher the degree of steel corrosion, the more serious the concrete spalling. All the corroded RC beams at different impact heights exhibit flexural failure characteristics. Steel corrosion has a certain influence on the impact resistance of reinforced concrete members. At the same drop hammer impact height, the peak impact force decreases and the mid-span displacement increases with the increase of steel corrosion rate. When the height of impact is 2.4 m, the peak value of impact force decreases to 29.0%. when corrosion rate is 30.0%, and the mid-span displacement increases to 15.7% when it is not corroded.

Keywords: Corrosion · Concrete beams · Impact load · Numerical simulation

1 Introduction

Reinforced concrete structures are often subjected to unexpected loads such as vehicle impact and rockfall impact in the process of use. In addition, reinforced corrosion often occurs in concrete structures near the sea and in some special industrial environments, resulting in degradation of material properties [1], which has a seriously affects on the safety and use of structures. Thus, it is increasingly important to study the impact resistance of corroded RC beams.

Fujikake et al. [2] studied the impact resistance of RC beams, and the results showed that the degree of local failure and the maximum mid-span deflection were important indexes to assess the damage degree of RC beams under impact loads. Zhao et al. [3] studied the bearing capacity of concrete beams by accelerating reinforcement corrosion test with electrochemical method, and according to the test results, obtained the

W. Guo and K. Qian (Eds.): GBCESC 2022, LNCE 211, pp. 762–770, 2023.
https://doi.org/10.1007/978-981-19-5217-3_76

calculation formula of concrete bond influence coefficient varying with reinforcement corrosion rate. Kishi et al. [4] conducted impact test on non-shear reinforced concrete beams, and the test results show that when the drop hammer impact velocity exceeds a certain value, the peak value of the reaction force on the support does not increase with the increase of impact velocity.

To analyze the influence of reinforcement corrosion on the impact resistance of concrete beams, finite element software ANSYS/LS-DYNA was used in this paper to simulate and analyze the no corrosion reinforced concrete beam structure completed by Fujikake et al. [2]. On the basis of verifying the correctness of the FEM, the influence of the impact height of drop hammer and the corrosion rate of reinforcement on the mechanical properties of concrete beams are further studied.

2 Establishment and Verification of Finite Element Model

2.1 Brief Introduction of Fujikaka Test

Fujikake et al. [2] studied the mechanical properties of RC beams under the impact of falling hammer. The size and reinforcement of the specimen was shown in Fig. 1, and the reinforcement of the specimen was shown in Table 1.

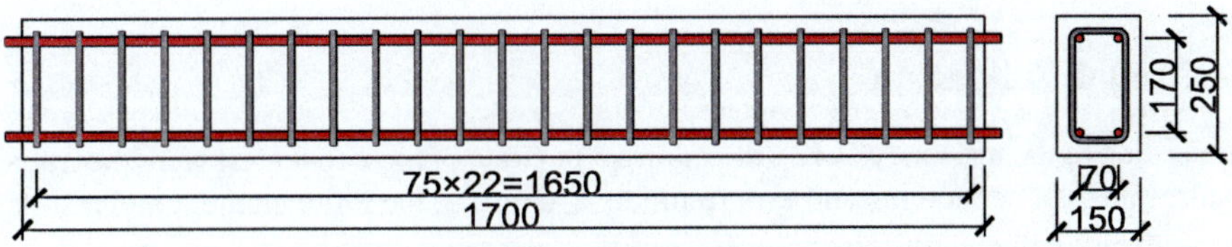

Fig. 1. Sample prototype diagram (Unit: mm)

Table 1. Test steel properties

Number	Upper reinforcement		Lower reinforcement	
	Amount	Diameter (mm)	Amount	Diameter (mm)
S2222	2	22	2	22

In this test, the drop hammer impact loading machine was used for impact loading. The drop hammer of 400 kg fell freely on the top surface of the mid-span reinforced concrete beams from four different heights. The corresponding impact heights of S2222 series reinforced concrete beams respectively were 0.3 m, 0.6 m, 1.2 m and 2.4 m respectively.

2.2 Finite Element Modeling

The study uses finite element software LS-DYNA to simulate. Qian et al. [5–11] used ANSYS/LS-DYNA software to model a refined FEM for reinforced concrete structures, and compared it with the experimental results, which shous that ANSYS/LS-DYNA could accurately predict various mechanical indexes of reinforced concrete structures.

In the establishment of the finite element model, SOLID164 element was used for concrete and BEAM161 element was used for reinforcement. The constitutive model used for concrete can effectively simulate the failure mode of concrete constitutive and RC structure under impact load. The constitutive model of steel bar was *MAT_PLASTIC_KINEMATIC. The finite element model is shown in Fig. 2.

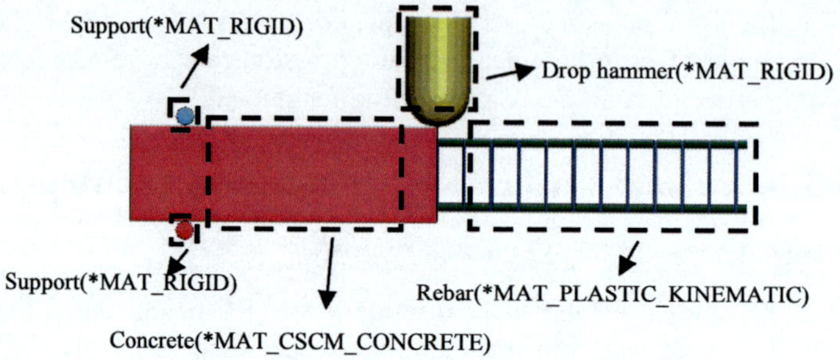

Fig. 2. RC beam finite element model

3 Model Validation

According to the test analysis [2], the main output data of the impact test were the impact loads, midspan deflections and failure mode. Therefore, the finite element and test data are compared and analyzed.

3.1 Comparison of Impact Response for S2222 Beams

Figure 3 shows the comparison of finite element and test impact load. When the impact height of drop hammer was 1.2 m and 2.4 m, the results of simulation and test were similar. Thus, the FEM can better predict the impact force time history curve of the specimen.

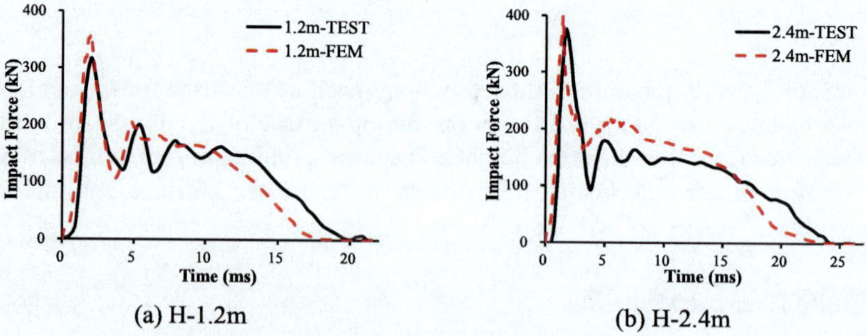

Fig. 3. Comparison of impact response for S2222

3.2 Comparison of Midspan Displacement for S2222 Beams

Figure 4 shown the comparison of midspan displacement curves between finite element simulation and test. When the impact height was 1.2 m and 2.4 m, the maximum difference between the calculated maximum midspan displacement and the test value was within 10%. It can be seen that the finite element model can better predict the midspan displacement of the specimen.

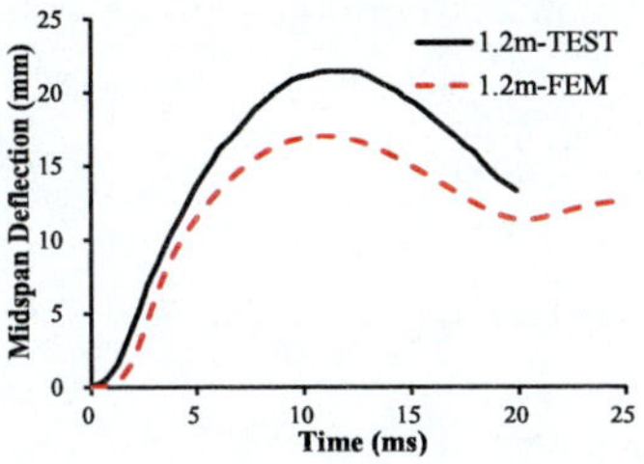

(a) H-1.2m

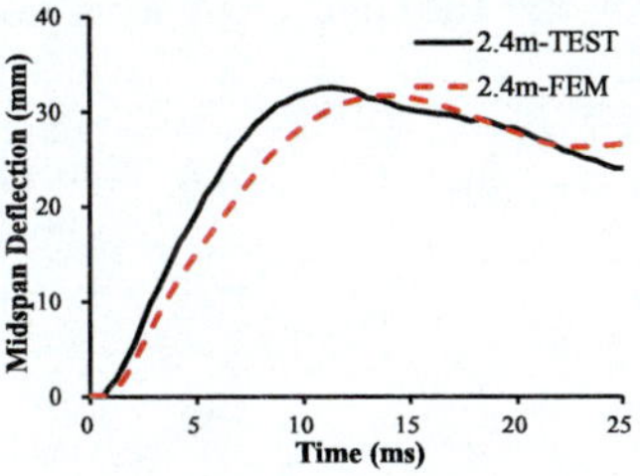

(d) H-2.4m

Fig. 4. Comparison of mid-span displacement time history curves for S2222

3.3 Comparison of Failure Modes for S2222 Beams

Figure 5 shown the comparison of failure modes between finite element and test. The figures shown that finite element could effectively simulate the crack development of the specimen and the failure of the concrete beam.

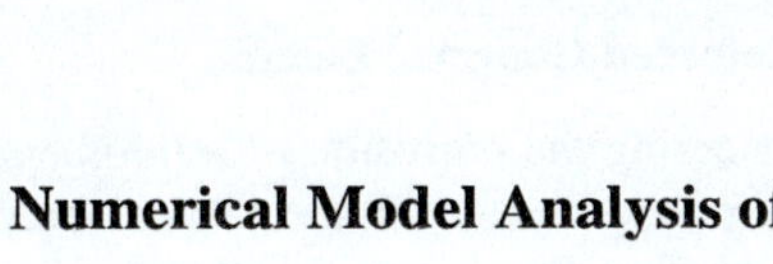

(a) H-1.2m

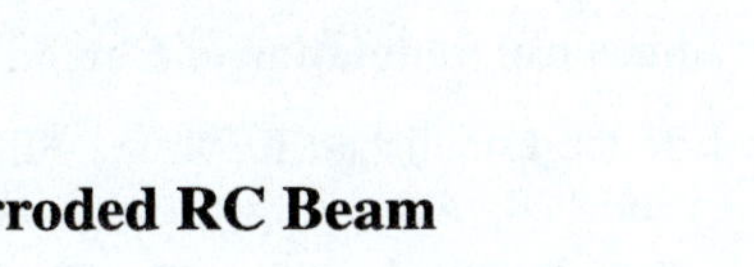

(b) H-2.4m

Fig. 5. Comparison of failure modes for S2222

4 Numerical Model Analysis of Corroded RC Beam

According to the verified finite element model, the influence of reinforcement corrosion rate on the impact resistance of concrete beams under different impact heights was studied.

4.1 Mechanical Properties of Steel and Concrete After Corrosion

Reinforcement corrosion is an inevitable problem in the use of reinforced concrete structures. It will reduce the service life and bearing capacity of concrete structures, and affect the safety and economy of structures. So it is necessary to research the impact resistance of reinforced concrete beams after corrosion.

The change of mechanical properties of reinforcement with the corrosion rate of reinforcement is mainly reflected in the reduction of reinforcement strength and ductility. Cairns et al. [12] gave the calculation formula of the reduction of yield strength, ultimate strength and ultimate elongation under the condition of reinforcement corrosion, as follows

$$f_y = \left(1.0 - \alpha_y \times Q_{corr}\right) f_{y0} \tag{1}$$

$$f_u = (1.0 - \alpha_u \times Q_{corr}) f_{u0} \tag{2}$$

$$\varepsilon_u = (1.0 - \alpha_1 \times Q_{corr}) \varepsilon_0 \tag{3}$$

where f_{y0} and f_y are the yield strength of reinforcement before and after corrosion; f_{u0} and f_u are the ultimate strength of reinforcement before and after corrosion; ε_0 and ε_u are the elongation of reinforcement before and after corrosion; α_y, α_u and α_1 are the empirical constants.

For the type of steel bar selected in this paper, the theoretical model proposed by Yuan et al. [13] was used to predict the bond strength between steel bar and concrete:

$$\tau_s^* = (1.0 - K_s \eta) \tau_s = \Phi_s \tau_s \tag{4}$$

$$\tau_{cr}^* = (1.0 - K_{cr} \eta) \tau_{cr} = \Phi_{cr} \tau_{cr} \tag{5}$$

$$\tau_u^* = (1.0 - K_u \eta) \tau_u = \Phi_u \tau_u \tag{6}$$

where τ_u and τ_u^* are the ultimate strength values of steel bars before and after corrosion; τ_{cr} and τ_{cr}^* are the splitting strength values before and after corrosion of reinforcement; τ_s and τ_s^* are the slip strength values of reinforcement before and after corrosion; Φ_u, Φ_{cr} and Φ_s are the degradation coefficient of bond strength; η Is the corrosion rate of reinforcement in the component; K_s, K_{cr} and K_u are the influence coefficient, which is related to the thickness of the protective layer and the diameter of the steel bar.

4.2 Numerical Simulation of Corroded Reinforced Concrete Beams

Based on the established FEM, the FEM considering the corrosion of reinforcement was established. According to formulas (3)–(5), the change of mechanical properties of reinforcement with corrosion rate of reinforcement was calculated. Finally, the theoretical model of bond degradation proposed by Yuan et al. [13] were used to simulate bond slip after corrosion of reinforcement, and the key word *CONSTRAINED_BEAM_IN_SOLID was used to simulate bond slip between concrete solid element and reinforced beam element [9].

4.3 Influence of Corrosion Rate of Reinforcement

Figure 6 shows the curves of impact force of corroded concrete beam changing with time. The impact force reaches the maximum value at the moment when the drop hammer contacts the specimen, and then the impact response fluctuates due to the gradual decrease of the relative velocity between the specimen and the drop hammer. When the drop hammer and the specimen reach the same speed, the impact force was basically unchanged. With the increase of corrosion rate, the peak of impact force tends to decreased. When the impact height of drop hammer was 1.2 m, the peak impact force decreased to 20.6% when the corrosion rate was 30.0%. When the drop hammer impact height was 2.4 m and the corrosion rat was 30%, the peak impact force decreased to 29.0%.

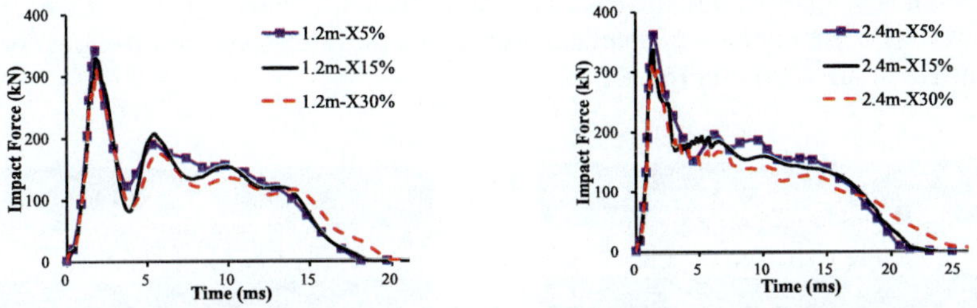

(a) H-1.2m (b) H-2.4m

Fig. 6. Comparison of impact response of different corrosion rates

Figure 7 shows the curves of midspan displacement at the bottom of corroded concrete beams changing with time. From the figures, we can know that when the impact height of the drop hammer was constant and only the corrosion rate of reinforcement was changed, the change curve of midspan displacement with time was consistent. At the same impact height, the maximum mid-span displacement and residual displacement increase with the increase of corrosion rate. Due to the rebound of the hammer head, the mid-span displacement time-history curve has a small amplitude waveform after the first half-sine curve. This is mainly due to the rebound of the drop hammer,

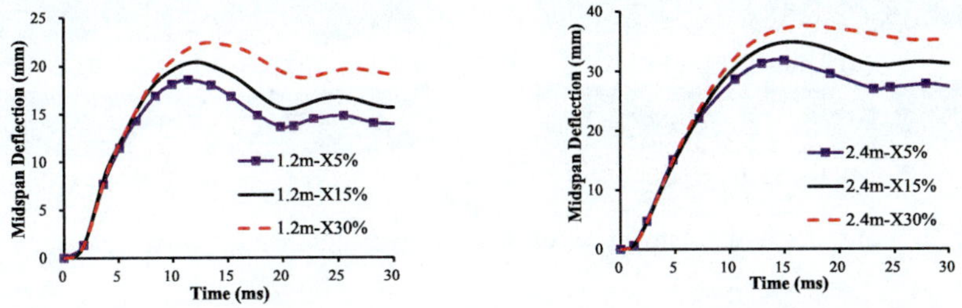

(a) H-1.2m (b) H-2.4m

Fig. 7. Comparison of mid span displacement time history curves with different corrosion rates which leads to a re-impact. Then the displacement tends to a constant value, namely residual displacement, which means that the specimen has a large plastic deformation after impact. When the impact height was 1.2 m, the mid-span displacement was 24.2% of that without corrosion when the corrosion rate was 30.0%. When the impact height was 2.4 m, the mid-span displacement with corrosion rate of 30.0% was 15.7% of that without corrosion.

Figure 8 and 9 shows the comparison of failure forms of different corrosion rates at different heights after falling hammer impact. By comparison, it can be found that the failure of the specimen all occurred near the impact point, the damage of the bottom of the mid-span beam appeared intensively, and the failure of steel bar and concrete appeared in different degrees.

When the impact height is 1.2 m, the concrete at the contact part between the drop hammer and the specimen was crushed under the impact action, and the concrete at the bottom was spalling and the steel bar was exposed. The degree of concrete spalling increased with the increase of steel corrosion rate. Failure modes of different corrosion rates at 1.2 m are shown in Fig. 8.

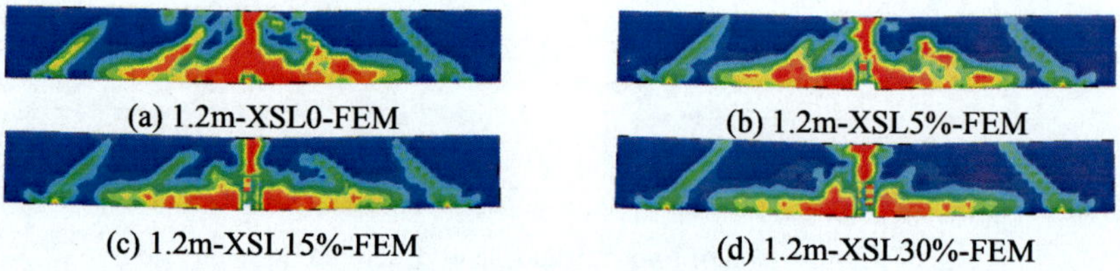

Fig. 8. Comparison of failure modes of steel bars with different corrosion rates at 1.2 m

When the impact height was 2.4 m, with the increase of the corrosion rate of reinforcement, the concrete at the bending crack at the middle of the span was spalling, and the spalling height was almost half of the height of the beam, indicating that both the impact height and the corrosion rate affect the failure mode of the concrete beam, showing a typical bending failure. Failure modes of different corrosion rates at 2.4 m are shown in Fig. 9.

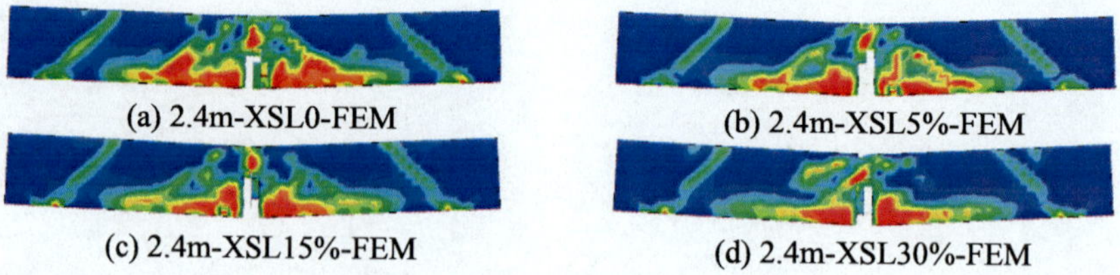

Fig. 9. Comparison of failure modes of steel bars with different corrosion rates at 2.4 m

5 Conclusions

According to the finite element analysis, the following conclusions can be drawn:

(1) Model established by finite element software ANSYS/LS-DYNA can better simulate the impact loads, midspan deflections and failure mode of concrete structure impact resistance test.
(2) In the drop hammer impact test, the reinforced concrete beams with different corrosion rates show typical bending failure characteristics under different impact heights.
(3) Under the same impact height, when the corrosion rate increases, the peak of impact force decreases gradually, the residual displacement and the maximum midspan displacement of the test beam under impact load increase.

Acknowledgement. The authors gratefully acknowledge the financial support provided by Guangxi Key Laboratory Project of New Energy and Building Energy Saving (Gui Ke Neng22-J-21-3), the Natural Science Foundation of Guangxi (No. 2021GXNSFFA196001), the Guangxi Science and Technology Program Project (Gui Ke AD21220071), the China Postdoctoral Science Foundation (2021M693814) and the Guangxi Young and Middle-aged Teacher Basic Ability Improvement Project (2021KY0254). Any opinions, findings and conclusions expressed in this paper do not necessary reflect the view of Guangxi Key Laboratory of New Energy and Building Energy Saving and Natural Science Foundation of Guangxi.

References

1. Xinbo, R.: Study on the diffussion performance of chloride ion in concrete under mechanical load and life prediction. Qingdao University of Technology (2012)
2. Fujikake, K., Li, B., Soeun, S.: Impact response of reinforced concrete beam and its analytical evaluation. J. Struct. Eng. **135**(8), 938–950 (2009)
3. Yuxi, Z., Weiliang, J.: Test study on bond stress-slip relationship of concrete and steel bar. J. Build. Struct. **23**(1), 32–37 (2002)
4. Kishi, N., Mikami, H., Ando, T.: Impact resistant behavior of shear-failure-type RC beams under falling-weight impact loading.In: Structures Under Shock and Impact VII, vol. 63, pp. 499–508 (2002)
5. Kai, Q., Zhi, L., Yunhao, W., Gang, W.: Load resisting mechanisms of assembled monolithic precast concrete beam-slab substructures to resist progressive collapse. J. Build. Struct. **42**(07), 183–193 (2021)
6. Zhi, L., Yuhao, W., Xiaofang, D., Kai, Q.: Behavior of precast concrete beam-slab substructures with welded connections to resist progressive collapse. J. Build. Struct. **41**(10), 121–128 (2020)
7. Yang, Y., Zhi, L., Longshan, X., Songyuan, G., Kai, Q.: Load resisting mechanism of precast structure under exterior column failure. Eng. Mech. **38**(04), 159–168 (2021)
8. Kai, Q., Zhi, L., Pan, H., Gang, W.: Progressive collapse mechanism of PC beam-slab substructure with bolted connections. J. Build. Struct. **41**(01), 173–180 (2020)
9. Xiaofang, D., Zhi, L., Yunhao, W., Xiaohui, Y., Kai, Q.: Experimental study on performance of prestressed concrete beam-column substructure against progressive collapse. J. Build. Struct. **40**(08), 71–78 (2019)

10. Kai, Q., Zhi, L., Yunhao, W.D.: Behavior of RC beam-slab structures to resist progressive collaps. Engineering Mechanics **36**(06), 239–247 (2019)
11. Kai, Q., Yunhao, W.: Application of ANSYS/LS-DYNA in Concrete Structure Engineering. China Machine Press, Beijing (2020)
12. Cairns, J., Plizzari, G.A., Du, Y., et al.: Mechanical properties of corrosion-damaged reinforcement. ACI Mater. J. **102**(4), 256–264 (2005)
13. Yingshu, Y., Suo, Y.: Deterioration of bond behavior of corroded reinforced concrete. Ind. Constr. **29**(11), 47–50 (1999)

GPU-Accelerated and Machine-Learning-Based Wind Turbine Damper Optimization

Shi Liu[1], Yize Wang[2(✉)], and Zhenqing Liu[2]

[1] China Southern Power Grid Technology Co., Ltd., Guangzhou 510000, Guangdong, China
[2] School of Civil and Hydraulic Engineering, Huazhong University of Science and Technology, Wuhan 430000, Hubei, China
wangyize@hust.edu.cn

Abstract. Wind turbine wake flow generally has larger turbulence intensity compared with free flow, which would result in larger damage to downstream wind turbines. However, there still has no research about wind turbine vibration considering wake effect. Meanwhile, although previous research had reduced wind turbine vibration by installing dampers, there is still no systematic global optimization method for wind turbine damper. The question that whether optimized damper without consideration about wake effect can be directly applied in wind turbines affected by wake or not is still not clear. To this end, this study first investigates wake effect on downstream wind turbine vibration. Then, tuned mass damper and rotational inerter double tuned mass damper are installed in wind turbine tower to control tower vibration. Innovatively, this study proposes a global optimization method for dampers based on radial basis function neural network and genetic algorithm, which is significantly accelerated by GPU acceleration technology. As well, wake effect on wind turbine dampers is studied by comparing optimized dampers with and without consideration of wake. Optimized dampers can reduce at most 44% tower bottom equivalent fatigue load. Numerical results can provide references for choosing damper and damper optimization in real engineering.

Keywords: Wind turbine · Vibration control · Wake effect

1 Introduction

Wind energy has been extensively exploited over the past tens of years. Wind turbine is utilized to harvest wind energy. For more energy output, wind turbine size grows rapidly, which causes excessive tower vibrations. Further, long-term vibration results in large tower bottom equivalent fatigue load (EFL), which is the primary reason for tower collapsing. Thus, investigating wind turbine tower vibration and how to control it get more and more important.

Generally, evaluation of tower bottom EFL includes: i) generating incoming wind speed history: previous research commonly adopted simplified analytical turbulence models. Those models are governed by statistical information of a real wind field, which can hardly satisfy N-S (Navier-Stokes) equations. Thus, wind turbine vibration calculated using those wind flows may be inconsistent with the reality. Consequently, this

W. Guo and K. Qian (Eds.): GBCESC 2022, LNCE 211, pp. 771–778, 2023.
https://doi.org/10.1007/978-981-19-5217-3_77

study utilizes large eddy simulation (LES) [1] to numerically calculate wind flow, which is simulated according to N-S equations. LES results are closer to the reality compared with analytical turbulence models. ii) Calculating load of wind turbine key parts caused by external wind load. During this process, blade element momentum method (BEM) [2] was generally adopted, which is also utilized in this study. iii) Structural dynamic analysis of wind turbine. There are generally two ways, i.e., finite element method (FEM)-based simulations and multi-degree-of-freedom (DOF)-based simplified simulations. Due to that FEM-based simulations involve amounts of DOFs, they are extremely time-consuming. Thus, they are not suitable for wind turbine damper optimization which requires to evaluate dozens of cases. This study uses multi-DOF-based simplified simulation method [3] to calculate wind turbine structural dynamics. iv) Calculating EFL. Based on numerical results of step iii, EFL can be calculated using rain flow counting and linear damage accumulating theory [3]. This study aims to control wind turbine tower fore-aft vibration. Thus, this study mainly evaluates wind turbine tower bottom EFL. In following discussions, wind turbine tower bottom EFL is simply called EFL.

To reduce EFL and further lengthen wind turbine fatigue life, Rezaei et al. [4] applied damper in wind turbine tower and found that damper can effectively reduce tower structural dynamics; meanwhile, EFL can also be reduced. Fitzgerald and Basu [5] installed active tuned mass damper (TMD) in wind turbine tower enabling that the TMD could almost always perform well when under various wind conditions. Subsequently, Dinh and Basu [6] installed multiple TMDs in wind turbine tower to further reduce tower vibration. They correspondingly optimized structural parameters of each TMD through parametric analysis. Sun [7] and Ghassempour et al. [8] also respectively reduced wind turbine tower vibration through installing dampers in it. However, previous research seldom considered upstream wind turbine wake effect on downstream wind turbine structural vibration and on vibration control. It is still not clear that whether optimized dampers without consideration about wake effect can be directly applied in wind turbines affected by wake or not. Meanwhile, previous research generally adopted empirical formulae or parametric analysis to optimized damper parameters, which can hardly be the global optimal parameter combination.

Therefore, this study controls wind turbine fore-aft vibration through installing dampers in it, as can be seen in Fig. 1. Both of TMD and rotational inerter double TMD (RIDTMD) are investigated. First, upstream wind turbine wake effects on downstream wind turbine vibrations are examined. It is cleared that how large a wind turbine spacing would result in unignorable wake effects. Next, damper parameters are respectively optimized with and without consideration of upstream wind turbine wake effects. Damper parameters are compared with each other to study wake effects on damper optimization. These comparisons can answer the question that whether optimized dampers without consideration of wake effects can be directly applied in wind turbines affected by wake flows or not. To systematically and globally optimize damper parameters, this study innovatively presents an optimization method based on radial basis function neural network (RBFNN) and genetic algorithm (GA), which is greatly accelerated by GPU (graphic process unit) acceleration technology. Numerical results can help engineers properly choose and optimize dampers in real engineering.

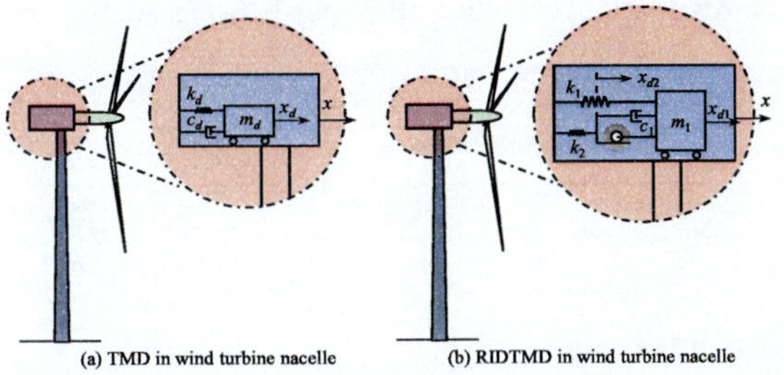

Fig. 1. (a) TMD in wind turbine nacelle; (b) RIDTMD in wind turbine nacelle.

2 PyGAOWT

This study presents a novel structural dynamic analysis software for wind turbine (PyGAOWT), which is developed on the basis of AOWT [9]. Figure 2 (a) and (b) present hardware layer and software layer of PyGAOWT, respectively. In Fig. 2 (a), it can be seen that a workstation generally consists of CPU (central process unit), host memory, GPU, and device memory. Data in these two different memories cannot directly exchange. They should be controlled by CPU. Meanwhile, CPU can call kernels on GPU to run computing tasks parallelly. Generally, GPU has thousands of times more computing cores compared with CPU. Thus, GPU is more appropriate when handling parallelable and computationally intensive tasks, which can make the program run thousands of times faster than that runs on CPU.

From Fig. 2 (b), it can be observed that wind turbine structural dynamic analysis includes following five steps: i) load wind speed time series; ii) calculate blade and tower loads using BEM theory; iii) run structural dynamic analysis of wind turbine; iv) post-process to calculate tower bottom moment time series; v) further post-process to calculate EFL using rain flow counting and linear damage accumulating theory.

Benefiting from GPU acceleration technology, PyGAOWT has significantly higher computing efficiency. As can be seen in Table 1, AOWT costs about 40.7 min to calculate one single case. The same case only costs 1.7 min in PyGAOWT. When there are 300 cases, PyGAOWT only cost 6.1 min, which is approximately 2000 times faster than AOWT.

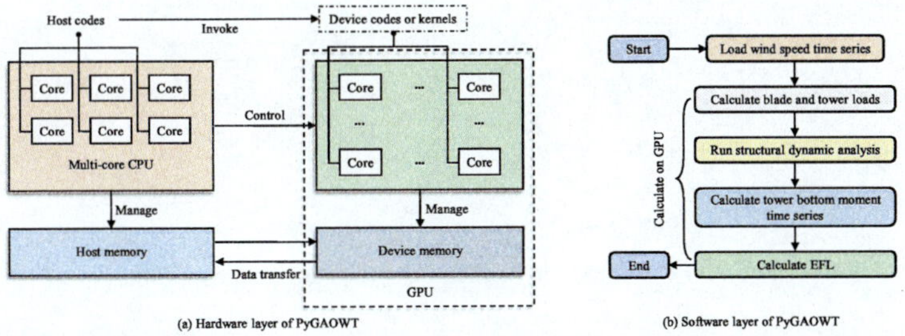

Fig. 2. (a) Hardware layer of PyGAOWT; (b) Software layer of PyGAOWT.

Table 1. Comparisons of computing efficiency between AOWT and PyGAOWT.

Number of cases	AOWT (min)	PyGAOWT (min)	AOWT/PyGAOWT
One single case	40.7	1.7	23.94
300 cases	12210	6.1	2000

3 Optimization Method

Due to that wind turbine is a time-varying structure, conventional empirical formulae for damper optimization is not suitable for wind turbine. Parametric-analysis-based optimization has poor accuracy and can hardly find the global optimal result. To improve optimization accuracy, parametric analysis generally requires to evaluate amounts of cases, which increases calculation time simultaneously.

Consequently, to solve the above-mentioned problems, this study proposes RBFNN- and GA-based optimization method, as can be seen in Fig. 3. Specifically, RBFNN is employed to set up a meta model of PyGAOWT to fast and accurately predict damper vibration control performance. First, many different damper configurations should be evaluated by PyGAOWT to set up a relationship between damper configuration and vibration control performance. Next, numerical results are inputted into RBFNN to set up the meta model. The attained meta model then can fast and accurately predict damper vibration control performance, as can be seen in Fig. 3 (a). Subsequently, GA is utilized to optimize damper parameters. In GA, a random population should be first generated. Each individual of the population represents one damper parameter combination. Their vibration control performances are respectively assessed using the above-mentioned meta model. Then, according to vibration control performances of every individuals, mutation, crossover, and selection occur in the original population to generate a new population, which is theoretically better than the original population. After numbers of iterations, GA can finally converge to the global optimal damper parameter combination, as shown in Fig. 3 (b). Detailed formulae of RBFNN and GA can be referred to the study of Liu et al. [3].

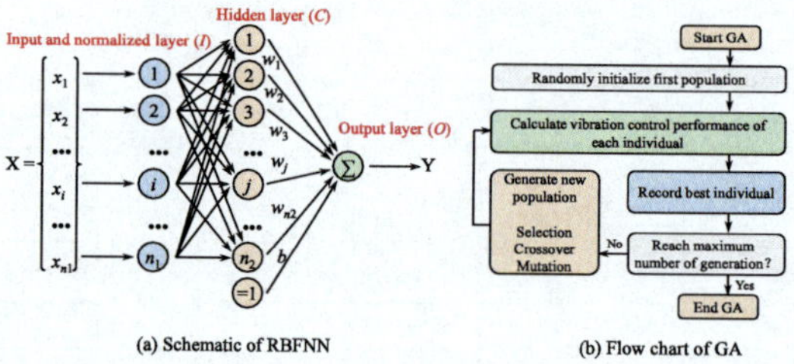

Fig. 3. (a) Schematic of RBFNN; (b) Flow chart of GA.

4 Wake Effects on Damper Optimization

Due to that this study aims to investigate wake effects on damper optimization, it is assumed that there have two wind turbines, which line up along wind propagation direction. TMD is installed on these two wind turbines and is optimized correspondingly. Four different wind turbine spacings are adopted, which are 2*D*, 4*D*, 6*D*, and 8*D*, respectively, with *D* as wind turbine rotor radius. Four different wind speed distributions (C1–4) [2] are considered.

First, wind turbine tower EFL are calculated for these four different spacings, when there has no TMD installed. EFLs are presented in Fig. 4. As can be seen in Fig. 4, EFLs calculated with consideration of wake effects are significantly larger than those calculated without consideration of wake effects. That is, upstream wind turbine wake effects would cause larger EFL of downstream wind turbine. As wind turbine spacing increases, EFL calculated with consideration of wake effects decreases, which exhibits smaller difference with that calculated without consideration of wake effects. Specifically, when wind turbine spacing is approximately 2*D*, EFL calculated with consideration of wake effects is approximately 1.2 times larger than that calculated without consideration of wake effects. However, when wind turbine spacing increases to 4*D*, EFL calculated with consideration of wake effects is only approximately 1.06 times larger than that calculated without consideration of wake effects. Consequently, it can be stated that when wind turbine spacing is approximately 2*D*, wake effects between the wind turbines should be encountered owing to that it can significantly increase wind turbine EFL.

Subsequently, TMD is applied in these two wind turbines. PyGAOWT and RBFNN-GA-based optimization method are utilized to globally optimize TMD structural parameters when under wind conditions C2–4 [2]. Table 2 lists optimized TMD parameters. From Table 2, it can be observed that optimized TMD parameters for wind turbines 1 and 2 are different. Possible explanation is that these two wind turbines are differently affected by wake effects. That is, wake effects will affect wind turbine damper optimizations, and damper optimized without consideration of wake effects can hardly be directly applied in wind turbines affected by wake.

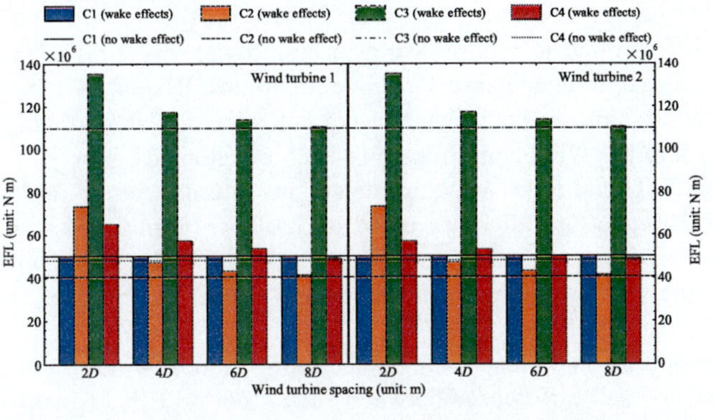

Fig. 4. Wind turbine ELFs without damper when wake effect is considered and not considered.

Figure 5 shows EFL reduction ratios of these two wind turbines after applying TMD. RBFNN-GA-optimized TMDs are compared with theoretically optimized ones. It can be observed that RBFNN-GA-optimized TMDs improve by 13% compared with theoretically optimized ones, and they can reduce at most 40% wind turbine tower EFL.

Table 2. Optimized TMD parameters when under wind conditions C2–4.

TMD				RIDTMD					
Case name	μ	ν	ξ	Case name	μ	μ_1	ξ	β	ν
C1	0.2100	1.1751	0.1769	C1	0.2100	1.6000	0.3946	0.9139	1.6000
C2	0.2089	1.0880	0.3108	C2	0.1886	1.2769	0.1402	0.4407	1.2576
C3	0.2100	0.9377	0.0882	C3	0.2100	0.6255	0.4390	0.9106	1.5999
C4	0.1998	0.9896	0.2253	C4	0.2100	0.4695	0.4149	0.1104	1.6000

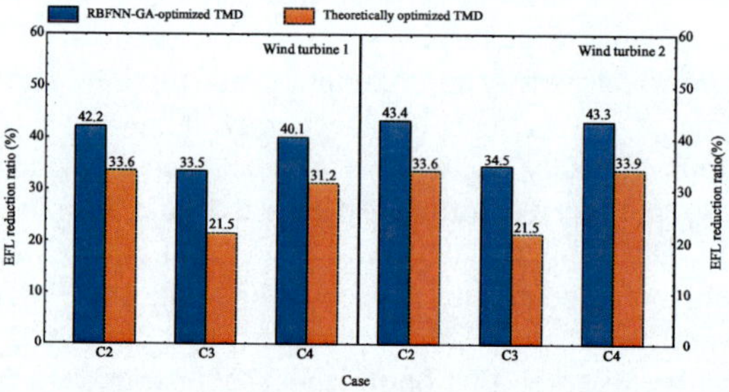

Fig. 5. EFL reduction ratio with RBFNN-GA-optimized and theoretically optimized TMD.

5 Comparison Between TMD and RIDTMD

RIDTMD is also applied to control wind turbine tower vibration in this section. Its parameters are also optimized using the above-mentioned RBFNN-GA-based optimization tool. In this section, wake effects are not considered. Dampers are optimized for only one wind turbine. Wind conditions C1–4 [2] are adopted again.

Optimized TMD and RIDTMD parameters are summarized in Table 3. Their corresponding EFL reduction ratios are listed in Table 4. From Table 4, it can be seen that wind turbine damper optimization is greatly affected by incoming wind conditions. Optimized dampers under different wind conditions are not generic. That is, damper optimized under C1 can hardly be utilized in C2, C3, and C4. Damper structural parameters are better to be optimized correspondingly according to local wind condition. TMD and RIDTMD exhibit small differences between each other on EFL reduction ratio. However, TMD performs a little better than RIDTMD. Optimized TMD and RIDTMD can at most reduce 44% wind turbine tower EFL when under real wind distributions.

Table 3. Optimized TMD and RIDTMD structural parameters.

TMD				RIDTMD					
Case name	μ	v	ξ	Case name	μ	μ_1	ξ	β	v
C1	0.2100	1.1751	0.1769	C1	0.2100	1.6000	0.3946	0.9139	1.6000
C2	0.2089	1.0880	0.3108	C2	0.1886	1.2769	0.1402	0.4407	1.2576
C3	0.2100	0.9377	0.0882	C3	0.2100	0.6255	0.4390	0.9106	1.5999
C4	0.1998	0.9896	0.2253	C4	0.2100	0.4695	0.4149	0.1104	1.6000

Table 4. EFL reduction ratios of optimized TMD and RIDTMD.

TMD		Wind condition				RIDTMD		Wind condition			
		C1	C2	C3	C4			C1	C2	C3	C4
Optimized TMD	C1	48.53%	48.97%	36.73%	30.36%	Optimized RIDTMD	C1	49.03%	48.98%	37.41%	30.68%
	C2	49.16%	49.12%	36.35%	31.03%		C2	48.10%	48.07%	38.74%	32.09%
	C3	44.35%	44.28%	41.17%	38.26%		C3	43.58%	44.55%	39.63%	39.08%
	C4	41.28%	42.23%	33.98%	44.21%		C4	40.96%	41.77%	34.60%	43.59%

6 Conclusion

In this study, TMD and RIDTMD are installed at top of wind turbine tower to reduce tower vibration. Novel PyGAOWT and RBFNN-GA-based optimization tool are presented. From numerical results, following conclusions can be drawn: i) TMD performs a little better than RIDTMD, which is more appropriate in wind turbine vibration control; ii) Wind condition greatly affect damper optimization result. In real engineering applications, wind turbine damper should be optimized correspondingly according to local wind distribution; iii) When wind turbine spacing is approximately $2D$, wake effects between them should be considered; iv) Optimized dampers reduce at mots 44% wind turbine tower EFL in this study.

References

1. Ishihara, T., Qian, G.: A new Gaussian-based analytical model for wind turbines considering ambient turbulence intensity and thrust coefficient effects. J. Wind Eng. Ind. Aerodyn. **177**, 275–292 (2018)
2. Liu, Z., Wang, Y., Hua, X., Zhu, H., Zhu, Z.: Optimization of wind turbine TMD under real wind distribution countering wake effects using GPU acceleration and machine learning technologies. J. Wind Eng. Ind. Aerodyn. **208**, 104436 (2021)
3. Liu, Z., Wang, Y., Nyangi, P., Zhu, Z., Hua, X.: Proposal of a novel GPU-accelerated lifetime optimization method for onshore wind turbine dampers under real wind distribution. Renew. Energy **168**, 516–543 (2021)

4. Rezaei, R., Fromme, P., Duffour, P.: Fatigue life sensitivity of monopile-supported offshore wind turbines to damping. Renew. Energy **123**, 450–459 (2018)
5. Fitzgerald, B., Basu, B.: Active tuned mass damper control of wind turbine nacelle/tower vibrations with damaged foundations. Key Eng. Mater. **569**, 660–667 (2013)
6. Dinh, V.N., Basu, B.: Passive control of floating offshore wind turbine nacelle and spar vibrations by multiple tuned mass dampers. Struct. Control. Health Monit. **22**, 152–176 (2015)
7. Sun, C.: Semi-active control of monopile offshore wind turbines under multi-hazards. Mech. Syst. Signal Process. **99**, 285–305 (2018)
8. Ghassempour, M., Failla, G., Arena, F.: Vibration mitigation in offshore wind turbines via tuned mass damper. Eng. Struct. **183**, 610–636 (2019)
9. Liu, Z., Li, Q., Ishihara, T., Peng, J.: Numerical simulations of fatigue loads on wind turbines operating in wakes. Wind Energy **23**, 1301–1316 (2020)

Influence of Openings on Fluctuating Wind Pressure Characteristics of Large-Span Cantilevered Roofs

Min Zhang[1,2], Pailin Fang[2], Qizhou Liu[1,2](✉), Dan Lu[3], and Dan Pan[2]

[1] GuangXi Key Laboratory of New Energy and Building Energy Saving, Guilin 541004, China
liuqz@glut.edu.cn

[2] College of Civil Engineering and Architecture, Guilin University of Technology, Guangxi 541004, China

[3] East China Architectural Design and Research Institute Co., Ltd., Shanghai, China

Abstract. Based on the rigid pressure measurement test of the stadium large-span cantilevered roof. The influence of openings on the fluctuating wind pressure characteristics of roof is studied. Firstly, compare the roof of no opening and opening in some aspects, such as fluctuating wind pressure coefficients, skewness and kurtosis, and probability density functions. Secondly, the two types of roofs are divided into non-Gaussian regions based on the cumulative probability curve. Finally, the Hermite moment model is used to calculate the peak factor to consider the non-Gaussian distribution characteristics of the roof. The results show that probability distributions of fluctuating pressures within the vicinity of the windward-side edge, which appear serious deviation and bulge compared with those of standard Gaussian distribution, and which present typical non-Gaussian characteristics. However, the opening can effectively slow down the flow separation phenomenon in the edge opening area, which weakens the non-Gaussian characteristics. The area located in the middle and rear of the roof with smooth airflow still obeys gaussian distribution after opening. At last, the reasonable values of peak factor of the two types of roofs are put forward.

Keywords: Cantilevered roof · Opening · Fluctuating wind load · Non-Gaussian characteristic · Wind tunnel test

1 Introduction

Large-span cantilevered roof, a typical form of large-span roof structures that are widely used in stadiums, gas stations and exhibition centers, etc., are characterized as aesthetic, light, flexible and extremely sensitive to wind loads [1]. The vulnerability of large-span roof structures to wind loads has well been recognized, and there are increasing evidences that wind action is the decisive loading condition for the structural design of large-span roof structures [2]. For large-span cantilevered roofs in particular, part of the roof can be easily lifted under strong winds due to the combined action of the high negative wind pressure on the upper surface and positive wind pressure on the lower surface of the roof.

W. Guo and K. Qian (Eds.): GBCESC 2022, LNCE 211, pp. 779–788, 2023.
https://doi.org/10.1007/978-981-19-5217-3_78

Efforts have been made towards the aerodynamic shape optimization of roofs in order to reduce the wind loads, with creating holes in selective positions proved to be one of the most economical and effective measures [3]. Despite the many in-depth discussions on the opening optimization for large-span roof with openings, most of the researches were based on the mean wind pressure characteristics with few on the characteristics of fluctuating wind pressure. It is imperative to estimate the extreme wind pressure on the roof surface with sufficient accuracy in order to ensure structural reliability.

At present, a peak factor based on Gaussian distribution postulation is easy to calculate and widespread applications in codes of practice, including the Chinese load code (GB50009-2012) [4], to calculate the extreme wind pressure on structures. Increasingly, studies indicate that the windward front areas of large-span roof affected by the separation vortex often show strong non-Gaussian (or non-G for short) characteristics. Ying et al. [5] separated non-G regions of the large-span flat roof based on statistics and revealed the formation mechanisms of the non-G characteristics. Min et al. [6] studied the high-order statistics and probability density distribution of wind pressure on large-span roof under different turbulence intensities and scales, and the findings indicated that the peak factor was closely related to the turbulence. Jie et al. [7] discussed the influence of non-G properties of large-span roof on the peak factor, and pointed out that the value of peak factor of wind-sensitive area of roof is much larger than that obtained by traditional methods. If the peak factor based on Gaussian hypothesis is still adopted, the extreme value of wind pressure on large-span roof will be underestimated, resulting in unsafe design. Some scholars have tried to improve the traditional methods for peak factor to determine the extreme wind pressure effectively and accurately. For example, Steven et al. [8] considered converting non-G process into Gaussian process, and proposed the calculation method for non-G peak factor based on the second-order Hermite series polynomial. Wei et al. [9] compared and analyzed various peak factor calculation methods, and proposed a new method based on the improved peak factor method. The proposed method has small calculation and good applicability in application, which is more convenient for engineering design.

In this paper, a stadium with large-span cantilevered roofs is used as an example. The synchronized pressure tests on the roof with and without openings were conducted in the wind tunnel. Based on the wind tunnel test results, the fluctuating wind pressure distribution characteristics for the roof with and without openings are compared and analyzed in terms of statistics, non-G region partition and Hermite moment-based non-G peak factor. The key findings are summarized in Conclusion.

2 Wind Tunnel Test

2.1 Model Design and Wind Pressure Measurement

The stadium consists of east stand and west stand, with the roof cantilevering up to 172 m and 260 m, respectively. The rigid pressure measurement model of the stadium with a length scale ratio of 1:250 was fabricated as shown in Fig. 1. The wind tunnel test was conducted in the CGB-1 wind tunnel laboratory. The test section's wide is 4 m, and the height is 2.8 m, the maximum wind speed is 20 m/s.

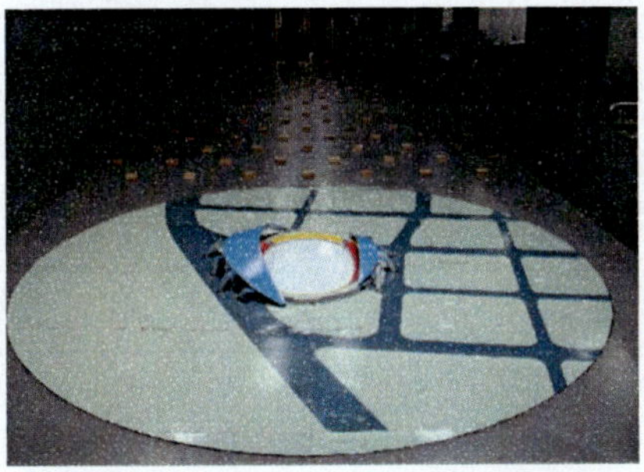

Fig. 1. Wind tunnel test model

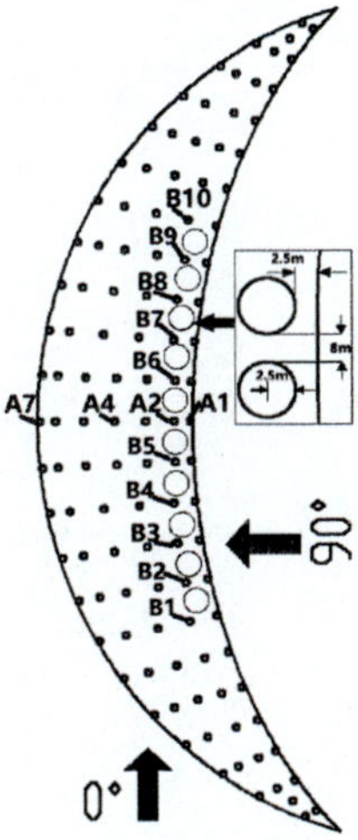

Fig. 2. Pressure taps and wind direction

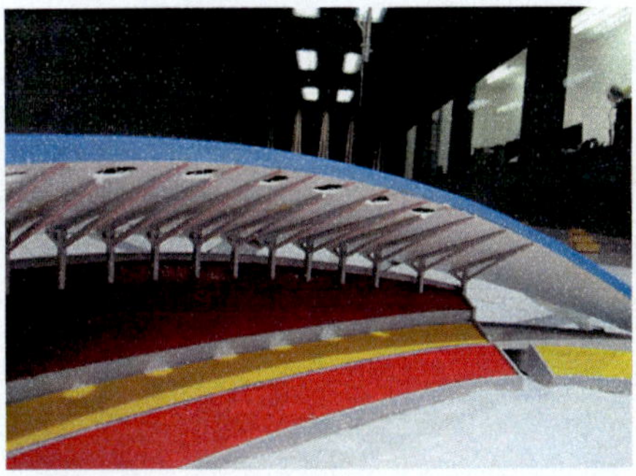

Fig. 3. Wind tunnel test model with openings

The plan view of the west roof is shown in Fig. 2. A total of 230 pressure taps were installed on the west roof of the stadium model and able to measure the wind pressure on the upper and lower surfaces synchronously. In the test, the sampling frequency of pressure taps was 312.5 Hz and the sampling duration was 32 s. The pressure measurements were repeated for a total of 24 wind directions from 0° to 360° at intervals of 15° by rotating the working table. Figure 2 shows the layout of pressure taps and the definition of wind direction. The test considered two conditions, i.e., with and without open holes in the roof. After the pressure measurements for the no opening condition, a total of 10 circular holes with a radius of 10 mm (i.e., equivalent to 2.5 m in the prototype scale and an opening ratio of 1.8%) were drilled in the west roof along the cantilever edge as shown in Fig. 2. Figure 3 shows the model with the openings in the roof. The pressure measurements on the roof with open holes were conducted the same as for the no opening condition.

2.2 Wind Field

In the test, the multi-functional simulator at the wind tunnel entrance and the ground roughness element are used to simulate the wind field in the atmospheric boundary layer of Class B landform specified in the Chinese code (GB50009-2012). The associated mean wind speed (U) and turbulence intensity (I_U) of the prescribed wind field can be well reproduced in the wind tunnel as shown in Fig. 4. Figure 5 shows the power spectrum of dimensionless wind speed measured at the reference height against those from several classic models.

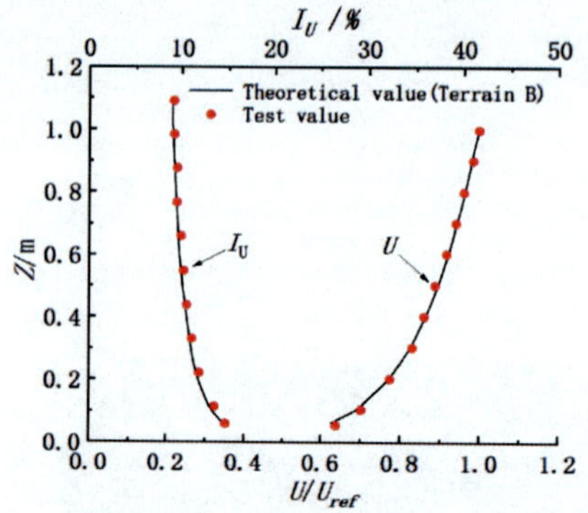

Fig. 4. Profiles of experimental mean wind velocity and turbulent intensity

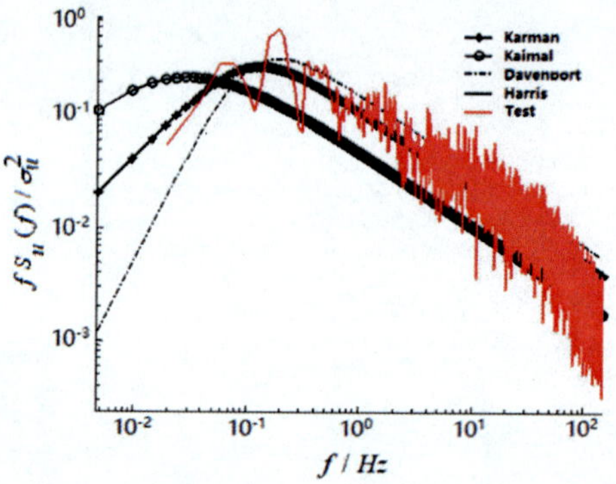

Fig. 5. Fluctuating wind speed spectrum

3 Results and Discussions

3.1 Fluctuating Wind Pressure Distribution Characteristics

In this paper, contours of fluctuating wind pressure coefficient are presented for the representative wind directions of 45° and 90° as shown in Fig. 6. It is clearly shows that the cantilevered end of the roof, when directly impacted by the approaching wind, demonstrates much larger fluctuating wind pressure coefficients than the rest parts. The presence of the openings is shown to greatly reduce the maximum fluctuating wind pressure coefficient by up to 38% from 0.52 to 0.32 for wind direction of 45° (Fig. 6a, b), or reduce the areas affected by significant fluctuating wind pressures as for wind direction of 90° (Fig. 6c, d). This has been ascribed to the high-speed upward airflow from the lower surface through the holes, disturbing the vortex around the upper surface and in consequence leading to alleviated airflow separation at the windward edge of the roof. In contrast with the windward edge, the fluctuating wind pressure coefficients at the other areas of the roof are seemingly to be little affected by the openings.

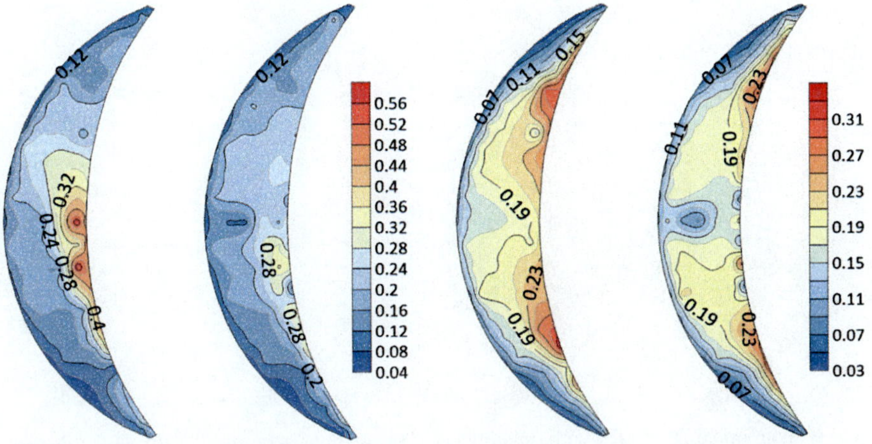

(a) without openings at 45° (b) with openings at 45° (c) without openings at 90° (d) with openings at 90°

Fig. 6. Contours of fluctuating wind pressure coefficients

3.2 Skewness and Kurtosis

In case of non-G pressure, high-order statistical moments, i.e., kurtosis γ_4 and skewness γ_3, are needed to describe the deviation and protrusion degree of the probability distribution of wind pressure signal, with $\gamma_3 = 0$ and $\gamma_4 = 3$ indicating standard Gaussian signals while $\gamma_3 \neq 0$ and $\gamma_4 > 3$ non-G ones. The skewness and kurtosis are obtained as follows:

$$\gamma_3 = n^{-1} \sum\nolimits_{i=1}^{n} \left[\left(C_{pi}(t) - C_{pi,mean} \right) / C_{pi,std} \right]^3 \tag{1}$$

$$\gamma_4 = n^{-1} \sum\nolimits_{i=1}^{n} \left[\left(C_{pi}(t) - C_{pi,mean} \right) / C_{pi,std} \right]^4 \tag{2}$$

where $C_{pi}(t)$ is the wind pressure coefficient for tap *i* on the model surfaces; $C_{pi,mean}$ and $C_{pi,std}$ are the mean and standard deviation, respectively, of the pressure coefficient at tap *i*; n is the total number of samplings at tap *i*.

Figure 7 and Fig. 8 show the contours of skewness and kurtosis of roof wind pressure, respectively. For wind direction of 45°, a few γ_3 values around the middle of the front windward edge are less than -1.0 and the γ_4 values are larger than 5.2. This is due to that the airflow separation occurred in the above-mentioned region, the fluctuating wind pressure presents obvious non-G characteristics under the action of separated vortex. On the other hand, the central and leeward parts of the roof with γ_3 values close to 0 and γ_4 values around 3 are likely to be Gaussian regions. The spatial distributions of skewness and kurtosis on the roof at 90° wind direction are similar to those at 45°, except that the non-G characteristics are obviously weaker than those at the 45° wind direction. After openings appear in the strong non-G regions of the roof, the skewness and kurtosis associated with the non-G characteristics are significantly reduced around the opening areas, while this phenomenon is less obvious for areas away from the openings.

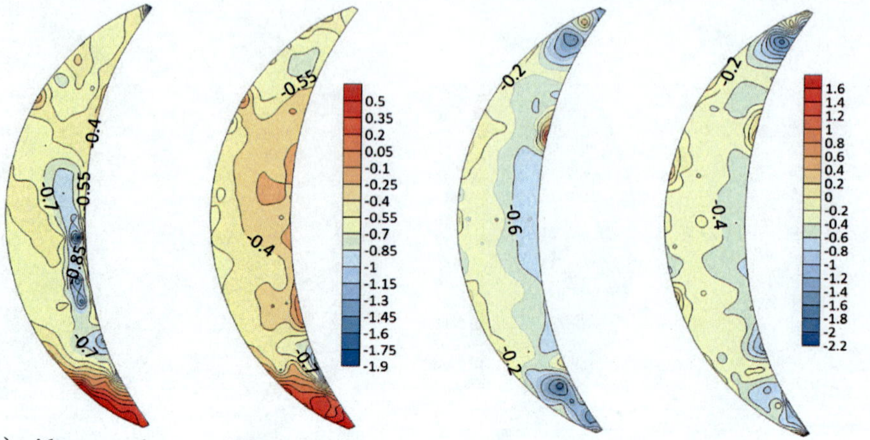

(a) without openings at 45° (b) with openings at 45° (c) without openings at 90° (d) with openings at 90°

Fig. 7. Contour of skewness of the fluctuating wind pressure

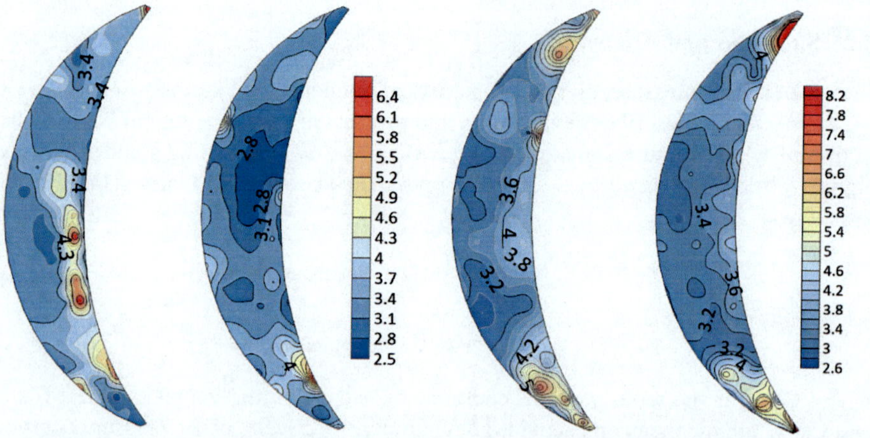

(a) without openings at 45° (b) with openings at 45° (c) without openings at 90° (d) with openings at 90°

Fig. 8. Contour of kurtosis of the fluctuating wind pressure

3.3 Probability Density Function

The probability density functions of the fluctuating wind pressure at taps A2, A4 and B1 (see Fig. 2) were compared with the standard Gaussian distribution curve for wind directions of 45°, as shown in Fig. 9.

As can be seen from Fig. 9, the probability densities of the fluctuating wind pressure at taps in the vicinity of the front windward edge, such as A2 and B1, deviate from the standard Gaussian distribution curve with noticeable negative skew and sharper bulge.

The non-G characteristics are relatively significant for the pressure at tap A2 under 45° wind direction, with the deviation degree attaining 38% and reduced to 15% after introduction of the openings (Fig. 9a, b). Meanwhile, the use of openings is found to almost eliminate the non-G characteristics of the wind pressure at tap B1 (Fig. 9f, h). In contrast, the tap A4 situated in the central part of the roof presents a Gaussian probability density function of fluctuating wind pressure regardless of openings (Fig. 9i–l).

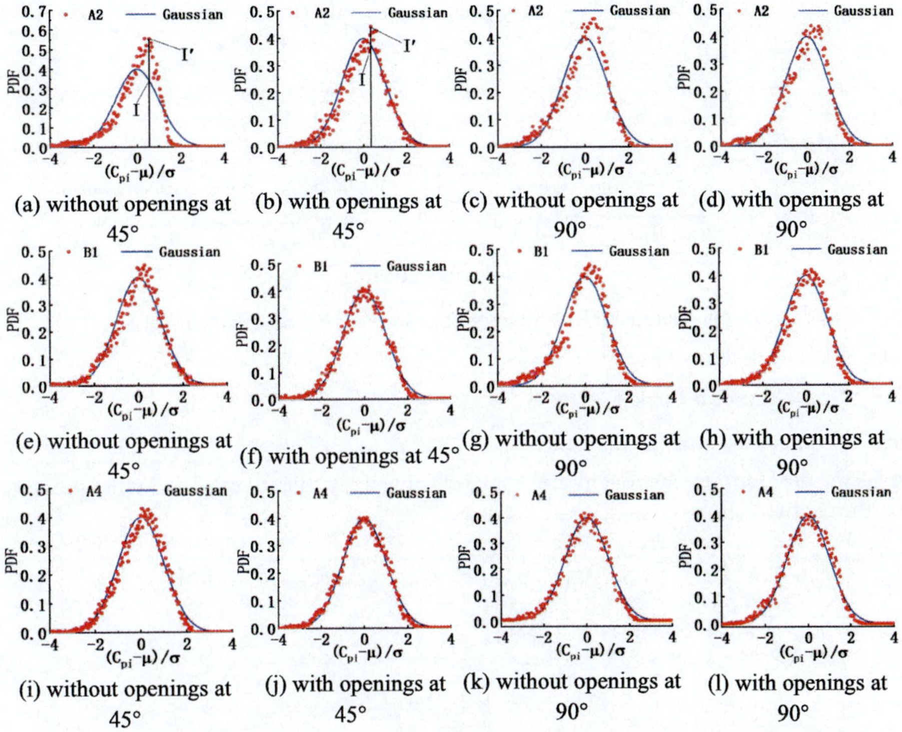

(a) without openings at 45° (b) with openings at 45° (c) without openings at 90° (d) with openings at 90°

(e) without openings at 45° (f) with openings at 45° (g) without openings at 90° (h) with openings at 90°

(i) without openings at 45° (j) with openings at 45° (k) without openings at 90° (l) with openings at 90°

Fig. 9. Probability density function of fluctuating wind pressure on typical pressure taps

3.4 Non-Gaussian Region Partitioning

Currently, the non-G region of fluctuating wind pressure on large-span roof is usually partitioned according to the range of coefficients of skewness and kurtosis. In this paper, a cumulative probability of 0.80 is employed to define the range of skewness and kurtosis values for non-G region partitioning as shown in Fig. 10. On the basis of this criteria, when the skewness and kurtosis of the wind pressure at the taps satisfy $|\gamma_3| > 0.39$ and $\gamma_4 > 3.48$, respectively, this region belongs to non-G region at 45° wind direction. Figure 10 provides the obtained threshold values of skewness and kurtosis between Gaussian and non-G regions of roofs with and without openings for wind directions of 45° and 90°.

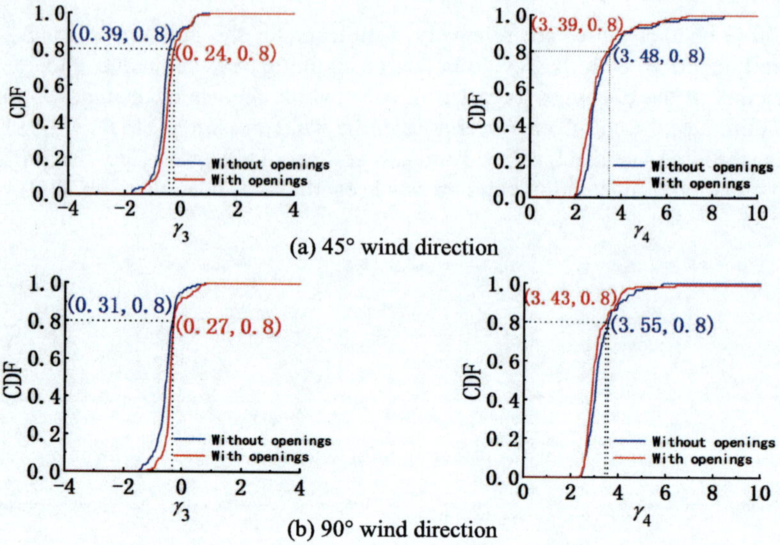

Fig. 10. Cumulative distribution functions of skewness and kurtosis.

3.5 Non-Gaussian Peak Factor

This paper has established the correlation between non-G process and Gaussian process by using the Hermite moment model, and derived the peak factor of non-G wind pressure on the roof as follows:

$$g = k\left\{\left(\beta + \frac{\gamma}{\beta}\right) + h_3\left(\beta^2 + (2\gamma - 1) + \frac{1.98}{\beta^2}\right) + h_4\left[\beta^3 + 3\beta(\gamma - 1) + \frac{3}{\beta}\left(\frac{\pi^2}{6} - \gamma + \gamma^2\right) + \frac{5.44}{\beta^3}\right]\right\} \tag{3}$$

$$h_3 = \frac{\gamma_3}{4 + 2\sqrt{1 + 1.5\gamma_4}} \tag{4}$$

$$h_4 = \frac{\sqrt{1 + 1.5\gamma_4} - 1}{18} \tag{5}$$

$$k = \frac{1}{\sqrt{1 + 2h_3^2 + 6h_4^2}} \tag{6}$$

where $\beta = \sqrt{2\log(\nu_0 T)}$; ν_0 is mean zero upcrossing rate; T is Observation time period; and $\gamma = 0.5772$ is Euler's constant.

Table 1. Peak factors of non-G pressure for selected taps

Wind direction	Taps	Without openings			With openings		
		γ_3	γ_4	g	γ_3	γ_4	g
45°	B1	−0.741	3.793	3.498	−0.338	3.316	3.145
	B3	−1.450	6.107	4.166	−0.495	3.729	3.567
	B4	−1.043	4.569	3.739	−0.425	3.365	2.803
	A2	−1.020	5.089	4.161	−0.432	3.565	3.824
90°	B4	−0.702	4.130	3.739	−0.587	3.528	3.351
	B5	−0.686	4.271	3.854	−0.414	4.118	3.437
	A2	−0.812	3.953	5.270	−0.606	3.514	3.705

Results of calculated peak factors for a few selected non-G taps along the windward edge are listed in Table 1 (the pressure taps subject to Gaussian characteristics after opening are not shown in the table). As can be seen, these values of peak factor are all considerably larger than 2.5 as currently stipulated in the Chinese code (GB50009-2012) based on Gaussian assumption. In other words, adopting a uniform peak factor of 2.5 is very likely to underrate the extreme wind pressure on the roof of the stadium. The utilization of the roof openings helps reduce the peak factors to some extent, but the values still surpass the expectation. Therefore, when the peak factor method is used to estimate the extreme value of roof wind pressure, the Hermite moment model theory can be used to consider the influence of non-G characteristics, otherwise it will acquire vulnerabilities to structural design.

4 Conclusion

The wind tunnel test and analysis based on a large-span cantilevered roof of a stadium with and without roof openings. The main conclusions were drawn as follows:

(1) The maximal negative pressure appears at the leading edge of roof, which is unfavorable to the cantilevered roof. The roof openings in the costal margin can disturb the vortex of airflow separation at the windward edge, leading to much reduced wind pressure at the cantilevered end and more favorable wind pressure distribution pattern.
(2) The windward edge of the roof is subjected to strong non-G pressure fluctuations based on the spatial distribution of skewness and kurtosis on the roof, while the central and leeward part of the roof is subjected to Gaussian pressure fluctuations. Both the non-G regions and non-G characteristics are reduced significantly as a result of introducing the openings.
(3) Although the peak factor of the opening roof in the non-G region is reduced, the non-G peak factor is much greater than the recommended peak factor of 2.5 based on Gaussian assumption. It is recommended that the non-G distribution areas of

large-span cantilevered roofs with and without openings can be confirmed by the Hermite moment model theory.

Acknowledgements. The authors acknowledge with thanks the financial assistance provided by the Natural Science Foundation of Guangxi (No. 2018GXNSFAA050145), the Guangxi S&T Base and Special Fund (No. Guike AD19110073) and Guangxi Key Laboratory Project of New Energy and Building Energy Saving (Gui Ke Neng 17-J-21-9). Any opinions expressed in this paper do not necessary reflect the view of Guangxi Key Laboratory of New Energy and Building Energy Saving and Natural Science Foundation of Guangxi.

References

1. Qinghua, Z.: Experimental study of wind pressure on a long-span canopy roof structure using a wind tunnel. Iranian J. Sci. Technol. Trans. Civ. Eng. **45**(1), 197–206 (2020)
2. Tiantian, L., Guirong, Y., Fangping, Y.: Dynamic structural responses of long-span dome structures induced by tornadoes. J. Wind Eng. Ind. Aerodyn. **190**(9), 293–308 (2019)
3. Jianfei, M.: Wind pressure characteristics and aerodynamic optimization of stadium grandstand's large-span cantilevered roof. Zhejiang University of Technology (2014)
4. Load Code for the Design of Building Structures: GB 50009-2012. China Architecture and Building Press, Beijing (2012)
5. Ying, S., Yue, W., Zhixing, L.: Non-Gaussian features of fluctuating wind pressures on long span roofs. Chin. Civil Eng. J. **40**(4), 1–5 (2007)
6. Min, L., Xinzhong, C., Qingshan, Y.: Characteristics of dynamic pressures on a saddle type roof in various boundary layer flows. J. Wind Eng. Ind. Aerodyn. **150**, 1–14 (2016)
7. Jie, S., Wei, X., Gang, H., Shuguo, L., Jian, T.: Non-Gaussian properties and their effects on extreme values of wind pressure on the roof of long-span structures. J. Wind Eng. Ind. Aerodyn. **184**, 106–115 (2019)
8. Steven, R.W.: Nonlinear vibration models for extremes and fatigue. J. Eng. Mech. **114**(10), 1772–1790 (1988)
9. Wei, L., Mingfeng, H., Wenjuan, L.: Peak factor method for non-Gaussian fluctuating pressures on long-span roofs. Build. Struct. **43**(15), 83–87 (2013)

Influence of Scour Depth on Pile Group with Complex Configuration in Clay

Zengliang Wang[1,2], Hang Zhou[1,2](✉), and Longyong Tong[1,2]

[1] College of Civil Engineering, Chongqing University, Chongqing 400045, China
zh4412517@163.com

[2] Key Laboratory of New Technology for Construction of Cities in Mountain Area, Chongqing University, Chongqing 400045, China

Abstract. The pile foundation in the current and marine environment will be affected by the scour of the current, which will cause the soil around the pile to be eroded, thereby reducing the bearing capacity of the pile. Many studies were pointed out that scour is one of the main reasons for bridge failure. However, A few studies have investigated the change in bearing capacity of laterally loaded pile groups under scour conditions. Based on the LIJIATUO Bridge 5# and 6# pile groups foundation and corresponding geological data on the JIALING River in Chongqing, a number of numerical studies are conducted to examine the lateral behavior of two pile groups with complex configurations under different scour conditions by using three-dimensional FEM model. The MCC model was adopted to describe the mechanical responses of soil. The formation process of the local scour hole around pile groups was simulated using the life and death element method. By comparing the bearing capacity of laterally loaded pile group and laterally loaded isolated pile under the same scour conditions. The results show that the increase in scour depth will reduce the bearing capacity of the laterally loaded pile group. The group effect of pile groups under lateral load will change with the varying scour depth. Among them, the scour depth has the most excellent effect on the group effect of the leading row piles.

Keywords: Local scour depth · Laterally loaded pile groups · Complex configuration · Numerical method · Pile group effect

1 Introduction

Bridges in oceans and rivers are generally subjected to scour caused by waves and currents. Scour can remove the soils around the bridge foundations, thus reducing the bridge foundation's axial and lateral bearing capacity. It became the main reason for the bridge failures, reduced its safety during service and caused significant economic losses [1]. Therefore, it is a crucial adjective to investigate the capacity of the bridge pile foundation considering scour.

Compared with the extensive research on scouring mechanisms, Only a few studies have looked into the consequences of local scour on pile groups and the infrastructure

W. Guo and K. Qian (Eds.): GBCESC 2022, LNCE 211, pp. 789–797, 2023.
https://doi.org/10.1007/978-981-19-5217-3_79

that supports them. The centrifuge model tests with various scour hole dimensions were conducted by Qi (2016) to study the effect of local scour on the lateral behavior of isolated pile in sand [2], and showed that general scour mainly affects the load eccentricity, and compared with the standard conditions, for a given depth, the formation of local scour holes increases the lateral stiffness of the soil. Ciancimino (2021) investigated the performance of piles through slow cyclic pushover considering local scour conditions, lateral monotonic and vertical tests using centrifuge tests [3]. It demonstrates that local scour has negligible impact on the pile's bearing capacity. However, local scour has a manifest influence on pile lateral bearing capacity, and the influence of general scour is even more severe. In addition, the formation of scour hole is similar to an unloading process, which changes the geo-stress state of the unscoured soil around the pile. Recently, several studies have focused on studying the effect of scouring on the laterally loaded isolated pile by considering changes in soil stress history caused by scouring [4], Li (2020) studied the evolution of the lateral response of a single pile under local scour and general scour by centrifuge tests, Chortis (2020) studied the scour type (general scour, local scour) and local scour depth on monopile *p-y* by centrifuge tests [5].

In summary, the influence of scouring on the performance of laterally loaded monopiles is the subject of current research. In practical engineering, the foundation of the bridge and infrastructure is mainly in the form of pile groups. However, there is currently no research concerning the effect of scouring on pile groups under lateral load.

The primary purpose of this paper is to understand the effect of scouring on pile groups under lateral load through 3D finite element analysis. According to extensive literature analysis, no data is available on the laterally loaded pile groups in clay with scouring. Firstly, a baseline finite element model is established, verified by the centrifuge test without scouring conditions. Secondly, according to the scour test, the simplified scour holes of the isolated pile and pile groups are obtained. Finally, the local scour hole formation process around the pile groups is simulated using the finite element software Abaqus. Then the lateral load is applied to the top surface of the pile group.

2 Baseline Finite Element Model

The centrifuge test conducted by Ilyas (2004) for laterally loaded single pile with no scour conditions is used to verify and set up the baseline FE model. The results can be used to verify the rationality of the finite element model [6], the centrifuge test was conducted at 70 g on the NUS, the model pile was a hollow aluminum tube, The details of the material parameters of the soil and model piles can be found in Ilyas (2004). The modified cam clay (MCC) constitutive model reflects the stress-strain relationship of soil and model pile is elastic material, the details of the MCC model can be found in Zhou (2017) [7]. The boundary of the model is set enough to avoid the boundary effects. The Coulomb friction law is used to reflect the behavior for the pile-soil interaction. The soil is modeled using the C3D8P element, and the C3D8R element is used for the model pile in the FE model. The baseline FE model mainly includes two analysis steps: the initial geostress balance and lateral loading analysis.

Figure 1 shows the load-lateral displacement relationship between the baseline FE model and centrifuge test at the pile head. The baseline FE model is in good agreement with the centrifuge test results. Therefore, the validity of the material parameters and the finite element model is proved.

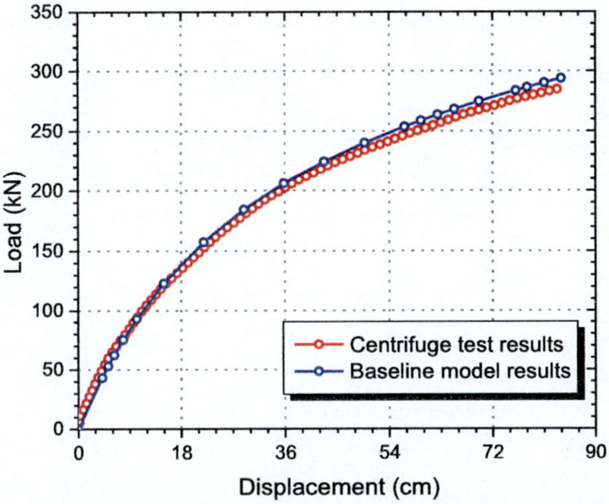

Fig. 1. Verification of the baseline FE model against the centrifuge test

3 FE Model of Laterally Loaded Pile Groups with Complex Configuration Considering Scour Effect

The model change analysis step is added after the initial geostress balance analysis step to simulate the formation progress of local scour holes around the pile group. Namely, three analysis steps are included in the FEM considering the scour effect: (i) the initial geostatic stress analysis; (ii) model change step, using the life-death element method in finite element software Abaqus to kill the soil elements within the scour hole to simulate the scour around the pile group; (iii) lateral loading analysis. Other parameters of the model are consistent with the baseline FE model. For the finite element model of the pile group, the top surface of the pile group is coupled to a point and a lateral displacement is applied to this point.

In this part, based on the LIJIATUO Bridge 5# and 6# pile groups foundation and corresponding geological data on the JIALING River in Chongqing, the FE models are established. The schematic diagram of the configuration of 5# and 6# pile groups of Lijiatuo Bridge is shown in Fig. 2. According to the pile group scour test [8]. The simplified forms of scour holes for two pile groups, as shown in Fig. 3.

According to the corresponding geological data, the riverbed soil near the Lijiatuo Bridge is clay, and the MCC constitutive model is used to characterize the stress-strain relationship. The soil parameters are shown in Table 1. The diameter of the pile is 1.6 m. The flexural rigidity $E_m I_m$ of the model pile is 922000 kNm^2. The Poisson's ratio of the pile is 0.2. The length of the pile $L = 28$ m. Due to the symmetry of the model and loading, as well as the time-saving of finite element calculation, half of the model is used for calculation. The schematic diagram of the FEM of the two pile groups is shown in Fig. 4.

Table. 1. Parameters of clays used in FE model

Parameter	Value
λ	0.122
κ	0.061
M	1.05
φ'	26.9
γ	19 kN/m^3
e_{sc}	0.78

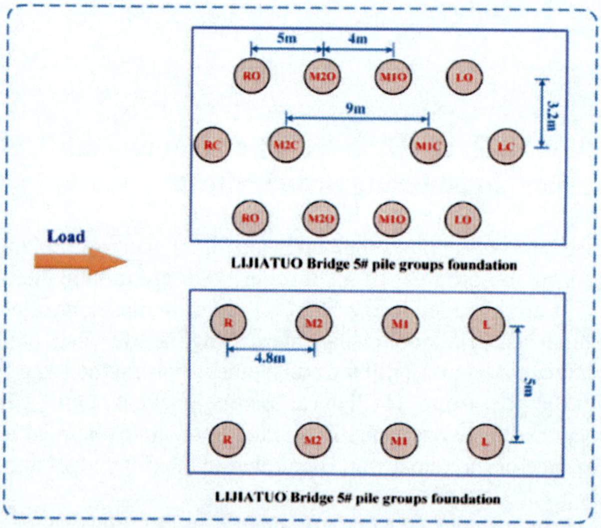

Fig. 2. Schematic of the configuration of Lijiatuo Bridge 5# and 6# pile groups foundation

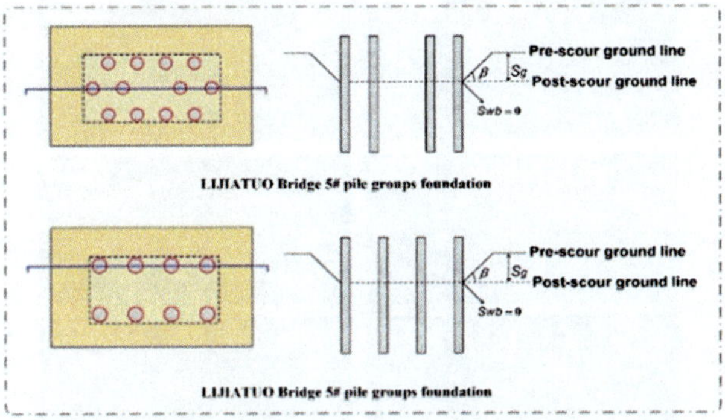

Fig. 3. Simplified form of scour hole of two pile groups

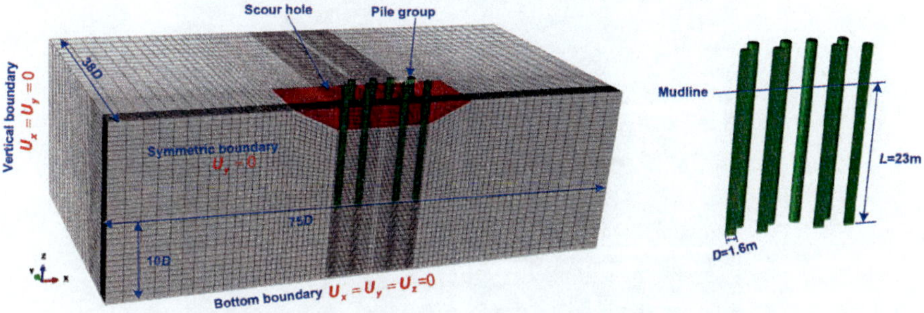

Fig. 4. Finite element model of pile group considering scour effect

4 Results and Analysis

Lin (2018) pointed out that scour depth significantly influences the lateral bearing capacity among all local scour dimensions [9]. In this part, for the two types of pile groups, three different scour depths (i.e., 2 m, 4.8 m, 6.4 m) are used to study the influence on their lateral bearing capacity while $\beta = 25°$. For each pile group model, both have a corresponding single pile model with the same scour conditions.

When the pile head lateral displacement is 0.32 m, the displacement profiles of pile groups and single piles with varying scour depths can been seen in Fig. 5. The load-lateral displacement relationships of 5# and 6# pile groups are shown in Fig. 6 and Fig. 7. When the lateral displacement is small, the local scour depth have the negligible influence on the lateral bearing capacity of the pile group, and since the lateral displacement of the pile group is 0.3D, the load-displacement relationship is approximately a straight line. With the increase of lateral displacement, the influence of local scour depth is gradually significant. Under the greater lateral displacement, the increase in local scour depth reduces the lateral load of pile groups. In addition, The reason is that the increase in the number of pile groups increases the pile-soil interaction and the shear stress area of the soil. The local scour depth reduces the effective buried depth of the pile group and increases the load eccentricity (Fig. 4).

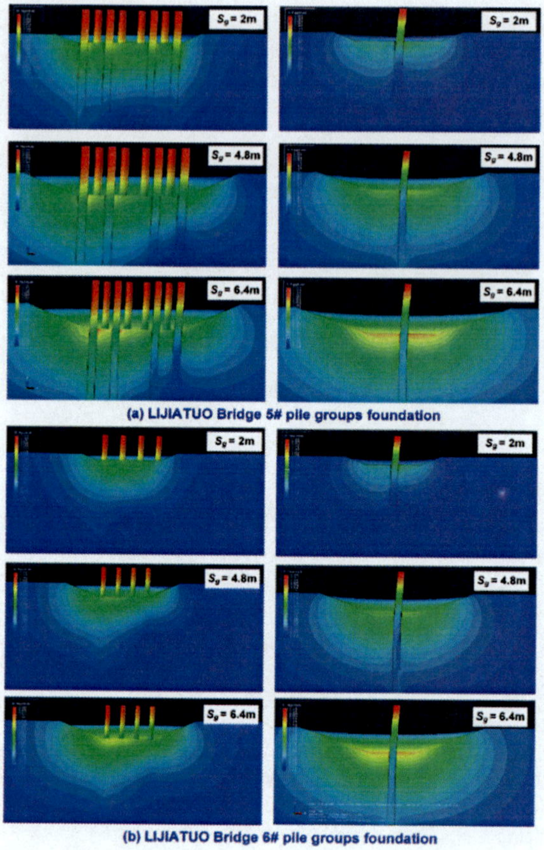

Fig. 5. Contours of soil displacement.

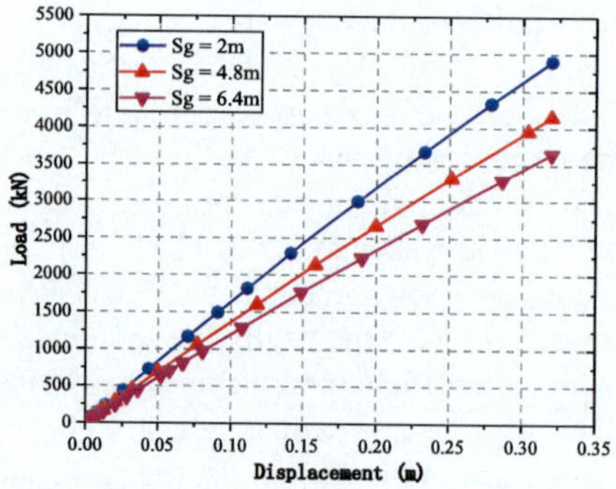

Fig. 6. Load-displacement relationship of 5# pile group under different scour depths

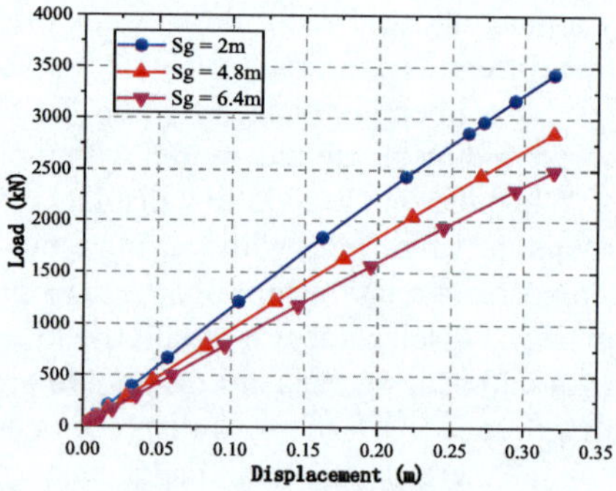

Fig. 7. Load-displacement relationship of 6# pile group under different scour depths

When pile groups are subjected to lateral loads, group effects (i.e., shallowing effect and edge effect) will appear due to the overlap of lateral soil resistance zones. Figure 8, 9 show the bending moment profiles of each piles-in-group under different scour depths for 5# and 6# pile groups when the pile head reaches a lateral displacement of 0.32 m, respectively. The depth reference in the figure is post-scour ground line. When the local scour depth is 2 m, the difference between the bending moments of different isolated piles is ignorable. The difference between the bending moments of each pile in the pile group gradually increases as scour depth increases. Among them, the maximum value of the bending moment for the piles in the LC pile and LO pile is more significant than other piles, and is gradually close to that of a single p9ile with the same conditions. In the loading direction, the remaining soil above the scour depth provides a more excellent lateral soil resistance for the leading row piles and a more tremendous passive earth pressure for the rear row. This phenomenon will be more evident as the scour depth increases. A similar phenomenon also appeared in the 6# pile group, as shown in Fig. 9.

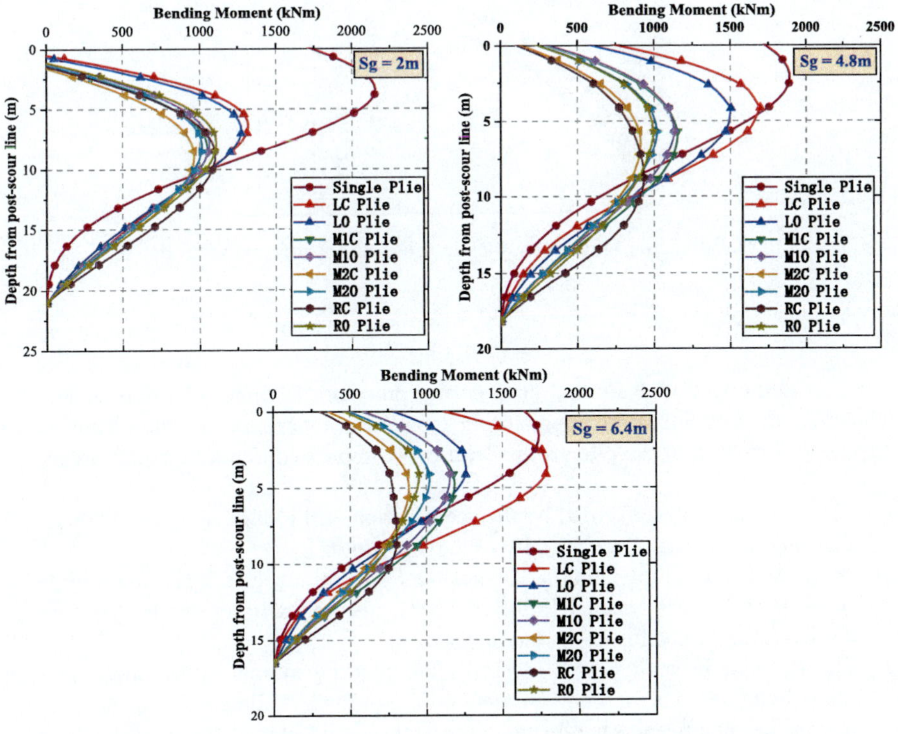

Fig. 8. Moment profile of 5# piles-in-group under different scour depths

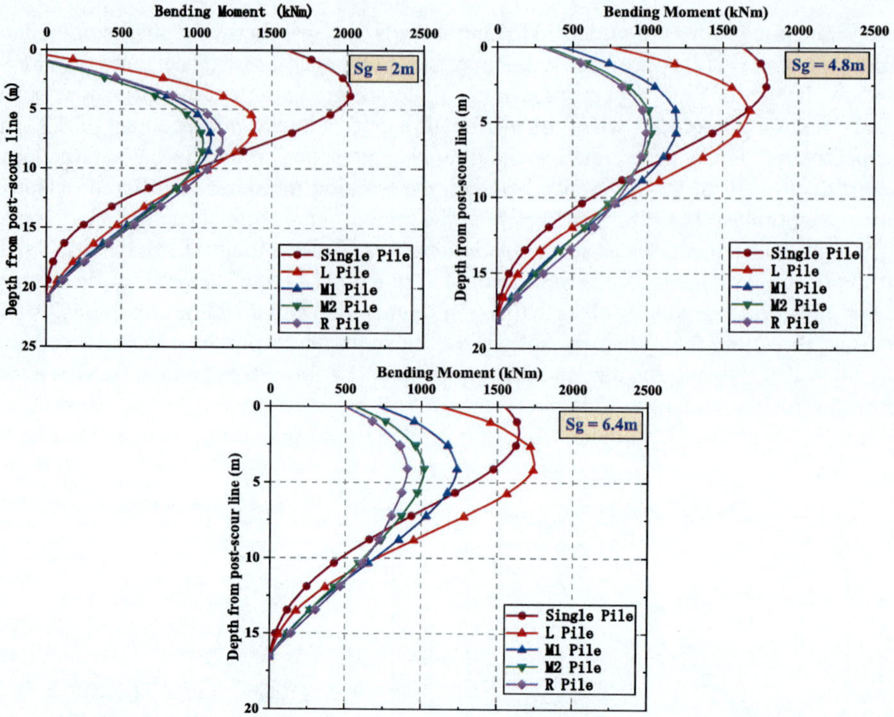

Fig. 9. Moment profile of 6# piles-in-group under different scour depths

5 Conclusions

Based on numerical analysis and engineering practice of Lijiatuo Bridge 5# and 6# foundation, the influence of the depth of the local scour pit around the pile on the lateral bearing performance of the pile groups with two complex configurations is studied.

(1) The local scour hole formed by the scour action will reduce the buried depth and thus increase the load eccentricity of the pile group.
(2) With the increase of the lateral displacement, the influence of the local scour depth on the lateral bearing capacity of the pile group is gradually significant. The greater the scour depth, the lower the lateral capacity.
(3) The increase in scour depth increases the bending moment difference between individual piles. The bending moment profile of the LO pile and LC pile gradually approaches that the single pile under the same conditions with the increase of local scour depth.

References

1. Fayun, L.: Experimental observations and evaluations of formulae for local scour at pile groups in steady currents. Mar. Georesour. Geotechnol. **35**(2), 245–255 (2017)
2. Wengang, Q.: Scour effects on p–y curves for shallowly embedded piles in sand. Géotechnique **66**(8), 648–660 (2016)
3. Ciancimino, A.: Experimental assessment of the performance of a bridge pier subjected to flood-induced foundation scour. Géotechnique, 1–18 (2021)
4. Yunjie, L.: Effects of scour-hole dimensions on lateral behavior of piles in sands. Comput. Geotech. **111**, 30–41 (2019)
5. Chortis, G.: Influence of scour depth and type on p–y curves for monopiles in sand under monotonic lateral loading in a geotechnical centrifuge. Ocean Eng. **197**, 106838 (2020)
6. Ilyas, T.: Centrifuge model study of laterally loaded pile groups in clay. J. Geotech. Geoenviron. Eng. **130**(3), 274–283 (2004)
7. Hang, Z.: A general semi-analytical solution for consolidation around an expanded cylindrical and spherical cavity in modified Cam Clay. Comput. Geotech. **91**, 71–81 (2017)
8. Amini, A.: Clear-water local scour around pile groups in shallow-water flow. J. Hydraul. Eng. **138**(2), 177–185 (2012)
9. Lin, C.: Evaluation of vertical effective stress and pile lateral capacities considering scour-hole dimensions. Can. Geotech. J. **56**(1), 135–143 (2019)

Finite Element Simulation Analysis of Double-Wall Steel Boxed Cofferdam for MaLuKou Zishui River Bridge on Guanzhuang-Xinhua Expressway

Lu Yi(✉)

CCCC Second Navigation Engineering Bureau Co., Ltd., Wuhan 430040, China
a15138051614@163.com

Abstract. Steel boxed cofferdam is a facility for underwater foundation construction, which has the advantages of high stiffness and simple construction. For the safe of steel boxed cofferdam, its overall mechanical properties need to be studied. Take the double-wall steel boxed cofferdam of MaLuKou Zishui River bridge for an example, the finite element model is established by MIDAS/civil. Its mechanical properties in different conditions are discussed and analyzed in this paper. Research shows: (1) The stress state of the double-wall steel boxed cofferdam is good, and the stress of the structural members and the bond stress of the bottom concrete meet the design requirements. (2) In the construction process, extreme stress occurs during pumping, which is the most dangerous working condition. (3) The stress of wallboard and steel pipe support is higher during pumping, and the pumping condition is the design control condition of wall and steel pipe support. (4) When pouring the bottom concrete, the stress of the bottom member is higher, which is the design control condition of the bottom plate. It is of great significance to the design and construction of the steel boxed cofferdam.

Keywords: Double-wall steel boxed cofferdam · Underwater construction · Finite element analysis · Mechanical property

1 Introduction

Steel boxed cofferdam is a helpful structure for building a deep-water high pile bearing platform. Its construction process is a series of continuous loading processes. And the spatial structure is complex, the construction period is long, the construction environment is complex, and the stress condition of the structure is difficult to determine [1–3]. Therefore, the mechanical analysis of its construction process to determine the mechanical performance of steel boxed cofferdam is one of the problems that need to be paid close attention to. The steel boxed cofferdam is divided into two forms: single wall and double wall, which mainly provides the construction environment for the underwater bearing platform [4].

W. Guo and K. Qian (Eds.): GBCESC 2022, LNCE 211, pp. 798–808, 2023.
https://doi.org/10.1007/978-981-19-5217-3_80

Since Jiujiang Bridge was built with steel boxed cofferdam in China, many underwater construction operations have been completed, but the design of steel cofferdam is conservative in practical projects. Or poor consideration of some details in the design, resulting in serious engineering accidents. Therefore, it is necessary to further research the construction of steel boxed cofferdam. This paper makes a theoretical analysis on the double-wall steel boxed cofferdam of MaLuKou Zishui River bridge, establishes the steel boxed cofferdam model by using the finite element software Midas/Civil, makes a mechanical analysis on the practical application of this technology, and provides guiding suggestions for the design and construction of the same type of steel boxed cofferdam [5].

2 Project Overview

MaLuKou Zishui River bridge is located in the middle and downstream reservoir area of the Zhexi reservoir in Hunan Province. The river channel of the Zhexi reservoir is U-shaped, and the maximum construction water depth is 68 m. The bridge adopts a double tower and double cable plane semi-floating system, with a total length of 940 m and a span arrangement of (60 + 160 + 500 + 160 + 60)m. The general layout of the bridge is shown in Fig. 1 [6]. 24 Bored Pile Group Foundations with a diameter of 2.8 m are adopted for the 6th Pier, and the plane size is 39.827 m × 22.8 m rectangular bearing platform with four corners chamfered. The hydrogeological conditions at the pier location are changeable, and the scouring is serious.

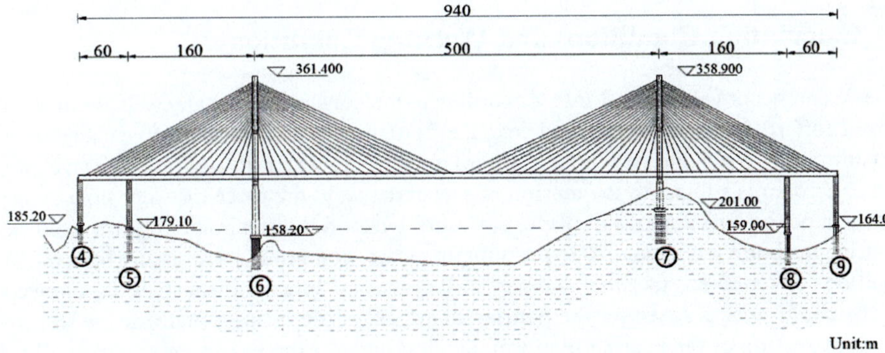

Fig. 1. General Layout of the MaLuKou Zishui River bridge

According to the hydrological environment of the 6th pier, the design and construction characteristics of the bearing platform, the project adopts a double-wall steel boxed cofferdam. The cofferdam is a large-scale non-implantation cofferdam, which is helpful for the construction of bearing platforms in deep water and high scouring. The height of the steel boxed cofferdam is depended on the construction water level and the elevation of the bearing platform. The designed dimension of the steel boxed cofferdam of the 6th pier of the bridge is 42.927 m × 25.900 m × 19.500 m, the distance between the inner and outer wall plates is 1.5 m, the thickness of the inner and outer wall plates is 6/8 mm, the height of the wall plate is divided into four Sects. (5.6 m, 6.5 m, 3.3 m and 4.1 m), the total weight of the steel boxed cofferdam is 1720t, and the thickness of C30 bottom concrete is 3.5 m. The construction site of the double-wall steel boxed cofferdam are shown in Fig. 2.

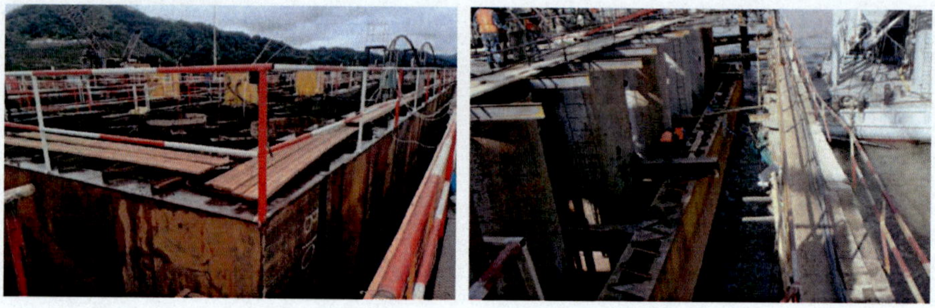

Fig. 2. Construction site

The steel boxed cofferdam is composed of the wallboard (beam), the bottom plate (beam), suspender, compartment plate, steel pipe support, and other components. All components are welded with each other [7–9].

3 Calculation Conditions and Working Conditions

The design water velocity is 1 m/s. According to the construction characteristics of steel boxed cofferdam, dividing the steel boxed cofferdam into the following five calculation conditions: ① When the steel boxed cofferdam entering water, in addition to its weight, the cofferdam is subject to the tension of suspenders and the force of water flow. After the steel boxed cofferdam is in place, the cofferdam and the steel pipe pile are fixed. ② Pouring the bottom concrete. ③ In the pumping water condition, the pressure bar system shall be converted at this time, and part of the internal support truss shall be removed to ensure the access space for the pouring materials of the bearing platform. ④ During the construction of the bearing platform, the first layer of the bearing platform shall be poured. At this time, the temporary steel pipe above the bottom concrete shall be cut off, and the pouring of the second layer of the bearing platform shall be carried out after removing part of the internal support. ⑤ Tower base and column construction.

The construction and calculation process of steel boxed cofferdam for the 6th Pier is shown in Fig. 3.

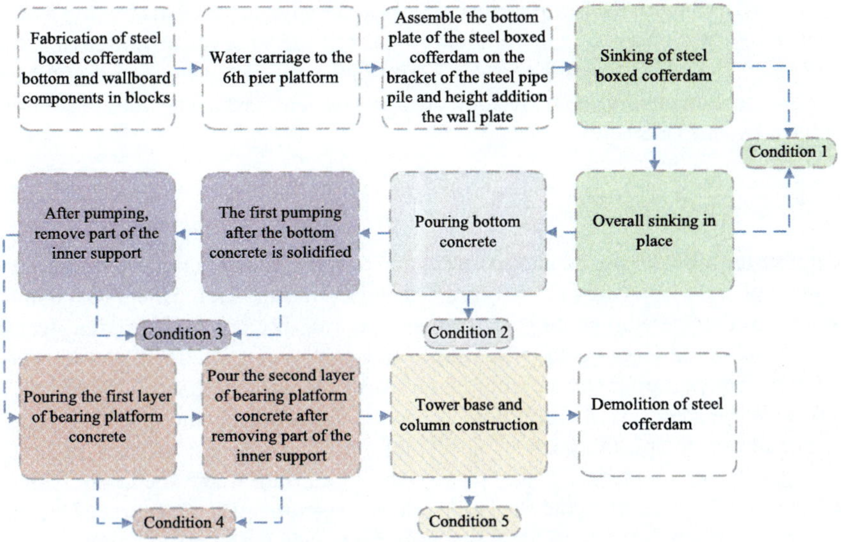

Fig. 3. Construction technology and calculation conditions

4 Calculation Model of Steel Boxed Cofferdam

At present, steel boxed cofferdam has been applied in the construction of a high pile bearing platform in deep water widely. The simplified two-dimensional finite element model can not fully grasp its overall mechanical properties. In this paper, a three-dimensional finite element model of double-wall steel boxed cofferdam is established by MIDAS/civil, and the stress under various working conditions is comprehensively analyzed. According to relevant research [10–13], most of the components of the double-wall steel boxed cofferdam have small deformation, and only a few components have stress redistribution due to plastic deformation. Therefore, the linear elastic constitutive relationship is adopted for the material of the steel boxed cofferdam in this paper.

4.1 Properties

Elastic modulus of steel is 2.0 × 105 Mpa, Poisson's ratio is 0.3, the density is 7850 kg/m^3; Modulus of elasticity of concrete is 3.0 × 104 Mpa, Poisson's ratio is 0.167, and the dry weight of concrete is 24 kN/m^3. As the steel boxed cofferdam is a temporary facility for the construction of an underwater bearing platform, considering the safety factor of the structure, the yield strength of steel is 215 Mpa.

4.2 Unit Type

The steel boxed cofferdam has many components and different shapes and sizes. In this paper, the finite element model is simplified according to the stress characteristics of each component. Combined with the element types provided by the finite element software MIDAS/civil, the beam element is adopted for the beam of the bottom plate, the support in the wallboard, and the internal support of the steel boxed cofferdam, which is simulated by Timoshenko theory [14]. The inner and outer wallboard and bottom plates of steel boxed cofferdam are simulated by plate elements. The bottom concrete is simulated by solid elements.

4.3 Boundary Condition

The constraint of steel boxed cofferdam under various construction conditions is relatively complex, so it can only be analyzed by certain simplification. Under the conditions of steel boxed cofferdam sinking in place and pouring bottom concrete, the steel pipe pile is consolidated at the center of the circle and rigidly connected with the steel boxed cofferdam. The pumping water condition increases the water pressure difference inside and outside the steel boxed cofferdam. To ensure the safe and smooth pumping of the steel boxed cofferdam, the analysis is divided into three stages: the first and second pumping and the completion of pumping. During pumping water, the bottom concrete constrains the displacement, the steel pipe pile is consolidated at the center of the circle, the internal support of the cofferdam is rigidly connected with the wall plate, and some wallboard and all bottom plates are restrained. When pouring the first and second layer bearing platform and tower column, the bottom concrete and the steel pipe pile shall be consolidated, and the contact between the wallboard and the bearing platform shall be consolidated. Due to the complexity of the construction process, this paper establishes multiple calculation models. The reason for the length, there are only some working condition steel boxed cofferdam calculation models are listed, as shown in Fig. 4.

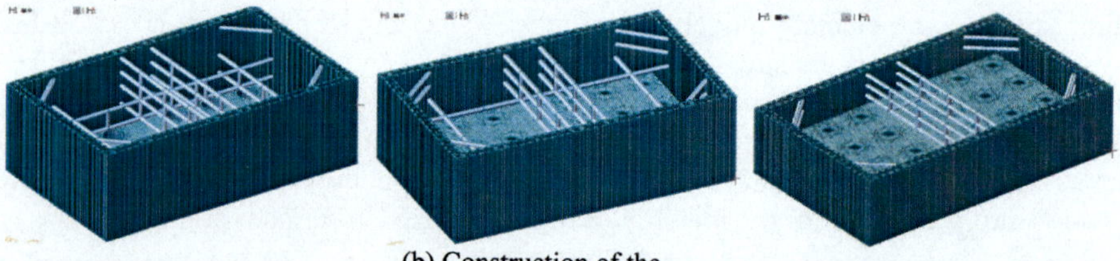

(a) Pumping water stage (b) Construction of the bearing platform (c) Tower column construction

Fig. 4. Calculation model of steel boxed cofferdam

4.4 Load Combination

In condition1, the most unfavorable load combination is considered as self-weight, the hydrostatic pressure of inner and outer wallboard, water flow force, bottom plate buoyancy, and wall buoyancy. When pouring the bottom concrete, the pressure of the concrete on the bottom plate and wallboard of the steel boxed cofferdam (pouring load) shall be additionally considered. The most unfavorable load combination after pumping water in condition3 is the same as that in condition1. At this time, the internal and external water pressure difference increases, so it is necessary to adjust the hydrostatic pressure of the internal and external wallboard. The most unfavorable load combination considered in conditions 4 and 5 is the same as that in conditions 2 and 1 respectively. According to the actual situation of the 6th pier, the wind load has little impact on the construction of steel boxed cofferdam in each stage, so the wind load is not considered here. Taking condition1 as an example, the load diagram of the steel box cofferdam is shown in Fig. 5.

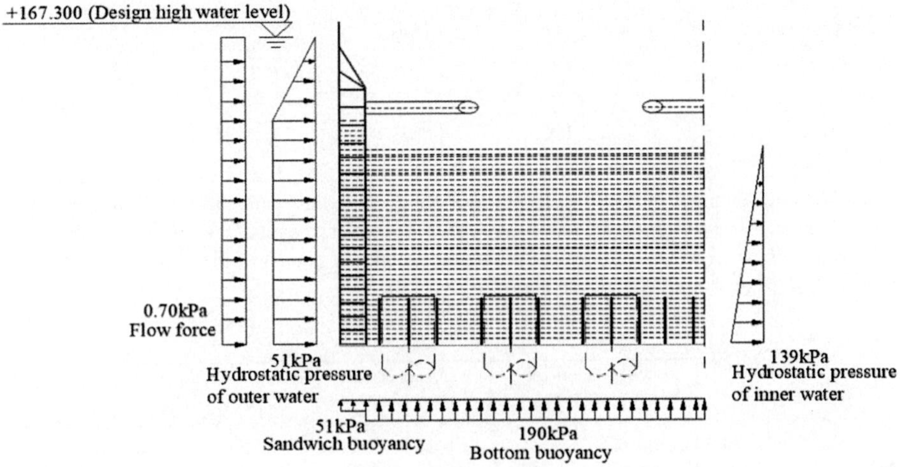

Fig. 5. Load diagram

4.5 Calculation Results and Analysis

The stress state of steel boxed cofferdam is a significant index affecting the smoothness of construction. The maximum stress value (combined stress) of some components under various conditions is shown in Table 1, and the stress variation curve of some components during construction is shown in Fig. 6.

Table 1. Maximum stress value of main components of the steel boxed cofferdam (Mpa)

Conditions	1	2	3	4	5
	Steel boxed cofferdam sinking	Bottoming concrete	Pumping water	Construction of bearing platform	Tower base and column construction
Wallboard element	187	160	203	202	179

(*continued*)

Table 1. (*continued*)

Conditions	1	2	3	4	5
	Steel boxed cofferdam sinking	Bottoming concrete	Pumping water	Construction of bearing platform	Tower base and column construction
Wall beam element	116	112	202	197	165
Steel pipe brace	24	44	185	123	72
Bottom plate element	41	182	97	157	157
Bottom beam element	165	190	63	93	53

Note: The wall element includes the inner and outer wallboard, compartment plates, and ring plates. The wall beam element includes the vertical secondary beam and a transverse brace. The bottom plate beam system includes the main and secondary beams of the bottom plate and the bottom plate edge sealing channel steel.

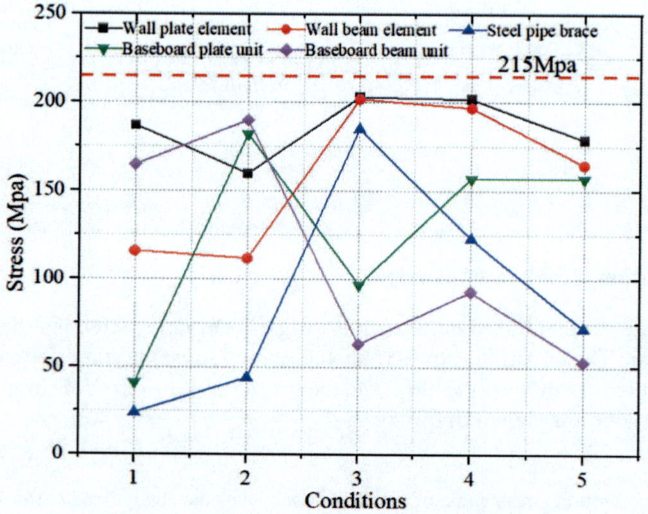

Fig. 6. Stress variation of some components

The stress of all components is less than 215 Mpa, meeting the design requirements. For length reasons, only the calculation results of condition5 are taken as an example, and the calculation results of some components are shown in Fig. 7.

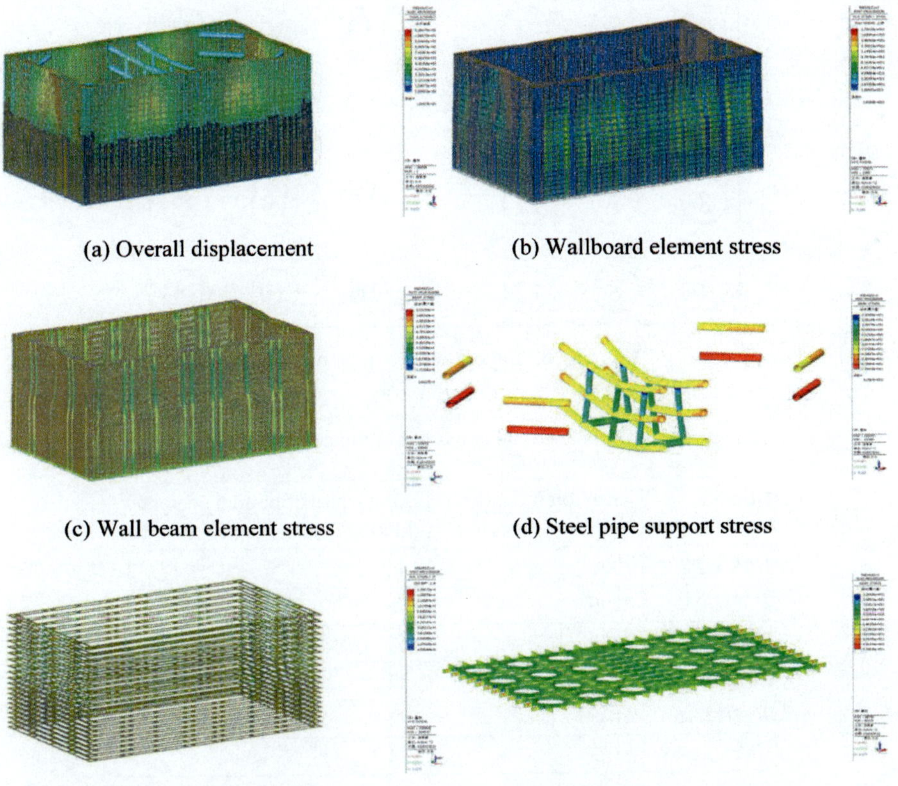

(a) Overall displacement (b) Wallboard element stress

(c) Wall beam element stress (d) Steel pipe support stress

(e) Horizontal ring plate beam stress (f) Bottom beam element stress

Fig. 7. Calculation results of some components of steel boxed cofferdam

The bond stress of the bottom concrete on the steel pipe pile can offset the small sinking force, avoid the safety problems such as the upward floating of the cofferdam and the inclination of the construction bearing platform, and is conducive to the stability and safety of the double-wall steel boxed cofferdam [15]. The core index of the thickness design of the bottom concrete is the bond stress (kPa) between the bottom concrete and the steel pipe pile, which is usually 80–120 kPa. In this paper, the bond stress is conservatively 150 kPa, which will be more conducive to the safe construction of the steel boxed cofferdam. The number of steel pipe piles of the 6th pier is shown in Fig. 8. When the thickness of C30 bottom concrete is 3.5 m, the bond stress is shown in Table 2.

Through the above analysis, the strength and stiffness of the double-wall steel boxed cofferdam under various construction conditions meet the design requirements. The double-wall steel boxed cofferdam has strong applicability when the section size of the components is appropriate. On the premise of ensuring safety, the construction of an underwater bearing platform can be completed successfully.

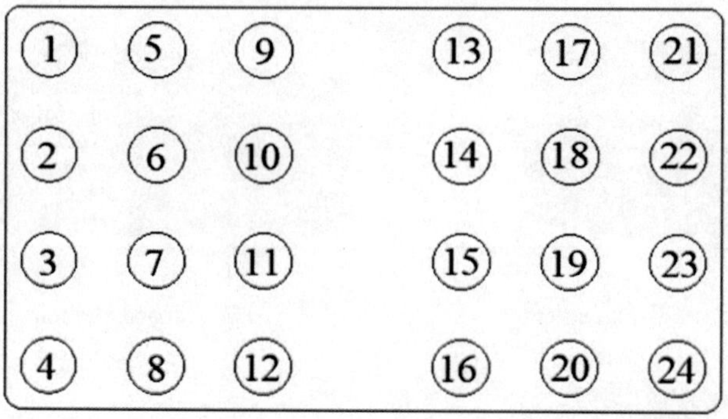

Fig. 8. Layout of steel pipe pile

Table 2. Bond stress of bottom concrete

Number	Reaction force (kN)	Bond stress of bottom concrete (kPa)
1/4/21/24	1936	146
2/3/22/23	2299	68
5/8/17/20	2285	67
6/7/18/19	2772	82
9/12/13/16	4148	122
10/11/14/15	4193	124

(1) During entering the water, the water level in the partition silo of the steel boxed cofferdam is higher than that outside the partition silo. The resulting internal and external water pressure difference makes the wallboard stress larger, and the deep-water pressure makes the beam element stress of the bottom plate of the cofferdam relatively larger.
(2) During pouring the bottom concrete, the vertical pressure of the bottom concrete also makes the stress of the bottom plate and beam element of the steel boxed cofferdam greater, but vertical water pressure.
(3) During pumping, the water level in the steel boxed cofferdam is low, the water pressure difference inside and outside the steel boxed cofferdam leads to the increase of the stress of the wall plate and beam element, the stress of the steel pipe support also reaches the maximum value. At this time, the bottom sealing concrete and the steel boxed cofferdam are stressed together, which restricts the steel boxed cofferdam.
(4) With the initial setting of the concrete and stressed together with the steel boxed cofferdam, the stress of the bottom plate and wallboard of the cofferdam is maintained in a certain state. Therefore, the change of component stress is relatively

small during the pouring of the bearing platform and the construction of the tower base and column.

(5) The thickness of the bottom concrete affects the strength of its holding and the bond stress with the steel pipe pile. According to the above calculation, the bond stress of the bottom concrete is less than 150 kPa, which meets the design requirements.

According to the above calculation, the stress of the wallboard, beam element, and steel pipe support of the steel boxed cofferdam reach the maximum during pumping water. Therefore, when designing the section of the wallboard and beam, the section size should be reasonably set after careful calculation of the pumping water condition to ensure structural safety. When pouring the bottom concrete, the concrete has not been initially set, and the vertical load on the bottom plate and beam system of the steel boxed cofferdam is the largest. After the initial setting, it will be stressed as a whole with the steel boxed cofferdam. The vertical displacement of the steel boxed cofferdam will reach the theoretical maximum before the initial setting of the bottom concrete, resulting in a relatively large bottom plate unit stress. Therefore, pouring the bottom concrete is significant in controlling the deformation of the steel boxed cofferdam, the design of steel boxed cofferdam bottom plate and beam system, the section size and spacing of main and secondary beams of the bottom plate shall be reasonably selected mainly according to this condition.

5 Conclusion

Based on the double-wall steel boxed cofferdam of MaLuKou Zishui River bridge, this paper carries out the finite element analysis of the construction of the steel boxed cofferdam of the 6th pier and studies its safety and overall mechanical properties. The main conclusions are as follows:

(1) The stress state of the double-wall steel boxed cofferdam is good, and the structural stress and the bond stress of the bottom concrete meet the design requirements.
(2) In each construction stage of the steel boxed cofferdam discussed in this paper, the stress of each component during pumping water is relatively large, which is the most unfavorable condition in the whole construction process.
(3) During pumping water, the stress of wall components and steel pipe support reaches the maximum value during construction, so the pumping condition is the design control condition of wall and steel pipe support. When pouring the bottom concrete, the bottom plate of the steel boxed cofferdam increases the vertical load of the concrete. At this time, the stress of the bottom plate element reaches the peak, so the condition of pouring the bottom concrete is the design control condition of the bottom plate components.
(4) In the absence of field measurement conditions, the use of finite element analysis can reflect the basic stress change of steel boxed cofferdam, which can be used as a reference for construction.

References

1. Guangyao, Y.: Research on key technology of double-wall steel-boxed cofferdam in deep water construction. Highway **65**(4), 221–224 (2020)
2. Hong, Z., Xianpeng, L.: Construction Technology of Deep Water High Pile Bearing Platform Foundation of Super Large Bridge. China Construction Industry Press, Beijing (2005)
3. Ziming, L.: Deep Water Foundation of Bridge. People's Communications Press, Beijing (2003)
4. Xiaoyong, O.Y., Huixing, R., et al.: Key Technology of Bridge Deep Water Foundation Construction-Construction Practice of South Tower Foundation Project of Sutong Bridge. People's Communications Press, Beijing (2006)
5. Minrun, H., Zanrong, L., et al.: Construction technology of single wall steel hanging box for large high pile bearing platform. World Bridge **40**(4), 30–33 (2012)
6. Yaobei, H., Xudong, S., et al.: Selection of main girder scheme for Malukou Zishui river bridge on Guanzhuang-Xinhua expressway. Bridge Constr. (6) (2019)
7. Xinli, L.: Positioning techniques for suspended steel box cofferdam of main Pier of Mingzhu Bay bridge in Guangzhou. Bridge Constr. (6) (2021)
8. Xinli, L.: Launching technology of double-wall steel hanging box cofferdam of Mingzhuwan bridge. World Bridge (6) (2019)
9. Qili, X.: Design of pylon cofferdams for navigational channel bridge of Pingtan Straits Rail-cum-Rode Bridge. Bridge Constr. (1) (2021)
10. Hai, D., Hongwei, X., et al.: Design and construction of double wall steel box cofferdam in highway bridge. Highw. Eng. (6) (2020)
11. Kang, X.: Research on Key Construction Technology of Double-Wall Steel Boxed Cofferdam for Cross-Sea Bridge. Hubei University of Technology (2019)
12. Xiaomin, L., Yian, S., et al.: Study on design and construction technology of large double-wall steel hanging cofferdam for deep-water foundation under tide difference. Constr. Technol. (5) (2021)
13. Fan, L., Jie, S.: The application of 3D finite-element analysis to the steel lifting Caission of Ma'anshan Changjiang River highway bridge. J. Anhui Instit. Archit. (Nat. Sci. Ed.) (5) (2010)
14. Yaodong, L., Tianqing, Y., et al.: Simulation calculation and some key technique for construction of double-wall steel boxed cofferdam. Bridge Constr. (02), 72–75 (2009)
15. China Railway Bridge Bureau Group Co., Ltd.: Anti floating and sinking steel casing, cofferdam and construction method: CN202110295651.8, 23 July 2021

Defect Detection and Damage Evaluation of Steel Rods Based on Ultrasonic Guided Waves

Han Zhu and Jing Zhang(✉)

Hefei University of Technology, Hefei, China
zhangj@hfut.edu.cn

Abstract. Structural defects caused by corrosion and fatigue of steel rods are the main problems in cable-supported engineering. Compared with the traditional ultrasonic detection methods, guided waves have attracted the attention of more and more researchers because of the one-place excitation, long-distance propagation, and wide-range detection. In this paper, the propagation characteristics and laws of the guided wave in the rod are theoretically studied, and the dispersion curve of the rod-shaped structure is deduced. Then, a large number of numerical simulation experiments are carried out for the steel rods with defects using ABAQUS software, and the optimal excitation mode and excitation frequency are selected. Finally, the relationship curve between the evaluation index and the defect depth can be obtained by calculation, which can be used to describe the damage of the steel rod quantitatively.

Keywords: Steel rod · Guided waves · Defect · Quantitative evaluation

1 Introduction

With the rapid development of China's economy, steel rods are widely used in engineering structures, such as bolts in rock and soil anchoring, booms in building pressure, and parallel wire bundles in stay cables. Those steel rods generally play the role of bearing and supporting in engineering. In the long-term service process, accidents could occur due to fatigue, rain erosion, impact, hidden defects in the production and construction, resulting in loss of supporting effect [1]. It will not only cause incalculable property losses but even lead to casualties. Therefore, for the structural normal working performance, it is important to use accurate, safe, and fast nondestructive testing methods for rods and assess the damage of defects.

The traditional nondestructive testing methods mainly include ultrasonic testing, magnetic flux leakage testing, eddy current testing [2], and so on. Most of those methods are point-scanning detection, and the detection instruments need to scan point by point along the test structure. Many human and material resources are required and long-distance detection cannot be performed [3]. In addition, this type of detection method requires the instrument to be close to the surface of the structural component, and it is not suitable for detecting structures that are buried or inconvenient for direct contacts, such as anchors in soil, steel bars in concrete, and parallel wire bundles in cables [4].

W. Guo and K. Qian (Eds.): GBCESC 2022, LNCE 211, pp. 809–815, 2023.
https://doi.org/10.1007/978-981-19-5217-3_81

Ultrasonic guided wave technique with a long propagation distance can solve this problem well. Due to the limitation of the boundary, the guided wave can only propagate inside the rod, so its energy attenuation is relatively small. After the guided wave is excited at a certain point, it will propagate along the extension direction and reflect and transmit at the discontinuous part of the medium. Finally, the waveform received by the receiving sensor contains all the information between the excitation and reception. Therefore, ultrasonic guided wave inspection only needs to excite a specific point on the structure to detect the entire structure, and the damage status can be obtained using signal analysis [5]. In this paper, based on the ultrasonic guided wave theory, the finite element software ABAQUS is used to simulate the steel rods with different defect depths numerically, and the propagation characteristics of ultrasonic guided waves in the rods with defects can be obtained and also the defect evaluation.

2 Basic Theory

2.1 Guided Waves in Rods

An infinite isotropic elastic circular rod and a cylindrical coordinate system are built and the parameters are shown in Fig. 1.

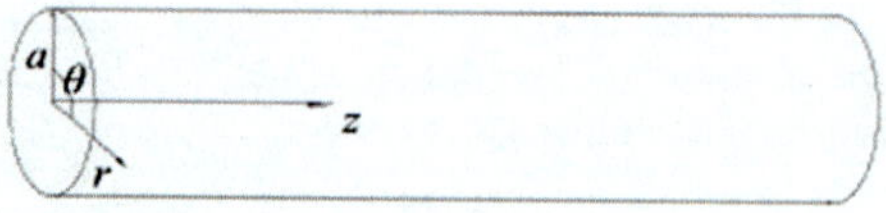

Fig. 1. Cylindrical-coordinate system.

There are three different modes of ultrasonic propagating along the z-direction in the rod, namely the Longitudinal mode, Torsional mode, and Bending mode [6] which are denoted as L(0, m), T(0, m), F(n, m), respectively. When the propagation of the guided wave is in the longitudinal mode, there are only axial and radial displacements, and the displacement of each point on the rod is symmetrical about the central axis of the rod [7]. The well-known Pochhammer [8] frequency equation can be obtained by solving the wave equation [9]. The equation expression can be expressed as follows:

$$\begin{aligned} \frac{2\alpha}{a}\left(\beta^2+k^2\right)J_1(\alpha a)J_1(\beta a)-\left(\beta^2-k^2\right)^2 J_0(\alpha a)J_1(\beta a) \\ -4k^2\alpha\beta J_1(\alpha a)J_0(\beta a)=0 \end{aligned} \tag{1}$$

where $\alpha^2=\frac{\omega^2}{c_L^2}-k^2$, $\beta^2=\frac{\omega^2}{c_T^2}-k^2$, ω is the frequency, k is the wavenumber and J is the Bessel function.

3 Numerical Simulation

Steel rods with different depths are simulated using the finite element analysis software ABAQUS. The relationship between defect depth and defect reflection is studied, and further quantitative analysis of defects is carried out to determine the damage degree of steel rods.

3.1 Establishment of the Finite Element Model.

The radius of the steel rod model is 9 mm, and the length is 1800 mm. Seven steel rods with different defect depths are established, and a steel rod without defects is established as a comparison. The model is a three-dimensional solid model, the primary element type is C3D8R, and the defect element type is C3D10M. Generally speaking, the element size should be no larger than 1/10 of the excitation guided wave wavelength [10], that is to say, it is necessary to ensure that there are at least ten elements within the length of a guided wave wavelength, the time step is generally not larger than the element size and the excitation guided wave speed ratio. After calculation, the element length of the model is selected as 2 mm, and the time step is set as 4.18×10^{-8} s. The steel rod model and meshing results are shown in Fig. 2.

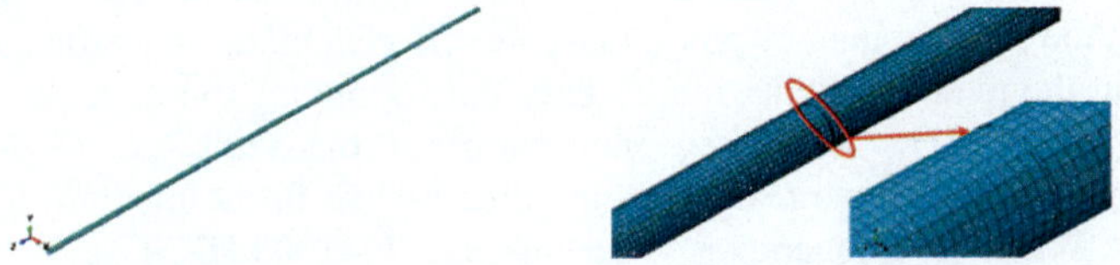

(a) steel rod model (b) Model meshing

Fig. 2. Schematic diagram of mesh division.

The L(0, 1) mode is used for detection, and an axial displacement load is applied to the left end face of the model to simulate the guided wave excitation. The reflection signal can be obtained by detecting the change of axial displacement with time. The excitation mode and the displacement signal receiving point are shown in Fig. 3.

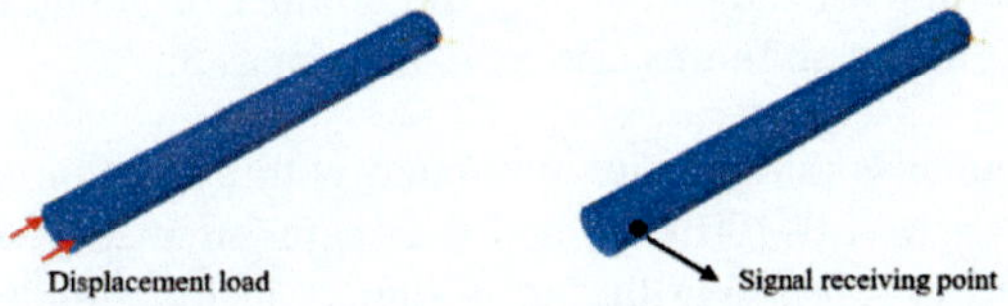

(a) Excitation point (b) Signal receiving point

Fig. 3. Excitation and signal receiving point.

3.2 Selection of Excitation Signal Selection of Excitation Signal

In this section, range of the excitation signal is determined by the dispersion curve [11], and the final selection is made by the reflection information of the excitation signal of different sizes. From the dispersion equation obtained, it can be found that the relationship between frequency and velocity is very complicated and cannot be seen intuitively from the equation. Therefore, this paper uses the DISPERSE software to draw the dispersion curve of the steel rod. The results are shown in Fig. 4.

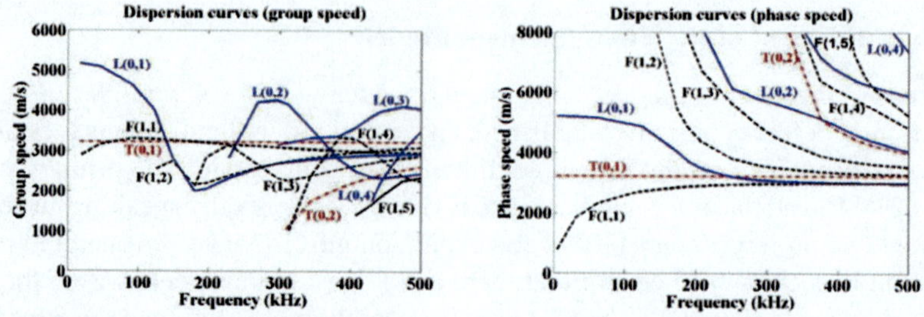

(a) Group velocity dispersion curve (b) Phase velocity dispersion curve

Fig. 4. Dispersion curves of guided waves in steel rods.

Figure 4 shows the dispersion curve from 0 Hz to 500 kHz. It can be seen from the figure that at the same frequency point, multiple different modes are propagating at their own speed in the medium. Other modes have corresponding cutoff frequencies except for L(0, 1) and F(1, 1). As the frequency increases, the number of modes increases, and the propagation characteristics of the guided wave in the medium become more complicated. When the frequency is in the range of 20–100 kHz, it can be seen that the dispersion of the L(0, 1) mode is relatively small, and the speed difference with the other two modes is relatively large, which is easy to identify and distinguish in a practical detection.

Signal Loading. The loaded signal function expression is:

$$f(x) = \begin{cases} 0.5\left[1 - \cos\left(\frac{2\pi f_c}{n}\right)\right]\sin(2\pi f_c), & 0 < t < \tau \\ 0, & t > \tau \end{cases} \tag{2}$$

where τ is the pulse time of the signal, n is the number of single tones, and f_c is the center frequency of the signal. In this article, n is taken as 5.

The steel rod model is numerically simulated with excitation signals of different frequencies in the range of 0–100 kHz, and the center frequency of the guided wave excitation is selected by observing the echo signals in the simulation results. After comparison, the center frequency is 50 kHz, and the 5-cycle sinusoidal signal modulated by the Hanning window is used as the excitation signal. The frequency spectrum of the signal is shown in Fig. 5.

3.3 Influence of Defect Depth on Inspection Results

Defects have various forms, and different types and sizes of defects affect the detection results. The study of the defect size helps to evaluate the damage degree of the steel rod. In this subsection, numerical simulations of steel rods with different defect depths are performed. The damage degree of the steel rod is explored by the amplitude of the reflection signal.

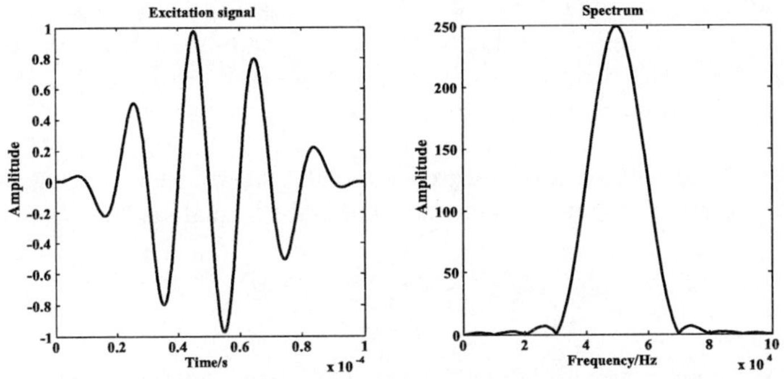

Fig. 5. Excitation signal and spectrogram.

Detection Results of Defects at Different Depths. The width of the defects in the finite element model of the steel rod established in this chapter are all 1 mm, and the depths are 3.6 mm, 4.5 mm, 5.4 mm, 6.3 mm, 7.2 mm, 8.1 mm, 9 mm. As a comparison, the steel rod does not have defects. The received echo signals are shown in Fig. 6.

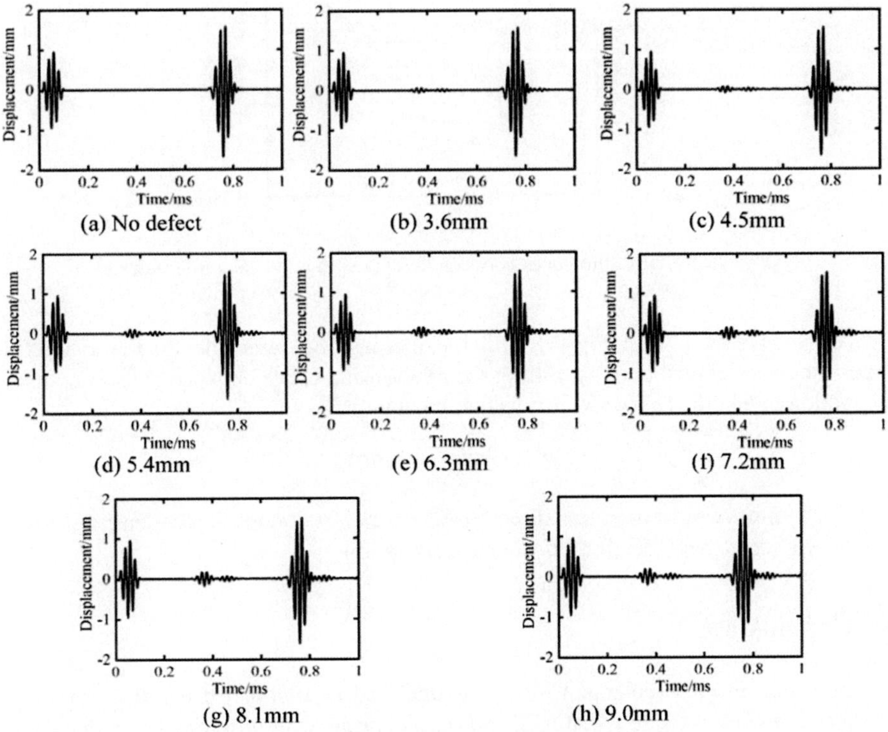

Fig. 6. Test results of defects at different depths.

It can be seen from the figure that with the increase of defect depth, the amplitude of defect echo also increases gradually, but the growth trend can not be seen from the curve alone. Therefore, it is necessary to define a defect damage evaluation system to quantitatively describe the damage of steel rods.

Defect Quantitative Evaluation Index. In order to determine the damage of steel rod, an evaluation index R is defined in this paper, and the formula is described as follows:

$$R = \frac{\sum_{i=1}^{n} |A_{a,i} - B_{1,i}|}{\sum_{i=1}^{n} |B_{1,i}|} \tag{3}$$

where $A_{a,i}$ represents the amplitude of defect echo of defective steel rod, and $B_{b,i}$ represents the signal amplitude of the steel rod with no defects simultaneously.

$A_{a,i}$ and $B_{1,i}$ corresponds to 0.3 ms–0.5 ms of the test result, where B corresponds to (a) in Fig. 6. Bring (b)–(h) in Fig. 6 into the formula to calculate the evaluation index R of different defect depths. Draw the relationship curve between defect depth and R, as shown in Fig. 7.

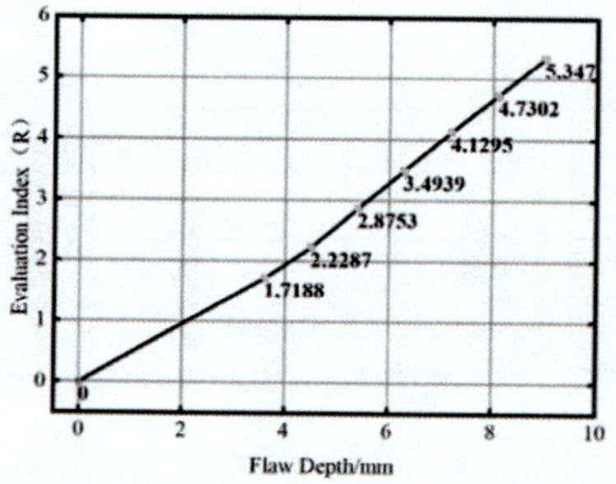

Fig. 7. Relationship curve between defect depth and evaluation index.

As can be seen from Fig. 7, with the increase of defect depth, the evaluation index increases almost linearly. Fitting the relationship curve between defect depth and evaluation index, the following results can be obtained:

$$R = 0.672x - 0.7012 \tag{4}$$

The formula can be used as a direct detection expression for defect depth detection, which provides a new idea for defect size description.

4 Conclusions

In this paper, the finite element model of steel rod is established based on the finite element analysis software ABAQUS, and the steel rods with different defect depths are tested by ultrasonic guided wave. The relationship between the defect depth of the steel rod and the amplitude of the reflection signal is analyzed, and the evaluation index *R* of lever defect is defined. Finally, the following conclusions are drawn.

According to the propagation characteristics of the guided wave in the steel rod, the dispersion curve of the guided wave is deduced, and then the frequency of the detected signal is selected according to the dispersion of the echo signal. Finally, the sinusoidal signal with the center frequency of 50 kHz modulated by the Hanning window in L(0, 1) mode is selected.

The damage evaluation index R is defined. The numerical simulation of steel rods with defects in different depths shows that the evaluation index *R* increases with the increase of defect depth, and the two are linearly correlated.

References

1. Cederquist, S.C.: Motor speedway bridge collapse caused by corrosion. Mater. Perform. **39**(7), 18–19 (2000)
2. Rehman, S., Ibrahim, Z., Memon, S.A.: Nondestructive test methods for concrete bridges: a review. Constr. Build. Mater. **107**(Mar. 15), 58–86 (2016)
3. Liu, Z.H., Zhao, J.C., Bin, W.U.: Application study on defect detection in steel strands by using high-order ultrasonic longitudinal guided waves. Eng. Mech. **28**(4), 214–220 (2011)
4. Ahmed, S.R., Saka, M.: A new ultrasonic angle-beam technique for sensitive evaluation of closed cracks. Ndt & E Int. **33**(4), 261–271 (2000)
5. Lindsay, R.B.: On waves propagated along the plane surface of an elastic solid (1885). Men of Physics Lord Rayleigh–the Man & His Work, pp. 155–163 (1970)
6. Viktorov, I.A.: Rayleigh and Lamb Wave – Physical Theory and Applications. Ultrasonic Technology (1967)
7. Chree, C.: A new solution of the equations of an isotropic elastic solid (1889)
8. Pochhammer, L.A.: Biegung des Kreiscylinders-Fortpflanzungs-Geschwindigkeit kleiner Schwingungen in einem Kreiscylinder (1876)
9. Stephens, R.W.B.: Sound waves in solids. Phys. Bull. **270**, 86–86 (2015)
10. Cao, X.L., Wu, L.M., Li, Z.M.: Behaviour of steel-reinforced concrete columns under combined torsion based on ABAQUS FEA. Eng. Struct. **209**, 109980 (2000)
11. Mehrabi, M., Hossein, S.M., Habibi, H., Kappatos, V.: An experimental technique for evaluating viscoelastic damping using ultrasonic guided waves. Ultrasonics **123**, 106707 (2022)

Study on Enhancing the Performance of Unglazed Solar Air Collector by Attached Jet

Jian Jin[1,2], Jianjun Hu[1,2(✉)], Gui Hu[1,2], and Shenghua Du[1,2]

[1] Hebei Province Low-Carbon and Clean Building Heating Technology Innovation Center, Yanshan University, Qinhuangdao 006004, China
kewei729@163.com

[2] Key Laboratory of Green Construction and Intelligent Maintenance for Civil Engineering of Hebei Province, Yanshan University, Qinhuangdao 066004, China

Abstract. This paper creatively designs a unglazed solar air collector using attached jet to enhance heat transfer. The collector enhances the heat exchange between air and Heat collector plate by creating a specific structure to block and reflect the jet and make it impact the heat collector plate. The structure can not only strengthen the heat exchange, solve the problems of insufficient heat exchange and heat loss of direct exposure to the environment of the unglazed heat collector, but also make up for the reduced effective heat collection area due to the opening on the heat collector plate. Through the simulation and analysis of the internal flow and heat transfer characteristics of the collector, it is found that the attached jet effect is obvious, which can better improve the above problems of the unglazed collector. The effects of slit width and baffle diameter on heat collection efficiency are analyzed by numerical simulation. Finally, the experiments showed that the collector's heat collection efficiency remained stable throughout the day, significantly improving the thermal performance of the traditional unglazed permeable solar collector.

Keywords: Solar air collector · Unglazed · Heat collection efficiency · Numerical simulation · Attached jet

1 Introduction

Flat plate solar air collector (**FPSAC**) is a solar thermal utilization device, which has the advantages of simple structure, low cost and convenient maintenance [1–3]. Based on the transparent cover plate, it is divided into glazed collector and unglazed collector. In order to improve the thermal performance of glazed collectors, scholars have proposed a variety of collectors, such as fins [4, 5], penetration [6, 7], reflux [8] and jet impact [9]. However, the cover plate will reflect sunlight and cause heat loss.

Therefore, an attached jet solar air collector (**AJSAC**) is designed in this paper, which will not reflect sunlight because there is no cover plate; By creating the corresponding structure to produce the attached jet to strengthen the heat exchange between the air and the collector plate, the problems of insufficient heat exchange and local high temperature

W. Guo and K. Qian (Eds.): GBCESC 2022, LNCE 211, pp. 816–825, 2023.
https://doi.org/10.1007/978-981-19-5217-3_82

of the FPSAC are improved; Baffles are arranged below the round holes, which solves the problem of reducing the effective heat collection area due to the opening of the heat collector plate.

2 Attached Jet Collector Model

2.1 Principle of Attached Jet Enhanced Heat Transfer

Under the sun, the heat collector plate and baffle absorb solar radiation and heat up rapidly. Under the action of the fan at the air outlet of the collector, the air outside the heat collector plate for heat exchange enters from the round hole of the heat collector plate and impacts the baffle, and then forms attached jet through the gap between the heat collector plate and the baffle to impact the lower surface of the heat collector plate, as shown in Fig. 1. During this period, the heat exchange between air and baffle and heat collector plate is strengthened due to jet impact.

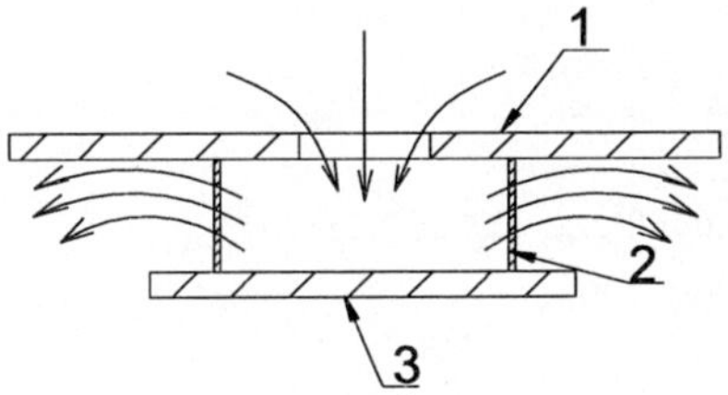

1.Heat collector plate 2.Support rod 3.Baffle

Fig. 1. Schematic diagram of attached jet enhanced heat transfer

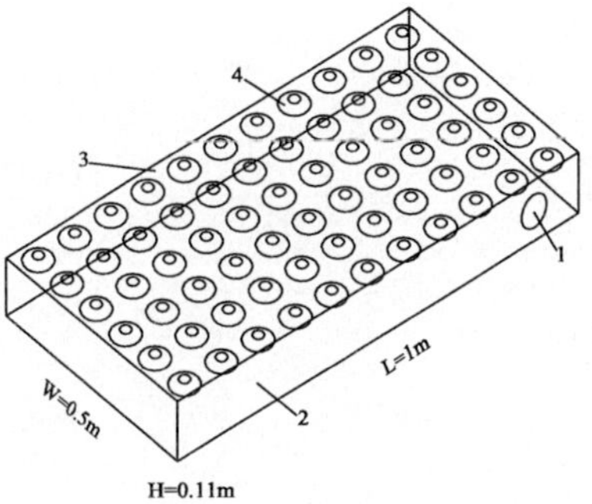

1.Air outlet 2.Periphery structure 3.Heat collector plate 4.Baffle

Fig. 2. Numerical model of attached jet collector

2.2 Geometric Model and Meshing

The original model of this paper is a AJSAC independently designed by the research group, and the net size of its cavity is 1.0 m × 0.5 m × 0.11 m, outlet size 0.06 m × 0.06 m. There are 66 round holes with a diameter of 25 mm evenly distributed with a transverse spacing of 85 mm and a longitudinal spacing of 90 mm on the Heat collector plate. There is a concentric circular baffle with a diameter of 60 mm at 6 mm directly below the round hole. The specific model is shown in Fig. 2.

This simulation uses gambit software for geometric modeling. Because the round holes and baffles on the heat collector plate in the model destroy the regularity of the geometric structure, the unstructured grid is used to divide the computational domain.

2.3 Determination of Model and Boundary Conditions

The internal flow of AJSAC is complex, especially the attached jet generated by baffle has a significant disturbance effect on the fluid. Realizable k-ε model can not only predict the negative pressure gradient flow, but also predict the dispersion rate of plane jet and circular jet more accurately. Therefore, the turbulence model in this paper selects realizable K-ε Two equation model.

In this paper, the S2S (Surface to Surface) radiation model is adopted, which is suitable for the case where only the radiation between surfaces is considered. The numerical model in this paper only considers the radiation heat transfer of Heat collector plate, baffle and boundary. The boundary conditions and material properties of the numerical model are shown in Table 1.

Table 1. Boundary conditions and material properties

Location	Boundary condition
Inlet	Temperature: 300 K Pressure inlet: 10 Pa
Outlet	Pressure outlet: 0 Pa
Heat collector plate	Heat Transfer Coefficient: 20 W/m^{2}k Free Stream Temperature: 300 K External Emissivity: 0.1 External Radiation Temperature: 300 K Wall Thickness: 1 mm Heat Generation Rate: 800000 W/m^3
Baffle(in)	Wall Thickness: 1 mm Heat Generation Rate: 800000 W/m^3
Baffle(out)	Coupled
Other walls	Heat Flux = 0

3 Simulation Analysis of Heat Collector

3.1 Heat Collection Efficiency

Heat collection efficiency (η_C) said in a certain time interval, the heat transfer working medium from the collector to take away the heat and the same time interval incident on the collector area of the ratio of solar energy, as shown in the formula (1).

$$\eta_c = \frac{mC_P(T_{out} - T_{in})}{AG} \tag{1}$$

where m – air flow of collector, kg/s; A - total area of absorber plate, m^2; G – total solar radiation intensity, W/m^2; T_{out} – outlet temperature of flowing medium, K; T_{in} – inlet temperature of flowing medium, K; C_P – specific heat of air at constant pressure.

3.2 Analysis of Flow and Heat Transfer Characteristics

Figure 3 shows the local flow field characteristics near two adjacent jet units in the AJSAC. It can be seen from the figure that the structure can form the expected flow form, that is, air enters from the round hole on the Heat collector plate, impacts the baffle, then reversely impacts the Heat collector plate, and forms a circumferential attached jet under the action of the slit between the Heat collector plate and the baffle. However, because the outflow direction of the left attached jet unit is opposite to the mainstream flow direction, an attachment vortex is formed on the left back side of the baffle and flow separation occurs. At the right back side of the baffle, the flow reattaches. Then, the reattachment fluid reacts with the downstream attached jet again to form a new attachment vortex until the outlet. Because the attached jet covers almost all the area under the Heat collector plate, the heat exchange area of the Heat collector plate is fully utilized, which is conducive to enhancing the thermal performance of the system. The vortex on the back side of the baffle was originally an unfavorable flow characteristic, but because the impinging jet on the front of the baffle was strong and the metal baffle was thin, this flow phenomenon would not have an adverse impact on the heat transfer of the baffle.

Figure 4 shows the temperature cloud diagram of the Heat collector plate of the AJSAC. It can be seen from the figure that the temperature at the edge of the opening of the Heat collector plate is the lowest, and then circularly distributed from the center to the four sides, gradually increasing, and the temperature around the opening of the Heat collector plate is significantly lower than that in the central areas of the four openings. This is because the intensity of the attached jet impinging on the eat collector plate will inevitably weaken with the increase of the distance between the air and the opening of the Heat collector plate, and the heat exchange intensity between the air and the Heat collector plate will also weaken. The low-temperature area around the opening of the absorber plate is large due to the action of attached jet, which can well prevent the formation of high-temperature area. So, it can improve the one-time heat transfer of permeable collector and the problem of local high temperature.

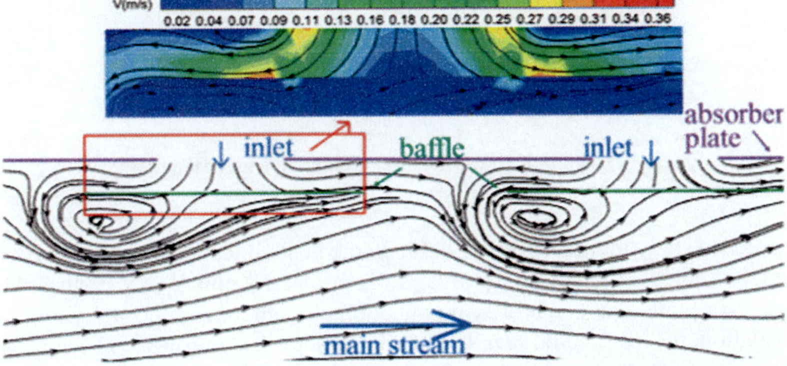

Fig. 3. Local streamline diagram of the attached jet collector

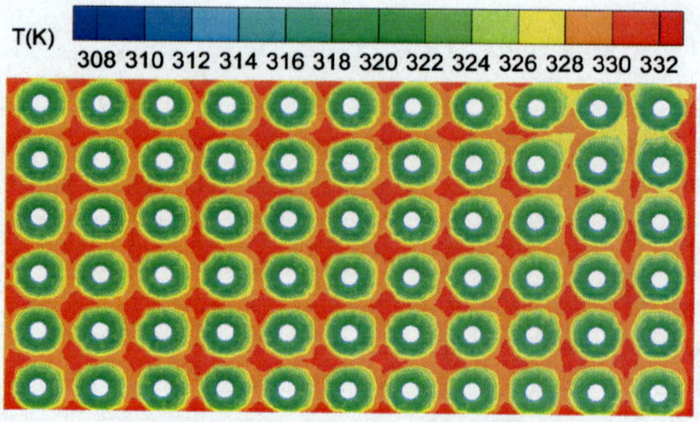

Fig. 4. Temperature cloud diagram of the Heat collector plate

3.3 Effect of Slit Width on Heat Collection Performance

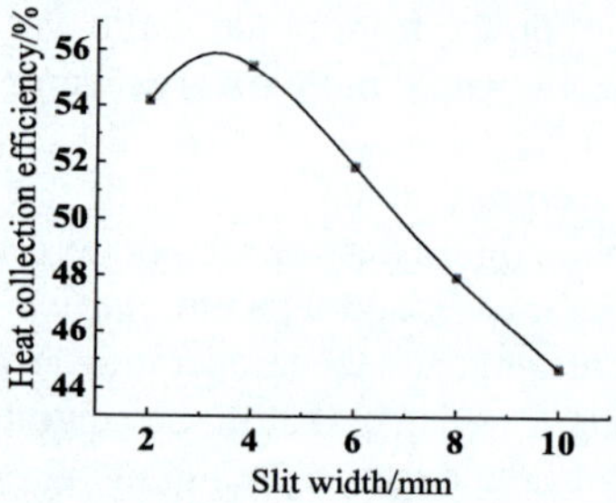

Fig. 5. Variation of heat collection efficiency with slit width

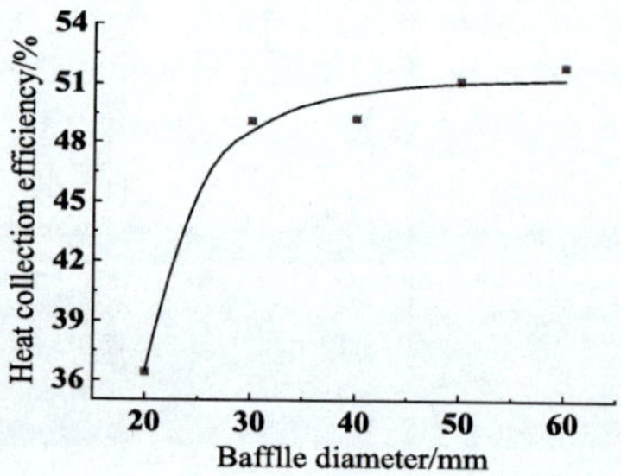

Fig. 6. Variation of heat collection efficiency with baffle diameter

In order to study the effect of slit width on the heat collection efficiency of AJSAC, this paper simulates the slit width of 2, 4, 6, 8, and 10 mm by the control variable method, and obtains the graph of heat collection efficiency with the slit width. During the simulation, the differential pressure of the collector is controlled to be constant. As shown in Fig. 5, the heat collection efficiency increases and then decreases as the slit width increases.

When the slit width increases from 2 mm to 4 mm, the resistance at the slit gradually decreases and the inflow air flow increases. At the same time, with the increase of the slit width, the jet range formed by the air after impacting the baffle also increases, so the heat exchange capacity is enhanced. When the slit width increases from 4 mm to 10 mm, the slit width is too wide, resulting in insufficient jet strength, so as to reduce the action range of attached jet and reduce the heat exchange capacity.

3.4 Effect of Baffle Diameter on Heat Collection Performance

Figure 6 shows the variation of the heat collection efficiency with the baffle diameter. During the simulation, the differential pressure of the collector is controlled as a constant. As shown in the figure, the heat collection efficiency increases with increasing baffle diameter in the range of 20 to 60 mm, and the growth rate decreases gradually. When the baffle diameter is smaller than the diameter of the circular hole, i.e., 25 mm, the baffle constrains the airflow less, and the amount of air forming the slit jet attached to the lower surface of the collector plate decreases, and part of the air cannot participate in the enhanced heat transfer in the form of the attached jet, resulting in the low collector efficiency; when the baffle diameter exceeds 25 mm, most of the air forms the attached jet, and the air volume no longer changes. However, the airflow constraint of the baffle is further enhanced, and as the diameter of the baffle increases, the area of the slit jet acting on the heat collection plate increases, and the heat collection efficiency increases. The closer the edges of the two adjacent baffles are, the closer the area of the slit jet is to the total area of the lower surface of the collector plate, so the growth rate of the heat collection efficiency gradually decreases and the heat collection efficiency gradually flattens out.

4 Experimental Verification

4.1 Experimental Process

In order to verify the performance of AJSAC under real meteorological conditions, a AJSAC test-bed is built in this paper. The overall structure of the test-bed is shown in Fig. 7. The inclination angle of the collector is 45° and the direction is due south.

See Table 2 for the main measuring equipment used. Omega hot wire anemometer is used to measure outlet flow. The differential pressure gauge is used to measure the outlet wind pressure. The solar radiation intensity meter TBQ-2 is used to detect the solar radiation intensity on the experimental day. When in use, the inclination angle is consistent with that of the collector, which is 45°. Agilent acquisition instrument 34972a is used to collect the temperature and solar radiation intensity of the measuring point in real time, and the data acquisition interval is 5 min. The hot wire anemometer is used to measure the ambient wind speed, and the acquisition interval is 15 min. In order to eliminate the influence of eddy current on the experimental results, the collector is operated by negative pressure extraction. The experimental data in this paper are taken from July 24, 2021, and the air volume of the collector is controlled to 39 m^3/h. Before the experiment, the collector will be exposed to sunlight for one hour to reduce the influence of thermal inertia of the system. The experiments started at 8 a.m. and ended at 16 p.m. for about 8 h. During the experiment, the experimental platform is always unobstructed by external objects.

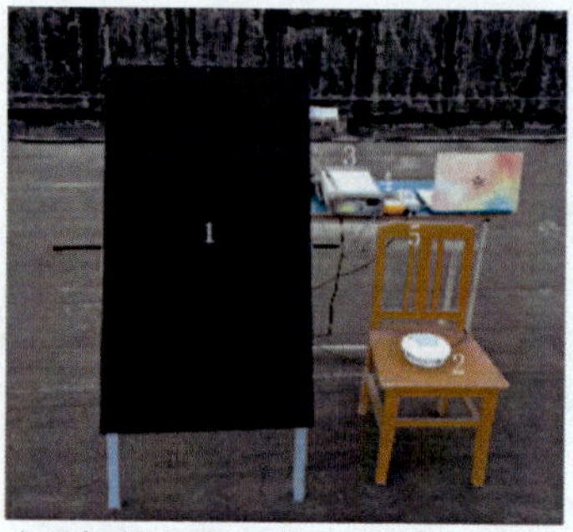

1. Attached jet solar collector 2. Solar radiation intensity meter 3. Agilent acquisition instrument 4. Differential pressure meter 5. Hot wire anemometer

Fig. 7. Test device diagram

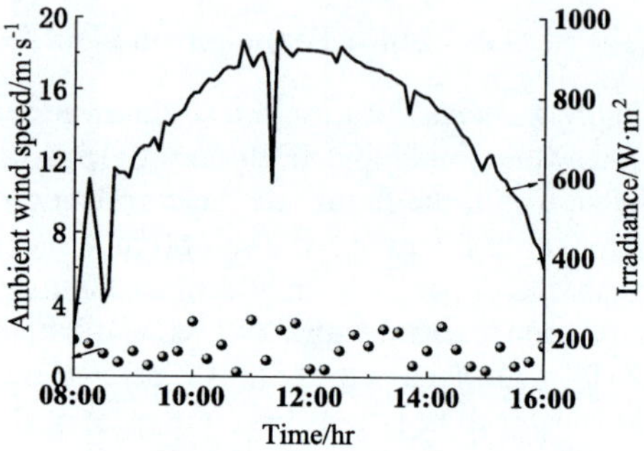

Fig. 8. Meteorological condition map

Table 2. Measuring devices and parameters

Equipment	Type	Parameters
Agilent acquisition instrument	34972A	Accuracy: 0.004%
The solar radiation intensity meter	TBQ-2	Sensitivity: 8.638 μVW·m^{-2} Spectrum: 280–3000 nm; Accuracy: <5%
Hot-wire anemometer	HHF-SD1	Working temperature: 0–50 °C Resolving power: 0.1 °C; Accuracy: ±0.8 °C
Thermocouple	K-Type	Range: −200–260 °C; Accuracy: ±0.75%

4.2 Thermal Performance Analysis

The meteorological conditions on the day of the test are shown in Fig. 8. The ambient wind speed changes little throughout the day and the value is small, fluctuating between 0–3 m/s. The solar illuminance curve is basically parabolic, and the maximum irradiance is 960 W/m^2. Due to the cloudy weather around 8:30 and 11:00, the solar irradiance curve fluctuates obviously during this period.

By processing the test data obtained by the measuring equipment, the operation of the AJSAC is shown in Fig. 9. Figure 9a is the analysis diagram of attached jet experiment data. It can be seen from the figure that the ambient temperature on the test day is 28.5 ± 3.5 °C. In the figure, 1, 2 and 3 are three measuring points on the collector plate that are farther and farther away from the center of the circular hole and are equally spaced. By comparing the temperatures of the three points, it can be seen that the farther away from the center of the circular hole, the higher the temperature on the collector plate, and the temperature difference between points 1 and 2 is large, and the temperature difference between points 2 and 3 is small and almost overlapped. It is concluded that due to the cooling effect of attached jet, there is an obvious temperature gradient around the circular hole, and the farther the temperature gradient is from the center of the circular hole, the smaller, that is, the closer the temperature curve is to overlap. The experimental results show that the attached jet can strengthen the heat transfer and improve the thermal performance of the collector, and the change trend of plate surface temperature is basically consistent with the simulation results.

Figure 9b shows the collector performance curve obtained by data processing. It can be seen from the figure that since the flow of the collector remains constant, the temperature difference between the inlet and outlet presents an obvious parabola, which is consistent with the change of solar irradiance, in which the maximum temperature difference can reach 16 °C.

Although the whole day heat collection efficiency curve of AJSAC fluctuates, its change trend is not parabolic, but basically stable throughout the day, roughly in a horizontal line. It is calculated that the average efficiency of the whole day is 48.1%. At the same time, this paper also increased the air volume to 93 m^3/h for the test, and calculated that the average heat collection efficiency of the whole day is 82.4%, which is 24.4% higher than the traditional unglazed solar air collector under the same air volume [10].

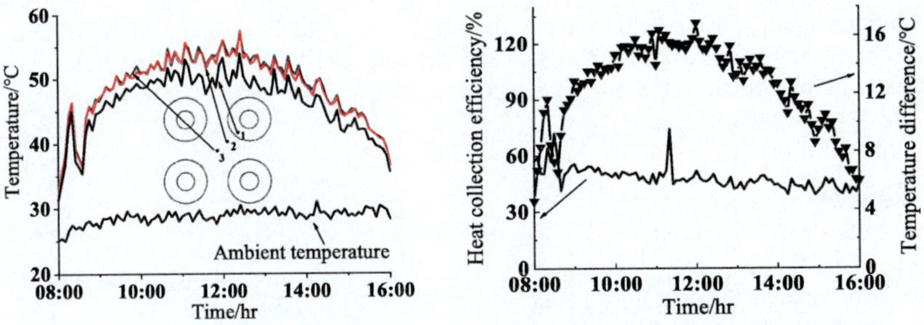

a. Attached jet analysis diagram b. Performance analysis diagram of collector

Fig. 9. Operation diagram of attached jet collector

It is an important feature that the heat collection efficiency of the unglazed collector remains stable throughout the day. Unlike the glazed collector, its heat collection efficiency is similar to the trend of solar irradiance in a parabola shape. The main reason is that the performance of glazed collector is limited by the reflection and incidence characteristics of transparent materials. For the unglazed collector, the solar radiation can be effectively absorbed when incident at any angle, which leads to the change of temperature difference at the inlet and outlet of the collector consistent with the change trend of solar irradiance. Combined with the definition of heat collection efficiency in formula (1), it is not difficult to understand why the heat collection efficiency of AJSAC can basically remain stable.

5 Conclusion

In this paper, the heat collection performance of AJSAC is numerically calculated and experimentally analyzed, and the following conclusions are drawn:

1) The reliability of the simulation model is verified. Through numerical simulation, it is found that the heat transfer enhancement effect of additional jet is obvious, which can improve the one-time heat transfer of permeable collector and the problem of local high temperature.
2) Through the simulation method, the relationship between the thermal efficiency of the collector and the two structural factors of slit width and baffle diameter is explored. The results show that the thermal efficiency first increases and then decreases with the increase of the slit width within the discussion range. The optimum slit width is 3 mm; With the increase of baffle diameter, the heat collection efficiency increases gradually.
3) The experimental results show that the additional jet can effectively improve the heat transfer strength of the collector, and its heat collection efficiency is relatively stable throughout the day, significantly improving the thermal performance of the conventional unglazed permeable solar air collector.

Acknowledgement. This research was supported by the Major Project of the National Natural Science Foundation of China (51890881), the Natural Science Foundation of Hebei Province, China (Grant numbers E2020203028) and the Science and Technology Research Project of Colleges and Universities in Hebei Province, China (Grant numbers ZC2016074).

References

1. Quanquan, L., Baoguo, L., Chuanhui, Z., et al.: Study on thermal efficiency of energy storage air solar collector. Therm. Power Eng. **34**(12), 116–121 (2019)
2. Salah, M.S., Jalal, M.J., Saleh, E.N.: Experimental and numerical analysis of double-pass solar air heater utilizing multiple capsules PCM. Renew. Energy **143**, 1053–1066 (2019)
3. Yalun, L., Baoguo, L., Shuqiang, S., et al.: Experimental research and analysis of air PCM solar heat pump heating system. Therm. Power Eng. **34**(08), 156–162 (2019)
4. Zainab, S., Ahmad, F., Zahari, I., et al.: Performance evaluation of absorber rod type solar air collector. IOP Conf. Ser.: Earth Environ. Sci. **268**, 012073 (2019)
5. Assadeg, J., Ali, H.A., Al-Waeli, A.F., et al.: Energetic and exergetic analysis of a new double pass solar air collector with fins and phase change material. Sol. Energy **226**, 260–271 (2021)
6. Junfeng, W., Feng, G., Liangze, Z., et al.: Study on thermal performance of guide plate permeable solar air collector. J. Solar Energy **34**(1), 162–166 (2015)
7. Goel, A.K., Singh, S.N.: Experimental performance evaluation of an impinging jet with fins type solar air heater. Environ. Sci. Pollut. Res. **28**(16), 19944–19957 (2021)

8. Ali, A., Gholamabbas, S., Habibollah, S.: An in depth evaluation of matrix, external upstream and downstream recycles on a double pass flat plate solar air heater efficacy. Therm. Sci. Eng. Prog. **21**, 100789 (2021)
9. Aboghrara, A.M., Baharudin, T., Alghoul, M.A., et al.: Performance analysis of solar air heater with jet impingement on corrugated absorber plate. Case Stud. Therm. Eng. **10**, 111–120 (2017)
10. Lixin, G., Shaozeng, S., Yuanfeng, W.: Experimental study on thermal performance of permeable unglazed solar air collector. Energy Saving Technol. **127**(30), 155–158 (2012)

Experimental Investigation on Stress-Strain Curve of Phosphogypsum Under Uniaxial Compression

Li Zhou[1,2](✉), Song Hu[1,2], Chao Yin[1], Yifeng Xu[1], and Qingyu Zou[1]

[1] College of Architecture and Urban Planning, Guizhou University, Guiyang 550025, China
Lzhou5@gzu.edu.cn

[2] Intelligent and Green Mountain Residence Engineering Research Center of Guizhou Province, Guiyang 550025, China

Abstract. To study the stress-strain relations of phosphogypsum (PG), 36 samples with different mix proportions were tested under uniaxial compression. The failure mode, measured curve, elastic modulus and Poisson's ratio were analyzed in detail. The results show that all samples showed similar failure modes, and all stress-strain curves can be divided into 3 stages. The strength of the sample increases with the reduction of the water-solid ratio, and the elastic modulus increases with the increase of the compressive strength. The mix proportion has little effect on the peak strain and Poisson's ratio of the PG sample. Furthermore, a stress-strain model was developed based on the mathematical fitting method, comparisons between the test and calculated curves showed that the proposed model has a good accuracy.

Keywords: Stress-strain curve · Phosphogypsum · Uniaxial compression

1 Introduction

Phosphogypsum (PG) is formed as a by-product of the phosphating industry, which is usually defined as industrial solid waste. Every year, the global annual output of the PG amounts to 200–300 million tons [1]. Unfortunately, only 15% of this waste is reused. The rest is released into the waste dump in the land surface, which has not been further treated. It is noteworthy that large amount of PG waste stacking could cause serious harm to the natural environment as well as human health. Globally, low utilization of PG has become a growing problem for many countries. Therefore, PG is tried to be applied in various fields, such as agricultural fertilizers [2] or soil stabilization amendments [3], cement production [4], and building materials [5].

In the field of building engineering, previous research works have been implemented to use PG in preparing various building products. Kumar prepared bricks using PG, lime and fly ash, the authors found that the new bricks have better tolerances and no efflorescence by comparing with the common bricks [6]. Ma et al. studied porous sound absorbing materials prepared with PG, the materials showed a good sound absorption

W. Guo and K. Qian (Eds.): GBCESC 2022, LNCE 211, pp. 826–835, 2023.
https://doi.org/10.1007/978-981-19-5217-3_83

property [7]. Wang et al. investigated the performance of foam insulation materials mixed with PG, the results showed that slag and fiber can effectively improve related properties [8]. Contreras et al. reported ceramic tiles prepared with PG as a admixture, the studies showed that the sintering behavior is influenced by the PG content. Besides, the apparent porosity and water absorption varies with the PG content [9]. Wu et al. investigated the seismic behavior concrete grid frame walls filled with PG, the research found that the bearing capacity of the wall can be effectively enhanced by the infill PG [10].

With the extensive application of PG in building engineering, more and more scholars have paid attention to the mechanical behaviors of PG materials. The strength, elastic modulus and Poisson's ratio of PG were investigated in detail [11, 12]. However, the research on the stress-strain relations of PG is rare, which greatly restricts the application of PG in engineering. Therefore, the present study aims the analysis of the stress-strain relations as well as the establishment of the stress-strain model of PG.

2 Materials and Test Methods

2.1 Materials and Sample Preparation

PG, cement, fiber and retarder were used in this work. PG was obtained from Guizhou phosphate chemical group, and the content of $CaSO_4{\bullet}0.5H_2O$ was 70.2%. P.O 32.5 Portland cement was used in mixtures. Polypropylene fiber was used in this work, and the fiber is 25 μm in diameter and 10 mm in length. The tensile strength of the fiber showed in product specification is 450 MPa. Retarder used in this study was procured from professional manufacturer of PG retarder, and the consumption of the retarder was determined according to the pouring time.

During the preparation of PG products, water-solid ratio is an important parameter [13]. In this study, the mixtures were composed of varying percentage of water and solid materials with constant weight. Also, the weight ratio of each solid material remains unchanged. Table 1 lists the mix proportions of this study.

Table 1. Mix proportions of mixtures.

Type	PG (%)	Cement (%)	Fiber (%)	Water-solid ratio	Retarder
A	90	10	0.2	0.7	Adopted
B	90	10	0.2	0.8	Adopted
C	90	10	0.2	0.9	Adopted
D	90	10	0.2	0.7	None
E	90	10	0.2	0.8	None
F	90	10	0.2	0.9	None

Weighed the solid materials in predetermined proportions according to Table 1, and mixed them well. Then, the water was added into the solid mixture and stirred for 1 min.

When the slurry was homogeneous, the slurry was poured into the mould of 100 mm × 100 mm × 300 mm, as illustrated in Fig. 1(a). For each mix proportion, 6 samples were manufactured. Some of the samples were shown in Fig. 1(b).

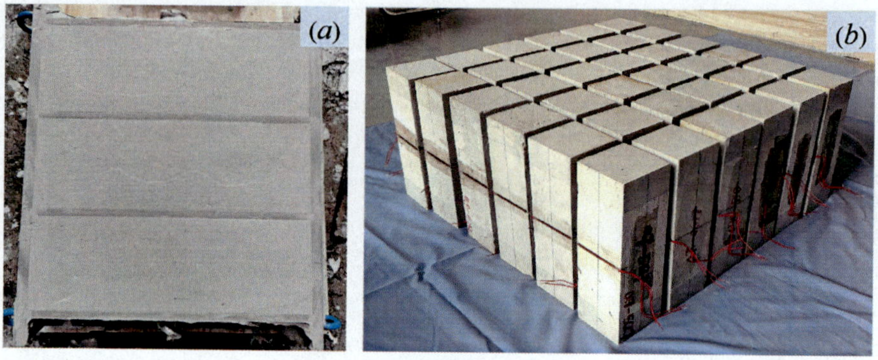

Fig. 1. Test samples of PG

2.2 Test Methods

The test sample was demoulded 4 h after pouring. Then, cured under indoor drying conditions until the sample was dried to a constant weight (the difference between two consecutive weighings is not more than 1% of the total weight). Before the formal test, the longitudinal strain gauge and transverse strain gauge were pasted in the middle of the side surface of the test sample, respectively. Then, uniaxial compressive tests were performed by using hydraulic universal testing machine, the maximum loading capacity of the machine is 100 kN. The test setup is shown in Fig. 2.

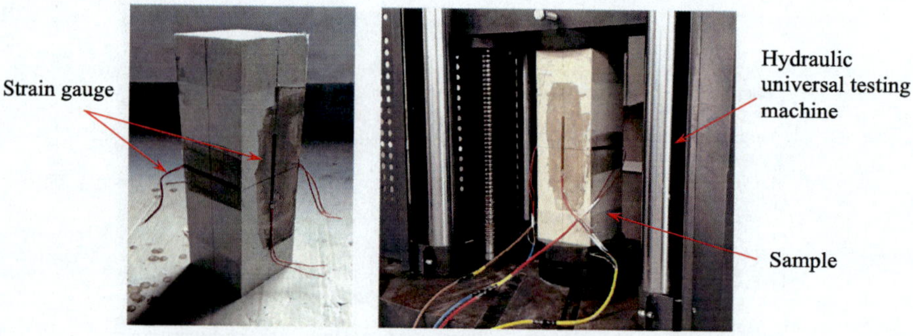

Fig. 2. Test setup of samples

3 Results and Discussion

3.1 Failure Characteristics

According to the test results, all samples showed similar failure modes, horizontal cracks appeared in the middle of the sample in early loading stage, this is due to the compression of micropores. Subsequently, the vertical cracks divided the sample into several small studs. Finally, the small studs were crushed under high pressure, resulting in the failure of the whole sample, as illustrated in Fig. 3.

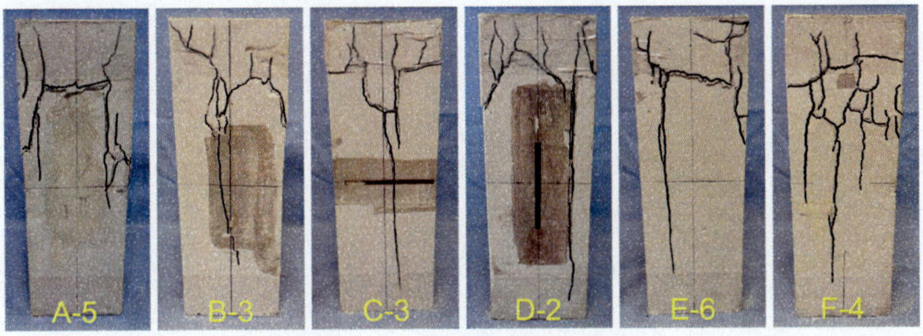

Fig. 3. Failure mode of samples

3.2 Stress-Strain Curves

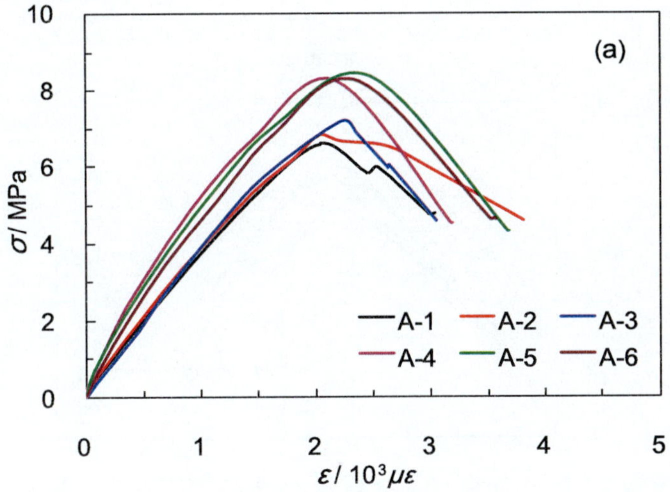

Fig. 4. Stress-strain curves of samples

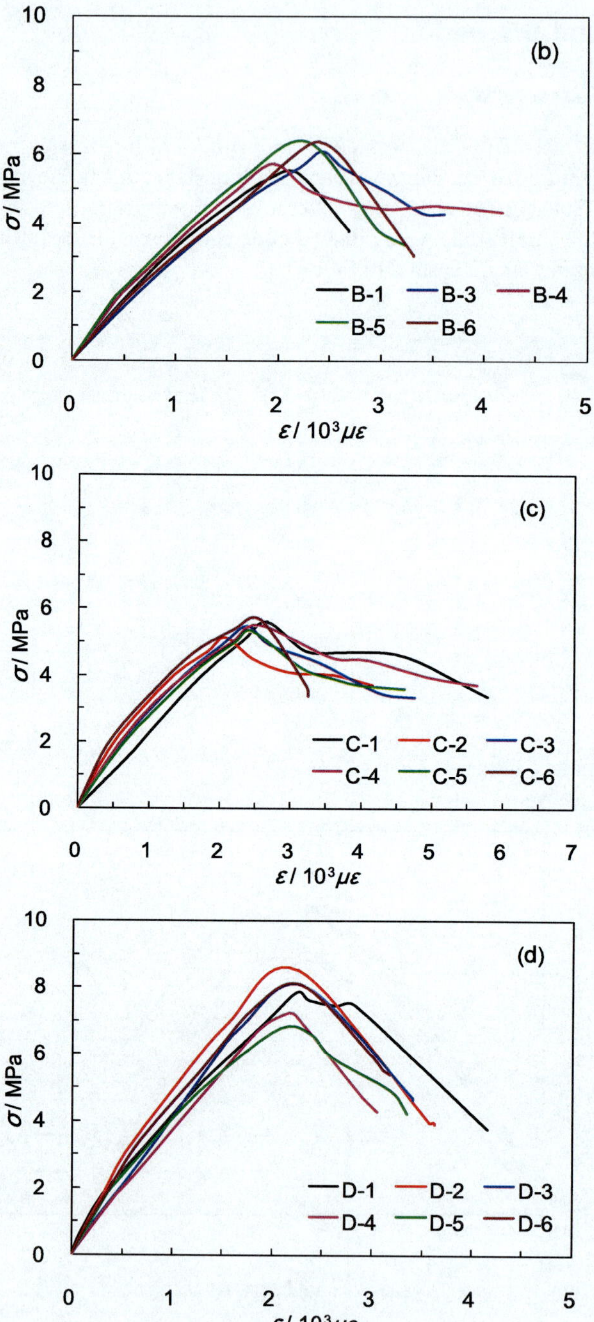

Fig. 4. continued

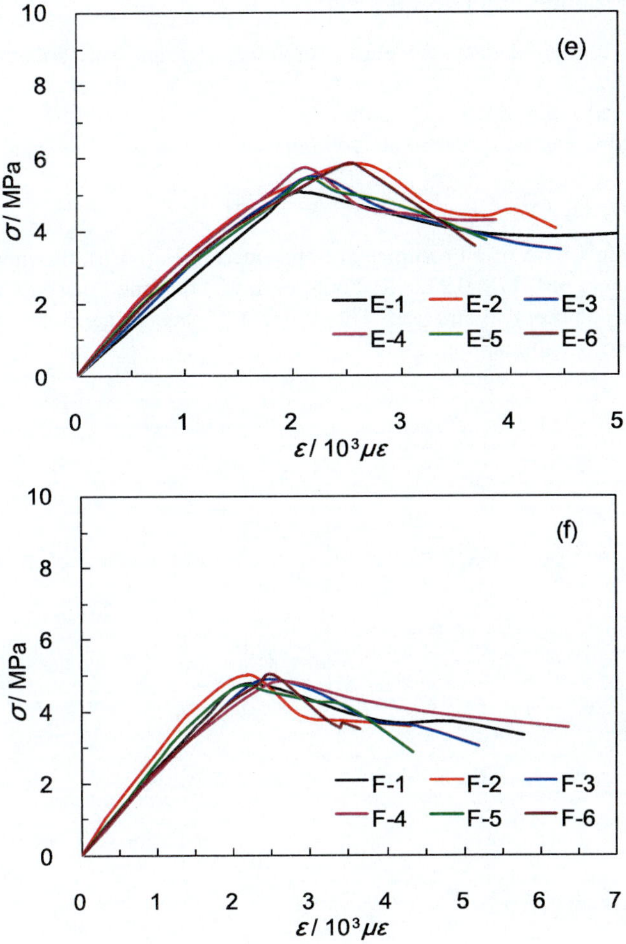

Fig. 4. continued

The stress and strain of each sample were obtained from the load sensor and strain gauges. Figure 4 displays the stress (σ) versus the strain (ε) curves for all samples under uniaxial compression. As can be seen, the test curves can be divided into three stages. The first stage is from the beginning of the loading to the yield state, and the curves are linear. The second stage is from the yield state to the peak state, and has a nonlinear characteristic. The third stage is the descending portion of the curve. Increasing the water-solid ratio can lead to an decrease in the compressive strength of the sample. When the water-solid ratio was 0.7 (Type A and D), 0.8 (Type B and E) and 0.9 (Type C and F), the average compressive strength was 7.65 MPa, 5.79 MPa and 5.14 MPa, respectively. The peak strain of all samples was in the range of 2222–2450 $\mu\varepsilon$, and the average value of peak strains was 2300 $\mu\varepsilon$.

3.3 Elastic Modulus and Poisson's Ratio

Based on the measured data, the elastic modulus (E) and Poisson's ratio (μ) were obtained, as illustrated in Fig. 5. As can be seen in Fig. 5(a), the elastic modulus of the sample increased as the compressive strength (f_c) increased. After fitting, the relationship between E and f_c can be approximately expressed as follows:

$$E = 749.4f_c - 747.8 \tag{1}$$

The Poisson's ratio of all samples are presented in Fig. 5(b), the measured values were in the range of 0.178–0.221. It should be noted that the Poisson's ratio changed little with the compressive strength, Therefore, 0.2 is recommended as the Poisson's ratio of PG during applications.

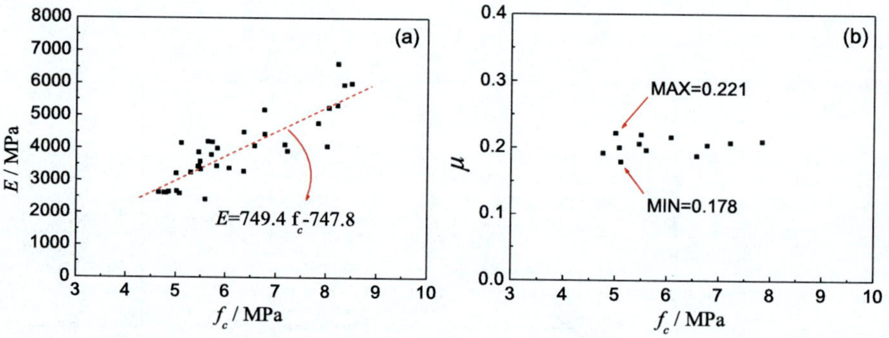

Fig. 5. Variations of E and μ with f_c

4 Stress-Strain Model

4.1 Method for Establishing the Model

For the convenience of the modeling, the stress-strain curves of all samples were normalized through the following relations.

$$x = \varepsilon/\varepsilon_0 \tag{2}$$

$$y = \sigma/f_c \tag{3}$$

Figure 6(a) presents the normalized curves of PG with different mix proportions. In general, the mix proportion affects the ascending stage of PG little. However, the descending stages of the curves showed great differences, those features were also shown in Fig. 4. According to the curve characteristics showed in Fig. 4 and Fig. 6(a), the typical curve of PG can be summarized, the typical curve includes an nonlinear ascending stage and a simplified linear descending stage, as illustrated in Fig. 6(b).

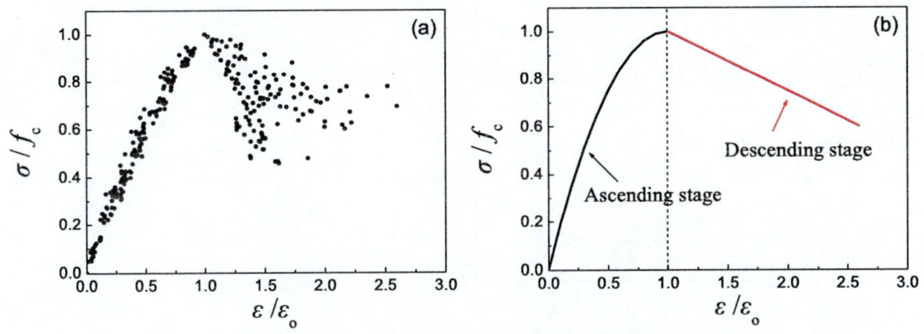

Fig. 6. Normalized curves and typical curve

For the ascending stage, the following relations should be conformed: $y_{x=0} = 0$; $y'' < 0$; $y_{x=1} = 1$; $y'_{x=1} = 0$. Hence, the mathematical expression of the ascending stage of the typical curve can be expressed as:

$$y = ax + (3 - 2a)x^2 + (a - 2)x^3 \quad (4)$$

$$a = E\varepsilon_o / f_c \quad (5)$$

For the descending stage, the curve is simplified to a linear form, so the mathematical expression of the descending stage can be expressed as:

$$y = 1 - b(x - 1) \quad (6)$$

By fitting the slopes of the descending stages of curves, parameter b can be obtained, and the expression is as follows:

$$b = 0.19f_c - 0.59 \quad (7)$$

4.2 Verification of Model

For samples with different mix proportions, the stress-strain curves were calculated by using the proposed model. Due to the variation in compressive strength and strain at peak stress of different samples, only partial samples were selected for validation, as illustrated in Fig. 7. As can be seen, the curves calculated by the proposed model have a good agreement with the test curves, which indicates that the proposed model has good accuracy. Besides, for sample with high water-solid ratio, the comparison results showed that the stiffness of the ascending stage showed in the test curve was slightly lower than that showed in the calculated curve, the possible reason is that high water-solid ratio leads to high porosity in the sample. When the sample suffered a high pressure, pores collapsed and the stiffness of the sample reduced.

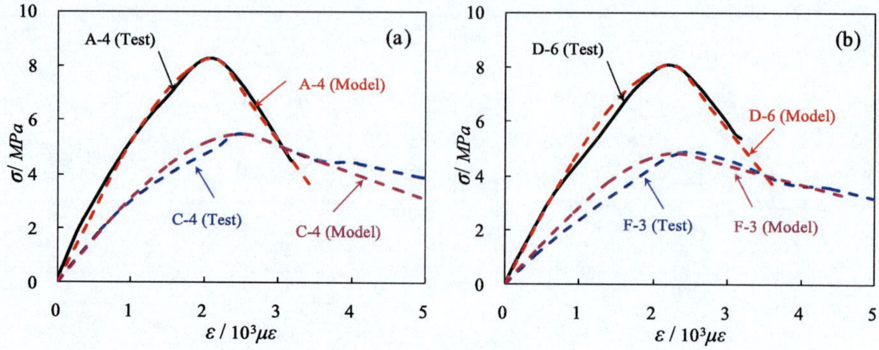

Fig. 7. Comparison between the calculated and test curves

5 Summary and Conclusions

In this study, the stress-strain relations of PG with different mix proportions were investigated, the failure mode, strength, elastic modulus and Poisson's ratio were summarized. Besides, a simplified model was developed for predicting the stress-strain curve of PG materials. According to the above research contents, the following conclusions can be drawn:

(1) PG samples with different mix proportions presented similar failure modes. The compression of micropores led to the development of horizontal cracks. The transverse expansion of the sample under compression led to the propagation of vertical cracks.
(2) The stress-strain curve of PG consisted of three stages, including the elastic stage, elastic-plastic stage and failure stage. The descending curve of the sample with low water-solid ratio is steeper than that of the sample with high water-solid ratio.
(3) The compressive strength of PG is greatly influenced by the water-solid ratio. The compressive strength is averagely increased by 12.6% and 48.8% respectively by decreasing the water-solid ratio from 0.9 to 0.8 and 0.7. Furthermore, the water-solid ratio has little influence on the peak strain of the sample.
(4) Decreasing the water-solid ratio or increasing the compressive strength can lead to increase the elastic modulus of the PG samples. By fitting the test data, a equation for predicting the elastic modulus of PG was developed. The study also showed that the mix proportion has no obvious effect on the Poisson's ratio of PG samples.
(5) A constitutive model was also proposed to represent the stress-strain relationship of PG samples based on mathematical fitting method. The proposed model was found to present a good correlation with the test results.

Acknowledgments. This research was under the support of the National Natural Science Foundation of China (No. 52068007), the Basic Research Program of Guizhou Province (No. QKHJC[2020]1Y417) and the Doctor Foundation of Guizhou University (No. GDRJ-HZ[2019]08).

References

1. Rafael, P.-L., Francisco, M., Carlos, R.C., et al.: Pollutant flows from a phosphogypsum disposal area to an estuarine environment: an insight from geochemical signatures. Sci. Total Environ. **553**, 42–51 (2016)
2. Olfa, H., Nelson, A., Ana Luísa, C., et al.: Phosphogypsum as a soil fertilizer: ecotoxicity of amended soil and elutriates to bacteria, invertebrates, algae and plants. J. Hazard. Mater. **294**, 80–89 (2015)
3. Gharaibeh, M.A., Rusan, M.J., Eltaif, N.I., Shunnar, O.F.: Reclamation of highly calcareous saline-sodic soil using low quality water and phosphogypsum. Appl Water Sci **4**, 223–230 (2014)
4. Shuhua, L., Lu, W., Baoying, Y.: Effect of modified phosphogypsum on the hydration properties of the phosphogypsum-based supersulfated cement. Constr. Build. Mater. **214**, 9–16 (2019)
5. Campos, M.P., Costa, L.J.P., Nisti, M.B., et al.: Phosphogypsum recycling in the building materials industry: assessment of the radon exhalation rate. J. Environ. Radioact. **172**, 232–236 (2017)
6. Kumar, S.: Fly ash-lime-phosphogypsum cementitious binder: a new trend in bricks. Mater. Struct. **33**, 59–64 (2000)
7. Baoguo, M., Zihao, J., Ying, S., et al.: Utilization of hemihydrate phosphogypsum for the preparation of porous sound absorbing material. Constr. Build. Mater. **234**, 117346 (2020)
8. Qiang, W., Yong, C., Junfeng, X.: Study on the improvement of the waterproof and mechanical properties of hemihydrate phosphogypsum-based foam insulation materials. Constr. Build. Mater. **230**, 117014 (2020)
9. Contreras, M., Teixeira, S.R., Santos, G.T.A., et al.: Influence of the addition of phosphogypsum on some properties of ceramic tiles. Constr. Build. Mater. **175**, 588–600 (2018)
10. Qing, W., Huagang, Z., Hongniao, C., et al.: Seismic behaviour of cast-in-situ phosphogypsum-reinforced concrete grid frame composite walls. Adv. Civ. Eng. **2019**, 1529137 (2019)
11. Degirmenci, N.: Utilization of phosphogypsum as raw and calcined material in manufacturing of building products. Constr. Build. Mater. **22**, 1857–1862 (2008)
12. Yichao, Z., Shaobin, D., Wanlin, W., et al.: Stress-strain relationship and seismic performance of cast-in-situ phosphogypsum. J. Appl. Biomater. Funct. Mater. **15**(Suppl. 1), 62–68 (2017)
13. Tian, T., Yun, Y., Zhihua, H., et al.: Utilization of original phosphogypsum for the preparation of foam concrete. Constr. Build. Mater. **115**, 143–152 (2016)

Numerical Simulation on Progressive Collapse Resistance of RC Beam-Slab-Column Substructures

Yi Zhang[1,2], Yirong Gao[2], Xiaolan Yuan[2], and Bing Liu[1,2](✉)

[1] GuangXi Key Laboratory of New Energy and Building Energy Saving, Guilin University of Technology, Guilin 541004, China
cnulb@glut.edu.cn

[2] College of Civil Engineering and Architecture, Guilin University of Technology, Guilin 541004, Guangxi, China

Abstract. In order to study the progressive collapse resistance (PCR) performance of unequal-span reinforced concrete (RC) spatial beam-slab-column substructures, numerical analysis based on the experimental study of the PCR performance of RC structures under central column failure completed by Qian was carried out. A finite element (FE) model was established by the high precision FE software ANSYS/LS-DYNA, and the accuracy of the model was verified by comparing with experimental results. Based on the verified FE model, the extended parameters of the model were analyzed, including the effects of concrete strength, slab thickness and slab reinforcement ratio on collapse resistance of RC spatial beam-slab-column substructures and mechanism of load redistribution. The FE analysis results showed that the increase of concrete strength would increase the first peak load of the structure but weaken the deformation capacity of the member. Appropriately increasing the slab thickness can significantly improve the bearing capacity of the structure. And with the increase of the reinforcement ratio at the bottom of the slab, the resistance to catenary mechanism was improved obviously, but the deformation capacity of the structure was weakened. In the whole process of RC frame collapse, the contribution of the slab to resistance was about 46% on average according to load redistribution analysis.

Keywords: RC beam-slab-column substructures · Pushdown method · FEM · Progressive collapse resistance · Resistance mechanism

1 Introduction

The progressive collapse refers to a continuous failure that is disproportionate to the initial failure after a small-scale initial failure of the structure caused by accidental factors [1]. There are quite a few reasons for the progressive collapse of the building structure, including human errors during design and construction, and the load caused by accidents outside the scope of design considerations. The collapses of the Ronan Point apartment in the United Kingdom in 1968 due to a gas explosion and the World Trade

W. Guo and K. Qian (Eds.): GBCESC 2022, LNCE 211, pp. 836–850, 2023.
https://doi.org/10.1007/978-981-19-5217-3_84

Towers in the United States in 2001 caused by aircraft strikes are two classic progressive collapse examples. The progressive collapse accident not only caused serious casualties and property losses, but also brought great social impacts. At the same time, it also caused the engineering community to attach great importance to the progressive collapse of the building structure and formulate corresponding regulations [2, 3] to prevent progressive collapse.

In order to study the various resistance mechanisms and failure conditions of building structures in the PCR, scholars at home and abroad have carried out a large number of studies through experiments, numerical simulations and theoretical analysis. Powell [4] used the principle of energy balance to analyze the PCR of a one-story two-span frame structure. The research results showed that under the failure condition of the central column, the gravitational potential energy of the structure gradually transformed into deformation energy and kinetic energy, and when the gravitational potential energy completely converted into deformation energy, the displacement of the structure reached the maximum value, and the kinetic energy of the frame was zero at this time. Zdenfk [5] explained the progressive collapse from the perspective of energy balance: when the kinetic energy generated by the impact on the outside of the structure exceeded the energy absorption limit of the structure itself, the progressive collapse of structure would appear. Li *et al.* [6] used pushdown method to study the PCR performance of welded PC beam-slab substructures under the side column failure scenario, and results showed that when the dynamic amplification factor is 2.0, the progressive collapse would not occur if a side column was removed instantaneously. Yu *et al.* [7] and Feng *et al.* [8, 9] took into account factors such as structural size, material strength and load to carry out uncertainty analysis and sensitivity analysis of the static resistance of RC frame structures to progressive collapse. The research results showed that the load of structure and the yield strength (YS) of steel bars were important factors affecting the capacity of the structure to resist progressive collapse. The experiment of non-beam floor slab carried by Zhang Fanzhen [10] showed that the damage of the structure was controlled by the film effect of the slab. Qian [11] *et al.* used frame beam longitudinal reinforcement ratio and transverse stirrup spacing as experimental variables to study the PCR of the remaining structure of four full-scale frame substructures after the center column was removed. The results showed that the failure mechanism of the structure was due to the formation of plastic hinges in the beam.

With the development of FE theory and the improvement of the function of computer, scholars are more inclined to perform nonlinear FE analysis on large concrete structures. It is possible for scholars to simulate the mechanical properties of various materials and analyze the bearing capacity of structures under different loads with the help of nonlinear FE analysis. At present, the FE analysis software commonly used includes ANSYS, MAC, ABAQUS and DIANA, etc. Lv Dagang *et al.* [12] used OpenSees to perform a nonlinear Pushdown analysis on reinforced concrete frame structures. The research results showed that the use of alternate load method and redundancy index could quantitatively evaluate the PCR of the structure under the failure of the main load-bearing members. Jun *et al.* [13, 14] used the Ls-Dyna display FE software to analyze the load conversion mechanism and bearing capacity of the beam-slab structure in the process of removing the side columns. Afterwards they analyzed the influence of boundary constraints, slab thickness and reinforcement to the load conversion mechanism and bearing capacity of the beam-slab structure. The research results showed that the bearing capacity of the structure under large deformation would be underestimated if floor slab was equivalent to the flange of T-beams or L-beams. Yu Xiaohui *et al.* [15] studied the influence of catenary effect on the PCR of RC frame structure by alternate load method, which turned out that after the central column is removed, the maximum value of the magnification factor should exceed 2.0 when considering the catenary effect.

Although a lot of researches have been carried out on the PCR of building structures, the current researches are mostly focused on the planar structure, only a few researches focused on the PCR of the space structure. However, in actual engineering, the building structure is mostly a spatial structure. Therefore, it is more practical to study the PCR of space structures. This paper used the large-scale dynamic numerical simulation software ANSYS/LS-DYNA to carry out the FE analysis of the progressive collapse test of the research group Qian [16], and expanded the parameter analysis on the basis of verifying the FE model of the test specimens. The influence of concrete strength, slab thickness, floor reinforcement ratio and floor load distribution mechanism on the PCR of RC beam-slab-column structure were studied by numerical simulation.

2 Establishment and Verification of FE Model

The prototype structure and research area of Qian [16] are shown in Fig. 1. The prototype structure is a three-story four-span frame structure with a height of 3.6 m. Longitudinal and transverse spans are 8.4 m and 6.0 m respectively. The properties were designed as follows: live load (LL): 3.0 kN/m^2, dead load (DL): 5.1 kN/m^2. Taking into account the limitations of the test conditions, the prototype structure was scaled by 1/4, and the test piece was designed according to the requirements of ACI318–08 [17] seismic design.

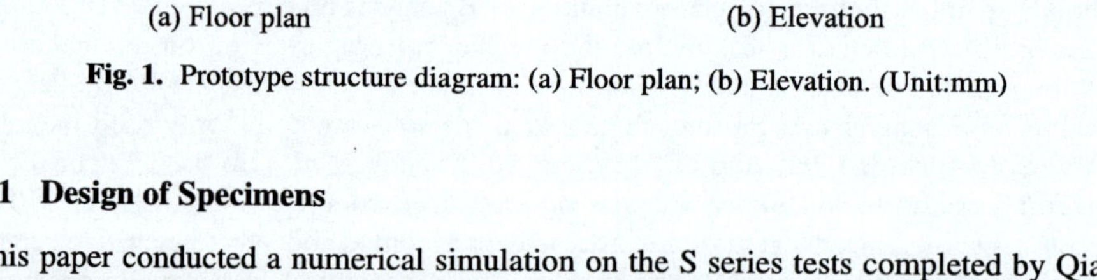

(a) Floor plan (b) Elevation

Fig. 1. Prototype structure diagram: (a) Floor plan; (b) Elevation. (Unit:mm)

2.1 Design of Specimens

This paper conducted a numerical simulation on the S series tests completed by Qian [16]. The test specimens of the S series were Tian-shaped space frame considering the effect of the slab, which included nine columns and twelve beams, and a 55-mm thick slab. The test specimens were all short stigma, and the side column stigma was partially strengthened by reinforcement. The size of S1 specimen column in S series was 200

× 200 mm. The spans of transverse and longitudinal beams were 1,500 and 2,100 mm respectively. Moreover, the cross-sectional dimensions of transverse and longitudinal beams were 140 × 80 mm and 180 × 100 mm, respectively. The size and structure of the S2 specimen was similar to those of the S1 specimen. The difference was that the transverse and longitudinal clear spans of the S2 specimen were both 1500 mm, and the cross-sectional dimensions of the transverse and longitudinal beams were both 140 × 80 mm. The concrete protective layer thicknesses of column, beam, and slab were 10, 7, and 7 mm (Table 1).

Table 1. Specimen properties

Test ID	Size type				Reinforcement type			
	Longitudinal beam	Transverse beam	Column	Slab	Longitudinal beam	Transverse beam	Column 1	Column 2
S1	A	B	C	D	4T10	4T10	4T13	8T16
S2	B	B	C	D	4T10	4T10	4T13	8T16

Note: Type A: Clear span = 1900 mm, section size is 180 mm × 100 mm; Type B: Clear span = 1300 mm, section size is 140 mm × 80 mm; Type C: section size is 200 mm × 200 mm; Type D: slab thickness is 55 mm

Since S2 specimen was similar to S1 specimen, this article only showed the size and the reinforcement diagram of S1 specimen (see Fig. 2). The target compressive strength of S1 and S2 concrete at age 28 days was 25 megapascals (MPa). The average compressive strength of S1 and S2 specimens are 21.4 MPa and 23.3 MPa, respectively. The material properties of the steel bars are shown in Table 2.

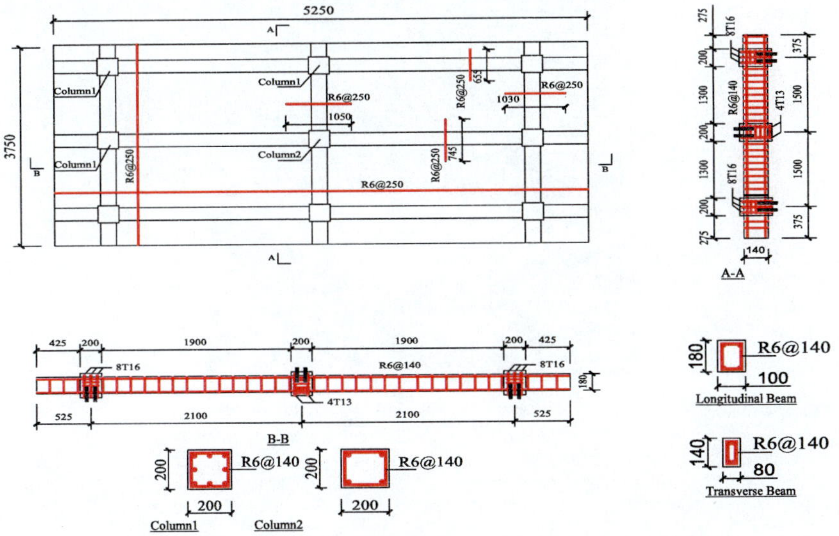

Fig. 2. Dimension and reinforcement details of S1. (Unit:mm)

Table 2. Material properties of reinforcements

Items	Yield strength (MPa)	Ultimate strength (MPa)	Yield strain/ (10^{-6})	Elasticity modulus (%)
R6	355	465	1910	17.5
T10	437	568	2273	13.1
T13	535	611	2605	11.6
T16	529	608	2663	14.3

Note: R6 represents plain bar with diameter of 6 mm; T10 represents deformed rebar with diameter of 10 mm. T13 represents deformed rebar with diameter of 13 mm. T16 represents deformed rebar with diameter of 16 mm

The loading device is shown in Fig. 3. In order to prevent out-of-plane instability of the test specimen, a limit device was installed under the hydraulic jack to ensure that the loading position of the jack would not change. Eight steel plates with a total weight of 380 kg were evenly suspended around the edges of the board to simulate the actual rotation constraints at the edge of the board. The side column was fixed on the steel column to simulate the fixed constraint of the side column. A downward concentrated load was applied to the upper surface of the center column Z2 by a hydraulic jack to simulate the collapse process of the center column under failure. The concentrated load was applied by displacement control load until the specimen failed and lost its bearing capacity.

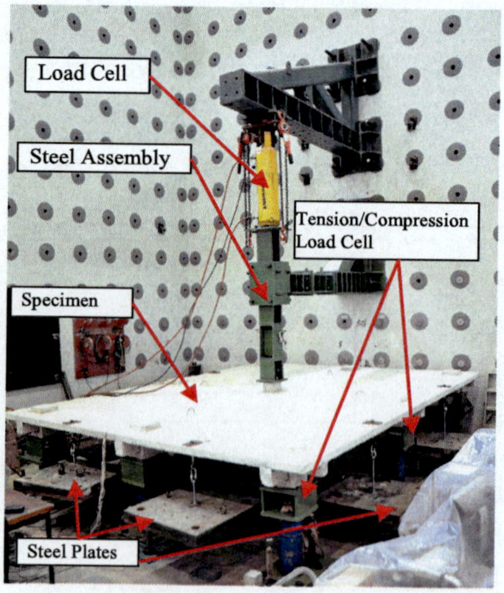

Fig. 3. Test setup

2.2 Modeling Process

The FE model was built by ANSYS/LS-DYNA for the simulation analysis. The SOLID164 element was used to simulate concrete. Among them, the SOLID164 element was a three-dimensional solid element that was defined by 8 nodes. This element could simulate the characteristics of the three-dimensional structure accurately for its default single-point integration algorithm. The BEAM161 element, which was a three-dimensional beam element defined by 3 nodes, was used to simulate steel reinforcement. It was often used to simulate steel reinforcement for its capacity to simulate axial tension and compression, biaxial bending and finite strain.

The FE model of the test specimen is shown in Fig. 4. The upper and lower surfaces of the side and corner column were all fixedly restrained. The concrete adopted *MAT_CSCM_CONCRETE model while the reinforcement adopted *MAT_PLASTIC_KINEMATIC material model. The coupling between reinforcement and concrete was defined through *CONSTRAINED_LAGRANGE_IN_SOLID and the rotation constraint of the slab was defined through the keyword *CONSTRAINED_LAGRANGE_IN_SOLID.

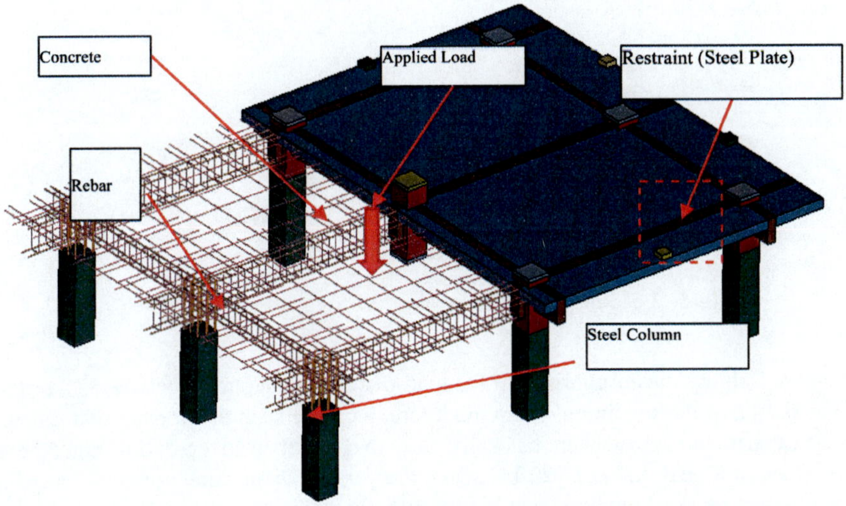

Fig. 4. Finite element model

2.3 Model Validation

Figure 5 shows the comparison of the load-displacement curves between the test and the simulation of S1 specimen. From the figure, it could be seen that the simulation curve and the experimental curve are basically close to each other. In the initial stage of loading, the simulated and experimental loads were in good agreement. In the test, when the vertical displacement was 47 mm, the first peak load was 115 kN, and when the vertical displacement was 36 mm in the simulation, the first peak load was 109 kN, 5% different from the test. Subsequently, the test load began to decrease, while the simulated load did not decrease significantly, staying stable relatively. The reason was that the constitutive relationship of the FE model was an optimization of the general situation, and there was still a deviation from the actual situation. Meanwhile, the micro-cracks of concrete emerged due to external factors in the process of production and curing, which was not considered in the simulation. Then, the catenary mechanism started to work, and the load rose again. The test reached the ultimate load of 169 kN when the vertical displacement was 210 mm. As a comparison, the simulation reached the ultimate load of 145 kN at 204 mm. Finally, when the vertical displacement was 305 mm, the load of experiment was stopped due to severe punching failure at the plate-column connection position.

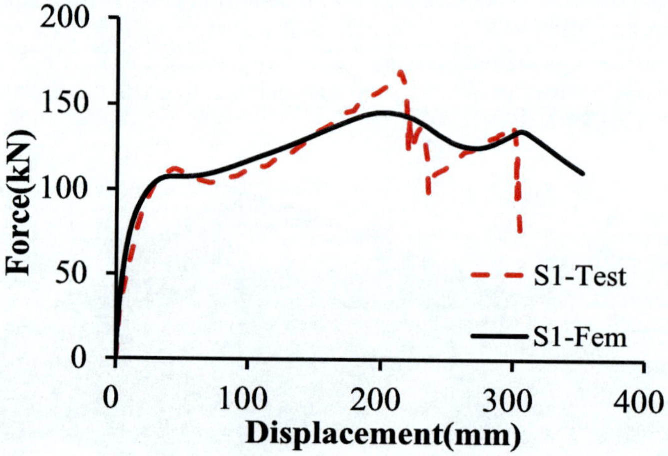

Fig. 5. Load-displacement curves of S1 from test and simulation

Figure 6 shows the comparison of the load-displacement curves between S2 test and the simulation of the specimen. From the figure, it can be seen that the simulation curve and the experimental curve were basically close to each other. In the initial stage, the test reached the first peak load of 123 kN when the vertical displacement was 42 mm, and the simulation reached the first peak load of 125 kN when the vertical displacement was 38 mm, 1.6% different from the test. Then due to the degradation of the compressive arch action, the load began to decrease. Subsequently, the specimen came to the catenary effect stage, and the load increased again. The test reached the ultimate load of 169 kN at 200 mm. As a comparison, the simulation reached the ultimate load of 181 kN at 224 mm. In the large deformation stage, there was much difference between the simulated load curve and the experimental load curve. It was because the failed element still existed in the model during the numerical simulation analysis instead of being removed, resulting in the simulated load being slightly larger than the experimental load curve.

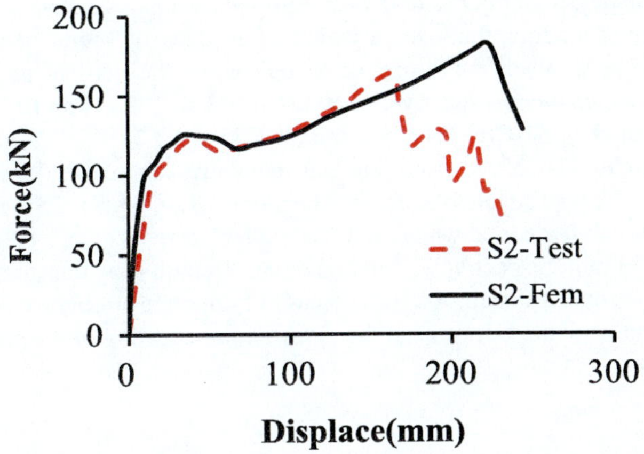

Fig. 6. Load-displacement curves of S2 from test and simulation

Figure 7 shows the failure modes of the tested and simulated specimens of S1. From the figure, we can see the cracks on the top slab in the test and simulation were all concentric ellipses with the center pillar as the center. Both the experimental and simulated slab bottom failure modes were concrete tensile failure, and the reinforcement at the bottom of the beam connected to the center column fractured.

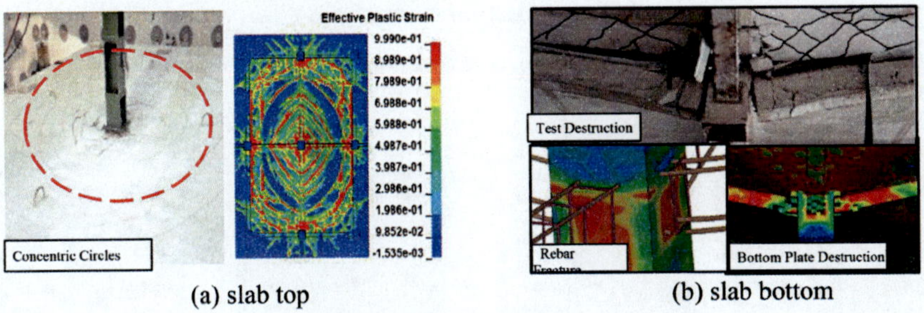

Fig. 7. Failure mode: (a) slab top; (b) slab bottom.

3 Influence of Concrete Strength

In order to further study the PCR performance of the unequal-span RC spatial structure under the failure condition of the center column, this section studies the influence of concrete with different strengths on the PCR performance of the RC beam-slab-column structure. The selected concrete strengths were 65 MPa, 55 MPa, 45 MPa and 25 MPa, and the model numbers were S1-C65, S1-C55, S1-C45 and S1-C25 respectively.

The load-displacement curve obtained from the FE analysis is shown in Fig. 8. At the initial stage of loading, the load curve increased linearly. Models S1-C65, S1-C55, S1-C45 and S1-C25 reached the first peak load when the vertical displacement was about 40 mm, which were respectively 149 kN, 139 kN, 129 kN and 109 kN. The first peak loads of models S1-C65, S1-C55, and S1-C45 were 36%, 27%, and 18% higher than that of model S1-C25, respectively. Subsequently, due to the development of the catenary effect, the load began to increase. The models S1-C65, S1-C55, S1-C45 and S1-C25 reached the ultimate load when the vertical displacements were 109 mm, 138 mm, 178 mm and 204 mm, respectively. The higher the strength was, the sooner the ultimate load reached. It could be seen that the increase of concrete strength will raise the first peak load of the structure but weaken the deformation capacity of the structure.

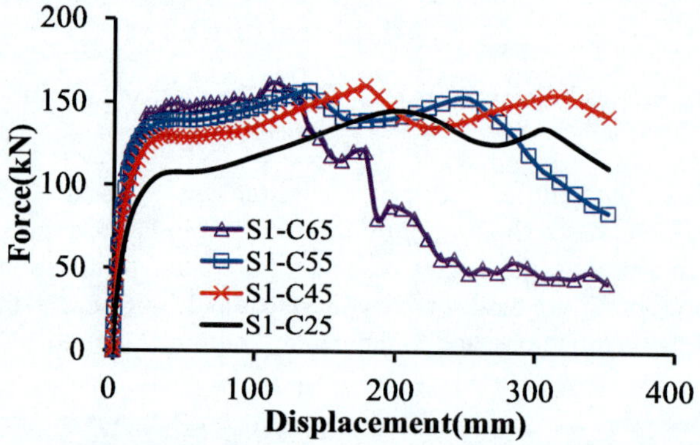

Fig. 8. Effects of concrete strength

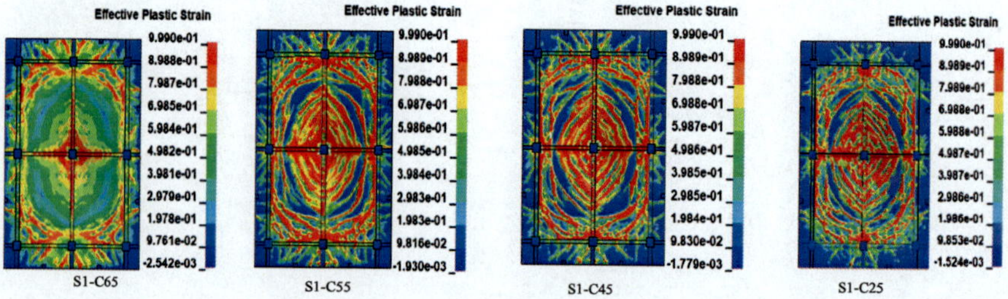

Fig. 9. Failure mode of vertical displacement is 300 mm

Figure 9 compared the failure modes of different concrete strength grades when the vertical displacement was 300 mm. It could be seen that as the strength of concrete increases, the damage of the slab decreases, and the stiffness of the members increases, so that the structure develops a higher bending mechanism and compression arch mechanism to resist collapse.

4 Influence of Concrete Slab Thickness

In actual projects, structural designers will adopt different thicknesses for the slabs in accordance with relevant design requirements, and the thickness of the slabs will also have a certain impact on the PCR performance of the structure. For this reason, this section studied the influence of slab thickness on the PCR performance of RC spatial structures with unequal spans under the failure condition of the center column. The slab thicknesses were 75 mm, 55 mm and 45 mm, and the corresponding model numbers were S1-B75, S1-B55 and S1-B45.

As shown in Fig. 10, the models S1-B75, S1-B55 and S1-B45 achieved the first peak loads at a vertical displacement of about 43 mm, which were 114 kN, 105 kN and 97 kN respectively. With the thickness of floor increased from 45 mm to 75 mm, the first peak load of the structure increased by 15%. Models S1-B75, S1-B55 and S1-B45 achieved ultimate loads which were 175 kN, 145 kN and 138 kN respectively when the vertical displacements were 242 mm, 204 mm and 195 mm. With the thickness of the floor slab increased from 45 mm to 75 mm, the ultimate load of the structure increased by 27%, the corresponding ultimate displacement increased by 24%. This was due to the increase in the thickness of the floor slab enlarged the height of the relative compression zone of the floor, improved the bending bearing capacity of the structure, thereby improved the PCR performance of the structure.

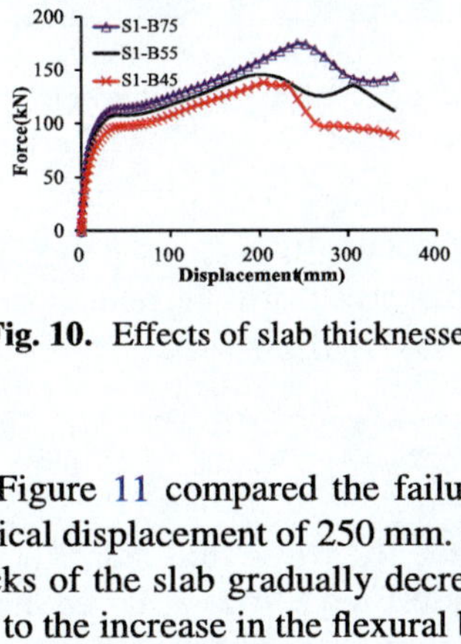

Fig. 10. Effects of slab thicknesses

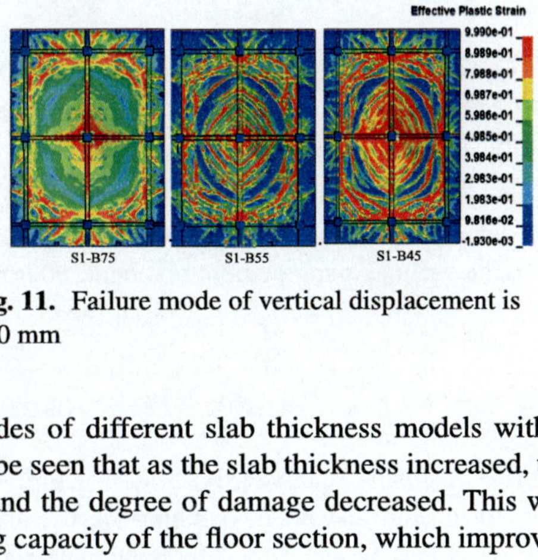

Fig. 11. Failure mode of vertical displacement is 250 mm

Figure 11 compared the failure modes of different slab thickness models with a vertical displacement of 250 mm. It can be seen that as the slab thickness increased, the cracks of the slab gradually decreased and the degree of damage decreased. This was due to the increase in the flexural bearing capacity of the floor section, which improved the progressive collapse bearing capacity of the structure.

5 The Influence of Slab Reinforcement Ratio

In this section, the influence of slab reinforcement ratio on the PCR performance of the structure was studied. Longitudinal steel bars with reinforcement ratios of 0.45%, 0.25%, and 0.17% were set on the slab, and the model numbers were S1-0.45%, S1-0.25%, and S1-0.17%, respectively.

The load-displacement curves of different reinforcement ratio models are shown in Fig. 12. At the initial stage of loading, the curves increased linearly and kept the same trend. At a vertical displacement of about 42 mm, models S1-0.45%, S1-0.25% and S1-0.17% achieved the first peak load, which were 115 kN, 109 kN and 10 3kN, respectively, and the first peak load of models S1-0.45% and S1-0.17% were 28% higher and 6% lower than the model S1-0.25% respectively. After that, as the load decreased slightly the structure entered the catenary mechanism, then the load began to increase again. Finally, models S1-0.45%, S1-0.25% and S1-0.17% achieved ultimate loads which were 157 kN, 145 kN and 139 kN respectively when the vertical displacements were 183 mm, 204 mm, and 220 mm. As the floor reinforcement ratio increased from 0.17% to 0.45%, the ultimate load of the structure increased by 13%, and the corresponding ultimate displacement reduced by 21%. For the RC frame, when the frame beam section size, reinforcement and slab thickness were unchanged, the bottom reinforcement ratio had an obvious impact on the resistance of the catenary mechanism. As the bottom reinforcement ratio increased, the resistance of the catenary mechanism correspondingly improved.

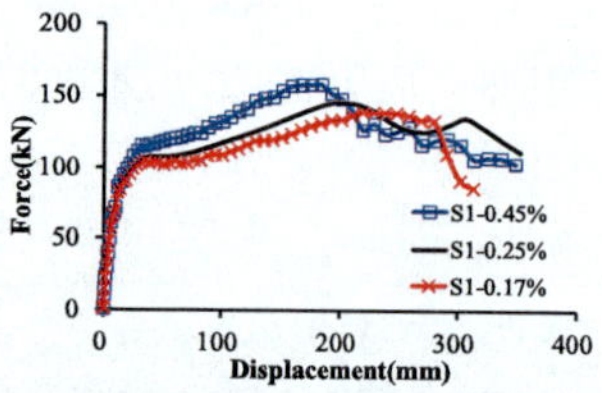

Fig. 12. Effects of slab reinforcement ratio

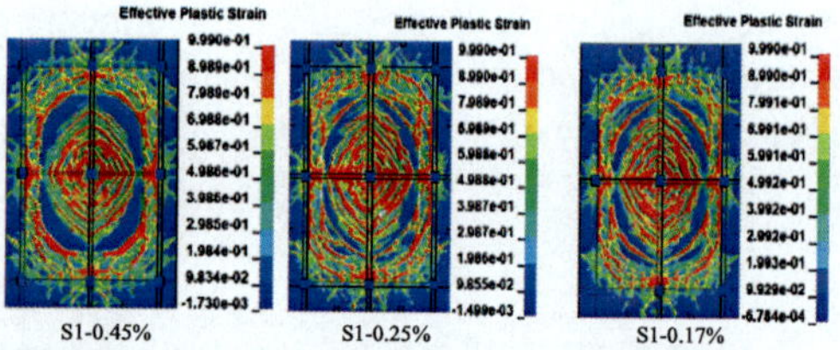

Fig. 13. Failure mode of vertical displacement is 250 mm

Figure 13 compared the failure modes of models S1-0.45%, S1-0.25% and S1-0.17% when the vertical displacement reached 250 mm. It can be seen that as the reinforcement ratio increased, the damage degree of the component was relatively weak.

6 Load Redistribution Mechanism

In order to further analyze the effect of the slab on the RC frame load redistribution mechanism and resistance mechanism, the slab in the S1 model was removed, and a pure frame FE model was established. The FE model of the pure frame structure is shown in Fig. 14.

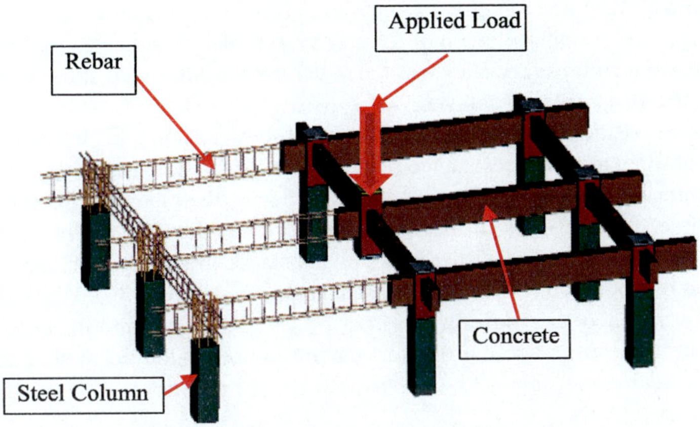

Fig. 14. Bare frame from numerical model

The load-displacement curves of the original model with slab (S1) and the pure frame (S1-L) are shown in Fig. 15. It can be seen that during the entire loading stage, the PCR of S1-L was less than that of S1. In the initial stage of loading, the initial stiffness of S1-L was 53 kN/m, and the initial stiffness of S1 was 85% higher than that of S1-L. When the vertical displacement was 48 mm, S1-L obtained the first peak load of 69 kN. Compared with S1-L, the first peak load and corresponding displacement of S1 were increased by 67% and 2% respectively. When the vertical displacement of S1 was 88 mm, the load reduced to the lowest value; and when the vertical displacement of S1-L was 198 mm, the load reduced to the lowest value, and the arching mechanism stage of S1-L was relatively long. It can be seen that the slab improved the bearing capacity of the beam mechanism. Due to the increase of the cross-sectional area and the tensile steel bars of the RC frame, the catenary mechanism got enhanced. Besides, the tensile film effect could improve the continuous collapse resistance of the member. After the failure of the steel bar in the beam, the steel bar in the plate bore the tensile force.

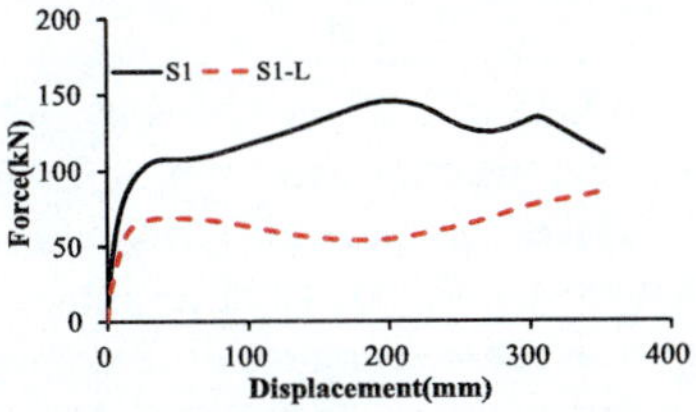

Fig. 15. Load-displacement curves from S1 and S1-L

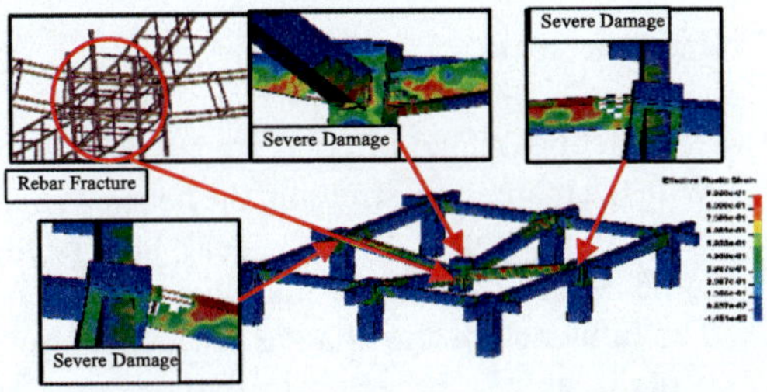

Fig. 16. Failure mode of bare frame from Finite element model

Figure 16 is a diagram of the failure mode of S1-L. It could be seen that compared with the S1 specimen, the concrete of S1-L spalled seriously at the bottom of the center column beam end and the top of the side column beam end. Both of the steel bars at the bottom of the center column beam of S1 and S1-L broke. It indicates that after adding the floor slab, the collapse resistance of the structure was improved, and the progressive collapse of the structure was delayed.

The decomposition of RC frame resistance is shown in Fig. 17. In the initial stage, the contribution of the slab resistance reached 60%. With the crushing of the concrete in the compression zone of the beam, the resistance of the beam began to decrease, at the same time the steel bars in the floor slab began to develop a film pulling effect, which made the contribution of the floor slab resistance gradually increase. The largest contribution of floor slab resistance at this stage accounted for 73%. When the vertical displacement reached 400 mm, the steel bar broke in the slab, and the catenary effect developed in the beam made the resistance contribution of the beam start to increase. When the structure was damaged, the beams and slabs contribute 47% and 51% of the resistance, respectively.

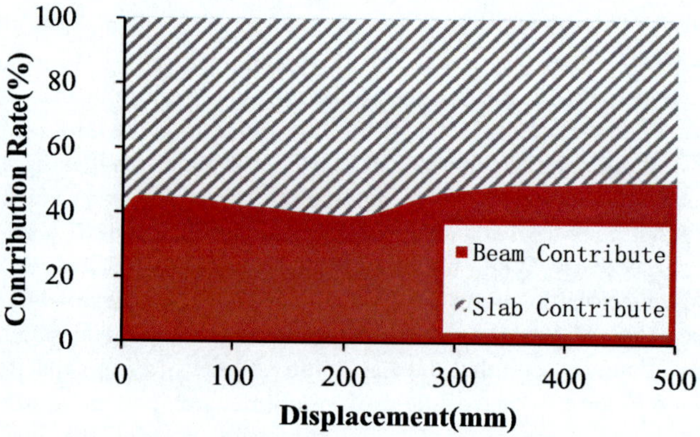

Fig. 17. Resistance capacity decomposition

7 Conclusion

1) The first peak load of concrete strength grade from C25 to C65 increases by 36%. With the increase in strength, the bearing capacity of the structure against progressive collapse increases. The first peak load of the frame with a concrete slab thickness of 65 mm is 27% higher than that of the frame with a slab thickness of 25 mm.
2) The influence of the change of the slab bottom reinforcement rate on the components is mainly reflected in the catenary mechanism, that is, as the slab bottom reinforcement rate increases, the resistance of the catenary mechanism will increase correspondingly, but the deformation ability of the structure is weakened.
3) For the RC frame structure, the slab can improve the resistance of the frame structure from the compression arch mechanism and catenary mechanism during the progressive collapse. The slab improves the ultimate bearing capacity of the RC frame by

up to 51%, and the average resistance contribution of the slab during the progressive collapse of the RC frame accounts for about 47%.

Acknowledgements. The authors gratefully acknowledge the financial support provided by Guangxi Key Laboratory Project of New Energy and Building Energy Saving (Gui Ke Neng22-J-21–4) and the Natural Science Foundation of Guangxi (No. 2021GXNSFFA196001).Any opinions, findings and conclusions expressed in this paper do not necessary reflect the view of Guangxi Key Laboratory of New Energy and Building Energy Saving and Natural Science Foundation of Guangxi.

References

1. Yi, L.: Research on the design method of RC frame structure resisting continuous collapse. Tsinghua University (2011). (in Chinese)
2. DOD: Design of buildings to resist progressive collapse. Washington DC: Department of Defense (2010)
3. GSA: Progressive collapse analysis and design guidelines for new federal office buildings and major modernization projects. Washington, DC: U.S. General Service Administration (2016)
4. Powell G.: Progressive collapse: Case studies using nonlinear analysis. In: Structures Congress, New York, pp. 1–14 (2005)
5. Zdenfk, P.B.: Mechanics of progressive collapse: learning from world trade center and building demolitions. J. Eng. Mech. **133**(3), 308–319 (2007)
6. Zhi, L., Yun Hao, W., Xiao Fang, D., Kai, Q.: Behavior of precast concrete beam-slab substructures with welded connections to resist progressive collapse. J. Build. Struct. **41**(10), 121–128 (2020)
7. Feng, C., Xie, C., Xu, J., et al.: Robustness quantification of reinforced concrete structures subjected to progressive collapse via the probability density evolution method. Eng. Struct. **202**, 109877 (2020)
8. Xiao Hui, X., et al.: Uncertainty and sensitivity analysis of reinforced concrete frame structures subjected to column loss. J. Perform. Constr. Facil. **31**(1), 1–14 (2017)
9. Ren, P., et al.: Experimental investigation of progressive collapse resistance of one-way reinforced concrete beam-slab substructures under a middle-column removal scenario. Eng. Struct. **118**, 28–40 (2016)
10. Fan Zhen, Z., Wei, J.Y.: Experimental study and analysis of the anti-collapse performance of beamless floor slabs. J. Hunan Univ. (Nat. Sci. Ed.) **37**(4), 1–5 (2010). (in Chinese)
11. Kai, Q., et al.: Experimental and analytical assessment on RC interior beam-column sub assemblages for progressive collapse. J. Perform. Constr. Facil. **26**(5), 576–589 (2011)
12. Da Gang, L., et al.: Robustness analysis of continuous collapse of structures based on the alternate load path Pushover method. J. Build. Struct. **31**(Supplement 2), 112–118 (2010). (in Chinese). https://doi.org/10.14006/j.jzjgxb.2010.s2.003
13. Jun, Y., et al.: Numerical investigation on structural behavior of RC beam-slab assemblies under an exterior column removal scenario. High Tech Concrete Where Technol. Eng. Meet **1**(1), 1252–1262 (2018)
14. Jun, Y., et al.: Numerical study of progressive collapse resistance of RC beam-slab substructures under perimeter column removal scenarios. Eng. Struct. **159**(1), 14–27 (2018)
15. Xiao Hui, Y., Kai, Q.: Analysis of continuous collapse resistance of reinforced concrete frame structure considering catenary effect. J. Build. Struct. **38**(4), 28–34 (2017). (in Chinese)

16. Kai, Q., et al.: Load-carrying mechanism to resist progressive collapse of RC buildings. J. Struct. Eng. **141**(2), 4014107–4014101 (2014)
17. ACI Committee 18-08. Building code requirements for structural concrete (ACI 318-08) and commentary(318R-08). Farmington Hills, MI, 433.ASCE/SEI (2008)

Effect of Vertical Load on Lateral Response of Single Piles in Coral Sand

Chunyan Wang[1,2] and Xuanming Ding[1,2(✉)]

[1] College of Civil Engineering, Chongqing University, Chongqing 400045, China
dxmhhu@163.com

[2] Key Laboratory of New Technology for Construction of Cities in Mountain Area, Chongqing University, Chongqing 400045, China

Abstract. Coral sand is distributed widely between 30 °S and 30 °N, and its mechanical properties are significantly different from general siliceous sand. To investigate the pile-soil lateral response in calcareous sand foundation under the vertical-horizontal load combinations, a series of effective numerical studies were conducted and analyzed. The lateral load-displacement curves, pile deformations, soil stress fields, and soil displacement fields were analyzed in detail. The combined loaded piles show a significant increase in the lateral capacities, or a corresponding reduction of pile displacements as compared to pure lateral loaded pile. In addition, the existence of vertical load can increase the stresses of sand around pile and weaken the rotation flow zone of sand near the pile tip.

Keywords: Coral sand · Vertical-horizontal load · Numerical analyses · Single pile · Hypoplastic constitutive model

1 Introduction

It is well recognized that vertical-horizontal load combinations are the most common load bearing form of pile foundation in offshore engineering, such as the vertical gravity combining with the horizontal load that results from winds, waves, or due to the impact of ships [1]. Noting that, the particle properties of coral sand show great differences with general sediment materials, such as the irregular shapes, low unit weights and high permeability [2]. Therefore, the design method based on the conventional sediments is inapplicable in coral sand foundations.

To date, numerous studies on the lateral pile-soil responses under vertical-horizontal combined load have been carried out [3]. On the one hand, once the lateral displacement occurs, P-Δ effect will be introduced, resulting in the reduction of the lateral capacity in some degrees or corresponding increase in the lateral deformation. On the other hand, it can be observed from in-situ tests [4] or numerical simulations [5] that the vertical compression, introduced by the vertical load, will reduce the pile deflections and bending moments. However, these researchers only put the focus on general siliceous sand and clay. It is necessary to conduct further studies on the pile-soil response under vertical-horizontal load combinations in coral sand foundations.

W. Guo and K. Qian (Eds.): GBCESC 2022, LNCE 211, pp. 851–859, 2023.
https://doi.org/10.1007/978-981-19-5217-3_85

Numerical approach can simulate the real stress and strain fields of foundation, which becomes the most common approach to explore the lateral pile-soil response. In this study, many effective numerical models were established to simulate the lateral pile-soil response in coral sand under different combined load levels and the mechanical mechanism of the variation of sand stiffness was revealed.

2 Material and Constitutive Model

Abaqus (a finite-element software) was employed to carried out the numerical simulation for pile-soil response in coral sand. Table 1 lists the main properties of the pile and soil used in this numerical analysis. The linearly elastic model and Hypoplastic (HP) constitutive model [6] was employed to characterize the stress-strain behavior of pile and coral sand respectively. The HP model can capture the reversible and irreversible deformations of coral sand, and reflect the pressure dependency and density dependency of sand behaviors. Dewi et al. [7] carried out a series of triaxial and oedometric tests on coral sand, and then calibrated the parameters of HP model for coral sand, which are summarized in Table 2. Therefore, the model parameters of coral sand suggested by Dewi et al. [7] were directly adopted in this numerical analysis.

Table 1. Main properties of piles and soil.

Pile details	Soil details
Diameter, D: 2 m Total length: 24 m Embedment length: 14 m Material of pile: Concrete C30 Young's modulus Ep: 30 GPa Unit weight: 24.5 kN/m^3 Poisson's ratio: 0.2	Type of soil: coral sand Unit weight: 14.7 kN/m^3 void ratio, *e*: 0.92 Particle density, ρ_s(g/cm^3): 2.82 Coefficient of uniformity, Cu: 3.434 Mean particle size: 0.397 mm Maximum dry density, $\rho_{d\,max}$(g/cm^3): 1.594 Minimum dry density, $\rho_{d\,min}$(g/cm^3): 1.244 Relative density: 70% Friction angle, φ: 34.3

Table 2. Model parameters of coral sand

φ_c	h_s	n	e_{d0}	e_{c0}	e_{i0}	α	β
40	100 MPa	0.08	0.769	1.267	1.45	0.25	1.8

3 Analysis Scheme

3.1 Building Model

In this numerical analysis, only one half of the pile-soil system was established. The interaction between sand and pile was set as hard contact in the normal direction and penalty friction in shearing direction, and the shear strength was defined with zero cohesive strength and two thirds of the friction angle. The vertical loads were applied prior to the horizontal loads and kept constant. After a lot of trial calculations with various mesh fineness and model sizes, as shown in Fig. 1, a suitable mensh fineness and model size (with 15, 19, and 30 times of the pile diameter in width, height, and length respectively) was found.

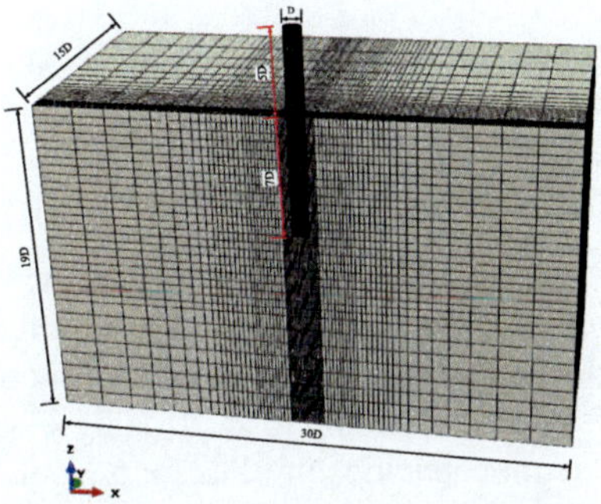

Fig. 1. Mensh fineness and model size.

3.2 Validation of the Finite-Element Model

Back prediction of test results is a good way to verify the validity of the 3D finite-element model. Chortis et al. [8] reported the results of centrifuge test for a rigid, open-ended monopile under vertical-horizontal load combinations. The prototype dimension of model pile is 1.8 m and 0.03 m in outer diameter and wall thickness, the bending stiffness and embedment length are 13.7 GN m^2 and 9 m, respectively. In the present numerical study, the sequences of load applications are the same as that in centrifuge tests. The distributed dead load was used to simulate constant vertical load of 3 MN in test, which was applied at pile head prior to lateral displacement.

The comparison for the calculated and measured relationships between lateral loads and displacements are shown in Fig. 2a, the predictions agree well with the test results except for the slight underestimation at the smaller displacement. Figure 2b shows the distributions of bending moments along embedment depth for a given lateral load (100 kN). The agreement between calculated and measured results validates the numerical model in simulating the pile-soil response under vertical-horizontal combined loading.

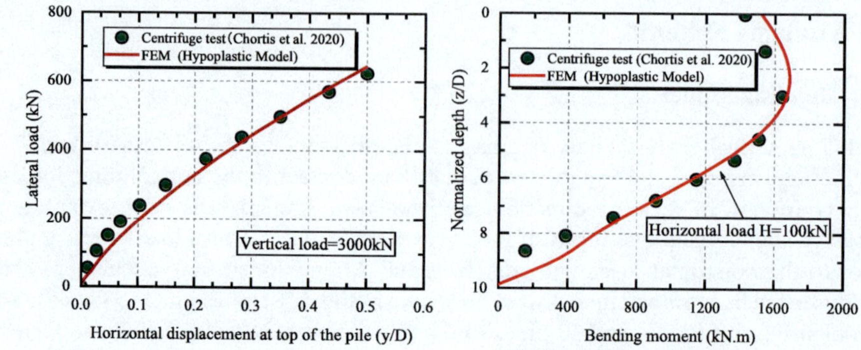

(a) Lateral load-displacement curves (b) Distribution of bending moment

Fig. 2. Comparisons of calculated and measured results.

4 Results and Discussions

4.1 Response of Pile

Figure 3(a) shows the evolution of lateral load-displacement relationships under different combined load levels. The lateral loads show a nonlinear growth trend with the increasing in the lateral displacements. These curves with combined loads are significantly higher than that with pure horizontal load. Broadly speaking, the applied vertical loads could improve the lateral capacities of piles for a given lateral displacement, and this phenomenon disagrees with the findings proposed by Karthigeyan and Rajagopal [5].

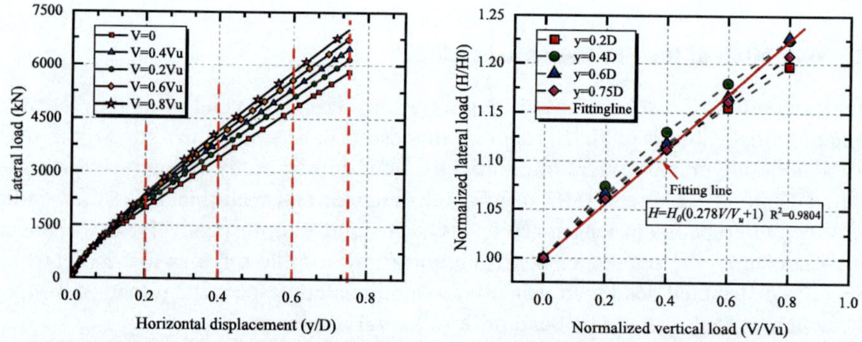

(a) Lateral load-displacement curves of piles (b) Variation of normalized lateral load

Fig. 3. Response of pile.

The variation of normalized lateral capacities under combined load conditions are shown in Fig. 3(b). Noting that the normalized lateral loads show a linear increase trend by increasing vertical load levels, which is not affected by the applied lateral displacement. For instance, when the applied vertical loads increased from 0 to 0.4 Vu and 0.8 Vu, the average lateral loads increased by approximately 11.9% and 21.4%, respectively. Therefore, the relationship between the normalized lateral loads and vertical load levels can be fitted as a linear curve, and the slope of the line is approximately 0.278 ($R^2 =$ 0.9804).

$$H = H_0\left(0.278\frac{V}{V_u} + 1\right) \tag{1}$$

where H is the lateral load under combined loading condition, H_0 is the pure lateral load, V is the applied vertical load, Vu is the ultimate capacity when the pile was subjected to pure vertical load.

Figure 4(a) shows the lateral displacements of piles under various combined load levels. It can be observed that the lateral displacements take place in the entire embedment depth and exhibit an increasing trend with the increasing of lateral loads. Moreover, it can be observed that some slight reverse displacements occur near the pile tip. These phenomena indicates that the sand in the entire embedment depth range has been mobilized to resist the applied lateral displacements, and the pile can be classified as the rigid short pile. In addition, the existence of vertical load significantly can reduce the lateral displacement of pile, and the magnitude of reduction becomes more remarkable with the increasing in vertical loads, which confirms the observation obtained from full-scale pile tests [4].

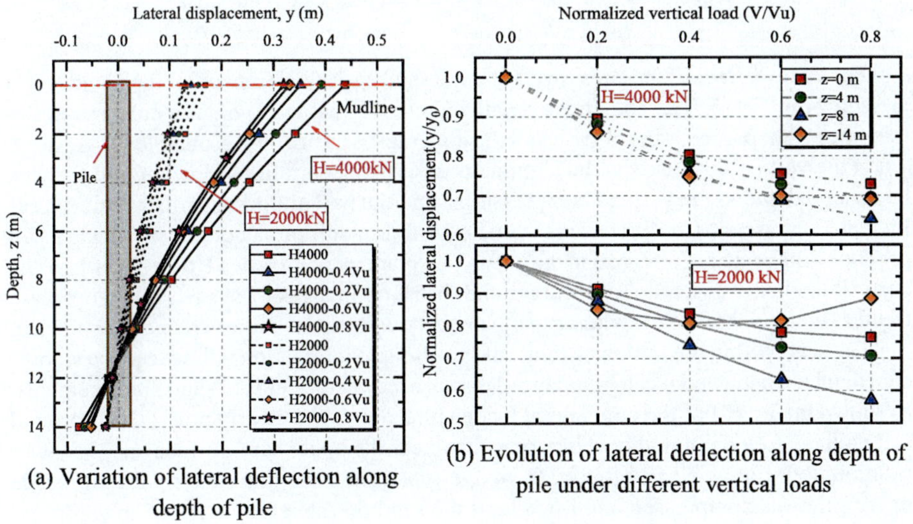

Fig. 4. Lateral deflection along depth of pile.

Figure 4(b) shows the evolution of lateral deflections along depth under different combined load levels. In the legend, y_0 represents the lateral displacement of pile under pure lateral load, *y* denotes the lateral displacements of piles under combined loading conditions. Increasing the pre-applied vertical load can result in the significant decrease in the pile displacements, and the magnitude of the reduction becomes more significant with the increase of depth except for pile tip. For instance, when the given lateral load was 4000 kN and the vertical load increased from zero to 0.8 Vu, the lateral displacement of pile was reduced by about 27% at ground elevation, and the reduction increased to 35% at the depth of 8 m.

4.2 Response of Sand

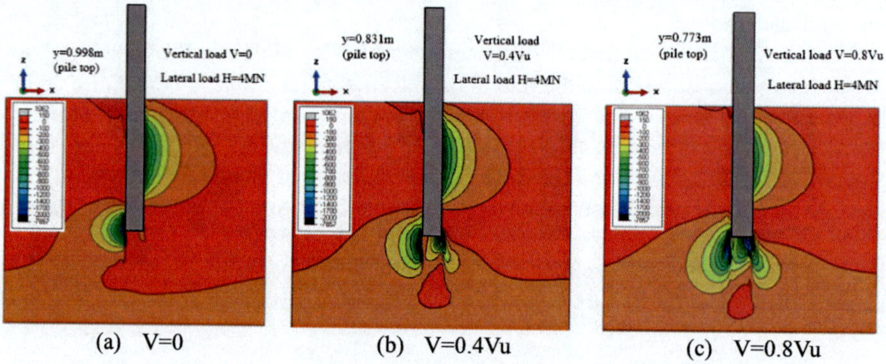

Fig. 5. Contour lines of the stress (σ_{xx}) of sand (unit: kPa)

The contour lines of the stress (σ_{xx}) of coral sand under different combined load levels are shown in Fig. 5. The soil stress mobilized by external loads mainly concentrates at the lower right part near the pile tip and the upper left part near the ground level. Besides, when the vertical load exists, the significant stress concentrations beneath pile tips can be observed due to the vertical compressions, and the magnitudes and distribution areas increase significantly by increasing vertical load. In addition, a certain sand range will produce soil resistance to resist the lateral displacement of pile. However, when the vertical loads increase, the magnitudes and distribution areas of stress show a decrease trend at the upper left part, whereas an increase trend can be observed near the pile tip.

The sand within the influence range of the pile-soil interaction will take place significant displacement, and it can be analyzed through the vector fields of sand displacements, as shown in Fig. 6. For the pure lateral loaded pile, due to the clearly pile rotation around a certain inflection point, the wedge-type flow near the upper sand surface and the lower rotation flow can be clearly found. In wedge-type flow zone, the sand on the right side of the pile was compressed and lifted, and the sand on the left side flows upwards and shows a tendency to fill the gap back the pile. However, the different flow mechanism was observed under combined load conditions. Referring to Fig. 6(b–c), the rotation zone cannot be found and the phenomenon of flowing upward on the left side also disappears. For a given vertical load of 0.8 Vu, the sand even shows a flow downward trend,

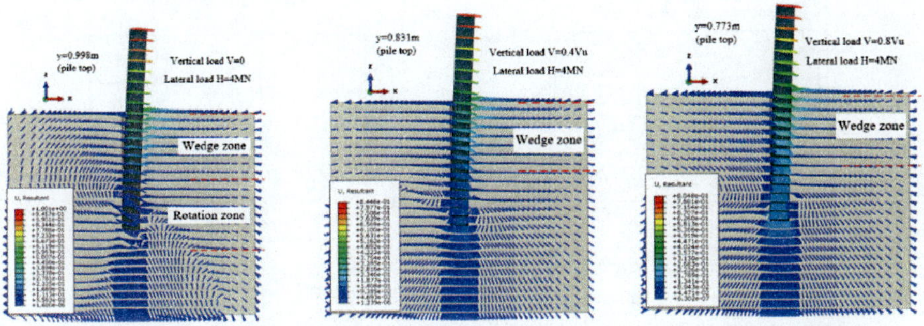

(a) V=0 (b) V=0.4Vu (c) V=0.8Vu

Fig. 6. Vector fields of sand displacement

resulting from the compression behavior beneath the pile tip and the friction between the sand and pile.

4.3 Response of Soil-Pile Interaction

To explain the mechanism of the improvement on the lateral bearing capacity resulting from vertical load, the lateral compressive stress (σ_{xx}) of sand in front of pile under different loading conditions are shown in Fig. 7, as well as the shear stress (σ_{xy}) of sand on the side of pile. The compressive stresses and shear stresses above the rotation point show an increase trend by increasing the lateral displacements or lateral loads. For a given lateral displacement, as shown in Fig. 7(a–b), the compressive stress shows an obvious increase trend hby increasing vertical loads, but the increase is not significant in shear stress. In addition, the growth of compressive stress is more regular and significant than that of shear stress. The existence of vertical load can increase confining stresses, which will lead to the increase of the stress levels on the pile-soil interfaces. Then, the larger compressive and shear stresses of the pile-soil interfaces will result in the greater shear strength of sand.

As shown in Fig. 7(c–d), for a given lateral load, the compressive stresses and shear stresses above rotation point show a significant decreasing trend with the increase in vertical load. It possibly resulted from that the increases of vertical loads can increase the sand shear strength, thereby reducing the pile-soil interaction stress, eventually resulting in the decrease of pile displacement.

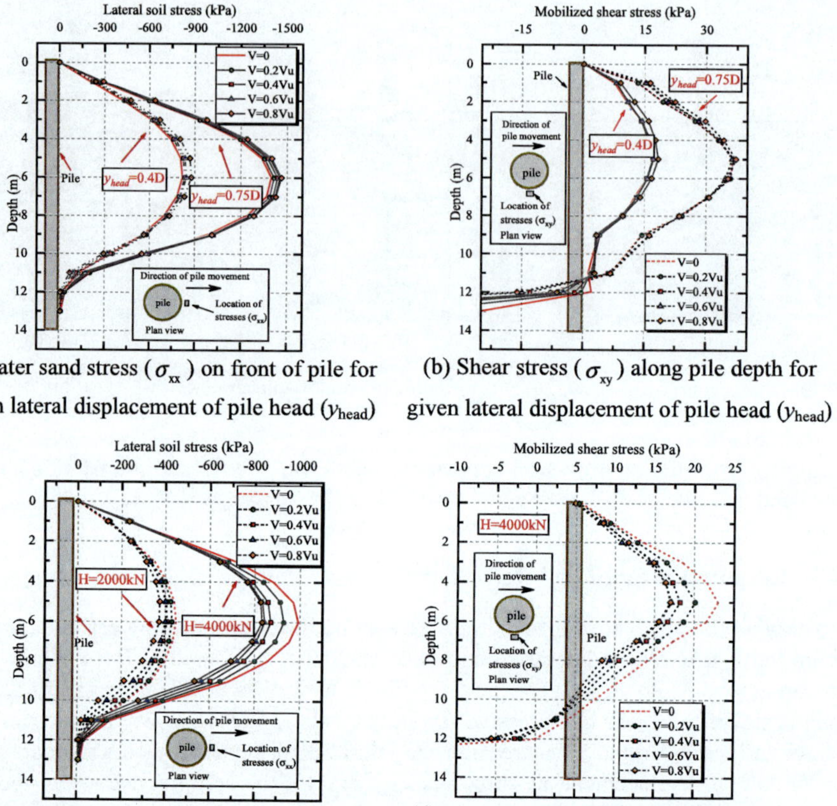

(a) Later sand stress (σ_{xx}) on front of pile for given lateral displacement of pile head (y_{head})

(b) Shear stress (σ_{xy}) along pile depth for given lateral displacement of pile head (y_{head})

(c) Later sand stress(σ_{xx}) on front of pile for given lateral load (H)

(d) Shear stress (σ_{xy}) along pile depth for given lateral load (H= 4000 kN)

Fig. 7. Variation of later sand stress (σ_{xx}) and shear stress (σ_{xy}) along pile depth under different loading condition.

5 Conclusion

This study addresses the pile-soil response under vertical-horizontal load combinations by numerical analyses. The key findings are summarized as follows:

1) The presence of vertical load can improve the lateral capacities of piles by as much as 21.4% in coral sand foundation.
2) The flow mechanism of sand changes with the vertical load level. The flow mechanism of sand under the pure lateral load consists of the upper wedge-type flow and the lower rotation flow. But the increase of vertical load can weaken the rotation flow zone of sand near the pile tip.
3) The existence of vertical load can lead to an increase in soil stress above the pile rotation point, thereby improves the soil resistances and subsequently reduces the lateral displacements of piles.

References

1. Wang, C., Liu, H., Ding, X., Wang, C., Ou, Q.: Study on the horizontal bearing characteristics of pile foundation in coral sand. Can. Geotech. J. **99**, 1928–1942 (2021)
2. Wang, C., Ding, X., Yin, Z., Peng, Y., Chen, Z.: Mechanical characteristics and particle breakage of coral sand under one-dimensional repeated loading. Acta Geotech. **17**, 3117–3130 (2021)
3. Li, Q., Prendergast, L., Askarinejad, A., Gavin, K.: Influence of vertical loading on behavior of laterally loaded foundation piles: a review. J. Mar. Sci. Eng. **8**, 1029 (2020)
4. Karasev, O., Talanov, G., Benda, S.: Investigation of the work of single situ-cast piles under different load combinations. Soil Mech. Found. Eng. **14**(3), 173–177 (1977)
5. Karthigeyan, S., Ramakrishna, V.V.G.S.T., Rajagopal, K.: Influence of vertical load on the lateral response of piles in sand. Comput. Geotech. **33**(2), 121–131 (2006)
6. von Wolersdor, P.A.: A hypoplastic relation for granular materials with a predefined limit state surface. Mech. Cohesive-frictional Mater.: Int. J. Exp. Model. Comput. Mater. Struct. **1**(3), 251–271 (1996)
7. Dewi, K., Vu, A.-T., Matsumoto, T.: Numerical study on behaviour of single piles in a coral sand (2016)
8. Chortis, G., Askarinejad, A., Prendergast, L.J., Li, Q., Gavin, K.: Influence of scour depth and type on p–y curves for monopiles in sand under monotonic lateral loading in a geotechnical centrifuge. Ocean Eng. **197**, 106838 (2020)

Study on the Effect of Pier Explosion on the Performance of RC Curved Girder Bridge with Double-Column Pier

Guohua Song, Sandeep Gharti Magar, Mingming Yang, and Jian Su(✉)

School of Civil Engineering, Zhengzhou University, Zhengzhou, People's Republic of China
sujian20210311@163.com

Abstract. A finite element (FE) model for reinforced concrete (RC) double-column pier curved continuous box girder bridge was established using ANAYS software. The FE modeling method was verified based on a pier explosion test. Considering two types of explosions including vehicle-borne explosion and knapsack explosion, the sensitivity of the columns in side span and mid-span under the explosion load was studied. The effect of the explosion location at different parts of the weak pier is also studied. The results show that the inner column of the side span was the most vulnerable column, and the weakest part for this column was the connection between column and bridge deck. The above research results provide a certain guiding significance for the design of double-column pier curved box girder bridge against explosion.

Keywords: Explosion · Finite element · Double-column pier · RC curved girder bridge

1 Introduction

The terrorist attacks in the past decades have shown that bridge structures have become an important target in terrorist attacks. Explosion of bridge structures will not only cause structural damage and casualties, but also, as an important lifeline infrastructure, will affect the resilience of the transportation network. A large number of studies have been carried out on the response of bridges under explosion loads, such as Geng et al. [1] tested the damage pattern of steel box girder scaled-down model under the action of explosion shock wave. Zong et al. [2] conducted tests on the explosion resistance of reinforced concrete pier columns and obtained the dynamic response of reinforced concrete columns. Liu [3] conducted tests on the damage mechanism of steel box girders under close explosion loading and obtained that parameters such as the thickness of the top plate can effectively improve the explosion resistance of steel box girders. In addition to tests, finite element simulation is also an important tool to study the explosion resistance of bridges [4–7].

It is worth noting that although there were a lot of explosion resistance studies on bridges, there was few studies focus on curved girder bridge. Actually, the geometric

W. Guo and K. Qian (Eds.): GBCESC 2022, LNCE 211, pp. 860–868, 2023.
https://doi.org/10.1007/978-981-19-5217-3_86

form and load distribution of curved girder bridges have obvious differences from those of straight bridges. To fill with this knowledge gap, this paper uses ANASY software to establish a numerical model of RC curved girder bridge with double column pier, and the effect of pier explosion on the performance of bridge was studied.

2 Introduction

2.1 Material Model

The RHT [8] damage material model was used for concrete materials. This model contains RHT strength model, damage model, p-α equation of state and erosion algorithm. It was provided can be well simulated the whole damage process of concrete from elasticity to failure well under explosion load. The J-C [9] material model was used for the steel reinforcement material, which is an ideal elastoplastic strengthening model that reflects the strain rate strengthening effect and the temperature rise softening effect, and is suitable for simulating the mechanical properties of the material with large strains, high strain rates, and high temperatures. Detailed material properties were shown in reference "RC curved girder bridge with double column pier under blast loading Dynamic Response and Safety Assessment".

2.2 Equations for Explosive and Air State

TNT explosives was assumed in this study, the explosive was simulated using the high-energy combustion model [10] with equation as

$$P = A[1 - \frac{\omega}{R_1 V}]e^{-R_1 V} + B_1[1 - \frac{\omega}{R_2 V}]e^{-R_2 V} + \frac{\omega \psi}{V} \tag{1}$$

where, ψ was the specific internal energy of explosives with value of 6.0×10^3 MJ/m^3; A and B were the material parameters with number of 373.8 and 3.747; R_1,R_2 and ω were constants with numbers of 4.15, 0.9, and 0.35, and the value of parameters were shown in Table 1.

Table 1. Explosive Material Parameters.

ρ/kg/m^3	A/GP_a	B/GP_a	R_1	R_2	ω
1630	3.738E+2	3.747	4.15	0.9	0.35

The air model is assumed to be an ideal gas and is simulated using a linear polynomial equation of state according to the Gama criterion [11], as

$$P_\alpha = (\gamma - 1)\frac{\rho}{\rho\alpha}E \tag{2}$$

where, E_0 is the initial internal energy, it was taking as 2.068×10^5J/m^3; γ was the specific heat ratio with value of 1.4; ρ_a was the air density with the value of 1.29kg/m^3.

2.3 Finite Element Simulation Verification

The experiment conducted by Tang [12] was used in this study for validation of FE simulation method. Detailed design and explosion case were listed in reference [12]. Figure 1 compares the simulation result and the experiment result. It was obvious from the figure that the method considered in this study can well simulate the concrete and steel damage caused by explosion.

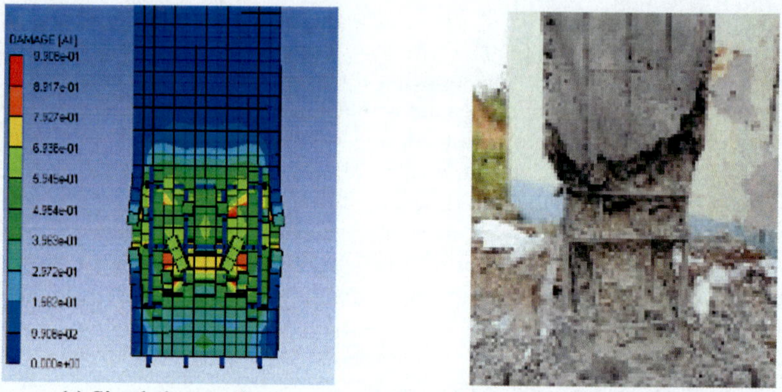

(a) Simulation result (b) Experiment result

Fig. 1. Comparison of simulation and experiment results.

3 Finite Element Model of RC Curved Girder Bridge

3.1 Geometric Dimensions

A RC double-column pier curved continuous box girder bridge with 4 spans in Chongqing was used as a prototype. Each span was 20 m. The concrete grade of the bridge was C40. The radius of curvature of the bridge was 140 m. Single box double chamber continuous box girder was used in this bridge. The width of the bottom plate was 7.5 m and the flange plate were 7.5 m and 0.9 m, respectively. The height of the box girder was 1.3 m, the thickness of the spacer at the abutment is 1.5 m, the thickness of the spacer at the pier was 2.2 m. Detailed typical cross section of the box girder was shown in Fig. 2.

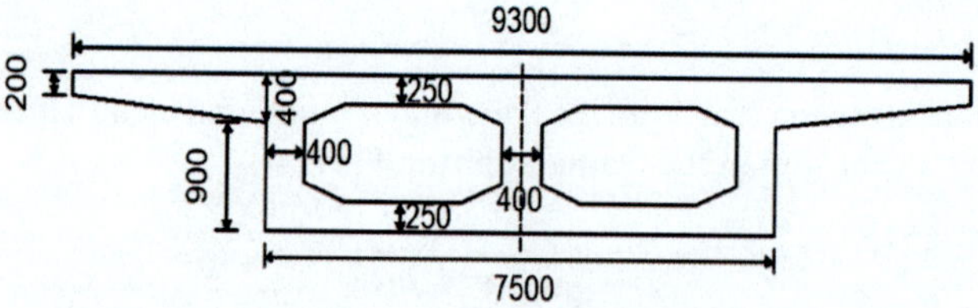

Fig. 2. Geometry diagram of box girder section (Unit: mm).

3.2 Bridge Finite Element Model

A combination of monolithic and separated modeling method [13] was used. The piers with were modeled using the separated model, and the box girders were modeled using monolithic model. SOLID164 [14] solid unit in ANSYS was used for both the main girders and piers, and the 3D solid model of the double-column pier curved box girder bridge based on the above modeling method, as shown in Fig. 3.

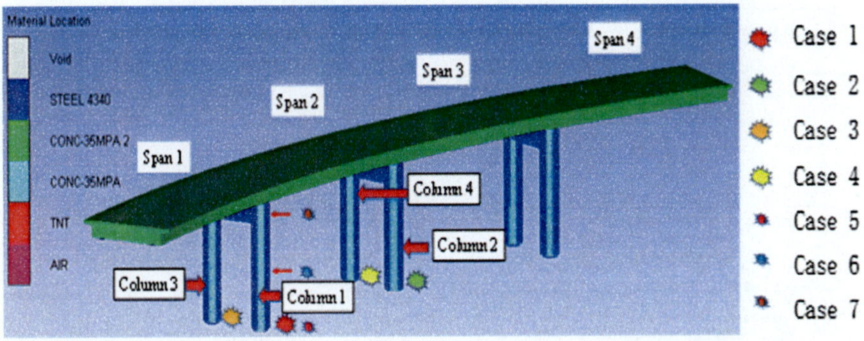

Fig. 3. 3D solid model of curved box girder bridge with double column pier.

4 Effect of Bridge Pier Explosion on Bridge Performance

4.1 Working Case Setting

According to the technical regulations of the highway bridge structural safety monitoring system in China, two kinds of TNT equivalents, 400 kg and 216 kg, are considered. The 400 kg was the simulated small vehicle explosion equivalent and 216 kg was the backpack explosion TNT equivalent. As shown in Fig. 3, for the inner and outer side columns for sides pier were numbered 1 and 3, and inner and outer middle piers for middle pier were numbered 2 and 4, respectively. A total of seven cases were considered as shown in Table 1, including 4 vehicle explosion cases and 3 knapsack explosion cases. Detailed explosive equivalent, explosive distance, explosive location and explosive height were shown in Table 2.

4.2 The Effect of Explosion on the Pier Bottom

The damage characteristics of cases 1 to 4 and phenomenon description were listed in Table 3. The bottom node of the span 2 was selected to judge the damage degree of bridge in this study. The response of the node displacement in the span under the action of four Cases were analyzed. Because the explosion response occurs in a very short time period. The time duration of the explosion action was assumed to 100 ms. After that, the bridge damage process under the action of self-weight was obtained. The whole calculation process was set to 500 ms, the results were shown in Fig. 4.

Table 2. Explosive Cases.

Case No	Type	Equivalent (kg)	Distance (m)	Location	Height (m)
1	Vehicle	400	0.1	bottom of column 1	0.5
2		400	0.1	bottom of column 2	0.5
3		400	0.1	bottom of column 3	0.5
4		400	0.1	bottom of column 4	0.5
5	Knapsack	216	0.1	bottom of vulnerable column	0.3
6		216	0.1	middle of vulnerable column	5.0
7		216	0.1	top of vulnerable column	9.5

As can be seen from Fig. 4, node displacement first rises and then falls for all Cases. The overall displacement of the bridge deck shows the wave shape if the bridge not collapse. The vertical displacement of the of Case 1 gradually increase and reaches about 360mm at 500ms, which was much larger than Cases 2, 3 and 4. The bridge was collapse due to the sudden increased load. In addition, the vertical displacement of the Cases 3 and 4 basically coincide, and they were smaller than Case 2. In the first 50ms, the displacement rises sharply, and the peak displacement of Case 2 reaches 22.5mm, which was larger than Cases 3 and 4. After 50ms, with the rapid weakening of the explosion load, the bridge gradually reaches its own equilibrium stability under the action of self-weight. From Table 2 and Fig. 4, it can be seen that the inner column of the side span was the most vulnerable column when the explosion load acts on the bottom of the pier.

4.3 The Effect of Different Explosion Height on the Vulnerable Column

The damage characteristics of Cases 5 to 7 and phenomenon description were listed in Table 4. The response of the vertical displacement in span 2 under the action of column bottom, middle and top explosion of three Cases was analyzed respectively, as shown in Fig. 5.

As can be seen in Fig. 5, when 216 kg TNT acts on the middle and bottom of the column, the vertical displacement of the midspan was small, and the change range was not large. The wave shaped displacement curves also occurred in the cases that bridge not collapse. Meanwhile, when the explosion site was located in the top of the column (9.6 m). The vertical displacement reached 390 mm in 500 ms, at which time the bridge may collapse damage. As can be seen above, when the explosive load in the most vulnerable column, and the top of column was weakest than other parts.

Table 3. Damage characteristics and phenomena description for Cases 1 to 4.

Case No.	Damage characteristics	Phenomenon description
1	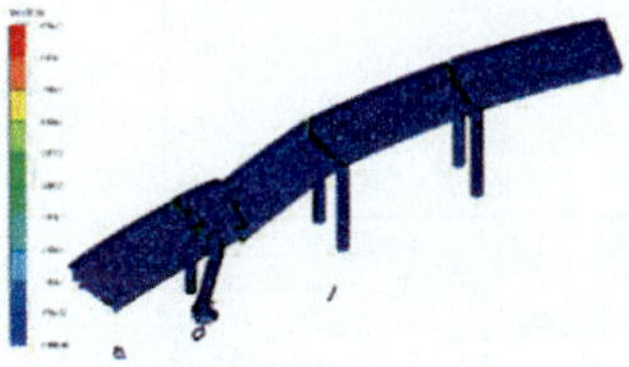	(1) At the moment of 0.8s, column 1 has been seriously inclined. The top of the box girder corresponding to the vertical direction of the diagonal crack at the bottom of the box girder in span 2 was greatly damaged. Oblique cracks occurs in span 1 and 2. (2) At the moment of 1.1s, the column 3 showed great drift due to the collapse of the box girder; (3) At the moment 1.4s, the bridge was completely collapsed.
2	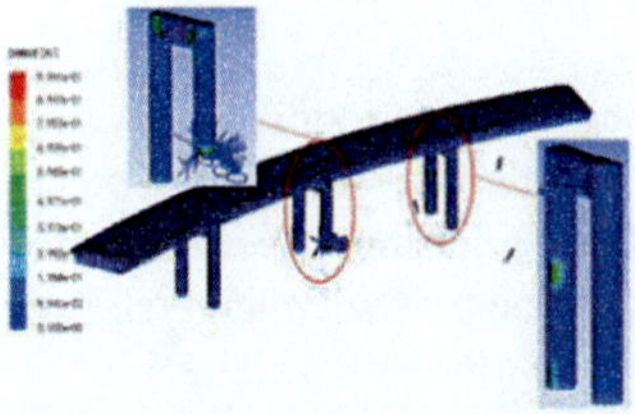	(1) Under the action of the explosion load, the concrete damage at the bottom of the inner side pier of Column 2 is more serious, and the reinforcement produces serious plastic deformation. (2) The damage to the connection part of the tie beam and the pier is more serious, and there is slight damage in the middle of the outer pier of Column 3 about 1.3m high, but the bridge as a whole has not collapsed and damaged.
3	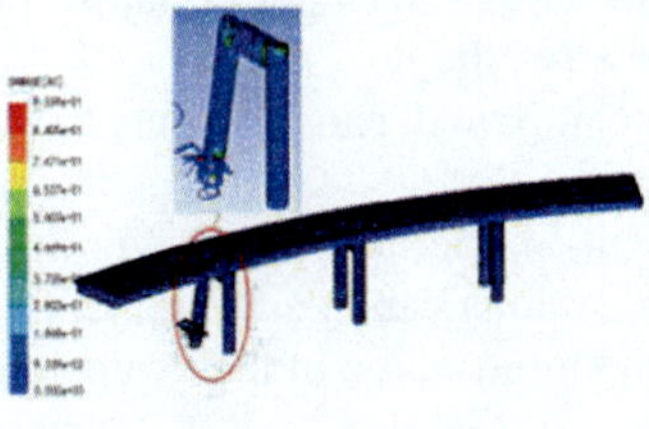	(1) After the explosion load, the bottom of the Column 3 produced serious damage. The bottom of the pier body has a large displacement in the direction of the bridge, the top of the pier and the tie beam connection part of the concrete torsional damage. (2) The material damage in the connection part of the tie beam and pier was more serious, and the outer side pier lost its bearing capacity, but the bridge as a whole did not collapse and damage.
4	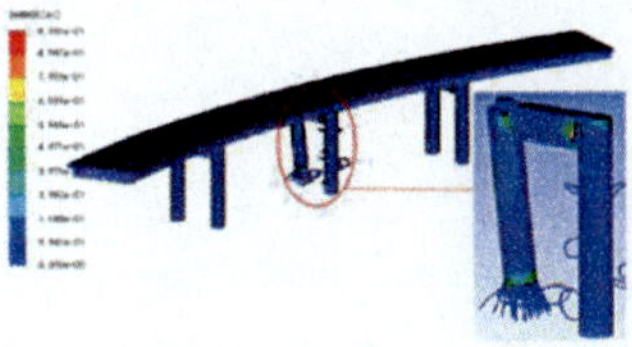	(1) Under the explosion load, the bottom of the Column 4 produced a large displacement in the direction of the bridge. (2) Column 4 and tie beam connection parts of the concrete damage is more serious, but the bridge as a whole did not collapse damage phenomenon.

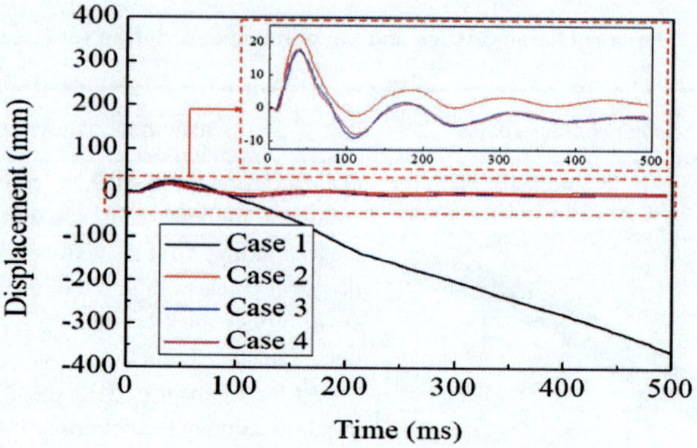

Fig. 4. Vertical displacement of the node in span 2.

5 Conclusions

In this study, a reinforced concrete curved girder bridge with four spans was investigated. Numerical simulations were carried out using finite element software. The failure mode and dynamic response of the curved girder bridge under the effect of explosion at the bottom of each pier and different locations of the same pier were analysis, and the main research results and conclusions are summarized as follows:

(1) When the explosion load acts on the pier, the bridge deck will first be upward displaced. After explosion, the bridge will have a downward vertical displacement under the action of self-weight. If the bridge does not collapse, the overall displacement of the bridge deck shows the wave shape.
(2) The explosive action in the top of the pier position will cause serious damage to the box girder concrete compares to other locations.
(3) The explosion response is extremely violent, in the near-explosive region of the concrete before and after 1ms that produce serious damage failure damage, while the reinforcement equivalent effect force under the action of the explosion load has increased significantly.

Table 4. Damage characteristics and phenomena description for Cases 5 to 7.

Case No.	Damage characteristics	Phenomenon description
5	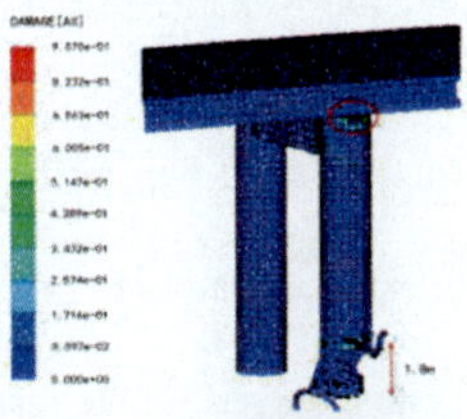	(1) Column 1 bottom explosion when the Column 1 damage height of about 1.8m, the damage range of steel reinforcement has a large plastic deformation.
6	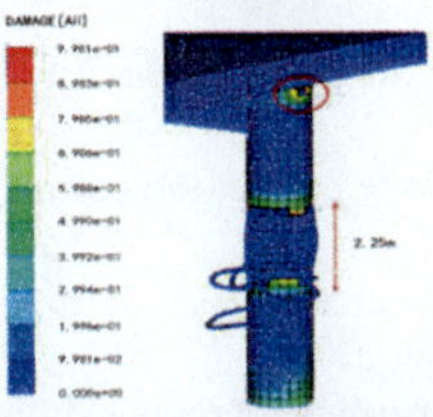	(1) Pier damage height in the column 1 explosion in about 2.25m, the height of damage is greater than the Column 1 bottom explosion pier damage height. (2) Some of the hoops fell out within the damage area, but none of the bridges collapsed.
7	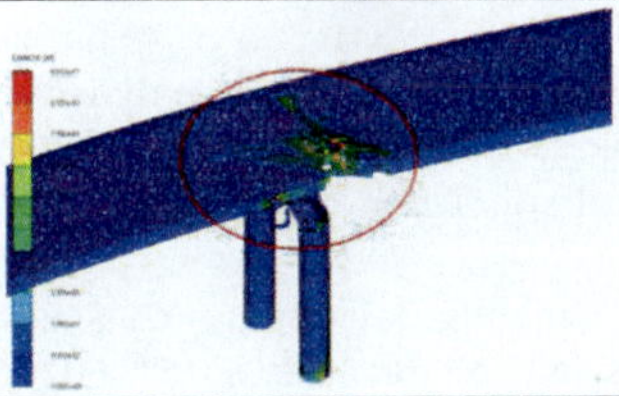	(1) Column 1 top explosion occurred when the Column 1 top concrete and explosives above the box beam occurred serious damage. (2) Column 1 top box girder section upper edge concrete failure damage, box girder under the action of self-weight to form a falling beam, the bridge collapse damage.

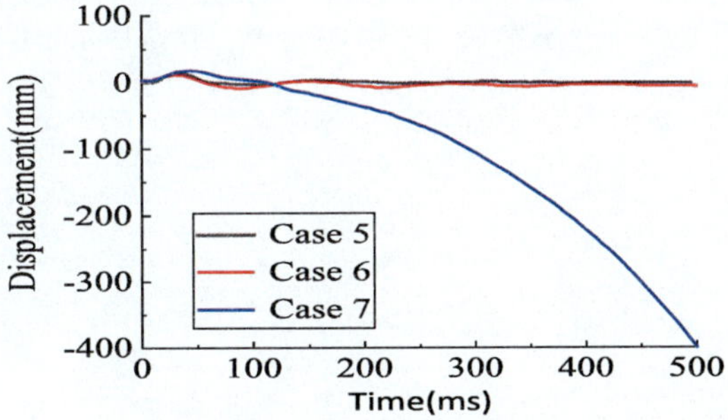

Fig. 5. Vertical displacement in span 2.

Acknowledgment. This work was supported by the National Natural Science Foundation of China (U1304515).

References

1. Shaobo, G., Yaling, L., Jianying, X.: Experimental study on the damage of steel box girder under the action of explosion shock wave of reduced scale model. **34**, 84-88 (2017)
2. Zhouhong, Z., Biao, T., Chao, G.: Explosion-resistant performance test of reinforced concrete pier column. **30**(9), 51–60 (2017)
3. Yaling, L., Shaobo, G., Yucun, L., Wangyue, W.: Experimental study on the influence of pavement layer on the blast resistance performance of steel box girder structure. **19**(15), 88–93 (2019)
4. Winget D G.: Analysis and Design of Critical Bridges Subjected to Explosion Loads. **131**(8), 1243–1255 (2005)
5. Tang, E.K.C., Hao, H.: Numerical simulation of a cable-stayed bridge response to explosion loads. **32**(10), 3180-3192 (2010)
6. Yi, Z., Agrawal, A.K., Ettouney, M.: Explosion load effects on highway. Bridges **30**(7), 1922–1937 (2013)
7. Jin, S., Lee, H.J.: Performance of cable-stayed bridge pylons subjected to explosion loading. **33**(4), 1133–1148 (2011)
8. Yuan, L., Jianyu, L., Yitao, H.: Study on the damage mechanism of reinforced concrete bridge deck slabs under blast loading. **34**(07), 291–300 (2021)
9. Johnson, G.R., Cook, W.H.: A constitutive model and data for metals subjected to large strains, high strain ratesand high temperatures. **21**, 541–548 (1983)
10. Yang, Y.B., Chang, K.C.: Extraction of bridge frequencies from the dynamic responses of a passing vehicle enhanced by the EMD technique. **332**(4–5), 718–739 (2009)
11. Jianhong, Y., Xi, Z., Zhenhua, Z.: Numerical simulation method of underwater explosion load. **33**(09), 18–23 (2011)
12. Zhouhong, Z., Biao, T., Chao, G., Lu, L., Minhong, L., Sujing, Y.: Explosion resistance performance test of reinforced concrete piers and columns. **30**(09), 51–60 (2017)
13. Mianjing, J.: Nonlinear Finite Element Analysis of Reinforced Concrete Structures. Shaanxi Science and Technology Press, Xi'an (1994)
14. Xiaojiang, S., Jianyu, S.: ANSYS/LS-DYNA Dynamic Analysis Methods and Engineering Examples. China Water Conservancy and Hydropower Press, Beijing (2006)

A Comparison Between PDEM-Based Approach and Linear-Regression-Based Approach in Seismic Fragility Assessment: Application into Low-Rise Frame Buildings

Xuyang Cao[1], Jigang Xu[2](✉), and Shizhi Chen[3]

[1] College of Civil and Transportation Engineering, Hohai University, Nanjing 210098, China
[2] College of Civil Engineering, Nanjing Tech University, Nanjing 211816, China
jgxu@njtech.edu.cn
[3] Highway College, Chang'an University, Xi'an 710064, China

Abstract. In this paper, a comparison between probability density evolution method (PDEM)-based approach and linear-regression-based approach in seismic fragility assessment is performed, and an application into low-rise frame building is conducted for implementation. The principles of the two fragility approaches are introduced first. The PDEM-based fragility approach has solid foundations in the reliability field and is expressed as a non-parametric form without predefined fragility shapes, while the linear-regression-based fragility approach is expressed as a classic parametric form, commonly under the assumption of lognormal distributions. Then a 3-span-4-storey reinforced concrete frame (RCF) building, which represents the existing buildings with large quantities in China, is used for seismic assessment by the two approaches. In general, the application shows the similar tendency of fragility curves for the two methods, and verifies the effectiveness of the non-parametric PDEM-based approach for seismic performance assessment. Comparatively, the PDEM-based approach omits the heavy computing burden as Monte Carlo simulation, and reflects the fragility characteristics with considerable accuracy as linear-regression-based approach, which turns on a new path for the fragility assessment scheme in the performance-based earthquake engineering.

Keywords: Seismic fragility · Earthquake engineering · Performance assessment · Probability analysis · Reinforced concrete frames · Pdem

1 Introduction

Seismic fragility is an effective tool for structural performance assessment in earthquake engineering. Since the Cornel et al. proposed the risk-based performance based earthquake engineering (PBEE) framework, extensive attention has been received by researchers all around the world to evaluate the seismic performance of engineering structures in light of the full probability theory.

W. Guo and K. Qian (Eds.): GBCESC 2022, LNCE 211, pp. 869–875, 2023.
https://doi.org/10.1007/978-981-19-5217-3_87

In the risk-based PBEE framework, the most important step is the probabilistic seismic fragility analysis (PSFA), which refers to the evaluation in the probability of structural responses exceeding different performance levels under the earthquake motions. This work describes the seismic behaviors of engineering structures quantitatively in a probabilistic aspect, and indicates the extent between the earthquake intensity and the structural damage from a macro sense. Choi et al. gave the seismic fragility of typical bridges in the Central United States, and provide the basis for the economic loss estimation as well as for retrofit prioritization. Ghosh and Padgett introduced the time-dependent seismic fragility of bridges and found that the seismic vulnerability significantly increased throughout the bridge lifetime with the ratio of 32%. Nielson and DesRoches analyzed the analytical fragility curves of structures considering the contribution of the major components. Park et al. introduced the seismic fragility procedure for low-rise unreinforced masonry structures in the United States, and the results are compared with those in HAZUS with lower variations. Lagaros et al. developed the seismic fragility analysis of geo-structures, and artificial neural networks is incorporated into the fragility framework to enhance the computational efficiency. Almaraz et al. carried out the full-scale dynamic tests of ceiling systems, and conclude the fragility curves that are appropriate for performance assessment and design. Ji et al. gave an analytical fragility framework for seismic evaluations of high-rise concrete buildings, and the accuracy of individual component was verified with the analytical models with more details. Seyedi et al. developed the seismic fragility surfaces of reinforced concrete structures, and the complete methodology was proposed through the damage calculation and dynamic analyses of multi-degree-of-freedom systems. Segura et al. performed the seismic fragility calculations of concrete gravity dams via meta-model, and a range of recommended parameters were given through the validation with safety thresholds. Cao et al. [1–5] adopted the classic parametric fragility equation to assess the existing buildings retrofitted with the novel external retrofitting techniques, and the fragility curves provided basis for the relevant performance design and optimization.

In this paper, a comparison between probability density evolution method (PDEM)-based approach and linear-regression-based approach in seismic fragility assessment is performed, and an application into low-rise frame buildings is conducted for implementation. The principles of the two fragility approaches are introduced first. The PDEM-based fragility approach has solid foundations in the reliability field and is expressed as a non-parametric form without predefined fragility shapes, while the linear-regression-based fragility approach is expressed as a classic parametric form, commonly under the assumption of lognormal distributions. Then a 3-span-4-storey reinforced concrete frame (RCF) building, which represents the existing buildings with large quantities in China, is used for seismic assessment by the two approaches. In general, the application shows the similar tendency of fragility curves for the two methods, and verifies the effectiveness of the non-parametric PDEM-based approach for seismic performance assessment. Comparatively, the PDEM-based approach omits the heavy computing calculations as Monte Carlo simulation, and reflects the fragility characteristics with considerable accuracy as linear-regression-based approach, which turns on a new path for the fragility assessment scheme in the performance-based earthquake engineering.

2 Principle of PDEM-Based Fragility Approach

As for the principle of PDEM-based fragility approach, the dynamic-motion balance equation of a structural system under earthquake excitation can be expressed as:

$$M\ddot{X}(\Theta, t) + C\dot{X}(\Theta, t) + KX(\Theta, t) = -M\ddot{x}_g(\Theta, t) \tag{1}$$

where M, C and K represent the mass matrix, damping matrix and stiffness matrix of structural system, respectively. $\ddot{x}_g(\Theta, t)$ represents the seismic earthquakes, and Θ represents the randomness of structural parameters. The theory of probability preservation is realized with the randomness depicted by Θ in the point selections in the PDEM-based method. In light of the principle of probability preservation, the generalized PDEM equation can be expressed as:

$$\frac{\partial p_{X\Theta}(X, \Theta, t)}{\partial t} + \dot{X}(\Theta, t) \cdot \frac{\partial p_{X\Theta}(X, \Theta, t)}{\partial X} = 0 \tag{2}$$

where t denotes the generalized time, and the initial condition for the dynamic system can be expressed as:

$$p_{X\Theta}(X, \Theta, t)|_{t=t_0} = \delta(X - X_0)p_\Theta(\Theta) \tag{3}$$

where δ denotes the Dirac function, and X_0 denotes the deterministic value at the initial time. With the above-mentioned equations, the probability density function (PDF) of the concerned variables can be acquired as:

$$p_X(X, t) = \int_{\Omega\Theta} p_{X\Theta}(X, \Theta, t) \cdot d\Theta \tag{4}$$

The generalized PDEM equation is then solved by a numerical procedure, and the representing points are needed to be given, according to the generalized F-Discrepancy method. After acquiring the PDF via PDEM approach, the cumulative distribution function (CDF) can be obtained through integration, and the fragility value can be also calculated, as the sum of CDF and fragility is 1. The acquired results are for a specific intensity level. After extending intensities to multiple levels and connecting the corresponding results, the PDEM-based fragility curves can be achieved.

3 Principle of Linear-Regression-Based Fragility Approach

As for the principle of linear-regression-based fragility approach, the results are commonly based on the incremental dynamic analyses (IDA). It is a classic parametric approach, and we assume the lognormal distributions for both structural capacity and demand. The corresponding fragility can be written as:

$$P_f = \Phi[\frac{lnS_{d|IM} - lnS_c}{\sqrt{\beta_{d|IM}^2 + \beta_c^2}}] \tag{5}$$

where P_f denotes the fragility value, $S_{d|IM}$ and S_c represent the medians of demand and capacity. S_c is taken as the thresholds according to FEMA-356, and $S_{d|IM}$ can be expressed as follows:

$$S_{d|IM} = a \cdot (IM)^b \tag{6}$$

The values of a and b can be acquired through the linear regression. β_c denotes the capacity uncertainty and is chosen as 0.2, $\beta_{d|IM}$ denotes the demand uncertainty and can be expressed as follows:

$$\beta_{d|IM} = \sqrt{\sum (\ln MIDR_i - \ln S_{d|IM})^2/(m-2)} \tag{7}$$

where $MIDR_i$ denotes the ith value of the maximum interstorey drift ratio (MIDR), and m denotes the total calculation points.

4 Application into Low-Rise Frame Buildings

The above-mentioned two fragility methods are applied and compared into low-rise frame buildings. Totally 22 earthquake ground motions recommended by FEMA695 are used, and the specific wave information can be available from Cao et al. [6, 7]. Totally 4 limit states are adopted herein, i.e., operational (OP), immediate occupancy (IO), life safety (LS) and collapse prevention (CP), and the engineering demand parameter (EDP) is adopted as the MIDR. The benchmark low-rise frame is designed as a 3-span-4-storey reinforced concrete frame building, which is located at a class-II site and the design group is type-I. As for the geometric dimensions and sectional details, the corresponding data can be available from Fig. 1.

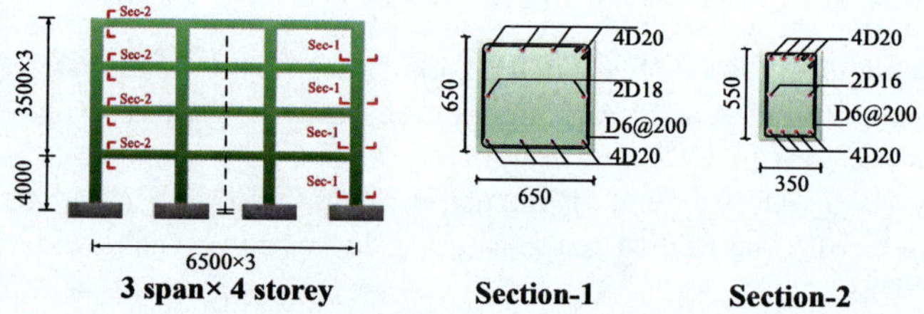

Fig. 1. The geometric dimensions and sectional details of low-rise frame buildings.

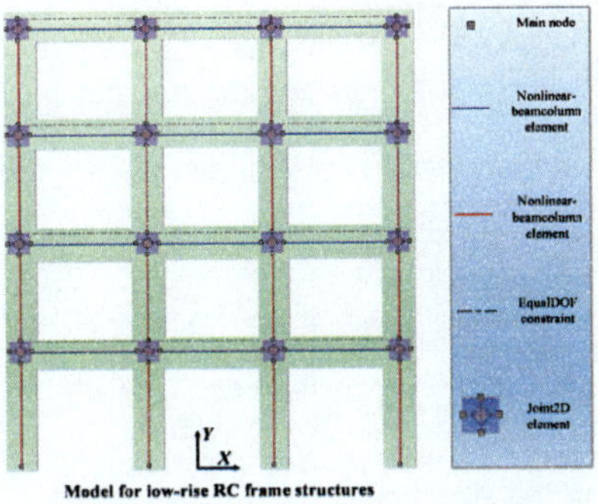

Fig. 2. The numerical model for low-rise frame buildings based on OpenSees.

Figure 2 presents numerical model for low-rise frame buildings based on OpenSees. The beams and columns are modelled by the force-based nonlinear beam-column elements, and EqualDOF constraints are adopted to limit the horizontal displacements of the two outer nodes on each floor, according to the rigid floor assumption. Besides, the Joint2D elements are chosen at the connections of beams and columns, in order to reflect the shear behaviors and bond-slip characteristics at the low-rise frame joints. The detailed modelling strategy in Fig. 2 can be available from Cao et al. [8–11].

With the design information and detailed application of the low-rise frame buildings, the PDEM-based fragility approach and linear-regression-based fragility approach as mentioned above are used to acquire the corresponding fragility curves. Figure 3 presents the CDF curves derived from the PDEM fragility approach (take the IM = 0.3 g and 1.1 g for example). From Fig. 3 we can obtain the CDF for different limit states under a specific intensity level, and the corresponding fragility can be given as 1-CDF. After connecting all the fragility values, as depicted in Fig. 4 in black solid lines, the PDEM-based fragility curves for the low-rise frame buildings are given. Figure 4 also shows the linear-regression-based fragility curves (i.e., in blue dotted lines), using the equations in Sect. 3. The results for all the OP, IO, LS and CP limit states are summarized and compared simultaneously, as illustrated in Fig. 4. In general, the application shows the similar tendency of fragility curves for the two methods, and verifies the effectiveness of the non-parametric PDEM-based approach for seismic performance assessment. Compared with the linear-regression-based approach, the PDEM-based approach omits the heavy computing calculations and reflects the fragility characteristics with more accuracy, which turns on a new path for the fragility assessment scheme in the performance-based earthquake engineering.

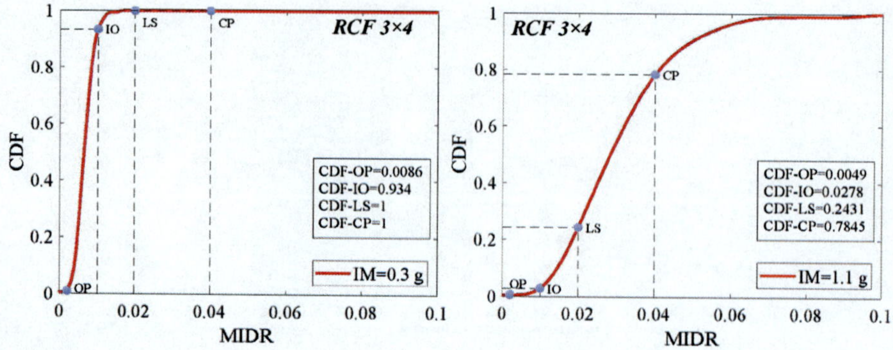

Fig. 3. The CDF curves derived from the PDEM fragility approach (take the IM = 0.3 g and 1.1 g for example).

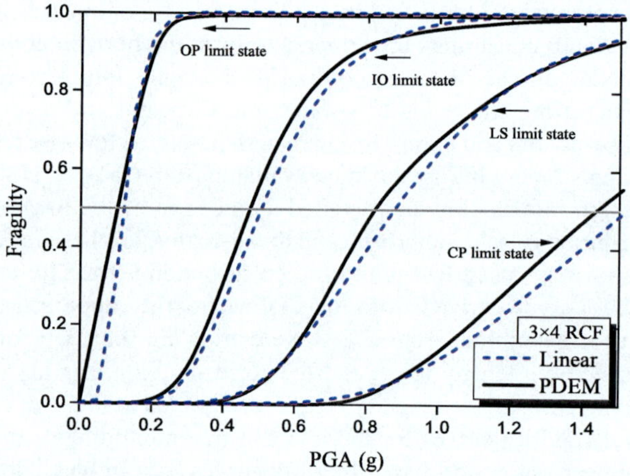

Fig. 4. The fragility curves for both PDEM-based and linear-regression-based approaches under all the OP, IO, LS and CP limit states.

5 Conclusion

A comparison between the PDEM-based approach and linear-regression-based approach in seismic fragility assessment is performed in this paper, and an application into low-rise frame buildings is conducted for implementation, from which the following conclusions may be drawn.

(1) The PDEM-based fragility approach has solid foundations in the reliability field and is expressed as a non-parametric form without predefined fragility shapes, while the linear-regression-based fragility approach is expressed as a classic parametric form, commonly under the assumption of lognormal distributions.

(2) The 3-span-4-storey reinforced concrete frame building, which represents the existing buildings with large quantities in China, is used for seismic assessment by the two approaches. In general, the application shows the similar tendency of fragility curves for the two methods, and verifies the effectiveness of the non-parametric PDEM-based approach for seismic performance assessment.
(3) Comparatively, the PDEM-based approach omits the heavy computing burden as Monte Carlo simulation, and reflects the fragility characteristics with considerable accuracy as linear-regression-based approach, which turns on a new path for the fragility assessment scheme in the performance-based earthquake engineering.

References

1. Cao, X.Y., et al.: Research on the seismic retrofitting performance of RC frames using SC-PBSPC BRBF substructures. Earthq. Eng. Struct. D. **49**(8), 794–816 (2020)
2. Cao, X.Y., et al.: Experimental and numerical study of outside strengthening with precast bolt-connected steel plate–reinforced concrete frame-brace. J. Perform Constr. Fac. **33**(6), 04019077 (2019)
3. Cao, X.Y., et al.: Pushover-based probabilistic seismic capacity assessment of RCFs retrofitted with PBSPC BRBF sub-structures. Eng. Struct. **234**, 111919 (2021)
4. Cao, X.Y., et al.: Reusing & replacing performances of the AB-BRB with thin-walled concrete-infilled steel shells. Thin Wall. Struct. **157**, 107069 (2020)
5. Cao, X.Y., et al.: Probabilistic seismic performance assessment of RC frames retrofitted with external SC-PBSPC BRBF sub-structures. J. Earthq. Eng. 1–24 (2021)
6. Cao, X.Y., et al.: Seismic performance upgrade of RC frame buildings using precast bolt-connected steel-plate reinforced concrete frame-braces. Eng. Struct. **195**, 382–399 (2019)
7. Cao, X.Y., et al.: Parametric investigation of the assembled bolt-connected buckling-restrained brace and performance evaluation of its application into structural retrofit. J. Build. Eng. **48**, 103988 (2022)
8. Cao, X.Y., et al.: Seismic performance improvement of existing RCFs using external PT-PBSPC frame sub-structures: Experimental verification and numerical investigation. J. Build. Eng. **46**, 103649 (2021)
9. Cao, X.Y., et al.: Experimental and theoretical investigations of the existing reinforced concrete frames retrofitted with the novel external SC-PBSPC BRBF sub-structures. Eng. Struct. **256**, 113982 (2022)
10. Cao, et al.: (2021) https://doi.org/10.21203/rs.3.rs-1026047/v1
11. Cao, X.Y.: https://doi.org/10.1109/ICHCESWIDR54323.2021.9656478

Study on the Fissure Propagation of Jointed Rock Slope Under Unloading Condition

Sheng Li[1](✉) and Chen Yang[2]

[1] Xi'an Jiaotong University, Xi'an 710049, China
Lisheng@stu.xjtu.edu.cn
[2] China University of Ming and Technology, Xuzhou 221116, China

Abstract. The rock mass unloading is closely related to the change of geological environment, such as river valley down-cutting, frequent rainfall and evaporation, and slope excavation. For the jointed rock slope with complicated structure, these changes may trigger the propagation of cracks, posing a serious threat to slope stability and construction safety. Thus, it is necessary to analyze the fissure propagation of jointed rock slope under unloading condition. Currently, traditional finite element method (FEM) has been widely used to analyze the failure of slope because of its rigorously theoretical system. However, for discontinuous problems such as crack propagation, the result in FEM often presents high stress concentration in crack tip (numerical singularity), causing calculation to abort. In this connection, this paper intended to introduce cohesive zone model (Theory of cohesive element) from fracture mechanics to deal with the numerical singularity, and meanwhile adopt Voronoi diagram algorithm for rapidly modelling the structural planes. Firstly, a jointed rock slope model was taken as an example, discontinuity network of which was generated by Voronoi diagram, and then 0-thickness cohesive elements were embedded into structural planes to establish a numerical model. Finally, the study compared and analyzed the crack evolution of slope under different unloading conditions by changing the geometry of slope.

Keywords: Fem · Jointed rock slope · Czm · Voronoi diagram · Crack propagation

1 Introduction

Jointed rock slope, a commonly encountered geological body in mining [1], roadway [2], and slope engineering [3], has always been one of the focuses in geotechnical discipline [4, 5]. The rock slope features discontinuous structural planes, a characteristic determining the weak stability and poor engineering properties of rock slope, making it prone to slide and even cause natural calamities. Therefore, it is significant to study the stability and failure mode of jointed rock slope for a construction safety.

At present, scholars have carried out relevant researches on the mechanical properties and failure mechanism of jointed rock slope by many methods, including in-situ test [6], indoor test [7], empirical formula estimation [8], making great progress in studying

W. Guo and K. Qian (Eds.): GBCESC 2022, LNCE 211, pp. 876–883, 2023.
https://doi.org/10.1007/978-981-19-5217-3_88

macro stability of the slope. However, the damage of many jointed rock masses begins with the development of microcracks, and complicated fissures in random distribution in slope mean it is hard to understand the mechanism between macro and micro failure by routine geological survey and test. In this connection, some researchers turn to numerical methods for further understanding the destruction of jointed rock slope. Currently, the finite element method (FEM) [9], discontinuous deformation analysis (DDA) [10], discrete element method (DEM) [11] and boundary element method (BEM) [12] have been mainly used to simulate the deformation and failure of jointed rock mass. Among them, the finite element method (FEM) boasts its rigorous theory system, in which the stress-strain relationship of elements is strictly considered, displaying comparative advantages over the deformation calculation of rock slope. Nevertheless, for such a typically discontinuous geological body, the calculation results in FEM usually present high stress concentration in crack tip (numerical singularity) as it is based on the continuity assumption. This will cause calculation to abort. Additionally, numerous jointed fissures mean a complex process modelling jointed rock slope. Therefore, this paper intended to introduce cohesive zone model (Theory of cohesive element) from fracture mechanics to deal with the numerical singularity, and meanwhile adopt Voronoi diagram algorithm for rapidly modelling the structural planes. Then, by changing the geometry of slope, the relationship between macro and micro failure mechanism under different unloading conditions was discussed.

2 Theory of Cohesive Element

The cohesive zone model (CZM) was introduced to simulate the cracking of ductile metal materials in 1960s [13], After which Hillerborg et al. proposed a cohesive crack model suitable for concrete [14]. At present, this method has been widely applied to analyze the fracture process of solids in many fields, such as crack propagation of hydraulic fracturing [15], gravity dam failure [16], Brazilian disk splitting test [17] and so on. Traction-separation (TS) criterion is a common damage law in CZM to describe the crack initiation and propagation, in which the bilinear constitutive model was used in this paper. In the initial stage, the relationship between stress and displacement is linear, so the constitutive equation is as follows:

$$\boldsymbol{t} = \begin{Bmatrix} t_n \\ t_s \end{Bmatrix} = \begin{bmatrix} k_{nn} & k_{ns} \\ k_{ns} & k_{ss} \end{bmatrix} \begin{Bmatrix} \delta_n \\ \delta_s \end{Bmatrix} = \boldsymbol{K\delta} \tag{1}$$

where $\boldsymbol{t}$ represents the stress vector of cohesive element; $\boldsymbol{\delta}$ the displacement component in each direction; $\mathbf{K}$ the stiffness matrix of cohesive element; t_n and t_s are the normal and tangential stresses. And then, with the increase of stresses, the cohesive element damage appears and stiffness degenerates gradually. The damage index D ~ (0, 1) is introduced to establish the degradation relationship of stresses in pre- and post- stiffness degradation. The normal and tangential stress components are written as:

$$\begin{cases} t_n = \begin{cases} (1-D)\hat{t}_n \ , \hat{t}_n \geq 0 \\ \hat{t}_n \ , \hat{t}_n < 0 \end{cases} \\ t_s = (1-D)\hat{t}_s \end{cases} \tag{2}$$

where $\hat{t}_n$ and $\hat{t}_s$ are initial normal and tangential stresses. The crack growth criterion of cohesive element consists of displacement and energy criterion. In this paper, the linear damage criterion of energy growth criterion was adopted, and the calculation formula of damage factor D is shown in Eq. (3).

$$D = \frac{d_m^f\left(d_m^m - d_m^0\right)}{d_m^m\left(d_m^f - d_m^0\right)} \tag{3}$$

where d_m^m is maximum effective relative displacement during loading; d_m^0 and d_m^f represent the effective relative displacements where the initial damage and complete failure of elements occur respectively. Additionally, the paper adopted the maximum stress criterion to simulate crack initiation, which is shown in Eq. (4).

$$\text{MAX}\left\{\frac{\langle t_n \rangle}{t_{n0}}, \frac{t_s}{t_{s0}}\right\} = 1 \tag{4}$$

where t_{n0} is the maximum tensile stress in normal direction and t_{s0} equals to the maximum shear stress in the tangential direction.

3 Numerical Model

The simplified geological model is shown in Fig. 1. As presented, the unloading area was divided into five parts (I-V), each of which was set at 5°. The number I-V represents the sequence of unloading, which may be initiated by weathering, collapse or other geological activities in nature. The slope was cut to a slope angle with 45° after unloading eventually. The cohesive element can be inserted in structural plane or in the whole model, but as the strength of structural plane is lower than that of the intact rock, the two-dimensional four nodes (COH2D4) cohesive element with 0-thickness was merely inserted into structural surface in this study, sharing nodes with solid element. The embedding method is shown in Fig. 2. For the generation of structural planes, one can find more about Voronoi diagram algorithm by the reference [18]. When it comes to boundary condition, the left and lower right side of rock slope were constrained horizontally, and the bottom was constrained horizontally and vertically. Additionally, the upper side and free face had no constraints. The Mohr Coulomb criterion was adopted as yield criterion and the CPE3 plane strain element was applied in solid element. The physical and mechanical parameters of model are shown in Table 1.

Apart from that, in situ stresses are of great significance to numerical results. Therefore, the paper had balanced in-situ stresses before performing numerical calculation, by which the model possessed initial in-situ stresses. Figure 3 presents the state of structural plane before unloading. In the picture, the value of SDEG represents the stiffness degradation rate of cohesive element. When SDEG equals to 0, the cohesive element falls under linear elastic deformation; When SDEG is between 0 and 1, the element stages into damage evolution; When SDEG is equal to 1, the element fails and crack begins to expand. After the equilibrium of in situ stresses, it can be seen from Fig. 3 that the structural plane of jointed rock slope appears stiffness degradation. Actually, some fissures should be in damage state in nature with geological activities and environment changes, so the numerical calculation started in this initial state.

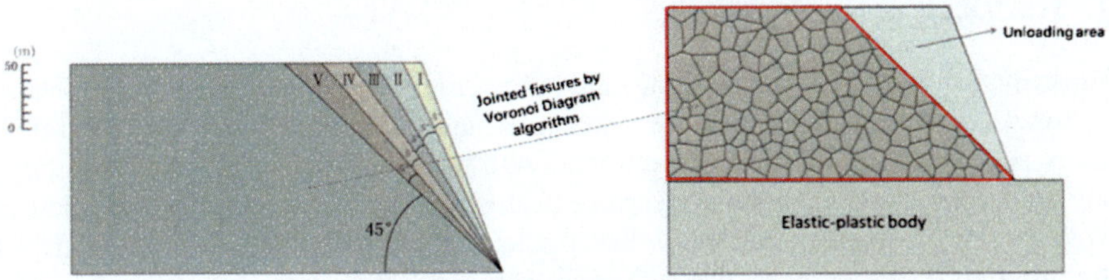

Fig. 1. The geological model.

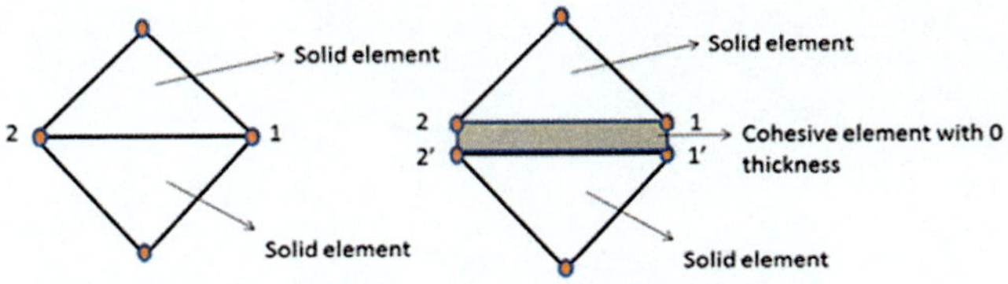

(a) Cohesive element not inserted (b) Cohesive element inserted

Fig. 2. The embedding method of cohesive element with thickness of 0

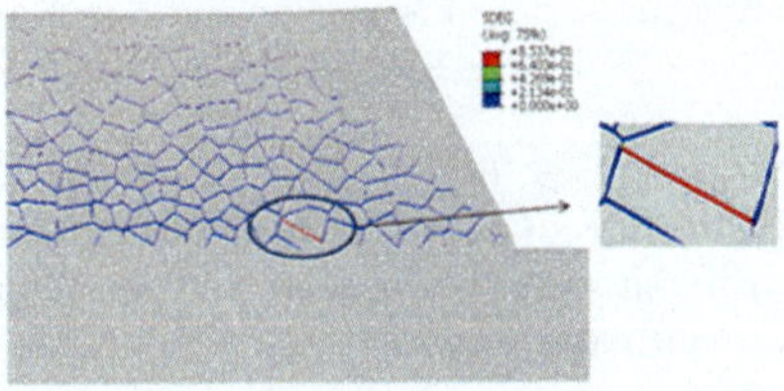

Fig. 3. The structural plane state before unloading

Table 1. The physical and mechanical parameters

	Elastic Modulus, E, GPa	Poisson's ratio μ	Internal friction angle φ, °	Cohesion c, kPa	Normal Stiffness, GPa·m^{-1}	Shear Stiffness, GPa·m^{-1}	Tensile Strength, kPa	Fracture Energy, N·m^{-1}
Intact rock	8	0.3	34.5	5.5×10^2	—	—	—	—
E-P body	36	0.22	52	1.5×10^3	—	—	—	—
Fissure	—	—	—	—	4	1.2	8.6×10^2	25

4 Results

The evolution law of unloading fissures in jointed rock slope under condition I is shown in Fig. 4 (1000 times magnification). In Fig. 4, the initially damaged structural plane firstly shows the trend of failure, and then with the redistribution of stress caused by unloading, the damaged structure surface begins to extend from one side. In addition, what can be obviously observed is that the left failure path expands continuously to the inside of the slope while the right one extending towards the free face, forming a V-shaped propagation.

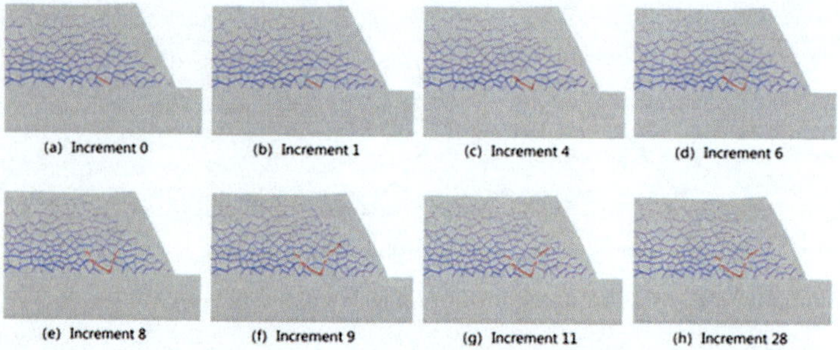

Fig. 4. The crack propagation of jointed rock slope under unloading condition I

For a clear observation about the failure mode of cracks under unloading, cracks in Fig. 4 were enlarged and analyzed, as shown in Fig. 5. It can be seen from Fig. 5 that the structural plane preliminarily shows an obvious trend of shear slip, a stage dominated by shear stress. And then with the gradual redistribution of stress field, the sliding displacement between rocks increases, resulting the structural plane fractured when the tensile-shear crack start to appear. Generally, the propagation speed of cracks expanding towards free surface is significantly higher than that developing into the inside, and the failure mode is mainly tensile-shear composite damage.

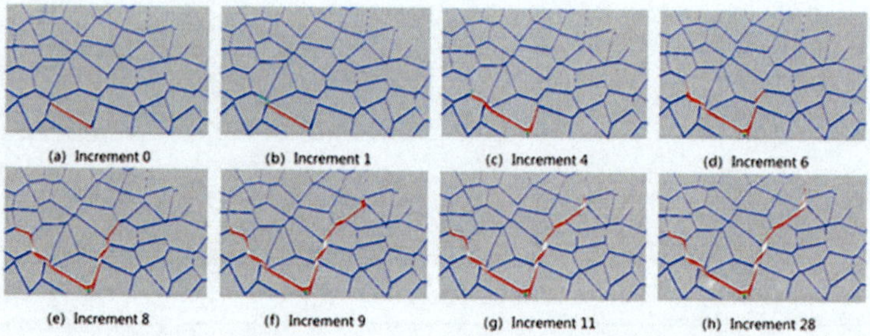

Fig. 5. The crack propagation of jointed rock slope under condition I

Figure 6 shows the evolution of fissures from condition I to condition V. As presented in Fig. 6, from condition I to condition II, the crack extending towards free face has no obvious change, but the other branch appears tensile cracks. When unloading to condition III, the crack propagation is relatively stagnant, so the calculation result was not listed. when unloading to condition IV, a new tension-shear fracture appears near the top of the right crack. Eventually, when unloading to condition V, the left crack tip (towards inside slope) opens under tensile stress, while the fissure around right crack (towards free face) tip is subjected to compressive stresses, with a closed trend. Generally, the crack propagation stays at the fastest speed in the initial stage of stress release, but when the unloading volume increases, the range of fracture propagation has no conspicuous change. To explore the influence of unloading volume on crack propagation, the angle of condition I was adjusted to 10°, 15°, 20° and 25° respectively. The revised unloading plan is shown in Fig. 7. According to numerical results, the crack propagation process in the initial angle of 5°,10°, 15°, 20° has few changes, so the results of 5°and 25° are merely shown for comparison in Fig. 8. It can be seen from the figure that when the initial angle is adjusted to 25°, the deformation of the slope is greater than that of 5°, with crack towards free face extending 4m and more than the initial angle of 5°, closer to the slope free face.

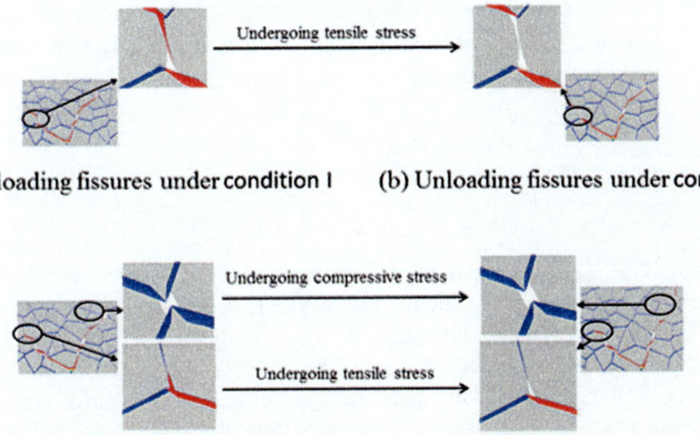

(a) Unloading fissures under condition I (b) Unloading fissures under condition II

(c) Unloading fissures under condition IV (d) Unloading fissures under condition V

Fig. 6. The evolution law of fissures under different unloading conditions

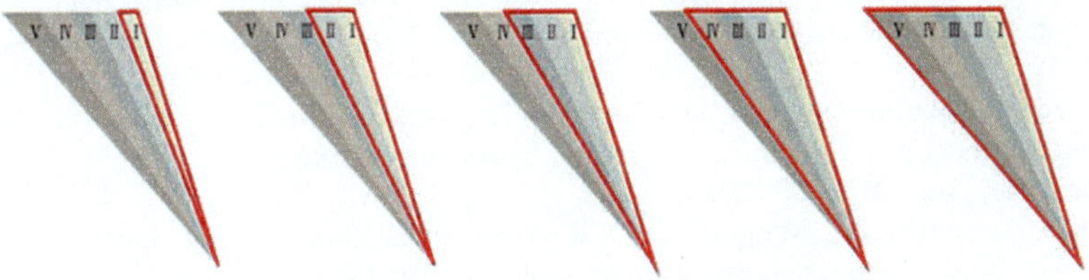

(a) 5° (b) 10° (c) 15° (d) 20° (d) 25°

Fig. 7. Five unloading conditions with different initial angles

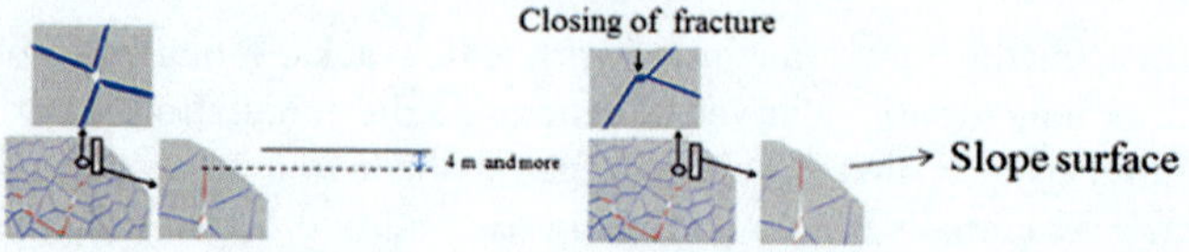

Fig. 8. The comparison of crack propagation between the initial angle of 5°and 25°

5 Conclusions

(1) During the whole unloading process, the structural plane is mainly subjected to tensile-shear failure, and the speed of crack propagation extending to free face is faster than that towards interior side of jointed rock slope.

(2) The crack expansion speed of jointed rock slope is the fastest in the initial stage of stress release, and stress state of micro crack will change when continuing to unload.

(3) when the initial angle is adjusted to 25°, the deformation of the slope is greater than that of 5°, with crack towards free face extending 4m and more than the initial angle of 5°. This indicates that the initial unloading volume has an influence on the deformation of jointed rock slope.

References

1. Lin, Y.D., Zhu, D.P., Deng, Q.L., He, Q.D.: Collapse analysis of jointed rock slope based on UDEC software and practical seismic load. Procedia Eng. **31**, 441–446 (2012)
2. Kim, D.H., Gratchev, I., Balasubramaniam, A.: Back analysis of a natural jointed rock slope based on the photogrammetry method. Landslides **12**(1), 147-154 (2015)
3. Li, S.H., Wang, J.G., Liu, B.S., Dong, D.P.: Analysis of critical excavation depth for a jointed rock slope using a face-to-face discrete element method. Rock Mech. Rock Eng. **40**(4), 331–348 (2007)
4. Jiang, M.J., Jiang, T., Crosta, G.B., Shi, Z.M., Chen, H., Zhang, N.: Modeling failure of jointed rock slope with two main joint sets using a novel DEM bond contact model. Eng. Geol. **193**, 79–96 (2015)
5. He, L., An, X.M., Ma, G.W., Zhao, Z.Y.: Development of three-dimensional numerical manifold method for jointed rock slope stability analysis. Int. J. Rock Mech. Min. Sci. **64**, 22–35 (2013)
6. Fernandez-Rodriguez, R., Gonzalez-Nicieza, C., Lopez-Gayarre, F., Amor-Herrera, E.: Characterization of intensely jointed rock masses by means of in situ penetration tests. Int. J. Rock Mech. Min. Sci. **72**, 92–99 (2014)
7. Che, A.L., Yang, H.K., Wang, B., Ge, X.R.: Wave propagations through jointed rock masses and their effects on the stability of slopes. Eng. Geol. **201**, 45–56 (2016)
8. Cai, M., Kaiser, P.K., Tasaka, Y., Minami, M.: Determination of residual strength parameters of jointed rock masses using the GSI system. Int. J. Rock Mech. Min. Sci. **44**(2), 247–265 (2007)

9. Lian, J.J., Li, Q., Deng, X.F., Zhao, G.F., Chen, Z.Y.: A numerical study on toppling failure of a jointed rock slope by using the distinct lattice spring model. Rock Mech. Rock Eng. **51**(2), 513–530 (2018)
10. Hatzor, Y.H., Bakun-Mazor, D.: Modelling dynamic deformation in natural rock slopes and underground openings with DDA: review of recent results. Geomechanics Geoengineering **6**(4), 283–292 (2011)
11. Regassa, B., Xu, N.X., Mei, G.: An equivalent discontinuous modeling method of jointed rock masses for DEM simulation of mining-induced rock movements. Int. J. Rock Mech. Min. Sci. **108**, 1–14 (2018)
12. Nazari, S.; Yarahmadi Bafghi, A.; Fatehi Marji, M.: Numerical modeling of crack propagation mechanism in jointed rock slopes using indirect BEM and DEM. In: Proceedings of the 49th U.S. Rock Mechanics/Geomechanics Symposium, pp. 2084–2094. American Rock Mechanics Association (ARMA), San Francisco, CA, USA (2015)
13. Barenblatt, G.I.: The mathematical theory of equilibrium cracks in brittle fracture. J. Adv. Applied Mech. **7**, 55–129 (1962)
14. Hillerborg, A., Modeer, M., Petersson, P.E.: Analysis of crack formulation and crack growth in concrete by means of fracture mechanics and finite elements. J. Cement Concrete Res. **6**, 773–782 (1976)
15. Wu, Z.J., Sun, H., Wong, L.N.Y.: A cohesive element-based numerical manifold method for hydraulic fracturing modelling with voronoi grains. Rock Mech. Rock Eng. **52**(7), 2335–2359 (2019)
16. Qin, X.N., Guo, J.J., Gu, C.S., Chen, X.D., Xu, B.: A discrete-continuum coupled numerical method for fracturing behavior in concrete dams considering material heterogeneity. Constr. Build. Mater. **305**, 124741 (2021)
17. Jiang, H.X., Meng, D.G.: 3D numerical modelling of rock fracture with a hybrid finite and cohesive element method. Eng. Fract. Mech. **199**, 280–293 (2018)
18. Encalada, Á., Barzola-Monteses, J., Espinoza-Andaluz, M.: A permeability-throat diameter correlation for a medium generated with delaunay tessellation and voronoi algorithm. Transp. Porous Media **132**(1), 201–217 (2020)

Determining and Estimating the Fracture Risk of Diaphragm Wall from Observed Lateral Deflection

Xiong Wang, Yue Pan(✉), Mingguang Li, and Jinjian Chen

Shanghai Key Laboratory for Digital Maintenance of Buildings and Infrastructure, Department of Civil Engineering, Shanghai Jiao Tong University, 800 Dongchuan Road, Shanghai, China

panyue001@sjtu.edu.cn

Abstract. Due to the great uncertainty and complexity in the underground environment, the investigation and analysis of the internal force of the diaphragm wall are a highly challenging problem. Aiming to achieve the safety estimation of foundation pit in deep excavation, this paper proposes a framework for the calculation of bending moment and determination of fracture risk of the diaphragm wall. Firstly, the five-point smoothed cubic smoothing algorithm is used to eliminate the adverse impact of data noise on the curve fitting. Secondly, the developed natural spline smoothing algorithm is used to implement the curve fitting by establishing the connection of piecewise approximation and smoothness. In addition, three case studies are analyzed to verify the effectiveness and applicability of the developed algorithm. Thirdly, the curvature is obtained by the second derivative of the fitted curve to calculate the actual bending moment with the elastic bending theory. Finally, the efficiency factor of bearing capacity (K) is calculated to investigate the behavior and determine the fracture risk level of the diaphragm wall. Meanwhile, a deep well excavation in Shanghai's Hard X-ray project is utilized as a case study for demonstration purposes. The contribution of this research lies in that the approach can be effectively applied to calculate the bending moment of the diaphragm wall and implement the risk assessment for deep excavation engineering.

Keywords: Bending moment · Fracture risk · Natural spline smoothing algorithm · Curvature · Risk assessment

1 Introduction

With the acceleration of urbanization, underground space exploitation tends to be larger, deeper, and more intensive [1]. The diaphragm walls have been widely used as the retaining structure to undertake the combined action of water and soil pressure, support axial force, and ground load. Therefore, the lateral deflection and bending moment of the diaphragm wall should be investigated and analyzed to evaluate the safety and stability of the foundation pit. There are many studies on lateral wall deflection, including statistical

W. Guo and K. Qian (Eds.): GBCESC 2022, LNCE 211, pp. 884–892, 2023.
https://doi.org/10.1007/978-981-19-5217-3_89

analysis, field monitoring, risk assessment, and real-time prediction [2]. However, the potential fracture risk caused by excessive bending moment has a more adverse impact and damage to the diaphragm wall, which lacks attention and systematic study. To keep the risk within a reasonable range, a primary task is to effectively determine the fracture risk level of the diaphragm wall in deep excavations.

The direct way is to obtain the efficiency factor of bearing capacity (denoted as K) by comparing the actual bending moment with the control value. Traditional monitoring schemes mainly concentrate on the lateral deflection and vertical movement of the diaphragm wall but lack effective and widely accepted monitoring techniques to measure bending moment [3]. Although the direct measurement is difficult, it can be calculated based on the elastic bending theory ($M = EI\phi$) with the longitudinal distribution of the lateral deflection [4]. There are several common methods to calculate the curvature ϕ and bending moment M, such as the three-point circle method [5], high-order polynomial curve fitting method [6], and cubic B-spline method [7, 8], etc. However, there is an inevitable problem that the calculated curvature curves will be oscillate or inconsistent with the objective laws, which is caused by two main aspects. On the one hand, certain noise exists in the data of lateral deformation collected from the monitoring instruments. On the other hand, the curve fitting process concentrates on the approximation to the original data with the least square method but ignores the requirements of curve smoothing.

In order to well handle the above-mentioned problems, the natural spline smoothing algorithm is used to fit the measured data from the longitudinal distribution of lateral wall deflection. The curvature can be obtained by the second derivative of the fitted curve, aiming to get a more accurate calculation of the actual bending moment. Then, the efficiency factor of bearing capacity (K) and corresponding fracture risk level of the diaphragm wall can be determined. The contribution of the research lies in that the developed algorithm implements the connection of segmented approximation and smoothness with a smoothing parameter (p) in curve fitting. The approach can be applied to calculate the bending moment of the diaphragm wall and provide convincing risk information for deep excavation engineering.

2 Methodology

In this study, a framework is developed to determine the bending moment and estimate the fracture risk of the diaphragm wall in deep excavation, as shown in Fig. 1. There are three main steps incorporated, as elaborated below.

2.1 Data Preprocessing

To eliminate the adverse impact of data noise on the curve fitting and keep the original variation of measured data simultaneously, the five-point smoothed cubic smoothing algorithm is used based on the least square method [9].

The basic idea of the cubical smoothing algorithm with five-point approximation is to use cubic polynomial function fitting five adjacent data points. The five equidistant nodes are marked as (x_{-2}, y_{-2}) (x_{-1}, y_{-1}), (x_0, y_0), (x_1, y_1), (x_2, y_2), and d is the equidistant value of x_i. The original nodes are converted into $t_n = n$ by taking $t = (x - x_0)/d$. Then, the least square method is used to determine the coefficients $a_k (k = 0, 1, 2, 3)$ of the polynomial function, as shown in Eqs. (1) and (2). By taking a derivative with respect to a_k respectively, and making it to 0, the algorithm is carried out in Eq. (3).

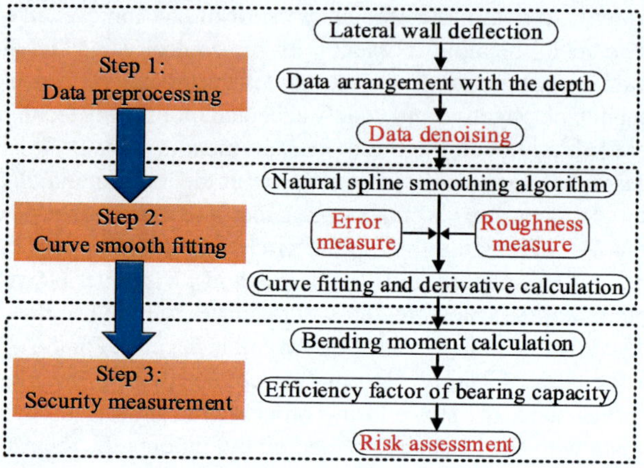

Fig. 1. The framework of determining the bending moment and estimating fracture risk.

$$y(t) = a_0 + a_1 t + a_2 t^2 + a_3 t^3 \tag{1}$$

$$\varphi(a_0, a_1, a_2, a_3) = \sum_{i=-n}^{n} [\sum_{j=0}^{3} a_j t_i^j - y_i]^2 \tag{2}$$

$$\begin{cases} \overline{y}_{-2} = (69y_{-2} + 4y_{-1} - 6y_0 + 4y_1 - y_2)/70 \\ \overline{y}_{-1} = (2y_{-2} + 27y_{-1} + 12y_0 - 8y_1 + 2y_2)/35 \\ \overline{y}_0 = (-3y_{-2} + 12y_{-1} + 17y_0 + 12y_1 - 3y_2)/35 \\ \overline{y}_1 = (2y_{-2} - 8y_{-1} + 12y_0 + 27y_1 + 2y_2)/35 \\ \overline{y}_2 = (-1y_{-2} + 4y_{-1} - 6y_0 + 4y_1 + 69y_2)/70 \end{cases} \tag{3}$$

Where $\overline{y}_i$ represents the improvement value of y_i.

2.2 Curve Smooth Fitting

The natural spline smoothing algorithm implements the connection of piecewise approximation and smoothness in curve fitting based on the cubic spline interpolation. Every piecewise curve is a cubic polynomial function to make the continuity of the first and second derivatives of the whole fitted curve.

The target function of the smoothing spline is depicted as Eq. (4), including the error measure for curve approximation and roughness measure for curve smoothness, respectively. The smoothing parameter p, specified as a scalar value between 0 and 1, is applied to determine the relative weight to place on the contradictory demands of approximation or smoothness. Due to that the shear force and bending moment in the top and bottom of the diaphragm wall should be zero, the boundary conditions are given as the Eq. (5). Finally, the curvature ϕ can be calculated in Eq. (6).

In the practical application, the smoothing parameter p is determined by comparing the global fitted curve and local six-degree polynomial curve with two established criteria. One is to keep the relative errors of maximum curvature values within 5%, and the other one is to keep the relative errors of the positions where the maximum curvature occurs within 10%.

$$\min\{p\sum_{i=1}^{n}[y(x_i)-y_i]^2+(1-p)\int[y''(x)]^2dx\} \tag{4}$$

$$y'(0)=y'(L)=y''(0)=y''(L)=0 \tag{5}$$

$$\phi=\frac{1}{\rho}=\left|\frac{y''(x)}{[1+y'(x)^2]^{3/2}}\right| \tag{6}$$

Where n is the number of entries of (x_i, y_i). $y(x_i)$ and y_i refer the fitted value and the original value corresponding tox_i, respectively. $y'(x)$ and $y''(x)$ denote the first and second derivatives of the function$y(x)$, respectively.

2.3 Security Measurement

The diaphragm wall can be regarded as the linear elastic structure during the normal operation period. Therefore, the elastic bending theory can be applied to calculate the bending moment, as depicted in Eq. (7). The efficiency factor of bearing capacity (K) of the diaphragm wall is calculated by comparing the actual bending moment and control value in Eq. (8). Additionally, the reliability-based risk assessment with load and resistance factors is carried out to ensure the safety and stability of the structures. Hence, the fracture risk level can be determined as follows: $K \leq 0.5$, low risk; $0.5 < K \leq 0.7$, medium risk; $K > 0.7$, high risk.

$$M=EI\phi \tag{7}$$

$$K=\gamma(M/M_u)\times 100\% \tag{8}$$

Where M and M_u is the actual bending moment and controlled value ($\text{N}\cdot\text{mm}^2/\text{m}$). EI is the flexural stiffness ($\text{N}\cdot\text{mm}^2$). $\varnothing$ is the curvature (mm^{-1}). γ is the load and resistance factor, which is equal to 1.35 in this study.

3 Algorithm Verification

Three case studies with complete monitoring data of lateral wall deflection and bending moment are analyzed to verify the effectiveness and applicability of the developed algorithm. These case studies are described in Table 1, including project area, soil condition, retaining structure, and depth.

Figure 2 describes the comparisons between the calculated values (resulting from the proposed algorithm) and the observed values (obtained from the literature). The relative error of bending moment is within 10% for all cases. The comparison results well demonstrate that the developed algorithm is effective and applicable to determine the bending moment from the observed lateral wall deflection.

Table 1. Descriptions of the case studies.

Case	Area	Soil condition	Retaining structure	Depth(m)
Case I [10]	London	Sand and gravels	Piled wall	20
Case II [11]	Taipei	Silty clay and sand	Diaphragm wall	35
Case III [12]	Hildenborough	Weald clay	Discrete piles	10

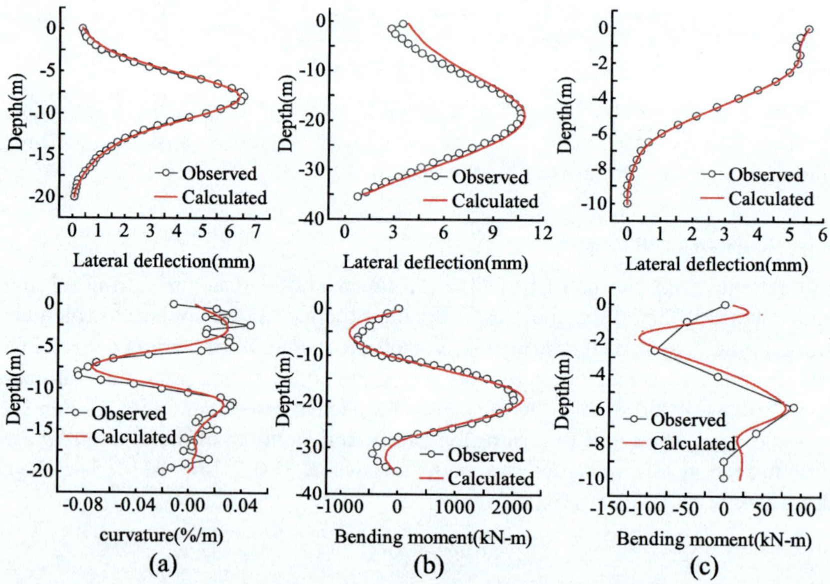

Fig. 2. Comparisons of the observed values (obtained from the literature) and calculated values (resulting from the proposed algorithm). (a) Case I; (b) Case II; (c) Case III.

4 Application

4.1 Project Profile

The proposed algorithm is applied to determine and estimate the fracture risk level of a deep well excavation in Shanghai's Hard X-ray project. The excavation covers an area of approximately 2750 m^2 (55 m-length × 50 m-width) with a depth of 40 m. An open-cut method is used to construct the pit with impermeable diaphragm walls and nine inner struts. The thickness and depth of diaphragm walls are 1.2 m and 82 m, respectively.

Due to the deep excavation and complex ground conditions, field monitoring is essential to investigate the behavior and estimate the potential fracture risk of the diaphragm wall. This project is utilized as a case study for demonstration purposes.

4.2 Data Collection and Description

The clinometer with maximum lateral wall deflection is extracted as the research target for the calculation of bending moment and risk analysis. The excavation data is collected from the BIM management system that records the layout of all measured points and field monitoring data. These extracted data are preprocessed by first removing the null and zero-values data.

Table 2 shows the typical excavation stages in the project. Figure 3 summarizes the lateral wall deflection and calculated curvature at four different construction stages. The wall develops typical deep-seated lateral deflection as excavation proceeds to deep layers. Compared with the original data (discrete points in Fig. 3(a)), the fitted curves based on the developed algorithm have an excellent performance. Different from the variation of lateral deformation, the longitudinal distribution of the curvature is complex with positive and negative values. The maximum value of curvature increases with the increase of excavation depth substantially. It is noticed that the locations of the maximum lateral deflection and maximum curvature value are basically at the same depth in different excavation periods.

Table 2. Typical excavation stages in the project.

Stage	Excavation depth	Date (mm/dd/yyyy)	Spent (d)	Duration
1	11 m	4/10/2020 to 4/30/2020	21	21st day
2	20 m	4/30/2020 to 5/23/2020	24	45th day
3	29 m	5/23/2020 to 6/23/2020	31	76th day
4	40 m	6/23/2020 to 8/7/2020	44	120th day

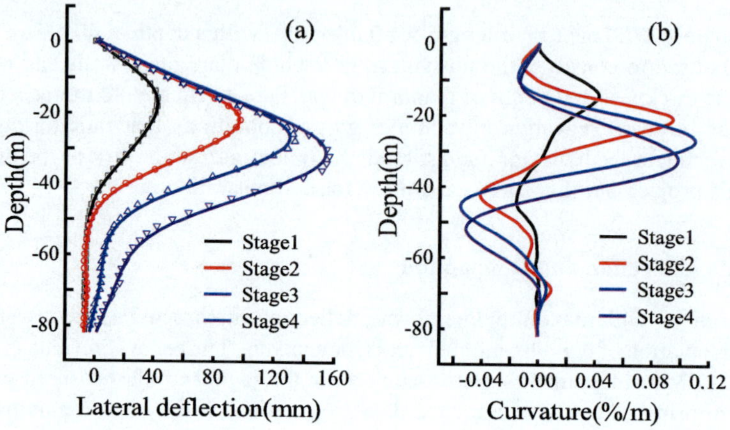

Fig. 3. Distribution of lateral wall deflection (a) and curvature (b) at different construction stages.

4.3 Risk Assessment

According to the design parameter, the flexural stiffness EI=4.68 × 10^{15}N · mm^2 and the control value of bending moment M_u=1.04 × 10^{10}N · mm. Consequently, the efficiency factor of bearing capacity (K) of the diaphragm wall in the whole excavation period can be calculated to estimate the fracture risk level.

Figure 4 plots the variation of K and the distribution of three risk levels. With the process of excavation, the factor K goes through three phases of sharp increase, steady development, and slow decline. The fracture risk level of the diaphragm wall in Stages 14 is low risk, low risk, high risk and medium risk, respectively. When the excavation reaches half of the design depth, the fracture risk of the wall increases significantly. The long-term high risk should be paid attention to by construction and technical personnel. When the excavation is complete, the overall support stiffness is improved due to the bottom slab construction, which reduces the fracture risk to some extent. Nonetheless, the diaphragm wall may be at the operation stage with cracks in the later construction process because it has experienced a long-term stage of high risk. Therefore, reasonable

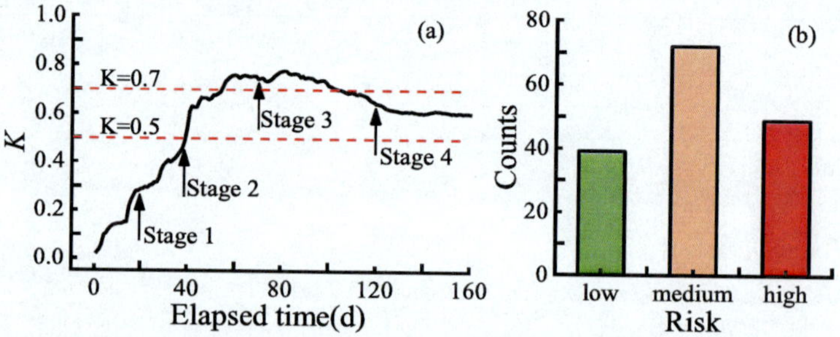

Fig. 4. The variation of K (a) and the distribution of three risk levels (b).

countermeasures should be taken to ensure the safety and stability of the diaphragm wall.

In conclusion, the proposed framework can be effectively applied to calculate the bending moment of the diaphragm wall and provide convincing risk information for deep excavation engineering.

5 Conclusions and Future Works

The study proposes a framework to calculate the actual bending moment and determine the fracture risk of the diaphragm wall from observed lateral deformation. The main achievements and conclusions are summarized as follows:

(1) The developed natural spline smoothing algorithm implements the connection of piecewise approximation and smoothness in curve fitting. The method can be used to obtain the longitudinal distribution of curvature value with excellent performance.
(2) The actual bending moment of the diaphragm wall can be calculated based on the elastic bending theory. And the efficiency factor of bearing capacity (K) is defined to investigate the behavior of the diaphragm wall and determine the fracture risk level.
(3) The proposed framework can be effectively applied to provide convincing risk information for deep excavation engineering.

Future works will focus on the variation of flexural stiffness (EI) due to that the retaining wall will not behave linearly and elastically when the strains rise above the threshold of concrete cracking [13, 14].

Acknowledgement. This work was substantially sponsored by the National Natural Science Foundation of China (No. 41727802), the Program of Shanghai Academic Research Leader (No. 20XD1422100), and the Shanghai Sailing Program (No. 22YF1419100).

References

1. Jinjian, C., Yunfei, Z., Mingguang, L., Suolin, W.: Novel excavation and construction method of an underground highway tunnel above operating metro tunnels. J. Aerosp. Eng. **28**(6), A4014003 (2015)
2. Mahsa, K., Ming, X., Tong, Q., Jennifer, N.: Evaluation of prediction methods for lateral deformation of GRS walls and abutments. J.of Geotechnical Geoenvironmental Eng. **143**(2), 06016022 (2017)
3. Guobin, L., Weidong, W.: Handbook for Excavation Engineering. China Architecture and Building Press, Beijing (2009)
4. Raul, F.: Internal forces of underground structures from observed displacements. Tunn. Undergr. Space Technol. **49**, 50–66 (2015)
5. Yujin, S., Xiao, X., Mingguang, L.: Long-term longitudinal deformation characteristics of metro lines in soft soil area. J. Aerosp. Eng. **31**(6), 04018080 (2018)

6. Chunge, W., Gaochao, Y., Wei, W., Jun, Z.: Deflection detection and curve fitting in three-roll continuous straightening process for LSAW pipes. J. Mater. Process. Technol. **255**, 150–160 (2018)
7. Hongbing, Z., Jinjian, C., Fan, F., Jianhua, W.: Deformation monitoring and performance analysis on the shield tunnel influenced by adjacent deep excavations. J. Aerosp. Eng. **30**(2), B4015002 (2017)
8. Nuodi, H., Li, H., Xi, H., Yang, Z., Limin, Z., Dirk, B.: B-spline-based corner smoothing method to decrease the maximum curvature of the transition curve. J. Manufacturing Science Eng.-Transa. ASME **144**(5), 054503 (2022)
9. Fei, W., Tenglong, Y., Chenxi, L., Yuanke, Z., Yifan, W., Jian, L.: Prediction of lower limb joint angle using sEMG based on GA-GRNN. In: IEEE International Conference on Cyber Technology in Automation, Control, and Intelligent Systems, pp. 1894–1899 (2015)
10. Hisham, M., Kenichi, S., Adam, P., Peter, J.B.: Performance monitoring of a secant-piled wall using distributed fiber optic strain sensing. J. Geotechnical Geoenvironmental Eng. **137**(12), 1236–1243 (2011)
11. Changyu, O., Juitang, L., Hornda, L.: Performance of diaphragm wall constructed using top-down method. J. Geotechnical Geoenvironmental Eng. **124**(9), 798–808 (1998)
12. Joel, S., William, P.: Monitoring and analysis of the bending behaviour of discrete piles used to stabilise a railway embankment. Géotechnique **57**(8), 663–677 (2007)
13. Restrepo, J.I., Rodriguez, M.E.: Stiffness modifiers to support the seismic design of reinforced concrete rectangular columns. J. Struct. Eng. **147**(10), 04021161 (2021)
14. Xuansheng, C., Haibo, L., Jiaxuan, S., Liang, M., Guoliang, L.: Flexural performance of prestressed beams with grouting material of various compactnesses. KSCE J. Civ. Eng. **24**(8), 2419–2434 (2020)

Hysteretic Performance of Self-centering Glulam Beam-To-Column Connections with Steel Core Panel Zone

Zheng Li[1(✉)], Fei Chen[1], Xiaofeng Sun[1], Zhan Shu[2], and Minjuan He[1]

[1] Department of Structural Engineering, Tongji University, Shanghai 200092, China
zhengli@tongji.edu.cn

[2] Department of Civil Engineering, Shanghai University, Shanghai 200444, China

Abstract. Resilient timber structures have arisen global attention due to the advantages in seismic resisting, of which the low repair cost and acceptable downtime are most favored to the public. Post-tensioned glued-laminated timber (PT-GLT) connections played a key role in assigning timber structures the seismic resilience. With the shortening of steel tendons, members connected to the PT-GLT connections are pulled back to their original position. However, there's still room for improvement. Based on the authors' previous experimental research on PT-GLT connections, the efficiency in concentrating damage was low. The connection was unable to entirely depend on the replaceable energy-dissipating elements to dissipate energy. The accumulative damage yielded to the permanent compression deformation on the timber column face, which was not recoverable This paper presents a preliminary study on the feasibility of a method for improving the PT-GLT connections. It was attempted to use a steel member in the intersection of the beam and column. The steel member provided a stiff foundation when the rocking occurred. A cyclic loading test of two PT-GLT connections was conducted. A comparison was made between the connections with and without the improvement method. It was proved that the method was beneficial to improve the connection efficiency in the energy dissipation, connection ductility, and stiffness.

Keywords: Post-tensioned connections · Steel-timber hybridization · Performance improvement

1 Introduction

The study of resilient timber structures is one of the hot spots within the research field of timber structures [1]. It is due to not only the increasing attention on environmentally friendly materials, but also the interest on building structures having resilience in resisting seismic attacks. Resilient structures can be repaired quickly after earthquakes. The repair costs are then significantly reduced. The potential crisis due to the people resettlement and the shutdown of lifeline structures (e.g., hospitals, power plants) can also be handled with acceptable economical costs.

To achieve resilience in laminated veneer lumber (LVL) structures, researchers [2] from the University of Canterbury first proposed to use rocking between LVL members.

W. Guo and K. Qian (Eds.): GBCESC 2022, LNCE 211, pp. 893–900, 2023.
https://doi.org/10.1007/978-981-19-5217-3_90

The rocking mechanism was accomplished with the help of post-tensioned (PT) steel tendons. Between the beam-column, the column-foundation, or the wall-foundation, members are clamped together after the post-tensioning of steel tendons. For LVL members connected to the rocking connection, experimental tests showed that they had the re-centering capacity after the cyclic loading. The hysteretic response of the rocking connection was in the flag shape. Numerical simulation and shaking table tests further validated the seismic resilience of LVL structures with rocking connections [2].

As to glued-laminated timber (GLT) structures, the authors of the research experimentally investigated the hysteretic performance of hybrid post-tensioned glued-laminated timber (PT-GLT) connections [3]. The GLT column and beam were assembled using PT steel tendons and angle dissipaters. In PT-GLT connections, angle dissipaters were intendedly taken as the replaceable members to be damaged and to dissipate energy. Based on research results from the resilient LVL structures [2], a 10 mm steel plate was inserted between the beam and column to avoid the local crushing in the column face. However, the local embedment in the column face was still observed after the test. The permanent compression deformation in the column face indicated that the connection was not efficient in damage concentration, which was also supported by the delay of the failure of angle dissipaters. In the test [3], the final failure of angle dissipaters was not observed until the loading drift ratio was up to 10%. The low efficiency of PT-GLT connections is partially due to the low strength of timber material perpendicular to the grain. In addition, as per [4], the large shear deformation in the shear panel of the beam-column connection also increased the connection flexibility, which was an obstacle for a large gap-opening in the beam-column interface. In this paper, the authors seek to make an improvement on the PT-GLT connections based on the philosophy of steel-timber hybridization [5, 6]. In the intersection of the beam and column, a steel member was used to eliminate the compression and shear deformation in the column, as shown in Fig. 1. The steel member was connected to the timber column with the glued-in rod technology. A preliminary experiment was conducted. With the comparison between PT-GLT connections with and without the improvement method, the effects of the improvement method were explored (Fig. 1).

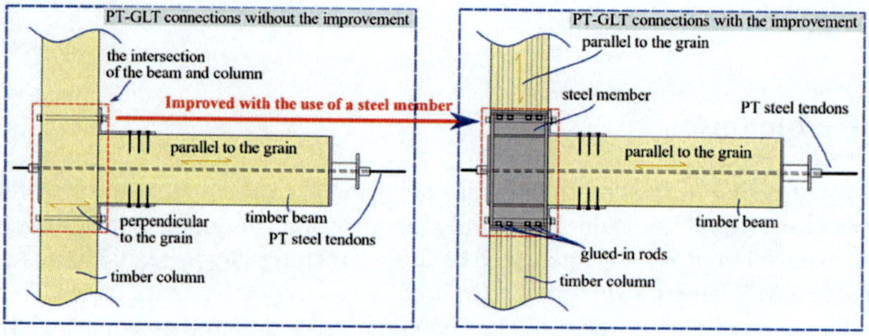

Fig. 1. The improvement method on PT-GLT connections

2 Experimental Program

2.1 Improvement Method and Specimen Details

The improvement method was based on the use of a steel member and the glued-in rod connection. As shown in Fig. 2, the steel member was bolted to the other two timber members before being compressed against the end of the beam. Steel tendons were then pre-tensioned to clamp the beam and column together. The improvement method was used in two connection specimens: Imp_S1 and Imp_S2 (Imp was the abbreviation of improvement). For brevity, they were also referred to as the improved connections hereafter. Except for the use of the steel member in the beam-column intersection, the two specimens had the same geometry and material features as the seven PT-GLT connection specimens (S1 through 7) of the authors' previous research [3]. The seven connections were referred to as the referenced connections (Fig. 2).

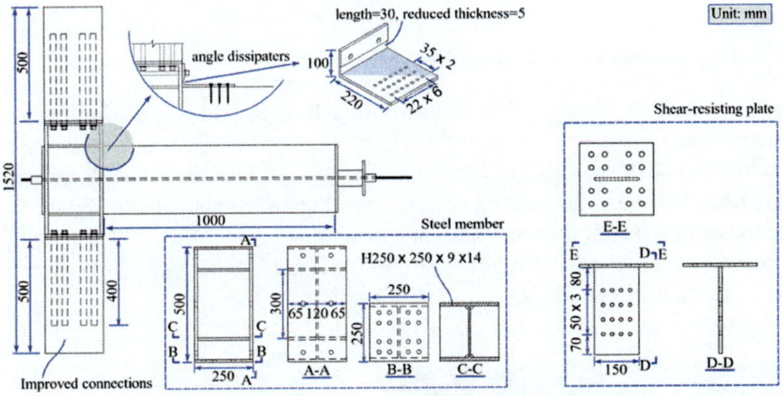

Fig. 2. Illustration on the improved connections

The initial PT-force of Imp_S1 and Imp_S2 was set as 60 kN, which was the highest initial PT force among all seven referenced connections. The initial PT force of S4 and S6 was 60 kN. The information of the two referenced connections S4 and S6 was compared against that of the improved connections Imp_S1 and Imp_S2, as listed in Table 1. The actual initial PT force of Imp_S1 and Imp_S2 was 54.87 kN and 76.20 kN, respectively. In addition, both the improved and referenced specimens used the GLT with a grade of GL 24 h. The angle dissipaters used the Q235 steel. Dissipaters of the improved connections were kept the same as those of the referenced connections. In the intendedly weakened part of angle dissipaters, the length and thickness was 30 mm and 5 mm, respectively.

Table 1. Basic information of the improved and the referenced specimens

Specimens	Img_S1	Img_S2	S4	S6
Initial PT force (kN)	54.87	76.20	56.52	60.71
Improvement method	w/	w/	w/o	w/o

Special concern was put on the design of glued-in rod connections in each improved connection. To ensure a reliable connection between the steel and timber members, each glued-in rod connection had a shear-resisting plate and 16 steel rods. Each steel rod had a slenderness ratio of 25 after being glued into the prefabricated hole with the two-component epoxy resin adhesive. The slenderness ratio was an extremely conservative value to ensure the ductile failure (e.g., the yielding of steel rods) in the glued-in rod connection. Before the test, each glued-in rod connection was fabricated and curved for a week before being used in the improved connections.

2.2 Loading Protocol and Testing Measurements

The loading setup is shown in Fig. 3 (a). For each improved connection, two ends of the column were fixed to a foundation beam. The top of the beam was hinged to a displacement-controlled actuator. The loading protocol was the CUREE protocol that was also used in the cyclic loading of referenced connections [3]. As shown in Fig. 3 (b), the maximum lateral displacement of the loading protocol was 60 mm, which corresponded to 7% drift ratio (60 mm over 850 mm equals 0.07, where 850 mm was the vertical distance between the actuator and the column) (Fig. 3).

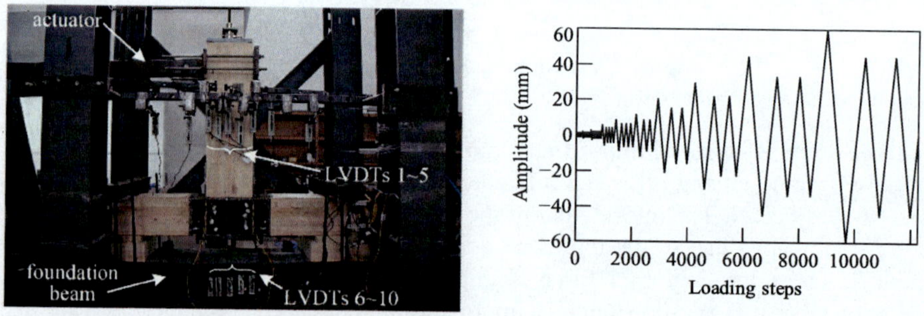

(a) (b)

Fig. 3. Experimental setup: (a) Measuring arrangements; (b) loading protocol

Ten linear variable differential transformers (LVDTs) were arranged in the beam-column interface. LVDT 1 through 5 were connected to the end of the beam and were linearly placed along with the height of the beam. The rest LVDTs were connected to the steel member. The difference of readings from LVDT i and LVDT $i + 5$, where i

ranges from 1 to 5, is the relative displacement between the beam and column Δ_i. The relative displacement Δ_1 and Δ_5 were used to calculate the connection rotation $\theta_{con} = (\Delta_5 - \Delta_1)/290$.

3 Experimental Results

3.1 Hysteretic Curves and Failure Modes

The moment-rotation curve of the two improved connections is shown in Fig. 4. The moment-rotation curve is in a flag shape. Within the test, the steel member in the middle of the column acted as a stiff foundation for the timber beam. As a result, the rocking mechanism was obtained in the beam-column interface. For specimen Imp_S1, the rocking started at the rotation of 0.0066 rad and -0.0064 rad along the positive and negative directions, respectively. The rocking of Imp_S2 occurred when the connection rotation reached over 0.007 rad (0.0076 rad in positive direction and -0.0072 rad in negative direction) (Fig. 4).

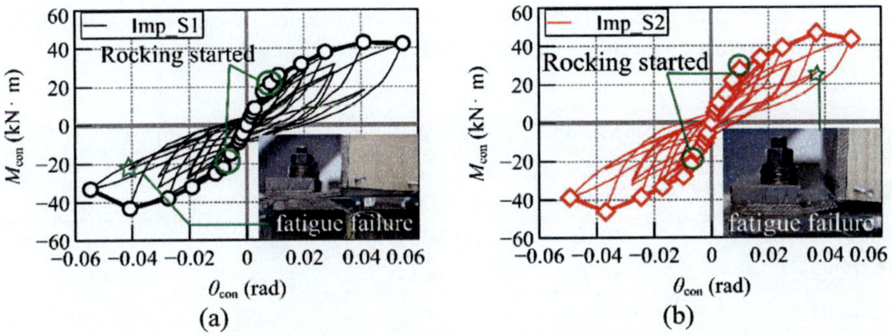

Fig. 4. Hysteretic response and failure modes of (a) Imp_S1; (b) Imp_S2

During the rocking between members, the gap-opening developed in the beam-column interface. Angle dissipaters were forced to deform due to the increase of the gap. The weakened part of angle dissipaters had the weakest resistance and was the first to yield. When the cyclic loading amplitude was up to 60 mm, the fatigue failure of the angle dissipaters was observed for both improved connections. As shown in Fig. 4, the fatigue cracking located near the fillet of the angle dissipaters. Even though the angle dissipaters of referenced connections also failed similarly, the corresponding lateral displacement was larger than 100 mm. The comparison between the time when angle dissipaters failed provide a preliminary insight on the benefits of the improvement method.

3.2 Effects of the Improvement Method

Strength, Ductility, and the Initial Stiffness. The comparison between skeleton curves of the improved connections (Imp_S1 and Imp_S2) and those of the referenced connections (S4 and S6) is shown in Fig. 5.

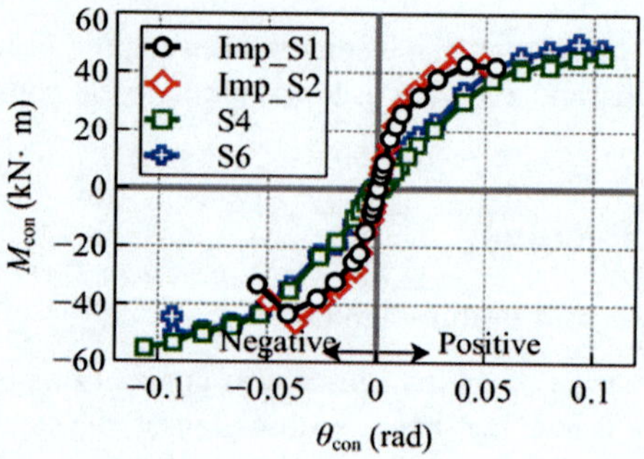

Fig. 5. Comparison on the skeleton curves of both improved and referenced connections

Different indexes were obtained from the skeleton curves of each connection, including the maximum moment strength M_{max}, the yielding moment M_y and the corresponding rotation θ_y (determined based on the Equivalent Energy Elastic–Plastic (EEEP) principle), the ductility μ (the ultimate connection rotation θ_{ult} over the yielding rotation θ_y), and the initial secant stiffness K_{ini} (the load factor for secant calculation was 0.1). The comparison of different indexes was listed in Table 2.

Table 2. Comparison on strength, ductility, and the initial stiffness

Specimens	Img_S1	Img_S2	S4	S6
M_{max} (kN·m)	43.01	46.32	50.51	50.39
M_y (kN·m)	35.77	38.53	44.87	44.89
θ_y (rad)	0.024	0.024	0.068	0.062
θ_{ult} (rad)	0.052	0.049	0.105	0.099
μ (rad/rad)	2.13	2.01	1.54	1.58
K_{ini} (kN·m/rad)	3208.49	3681.27	888.23	1481.32

The improvement method has no obvious effects on the moment capacity of PT-GLT connections. The average moment capacity of Imp_S1 and Imp_S2 is 44.67, which was 89% of the average moment capacity of two referenced connections. The minor discrepancy might be due to variance in the material property. As to the yielding moment, the improved connections had a lower yielding moment than the referenced connections. For rocking connections, the yielding point is also referred to as the "knee point", where the geometrical non-linearity occurs instead of the material plasticity. As a result, the lower-yielding point of the improved connections indicated that they are easier to develop a gap between the beam-column than the referenced connections. For referenced connections, the compression deformation on the timber column face acted as an obstacle to the development of the beam-column gap. With using the steel member as the improvement, the improved connections had a stiff foundation when the beam started to rock.

In addition, the improvement method also increased the ductility and the initial secant stiffness of the PT-GLT connections. The average ductility of two improved connections is 2.07, which is 1.56 times more than the average ductility of referenced connections. The average stiffness of Imp_S1 and Imp_S2 is 3444.88 kN·m/rad, which is 2.91 times more than the average stiffness of S4 and S6.

Energy Dissipation and Stiffness Degradation

For the two improved connections, the accumulative energy dissipation along the loading cycles was calculated, as shown in Fig. 6 (a). To make a fair comparison, the accumulative energy dissipation of the referenced connections S4 and S6 was also plotted in Fig. 6 (a). Within the first 20 loading cycles, both improved and referenced connections dissipated tiny energy. After that, the accumulated energy dissipation of two improved connections started to increase. While the accumulated energy dissipation of two referenced connections was still in the plateau. Upon the finish of the cyclic loading, Imp_S1 and Imp_S2 dissipated 9.66 kJ and 9.30 kJ, respectively. For connections without the improvement, the accumulated energy dissipation was 2.67 kJ and 2.77 kJ for S4 and S6, respectively. The improvement method increased the efficiency of PT-GLT connections in dissipating energy. With the improvement method, it is easier for PT-GLT connection to develop a gap in the beam-column interface. Angle dissipaters engage in dissipating energy at the onset of gap-opening. As a result, the accumulated energy dissipation of improved connections was 3.49 times more than that of referenced connections.

The degradation of stiffness of all four connections was shown in Fig. 6 (b). For improved connections, the initial secant stiffness was much higher than that of referenced connections. With the increase of the loading drift, the secant stiffness decreased due to the yielding of angle dissipaters. While for referenced connections, the secant stiffness of referenced connections increased with the loading drift within the initial loading stage. During this stage, the compression deformation on the column face developed. The timber column face was compressed gradually until a limit of the compression deformation was reached. Then the gap developed between the beam and column of referenced connections. Due to the existence of the soft timber column, the stiffness of

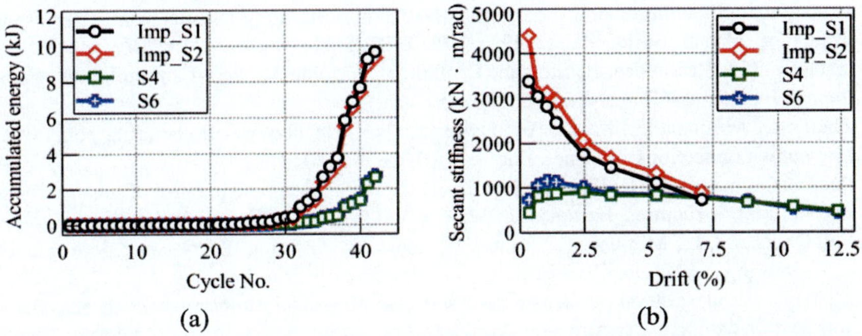

Fig. 6. Comparison between improved and referenced connections: (a) energy dissipation; (b) stiffness degradation

referenced connections was always lower than that of improved connections no matter how the degradation developed (Fig. 6).

4 Conclusions

This study presents a preliminary experimental study on a method to improve the hysteretic performance of PT-GLT connections. The following conclusions can be drawn:

- The rocking in the PT-GLT connections is easy to happen with the use of the steel member in the intersection of beam and column. Due to the easy rocking mechanism, steel angles of improved connections started to dissipate energy earlier than those of referenced connections.
- The improvement method concentrated the accumulative damage to angle dissipaters, which were intendedly designed as replaceable members. The final failure of two improved connections was the fatigue failure of angle dissipaters.
- The efficiency of dissipating energy of PT-GLT connections was increased with the help of the improvement method. Upon the same loading cycles, improved connections dissipated 3.49 times more than the energy accumulatively dissipated by referenced connections.
- The improvement method yielded to the increase of the connection ductility and stiffness. The ductility and initial secant stiffness of improved connections were 1.56 and 2.91 times more than those of referenced connections.

Acknowledgement. The authors gratefully acknowledge National Natural Science Foundation of China (Grant No. 51878476) and Shanghai Rising-Star Program (Grant No. 21QA1409300) for supporting this research.

References

1. Xiaofeng, S., Minjuan, H., Zheng, L.: Novel engineered wood and bamboo composites for structural applications: state-of-art of manufacturing technology and mechanical performance evaluation. Constr. Build. Mater. **249**, 118751 (2020)
2. Gabriele, G., Alessandro, P., Stefano, P., Shiling, P., John, V.: Pres-Lam buildings: state-of-the-art. J. Structural Eng. **146**(6), 04020085 (2020)
3. Zheng, L., Minjuan, H., Kangli, W.: Hysteretic performance of self-centering glulam beam-to-column connections. J. Struct. Eng. **144**, 04018031 (2018)
4. Patricio Quintana, G., David, M.C., Ramiro, B.: State of the art and practice of seismic-resistant hybrid timber structures. European J. Wood Wood Products **79**(1), 5–28 (2021)
5. Fei, C., Zheng, L., Minjuan, H., Yanbo, W., Zhan, S., Guirong, H.: Seismic performance of self-centering steel-timber hybrid shear wall structures. J. Building Eng. **43**, 102530 (2021)
6. Zheng, L., et al.: Lateral performance of self-centering steel–timber hybrid shear walls with slip-friction dampers: experimental investigation and numerical simulation. J. Struct. Eng. **147**, 04020291 (2021)

Behavior of CFRP-Confined Basalt Fibre Reinforced Recycled Aggregate Concrete

Yi Xu[1], Jiangfeng Dong[1,2](✉), and Qingyuan Wang[2]

[1] School of Architecture and Environment, Sichuan University, Chengdu 610065, China
dongjf@scu.edu.cn

[2] Failure Mechanics and Engineering Disaster Prevention and Mitigation Key Laboratory of Sichuan Province, Sichuan University, Chengdu 610065, China

Abstract. The research findings of an experimental study on the axial compressive behavior of carbon fibre reinforced polymer (CFRP) confined basalt fiber (BF) reinforced recycled aggregate concrete (RAC) are presented in this paper. Six CFRP confined BF reinforced RAC (BFRC) short columns were tested under axial compression. The effects of the recycled concrete aggregate (RCA) replacement ratio and CFRP constraint form were studied. The failure mode, stress-strain curve, finite element model and ultimate bearing capacity prediction are analyzed in detail. The results show that the failure modes of unconfined BFRC are similar, while those of CFRP-confined BFRC are different. CFRP constraints can effectively improve the bearing capacity and deformation capacity of BFRC square columns, and the degree of improvement is positively correlated with the thickness and the confined area of CFRP. Based on the experimental data, a simple prediction model of ultimate bearing capacity is established. The model has good accuracy.

Keywords: CFRP constraint · BFRC · Axial compression properties

1 Introduction

In recent years, with the development of real estate and infrastructure in China, the demolition of a large number of old buildings has produced a large amount of construction waste. At present, the annual output of construction waste in the world is 16 billion tons, and that in China is about 2.4 billion tons [1]. The recycling of construction waste is conducive to environmental protection and the conservation of natural resources. It is a promising way to realize environmental protection and sustainable construction. The research and use of recycled aggregate concrete will effectively realize the recovery and reuse of construction waste resources. The existed studies have shown that due to the defects of RAC [2], its strength [3] and durability [4] are often lower than normal aggregate concrete (NAC), which greatly restricts its application.

In order to expand the application scope of RAC, many scholars have found that adding fiber materials [5] and imposing external constraints [6] can effectively improve the mechanical properties of RAC components. Fiber reinforced polymer (FRP), as a

W. Guo and K. Qian (Eds.): GBCESC 2022, LNCE 211, pp. 901–909, 2023.
https://doi.org/10.1007/978-981-19-5217-3_91

new lightweight fiber material, has the advantages of high strength, good durability and corrosion resistance and has been used by many scholars to confined RAC to improve its mechanical properties in recent years. At present, scholars home and abroad have studied using flax fiber [7], carbon fiber [8] and glass fiber [9, 10] to confine RAC. The results show that the strength and deformation capacity of the specimens have been significantly improved. Xie [11] studied the effects of RCA replacement rate, section form, recycled concrete strength and carbon fiber instead of basalt fiber on the axial compression performance of members. Chen, et al. [12] studied the behavior of CFRP-confined RAC under axial compression and found that FRP-confined RAC can be reasonably approximated by existing stress-strain models for FRP-confined NAC.

Previous studies mainly focused on FRP confined RAC, but few on FRP confined fiber reinforced RAC. Therefore, the axial compression test of six CFRP confined BFRC square short columns was carried out to acquire the stress-strain curves of specimens under uniaxial compression in this study. The effect of CFRP constraint form and replacement rate of RCA on mechanical properties of BFRC was studied. Finally, a simple prediction model of ultimate bearing capacity is established by combining the test results with the finite element simulation results of ABAQUS.

2 Materials and Test Methods

2.1 Materials and Sample Preparation

Ordinary Portland Cement (namely 32.5 grade) was used in the study. The RCA used in this study is made by mechanical crushing and manual screening of the concrete waste from collapsed buildings after the earthquake. The material properties of the RCA and the natural coarse aggregates (NCA) are presented in Table 1. Due to the high water-absorption of RCA, it was pre-soaked before mixing to minimize its impact on the workability and water cement ratio of concrete. The sand used in this test was natural coarse sand, and the water used in the test was ordinary city tap water in Chengdu.

Table 1. Material properties of NCA and RCA.

Coarse aggregate	Grading (mm)	Bulk density (kg/m^3)	Apparent density (kg/m^3)	Water absorption (%)	Crushing index (%)
Natural	2.36–19.00	1382	2347	0.25	11.58
Recycled	2.36–19.00	1219	2172	2.25	13.42

The length of BF is between 15 and 19 mm, with 13 μm in diameter and the content is 4 kg/m^3. The tensile strength and elastic modulus are 2000 MPa and 93 GPa respectively. The poisson ratio, elastic modulus and yield strength of the CFRP wraps used in the test are 0.2, 242 GPa and 2400 MPa. Mix design was shown in Table 2.

Table 2. Mix proportions.

Specimen I.D.	RCA ratio (%)	W/C (%)	Unit weight (kg/m^3)					Slump (mm)
			Water	Cement	Sand	NCA	RCA	
BR0	0	0.53	180.5	340.6	653.6	1215.7	0	31.9
BR50	50	0.53	180.5	340.6	648.4	609.6	609.6	32.7
BR100	100	0.53	180.5	340.6	632.8	0	1186.4	30.1

The specimen number and specific parameters are shown in Table 3. For the naming of the specimens, the "S" represents square section, "BR" represents BFRC, the following number of "0", "50" and "100" represent the replacement rate of RCA 0%, 50% and 100% respectively. In addition, the last two letters of "W1" represents one layer fully confined by CFRP, "B1" and "B2" represent the specimens with one or two layers partially confined by CFRP, respectively. The specific dimensions and reinforcing arrangements of the specimens are shown in Fig. 1.

Table 3. Specimen details.

Specimens	RP (%)	*D* (mm)	*L* (mm)	The constrained height of CFRP (mm)	The thickness of CFRP (mm)
SBR0	0	100	400	/	/
SBR50	50	100	400	/	/
SBR100	100	100	400	/	/
SBR50B1	50	100	400	300	0.167
SBR50B2	50	100	400	300	0.334
SBR50W1	50	100	400	300	0.167

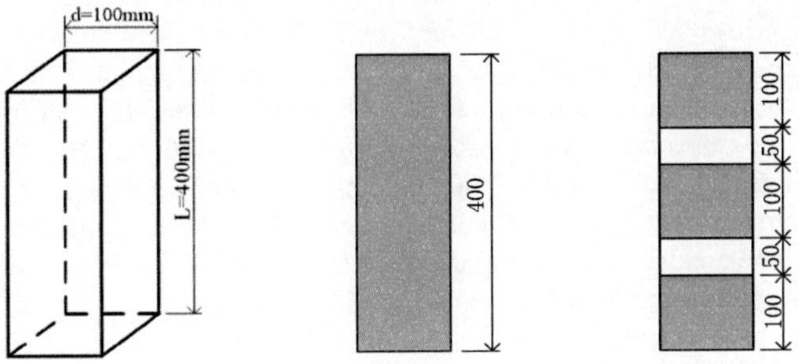

Fig. 1. The reinforcing arrangements for the specimens.

2.2 Test Methods

The test specimens were loaded on a 500t hydraulic testing machine. Before loading, fiber strain gauges were pasted in the middle of the specimen along the axial and circumferential directions to accurately measure the deformation. During the uniaxial compression test, three displacement meters are used to connect the three corners of the upper cover plate respectively, so as to measure the vertical displacement of each corner during loading. The actual layout is shown in Fig. 2. The loading scheme is graded loading, and the loading speed is 0.5 KN per second. Load 5 KN at each stage and keep the load for 3 min. After the instrument value is stable, record the test data, including vertical displacement, hoop strain, axial strain and specimen damage.

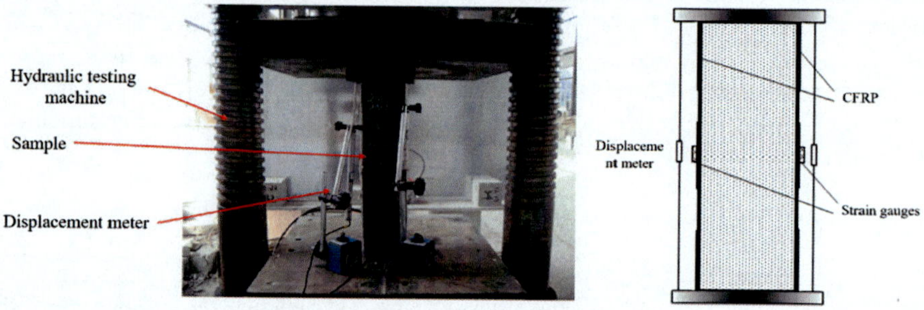

Fig. 2. Test equipment and instruments.

3 Results and Discussion

3.1 Unconfined Specimens

The compressive strength test of the unconfined specimen is conducted and the results are shown in Table 4. It can be seen that the unconfined compressive strength of BFRC increases with the escalation of the replacement rate of RCA, and the increase range is within 6.23%. However, previous relevant research results show that the compressive strength of RAC decreases with the escalation of the replacement rate of RCA [13, 14]. These research conclusions are contrary to the results of this test. Through research and analysis, the reasons for the improvement of the strength of BFRC are as follow: The recycled aggregate obtained by machine crushing and manual screening has higher strength [15]. In addition, it can be found that the ultimate strain of unconfined specimen increases with the escalation of the replacement rate of RCA. This test result is consistent with other relevant research conclusions [11, 16], which presents that the increase of RCA will lead to the increase of strain-softening in BFRC under compression. This is attributed to the lower elastic modulus of RCA than NCA, which will reduce the stiffness of BFRC. It is also manifested that the elastic modulus of BFRC in the table decreases with the growth of the replacement rate of RCA.

Table 4. The test results of unconfined specimens.

Concrete mix	f'_{co} (MPa)	E_c (GPa)	ε_{co} (%)
SBR0	30.5	27.8	0.12
SBR50	30.6	26.3	0.19
SBR100	32.4	20.8	0.20

3.2 Failure Mode

The typical failure models of unconfined and confined specimens are shown in Fig. 3. It can be observed that the failure mode of the unconfined specimen is crack initiation in the middle. With the increasing load, the crack extends to the upper and lower parts, and finally runs through the whole specimen, resulting in failure. The failure models of SBR50B1 and SBR50B2 is that the BFRC of the unconfined part is crushed and damaged, and the crushed part mainly occurs in the upper unconfined part of the specimen. This is because the BFRC of the unconfined part is not in the three-dimensional compression state, and the compressive strength is relatively lower. The failure model of SBR50W1 is the tearing failure of CFRP wrap, which leads to the sudden failure of the specimen with the increase of load. The failure part is concentrated at or near the corner of the specimen, and the main reason for it is the stress concentration at the corner of the square column.

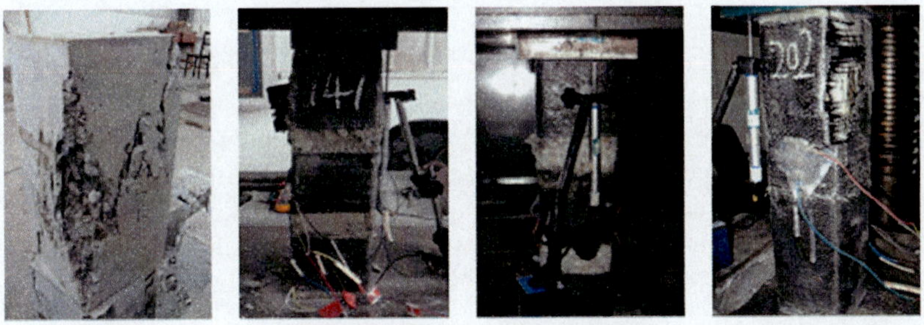

Fig. 3. Failure mode of samples.

3.3 Stress-Strain Response

Figure 4 displays the stress-strain curves for all samples under uniaxial compression. It can be seen that the specimen with higher replacement rate of RCA shows better strength and deformation capacity, but SBR50 has a hoop strain lower than SBR0. The reason for this phenomenon may be that the hoop strain gauge is not firmly bonded to the surface of the specimen. As can be observed from Fig. 4 (b), CFRP confined BFRC short columns show significantly better strength and deformation capacity, with an average increase of

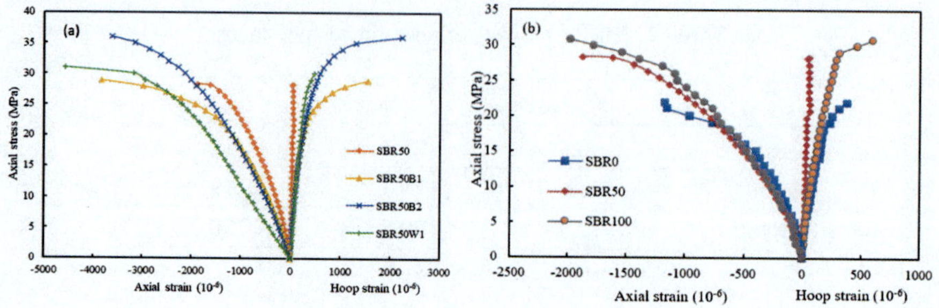

Fig. 4. Stress-strain curves of samples.

15.4% and 113.8% respectively. This is because CFRP exerts a hoop constraint on the compressed BFRC to make it in a three-dimensional compression state and improve its compressive strength. The restraint of CFRP makes the BFRC entering the plastic stage soften more fully, so that the axial and hoop ultimate strains have been greatly improved.

Table 5. The test results of confined specimens.

Concrete mix	Ultimate strength		Ultimate strain	
	σ (MPa)	r_σ (%)	ε (10^{-6})	r_ε (%)
SBR50B1	29.8	4.9	3842	104.8
SBR50B2	36.0	26.8	3626	93.3
SBR50W1	32.5	14.4	4565	143.4

Table 5 displays the ultimate strength and the corresponding ultimate strain of the specimens with 50% and 100% RCA replacement ratio and different FRP reinforcing arrangement. It can be seen that the increasing of the CFRP thickness (B2) and the CFRP constraint area (W1) can improve the constraint effect of CFRP. The fully wrapped constraint form (W1) can fully soften the internal BFRC and improve its deformation performance.

4 Ultimate Bearing Capacity of CFRP-Confined BFRC

The ultimate bearing capacity model in this paper is based on the model in [17], which is modified and extended combined with the test results and finite element results. The ultimate bearing capacity (N_u) model of CFRP-confined BFRC square short columns is as follows:

$$N_u = \left[\gamma f'_{co} + \alpha(f_{lu} - f_{lo})\right] A_s \tag{1}$$

where f'_{co}, f_{lu} and f_{lo} are in unit MPa. They are peak axial compressive stress of unconfined BFRC, threshold confining pressure and nominal lateral confining pressure at ultimate respectively. α is axial strength enhancement coefficient, which is 3.6 [16] and A_s is compression area. γ is parameter in compressive strength expression, which is calculated by the following formula:

$$\gamma = 1 + 0.0058\frac{K}{f'_{co}} \tag{2}$$

$$K = \frac{2hE_f t_f}{HD} \tag{3}$$

where K is lateral confinement stiffness and E_f is elastic modulus of CFRP, both are in unit MPa. h and H are the height of CFRP wrap and height of specimen. t_f is the thickness of CFRP and D is diameter of BFRC core.

$$f_{lo} = K\varepsilon \tag{4}$$

$$\varepsilon = \left(0.43 + 0.009\frac{K}{f'_{co}}\right)\varepsilon_{co} \tag{5}$$

$$f_{lu} = K\varepsilon_h \tag{6}$$

$$\varepsilon_h = K_{\varepsilon,f}\varepsilon_f \tag{7}$$

where ε_{co} is axial strain of unconfined concrete at f'_{co}, ε is hoop strain of CFRP-confined BFRC at peak axial compressive stress and ε_h is hoop rupture strain of CFRP shell. $K_{\varepsilon,f}$ and ε_f are hoop strain reduction factor of CFRP and ultimate tensile strain of CFRP respectively.

According to the ultimate bearing capacity model in this paper, the ultimate bearing capacity of SBR50B1, SBR50B2 and SBR50W1 are calculated. The comparison between the test results and the calculated results of the model is shown in the Table 6, a good agreement and a maximum of 5% error between the test result and calculation result.

Table 6. The comparison between the test results and the calculated results.

Specimen	Test result (KN)	Calculated result (KN)	Error (%)
SBR50B1	298.4	313.3	5.0
SBR50B2	360.0	342.3	4.9
SBR50W1	324.8	323.0	0.6

5 Conclusions

This paper introduces, summarizes and analyzes the test results of axial compression performance of BFRC square short columns restrained by CFRP. The following conclusions can be drawn:

(1) For unconfined BFRC, its compressive strength and ultimate strain will increase with the replacement rate of RCA, while its stiffness and elastic modulus are opposite.
(2) The failure model of unconfined BFRC is similar to the unconfined NAC failure model. The failure model of BFRC under half package constraint is the failure caused by crushing of unconstrained part of BFRC, and the failure model of BFRC under full package constraint is the sudden failure caused by tearing of CFRP wraps.
(3) CFRP restraint can effectively improve the compressive strength and deformation capacity of BFRC square short columns. The thicker the CFRP, the larger the restraint area, and the greater the improvement of compressive strength and deformation capacity. Compared with half package, full package can significantly improve the deformation capacity.

References

1. Kaidong, X., Jina, W., Zhixin, L., et al.: Systematized study on application technology indicators of recycled aggregate reproduced by masonry construction waste. Bull. Chin. Ceram. Soc. **39**(09), 2905–2911 (2020)
2. Don, J.M.S., Pedro, A.G.: Study on the influence of attached mortar content on the properties of recycled concrete aggregate. Constr. Build. Mater. **23**(2), 872–877 (2009)
3. Wai Hoe, K., Mahyuddin, R., Kenn Jhun, K., et al.: Influence of the amount of recycled coarse aggregate in concrete design and durability properties. Constr. Build. Mater. **26**, 565–573 (2012)
4. Ciro, F., Carmine, L., Enzo, M., et al.: Mechanical and durability performance of sustainable structural concretes: An experimental study. Cement Concr. Compos. **71**, 85–96 (2016)
5. Wisal, A., Lim, C.W.: Production of sustainable and structural fiber reinforced recycled aggregate concrete with improved fracture properties: A review. J. Cleaner Prod. **279**, 123832 (2021)
6. Jianliang, Z., Tao, Y., Jinguang, T.: Stress-strain behavior of FRP-confined recycled aggregate concrete. J. Compos. Constr. **19**(3), 04014054 (2015)
7. Bin, Y., Liang, H., Libo, Y., et al.: Behavior of flax FRP tube encased recycled aggregate concrete with clay brick aggregate. Constr. Build. Mater. **136**, 265–276 (2017)
8. Chang, G., Liang, H., Libo, Y., et al.: Behavior of glass and carbon FRP tube encased recycled aggregate concrete with recycled clay brick aggregate. Compos. Struct. **155**, 245–254 (2016)
9. Jianzhuang, X., Yijie, H., Jie, Y., et al.: Mechanical properties of confined recycled aggregate concrete under axial compression. Constr. Build. Mater. **26**(1), 591–603 (2012)
10. Jianzhuang, X., Jie, Y.: On recycled concrete confined by GFRP tube under axial compression. J. Tongji Univ. (Nat. Sci.) **37**(12), 1586–1591 (2009)
11. Tianyu, X., Togay, O.: Behavior of recycled aggregate concrete-filled basalt and carbon FRP tubes. Constr. Build. Mater. **105**, 132–143 (2016)
12. Guangming, C., Yanhong, H., Tao, J., et al.: Behavior of CFRP-confined recycled aggregate concrete under axial compression. Constr. Build. Mater. **111**, 85–97 (2016)
13. Gonzalez-Fonteboa, B., Martinez-Abella, F., Lopez Diego, C., et al.: Stress–strain relationship in axial compression for concrete using recycled saturated coarse aggregate. Constr. Build. Mater. **25**(5), 2335–2342 (2011)
14. Ulloa, V.A., García-Taengua, E., Pelufo, M.J., et al.: New views on effect of recycled aggregates on concrete compressive strength. ACI Mater. J. **110**(6), 1–10 (2013)

15. Bozhi, Z., Sheliang, W., Bo, Z., et al.: Experimental analysis of the basic mechanical properties of recycled concrete. Concrete **7**, 4–6 (2011)
16. Qiong, L., Jianzhuang, X., Zhihui, S.: Experimental study on the failure mechanism of recycled concrete. Cem. Concr. Res. **41**(10), 1050–1057 (2011)
17. Togay, O., Jian, C.L.: Axial compressive behavior of FRP-confined concrete: Experimental test database and a new design-oriented model. Compos. Part B Eng. **55**, 607–634 (2013)

Influence of Hammer Geometry on Impact Performance of RC Beams

Tianqi Xue[2], Xiaolan Yuan[2], Xin Zhang[3], Zhi Li[1,2], and Xiaofang Deng[1,2](✉)

[1] GuangXi Key Laboratory of New Energy and Building Energy Saving, Guilin 541004, Guangxi, China
XiaofangDeng@126.com

[2] College of Civil Engineering and Architecture, Guilin University of Technology, Guilin 541004, Guangxi, China

[3] School of Information Engineering, Xuzhou College of Industrial Technology, Xuzhou 221140, Jiangsu, China

Abstract. Finite element models were created based on the impact test of RC beams to examine the influence of hammer curvature radius on the resistance of RC beams under drop weight load. The influence of varied hammer curvature of drop weight on the dynamic response of RC beam under impact load was investigated using a confirmed finite element model. The findings showed that the FE model accurately predicted the dynamic response of an RC beam under impact load. It is also discovered that increasing the curvature radius of the hammer increases the impact loading rate and the peak of impact force. The peak value of impact force recorded by the drop hammer is more stable at varying beginning inclination degrees of the drop weight when the curvature of the hammer is not zero.

Keywords: Rc beam · Impact resistance · Numerical simulation · Curvature radius

1 Introduction

Reinforced concrete (RC) structures will inevitably encounter extreme loads during life cycle, which will cause irreparable damage to the structure and make it completely lose its normal function. As an important bearing member of the building structure, the reaction of beam under impact load will have a significant influence on the building structure's safety performance. Many studies have been conducted to evaluate the dynamic behavior of concrete beams that have been subjected to impact loads [1, 2].

Kazunori F. et al. [3] conducted impact tests on I-section beams, and verified applicability of the theoretical calculation model of two-degree of freedom. Huawei L. et al. [4, 5] studied the impact resistance of beams by finite element method, mainly considering the influence of impact sensor position on measurement accuracy. And a calculation method based on the developed 2DOF analytical model is proposed to analyze the impact

W. Guo and K. Qian (Eds.): GBCESC 2022, LNCE 211, pp. 910–917, 2023.
https://doi.org/10.1007/978-981-19-5217-3_92

response of concrete beams. However, in most of previous studies, the curvature radius of drop weight is different [6–8].

Although it is well known that the curvature radius of a drop weight has a significant impact on RC beam impact resistance. As a result, a systematic investigation is required to measure the impact performance of RC beams as a function of drop weight curvature radius in order to determine the response characteristics of RC beams under impact load. The ANSYS/LS-DYNA is used to develop a series of RC beams in this article, and the impact of a drop hammer with varying curvature radius on RC beams under impact load is investigated.

2 Numerical Model Development

2.1 Introduction of Test

Impact tests by Fujikake K et al. [9] examined the resistance of 12 RC beams. The RC beam section size is 250 mm × 150 mm, the length is 1700 mm, and the effective span is 1400 mm. The reinforcement of different specimens was shown in Table 1, and the dimensions was shown in Fig. 1.

Table 1. Table of reinforcement for specimens

Designation	Compression side		Tension side	
	Number	Rebar diameters (mm)	Number	Rebar diameters (mm)
S1616	2	16	2	16
S1322	2	13	2	22
S2222	2	22	2	22

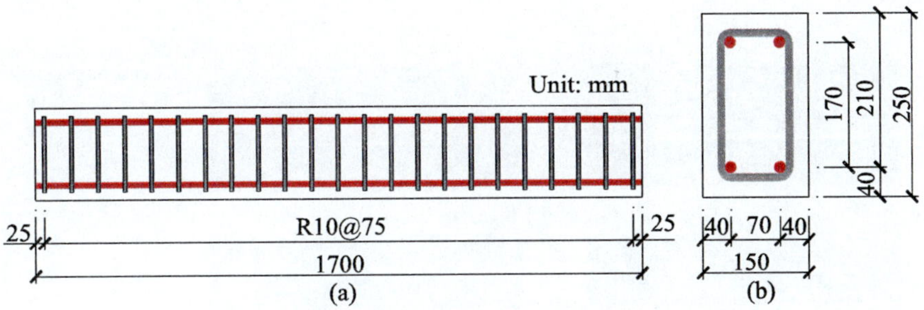

Fig. 1. Dimensions of the specimen.

In this test, a 400 kg drop weight is dropped freely from various heights. A hemisphere with a radius of 90 mm serves as the hammer head.

2.2 Numerical Model

For the RC beams under impact load described by Fujikake K et al. [9], a finite element (FE) model was generated with LS-DYNA. S2222 series tests were chosen for modeling in this investigation, and the model is presented in Fig. 2. RC beam model.

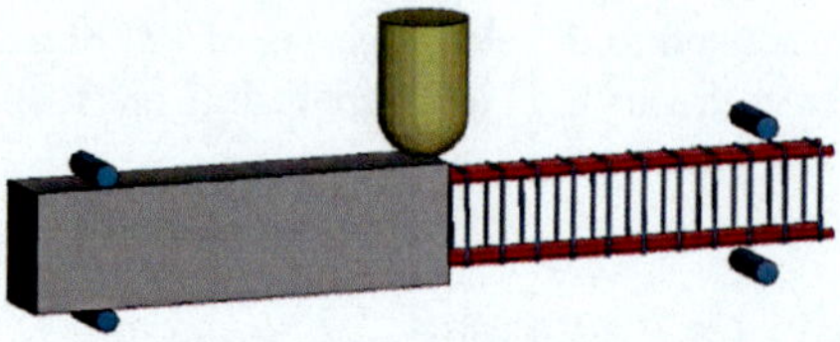

Fig. 2. RC beam model.

Material model *MAT CSCM CONCRETE is the constitutive model for concrete, whose constitutive model is characterized by the application of the maximum principal strain failure criterion, which can simulate the failure scenarios of RC structures under impact load effectively [10]. The material constitutive model used by the rebar is *MAT_PLASTIC_KINEMATIC, whose constitutive model relationship is bilinear, similar to that of the rebar. Parameters required by the two material models were calculated in reference.

3 Verification of Numerical Model

3.1 Failure Scenarios

The comparison of failure scenarios between model and test is shown in Fig. 3. The failure scenarios predicted by the FE model are nearly identical to those predicted by the test, implying that the model can better predict RC beam failure scenarios under impact loads.

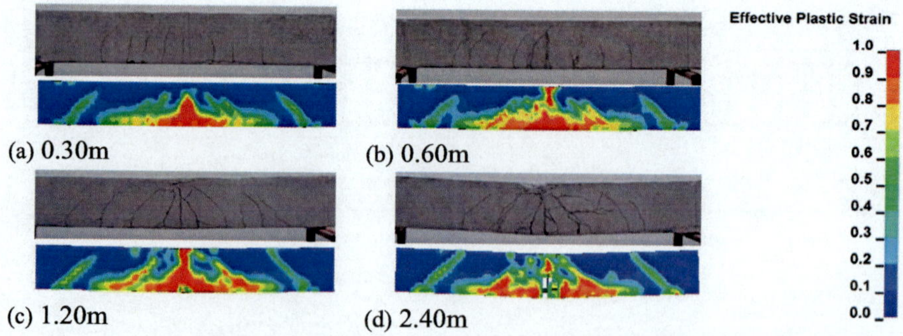

Fig. 3. Comparison of failure scenarios for S2222.

3.2 Impact Force

Figure 4 shows that the FE model's projected impact force time history curves are in well agreement with the test, the variation trend is similar.

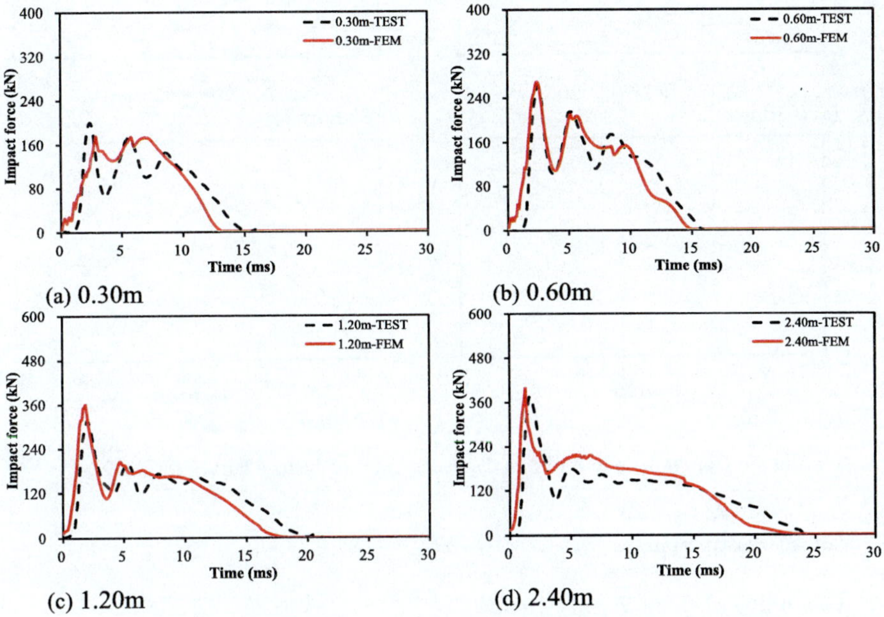

Fig. 4. Comparison of impact force time history curves for S2222.

3.3 Midspan Displacement

The midspan displacement predicted by FE model are somewhat lower than the test, as seen in Fig. 5. Which is because the material models used in the numerical model does not consider the initial defects, resulting in slightly higher overall structural stiffness than tested.

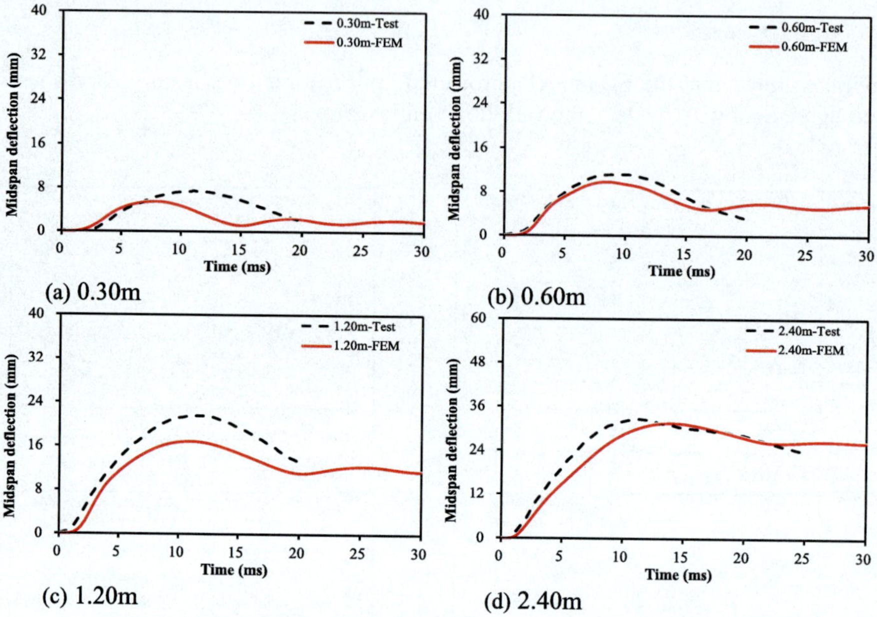

Fig. 5. Comparison of midspan displacement time history curves for S2222.

4 Parametric Study

4.1 Geometry of Drop Weight Head

There are five types of drop hammer shown in Fig. 6. Different head geometries, each having a distinct curvature radius of the hammer heads, named R90, R250, R500 and FLAT, with curvature radius of 90 mm, 250 mm, 500 mm, and infinite. The 400 kg drop weight impacts beam at a speed of 4.85 m/s.

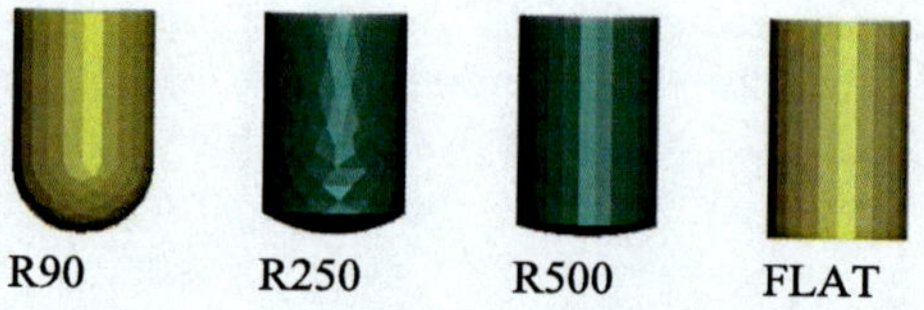

Fig. 6. Different head geometries.

Impact force time history curves for different head geometries, as shown in Fig. 7. Because of the complex interaction between the drop hammer and the reinforced concrete beam, the second peak impact force is observed in each specimen. Because the first peak is substantially bigger than the second, the first dictates the beam's performance. As a result, our research focuses solely on the dynamic response of beams during the initial hit.

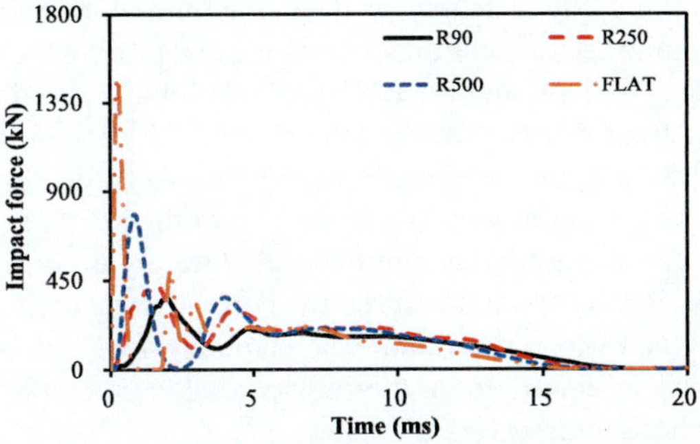

Fig. 7. Impact force curves for different head geometries.

It can be seen from Fig. 7 that as the radius of curvature of hammer head increases, the slope of loading section of impact force time history curve increases and the loading rate increases. R90 has the lowest impact peak value, 358.88 kN, as can be observed from the data. The highest value of impact force grows as the radius of curvature increases. When the radius of curvature of the hammer approaches infinity, the maximum impact force associated with the hammer is 1478.44 kN, which is 4.13 times that when the radius of curvature is 90 mm. With the increase in hammer curvature radius, the peak arrival time decreases from 2.00 ms to 0.30 ms. That is, the higher the radius of hammer curvature, the greater the maximum of force and the quicker it comes.

4.2 Initial Angle of Hammer

During the impact test, hammer is dropped on the RC beam at a predetermined height. A guide rail is used to regulate the movement of the drop hammer and guarantee that it meets the RC beam at the prescribed impact location and does not fall to the ground after impact. However, the initial gap and friction difference between the drop hammer and the guide rail will lead to the initial angle of the drop. The beginning inclination angles (0°, 0.5°, 1°, and 2°) were investigated in order to evaluate the impact force of the drop hammer head with varying curvature radius at different inclination degrees. The

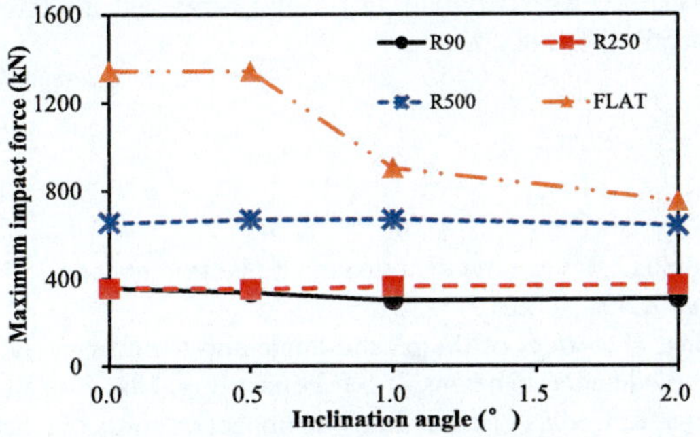

Fig. 8. Comparisons of the maximum impact force for impact inclination angle.

inclination angle in this study is defined as the angle formed by the normal vector of the beam and the drop hammer's center axis. At various beginning inclination degrees, the drop weight was set at 400 kg, and the RC beam was hit at a speed of 4.85 m/s.

As can be seen from the Fig. 8, the peak for FLAT decreases significantly from 1343.56 kN to 751.37 kN, and the maximum impact force decreases by 44.08%. By contrast, the effect of the initial angle on the peak of others are negligible, because the drop hammer's small starting tilt has minimal influence on its contact area.

Finally, the peak impact force measured by a drop hammer with a certain curvature head is more stable at various drop hammer beginning inclination angles. The effect of the drop hammer's initial angle on the impact load can be efficiently decreased by utilizing a hammer head with a certain curve.

5 Conclusions

The FE model was developed by ANSYS/LS-DYNA, and the accuracy of model was confirmed by comparing experimental findings. The following are the results of the extended parameter analysis of the FE model.

(1) Refined finite element model can accurately simulate the failure modes, impact forces, and midspan displacement of RC beams subjected to impact.
(2) The peak of force grows as curvature radius of the drop hammer head grows, and the arrival time of peak is also delayed.
(3) At various beginning inclination angles of the drop weight, the peak of impact force recorded by the hammer with curvature higher than zero is more stable. By utilizing a hammer head with a curvature higher than zero, the impact force of the starting angle of the drop weight can be lowered.

Acknowledgement. The authors are appreciative for the financial support by Guangxi Key Laboratory Project of New Energy and Building Energy Saving (Gui Ke Neng22-J-21–3), the Natural Science Foundation of Guangxi(No.2021GXNSFFA196001), the Guangxi Science and Technology Program Project (Gui Ke AD21220071), the China Postdoctoral Science Foundation (2021M693814) and the Guangxi Young and Middle-aged Teacher Basic Ability Improvement Project (2021KY0254). Any thoughts, findings, or conclusions presented in this study are not necessarily those of Guangxi Key Laboratory of New Energy and Building Energy Saving and Natural Science Foundation of Guangxi.

References

1. Thong, M.P., Hong, H.: Review of concrete structures strengthened with FRP against impact loading. Structures **7**, 59–70 (2016)
2. Thong, M.P., Hong, H.: Effect of the plastic hinge and boundary conditions on the impact behavior of reinforced concrete beams. Int. J. Impact Eng. **102**, 74–85 (2017)
3. Fujikake, K., Senga, T., Ueda, N., et al.: Study on impact response of reactive powder concrete beam and its analytical model. J. Adv. Concr. Technol. **4**(1), 99–108 (2006)

4. Huawei, L., Wensu, C., Hong, H.: Factors influencing impact force profile and measurement accuracy in drop weight impact tests. Int. J. Impact Eng. **145**, 103688 (2020)
5. Huawei, L., Wensu, C., Thong, M.P., et al.: Analytical and numerical studies on impact force profile of RC beam under drop weight impact. Int. J. Impact Eng. **147**, 103743 (2021)
6. Chen, Y., May, I.M.: Reinforced concrete members under drop-weight impacts. Proc. Inst. Civ. Eng. Struct. Build. **162**(1), 45–56 (2009)
7. Yan, Q., Sun, B., Liu, X., et al.: The effect of assembling location on the performance of precast concrete beam under impact load. Adv. Struct. Eng. **21**(8), 1211–1222 (2018)
8. Kishi, N., Mikami, H.: Empirical formulas for designing reinforced concrete beams under impact loading. ACI Struct. J. **109**(4), 509–519 (2012)
9. Fujikake, K., Li, B., Soeun, S.: Impact response of reinforced concrete beam and its analytical evaluation. J. Struct. Eng. **135**(8), 938–950 (2009)
10. Kai, Q., Yun-Hao, W., Feng, F., et al.: Numerical evaluation of the reliability of using single-story substructures to study progressive collapse behaviour of multi-story RC frames. J. Build. Eng. **33**, 101636 (2021)

Component-Based Simulation on Progressive Collapse of a 2D-Framed Steel Structure Under Debris Impact

Yufang Yuan(✉), Kang Chen, and Bo Yang

Chongqing University, No. 174 Shazheng Street, Shapingba, Chongqing, China
yufanyuan0930@163.com

Abstract. In this study, component-based models for FP joints and WUF-B joints were proposed and validated against impact tests. Then a steel-framed structure with 10-floor designed by NIST was studied by removing column method with the application of the models. The locations of the removed columns, including different floors and bays, and the number of the removed columns were considered as variables. The simulation results showed that removal of a column located at a lower floor, contrasted with that at a higher floor, can cause a larger increase of axial forces of the columns at the same floor with the removed ones. The vertical displacement at the removed columns decreased with the floor number decreasing. Since the gravity bay was more vulnerable than the moment bay, column removal at the transition area would lead to the greatest change of axial force occurring at the column belonging to the adjacent moment bay. In addition, a sufficient lumped mass was required to trigger the progressive collapse of structures.

Keywords: Component-based model · Progressive collapse · Multi-story steel structures · Simulation

1 Introduction

Progressive collapse is a consequent reaction caused by the failure of local structural components of building structures subjected to accidental load, which leads to the disproportionately large part collapse of the whole structures. The unexpected collapse of the Ronan Point apartment in the U.K. firstly caught the attention of researchers on the robustness of structures. Subsequently, such research area drew an increasing amount of attention after several catastrophic events, including the collapse of the Murrah Federation building and the World Trade Center. Both resulted in serious safety problems and heavy economic losses. Recently, the collapse incidents of Jiaxin hotel in Quanzhou, China in 2020 and a high-rise apartment in Korea in 2022 indicated that further efforts are needed to explore the mechanism of progressive collapse and to propose reliable design guidelines to reduce the probability of occurrence of similar incidents and mitigate the extent of the influence of progressive collapse.

Till date, the performance of the steel structures subjected to progressive collapse have been studied through both experimental investigations and numerical simulations.

W. Guo and K. Qian (Eds.): GBCESC 2022, LNCE 211, pp. 918–925, 2023.
https://doi.org/10.1007/978-981-19-5217-3_93

For the experimental study, testing a full-scale building could be costly and require numerous technical supports. Hence, the majority of experimental research studies were carried out through tests on sub-assemblies to figure out the effect of different types pf joints and slabs on the anti-progressive collapse performance of structures. Yang [1] and Dinu [2] tested the structural performance of various steel beam-column connections under a quasi-static central-column-removal scenario, which pointed out that catenary action can be developed to withstand the applied external loading during the large displacement stage. Furthermore, the rotation capacity and tensile strength of a beam-column connection can be a key factor affecting the progressive collapse resistance of a structure. Besides quasi-static experiments, some impact tests of joints were carried out by Wang [3] and Wang [4]. They suggested that when evaluating impact resistance, the energy absorption capability of a structure was essential and the good ductility rather than high strength played a significantly vital role in the development of catenary action. Wang [5] conducted quasi-static tests of two full-scale single-story two-bay by one-bay steel-concrete composite floor system to evaluate the influence of the slab. The tests showed that tensile membrane action was insensitive to the horizontal boundary conditions, while the development of catenary effect was highly affected.

As for the simulation aspect, both micro-models and macro-models were adopted. Khandelwal [6] proposed macro-models with the aid of the component-based method to simulate the structural responses of joints in a 10-story two-dimensional framed steel structure to investigate the ductility demand of joints against progressive collapse. Sadek [7] and Main [8] developed macro-models for welded unreinforced flanged-bolted web (WUF-B) connection and single plate shear (fin plate, FP) connection, which were calibrated with corresponding micro-models and experimental results. Li [9] and Wang [10] applied the aforementioned macro-models to assess the anti-progressive collapse performance of three-dimensional framed steel buildings with composite slabs.

Above all, a large proportion of researchers placed their focus on investigating the progressive collapse at the component-level or substructure-level, not only in the experimental aspect, but also in the numerical aspect. Furthermore, the target structures were normally with few floors and normally the removed columns were located at the base floor. This study intends to investigate the mechanism of progressive collapse of multi-floor planar steel-framed structures under alternate path method. Due to the limitation of the feasibility of the experiment and consideration of economic cost, the numerical simulation with the component-based joint modelling method is adopted. The whole study can be divided into three parts:

(1) To propose component-based joint models;
(2) To verify the model with previous impact experiments of substructures;
(3) To model a two-dimensional steel-framed structure and to carry out relevant simulations.

2 Numerical Model of Joints

2.1 Component-Based Model of Joints

The models shown in Fig. 1 were component-based connection models for FP joints and WUF-B joints, respectively.

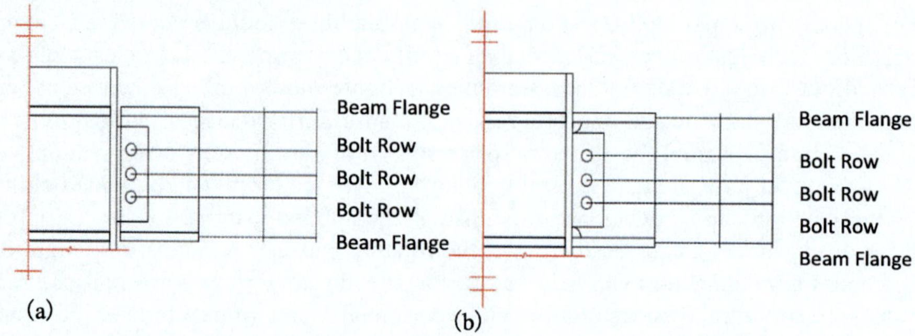

Fig. 1. Component-based models: (a) FP connection; (b) WUF-B connection

Beam Flange Spring. For FP joints, the springs for beam flange will not take effect unless the beam flange get in touch with column flange. For WUF-B joints, the corresponding springs are modelled to support the beam flange in case the penetration occurs and provide tensile force if the beam flange is in tension.

Bolted Spring. The ultimate strength $R_{n,br}$ of bolt in bearing is calculated by the equation in AISC 360–10 as follows:

$$R_{n,br} = 1.5\left(L_e - \frac{d_b}{2}\right)t\sigma_u \leq 3td_b\sigma_u \tag{1}$$

where L_e is the end distance from the center of a bolt hole to the edge of the fin plate, d_b is the nominal diameter of the bolt, and t is the thickness of the plate.

According to Rex and Easterling [11, 12], the stiffness of bolt in bearing k_i and force-displacement relationship of bolts in bearing can be determined.

2.2 Verification of Joint Model

The proposed joint model for WUF-B connection was established by LS-DYNA and the results were verified by the experiments carried out by Wang [4]. The impact loading history curve and the displacement history curve of the mid-span of the beam are shown in Fig. 2, respectively. The impact loading and the displacement of the mid-span of the beam of the simulated models satisfied well with the experimental results.

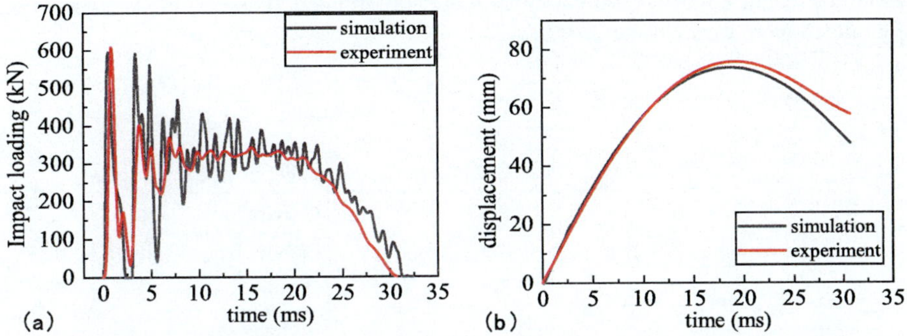

Fig. 2. (a) Impact loading history curves of simulation and experiments; (b) Displacement history curves of simulation and experiments.

3 Modeling of the 2D Framed Steel Structure

Having validated the joint model and relevant modelling technique, a two-dimensional 10-story framed steel structure was modeled. The prototype structure (See Fig. 3) was referred to Khandelwal [6].

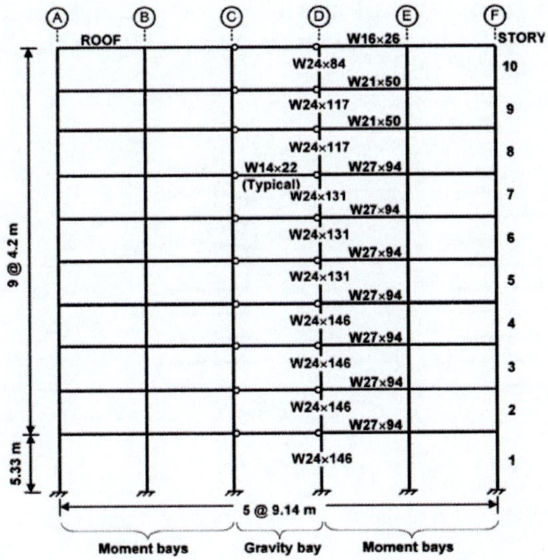

Fig. 3. The prototype structure of the 10-story steel frame

Numerical analysis of progressive collapse of the structure was conducted under column-removal scenarios by LS-DYNA. The locations of removed columns, including different floors and bays, and the number of the removed columns were considered as variables. Eight column removal cases were investigated. The column removal cases conducted were summarized in Table 1.

Table 1. Column removal cases

Column removal case	Column removed	Column type
1	9D	Gravity bay
2	7D	Gravity bay
3	3D	Gravity bay
4	3E	Moment interior
5	3F	Moment corner
6	7DE	/
7	6DE	/
8	5DE	/

3.1 Removal of Columns at Different Floors

Columns at different floors were selected in case 1, case 2, and case 3. The failure modes of the whole structures for those 3 cases were shown in Fig. 4. Other detailed information was discussed as follows.

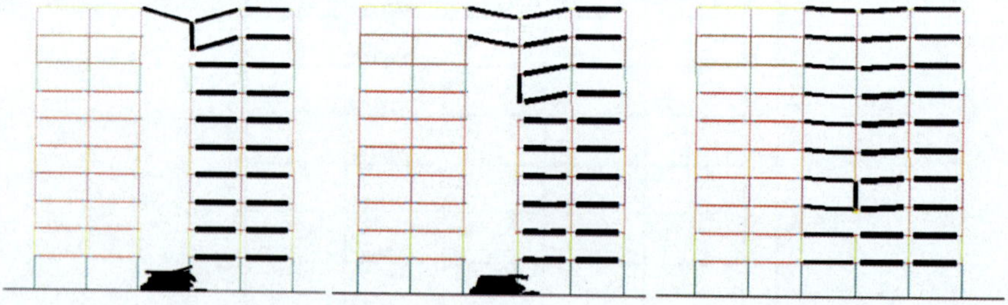

(a) case 1: 9D (b) case 2: 7D (c) case 4: 3D

Fig. 4. The failure mode of the whole structure for different column removal cases

For case 1, when column 9D was abruptly removed, the node of the corresponding column's top gradually moved downwards and finally experienced a vertical displacement of 2020 mm. The FP joints in the gravity bay of beam 9CD fractured and impact on the lower beam, causing the progressive collapse of the gravity bay except for the beam 10CD. The axial force of the remained columns on the 9th floor can be shown in Fig. 5(a). In general, the axial forces of column 9A and 9F were decreasing while the one of column 9E was increasing. Specifically, the maximum axial force occurred at column 9E, increasing from 301 kN to 524 kN, which was 1.74 times. In contrast, the minimum axial force occurred in column 9A and 9F, decreasing from 159 kN to 105 kN.

As presented in Fig. 5(b), a slightly different result was observed in case 2: removing column 7D. The axial force in column 7A was increasing while that in 7C was decreasing. The maximum axial force still appeared at column E, changing from 698 kN to 1520 kN, which was around 2.18 times. The node of the corresponding column's top ultimately reached a deflection of 1440 mm. In this case, the FP joints in the gravity bay of beam 7CD and 8CD fractured firstly and impact on the lower beam, inducing the occurrence of progressive collapse of gravity bay from floor 8 to floor 1.

It can be obtained from Fig. 5(c) that the vertical displacement at the removed column 3D ultimately reached 454 mm. The maximum axial force still appeared at column E, changing from 1480 kN to 4160 kN, which was almost 3 times. In addition, removing column 3D did not lead to the progressive collapse of the gravity bay.

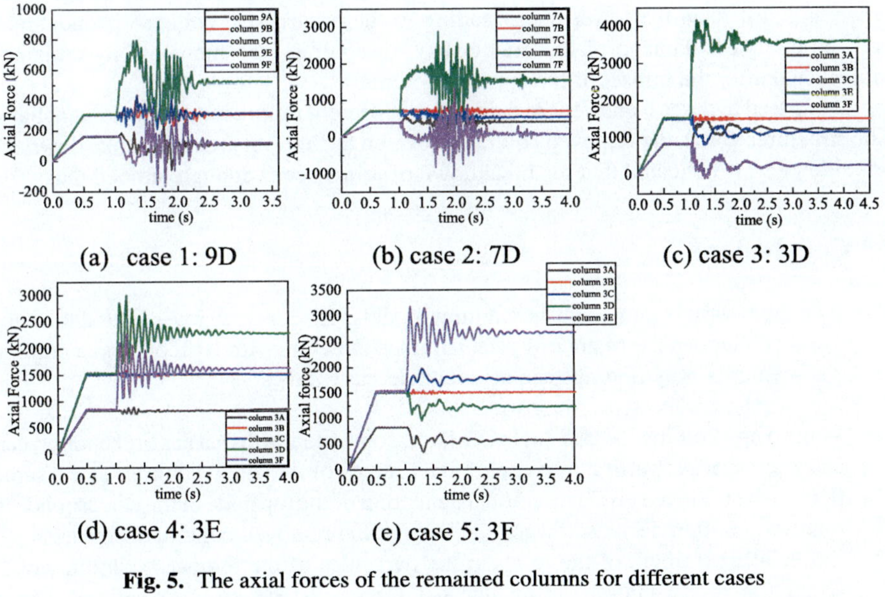

Fig. 5. The axial forces of the remained columns for different cases

3.2 Removal of Columns at Different Bays

Columns belonging to different bays were selected in case 3, case 4, and case 5. No failure of any other components occurred among all three cases.

The vertical displacements of the top of the corresponding column in case 3, case 4, and case 5 were 454 mm, 53.1 mm, and 88 mm, respectively. For case 4, the removal of 3E did not trigger any variation of the axial forces in columns belongs to the left moment bay in the third floor. In addition, the axial forces in columns adjacent to the removed column 3E, both endured a growth with a nearly same value around 770 kN. However, for case 3, the adjacent column 3C experienced a decrease in axial force while another adjacent column 3E experienced a significant increase in axial force. It indicated that the relative strength of the joints at the two sides of the removed column would affect the redistribution of the internal forces. To be more specific, a larger proportion of internal forces will transfer to the side with higher strength.

3.3 Number of Removed Columns

Based on previous simulations, it was found that removing single column can hardly generate a significant collapse on account of the redistribution of the internal forces, catenary action, and Vierendeel action. Hence, the following cases were performed.

Removal of 6DE and 7DE resulted in almost identical responses. The upper structures moved downwards immediately when the corresponding columns were removed, which finally caused the fracture of the springs of adjacent beam-column joints in the gravity bay after they experienced enough deformation. Then the beams in the upper region fell down and struck on the lower ones, leading to the progressive collapse phenomenon and the complete damage of the gravity bay. In contrast, the moment bay remained undamaged after the impact of the upper structures.

Compared to cases 6 and 7, case 8 displayed different failure modes in moment bays. The structures above the removed column stroke on the bottom area and broke down the left side joint. It indicated that the breakdown of joints needs enough lumped mass.

4 Conclusions

Based on the verified component-based joint models, this research investigated the anti-progressive collapse performance of a steel-framed structure with 10 floors under column removal scenarios. Key findings can be concluded as follows:

(1) Removing a column located at a lower floor, contrasted with that at a higher floor, can cause a larger proportion increase of axial forces of remained columns at the same floor of the removed one. The vertical deflection of the top node of the corresponding removed column decreased when the floor of the removed column decreased.
(2) The relative strength of the joints at the two sides of the removed column would affect the redistribution of the internal forces. To be more specific, a larger proportion of internal forces would transfer to the side with higher strength.
(3) Sufficient lumped mass or an adequate number of removed columns were required to trigger the disproportionated collapse of the structure.

Progressive collapse is accompanied with dynamic response of the structural members while the component-based models for joints have not considered the strain rate effect so far. Hence, dynamic behavior deserves to be further considered in the future.

References

1. Bo, Y., Kang Hai, T.: Experimental tests of different types of bolted steel beam–column joints under a central-column-removal scenario. Eng. Struct. **54**, 112–130 (2013)
2. Florea, D., Ioan, M., Dan, D.: Experimental testing and numerical modelling of steel moment-frame connections under column loss. Eng. Struct. **151**, 861–878 (2017)
3. Haitao, W., Jingsi, H., Yanzhi, L., et al.: Dynamic performance of composite beam-column connections subjected to impact loadings. J. Constr. Steel Res. **178**, 106498 (2021)
4. Hao, W., Kang Hai, T., Bo, Y.: Experimental tests of steel frames with different beam-column connections under falling debris impact. J. Struct. Eng. **146**(1), 04019183 (2020)
5. Wei, W., Junjie, W.: Progressive collapse tests and failure mechanism of large-scale steel-concrete composite floor system. J. Build. Struct. **42**(5), 82–91 (2021)
6. Khandelwal, K., El-Tawil, S., Kunnath, S.K., Lew, H.S.: Macromodel-based simulation of progressive collapse: steel frame structures. J. Struct. Eng. **134**(7), 1070–1078 (2008)
7. Sadek, F., El-Tawil, S., Lew, H.S.: Robustness of composite floor systems with shear connections: modeling, simulation, and evaluation. J. Struct. Eng. **134**(11), 1717–1725 (2008)
8. Main, J.A., Sadek, F.: Robustness of steel gravity frame systems with single-plate shear connections. NIST Technical Note 1749. National Institute of Standards and Technology, Gaithersburg (2013)
9. Honghao, L.: Modeling, behavior and design of collapse-resistant steel frame buildings. PhD Thesis, University of Michigan (2013)
10. Junjie, W., Wei, W.: Theoretical evaluation method for the progressive collapse resistance of steel frame buildings. J. Constr. Steel Res. **179**, 106576 (2021)
11. Rex, C.O., Easterling, W.S.: Behavior and modeling of a single plate bearing on a single bolt. Virginia Polytechnic Institute and State University, Blacksburg, V.A., U.S. (1996)
12. Rex, C.O., Easterling, W.S.: Behavior and modeling of a bolt bearing on a single plate. J. Struct. Eng. **129**(6), 792–800 (2003)

Analysis of Mechanical Properties of Turbogenerator Foundation Based on ANSYS

Mingyi Tang, E Wenqi(✉), and Cong Zeng

Northeast Electric Power University, Jilin 132012, China
zc_1113@126.com

Abstract. As a working platform supporting the normal operation of the generator set, the turbo-generator foundation, its dynamic characteristics are very important to the safe operation of the whole set. In this paper, a finite element model was established according to the measured size of the turbine generator foundation, and the finite element software ANSYS was used to perform a static analysis on the foundation of the turbo-generator, and the deformation and stress cloud diagrams were obtained. The results shown that the local stress concentration phenomenon occurred in the foundation, which leaded to the high tensile stress at this stress position of the foundation. Then, the modal analysis of the foundation was performed to obtain the natural vibration characteristics of the foundation. Then the time-history analysis and calculation of the foundation structure under earthquake action were carried out. The results shown that the foundation of the turbo-generator met the seismic requirements, and the earthquake action had little influence.

Keywords: Steam turbine · ANSYS · Static analysis · Modal analysis · Seismic performance · Dynamic time history analysis

1 Introduction

In the power plant project design, in addition to the design of the factory building, it also includes the design of many equipment foundations. In the design of the equipment foundation, the design of the turbine generator base is complicated due to its complex force and irregular shape. The basis of power equipment is the difficulty and key point of design [1]. The turbo-generator foundation is the core equipment foundation of the main building, and its seismic performance plays a vital role in ensuring the safety of the turbo-generator and other equipment during earthquakes [2]. The dynamic characteristics of the turbine foundation directly affect the normal operation of the unit, so it is particularly important to study the dynamic characteristics of the turbine foundation [3].

Through ANSYS modeling, people can understand the deformation and stress distribution laws of components or structure. Scholars at home and abroad attach importance to the use of ANSYS analysis results to test the accuracy of theoretical methods [4]. In this paper, the dynamic calculation and analysis of a turbo-generator foundation is carried out by the finite element ANSYS software, and the mechanical properties of the equipment foundation are tested by analyzing the simulation results.

W. Guo and K. Qian (Eds.): GBCESC 2022, LNCE 211, pp. 926–933, 2023.
https://doi.org/10.1007/978-981-19-5217-3_94

2 The Establishment of Finite Element Model

2.1 Model Overview

The research object is the foundation of a new thermal power station steam turbine generator, and a finite element model is established according to the measured size of the foundation of the steam turbine. The foundation consists of a platform, a short column and a bottom plate. The foundation platform is supported by six foundation short columns, and the foundation bottom plate is under the short columns. The size of the base plate is 1980 mm × 9400 mm × 1350 mm, the height of the column is 4550 mm, and the interface size is 1200 mm × 1200 mm. The interface size of the upper platform is 18000 mm × 8000 mm × 1800 mm. SolidWorks three-dimensional design software was used to establish a finite element model, and the finite element model is shown in Fig. 1.

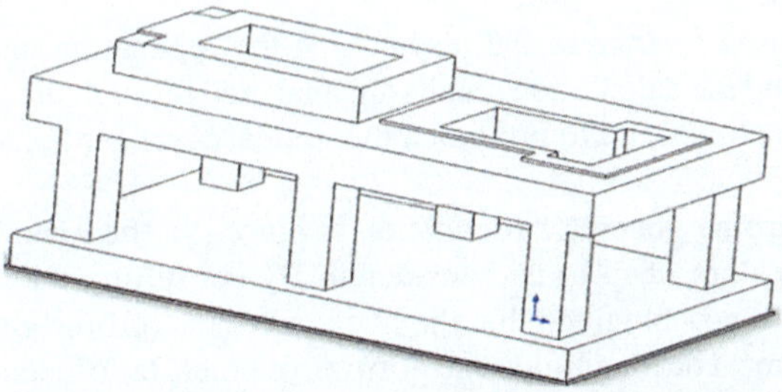

Fig. 1. Turbo generator basic finite element model

2.2 Element Selection and Material Parameter Setting

The model used solid185 solid elements, which were used to construct three-dimensional solid structures. The element was defined by 8 nodes, each with 3 degrees of freedom for translation along the XYZ directions. The element has super elasticity, stress toughening, creep, large deformation and large strain capacity. The solid185 unit is more advanced and advantageous for concrete model analysis, and the calculation convergence is also very strong. The material parameters obtained through experimental research and literature specifications are shown in Table 1.

Table 1. C40 concrete material parameter table

Material type	Modulus of elasticity (pa)	Density (kg/m^3)	Poisson's ratio
C40 concrete	C40 concrete	2500	0.2

2.3 Meshing and Setting of Boundary Conditions

Import the 3D model for analysis created by SolidWorks software into Hyper Mesh, and set material properties in it. Then meshing was carried out, and the meshing adopted the Hex Dominant (hexahedron) method.

The bottom plate of the turbine generator foundation is buried deep in the soil and connected to the pile foundation, so the foundation stability is good. Both the soil and the pile foundation have a constraining effect on the turbo-generator foundation. Therefore, when the basic model was established, the bottom of the foundation was set to be fully constrained in three directions of XYZ, and the automatic degrees of all three directions of XYZ were restricted.

3 Static Analysis of Turbo Generator Foundation

3.1 Static Load Loading Arrangement

Static analysis is used to analyze the response of the foundation under the action of a given static load, and the calculation results are analyzed through the displacement cloud map and the stress cloud map to provide a theoretical basis for the safety performance of the foundation.

Because the turbine generator foundation is placed in the main building, the wind and snow load does not need to be considered. By consulting the design rules of the main powerhouse, it was obtained that the live load of the turbine generator foundation platform is 15 KN/m^2.The magnitude and position of other static loads were set according to the foundation load diagram. The base static loading position of the turbo-generator is shown in Fig. 2.

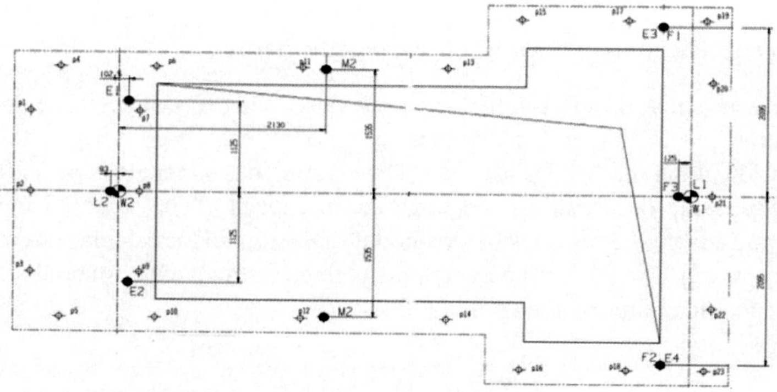

Fig. 2. Turbine foundation load layout

The generator load layout diagram is shown in Fig. 3. The static loading point, size and direction are shown in Table 2.

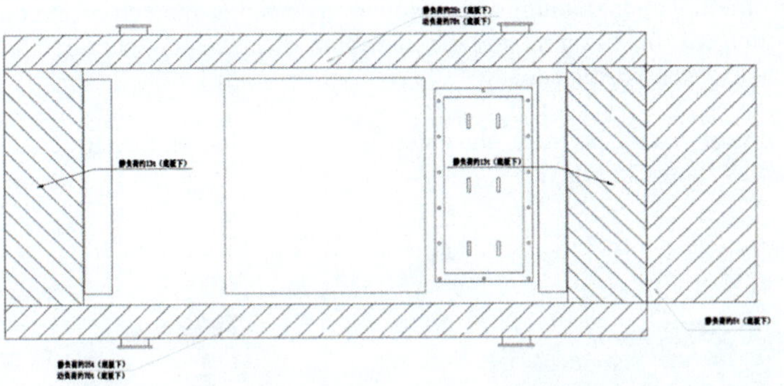

Fig. 3. Generator foundation load layout

Table 2. Loading data of steam turbine foundation

Load	Rotor	Cat paw	Axle-bearing	Guide key	Chassis load	Valve
Point of action	W1/W2	E1 E2/E3 E4	L1/L1	F1/F2	P1–P23	M2
Size (KN)	304/270	791/1776	207/165	339/339	15	75
Direction	−Fy/−Fy	−Fy/−Fy	−Fy/−Fy	−Fx/−Fx	−Fy	−Fy

3.2 Static Cloud Map Analysis

Displacement Contour Analysis. After the calculation and solution, the displacement cloud diagram was obtained as shown in Fig. 4.

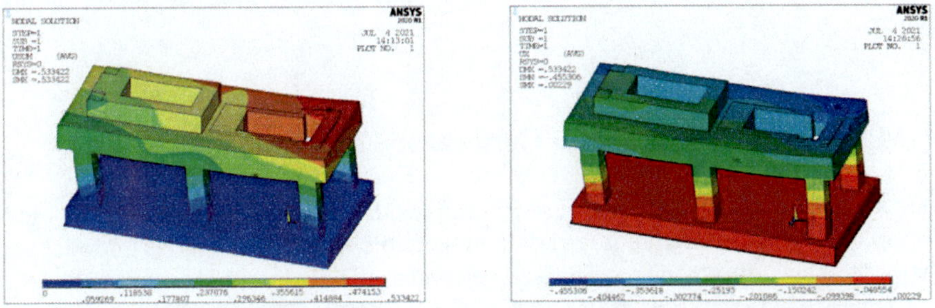

(a)Maximum displacement contour (b)Displacement contour in X direction

Fig. 4. Static analysis displacement contour

The following conclusions can be drawn from the comparative analysis of the displacement nephograms:

Through the cloud map of the overall displacement of the turbo-generator foundation, it could be seen that the overall displacement of the turbo-generator foundation is relatively small. The maximum displacement occurred at the end of the turbine foundation, which was 0.5334 mm, and the minimum displacement occurred at the bottom plate, with almost no displacement.

Stress Contour Analysis. After the calculation and solution, the stress cloud diagram is obtained as shown in Fig. 5.

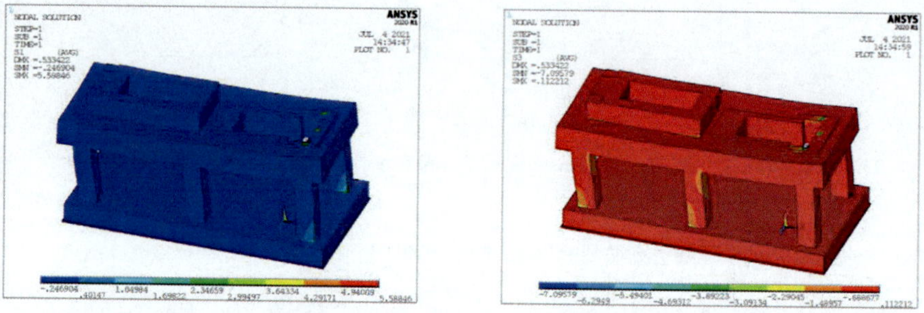

(a)First principal stress cloud map (b)Third principal stress cloud map

Fig. 5. Static analysis stress contour

The following conclusions could be drawn from the comparative analysis of the stress nephograms:

Through the analysis of the stress nephogram, the maximum stress generated by the foundation was generated at the end of the turbine generator foundation, the action points were E2, F4, the maximum tensile stress was 5.5884 Mpa, the maximum compressive stress was 7.0957 Mpa, and the tensile stress at the root of the foundation short column was also reached 1.0498 Mpa. The stress distribution was uneven and stress concentration occurred.

4 Modal Analysis of Turbo Generator Foundation

The modal analysis method can be used to find out the characteristics of the main modes of each order of the structure in a certain susceptible frequency range [5]. Modal analysis is the basis of dynamic analysis. The modal of the structure can directly reflect the stiffness of the structure itself and the characteristics of the structural arrangement, which is the basis of the seismic analysis of the structure [6]. In order to avoid resonance, the natural frequency of the foundation should avoid the disturbance frequency of the machine, so the vibration characteristics (natural frequency and mode shape) of the structure should be determined by modal analysis. In this simulation, the first 6-order modes were extracted by the Lanzos method, and the basic natural frequency and formation of the turbo-generator were obtained. The frequencies and periods are shown in Table 3.

Table 3. Turbine power generation and basic natural frequency and period

Formation	1 order	2 order	3 order	4 order	5 order	6 order
Natural frequency (HZ)	9.1393	9.4771	11.822	34.745	40.281	40.368
Natural period (s)	0.1094	0.1055	0.0846	0.0288	0.0248	0.0247

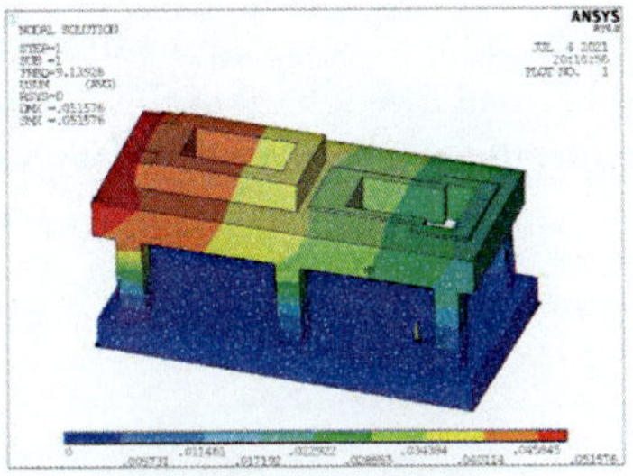

(a)first mode shape

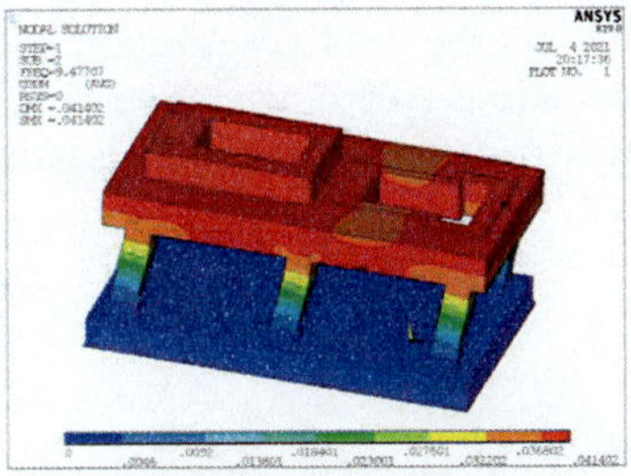

(b)second mode shape

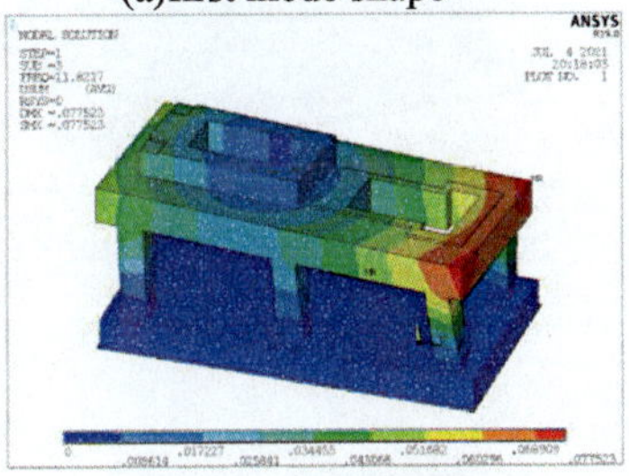

(c)third mode shape

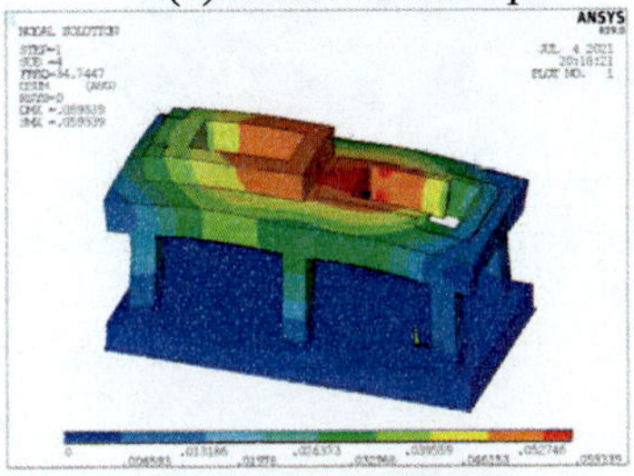

(d)fourth mode shape

Fig. 6. Foundation first four formations

For base structure frequency and modal shape analysis (Fig. 6):

The first three orders of the mode were analyzed. The first order mode swung left and right along the horizontal direction, the second order mode vibrated along the longitudinal direction of the foundation, the deformation gradually increased from the bottom plate to the foundation platform, and the third order mode shape was torsion of the foundation structure deformed. After the third order, the frequency increased and was not considered.

5 Earthquake Time History Analysis

The response spectrum analysis method can only approximate the maximum value of the dynamic response of the structure [7]. However, the time history analysis method is a relatively sophisticated method, which can not only consider the internal force redistribution after the structure enters plasticity, but also record the entire process of the structural response, and the calculation results are closer to the real dynamic response of the structure [8].

In this study, the El-Centro wave of magnitude 8 commonly used in seismic analysis of structures was used as the input [9]. Since the vertical seismic load has little influence on the dynamic response of the entire foundation structure, in general, only the horizontal seismic action was considered, so the seismic action was calculated separately in the two horizontal directions of the foundation. They correspond to the X direction (horizontal) and Z direction (horizontal) of the overall coordinate system of the ANSYS interface.

According to the "Seismic Code", the peak value of the frequent and rare earthquakes of magnitude 8 earthquakes was adjusted, and the peak value of the El-Centro wave was adjusted, and the peak acceleration was 0.7 m/s^2 and 4 m/s^2 respectively. Structural damping ratio is 0.05. Then the seismic wave was input through the ANSYS command flow mode to determine the spectral value. The specific cloud map was shown in Fig. 7.

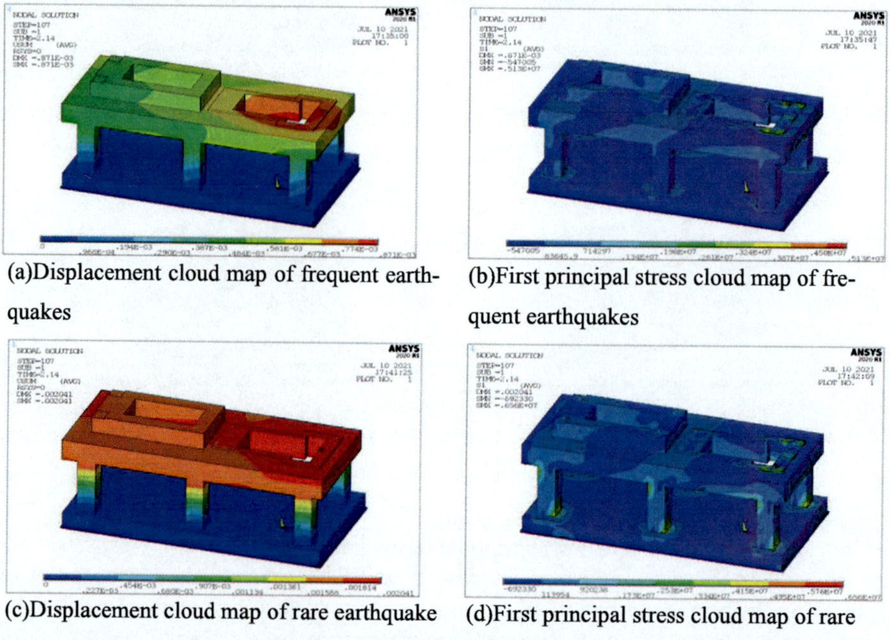

(a)Displacement cloud map of frequent earthquakes

(b)First principal stress cloud map of frequent earthquakes

(c)Displacement cloud map of rare earthquake

(d)First principal stress cloud map of rare earthquake

Fig. 7. Seismic analysis cloud map

The seismic analysis of the foundation structure was carried out through the above figure: Under the action of horizontal frequent earthquakes and rare earthquakes, the frequent earthquakes had little impact on the foundation of the turbo-generator. The maximum displacement, tensile stress and compressive stress all occurred at the end of the turbine foundation, and their sizes were 2.1050 mm, 6.8100 Mpa, and 12.4000 Mpa, respectively.

6 Conclusion

In this paper, the following conclusions can be drawn from the static analysis, modal analysis and time history analysis of the turbo-generator foundation under static load through ANSYS:

Under the action of static load, the loads at the action points E3, E4 and W1 at the end of the steam turbine foundation were relatively large, and the maximum deformation and stress of the foundation occurred here. It could be seen from the stress cloud diagram that stress concentration occurred here, resulting in the tensile stress at the E3 action point reaching 5.5884 Mpa, and the tensile stress at the root of the foundation short column here also reaching 1.0498 Mpa.

The modal analysis method extracts the fundamental mode shape and frequency, which could effectively avoid structural resonance and optimize structural characteristics. Through modal analysis, the basic natural frequency and mode shape of the turbo-generator were obtained, the basic natural frequency was 9.1393 Hz, and the period was 0.1094 s.

The basic structure could intuitively study the stress distribution and displacement deformation through the time history analysis method. Through the cloud map analysis, it could be seen that the maximum displacement and stress occurred at the end of the turbine foundation, and certain attention should be given to the design.

References

1. Chongshu, G.: Brief analysis of turbogenerator foundation design. Power Station Syst. Eng. **36**(01), 57–59 (2020)
2. Baoqi, W., Hua, M., Xiyuan, Z., Zebing, D.: Experimental research on seismic performance of large turbogenerator base structure. Electr. Power Constr. **08**, 13–17 (2008)
3. Hongmei, Y., Rui, L., Yinfang, L.: Basic dynamics analysis of turbogenerator based on ANSYS. Shanxi Archit. **45**(04), 79–80 (2019)
4. Jialai, Z., Xianzhun, L.: ANSYS finite element static analysis under the load of cantilever I-beam. Shanxi Archit. **44**(06), 41–42 (2018)
5. Hongye, W., et al.: Research on the seismic performance of ancient city walls based on ANSYS. J. Hebei Inst. Archit. Eng. **38**(01), 30–35 (2020)
6. Jiufeng, Z.: Analysis of ferris wheel modal and earthquake response spectrum based on ANSYS. Mod. Mach. **06**, 65–68 (2018)
7. Lin, W.H., de la Llera, C.: Accidental torsion in buildings: Analysis versus earthquake motions. J. Struct. Eng. **127**(5), 475–481 (2001)
8. Kan, C.L., Chopra, A.K.: Tortional coupling and earthquake response of simple elastic and inelastic systems. J. Struct. Div. **107**(4), 1569–1588 (1981)
9. Yong, D., Lanfeng, Y., Zeping, X., Jiangping, X.: Dynamic response analysis of boom-type tower crane structure under earthquake excitation. Mod. Manuf. Eng. (01), 75–80 (2018)

Mechanical Properties of Spray-Based 3D Printed Micro-cable Reinforced Concrete

Ji Zhou, Guanyu Hou, Xiongfei Liu(✉), Qi Li, Li Wang, and Jixiang Li

School of Civil Engineering and Transportation, Hebei University of Technology, Tianjin 300401, China
liuxfking@foxmail.com

Abstract. Spray-based 3D (S-3D) printing is an intelligent architectural technology by spraying concrete layer by layer, which can solve the problem of collaborative construction between 3D printing and reinforcement. This paper studies the feasibility of S-3D printed concrete with reinforcement. Effects of micro-cables diameters with 0.6, 0.8 and 1.0 mm on the flexural properties of S-3D printed micro-cable reinforced beams are considered. In addition, the interfacial bonding properties between micro-cable and concrete is observed by scanning electron microscope (SEM). The results showed that the ultimate flexural strength of S-3D printed micro-cable reinforced beam is improved by 43.4%–114.1% with the ductility improved by 58.4%–315.7%, respectively, which benefits the firmly bonded interface between micro-cable and concrete. This paper provides an effective method to improve the cooperation between 3D printing and reinforcement.

Keywords: Spray-based 3D printing · Micro-cable reinforced concrete · Flexural properties

1 Introduction

In light of the features of model-free, flexibility and intelligence [1–3] of 3D printing concrete technology, it has been widely used in the engineering construction [4, 5]. Nevertheless, there are still some limitations for 3D printing, such as low interface bonding strength [6, 7], unable to print in multi-angle directions [8, 9]. In addition, in the process of 3D printed concrete with reinforcement, the space limitation existed in the pre-positioned steel cage [10, 11] greatly restricts the engineering application of 3D printing in large-scale structures.

Spray-based 3D (S-3D) printing concrete technology [19] has the characteristics of high-speed wind pressure assistance, flexible printing in horizontal, vertical and top directions [12]. The inter-layer strength of S-3D printing concrete is higher than that of extruded 3D printing concrete [13]. In addition, the angle between nozzle and jet surface is not limited, which can build complex special-shaped reinforced structures [14]. Although, 3D printed reinforced concrete has been widely studied [15–17], which shows

Ji Zhou has passed away before the publication of this book.

W. Guo and K. Qian (Eds.): GBCESC 2022, LNCE 211, pp. 934–942, 2023.
https://doi.org/10.1007/978-981-19-5217-3_95

obviously inconvenience in construction. It can be seen that compared with extrusion-based 3D printing technology, S-3D printing technology can cooperated better with steel cage, and can effectively improve the efficiency and speed to build reinforcement concrete. Researches on S-3D printing technology mainly focused on materials [12, 18, 19] and printing parameters [13]. At present, the research on the integration of enhancers in S-3D printing is limited. Studies have shown that S-3D printing has better adhesion between concrete and reinforcement [14]. In order to prevent the problem of weak adhesion between concrete and reinforcement. The distance between the nozzle and the reinforcement and the included angle with the vertical direction should be adjusted appropriately [20]. Although problems such as space collision and complex structure are still difficult to overcome, it can be improved by using S-3D printing technology.

This paper studies the feasibility of S-3D printed concrete with reinforcement, which expands the application of S-3D printing in engineering fields. Effects of micro-cables diameters with 0.6, 0.8 and 1.0 mm on the flexural properties of S-3D printed micro-cable reinforced beams are considered. In addition, the interface bonding properties between micro-cable and concrete is observed by scanning electron microscope (SEM). Therefore, S-3D printing technology provides an effectively method to realize 3D printed reinforcement concrete structure.

2 Materials

2.1 S-3D Printing System

The S-3D printing system consists of a 6-axis robotic arm, a spray gun, a material feeder, an air compressor, as shown in Fig. 1 (a). The S-3D printing process starts with putting the prepared concrete into the feeder, it is then pumped through the material pipe to the spray gun nozzle. The air compressor provides air pressure of 70 kPa at the nozzle to make the material sprayed out with a high speed. The printing path is designed by the SprutCAM15 software, and controlled by the robotic arm. The printing range is 2 m (X) × 2 m (Y) × 1.5 m (Z), and the moving speed of the robotic arm is 30 mm/s.

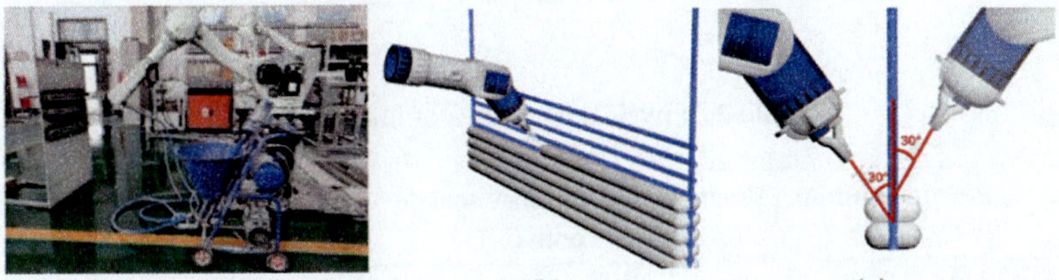

(a) (b) (c)

Fig. 1. (a) S-D printing system; (b) process of S-3D printed micro-cable reinforced concrete; (c) S-3D printing angle.

2.2 Raw Materials

The S-3D printing concrete consists of ordinary Portland cement (P·O 42.5), silica fume, I-grade fly ash and quartz sand. The particle size distribution of the material is shown in Fig. 2 with the chemical composition shown in Table 1. Appropriate amount of silica fume and fly ash are added to reduce the dynamic yield stress and increase the static yield stress to meet the rheological properties of S-3D printable concrete. At the same time, an appropriate amount of polycarboxylate is added to reduce the water-cement ratio. The water reducing rate of polycarboxylate superplasticizer is 18–25%.

The micro-cable is SUS304 type with diameters of 0.6, 0.8, and 1 mm are used, and the details are shown in Table 2.

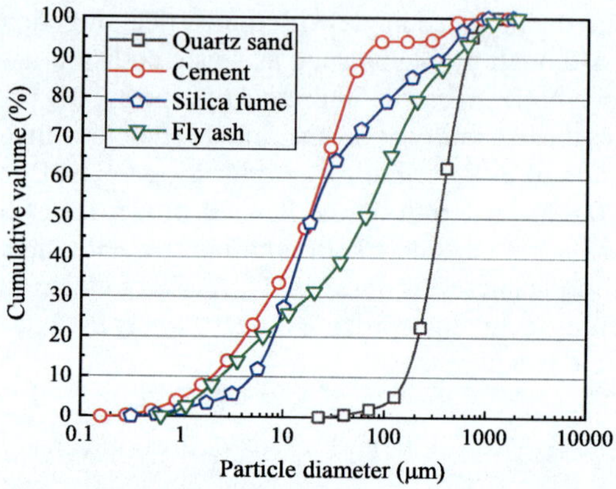

Fig. 2. Particle size distribution of raw materials

Table 1. Chemical composition of raw materials

Components	Al_2O_3	SiO_2	CaO	Na_2O	MgO	K_2O	Fe_2O_3	TiO_2	LOI
Cement	18.90	63.80	1.33	3.61	1.57	4.92	4.85	0.33	3.50
Silica Fume	0.79	91.0	1.40	0.97	2.37	1.13	1.13	–	4.14
Fly ash	19.10	61.40	1.49	1.80	4.67	6.64	6.64	0.65	4.12

Table 2. Physical properties of micro-cable

Grade	Elastic modulus (GPa)	Fracture elongation	Ultimate tensile force (N)	No. of strands	Diameter (mm)
SUS304	192	40%	880	7 × 7	0.6/0.8/1

2.3 Material Preparation

Preparation of Micro-cable Mesh. As is shown in Fig. 1(b), the two ends of the micro-cable are fixed on two vertical columns. Then micro-cables are arranged at intervals of 7 mm along the height direction with pre-stress applied at both ends of the cables to ensure the micro-cables horizontal. Types of micro-cable include D0.6, D0.8 and D1.0.

S-3D Printed Micro-cable Reinforced Concrete. Cement, quartz sand, fly ash and silica fume are put into the mixer and stirred at low speed for 3 min to obtain a well-mixed dry powder mixture. Water and water reducer are mixed for 30 s and then poured into the mixer. After that, the mixture is stirred at low speed for 1.5 min and then stirred at high speed for 3 min. Finally, the well-mixed concrete is added into the S-3D printing system. The concrete mix ratio is shown in Table 3.

Table 3. S-3D printed concrete

OPC	Fly ash	Silica fume	W/B	Quartz sand	HRWRA
0.90	0.05	0.08	0.35	0.60	0.30

During the process of S-3D printing, in order to ensure each layer of printed concrete can be stably laid with micro-cable, the printing thickness of concrete ought to be regulated in advance. Each printed concrete layer thickness is designed as 7 mm with the micro-cables spaced 7 mm apart. The left and right sides of the micro-cable mesh are alternately performed, and angle between the jet axis and the plane of the micro-cable mesh is set as 30°,as is shown in Fig. 1(b) and (c). D0.6, D0.8 and D1.0 are used to represent the S-3D printed micro-cable reinforced beams, and D0 is the beam without micro-cable. After the sample is printed, the mold is demolded after curing at room temperature for 1 d, and then placed in a curing room with a temperature of 20 ± 5 °C and a relative humidity of 95 ± 5% for 28 d.

2.4 Test Methods

A 430 mm × 40 mm × 35 mm beam is cut from the S-3D printed specimen for four-point bending test. The four-point bending test adopts the MTS servo hydraulic universal testing machine with a range of 500 kN and the loading rate of 0.1 mm/min. After the test, the fractured part of the sample is taken to test the bonding morphology of the interface between the micro-cable and the concrete by SEM.

3 Results and Discussion

3.1 Fractured Morphology of S-3D Printed Micro-cable Concrete Beam

It can be seen from Fig. 3 that the micro-cable changes the fractured morphology of S-3D printed micro-cable reinforced concrete beam. D0 and D0.6 beams are broken at the center position, which is the same as the traditional concrete beam with the brittle performance. Under the action of four-point bending test, the crack of the printed D0.8 beam extends from the right support to the upper fulcrum, which is typical shear pressure damage. The fracture cracks of the printed D1.0 beam are at the left support, and there is no damage in the pure bending section of the beam, showing a typical cable-tension damage.

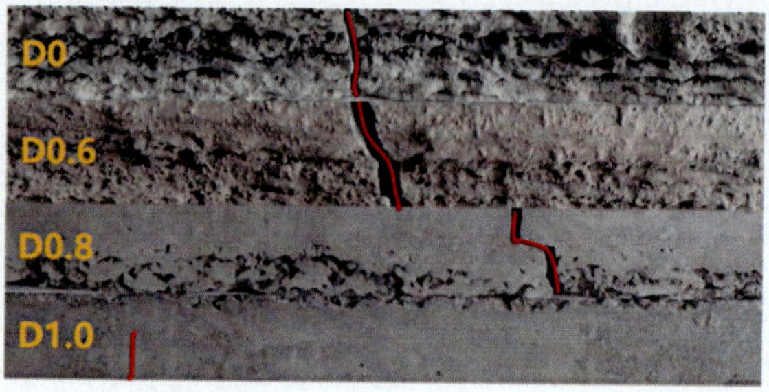

Fig. 3. Failure morphology of beam by spray-based 3D printing

3.2 Mechanical Properties of S-3D Printed Micro-cable Concrete Beam

The four point bending load-displacement relationship of S-3D printed micro-cable concrete beam is shown in Fig. 4.

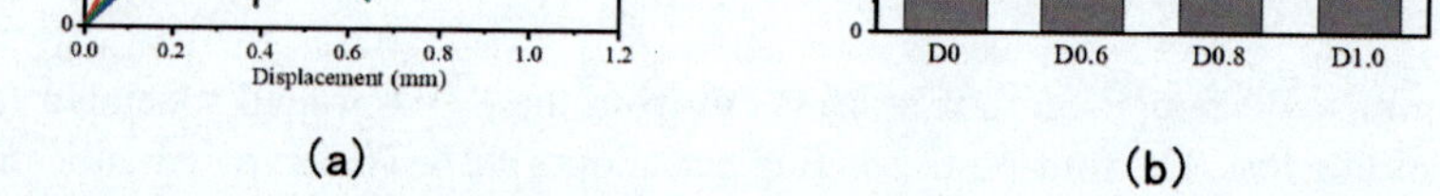

Fig. 4. (a) Curves of flexural load to displacement; (b) Ultimate flexural strength of S-3D printed micro-cable concrete beam.

As shown in Fig. 4, micro-cable can significantly improve the flexural strength and toughness of the S-3D printed concrete. Although the failure modes are all brittle fracture, the ductility is significantly improved after the addition of micro-cable. The ductility of D0.8 maximum increases by 152.3% compared to that of D0.

As is shown in Fig. 4 (b), the ultimate flexural strength of S-3D printed D0 beam is 2.5 MPa, and the ultimate flexural strength of S-3D printed micro-cable concrete beams with different diameters is 3.59 MPa, 4.74 MPa and 5.35 MPa, which respect increased by 43.4%, 89.7% and 114.1% than that of D0. The interface between S-3D printed concrete and the micro-cable shows excellent wrapping, and the reinforcement has not been pulled out. When the sample breaks, the micro-cable and the concrete break almost at the same time, showing a typical brittle fracture. The improvement of reinforcement diameter keeps pace with the improvement of bending resistance. Therefore, although the micro-cable plays a significant role in resisting load before fracture. Compared with D0, the load of D0.6 rises sharply and drops instantly at the peak point, losing the carrying capacity, which may be due to the better wrapping property of the small diameter micro-cable and concrete, and the synchronous bearing tension. Compared with D0.8 and D1.0 diameter, the peak point appeared in the curve rising process, indicating that the micro-based plays a role in resisting stress in the damaging process of concrete under load. The deflection of D0.8 increased by 143.5% and the deformation ability is the optimum one. Combined with the failure topography in Fig. 3, D0.8 conforms to the damage characteristics of shear compression damage.

3.3 Fracture Energy of S-3D Printed Concrete

Micro-cable mesh greatly increases the ductility of the S-3D printed sample. The energy required for crack propagation per unit area of the printed beam before the sample breaks or does not bear force is measured by fracture energy w. The fracture energy w is calculated by Eq. (1):

$$w = \int_0^{\varepsilon} \sigma(\varepsilon) d\varepsilon \tag{1}$$

where, ε is the strain, σ indicates the stress corresponding to strain ε. The higher the tensile strength and ductility of the material shows the greater fracture energy.

As is shown in Fig. 5, the energy absorption of printed beams in each group are different. As the diameter of micro-cable increases, the flexural performance of samples improves. Compared with D0, the fracture energy of samples D0.6 and D0.8 increases by 58.4% and 315.7% respectively. When the reinforcement ratio is too high, the fracture mode of the sample changes; when the diameter of micro-cable increases to 1 mm, the fracture energy w decreases by 24.8% compared with that of D0.8.

The flexural toughness of S-3D printed beam is improved by micro-cable. The flexural strength of D1.0 is higher than that of D0.8, but the increase of ductility is not as large as that of D0.8, which indicates that there is an optimal point for the diameter of micro-cable. When the diameter of micro-cables increases, the fracture toughness and ductility are not significantly improved [11]. 0.8 mm diameter of micro-cable has the best effect on ductility. When the diameter is less than 0.8 mm, it will cause brittle fracture due to less reinforcement. While, if the steel cable diameter up to 1 mm, on the surface of the steel back to spray insufficient part gelling material extrusion dense and causes part of micro-cable facing away from the jet port, resulting in rapid destruction of the sample.

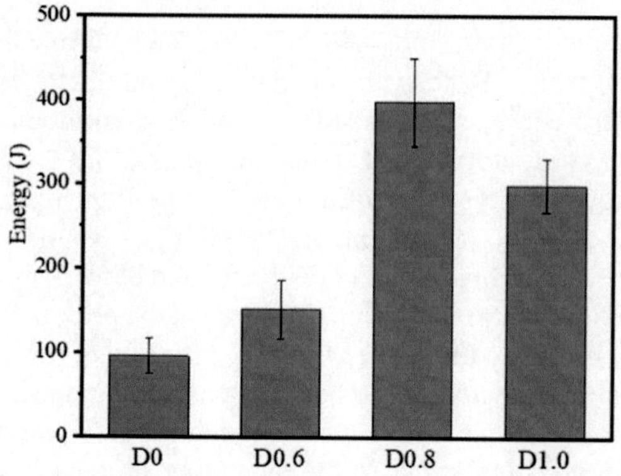

Fig. 5. The fracture energy of the S-3D printed concrete

3.4 Microscopic Properties

Figure 6 shows the SEM morphology of the fracture site at different magnifications. The section is polished before the test, and it is observed that there are no obvious bubbles and boundary segmentation between the micro-cable and the cementitious material. Meanwhile, the groove formed by the micro-cable being pulled out after being stressed is not found in the SEM test. There is no obvious damage phenomenon between micro-cable and concrete interface. The multi-angle and multi-direction printing method of spray-based 3D printing effectively wraps the micro-cable. The close combination of the cementitious material and micro-cable can improve the shear force. When subjected to external load, the shear force improvement brought by the dense connection between micro-cable and concrete is one of the main reasons to improve the flexural strength. Meanwhile, the micro-cable is not pulled out during loading, which plays a role in bridging the micro-cracks of concrete, improving the ductility of the sample.

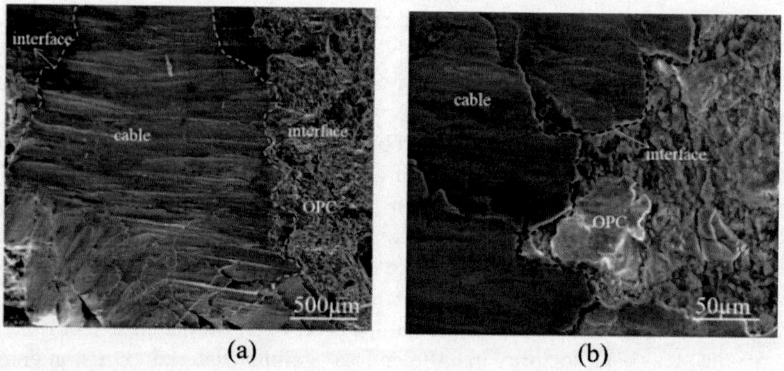

Fig. 6. SEM photos of micro-cable concrete with different multiples. (a) 50 times; (b) 500 times.

4 Conclusion

In this article, the mechanical properties of S-3D printed micro-cable reinforced concrete are comprehensively studied with the interface micromorphology analyzed by SEM. The main conclusions are:

- S-3D printing can effectively solve the problem of co-arrangement of 3D printing and reinforcement. S-3D printed reinforcement concrete shows excellent flexibility and strength performance.
- 0.8 mm micro-cable has good ductility in S-3D printed reinforcement concrete, and the fracture energy of D0.8 is 315.7% higher than that of D0.
- The interface between micro-cable and concrete is dense, which plays a role in bridging micro-cracks of concrete, thus improves the ductility and flexural properties of S-3D printed reinforcement concrete.

Acknowledgement. The authors would like to acknowledge the financial supports from the National Natural Science Foundation of China (Nos. 51908182 and 52078181), Natural Science Foundation of Hebei (No. E2020202043, and E2022202041), National Key Laboratory Foundation of Science and Technology on Materials under Shock and Impact (No. 6142902200304).

References

1. Flavio, C., Shadi, N., Helean, B.: An automated system for 3D printing functionally graded concrete-based materials. Addit. Manuf. **33**, 101146 (2020)
2. Leo, D., Nicolas, D., Romain, M.: Flow-based pultrusion of continuous fibers for cement-based composite material and additive manufacturing: rheological and technological requirements. Comp. Struct. **262**, 113564 (2021)
3. Li, L.G., Xiao, B.F., Fang, Z.Q.: Feasibility of glass/basalt fiber reinforced seawater coral sand mortar for 3D printing. Add. Manuf. **37**, 101684 (2021)
4. Gieljan, V., Wouter, D., Emad, S.: 3D printing of a post-tensioned concrete girder designed by topology optimization. Autom. Constr. **112**, 103084 (2020)
5. Eric, L.K., Megan, A.K., Michael, P.C.: Development of the construction processes for reinforced additively constructed concrete. Add. Manuf. **28**, 39–49 (2019)
6. Yiwei, E., Mingyang, L., Dong, Z.: Investigation of interlayer adhesion of 3D printable cementitious material from the aspect of printing process. Cem. Concr. Res. **143**, 106386 (2021)
7. Xuanting, L., Bohua, S.: The influence of interface on the structural stability in 3D concrete printing processes. Addit. Manuf. **48**, 102456 (2021)
8. Shravan, M., Sayanthan, R., Jay, S.: Technologies for improving buildability in 3D concrete printing. Cement Concr. Compos. **122**, 104144 (2021)
9. Bos, F.P., Kruger, P., Lucas, S.S.: Juxtaposing fresh material characterisation methods for buildability assessment of 3D printable cementitious mortars. Cement Concr. Compos. **120**, 104024 (2021)
10. Viktor, M., Richard, B., Harald, K.: Integrating reinforcement in digital fabrication with concrete: a review and classification framework. Cement Concr. Compos. **119**, 103964 (2021)

11. Jian Hui, L., Biranchi, P., Quang-Cuong, P.: Improving flexural characteristics of 3D printed geopolymer composites with in-process steel cable reinforcement. Constr. Build. Mater. **178**, 32–41 (2018)
12. Bing, L., Weiping, Z., Yiwei, W.: Study of MgO-activated slag as a cementless material for sustainable spray-based 3D printing. J. Clean. Prod. **258**, 120671 (2020)
13. Harald, K., Hans-Werner, K., Norman, H.: Influence of process parameters on the interlayer bond strength of concrete elements additive manufactured by Shotcrete 3D printing (SC3DP). Cem. Concr. Res. **134**, 106078 (2020)
14. Inka, D., Niklas, F., Dirk, L.: The effect of accelerator dosage on fresh concrete properties and on interlayer strength in shotcrete 3D printing. Materials **13**(2), 374 (2020)
15. Freund, N., Dressler, I., Lowke, D.: Studying the bond properties of vertical integrated short reinforcement in the shotcrete 3D printing process. In: Bos, F.P., Lucas, S.S., Wolfs, R.J.M., Salet, T.A.M. (eds.) DC 2020. RB, vol. 28, pp. 612–621. Springer, Cham (2020). https://doi.org/10.1007/978-3-030-49916-7_62
16. Hack, N., Lindemann, H., Kloft, H.: Adaptive modular spatial Structures for Shotcrete 3D printing. In: Intelligent and Informed-Proceedings of the 24th CAADRIA Conference, vol. 2, pp. 363–372 (2019)
17. Biranchi, P., Suvash Chandra, P., Ming, J.T.: Anisotropic mechanical performance of 3D printed fiber reinforced sustainable construction material. Mater. Lett. **209**, 146–149 (2017)
18. Bing, L., Ye, Q., Mingyang, L.: Designing spray-based 3D printable cementitious materials with fly ash cenosphere and air entraining agent. Constr. Build Mater. **211**, 1073–1084 (2019)
19. Xiongfei, L., Qi, L., Jixiang, L.: Shrinkage and mechanical properties optimization of spray-based 3D printed concrete by PVA fiber. Mater. Lett. **319**, 132253 (2022)
20. Taylor, M., Jay, S.: Mesh reinforcing method for 3D concrete printing. Autom. Constr. **109**, 102992 (2020)

The Distributed Hybrid Test and Validation of Multi-substructural Space Steel-Braced Frame Models Based on OpenFresco

Tengfei Li(✉)

School of Civil Engineering, Xi'an University of Architecture and Technology, Xi'an 710055, China
litengfei0217@163.com

Abstract. The distributed hybrid test (DHT) is a new structural seismic test method, which is created to accommodate the increasing size and complexity of the test structure. Based on the OpenFresco platform, this paper introduces the principle and system components of the DHT for multi-substructures, and describes the modeling process, parameter selection and test system network construction process of the DHT model in the finite element analysis software OpenSees Navigator. A multi-story space steel-braced frame model is used as the research object to conduct the DHT, and the good loading accuracy and communication performance of the test system are proved. OpenSees Navigator was used to establish pure numerical models of the global structure respectively, and the analytical results of both were compared with the results of DHTs to prove the applicability and accuracy of the proposed test method.

Keywords: Distributed hybrid test · Multi-substructures · Space steel-braced frame

1 Introduction

The conventional seismic test methods mainly include the cyclic loading test, seismic simulated shaking table test, pseudo-dynamic test, and substructure pseudo-dynamic test [1]. The cyclic loading test and the pseudo-dynamic test use the method of static loading to load the specimen, which is a static test. The seismic simulated shaking table test is to apply seismic excitation or simulated seismic excitation to the structural model to study the seismic response of the structure, so it belongs to the dynamic test. With the development of high-tech industry on the Internet, computer technology has a huge role in promoting the study of numerical simulation and physical test, and the distributed hybrid test (DHT) methods proposed based on the substructure pseudo-dynamic test have also emerged one after another and gradually been studied and applied by the majority of researchers.

In the DHT, a global structure model is divided into several substructures according to the complexity of the forces, where the complex structure parts are placed in different

W. Guo and K. Qian (Eds.): GBCESC 2022, LNCE 211, pp. 943–950, 2023.
https://doi.org/10.1007/978-981-19-5217-3_96

laboratories as the test substructures and the rest of the model is placed in the finite element program as a numerical model, and the DHT platform (e.g. OpenFresco [2]) is used as a transfer station between the different substructures to complete data transfer and collaborative work to obtain the global structure's dynamic response during earthquakes. The Network for Earthquake Engineering Simulation (NEES) has been pioneered to conduct the DHT research in the USA [3]. In Europe, a dedicated fiber network was used to connect a geotechnical centrifuge and a structural dynamic load equipment for communication between different systems [4]. Bousias et al. investigated the impact of remote host distance on the accuracy and performance of a hybrid simulation of a realistic three-span reinforced concrete bridge case between European and North American partners [5]. In Japan, the world's largest earthquake simulation shaking table E-Defense was established and actively networked with NEES to conduct experimental research [6]. In Korea, a research project called KOrean Construction Engineering Development (KOCED) was initiated to establish a virtual structural laboratory using web-based technology [7]. The National Center for Research on Earthquake Engineering (NCREE) in Chinese Taipei developed the Internet-based Simulation for Earthquake Engineering (ISEE) and conducted several experimental research on DHTs [8]. In China Mainland, a research project called Modern Collaborative Test Method and System for Dynamic Simulation of Structures was carried out [9].

At present, most DHT studies mainly select single-degree-of-freedom members such as columns and piers as test substructures, and the research on DHT methods for large and complex structures is still insufficient. Based on the OpenFresco test platform, this paper introduces the principle and system composition of DHT for multiple substructures, and describes the modeling process, parameter selection and test system network construction process of DHT model in OpenSees Navigator. A multi-story steel-braced space frame model is used as the research object to conduct DHTs to verify the loading and communication performance of the test system. Finally, OpenSees Navigator is used to build a full-structure pure numerical model for simulations to verify the applicability of the DHT by comparing the simulation results with the DHT results.

2 The DHT Model

2.1 A DHT System

In this study, a DHT system is established as shown in Fig. 1, which mainly includes OpenSees Navigator, a general finite element Software for DHT model buildings and finite element analysis, OpenFresco, a network test platform for data communication, a local MTS test loading system and a remote MTS test loading system. Among them, communication between OpenFresco and the remote test system is required via Internet. A typical DHT dynamic equation can be expressed as Eq. (1). During the test, the DHT model is calculated by OpenSees Navigator to obtained the target displacement of the test substructures, and the displacement commands are sent to the local and remote MTS test loading system through the OpenFresco, respectively.

$$(M_{\mathrm{n}}\ddot{X}_i + C_{\mathrm{n}}\dot{X}_i + K_{\mathrm{n}}X_i) + (M_{\mathrm{t}}\ddot{X}_i + C_{\mathrm{t}}\dot{X}_i + K_{\mathrm{t}}X_i) = F_i \tag{1}$$

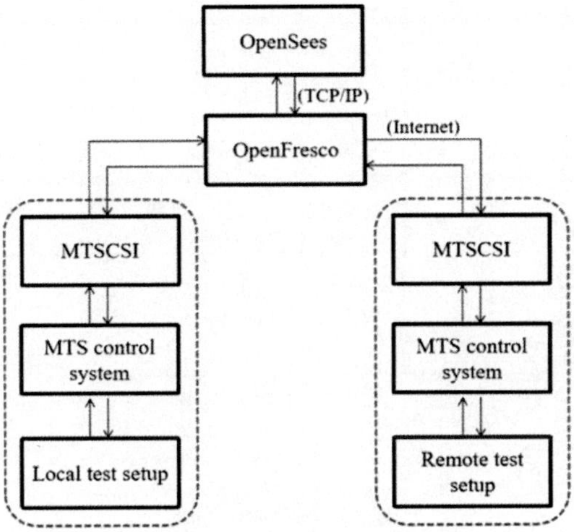

Fig. 1. A DHT system

where M_{n} and M_{t} are the mass matrices of the numerical substructure and the test substructure, respectively. C_n and C_t are the damping matrices of the numerical substructure and the test substructure, respectively. K_n and K_t are the stiffness matrices of the numerical substructure and the test substructure, respectively. $\ddot{X}_i$, $\dot{X}_i$, and X_i are the acceleration, velocity, and displacement matrices of the substructures, respectively. F_i is the seismic force.

2.2 The Prototype and Substructures

Figure 2 shows the prototype structure, which is a three-story, five-span space steel braced frame. The span of the structure is 5650 mm and the story height is 3600 mm. The global structure is divided into three parts. The three-story frame at second span is taken as the test substructure 1 for loading in the local laboratory, the bottom-story frame at fourth span is taken as the test substructure 2 for loading in the remote laboratory, and the rest of the structure is simulated in OpenSees Navigator as a numerical substructure. Considering the limitation of laboratory equipment size, the prototype structure was scaled down to 1/2 scale in the actual DHT modeling, and the span of the structure after scaling down is 2825 mm and the story height is 1800 mm.

2.3 The Numerical Substructure Model

The fiber model element provided in OpenSees Navigator is used to build the numerical substructure model. The material of element is selected as Steel02. The damping is selected as Rayleigh damping. A total number of 1000 steps are performed for the El Centro wave, and the time integration step is set to 0.01s.

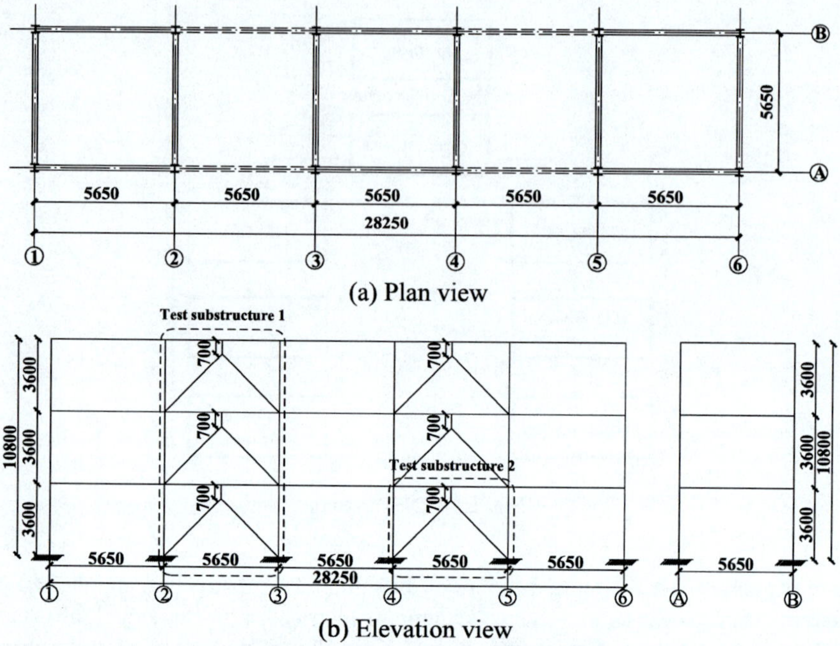

Fig. 2. The prototype and substructures

2.4 Parameter Settings in OpenFresco

The test loading devices for the two test substructures are shown in Fig. 3. Where test substructure 1 is loaded with three actuators for translational loading and test substructure 2 is loaded with two actuators for translational loading.

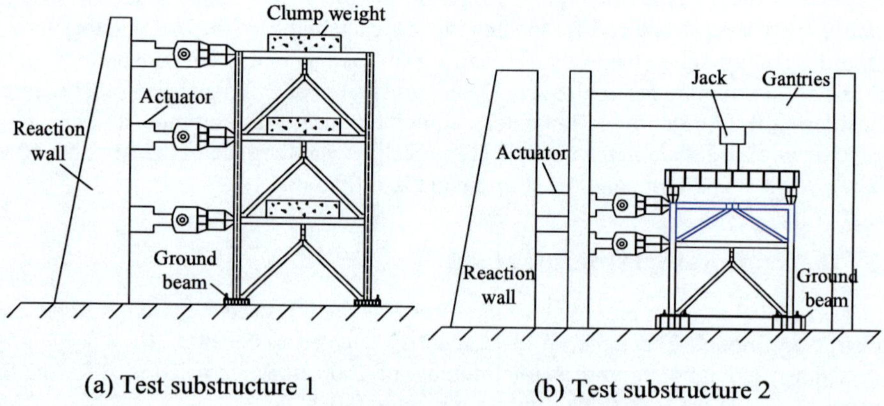

Fig. 3. Substructure test loading devices

In order to achieve equivalent communication between the test substructure loading and the OpenFresco platform. First, the test substructure 1 corresponds to a generic Test Element that can be defined with multiple degrees of freedom, and the test substructure 2 corresponds to a beamColumn Test Element that can simulate two-dimensional 3 degrees of freedom or three-dimensional 6 degrees of freedom. Second, the Test Control module in OpenFresco is utilized to implement the communication between the Test Element and the test loading control system. The test substructure 1 is loaded with three actuators and the test substructure 2 is loaded with two actuators, and the parameters of test substructure 1 (Command 1, Feedback 1, MtsCsi 1) and test substructure 2 (Command 2, Feedback 2, MtsCsi 2) are set using Control Point commands [10]. Figure 4 shows the control parameter settings for test substructure 1.

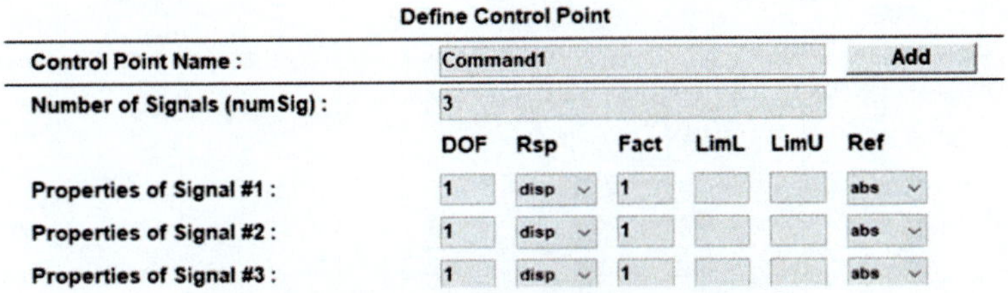

(a) Command 1 for test substructure 1

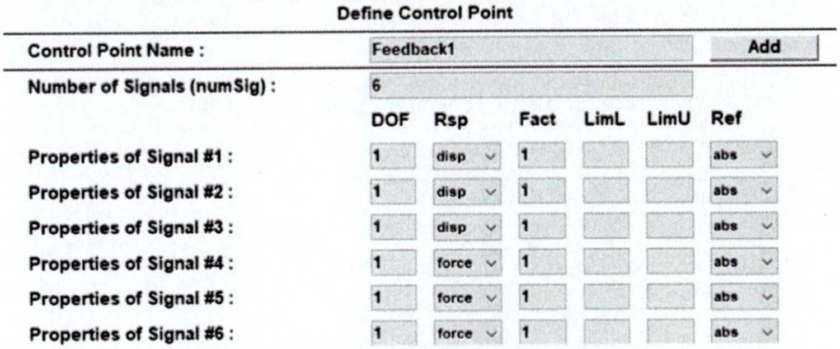

(b) Feedback 1 for test substructure 1

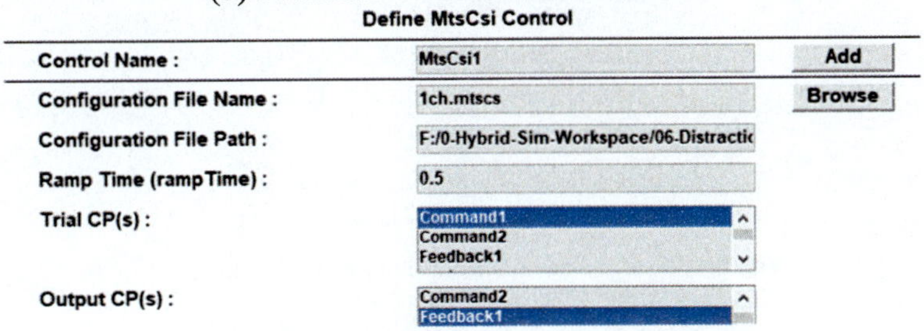

(c) MtsCsi control for test substructure 1

Fig. 4. Test control settings

3 DHT Result Analysis

3.1 DHT Results

Figure 5 shows the displacement time-history curves of the DHT model under the action of El Centro wave with peak acceleration of 0.120 g. It can be seen that the three curves are the same in trend, and the first vibration type deformation is dominant. Figure 6 shows the seismic force distribution of each story of the test substructure 1 under each load condition. It can be seen that the seismic force distribution is a standard inverted triangle distribution, in line with the shear deformation distribution law of the global structure.

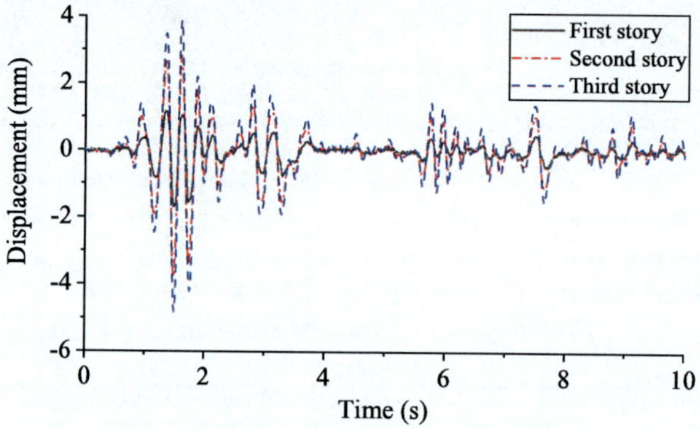

Fig. 5. Displacement time-history curves

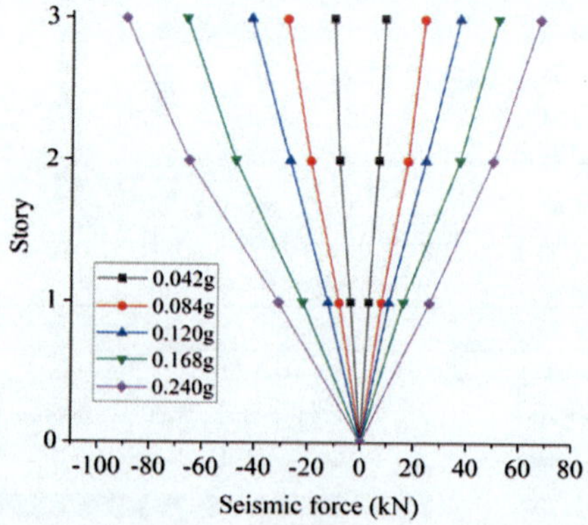

Fig. 6. Seismic force distribution of test substructure 1

3.2 Comparison of DHT Results and Pure Numerical Simulation Results

Figure 7 shows the roof displacement and base shear time-history curves of the DHT under the action of El centro wave with acceleration peak of 0.240 g and the comparison with that of the pure numerical simulation. It indicates that the DHT and simulation results basically agree with each other, and the relative errors of both at the peak do not exceed 15%.

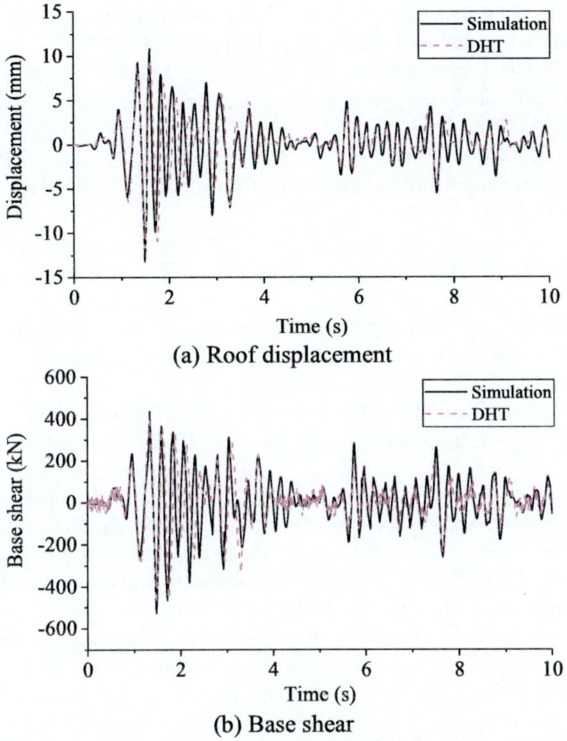

Fig. 7. Comparison of DHT results and pure numerical simulation results

4 Conclusions

Based on the openfresco test platform, this paper introduces the principle and system components of the DHT for multiple substructures, and describes the modeling process in openSees navigator and the parameter settings in openFresco. A multi-story steel braced space frame model is used as the research object to conduct the DHT. The displacement time-history curves and seismic force distribution analyses show that the DHT can be used to simulate the seismic response of the space steel-braced frame model. The comparison of the DHT results and pure numerical simulation results shows that the developed DHT model has a sufficiently high precision.

Acknowledgement. The work was supported in part by the National Natural Science Foundation of China (No. 51178382).

References

1. Stavros, D., Stephen A.M.: Development of substructuring techniques for on-line computer controlled seismic performance testing. Report No.UCB/EERC-85/04. Earthquake Engineering Research Center, University of California, Berkeley (1985)
2. Bozidar, S., Gilberto, M., Stephen A.M.: Event-driven control system for geographically distributed hybrid simulation. J. Struct. Eng. **132**(1), 68–77 (2006)
3. Stephen, M.: White Paper: Towards a Vision for the NEES Collaboratory, USA: Reports of NEES Consortium Development project. Consortium of Universities for Research in Earthquake Engineering (2002)
4. Mobin, O., Ignacio Lamata, M., Matt S.D., et al.: Geographically distributed hybrid testing & collaboration between geotechnical centrifuge and structures laboratories. Earthquake Eng. Eng. Vibr. **17**(1), 53–71 (2018)
5. Stathis, B., Anastasios, S., Oh-Sung, K., et al.: Intercontinental hybrid simulation for the assessment of a three-span R/C highway overpass. J. Earthquake Eng. **23**(7), 1194–1215 (2019)
6. Eiichi, W., Chung Bang, Y., Kunitomo, S., et al.: On line interactive testing between KAIST and Kyoto University. In: Proceeding of the 14th KKNN Symposium on Civil Engineering, Kyoto, Japan, pp. 369–374 (2001)
7. Jae-Kwan, K.: KOCED collaboratory program. In: Proceedings of the 2004 ANCER Annual Meeting: Network of Young Earthquake Engineering Researchers and Professionals, Hawaii, USA (2004)
8. Yuan-Sen, Y., Kung-Juin, W., Shiang-Jung, W., et al.: Networked pseudo-dynamic testing part I: database approach. In: Proceedings of the 13th World Conference on Earthquake Engineering, Vancouver, BC, Canada, pp. 1910–1919 (2004)
9. Yan, X., Yurong, G., Pingsheng, Z., et al.: Networked pseudodynamic testing of bridge pier and precast pile foundation. Eng. Struct. **38**, 32–41 (2012)
10. Andreas, S., Hong K.K., Yoshikazu, T., et al.: OpenFresco Command Language Manual. The Regents of the University of California, CA, USA (2009)

Kalman Filter Based Adaptive Control for Real-Time Hybrid Simulation with Nonlinear Substructure

Wei Huang[1] and Xizhan Ning[1,2(✉)]

[1] College of Civil Engineering, Huaqiao University, Xiamen 361021, China
xzning@hqu.edu.cn

[2] Key Laboratory for Intelligent Infrastructure and Monitoring of Fujian Province, Huaqiao University, Xiamen 361021, China

Abstract. Real-time hybrid simulation (RTHS) plays an important role in obtaining the structural responses under dynamic loads. However, the nonlinearities in physical testing system often induce time-delay, which is detrimental to RTHS. To deal with this issue, a Kalman filter based adaptive delay compensation (KF-ADC) method had been proposed based on the discrete system model, whose paramcters is estimated by Kalman filter (KF) algorithm. The delay compensation method was validated through a Benchmark problem in RTHS. This study intends to further investigate the performance of KF-ADC dealing with nonlinear physical substructures (PS). Virtual RTHSs are carried out on PS with bilinear and Bouc-wen restoring force characteristics, respectively. The results of virtual RTHSs reveal that the KF-ADC has an excellent tracking performance for nonlinear physical substructure and is beneficial for improving the accuracy and stability of RTHS.

Keywords: Real-time hybrid simulation · Adaptive delay compensation · Kalman filter

1 Introduction

A huge number of efforts had been paid to obtain the structural seismic performance, such as the quasi-static testing, shaking table testing, and hybrid simulation (HS) [1]. Due to the effectiveness and economics in acquiring the structural dynamics, HS, especially the real-time HS (RTHS), has received widespread attention.

However, the accuracy and stability of RTHS are often affected by the time-delay, which is usually induced by the nonlinearities of physical testing system. In the past almost two decades, many delay compensation methods or control strategies had been proposed. Polynomial extrapolation time-delay compensation method was first proposed by Horiuchi et al. [2], which is the most widely used time-delay compensation method in RTHS. Darby et al. [3] proposed a more accurate time-delay estimation method based on linear feedback theory, and verified the performance of the method for linear

W. Guo and K. Qian (Eds.): GBCESC 2022, LNCE 211, pp. 951–959, 2023.
https://doi.org/10.1007/978-981-19-5217-3_97

and nonlinear systems. Chen et al. [4] proposed an adaptive inverse control compensation method using an error tracking indicator to adapt compensation parameters. Then Phillips and Spencer proposed a model-based feedforward - feedback actuator control [5]. In addition, Chae et al. [6] proposed an adaptive time series compensation scheme, which updates the coefficients of the system transfer function by online real-time linear regression analysis. Zhou and Li [7] proposed an improving model-based compensation method considering the error between the identified transfer function and the control plant.

For practical application, Ning et al. [8] proposed an adaptive delay compensation method based on the discrete physical testing system model. Subsequently, the Kalman filter was introduced to improve the parameter estimation method by Ning et al. [9], which is termed as KF-ADC. This study intends to further investigate the performance of the proposed KF-ADC method in dealing with nonlinear substructure.

2 Kalman Filter-Based Adaptive Control Method

Block diagram of real-time hybrid simulation with KF-ADC is shown in Fig. 1. As can be seen in this figure, the KF-ADC consists of a feedforward controller with adjustable parameters and an online parameter estimator using Kalman filter algorithm. The transfer system and PS can be collectively seen as system model. The parameter estimator can rapidly calculate the system model by the relationship between on the command displacement d^c and the measured displacement d^m. Then, the estimated parameters are forwarded to the feedforward controller.

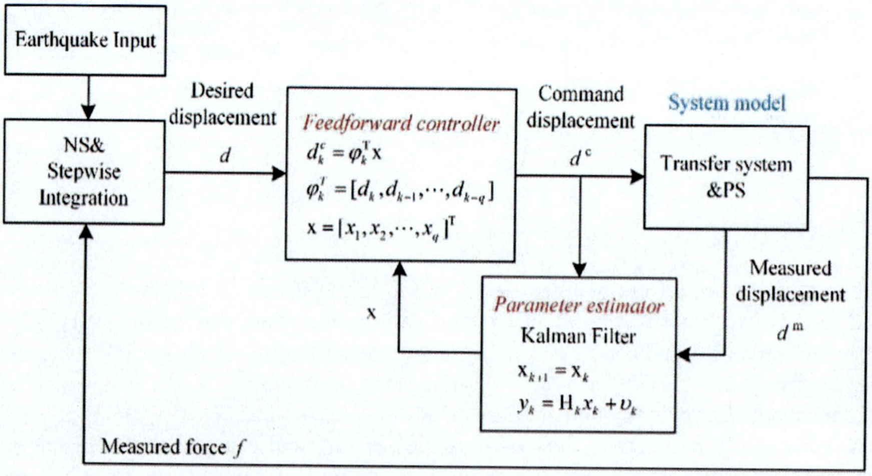

Fig. 1. Block diagram of real-time hybrid simulation with KF-ADC.

In the proposed control strategy, the feedforward controller is determined by the discrete inverse model of the system model, which is formulated as a difference equation, namely

$$d_k^c = \boldsymbol{\varphi}_k^T \mathbf{x} \tag{1}$$

where $\boldsymbol{\varphi}_k^T = [d_k, d_{k-1}, \cdots, d_{k-q}], \mathbf{x} = [x_1, x_2, \cdots, x_q]^T$. d and d^c are the desired displacement and the command displacement, respectively; x are the parameters of system model; $\mathbf{x}$ is the state vector which is consists of the parameter x. It should be noted, for application convenience, the measured displacements d^m are replaced by the desired displacement d, as shown in Eq. (1).

The parameters x in Eq. (1) are estimated and updated by the estimator, which can be written in the state-space format, namely

$$\begin{aligned} \mathbf{x}_{k+1} &= \mathbf{x}_k \\ y_k &= \boldsymbol{H}_k \mathbf{x}_k + \upsilon_k \end{aligned} \tag{2}$$

where $\mathrm{H}_k = [d_k^m, d_{k-1}^m, \cdots, d_{k-q}^m]$, y_k is the displacement command at the k-th time step, and υ is the measurement noise that satisfies normal distribution

$$\boldsymbol{E}\upsilon_k = 0, \boldsymbol{E}\upsilon_k \upsilon_k^{\mathrm{T}} = R \tag{3}$$

where R is the measurement noise covariance.

3 RTHS with Nonlinear Physical Substructure

3.1 Overview of the Simulation

The validation simulations are carried out using the Benchmark problem in RTHS by Silva et al. [10]. A three-story two-bay moment-resisting steel frame structure is divided into a physical substructure with one-bay one-story and a numerical substructure with the rest. The structure is modeled by a three degree of freedom equation of motion, where only the horizontal DOF is considered. The PS and the NS are associated together by a transfer system, which is represented by transfer functions as shown in Fig. 2. It should be noted, the parameters in the figure are estimated employing the experimental data, and uncertainties in the parameters are accounted for by using random variable with a

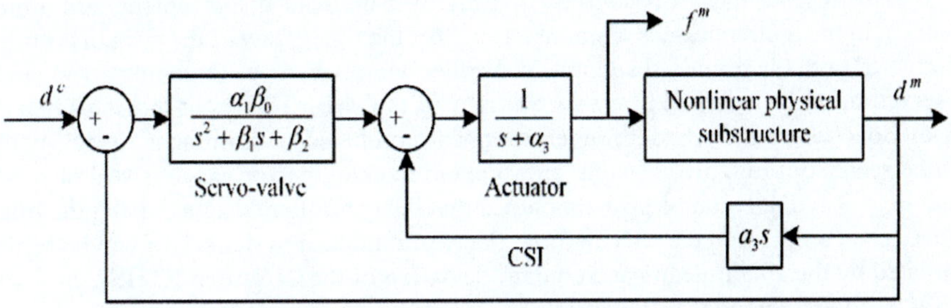

Fig. 2. Block diagram of the transfer system in the Benchmark problem.

Table 1. Parametric values of the transfer system.

Parameter	Component	Nominal value	Standard deviation	Units
$\alpha_1\beta_0$	Servo-valve	2.13×10^{13}	–	m·Pa·s^{-1}
α_2	CSI	4.23×10^6	–	m Pa
α_3	Actuator	3.3	1.3	1/s
β_1	Servo-valve	425	3.3	unitless
β_2	Servo-valve	10×10^4	3.3×10^3	1/s

normal distribution of each parameter. The parameters used in this study are listed in Table 1.

In this study, the reference floor mass and modal damping ratio are 1300 kg and 3%, respectively. Two types of nonlinear physical substructure are considered, one is represented by a bilinear model (Kz $= 1.1688 \times 10^6$ N/m, $d_y = 0.0025$ m, $b = 0.45$) and the other is represented by a Bouc-wen model ($k = 1.19 \times 10^6$ N/m, $\alpha = 0.02$, $\beta = 300$, $\gamma = 300$, $n = 1.05$). The external excitation is El-Centro earthquake record, which is scaled to 0.3g. A total of 21 virtual RTHS were carried out, one is for transfer system with nominal value, and the others are with perturbed parameters. Pure numerical simulation with ideal transfer system is served as the reference solution.

The adaptive controller designed by Ning et al. [9] is used in this study, which is given by:

$$d_k^c = x_1 d_k + x_2 d_{k-1} + x_3 d_{k-2} + x_4 d_{k-3} \tag{4}$$

The initial value $\mathbf{x}_0$, measurement noise variance R, and the initial covariance matrix are given by $[11.2458\ -21.0426\ 14.6847\ -3.9078]^T$, 1×10^{-4}, and diag([11.2458^2 21.0426^2 14.6847^2 3.9078^2]), respectively.

3.2 Analysis of Simulation Results

The displacement time histories of the first story for the nominal case with bilinear model and Bouc-wen model are shown in Fig. 3 and Fig. 4, respectively. The reference solutions are also provided in these two figures.

It is seen from these two figures that the measurement displacements are almost identical to the reference ones, demonstrating that the time-delay of the system is almost eliminated with the proposed adaptive controller. However, a peak tracking error can be observed from the enlarged figures, especially in Fig. 4(b). The result indicates that the PS of Bouc-wen model had stronger nonlinear compared with bilinear model, which can be seen from Fig. 5(b). Tough, an excellent tracking performance is still observed from Figs. 3 and 4. It can be also concluded that the controller designed using the linear system can compensate the varying time-delay of nonlinear system. This can be further validated by the small mean and standard deviation of the 21 virtual RTHSs, as shown in Table 2 and Table 3, respectively.

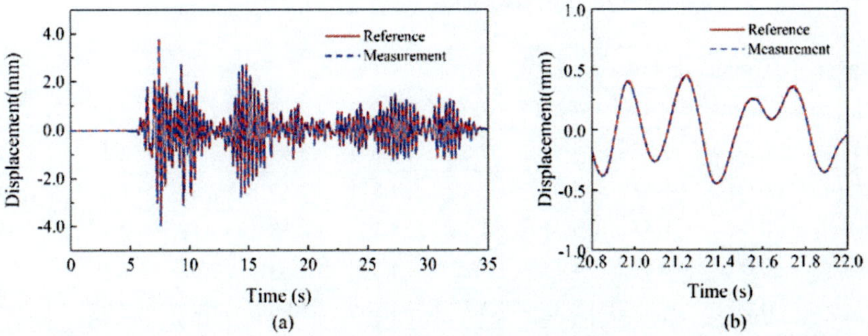

Fig. 3. Displacement time histories of the first floor for bilinear model. (a) Overview; (b) Enlarged.

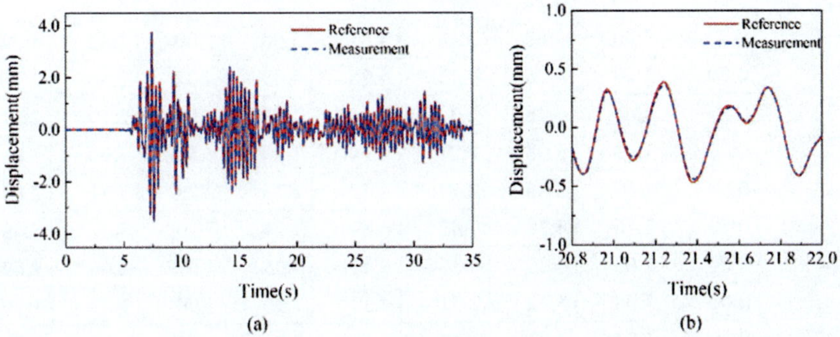

Fig. 4. Displacement time histories of the first floor for Bouc-wen model. (a) Overview; (b) Enlarged.

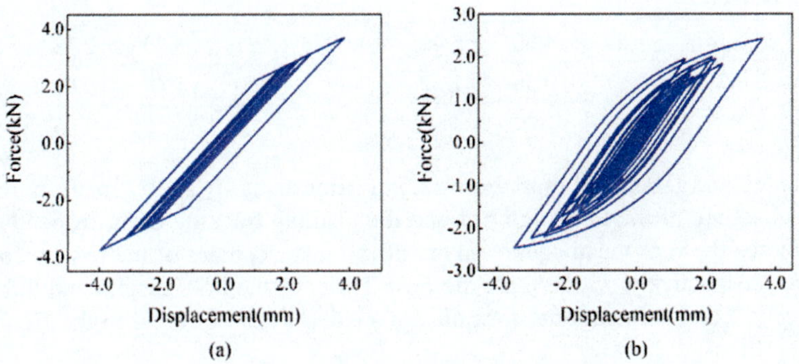

Fig. 5. Hysteresis relationship of two nonlinear physical substructures. (a) bilinear model; (b) Bouc-wen model.

Table 2. Virtual RTHS evaluation criteria for bilinear model.

Data type	Evaluation criteria								
	J_1(ms)	J_2(%)	J_3(%)	J_4(%)	J_5(%)	J_6(%)	J_7(%)	J_8(%)	J_9(%)
Nominal	0.20	1.52	4.47	3.55	4.84	3.41	3.40	2.13	2.14
Perturbed	0.00	1.52	4.96	2.88	4.82	2.67	2.67	1.69	1.69
	0.00	1.55	5.46	2.48	4.75	2.18	2.18	2.15	2.06
	0.00	1.48	4.74	2.72	4.52	2.53	2.52	1.58	1.57
	0.00	1.83	5.33	2.37	4.14	1.84	1.83	2.00	1.97
	0.00	1.51	4.76	3.19	4.81	3.06	3.06	2.02	2.01
	0.00	1.55	4.91	2.34	4.29	2.03	2.03	1.75	1.67
	0.50	1.79	4.22	5.11	5.31	4.93	4.90	3.51	3.52
	0.00	1.50	4.88	3.18	4.87	3.00	3.00	1.83	1.82
	0.20	1.63	4.26	4.15	4.98	4.03	4.02	2.66	2.70
	0.00	1.64	5.48	2.32	4.50	1.94	1.94	2.19	2.11
	0.00	1.63	5.08	2.21	4.34	1.80	1.80	1.90	1.81
	0.00	1.89	5.86	2.77	4.27	2.25	2.23	2.71	2.64
	0.70	2.16	3.78	7.42	6.09	7.24	7.20	5.47	5.49
	0.20	1.63	4.30	4.15	4.91	4.05	4.03	2.66	2.69
	0.00	1.58	5.12	2.21	4.32	1.90	1.90	1.87	1.79
	0.50	1.84	3.97	5.42	5.18	5.24	5.21	3.77	3.79
	0.20	1.65	4.18	3.89	4.71	3.74	3.72	2.46	2.49
	0.00	1.53	4.96	2.75	4.74	2.52	2.52	1.56	1.55
	0.00	1.56	5.35	2.33	4.60	2.01	2.01	2.09	2.00
Mean	0.14	1.66	4.80	3.46	4.79	3.21	3.20	2.48	2.46
Standard	0.23	0.17	0.56	1.42	0.49	1.50	1.48	0.98	1.00

In Table 2 and Table 3, $J1$ represents tracking time delay; J_2 and J_3 are the normalized root mean square of the tracking error and the peaking tracking error, respectively; J_4 to J_6 denotes the root mean square errors of the displacement of the first, second and third floor, respectively. J_7 to J_9 are the peak errors of the first, second and third floor, respectively. The computational formulas of J_1 to J_9 can be found in the Benchmark problem [10].

As shown in Table 2, the values of evaluation criteria are mostly less than 5%. The nominal values of evaluation criteria are almost same with the mean values of ones. In addition, the standard values of J_1, J_2, and J_3 are less than 1%. Overall the table suggests that the proposed delay compensation method has an excellent tracking performance and exhibits a strong robustness.

In comparison with Table 2, the values of evaluation criteria in Table 3 are much larger. And, some perturbed cases are more than 15%, which demonstrates that the physical substructure with Bouc-wen model has stronger nonlinear than that with bilinear model. However, the standard values of J_1 to J_3 are less than 1%, meanwhile the mean values of J_1 to J_3 are almost identical with the nominal values of ones. It demonstrates that the performance of KF-ADC is acceptable in solving variant time-delay problem of stronger nonlinear system.

Table 3. Virtual RTHS evaluation criteria for Bouc-wen model.

Data type	Evaluation criteria								
	J_1(ms)	J_2(%)	J_3(%)	J_4(%)	J_5(%)	J_6(%)	J_7(%)	J_8(%)	J_9(%)
Nominal	0.00	2.23	5.86	3.22	5.34	2.61	2.62	2.02	2.03
Perturbed	0.00	2.91	7.53	4.80	5.67	3.56	3.54	4.08	4.05
	0.00	2.34	6.31	3.17	5.08	2.34	2.32	2.92	2.93
	0.00	2.24	5.87	15.70	9.14	15.47	15.52	9.06	8.97
	0.00	2.27	6.44	14.23	8.34	14.26	14.31	8.37	8.29
	0.00	2.19	5.83	14.94	8.79	14.80	14.85	8.72	8.63
	0.00	2.24	5.44	8.48	4.83	8.07	8.10	4.60	4.56
	0.00	2.83	6.93	12.10	9.13	11.00	11.00	8.06	7.98
	0.00	2.42	6.75	7.51	5.31	7.65	7.68	4.32	4.29
	0.00	2.48	6.94	15.20	9.02	15.55	15.60	9.08	9.00
	0.00	2.47	6.82	5.58	5.55	5.50	5.53	3.03	3.01
	0.00	2.72	7.49	9.37	5.55	9.70	9.74	5.64	5.59
	0.20	2.24	5.38	9.34	5.29	8.93	8.97	5.12	5.07
	0.00	2.59	7.02	4.64	5.88	3.65	3.63	4.02	4.03
	0.00	2.59	6.89	4.13	5.46	3.92	3.94	2.40	2.35
	0.50	2.53	4.80	5.96	5.69	6.48	6.47	5.23	5.17
	0.00	2.37	6.72	9.03	5.26	9.14	9.18	5.27	5.23
	0.00	2.22	5.75	9.60	7.71	9.52	9.51	7.25	7.14
	0.50	2.43	4.85	8.25	6.50	8.90	8.90	6.53	6.41
	0.50	2.35	4.72	6.15	5.96	6.54	6.53	5.47	5.41
Mean	0.12	2.45	6.13	9.62	7.00	9.51	9.53	6.25	6.20
Standard	0.23	0.21	0.99	5.15	2.58	5.44	5.46	3.08	3.06

The evaluations of the estimated system model parameters for two nonlinear physical substructures are respectively shown in Fig. 6 and Fig. 7. As shown in these two figures, the variation trends of the system model parameters for two physical substructures are almost identical. Actually, in these two graphs, the convergence trend of system model parameters is not obvious. It is because that the compensator is designed using linear system and is used to address the variant time-delay of a higher-order nonlinear system. Combined with the above analysis, the performance of the proposed KF-ADC still can be acceptable.

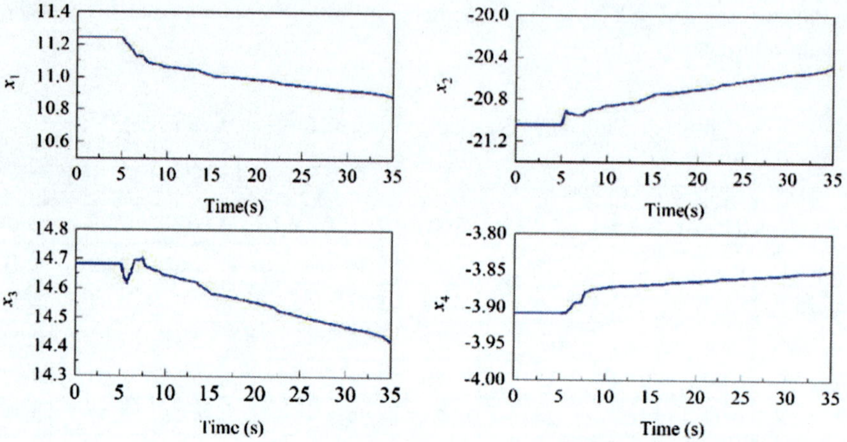

Fig. 6. Evaluations of the estimated system model parameters for bilinear model.

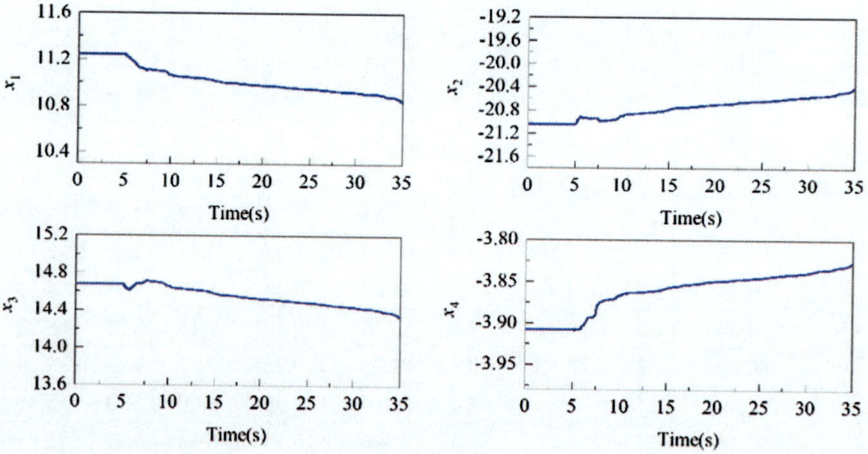

Fig. 7. Evaluations of the estimated system model parameters for bouc-wen model.

4 Conclusion

A Kalman filter-based adaptive control strategy is proposed to deal with the variant time-delay problem of RTHS with nonlinear physical substructures. In this method, a parameterized discrete inverse model of the physical testing system is served as the feedforward controller and a Kalman filter is used to estimate and update the parameters by the relationship between the measurement displacement and command displacement. A total of 42 vRTHSs are carried out to investigate the capability of KF-ADC in the benchmark problem. Conclusions can be addressed as follow:

(1) The evaluation criteria for the nominal cases are almost identical to that for the perturbed cases and that the standard values of J_1 to J_3 are less than 1%. It demonstrates that KF-ADC has a brilliant compensation performance and strong robustness for nonlinear system.
(2) Discrepancies in amplitude can be observed from the displacement time histories, especially in the case of PS with Bouc-wen model. The system model parameters cannot converge to constant values for the cases with nonlinear substructures. It is because that the compensator is designed by linear discrete model and is used to address the variant time-delay of nonlinear system. Hence, the accuracy of RTHS with KF-ADC is acceptable.

References

1. Horiuchi, T., Inoue, M., Konno, T., et al.: Real-time hybrid experimental system with actuator delay compensation and its application to a piping system with energy absorber. Earthquake Eng. Struct. Dynam. **10**(28), 1121–1141 (1999)
2. Nakashima, M., Kato, H., Takaoka, E.: Development of real-time pseudo dynamic testing. Earthquake Eng. Struct. Dyn. **21**(1), 79–92 (1992)
3. Darby, A.P., Williams, M.S., Blakeborough, A.: Stability and delay compensation for real-time substructure testing. J. Eng. Mech. **128**(12), 1276 (2002)
4. Cheng, C., Ricles, J.M.: Tracking error-based servohydraulic actuator adaptive compensation for real-time hybrid simulation. J. Struct. Eng. **136**(4), 432–440 (2010)
5. Phillips, B.M., Spencer, B.F.: Model-based feedforward-feedback actuator control for real-time hybrid simulation. J. Struct. Eng. **139**(7), 1205–1214 (2013)
6. Chae, Y., Kazemibidokhti, K., Ricles, J.M.: Adaptive time series compensator for delay compensation of servo-hydraulic actuator systems for real-time hybrid simulation. Earthquake Eng. Struct. Dyn. **42**(11), 1697–1715 (2013)
7. Zihao, Z., Ning, L.: Improving model-based compensation method for real-time hybrid simulation considering error of identified model. J. Vib. Control **27**(21–22), 2523–2535 (2021)
8. Xizhan, N., Zhen, W., Chunpeng, W., Bin, W.: Adaptive feedforward and feedback compensation method for real-time hybrid simulation based on a discrete physical testing system model. J. Earthquake Eng. (2020)
9. Xizhan, N., Zhen, W., Bin, W.: Kalman filter-based adaptive delay compensation for benchmark problem in real-time hybrid simulation. Appl. Sci. **10**(20), 7101 (2020)
10. Silva, C.E., Gomez, D., Maghareh, A., Dyke, S.J., Spencer, B.F.: Benchmark control problem for real-time hybrid simulation. Mech. Syst. Signal Process. **135**, 106381 (2020)

Comparative Analysis of Tunnel Structures with Different Sections in Open Excavation and Buried Section of Urban River Passage

Hongyu Guo[1](✉), Helin Fu[2], Guijun Luo[1], Hongwei Zhang[1], and Kaiyuan Zhang[2]

[1] China Construction Fifth Engineering Division Corp., Ltd., Changsha, Hunan, China
skycboy@163.com
[2] Central South University, Changsha, Hunan, China

Abstract. Based on the engineering background of the cross-river tunnel in the Rongjiang New District of Ganzhou, this paper uses Midas Gtx to establish three different tunnel cross-section structures: rectangle, polyline, and arch, and compare and analyze their ability to withstand bending moments and shear forces. Deformation, material consumption. Finally, the best section form is obtained. Studies have shown that under the conditions of large-section tunnels, arched tunnels have better mechanical performance. The maximum bending moment that arched tunnels can withstand is 47% of rectangular tunnels and 54% of broken-line tunnels. The structural deformation of the arch tunnel is the smallest, the maximum deformation of the rectangular tunnel is 19.97 mm, the maximum deformation of the broken line tunnel is 18 mm, and the maximum deformation of the rectangular tunnel is 15.6 mm. At the same time, the arched section structure is more economical in terms of material consumption. The concrete consumption of the polyline structure tunnel is 1.44 times that of the arch tunnel, and the concrete consumption of the rectangular structure is 1.67 times that of the arch structure. Under the same supporting conditions during construction, the safety reserve of the arch tunnel is higher.

Keywords: Open cut section · Section form · Numerical simulation · Force and deformation · Stability

1 Preface

According to the construction experience at home and abroad, bridges and tunnels have their own advantages when crossing rivers. However, considering the construction difficulty and the impact on the river navigation capacity after completion, the tunnel scheme in some areas is better than the bridge scheme [1]. When the tunnel is selected as the river crossing scheme, the tunnel is usually divided into open excavation and concealed excavation. As the transition section connecting the tunnel at the bottom of the river and the existing road traffic, the concealed excavation part needs to play a good transition

W. Guo and K. Qian (Eds.): GBCESC 2022, LNCE 211, pp. 960–973, 2023.
https://doi.org/10.1007/978-981-19-5217-3_98

role in the continuity of the structural section. The section geometry of open cut tunnel has a great impact on itshydrodynamic characteristics, structural performance and service level. Therefore, the section geometry has become the primary factor of tunnel optimization [2]. Which channel section structure is more reasonable in the open cut section of the river crossing channel remains to be studied in China.

Some scholars have studied that due to the characteristics of shape, rectangular tunnel shows different stress characteristics from circular shield tunnel [3]. In terms of stress, shallow buried rectangular tunnel is mainly affected by structural self weight and load [4]. The maximum bending moment of sub rectangular tunnel first decreases and then increases with the increase of radius R. the optimal shape of sub rectangular tunnel is determined by the minimum of the respective maximum bending moment mmax corresponding to different radius r [5]. The larger the span of the broken line tunnel, the larger the size of the structural section, the lower the utilization rate of the section and the greater the amount of concrete [6]. The arch tunnel section can effectively control the deformation amount and deformation rate of surrounding rock, especially the horizontal convergence deformation [7]. The stress area of the arch crown of the arch tunnel with relatively flat shape is large, and the part with large stress moves down from the arch waist to near the side wall [8]. Through the deformation monitoring measurement and data analysis of the straight wall section and curved wall section of the large section soft rock tunnel, Hu Xiaoming and others proposed that the curved wall tunnel is more reasonable in terms of stability and safety than the straight wall tunnel [9]. Sun Ping et al. Proposed that the equivalent stress value of three center circular lining is almost the same as that of circular lining, but the vault subsidence displacement and inverted arch uplift displacement are less than that of circular lining [10]. Chang Jianqiang and others analyzed the stress field and displacement field of three different cross-section tunnels such as horseshoe, circular arch and ellipse through finite element software, and put forward that horseshoe tunnel has more advantages in controlling surrounding rock deformation [11].

Starting from the river crossing channel in Rongjiang new area, Ganzhou, this paper will determine the section form of the open cut section of the urban river crossing channel under the condition of large section by comparing the stress performance, material use and construction stability of arch, rectangular and broken line tunnels.

2 Project Overview

The Rongjiang tunnel in Ganzhou Rongjiang new area generally runs north-south and crosses Zhangjiang River, mainly connecting Fenggang area and Rongjiang new area. It is an important high-capacity channel between the two areas. The total length of the tunnel is about 1.74 km, and the buried section is 949.5 m.

2.1 Geological Overview

The site stratum is mainly composed of artificial fill (q4ml), quaternary Holocene alluvium (q4al) and upper Pleistocene eluvial diluvium (q3edl). The geological profile is shown in Fig. 1. The physical property indexes of each layer of soil are shown in Table 1.

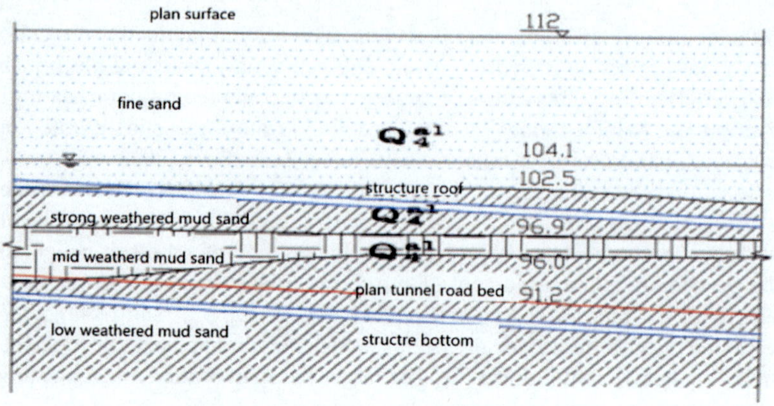

Fig. 1. Geological cross section (unit: m)

Table 1. Physical property index of soil layer

Soil layers	Density (KN/m^3)	Elastic modulus (MPa)	Poisson's ratio	Cohesion (KPa)	Friction angle (°)
Fine sand	18	7.3	0.35	30	18
Strong weathered mud sand	21.5	120	0.25	50	25
Mid weathered mud sand	25	146	0.2	146	43.8
Mid weathered sand	22	1.63	0.35	23.7	11.5

2.2 Conceptual Design

The river crossing channel is mainly three lane highway. The section net width of three lane highway tunnel is 12.5 m and the net height is 4.5 m. The most unfavorable section is the section with the largest buried depth in the open cut section of the tunnel. On this basis, three different tunnel section shapes are designed:

Arched Tunnel: The full section of arched tunnel adopts 1.1 m thick lining structure. The outline length of the structure is 15.45 m and the height is 10.8 m. The detailed drawing is shown in Fig. 2 (a).

Broken Line Tunnel: The broken line section at the top of the broken line tunnel and the lining structure of the side wall are 1.3 m thick and the bottom plate is 1.9 m thick.

The outline length of the structure is 15.1 m and the height is 10.15 m. As shown in Fig. 2 (b).

Box Tunnel: The thickness of the top plate of the box tunnel is 1.8 m, the thickness of the side wall is 1.5 m and the thickness of the bottom plate is 1.8 m. The outline length of the structure is 15.5 m and the height is 8.5 m. As shown in Fig. 2 (c).

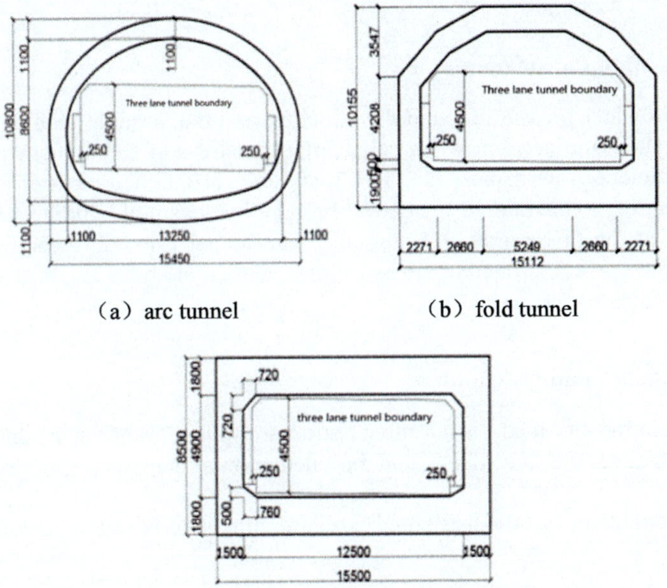

(a) arc tunnel (b) fold tunnel

(c) rectangle tunnel

Fig. 2. Structural section of open cut and buried section of tunnel (unit: mm)

3 Calculation Model

3.1 Calculation Model

The limit state method is more suitable for the design of urban open-cut tunnel when the calculation of tunnel structure is carried out. According to the characteristics of the analysis problem, Midas GTS is used to analyze three kinds of tunnel structures with load structure model. When the load-structure model is used, the length of tunnel is 1 m, and the tunnel structure is simplified as plane strain problem. The beam element is used to simulate the structure, and the foundation spring is used to simulate the ground force. The tunnel calculation model is shown in Fig. 3.

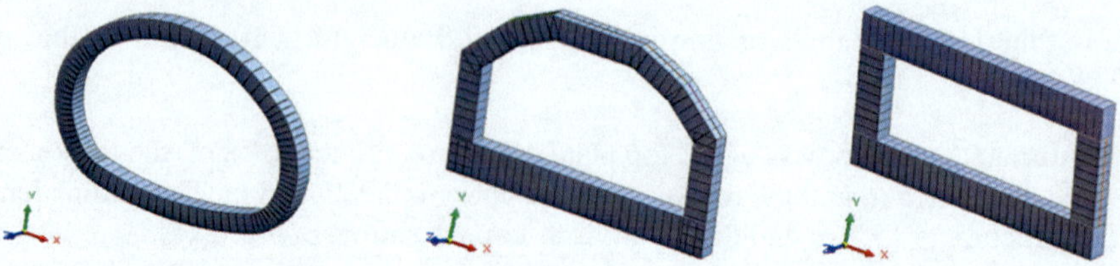

Fig. 3. Structural calculation model of open-cut and buried section of tunnel

3.2 Calculation Parameters

The soil and water pressure around the structure and the variable load on the ground should be taken into account when calculating the stress of the tunnel structure. The partial coefficient of permanent load is 1.3, and the partial coefficient of variable load is 1.5. According to the unified standard for reliability design of building structures, the structure is of the first degree of importance and its corresponding coefficient is 1.1. The parameters of C45 concrete are as follows: elastic modulus Ec, bulk density $\gamma = 25\ \text{kN/m}^3$, Poisson's ratio $\mu = 0.2$.

3.3 Structural Load Calculation

When calculating the load on the tunnel structure, put the bottom of three different tunnel structures at the same depth, and the calculation of corresponding structures is as follows.

The vertically uniformly distributed earth pressure of surrounding rock is calculated as follows:

$$q = \gamma h \tag{1}$$

where: γ-surrounding rock density, KN/m^3; h-tunnel buried height, m.

The horizontal uniformly distributed pressure of surrounding rock is calculated according to Rankine formula, i.e.

$$e = \lambda q \tag{2}$$

where: λ-Lateral pressure coefficient of corresponding soil layer.

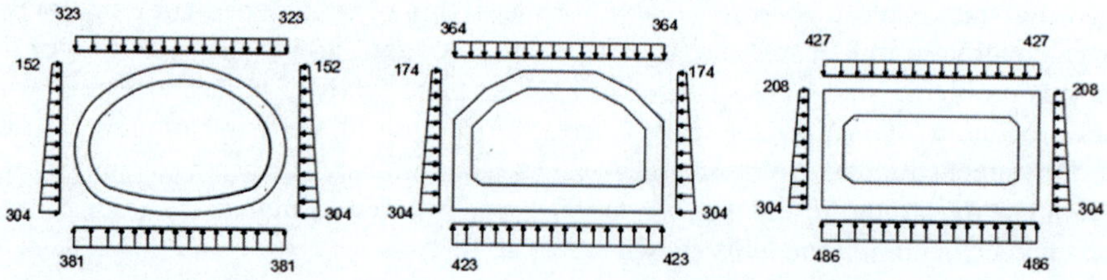

Fig. 4. Structural load diagram of tunnel open excavation and buried section (unit: kN/m)

During the calculation, the corresponding partial coefficient of load shall be considered at the same time. After calculation, the loads borne by three different tunnel structures are shown in Fig. 4.

4 Result Analysis

Through numerical analysis, the stress and deformation results of the main structure of the tunnel with three different cross-sections are obtained, which are analyzed according to the results of the following aspects of the key nodes (1–5) (Fig. 5).

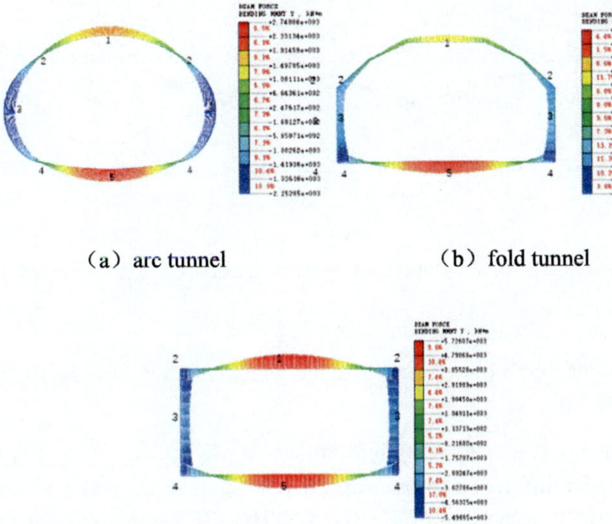

(a) arc tunnel (b) fold tunnel

(c) rectangle tunnel

Fig. 5. Structural bending moment results of open cut and buried section of tunnel

4.1 Comparative Analysis of Stress

Structural Bending Moment Analysis. According to the tunnel calculation results, the maximum bending moment of arch tunnel MS = 2.74 × 103 MPa is at the arch bottom, and the bending moment M5 = 4.95 × 103 MPa is at the arch bottom (No. 5 node); The bending moment of the broken line tunnel M4 = 4.7 × 103 MPa is at the middle of the bottom plate (No. 5 node) and at both ends of the bottom (No. 4 node); The bending moment of the rectangular tunnel M5 = 5.5 × 103 MPa is at the middle of the bottom plate (No. 5 node) and M4 = 5.72 × 103 MPa at both ends of the bottom (No. 4 node).

Figure 6 shows the comparison of bending moments of key nodes of three structures. According to the comparison chart, the total variation trend of bending moment of each node of the three structures is similar. Each structure has a large bending moment in the middle of the plate and wall (nodes 1, 3 and 5), and a relatively small bending moment at the fulcrum on both sides (nodes 2 and 4). According to the comparison, the bending moment at the same position of each section: the bending moment of arch structure is the largest, followed by broken line structure, and the arch shape is relatively the smallest.

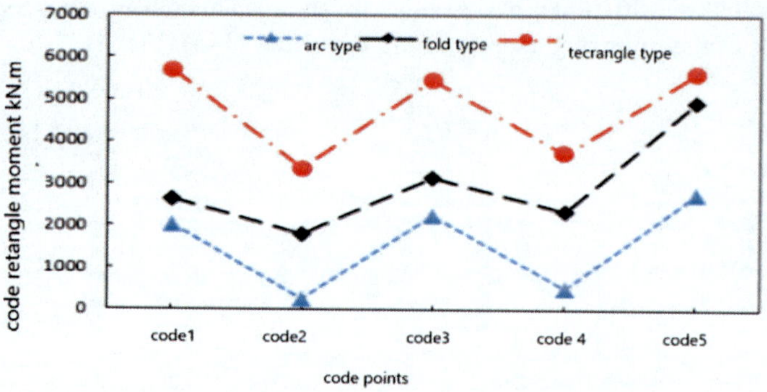

Fig. 6. Comparison diagram of structural bending moment of open excavation and buried section of tunnel

The above shows that the arch structure has the advantage of reducing the bending moment of the tunnel.

Structural Shear Analysis. The following Fig. 7 shows the shear calculation results of tunnel structure under different section forms: according to the tunnel calculation results, the maximum shear of arch tunnel F4 = 1.25 × 103 kN appears at both sides of arch bottom (No. 4 node), and the maximum value is; The maximum shear force of broken line tunnel appears at both ends of the bottom plate (No. 4 node), and the maximum value is,; The maximum shear force of rectangular tunnel is at both ends of the bottom plate (No. 4 node), and the maximum value is.

Figure 8 shows the comparison of shear forces at key nodes of three structures. According to the comparison diagram, the shear force of the three structures is small in the middle of the slab and wall (nodes 1, 3 and 5), and the shear force is large at the fulcrum on both sides (nodes 2 and 4). According to the comparison chart, the shear value of arch structure and broken line structure is much greater than that of arch structure. The arch structure will have large shear force at nodes 2 and 4. Compared with the bending moment diagram, the maximum shear area of the arch structure F4 = 2.86 × 103 kN is just staggered with the maximum bending moment area. The larger shear force of broken line tunnel and rectangular tunnel appears at the end of each slab (nodes 2 and 4), and the larger shear force of broken line structure F4 = 3.37 × 103 kN at the end of bottom slab (node 4). For rectangular structure, there will be large shear at the bottom

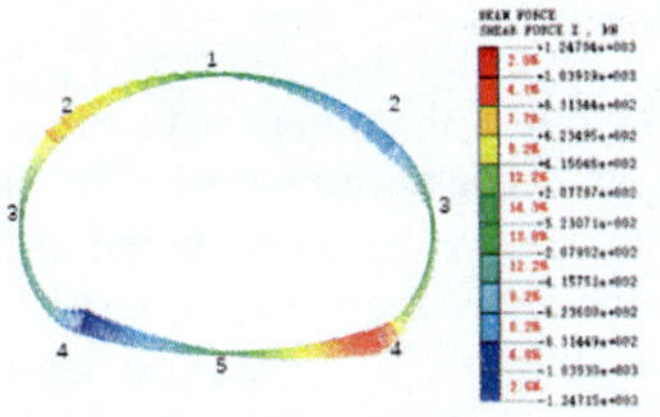

(a) arc tunnel

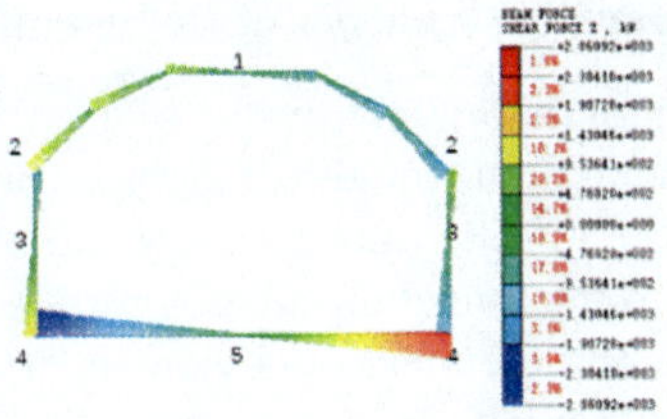

(b) fold tunnel

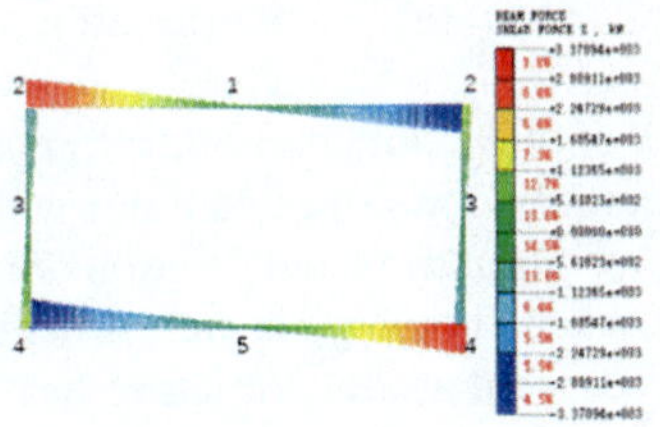

(c) rectangle tunnel

Fig. 7. Structural shear force diagram of open cut and concealed buried section of tunnel

plate and top plate end (nodes 2 and 4). Combined with the bending moment diagram of rectangular tunnel, these two places are also areas with large bending moment, which leads to poor stress in these places and is unfavorable to the structure, which usually increases the reinforcement of the structure.

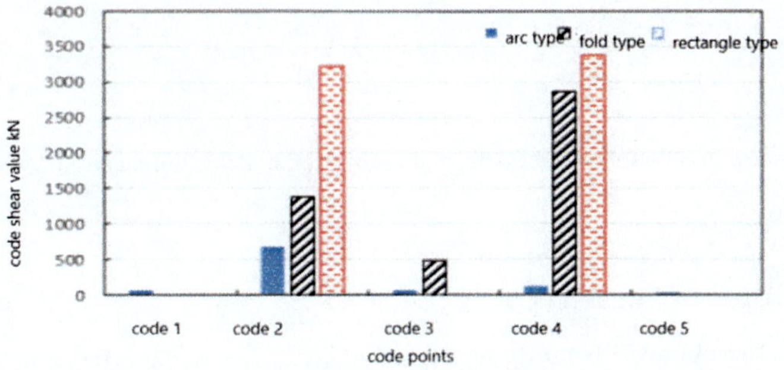

Fig. 8. Comparison diagram of structural shear force in open cut and concealed buried section of tunnel

4.2 Comparative Analysis of Deformation

The maximum deformation value of each section of the tunnel structure *f1* = 15.6 mm is concentrated in the middle of the plate or wall. The deformation of key nodes of each section of the tunnel structure under different section forms is shown in Fig. 9 below. According to the tunnel calculation results, the arch tunnel has the largest deformation at the vault, and the maximum value is *f2* = 18 mm; The maximum deformation *f3* = 19.97 mm at the middle of the tunnel floor is the fold line; The maximum deformation of the rectangular tunnel is now in the middle of the roof, and the maximum value is. The structural deflection is checked according to the code for design of highway tunnels: when the span exceeds 8 m, the maximum deformation of the tunnel structure does not exceed *l0*/400, and the results show that the deformation of the three tunnel structures meets the requirements.

Overall, the deformation of the middle part of the top plate and bottom plate of each structure is large, while the deformation value of the side wall is relatively small. From the comparison of deformation results at the same position of tunnels with different shapes, the order of deformation from small to large is arch structure, broken line structure and rectangular structure. This shows that arch structure has great advantages in bearing deformation, while rectangular structure will have large deflection in the middle of the plate due to its own span and other reasons.

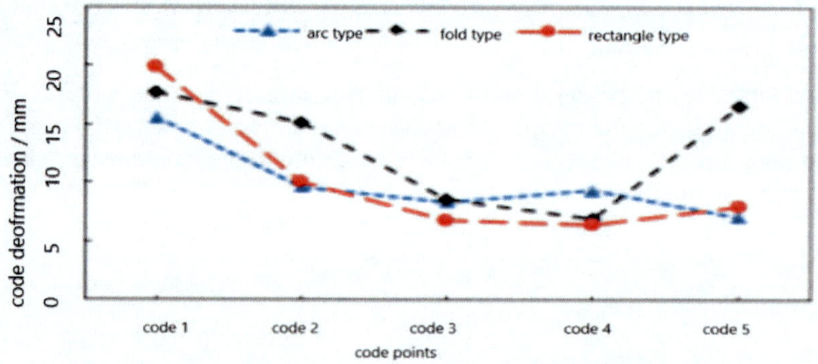

Fig. 9. Comparison of deformation of each node of tunnel structure

4.3 Comparison of Material Use

According to the obtained bending moment and shear force, the reinforcement of three tunnels with different sections is carried out and the quantities are calculated. According to the code for design of concrete structures, the reinforcement calculation under the serviceability limit state is carried out for the results of three different sections. HRB400 reinforced concrete structure is adopted. Combined with the content of the specification, the crack width limit of waterproof concrete structure is less than 0.2 mm. Under the condition of strictly limiting the crack width, the reinforcement results of the three schemes are shown in Table 2.

Table 2. Reinforcement calculation of tunnel structure.

Structure position		Stress type					
		Bending moment (KN • m)			Shear force (KN)		
		Arc type	Rectangle type	Fold type	Arc type	Rectangle type	Foldtype
Roof	Bearing	1151	3355	1791	850	2901	815
	Mid span	2009	5719	2636	0	0	0
Bottom	Bearing	641	5485	4704	1012	3371	2861
	Mid span	2714	5642	4954	0	0	0
Lateral wall	Top support	1151	5485	3130	850	986	834
	Bottom support	641	4863	3862	1012	1091	1184
	Mid span	2245	3734	2350	107	93	394

Table 3. Reinforcement calculation of tunnel structure.

Structure position		Stress type			
		Reinforcement			
		Arc type	Arc type	Rectangle type	Fold type
Roof	Bearing	1151	28@150 + 28@100	36@100 + 36@100	28@150 + 28@100
	Mid span	2009	40@150 + 40@100	40@100 + 40@100	36@150 + 36@100
Bottom	Bearing	641	28@150 + 28@100	36@100 + 36@100	40@150 + 40@100
	Mid span	2714	40@150 + 40@100	40@100 + 40@100	40@150 + 40@100
Lateral wall	Top support	1151	40@150 + 40@100	40@100 + 40@100	40@150 + 40@100
	Bottom support	641	28@150 + 28@100	40@100 + 40@100	40@150 + 40@100
	Mid span	2245	40@150 + 40@100	36@100 + 36@100	36@150 + 36@100

The concrete consumption of three tunnel structures with a width of 1 m is calculated according to the tunnel section, as shown in Table 3. The three tunnels with different sections are calculated according to the width of 1 m. Combined with the reinforcement calculation in Table 2, the material consumption of each meter width is shown in Fig. 10.

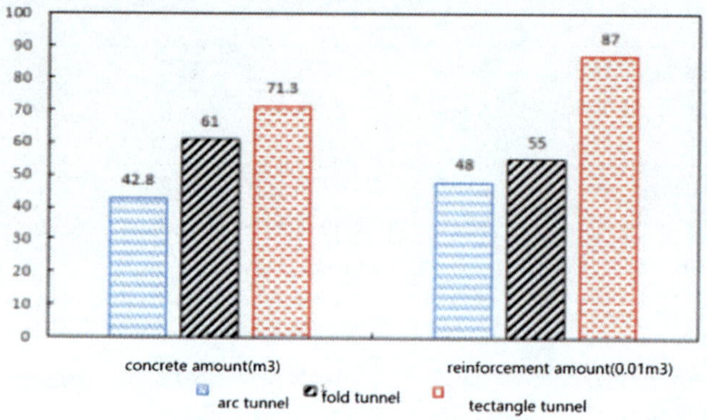

Fig. 10. Material consumption table

According to the calculation results in Fig. 10, the concrete consumption of arch tunnel is (42.8 m^3), that of broken line tunnel is (61 m^3), and that of rectangular tunnel is (71.3 m^3). The concrete consumption of broken line tunnel is 1.44 times that of arch tunnel, and that of rectangular structure is 1.67 times that of arch structure. The concrete consumption of arch tunnel will be greatly saved.

The corresponding reinforcement amount of arch tunnel, broken line tunnel and rectangular tunnel increases in turn. Compared with broken line tunnel and rectangular tunnel, the reinforcement of arch tunnel decreased by 12.7 and 36.8 respectively.

4.4 Construction Stability

During tunnel construction, due to the increase of roof thickness and reinforcement of broken line tunnel and rectangular tunnel, the force on the formwork support system will be greater when the tunnel is erected. It will have an adverse impact on stability. The construction stability of tunnels with different sections under the same support system is analyzed. All three tunnel structures adopt φ60.2 × 3.2 mm Q235 turnbuckle steel pipe support. The layout spacing of steel pipe support is 90 cm longitudinally × Horizontal 60 cm, cross bar step distance 150 cm. Longitudinal and horizontal vertical inclined rods are fully distributed, four layers of inclined rods are set at the top of the horizontal, and the bottom layer is set with inclined rods.

The vertical pole of the support is subjected to the load transmitted by the longitudinal I-beam (longitudinal distribution beam). Each vertical pole bears the load within the range of 60 cm × 90 cm.

The loads on the vertical pole are:

- Self weight of concrete: Fa = 0.6 × 0.9 × h×γ c, h is the thickness of structural roof slab, concrete density γC = 26 KN/m^3.
- Weight of formwork and primary and secondary keels: Fb is calculated as 5% of the self weight of concrete.
- Concrete pouring, construction personnel and equipment load: Fc = (2.5 + 2 + 2) KN/m^2 × 0.6 m × 0.9 m = 3.51 KN.

The load combination of a single vertical pole is: N = 1.3 × (Fa + Fb) + 1.5 × Fc. Pole stability calculation:

$$\sigma = \frac{N}{\phi A} \leq [\sigma] \tag{3}$$

where: A-Cross sectional area of vertical pole: A = 572 mm^2.

[σ] - Allowable stress of steel, here [σ] = 215 MPa.

Φ - The stability coefficient of axial compression member is determined by the slenderness ratio of vertical rod λ determine.

Slenderness ratio of vertical pole λ = l_0/i.

l_0 - Calculated length of support pole, l_0 = kμl.

k - Additional coefficient of effective length, k = 1.15.

μ - The length coefficient of the vertical pole of the support, and the two ends of the vertical pole are fixed, μ = 0.5

i - Slewing radius of support pole, i = 20.1 mm.

According to Appendix D of safety technical specification for socket and turnbuckle steel pipe support in building construction, the table is found φ = 0.934. The calculation results are shown in Table 4.

Table 4. Reinforcement calculation of tunnel structure

Tunnel type	Construction difficulty	Plate thickness h(m)	Fa(KN)	N	Pole stress σ (MP)a	Allowable stress [σ] (MPa)	Pole stress/critical stress
Arc tunnel	Inconvenient erection of tunnel slope support	1.10	15.44	26.35	49.31	215.00	0.23
Fold tunnel	Inconvenient erection of tunnel vault support	1.30	18.25	30.18	56.49	215.00	0.26

(continued)

Table 4. (*continued*)

Tunnel type	Construction difficulty	Plate thickness h(m)	Fa(KN)	N	Pole stress σ (MP)a	Allowable stress $[\sigma]$ (MPa)	Pole stress/critical stress
Rectangle tunnel	Regular structure and convenient support erection	1.80	25.27	39.76	74.42	215.00	0.35

According to the calculation results in Table 4, the stress of the vertical pole of rectangular tunnel is the largest in three cases, and the stress value is 74.42 MPa; The stress of the vertical pole of the arch tunnel support is the smallest, and the stress value is 49.31 MPa. According to the ratio of neutral bar stress/critical stress of the support system: arch, broken line and rectangle decrease in turn. This shows that in terms of safety reserve of support system, arch tunnel is the best, broken line tunnel is in the middle, and rectangular tunnel is the lowest. Therefore, the stability of arch structure is better in terms of support stability.

5 Conclusion

The bending moment, shear force and deformation of different tunnel sections are calculated by Midas GTS. The corresponding amount of reinforcement and concrete is obtained through corresponding calculation, and the following conclusions are obtained:

(1) The concrete and steel consumption of the tunnel with rectangular section structure is large, which is 1.5 times that of the arch section structure. Therefore, the internal force of the whole tunnel structure is also large, which is not conducive to the stress of the structure. At the same time, due to the span, the deflection of the rectangular tunnel is also large, and the maximum deflection is. For the area with large buried depth, a cavity layer needs to be set to ensure the stress of the structure.
(2) Compared with rectangular and arched structures, the broken line tunnel section is in the middle in terms of mechanical performance, material use and section utilization. The broken line structure will have the disadvantage of inconvenient erection of formwork support system during construction.
(3) Arch tunnel has better mechanical performance, and the internal forces such as bending moment are relatively small. The maximum bending moment is only 47% of the maximum bending moment of rectangular tunnel. The thickness of the structure and the reinforcement can be reduced accordingly. Formwork trolley can be used during construction, with the best safety and convenience. Compared with broken line tunnel and rectangular tunnel, the reinforcement amount is reduced by 12.7 and 36.8 respectively. It can achieve good economic results. Therefore, arch tunnel is more suitable for large cross-section urban river crossing channel in open excavation and concealed buried section.

References

1. Mengshu, W.: Comparison and selection of the West Bridge Scheme of the East Tunnel and the whole tunnel scheme of the Shenzhen Guangzhou middle corridor. In: National Underwater Tunnel Construction and Management Technology Exchange Meeting. China Highway Society (2015)
2. Wang, Q., Xin, Z., Jiang, B.: Comparative experimental study on mechanical mechanism of combined arches in large section tunnels. Tunn. Undergr. Space Technol. **99**, 103386 (2015)
3. Wei, S., Linxing, G., Zhuyin, W.: Stress analysis and engineering application of large section rectangular shield tunnel. Tunnel Constr. **35**(10), 1028–1033 (2015)
4. Xiaoduo, O., Hongying, W., Songyang, H.: Construction monitoring and mechanical characteristics analysis of super large section box open cut tunnel. J. Civ. Eng. **07**, 133–140 (2013)
5. Dianchun, D., Daniel, D., Ngocanh, D.: Lining performance optimization of sub-rectangular tunnels using the hyperstatic reaction method. Comput. Geotech. **117**(1), 103279.1–103279.16 (2020)
6. Daxi, F.: Study on section form of urban large section open cut tunnel. Tunnel Constr. **08**, 1026–1031 (2017)
7. Xianmin, H., Minglei, S., Wenjiang, L.: Optimization of tunnel section shape and support parameters under complex conditions, Geotech. Mech. (S1), 725–731 (2011)
8. Venus, S.., Xibo, H.: Stability analysis of surrounding rock in super large tunnel construction stage, Railway Constr. **58**(11), 79–82 (2018)
9. Xiaoming, H., Wenling, J., Peng, P.: Comparative analysis of section forms of a large section tunnel. West. Transp. Technol. **06**, 74–77 (2016)
10. Ping, S., Yuan, X., Kunming, M.: Analysis and optimization of tunnel lining structure. J. Jinling Univ. Sci. Technol. **02**, 29–33 (2019)
11. Jianqiang, C., Jihua, Z., Jubing, Z.: Research on tunnel section shape optimization and support technology. West. Transp. Technol. **12**, 76–80 (2019)

Study on Flexural Behavior of Coral Concrete Beams Reinforced with BFRP Bars

Yi Zhang[1,2], Baikun Chen[1], Xiaolan Yuan[1], and Bing Liu[1,2](✉)

[1] College of Civil Engineering and Architecture, Guilin University of Technology, Guilin 541004, Guangxi, China
cnulb@glut.edu.cn

[2] Guangxi Key Laboratory of New Energy and Building Energy Saving, Guilin 541004, Guangxi, China

Abstract. Coral concrete structure reinforced with fiber reinforced polymer (FRP) bars is a new type of structure with broad application prospect in island construction. To this end, the failure pattern, midspan deflection, concrete strain, basalt fiber reinforced polymer (BFRP) bars strain and crack characteristics of coral concrete beams reinforced with BFRP bars (CCBRB)were analyzed through the three-point flexural tests. The test results showed that the CCBRB have two failure modes: flexural failure and shear failure, and the strain of mid-span concrete conforms to the plane section assumption. The flexural behavior of CCBRB is similar to that of ordinary concrete beams reinforced with BFRP bars (OCBRB). Reinforcement ratio increased from 1.22% to 2.34%, the cracking load and ultimate load of CCBRB increased by 33% and 25% respectively, and the mid-span deflection, crack width and average crack spacing decrease by 30%, 50% and 66% respectively. The cracking load of all specimens is 15–20 kN, about 10% of the ultimate load. CCBRB has a certain deformation recovery ability. The flexural capacity of CCBRB is calculated by the formula derived by Wang et al. The calculated value is closer to the experimental value.

Keywords: Coral concrete · BFRP bars · Flexural behavior · Failure pattern · Reinforcement ratio

1 Introduction

Island construction is the key to promoting the strategy of maritime power, but it is faced with problems such as lack of traditional materials and inconvenient transportation. In view of this situation, domestic and foreign scholars have conducted systematic experimental studies on coral concrete and found that it is a new building material suitable for island construction. However, the chloride ion content in coral concrete is too high and the service environment is savage, which leads to serious corrosion of reinforcement and serious degeneration of structural durability. In order to solve this problem, scholars proposed to use FRP bars to replace steel bars to form a new type of FRP reinforced with coral concrete structure. Due to the good corrosion resistance of FRP bars, FRP bars are not easily corroded in coral concrete structures, which greatly improves the durability

W. Guo and K. Qian (Eds.): GBCESC 2022, LNCE 211, pp. 974–983, 2023.
https://doi.org/10.1007/978-981-19-5217-3_99

of the structure. BFRP rebar is a new type of FRP rebar with high performance-price ratio, and its application prospect is more extensive. Beam is a basic transfer member in reinforced concrete structures, which may be suffer the load caused by local collapse [1–15]. Therefore, this paper selects BFRP rebar as FRP rebar to form CCBRB, and then studies its static flexural behavior. The research results can provide experimental basis and technical reference for promoting the application of coral concrete structure reinforced with BFRP in island construction.

2 Experiment

2.1 Specimens Design

Three CCBRB and one OCBRB were made in this test. The test variables include reinforcement ratio and concrete type. The total length of the beam is 1800 mm, the section size is 150 mm × 250 mm, and the thickness of the concrete protective layer is 15 mm. The dimensions and reinforcement details of the test beams are shown in Fig. 1, and the specific reinforcement parameters for each beam are shown in Table 1.

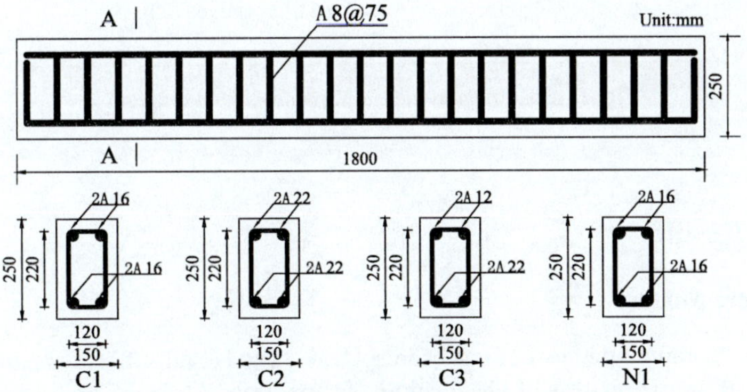

Fig. 1. Specimens details

Table 1. Specimen information

Designation	Compression side	Tension side	ρ/%	ρ'/%
C1	2ϕ16	2ϕ16	1.22	1.22
C2	2ϕ22	2ϕ22	2.34	2.34
C3	2ϕ12	2ϕ22	2.34	0.68
N1	2ϕ16	2ϕ16	1.22	1.22

Note: C stands for coral concrete and N stands for ordinary concrete

2.2 Test Loading and Measurement

Figure 2 is the loading and measuring point layout diagram. MTS with a range of 1000 kN was used to carry out mid-span single point loading, and graded monotone loading was carried out on the specimen according to GB/T50152-2012 standard [16]. The specimen was loaded 5kN in each step before cracking and 5mm in each step after cracking. The measurement mainly includes load, deflection, longitudinal reinforcement strain, stirrup strain, concrete strain and crack width. The data are collected by DH3816N static stress and strain testing and analysis system.

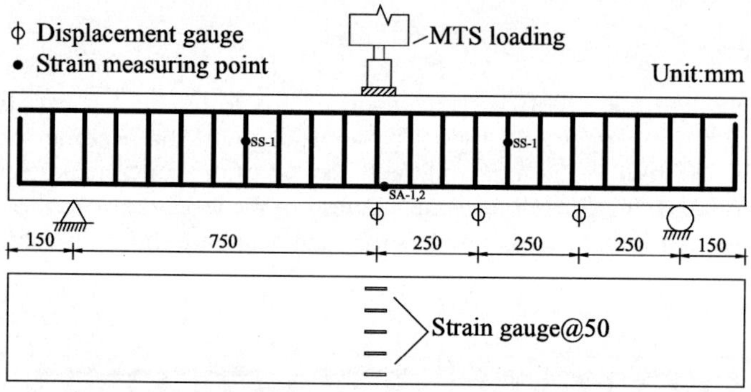

Fig. 2. Loading and measuring point layout diagram

3 Test Results

3.1 Failure Modes

Figure 3 is the failure patterns of test beams. C1, C3 and N1 all suffered flexural failure. The flexural failure process of C3 specimen is analyzed. At the beginning of loading, the specimen is in full section working state. As the load increases, the vertical cracks first appeared in the middle of the span, and then the cracks began to expand rapidly. The number of cracks continued to increase and the vertical cracks evolved into flexural cracks. The compression zone of concrete is shrinking, and the mid-span deflection increases rapidly after reaching 80% of the ultimate load. Finally, the specimen was damaged because the concrete in the compression zone reached the ultimate compressive strength. Shear failure occurred in C2 because the stirrup reached its ultimate tensile strength. One is due to the large reinforcement ratio of C2, the other is due to the strength reduction of thermosetting BFRP stirrups caused by reprocessing, resulting in stress concentration. The crack distribution of C1 and N1 is basically similar, but C1 has more small branch cracks. CCBRB with large reinforcement ratio have denser crack spacing.

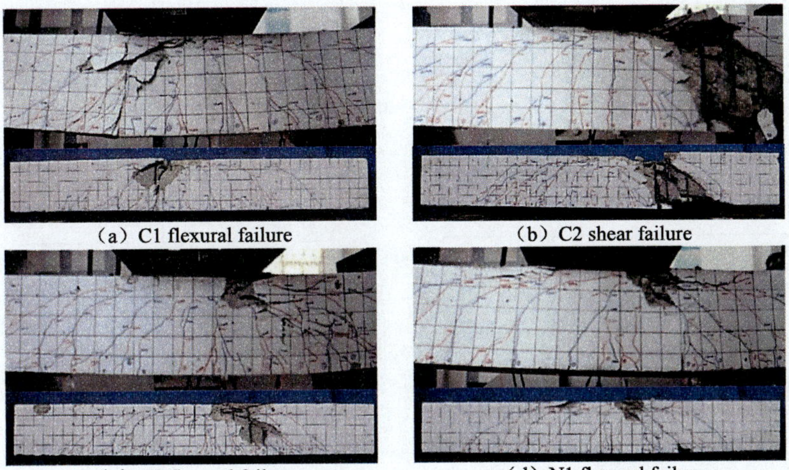

(a) C1 flexural failure (b) C2 shear failure

(c) C3 flexural failure (d) N1 flexural failure

Fig. 3. Failure pattern of specimens

3.2 Concrete Strain

Figure 4 is the concrete strain at the mid-span section under different loads. The strain values of the measuring points on the mid-span section of all specimens are linearly related to the distance from the point to the neutral axis, which indicated that the CCBRB meet the plane section assumption under normal design conditions.

3.3 Load-Deflection Curve

Figure 5 is the load-deflection curves of all specimens, showing bilinear characteristics before reaching the peak load. The specimen is in elastic working stage at the beginning of loading and the curve is approximately linear. After cracking load, the curve appears inflection point and the deflection growth rate gradually accelerates. After 80% of the peak load, the slope of the curve further decreases. The descending section of C1 and C3 curves is gentle, indicating that the CCBRB with these reinforcement ratios can continue to bear the load after reaching the ultimate load and has good ductility. After the ultimate load of C2, the bearing capacity decreased sharply, indicating that its ductility is not good. Under the same load, the deflection of C2 is about 30% smaller than that of C1, indicating that increasing the ratio of reinforcement can inhibit the development of CCBRB deflection.

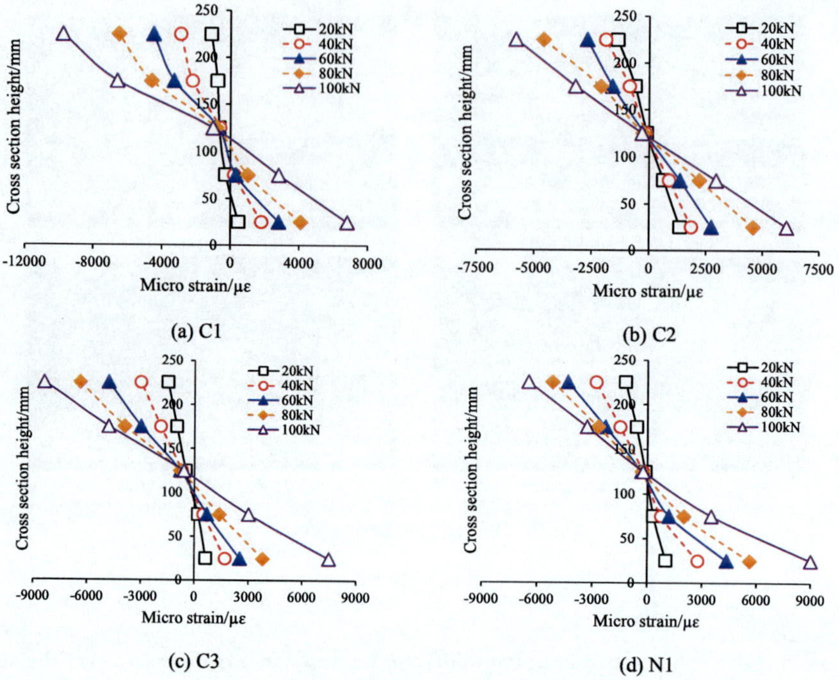

Fig. 4. Concrete strain-section depth relationships

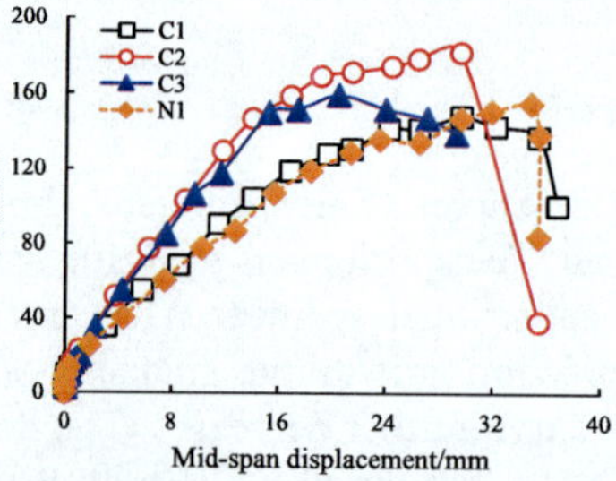

Fig. 5. Load-deflection relationships

Figure 6 is the cracking load and ultimate load of all specimens. The initial crack load of all specimens is 15–20 kN, which is about 10% of the peak load. The cracking load is not significantly affected by the reinforcement ratio, because it mainly depends on the ultimate tensile strength of concrete. The ultimate load of C2 is 25% and 13% higher than that of C1 and C3, respectively. It shows that the peak load of CCBRB can be improved by increasing the reinforcement ratio and applying large diameter compressive reinforcement, because this can improve the flexural rigidity of the specimens, and then improve the ultimate flexural capacity. The difference of ultimate flexural capacity between C1 and N1 is 5%, and the initial cracking load is the same. It shows that when

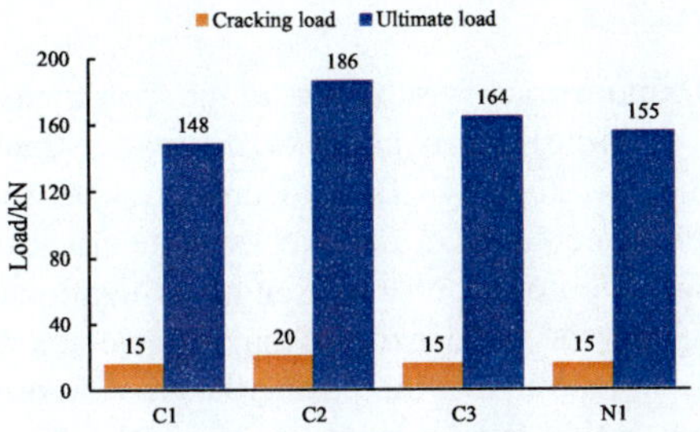

Fig. 6. Cracking load and ultimate load

the reinforcement ratio is identity, the flexural bearing capacity of CCBRB and OCBRB is similar.

3.4 Strain Analysis of BFRP Bars

Figure 7 shows the load-longitudinal strain curve, which is the average of SA-1 and SA-2. The curve trend is similar to its load-deflection curve. The strain of longitudinal reinforcement is similar before cracking, because the tensile reinforcement and concrete bear the load together, and the stress is small. The slope of the curve decreases obviously after cracking, because the stress redistribution occurs in the cross section, and the BFRP bars need to bear more load, so the strain growth rate of the longitudinal reinforcement is accelerated. Under the same load, the longitudinal strain of C2 is about 50% and 20% smaller than that of C1 and C3, respectively. It is further verified that increasing the reinforcement ratio can reduce the deflection of CCBRB.

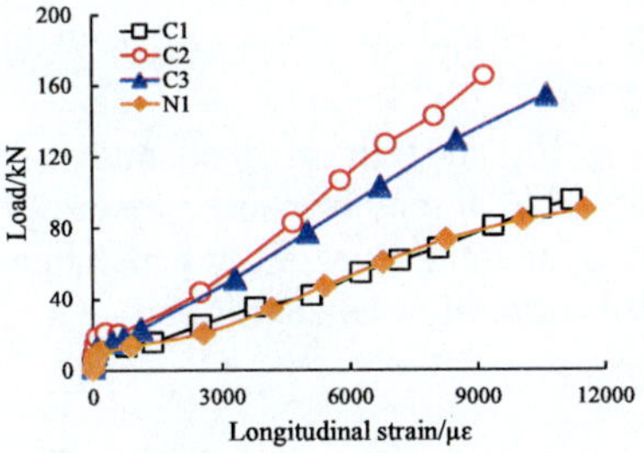

Fig. 7. Load-reinforcement strain curves

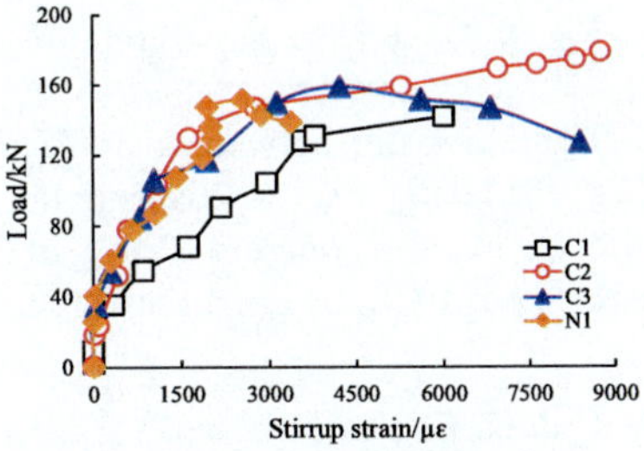

Fig. 8. Load-stirrup strain curves

Figure 8 is the load-stirrup strain curve at the SS-1 strain measuring point. Before the formation of inclined cracks, the stirrup strain is zero, and most of the shear force is resisted by concrete. After the formation of oblique cracks, concrete and stirrup resist shear force together. The curve shows an obvious inflection point, and the strain growth rate of stirrups has accelerated. The strain value C2 is the largest, followed by C3, C1 and N1, which is consistent with failure mode.

3.5 Crack Characteristics

Figure 9 is the load-maximum crack width curve of the specimens. Maximum crack width C2 corresponding to the same load is about 50% smaller than that of C1, indicating that the maximum crack width of CCBRB decreases with increasing reinforcement ratio. Increasing the reinforcement ratio can resist large cracking stress and effectively inhibit the development of cracks, but the strength utilization rate of BFRP bars will decrease accordingly. Under the same load, the maximum crack width of C2 is about 25% smaller than that of C1, indicating that increasing the area of compression reinforcement contributes to inhibiting crack development but has little effect. The maximum crack width at ultimate load is 1.4 mm for C1, 1.0 mm for C2, 0.9mm for C3 and 1.2 mm for N1. The maximum crack width of concrete beams reinforced with reinforced FRP should be strictly controlled during service. ACI 440.1 R-15 [17] limit structure maximum allowable crack width is 0.5 mm. If the load corresponding to 0.5 mm is used as the control standard, C1 and N1 are only 30% ~ 35% of the ultimate load, while C2 and C3 are 60%– 65% of the ultimate load. This should be considered in design use. The residual crack width decreases after unloading, indicating that CCBRB have a certain deformation recovery ability.

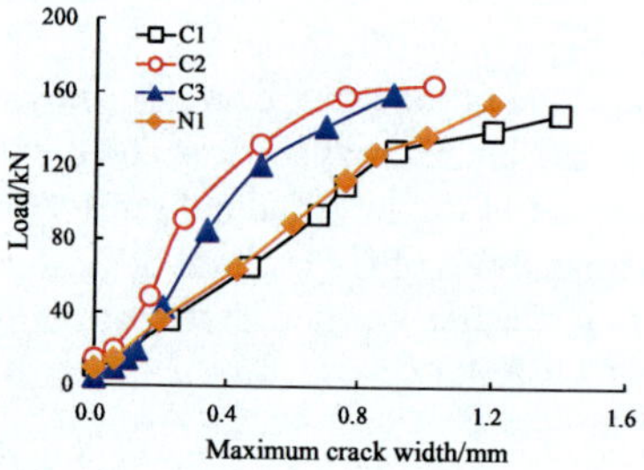

Fig. 9. Load-max crack width curves

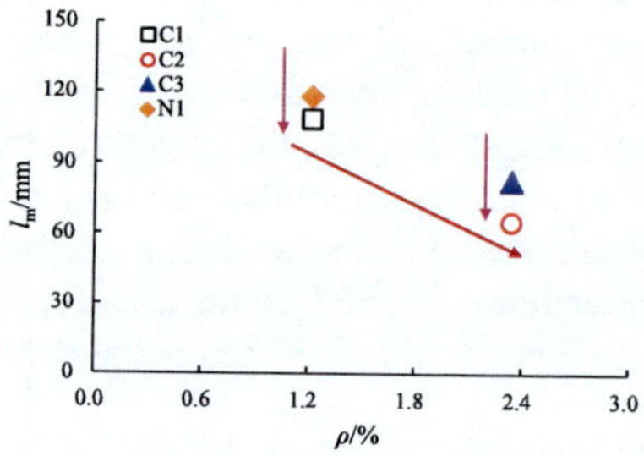

Fig. 10. $l_m - \rho$ curves

The relationship between average crack spacing (l_m) and reinforcement ratio is shown in Fig. 10. The l_m f C1 is 20% smaller than that of N1, it may be caused by the greater brittleness of coral concrete. The l_m of C2 is 66% smaller than that of C1, which indicates that the l_m of CCBRB decreases with the increase of reinforcement ratio.

3.6 Calculation of Flexural Capacity

There is no code for calculating flexural capacity of CCBRB. Literature [18], GB 50608-2010 [19] and ACI 440.1R-15[17] code of FRP reinforced concrete beam flexural capacity calculation formula as shown in Table 2. Theoretical and experimental values are shown in Table 3. The results obtained by various calculation methods are less than the actual test results. The results calculated by ACI 440 1R-15 [17] code are most conservative, about 30% lower than the experimental values. The calculated value of GB 50608-2010 [19] code is closer to the experimental value than that of ACI 4401R-15 [17] code. The main reason is that considering the safety degree, the design value of materials is used in the specification formula. The reason that the results calculated by the derived formula in the Literature [18] are closer to the experimental values is that the contribution of the compression bars is considered.

Table 2. Comparison between calculated and experimental values

No	P_u	M_u	Literature [18]		GB 50608–2010 [19]		ACI 440 1R-15 [17]	
			M_c	M_c/M_u	M_c	M_c/M_u	M_c	M_c/M_u
C1	148	55.5	47.2	0.81	44.9	0.73	41.1	0.67
C3	164	61.5	53.5	0.84	49.2	0.76	46.1	0.71
N1	155	58.1	48.2	0.82	45.9	0.74	44.2	0.69

Note: P_u is ultimate load (kN), M_u is experiment value, M_c is theoretical value (kN.m)

Table 3. Calculation formula

Category	Formula
Literature [18]	$M = \alpha_1 \beta_1 f_c bx\left(h_0 - \frac{\beta_1 x}{2}\right) + E_f \varepsilon_{f'} A'_f (h_0 - a'_s)$ $\varepsilon_f = \varepsilon_{cu} \frac{(h_0 - x_0)}{x_0} \left(\varepsilon_{f'} = \frac{\varepsilon_{cu}(x_0 - a'_s)}{x_0}\right)$ $\alpha_1 \beta_1 f_c bx + E_f \varepsilon_{f'} A'_s = E_f \varepsilon_f A_f$ $x = \beta_1 x_0 (\rho_f \geq \rho_{fb})$
GB 50608-2010 [19]	$M = f_{fe} A_f (h_{0f} - x/2)\ \rho_{fb} = \frac{\alpha_1 f_c \beta_1 \varepsilon_{cu}}{f_{fd}(\varepsilon_{cu} + f_{fd})/E_f}$ $f_{fe} = f_{fd}(\rho_f/\rho_{fb})^{-0.5} (\rho_f \geq 1.5\rho_{fb})\ x = \frac{f_{fe} A_f}{f_c b}$
ACI 440 1R-15 [17]	$M_n = \rho_f f_f \left(1 - 0.59 \frac{\rho_f f_f}{f'_c}\right) bd^2$ $f_f = \sqrt{\frac{(E_f \varepsilon_{cu})^2}{4} + \frac{0.85 \beta_1 f'_c E_f \varepsilon_{cu}}{\rho_f}} - 0.5 E_f \varepsilon_{cu}$ $\rho_{fb} = 0.85 \beta_1 \frac{f'_c E_f \varepsilon_{cu}}{f_{fu}(E_f \varepsilon_{cu} + f_{fu})} (\rho_f \geq \rho_{fb})$

4 Conclusions

(1) There are two failure patterns of CCBRB: flexural failure and shear failure. The types of concrete (ordinary concrete and coral concrete) have little effect on the ultimate bearing capacity of BFRP reinforced concrete beams. Their cracking loads are the same, and the difference in ultimate loads is only 5%.

(2) The cracking load of CCBRB is 15–20 kN, which is about 10% of the ultimate load. Increasing reinforcement ratio within a certain range can reduce deflection and improve ultimate bearing capacity.
(3) The maximum crack width and average crack spacing of CCBRB are greatly affected by the reinforcement ratio. If the load corresponding to 0.5 mm is used as the control standard, C1 and N1 are only 30%–35% of the ultimate load, while C2 and C3 are 60%–65% of the ultimate load. CCBRB have a certain deformation recovery ability.
(4) The flexural capacity of CCBRB is calculated by ACI 440 1R-15 [17] and GB 50010-2010 [19], and the results are about 30% lower than the experimental values. The results calculated by the derived formula in the literature [18] are closer to the experimental values.

Acknowledgement. The work was supported by NSFC project 52022024, Major Scientific Research Project of Guangxi (Gui Ke Neng22-J-21-4, No.2021GXNSFFA196001, Gui Ke AD21220071, 2021KY0254), and the China Postdoctoral Science Foundation (2021M693814).

References

1. Qian, K., Li, Z., Weng, Y., et al.: Behavior of RC beam-slab structures to resist progressive collapse. Eng. Mech. **36**(06), 239–247 (2019)
2. Li, Z., Weng, Y., Deng, X., et al.: Behavior of precast concrete beam-slab substructures with welded connections to resist progressive collapse. J. Build. Struct. **41**(10), 121–128 (2020)
3. Qian, K., Li, Z., Weng, Y., et al.: Load resisting mechanisms of assembled monolithic precast concrete beam-slab substructures to resist progressive collapse. J. Build. Struct. **42**(07), 183–193 (2021)
4. Qian, K., Li, Z., He, P., et al.: Progressive collapse mechanism of PC beam-slab substructure with bolted connections. J. Build. Struct. **41**(01), 173–180 (2020)
5. Deng, X., Li, Z., Weng, Y., et al.: Experimental study on performance of prestressed concrete beam-column substructure against progressive collapse. J. Build. Struct. **40**(08), 71–78 (2019)
6. Yu, Y., Li, Z., Xiao, L., et al.: Load resisting mechanism of precast structure under exterior column failure. Eng. Mech. **38**(04), 159–168 (2021)
7. Qian, K., Luo, D., He, S., et al.: Progressive collapse performance of reinforced concrete frame structures subjected to ground two-column removal scenario. J. Build. Struct. **39**(01), 61–68 (2018)
8. Qian, K., Li, X., Yuan, X., et al.: Behavior of multi-story steel frame considering influence of composite slab to resist progressive collapse. J. Build. Struct. **42**(S2), 11–21 (2021)
9. Yu, X., Qian, K., Lü, D.: Progressive collapse capacity analysis of reinforced concrete frame structures considering catenary action. J. Build. Struct. **38**(04), 28–34 (2017)
10. Yu, X., Qian, K., Lü, D.: Probabilistic assessment of structural resistance of RC frame structures against progressive collapse using random Pushdown analysis. J. Build. Struct. **38**(02), 83–89 (2017)
11. Li, Z., Weng, Y., Deng, X., et al.: Tests and numerical simulation for anti-progressive collapse of RC frame reinforced with steel braces. J. Vib. Shock. **41**(05), 173–181 (2022)
12. Qian, K., Huang, Z., Li, H., et al.: Numerical study on progressive collapse performance of infilled RC frame. J. Build. Struct **41**(S1), 221–229 (2020)

13. Qian, K., Li, B., Ma, J.X.: Load-carrying mechanism to resist progressive collapse of RC buildings. J. Struct. Eng. **141**(02), 04014107 (2015)
14. Qian, K., Liu, Y., Yang, T., et al.: Progressive collapse resistance of posttensioned concrete beam-column subassemblages with unbonded posttensioning strands. J. Struct. Eng. **144**(01), 04017182 (2018)
15. Qian, K., Li, B., Yang, T., et al.: Progressive collapse resistance of posttensioned concrete beam-column subassemblages with unbonded posttensioning strands. J. Struct. Eng.**144**(01), 04017182 (2018)
16. GB 50152–2012. Standard for test method of concrete structures. China Architecture & Building Press (2012)
17. ACI 440 1R-15. Guide for the design and construction of structural concrete reinforced with fiber reinforced polymer bars (2015)
18. Wang, X., Ma, H., Liu, X.: Calculation formula for bending capacity of concrete beams with FRP bars. J. Shenyang Univ. **33**(03), 252–257 (2020)
19. GB 50608–2010, Technical specification for application of fiber reinforced composite materials in construction projects. China Planning Press (2010)

Stochastic Model of Typhoons Along the China Coast and Its Application into Risk Assessment of Sea-Crossing Bridge Under Multiple Hazards

Kai Wei[1](✉), Zhonghui Shen[2], Xi Zhong[1], and Zhiyu Bai[1]

[1] Department of Bridge Engineering, Southwest Jiaotong University, Chengdu, China
kaiwei@swjtu.edu.cn

[2] Department of Hydraulic Engineering, Tsinghua University, Beijing, China

Abstract. Typhoon-induced strong wind, extreme wave, and storm surge pose a significant threat to the structural and operational safety of a sea-crossing bridge. However, historical typhoons are usually insufficient in determining extreme hazards. Stochastic modeling of typhoons can expand the TC database and has been widely used in the risk assessment of TC-induced hazards. This paper introduced a full-track stochastic model to generate synthetic TCs based on historical TCs and climatology data along the China coast. A variable geo-graphical sampling approach is used to capture spatial heterogeneity. The track model is derived using the Cartesian coordinate system and considers the statistical characteristics of TC lysis. The one-dimensional ocean model is included to consider the ocean feedback on the typhoons. A catalog of 10,000-year synthetic typhoon events around the China coast is then generated by the developed stochastic typhoon model. The practical applications of the developed model are illustrated by an example sea-crossing cable-stayed bridge. The wind, wave, storm surge and current at bridge site under all the selected synthetic typhoon events are simulated by SWAN+ADCIRC. The joint probability of the typhoon-induced extreme conditions is established using the Copula theory. The environmental contour with a given return period is then obtained by the inverse first order reliability method. The stochastic dynamic responses of bridges during typhoons are finally investigated based on the obtained environmental contour under typhoons. This paper provides an insight into the risk assessment of the sea-crossing bridge under typhoon-induced extreme conditions in a life-cycle context.

Keywords: Typhoon · Sea-crossing bridge · Stochastic models

1 General Instructions

Typhoon is a destructive low-pressure weather system formed in tropical or subtropical oceans. And it poses a serious threat to the safety of sea-crossing bridges due to accompanied strong winds, huge waves, currents. The design of sea-crossing bridges, ports, offshore wind turbines, cofferdams and so on should ensure the safety under extreme conditions induced typhoons [3, 4]. Due to the limitation of historical typhoons, the

W. Guo and K. Qian (Eds.): GBCESC 2022, LNCE 211, pp. 984–991, 2023.
https://doi.org/10.1007/978-981-19-5217-3_100

traditional probability model extrapolation cannot meet the estimation of the long return period of extreme environmental factors in the sea area of all interest points. At the same time, extrapolating is sensitive to the tail of the probability distribution, and it exists obvious uncertainty. Compared with historical typhoons, the full-path stochastic simulation of typhoons can expand the typhoon database, and has been widely used in typhoon disaster insurance and other fields that require high accuracy of probability distribution calculation, as well as in wind engineering specifications [5]. The typhoon design wind speed of ASCE 7-10 is determined based on the hurricane model proposed by Vickery et al. [6] At the same time, the integration of typhoon full-path model and global climate model is a widely used downscaling method to study the impact of climate change on typhoon activities, which has important potential for assessing the future typhoon disaster risk and structural toughness [7].

Vickery et al. [6] first developed the artificial technology for synthesizing hurricanes. However, these studies usually divide the geographical space evenly, ignoring the spatial heterogeneity of typhoons. Uniform grid division leads to too much data in some areas, while some areas lack sufficient data for reliable regression analysis. Meanwhile, in the process of typhoon crossing the marine area, it will cause substantial disturbance to the water and increase the depth of marine mixed layer. When the depth of the ocean mixing layer increases, the ocean surface temperature decreases and the heat supplement of the typhoon decreases. This process is called ocean feedback. Therefore, ocean feedback to some extent inhibits the unrealistic development of typhoon [6]. Considering the in-sufficient consideration of spatial heterogeneity and ocean feedback of typhoons in the previous typhoon full-path model, this paper carried out the stochastic modeling of typhoon full-path considering spatial heterogeneity and ocean feedback. At the same time, three important application practices of typhoon full-path stochastic simulation technology are described: (1) Typhoon disaster map is the basis of offshore typhoon disaster assessment, risk analysis, bridge location and wind resistance design. The typhoon disaster along the eastern coast of China is calculated by using typhoon full-path simulation technology; (2) Based on the long-term numerical simulation of typhoon and wind, wave and tide under typhoon, the multi-dimensional joint probability model of wind and wave is established by using Pair-Copula theory. (3) In order to explore the dynamic disaster mechanism of sea-crossing bridge under typhoon, the stochastic dynamic analysis method of bridge under typhoon wind-wave-current-tide is established, and the dynamic response characteristics of sea-crossing bridge under typhoon are studied.

2 Full Path Stochastic Model of Typhoon Considering Spatial Heterogeneity and Ocean Feedback

Based on the time discretization thought of Vickery et al. [6] the typhoon full path model established in this paper is as follows.

Path propagation model:

$$x_{i+1} = x_i + \Delta x_i \tag{1}$$

$$y_{i+1} = y_i + \Delta y_i \tag{2}$$

$$\Delta x_i = a_0 + a_1 \Delta x_{i-1} + a_2 \Delta y_{i-1} + \varepsilon_{\Delta x_i} \tag{3}$$

$$\Delta y_i = b_0 + b_1 \Delta y_{i-1} + b_2 \Delta x_{i-1} + \varepsilon_{\Delta y_i} \tag{4}$$

Termination model:

$$P_i = \frac{\sum_j \Theta_j e^{-d_j^2/2L^2}}{\sum_j e^{-d_j^2/2L^2}} \tag{5}$$

Strength regression model:

$$\ln(I_{i+1}) = c_0 + c_1 \ln(I_i) + c_2 T_{s,i} + c_3 (T_{s,i+1} - T_{s,i}) + c_4 V_s + \varepsilon_I \tag{6}$$

Land filling model:

$$\Delta p(t) = \Delta p_0 \times \exp(-d \cdot t) \tag{7}$$

$$d = d_0 + d_1 \Delta p_0 + \varepsilon_d \tag{8}$$

The kilometer posts (KPs) along the coast of China are used to validate the established typhoon full path model. In this paper, the characteristic parameters of historical typhoon and random simulated typhoon in the nearshore are compared, including annual incidence, moving speed, moving direction, minimum distance from typhoon to station and maximum central pressure difference in center, as shown in Fig. 1. At the same time, in order to illustrate the advantages of considering geographical spatial heterogeneity, this paper also calculates the simulation results of path characteristics using uniform grids (5° × 5°).

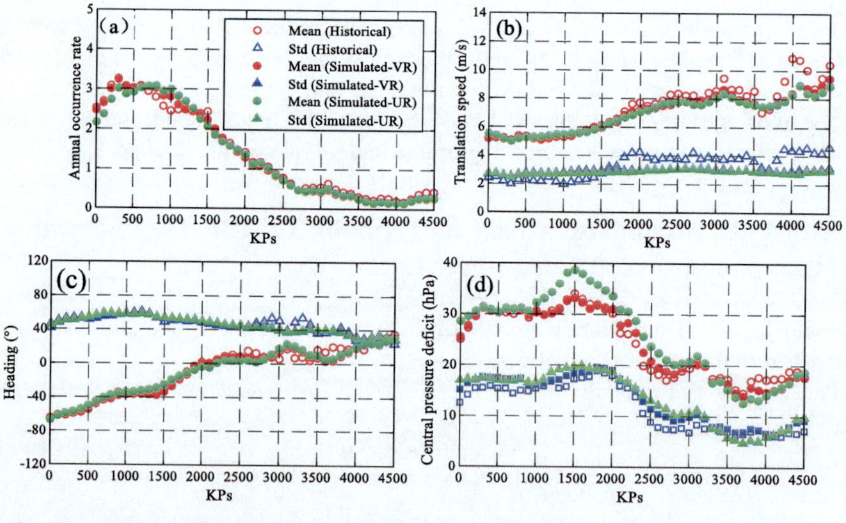

Fig. 1. The statistical characteristics of TCs at each KP; (a) annual occurrence rate λ; (b) translation speed c (m/s); (c) heading θ (°); (d) central pressure deficit Δpc (hPa).

As shown in Fig. 1, the model established in this paper can well capture the statistical characteristics of coastal typhoons. At the same time, compared with the uniform geographical space for regression, the path characteristics obtained by using the variable geographical space based on the density around the target point are closer to the historical path characteristics.

In order to illustrate the effect of ocean feedback on the strength model, Fig. 2 shows the time evolution of typhoons in the whole life cycle with and without ocean feedback. As shown in Fig. 2, considering the ocean feedback can effectively control the development of typhoon intensity. At the same time, the time development of typhoon in the whole life cycle simulated by the intensity model in this paper is in good agreement with the historical results, and has certain credibility.

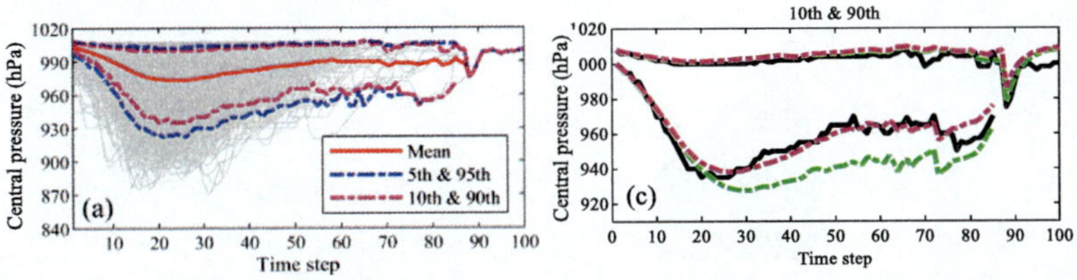

Fig. 2. The temporal development of the central pressure during the total lifetime of TCs.

3 Design Parameters of Extreme Environment for Sea-Crossing Bridges

3.1 Typhoon Disaster Along the East Coast of China

Based on the established typhoon full-path modeling technology, the typhoon wind disaster maps (50-year, 100-year, 200-year typhoon wind speed) along the eastern coast of China are calculated, as shown in Fig. 3.

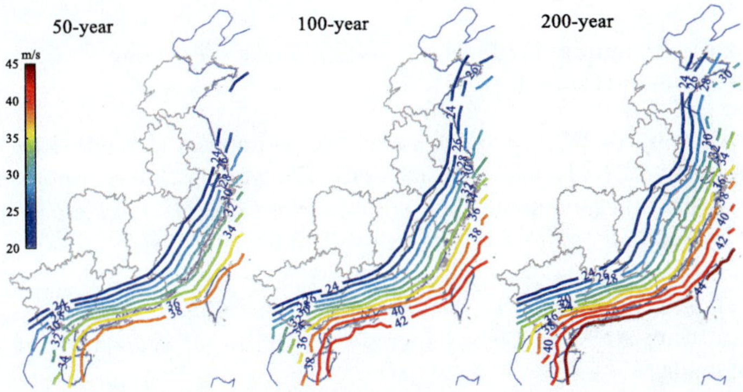

Fig. 3. Wind hazard maps of the coastal region of China: a) 50-year; b) 100-year; and c) 200-year.

3.2 Multidimensional Joint Probability Model of Wind and Wave

Based on the established typhoon full-path modeling technology and the numerical model of wave and storm surge under typhoon (SWAN+ADCIRC wave-current coupling model), the environmental factors of wind speed and wave height under long-term typhoon affecting Pingtan Bridge are calculated. At the same time, based on Pair-Copula theory and inverse first-order reliability method, the multidimensional joint probability model of typhoon wind and wave and the environmental equivalent surface (recurrence period) model are calculated, as shown in Fig. 4.

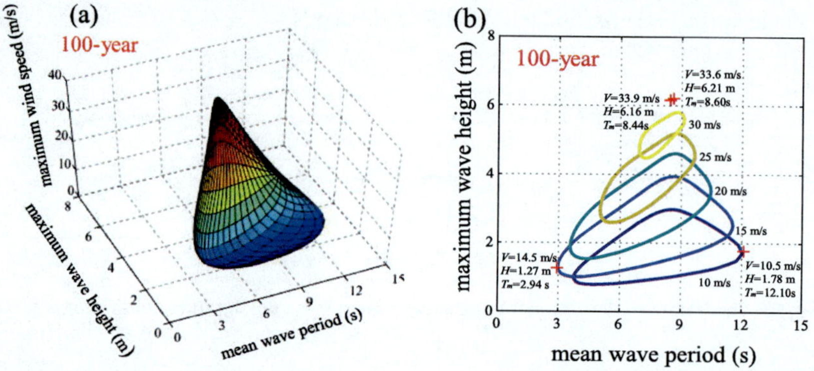

Fig. 4. Three-dimensional probability and recurrence period model of maximum wind speed, maximum wave height and average wave period of typhoon: a) Three-dimensional probability and recurrence period model of maximum wind speed, maximum wave height; and b) Average wave period of typhoon

Three-dimensional Pair-Copula model:

$$f(x_1, x_2, x_3) = c_{12}(F_1(x_1), F_2(x_2)) \cdot c_{23}(F_2(x_2), F_3(x_3)) \cdot \\ c_{13|2}\big(F_{1|2}(x_1|x_2), F_{3|2}(x_3|x_2)\big) \cdot f_1(x_1) \cdot f_2(x_2) \cdot f_3(x_3) \tag{9}$$

3.3 Stochastic Dynamic Analysis of Bridge Under Typhoon Wind-Wave-Current-Tide

Using typhoon database and numerical simulation technology (SWAN+ADCIRC wave-current coupling model) of wind, wave and tide under typhoon, the environmental parameters under typhoon are obtained. The finite element model of sea-crossing bridge is established by ANSYS.

(1) *Finite element model of sea -crossing bridge.* The bridge is modeled by finite element software ANSYS. Figure 5 shows a diagram of typhoon wind, wave and current loads.

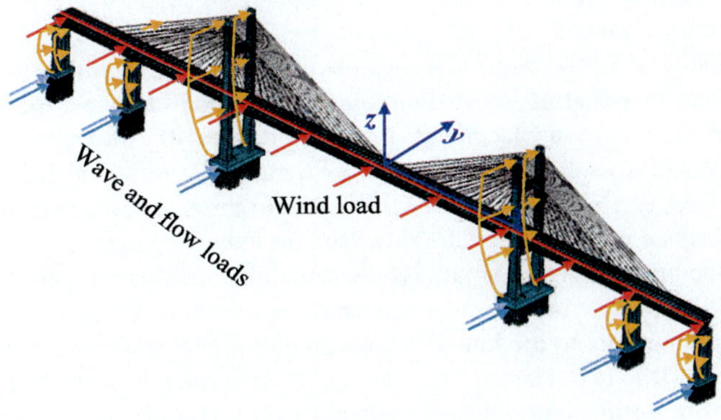

Fig. 5. Finite element model of the bridge

(2) *Study case* One synthetic typhoon was selected to analyze the random vibration characteristics of sea-crossing bridge under typhoon.

In order to describe the random vibration characteristics of the bridge during typhoon 'Aere' landing, the random dynamic responses of the bridge at *t1*, *t2*, *t3* and *t4* are calculated, and the simulation time is 10 min.

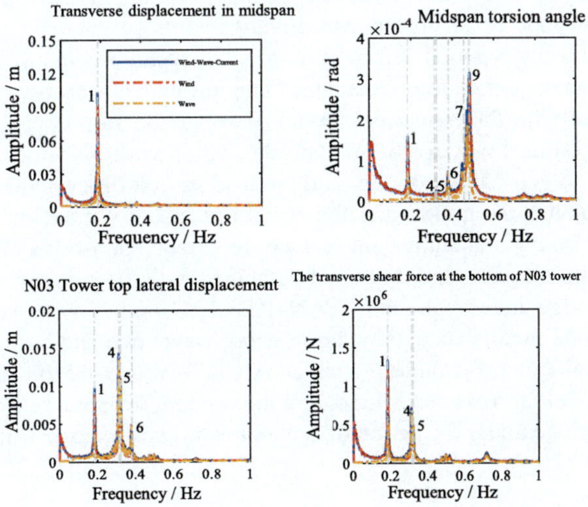

Fig. 6. The dynamic response amplitude spectrum of the: a) Mid-span lateral displacement; b) Mid-span torsion angle; c) Transverse displacement of tower top; and d) Transverse Shear Force of Tower Bottom.

In order to illustrate the dynamic response characteristics of bridges during typhoon landing, taking time *t2* as the representative, Fig. 6 shows the response amplitude spectra of B1, B2, B3 and B4 under the action of single wind load, single wave load and wind-wave-current disasters.

According to Fig. 6, the wind load has obvious influence on the mid-span lateral displacement and torsional angle of the main beam, while the wave load has a great contribution to the lateral displacement of the top of the N03 tower, which is mainly related to the double-column bridge tower of Pingtan Bridge. According to the CFD numerical analysis carried out by Li et al. [8], due to the interference of the upstream tower, the resistance coefficient of the downstream tower is negative, and the overall wind load on the bridge tower is small. At the same time, compared with the first-order vibration, the high-order vibration caused by waves is obvious. Under single wave load, the modes that contribute to the lateral displacement of N03 tower top are mainly 4–6 orders (>0.3 Hz). Due to the strong wind action of typhoon, the wind wave component of wave spectrum during typhoon is significant (>0.1 Hz), Therefore, for long-span bridges with significant participation of high-order vibration, ignoring the variation of wave spectrum during typhoon (significant wind wave component) will undoubtedly underestimate the dynamic response of bridges.

4 Conclusions

(1) Considering the spatial heterogeneity of typhoon and ocean feedback can effectively improve the stochastic simulation accuracy of typhoon, and for statistical intensity model, ocean feedback can control the development of typhoon intensity due to the balance between storm and ocean; the established typhoon path model and intensity model show good performance respectively;
(2) According to the typhoon full path model, the typhoon wind disaster map along the eastern coast of China is calculated. The full-path stochastic model of typhoon is combined with the numerical model of wave and storm surge under typhoon (SWAN+ADCIRC model) to simulate the wind, wave and tide under long-term typhoon. The Pair-Copula theory and the inverse first-order reliability method are used to establish the multidimensional joint probability model of wind and wave and the environmental equivalent surface (recurrence peri-od) model.
(3) The typhoon full-path random model combined with the numerical model of wave and storm surge under typhoon (SWAN+ADCIRC model) was used to simulate the environmental factors such as typhoon wind, wave, tide and current. At the same time, the finite element models such as ANSYS were used to establish the finite element model of cross-sea bridge, and the random dynamic response analysis of cross-sea bridge under the action of typhoon wind-wave-current-tide was realized.

Acknowledgements. The authors would like to acknowledge the support from the National Natural Science Foundation of China (Grant No. 51978578).

References

1. Kai, W., Hasan, I., Shunquan, Q.: Parametric wave spectrum model for typhoon-generated waves based on field measurements in nearshore strait water. J. Offshore Mech. Arct. Eng. **143**(5), 051201 (2021)
2. Kai, W., Zhonghui, S., Zilong, T., Shunquan, Q.: Trivariate joint probability model of typhooninduced wind, wave and their time lag based on the numerical simulation of historical typhoons. Stoch Environ. Res. Risk A **35**(2), 325–344 (2021)
3. Zhigang, H., David, V., Peter, R.: Long-term hurricane risk assessment and expected damage to residential structures. Reliab. Eng. Syst. Saf. **74**(3), 239–249 (2001)
4. Abdullahi, M., Yue, L., Mark, G.: Evaluating system reliability and targeted hardening strategies of power distribution systems subjected to hurricanes. Reliab. Eng. Syst. Saf. **144**, 319–333 (2015)
5. Weihua, F., Wei, L.: Summarization of typhoon wind field models for disaster risk assessment. Geosci. Progr. **32**(6), 852–867 (2013)
6. Peter, J., Dhiraj, W., Lawrence, A., Francis, M.: U.S. hurricane wind speed risk and uncertainty. J. Struct. Eng. **135**(3), 301–320 (2009)
7. Kerry, E., Ragoth, S., John, W.: Hurricanes and global warming: results from downscaling IPCC AR4 simulations. Bull. Am. Meteorol. Soc. **89**(3), 347–368 (2008)
8. Yongle, L., Chen, F., Kai, W., Guanghao, Z., Haojun, T.: Frequency domain dynamic analyses of freestanding bridge pylon under wind and waves using a copula model. Ocean Eng. **183**, 359–371 (2019)

Comparison of Stability of Four Numerical Integration Methods Under Negative Stiffness

Fantao Meng[1(✉)], Jianfeng Zhao[2], and Xingqun Ruan[1]

[1] Shandong Huake Planning and Architectural Design Co., Ltd., Liaocheng 252000, China
super20030@163.com
[2] Qingdao University of Technology, Qingdao 266520, China

Abstract. Numerical Integration Algorithm plays a very important role in real-time hybrid experiments. Based on the Newmark ($\gamma = 1/2$, $\beta = 1/4$) average constant acceleration algorithm, the stability analysis equations of Chang Algorithm, CR Algorithm and TL Algorithm under the condition of negative stiffness were derived by using the principle of positive stiffness analysis. The stability of the four algorithms under the condition of negative stiffness was analyzed by Matlab, and the stability characteristics of the four algorithms were studied by changing the integral time interval, damping ratio and stiffness ratio. The research findings indicate that when stiffness is negative, the Newmark average constant acceleration algorithm ($\gamma = 1/2$, $\beta = 1/4$) is no longer unconditionally stable, whearas the other three algorithms are all unconditionally stable. The Chang Algorithm has the best stability, while the CR algorithm is nearly as stable as the TL algorithm. It is suggested that while performing hybrid testing with probable negative stiffness, the numerical integration algorithm should be carefully chosen.

Keywords: Integral algorithm · Negative stiffness · Unconditional stability · Damping · Time interval

1 Introduction

Stability of the Algorithm is important in real-time hybrid testing due to the high cost of loading tests [1]. As a result, the step-by-step integration algorithm becomes a critical area of research for hybrid testing. Due to the fact that many structures or materials exhibit the phenomenon of increasing deformation but decreasing load after reaching the yield point, when the incremental stiffness is used to solve the dynamic equation, the restoring force skeleton curve will show the phenomenon of negative stiffness. Cheng Minxian [2] studied the convergence and stability of numerical integration under negative yield stiffness, using Newmark-β average constant acceleration method [3] and Houbolt method as examples. The convergence and stability of the central difference method, z-transform method, wilson-θ method and two-step Adams explicit method are then investigated [4]. The central difference method, z-transform method and two-step Adams explicit method are all unconditionally stable, whereas the Wilson-θ method is conditionally stable. Wu [5] and Li [6] explored the convergence and stability of Newmark-β method when

W. Guo and K. Qian (Eds.): GBCESC 2022, LNCE 211, pp. 992–1000, 2023.
https://doi.org/10.1007/978-981-19-5217-3_101

the stiffness is negative. With the development of hybrid experiment technology, many scholars have proposed various numerical integration algorithms, including the Chang method [7–12], the CR method [13–17], and the TL method [18, 19]. The convergence and stability of the above algorithms are studied under positive stiffness conditions, but not under negative stiffness conditions. The stability under negative stiffness conditions is explored in this study using the three methods mentioned above and the Newmark-β method. The stability of the four algorithms under the condition of negative stiffness was analyzed by Matlab, and the stability characteristics of the four algorithms were studied by changing the integral time interval, damping ratio and stiffness ratio. The research findings indicate that when stiffness is negative, the Newmark average constant acceleration algorithm ($\gamma = 1/2$, $\beta = 1/4$) is no longer unconditionally stable, whereas the other three algorithms are all unconditionally stable. The Chang Algorithm has the best stability, while the CR algorithm is nearly as stable as the TL algorithm.

2 Introduction of Four Numerical Integration Algorithms

2.1 Newmark-β Integral Algorithm

In structural dynamic calculation, Newmark (Newmark 1959) method is widely used. The analysis of the Algorithm shows that the parameters γ and β have a great influence on the characteristics of the algorithm. When $\beta \geq (0.5 + \gamma)^2/4$, $\gamma \geq 1/2$, the algorithm is unconditionally stable. In the Newmark-β Integration Algorithm family, the displacement and velocity expression as follows (1) and (2):

$$d_{i+1} = d_i + \Delta t v_i + (\frac{1}{2} - \beta)\Delta t^2 a_i + \beta \Delta t^2 a_{i+1} \tag{1}$$

$$v_{i+1} = v_i + (1-\gamma)\Delta t a_i + \gamma \Delta t a_{i+1} \tag{2}$$

β and γ have different values corresponding to different numerical characteristics in Eq. (1) and Eq. (2).

2.2 Chang Algorithm

By using the Newmark average constant acceleration method, which is iterated once by Newton method and ignores the influence of the external load term, the unconditionally stable explicit Chang Algorithm can be obtained as shown in Eq. (3) and (4). The velocity expression of Chang method is consistent with Newmark average constant acceleration method, implying that it has an implicit velocity expression. In addition, it should be noted that the load term in Eq. (4) cannot be ignored, and it can be proved that omitting the load term in Eq. (4) results in a loss of second order velocity accuracy for linear systems.

$$d_{i+1} = d_i + \beta_1 \Delta t v_i + \beta_2 \Delta t^2 a_i \tag{3}$$

$$v_{i+1} = v_i + (1 - \gamma)\Delta t a_i + \gamma \Delta t a_{i+1} \tag{4}$$

Chang's method: $\beta_1 = \frac{1+2\gamma\xi\Omega_0}{1+2\gamma\xi\Omega_0+\beta\Omega_0^2}$, $\beta_2 = \frac{\frac{1}{2}-2(\beta-\frac{1}{2}\gamma)\xi\Omega_0}{1+2\gamma\xi\Omega_0+\beta\Omega_0^2}$.

2.3 CR Algorithm

On the basis of discrete control theory, it is found that the pole characteristic roots of the transfer function are the same as those of the Newmark average constant acceleration method when the dynamic equilibrium equation is combined with Eqs. (5) and (6). Chen and Riles proposed CR algorithm [13, 14]. Under the condition of positive stiffness, because the Newmark average constant acceleration method is unconditionally stable, the obtained CR method is also unconditionally stable. It is not difficult to demonstrate that the accuracy of this algorithm is lower than second order.

$$d_{i+1} = d_i + \Delta t v_i + \alpha \Delta t^2 a_i \tag{5}$$

$$v_{i+1} = v_i + \alpha \Delta t a_i \tag{6}$$

where, $\alpha = \frac{1}{1+2\gamma\xi\Omega_0+\beta\Omega_0^2}$.

2.4 TL Algorithm

Based on the concept of CR algorithm and discrete control theory, Tang proposed an unconditional stable TL algorithm in 2016 [18] as shown in Eq. (7) and Eq. (8), and the accuracy of the algorithm is also lower than second order, as compared to CR algorithm.

$$d_{i+1} = d_i + \alpha \Delta t v_i + \beta_1 \Delta t^2 a_i \tag{7}$$

$$v_{i+1} = v_i + \Delta t a_i \tag{8}$$

where, $\alpha = \frac{1}{1+2\gamma\xi\Omega+\beta\Omega^2}$, $\beta_1 = \frac{-8\gamma\xi^2-4\beta\xi\Omega+1+2\gamma}{2(1+2\gamma\xi\Omega+\beta\Omega^2)}$.

The parameters in the above equations: ξ damping ratio, $\omega = \sqrt{k_0/m}$, $\Omega = \omega\Delta t$. Introduction of four stability analysis method with negative stiffness

3 Stability Analysis Method with Negative Stiffness

By linearizing the structure's load-displacement curve, the stability of the integral technique can be explored using a single-degree-of-freedom linear system. For a system with a single degree of freedom, the dynamic equilibrium equations are as follows:

$$u''_{t+\Delta t} + 2\xi\omega u'_{t+\Delta t} + \eta\omega^2 u_{t+\Delta t} = -u''_{g_{t+\Delta t}} \tag{9}$$

where, $u''_{t+\Delta t}$, $u'_{t+\Delta t}$, $u_{t+\Delta t}$, are represented acceleration, velocity and displacement of the structure at time i respectively, $u''_{g_{t+\Delta t}}$ the acceleration of the ground motion at the time $t + \Delta t$, ξ the damping ratio, ω the natural frequency, $\eta = k_1/k_0$, η the stiffness coefficient of the structure, the elastic stage when $\eta = 1.0$, after yielding, $\eta < 0$, k_1 the negative stiffness, k_0 the initial stiffness. Equation (9) can be rewritten in the following recursive form:

$$u_{t+\Delta t} = A u_t + L g''_{t+\Delta t} \tag{10}$$

where, $u_{t+\Delta t} = [u''_{t+\Delta t}\ u'_{t+\Delta t}\ u_{t+\Delta t}]^{\mathrm{T}}$, $u_t = [u''_t\ u'_t\ u_t]^{\mathrm{T}}$.

Matrix A and vector L are integral approximation operator and load operator respectively. When $\eta \neq 1$, the expression (9) can be rewritten as follows:

$$u''_{t+\Delta t} + 2\frac{\xi}{\sqrt{\eta}}(\sqrt{\eta}\omega)u'_{t+\Delta t} + (\sqrt{\eta}\omega)^2 u_{t+\Delta t} = -u''_{g t+\Delta t} \tag{11}$$

Thus, the Matrix corresponding to Eq. (11) can be obtained by using the standard form of the known Matrix A as follows:

$$\begin{bmatrix} -(\frac{1}{2}-\beta)\varphi - 2(1-\gamma)\phi & \frac{1}{\Delta t}(-\varphi - 2\phi) & -\frac{\varphi}{\Delta t^2} \\ \Delta t[(1-\gamma) - (\frac{1}{2}-\beta)\gamma\varphi - 2(1-\gamma)\gamma\phi] & 1-\gamma\eta - 2\gamma\phi & -\frac{\gamma\varphi}{\Delta t} \\ \Delta t^2[(\frac{1}{2}-\beta) - (\frac{1}{2}-\beta)\beta\varphi - 2(1-\gamma)\beta\phi] & \Delta t(1-\beta\varphi - 2\beta\phi) & 1-\beta\varphi \end{bmatrix} \tag{12}$$

where, $\varphi = \frac{1}{\beta + \frac{1}{\eta\Omega^2} + \frac{2\gamma\xi}{\eta\Omega}}$, $\phi = \frac{\xi\varphi}{\eta\Omega}$.

The three eigenvalues of A can be obtained from the Characteristic polynomial as follows:

$$\lambda_1 = 1 - \frac{1}{4}(1+2\gamma)\varphi - \phi + \sqrt{-(1+\frac{\varphi}{2} - \gamma\varphi - 2\phi) + [-1 + \frac{1}{4}(1+2\gamma)\varphi + \phi]^2}$$

$$\lambda_2 = 1 - \frac{1}{4}(1+2\gamma)\varphi - \phi - \sqrt{-(1+\frac{\varphi}{2} - \gamma\varphi - 2\phi) + [-1 + \frac{1}{4}(1+2\gamma)\varphi + \phi]^2}$$

$$\lambda_3 = 0$$

Based on the above analysis results, the stability under negative stiffness is analyzed by the following equation:

$$\frac{\rho(\mathrm{A})}{e^{\mu\Delta t}} = \frac{\max(|\lambda_1|, |\lambda_2|, |\lambda_3|)}{e^{\mu\Delta t}} \leq 1 \tag{13}$$

where, $\rho(\mathrm{A})$ is the spectral radius of Matrix A, $\mu = (-\xi + \sqrt{\xi^2 - \eta})\omega$.

For the Chang Algorithm, the CR Algorithm, and the TL Algorithm, the Characteristic polynomial of negative stiffness can be obtained by following these steps:

$$\lambda(\lambda^2 - A_1\lambda + A_2) = 0 \tag{14}$$

According to the characteristics of the algorithm, one of the three roots is zero, and the other two roots need to be obtained by solving the linear equation. Therefore, according to Eq. (14), the A1 and A2 of the three algorithms with negative stiffness can be obtained:

$$A_{1Chang} = 2 - \frac{2\xi\Omega_0}{1+2\gamma\xi\Omega_0} - \frac{\{(\gamma+\frac{1}{2}) - 2[\beta - \gamma(\gamma+\frac{1}{2})]\xi\Omega_0\}\eta\Omega_0^2}{(1+2\gamma\xi\Omega_0)(1+2\gamma\xi\Omega_0+\beta\Omega_0^2)}$$

$$A_{2Chang} = 1 - \frac{2\xi\Omega_0}{1+2\gamma\xi\Omega_0} - \frac{\{(\gamma-\frac{1}{2}) - 2[\beta - \gamma(\gamma-\frac{1}{2})]\xi\Omega_0\}\eta\Omega_0^2}{(1+2\gamma\xi\Omega_0)(1+2\gamma\xi\Omega_0+\beta\Omega_0^2)}$$

$$A_{1CR} = 2 - \frac{2\xi\Omega_0 + \eta\Omega_0^2}{1 + 2\gamma\xi\Omega_0 + \beta\Omega_0^2}$$

$$A_{2CR} = 1 - \frac{2\xi\Omega_0}{1 + 2\gamma\xi\Omega_0 + \beta\Omega_0^2}$$

$$A_{1TL} = 2 - \frac{4\gamma\xi\Omega_0 + \eta\Omega_0^2 + 8\gamma\xi^2\Omega_0^2(1-\eta) + \gamma\xi\Omega_0^3(1-\eta)}{1 + 2\gamma\xi\Omega_0 + \beta\Omega_0^2}$$

$$A_{2TL} = 1 - \frac{4\gamma\xi\Omega_0 + 8\gamma^2\xi^2\Omega_0^2(1-\eta) + \gamma\xi\Omega_0^3(1-\eta)}{1 + 2\gamma\xi\Omega_0 + \beta\Omega_0^2}$$

4 Stability Analysis of algorithms with negative stiffness

4.1 Effect of Parameter β on Algorithm Stability

In the analysis of positive stiffness, $\Delta t/T = 1/10 - 1/20$,where T is the period of structure. To assist in the analysis, the example of negative stiffness is included. Figure 1 shows the variation of spectral radius of four algorithms with parameter β and negative stiffness coefficient η when $\gamma = 1/2$, $\Delta t/T = 1/10$, $\xi = 0.05$. The stability of Chang algorithm is better than that of CR Algorithm and TL algorithm. The Newmark-β algorithm is conditional stable and has local peak value. When $\gamma = 1/2$, $\Delta t/T = 1/20$, $\xi = 0.05$, the Newmark-β algorithm is conditionally stable, the CR algorithm is almost identical to the TL algorithm, and the Chang method is the most stable (Fig. 2).

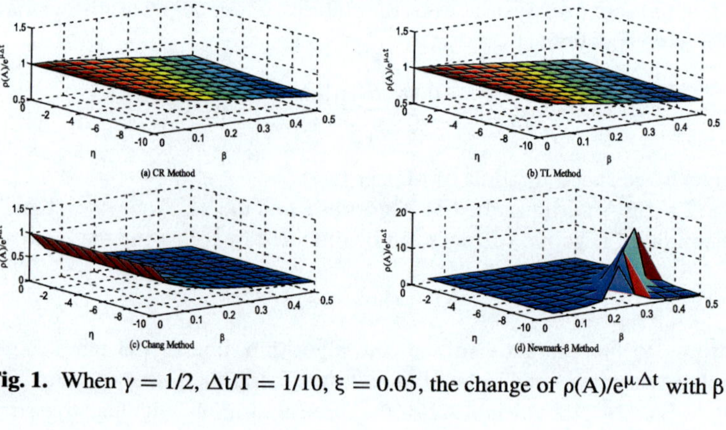

Fig. 1. When $\gamma = 1/2$, $\Delta t/T = 1/10$, $\xi = 0.05$, the change of $\rho(A)/e^{\mu\Delta t}$ with β

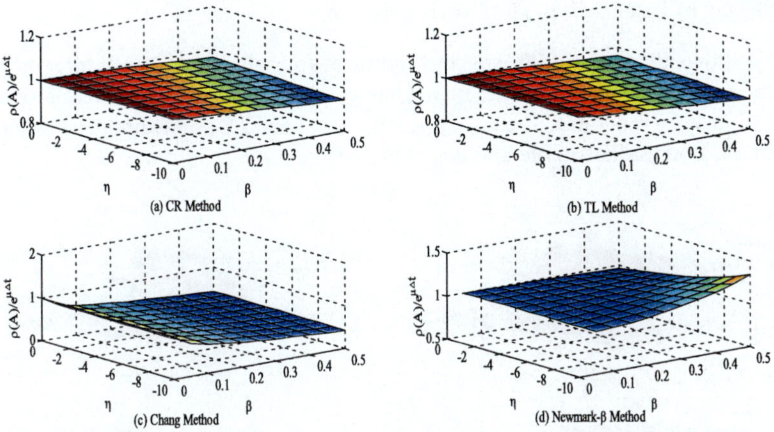

Fig. 2. When $\gamma = 1/2$, $\Delta t/T = 1/20$, $\xi = 0.05$, the change of $\rho(A)/e^{\mu\Delta t}$ with β

4.2 Effect of Damping

Figure 3 shows the variation of spectral radius of four algorithms with damping ratio ξ and negative stiffness coefficient η when $\gamma = 1/2$, $\beta = 1/4$, $\Delta t/T = 1/20$. The stability of Chang algorithm is better than that of CR Algorithm and TL Algorithm, but Newmark-β algorithm's stability is conditional.

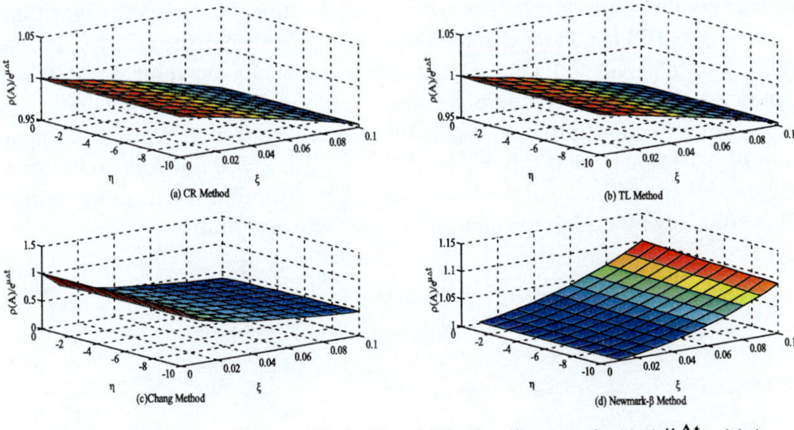

Fig. 3. When $\gamma = 1/2$, $\beta = 1/4, \Delta t/T = 1/20$, the change of $\rho(A)/e^{\mu\Delta t}$ with ξ

4.3 Effects of Integration Time Intervals

Figure 4 shows the variation of spectral radius of four algorithms with integral interval $\Delta t/T$ and negative stiffness coefficient η when $\gamma = 1/2$, $\beta = 1/4$, $\xi = 0.05$. The stability of Chang Algorithm is better than that of CR Algorithm and TL algorithm. The Newmark-β algorithm is conditional stable and has many peaks.

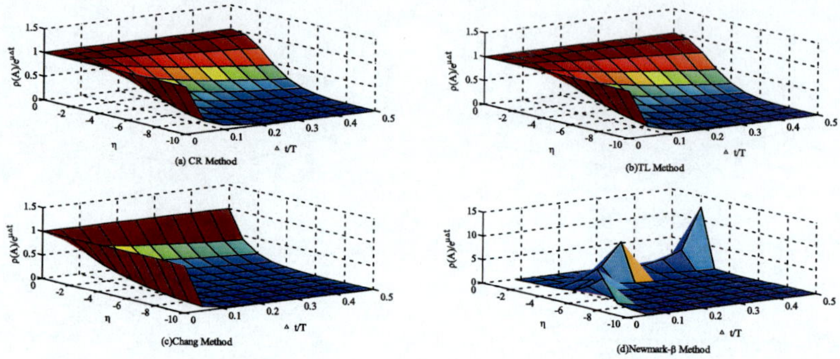

Fig. 4. When $\gamma = 1/2$, $\beta = 1/4$, $\xi = 0.05$, the change of $\rho(A)/e^{\mu \Delta t}$ with $\Delta t/T$

From the above analysis, it can be concluded that the parameters, damping ratio and time interval have some influence on the stability of the algorithm when stiffness is negative. Among the four methods, when $\gamma = 1/2$, $\beta = 1/4$, Newmark-β algorithm is no longer unconditionally stable, but conditionally stable. Newmark-β algorithm ($\beta \geq (0.5 + \gamma)^2/4$, $\gamma \geq 1/2$) is no longer unconditionally stable when the stiffness is negative. The stability of the algorithm can be guaranteed only when the value of $\Delta t/T$ is within a certain range. In order to compare the stability of the four algorithms under negative stiffness conditions more directly, the stability of the four algorithms is compared as shown in Fig. 5 when $\gamma = 1/2$, $\beta = 1/4$, $\Delta t/T = 1/10$, $\xi = 0.05$. As can be seen from Fig. 5, the stability of Chang method is optimal, CR algorithm is almost the same as TL algorithm, and Newmark-β is the conditional stability algorithm.

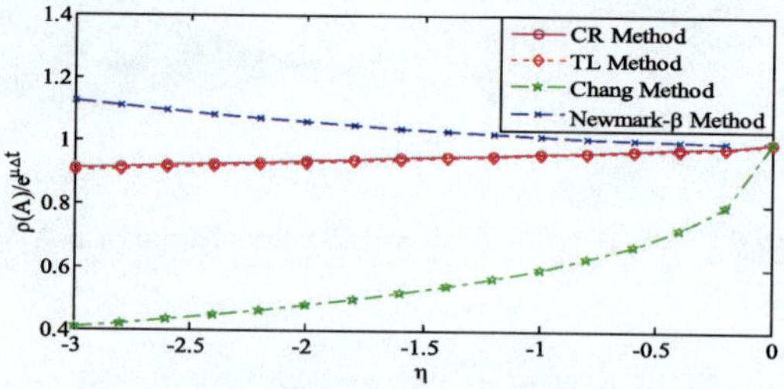

Fig. 5. The change of $\rho(A)/e^{\mu \Delta t}$ with η

5 Conclusion

For CR algorithm, TL algorithm, Chang algorithm and Newmark-β algorithm ($\gamma = 1/2$, $\beta = 1/4$), the stability analysis is carried out and the equation is derived under the condition of negative stiffness. The relationships between the algorithm's stability and the variation of the integral parameter β, the integral time interval and the damping ratio are established. With negative stifness, the Newmark-β algorithm ($\beta \geq (0.5 + \gamma)^2/4$, $\gamma \geq 1/2$) is no longer unconditionally stable. Among the three explicit algorithms, the Chang method has the best stability, while the CR algorithm is basically consistent with the TL algorithm. It is suggested that while performing hybrid testing with probable negative stiffness, the numerical integration algorithm should be carefully chosen.

References

1. Bin, W., Zhen, W., Guoshan, X., et al.: Research and application progress in hybrid testing of engineering structures. Eng. Mech. **39**(1), 1–20 (2022)
2. Minxian, C.: Convergence and stability of direct integration in case of negative-yield-stiffness. J. Theor. Appl. Mech. **20**(1), 31–40 (1988)
3. Newmark, N.M.: A method of computation for structural dynamics. Journal of the Engineering Mechanics Division, ASCE **85**(3), 67–94 (1959)
4. Minxian, C.: Convergence and stability of several step-by-step integration methods in structural dynamic analysis in case of negative stiffness. Eng. Mech. **6**(2), 35–47 (1989)
5. Yunfang, W.: Convergence and stability of Newmark method in case of negative stiffness. J. Chongqing Jianzhu Univ. **23**(1), 21–24 (2001)
6. Shuang, L., Changhai, Z., Hongbo, L., Lili, X.: On the stability of Newmark integration method for structure with negative stiffness. J. Harbin Inst. Technol. **43**(8), 1–5 (2011)
7. Chang, S.: Explicit pseudodynamic algorithm with unconditional stability. J. Eng. Mech. **128**(9), 935–947 (2002)
8. Chang, S.: Enhanced unconditionally stable explicit pseudodynamic algorithm. J. Eng. Mech. **133**(5), 541–554 (2007)
9. Chang, S.Y.: Assessments of dissipative structure-dependent integration methods. Struct. Eng. Mech. **62**(2), 151–162 (2017)
10. Chang, S.Y.: Family of structure-dependent explicit methods for structural dynamics. J. Eng. Mech. **140**(6), 06014005 (2014)
11. Chang, S.-Y.: A dual family of dissipative structure-dependent integration methods for structural nonlinear dynamics. Nonlinear Dyn. **98**(1), 703–734 (2019). https://doi.org/10.1007/s11071-019-05223-y
12. Chang, S.-Y.: Non-iterative methods for dynamic analysis of nonlinear velocity-dependent problems. Nonlinear Dyn. **101**(2), 1473–1500 (2020). https://doi.org/10.1007/s11071-020-05836-8
13. Chen, C., Ricles, J.M.: Development of direct integration algorithms for structural dynamics using discrete control theory. J. Eng. Mech. **134**(8), 676–683 (2008)
14. Chen, C., Ricles, J.: Stability analysis of direct integration algorithm applied to nonlinear structural dynamics. J. Eng. Mech. **134**(9), 703–711 (2008)
15. Gui, Y., Wang, J.-T., Jin, F., Chen, C., Zhou, M.-X.: Development of a family of explicit algorithms for structural dynamics with unconditional stability. Nonlinear Dyn. **77**(4), 1157–1170 (2014). https://doi.org/10.1007/s11071-014-1368-3
16. Fu, B., Feng, D., Jiang, H.: A new family of explicit model based integration algorithms for structural dynamic analysis. Int. J. Struct. Stab. Dyn. **19**(6), 1950053 (2019)

17. Li, J., Yu, K.: Noniterative integration algorithms with controllable numerical dissipations for structural dynamics. Int. J. Comput. Methods **16**(07), 1850111 (2019)
18. Tang, Y., Lou, M.: New unconditionally stable explicit integration algorithm for real-time hybrid testing. J. Eng. Mech. **143**(7), 04017029 (2017)
19. Tang, Y., Ren, D.: New family of explicit structure-dependent integration algorithms with controllable numerical dispersion. J. Eng. Mech. **147**(3), 04021001 (2021)

Study on Phase Change Process of Heat Source for a Water Source Heat Pump

Junhua Wu[1(✉)] and Nuo Wang[2]

[1] Laboratory of Green Construction and Intelligent Maintenance for Civil Engineering of Hebei Province, Yanshan University, Qinhuangdao 066004, China
td_wjh@163.com
[2] Hebei Province Low-Carbon and Clean Building Heating Technology Innovation Center, Qinhuangdao 066004, Hebei, China

Abstract. Allowing water phase change to release latent heat to provide heat for water source heat pump system can increase the running time of heat pump unit and reduce the use of auxiliary heat source. When water freezes and releases latent heat to provide heat energy for heat pump unit, its heat transfer process is more complex. This paper mainly studied heat transfer characteristics of ice-water phase change for a spiral coil heat exchanger. Firstly, the phase change process of water was studied experimentally. Then, the numerical model was established based on the experiments and the model was verified by the experimental data. Finally, the temperatures and liquid fraction in the heat exchanger container were studied by the verified numerical model.

Keywords: Water source heat pump · Phase change · Coil heat exchanger

1 Introduction

The problem of energy shortage and the contradiction between energy and environment have attracted more and more attention of researchers. In the field of winter heating for buildings, the use of renewable energy can reduce the dependence on traditional energy [1]. Therefore in northern China, heat pump systems that uses renewable energy are more and more widely adopted. The water source heat pump (WSHP) uses surface water as heat source for the unit. The buildings heated by the WSHP are often near the sea, river or lake. These buildings are often far from the centre of city and cannot be provided with central heating. So they are more suitable to adopt decentralized heating, such as heat pump heating system. Recently, air source heat pump (ASHP) is widely used. Compared with WSHP heating system, ASHP has high operation noise and high defrosting energy consumption. When the outdoor air temperature is below 0 °C, the unit need defrosting and the energy consumption is increasing [2]. For the WSHP, the water temperature is an important parameter. In winter, the water temperature is lower than 4 °C for a comparatively long period in north of China. When the water temperature is low, it is easy to cause the water icing and the heat pump unit will be damaged. In this case, the auxiliary heat source will be turned on. If the phase change of water is allowed,

W. Guo and K. Qian (Eds.): GBCESC 2022, LNCE 211, pp. 1001–1008, 2023.
https://doi.org/10.1007/978-981-19-5217-3_102

the running time of the system will increase and the service time of auxiliary heat source can be reduced. However, the fluid entering the unit has a lower freezing point, and the unit's operating efficiency will be reduced. In this paper we will study the phase change process of heat source under a reasonable coefficient of performance (COP) for a WSHP system.

The melting point of water is low, but its specific heat is large and the phase change process has an approximately constant temperature. In recent years, there have been some researches on the heat pump heating system with the phase change heat of water as heat source [3–9]. These researches mainly focused on the extracting the phase change latent heat of water by the experimental devices, the methods of rapid and effective deicing and ensuring the continuous heat collection operation of the system. This paper mainly carried out experimental and simulating researches on a WSHP system with a designed spiral coil heat exchanger, calculated the COP of the system, and studied the influence of intermittent operation on phase change.

2 Experimental Setup

For the heat source of WSHP heating system, a series of experiments were carried out and relevant laws of phase change of water were analyzed. The experimental system consisted of heat pump unit, heat exchange tank, two containers and fan coil. The schematic diagram of the experimental system is shown in Fig. 1. It was built in a public space near the South Gate of No. 1 Jinmeng Bay, Qinhuangdao harbor area. The area of building is about 100 m^2 and the experiments were done on January 3, 2021.

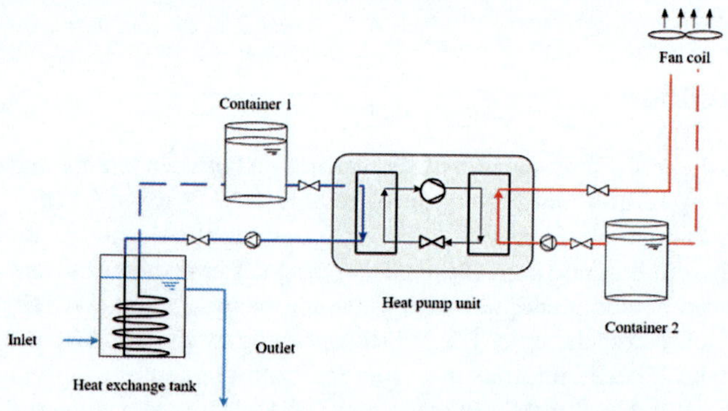

Fig. 1. Schematic diagram of experimental water source heat pump system.

In the heat exchange tank, water exchange heat with glycol solution through a spiral coil heat exchanger. The instrument parameters used in the experiment are shown in Table 1.

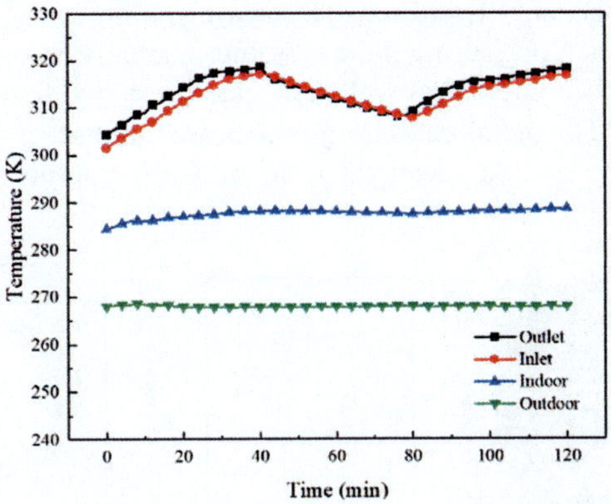

Fig. 2. Schematic diagram of system operating parameters.

In the experiments, the air temperatures of indoor and outdoor, inlet and outlet of water for the users are measured and shown in Fig. 2. The heat pump unit delivered heat to room by the fan coil. When the unit was running, the water temperature of outlet was rising. When the temperature arrived at 318 K which is a set value, the unit stopped running. Then the water temperature begun to drop because of the continuous heat demand for the building. When the temperature arrived at 308 K, the unit started again. The heat pump unit run intermittently.

Table 1. The parameters of the experimental apparatus.

Apparatus	Parameter	Value
Fan coil	FP-102BG	Heating capacity 4.6–4.8 kW
Heat pump unit	WB-KX10s	Heating capacity 7.2 kW
Heat exchange tank	2.1 Printing area	380 mm × 380 mm × 450 mm
Spiral coil	316 Stainless steel	r = 16 mm, D = 250 mm
Pipe	PE	DN25
Water pump	FUJ-10601	Maximum lift 30 m/25 m
Container 1	Stainless steel	d = 500 mm, H = 700 mm
Container 2	Stainless steel	d = 500 mm, H = 700 mm

A spiral coil heat exchanger (SCHE) was placed in the heat exchange tank. The SCHE is made of stainless steel pipe with a diameter of 16 mm and a thickness of 1 mm. The winding diameter of the SCHE was 250 mm. The number of inner and outer turns of the double-layer spiral coil was 20. The total length of the SCHE was about 30 m. Heat exchange between water out of the coil and 40% ethylene glycol in the coil. In the experiment, the water flow was set as 210 L/h. The SCHE was placed in the center of the heat exchange tank. Figure 3 shows the layout of temperature measuring points of

SCHE in heat exchange tank. There are 10 measuring points arranged in this experiment. The temperature of each measuring point was automatically recorded by multi-channel temperature recorder. Temperatures were collected every two minutes and the data was automatically saved. The metal measuring probes are thermal resistance PT100 and the accuracy is 0.1 K. The specific locations of the measuring points are shown in Fig. 3.

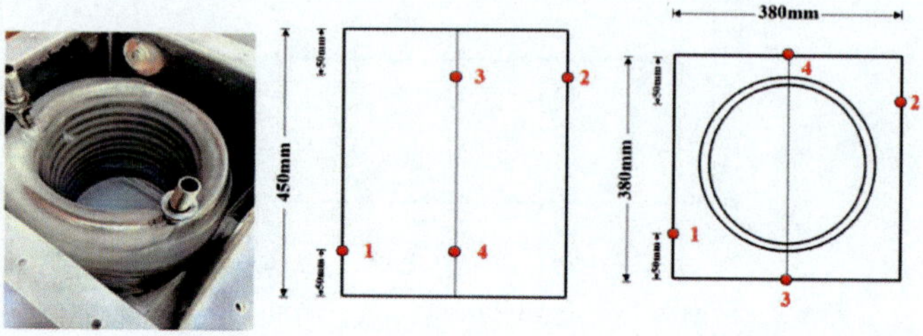

Fig. 3. Schematic diagram of measuring points in heat exchanger.

The power consumption of heat pump unit and water pump was measured with a power consumption meter. The measurement range is 0–9999 kWh and recorded manually at an interval of 5 min. The total power consumption was 4.5 kWh.

3 Numerical Simulation

The numerical model of heat exchange tank was established by using ANSYS software. The numerical simulation was carried out under the following assumptions:

- The wall thickness of heat exchange tank is set to 0, neglecting the heat exchange losses between the outer wall of heat exchange tank and the outside.
- As the phase change medium, water was set to be homogeneous;
- The "enthalpy porosity" method regarded the paste region (partially solidified region) as a porous medium [10].

Based on these assumptions, the model of SCHE in heat exchange tank was established. The model was used to simulate the temporal and spatial changes of water temperature, ethylene glycol temperature and liquid velocity. The 3D numerical model is shown in Fig. 4. The SCHE was simplified to a single-layer model.

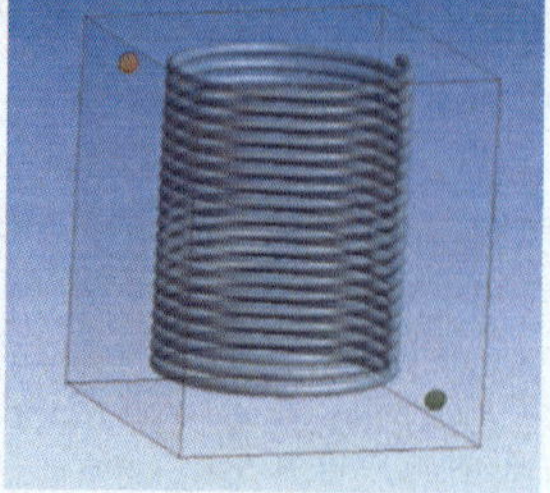

Fig. 4. 3D numerical model.

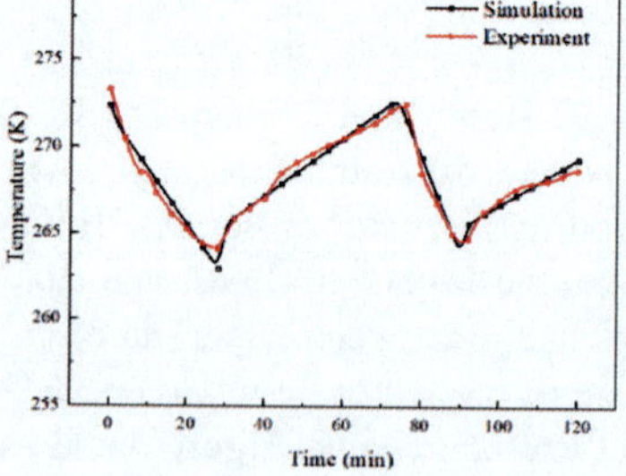

Fig. 5. Ethylene glycol inlet temperature.

Table 2. Physical parameters of three materials.

Material name	Stainless steel	Water	Ethylene glycol
Type	Solid	Fluid	Fluid
Density (kg/m^3)	8030	998.2	1111.4
Specific heat capacity (J/(kg·K))	502.48	4182	2415
Thermal conductivity (W/(m·K))	16.27	0.6	0.252
Viscosity (Pa·s)	–	0.001003	0.0157
Latent heat (kJ/kg)	–	336	187
Phase change temperature (K)	–	273.15	247.85

The parameters of the stainless steel, water and ethylene glycol are shown in Table 2.

Corresponding to the experiment, inlet temperature of water was set as 279.75 K and the inlet temperature of ethylene glycol was programmed by UDF, as shown in Fig. 5. The inlet flow rate is 0.52 m/s. Except the basic equations, such as the continuity equation, momentum equation and energy conservation equation, the solidification melting model is used in the numerical simulation. The time step is set to 60 s and the simulation step is set to 21, 15 and 24 respectively according to the experiment.

4 Results and Discussion

As the experimental results, the water temperature is shown in Fig. 6. As can be seen from Fig. 6, the water temperature of outlet has an obvious fluctuation although the inlet temperature is almost constant. The temperature decreases when the unit is running and recovers when the unit stops running.

4.1 Calculation of Results and Analysis

COP is an important parameter for the heat pump system. The heating capacity of the heat pump unit is calculated as follows:

$$Q = V_{\mathrm{w}} \rho_{\mathrm{w}} c_{\mathrm{w}} \left(T_{\mathrm{w,out}} - T_{\mathrm{w,in}}\right) \tag{1}$$

where Q is the heating capacity of the heat pump unit, V_{w} is flow of hot water, ρ_{w} is density of water, c_{w} is specific heat capacity of water, $T_{\mathrm{w,in}}$ and $T_{\mathrm{w,out}}$ are the inlet and outlet temperature of hot water.

The experiment lasted two hours. When the water temperature on the side of user arrived at 45 °C, the unit stopped. When the temperature decreased to 35 °C, the unit opened again. The calculation formula for COP of heat pump system is:

$$COP_{\mathrm{s}} = \frac{Q}{\sum W} \tag{2}$$

where COP_s is performance coefficient of heat pump system, and $\sum W$ is total power consumption. The total power consumption is 4.5 kW · h which recorded by power meter. So the COP_s of the heat pump system is calculated as 3 under the experimental conditions.

4.2 Validation of the Model

Temperature comparisons between experiment and simulation for outlet of ethylene glycol, outlet of water, point 3 and point 4 are shown in Fig. 7, 8, 9, 10, respectively.

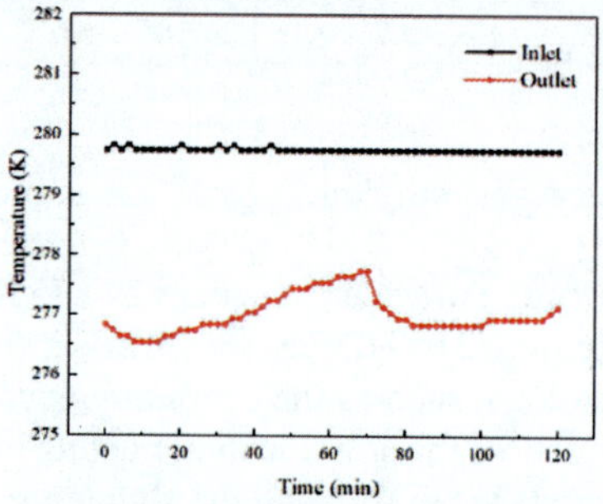

Fig. 6. Water temperature.

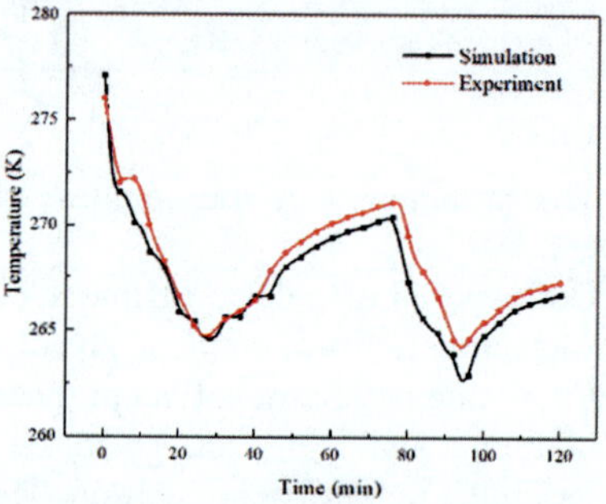

Fig. 7. Outlet temperature of ethylene glycol solution.

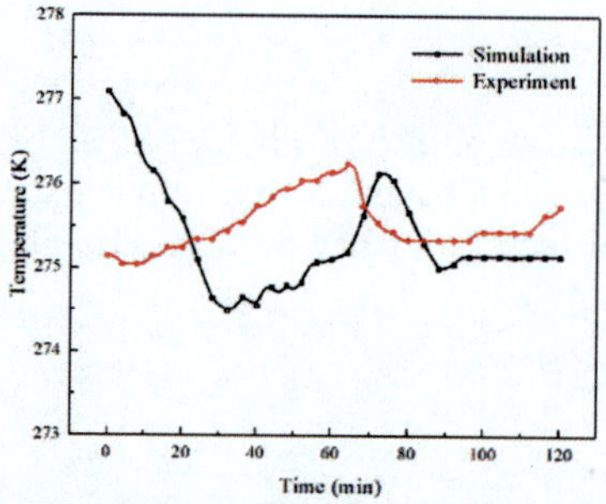

Fig. 8. Outlet temperature of water.

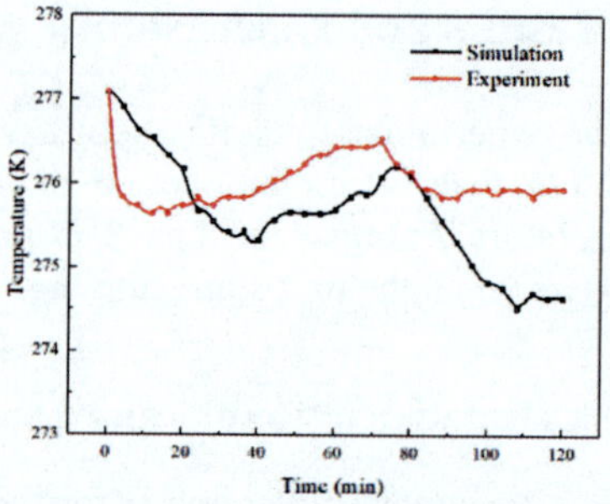

Fig. 9. Temperature of point 3.

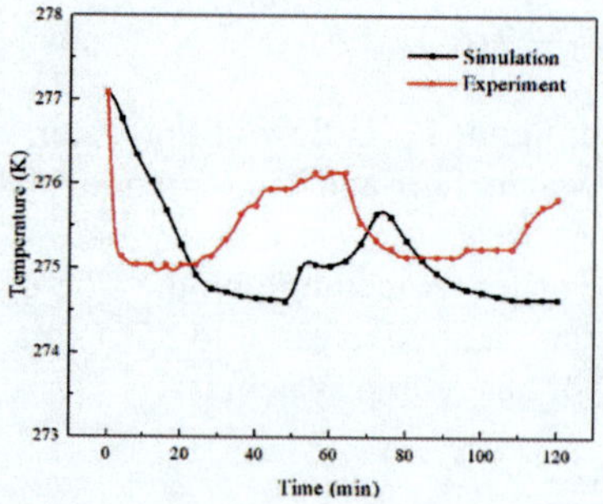

Fig. 10. Temperature of point 4.

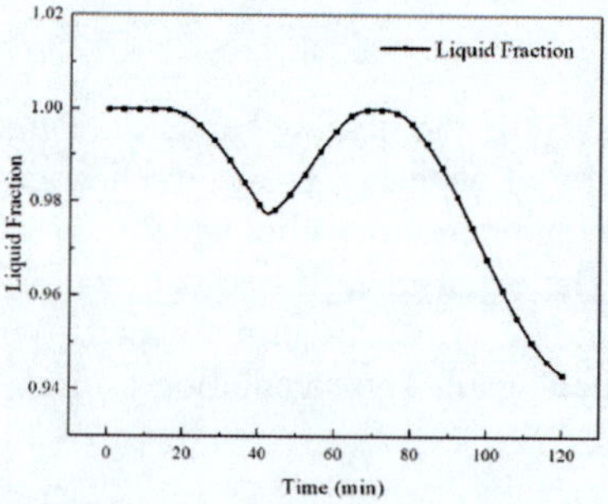

Fig. 11. Liquid fraction of water-ice.

There is a main data residual checking method: average relative error α [11]:

$$\alpha = \frac{1}{N}\sum_{k=1}^{N}\left|\frac{\varepsilon(k)}{x_0^{(0)}(k)}\right| \tag{3}$$

where $\varepsilon(k)$ stands for the absolute degree of incidence between the raw data $x_0^{(0)}(k)$ and the simulated value $\hat{x}_0^{(0)}(k)$, $\varepsilon(k) = x_0^{(0)}(k) - \hat{x}_0^{(0)}(k)$. The average relative errors of water outlet, measuring point 3 and measuring point 4 are 0.5%, 0.62% and 0.36% respectively. Therefore, it is considered that the numerical model can be verified by experiments.

4.3 Phase Change

As a result of simulation, Fig. 11 shows the liquid fraction of water-ice in the heat exchange tank. There were two heat release processes. The ice formed in the first heat release process and melted into water during the time when the unit was out of operation. However, in the second heat release process, the solid ice formed too much. Increased thermal resistance resulted in insufficient heat transfer. The unit started the self-protection program and stopped running.

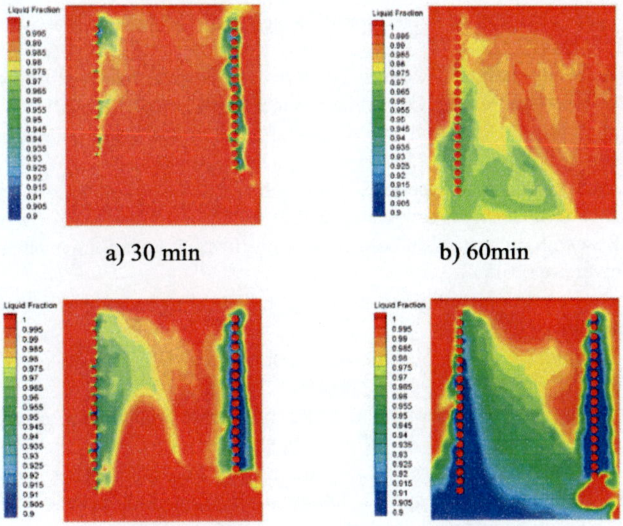

a) 30 min b) 60min

c) 90 min b) 120 min

Fig. 12. Liquid fraction at different moments.

The cloud charts of liquid fraction at different moments are shown in Fig. 12. The ice forms on the surface of coil heat exchanger. Because the water flowing through the heat exchange tank which has an impact on the ice forming, the bottom of the icing area is deformed.

5 Conclusions

In order to extend the heating time of WSHP and reduce the use of auxiliary heat source, a heat source with phase change for the unit was adopted in this paper. The phase transformation of heat source was tested and simulated. Firstly, the water temperature of outlet had an obvious fluctuation although the inlet temperature was almost constant during the experiment. Besides that, the temperature decreased when the unit was running and recovered when the unit stopped running. The COP of the heat pump system was calculated as 3 under the experimental conditions. Then the numerical model was built and verified by experiments. The numerical simulating results of liquid fraction curves and cloud charts were analyzed. The results shown that intermittent operation of WSHP unit helped to melt the ice. The cloud charts of liquid fraction at different moments in the heat exchange tank were helpful to understand the phenomenon of icing and guide the structure design of heat exchanger.

Acknowledgement. The research was funded by Natural Science Foundation of Hebei Province (E2019203495).

References

1. Fengli, L.: Application of seawater source heat pump in Qinhuangdao area. Dissertation, Tianjin University (2012)
2. Fazilati, M.A., Alemrajabi, A.A.: Phase change material for enhancing solar water heater, an experimental approach. Energy Convers. Manage. **71**(08), 138–145 (2013)
3. Yanqiu, Z., Demin, S.: Research on core technology of ice water source heat pump. Sci. Technol. Innov. Appl. **08**(03), 155–156+159 (2020)
4. Jianwei, Z.: Study on performance and ice strength of extraction solidification heat exchanger. Dissertation, Harbin Business University (2019)
5. Jishen, Z.: Research on extraction cold water solidification heat pump system. Dissertation, Qingdao University (2017)
6. Jingxin, G.: Research on solar ice melting heat storage heating heat pump system. Dissertation, Yanshan University (2019)
7. Zhibin, L., Ronghua, W., Hao, Y.: Operation test of water phase variable energy heat pump heating system. HVAC **49**(05), 96–100 (2019)
8. Xiao, W.: Simulation and experimental research on cold water phase change energy heat pump system. Dissertation, Qingdao University (2019)
9. Yuanbo, Y.: Operation characteristics and engineering experimental research of cold water phase change energy heat pump system. Dissertation, Qingdao University (2020)
10. Yingxiang, Z.: Study on unsteady state of low temperature phase change heat source of water source heat pump. Dissertation, Yanshan University (2021)
11. Jie, Z., Yiqun, T.: Centrifuge experimental study of thaw settlement characteristics of mucky clay after artificial ground freezing. J. Eng. **190**, 98–108 (2015)

Bidirectional Hybrid Simulation of a Six-Story Full-Scale Frame Specimen with Incomplete Boundary Conditions

Shangzhang Wang[1,2], Ge Yang[1,2,3], Liang Tang[4,5], Zhen Wang[1,2], Bin Wu[1,2(✉)], Yanfang Liang[4,5], Zhen Tian[6], and Jiajun Xiao[1,2]

[1] Sanya Science and Education Innovation Park, Wuhan University of Technology, Sanya, China
wub@whut.edu.cn

[2] School of Civil Engineering and Architecture, Wuhan University of Technology, Wuhan, China

[3] Key Laboratory of Earthquake Engineering Simulation and Seismic Resilience of China Earthquake Administration, Tianjin University, Tianjin, China

[4] China Construction Industrial Engineering and Technology Research Academy Co., Ltd., Beijing, China

[5] China State Construction Technical Center, Beijing, China

[6] The Civil Engineering Group Corporation of China Second Engineering Bureau. Ltd., Beijing, China

Abstract. It is imperative to carry out full-scale bidirectional hybrid simulations to accurately reproduce the responses of complex structures under earthquake excitations. To study the seismic performance of a six-story steel frame structure, a bidirectional hybrid simulation was carried out on a six-story, one-bay one-span experimental substructure by HyTest. In the test, only three degrees of freedom in the floor plane were considered. Four large-ton actuators were used for loading onto each floor. In this case, there was a problem with redundant actuator control. The control mode of "three displacements and one force" was used to complete the bidirectional test study, in which the redundant actuator is operated in force control mode. Furthermore, the "table-type" experimental substructure has a strong problem of incomplete boundary conditions. Thus, a hybrid simulation method with incomplete boundary conditions based on restoring force correction was proposed. The numerical results show that this method can effectively solve the problem of incomplete boundary conditions well and is a promising method.

Keywords: Bidirectional hybrid simulation · Redundant actuator control · Incomplete boundary conditions · Restoring force correction · Six-story structure

1 Introduction

In the field of earthquake engineering, the main methods to explore structural damage and failure behavior are numerical analysis, physical experiments, and hybrid simulation (HS). For large and complex structures, numerical analysis is difficult to accurately

W. Guo and K. Qian (Eds.): GBCESC 2022, LNCE 211, pp. 1009–1021, 2023.
https://doi.org/10.1007/978-981-19-5217-3_103

simulate their mechanical behavior, physical experiment is expensive and requires the high performance of hardware equipment. HS can economically and effectively complete large-scale or full-scale experiments by connecting numerical analysis with physical experiments through online interaction [1]. In HS, the part of the structure that is difficult to numerically analyze and has complex nonlinear behavior is treated as the experimental substructure, and the rest part is treated as the numerical substructure. Finally, these two parts are coordinated to complete the simulation together.

Because the mass and stiffness of the structure or the excitation are not along the direction of the main axis of the structure, the structure produces responses in multiple directions. In the unidirectional HS, the damage and failure of the earthquake are often underestimated, and ignoring the torsional deformation will reduce the safety guarantee. Thewalt and Mahin [2] conducted the first bidirectional pseudo-dynamic test using the linear transformation relationship between the centroid and the actuators. Molina et al. [3] conducted a large-scale bidirectional pseudo-dynamic test on a full-scale three-story building affected by strong earthquake motions for the first time. Liu and Chang [4] explained the analytical geometric relationship between a rigid specimen, actuators, and reaction frame, and conducted bidirectional pseudo-dynamic tests under horizontal and vertical seismic excitation. Cai et al. [5] used NetSLab to complete the pseudo-dynamic test of a multi-span continuous bridge in multiple laboratories. Zhou et al. [6] discussed the problem of multi degrees of freedom (DOFs) synchronous control and verified the advantages of the modified Newton Raphson iterative method combined with PI control in the form of numerical simulation and experiment.

One of the key scientific issues in HS technology for engineering structures is the realization of experimental boundary conditions. For a simple experimental substructure subjected to single-directional loading, only one actuator is required for the control point and the boundary conditions are easy to reproduce. However, for the experimental substructure with multi DOFs boundary conditions, complete realization of all the boundary conditions would require many actuators and may exceed the loading capability of the laboratory. Based on the assumption that the position of the inflection point is fixed, an overlapping technique of substructure boundaries was used. To weaken the influence of the assumption, the bottom one and a half stories were taken as the experimental substructure, and the boundary position of the substructure is extended to the numerical substructure [7]. Under the constraints of laboratory conditions, a weakly coupled hybrid simulation method is proposed [8]. For the DOFs that can be loaded, the corresponding restoring force is provided by the experimental substructure, while for the DOFs that can't be loaded, the corresponding restoring force is provided by the numerical model of the experimental substructure. Finally, taking the plane frame as an example, the experimental substructure only completes two translational DOFs, while ignoring the rotational DOF. To obtain the corresponding rotational restoring force, a numerical model with complete boundary conditions is established for the experimental substructure to control three DOFs; finally, the two translational DOFs and one rotational DOF obtained are combined to form the restoring force vector of the experimental substructure with complete boundary conditions. Different from this method of directly combining experimental restoring force and numerical restoring force, an online numerical simulation method based on model update [9] considers that the constitutive parameters of the experimental structure are invariant whether it is a complete boundary or an incomplete boundary. Therefore, it establishes a numerical model for the overall structure, and only loads the main boundary conditions for the experimental substructure. The parameters of the experimental substructure are identified by the test data and used to update the numerical model of the overall structure with the same parameters as the experimental substructure. It not only reduces the influence of boundary conditions but also improves the accuracy of the numerical model.

Literature reviews show that: To consider the problem of bidirectional HS in complex structures, one is to use the overlapping technology to expand the scope of the substructure, but the expansion of the substructure may bring great difficulties to the laboratory, and the other is to compensate the incomplete boundary conditions by the numerical model. According to the second method, a hybrid simulation method with incomplete boundary conditions based on restoring force correction is proposed. This method not only does not have a stability problem, such as the Unscented Kalman Filter (UKF) algorithm [10] but also is expected to meet the requirements of real-time loading [11, 12].

Therefore, a bidirectional hybrid simulation of a six-story full-scale frame was carried out in this paper. In Sect. 2, the basic structural information and the adopted redundant control method, and the hybrid test method of incomplete boundary conditions are mainly introduced; The third section is the theoretical analysis and numerical verification. The conclusion is summarized in the last section. The work done in this paper is mainly to prepare for the next experiment.

2 Overview of the Six-Story Structure and Test Method

2.1 Structural Information

The structural prototype is a six-story, seven-bay three-span steel frame structure with a height of 18.3 m (the height of the first story is 3.3 m and the other stories are 3.0 m), and the length and width of the floor plan are 42 m and 18 m respectively (the horizontal and longitudinal spans are 6.0 m). The structural plane layout is shown in Fig. 1, and the vertical layout is shown in Fig. 2. The building is planned to be built in the area of Grade 8, the site category is class II, the design earthquake is divided into group II, the load is selected according to the current load code in China, and the structural design is completed by using the building structure design software PKPM. After section selection and checking calculation, the column section of the first floor is HW414 mm × 405 mm × 18 mm × 28 mm. The column section on the second floor is HW400 mm × 400 mm × 13 mm × 21 mm. The others are HW350 mm × 350 mm × 12 mm × 19 mm. Q235B hot-rolled section steel and HW340 mm × 250 mm × 9 mm × 14 mm is selected as the frame beam section of all floors. The section of the secondary beam is HW294 mm × 200 mm × 8 mm × 12 mm. The material of column and beam is Q355B hot-rolled section steel and Q235B hot-rolled section steel respectively. The floor is 100 mm thick profiled steel sheet composite slabs, and the concrete strength grade is C30. One horizontal and longitudinal span in the central area of the prototype structure is selected as the experimental substructure, and the rest is used as the numerical substructure, as shown in Fig. 1.

In the test, the Northridge (NS) seismic wave was selected, the peak acceleration of ground motion was adjusted to 200 gal, and the Northridge (WE) seismic wave was selected, and the peak acceleration of ground motion was adjusted to 170 ga. The seismic wave duration is 34.8 s, and the seismic wave is shown in Fig. 3.

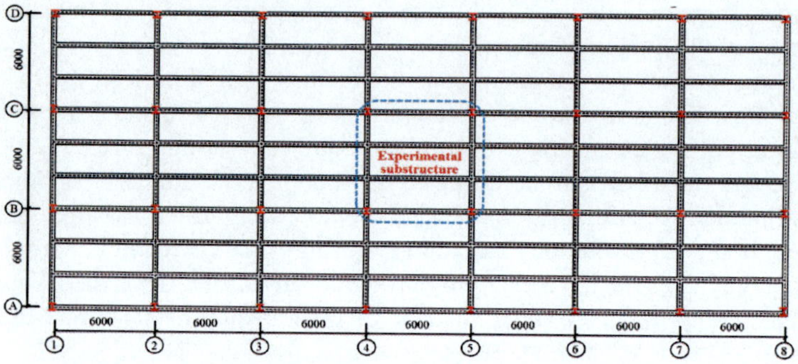

Fig. 1. The plane layout of the prototype structure (Unit: mm)

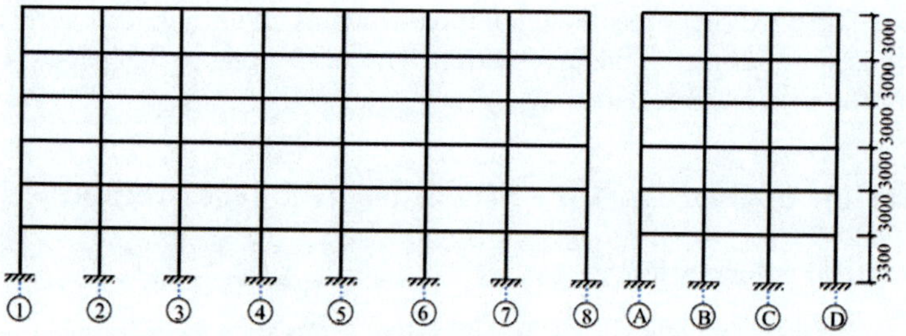

Fig. 2. Elevation of prototype structure (Unit: mm)

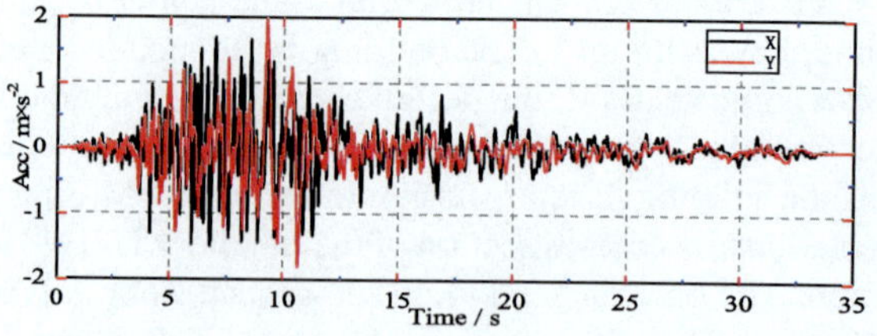

Fig. 3. Seismic waves

2.2 The Redundant Actuator Control Scheme

The physical specimen is constructed and fabricated in the two-way reaction wall in the structural laboratory of China State Construction Engineering Corporation (CSCEC), as shown in Fig. 4. The two-way reaction wall is 25.5 m high, 50.0 m on the long side, and 18.5 m on the wide side. It is equipped with high-performance loading equipment and hybrid simulation software HyTest [13], which can provide a guarantee for the success of the experiment [14].

The base of the specimen is anchored to the reaction floor by prestressed steel rods to prevent the specimen from slipping by static friction. Four actuators are used on each floor to control the loading of two translational and one rotational DOFs at the centroid [15, 16]. The loading capacity ranges of the actuators in the x-direction and y-direction of one floor are ±1000 kN and ±500 kN, respectively, and the loading capacity ranges of the other floors of actuators are shown in Table 1.

To achieve the loading target more accurately, the outer-loop displacement control algorithm is used [17]. This algorithm needs to place four displacement sensors on each floor to measure the displacement of the floor centroid. As shown in Fig. 5, the front and rear ends of the actuators and displacement sensors are made of hinged joints [18].

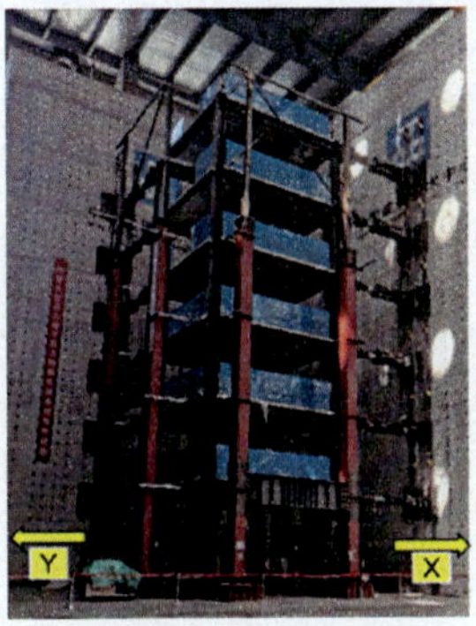

Fig. 4. Photographs at CSCTC

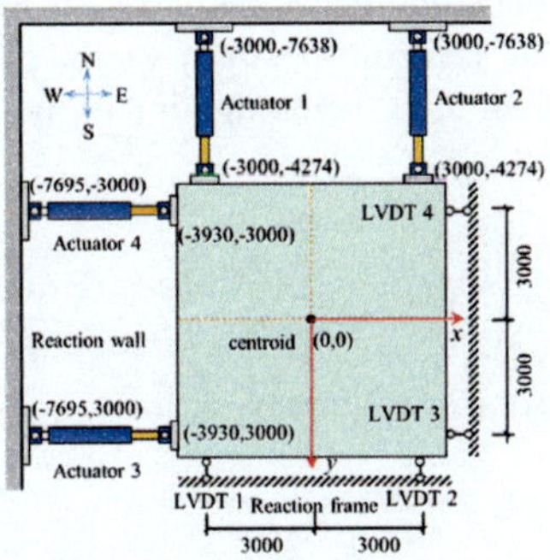

Fig. 5. Layout of specimen

Table 1. Loading capacity of each floor actuators

Num of floor	Two actuators in the x-direction	Two actuators in the y-direction
6	1500 kN	2000 kN
5	1000 kN	2000 kN
4	1000 kN	1500 kN
3	1000 kN	1500 kN
2	1000 kN	1000 kN
1	1000 kN	500 kN

It can be seen that the number of actuators used in this experiment is much more than the number of DOFs to be controlled by the specimen, so the loading system is redundant. The redundant system can effectively reduce test failures, and make it possible for a small system to do big things. However, if the control is not good, there will be internal forces in the system, which will damage the structure and equipment. the redundant control method proposed by Wang et al. [15] was adopted, which is recorded as "three displacements and one force".

The displacement-controlled actuator elongation is obtained by geometric motion. When the displacement command at the centroid of the specimen is $\mathbf{d}_{\mathrm{com}} = \left[x_{\mathrm{d}}\ y_{\mathrm{d}}\ \theta_{\mathrm{d}}\right]^{\mathrm{T}}$,

$$d_{\mathrm{a}}^{i} = \left\|\mathbf{A}_{\mathrm{a}}^{i} - \mathbf{A}_{\mathrm{ae}}^{i}\right\|_2 - \left\|\mathbf{A}_{\mathrm{a0}}^{i} - \mathbf{A}_{\mathrm{ae0}}^{i}\right\|_2 = \left\|\mathbf{T}\cdot\left(\mathbf{A}_{\mathrm{a0}}^{i} - \mathbf{C}_{\mathrm{c0}}\right) + \mathbf{C}_{\mathrm{c}} - \mathbf{A}_{\mathrm{ae}}^{i}\right\|_2 - \left\|\mathbf{A}_{\mathrm{a0}}^{i} - \mathbf{A}_{\mathrm{ae0}}^{i}\right\|_2 \tag{1}$$

where superscript i is the actuator i; Subscript a and ae are represented the position vector of the connection point between the actuator and the specimen and the position vector of the fixed point between the actuators and the reaction wall when the displacement command is completed respectively. Subscript a0 and ae0 represent the position vector of the connection point between the actuator and the specimen and the position vector of the fixed point between the actuator and the reaction wall when the specimen is in the initial state respectively. $\mathbf{C}_{\mathrm{c0}}$ is the initial position vector of the specimen centroid. $\mathbf{C}_{\mathrm{c}}$ is the centroid position vector of the specimen after completing the displacement command. d_{a} is the elongation of the actuator. $\mathbf{A}$ is the actuators position vector; $\|\cdot\|_2$ is 2-norm;

$$\mathbf{T} = \begin{pmatrix} \cos\theta_d & \sin\theta_d \\ -\sin\theta_d & \cos\theta_d \end{pmatrix}$$

the above Eq. (1) can be written as

$$d_{\mathrm{a}}^{i} = g^{i}(\mathbf{d}_{\mathrm{com}}) \tag{2}$$

the above formula indicates that when the centroid displacement of the specimen is $\mathbf{d}_{\mathrm{com}}$, the elongation and displacement feedback of actuators i shall be d_{a}^{i}.

The force of the redundant actuator is obtained by the dynamic equation. The mechanical relationship between the force of the actuator i and the restoring force of the specimen is

$$\mathbf{r} = \mathbf{A}\cdot\mathbf{F} \tag{3}$$

where, $\mathbf{F} = \left[f_1\ f_2\ f_3\ f_4\right]^{\mathrm{T}}$, $\mathbf{r} = \left[r_x\ r_y\ r_\theta\right]^{\mathrm{T}}$, $\mathbf{A}_{\mathrm{a}}^{i} = \left[x_{\mathrm{a}}^{i}\ y_{\mathrm{a}}^{i}\right]^{\mathrm{T}}$,$\mathbf{A}_{\mathrm{ae}}^{i} = \left[x_{\mathrm{ae}}^{i}\ y_{\mathrm{ae}}^{i}\right]^{\mathrm{T}}$, $\mathbf{E}_{\mathrm{a}}^{i} = \mathbf{A}_{\mathrm{a}}^{i} - \mathbf{A}_{\mathrm{ae}}^{i}$, $\mathbf{C}_{\mathrm{c}} = \left[x_{\mathrm{c}}\ y_{\mathrm{c}}\right]^{\mathrm{T}}$,

$$\mathbf{A} = \begin{bmatrix} \frac{x_a^1 - x_{ae}^1}{\|E_a^1\|_2} & \frac{y_a^1 - y_{ae}^1}{\|E_a^1\|_2} & \frac{(x_a^1 - x_{ae}^1)\cdot(y_c - y_a^1) - (y_a^1 - y_{ae}^1)\cdot(x_c - x_a^1)}{\|E_a^1\|_2} \\ \frac{x_a^2 - x_{ae}^2}{\|E_a^2\|_2} & \frac{y_a^2 - y_{ae}^2}{\|E_a^2\|_2} & \frac{(x_a^2 - x_{ae}^2)\cdot(y_c - y_a^2) - (y_a^2 - y_{ae}^2)\cdot(x_c - x_a^2)}{\|E_a^2\|_2} \\ \frac{x_a^3 - x_{ae}^3}{\|E_a^3\|_2} & \frac{y_a^3 - y_{ae}^3}{\|E_a^3\|_2} & \frac{(x_a^3 - x_{ae}^3)\cdot(y_c - y_a^3) - (y_a^3 - y_{ae}^3)\cdot(x_c - x_a^3)}{\|E_a^3\|_2} \\ \frac{x_a^4 - x_{ae}^4}{\|E_a^4\|_2} & \frac{y_a^4 - y_{ae}^4}{\|E_a^4\|_2} & \frac{(x_a^4 - x_{ae}^4)\cdot(y_c - y_a^4) - (y_a^4 - y_{ae}^4)\cdot(x_c - x_a^4)}{\|E_a^4\|_2} \end{bmatrix}^{\mathrm{T}}$$

the objective function is

$$\min \mathrm{h}(\mathbf{F}) = \sum_{i=1}^{n} \left(\frac{f_i}{f_i^m} \right)^2 \tag{4}$$

where, f_i^m is the maximum loading capacity of actuator *i*. For solving the above problem, the Lagrange multiplier method is used, is that

$$L(\mathbf{x}) = \mathrm{h}(\mathbf{F}) + \boldsymbol{\lambda} \cdot (\mathbf{r} - \mathbf{A} \cdot \mathbf{F}) \tag{5}$$

where, $\boldsymbol{\lambda} = \left[\lambda_1 \; \lambda_2 \; \lambda_3 \right]$, $\mathbf{x} = \left[\mathbf{F} \; \boldsymbol{\lambda} \right]$.

When the partial derivative of the Lagrange function to each variable is 0, the force value of the optimized actuators can be obtained

$$\frac{\partial L(\mathbf{x})}{\partial \mathbf{x}} = 0 \tag{6}$$

2.3 Hybrid Simulation Method with Incomplete Boundary Conditions

The proposed method requires the numerical models of the overall structure and the experiment substructure, the latter is called the surrogate model. The surrogate model has the same boundary conditions as the experiment substructure and is used to correct the restoring force of the structure. Taking the structure mentioned in Sect. 2 as an example, the implementation process of this method is shown in Fig. 6.

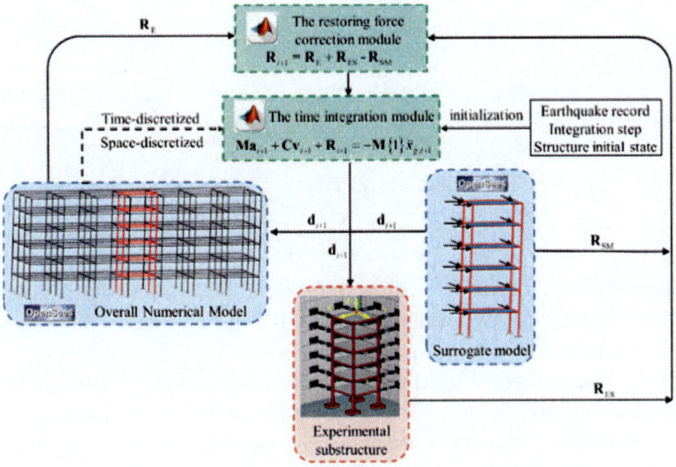

Fig. 6. The basic framework of the proposed method

Step1: First, the overall structure is discretized in the time and space domains to obtain the structural dynamic equation, which is

$$\mathbf{Ma}_{i+1} + \mathbf{Cv}_{i+1} + \mathbf{R}_{i+1} = -\mathbf{M}\{1\}\ddot{x}_{g,i+1} \tag{7}$$

where, $\mathbf{M}$, $\mathbf{C}$ are the mass and damping matrices of the structure, $\mathbf{a}$, $\mathbf{v}$ are acceleration and velocity vectors for the structure, $\mathbf{R}$ is the restoring force vector of the structure, $\{1\}$ is a vector whose elements are 1; The subscript $i+1$ is the $i+1$th time step, and $\ddot{x}_{g,i+1}$ is the seismic acceleration corresponding to the $i+1$th time step. The discrete structural dynamic equation is established by MATLAB, which is called the time integration module.

Step2: Using numerical integration algorithms such as the central difference algorithm, the dynamic equation of the structure are solved, and the displacement vector $\mathbf{d}_{i+1}$ of the structure at the $i+1$th time step is obtained.

Step3: Send the displacement vector $\mathbf{d}_{i+1}$ to finite element software such as OpenSees, calculate the restoring force $\mathbf{R}_{\mathrm{E}}$ of the overall structure model and the surrogate model restoring force $\mathbf{R}_{\mathrm{SM}}$, and send the obtained restoring force to the MATLAB, which is called the restoring force correction module.

Step4: The displacement vector $\mathbf{d}_{i+1}$ is sent to the loading system of the experimental substructure. The restoring force $\mathbf{R}_{\mathrm{ES}}$ of the specimen is obtained after the physical loading is completed and sent to the restoring force correction module.

Step5: The restoring force correction module completes the structural restoring force vector corresponding to the $i+1$th time step in the form of $\mathbf{R}_{i+1}=\mathbf{R}_{\mathrm{E}}+\mathbf{R}_{\mathrm{ES}}-\mathbf{R}_{\mathrm{SM}}$ and substitutes it into Step1 to solve the displacement vector of the next time step.

Repeat this until the end of the test.

The central difference algorithm was adopted, which is an explicit integration algorithm and has high computational efficiency.

$$\dot{\mathbf{x}}_i=(\mathbf{x}_{i+1}-\mathbf{x}_{i-1})/(2\Delta t) \tag{8}$$

$$\ddot{\mathbf{x}}_i=(\mathbf{x}_{i+1}-2\mathbf{x}_i+\mathbf{x}_{i-1})/\Delta t^2 \tag{9}$$

substitute the central difference method formula (2) and formula (3) into the structural dynamic equation formula (1) to obtain the explicit expression of the displacement.

$$\mathbf{x}_{i+1}=\frac{-\mathbf{M}\ddot{\mathbf{x}}_g-\mathbf{R}_{\mathrm{N}}(\mathbf{x}_i)-\mathbf{R}_{\mathrm{E}}(\mathbf{x}_i)+\frac{2\mathbf{M}}{\Delta t^2}\mathbf{x}_i-\left(\frac{\mathbf{M}}{\Delta t^2}+\frac{\mathbf{C}}{2\Delta t}\right)\mathbf{x}_{i-1}}{\frac{\mathbf{M}}{\Delta t^2}+\frac{\mathbf{C}}{2\Delta t}} \tag{10}$$

where the time interval Δt is 0.02 s. Since only two translational DOFs are considered, the DOFs involved are only two translational DOFs in one plane. The following is the same and will not be described again.

The numerical substructure used is modeled by OpenSees. The floor of the model is assumed to be rigid, and the material of the components is simulated by steel01. The values of its constitutive parameters are shown in Table 2. The component elements are simulated by the nonlinearBeamColumn element, and the columns consider the P-Delta effect. The load and mass values of the frame structure are shown in Table 3.

Table 2. Material information

element	Yield strength Fy/MPa	Elastic modulus Es/GPa	strain hardening rate bs
Column	355	206	0.008312
Beam	235	206	0.008312

Table 3. Load and mass information

Number of floor	Loads on side members / N × m−1	Loads on middle members /N × m^{-1}	Mass / Ton
1	−13640.28	−27280.56	701.5
2	−13486.67	−26973.33	693.6
3–5	−13418.61	−26837.22	690.1
6	−12018.61	−24037.22	618.1

3 Numerical Analysis of Bidirectional Hybrid Simulation Method

3.1 Numerical Simulation and Verification

The research object of this paper is the experimental substructure, and it is assumed that the numerical substructure model is accurate. Due to the model error between the surrogate model and the experimental substructure, there is a deviation between the constitutive material value of the surrogate model and the experimental substructure, which indicates that the model is inaccurate.

To fully consider the two influencing factors of the HS, four types of Working conditions are set. Type 1 is the overall structure simulation. The boundary conditions of this condition are complete and the model is accurate. The results of this condition are recorded as ref as the reference solution. Type 2 is the conventional HS simulation, which divides the structure into the experimental substructure and numerical substructure. Under this working condition, there are incomplete boundary conditions, and the results can be used as the boundary influence degree, which is recorded as BC1ME0 (BC1 indicates that there is the boundary condition problem, Me0 indicates that the experimental substructure model is accurate); Type 3 is the overall structure simulation. Although the boundary conditions of this working condition are complete, there is a model error. The results are recorded as BC0ME1 as the influence degree of the model (BC0 indicates that there is no boundary condition problem in this working condition, ME1 indicates that the model of the experimental substructure is inaccurate and there is an error). The last one is the proposed method, recorded as NewMethod. The surrogate model established and the surrogate model in the overall structure model have the same model error. The values of constitutive materials under the above four working conditions are shown in Table 4. The seismic wave and the peak acceleration of ground motion previously mentioned are still used.

Table 4. The values of four working conditions

Working condition	Yield strength\MPa		Elastic modulus\GPa	Strain hardening rate
	Column	Beam		
Ref	355	235	206	0.008312
BC1ME0	355	235	206	0.008312
BC0ME1	500	350	260	0.020
NewMethod	500	350	260	0.020

3.2 Simulation Results

To more clearly show the characteristics of the above four working conditions, it can be used that the root mean square error (RMSE) curves varying with time as the judgment standard,

$$\mathrm{RMSE}^j = \sqrt{\frac{\sum_{i=1}^{j}\left(d^i_{\mathrm{simu}} - d^i_{\mathrm{ref}}\right)^2}{\sum_{i=1}^{j}\left(d^i_{\mathrm{ref}}\right)^2}} \tag{11}$$

where RMSE^j represents the RMSE corresponding to the jth time step, the subscript s is the simulation value of the structure, and subscript ref is the reference solution of the structure.

Since the results of each floor are similar, only the bidirectional RMSE curve of the top floor of the structure with time is drawn, as shown in Fig. 7. Obviously, in both directions, the proposed method is in the highest agreement with the reference solution, and the maximum RMSE is only 0.03. The simulation results of the other two methods are only significantly different from the reference solution, and the maximum RMSE is up to 60%. The above also shows that there are two kinds of errors in the HS, namely, the error of the surrogate model and the incomplete boundary conditions. To improve the accuracy of the HS, it can not only improve the source of error on the one hand.

Figure 8 shows the comparison results of top floor displacement and base shear curve of the four methods. It can be seen that the hysteretic curves of the proposed method are in good agreement with the reference solution in two directions, which further illustrates the superior performance of this method.

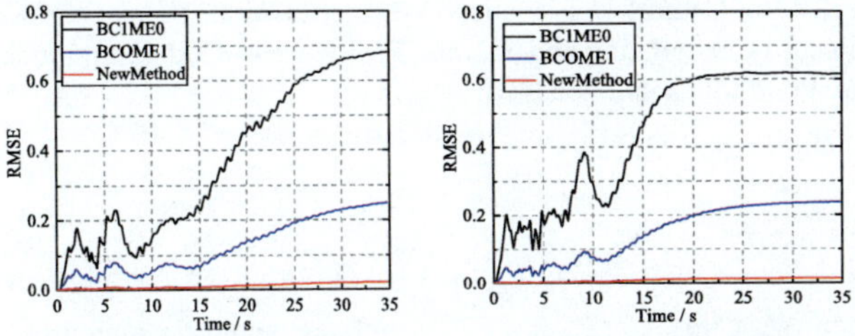

RMSE in the x-direction RMSE in the y-direction

Fig. 7. Bidirectional RMSE curves of the top floor of the structure

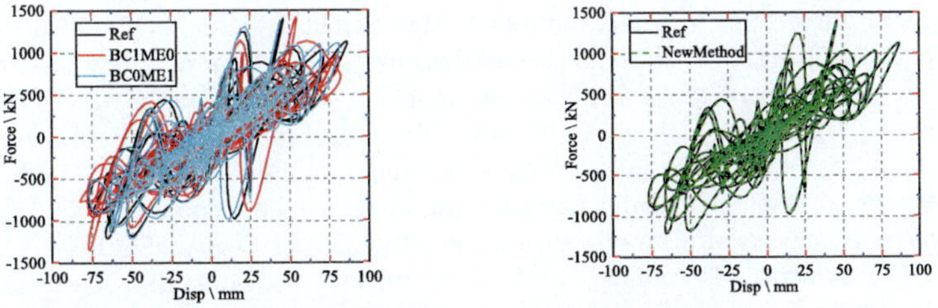

Bottom shear force-top displacement hysteresis curve in X-direction

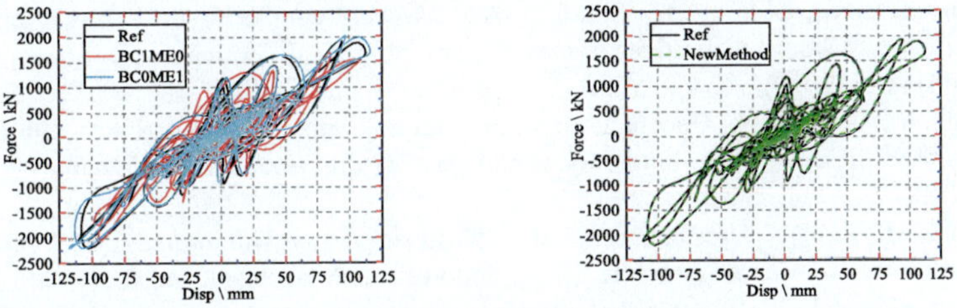

Bottom shear force-top displacement hysteresis curve in Y-direction

Fig. 8. The bidirectional hysteretic curve of the structure

4 Conclusions

To explore the seismic performance of large and complex structures, bidirectional HS, as a cost-effective scheme, was used in the six-story full-scale steel frame. Firstly, the model information of this structure and the process of implementing bidirectional HS were introduced. Furthermore, for the problem of incomplete boundary conditions, a hybrid simulation method with incomplete boundary conditions based on restoring force correction was proposed. Through the numerical analysis, the effectiveness and improvement of the proposed method can be seen, and this method is promising in real-time hybrid simulation with incomplete boundary conditions.

Acknowledgment. Financial support by the Hainan Special PhD Scientific Research Foundation of Sanya Yazhou Bay Science and Technology City(Grant No.HSPHDSRF-2022-03-014), National Natural Science Foundation (Grant No. 52008320, 51878525), the Fundamental

Research Funds for the Central Universities (WUT:2021IVA031), and the Key Laboratory of Earthquake Engineering Simulation and Seismic Resilience of China Earthquake Administration (open project) are gratefully acknowledged.

References

1. Bin, W., Zhen, W., Guoshan, X., et al.: Research and application progress in hybrid testing of engineering structures. Eng. Mech. **9**(1), 1–20 (2020)
2. Mercan, O., Ricles, J.M., Sause, R., et al.: Kinematic transformations for planar multi-directional pseudodynamic testing. Earthq. Eng. Struct. Dynam. **38**(9), 1093–1119 (2019)
3. Molina, F., Verzeletti, G., Magonette, G., et al.: Bi-directional pseudo dynamic test of a full-size three-storey building. Earthquake Eng. Struct. Dynam. **28**(12), 1541–1566 (1999)
4. Liu, G.Y., Chang, S.Y.: Bi-axial pseudo dynamic testing. In: Proceedings of the 12th World Conference on Earthquake Engineering, Auckland, New Zealand, January 2000
5. Xinjiang, C., Shizhu, T., Dapeng, W., et al.: Networked collaborative pseudo-dynamic testing of a multi-span bridge based on NetSLab. Earthq. Eng. Eng. Vib. **8**(3), 387–397 (2009)
6. Huimeng, Z., Tao, W., Chunbo, D., et al.: Multi-degree-of-freedom force-displacement mixed control strategy for structural testing. Earthq. Eng. Struct. Dynam. **50**, 354–374 (2021)
7. Tao, W., Gilberto, M., Andres, J., et al.: Performance evaluation of a distributed hybrid test framework to reproduce the collapse behavior of a structure. Earthq. Eng. Struct. Dynam. **41**(2), 295–313 (2012)
8. Georgios, G., OhSung, K., Shamim, A.S.: Weakly coupled hybrid simulation method for structural testing: theoretical framework and numerical verification. J. Struct. Eng. **146**(2), 04019196 (2020)
9. Bin, W., Xizhan, N., Guoshan, X., et al.: Online numerical simulation: a hybrid simulation method for incomplete boundary conditions. Earthq. Eng. Struct. Dynam. **47**(4), 889–905 (2018)
10. Zhu, M., Bin, W., Oreste, S.B., et al.: Hybrid simulation with online model updating: application to a reinforced concrete bridge endowed with tall piers. Mech. Syst. Signal Process. **123**, 533–553 (2019)
11. Ning, L., Zihao, Z., Zhongxian, L.: Stability prediction for real-time hybrid simulation with different physical and numerical substructure discretization using asynchronous multirate simulation. J. Eng. Mech. **147**(11), 04021086 (2021)
12. Zhenyun, T., Matt, D., Zhenbao, L., et al.: The performance of delay compensation in real-time dynamic substructuring. J. Vib. Control **24**(21), 5019–5029 (2018)
13. Ge, Y., Bin, W., Ge, O., et al.: HyTest: platform for structural hybrid simulations with finite element model updating. Adv. Eng. Softw. **112**, 200–210 (2017)
14. Yungui, L., Jianyun, S., Pengfei, S., et al.: New facilities in laboratory of CSCEC for large and full scale structure test. In: 6th International Conference on Advances in Experimental Structural Engineering and the 11th International Workshop on Advanced Smart Materials and Smart Structures Technology. University of Illinois, Urbana-Champaign (2015)
15. Zhen, W., Qiyang, T., Pengfei, S., et al.: Performance validation and application of a mixed force-displacement loading strategy for bi-directional hybrid simulation. Smart Struct. Syst. **26**(3), 373–390 (2020)
16. Shangzhang, W., Ge, Y., Bin, W., et al.: A redundant actuator control method for hybrid simulation based on the interior point and the Lagrange multiplier. J. Vibr. Shock **40**(12), 23–30+37 (2021)

17. Guoshan, X., Zhen, W., Bin, W., et al.: Pseudo dynamic tests with substructuring of a full-scale precast box-modularized structure made of reinforced concrete shear walls. Struct. Des. Tall Special Build. **26**(16), e1354 (2017)
18. Bin, W., Qiyang, T., Pengfei, S., et al.: Substructure modeling and loading-control techniques for the test of a full-scale spatial RC frame with buckling-restrained braces subjected to bidirectional loading. J. Struct. Eng. **147**(9), 04021136 (2021)

Check for updates

Main-Aftershock Modeling According to Measured Records

Zixin Liu[1(✉)], Yunmu Jiang[2], Zhangjun Liu[2], and Xinxin Ruan[2]

[1] Key Laboratory of Building Collapse Mechanism and Disaster Prevention, Institute of Disaster Prevention, China Earthquake Administration, Sanhe 065201, Hebei, China
liuzixin@cidp.edu.cn

[2] School of Civil Engineering and Architecture, Wuhan Institute of Technology, Wuhan 430074, Hubei, China

Abstract. At first, the coherence between main and aftershock measured records is systematically studied, and a Fourier trigonometric series model for the coherence function is proposed on the basis of this. After that, the parameters of evolutionary power spectrum density (EPSD) function model of main-aftershock ground motion are determined using the best square approximation. At last, to realize the dimension-reduction modeling of main-aftershock ground motion, it is regarded as a one-dimensional two-variate (1D-2V) stochastic vector process simulated by the POD-based dimension-reduction representation method. Thus, the dimension-reduction modeling for main-aftershock ground motion according to measured records is achieved with merely two elementary random variables. Numerical results show that the representative samples are consistent well with the measured records in terms of the coherence and the response spectrum, which demonstrates the correctness of the proposed main-aftershock model and the effectiveness of the dimension-reduction modeling method. It provides a positive way for the accurate dynamic response and reliability analysis of engineering structures subjected to main-aftershocks.

Keywords: Main-aftershock ground motion · Measured records · Coherence function · EPSD model · Dimension-reduction modeling

1 Introduction

The research shows that after a main earthquake, it is often accompanied by multiple aftershocks [1–3]. Not only the number of aftershocks is large, but also some of them have large magnitudes, which becomes the main factor of secondary damage of structures. However, due to the limited number of measured main-aftershock ground motion records and the requirement to process the measured records to satisfy structural seismic analysis, the generation of main-aftershock ground motion time-histories by artificial synthesis method has attracted extensive attention in recent years.

Generally speaking, the synthetic methods of main-aftershock ground motions can be roughly divided into deterministic methods and stochastic methods. For the deterministic methods, Hatzigeorgiou et al. [4] used the repetition method to construct the

W. Guo and K. Qian (Eds.): GBCESC 2022, LNCE 211, pp. 1022–1029, 2023.
https://doi.org/10.1007/978-981-19-5217-3_104

main-aftershock sequence. This method assumes that the characteristics of the main-aftershocks are consistent, which is obviously inconsistent with the actual situation. Li et al. [5] proposed the random combination method to construct the main-aftershock sequence, that is, a motion record is randomly selected from the main and aftershock record database to form the main-aftershock sequence. However, the main-aftershock sequence obtained by this method does not have randomness in essence. For the random method, Hu et al. [6] used the amplitude modulation filtering white noise method to generate the main-shock process, and then generated the aftershocks according to the branch aftershock sequence method. However, this method belongs to the conventional Monte Carlo method, which cannot accurately describe the main-aftershock process at the probability level. Jiang et al. [7] established the relevant structure of main-aftershock parameters based on Copula function, and introduced the random function of orthogonal random variables, so as to realize the efficient simulation of aftershocks with the given main shock parameters. However, the time-frequency fully non-stationary evolutionary power spectrum model used in this study was coupled in time-frequency domain, errors were inevitably introduced in the identification of main-aftershock parameters. Meanwhile, the study regarded the main-aftershock type ground motion as two independent random processes, and did not consider the coherence between the main and aftershocks. As a result, two simulation processes were required to obtain the representative time-histories of the main-aftershock type ground motion. To sum up, it is urgent to develop an integrated and efficient method for simulating main-aftershock ground motions at the probabilistic level.

Under such background, the innovations of this study include the dimension-reduction simulation of the main-aftershock type ground motion which is regarded as a random vector process, and the proposed coherence function as well as the evolutionary power spectrum density (EPSD) model of the main-aftershock earthquakes. Firstly, according to the measured records, the coherence function model of main-aftershocks is established, which effectively considers the coherency between the main and aftershocks in the frequency level. Secondly, according to the energy curve and response spectrum of ground motion, the parameters of the EPSD model of main-aftershocks are identified, and the values of all parameters corresponding to different soil sites are given. Finally, combined with the random function method based on proper orthogonal decomposition (POD) [8, 9], the dimension-reduction modeling of main-aftershock vector process is realized. Numerical examples verify the correctness of the main-aftershock coherence function model and the effectiveness of the parameter identification method.

2 The Coherence Function Model of Main-Aftershocks

It is well known that local sites have a great influence on the spectral characteristics of ground motions, and the main shocks and aftershocks recorded by the same station occurred in the same local site. From this, it is inferred that there may be some kind of coherence in the spectral characteristics of the main and aftershocks. To this end, according to the selected 488 sets of measured strong motion records, a statistical analysis of the coherence between the main shocks and aftershocks was carried out. For simplicity, considering both the main shocks and aftershocks as stationary processes, the (lagged) coherence function is defined as:

$$\gamma_{ma}(\omega) = \left| \frac{S_{ma}(\omega)}{\sqrt{S_m(\omega) \cdot S_a(\omega)}} \right| \tag{1}$$

where $S_{ma}(\omega)$ denotes the cross-PSD of main-aftershocks, $S_m(\omega)$ and $S_a(\omega)$ are the auto-PSDs of the main and aftershocks, respectively.

It can be seen from the analysis that in the frequency range 0–300 rad/s commonly used in engineering, the coherence of main-aftershocks generally shows a decreasing trend with the increase of frequency. To this end, the first three-order Fourier series model can be used to fit the coherence function of the main aftershocks, following that:

$$\overline{\gamma}_{ma}(\omega\,; \boldsymbol{\lambda}_\gamma) = A + \sum_{i=1}^{3} [B_i \sin(i \times D \times \omega) + C_i \cos(i \times D \times \omega)] \tag{2}$$

where $\boldsymbol{\lambda}_\gamma = (A, B_1, C_1, B_2, C_2, B_3, C_3, D)$ denotes the parameter vector of the Fourier series model.

According to the principle of the best square approximation, with $\gamma_{ma}(\omega)$ as the target value, the parameter vector $\boldsymbol{\lambda}_\gamma$ of the Fourier series model can be identified, and then the values of the model parameters can thus be obtained. The fitting result of the Fourier series model and the coherence function of the measured main-aftershocks is shown in Fig. 1.

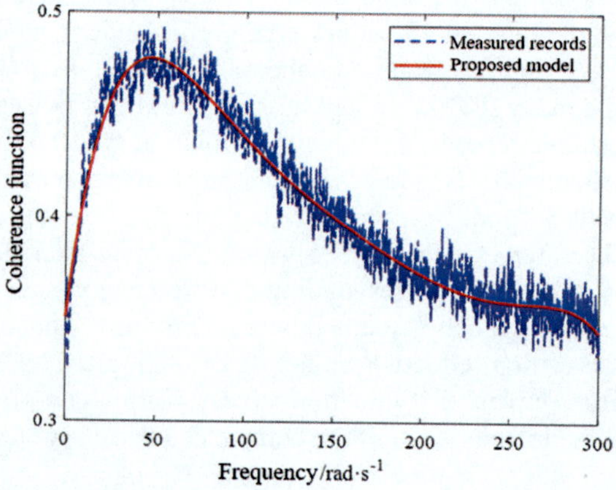

Fig. 1. Fitting result between the proposed coherence function model and measured records

3 The EPSD Parameter Identification for Main-Aftershocks

For simplicity, the EPSD function form of the intensity modulation is used [10]:

$$S_J(\omega, t;\, \boldsymbol{\lambda}_J) = \left|q(t;\, \boldsymbol{\lambda}_q)\right|^2 S(\omega;\, \boldsymbol{\lambda}_S) \tag{3}$$

where $S_J(\omega, t; \boldsymbol{\lambda}_J)$ denotes the one-sided EPSD function of non-stationary ground motion, $q(t; \boldsymbol{\lambda}_q)$ denotes the intensity modulation function, and $S(\omega; \boldsymbol{\lambda}_S)$ denotes the one-sided PSD function of stationary ground motion, respectively.

For specific, the intensity modulation function $q(t; \boldsymbol{\lambda}_q)$ adopts the three-stage model proposed by Amin-Ang [11]:

$$q(t; \boldsymbol{\lambda}_q) = \begin{cases} t^2/t_1^2 & 0 \le t \le t_1 \\ 1 & t_1 \le t \le t_2 \\ \mathrm{e}^{-\alpha(t-t_2)} & t_2 \le t \end{cases} \tag{4}$$

where $\boldsymbol{\lambda}_q = (t_1, t_2, \alpha)$ denotes the parameter vector of intensity modulation function $q(t; \boldsymbol{\lambda}_q)$. For the one-sided PSD function of stationary ground motion, the classical Clough-Penzien model [12] is used:

$$S(\omega; \boldsymbol{\lambda}_S) = 2 \times \frac{\omega_g^4 + 4\xi_g^2\omega_g^2\omega^2}{(\omega^2 - \omega_g^2)^2 + 4\xi_g^2\omega_g^2\omega^2} \times \frac{\omega^4}{(\omega^2 - \omega_f^2)^2 + 4\xi_f^2\omega_f^2\omega^2} \times S_0 \tag{5}$$

in which

$$S_0 = \frac{A_{\max}^2}{\bar{r}^2\omega_e}, \ \omega_e = \frac{1}{S_0}\int_0^\infty S(\omega; \boldsymbol{\lambda}_S)\mathrm{d}\omega \tag{6}$$

where ω_g and ξ_g denote the superior circular frequency and damping ratio of the site soil, ω_f and ξ_f denote the superior circular frequency and damping ratio of the bedrock, respectively. In general, there exists $\omega_f = 0.1\omega_g$ and $\xi_f = \xi_g$. $A_{\max}$ denotes the mean value of the peak ground acceleration, and $\bar{r}$ denotes the peak factor, respectively. Thus, $\boldsymbol{\lambda}_S = (\omega_g, \xi_g, \bar{r})$ is the parameter vector of one-sided PSD function of stationary ground motion $S(\omega; \boldsymbol{\lambda}_S)$.

According to the measured strong motion records of main-aftershocks in different site categories, the best square approximation is used to identify the parameters of the EPSD model of the main-aftershock vector process, and the mean value of each parameter is obtained. The recommended values of EPSD model parameters are shown in Table 1.

Table 1. Recommended values of main-aftershock evolutionary power spectrum parameters for different soil site classifications.

Earthquake	Soil site	Paremeters				
		t_1 (s)	t_2 (s)	α	ω_g(rad/s)	ξ_g
Main shock	I_1	3.51	6.93	0.14	16.25	0.47
	II	2.98	7.19	0.15	13.63	0.42
	III	1.66	8.06	0.11	10.21	0.44
Aftershock	I_1	4.11	6.05	0.16	18.01	0.40
	II	2.17	6.44	0.20	15.13	0.39
	III	1.09	6.99	0.14	13.21	0.46

4 Dimension-Reduction Modeling of Main-Aftershock Vector Process Based on POD

For the main-aftershock type ground motion process, it can be regarded as a 1D-2V (one-dimensional two-variate) random vector process with zero mean, so it can be simulated by the proper orthogonal decomposition (POD) method. The EPSD matrix $\boldsymbol{S_J}(\omega,\ t)$ of the non-stationary main-aftershock vector process can be written as:

$$\boldsymbol{S_J}(\omega, t) = \begin{bmatrix} S_{\mathrm{m}}(\omega, t) & S_{\mathrm{ma}}(\omega, t) \\ S_{\mathrm{am}}(\omega, t) & S_{\mathrm{a}}(\omega, t) \end{bmatrix} \tag{7}$$

which can be further decomposed as the following sense:

$$\boldsymbol{S_J}(\omega, t) = \boldsymbol{D}(\omega, t)\boldsymbol{\gamma}(\omega)\boldsymbol{D}^{*T}(\omega, t) \tag{8}$$

The eigen decomposition of the coherence function $\boldsymbol{\gamma}(\omega)$ matrix can be obtained:

$$\begin{cases} \boldsymbol{\gamma}(\omega) = \boldsymbol{\Psi}(\omega)\boldsymbol{\Lambda}(\omega)\boldsymbol{\Psi}^{*\mathrm{T}}(\omega) \\ \boldsymbol{\Psi}^{*\mathrm{T}}(\omega)\boldsymbol{\Psi}(\omega)=\mathbf{I} \end{cases} \tag{9}$$

where $\boldsymbol{\Psi}(\omega)$ denotes the eigenvector matrix, and $\boldsymbol{\Lambda}(\omega)$ denotes the eigenvalue diagonal matrix, respectively. If we define: $\boldsymbol{\psi}_i(\omega_k) = \boldsymbol{\chi}_i(\omega_k) + \mathrm{i}\boldsymbol{\mathcal{Z}}_i(\omega_k)$. Then, according to a series of derivations, the POD formula of the r-th component process of the main-aftershock vector process can be expressed as:

$$J_r(t) \approx \sqrt{2}\sum_{i=1}^{2}\sum_{k=1}^{N}\sqrt{S_{J_r}(\omega_k,t)\Lambda_i(\omega_k)\Delta\omega} \times \{\chi_{ri}(\omega_k)[R_{ik}\cos(\omega_k t) - I_{ik}\sin(\omega_k t)]$$
$$-\mathcal{Z}_{ri}(\omega_k)[R_{ik}\sin(\omega_k t) + I_{ik}\cos(\omega_k t)]\} \tag{10}$$

where R_{ik} and I_{ik} are orthogonal random variables with zero mean, satisfying the following basic conditions:

$$E[R_{ik}] = E[I_{ik}] = 0,\ E[R_{ik}I_{sl}] = 0,\ E[R_{ik}R_{sl}] = E[I_{ik}I_{sl}] = \frac{1}{2}\delta_{is}\delta_{kl}, \quad i, s = 1, 2;\ k, l = 1, 2, \cdots, N \tag{11}$$

In this paper, the idea of dimension-reduction based on random function is introduced, and the orthogonal random variables are defined as the functional form of the elementary random variables [13]:

$$R_{ik} = \sqrt{2}\cos(s \times \Theta_1 + \pi/4) \cdot \sin(l \times \Theta_2 + \pi/4), \quad I_{ik} = \sqrt{2}\sin(s \times \Theta_1 + \pi/4) \cdot \cos(l \times \Theta_2 + \pi/4) \tag{12}$$

where $i, s = 1, 2;\ k, l = 1, 2, \cdots, N$, $\Theta_i(i = 1, 2)$ denotes the elementary random variables independent of each other and uniformly distributed over the interval $(0, 2\pi]$.

Obviously, the random function form defined by Eq. (12) fully satisfies the basic conditions of Eq. (11). In this way, just two elementary random variables are needed to accurately simulate the main-aftershock vector process, and the representative time-histories generated have assigned probability, and the set of all representative time-histories constitutes a complete probability. It lays a solid foundation for realizing the refined dynamic response analysis and dynamic reliability evaluation of complex structures in combination with the probability density evolution method.

The simulation results are shown in Fig. 2 and Fig. 3. As presented in the Fig. 2, the representative time-histories generated have obvious characteristics of main-aftershock sequence type ground motions, that is, the main shock and aftershocks show obvious differences in duration, frequency spectrum and amplitude. Figure 3 displays the comparison of the simulated value and the corresponding target of the coherence upon the main-aftershocks. It can be seen that the representative time histories of the main aftershocks have good coherence and are consistent with the measured records, which proves the correctness of the coherence function model proposed in this paper. What's more, in case that there are just more than one hundred representative time-histories, the standard deviation error is less than 5%, which can meet the engineering requirements and verify the remarkable accuracy possessed by the dimension-reduction method.

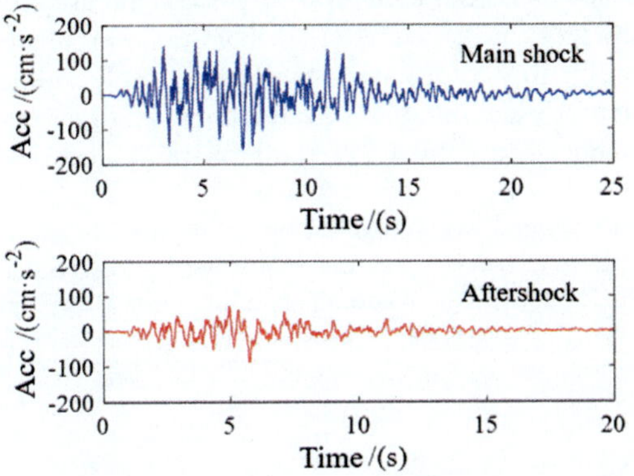

Fig. 2. The representative time-histories of main-aftershock simulated by the dimension-reduction simulation

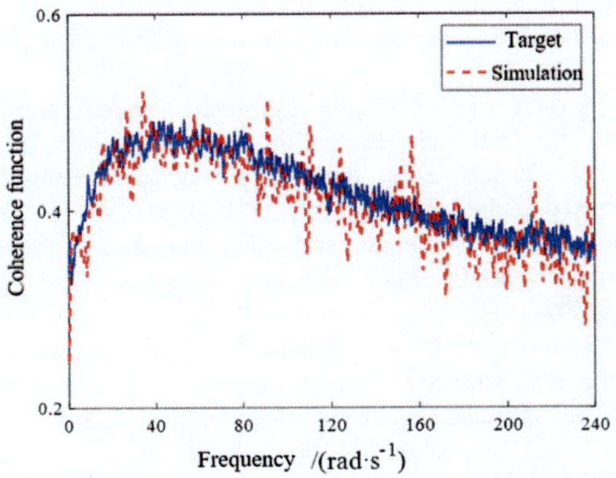

Fig. 3. Comparison of the simulated value and the corresponding target of the coherence upon the main-aftershocks

5 Conclusions

The measured strong motion records show that there is obvious coherence between the main and aftershocks, and the coherence generally shows a decreasing trend with the increase of the frequency. Accordingly, this paper proposes a Fourier series model for simulating the coherence of main-aftershocks, and verifies the validity of the coherence function model by comparing it with the measured records. Considering the main-aftershock as a vector process, the integrated and efficient simulation of the main-aftershock sequence type ground motion is realized. With the help of this method, the representative time-histories of the main shock and aftershocks can be generated at once, instead of simulating the main and aftershocks as two random processes separately, which not only reflects the coherence between the main and aftershocks, but also greatly improves the computational efficiency. It is worth mentioning that the dimension-reduction method based on random function only needs hundreds of representative time histories to reflect the probability characteristics of the vector process of main and aftershocks on the full probability information, which lays a solid foundation for the refined dynamic response analysis and reliability evaluation of engineering structures under the action of main-aftershocks combined with the probability density evolution theory.

Acknowledgement. This work was supported by the Natural Science Foundation of Hebei Province [Grant No. E2021512001], and the Fundamental Research Funds for the Central Universities [Grant No. ZY20210210]. These financial supports are gratefully acknowledged.

References

1. Zhang, Y., Zhi, X., Fan, F.: Study of the correlations between main shocks and aftershocks and aftershock synthesis method. Earthq. Eng. Eng. Vib. **18**(4), 759–775 (2019). https://doi.org/10.1007/s11803-019-0535-4
2. Zhai, Y.F., Zhang, L.J., Bi, Z.H., et al.: Seismic performance evaluation of AAR-affected concrete gravity dams under main aftershock sequence. Soil Dyn. Earthq. Eng. **157**, 107258 (2021)
3. Mun, C., Song, J.: Bayesian-network-based risk modeling and inference for structures under a sequence of main and aftershocks. Earthq. Eng. Struct. Dynam. **51**(5), 1058–1075 (2022)
4. Hatzigeorgiou, G.D., Beskos, D.E.: Inelastic displacement ratios for SDOF structures subjected to repeated earthquakes. Eng. Struct. **31**(11), 2744–2755 (2009)
5. Li, Q., Ellingwood, B.R.: Performance evaluation and damage assessment of steel frame buildings under main shock-aftershock earthquake sequences. Earthquake Eng. Struct. Dyn. **36**(36), 405–427 (2007)
6. Hu, S., Gardoni, P., Xu, L.: Stochastic procedure for the simulation of synthetic mainshock-aftershock ground motion sequences. Earthq. Eng. Struct. Dyn. **47**(11), 2275–2296 (2018)
7. Jiang, Y.M., Ruan, X.X., Liu, Z.J.: Dimension-reduction simulation of main aftershock type ground motion process. J. Vibr. Shock **40**(24), 282–292 (2021). (in Chinese)
8. Yuen, K.V., Mu, H.Q.: Peak ground acceleration estimation by linear and nonlinear models with reduced order Monte Carlo simulation. Comput.-Aided Civ. Infrastr. Eng. **26**(1), 30–47 (2011)

9. Liu, Z.X., Liu, Z.J., Ruan, X.X., et al.: Spectral representation-based dimension reduction for simulating multivariate non-stationary ground motions. Soil Dyn. Earthq. Eng. **114**, 313–325 (2018)
10. Liu, Z.J., Liu, Z.X.: Simulation of fully non-stationary ground motion based on seismic design response spectrum. J. Sound Vib. **30**(3), 457–465 (2017). (in Chinese)
11. Amin, M., Ang, A.H.S.: Nonstationary stochastic models of earthquake motions. J .Eng. Mech. Div. **94**(2), 559–584 (1968)
12. Clough, R., Penzien, J.: Dynamic of Structures. McGraw-Hill Inc, New York (1993)
13. Liu, Z.J., Liu, Z.X., Ruan, X.X., et al.: POD-based dimension reduction representation of stochastic ground motion fields. Sci. Sinica (Technol.) **49**(5), 103–111 (2018). (in Chinese)

Study on Mechanical Performance of Column Base Joint with Slip-Friction Arc Endplates

Chengyu Li[1,2](✉) and Zimeng Lai[1]

[1] School of Urban Construction, Wuhan University of Science and Technology, Wuhan, China
lee_chengyu@163.com

[2] Institute of High Performance Engineering Structure, Wuhan University of Science and Technology, Wuhan, China

Abstract. In order to study the mechanical performance of column base joint with slip-friction arc endplates, the finite element simulation analysis of this joint is carried out. By changing the radius of arc endplate, axial compression ratio, and other parameters, the effect of different parameters on the bending capacity, joint stiffness, and bolt group force distribution of the joint is analyzed. The results show that the slip behavior of the column base joint is the best when the arc endplate radius is 0.7H (H is the height of the column section); This column base joint is semi-rigid and has good rotation capability. Based on the force distribution of the bolt group in the finite element analysis, the neutral axis of the bolt group is assumed to be located at the outermost row of bolts. Through theoretical analysis, the calculation formula for the sliding threshold of this joint is obtained. Compared with the simulated value, the two are in good agreement by using a correction coefficient of 0.75.

Keywords: Column base joint · Arc endplates · Sliding friction · Force analysis

1 Introduction

Steel structures are increasingly used in high seismic areas due to their excellent performance and good performance in previous earthquakes. However, a seismic damage survey showed that steel joints suffered different degrees of damage in the earthquake, making post-earthquake repair difficult and costly. Friction connections are one of the most commonly used low damage connections and have been investigated by many scholars [1]. The test of the low damage sliding hinged (SHJ) column base joint shows that this joint has no obvious damage after a strong earthquake [2]. Tests on a capped sliding friction self-resetting foot node showed that the bolts at the flange of the node could slip and dissipate energy along the long circular bolt holes, thus achieving no loss of energy in the structure [3]. Elena et al. [4] discussed a new type of self-centering undamaged steel column base joint and its beneficial effect in flexural steel frames (MRFs). Li et al. [5] carried out a parametric simulation on the hysteretic behavior of the column base joint with slip-friction arc endplates. It showed that the joint can use friction slip to consume seismic energy and achieve the design purpose of damage control.

W. Guo and K. Qian (Eds.): GBCESC 2022, LNCE 211, pp. 1030–1038, 2023.
https://doi.org/10.1007/978-981-19-5217-3_105

In this paper, finite element analysis is carried out on the static performance of rectangular steel tube column base joint with slip-friction arc endplates. It studies the effects of different arc plate radius, axial pressure ratios on the bending capacity, joint stiffness, and bolt group force performance. On this basis, the theoretical force analysis of the column base joint is carried out. The formula for calculating the slip force of the joint is proposed and compared with the simulated values.

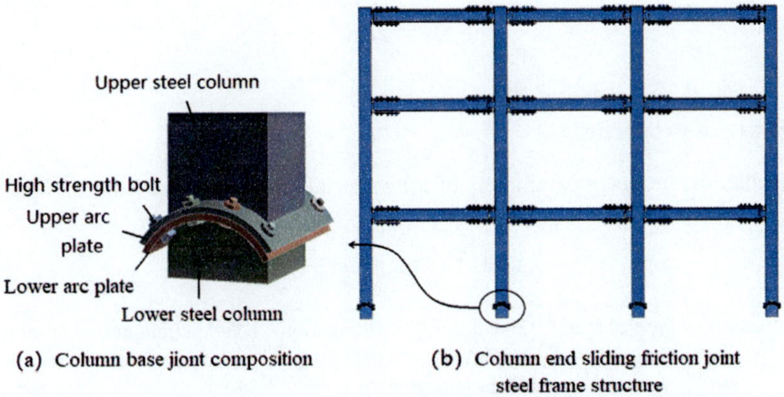

Fig. 1. Rectangular steel tube column base with slip-frictionarc endplates

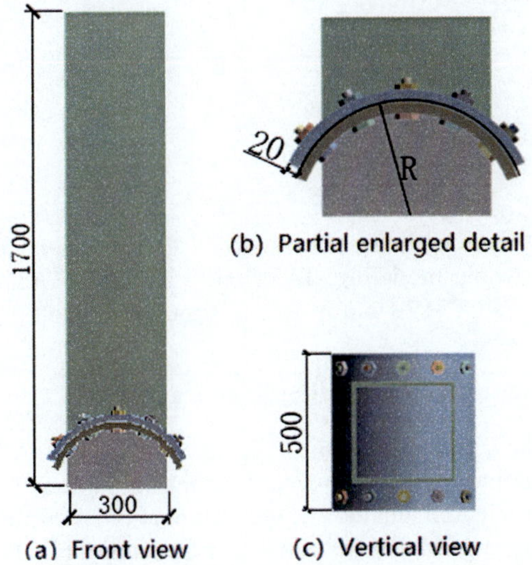

Fig. 2. Dimensions of the column base joint

2 Finite Element Analysis of Mechanical Behavior of Joints

2.1 Finite Element Model

The rectangular steel tube column base joint with slip-friction arc endplates is shown in Fig. 1. Dimensions of the joint is shown in Fig. 2. The slip length of the long circular hole of the lower arc plate is determined by Formula (1). Where: φ is the joint corner radian, this paper takes $\varphi = 0.05$; R is the arc plate radius.

$$B = 2\varphi R \tag{1}$$

2.2 Material Properties

The mechanical property parameters of the material of the joint are shown in Table 1.

Table 1. Material property.

Component	Type of steel	Elastic modulus $E/(N/mm^2)$	Elongation (%)	Yield strength $f_y/(N/mm^2)$	Ultimate strength $f_u/(N/mm^2)$
Steel column	Q355	2.06×10^5	21	355	560
Arc plate	Q355	2.06×10^5	21	355	560
High strength bolt	45# steel	2.06×10^5	12	640	800

2.3 Unit Type

The simulation analysis is carried out by ANSYS finite element software. SOLID185 unit is used for the whole modeling. TARGE170 and CONTA174 units are used to simulate the contact. PRETS179 unit is used to simulate the bolt.

2.4 Boundary Conditions and Loading Regimes

The bottom end of the column base joint is set as the fixed end. All contacts are set as frictional contacts, adopting the incremental Lagrange algorithm.

This paper adopts a loading method of the combination of displacement loading and concentrated force loading, which was carried out in 3 steps. Firstly, apply bolt preload P; followed by the axial pressure N on the column top; finally, apply the horizontal displacement load on the left side of the column top.

2.5 M-θ Curve

Figure 3 gives the M-θ curve of single column base joint under different parameters.

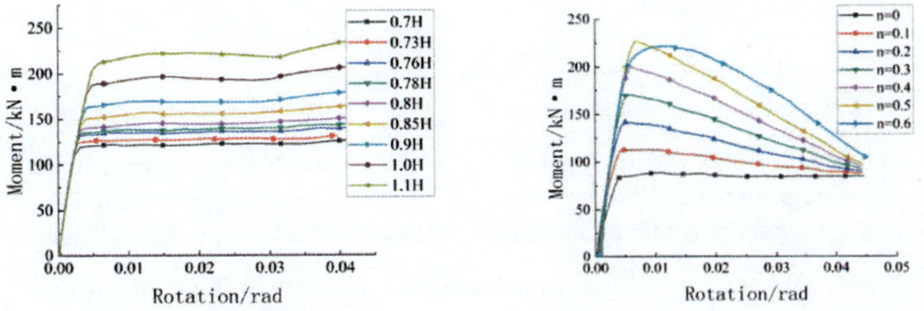

(a) M-θ curves of joints with different arc plate radius

(b) M-θ curve under different axial compression ratio

Fig. 3. Moment rotation curves of column base joints with different parameters

It can be seen: (1) The radius of the arc plate has a significant impact on the steel column base joint with slip-friction arc endplates. With the increase of the arc plate radius, the flexural bearing capacity of the column base joint gradually increases. But the fluctuation of the curve sliding section becomes more significant, and the joint may not rotate. (2) When n $\leq$ 0.5, the bearing capacity of the column base joint increases with the increase of the axial compression ratio. Due to the P $-$ Δ effect, there is a falling section in the M-θ curve, and the slope of the falling section increases with the increase of the axial compression ratio. When the axial compression ratio is too large, with the increase of axial compression ratio, the additional bending moment of the joint increases, the steel column buckles, and the flexural capacity of the joint decreases.

2.6 Joint Stiffness

A simple and effective method was proposed for classifying nodes into rigid, semi-rigid, and hinged connections [6]. The method is as follows: The classification criteria for the connection is defined as follows: the joint is a rigid connection when $\overline{m} > 0.7, \overline{m} > 2.5\overline{\theta}$; the joint is a semi-rigid connection when $0.2 < \overline{m} < 0.7$, $0.5\overline{\theta} < \overline{m} < 2.5\overline{\theta}$; the joint is a hinged connection when $\overline{m} < 0.2$, $\overline{m} \leq 0.5\overline{\theta}$. Where $\overline{m} = M/M_p$, $\overline{\theta} = \theta/\theta_p$, $\theta_p = M_p/(EI/L_p)$; M is the actual moment; M_p is the plastic moment; θ is the actual angle; θ_p is the plastic angle; E is the modulus of elasticity; I is the section moment of inertia; Lp is the calculated length of the column.

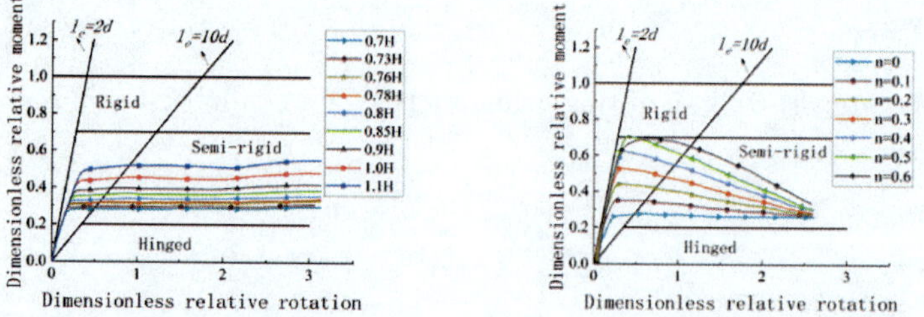

(a) Joint stiffness with different arc plate radius

(b) Joint stiffness under different axial compression ratio

l_e is the reference length of the column; d is the section height of the column

Fig. 4. Connection stiffness of column base joint with different parameter

The M-θ curves obtained from the simulation are dimensionless and listed in Fig. 4. It can be seen that (1) The joint stiffness increases as the radius of the arc plate increases, and they all belong to semi-rigid joints with good rotational capacity (see Fig. 4a). (2) The joint stiffness increases continuously with the increase of axial compression ratio. When n = 0–0.4, the joints are semi-rigid joints with good rotation capacity; when n = 0.5, the joints are close to rigid joints (see Fig. 4b).

2.7 Bolt Stress Analysis

The structure and stress of this column base joint are symmetrical about the horizontal centerline, so one side bolt is taken to study the stress of the bolt. It can be seen from Fig. 5 that when the axial compression ratio is equal to 0, the internal force of bolts from 1 to 5 shows a decreasing trend. With the increase of arc plate radius, the bolt tension increases continuously. When R = 0.9H, the tensile force of outermost bolts reaches the maximum tensile strength of bolts. Therefore, it is more appropriate to select a smaller arc plate radius R = 0.7h for the column base joint.

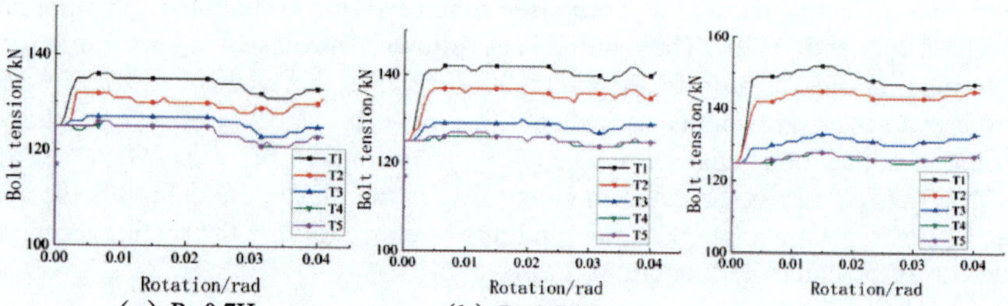

(a) R=0.7H (b) R=0.76H (c) R=0.8H

Fig. 5. Force of bolt group in joints with different radius of arc endplate theoretical analysis of the joint

According to the analysis of bolt stress in Sect. 2.7, this paper makes the following assumptions: for the column base joint, it is assumed that the neutral axis is at the outermost side of the bolt group. Under the bending moment, the stress of the bolt group is linearly distributed, and the stress of the outermost row bolt is the largest. The calculation formula of the maximum tension when the bending moment of the high-strength bolt group is shown in Eq. (2):

$$N_M = \frac{My_i}{\sum y_i^2} \tag{2}$$

where: y_i is the distance from any bolt to the neutralization axis.

When the sliding friction column base joint is subjected to the combined action of bending moment, shear force, and axial pressure, take the isolator for stress analysis.

Decompose the forces N_M, V_F, and N_N along the radial and tangential directions of the arc plate, as shown in Fig. 6.

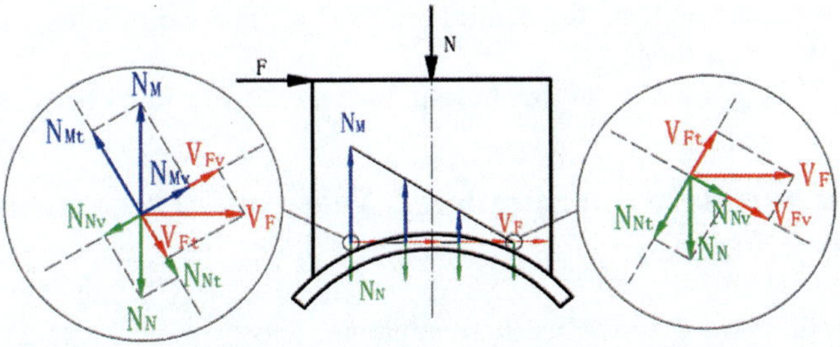

Fig. 6. Force analysis of the column base joint under bend-shearing-pressure combination

N_M can be calculated by Eq. (1), and the shear force and axial pressure are distributed evenly: $V_F = \frac{F}{s}, N_N = \frac{N}{s}$ (s is the number of bolts).

Extrusion force between left arc plates of symmetry axis:

$$C = P - N_{Mt} + V_{Ft} + N_{Nt} \tag{3}$$

Tangential shear force along arc plate:

$$V = N_{Mv} + V_{Fv} - N_{Nv} \tag{4}$$

Extrusion force between arc plates on the right side of symmetry axis:

$$C = P - N_{Mt} - V_{Ft} + N_{Nt} \tag{5}$$

Tangential shear force along arc plate:

$$V = V_{Fv} + N_{Nv} - N_{Mv} \tag{6}$$

Total extrusion force between two arc plates:

$$\Sigma C = sP + \Sigma N_{Nt} - \Sigma N_{Mt} \tag{7}$$

Total friction between arc plates:

$$\Sigma F_f = \mu(sP + \Sigma\frac{N}{s} \cdot \cos\alpha_i - \frac{My_i}{\sum y_i^2} \cdot \cos\alpha_i) = \mu sP + \mu\Sigma\frac{N}{s} \cdot \cos\alpha_i - \mu\frac{My_i}{\sum y_i^2} \cdot \cos\alpha_i) \quad (8)$$

Total shear force along the tangential direction of arc plate:

$$\Sigma V = \Sigma\frac{F}{s} \cdot \cos\alpha_i + \sum_{j=1}^{\frac{i}{2}} \frac{My_j}{\sum y_i^2} \cdot \sin\alpha_j - \sum_{j=\frac{i}{2}+1}^{i} \frac{My_j}{\sum y_i^2} \cdot \sin\alpha_j \quad \text{(i is an even number)} \quad (9)$$

$$\Sigma V = \Sigma\frac{F}{s} \cdot \cos\alpha_i + \sum_{j=1}^{\frac{i-1}{2}} \frac{My_j}{\sum y_i^2} \cdot \sin\alpha_j - \sum_{j=\frac{i}{2}+3}^{i} \frac{My_j}{\sum y_i^2} \cdot \sin\alpha_j \quad \text{(i is an odd number)} \quad (10)$$

where: μ is the contact surface friction coefficient; P is the preload force in the bolt; α is the angle between the bolt and the neutral axis; s is the number of bolts; i is the number of bolts on one side of the joint.

When $\sum V = \sum F_f$, the column base joint has relative sliding between arc plates.

3 Comparison Between Simulated Value and Theoretical Value of Sliding Force

Table 2 shows the comparison between the simulated and theoretical values of the column base joint under different axial pressure ratios. It can be seen that the simulated values of the joint with eight bolts and ten bolts are between 75% and 80% of the theoretical values under different axial pressure. The reason for the difference between the two values is that the theoretical calculation considers the starting slip force of the column base joint in an ideal state. Practically, the bolt is subjected to prying force, which increases the tensile force on the bolt and reduces the extrusion force between arc plates, resulting in the reduction of friction force to be overcome for joint sliding and the reduction of joint sliding force. Therefore, a correction factor of 0.75 is introduced to carry out the theoretical bending capacity calculation. However, the accuracy of this factor remains to be verified by test.

Table 2. Comparison between the simulated value and the theoretical value.

Number of bolts	Axial compression ratio	Simulated value of sliding force F′/kN	Theoretical value of sliding force F/kN	F′/F
8	0	52.0	64.4	0.81
	0.1	66.2	85.4	0.78
	0.2	83.5	106.4	0.78
	0.3	100.5	127.5	0.79
	0.4	117.8	148.5	0.79

(continued)

Table 2. (*continued*)

Number of bolts	Axial compression ratio	Simulated value of sliding force F′/kN	Theoretical value of sliding force F/kN	F′/F
10	0	64.4	81.6	0.79
	0.1	79.3	103.4	0.77
	0.2	95.6	125.0	0.76
	0.3	112.9	147.0	0.77
	0.4	127.8	168.5	0.76

4 Conclusion

In this paper, the theoretical analysis and finite element analysis of the rectangular steel tube column base joint with slip-friction arc endplates are carried out, and the following conclusions are obtained:

(1) The radius of the arc plate greatly influences the slip performance of the column base joint. With the increase of the radius of the arc plate, the tensile force on the bolt increases. Therefore, according to the stress of the bolt group, when the radius of the arc plate is $r = 0.7H$ (H is the height of the column section), the slip performance of the column base joint is better.

(2) With the increase of axial compression ratio ($n = 0.1$–0.3), the bearing capacity and joint stiffness of the rectangular steel tube column base joint with slip-friction arc endplates increase.

(3) The theoretical calculation formula in this paper is based on the ideal conditions. Considering the influence of bolt prying force, $P - \Delta$ effect, and other factors, a correction coefficient of 0.75 is introduced in this paper. Compared with the simulated and theoretical value of column base joint sliding force, the two are in good agreement.

Acknowledgement. This research was funded by the National Natural Science Foundation of China, grant number 51878522.

References

1. Chouery, K.E., et al.: State-of-the-art review of symmetric and asymmetric friction connections: seismic behavior and design methods. Eng. Mech. **38**(5), 22–37 (2021)
2. Borzouie, J., et al.: Experimental studies on cyclic performance of column base strong axis aligned asymmetric friction connections. J. Struct. Eng. 04015078 (2015)
3. Santiago, A., Silva, L.S.D., et al.: Experimental behavior of base plate joints equipped with self-centering system and friction dampers. In: Eurosteel 2017 (2017)

4. Elettore, E., et al.: Design and analysis of a seismic resilient steel moment resisting frame equipped with damage-free self-centering column bases. JCSR **179**(6), 106543 (2021)
5. Chengyu, L., Qi, L., Gongwen, L.: Seismic behavior of steel column base with slip-friction connections. Materials **13**(18), 3986 (2020)
6. Bjorhovde, R., Colson, A., Brozzetti, J.: Classification system for beam-to-column connections. J. Struct. Eng. **116**(11), 3059–3076 (1990)

Effect of Tung Oil on Mechanical Behavior of Silica Sand

Ke Chen and Sérgio D. N. Lourenço(✉)

The University of Hong Kong, Pok Fu Lam, Hong Kong S.A.R., China
lourenco@hku.hk

Abstract. Cemented sands have great potential as construction materials compared with clean sand due to the higher strength, but they have insignificant environmental benefits such as high embodied energy and carbon footprint. To provide an environmentally friendly, cost effective and reliable method to create an engineered soil, this study investigated on how Tung oil affects the mechanical behavior (strength and stress-strain behavior) of silica sand. Sand is mixed with a certain concentration (5%) of Tung oil. Heating could accelerate the equilibration of Tung oil-sand mixture and may contribute to the increase of mechanical strength compared with those at normal temperature condition. Therefore, conventional unconfined uniaxial compression tests are performed on sample equilibrated at 60 °C for different durations of heating. The Tung oil-sand mixture was found to exhibit (1) brittle failure under compression, (2) shorter time of equilibration and increasing compressive strength with longer duration of heating within 14 days.

Keywords: Silica sand · Tung oil · Mechanical behavior

1 Introduction

Soil is a natural granular material that exhibits low strength compared with other man-made building materials such as steel and concrete. For decades, various chemical methods of soil stabilization have been utilized to enhance the soil strength and stabilize the ground. For example, cement, lime, fly ash, bentonite and blast furnace slag are widely used as stabilizers for years. However, some of the abovementioned conventional chemical stabilization techniques have a high embodied energy and may lead to environmental pollution [1]. Therefore, attempts of applying environmental-friendly stabilizers to weak sands should be called for.

As a possible ecofriendly coating, Tung oil has the advantage of its natural origin, commercial availability, and low cost (e.g., 7 US$/liter). Historically, Tung oil has been used as a waterproof finish for wood in a range of applications, from boats to furniture. Recently, Tung oil has been used to impart water repellency to soil [2], and also as a self-healing material in polymers, concrete and steel [3, 4]. Most importantly, Tung oil could dry and harden by reacting with atmospheric oxygen [5]. Given this nature, preliminary investigations have been conducted on the effect of Tung oil on mechanical strength of soil. [6] attempted to mix both Tung oil and sticky rice juice to increase the strength of

W. Guo and K. Qian (Eds.): GBCESC 2022, LNCE 211, pp. 1039–1044, 2023.
https://doi.org/10.1007/978-981-19-5217-3_106

silt by using the direct shear box test. The optimum mass ratio of Tung oil, rice juice and soil were established until the maximum value of both friction angle and cohesion were obtained. However, the normal stress level and the corresponding stress-strain relationship were not clear. [7] found that hardened Tung oil bonded particles in natural condition and enhanced the unconfined compressive strength of CDG (i.e., Completely Decomposed Granite). Also, Tung oil has a time-dependent hardening behavior. With the time of equilibrium prolonged, the compressive strength increased. In order to shorten the time required for Tung oil to dry and harden, heating under appropriate temperature and duration proved to be effective [8]. However, the sole effect of Tung oil on stress-strain behavior of soil, which is important to any material for analyzing its stability as part of an engineering system, remains unexplored.

The development of a hardening film of Tung oil and its effect on sand depends on the environmental conditions such as temperature. During heating, the Tung oil coating on sand undergoes a rapid crystallization process, producing greater water repellency [8]. Therefore, it is estimated that a relatively high temperature (60 °C) could accelerate the hardening of Tung oil and the bonding effect on sand particles. On the other hand, an appropriate heating temperature will not induce evident loss of Tung oil due to evaporation.

In this study, the temperature effect of Tung oil on mechanical strength is explored by means of unconfined uniaxial compression tests. Since heating could accelerate the drying and hardening of Tung oil as already proved by [8], in this study the effect of heating duration was investigated. Samples were compacted in molds and placed in the oven at 60 °C for up to 28-days heating, and finally removed for testing. The effect of heating on compressive strength and stress-strain behavior of sand-Tung oil mixtures was then explored. [8] investigated the aggregate tensile strength of Tung oil treated CDG at different heating conditions. This study concentrated on compressive strength of Tung oil treated silica sand and the effect of duration of heating. Note that the heating time was set longer, from 1 day to 28 days at a specific temperature.

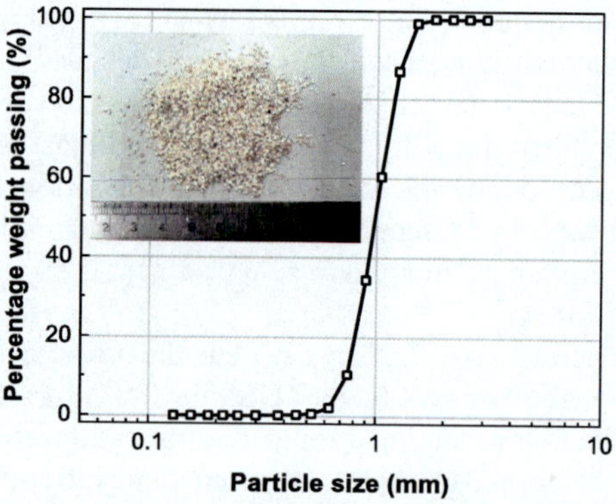

Fig. 1. Particle size distribution of the silica sand used in this study.

2 Material, Testing Program and Procedures

2.1 Material Tested

Laboratory tests were performed on clean silica sand and silica sand mixed with Tung oil with a concentration of 5%. The commercially available silica sand ranges from 0.60 mm to 1.18 mm and can be described as a clean quartzitic sand with a uniform grading. The gradation curve and the image of silica sand is shown in Fig. 1. The characteristics of the silica sand used are given in Table 1. As for information of Tung oil, details could be referred to [7].

Table 1. Physical properties of silica sand.

Properties	Silica sand
Specific gravity	2.65
Uniformity coefficient, C_u	1.40
Curvature coefficient, C_c	1.03
Minimum void ratio, e_{min}	0.62
Maximum void ratio, e_{max}	0.96

2.2 Testing Program and Procedures

Specimens with Tung oil concentration of 5% were selected for unconfined uniaxial compression tests. Sand and Tung oil were mixed and compacted in a cylindrical mold with dimensions 50 mm in diameter by 100 mm in height at the same relative density of 70%. After compaction, these specimens were placed in a temperature-controlled oven and equilibrated at 60 °C for 1, 3, 5, 7, 14 and 28 days. After heating, the specimens were cooled at ambient temperature (20–25 °C) for 1 day and tested in a universal compression machine (TriSCAN-50) in a deformation-controlled condition with the displacement rate of 1.0 mm/min. The schematic diagram of the apparatus used in this study was plotted in Fig. 2.

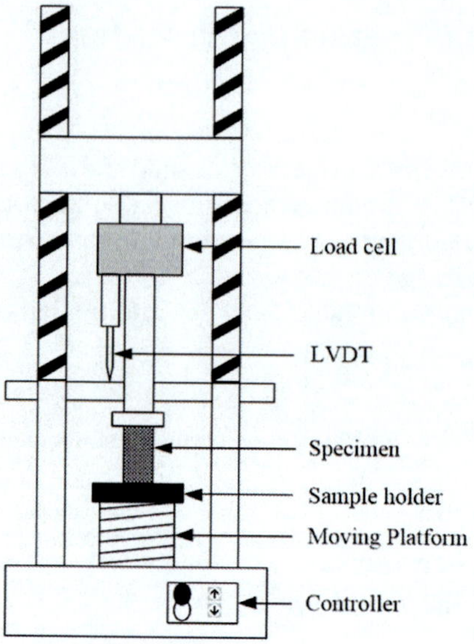

Fig. 2. The schematic diagram of the compression machine.

3 Results and Discussion

3.1 Unconfined Compression Strength

To explore the effect of heating duration on the compressive strength of Tung oil-sand mixture, unconfined uniaxial compression tests are conducted, and the relationship of compressive strength and the duration of heating is shown in Fig. 4. Basically, with increasing time of heating, the compressive strength increases from the original cohesionless state (0–14 days). The slight decrease on the 28th day may be due to the excessive melting and loss of Tung oil during a long time of heating. It should be noted that normally cemented sand is defined as having compressive strength between 100 to 400 kPa. Even with only 1day of heating, the Tung oil enhanced sand will own a compressive strength larger than 200 kPa.

3.2 Stress-Strain Behavior

The stress-strain behavior of sand-Tung oil mixture under different durations of heating are plotted in Fig. 3 to present the pre-failure behavior, peak strength and failure mode of this chemically cemented sand. Generally, when the duration of heating ranges from 1 day to 14 days, both the stiffness and the peak strength increase rapidly. After reaching the peak, the compressive stress will fall swiftly due to brittle failure. As mentioned in the previous section, 28 days of heating will cause melting and evaporation of Tung oil. Also, the excessive loss of Tung oil will largely weaken the contribution of the boding effect between Tung oil and sand particles, leading to decreasing stiffness and peak strength of sand-Tung oil mixture. One can roughly draw the conclusion from both Fig. 2 and Fig. 3 that 14 days of heating is the optimum time of equilibration for sand-Tung oil mixtures.

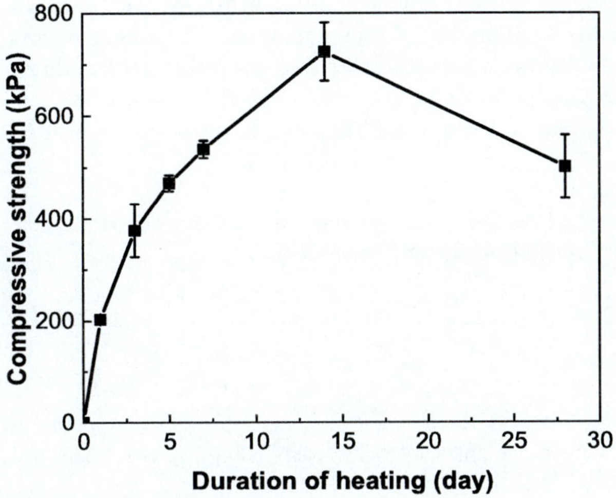

Fig. 3. Effect of different duration of heating on the unconfined uniaxial compressive strength of sand-Tung oil mixture under 60 °C.

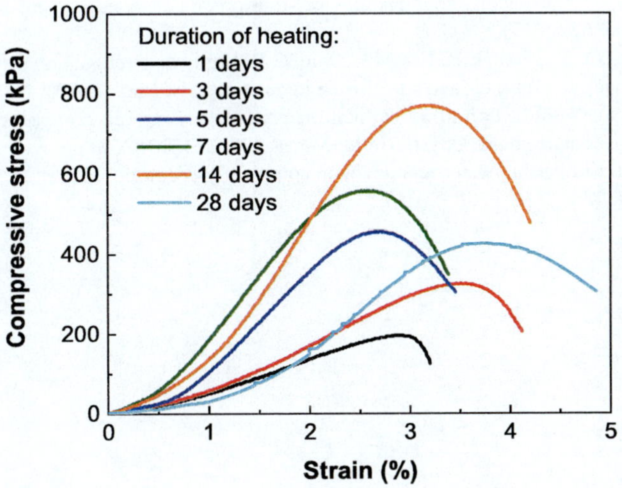

Fig. 4. Effect of different duration of heating on the stress-strain behavior of sand-Tung oil mixture under 60 °C.

4 Conclusions

This study proposes a possible method to improve the mechanical strength of sand by adding Tung oil to sand and heating at a relatively high temperature under different durations of heating. By conducting uniaxial compression tests on Tung oil-sand mixture, the following conclusions have been drawn:

(1) Tung oil requires at least 6 days to equilibrate in normal temperature, but heating can accelerate the equilibration. With longer time of heating, the compressive strength of Tung oil-sand mixture increases fast, but excessive heating could reduce the compressive strength.
(2) The stress-strain curves of sand-Tung oil mixtures show a brittle failure mode.

Acknowledgement. This work was supported by a Collaborative Research Fund from the Research Grants Council Hong Kong (C6006–20 GF).

References

1. Gowthaman, S., Kazunori, N., Satoru, K.: A state-of-the-art review on soil reinforcement technology using natural plant fiber materials: past findings, present trends and future directions. Materials **11**(4), 553 (2018)
2. Steenhuis, T.S., Hunt, A.G., Parlange, J.Y., Ewing, R.P.: Assessment of the application of percolation theory to a water repellent soil. Soil Res. **43**(3), 357–360 (2005)
3. Samadzadeh, M., Boura, S.H., Peikari, M., Ashrafi, A., Kasiriha, M.: Tung oil: An autonomous repairing agent for self-healing epoxy coatings. Prog. Org. Coat. **70**(4), 383–387 (2011)
4. Chen, Y., et al.: Self-healing coatings for steel-reinforced concrete. ACS Sustain. Chem. Eng. **5**(5), 3955–3962 (2017)
5. Tang, X.W., Yu, Y., Zhou, L.P., Fei, M.L., Sun, G.P.: Study on strength and permeability of silt soils improving by Tung oil and sticky Rice juice. Adv. Civ. Eng. (2020)
6. Lin, H., et al.: Stabilization of an earthen material with Tung oil: compaction, strength and hydrophobic enhancement. Constr. Build. Mater. **290**, 123213 (2021)
7. Lin, H., et al.: Imparting water repellency in completely decomposed granite with Tung oil. J. Clean. Prod. **230**, 1316–1328 (2019)

A Study on Wind Resistance of Steel Arch Pedestrian Bridge

Jiahua Xian[1], Junfeng He[1], Shaocai Zhu[1], and Minting Zhong[2(✉)]

[1] China-Singapore Guangzhou Knowledge City Fiscal Invested Construction Project Management Center, Guangzhou 510700, China
[2] School of Civil Engineering, Guangzhou University, Guangzhou 510006, China
2111916048@e.gzhu.edu.cn

Abstract. Pedestrian bridge is an important part of urban landscape. On the basis of satisfying the function of pedestrian traffic, it gradually steps forward to the comprehensive requirements of light structure, unique shape, large span, high bearing capacity and convenient installation. With the increase of bridge span, bridge structure becomes more portable and flexible. The wind resistance of bridge structures is increasingly prominent, especially for the flutter stability and eddy stability of bridges, which have become the key factors directly affecting the further growth of bridge span. Taking a steel arch pedestrian bridge in Guangzhou as an example, the flutter stability and eddy stability are investigated. The result shows that the bending-torsional coupling critical wind speed of flutter and separation flow flutter is less than the critical wind speed of flutter, suggesting that the flutter will not occur in the design reference period.

Keywords: Steel arch footbridge · Wind resistance · Flutter stability

1 Introduction

In recent years, with the continuous development of the national economy, China's urban traffic has also been rapidly improved, and the traffic flow and people flow are increasing rapidly. Urban traffic problem has become one of the important problems in urban modernization. Especially in the process of old city reconstruction, the phenomenon of people and cars competing for roads is very serious. In order to alleviate this contradiction, the construction of pedestrian overpass is one of the feasible methods at present [1, 2]. With the development of pedestrian overpass and the needs of urban planning, the span of pedestrian overpass is increasing, the structural mass is lighter and lighter, the structural stiffness is smaller and smaller, and the structural damping is lower and lower, which leads to greater and greater sensitivity to wind-induced action [3, 4]. In case of strong wind, it may be damaged due to poor elastic deformation or improper fixation. At the same time, the damage may be accompanied by the falling of components, which may endanger personal safety and cause extremely serious consequences. The wind resistance of pedestrian overpass is greatly affected by flutter stability [5].

W. Guo and K. Qian (Eds.): GBCESC 2022, LNCE 211, pp. 1045–1052, 2023.
https://doi.org/10.1007/978-981-19-5217-3_107

A large number of scholars have studied the wind vibration problems such as flutter vibration of the bridge. Chen et al. [6] Based on a steel truss pedestrian bridge under construction, conducted wind tunnel test on the vortex vibration performance and flutter performance of the main beam by using the wind tunnel test method. It is found that there is obvious vortex induced resonance in the asymmetric steel truss curved pedestrian bridge, and tuned mass damper (TMD) is used It can effectively reduce the amplitude of vortex vibration response of steel truss pedestrian bridge. Taylor et al. [7] studied the dynamic characteristics and aeroelastic stability of a pedestrian bridge on a highway in northern England by using the discrete vortex method, and improved the flutter stability of the bridge section by adjusting the number and angle of railing plates, so as to meet the requirements of bridge flutter stability. Cammelli et al. [8] conducted a wind tunnel test on a pedestrian bridge with a span of 115 m in Britain. The test found that there was an obvious vortex galloping coupling phenomenon of the bridge, resulting in a large wind-induced vibration response of the bridge within the design wind speed range. Finally, it was suggested to reduce the wind-induced vibration response of the bridge structure by changing the stiffness, mass and damping of the main beam. Pirner et al. [9] conducted an aeroelastic model test on a pedestrian suspension bridge with a span of 252 m. The results show that the bridge has vortex induced resonance in the wind speed range of 15–17.5 m/s, and dampers are set at the railing to control vortex vibration. The test results show that flutter instability is not observed. Wu et al. [10] Increased the frequency of each key vibration mode of pedestrian suspension bridge by setting wind resistant cables. The setting of wind resistant cable can significantly enhance the static wind stability of pedestrian suspension bridge and improve its torsional divergence critical wind speed. Li et al. [11] Studied that the transverse bending, vertical bending and torsional vibration mode frequencies of stiffeners are greatly improved by arranging wind resistant cables at different positions. At the same time, setting wind resistant cables at different positions is different to improve the vibration mode frequency amplitude of bridge structure.

To sum up, the wind resistance performance of pedestrian bridge structure is a key concern of pedestrian bridge. Based on a steel arch pedestrian bridge, this paper tests whether the wind resistance performance of the bridge meets the requirements through the calculation method in the Wind-Resistant Design Specification for Highway Bridges.

2 Project Overview

2.1 Overview of Pedestrian Overpass

The engineering background of this paper is a pedestrian overpass located in Guangzhou, Guangdong, China. The bridge is a steel arch bridge. The main bridge is arranged as 144.5 m. Three branches with spans of 4.064 m, 19.0 m and 15.128 m are rigidly connected with the main bridge as a whole. The Midas civil finite element model of the overall bridge is shown in (Fig. 1).

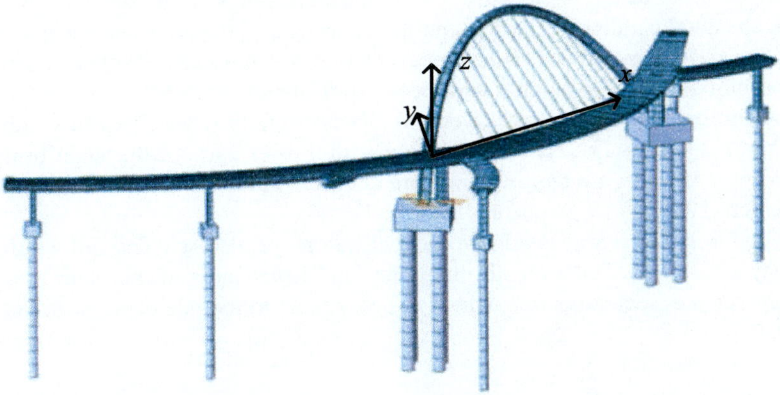

Fig. 1. Midas civil finite element model of the pedestrian overpass

The arch axis of the arch rib is a quadratic parabolic equation ($y = -23.2x^2/292$), the calculated span is 58 m, the rise height is 23.2 m, the calculated rise span ratio is 1/2.5, and the overall camber of the arch rib is 20°. The section of the arch rib is a regular pentagon, and the section from the arch foot to the arch crown is variable. It is controlled according to the radius r of a regular pentagon inscribed circle. R = 1.0 m at arch foot and R = 0.6 m at arch crown. The plate thickness is 30 mm.

The steel main beam section adopts a single box and single chamber section, with beam height of 1.2 m, beam width of 4–6 m, the top plate thickness of 12–16 mm, the bottom plate thickness of 12–22 mm, the web plate thickness of 12–14 mm, and the spacing of top and bottom plate stiffeners is controlled as 350mm. The cross section is shown in (Fig. 2).

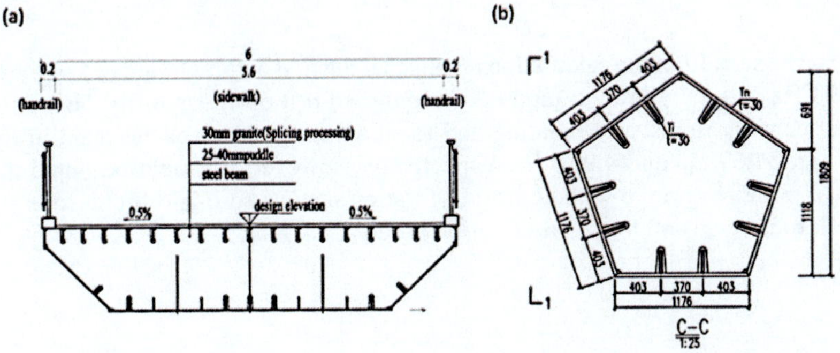

Fig. 2. Cross section of footbridge (a) Main bridge section (b) arch rib interface

Replaceable suspenders and slings are used for the bridge, and a total of 13 slings are set for the whole bridge. The slings are symmetrically arranged along the main arch rib and main span. The spacing on the arch axis of the arch rib is 2.5 m and the spacing on the centerline of the main beam is 3.0 m. Single slings are used. The sling adopts HDPE sheathed parallel steel wire, and the upper end of the arch suspender cable is a lifting lug and the lower end is a tensional anchor box. Considering fatigue, hoisting and replaceability, the design safety factor of the sling is more than 3.0.

In addition, 4 stairways and 2 vertical elevators are also set. The full width of the stairway is 3.5/4 m. The arch rib, the upper and lower parts of the main bridge and the stairway are steel structures, which are connected to the pile cap cast-in-place pile foundation.

In addition, 4 stairways and 2 vertical elevators are also set. The full width of the stairway is 3.5/4 m. The arch rib, the upper and lower parts of the main bridge and the stairway are steel structures, which are connected to the pile cap cast-in-place pile foundation.

2.2 Structural Dynamic Characteristic Analysis

The loads such as deck pavement, pedestrian railings and other auxiliary structures on the bridge are transformed into mass. The first, third, fourth and ninth order formations of the first 10 natural vibration frequencies and vibration modes of the bridge are shown in Table 1, and the corresponding vibration mode diagram is given in (Fig. 3).

Table 1. Natural frequency and mode shape of bridge.

Order	Frequency (Hz)	Period/(s)	Mode description
1	1.498	0.668	Arch rib side inclination
3	3.091	0.324	Vertical bending of main bridge
4	3.269	0.306	Arch rib bending and main beam transverse bending
9	4.934	0.203	Arch rib inclination and main bridge torsion

It can be seen from the calculation results that the first natural frequency of the bridge is 1.498 Hz and the vibration mode is the outward roll of the arch rib. The first-order vertical bending, transverse bending and torsional frequencies of the main bridge are 3.091 Hz, 3.269 Hz and 4.934 Hz respectively. However, it should be noted that the bridge is a curved bridge with complex vibration modes, so it is difficult to have pure vertical, transverse and torsional vibration modes.

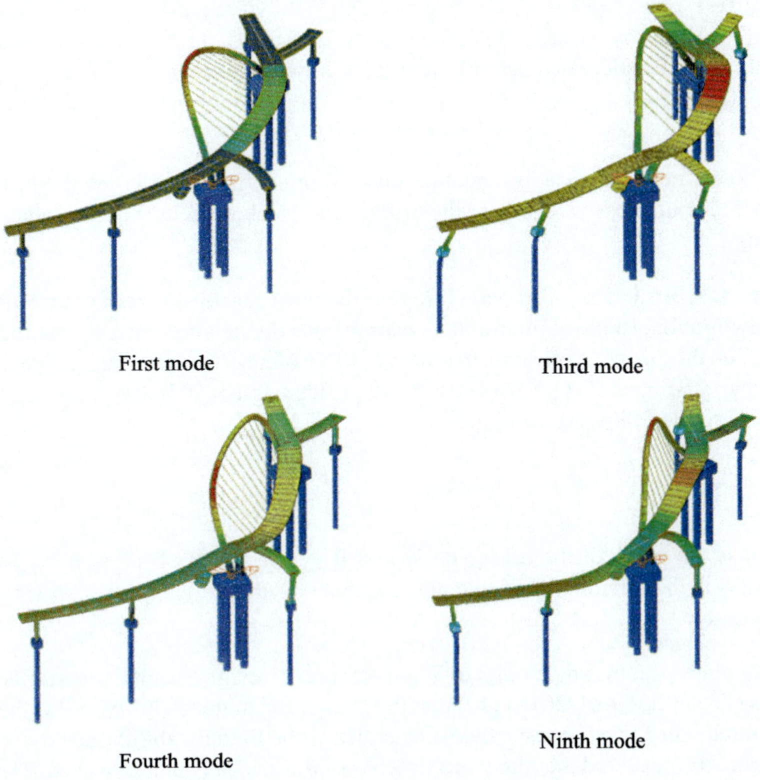

Fig. 3. Bridge vibration mode diagram.

3 Stability Analysis

3.1 Flutter Stability Checking Calculation

Design Reference Wind Speed The design reference period and design service life of the bridge are 100 years. Referring to the national basic wind speed value and basic wind speed distribution map in Appendix A of Wind-Resistant Design Specification for Highway Bridges (JTG/T 3360-01-2018) [12], the basic wind speed in Guangzhou under 100-year return period is $V_{s10} = 31.3$ m/s.

This bridge is a pedestrian overpass among urban bridges. The bridge site belongs to the dense area of medium and high-rise buildings. The surface roughness is considered class D, so the power index is taken as 0.3. According to the design drawings, the deck elevation is 49.046 m and the surface elevation is 42.400 m, so the height from the deck to the ground is 6.646 m. Design reference wind speed at bridge deck:

$$V_s = V_{s10} \cdot Z/10^{0.3} \tag{1}$$

Flutter Test Wind Speed. According to article 6.3.8 of Wind-Resistant Design Specification for Highway Bridges (JTG/T 3360-01-2018) [12], the flutter test wind speed of the main beam is calculated according to the following formula:

$$[V_{cr}] = 1.2 \cdot \mu_f \cdot \mathrm{V_d} \tag{2}$$

where μ_f is the wind speed fluctuation correction coefficient, it is related to the span and surface roughness category of the bridge, and is taken as 1.49 in the calculation of the bridge.

Flutter Stability Index. The main bridge is the most sensitive part of the whole bridge to wind vibration, so the main bridge is mainly considered when considering the impact of flutter on the bridge. According to article 6.3.1 of Wind-Resistant Design Specification for Highway Bridges (JTG/T 3360-01-2018) [12], the flutter stability index is calculated according to the following formula:

$$I_f = \frac{[V_{cr}]}{f_t \cdot B} \tag{3}$$

where B is the width of the bridge deck, and B = 6 m for this bridge, B' = 1.8 m for arch rib; f_t is the torsional fundamental frequency of the bridge, f_t = 4.96 Hz is taken for this bridge.

It is pointed out in article 6.3.3 of Wind-Resistant Design Specification for Highway Bridges (JTG/T 3360-01-2018) [12] that the greater the flutter stability index, the higher the requirements for wind resistance. Generally, if the flutter stability index is less than 2.5, it can be considered that the wind stability of the bridge is in a safe state. The flutter stability index calculated by the bridge is 1.6 < 2.5, so it can be preliminarily judged that the wind resistance stability of the pedestrian overpass has sufficient guarantee through the flutter stability index.

Checking Calculation of Flexural Torsional Coupling Flutter. For flexural torsional coupled flutter, the critical flutter wind speed can be estimated according to the van der put formula commonly used in Engineering (article 6.3.4 of Wind-Resistant Design Specification for Highway Bridges (JTG/T 3360-01-2018)) [12]:

$$V_{cr1} = \eta_s \cdot \eta_\alpha \cdot [1 + (\varepsilon - 0.5)\sqrt{0.72\mu \cdot r/b}] \cdot \omega_b \cdot b \tag{4}$$

where η_s is the influence coefficient of the section shape of the main beam, which is taken as 0.7 for the beam and 2.5 for the arch rib; η_α is the angle of attack effect coefficient, 0.8 for this bridge; b is the half-width of the bridge deck, 3 m for this bridge, 0.9 for the arch rib; r is the inertia radius, 1.82 m for this bridge; ε is the twist frequency ratio, $\varepsilon = \frac{f_t}{f_b}$, $\mathrm{f_t}$ is the fundamental frequency of symmetric torsion and $\mathrm{f_b}$ is the fundamental frequency of symmetric vertical bending, ε = 1.596; μ is the ratio of beam mass to air density, the mass per unit length of the bridge is m = 2285.35 g/m, so; is the circular frequency of the fundamental frequency of vertical bending of the bridge, $\omega_b = 2\pi f_b = 19.41\ \mathrm{rad}/s$.

Since main bridge $Vcr1$ = 196.4 m/s > $[Vcr]$ = 49.51 m/s; the arch rib $Vcr1$ = 329.7 m/s > $[Vcr]$ = 49.51 m/s, the flexural torsional coupling flutter of the pedestrian overpass meets the requirements. Comparing the flexural torsional coupling flutter of the main bridge and arch rib, more attention should be paid to the main bridge.

Checking Calculation of Torsional Flutter of Separated Flow. The critical wind speed of separated flow torsional flutter is estimated by Herzog formula in article 6.3.4 of Wind-Resistant Design Specification for Highway Bridges (JTG/T 3360-01-2018) [12]:

$$V_{cr2} = \eta_s \cdot \eta_\alpha \cdot T_h^{-1} \cdot f_t \cdot B \tag{5}$$

where T_h^{-1} is the reciprocal of Theodosia' number, which is calculated according to the following formula:

$$T_h^{-1} = 2.5\sqrt{\mu\frac{r}{b}} \tag{6}$$

Since main bridge $Vcr2 = 114.6$ m/s > $[Vcr] = 49.51$ m/s; the arch rib $Vcr2 = 419.2$ m/s > $[Vcr] = 49.51$ m/s, the shunt coupling flutter of the pedestrian overpass meets the requirements.

4 Conclusion

Through the analysis and calculation of the flutter stability of the pedestrian overpass, the following conclusions can be drawn: the critical wind speed of the arch rib is significantly greater in comparison to the main beam, meaning that more flutter attention should be paid to the main beam in the same type of flutter stability analysis. The critical wind speeds of flexural torsional coupling flutter and separated flow flutter of the main beam are 196.4 m/s and 114.6 m/s respectively, significantly higher than the flutter test wind speed of 49.51 m/s. Thus, the pedestrian overpass will not flutter during the design reference period, meeting the safety requirements.

References

1. Haijun, Z., Shaodong, L.: The Design of Pedestrian Overpass in the City. Fujian Architecture & Construction (2014)
2. Weidong, J., Xiaowei, X., Tengfei, Z.: Rethinking on the Design of Urban Pedestrian Overpass–Designing of Pedestrian Overpass on the front Square of Carrefour on Lishan. Chinese & Overseas Architecture (2013)
3. Penghui, L., Zili, H., Jianye, Z., Jing, Y.: Experimental study on dynamic performance of railway pedestrian overpass under high speed train fluctuating wind action. Railway Eng. (2019)
4. Pan, T.C.: Vibration of pedestrian overpass. J. Perform. Constr. Facil. **6**, 34–45 (1992)
5. Ivanovi, S., Pavic, A., Reynolds, P.: Vibration serviceability of footbridges under human-induced excitation: a literature review. J. Sound Vib. **279**, 1–74 (2005)
6. Hua, B., Shuaiyu, C., Hui, Z., Jiawu, L.: The rapid evaluation of bridge flutter stability and its application

7. Taylor, I.J., Vezza, M., Salisbury, I.: Numerical investigation of the effects of pedestrian barriers on aeroelastic stability of a proposed footbridge. J. Wind Eng. Ind. Aerodyn. **96**, 2418–2437 (2008)
8. Cammelli, S., Bagnara, A., Navarro, J.G.: VIV-galloping interaction of the deck of a footbridge with solid parapets. Procedia Eng. **199**, 1290–1295 (2017)
9. Pirner, M.: Aeroelastic characteristics of a stressed ribbon pedestrian bridge spanning 252 m. J. Wind Eng. Ind. Aerodyn. **53**, 301–314 (1994)
10. Changqing, W., Zhitian, Z., Xiaobo, W.: Influences of wind cables on dynamic properties and aerostatic stability of pedestrian suspension bridges. Bridge Constr. **47**, 77–82 (2017)
11. Maoqi, L.: Research on the influence of wind-resistant cables layout on dynamic characteristics & static wind stability of double tower singlespan suspension bridge. China Municipal Eng. 8–11+112–113 (2021)
12. JTG/T 3360-01-2018 Wind-Resistant Design Specificationfor Highway Bridges. China Communications Press Co., Ltd., Beijing, China (2018)

Research on Impact Resistance of H-shaped Steel Beam

Yunhao Weng[1(✉)], Shihong Li[2], Xiaolan Yuan[2], and Tianqi Xue[2]

[1] Guangxi Key Laboratory of New Energy and Building Energy Saving, Guilin 541004, China
wengyh@st.gxu.edu.cn

[2] College of Civil Engineering and Architecture, Guilin University of Technology, Guilin 541004, Guangxi, China

Abstract. This paper established a fine finite element model based on the impact test of H-shaped steel beams completed by Huo Jingsi et al. From Hunan University, in order to verify the accuracy of the finite element model. The paper used the validated finite element model for parametric analysis to study the impact of hammer geometry and boundary conditions on the impact resistance of steel beams. According to the parameter analysis: changing the hammer head geometry leads to changing the contact stiffness, so the impact force time-history curve changes; when the steel beam boundary condition is changed from hinged to rigid connection, the bearing on the beam is weakened, so the impact force is greatly reduced, the beam span deflection increases.

Keywords: H-shaped steel beam · Impact resistance · Numerical simulation · Failure mode · Impact condition

1 Introduction

Because of the large number of casualties and huge economic losses caused by the "911 terrorist attack", people are increasingly concerned about the impact damage and continuous collapse of buildings [1–3]. In order to reasonably evaluate the impact resistance of buildings, it is necessary to systematically study the impact resistance of buildings [4–6] in engineering construction.

There are many impact test studies on building components, among which Fujikake et al. [7] studied the impact resistance of reinforced concrete beams, and proposed the impact resistance method for beams under bending failure. According to the impact test of reinforced concrete beams. Bin et al. [8] carried out low-speed impact test on rectangular steel plate by hemispherical hammer, studied the plastic deformation of steel plate under impact, and proposed a linear constitutive equation that can describe the true stress-strain curve at high strain rate.

Pengcheng C et al. [9] established a finite element model of H-shaped steel beams under lateral impact to study the impact resistance of H-shaped steel beams. Thairy et al. [10] established a finite element model to analyze the dynamic response and failure mode of axial pressure, impact velocity, impact mass and impact position on H-shaped steel

W. Guo and K. Qian (Eds.): GBCESC 2022, LNCE 211, pp. 1053–1061, 2023.
https://doi.org/10.1007/978-981-19-5217-3_108

column subjected to lateral impact The study showed that the thickness of the web and flange have an impact on the peak values of the impact force. Previous research parameters of impact H-beams are mainly reflected in: impact velocity, impact mass and width and thickness ratio. In order to further study the change of the impact resistance of H-shaped steel beam under different impact conditions, this paper will study the change of the hammer head geometry and boundary conditions.

2 Numerical Model Development

2.1 Brief Introduction of Test

Jingsi H et al. [11] from Hunan University studied the impact resistance of double-span restrained steel beams under upper concentrated loads. The prototype is shown in Fig. 1, and the properties of the steel are shown in Table 1. This paper added steel plates at the upper flange and hinge position at the impact point. Four steel beams with the same geometric parameters are impacted. Among them, HR4–3, HR4–6, HR4–8 correspond to the impact of 450 kg, the impact heights are 3.2 m, 6.0 m, 8.0 m, HR5–6 corresponds to the impact of 575 kg, the impact height is 6.0 m.

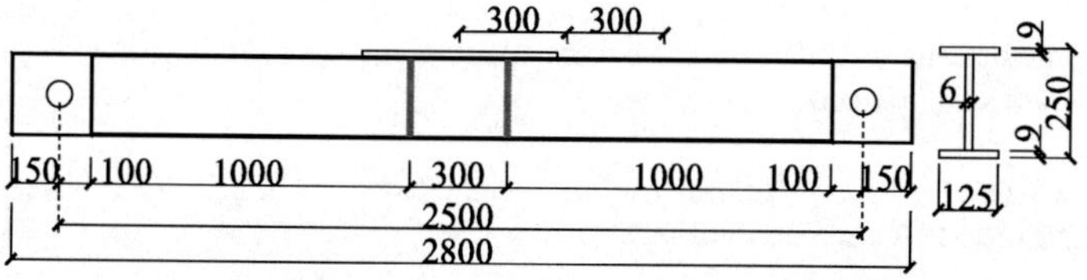

Fig. 1. Sample prototype diagram (Unit:mm)

Table 1. Characteristics of steel profiles

Material location	Thickness	Yield strength	Compressive strength	Elastic modulus	Elongation
	mm	f_{ys}/MPa	f_u/MPa	E_s/MPa	%
Wing edges	9	239.1	390.0	1.82×10^5	20.6
Web	6	251.1	398.9	1.89×10^5	18.6

2.2 Numerical Model

This study establishes the above experiment model by ABAQUS. The element type used for steel beams, additional steel plates and steel shafts adopts C3D8R, an eight-node solid element. The hammer head is made of high strength steel, and the hammer head is defined as a rigid body using the rigid element R3D4. 20 mm is used as the mesh size of the finite element model. The H-beam is made of Q235 grade low carbon steel with a density of 7.85 g/cm^3, an elastic modulus of 200 GPa and a Poisson's ratio of 0.3. The finite element model of the steel beam is shown in Fig. 2.

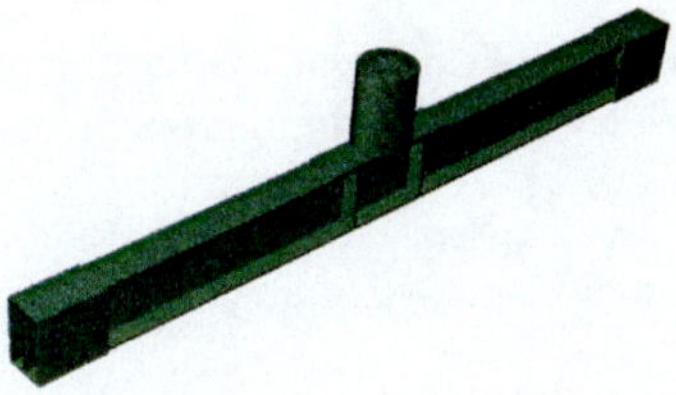

Fig. 2. Steel beam finite element model

3 Model Verification

According to the experimental analysis [12–15], the main output data in the impact test are the impact force and mid-span displacement, so this paper compares and analyzes the experimental data and the simulated data.

3.1 Contrast of Impact Time History Curves

Figure 3 shows the comparison of the impact force time history curve and the simulated impact force. It can be found that the results of the two curves are relatively consistent.

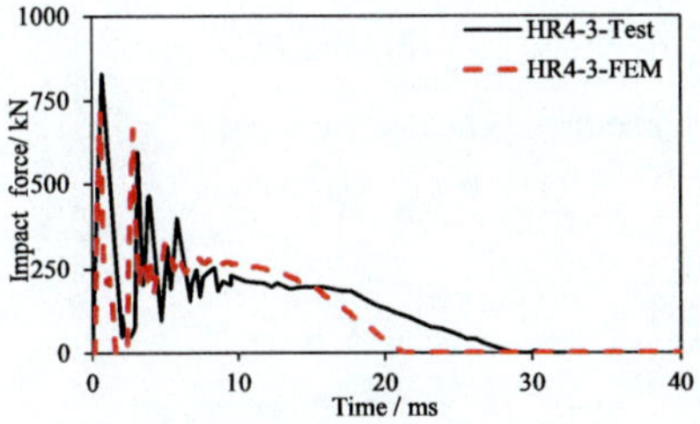

(a) HR4-3

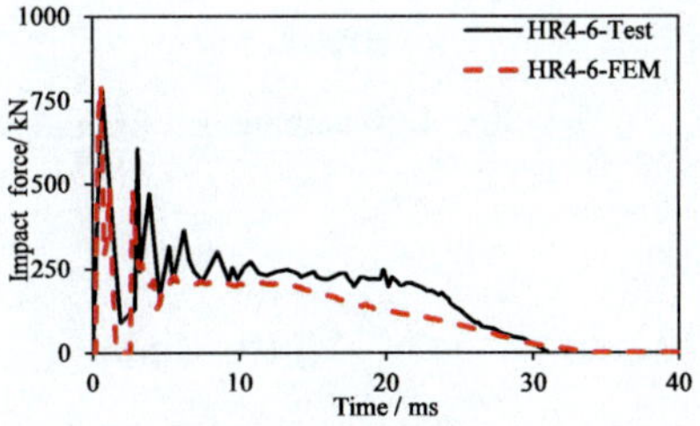

(b) HR4-6

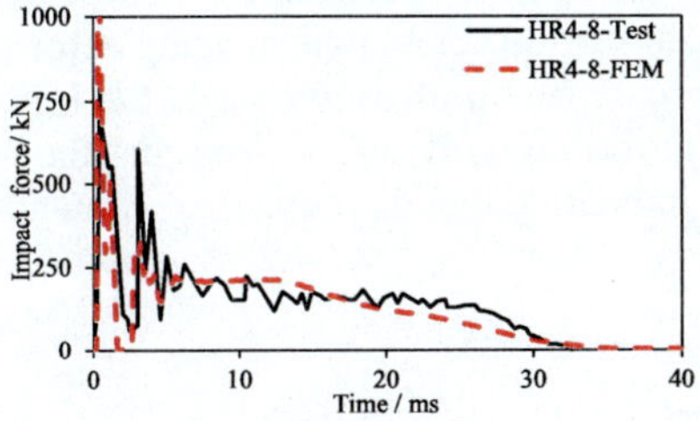

(a) HR4-8

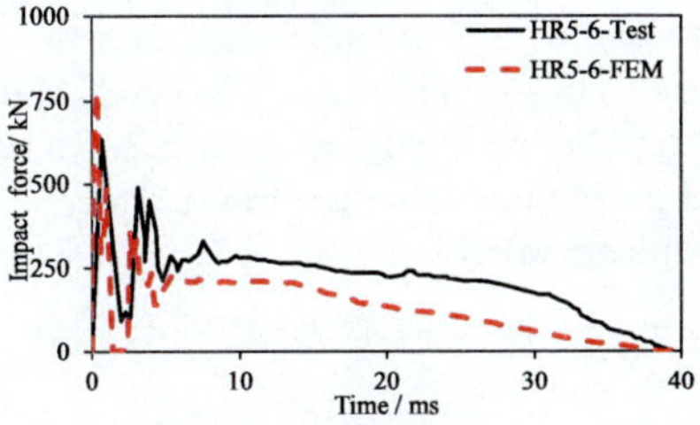

(b) HR5-6

Fig. 3. Comparison of impact force time history curves

The change trends of the impact force in the elastic-plastic stage, the stable stage and the falling stage are the same. The finite element model can well predict the impact force during the test.

3.2 Contrast of Impact Time History Curves

Comparing the time history curve of mid-span displacement in Fig. 4, it can be seen that the change trend of the test curve and the simulation curve is consistent.

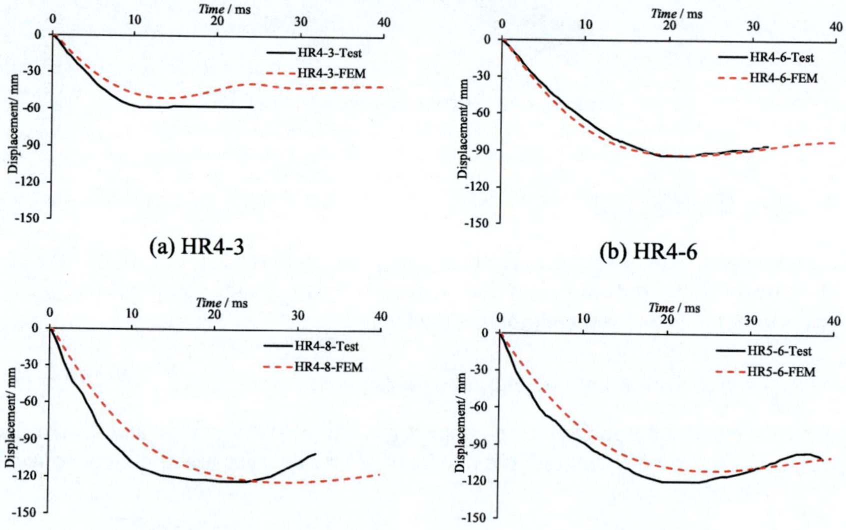

(a) HR4-3

(b) HR4-6

(c) HR4-8

(d) HR5-6

Fig. 4. Comparison of mid-span displacement time history curves

3.3 Comparison of Failure Modes

The post-impact test and simulated failure mode are compared in Fig. 5, with all flanges showing local buckling near the plastic hinge. The HR4–3 specimen showed bending deformation on both sides of the upper reinforced steel plate, and the bending deformation of HR4–6, HR4–8, HR5–6 was more obvious. Due to the increase of the impact energy, the residual deflection also increased accordingly. It can be seen that the test and the simulated failure mode are consistent, indicating that the numerical simulation results are reliable.

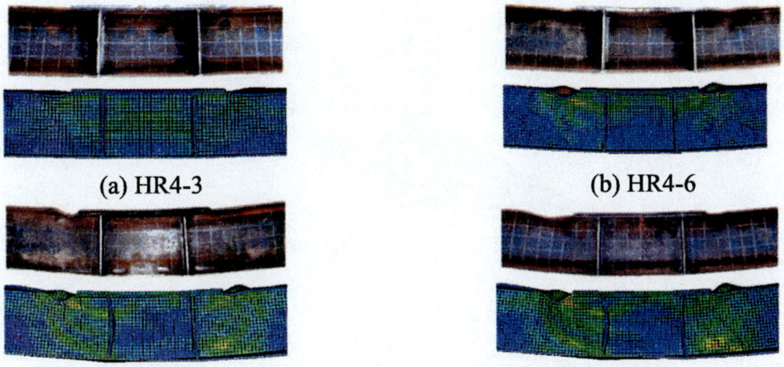

(a) HR4-3 (b) HR4-6

(c) HR4-8 (d) HR5-6

Fig. 5. Comparison of failure modes between test and simulation of steel beams

4 Extended Parameter Analysis

In order to study the effect of different impact conditions on the impact resistance of steel beams, this paper will analyze the hammer head geometry and change the boundary conditions.

4.1 Hammer Head Geometry

According to the HR4–3 model, it is assumed that the shape of the hammer head is a cylinder and contacts a circular plane. In order to study the impact resistance under different hammer surfaces, the hammer head is set as a hemispherical hammer head as shown in Fig. 6. It can be seen from Fig. 7 to Fig. 8 that when the shape of the hammer head is changed, the impact force curve changes significantly. At this time, the peak value of the impact force decreases by 27.0%, while the mid-span displacement does not change significantly.

According to the research of Pengcheng J et al. [16], shows that the inertial force of particles in the impact area of the impact component makes the impact force reach its peak rapidly,after comparing the hemispherical surface and the flat hammer surface, the flatter the contact surface between the drop weight and the specimen, the greater the contact stiffness, the greater the acceleration of the specimen during impact, the longer the speed of the beam exceeds the drop hammer, and the more kinetic energy dissipated in the inertial resistance stage. It can be seen from the above analysis that after changing the geometry of the hammer head, the contact stiffness changes, which affects the energy dissipation in the inertial resistance stage. At this time, only the time history curve of the impact force changes significantly.

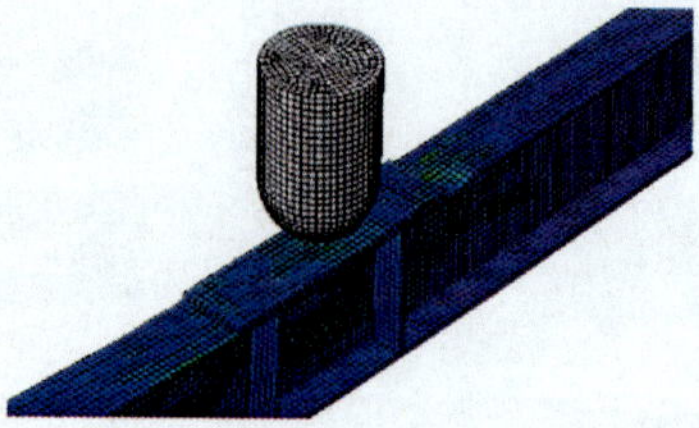

Fig. 6. Hemispherical hammer model

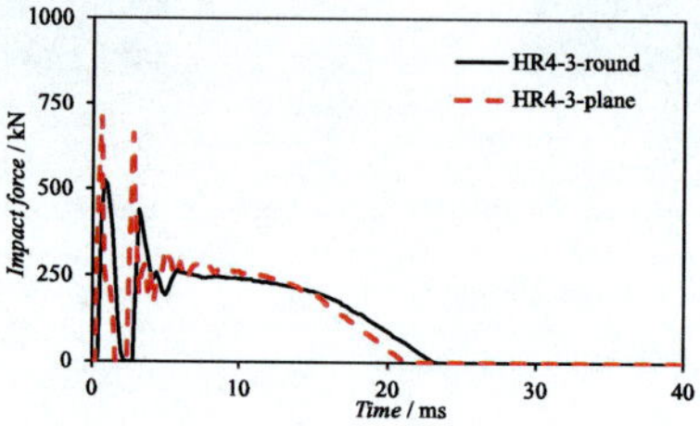

Fig. 7. Impact force time history curves of different contact surfaces (Note: The round is a semicircular sphere surface, and the plane is a plane surface)

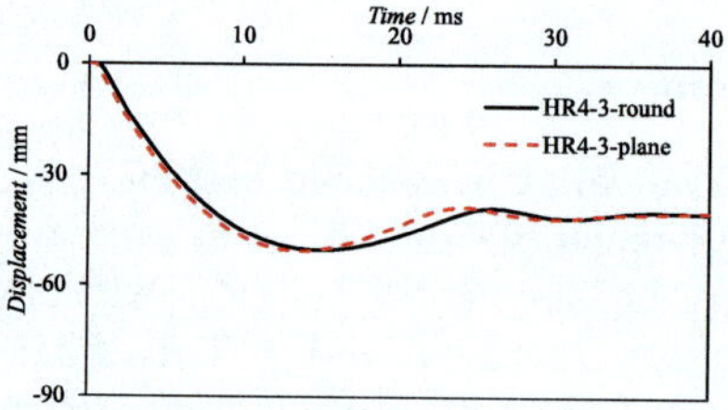

Fig. 8. Mid-span displacement time history curves of different contact surfaces

4.2 Change Boundary Conditions

According to the research of Juanlin C et al. [17], under different boundary conditions, the hot-rolled H-shpaed steel beam is affected by the lateral impact load. Under the constraint boundary condition, the specimen is mainly bent and deformed as a whole. The buckling deformation is dominant, indicating that the boundary conditions have a significant impact on the impact force time history curve. Under concentrated load when the steel beam is at both ends, the two ends of the beams between the two inflection points (the length accounts for half of the total span of the beam) are set to be hinged and fixed, which is the HR4–3 model boundary condition. The boundary conditions are changed to the actual steel beam length of 5000 mm, and the two ends are rigidly connected. After comparing the impact force time-history curve in Fig. 9 with the mid-span displacement time history curve in Fig. 10, it is found that when the boundary conditions are changed, the impact force decreases as a whole. Among them, the peak value of the impact force decreases by 8.4%, while the mid-span displacement value increases by 22.7%.

When the boundary condition is changed to full-length rigid connection, the time for the impact force to be transmitted to the support increases, and the response time to the inertial load in the elastic-plastic stage is prolonged. Since the restraint force of the bearing on the steel beam is weakened, the impact force value decreases. According to the comparison of the deflection curves in Fig. 11, it can be seen that after changing the boundary conditions, the overall deflection of the steel beam shows an increasing trend. When the boundary condition of the steel beam is changed from hinged connection to rigid connection, because the restraint force of the support on the beam body is weakened, the impact force is greatly reduced, so the beam span deflection increases.

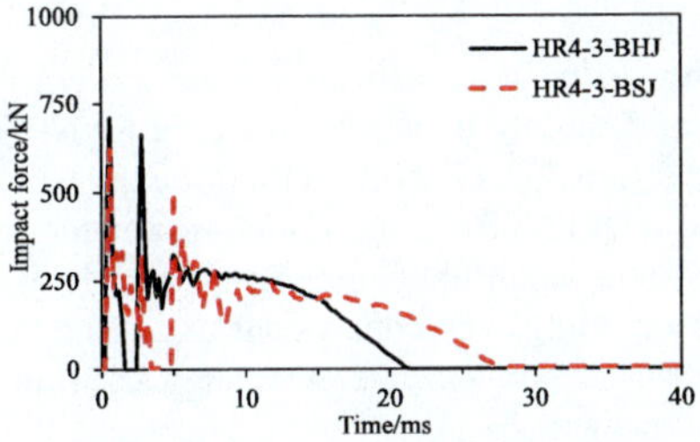

Fig. 9. Impact force time history curves of changing boundary conditions (Note: BHJ is hinged at both ends, BSJ is rigid at both ends of the full length.)

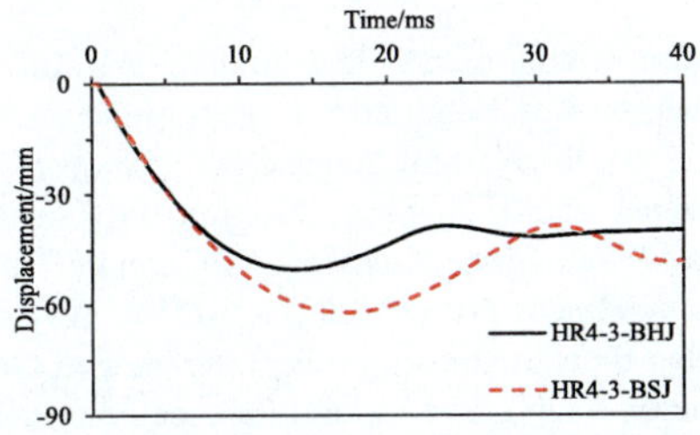

Fig. 10. Mid-span displacement time history curves of changing boundary conditions

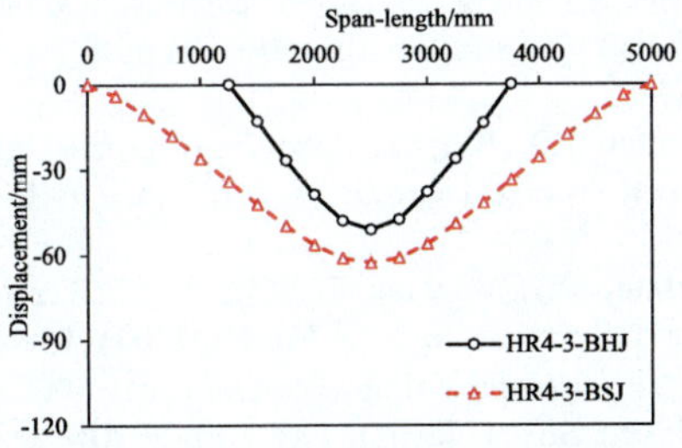

Fig. 11. Deflection curves of changing boundary conditions

5 Conclusions

This paper compares the test and simulation results to verify the accuracy of the model by establishing ABAQUS finite element simulation model, which is subjected to load redistribution and internal force change, the following conclusions are obtained after analysis:

- Through test, according to the impact force time history curve, time history curve of mid-span displacement for H-shaped beam, the impact force and the final failure, the ABAQUS finite element simulation and the test results are consistent under the impact load.

- When the hammer head geometry is changed, the time history curve of the impact force changes obviously, while the time history curve of the mid-span deflection and the reaction force have no obvious change, the change of the contact stiffness affects the energy dissipation in the inertia resistance stage, so the impact force time history curve changes.
- When the boundary condition of the steel beam is changed from hinged connection to rigid connection, the restraint force of the support on the beam is weakened, hence the impact force is greatly reduced, and the beam span deflection increases.

Acknowledgement. The authors gratefully acknowledge the financial support provided by Guangxi Key Laboratory Project of New Energy and Building Energy Saving (Gui Ke Neng22-J-21–3), the Natural Science Foundation of Guangxi (No.2021GXNSFFA196001), the Guangxi Science and Technology Program Project (Gui Ke AD21220071), the China Postdoctoral Science Foundation (2021M693814) and the Guangxi Young and Middle-aged Teacher Basic Ability Improvement Project (2021KY0254). Any opinions, findings and conclusions expressed in this paper do not necessary reflect the view of Guangxi Key Laboratory of New Energy and Building Energy Saving and Natural Science Foundation of Guangxi.

References

1. Kai, Q., Zhi, L., Yunhao, W., Gang, W.: Load resisting mechanisms of assembled monolithic precast concrete beam-slab substructures to resist progressive collapse. J. Build. Struct. **42**(07), 183–193 (2021)
2. Zhi, L., Yuhao, W., Xiaofang, D., Kai, Q.: Behavior of precast concrete beam-slab substructures with welded connections to resist progressive collapse. J. Build. Struct. **41**(10), 121–128 (2020)
3. Yang, Y., Zhi, L., Longshan, X., Songyuan, G., Kai, Q.: Load resisting mechanism of precast structure under exterior column failure. Eng. Mech. **38**(04), 159–168 (2021)
4. Kai, Q., Zhi, L., Pan, H.: Progressive collapse mechanism of PC beam-slab substructure with bolted connections. J. Build. Struct. **41**(01), 173–180 (2020)
5. Xiaofang, D., Zhi, L., Yunhao, W., Xiaohui, Y., Kai, Q.: Experimental study on performance of prestressed concrete beam-column substructure against progressive collapse. J. Build. Struct. **40**(08), 71–78 (2019)
6. Kai, Q., Zhi, L., Yunhao, W., Xiaofang, D.: Behavior of RC beam-slab structures to resist progressive collapse. Eng. Mech. **36**(06), 239–247 (2019)
7. Fujikake, K., Li, B., Soeun, S.: Impact Response of reinforced concrete beam and its analytical evaluation. J. Struct. Eng. **135**(8), 938–950 (2009)
8. Kun, L., et al.: Assessment of material strain rate effects on square steel plates under lateral dynamic impact loads. Ships Offshore Struct. **13**(12), 217–225 (2018)
9. Pengcheng, C., Xin, C., Lei, G., et al.: Impact resistance of H-shaped steel beams with different width and thickness ratios. Explosion ShockWaves **38**(06), 1378–1385 (2018)
10. Al-Thairy, H., Wang, Y.C.: A numerical study of the behaviour and failure modes of axially compressed steel columns subjected to transverse impact. Int. J. Impact Eng. **38**(8–9), 732–744 (2011)

11. Jingsi, H., Jinqing, Z., Baisheng, C., et al.: Dynamic behaviors of hot-rolled steel beams under drop weight impact loading. J. Build. Struct. **32**(012), 242–249 (2011)
12. Kai, Q., Zhi, L., Yunhao, W., Gang, W.: Load resisting mechanisms of assembled monolithic precast concrete beam-slab substructures to resist progressive collapse. J. Build. Struct. **42**(07), 183–193 (2021)
13. Kai, Q., Song-Yuan, G., Shi-Lin, L., Feng, F., Jun, Y.: Effects of loading regimes on the structural behavior of RC beam-column sub-assemblages against disproportionate collapse. Eng. Struct. **251**, 113470 (2022)
14. Yun-Hao, W., Kai, Q., Feng, F., Qin, F.: Numerical investigation on load redistribution capacity of flat slab substructures to resist progressive collapse. J. Build. Eng. **29**, 101109 (2020)
15. Kai, Q., et al.: Robustness of post-tensioned concrete beam-column subassemblies under various column removal scenarios. J. Struct. Eng. **148**(5), 04022032 (2022)
16. Pengcheng, J., Hao, W., Qin, F.: Low-velocity lateral impact analyses of UHPC member based on CSC model. J. Build. Struct. **2019**, 0703 (2020)
17. Juanling, C., Zhaosheng, G., Rui, W., et al.: Test for behavior of a hot rolled H-shaped steel column under lateral impact. J. Vib. Shock **33**(18), 133–139 (2014)

Segmental Prefabrication and Assembly Technology of Composite Box Girder with Corrugated Steel Webs

Huijun Shen[1,2](✉), Hehui Zheng[1,2,3], Fei Tian[1,2,3], and Hao Dai[1,2]

[1] CCCC Second Harbour Engineering Co., Ltd, Wuhan 430040, China
shenhuijun@seu.edu.cn
[2] Key Laboratory of Large-Span Bridge Construction Technology, Wuhan 430040, China
[3] CCCC Highway Bridge National Engineering Research Centre Co. Ltd, Wuhan 430040, China

Abstract. The three approach bridges of the Jiangxinzhou Yangtze River Bridge in Nanjing are prestressed concrete box girder bridges with corrugated steel webs, which is the first in the world to adopt the process of short-line matching prefabrication and cantilever assembly by bridge erector. A series of new technologies and equipment are proposed for the construction process from the manufacturing of corrugated steel webs to the matching prefabricated and cantilever assembly of composite beams, such as a special matching prefabricated template system, manufacturing of corrugated steel webs in yard by short-line matching method, technological process of molding with partial mode for steel cage and high precision cantilever assembly by bridge erector. On this basis, combined with field measurement and numerical simulation, the mechanical behavior of segmental precast assembled box girder with corrugated steel webs during construction is revealed.

Keywords: Box girder with corrugated steel web · Segmental prefabrication and assembly · Short-line matching method · Construction monitoring

1 Introduction

In recent years, composite girder bridge with corrugated steel webs has been widely used in China due to its light weight, high prestressed efficiency, good durability and beautiful shape (Chen et al. 2018). At present, this type of bridge basically adopts construction technology such as cantilever casting and bracket cast-in-place (Deng et al. 2019). With the development of bridge industrialization, the segmental prefabrication technology with the advantages of energy saving, environmental protection, high efficiency and durability has been promoted in concrete bridges (Zhang et al. 2011). Introducing this technology into the construction of composite girder bridges with corrugated steel webs can not only significantly improve the construction quality and speed up the construction, but also improve the mechanical performance and seismic performance and reduce construction costs (Zhou et al. 2019; Shao et al. 2010).

W. Guo and K. Qian (Eds.): GBCESC 2022, LNCE 211, pp. 1062–1069, 2023.
https://doi.org/10.1007/978-981-19-5217-3_109

There are three approach bridges in Nanjing Jiangxinzhou Yangtze River Bridge, Crossing Lixin Road Bridge with spans of (31 + 46 + 31) m, Crossing Fengzi River Bridge (Fig. 1) with spans of (38 + 68 + 38) m, and Riverside Avenue Bridge with spans of (41 + 78 + 45) m, which is the first in the world to adopt the process of short-line matching prefabrication and cantilever assembly by bridge erector (Zhang et al. 2017; Li et al. 2015). It is divided into 314 precast segments with the length of 2.4 m and 3.2 m, which are divided into single box with one chamber and single box with double chamber. The corrugated steel web connected with concrete top plates by double PBL keys, and connected with concrete bottom plates by embedded form. The segments of corrugated steel webs are connected by butt welding.

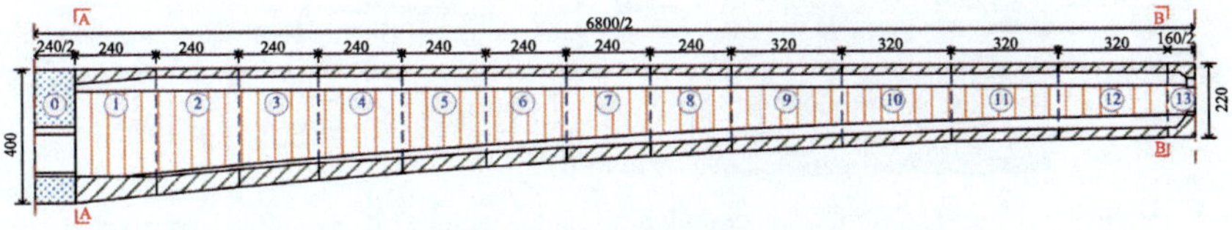

(a) Segmentation diagram of 1/2 mid-span beam

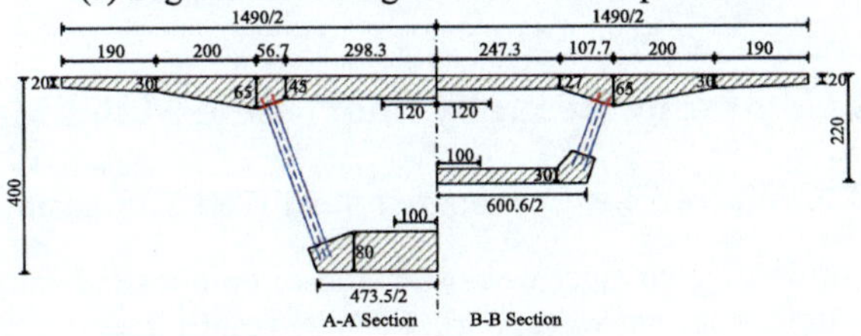

(b) Cross-section of beam

Fig. 1. Structure of Crossing Fengzi River Bridge (unit: cm).

2 Overall Construction Technology

After the corrugated steel webs are manufactured in the prefabricating yard of steel structure, they are transported to the prefabricating yard of concrete components. The short line matching process is used to precast the segment of composite beam step by step. The segment is transported to the construction site after the storage period meets the requirements. The substructure is constructed while the beam segment is prefabricated. And then segments 1# of two middle piers are hoisted and 0# segments are casted. The composite box beam segments with corrugated steel webs are cantilever assembled by bridge-craft, and the side span and middle span closure are completed in turn. The structure system transformation of continuous beam is completed after the temporary consolidation constraint of pier-beam is removed.

3 Manufacturing of Corrugated Steel Webs in Yard

The manufacturing alignment of beam segments with corrugated steel web precast by the short-line matching method should meet the requirements of positioning and installation in the pedestal (Fig. 2). That is, both ends of the first corrugated steel web segment should be perpendicular to the fixed or floated end template, and the standard corrugated steel web segments should be perpendicular to the fixed end template at the end template side (the precast direction). After the corrugated steel web segments are manufactured, half-span or full-span preassembly should be carried out, and temporary matching parts should be installed between the corrugated steel webs.

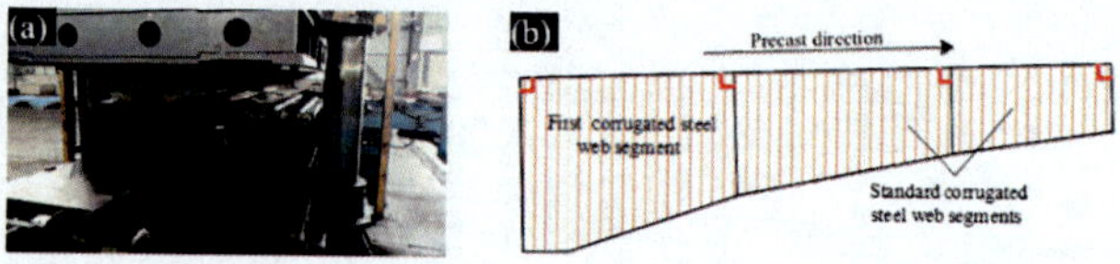

Fig. 2. Manufacturing segments of corrugated steel webs: (a) manufacturing in yard; (b) manufacturing schematic.

4 Precast of Composite Beam by Short-Line Matching Method

4.1 Molding of Reinforcement-Corrugated Steel Web Composite Skeleton

The reinforcement-corrugated steel webs skeleton of composite beam segment can be placed into the template in two ways: partial (Fig. 3) and integral. Combined with the connection type and reinforcement layout characteristics in the project, BIM technology is used for virtual assembly, and finally the partial mode in the precast pedestal is selected.

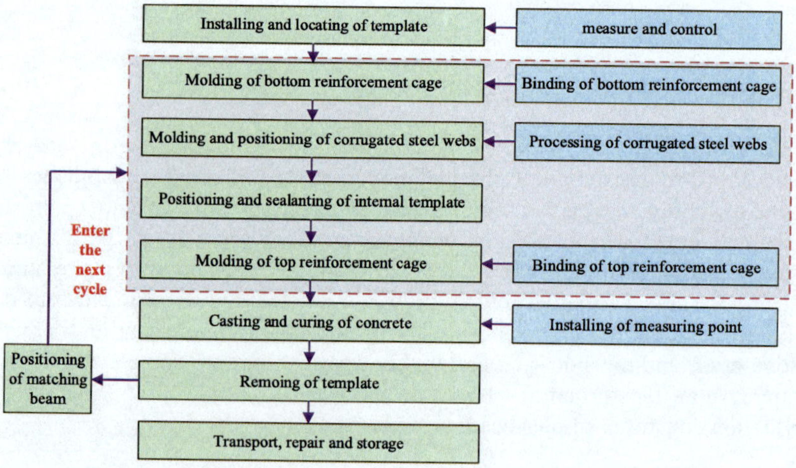

Fig. 3. Technological process of molding with partial mode.

4.2 Design of Special Precast Template System for Short-Line Matching

The design of precast template system for composite beam with corrugated steel webs must to be consider the following issues: (a) the fixed end template should be set movable blocks in the web area to facilitate the placement of corrugated steel webs into the template (Fig. 4); (b) fixed or floated end template and side template should be arranged for three-dimensional positioning tools of corrugated steel webs; (c) the measures to prevent slurry leakage should be set at the junction of corrugated steel webs and the top plate and bottom of box girder during concrete casting.

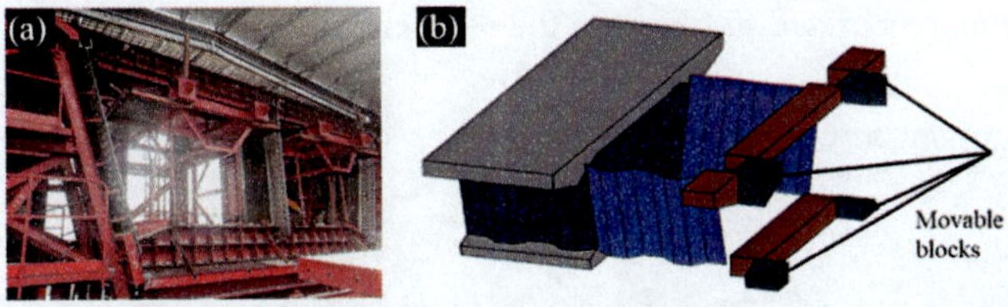

Fig. 4. Design of special precast template system: (a) special precast template; (b) movable blocks at the fixed end template.

4.3 Installation and Positioning of Corrugated Steel Webs in the Template

The positioning of corrugated steel webs in the template includes the positioning on the side of the matching beam and the positioning on the side of the fixed end template (Fig. 5). Since the corrugated steel web segments have been pre-assembled according to the manufacturing line before leaving the yard, and temporary matching connectors are set between adjacent segments, the corrugated steel webs of standard segments except the starting segment can be adjusted after connecting with the matching beam through the temporary matching connectors. The temporary matching connectors should be closely pasted and combined with punch pins to limit the displacement of the corrugated steel webs in the casting process. At the side of fixed end template, the corrugated steel webs can be accurately positioned by means of three-dimensional positioning tool, auxiliary jack, chain hoist, etc., and fixed by limiting baffles and ear socket for avoiding leakage.

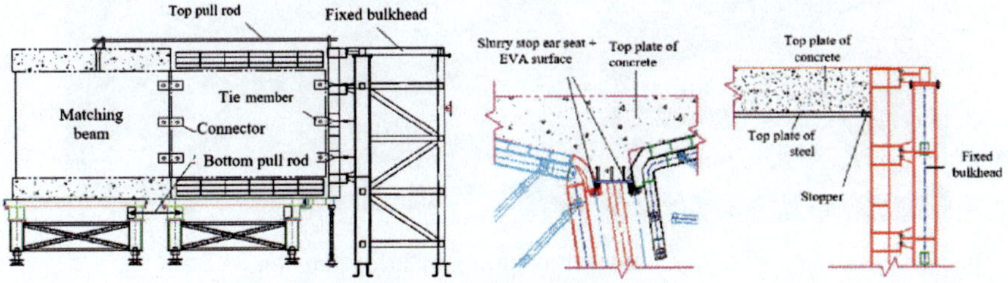

Fig. 5. Positioning of corrugated steel webs in the template.

5 Cantilever Assembly by Bridge Erector

5.1 High Precision Installation of Segment 1#

Segment 1# is installed with a 200t crawler crane. Four three-dimensional jacks are used to accurately adjust the position of segment 1#. Then, the beam is clamped with the prepared wedge pad at the bottom of the beam to ensure that the beam does not fall back after the return. By inserting thin steel plate to keep close between the limit block and the box girder. Two JL32 finely-rolled threaded bars were used to temporarily anchor the beam and the pier-side brackets. The segment 1# was fixed by the temporary prestress at the top and bottom of the beam to resist the influence of the lateral pressure for the segment 0# which is casted in-place on the segment 1# (Fig. 6).

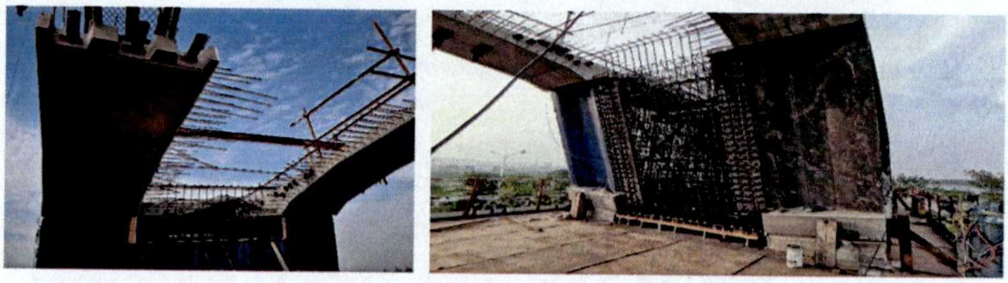

Fig. 6. Segment 1# installed with a 200t crawler crane

5.2 Cantilever Assembly for Standard Segments

From segment 2# to the maximum cantilever segment are installed by bridge erector (Fig. 7). Due to the different supporting conditions of segments 1# and 2#, the deformation between adjacent segments will inevitably be different. Therefore, it is necessary to analyze the specific value of deformation difference in advance, and adjust it by controlling the tensile value of transverse prestress in the box girder roof. The main installation process by bridge erector for standard segments includes pre-assembly, smearing epoxy adhesive, formal assembly, temporary prestress tension, permanent prestress tension and welding between segments of corrugated steel web.

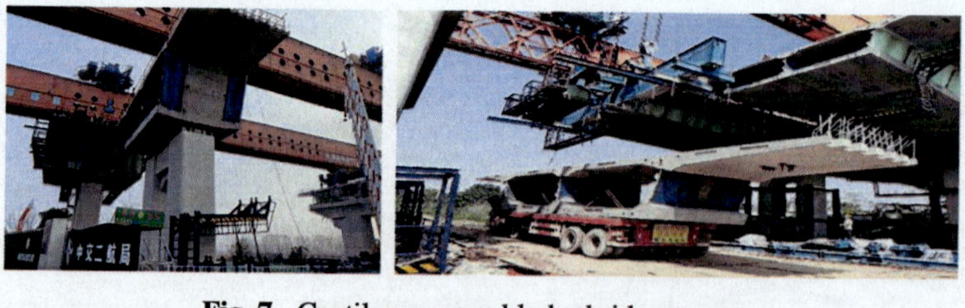

Fig. 7. Cantilever assembly by bridge erector.

6 Construction Monitoring During Installation Stage

The stress measuring point is located at the joint section of segment 1# and segment 2# and the middle section of segment 2# (Fig. 8 and Fig. 9). The monitoring contents include the stress-strain of concrete roof and floor (concrete longitudinal strain gauge) and corrugated steel web (three-way steel strain rosette). Due to the influence of steel bucket, the arrangement of two cross-section concrete strain gauges is not exactly the same.

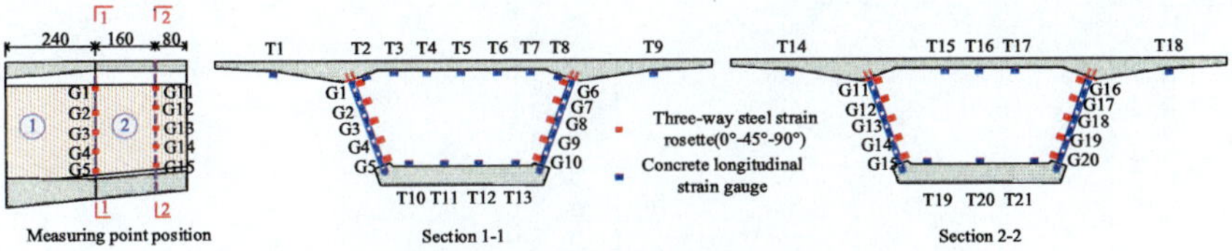

Fig. 8. Measuring point layout schematic (unit: cm).

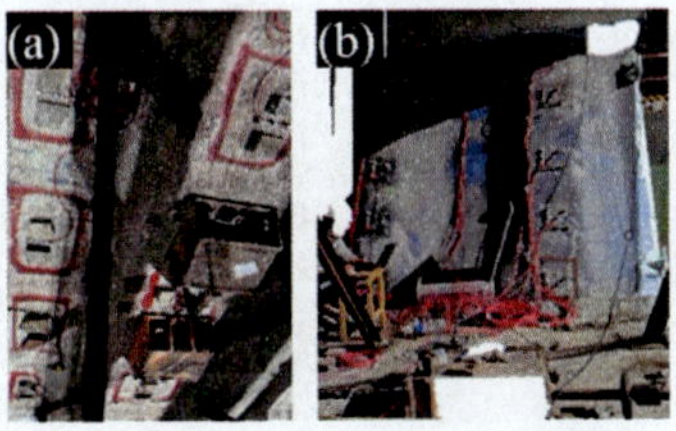

Fig. 9. Site measuring point layout: (a) concrete longitudinal strain gauge; (b) three-way steel strain rosette.

The spatial finite element model of the main girder with corrugated steel webs is established based on Midas FEA (Fig. 10). The concrete is simulated by solid element, the corrugated steel webs and their double PBL shear connectors are simulated by plate element, and the studs are simulated by rod element. The stud nodes are coupled with the surrounding concrete nodes in three directions, without considering the slip between them (Zhou et al. 2019). As the main girder is always in elastic stage during construction, concrete and steel are elastic constitutive (Zhang et al. 2018).

Studies have shown that even if the epoxy resin adhesive and concrete are flat joints, the bonding force can also reach more than 1.5 MPa (Shen et al. 2019). Through trial calculation, the tensile stress of the concrete in this bridge is within 1 MPa during the construction, so there is no cracking at the joint. Therefore, the concrete elements at the joints are connected by the nearest node.

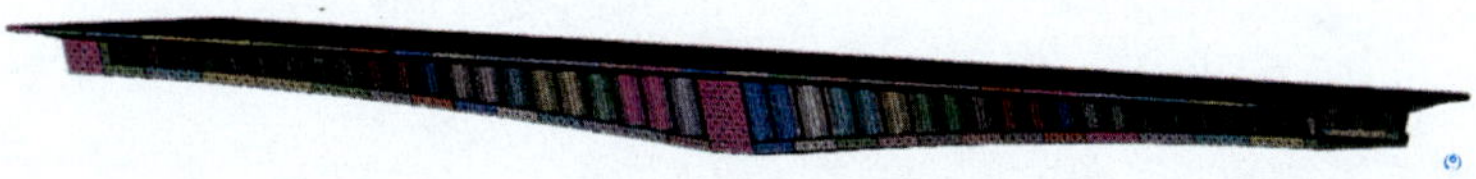

Fig. 10. Spatial finite element model of 1/2 main girder with corrugated steel webs.

The comparison between the measured and theoretical values of the longitudinal stress for the joint section concrete is shown in Fig. 11. It can be seen that the longitudinal stress of the concrete at the top has obvious shear lag phenomenon, and the stress distribution of the concrete at the bottom is relatively uniform. Overall, the longitudinal stress of the measured value is in good agreement with the theoretical value for the concrete.

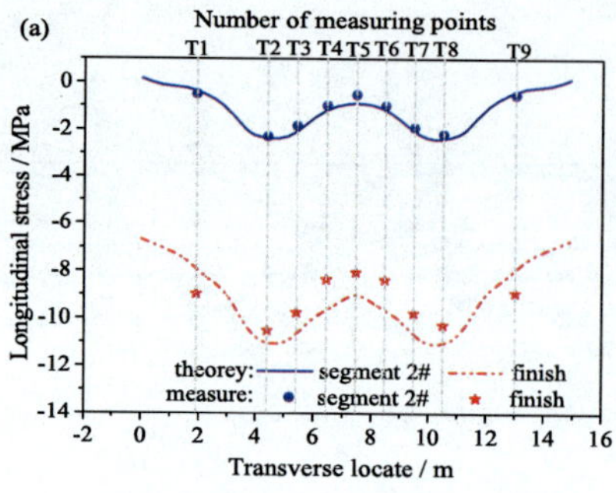

Fig. 11. The longitudinal stress of the concrete: (a) top; (b) bottom.

The comparison between the measured and theoretical values of shear stress and longitudinal stress for corrugated steel webs is shown in Fig. 12. It can be seen that the shear stress of corrugated steel webs is uniformly distributed along its height direction, and the longitudinal stress is small except for the large area near the contact with concrete. The result shows that corrugated steel webs mainly bear shear stress, and almost do not bear longitudinal stress, which is in line with the mechanical mechanism of corrugated steel webs. Overall, the shear stress and longitudinal stress of measured values are in good agreement with the theoretical values for corrugated steel webs.

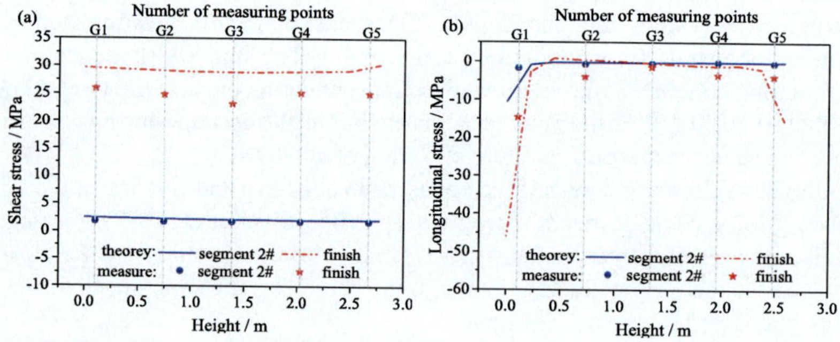

Fig. 12. The shear stress and longitudinal stress for corrugated steel webs: (a) shear stress; (b) longitudinal stress.

7 Conclusions

The combination of segmental precast assembly technology and composite beam with corrugated steel web can maximize the respective advantages of structure and construction technology, such as light-weight, beautiful shape, excellent mechanical performance and seismic performance of joints, easy quality control, and fast construction speed, which has good promotional value. On July 18, 2020, the main structure of the north approach bridge of Jiangxinzhou Yangtze River Bridge in Nanjing is connected, which marks that all 314-segmental beams of three box girder bridges with corrugated steel webs are successfully assembled. During the implementation of the project, a special matching prefabricated template system was developed, which can meet the requirements of rapid molding of corrugated steel webs, high-precision positioning and matching connection, and high-quality casting of steel-concrete joints. The application of technological process of molding with partial mode for steel cage can not only ensure the construction accuracy but also improve the construction efficiency. Combined with field measurement and numerical simulation, the mechanical behavior of segmental precast assembled box girder with corrugated steel webs during construction is revealed. The innovative practice of this project can provide new ideas for the assembly construction of medium-span composite girder bridges.

References

Chen, Y., Dong, J., Xu, T.: Composite box girder with corrugated steel webs and trusses: a new type of bridge structure. Eng. Struct. **166**(July 1), 354–362 (2018)

Deng, W., Liu, D., Xiong, Y., et al.: Experimental study on asynchronous construction for composite bridges with corrugated steel webs. J. Constr. Steel Res. **157**(June), 93–102 (2019)

Zhang, H., Xigang, Z., Feng, D., et al.: Research on new technology of segment prefabricated assembled bridge by short line matching method. Highway **02**, 76–82 (2011). (in Chinese)

Zhou, M., Liu, Y., Deng, W., et al.: Transverse analysis of full-scale precast segmental box girder segments with corrugated steel webs: experimental tests and FE modelling. Eng. Struct. **187**(May 15), 231–241 (2019)

Shao, X., Wang, H., Zhao, H., et al.: Experimental study on multicantilever prestressed composite beams with corrugated steel webs. J. Struct. Eng. **136**(9), 1098–1110 (2010)

Zhang, H., Hehui, Z., Ming, C.: Tests of segmental precasting and assembling technology for composite box girder bridge with corrugated steel webs. Bridge Constr. **47**(1), 82–87 (2017). (in Chinese)

Li, J., Wu, H., Chen, H.: Analysis of steel web shear capacity of variable section continuous system composite box beam bridge with corrugated steel webs. Bridge Constr. **45**(1), 79–84 (2015). (in Chinese)

Zhang, B., Chen, W., Xu, J.: Mechanical behavior of prefabricated composite box girders with corrugated steel webs under static loads. J. Bridge Eng. **23**(10), 04018077 (2018)

Shen, Y., Peng, C., Lisheng, C., et al.: Shear strength of keyed joints in segmental precast concrete bridges. J. Tongji Univ. (Nat. Sci.) **47**(10), 1414–1420 (2019). (in Chinese)

Check for updates

Correlation of Spectral Shapes of Horizontal and Vertical Ground Motions at the Same Period Based on Copula Function

Xiaolei Wang[1,2], Zixu Zhao[1], and Ruiguang Zhu[3](✉)

[1] School of Civil Engineering, Shenyang Jianzhu University, Shenyang, China
[2] Laboratory of Earthquake Disaster Prevention and Risk Assessment, Institute of Disaster Prevention, Sanhe, China
[3] School of Civil Engineering and Mechanics, YanShan University, Qinhuangdao, China
Zrg179@163.com

Abstract. Correlation and joint distribution model of spectral shapes of horizontal and vertical ground motions are important for the study of some applications. This study investigates the correlation and joint distribution model of the epsilons between vertical ground motions and associated horizontal ground motions at the same period via copula technique. The correlation of spectral shapes of horizontal and vertical ground motions based on four sets of horizontal and vertical NGA-West2 GMPEs are respectively studied, and a pair of horizontal and vertical GMPEs is selected for illustration. The joint probability distributions between horizontal ground motions epsilons and vertical ground motions epsilons via copula functions have been built, and then the conditional mean epsilons of vertical ground motions are calculated to verify the accuracy of the correlation of spectral shapes of horizontal and vertical ground motions. It has finally been found that epsilons of horizontal and vertical ground motions at the same period follow bivariate normal distribution. The correlation and joint distribution model of spectral shapes of horizontal and vertical ground motions from this study could be used as a theoretical basis for the site-specific spectrum generation in the future.

Keywords: Correlation coefficient · Joint distribution · Copula function · Horizontal and vertical ground motions · Spectral shape

1 Introduction

Earthquakes will give rise to horizontal and vertical ground motions on the ground. Horizontal ground motions are the main component of ground motions, which have a significant impact on the seismic resistance of structures. Meanwhile, vertical ground motions are also needed to predict actual seismic responses of specific structures [1–3], which even lead to destructions of these structures. Epsilon is an important feature of spectral shapes of ground motions, and past studies have indicated that the epsilon affects structural responses [4]. It is necessary to consider the influence of the epsilons of vertical ground motions on the structure.

W. Guo and K. Qian (Eds.): GBCESC 2022, LNCE 211, pp. 1070–1083, 2023.
https://doi.org/10.1007/978-981-19-5217-3_110

The epsilons of vertical ground motions are not the same as those of associated horizontal ground motions. However, similar to the other seismic features, epsilons of horizontal and vertical ground motions should be correlated due to certain commonalities in the seismic features. In light of this reason, the epsilon of the vertical ground motion could be predicted conditional on that of the horizontal ground motion. Some researchers [5–9] have verified that the joint probability distribution of horizontal ground motion intensity measures, and the joint probability distribution of epsilons of mainshock and aftershock ground motions all follow the multivariate normal distribution [10]. The epsilons between horizontal and vertical ground motions were assumed to jointly follow a multivariate normal distribution [11]. However, as far as authors know, the joint probability distributions of epsilons of horizontal and vertical ground motions have not been validated by statistics in the public documents.

The main objective of this study aims to develop the dependence modeling of epsilons of horizontal ground motions and those of vertical ground motions. To do this, firstly four horizontal and vertical NGA-West2 GMPEs (including horizontal CY14 [12] model and associated vertical CY13 [13] model, horizontal CB14 [14] model and associated vertical BC16 [15] model, horizontal BSSA14 [16] model and associated vertical SBSA16 [17] model, and horizontal ASK14 [18] model and associated vertical GKAS17 [19] model) and associated horizontal and vertical ground motions have all been selected. As an illustration, correlation by the vertical SBSA16 [17] model and associated horizontal BSSA14 [16] model are detailedly exemplified in this paper. Then the copula technique is used to build the joint probability distribution of epsilons of horizontal ground motions and those of vertical ground motions. It is found that epsilons of horizontal and vertical ground motions at the same period follow bivariate normal distribution.

2 Correlation of Spectral Shapes of Horizontal and Vertical Ground Motions Based on Copula Function

2.1 A Simple Procedure to Build Correlation of Spectral Shapes of Horizontal and Vertical Ground Motions based on Copula Function

Step 1: Determine the Target ε at Every Period and the Associated M and R. The *SA* values of horizontal and vertical ground motions at every period should be identified. Then once a GMPE is specified, the values of $\varepsilon_{\rm H}(T_i)$ and $\varepsilon_{\rm V}(T_i)$ could be computed for every period. The normalized residual (epsilon or ε) at the period of interest T^* could be expressed as:

$$\varepsilon(T^*) = \frac{\ln SA(T^*) - \mu_{\ln SA}(M, R, T^*)}{\sigma_{\ln SA}(T^*)} \tag{1}$$

in which ln$SA(T^*)$ denotes the log of the spectral acceleration (*SA*) of interest periods T^*; logarithmic mean and standard deviation of *SA* at all periods of interest T^* and its associated magnitude (*M*) and source-to site distance (*R*) of the target earthquake, which are denoted as $\mu_{\ln SA}(M,R,T^*)$ and $\sigma_{\ln SA(T^*)}$, could be computed via a specific GMPE.

Step 2: Construct Correlation of Spectral Shapes of Horizontal and Vertical Ground Motions Based on Copula Function. Correlation of spectral shapes of horizontal and vertical ground motions could be calculated by joint probability density function (PDF) of $\varepsilon_H(T_i)$ and $\varepsilon_V(T_i)$, $f[\varepsilon_V(T_i), \varepsilon_H(T_i)]$, which is the core and important of the whole calculation process.

The joint PDF $f[\varepsilon_V(T_i),\varepsilon_H(T_i)]$ could be determined by the copula technique. The copula technique has been used by Zhu et al. [10] for the joint distribution of mainshock epsilons and aftershock epsilons, which has the advantage of describing the functions that join together 1D distribution functions to form multivariate distribution functions inherently. Moreover, the copula technique could handle the cases of complex and non-linear correlations [19]. According to the copula technique, the joint PDF of multivariate data could be formed from their marginal probability distributions. Let $F[\varepsilon_V(T_i),\varepsilon_H(T_i)]$ be the joint cumulative distribution function (CDF) of $\varepsilon_H(T_i)$ and $\varepsilon_V(T_i)$ with the corresponding marginal CDFs of $F[\varepsilon_H(T_i)]$ and $F[\varepsilon_V(T_i)]$, respectively; then there exists a copula C such that

$$F[\varepsilon_V(T_i), \varepsilon_H(T_i)] = C\{F_V[\varepsilon_V(T_i)], F_H[\varepsilon_H(T_i)]\} = C(u_V, u_H) \tag{2}$$

in which $u_V = F_V[\varepsilon_V(T_i)]$ and $u_H = F_H[\varepsilon_H(T_i)]$. From a suite of optional copula functions, the optimum copula could be determined by fitting the empirical copula samples. Based on Eq. (2), the joint PDF $f[\varepsilon_V(T_i),\varepsilon_H(T_i)]$ could be calculated by taking the derivative of $F[\varepsilon_V(T_i),\varepsilon_H(T_i)]$, yielding

$$f[\varepsilon_V(T_i), \varepsilon_H(T_i)] = D(u_V, u_H) \times f_V[\varepsilon_V(T_i)] \times f_H[\varepsilon_H(T_i)] \tag{3}$$

in which $f_H[\varepsilon_H(T_i)]$ and $f_V[\varepsilon_V(T_i)]$ are respectively marginal PDFs of $\varepsilon_H(T_i)$ and $\varepsilon_V(T_i)$, and $D(u_V, u_H)$ is the density function of the copula function $C(u_V, u_H)$, which could be expressed as:

$$D(u_V, u_H) = \partial^2 C(u_V, u_H)/\partial u_V \partial u_H \tag{4}$$

Step 3: Verify the Accuracy of the Correlation of Spectral Shapes of Horizontal and Vertical Ground Motions. As the correlation of spectral shapes of horizontal and vertical ground motions is calculated, it can be used to calculate the conditional mean epsilon of vertical ground motion, *given* $\varepsilon_H(T_i)$. This verifies the accuracy of the correlation of spectral shapes of the horizontal and vertical ground motions. The conditional mean epsilon $\mu_{\varepsilon_V(T_i)|\varepsilon_H(T_i)}$ could be computed by

$$\mu_{\varepsilon_V(T_i)|\varepsilon_H(T_i)} = E[\varepsilon_V(T_i)|\varepsilon_H(T_i)] = \int_{-\infty}^{+\infty} \varepsilon_V(T_i) \times f[\varepsilon_V(T_i)|\varepsilon_H(T_i)]d\varepsilon_V(T_i) \tag{5}$$

in which $f[\varepsilon_V(T_i)|\varepsilon_H(T_i)]$ is the conditional probability density function (PDF) of $\varepsilon_V(T_i)$, given $\varepsilon_H(T_i)$, could be defined as

$$f[\varepsilon_V(T_i)|\varepsilon_H(T_i)] = \frac{f[\varepsilon_V(T_i), \varepsilon_H(T_i)]}{f_H[\varepsilon_H(T_i)]} \tag{6}$$

Substituting Eqs. (6) and (3) into Eq. (5), then the conditional mean epsilon of the vertical ground motion could be calculated by

$$\mu_{\varepsilon_V(T_i)|\varepsilon_H(T_i)} = \int_{-\infty}^{+\infty} \varepsilon_V(T_i) \times D(u_V, u_H) \times f_V[\varepsilon_V(T_i)]d\varepsilon_V(T_i) \tag{7}$$

3 Correlation Example for PEER NGA-WEST2 Database

3.1 Horizontal-Vertical Ground Motions

The NGA-West2 database is a valuable source to select horizontal and vertical ground motions. Four sets of horizontal and vertical NGA-West2 GMPEs are respectively selected and studied, and the horizontal BSSA14 model and associated vertical SBSA16 model are detailedly exemplified in this study. A set of 15862 HGM-VGM sequences of 336 HGM-VGM earthquakes from BSSA14 and SBSA16 model have been used. It has been selected with the similar criteria. The other detailed specific criteria used for ground-motion selection could be found in the Boore et al. [16] and the Stewart et al. [17].

Some seismological information such as the magnitudes of HGM-VGM ranging from 3.01 to 7.9, their rupture distances are located in the scope between 0 and 300 km, and spectral acceleration values of horizontal ground motion records and associated vertical ground motion records are compared at the 0–10 s period ranges, which have been shown in Fig. 1a and Fig. 1b.

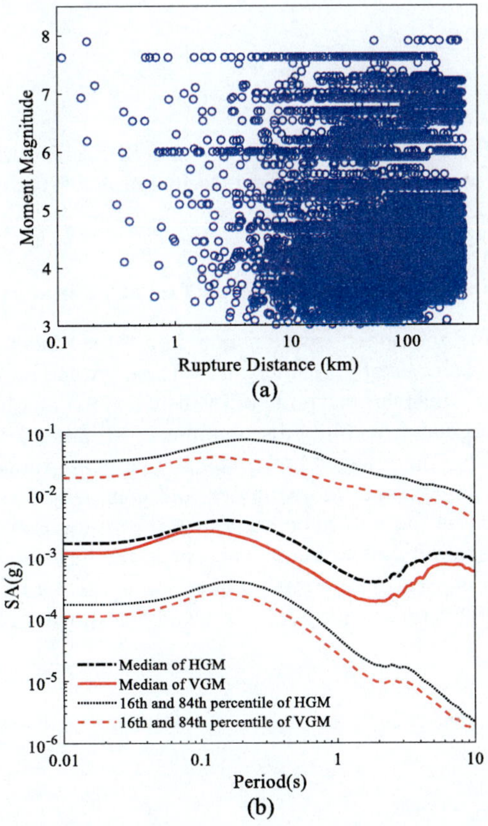

Fig. 1. The selected HGM-VGM sequences: (a) The magnitude–distance distribution; and (b) The median and percentile response spectra

As defined in the NGA-West2 database, resulting in a period-varying dataset (as described by Abrahamson et al. [20]), the number of the available HGM-VGM seismic sequences is reduced at long periods, which is shown in Fig. 2.

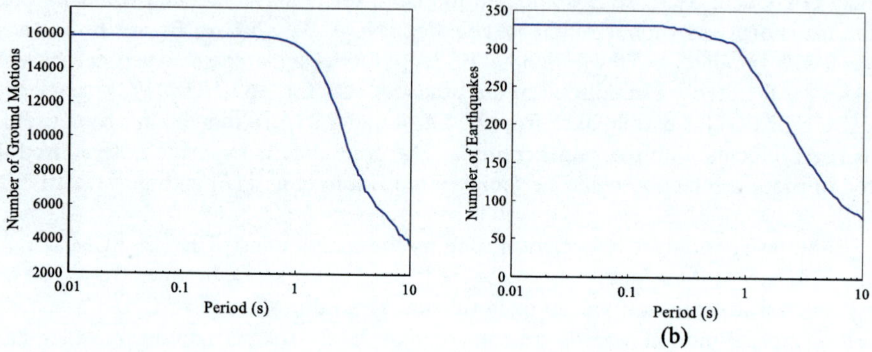

Fig. 2. The number of the HGM-VGM data: (a) the number of ground motions; and (b) the number of earthquakes.

3.2 Calculation of ε_H and ε_V

The values of $\varepsilon_H(T_i)$ and $\varepsilon_V(T_i)$ of BSSA14 and SBSA16 GMPE model could be computed for every period via Eq. (1). The marginal probability distributions of ε_H and ε_V would be discussed in the next section.

3.3 Marginal Probability Distribution Modeling of ε_H and ε_V

The correlation coefficient model between ε_H and ε_V is the basis of this research on the correlation of spectral shapes of horizontal and vertical ground motions. To develop the joint PDF of ε_H and ε_V, the marginal probability distributions should be firstly identified. In this study, the marginal probability distributions of ε_H and ε_V are derived by fitting the data calculated from the selected horizontal and vertical ground motion sequences. According to the histogram, normal, Gumbel, and generalized extreme value (GEV) distribution are taken as the null hypothesis of the Kolmogorov-Smirnov (K-S) test. If more than one probability distribution is accepted, the bayesian information criteria (BIC) principle [21] is used to determine which one is the optimum distribution. The histograms and the PDFs for ε_H and ε_V at 8 s have been shown in Fig. 3.

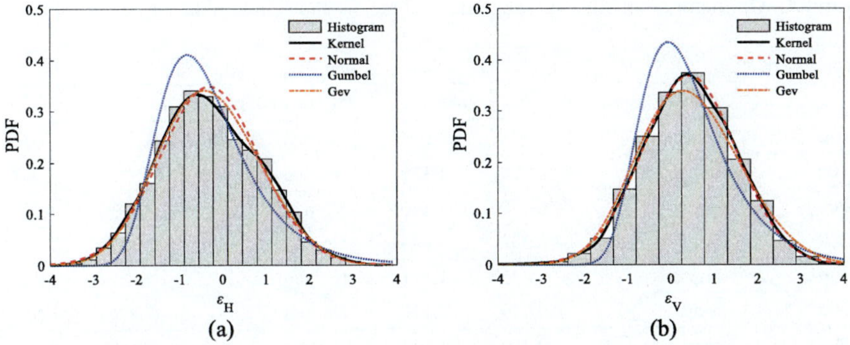

Fig. 3. The histograms and the PDFs for ε_H and ε_V at different periods: (a) ε_H at 8 s; (b) ε_V at 8 s.

Optimum probability distributions and means and standard deviations of ε_H and ε_V have been listed in Table 1. The optimum probability distributions of ε_H and ε_V follow the normal distribution at most periods (except for a long period after 2 s of ε_H, in which optimum probability distributions follow the generalized extreme value (GEV) distribution. The means of ε_H and ε_V are approximately equal to zero at the whole periods, and the standard deviation of ε_H and ε_V are approximately equal to one at the whole periods, the reason of which is although the selected ground motions are selected according to the selection criteria of the Boore et al. [16] and the Stewart et al. [17], the selected horizontal and vertical ground motion records could not be exactly the same as the records selected by the Boore et al. [16] and the Stewart et al. [17], and it is inevitable to be biased. Based on the above discussion, to avoid introducing additional model errors, it is reasonable for the distribution of ε_H and ε_V to follow a normal distribution or even to follow a standard normal distribution. The effect of using different distributions of ε_H and ε_V would be discussed in the last section.

3.4 Joint Distribution of ε_H and ε_V Using Copula Functions

Three different correlation coefficient models (including Pearson linear correlation coefficient model, Kendall rank correlation coefficient model, and Spearman rank correlation coefficient model) are selected to study the correlation between ε_H and ε_V in this study. The specific values of the above correlation coefficients between ε_H and ε_V at different periods have been listed in Table 2, and the trend graph of three correlation coefficient models has been shown in Fig. 4. It could be seen that the Kendall rank correlation coefficient is lower than Pearson linear correlation coefficient and Spearman rank correlation coefficient at any period, indicating that the consistency between ε_H and ε_V is relatively weak. The Pearson linear correlation coefficient and the Spearman rank correlation coefficient are not much different. The Pearson linear correlation coefficient is higher at the short period, but the correlation coefficient is lower at the long period. Pearson linear correlation coefficient is mostly between 0.7 and 0.8, indicating that its linearity is relatively strong, In addition, the correlation between ε_H and ε_V is the strongest at the period 0.075 s, and then the correlation gradually decreases as the period increases.

Table 1. Optimum probability distributions and means and standard deviations of ε_H and ε_V

Period(s)	ε_H			ε_V		
	Optimum type	μ	σ	Optimum type	μ	σ
0.010	Normal	0.135	1.096	Normal	0.012	1.013
0.100	Normal	−0.028	1.118	Normal	−0.027	1.007
0.200	Normal	0.138	1.146	Normal	−0.019	1.028
0.300	Normal	0.332	1.131	Normal	0.020	1.034
0.400	Normal	0.435	1.121	Normal	0.024	1.021
0.500	Normal	0.470	1.116	Normal	−0.008	1.020
1.000	Normal	0.269	1.091	Normal	0.113	0.999
2.000	GEV	0.042	1.076	Normal	0.272	1.011
3.000	GEV	−0.135	1.041	Normal	0.309	0.997
5.000	GEV	−0.435	1.067	Normal	0.276	1.015
10.000	GEV	−0.204	1.181	Normal	0.439	1.124

Table 2. Correlation coefficients of the epsilons of the HGM-VGM at different periods

Period(s)	Correlation coefficient			Period(s)	Correlation coefficient		
	Pearson	Kendall	Spearman		Pearson	Kendall	Spearman
0.010	0.849	0.653	0.838	0.500	0.784	0.580	0.770
0.020	0.850	0.654	0.839	1.000	0.802	0.607	0.798
0.050	0.857	0.663	0.846	2.000	0.810	0.618	0.811
0.100	0.858	0.660	0.846	3.000	0.789	0.593	0.786
0.200	0.836	0.636	0.825	5.000	0.763	0.574	0.765
0.300	0.802	0.596	0.787	10.000	0.671	0.514	0.705

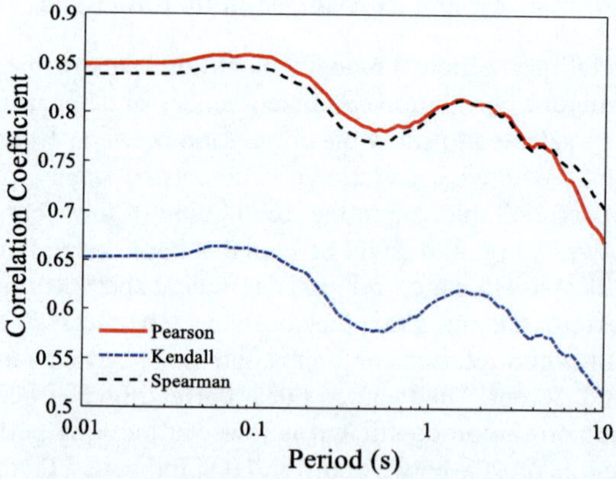

Fig. 4. The trend of correlation coefficient of the epsilon of the HGM-VGM sequences at different periods.

The scatter plots of ε_{H} and ε_{V} at 1 s and 10 s have been shown in Fig. 5. It could be observed that the data of ε_{H} and ε_{V} shows a symmetrical property concerning linear, and the epsilons of the vertical ground motions are obviously normally correlated to those of their corresponding horizontal ground motions.

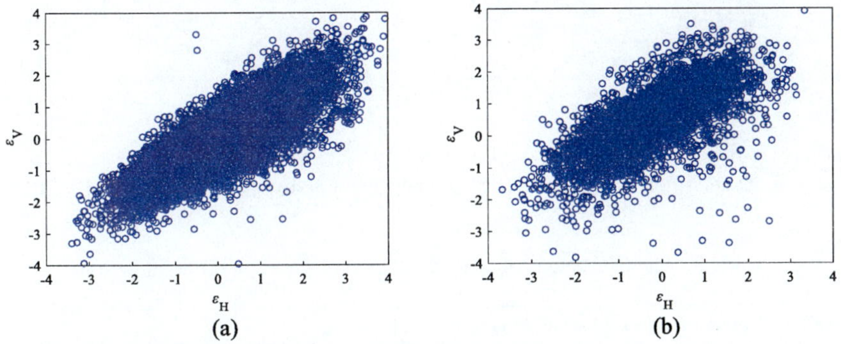

Fig. 5. Scatter plots of ε_{H} and ε_{V} at some interesting periods: (a) $T = 1$ s; (b) $T = 10$ s.

As the marginal probability distributions are identified, the final ingredient to complete the joint probability distribution is Copula function of ε_{H} and ε_{V}. To identify a suitable copula of ε_{H} and ε_{V} more formally, the candidate Copula function should be selected via the empirical cumulative distribution sample of ε_{H} and ε_{V}. The empirical cumulative distribution sample of ε_{H} and ε_{V} are represented by $U(\varepsilon_{\mathrm{H}})$ and $V(\varepsilon_{\mathrm{V}})$, respectively. The corresponding scatter plots of $U(\varepsilon_{\mathrm{H}})$ and $V(\varepsilon_{\mathrm{V}})$ are shown in Fig. 6, which provides an insight for a suitable copula selection. It could be observed that the data of $U(\varepsilon_{\mathrm{H}})$ and $V(\varepsilon_{\mathrm{V}})$ show a symmetrical property with respect to the diagonal lines of a unit square, and the upper and lower tails might be related. Therefore, seven frequently used symmetrical bivariate copulas are fitted using the maximum-likelihood method, which includes Gaussian copula, t copula, Plackett copula, Frank copula, No. 16 copula, Gumbel copula. The above selected copula models could not only reflect symmetry but also capture the correlation between the upper and lower tails. To measure the goodness fit of the copula function, the BIC principle is adopted. A copula associated with the smallest BIC values is considered for the best-fit copula. The best-fitting results of the considered copula models have been listed in Table 3, and the best-fit copula is highlighted in bold. It has been shown that the t copula has fewer BIC values than the other copulas at most periods and Frank is the optimal copula function only at the long periods. It is noteworthy that the Gaussian copula and t copula have almost identical BIC values. The effect of copula functions on the correlations of spectral shapes of horizontal and vertical ground motions is investigated in the next section.

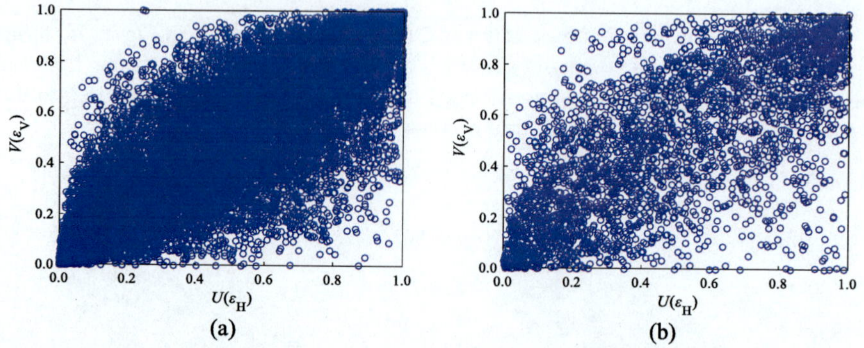

Fig. 6. Scatter plots of $U(\varepsilon_H)$ and $V(\varepsilon_V)$ at different periods: (a) $T = 1$ s; (b) $T = 10$ s.

Table 3. The BIC values of different copula functions at different periods

Period(s)	Optimum copula	BIC of different copula functions						
		Gaussian	Plackett	Frank	No. 16	t	Gumbel	Clayton
0.010	t	−2488	−2543	−2564	47086	**−2580**	−2315	−1839
0.030	t	−20233	−19665	−19035	23310	**−20823**	−18578	−17870
0.050	t	−20974	−20497	−19708	20710	**−21639**	−19794	−17715
0.100	t	−20906	−19974	−19484	21503	**−21302**	−19032	−18237
0.300	t	−16329	−15318	−15022	48518	**−16599**	−14585	−14411
0.500	t	−15145	−14345	−14016	58688	**−15454**	−13864	−12672
1.000	t	−15599	−14907	−14764	40506	**−15835**	−14440	−12455
2.000	t	−12240	−11670	−11752	27195	**−12446**	−11452	−9466
5.000	t	−4615	−4551	−4536	21362	**−4799**	−4466	−3510
10.000	Frank	−2000	−2126	**−2191**	35798	−2070	−1797	−1595

The minimum of BIC is the best-fit copula and is highlighted in bold

3.5 Conditional Mean Epsilon of Vertical Ground Motions

Once the joint probability distribution of ε_H and ε_V is determined, the conditional mean epsilon of vertical ground motions could be calculated by Eq. (5). The scatter plots of conditional mean epsilon of vertical ground motions, $\mu_{\varepsilon_V(T_i)|\varepsilon_H(T_i)}$, at the periods 1 s and 7.5 s has been shown in Fig. 7. It has been observed that $\mu_{\varepsilon_V(T_i)|\varepsilon_H(T_i)}$ calculated by the copula technique could reflect the linear as well as nonlinear correlation between ε_H and ε_V.

To illustrate the necessity to consider the dependency between ε_H and ε_V, the conditional PDFs and CDFs of ε_V, respectively based on the independence assumption and the optimum copula at 1 s have been shown in Fig. 8. It could be shown that the independence assumption leads to biased evaluations of the probability distribution of ε_V, whereas the conditional PDF and CDF based on the copula function could better match the recorded data. The result illustrates that the copula-based joint probability distribution of ε_H and ε_V is useful and accurate to predict ε_V values conditioned on the corresponding ε_H values.

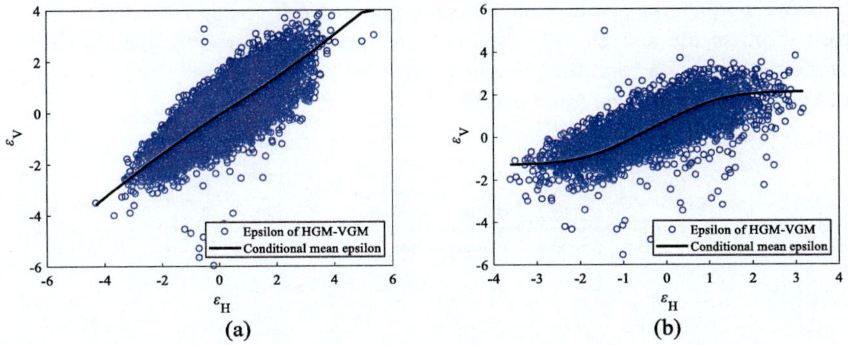

Fig. 7. The conditional mean of ε_V, given ε_H at different conditional periods: (a) $T = 1$ s and (b) $T = 7.5$ s.

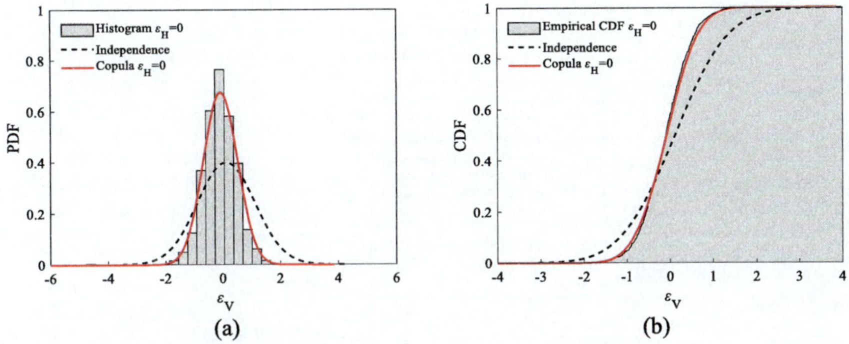

Fig. 8. The conditional PDFs and cumulative distribution functions (CDF) of ε_V, given ε_H, at different periods: (a) conditional PDF at 1 s, (b) conditional CDF at 1 s.

The mean and standard deviation of the predicted ε_V using the optimal Copula function given conditional ε_H has been shown in Fig. 9a. The optimal copula follow the Frank Copula distribution at the long period after 7.5 s, and the optimal copula follow the t Copula distribution at other periods, which have been shown in Table 3. The reason of the above results is that although the selected ground motions are selected according to the selection criteria by Boore et al. [16] and the Stewart et al. [17], the selected horizontal and vertical ground motion records could not be exactly the same as the records selected by the Boore et al. [16] and the Stewart et al. [17], and it is inevitable to be biased. Meanwhile, if the degree of freedom parameter of the t Copula function approaches infinity, its limit is the Gaussian Copula function. Hence, it could be a reasonable option to use t or Gaussian copula function at all periods when generating $f[\varepsilon_V(T_i), \varepsilon_H(T_i)]$. The mean and standard deviation of ε_V predicted using t or Gaussian Copula functions given conditional ε_H have been shown in Fig. 9b. It could be seen that the mean value of the predicted value of ε_V almost entirely coincides with the mean value of the ε_V from recorded ground motion, but the standard deviation of the predicted value is smaller than the standard deviation of the value from recorded ground motion.

The above predicted ε_V values have been computed only using the conditional mean information, so the discrete information could not be reflected for the above results. Therefore, the copula conditional mean of ε_V could accurately predict the mean of ε_V, given conditional ε_H at the same period.

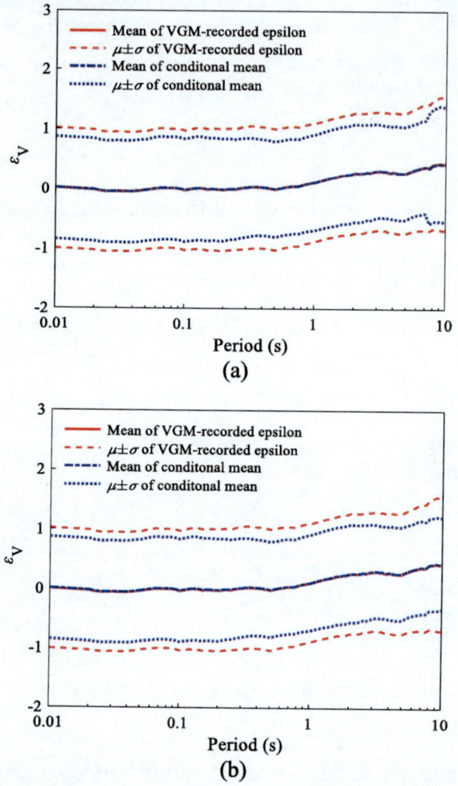

Fig. 9. Mean and standard deviation of the predicted ε_V using different Copula conditional mean: (a) optimal Copula conditional mean; (b) t or Gaussian copula conditional mean.

Based on the above discussion. Hence, it is reasonable to replace the optimal marginal distributions of ε_H and ε_V with the normal distribution. If the marginal distributions of two variables are normally distributed and joint probability distribution between them could be built by Gaussian or t copula function, then the joint distribution of these two variables follow a bivariate normal distribution. In this case, the conditional mean and conditional standard deviation of ε_V given conditional ε_H could be calculated by:

$$\mu_{\varepsilon_V(T_i)|\varepsilon_H(T_i)} = \mu_{\varepsilon_V(T_i)} + \rho[\mu_{\varepsilon_V(T_i)}, \mu_{\varepsilon_H(T_i)}] \cdot \frac{\sigma_{\varepsilon_V(T_i)}}{\sigma_{\varepsilon_H(T_i)}} \cdot [\varepsilon_H(T_i) - \mu_{\varepsilon_H(T_i)}] \quad (8)$$

$$\sigma_{\varepsilon_V(T_i)|\varepsilon_H(T_i)} = \sqrt{1 - \rho^2[\varepsilon_V(T_i), \varepsilon_H(T_i)]} \cdot \sigma_{\varepsilon_V(T_i)} \quad (9)$$

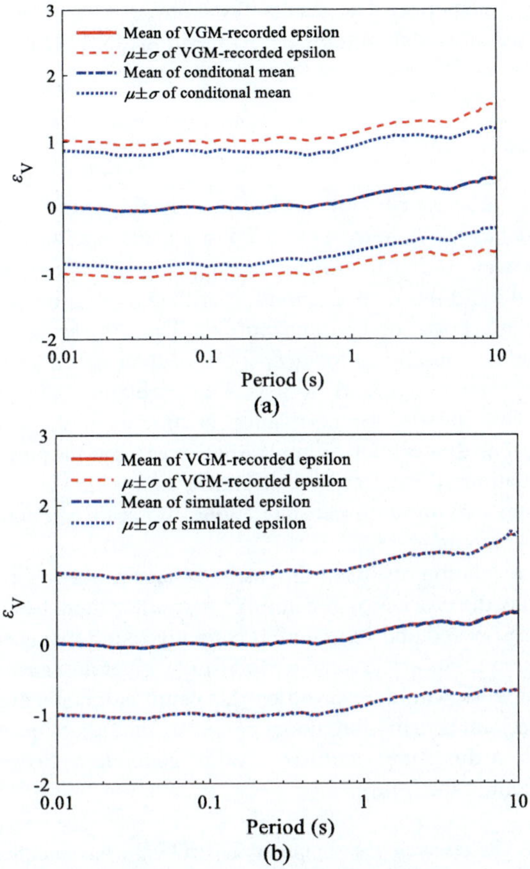

Fig. 10. Different ways to prove conditional mean follow bivariate normal: (a) Conditional mean of ε_V using normal marginal distribution and bivariate Gaussian joint distribution; (b) ε_V obtained by Monte Carlo simulation with conditional mean and deviation.

where $\mu_{\varepsilon_V(T_i)}$, $\mu_{\varepsilon_H(T_i)}$, $\sigma_{\varepsilon_V(T_i)}$ and $\sigma_{\varepsilon_H(T_i)}$ are the means and standard deviations of ε_V and ε_H at the same period T_i; $\rho[\varepsilon_V(T_i),\varepsilon_H(T_i)]$ is the correlation coefficient between ε_V and ε_H at the period T_i. The calculation method of ε_V given conditional ε_H could be simplified. It is clear that the conditional mean of ε_V still coincides with the mean of the recorded value, and there is still a specific difference between standard deviations of predicted values and those of the recorded values, which have been shown in Fig. 10a. Therefore, an accurate conditional mean of ε_V could be predicted via the above simplified method, which is proposed in this paper (see Eq. (8) and Eq. (9)). In addition, given conditional ε_H, the predicted value of ε_V could be generated by Monte Carlo simulation. The mean and standard deviation of ε_V obtained via Monte Carlo simulation given conditional ε_H have been shown in Fig. 10b. It is clear that the mean and standard deviation of the simulated value of ε_V basically coincides with the mean and standard deviation of the recorded value of ε_V. The result illustrates that ε_H and ε_V follow a bivariate normal distribution at the same period. The results based on three other horizontal and vertical NGA-West2 GMPEs is similar with the above-mentioned results based on BSSA14 and SBSA16 model.

4 Conclusions

The correlation and joint distribution is illustrated using ground motions selected from the NGA-West2 database. A new method for study correlation of spectral shapes of horizontal and vertical ground motions is presented in this study. This method adopts copula functions to build the joint distribution of the epsilons between vertical ground motions and associated horizontal ground motions. Then, the accuracy of the correlation is verified by using the conditional mean epsilons of the vertical ground motions, given the epsilons of the horizontal ground motions. The correlation between ε_H and ε_V varies with the periods. Specifically, the correlation coefficient is the strongest around the 0.075 s period, and then the correlation gradually decreases as the period increases. When investigating correlations, the marginal distributions of ε_H and ε_V and the corresponding copula functions are varying with periods (Either the optimal copula, the t copula or Gaussian copula is recommended).

Although the correlation results of the research have proven that such a treatment is accurate and reliable, the process of generating the result is more troublesome. A simplified assumption is proposed and validated. It is finally found that epsilons of horizontal and vertical ground motions at the same period follow bivariate normal distribution, and a lot of calculation work would be saved by this result and convenient for application. The correlation and joint distribution model of spectral shapes of horizontal and vertical ground motions from this study could be used to generate a theoretical basis for the site-specific spectrum in the future.

Acknowledgement. The research was supported in part by the National Science Foundation of China (under Grant Number 51908379), Hebei Key Laboratory of Earthquake Disaster Prevention and Risk Assessment (Grant No. FZ213206), Doctoral Scientific Fund Project of the Yanshan University (BL18053) and Postdoctoral Research Project of Hebei Provincial Department of Human Resources and Social Security (Personnel Department) (B2019003025).

References

1. Gu, Z.-Y., Wang, S.-G., Liu, W.-Q., Du, D.-S., Xu, W.-Z.: Experimental research and numerical simulation of a large-span isolated structure considering multi-dimensional input effects. Int. J. Steel Struct. **17**(4), 1583–1595 (2017). https://doi.org/10.1007/s13296-017-1224-7
2. Chen, Y., Kun, C., Larkin, T., Chouw, N.: Impact of vertical ground excitation on a bridge with footing uplift. J. Earthq. Eng. **20**(7), 1035–1053 (2016)
3. Hwang, J.S., Hsu, T.Y.: Experimental study of isolated building under triaxial ground excitations. J. Struct. Eng. **126**(8), 879–886 (2000)
4. Jack, W.B., Allin, C.: A vector-valued ground motion in tensity measure consisting of spectral acceleration and epsilon. Earthq. Eng. Struct. D **34**, 1193–1217 (2005)
5. Brendon, A.B.: A ground motion selection algorithm based on the generalized conditional intensity measure approach. Soil Dyn. Earthq. Eng. **40**, 48–61 (2012)

6. Brendon, A.B.: A generalized conditional intensity measure approach and holistic ground motion selection. Earthq. Eng. Struct D **39**(12), 1321–1342 (2010)
7. Brendon, A.B.: Empirical correlations between peak ground velocity and spectrum-based intensity measures. Earthq. Spectra **28**(1), 17–35 (2012)
8. Athanasios, N.P., Mohsen, K., Paolo, B.: Correlation of spectral acceleration values of mainshock-aftershock ground motion pairs. Earthq. Spectra **35**(1), 39–60 (2019)
9. Nirmal, J., Jack, W.B.: Statistical tests of the joint distribution of spectral acceleration values. Bull. Seismol. Soc. Am. **98**(5), 2231–2243 (2008)
10. Zhu, R.G., Lu, D.G., Yu, X.H., Wang, G.Y.: Conditional mean spectrum of aftershocks. Bull. Seismol. Soc. Am. **107**(4), 1940–1953 (2017)
11. Neal, K., Kishor, J., Nicolas, L., Baker, J.W.: Selecting three components of ground motions from conditional spectra for multiple stripe analyses. In: 17th World Conference on Earthquake Engineering, Sendai, Japan (2020)
12. Chiou, J., Youngs, R.R.: Update of the Chiou and Youngs NGA model for the average horizontal component of peak ground motion and response spectra. Earthq. Spectra **30**(Special I), 1117–1153 (2014)
13. Chiou, J., Youngs, R.R: NGA-West2 Ground motion prediction equations for vertical ground motions. In: PEER Report 2013/24. Pacific Earthquake Engineering Research Center, Headquarters at the University of California, Berkeley (2013)
14. Kenneth, W.C., Yousef, B.: NGA-West2 ground motion model for the average horizontal components of PGA, PGV, and 5% damped linear acceleration response spectra. Earthq. Spectra **17**, 3615–3635 (2014)
15. Yousef, B., Kenneth, W.C.: Vertical ground motion model for PGA, PGV, and linear response spectra using the NGA-West2 database. Earthq. Spectra **32**(2), 979–1004 (2016)
16. David, M.B., Jonathan, P.S., Emel, S., Gail, M.A.: NGA-West2 equations for predicting PGA, PGV, and 5% damped PSA for shallow crustal earthquakes. Earthq. Spectra **30**(3), 1057–1085 (2014)
17. Jonathan, P.S., David, M.B., Emel, S., Gail, M.A.: NGA-West2 equations for predicting vertical-component PGA, PGV, and 5%-damped PSA from shallow crustal earthquakes. Earthq. Spectra **32**(2), 1005–1031 (2016)
18. Ronnie, K., Norman, A.A., Walter, J.S.: Nonlinear horizontal site response for the NGA-West2 project. Earthq. Spectra **30**, 1223–1240 (2014)
19. Zeynep, G., Ronnie, K., Norman, A.A., Walter, J.S.: Ground motion prediction equations for the vertical ground motion component based on the NGA-W2 database. Earthq. Spectra **33**(2), 499–528 (2017)
20. Norman, A.A, Walter, J.S., Ronnie, K.: Update of the AS08 ground-motion prediction equations based on the NGA-West2 data set. PEER Report 2013/04. Pacific Earthquake Engineering Research Center, University of California, Berkeley (2013)
21. Ken, A., DeWayne, D., Teri, P.: Model selection for ecologists: the worldviews of AIC and BIC. Ecology **95**(3), 631–636 (2014)

Hydrophobicity Assessment of Sands Treated with Tung Oil

Xin Xing and Sérgio D. N. Lourenço(✉)

Department of Civil Engineering, The University of Hong Kong, Hong Kong SAR, China
lourenco@hku.hk

Abstract. Hydrophobic (or water repellent) soils have been reported worldwide. Soil hydrophobicity shows significant effects on water infiltration, water movement, evaporation, and soil strength. The unique properties of hydrophobic soils have been recognized by geotechnical engineers. They have been considered as novel construction materials for ground infrastructure. Synthetic hydrophobic soil was then proposed as it can be more economic, sustainable, and durable. Artificially inducing soil hydrophobicity can be obtained by using hydrophobizing agents, e.g. silane compounds and fatty acid. This paper presents a study of synthetic hydrophobic sand treated with Tung oil, a natural vegetation oil. The soil hydrophobicity was assessed through varied methods (contact angles, water drop penetration time and water entry pressure). Relationships between different hydrophobicity indicators and the amount of the applied Tung oil were investigated. The most efficient concentration for applying Tung oil to sand was proposed.

Keywords: Hydrophobicity · Geosynthetics · Water entry pressure

1 Introduction

Hydrophobic (or water repellent) soils have been reported worldwide for the past century [1]. Natural soil hydrophobicity originates from attachments (e.g. organic polymers and waxes) on soil particles [2]. Previous studies have revealed that soil hydrophobicity has significant effects on water infiltration, water movement, evaporation, and soil strength [3–7]. It has been reported that even a small percentage of hydrophobic sand particles in wettable soil has the ability to affect water absorption and infiltration [8, 9].

The importance of hydrophobic soils is recognized in the geotechnical engineering field. It has been studied as a promising material applied in ground infrastructure, e.g. hydraulic barriers for embankments, protection surfaces for horse racing tracks, and landfill cover [10–12]. Synthetic hydrophobic soil, which can be economic and sustainable, then becomes an alternative as a construction material [13]. Soil hydrophobicity can be artificially induced by using hydrophobizing agent, e.g. polydimethylsiloxane, stearic acid, wax, and Tung oil. Tung oil is a natural vegetation oil, being low cost, clean, and durable. It has been considered a waterproofing material and healing material

W. Guo and K. Qian (Eds.): GBCESC 2022, LNCE 211, pp. 1084–1090, 2023.
https://doi.org/10.1007/978-981-19-5217-3_111

applied in wood protection and reinforced concrete [14, 15]. The potential use of Tung oil in soils has also been studied recently [13].

The two commonly used methods for quantifying soil hydrophobicity are measuring contact angle (CA) and water drop penetration time (WDPT). When a droplet is placed on a flat surface, the contact lines among the three phases (i.e. solid, liquid, vapor) formed the CA. It depicts the interaction of the interfacial energies. On flat solid surfaces, the relationship among these energies can be described using Young's equation (Eq. 1).

$$\cos\theta_Y = (\gamma_{sv} - \gamma_{sl})/\gamma_{lv} \tag{1}$$

where θ_Y is Young's contact angle, γ represents the interfacial energies between different phases. If CA > 90°, the surface is hydrophobic (low wettability). While if the CA < 90°, the surface is hydrophilic (high wettability) [16]. WDPT is defined as the time required for a drop of water to fully infiltrate soil. It quantifies the persistence of soil hydrophobicity.

Besides CA and WDPT, water entry pressure (WEP), although less documented, is also an important indicator to quantify soil hydrophobicity [17]. WEP is defined as the pressure needed for water to break through the soil matrix. It quantifies the soil hydrophobicity at a macro scale.

This paper presents a study of synthetic hydrophobic sand treated with Tung oil. Different methods for assessing soil hydrophobicity were adopted. The relationship between the soil hydrophobicity and the amount of the applied Tung oil was investigated.

2 Materials and Methods

2.1 Materials

Xiamen ISO standard sand was used in this study. The sand is silica-based (silica proportion ≥98%) with particle sizes ranging from 63 μm to 2000 μm. The sand has been sieved into three fractions (μm) for the tests in this study: fine (63–300), medium (300–1180), and coarse (1180–2000), and denoted as F, M, and C respectively hereinafter. The particle characteristics of the sand were determined using a dynamic image analyzer, QicPic™ (Sympatec GmbH, Clausthal-Zellerfeld, Germany). The maximum and minimum dry density of the sand in different fractions was tested following the procedures in [18].

The ISO sand was rendered hydrophobic by mixing with natural Tung oil which was transparent with an amber color [13]. The amount (concentration) of Tung oil applied to sand was determined using mass ratio (the mass of Tung oil to the mass of sand). Concentrations used in this study ranged from 0.01% to 0.5%. According to [13], the hydrophobicity of Tung oil-treated soil can be enhanced (higher CA) and exhibit stronger persistence (longer WDPT) through heating. The samples used in this study were oven heated under 200 °C for 2 h immediately after being mixed with Tung oil.

2.2 Hydrophobicity Assessment

The sessile drop method was adopted in this study to measure CA. Following procedures in [19], a double-sided adhesive tape was adhered to a laboratory microscope glass slide to help fix the sand particles to a monolayer structure. A drop of 10 μl of deionized water was dispensed from a micro-syringe and put on the soil surface. The videos of the droplet touching the particles were recorded. The CA was determined with image processing techniques aided by ImageJ software. The CA was determined from the initial contact moment of droplet and sand particles for this study. The WDPT was measured by placing 6 drops (65 μl for each) on the samples and documenting the time for complete entry of the drops into the sand.

The WEP was measured using the water ponding method [17]. The sand was filled in a transparent acrylic tube with a moisture sensor inserted from the bottom. The sand was compacted to a relatively dense state (relative density = 0.8). Water was added atop the sand column at a constant flow rate until infiltration occurred when a certain water head was reached. A data logger, connected to the moisture sensor, reported the breakthrough moment. The water head at the onset of infiltration was recorded for calculating the WEP. The gravitational acceleration is 9.8 m/s^2.

3 Results and Discussion

3.1 Particle Size Distribution

As the soil aggregation effect caused by the addition of Tung oil has been addressed in the previous studies, and the change of particle size may lead to a significant change in WEP, it is important to compare the particle size change before hydrophobicity assessment. The particle size distributions of sand samples treated with different amounts of Tung oil are shown in Fig. 1. It can be found that the change of particle size with the concentration

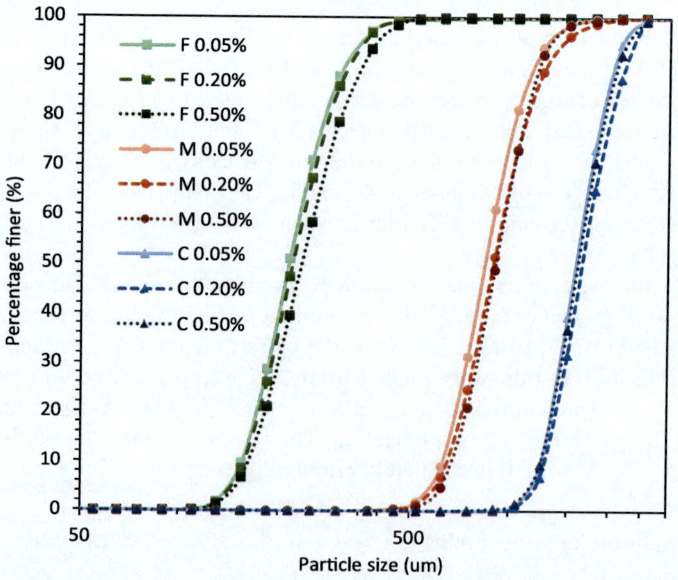

Fig. 1. Particle size distributions of sand samples treated with different amounts of Tung oil. F, M, and C represent fine, medium, and coarse-grained sand respectively, followed by the concentration of Tung oil applied.

of additional Tung oil is minor. Only fine sand exhibited a slight increase in particle size. For medium sand and especially coarse sand, the curves almost overlap.

3.2 Contact Angle

The relationship between CA and Tung oil concentration is illustrated in Fig. 2. A similar tendency was observed for the sand of different particle sizes: CA increased with the concentration of the applied Tung oil at first, then reached the peak when 0.05% Tung oil was applied. For higher concentrations of Tung oil (>0.05%), the CA decreased. The values of the CAs for different particle sizes did not differ much. The peak values of the CA for fine, medium, and coarse-grained sand are 97.2°, 99.3°, and 95.2° respectively.

The decrease of the CA when a relatively higher concentration of Tung oil was applied can be explained by the Tung oil molecular structure. Tung oil, like other fatty acids, contains chains with two ends: a hydrophobic end ($-CH_3$) and a hydrophilic end (-COOH or -COO-). When a small amount of Tung oil was mixed with the sand, the hydrophilic ends adhered to the sand particle surface and formed hydrophobic coating (with hydrophobic ends oriented outward). CA increased with the Tung oil concentration. The CA reached a peak when the coating is optimum, as a monolayer of the chains adhered to the particle surface. With more Tung oil applied, some hydrophilic ends are exposed outward the particle surface, and the CA started to decrease. This mechanism has been proposed elsewhere [20]. The major soil property that controls the concentration corresponding to the peak CA is the specific area of soil grains. A higher concentration is usually required to reach the peak CA for soils with a larger specific area [21].

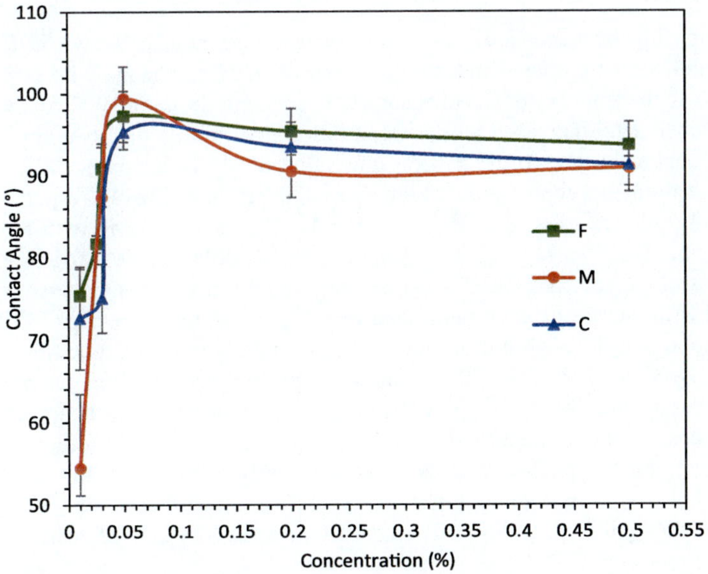

Fig. 2. The relationship between contact angle (CA) and concentration of Tung oil. F, M, and C represent fine, medium, and coarse-grained sand respectively

3.3 Water Drop Penetration Time

The results of WDPT tests are summarized in Table 1. The time required for water drops completely penetrate sand varied for different particle sizes and concentrations of Tung oil. However, some of the WDPT for the sand of different particle sizes lies in a certain range, and a pattern can be addressed here. Generally, the WDPT increases with the concentration of Tung oil in this study. For all the sand samples, despite different particle sizes, 0.01% addition of Tung oil rendered WDPT within 2 s to 5 s. 0.03% addition of Tung oil rendered WDPT within 90 s to 590 s. While all the WDPT was above 3600 s if treated with more than 0.05% Tung oil (0.05%–0.5% in this study).

Table 1. The summary of the water drop penetration time (WDPT) test results (for all the three particle size ranges (μm): 63–300, 300–1180, and 1180–2000).

Concentration (%)	0.01	0.03	0.05–0.50
WDPT (s)	2–5	90–590	>3600
Persistence rating	None	Moderate	Extreme

According to [22], the persistence of the hydrophobicity can be rated and described as none (wettable), moderate, and extreme for sand treated with 0.01%, 0.03%, and 0.05%–0.50% respectively.

3.4 Water Entry Pressure

The relationship between WEP and Tung oil concentration is shown in Fig. 3. For medium and coarse-grained sand, the tendency of WEP to change with concentration was similar to the tendency of CA to change with concentration. The WEP of medium and coarse-grained sand increased with concentration first, until a peak value was reached at 0.05% concentration. Above 0.05%, the value of WEP decreased. The maximum WEP for medium and coarse-grained sand was 0.38 kPa and 0.22 kPa respectively when 0.05% Tung oil was used. For fine sand, the WEP kept increasing with the Tung oil concentration, from 0.59 kPa (0.03% Tung oil) to 1.14 kPa (0.50% Tung oil).

The correlation between WEP and concentration for medium and coarse sand can be explained with the mechanism mentioned in 3.2. For fine sand, the peak WEP was not reached when 0.05% Tung oil was applied. The concentration corresponding to the peak WEP was within 0.20% and 0.50%. This observation agrees with the previous finding that higher concentrations are usually required to reach the peak CA for soils with larger specific areas, as fine-grained sand has a larger specific area than the other two.

However, the CA of fine sand had reached a peak at the concentration of 0.05%. This might be due to the aggregation in fine sand (Fig. 1), with the extent of aggregation having a more significant effect in CA measurement but less in WEP tests.

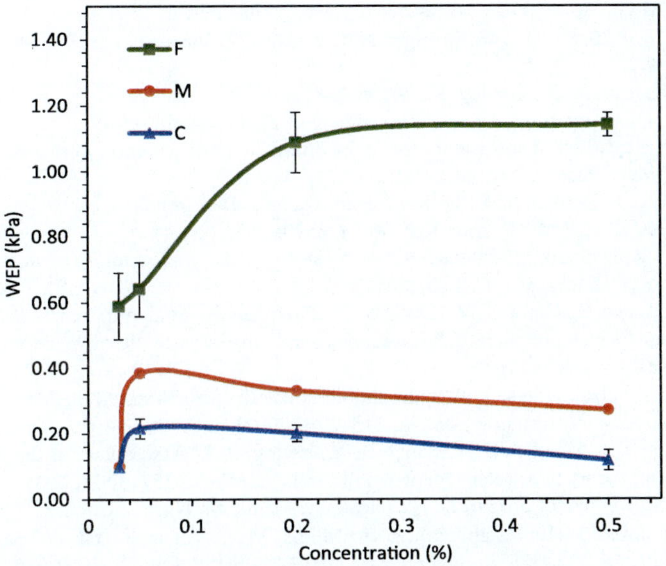

Fig. 3. The relationship between water entry pressure (WEP) and concentration of Tung oil. F, M, and C represent fine, medium, and coarse-grained sand respectively

4 Conclusion

This paper presents a study of synthetic hydrophobic sand treated with Tung oil. Xiamen ISO sand, pre-sieved to three particle size ranges, was mixed with Tung oil (concentration range: 0.01%–0.50%) and heated to render hydrophobic. The soil hydrophobicity was assessed with CA, WDPT, and WEP. Based on the results from WDPT measurement, the persistence of the hydrophobicity can be rated as none (wettable), moderate, and extreme for all sand treated with 0.01%, 0.03%, and above 0.05% respectively. CA and WEP increased with Tung oil concentration until the peak values were reached. The Tung oil concentration corresponding to the peak was 0.05% for medium and coarse-grained sand and might be higher for fine-grained sand (around 0.50%). Overall, 0.05% Tung oil is suggested to use for medium and coarse-grained sand to achieve maximum efficiency. For fine-grained sand, a larger amount of Tung oil is required (concentration around 0.50%).

Acknowledgement. This work was supported by the Collaborative Research Fund from the Research Grants Council Hong Kong (C6006-20GF). The help with the experimental work from Mr. Ho Yan Chun Michael is much appreciated.

References

1. De Bano, L.: Water repellency in soils: a historical overview. J. Hydrol. **231**, 4–32 (2000)

2. Wallis, M.G., Horne, D.J.: Soil water repellency. In: Stewart, B.A. (eds.) Advances in Soil Science, vol. 20, pp. 91–146. Springer, New York(1992). https://doi.org/10.1007/978-1-4612-2930-8_2
3. Fink, D., Frasier, G., Cooley, K.: Water harvesting by wax-treated soil surfaces: progress, problems, and potential. Agric. Water Manag. **3**(2), 125–134 (1980)
4. Wallach, R.: Effect of soil water repellency on moisture distribution from a subsurface point source. Water Resour. Res. **46**(8) (2010)
5. Ritsema, C.J., Dekker, L.W.: How water moves in a water repellent sandy soil: 2. Dynamics of fingered flow. Water Resour. Res. **30**(9), 2519–2531 (1994)
6. Bardet, J.P., Jesmani, M., Jabbari, N.: Permeability and compressibility of wax-coated sands. Géotechnique **64**(5), 341–350 (2014)
7. Hall, D., Jones, H., Crabtree, W., Daniels, T.: Claying and deep ripping can increase crop yields and profits on water repellent sands with marginal fertility in southern Western Australia. Soil Res. **48**(2), 178–187 (2010)
8. Bauters, T., et al.: Physics of hydrophobic soils. In: Soil Water Repellency: Occurrence, Consequences and Amelioration, pp. 215–224 (2003)
9. Steenhuis, T.S., Hunt, A.G., Parlange, J.-Y., Ewing, R.P.: Assessment of the application of percolation theory to a water repellent soil. Soil Res. **43**(3), 357–360 (2005)
10. Zheng, S., Lourenço, Sérgio. D. N., Cleall, P., Millis, S., Ng, A., Chui, T.: Synthetic water repellent soils for slope stabilization. In: Mikoš, M., Arbanas, Ž, Yin, Y., Sassa, K. (eds.) WLF 2017, pp. 523–528. Springer, Cham (2017). https://doi.org/10.1007/978-3-319-53487-9_61
11. Zheng, S., Xing, X., Lourenço, S.D., Cleall, P.J.: Cover systems with synthetic water-repellent soils. Vadose Zone J. **20**, e20093 (2020)
12. Choi, Y., Choo, H., Yun, T., Lee, C., Lee, W.: Engineering characteristics of chemically treated water-repellent kaolin. Materials **9**(12), 978 (2016)
13. Lin, H., et al.: Imparting water repellency in completely decomposed granite with Tung oil. J. Clean. Prod. **230**, 1316–1328 (2019)
14. Samadzadeh, M., Boura, S.H., Peikari, M., Ashrafi, A., Kasiriha, M.: Tung oil: an autonomous repairing agent for self-healing epoxy coatings. Prog. Org. Coat. **70**(4), 383–387 (2011)
15. Chen, Y., et al.: Self-healing coatings for steel-reinforced concrete. ACS Sustain. Chem. Eng. **5**(5), 3955–3962 (2017)
16. Goebel, M.O., Bachmann, J., Reichstein, M., Janssens, I.A., Guggenberger, G.: Soil water repellency and its implications for organic matter decomposition–is there a link to extreme climatic events? Glob. Change Biol. **17**(8), 2640–2656 (2011)
17. Wang, Z., Wu, L., Wu, Q.: Water-entry value as an alternative indicator of soil water-repellency and wettability. J. Hydrol. **231**, 76–83 (2000)
18. British Standard, BS 1377-4-1990: Methods of test for Soils for civil engineering purposes-Part 4: Compaction-related tests. British Standard Institute, London, UK (1990)
19. Bachmann, J., Woche, S., Goebel, M.O., Kirkham, M., Horton, R.: Extended methodology for determining wetting properties of porous media. Water Resour. Res. **39**(12) (2003)
20. Wijewardana, N.S., Kawamoto, K., Moldrup, P., Komatsu, T., Kurukulasuriya, L.C., Priyankara, N.H.: Characterization of water repellency for hydrophobized grains with different geometries and sizes. Environ. Earth Sci. **74**(7), 5525–5539 (2015). https://doi.org/10.1007/s12665-015-4565-6
21. Lin, H., Yan, C., Lourenço, S.: Fatty acids as hydrophobizing agents for granular materials. In: 16th Asian Regional Conference on Soil Mechanics and Geotechnical Engineering (16ARC) (2019)
22. Doerr, S., Shakesby, R., Dekker, L., Ritsema, C.: Occurrence, prediction and hydrological effects of water repellency amongst major soil and land-use types in a humid temperate climate. Eur. J. Soil Sci. **57**(5), 741–754 (2006)

Study on Essential Maintenance of Long-Span Bridge in Railway

Xin Jiang[1,2], Zhouyu Zhang[1], Peng Wang[3](✉), Xinlin Ban[2,4], Min Huang[1], and Jichao Guo[1]

[1] Wuhan Institute of Technology, Wuhan 430074, Hubei, China
[2] State Key Laboratory for Track Technology of High-Speed Railway, China Academy of Railway Sciences, Beijing 100081, China
[3] China Railway 11 Bureau Group Co., Ltd., Wuhan 430061, Hubei, China
xjiang11@wit.edu.cn
[4] Railway Engineering Research Institute, China Academy of Railway Sciences Co., Ltd., Beijing 100081, China

Abstract. Bridge is one of the most critical structures on the railway. Steel structure is widely applied in the long-span bridges crossing natural obstacles such as canyons, rivers, and lakes. The focus of the long-span bridge is gradually diverted from design to maintenance. The points of maintenance of a long-span railway bridge in operation are discussed in this paper. The basic principles of maintenance are proposed. In this paper, spherical bearings, suspenders, and bolted connections are defined as key components for long-span railway steel bridges. The inspection strategies and typical diseases of spherical bearings are introduced. The testing methods and contents of different types of suspenders are listed. Numerical control constant torque electric wrenches are recommended for inspecting high-strength bolts, and the delayed fracture factors of high-strength bolts are discussed. The essential maintenance in this study is recommended for long-span bridges in railway.

Keywords: Structure safety · Long-span bridge · Railway · Inspection and maintenance

1 Introduction

Bridge is one of the most critical structures on the high-speed railway (HSR). By the end of 2020, China's railway mileage has reached 146000 km, including 38000 km of HSR [1]. As shown in Fig. 1, China's railway mileage has increased at a rate of about 4% each year for the last five years. With the continuous construction of China's HSR, a large number of bridges have been built. In addition to the standard span concrete beam bridge which is dominant in number, special long-span steel bridges such as arch bridges or cable-stayed bridges are often used when the line crosses natural cuttings such as canyons, rivers and lakes.

Most long-span steel bridges of China's railway have been in service over ten years. But only a few long-span railway bridges are equipped with health monitoring systems.

W. Guo and K. Qian (Eds.): GBCESC 2022, LNCE 211, pp. 1091–1098, 2023.
https://doi.org/10.1007/978-981-19-5217-3_112

Nowadays, manual inspection is one of the main ways of bridge maintenance. When bridge inspectors have inspected HSR bridges for damage, they are in accordance with "Manual of maintenance and repair of bridges and tunnels on High Speed Railway" [2]. The specification is more appropriate for conventional simply supported girders with spans of 24 m and 32 m.

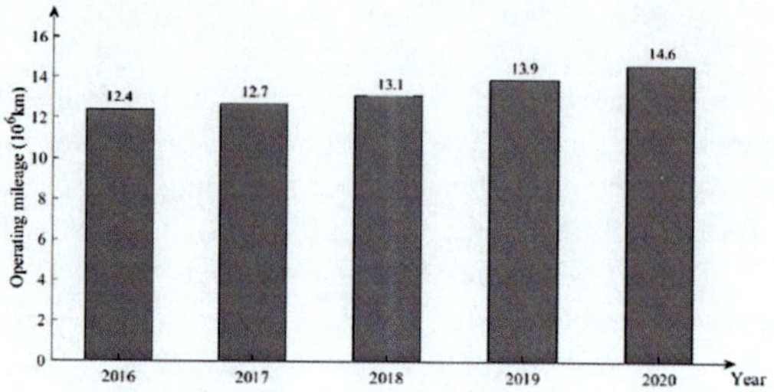

Fig. 1. Railway operating mileage in China from 2016 to 2020

Many scholars studied the maintenance evaluation of long-span bridges. Huang [3] proposed a series of bridge condition assessment and health management methods based on fuzzy theory and the analytic hierarchy process. Yang [4] used prognostics and health management (PHM) technology to monitor HSR bridges and realize structure disease diagnosis and prediction. Jiang [5] applied PHM technology to HSR bridge monitoring, combining health monitoring with intelligent manual inspection to create a data analysis and decision-making system based on the building information model (BIM). Many researchers focused on the evaluation of the service state of the bridge, but paid little attention to the evaluation of the key members of the bridge.

As for the durability of long-span steel bridges, corrosion and fatigue are the main two effects. The behavior of long-span bridges is greatly influenced by train running with high-speed. Other influencing factors refer to the environmental factors such as wind, temperature, humidity, and so on. Field investigation reveals local structural defects such as bearing disease [6], unclear state of suspender [7] and occasional bolt fracture [8] occur in long-span steel bridges. These damages may significantly weaken the durability. Therefore, timely maintenance is required.

In this paper, it is discussed how to maintain the key members of long-span bridges in service. The basic principles of maintenance are proposed. Spherical bearings, suspenders and bolted connections are the characteristic components of long-span steel bridges. The inspection strategies and typical diseases of spherical bearings are introduced. The testing methods and contents of different types of suspenders are listed. For the inspection of high-strength bolts, numerical control constant torque electric wrenches are recommended. And the delayed fracture factors of high-strength bolts are discussed. The maintenance studies are useful for inspecting and repairing long-span steel bridges.

2 Essential Maintenance of Long-Span Bridges

The key members of the long-span steel bridges are concerned. The maintenance work of concrete bridges is concerned with the cracks in the structure, while the maintenance work of steel bridges is focused on the corrosion and fatigue damage of steel structures. There is sufficient experience in the maintenance of corrosion and fatigue diseases of steel bridges. For long-span steel bridges, more attention is paid to the maintenance of key members. In this paper, the inspection and repair of spherical bearings, suspenders and high-strength bolts are analyzed and discussed.

The following are the basic principles for maintaining long-span steel bridges: (1) Long-span bridges must maintain structural safety throughout their service life. (2) Bridges must be maintained in the most cost-effective way over the long term. (3) As the bridge is irreplaceable, the maintenance work that interrupts the line must be avoided.

2.1 Maintenance of Spherical Bearings

Spherical bearing is a kind of bridge bearing developed on the basis of the basin rubber bearing and steel bearing. The bearing is composed of the bedplate, spherical crown lining board, lower bedplate, plane antifriction plate, spherical antifriction plate and so on. It has the characteristics of reliable and stable performance. Spherical bearings, basin bearings and other steel bearings account for more than 90% of the long-span bridge bearings in China [6]. The spherical bearing structure is shown in Fig. 2. As a significant member transferring loads from the superstructure to the pier or the abutment, the health status of the spherical bearing shall be controlled in good condition. However, it is rare to find appropriate suggestion on inspection and repair.

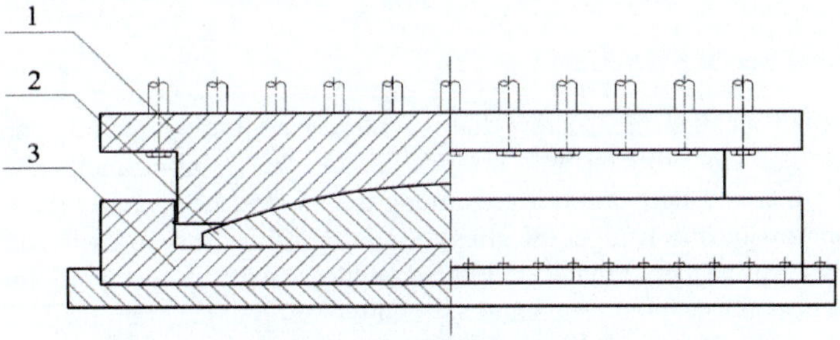

1. bedplate; 2. spherical crown lining board; 3. lower bedplate

Fig. 2. Spherical bearing

Diseases of Spherical Bearings. The main diseases of steel bearings include corrosion, deformation and wear of bedplate, and aging. Similar to ordinary steel structures, the main reason for the corrosion of spherical bearings is related to the environment. As shown in Fig. 3, corrosion develops rapidly once the protective coating falls off. And the bearing disengagement caused by the deformation of the bearing is also a common disease. The differential settlement of the bridge structure is the primary cause of bearing deformation and disengagement. The liquefaction and leakage of bearing rubber sealing ring is also a typical disease. The disease of rubber ring is related to the aging of rubber caused by temperature and long-term pressure.

Inspection of Spherical Bearing. Bearing inspection includes regular inspection and special inspection. Regular inspection refers to bearing appearance inspection, including steel cracks and bearing deformation, corrosion, structural wear and temperature. The frequency of regular inspection is once a month. And the special inspection is based on the results of regular inspection. If a damaged bearing is detected, a special inspection will be conducted. Some inspection suggestions are as follows. (1) The bearing should be repaired as soon as possible when there are cracks or abnormal deformation. (2) Bearing corrosion is a common problem that should be removed as soon as it is found. (3) Bedplate wear and displacement angle should be checked or monitored regularly. (4) The movable bearings need to be checked whether its displacement direction is consistent with the temperature change.

In recent years, a series of new monitoring technologies are applied to steel bearings. Dynamic measurement methods are widely used and include the use of strain sensors, laser displacement sensors, and angle sensors. There is also a quasi-static method for monitoring the displacement of bearings. This method refers to the video monitoring of preset marks on spherical bearings. According to the comparison of the preset marks, the position change of the marks relative to the monitoring point is obtained, and then the displacement change of the spherical bearing is displayed. In practical application, two-dimensional codes are usually used as preset markers. The maintenance personnel may also scan the two-dimensional code to get the production and maintenance records of the bearing. This method automatically monitors the appearance damage and displacement information of the bearing.

2.2 Maintenance of Suspenders

Suspender is a special steel structure that is widely used in long-span tied arch bridges or concrete-filled steel tubular arch bridges. Taking the Beijing-Shanghai HSR as an example, there are 11 tied arch bridges among the 15 steel bridges. As the force transmission component of a bridge, the stress condition of the suspender is complex. The suspender is sensitive to external load, which produces vortex-induced vibration and galloping vibration at low wind speed and vertical vibration under train load. The internal force of the suspender caused by the load is constantly changing, resulting in material fatigue. And there is also the corrosion problem of steel components. If the diseases are not inspected in time, it may adversely impact the safety and reliability of the structure. The HSR bridges are in good condition because the HSR has been well-maintained. But suspender diseases of common tied arch bridges do occur on occasion. For example, the suspender fracture accidents of Xiaonanmen bridge in Sichuan [9] and Qingchuan bridge in Wuhan [10]. As shown in Fig. 4, fatigue failure occurred on short suspender of Egongyan railway transit bridge in Chongqing [11]. These typical diseases are referred to in the maintenance of HSR long-span bridges.

Diseases of Suspenders. The suspender belongs to the key force transmission component in the bridge structure. The common diseases of suspenders include corrosion, fatigue cracking and hydrogen embrittlement fracture. The main cause of suspender corrosion is the failure of protective measures. Under the excitation of wind and rain in the natural environment, the suspenders show obvious vibration. Vibration may increase suspender tension and aggravate suspender fatigue failure. To date, the inspection of the suspender is mostly concerned with appearance, and there is a lack of inspection guidance for the internal force and vibration response of the suspender.

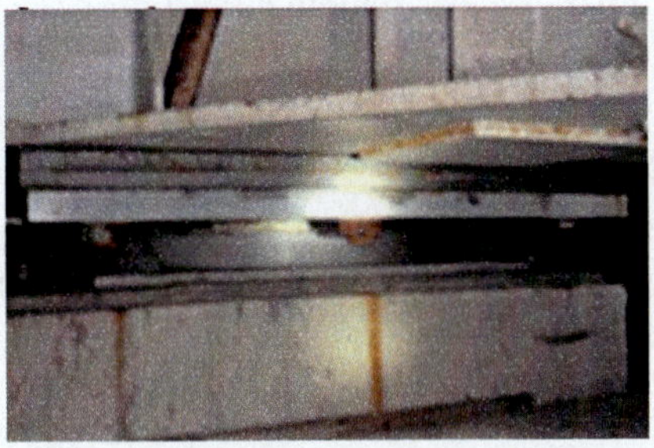

Fig. 3. Corrosion of spherical bearing

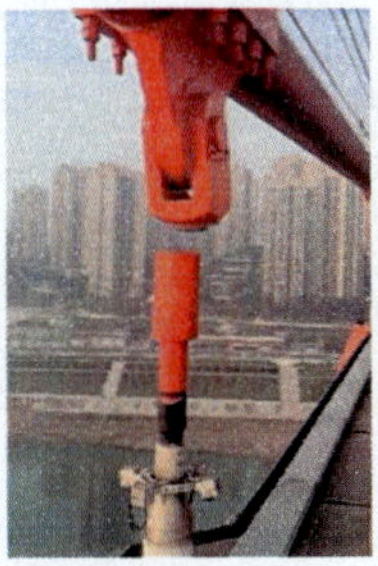

Fig. 4. Fatigue failure of short suspender [11]

Inspection of Suspenders. In suspender maintenance, it is more important to find diseases. Disease inspection includes appearance detection, long suspender vibration detection, short suspender fatigue detection, and suspender stress detection. The specific inspection measures are as follows.

(1) Suspenders appearance inspection is similar to steel structures appearance inspection, such as inspection of painting, cracks, bolts and so on.
(2) The vibration sensor is used to observe the longitudinal, transverse, and vertical vibration of the long suspenders. Among them, the measuring point of longitudinal and transverse vibration should be close to the middle of the suspender. The vertical vibration inspection should be at the end of the suspender near the bridge deck. For the excessive cable vibration found in the inspection, the force monitoring shall be carried out and the vibration shall be controlled in time.
(3) The fatigue failure of short suspenders is due to the disease under the action of impulse stress caused by cyclic load. This is due to the gradual decrease of the bearing area caused by the formation of local defects and cracks, eventually resulting in the brittle fracture. Compared with the long suspender, the change of the stress amplitude of the short suspender under the train load is much larger. Attention should be paid to the fatigue failure of the short suspender. Therefore, the fatigue inspection of short suspenders focuses on the stress concentration of joints, welds and so on.
(4) The stress state of the suspender is an important basis for judging whether the bridge is in a normal working state. Check the stress state of each suspender, using the normal suspender stress as a reference value, to determine whether the suspender or other structural members are damaged.

2.3 Maintenance of Bolted Connections

Welding, riveting and bolting are the main connections in steel structures. As for the long-span steel bridges in railway, high-strength bolts are widely applied. Corrosion and fracture of bolts are the typical diseases. Especially, the latter may result in the failure of the connection, threatening the safety of the structure. To date, bolt fracture due to fatigue has been analyzed while delayed fracture was seldom discussed.

Disease of Bolted Connections. The stress state of a high-strength bolt is complex. The bolts are under the action of high frequency and high amplitude dynamic cyclic load for a long time. Microcracks at the thread edges of bolts are common. In the case of rainwater intrusion, the bolts undergo a chemical reaction with the surrounding air to gradually rust or even break off. As shown in Fig. 5, the brittle fracture phenomenon is called delayed fracture, which is the hydrogen-induced embrittlement of high-strength materials caused by environment and stress. Lu [12] studied the delayed fracture of high-strength bolts. When the tensile strength of steel exceeds 1200 MPa, delayed fracture is common, especially in wet or corrosive environments. The sensitivity of delayed fracture increases as material strength increases. Besides, high-strength bolts belong to notched parts, which are more prone to cracking at the notch. Failure bolts mainly fracture at the root of the nut and thread, both of which are under stress concentration.

The causes of delayed fracture of high-strength bolts include material defects, adverse effects of the climate and environment, and high stress states [8]. (1) There is metalloid impurity in the materials, and the invasion of various elements in material smelting and processing leads to the initial defects of bolts. New defects may also occur in the manufacturing process. All of this may contribute to delayed fracture of high-strength bolts. (2) Most of the existing HSR lines in China are in monsoon climate. The environment is humid and rainy, with a large temperature difference between day and night. Such environmental and climatic conditions have a great impact on the weather resistance of high-strength bolts and they are prone to corrosion. It is one of the reasons for the fracture of high-strength bolts. (3) During the construction of high-strength bolts, surface lubricant is often used to treat bolts to reduce direct contact with metal to prevent wear. However, the surface lubricant also affects the control of bolt pretension, which is easy to cause the discretization of torque coefficient and the problem of super tighten. Excessive stress on high-strength bolts leads to bolt diseases.

Inspection of Bolted Connections. The fracture failure of steel bridge high-strength bolts leads to the degradation of joint connection performance. This may reduce the structural stiffness and cause great safety risks. Figure 6 shows the replacement of fractured bolted connection in a truss member of a railway bridge. Li [13] tried to monitor the bolt fracture by arranging fiber brag grating sensors in the bolted connections. However, this method may not accurately locate the damaged bolts. Therefore, the inspection and maintenance of steel bridge bolts in service is very necessary.

In terms of inspection equipment, new equipment such as numerical control constant torque electric wrenches and portable torque testers has been developed [14]. The wrench contains a high-precision torque sensor and an angle sensor, which automatically display and store the test results. The wrench showing output torque is suitable for the tight-fitting inspection method and has been tried out on bridge projects [15]. Pan [16] developed a bolted connection construction management system, which is composed of a numerical control constant torque electric wrench and an application system. The application of the management system realizes effective management and control of the whole process of bolt construction. The workload of manual recording and data analysis is greatly reduced, and the full element management of the bolt life cycle process is realized. The use of electric wrench improves the accuracy of inspection results, reduces the influence of human factors, and reduces the intensity of maintenance work.

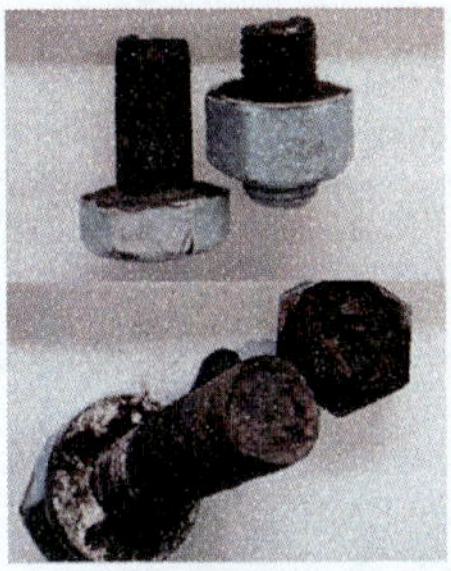

Fig. 5. Delayed fracture of bolts

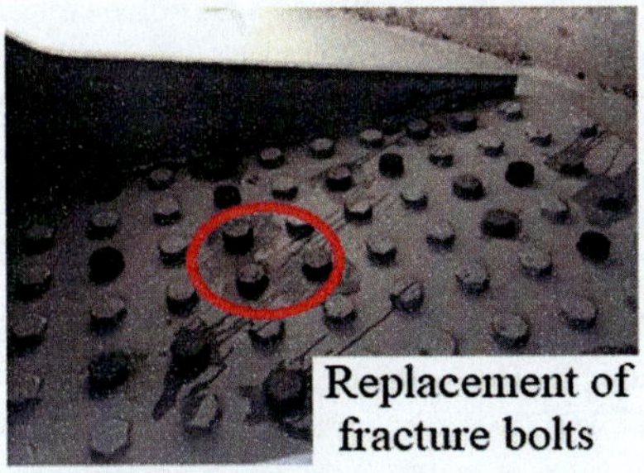

Fig. 6. Disease of bolted connections

3 Summary and Discussion

In this paper, it is discussed how to maintain the key members of long-span railway bridges in service. The basic principles of maintenance are proposed. Spherical bearings, suspenders and bolted connections are discussed as representative parts of long-span bridges. The three points listed below should be studied for better bridge maintenance.

(1) Informatization of maintenance work. A large amount of maintenance data is generated in the process of bridge maintenance. The storage of information and the uploading of maintenance data significantly help analyze the health status of bridges.
(2) Fusion of data from maintenance and monitoring. It is critical to evaluate the bridge's disease and health status in combination with data from maintenance and monitoring, which guide bridge maintenance.
(3) Portable inspection equipment. The quality of bridge maintenance depends on inspection work. Maintenance work can be performed efficiently using portable devices.

Acknowledgment. This project was supported by the foundation of State Key Laboratory for Track Technology of High-speed Railway (2020YJ216), China Academy of Railway Sciences, the science and technology research project of Department of Education of Hubei Province (B2019052), the science foundation of Wuhan Institute of Technology (K201961) and Graduate Innovative Fund of Wuhan Institute of Technology (CX2021109).

References

1. The Ministry of Transport of the People's Republic of China. Statistical Bulletin of Transport Industry Development in 2020 (2021)
2. The Ministry of Railways of the People's Republic of China. Manual of Maintenance and Repair of Bridges and Tunnels on High Speed Railway. China Railway Publishing House, Beijing (2011)
3. Qiao, H., Haihong, T., Yuan, R.: Study of assessment theory of long-span bridges based on fuzzy theory. J. Highw. Transp. Res. Dev. **27**(1), 62–66 (2010)
4. Huaizhi, Y.: Elementary discussion on application of PHM system for maintenance and repair in large bridges of high speed railway. Railw. Eng. **2017**(6), 12–16+35 (2017)
5. Xin, J., Zhao, X., Hu, S.: A PHM system for bridge monitoring on high speed railway. In: 2017 International Conference on Sensing, Diagnostics, Prognostics, and Control (SDPC), pp. 371–376 (2017)
6. Jiaming, D., Xiaoqiu, Z., Qiujun, S.: Analysis on influential factors of corrosion resistance and weatherability of railway bridge bearing and strengthening measure study. Railw. Eng. **2017**(6), 36–40 (2017)
7. Yupu, S., Hanyong, L.: Fatigue test and stress analysis of threaded connection of arch bridge stainless steel suspender. Key Eng. Mater. **985**(905), 541–544 (2011)
8. Suhua, Z.: Fracture analysis and maintenance suggestions of high strength bolts of steel truss girder of super large bridge of high-speed railway. Shanghai Railw. Sci. Technol. **156**(04), 79–80+89 (2016)
9. Jiandong, W.: Urgent reinforcement and restoration of Xiaonanmen Bridge in Yibin City. Highway **2003**(4), 34–38 (2003)
10. Jinghua, F., Lixin, C., Desheng, J.: Influence of ties broken on arch bridge health state. J. Wuhan Univ. Technol. **28**(10), 61–65 (2006)
11. Fracture failure of short suspender of Egongyan railway transit bridge. https://www.sohu.com/a/517719621_682294. Accessed 19 Jan 2022
12. Chunzhen, L.: Study on delayed fracture of high strength bolts for bridges. Master thesis of China Academy of Railway Sciences (2019)
13. Junxin, L., Zhihua, L., Xiaojun, Q.: Model test on monitoring for bolts fracture and dropping in railway steel bridge based on fiber Bragg grating sensor technology. Railw. Eng. **60**(1), 40–43 (2020)
14. Wallace, P.: Energy torque and dynamics in impact wrench tightening. J. Manuf. Sci. Eng. **137**(2), 24503 (2015)
15. Xuewen, L.: A summarized account of construction techniques for main bridge of Nanjing Dashengguan Changjiang river bridge on Beijing-Shanghai high-speed railway. Bridge Constr. **2010**(4), 1–4 (2010)
16. Yongjie, P.: Construction management system of high strength bolt for life cycle management. Railw. Eng. **60**(4), 115–119 (2020)

Load-Path Method for Design of Disturbed Regions in Concrete Structures

Jianlin Wang[1], Zhiqiang Wang[1], and Zhiqi He[2](✉)

[1] Zhejiang Hang-Shao-Yong Expressway Co., Ltd., Hangzhou 310014, China
[2] School of Civil Engineering, Southeast University, Nanjing 211189, China
z.he@seu.edu.cn

Abstract. This study develops a quantitative load path method for the shape selection of the strut-and-tie models in disturbed regions (D-regions) of structural concrete. Firstly, quantitative load paths are generated in two ways. One approach is to develop a computer-based tool to plot principal stress trajectories seen as load paths. Another approach is to give mathematic descriptions of the load paths based on their geometric boundary conditions in some typical D-regions. Then, the characteristics of the load paths are explored. With the help of the quantitative load paths, the shape selection of the strut-and-tie models can be significantly facilitated. The suitability and efficiency of the proposed methodology are demonstrated in three types of D-regions, including deep beams, bottle-shaped struts, and dapped beams.

Keywords: Structural concrete · Disturbed regions · Load paths · Strut-and-tie models

1 Introduction

Since the late 1980s, the strut-and-tie model (STM) has gradually been regarded as a rational tool for the design of disturbed regions (D-regions) in concrete structures subject to complex states of stress. The STM is a truss model consisting of struts and ties that represent the load paths in the concrete structure [1]. Despite the fact that the STM has been included in many design provisions, there is still reluctance for practitioners. Putting aside its immaturities, the STM has also been depicted by many researchers as an "art" for engineers with sound engineering judgment and design experience. Many challenges still exist in making this conceptually powerful method into a general and practical design tool.

In designing with the STM, one of the big challenges is the shape selection of the model. Experimental results validated that the STM provided a conservative design in the ultimate limit state even when unfavorable shapes were selected [2, 3]. However, a poor selection of the shape could result in ineffective structural performance during service states [4, 5]. In practice, the failure accidents of D-regions happen rarely but can be dramatic, and the cracking problems frequently occur in regions such as end anchorages, pier caps, and diaphragm walls. Therefore, a direct criterion for evaluating

W. Guo and K. Qian (Eds.): GBCESC 2022, LNCE 211, pp. 1099–1106, 2023.
https://doi.org/10.1007/978-981-19-5217-3_113

the shape selection of the STM should be focused on good structural performances, including ultimate strength and serviceability. Schlaich et al. proposed that a good STM can be developed by orienting the geometry of the model along the force paths indicated by the elastic stress field [6]. Under this criterion, both serviceability and ultimate limit states are expected to be satisfied. Two useful modeling methods were also presented by Schlaich et al. [6]. One solution is to arrange the configuration of the STM according to the direction of the principal stress trajectories. The other solution is the "load path method" if no elastic analysis is available, which was highlighted and advocated by some literature [7, 8].

Although the above two modeling methods are powerful and widely used, some defects still remain as follows: (1) both methods are somewhat sophisticated due to the requirement to do hand-based procedures [9, 10]; (2) The load path method requires some intuition and experience on how the forces flow within the D-regions, which is hard to be traced in a complex D-region; and (3) due to their hand-based nature, it is difficult to obtain a quantitatively determined solution for a given problem. To overcome these defects, this study presents two potential solutions for the shape selection of the STM: (1) a computer-based design tool to plot the load paths automatically; and (2) mathematic descriptions of the load paths in some typical D-regions. These two methods feature to provide a quantitative approach to the design of structural concrete.

2 Automatic Generation of Load Paths

As mentioned previously, principal stress trajectories originating from the finite element analysis (FEA) are useful in laying out a STM. Until now, most general-purpose finite element programs have the function of doing vector plots indicating the directions and relative magnitudes of principal stresses at nodes or element centers, but they usually cannot generate continuous trajectories, resulting in the designers having to draw the stress trajectories by hand using the vector plots as a backdrop. For this reason, this study develops a tool for the automatic generation of the stress trajectories, which is based on the FEA software ANSYS and its parametric design language. Similar computer-based tools were developed by Kelly et al., Hörmann, and some others [11, 12].

Figure 1 shows how a principal compressive stress trajectory can be automatically drawn in an end anchorage. The first step is to perform the regular finite element solution to obtain the stresses at all nodes. The second step is to select a start point (P_0) of the stress trajectory. The start point P_0 is placed on the loading point. After that, the values of two-dimensional stresses (σ_x, σ_y, and τ_{xy}) at P_0 need to be determined. The third step is to calculate the direction of the principal compressive stress at P_0, which can be done using the Mohr's circle:

$$\tan 2\alpha = \frac{2\tau_{xy}}{\sigma_x - \sigma_y} \tag{1}$$

The fourth step is to find the second point (P_1). The step length is taken as s, and the location of P1 can be expressed as

$$x_1 = x_0 + s \cdot \cos \alpha \tag{2a}$$

$$y_1 = y_0 + s \cdot \sin \alpha \tag{2b}$$

Repeating the above procedures, a set of points can be located until the endpoint is outside the domain. Then, the stress trajectory can be drawn by linking these points with straight lines. Finally, a tuft of stress trajectories can be generated by setting different start points.

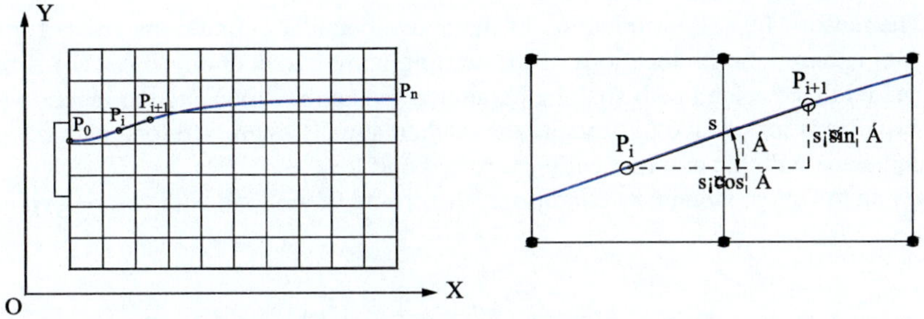

Fig. 1. Procedures for the automatic drawing of a stress trajectory.

3 Applications of Load Path Method

3.1 Deep Beams

Reinforced concrete deep beams are one of the typical D-regions, which are usually included in a design code to show the design procedures and fundamental provisions of the strut-and-tie method. For a simple deep beam subjected to a uniform load on the top, the distributed load can be simplified as two equivalent point loads. As shown in Fig. 2(a), automatically generated load paths are plotted for deep beams with different span-depth ratios. At the same time, mathematic expressions for the load paths of deep beams with two symmetrical point loads are also derived in the following.

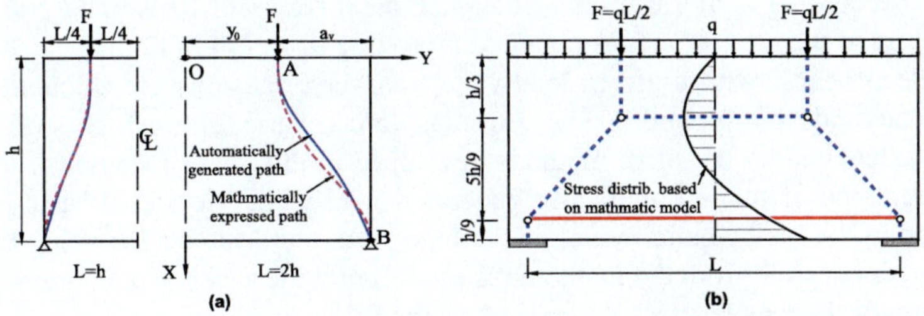

Fig. 2. Deep beams: (a) load paths; (b) strut-and-tie model.

Similarly to the anchorage zone, five boundary conditions of the load path are established as follows

$$y|_{x=0} = y_0, \quad y|_{x=h} = y_0 + a_v \tag{3a}$$

$$\left.\frac{dy}{dx}\right|_{x=0} = 0, \quad \left.\frac{dy}{dx}\right|_{x=h} = 0 \tag{3b}$$

$$\left.\frac{d^2y}{dx^2}\right|_{x=0} = 0 \tag{3c}$$

in which, y_0 is the distance between the concentrated load and the center line of the beam; and a_v is the shear span.

The reasons for the introduction of the above boundary conditions are as follows: Eq. (3a) is based on the locations of the starting point and endpoint; Eq. (3b) is based on the fact that the load path shall be parallel to the applied load and the reaction load; and Eq. (3c) is introduced to promote the mathematically expressed load path having a similar shape with the automatically generated one.

Using the above boundary conditions, the profile of the load path can be expressed by

$$y = \frac{-3a_v}{h^4}x^4 + \frac{4a_v}{h^3}x^3 + y_0 \tag{4}$$

The transverse stresses along the middle section can be expressed as

$$\sigma_{trans} = -\frac{F}{t}\frac{d^2y}{dx^2} = 12a_v x\frac{3x - 2h}{h^4 t}F \tag{5}$$

Finally, the transverse tension can be obtained as

$$T = \int_{2h/3}^{h} \sigma_{trans} t dx = \frac{16a_v}{9h}F \tag{6}$$

As shown in Fig. 2(a), the mathematically expressed load paths almost coincide with the automatically generated ones. According to the transverse stress distribution based on Eq. (5), the locations of the horizontal strut and tie can be determined, as shown in Fig. 2(b). According to this strut-and-tie model, the force in the tie is $9FL/(20h)$, which is very close to the term $4FL/(9h)$ obtained by setting $a_v = L/4$ in Eq. (6).

For deep beams with uniform loads, the transverse tensions are calculated using different methods and plotted in Fig. 3 to make a comparison. It can be seen that the tensions calculated by the three methods developed in this study (Method I: using the maximum slope of the automatically generated load path; Method II: using Eq. (6) derived from the mathematic model of the load path; and Method III: using the strut-and-tie model derived from the mathematic model) provide acceptable agreements with the integral of the tensile stresses obtained by the FEA.

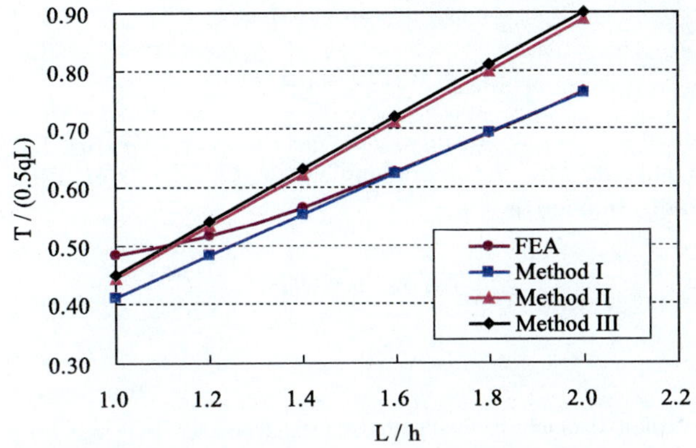

Fig. 3. Transverse tensions in deep beams with uniform loads.

3.2 Bottle-Shaped Struts

A bottle-shaped strut is a typical strut in D-regions [13]. For a square concrete panel (aspect ratio = 1) shown in Fig. 4(a), the automatically generated stress trajectories clearly show the bottle-shaped profile of the compression dispersion. Two load paths of the panel are also plotted in Fig. 4(b). To maintain equilibrium in the curved load paths, transverse compression and tension are developed. The transverse tension may cause brittle splitting failure to the bottle-shaped strut. Thus, web reinforcement should be placed to resist it [14]. As shown in Fig. 4(c), a bottle-shaped strut can be modeled by a strut-and-tie model to reflect the transverse tension [15]. The dispersion angle (α) or the deviation distance (*e*) of the load path is a critical parameter in constructing the model. As listed in Table 1, using the computer-based tool developed in this study, a parametric analysis is conducted to determine the deviation distances of the load paths in square panels with different bearing plate widths.

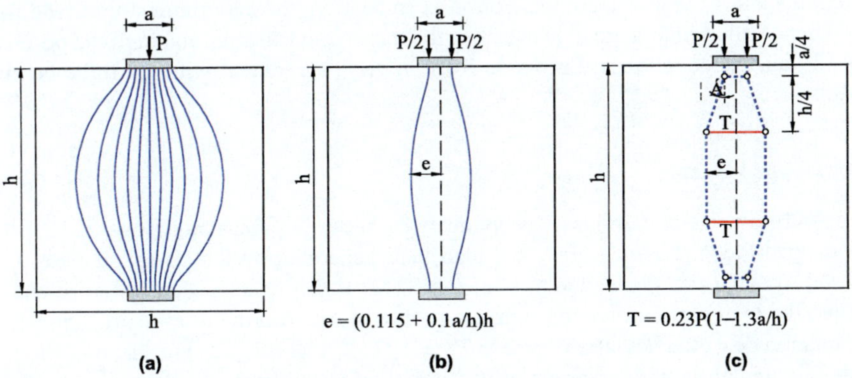

Fig. 4. Bottle-shaped strut: (a) stress trajectories; (b) load paths; (c) strut-and-tie model.

The fitted deviation distance can be expressed as

$$e = (0.115 + 0.1a/h)h \tag{7}$$

The horizontal ties can be placed at a distance of $(a + h)/4$ from the bearing plates. Therefore, the tangent of the dispersion angle is $\tan\alpha = (4e - a)/h$. Finally, the force in a single tie can be obtained as

$$T = \frac{P}{2}\tan\alpha = 0.23P\left(1 - \frac{1.3a}{h}\right) \tag{8}$$

Table 1. Deviation distances of load paths and transverse tensions in bottle-shaped struts.

a/h	e/h	The proportion of total transverse tension to the applied load ($2T/P$)			
		FEA	Max. slope of load path	STM [Eq. (8)]	Eurocode 2 [Eq. (9)]
0.1	0.125	0.438	0.422	0.400	0.430
0.2	0.135	0.343	0.336	0.340	0.360
0.3	0.145	0.271	0.255	0.281	0.290
0.4	0.155	0.213	0.209	0.221	0.220

On the other hand, Eurocode 2 (CEN 2004) also provides an equation to calculate the tensile force in a bottle-shaped strut with full discontinuity [16]. The equation in Eurocode 2 Part 1–1 can be rewritten as

$$T = 0.25P(1 - 1.4a/h) \tag{9}$$

It can be seen that Eq. (8) and Eq. (9) have very similar expressions. The proportions of total transverse tension to the applied load are calculated using different methods and listed in Table 1 to make a comparison. As can be seen, the maximum slopes read from the computer-based tool are able to reflect the transverse tensions, and both the proposed Eq. (8) and the equation in Eurocode 2 are in good agreement with the finite element results.

3.3 Dapped Beams

Dapped beams are usually used as connections between precast components [17]. For the dapped beam shown in Fig. 5, a load path generated by the computer-based tool is used to assist the shape selection of the STM. It can be seen that in the re-entrant corner, the load path has an approximate profile of one-fourth concave arc. Therefore, the magnitude of the deviation force is $T_R = \sqrt{2}\,P$. According to the shape of the load path, a diagonal tie with an angle of 45° is placed to reflect the deviation force. Other parts of the STM can be developed following the classic truss in a reinforced concrete beam. The diagonal tie follows the principal stress pattern closely, and the diagonal reinforcement provides better control of cracking than other reinforcing details which utilize vertical reinforcement [18]. However, the diagonal reinforcement is slightly more difficult to place and anchor properly, so that vertical and horizontal reinforcements are usually also placed to balance partial or total of the diagonal tension.

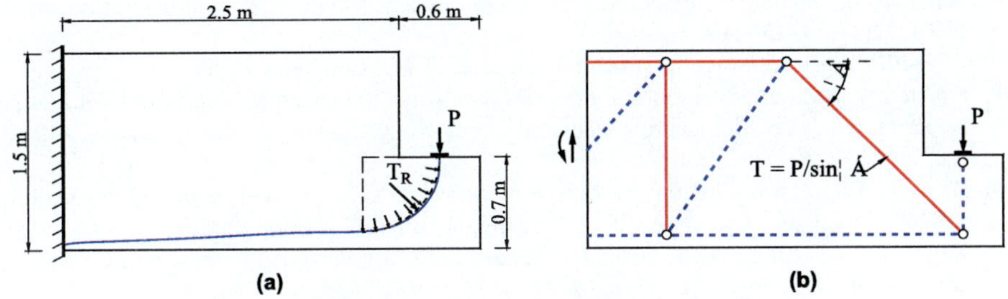

Fig. 5. Dapped beam: (a) load path; (b) strut-and-tie model.

4 Conclusions

A quantitative load path method for the shape selection of the strut-and-tie model was developed in this study. It was achieved through the automatic generation of the load path using a computer-based design tool or the mathematic description of the load path. The quantitative characteristics of the load path are useful in the STM shape selection. The proposed method was successfully applied to three types of D-regions. The main findings of this study are:

(1) The principal compressive stress trajectory starting from the applied point load must end at the support and thus can be seen as the load path.
(2) Equations for the transverse forces in bottle-shaped struts and deep beams are developed, which can be applied in the crack-control design of these D-regions.
(3) The profile of the load path in the dapped beam reveals the diagonal tension mechanism, leading to a diagonal tie in the strut-and-tie model.

Acknowledgement. The work was supported in part by NSFC project 52178116 and Transportation Science and Technology Project of Zhejiang Province, China (No. 2021048).

References

1. Liang, Q.Q., Uy, B., Steven, G.P.: Performance-based optimization for strut-tie modeling of structural concrete. J. Struct. Eng. **128**(6), 815–823 (2002)
2. Ley, M.T., Riding, K.A., Bae, S., Breen, J.E.: Experimental verification of strut-and-tie model design method. ACI Struct. J. **104**(6), 749–755 (2007)

3. Kuchma, D., Yindeesuk, S., Nagle, T., Hart, J., Lee, H.H.: Experimental validation of strut-and-tie method for complex regions. ACI Struct. J. **105**(5), 578–589 (2008)
4. Yindeesuk, S.: Design and analysis of complex D-regions in reinforced concrete structures. Ph.D. Dissertation, University of Illinois at Urbana-Champaign, Urbana, IL (2009)
5. Park, J.W., Yindeesuk, S., Tjhin, T., Kuchma, D.: Automated finite-element-based validation of structures designed by the strut-and-tie method. J. Struct. Eng. **136**(2), 203–210 (2010)
6. Schlaich, J., Schäfer, K., Jennewein, M.: Toward a consistent design of structural concrete. PCI J. **32**(3), 74–150 (1987)
7. CEB-FIP: Practical design of structural concrete. SETO, London, UK (1999)
8. Schäfer, K.: Strut-and-tie models for the design of structural concrete. Notes from Workshop, Department of Civil Engineering, National Cheng Kung University, Tainan, Taiwan, 140 pp (1999)
9. Reineck, K.H.: Examples for the design of structural concrete with strut-and-tie models. SP-208, American Concrete Institute, Farmington Hills, MI, 242 pp (2002)
10. Tjhin, T.N., Kuchma, D.A.: Computer-based tools for design by strut-and-tie method: advances and challenge. ACI Struct. J. **99**(5), 586–594 (2002)
11. Kelly, D.W., Hsu, P., Asudullah, M.: Load paths and load flow in finite element analysis. Eng. Comput. **18**(1–2), 304–313 (2001)
12. Hörmann, A.: Determination of optimized strut and tie models based on force flow as an application of platform independent remote process interfacing. Ph.D. Dissertation, Technische Universität München, Germany (2006)
13. American Concrete Institute (ACI): Building code requirements for structural concrete (ACI 318-08) and commentary (ACI 318R-08). ACI 318-08, Farmington Hills, MI (2008)
14. Sahoo, D.K., Singh, B., Bhargava, P.: Minimum reinforcement for preventing splitting failure in bottle-shaped struts. ACI Struct. J. **108**(2), 206–216 (2011)
15. Brown, M.D., Bayrak, O.: Minimum transverse reinforcement for bottle-shaped struts. ACI Struct. J. **103**(6), 813–821 (2006)
16. European Committee for Standardization (CEN): Design of concrete structures—part 1-1: general rules and rules for buildings. BS EN 1992-1-1:2004, Eurocode 2, London, UK (2004)
17. Nanni, A., Huang, P.C.: Validation of an alternative reinforcing detail for the dapped ends of prestressed double tees. PCI J. **47**(1), 38–49 (2002)
18. Mattock, A.H., Theryo, T.S.: Strength of precast prestressed concrete members with dapped ends. PCI J. **31**(5), 58–75 (1986)

A Study on the Design Strategy of Living Space Renovation Under the Background of "Aged Caring at Home"

Feng Wang[1](✉), Xiangyun Wang[1], and Jie Liu[2]

[1] Shandong Youth University of Political Science, Jinan 31699, Shandong, China
160109@sdyu.edu.cn
[2] Qilu Normal University, Wenbo Zhangqiu 2, Jinan, China

Abstract. The aging of the population has become a problem that cannot be ignored in China. Due to the current economic development in China, it is not possible to copy the aging system of western developed countries for the time being, and the "aging in place" model will become the first choice for most of the elderly in China. This paper summarizes the current situation of the elderly living space and proposes the principles of aging-friendly renovation applicable to the current development situation in China, and finally proposes the strategy of living space renovation in the context of aging in place.

Keywords: Aging · Home care · Space transformation · Design strategy

1 Research Background

According to the latest data of the 7th Population Census released by the National Bureau of Statistics of China, China's population is aging further and the number of elderly people will remain high for a long period of time, among which those aged 60 to 70 account for more than half of China's total elderly population (Fig. 1).

At present, the construction of China's elderly system is not perfect, and the capacity of elderly institutions is far from matching the current number of elderly people, and it is impossible to change the status quo of "beds are hard to find" in elderly institutions in a short time. Therefore, the majority of elderly people will adopt the form of home care. According to the relevant research, the demand for aging-appropriate renovation of living space is generally higher among the aged 60 to 70 years old due to the comprehensive factors such as knowledge, economy, skills and physical status.

W. Guo and K. Qian (Eds.): GBCESC 2022, LNCE 211, pp. 1107–1115, 2023.
https://doi.org/10.1007/978-981-19-5217-3_114

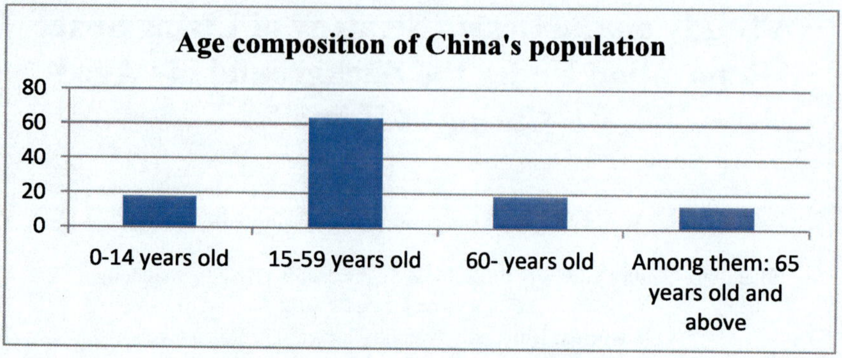

Fig. 1. Age composition of China's population.

2 The Current Situation of Living Space Designs

There are many forms of existing residences in China, and different types of building structure forms are adopted. The choice of building structure form directly affects the difficulty of renovating residences, which will eventually be reflected in the difference of renovation costs. Most of the residences built in the early period are brick and concrete structures, which belong to the wall bearing system, and the transformation cost of the original walls is very high; as the urbanization process in China accelerates, the construction technology is also updated, and the later residences mostly adopt frame structures, steel structures and other column bearing systems, and the walls are mostly non-load-bearing structures, which only provide the enclosure function, and the wall transformation cost of space is relatively low, and the difficulty of space layout transformation is relatively low.

At the same time, limited by the timeliness of the design, the elderly now live in the housing age of about 10 years, the corresponding residential interior design is mostly about 15 years ago, the corresponding space and facilities design to the characteristics of the middle-aged owners at that time as the entry point, with the owners into the aging stage, the space and equipment and facilities design is gradually unable to meet the current needs of the owners.

3 Residential Space Renovation Design Principles

3.1 Systematic Principle

The "aging-in-place" model is a systematic project. The home cannot be arbitrarily separated from the outside world, and the design of the aging-friendly indoor space cannot be separated from each other [1].

The community service is incorporated into the home aging service system, and it becomes a bridge between the home and the recreation hospital, so as to build a "hospital-community-home" linkage aging system. Professional, systematic and intelligent transformation of the home care system can effectively improve the efficiency of China's "home care" model.

Interior space design should also be centered on elderly residents and analyze the space from a systematic perspective. Traditional interior space design mostly takes the use function of a single space as the entry point for functional division and space arrangement, and gives less thought to the traffic space between spaces [2]. The elderly people have complex behavior trajectories at home, and their physical functions are declining to different degrees, so designers should systematically grasp the whole living space from a global perspective. Reconstruction has selection strategy.

With the development of China's economy, there is a relatively clear development of China's residential flat. From the earliest "tongzi building" to "unit house" and later "wide hall", "square hall" and so on. The vast majority of China's elderly population will also live in these apartments for a long time.

The change of house type has also directly affected the change of residential building structure. The mainstream structure selection has changed from the earliest brick and concrete structure to the frame structure with more flexible layout. The change of structural form directly affects the reconstruction cost of housing [3]. The bearing capacity of brick and concrete structure is mainly composed of board and wall, and the frame structure is composed of beam, board and column. The floor load of the former is transferred from the floor layer to the lower wall. Due to the load-bearing of the wall, the economic cost of the reconstruction and reinforcement of the wall is significantly higher than that of the frame structure residence if the wall part is reformed in the process of reconstruction, such as opening the lighting window hole and expanding the door hole.

In the initial stage of the renovation of suitable for aging living space, careful survey and analysis should be made to calculate the economic cost of the renovation.

3.2 Principle of Economy

With the low per capita wealth of China's elderly population today, the economics of home renovation for the aging population is also particularly important. The renovation process should ensure that the needs of the elderly are met and that unnecessary economic waste is minimized. For renovation projects that are difficult and time-consuming, comprehensive consideration should be given to avoid excessive waste of resources.

3.3 Age-Appropriate Principle

When the space is used by the elderly, the principle of aging appropriateness should be grasped at all times. The aging-friendly principle mainly includes two connotations: safety and preventive renovation and physical and mental health renovation. Safety prevention is the most important part of the aging-friendly living space renovation. As the living space is the longest space used by the elderly in their daily life, the safety of the space environment is also the most important part of the aging-friendly renovation. At the same time, the concern for physical and mental health should not be neglected. With the decline of physical functions, the psychology of the elderly group will more or less produce some changes, so the space transformation should be based on the physiological and psychological perspective of the elderly, from the large space scale relationship, space color selection, to the small space furnishing design and interface material application. For example, the net width of often interior doors should meet the size of wheelchair passage after installation, and the main traffic space should meet the size required for wheelchair turning. In the face of visual decline, we should provide a more comfortable and clear visual environment for the elderly through color adjustment and moderate increase of signs; in the face of tactile decline, we should reasonably set up leaning handrails in the residential traffic space, and the materials of handrails are usually rougher and more stimulating to the tactile senses of the elderly such as palms. The creation of auditory environment, through the opening of indoor communication windows, the removal of unnecessary partition walls and other ways to increase the sound penetration of different spaces …… reduces the psychological gap between the elderly due to the decline of physical functions caused by the reduction of control of space, and finally achieve the purpose of promoting the physical and mental pleasure of the elderly [4].

4 Design Strategy of Living Space Renovation

4.1 Home and Community Hospital Interlinking Strategies

According to the China Development Report 2020: Development Trends and Policies for China's Aging Population, released by the China Development Foundation, the report predicts that China will enter an aging society from an aging society around 2022, i.e., the number of people aged 65 and above will account for more than 14% of China's total population. This national social problem can no longer be ignored. Therefore, the elderly at home should take the initiative to dovetail with the medical and rehabilitation system, and plan at the national, city and community levels to establish a multi-level and systematic elderly care system.

As an important part of the elderly care system, we should establish system service awareness during the process of living space renovation, arrange the indoor intelligent rehabilitation system in advance before the community elderly care service is completely built, and take the initiative to connect with the community, institutions, hospitals and other elderly care service departments by means of Internet signal transmission [5].

4.2 System Layout Strategy

The space design for the elderly cannot follow the traditional space design method, which takes the use of a single space as the entry point. Older people have been at home for a long time, and their behavioral trajectories are more complicated than those of young people, and their physical functions are weakened to different degrees (Figs. 2 and 3).

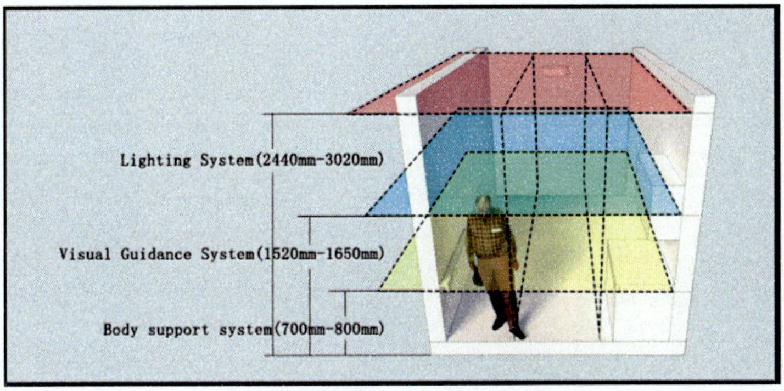

Fig. 2. System layout strategy.

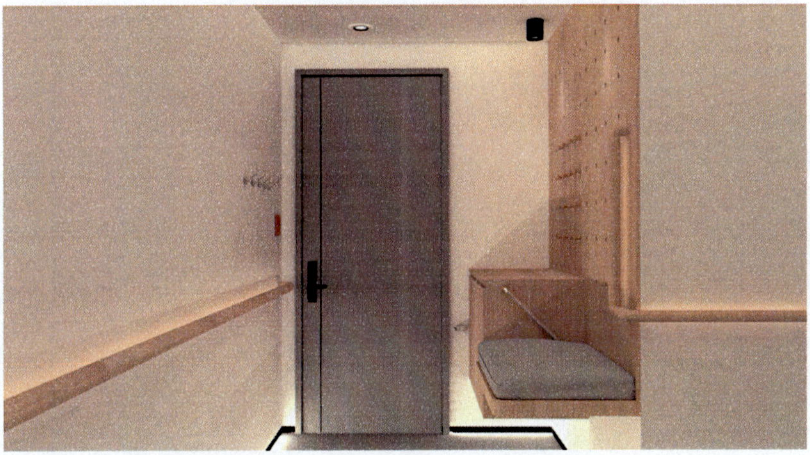

Fig. 3. Age-appropriate entrance hall

The design should take the behavioral characteristics of the elderly as the entry point and build several systems for the home space to jointly maintain the space use needs of the elderly. For example, in the "home support system", a more comfortable support height for the elderly is selected as the reference plane, and on this level, the use of open handrails, furniture platforms of appropriate height and landscape ornaments, etc. together to build a complete "support system". Another example is the lighting system, space guidance system, etc. (Fig. 4).

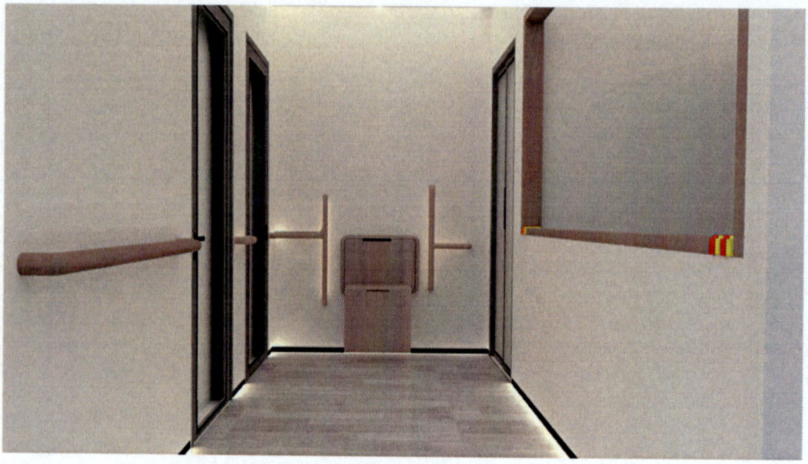

Fig. 4. Home support system

4.3 Choice Based Retrofit Strategy

With China's economic development, there is a clear lineage of development and change in China's residential house plan. From the earliest "silo" to "flat" and then to the later "wide hall" and "square hall". The vast majority of China's aging population will also live in these types of households for a long time. The change of household type also directly affects the change of residential building structure, and the mainstream structure selection has changed from the earliest brick-and-mortar structure to the more flexible frame structure. The change of structural form directly affects the cost of housing renovation. The load-bearing of brick-and-mortar structure is mainly composed of slabs and walls, while frame structure is composed of beams, slabs and columns. The floor load of the former is transferred from the floor level to the wall below, and due to the load-bearing of the wall, the economic cost of the corresponding renovation and reinforcement is obviously higher than that of the frame structure housing if there is a renovation of the wall part during the renovation process, such as the opening of light window holes and the expansion of doorways (Fig. 5).

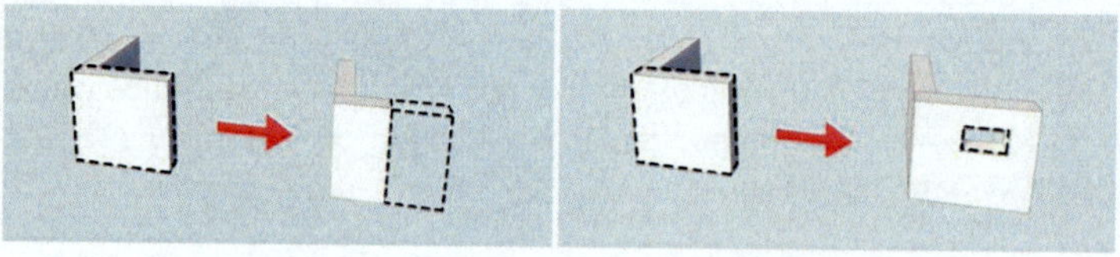

Fig. 5. Choice based retrofit strategy

At the same time, interior renovation should be carried out in a relatively low cost way under the premise of ensuring safety and environmental protection, and "small demolition and modification" should be used instead of "big demolition and modification"; low-cost decoration methods and materials should be chosen as much as possible. For example, for the sake of the line of sight through, the wall can be demolished to adopt a partial wall transformation, only in part to dig out the window hole required for the line of sight instead of the entire wall demolition; and such as the treatment of the wall, the use of latex paint instead of wallpaper, choose to meet the lighting needs of the down lights instead of luxury crystal lamps and other ways.

4.4 Strategies to Induce Change with Safety in Mind

Safety is the top priority in the renovation of living space for the elderly, and the renovation design needs to eliminate all safety hazards and pursue the ultimate safety strategy of "level", "smooth" and "reach". "Flat" means unified indoor horizontal elevation; "smooth" means to ensure the smooth passage of the elderly in the living space; "up" means the clarity and accessibility of the sight line (Fig. 6).

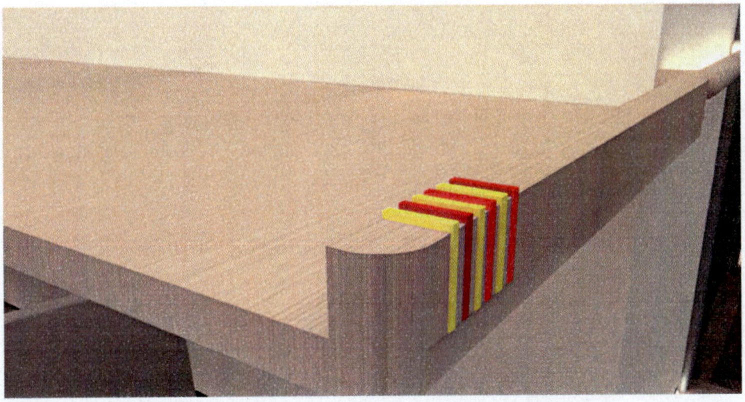

Fig. 6. Aging-friendly tips at the end of the handrail

According to the data of China's disease monitoring system, falls have become the first direct cause of death for people over 65 years old in China. In the renovation process, we focus on the details of the space that may induce the elderly to fall. For example, the height difference in the design of daily bathroom space is common. The conventional design of bathroom elevation is about 30 mm lower than the outdoor elevation for blocking the short time water in the bathroom due to cleaning and showering. Due to the decline of motor function of the elderly, a small height difference, material and even color changes may lead to the loss of balance of the elderly triggered by falls, resulting in greater physical injury. So in this kind of space detail processing, adopt a more suitable way for the elderly, adopt the form of long drainage grate at the entrance of the bathroom, maintain the bathroom plane, drainage grate, bathroom outside the traffic space plane at the same level height; according to the daily behavior trajectory of the elderly, smooth path space; finally, through reasonable line of sight planning design, lighting design to ensure that the elderly achieve access, facilitate The overall and detailed grasp of the space by the elderly improves the overall safety.

However, the ultimate safety strategy is not the same as the arbitrary "complete barrier-free", moderate "induced barriers", that is, under the premise of absolute safety to increase the induced, non-command "barriers" For example, the design of the route from "point-to-point" straight line design to more selective "migratory" route design, the elderly actively choose different traffic paths; another example, in the main daily traffic routes, through the means of For example, the same traffic handrail, use different handrail material, choose the handrail surface material with higher friction coefficient. Adding some "induced barriers" to the design of the living space for the elderly is conducive to the sensory mobilization and participation of the elderly, which can help to awaken their sensory perception and slow down the degeneration of physical functions. Finally, the elderly can adapt to the home life as soon as possible, and maintain the semi-self-care and self-care state of the elderly life as much as possible [6].

5 Conclusion

At this stage, there is still a lot of room for improvement in the design of living space renovation for the elderly. Designers should pay attention to the physical and mental needs of the elderly in the design process, take the initiative to meet the needs of appropriate aging transformation, and follow the space design strategy of system integration, transformation selection, and the co-existence of safety and "obstacles". This will help to build a systematic space for aging in place.

References

1. Muller, A., Missotten, P., Adam, S.: Transforming nursing home culture: opinions of older people. A cross-sectional study in Belgium. J. Aging Stud. **61** (2022)
2. Fu, J., et al.: A modified conceptual framework for peer relationship among community-dwelling older people in China: a qualitative study. Health. Soc. Care Commun. (2022)
3. Blay, L., Schulz, A.J.: The relationship of built environment to health-related behaviors and health outcomes in older adults community residents in a middle income country. J. Public Health Res. **4**(2), 548 (2015)
4. Hamdy, M., Carlucci, S., Hoes, P.: The impact of climate change on the overheating risk in dwellings—a Dutch case study. Build. Environ. **122**, 307–323 (2017)

5. Singelenberg, J.: Integrated service areas: an innovative approach to housing, services and supports for older persons ageing in place. J. Commun. Appl. Soc. Psychol. **24**(1), 69–73 (2014)
6. Guo, K.: The US long term care system: development and expansion of naturally occurring retirement communities as an innovative model for aging in place. Aging Int. **37**, 210–227 (2011)

Experimental Study on Bending Toughness of Engineered Cementitious Composites

Huiqing Xue(✉)

Beijing Municipal Engineering Research Institute, Beijing 100037, China
xhqmoon@126.com

Abstract. Through the three-point bending test of ECC, the bending toughness of ECC is studied, and the load deflection curve is obtained to evaluate the bending mechanical properties of ECC. The conclusions are as follows: after the initial crack, the load of ECC continues to rise until the ultimate load, showing good strain hardening characteristics, but the increase of deflection corresponds to the opening of a single crack, and there is no multi crack cracking; The flexural performance of ECC of ordinary cement series is better than that of ECC of composite cement series. The composite preparation of cement has a certain impact on the flexural performance of ECC. The amount of fiber has a certain influence on the flexural performance of ECC, especially the ultimate deflection. The variety of PVA fiber has little influence on the bending performance of ECC, which is generally controlled at 5%. Under the same dosage, the influence of domestic PVA fiber and imported PVA fiber on the bending performance of ECC is basically the same.

Keywords: Bending toughness · ECC · Crack resistance and toughening

1 Preface

Lately, as the most widely used building material in the world, the performance of concrete has been improved, but the shortcomings of concrete which is difficult to control, such as low tensile strength, brittle and easy to crack, crack width after cracking, limiting the application of reinforced concrete structure in infrastructure. Therefore, especially for the national economy and the people's livelihood major structural engineering, higher requirements are put forward in the aspects of crack resistance, crack limitation and toughening. It is particularly important to find a new type of structural material which can effectively limit the generation and development of cracks [1–3].

A large number of researches and applications have been made on the Engineered Cementitious Composites (ECC). One of the main differences between ECC and traditional fiber-reinforced cement-based composites is that the composition of each component of ECC is based on micromechanics design [4, 5]. ECC was first proposed by Professor Li VC of the University of Michigan, using the basic principles of micromechanics and fracture mechanics. It is a new type of engineering cement-based composite material with strain hardening characteristics and multi crack cracking characteristics

W. Guo and K. Qian (Eds.): GBCESC 2022, LNCE 211, pp. 1116–1124, 2023.
https://doi.org/10.1007/978-981-19-5217-3_115

based on the principle of microphysics and mechanics. The volume ratio of the composite material is only about 2% (mainly synthetic fiber) More than 3% of the tensile strain capacity can be obtained with the addition [6, 7]. In view of the superior properties of ECC materials, this paper studies the bending toughness of ECC materials through three-point bending test of ECC materials, and obtains the load deflection curve, which is used to evaluate the bending mechanical properties of ECC materials.

2 Test Materials and Mix Proportion

2.1 Test Materials

(1) **PVA fiber characteristics**
As a new type of fiber, PVA fiber has high tensile strength, elastic modulus, good acid-base resistance, good chemical compatibility with Portland cement, good hydrophilicity. And there is a good interfacial bonding force with cement substrate, mainly because the non-circular and irregular cross section of PVA fiber helps to increase the bonding surface between the fiber and the matrix, and the -C-OH group in the molecular structure of PVA fiber and the -OH group in cement hydrate can form a basic bond.

Imported PVA fiber is mainly used in the test, which is provided by Kuraray company of Japan, represented by I; Domestic fiber is provided by Beijing Rongnaier Engineering Materials Co., Ltd., represented by D, as shown in Fig. 1, and the material properties of PVA fibers are shown in Table 1.

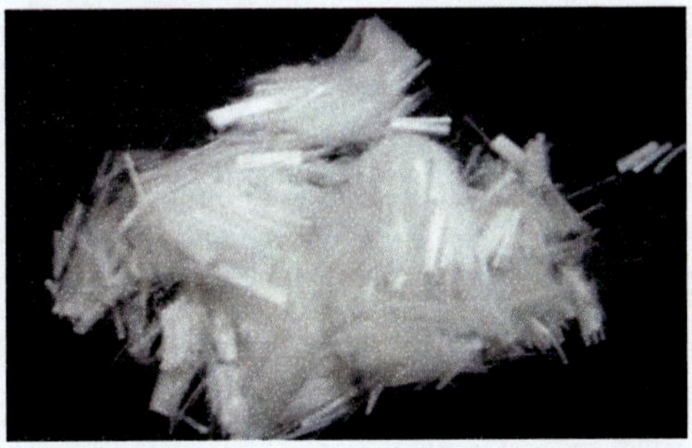

Fig. 1. PVA fibers

Table 1. Properties of PVA fiber materials

Number	Origin	Diameter/μm	Length/mm	Density/(g/cm^3)	Tensile strength/MPa	Modulus of elasticity/GPa	Ultimate elongation/%
I	Japan	39	12	1.3	1620	42.8	6
D	Beijing	21	12	1.3	1860	35	7

(2) **Physical properties of cement**
The special cement is 42.5 grade sulphoaluminate cement produced by Tangshan polar bear building materials Co., Ltd. and the ordinary cement is 42.5 grade cement produced by Beijing Jinyu company. The physical properties of cement is shown in Table 2.

Table 2. Physical properties of cement

Cement variety	Compressive strength /MPa			Flexural strength /MPa		
42.5 Sulphoaluminate cement	1d	3d	28d	1d	3d	28d
	44.8	59.6	70.4	7.2	7.4	8.7
	Initial setting time: 26 min; Final setting time: 38 min Density: 2.88 g/cm^3 Specific surface area: 4860 cm^2/g					
42.5 Ordinary portland cement	1d	3d	28d	1d	3d	28d
	-	27.2	54.0	-	5.8	8.0
	Initial setting time: 2 h 44 min; Final setting time: 3 h 48 min Density: 3.02 g/cm^3 Specific surface area: 4228 cm^2/g					

2.2 Test Mix Proportion

The material mix proportion is shown in Table 3. Special cement is 42.5 grade sulphoaluminate cement, ordinary cement is P.O42.5 grade ordinary portland cement; Quartz sand specification is 100–200 mesh; Fly ash is grade I low calcium ash; Specific surface area of silica fume is 20000 m^2/kg; Expansion agent and water reducer are provided by Wuhan Sanyuan building materials company; Calcium gluconate is used as retarder. Water reducing agent is 10.6 kg/m^3, Swelling agent is 133 kg/m^3.

Table 3. Mix proportion of base material Unit: kg/m^3

Number	Speacial cement	Ordinary portland cement	Quartz sand	Fly ash	Silica fume	Water	Fiber	Retarder
ECC1	311	623	331	313	78	424	26-I	0.66
ECC2	311	623	331	313	78	424	26-D	0.66
ECC3	311	623	331	313	78	424	19.5-I	0.66
ECC4	0	934	331	313	78	431	26-I	0

2.3 Mixing Process

1) Prepare the base material: first pour the cement, quartz sand, etc. into the mixer, mix them evenly at low speed, then dissolve the water reducer into the water, and then pour them into the mixer for mixing for 5 min to obtain the base material with uniform flow.
2) Adding fiber: adding fiber by hand while stirring at low speed to ensure uniform distribution of fiber without agglomeration.
3) Pouring and maintenance: all test pieces shall be poured in two layers, first poured in 2/3, vibrated for 1–2 min to ensure the material vibrates compactly, then poured in the remaining materials in the same way, finally, the surface shall be trowelled and covered with a plastic film to prevent water evaporation. After the test piece is formed for 24 h, the formwork shall be removed and put into the standard curing room for curing to the required age. The test piece shall be taken out for drying preparation test for 3 h before the test.

2.4 Test Method

400 mm × 100 mm × 100 mm rectangular specimen is used for three-point bending test. The bending test was carried out on ToniNORM2000 material testing machine with servo control function. The three-point bending loading method was adopted, and the span was 350 mm. The linear displacement sensor (LVDT) was used to measure the mid span deflection of the specimen and control the loading rate, which was 0.10 mm/min. During the loading process, the computer can automatically collect the loading time, mid span load value, mid span deflection and other data, and the collection frequency is 20 times/second. The load value can be transformed into the equivalent bending stress at the bottom of the beam, and the relationship between the bending stress and the mid span deflection can be established to evaluate the bending mechanical properties of ECC.

3 Test Results and Analysis

3.1 Test Phenomenon and Failure Mode

The cracking picture measured by the test is shown in Fig. 2.

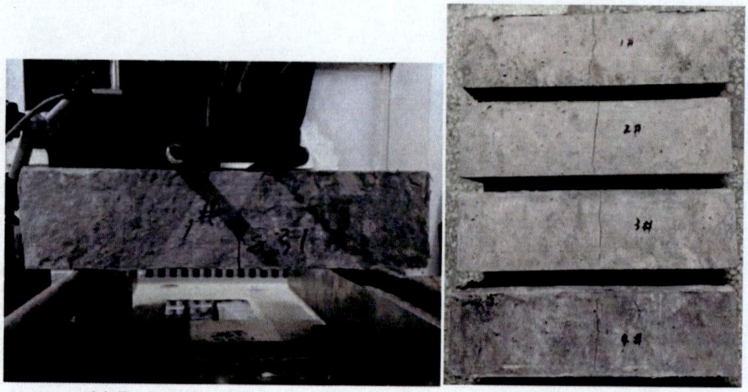

(a) Bottom cracking (b) Side cracking

Fig. 2. Cracking picture

As shown in Fig. 2, the crack patterns measured by bending toughness test are all cracked from the bottom of the test piece, and there is only one crack.

3.2 ECC Load Deflection Curve Characteristics

(1) **ECC load deflection curve**

The full load deflection curve of ECC is shown in Fig. 3.

It can be seen from Fig. 3 that the load of ECC continues to rise after the initial crack until the ultimate load, showing good strain hardening characteristics, but the increase of its deflection corresponds to the opening of a single crack, and there is no multi crack phenomenon. After the peak value, the bending curves of individual specimens vibrated, which was mainly due to the damage and spelling of the base materials on both sides of the cracks caused by PVA fiber during the pulling out process, resulting in accessory cracks, rather than multiple cracks, which was consistent with the crack diagram in front.

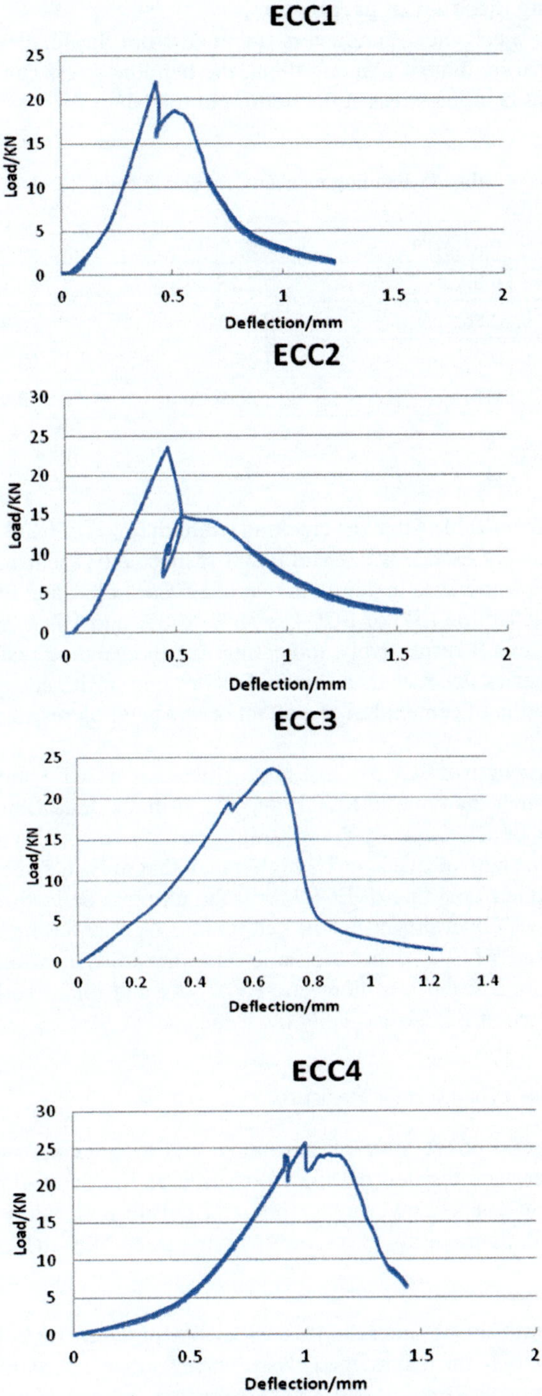

Fig. 3. ECC load deflection curve

(2) **ECC bending mechanical parameters**
The bending mechanical parameters obtained from the load deflection curve of ECC are listed in Table 4. Among them, the bending stress can be obtained from the equivalent bending stress at the bottom of the beam.

Table 4. Bending mechanical parameters of ECC

编号	$\sigma_{fc,f}$/MPa	σ_{fl}/MPa	d_{fl}/mm
ECC1	0.82	2.25	0.47
ECC2	0.86	2.35	0.49
ECC3	0.98	2.38	0.71
ECC4	1.13	2.59	0.98

It can be seen from Table 4 that the cracking strength ($\sigma_{fc,f}$) of ECC4 is 15%, 32% and 38% higher than that of ECC3, ECC2 and ECC1 respectively, the bending strength (σ_{fl}) of ECC4 is 9%, 10% and 15% higher than that of ECC3, ECC2 and ECC1 respectively, and the ultimate deflection (d_{fl}) of ECC4 is 38%, 100% and 109% higher than that of ECC3, ECC2 and ECC1 respectively, indicating that the bending resistance of ECC of common cement series is better than that of ECC of composite cement series, and the composite preparation of cement has one effect on the bending resistance of ECC Fixed impact.

The cracking strength of ECC3 is 20% higher than that of ECC1, the bending strength of ECC3 is 6% higher than that of ECC1, and the ultimate deflection of ECC3 is 51% higher than that of ECC1.

The cracking strength of ECC2 is 5% higher than that of ECC1, the bending strength of ECC3 is 5% higher than that of ECC1, and the ultimate deflection of ECC3 is 5% higher than that of ECC1. It shows that the variety of PVA fiber has little influence on the bending performance of ECC under the same ratio, generally controlled at 5%. Under the same dosage, the influence of domestic PVA fiber and imported PVA fiber on the bending performance of ECC is basically the same.

3.3 Mechanism of Fiber Crack Resistance

PVA fiber has a good effect of crack resistance and toughening, which can significantly improve the crack resistance and deformation ability of ECC materials, that is, ECC achieves multi-crack growth under uniaxial tensile load, with strain hardening characteristics. PVA fiber mainly plays the following three roles in ECC materials:

(1) Crack resistance
Fiber can prevent the generation and expansion of microcracks in cement matrix, which exists in both the unhardened plastic stage and the hardening stage of cement matrix. The tensile strength of the cement matrix is very low within 24 h after

pouring. If it is in a constrained state, a large number of cracks are easily generated when the moisture contained in the cement matrix evaporates rapidly. The fibers evenly distributed in the cement matrix can bear the tensile stress caused by plastic shrinkage, so as to prevent or reduce the generation of cracks. When the cement matrix is hardened, if it is in a restrained state, and the tensile force caused by drying shrinkage exceeds its tensile strength due to the change of surrounding temperature and humidity, a large number of cracks are easily generated, in this case, the fiber can also prevent or reduce the generation of cracks.

(2) Enhancement

The tensile strength of cement matrix is not only low, but also difficult to guarantee due to the existence of internal defects.

(3) Toughening effect

Under the load, even if the cement matrix cracks, the fiber can cross the cracks to bear the tensile stress and make the composite have certain ductility, which also means that the composite can have certain toughness.

4 Conclusion

Through the bending toughness test of ECC material, the following conclusions are obtained:

(1) After the initial crack, the load of ECC continues to rise until the ultimate load, showing good strain hardening characteristics, but the increase of its deflection corresponds to the opening of a single crack, and there is no multi crack phenomenon. After the peak value, the bending curves of individual specimens vibrated, which was mainly due to the damage and spalling of the base materials on both sides of the cracks caused by PVA fiber during the pulling out process, resulting in accessory cracks, rather than multiple cracks, which was consistent with the crack diagram in front.
(2) The flexural performance of ECC of common cement series is better than that of composite cement series. The composite preparation of cement has a certain influence on the flexural performance of ECC.
(3) The fiber content has a certain influence on the bending performance of ECC, especially the ultimate deflection.
(4) The variety of PVA fiber has little influence on the bending resistance of ECC, which is generally controlled at 5%. Under the same dosage, the influence of domestic PVA fiber and imported PVA fiber on the bending resistance of ECC is basically the same.

References

1. Chuanying, C.: Research on Concrete Raw Materials for the Three Gorges Project. China Water Conservancy and Hydropower Press, Beijing (1999)

2. Luoshu, G., Chunpu, L.: Durability and Damage Repair of Concrete. China Construction Industry Press, Beijing (1990)
3. Rongxi, S., Qi, C., Qinghai, L.: New Fiber Reinforced Cement-Based Composite. China Building Materials Industry Press, Beijing (2004)
4. Zhongwei, W.: The future of fiber reinforced cement-based materials. Concr. Cem. Products **1**, 5–6 (1999)
5. Chengxu, G.: Research and Development of High Toughness and Low Shrinkage Fiber Reinforced Cement-Based Composite. Department of civil engineering, Tsinghua University, Beijing (2008)
6. Li, V.C.: Research progress and application of high ductility fiber reinforced cement-based composite. J. Silicate **35**(4), 531–536 (2007)
7. Li, V.C.: Advances in ECC Research. ACI Special Publication on Concrete: Material Science to Applications, SP206-23: 373-400 (2002)

Near-Fault Pulse Ground Motion Decomposition Based on Improved Wavelet Multi-scale Analysis Method

Longlong Zhou[1,2], Huiguo Chen[1](✉), Yixin Lei[1], and Xiaorui Yang[2]

[1] Army Logistics Academy of PLA, Chongqing, China
chenhg_mail@163.com
[2] PLA Joint Logistic Support Force, Wuhan, China

Abstract. Near-fault pulse ground motion is a kind of special destructive ground motion. Domestic and foreign scholars usually decompose it into different frequency components in the research process. To accurately decompose the ground motion records, this paper improves the wavelet multi-scale analysis method and obtains the low-frequency and high-frequency components that can decompose the near-fault pulse-type ground motion into specific values with the boundary frequency. The effectiveness and applicability of this method are verified by time-domain and frequency-domain analysis.

Keywords: Pulse ground motion · Multi-scale analysis · Improved application · Time-domain · Frequency-domain · Seismic decomposition

1 Introduction

Compared with other ground motions, the near-fault pulse ground motion has greater destructive force on civil high-rise buildings, military large-span buildings, high sidewall special engineering, and other engineering structures [1–3]. At present, the hybrid model method can integrate the characteristics of low-frequency pulse and high-frequency random components of near-fault pulse ground motion to a certain extent, and simulate ground motion more completely. To further study the hybrid simulation method and obtain the simulation method that can simulate the closest to the actual seismic records, scholars at home and abroad have carried out a lot of research on pulse separation and high and low-frequency separation methods. Makris [4], Alavi [5], Bozorgnia [6] and Xu Longjun [7] used simple geometric functions such as sine and cosine, rectangle, triangle, and so on to achieve the purpose of pulse separation by simplifying the velocity pulse, which made a useful exploration in the study of equivalent ground motion pulse model. Fourier analysis can realize the conversion of signal in time-domain and frequency-domain by Fourier transform, and the decomposition of high and low-frequency components can be realized by low-pass filtering technology. However, the implementation of these two methods depends on many subjective factors (the selection of equivalent pulse model, pulse period, etc.), resulting in a certain difference between the equivalent

W. Guo and K. Qian (Eds.): GBCESC 2022, LNCE 211, pp. 1125–1135, 2023.
https://doi.org/10.1007/978-981-19-5217-3_116

pulse and the original pulse. Baker [8] used continuous wavelet transform to separate near-fault ground motion pulses. Firstly, the continuous wavelet transform is performed on the ground motion data, and then the ground motion pulse is obtained according to the wavelet coefficient, and the pulse is filtered to obtain the remaining part. Ghahari [9] uses moving average filtering to decompose near-fault pulse-type ground motion into low-frequency pulse parts and high-frequency parts. To decompose the ground motion more accurately, the multi-scale analysis method in wavelet analysis is introduced and improved in this paper. The actual near-fault pulse-type ground motion record is decomposed by the improved multi-scale analysis method to obtain multiple sets of high and low-frequency components, which can provide the basis for the study of non-stationary characteristics and artificial synthesis of near-fault pulse-type ground motion.

2 Multi-scale Analysis

Multi-scale analysis, also known as multi-resolution analysis, was first proposed by Mallat in 1989. The idea of multi-scale analysis is that in the function space, the function can be regarded as the limit of a series of approximate functions, and the function is approximated step by step at different scales. Each approximation is a smooth result, and the approximate function will be more and more refined.

Taking the high-frequency part of decomposition in the multi-scale analysis as W_n and the low-frequency part of decomposition as V_n, then W_n is the orthogonal complement of V_n in V_{n+1}, namely:

$$V_m \perp W_n,\ V_n \oplus W_n = V_{n+1},\ n \in Z \tag{1}$$

At this time, the subspace V_0 of multi-scale analysis can be approximated by finite subspaces, namely:

$$V_0 = V_1 \oplus W_1 = V_2 \oplus W_2 \oplus W_1 = \cdots = V_N \oplus W_N \oplus \cdots \oplus W_2 \oplus W_1 \tag{2}$$

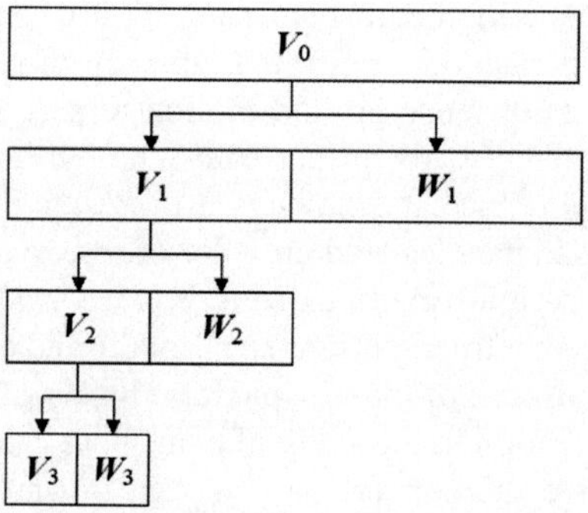

Fig. 1. Multi-scale analysis decomposition diagram.

As shown in Fig. 1, if any signal $f(x) \in V_0$ is taken, it can be decomposed into high-frequency component W_n (details) and low-frequency long-period component V_n

(approximation) through multi-scale analysis, and then the low-frequency component can be decomposed gradually until approaching the target scale. If the original signal $x(t)$ is taken and its total frequency band is $(0 \sim \omega)$, the frequency space is divided into two subspaces for the first time: low-frequency component V_1 (frequency band $(0 \sim \omega/2)$) and high-frequency component W_1 (frequency band $(\omega/2 \sim \omega)$). The second decomposition of the low-frequency component V_1 can get low-frequency V_2 (band $(0 \sim \omega/4)$) and high-frequency (band $W_2(\omega/4 \sim \omega/2)$)... By analogy, if any signal function $x(t) \in V_0$ is taken, $a_n \in V_n$ is denoted as the approximation of the original signal, and $d_n \in W_n$ is the error of approximation, then Eq. (2) can be expressed as:

$$a_0 = a_1 + d_1 = a_2 + d_2 + d_1 = \cdots = a_n + d_n + \cdots + d_2 + d_1 \quad (3)$$

where a_0 represents the reconstruction after signal decomposition.

3 Improvement of Multi-scale Analysis Method

In the process of decomposition and reconstruction of the signal, the purpose is to obtain the signal of any frequency band. In the description of the above multi-scale analysis method, the boundary frequency is set to f. If any original signal is decomposed step by step, the low-frequency component V_n (frequency band $(0 \sim \omega/2^n)$) and the high-frequency component $W_n \oplus \cdots \oplus W_2 \oplus W_1$ (frequency band $(\omega/2^n \sim \omega)$) close to the target boundary frequency can be obtained. However, when the difference between the boundary frequency $\omega/2^n$ and the target boundary frequency f is relatively large, and the target accuracy is not reached, the error between the boundary frequency $\omega/2^{(n+1)}$ and the target boundary frequency f is increasing after the low-frequency component is decomposed again. At this time, the boundary frequency $\omega/2^n$ obtained by multi-scale decomposition cannot be infinitely close to the target boundary frequency f, and the error is large. Therefore, we improve the multi-scale analysis method to decompose the signal which is infinitely close to the target frequency band.

The specific steps of the improved multi-scale analysis method are as follows:

1. Determining the Error Precision of Obtaining Frequency Band and Target Frequency Band;
2. After n-scale decomposition of the original signal, $d_1,\ d_2 \cdots d_n,\ a_n$ is obtained after the boundary cut-off frequency is first close. The signals with different frequency ranges can be obtained by reconstructing each component. Then $a_0 = a_n + d_n + \cdots + d_2 + d_1$, where a_0 represents the reconstructed signal.

Determine whether the obtained frequency band meets the accuracy requirements. When not satisfied, let $ai_0 = a_0 - a_n - a1_n - \cdots - ai_n$, where $i = r - 1$, r is the number of multi-scale analyses. The sequence ai_0 is decomposed by k scale to get $di_1,\ di_2 \cdots di_k$, ai_k. The reconstructed signal at this time is denoted as $a_0', a_0' = ai_n + di_n + \cdots + di_2 + di_1$. After multiple deductions of low-frequency components, and then multi-scale analysis, the obtained bound frequency is infinitely close to the target frequency. When meeting the accuracy requirements, $x = a_n + a1_n + \cdots + ai_n$ is the low-frequency signal to meet the target frequency band.

At this time, the subspace V_0 after multi-scale analysis can be approximated by finite subspaces, namely:

1st time: $V_0 = V_1 \oplus W_1 = V_2 \oplus W_2 \oplus W_1 = \cdots = V_N \oplus W_N \oplus \cdots \oplus W_2 \oplus W_1$

2nd: $V_0 = V_N \oplus V1_1 \oplus W1_1 = V1_2 \oplus W1_2 \oplus W1_1 = \cdots = V_N \oplus V1_K \oplus W1_K \oplus \cdots \oplus W1_2 \oplus W1_1$

The rth:

$$\vdots$$

$$V_0 = V_N \oplus V1_K \oplus \cdots \oplus Vr_K \oplus Vr_1 \oplus Wr_1 = V_N \oplus V1_K \oplus \cdots \oplus Vr_K \oplus Vr_2 \oplus Wr_2 \oplus Wr_1 = \cdots = V_N \oplus V1_K \oplus \cdots \oplus Vr_K \oplus Vr_{KK} \oplus W_{KK} \oplus \cdots \oplus W_2 \oplus W_1$$

In each multi-scale decomposition, the decomposition order (scale) is different, where the second decomposition scale is K, and the rth decomposition scale is KK.

4 Verification of Actual Ground Motion Decomposition

Near-fault pulse-type ground motion includes a low-frequency pulse component and a high-frequency random component. Tian Yuji et al. [10] found that the near-fault velocity pulse is a low-frequency component less than 1 Hz after analyzing 11 ground motions and 28 seismic records. The near-fault pulse ground motion can be obtained by simulating the low-frequency pulse component and the high-frequency component respectively. Therefore, to obtain the low-frequency pulse component and high-frequency random component, the boundary frequency of near-fault pulse-type ground motion decomposition is 1 Hz, and the error range is 0.1 Hz.

A large number of studies have shown that the concept of a near-field region defined by fault distance is clear and the physical meaning is clear [11–14]. Zhao et al. [15] proposed that the velocity pulse effect is more significant when the fault distance is small. To verify the validity and applicability of the decomposition method, this paper sets three ranges ($0 < M \leq 20$ km, $20 < M \leq 40$ km, $40 < M \leq 60$ km) according to the fault distance and selects three actual near-fault pulse ground motion records in different fault distance ranges as the data of this analysis, such as Table 1.

Table 1. Three near-fault pulse seismic records

Earthquake name	Station	Direction	Moment magnitude	Site classification	Distance from the fault/km
Imperial Valley	EC County Center FF	92	6.53	CD	7.6
Landers	Yermo Fire Station	360	7.28	CD	26.3
Chi-Chi, Taiwan	TCU017	E	7.62	AB	52.2

4.1 Determination of Decomposition Times and Decomposition Layers in Multi-scale Analysis

The multi-scale analysis of impulse ground motions at EC County Center FF station in 1979 Imperial Valley ($M_W = 6.53$) earthquake was carried out. The sampling time of the ground motion is 0.005 s, and it is resampled to make the sampling time of 0.005 s become 0.02 s, and the sampling frequency is 1/0.02 = 50 Hz. According to the Nyquist theory, the original signal frequency is $\leq$50/2 = 25 Hz, so its frequency range is [0, 25 Hz]. It is decomposed into five layers, taking *ai* as the low-frequency part of each layer and *di* as the high-frequency part, then the frequency bands of each layer are decomposed as follows:

a1: 0–12.5 Hz	d1: 12.5–25 Hz
a2: 0–6.25 Hz	d2: 6.25–12.5 Hz
a3: 0–3.125 Hz	d3: 3.125–6.25 Hz
a4: 0–1.563 Hz	d4: 1.563–3.125 Hz
a5: 0–0.781 Hz	d5: 0.781–1.563 Hz

At this time, the error is 1 Hz–0.781 Hz = 0.219 Hz > 0.1 Hz, the second multi-scale decomposition is needed, so the a5:0–0.781 Hz segment is retained, and the remaining signal is decomposed again. The decomposition layer is 7 layers. Take ali to represent the low-frequency part of each layer, dli to represent the high-frequency part, and divide the frequency bands of each layer as follows:

al1: 0.781–12.891 Hz	dl1: 12.891–25 Hz
al2: 0.781–6.836 Hz	dl2: 6.836–12.891 Hz
al3: 0.781–3.809 Hz	dl3: 3.809–6.836 Hz
al4: 0.781–2.295 Hz	dl4: 2.295–3.809 Hz
al5: 0.781–1.538 Hz	dl5: 1.538–2.295 Hz
al6: 0.781–1.160 Hz	dl6: 1.160–1.538 Hz
al7: 0.781–0.971 Hz	dl7: 0.971–1.160 Hz

The boundary frequency error is 1 Hz–0.971 Hz = 0.029 Hz < 0.1 Hz, which meets the accuracy requirements.

4.2 Decomposition and Reconstruction

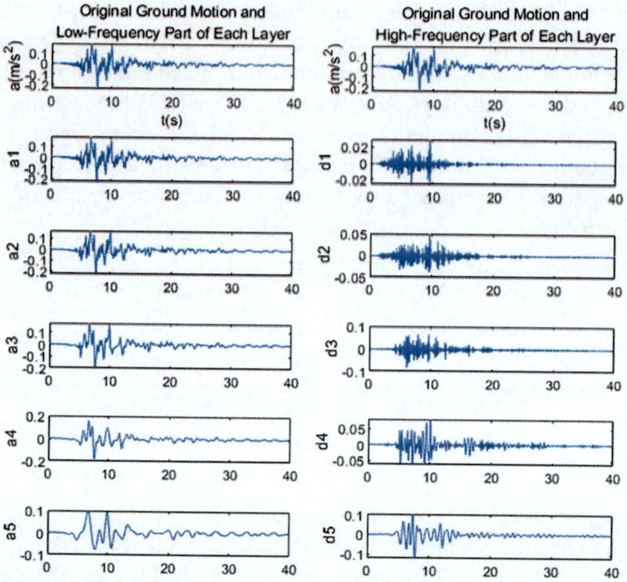

Fig. 2. Frequency band signals of ground motion decomposed by five layers wavelet.

Figure 2 shows the original signal and the high and low-frequency parts of each layer after the 5-layer wavelet decomposition of the 92° directional acceleration time history of EC County Center FF station. Figure 3 shows the adjusted signal (the original signal minus the bottom low-frequency signal) after 7-layer wavelet decomposition and the high and low-frequency parts of each layer. In the wavelet decomposition, the db8 wavelet is used, which can not only decay rapidly in the frequency domain, but also has good locality in the time domain, and can detect small information more easily. According to the Mallat algorithm, taking Fig. 2 as an example, there is a recursive relationship between the high-frequency part and the low-frequency part of each layer. After two multi-scale decompositions, the original ground motion signal can be reconstructed by the sum of the low-frequency component $a5$ at the first decomposition, the low-frequency component $a17$ at the second decomposition, and the high-frequency component $d11$–$d17$. Figure 4 shows the comparison between the reconstructed ground motion and the original ground motion by using $a5$, a17, and $d11$–$d17$, and the error between them is obtained. Figure 5 shows the final target low-frequency component ($a5 + a17$) after decomposition and reconstruction, and its frequency band is (0–0.971 Hz). High-frequency component ($d11 + d12 + d13 + d14 + d15 + d16 + d17$), the frequency band is (0.971 Hz–25 Hz).

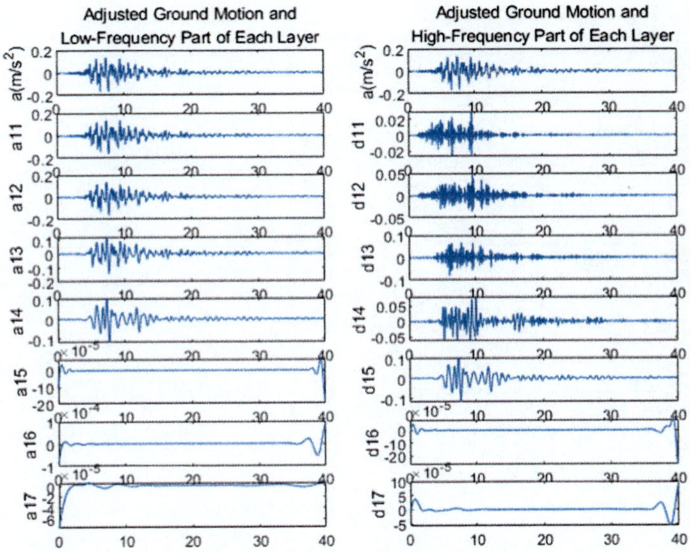

Fig. 3. Frequency band signals of ground motion decomposed by seven layers wavelet.

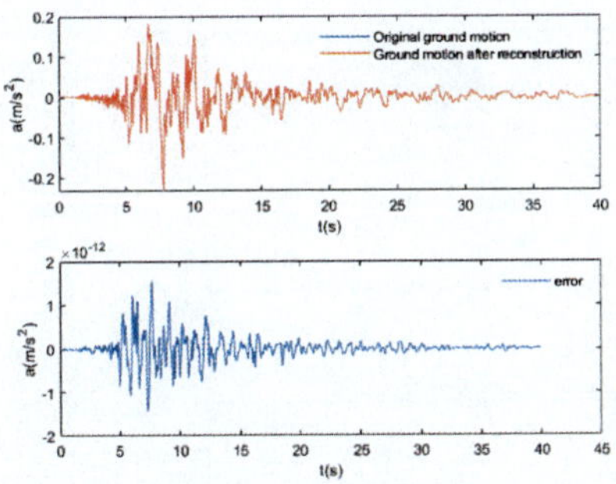

Fig. 4. Comparison and error between the two after reconstruction.

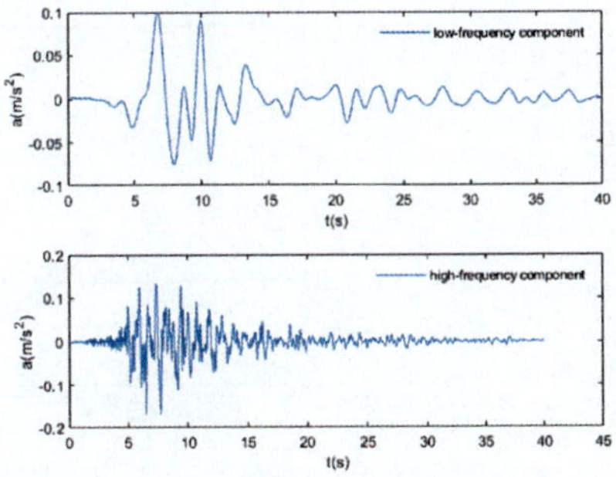

Fig. 5. Low-frequency and high-frequency components after decomposition and reconstruction.

4.3 Effectiveness Analysis

To determine the effectiveness of the multi-scale analysis method, the following two aspects are discussed in time-domain and frequency-domain.

Time-Domain Analysis. Figure 4 is the comparison of the acceleration time history of the original ground motion and the reconstructed ground motion and the corresponding error. As the figure shows, the reconstructed ground motion acceleration time history curve can basically coincide with the original ground motion, and the error is very small.

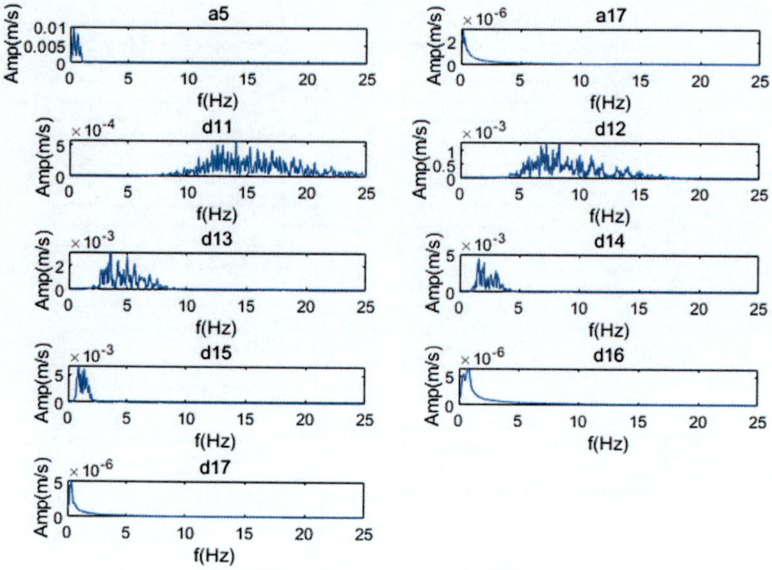

Fig. 6. Fourier amplitude spectrum of different frequency bands.

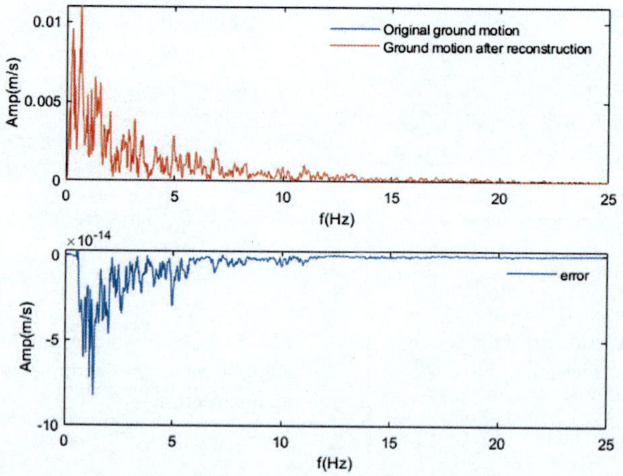

Fig. 7. Comparison of Fourier amplitude spectrum of original ground motion and reconstructed ground motion and their error.

Frequency-Domain Analysis. Figure 6 shows the Fourier amplitude spectrum of each frequency band after multi-scale decomposition. It can be seen from its distribution characteristics that it can well correspond to the expected frequency band. Figure 7 shows that the amplitude spectrum of the reconstructed seismic signal is similar to the original seismic signal, and the error magnitude is very small.

Therefore, the improvement of the multi-scale analysis method is also effective in time-domain and frequency-domain.

4.4 Other Instance Validation

After the analysis of the above-ground motions, the effectiveness of the improved wavelet multi-scale analysis method is determined. To verify the applicability of the method, the near-fault pulse ground motions (LANDERS_YER360, CHICHI_TCU017-E) in other ranges (fault distances are $20 < M \leq 40\,\text{km}$, $40 < M \leq 60\,\text{km}$) in Table 1 are selected for analysis, and the corresponding high-frequency components and low-frequency components of the target are obtained.

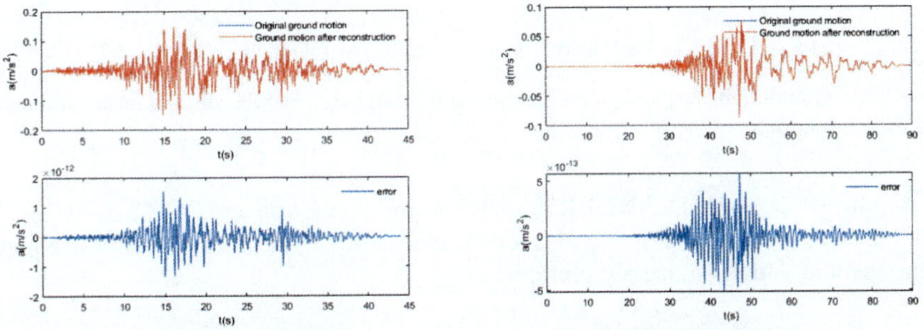

(a) LANDERS_YER360 (b) CHICHI_TCU017

Fig. 8. Comparison and error of original ground motion and reconstructed ground motion in time-domain.

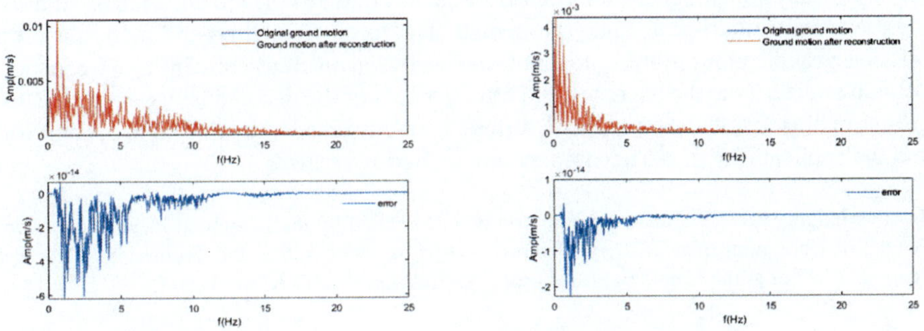

(a) LANDERS_YER360 (b) CHICHI_TCU017-E

Fig. 9. Comparison and error of original ground motion and reconstructed ground motion in frequency-domain.

It can be seen from Figs. 8 and 9 that the LANDERS_YER360 ground motion and CHICHI_TCU017-E ground motion within different fault distances are decomposed by the improved wavelet multi-scale analysis method. The reconstructed ground motion has good consistency with the actual ground motion in the time-domain and frequency-domain. It is proved that the improved multi-scale analysis method is applicable to impulse ground motion.

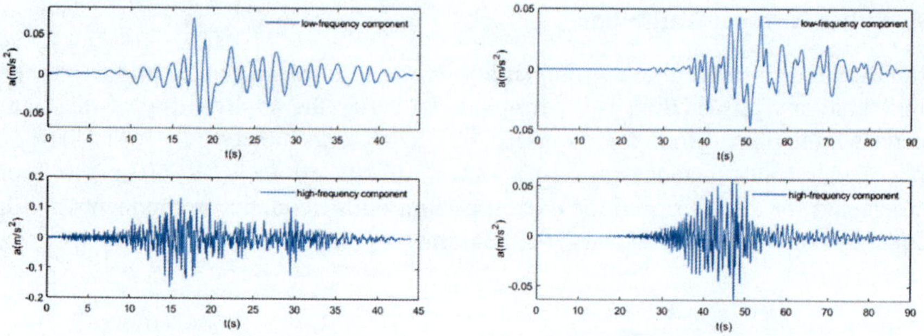

(a) LANDERS_YER360 (b) CHICHI_TCU017-E

Fig. 10. Low and high-frequency components of the original groundmotion after decomposition and reconstruction.

Figure 10 shows that LANDERS_YER360 ground motion and CHICHI_TCU017-E ground motion are decomposed and reconstructed to obtain the final target low-frequency component and high-frequency component.

5 Conclusions

This paper introduces the multi-scale analysis method in wavelet analysis and proposes an improved wavelet multi-scale analysis method by improving it. A series of high and low-frequency components with a cutoff frequency of 0.971 Hz are obtained by improving the wavelet multi-scale analysis method. The reconstructed ground motion and the original ground motion are compared and analyzed in time-domain and frequency-domain to determine the accuracy and effectiveness of the decomposition method. Two groups of pulse ground motions with different fault distances are selected to decompose, and the applicability of the decomposition method is verified.

Acknowledgments. This financial supports were provided by the Basic and Frontier Research Project of Chongqing (cstc2019jcyj-msxmX0218) and The Science and Technology Research Program of Chongqing Municipal Education Commission (KJZD-K201912903).

References

1. Xinle, L., Xi, Z.: Study on equivalent velocity pulse of near-fault ground motions. Acta Seismol. Sin. **26**(6), 634–643 (2004). (in Chinese)
2. Junfeng, J., Xiuli, D., Qiang, H.: The state-of-art review of near-fault earthquake ground motion characteristics and effects on engineering structures. J. Build. Struct. **36**(1), 1–12 (2015). (in Chinese)

3. Makris, N.: Rigidity-plasticity-viscosity: can electrorheological dampers protect base-isolated structures from near-source ground motions. Earthq. Eng. Struct. Dyn. **26**(5), 571–91 (1997)
4. Ze, Y., Jian, S., Xianting, J., Zhenghua, Z.: Simulation of near-fault ground motion of Jinghe M_S6.6 earthquake. J. Disaster Prev. Mitig. Eng. **40**(06), 1024–1036+1052 (2020). (in Chinese)
5. Krawinkler, H., Alavi, B.: Development of an improved design procedure for near-fault ground motions. Oakland C A, SMIP 98 seminar on utilization of strong motion data (1998)
6. Bozorgnia, Y., Mahin, S.A.: Ductility and strength demands of near-fault ground motions of the Northridge earthquake. In: Earthquake Engineering Research Institute, Seattle, Proceedings of the 6th US National Conference on Earthquake Engineering (1998)
7. Xu, L.J., Rodriguez-Marek, A., Xie, L.L.: Design spectra including effect of rupture directivity in near-fault region. Earthq. Eng. Eng. Vib. **5**(2), 2-1 (2006)
8. Baker, J.W.: Quantitative classification of near-fault ground motions using wavelet analysis. Bull. Seismol. Soc. Am. **97**(5), 1486–1501 (2007)
9. Ghahari, S.F., Jahankhah, H., Ghannad, M.A.: Study on elastic response of structures to near-fault ground motions through record decomposition. Soil Dyn. Earthq. Eng. **30**(7), 536–546 (2010)
10. Yuji, T., Qingshan, Y., Mingqi, L.: Simulation method of near-fault pulse-type ground motion. Acta Seismol. Sin. **29**(1), 77–84 (2007). (in Chinese)
11. Chopra, A.K., Chintanapakdee, C.: Comparing response of SDF systems to near-fault and far-fault earthquake motions in the context of spectral regions. Earthq. Eng. Struct. Dyn. **30**(12), 1769–1789 (2001)
12. Bray, J.D., Rodriguez-Marek, A.: Characterization of forward-directivity ground motions in the near-fault region. Soil Dyn. Earthq. Eng. **24**(11), 815–828 (2004)
13. Liao, W.I., Loh, C.H., Wan, S.: Earthquake response of RC moment frames subjected to near-fault ground motions. Struct. Des. Tall Build. **10**(3), 219–229 (2001)
14. Liwei, Y., Jianwen, C., Tianji, Z.: Analysis of near-field strong motion characteristics of the Yangbi M6.4 earthquake in Yunnan province in 2021. World Earthq. Eng. **37**(04), 81–90 (2021). (in Chinese)
15. Weifeng, Z., Songwu, T., Jing, Z.: Shear resistant performance of RC columns subjected to vertical near-fault velocity pulse-like earthquake actions. J. Chongqing Univ. **36**(6), 103–111 (2013). (in Chinese)

Smart City

A Comparative Study of Point Cloud Mapping Algorithms Towards Heterogeneous Traffic Scenarios

Jiazhong Zhang, Shuai Wang(✉), Xiaojun Tan, and Minghao Chen

School of Intelligent Systems Engineering, Sun Yat-sen University, Shenzhen 518107, China
wangsh368@mail.sysu.edu.cn

Abstract. In recent years, the traditional automobile industry has paid an increasing attention to autonomous driving. For autonomous driving, the point cloud map contains abundant information of environment and can support various algorithms such as intelligent vehicle positioning and moving obstacle detection. Therefore, it is significant to make an accurate point cloud map for driving tasks. However, the actual road conditions are complicated in practice, especially in the underground scenes where GPS is unavailable like tunnels. Meanwhile, common point cloud mapping algorithms cannot be directly applied to long and narrow underground tunnels on the grounds that the lack of feature information, where the quality of point cloud is degenerated. Dealing with the problems above, this paper adopts a 64-line mechanical lidar and applies different point cloud mapping algorithms to construct maps, so as to explore the effect of different point cloud mapping algorithms in heterogeneous traffic scenes. The main work is as follows. Firstly, point cloud data collecting platform system has been established on a vehicle. The information interaction between the system and the environment is achieved according to the requirements of specific scenes. Then data are collected in tunnels and conventional roads, and mapped via different algorithms. Empirical validations indicate that in the scene of underground scenes which lacks of geometric structure and point cloud degradation (e.g. tunnels), Lightweight and Ground-Optimized Lidar Odometry and Mapping based on scan context (SC-LeGO-LOAM) can achieve better point cloud map over normal distribution transformation (NDT).

Keywords: Autonomous driving · Autonomous driving maps · Point cloud map · Heterogeneous traffic scenes

1 Introduction

As human civilization and technology improved by leaps and bounds, cars have played an important role in daily life, which is indispensable for the society. The attached problems like traffic safety, congestion and energy consumption have also been intensively studied. Developing automatic driving technology has become an indispensable way to improve transportation efficiency and relieve congestion.

The autonomous driving map is a key concern of autonomous vehicles. In the existing function modules of intelligent connected vehicles, autonomous driving map is not

W. Guo and K. Qian (Eds.): GBCESC 2022, LNCE 211, pp. 1139–1150, 2023.
https://doi.org/10.1007/978-981-19-5217-3_117

only used to support macro-lane-level path planning [1], but also provides accurate prior information for local trajectory planning [2], working as a prerequisite for active map-based positioning [3]. Compared with the traditional navigation map, autonomous driving map is characterized by precise and complete information description, which the surrounding environment information can be catched accurately and reliably, and the cost of environmental perception be reduced effectively. It is crucial for autonomous driving. In order to achieving high precision positioning using point clouds, accurate pose estimation and multi-source information association are two core problems.

Simultaneous localization and mapping (SLAM) refers to the situation when autonomous cars are located in a thoroughly unexplored scenario, moving in an unknown environment while proceeding incremental mapping of the environment. The interconnection between localization and constitution causes complexity to the SLAM problem. The accurate positioning of autonomous driving relays on the map which includes the consistency of the surrounding environment. Similarly, the construction of a highly consistent environment map is based on the premise of positioning with accurateness as well [4]. SLAM can be divided into laser SLAM and visual SLAM. Visual sensor can be applied to indoor and outdoor environment, and visual sensor can perceive and obtain rich environmental information in use. Many scholars have carried out research on visual SLAM, but visual sensor has the disadvantage of being easily interfered by light. In contrast, lidar is not affected by light and has high robustness. Up to now, lidar-based real-time positioning and mapping is still a popular technology in autonomous driving [5].

In real traffic scenes, the characteristics of road conventional traffic scenes are quite different from those of tunnels. Tunnels are usually constructed of artificial materials and may possess several kilometers' length, which are characterized by lack of geometric structure and may cause degradation to the point cloud. Accurate positioning and environment map construction, path planning and motion control are the key technologies to improve the reliability and intelligence degree of autonomous driving in diversified scenes. Path planning and obstacle avoidance require environmental map and precise positioning. In order to guarantying vehicles are able to obtain accurate environmental maps in most scenes, it is significant to explore the mapping effects of different cloud map generation algorithms in different traffic scenes. The analytic result may contribute to improving the positioning accuracy and the robustness of autonomous driving.

Most studies are based on conventional roads using a single algorithm, and few studies have touched upon the performance comparison of different approaches for constructing point cloud maps. This paper collects point cloud data of conventional roads and tunnels bases on lidar. NDT algorithm uses probability model to match by calculating probability density rather than matching the features of points, SC-LeGO-LOAM added loopback detection based on Scan Context on the basis of LeGO-LOAM, the features of lines and surfaces are matched in the calculation [6]. These two algorithms belong to two different registration modes. Therefore, NDT algorithm and SC-LeGO-LOAM algorithm are used to make point cloud maps of conventional roads and tunnels, respectively. By comparing the production effects of the two algorithms in the same environment, the above two algorithms are objectively analyzed. The poor mapping effect in the scene of tunnel where lacking geometric structure and degenerated point cloud of NDT algorithm has showed in the experimental results. Comparatively, SC-LeGO-LOAM can get a better mapping effect.

The frame of this paper is as follows. The second segment introduces the registration of traditional point cloud algorithms Iterative Closest Point (ICP) and NDT. After theoretical analysis of ICP and NDT algorithms, NDT algorithm is selected as one of the algorithms in this experiment. Section 3 introduces Lidar Odometry and Mapping (LOAM), Lightweight and Ground-optimized Lidar Odometry and Mapping (LeGO-LOAM) and SC-LeGO-LOAM, where SC-LeGO-LOAM is selected as one of the algorithms in this paper. In the fourth section, NDT algorithm and SC-LeGO-LOAM algorithm are selected into experimental platform, and the above two algorithms are used to make point cloud maps for the conventional road and off-campus tunnel environment. The above two algorithms are analyzed through the making effect of point cloud map. The summary of this work is given in Sect. 5.

2 Traditional Point Cloud Mapping Algorithm

2.1 ICP Algorithm

ICP is the point cloud registration algorithm. the correspondence of points set is determined first, and the optimal transformation of rigid body is calculated repeatedly until the convergence condition is satisfied, which is essentially the optimal matching based on the least square method [7]. The primary target of ICP is to solve the spatial transformation between source P and target Q in the overlapping region across two data sets, which is aimed at minimizing the distance between of P and Q [8]. Let $E(R, t)$ be the objective function. The error of source P and target Q under the transformation matrix (R, t), which is expressed by $E(R, t)$. where R is the rotation matrix and t is the translation matrix [9]. Then the optimal transformation matrix can be transformed into the optimal solution (R, t) satisfying $minE(R, t)$ as following formula:

$$E(R,t) = \Sigma_{i=1}^{n} \|Q_i \cdot R + t\|^2 \quad (1)$$

The optimal transformation matrix is obtained under minimizing the objective function. The explicit step is illustrated following:

1. Each nearest point in source is calculated into the target;
2. Use the Singular Value Decomposition (SVD) to calculate the rotation matrix R and shift matrix T and minimize the objective function $E(R, t)$;
3. Calculating rotation matrix R and translation matrix T to rotate and translate the point set P into the new corresponding points set P' is obtained;
4. Calculating the average distance d between the new points set P' and the objective points cloud Q:

$$d = \frac{1}{n} \Sigma_{i=1}^{n} \|P_i' - Q_i\| \quad (2)$$

5. Set a maximum iteration, calculate the average distance D under the limitation of maximum iteration until it satisfies the threshold or the iterations reach the predefined number.

2.2 NDT Algorithm

NDT is a statistical model. If a set of random vectors satisfies a normal distribution, the probability density function [10] as expressing as follows:

$$P(x) = \frac{1}{(2\pi)^{\frac{D}{2}}\sqrt{|\Sigma|}} e^{\frac{\left(x - \vec{u}\right)^T \Sigma^{-1}\left(x - \vec{\mu}\right)}{(2\sigma)^2}} \tag{3}$$

where D is the dimension, $\vec{\mu}$ is the mean vector, Σ is the covariance matrix of the random vector. NDT algorithm can describe the scattering of point cloud by way of probability, which can effectively reduce the time required for registration [11]. The explicit step of algorithm is illustrated in following steps:

1. The point of target is scattered to each grid. The space where the laser point cloud is located is rasterized by a cube so that the laser point cloud is located in the corresponding cube;
2. Scattering the objective points cloud, then normal distribution probability density function in each grid is calculated subsequently. It is generally guaranteed that there are at least 5 points in the grid to avoid unsuitable sizes;
3. The initial coordinate transformation parameters of source points are determined with respect to objective point; This step provides the initial value close to the optimal value for the next iteration of transformation parameters;
4. The source point cloud performs the initial transformation of coordinate, then calculates the probability in the target grid. According to the (R, t) computed precedently, the coordinates of the source point cloud are converted to the objective grid. The converted source point cloud gets coordinated corresponding to the normal distribution probability density function of each voxel grid, then the probability of laser point coordinates after transformation is calculated.
5. The probability product of each point cloud is taken as the target likelihood function. The likelihood function with the maximum probability product is the optimal coordinate transformation, as shown in the equation:

$$-log\ \varphi = -\Sigma_{k=1}^{n} log\left(P\left(X'\right)\right) \tag{4}$$

where $-log\varphi$ is the negative logarithm form of the maximum likelihood function, which is used to facilitate the derivation later. Finally, the Newton iteration schema is used to acquire the optimal transformation parameters to complete point cloud registration.

2.3 Comparation of ICP Algorithm and NDT Algorithm

ICP, the algorithm that able to get registration precisely and effectly. It processes the data of points without feature extraction and segmentation. It can exhibit a good convergence ability with a good initial value. Compared with NDT algorithm, algorithm of ICP spends a large quantity of calculation on finding quantities of interrelated points, and the convergence region is small; the running speed is thus slow. Meanwhile, seeking the ICP algorithm adopts the assumption that the distance of Euclidean which can be obtained from the nearest point is corresponding to the point, which may result in wrong corresponding points. Comparably, the NDT algorithm has better robustness and an improved efficiency [12]. Therefore, this paper chooses NDT algorithm as one of the point cloud mapping algorithms.

3 Point Cloud Mapping Base on SC-LeGO-LOAM

SC-LeGO-LOAM adds loopback detection based on the Scan Context on the basis of LeGO-LOAM. Scan Context with the loopback detection, which was proposed from KAIST University in Korea, the author named Giseop Kim and Ayoung Kim [13]. The global descriptor with the main feature of non-histogram is proposed to help search "context" (current/previous data) in a quick and effective manner. Scan Context, for localization using a single 3D scan is an advanced spatial descriptor with a matching algorithm.

3.1 LOAM

LOAM is a 3D laser SLAM algorithm proposed by Ji Zhang et al., school of Robotics, Carnegie Mellon University, which has a good performance and ranks in the top of the accuracy of KITTI [14] odometer list. LOAM algorithm is a real time, low drift laser odometer and mapping method. LOAM divides the scanning and matching process of the laser odometer into two parts to recognize the relationship between point clouds at different moments. The high-frequency part adopts the "scanning-scanning" matching method and the low-frequency part adopts the "scanning-map" matching method. By using these two parts together, LOAM makes better use of computing resources and balance the problem of computing accuracy and efficiency. Figure 1 illustrates thc system framework of LOAM.

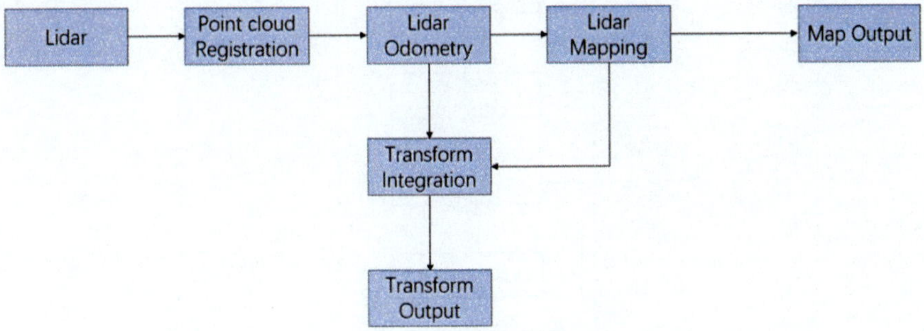

Fig. 1. Flow diagram of LOAM algorithm system

Meanwhile, LOAM uses curvature features to registrate the point cloud which get from the Lidar. Traditional point cloud registration method is based on ICP. Generally, these methods can obtain better precision theoretically, but the calculation process consumes a lot of time, LOAM selects points with large curvature as edge points and points with small curvature as plane points, By calculating the interconnection between edge points and the interrelated edge, together with the interconnection between plane points and the interrelated plane, the position and pose transformation relationship of two-point cloud is established.The amount of points can be reduce effectively involved in the operation while ensuring accuracy by using LOAM.

3.2 LeGO-LOAM

LeGO-LOAM is an algorithm optimized and improved by TixiaoShan et al. on the basis of LOAM [15]. Because there are sundry objects on the ground, the accuracy of point cloud map may be affected if the point features on the ground in LOAM cannot meet the actual requirements. LeGO-LOAM introduces point cloud segmentation on the basis of LOAM to separate ground information and then carries out subsequent processing such as feature extraction for the segmented information. Different from LOAM, LeGO-LOAM introduces two-step LM optimization [16], which first optimizes plane point feature pairs in vertical direction, roll direction and pitch direction, and then optimizes edge points and other directions. The calculation budget is less than that of LOAM. And loopback detection mechanism is added and the graph optimization is used in LeGO-LOAM, LeGO-LOAM has a higher practical value. The system architecture of LeGO-LOAM is shown in Fig. 2. Due to the limitation of lidar hardware frequency, the neighboring frame clouds often differ greatly when the vehicle is moving fast. In order to overcome the problem that the scanning effect of laser odometer is poor when the position changes sharply, more high-frequency information is generally introduced as the initial value of scanning matching.

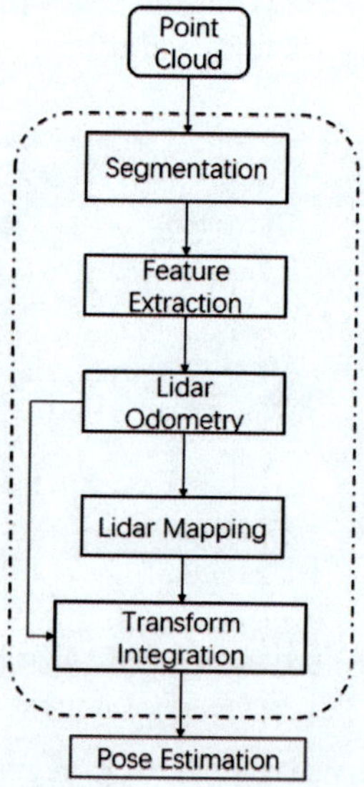

Fig. 2. Flow chart of LeGO_LOAM system

3.3 SC-LeGO-LOAM

It is necessary to solve the problems of descriptor rotation invariance and noise processing in lidar localization. First, the rotation invariance of descriptor does not change with changing viewpoints. Secondly, normals are lot of uncertain noisy due to the resolution of point cloud is easily fluctuate with distance, and noise processing is an unavoidable problem. At present, the mainstream approach is to use histogram to solve the two problems mentioned above. Considering the histogram method can only provide the random index of the surrounding environment, it cannot describe the detailed structure of the scene quickly and effectively, descriptors have some problems such as potential false positives and poor differentiability for location recognition.

The design method encodes all point clouds in 3d scanning as a matrix, and describes self-centered 2.5D information. Its main characteristics include:

- The encoding function of efficient bin. On the premise of point cloud density and normal invariance, the bin encoding function is used to identify the position more efficiently [13, 17].
- The internal structure of point cloud is preserved. Point cloud of bin determines the value of each element of the matrix. If the relative geometry of a point is described as a histogram [18], the absolute position of the point is lost. Therefore, by avoiding the adoption of histogram to preserve the complete inside structure of point cloud, the method aligns the viewpoints of query scan with candidate scan and improves the discrimination ability.
- Efficient two-stage matching algorithm. Scan Context provides a rotating invariant descriptor by searching the first nearest neighbor, and combines the rotating invariant descriptor with pairwise comparability score at the level. This operation helps avoiding cyclic detection in all databases, it can thus obtain a viable search time.

4 Experimental Evaluation

So as to validate the effectiveness of NDT algorithm and SC-LeGO-LOAM algorithm in making point cloud map under different traffic scenarios, a large number of real vehicle multi-scene data have been collected. The hardware environment of the experiment is Intel i7 processor with 16 GB memory, and the system environment is a laptop installed with Ubuntu18.04 and Ros Melodic robot operating system. The experimental platform is The Great Wall VV6 vehicle (as shown in Fig. 3), and a single 64-line mechanical lidar OS1-64 is installed on the car roof. The experimental environments were conventional driveway in campus and Luntou tunnel in Guangzhou.

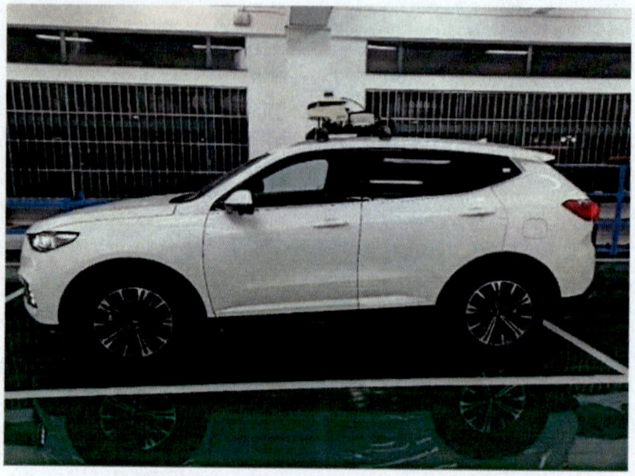

Fig. 3. The experimental platform

After collecting data in different scenes, NDT algorithm and SC-LeGO-LOAM algorithm are implemented to make point cloud map separately. Figure 4 shows the point cloud map made by NDT algorithm for conventional roads on campus.

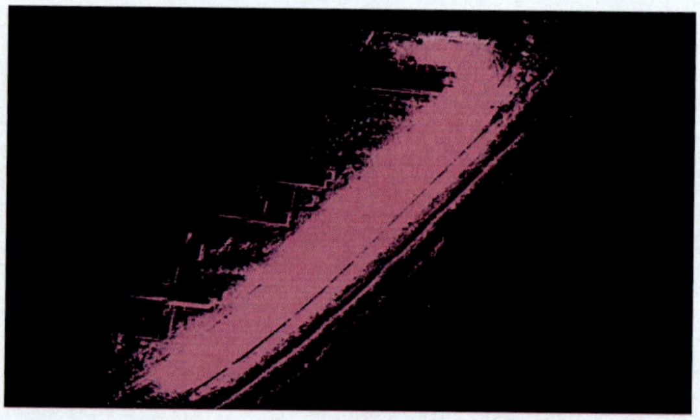

Fig. 4. Point cloud map made by NDT algorithm for conventional road

As it illustrated in Fig. 4, NDT algorithm can clearly reconstruct the contour of conventional roads with dense point clouds. Figure 5 shows the point cloud map made by SC-LeGO-LOAM algorithm for conventional roads on campus.

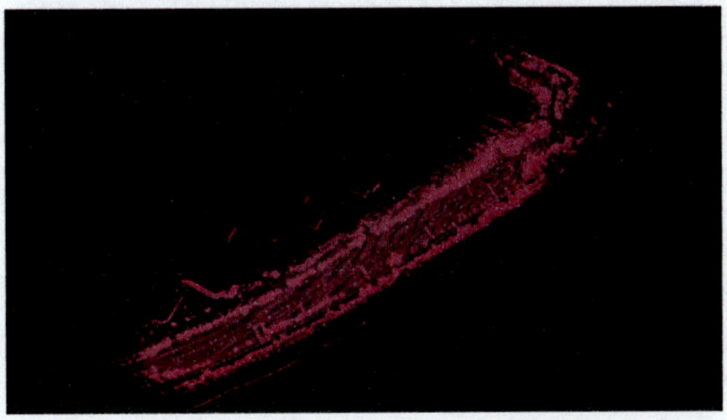

Fig. 5. Point cloud map made by SC-LeGO-LOAM algorithm for conventional road

As given in Fig. 5, SC-LeGO-LOAM algorithm can reproduce the contour of conventional lanes on campus. Compared with Fig. 4, the point cloud in Fig. 5 is relatively sparse due to its high filter. Figure 6 shows the point cloud map of Luntou Tunnel in Guangzhou made by NDT algorithm.

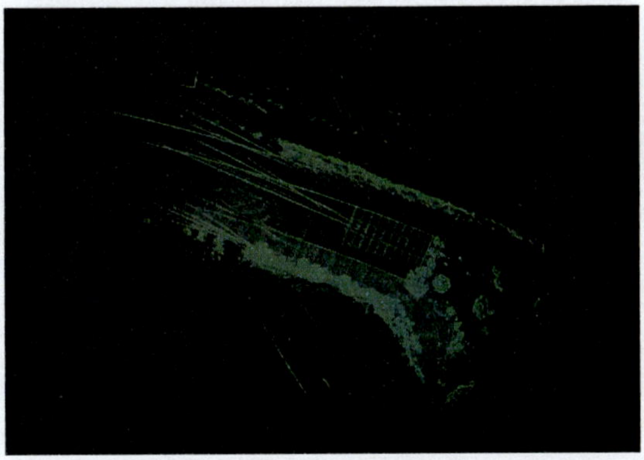

Fig. 6. Point cloud map of tunnel in Guangzhou made by NDT algorithm

As illustrated from Fig. 6 that the tunnel point cloud map made by NDT algorithm could not clearly reconstruct the contour of the tunnel, and serious double shadow and noise appeared. Figure 7 shows the tunnel point cloud map made by SC-LeGO-LOAM.

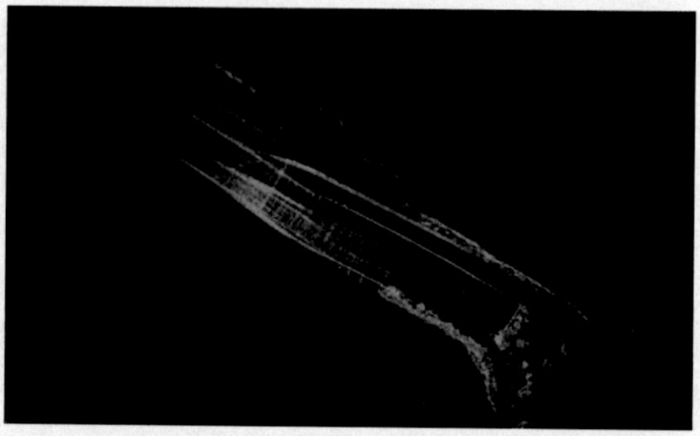

Fig. 7. Point cloud map of tunnel made by SC-LeGO-LOAM algorithm

Figure 7 shows that compared with NDT algorithm, SC-LeGO-LOAM algorithm holds better point cloud map in tunnel scenes and can clearly reproduce the outline of the tunnel. Figure 4 and Fig. 5 shows that both NDT algorithm and SC-LeGO-LOAM algorithm can clearly show the outline of the conventional road. Tables 1 and 2 list the specific data of the above four groups of point cloud maps.

Table 1. Campus driveway

Algorithm	The number of points in a point cloud map	Point cloud map size /M
NDT	30756443	1126.4
SC-LeGO-LOAM	594430	22.4

Table 2. Luntou Tunnel in Guangzhou

Algorithm	The number of points in a point cloud map	Point cloud map size /M
NDT	436979	15.8
SC-LeGO-LOAM	129224	5

5 Conclusion

This paper explores the effect of different point cloud registration algorithm in heterogeneous traffic scenarios. Firstly, the required experimental scenes are selected, and the experimental data acquisition platform and data processing platform are constructed, which are based on the actual requirements of the scene and experiment. A great amount of laser point cloud data is collected using lidar mounted on real vehicles in real traffic environments. In the comparative analysis towards the traditional mapping algorithm, NDT algorithm is selected as one of the experimental algorithms in this paper, and the

NDT algorithm and SC-LeGO-LOAM algorithm are used to make the point cloud maps of campus conventional lanes and tunnels.

Based on the actual point cloud mapping effect, it can be concluded that in conventional roads, both NDT algorithm and SC-LeGO-LOAM can clearly reproduce the outline of the road, and have a good mapping effect. In the scene of tunnel lacking of geometric structure and point cloud degradation like tunnels, NDT algorithm cannot make point cloud map and the cartographic effect is poor. Possible criteria can be obtained in this work to guide the algorithm determination task for making point cloud maps.

Acknowledgment. This work was supported by Guangdong Basic and Applied Basic Research Foundation under Grant 2021A1515110543.

References

1. Liu, C., Jiang, K., Xiao, Z., et al.: Lane-level route planning based on a multi-layer map model. In: 20th International Conference on Intelligent Transportation Systems (ITSC), pp. 1–7. IEEE (2017)
2. Chu, K., Lee, M., Sunwoo, M.: Local path planning for off-road autonomous driving with avoidance of static obstacles. IEEE Trans. Intell. Transp. Syst. **13**(4), 1599–1616 (2012)
3. Wolcott, R.W., Eustice, R.M.: Visual localization within lidar maps for automated urban driving. In: 2014 IEEE/RSJ International Conference on Intelligent Robots and Systems, pp. 176–183. IEEE (2014)
4. Wurm, K.M., Stachniss, C., Grisetti, G.: Bridging the gap between feature and grid-based SLAM. Robot. Auton. Syst. **58**(2), 140–148 (2010)
5. Thrun, S., Burgard, W., Fox, D.: Probabilistic Robotics. The MIT Press, Cambridge (2005)
6. Rusinkiewicz, S.: Efficient variants of the ICP algorithm. In: Proceedings of the Third International Conference on 3-D Digital Imaging and Modeling, pp. 145–152 (2010)
7. Zhiquan, Z., Fei, L., Jingjing, Q., Bing, K., Hui, J., Yujie, Z.: Hybrid scanning matching algorithm based on tight coupling of IMU and lidar. Comput. Syst. Appl. **30**(11), 203–209 (2021)
8. Qingshan, W., Jun, Z., Yuansheng, L., Xinchen, Z.: Point cloud registration algorithm based on combination of NDT and ICP. Comput. Eng. Appl. **56**(07), 88–95 (2020)
9. Segal, A., Haehnel, D.,Thrun, S.: Generalized-ICP. Robot.: Sci. Syst. **2**(4) (2009)
10. Biber, P., Strasser, W.: The normal distributions transform: a new approach to laser scan matching. In: Proceedings of 2003 IEEE/RSJ International Conference on Intelligent Robots and Systems, pp. 2743–2748 (2003)
11. Magnusson, M., Lilienthal, A., Duckett, T.: Scan Registration for Autonomous Mining Vehicles Using 3D-NDT: Research Articles, pp. 803–827. Wiley (2007)
12. Surachet, S., Chi-Ming, L., Syun, T., Guang-Je, T., Kai-Wei, C.: Performance analysis of 3D NDT scan matching for autonomous vehicles using INS/GNSS/3D LiDAR-SLAM integration scheme. In: 2021 IEEE International Symposium on Inertial Sensors and Systems (INERTIAL) Inertial Sensors and Systems (INERTIAL), pp. 1–4 (2021)
13. Giseop, K., Ayoung, K.: Scan context: egocentric spatial descriptor for place recognition within 3D point cloud map. In: IEEE/RSJ International Conference on Intelligent Robots and Systems (IROS), Madrid (2018)
14. Zhang, J., Singh, S.: LOAM: lidar odometry and mapping in real-time. In: Proceedings of Robotics: Science and Systems (2014)

15. Zhang, J., Singh, S.: Low-drift and real-time lidar odometry and mapping. Auton. Robot. **41**(2), 401–416 (2017)
16. Shan, T., Englot, B.: LeGO-LOAM: lightweight and ground-optimized lidar odometry and mapping on variable terrain. In: Proceedings of the 2018 IEEE/RSJ International Conference on Intelligent Robots and Systems (IROS), pp. 4758–4765 (2018)
17. He, L., Wang, X., Zhang, H.: M2DP: a novel 3D point cloud descriptor and its application in loop closure detection. In: Proceedings of the IEEE/RSJ International Conference on Intelligent Robots and System, pp. 231–237 (2016)
18. Himstedt, M., Frost, J., Hellbach, S., Böhme, H.J., Maehle, E.: Large scale place recognition in 2D LIDAR scans using geometrical landmark relations. In: Proceedings of the IEEE/RSJ International Conference on Intelligent Robots and System, pp. 5030–5035 (2016)

Assessment of Urban Bikeability with Structural Equation Model

Yijun Liu, Aitong Pang, and Gege Jiang(✉)

Sun Yat-Sen University, Shenzhen, China
jianggg@mail.sysu.edu.cn

Abstract. The concept of bikeability provides a basic framework for evaluation of cycling environment, which has great significance to promote bicycling as a low-carbon travel mode. This paper aims to propose a method for assessing urban bikeability by identifying the key characteristics influencing the willingness to use bicycle. An online questionnaire is designed for participants to score the indicators according to their cycling experiences and preferences. We adopt the structural equation model to divide the 15 indicators into five latent variables for information systemization. Based on the result, we address the relationship between the indicators and satisfaction level for cycling. Then, this paper develops an evaluation model to evaluate urban bikeability.

Keywords: Bikeability · Structural equation model · Cycling experience

1 Introduction

With the rapid economic development and urbanization, the motor vehicles in most cities have reached saturation, which has led to serious urban traffic problems. As an important component of the urban transportation system, bicycling has a great significance in energy conservation and emission reduction to build a low-carbon city. Therefore, it is of vital importance to promote the use of bicycles by building the cycling-friendly environment.

Bikeability is a broad concept related to many factors and it derives from the notion of walkability [1]. It can be used for illustrating the extent of a safe cycling environment [2, 3]. Comfort and safety are the most common factors to explain the notion of bikeability [4]. For example, Ehrgott took the suitability of the bicycling route into consideration [5], while Stinson and Bhat focused on safety and comfort [6]. In the field of urban planning, bikeability is often identified as factors to achieve a specific level of cycle service, such as connectivity and objectivity. In order to find a clear definition of bikeability, Manum established a relationship among street characteristics, bicycle flows and speeds by applying the space syntax analysis [7].

Bikeability was also regarded as an indicator to predict the dense of cyclists or the ratio of cycling in a given mode [8]. To evaluate the level of suitability of urban environments for bicycling, Krenn and Titze introduced the concept of bikeability through the observation of cycleway infrastructure and land use [9]. Furthermore, Gholamialam

W. Guo and K. Qian (Eds.): GBCESC 2022, LNCE 211, pp. 1151–1157, 2023.
https://doi.org/10.1007/978-981-19-5217-3_118

and Matisziw assessed bikeability by characterizing a road segment, including the level of traffic stress and route length [10].

Previous research mainly focused on describing bicycle characteristics and improving the comprehensive environment of cycleways through various evaluation methods. However, few studies demonstrate bikeability from the satisfaction of cyclists under specific travel purpose. This work proposes a modeling framework regarding to this issue. First, a survey is conducted to determine the indicators affecting the cycling experience for different purposes. Second, we present a cycling satisfaction evaluation method via the structural equation model to assess the urban bikeability. Third, we investigate the attitudes of cyclists to the overpass and underpass, which explores cyclists' preferences under the purposes for shopping.

2 Data

In this study, participants completed an online questionnaire. According to previous literature and empirical experiences, bicycles are mainly used for commuting, shopping, and leisure activities. Therefore, we classified the questionnaire according to the above three purposes and analyze them separately. For better illustration and explanation, each question is accompanied with some pictures. See Fig. 1 as an example, which presents the scenarios of signalized intersection and isolation facilities. The questionnaire is divided into two parts: 1) the personal characteristics of respondents, such as age, gender, education background; 2) the evaluation of cycling-related factors, such as frequency, time and purpose. The questionnaire is designed following the principles of State Preferences (SP) method.

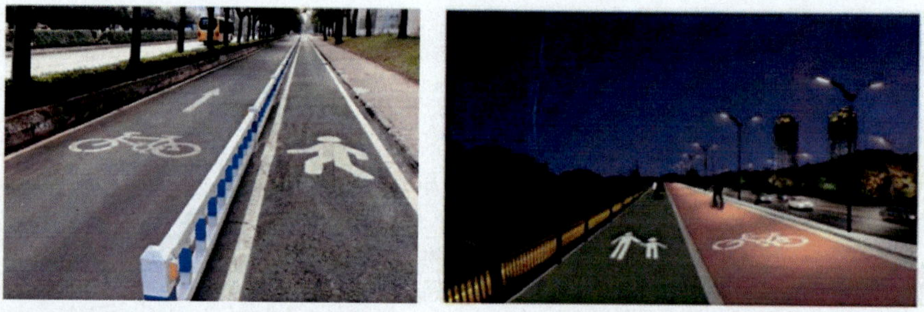

a) Setting of isolation facilities b) Setting of lighting facilities

Fig. 1. Illustration of scenarios.

The survey provides a detailed account of the factors that can influence cycling for different travel purposes. This study collects 492 results, most of the participants age between 18 to 30. It shows that the young people account for the majority of cyclists who ride at a high frequency (more than four times per week). We further analyze the influencing factors of bikeability under the purpose of shopping based on 157 respondents. The comparison between different traveling purposes will be discussed in our future study.

3 Methodology

The aim of the survey is to measure the respondents' preferences about infrastructural conditions under travel purpose for shopping. The methodology of the paper is unfolded in the following steps: 1) It analyzes the factors influencing the cyclists' preferences and constructs the evaluation model by calculating the Likert scale, which describes the suitability of cycling; 2) The structural equation model (SEM) was selected to evaluate bikeability by calculating the value of satisfaction. With the structural equation model, this paper work through the regression analysis to identify the relationship and interaction among indicators. The primary reference factors for cyclists to choose to travel by bicycle under different conditions will also be addressed.

Firstly, a series of Likert scale was used to measure the following factors on a scale from 0 (the lowest level of preferences of respondents for cycling experience) to 6 (the highest level of preferences of respondents for cycling experience). As Table 1 shows, this Likert scale is composed of 15 observation variables classified into five factors: Security, Comfort, Connectivity, Environment, Satisfaction. The structure of the model is depicted in Fig. 2, which illustrated the inter relationship among different variables.

Table 1. Indicators of model

Latent variable	Critical ratio	Observation variable
Security (F1)	5.479 ***	Interactions with other road users (V1)
		Setting of isolation facilities (V2)
		Setting of lighting facilities (V3)
Comfort (F2)	5.481 ***	Width of cycleways (V4)
		Flatness of cycleways (V5)
		Setting of shelter facilities (V6)
Connectivity (F3)	4.895 ***	Obstacles in cycleways (V7)
		Overpass/underpass in travel routes (V8)
		Waiting for traffic lights when crossing (V9)
Environment (F4)	4.958 ***	Planting trees around the travel routes (V10)
		Fumes and noise around the travel routes (V11)
		Streetscape around the travel routes (V12)
Satisfaction (F5)	–	Evaluation of cycling experiences (V13)
		Usage intention of cycling (V14)
		Recommendation intention of cycling (V15)

In Table 1, the values of critical ratio are all greater than 1.96, which means the model meets the requirements of validity and corresponding reliability. The asterisks in table mean the probability of getting a critical ratio in absolute value is less than 0.001.

In other words, the variance estimates for the latent variables are significantly different from zero at the 0.001 level (two-tailed).

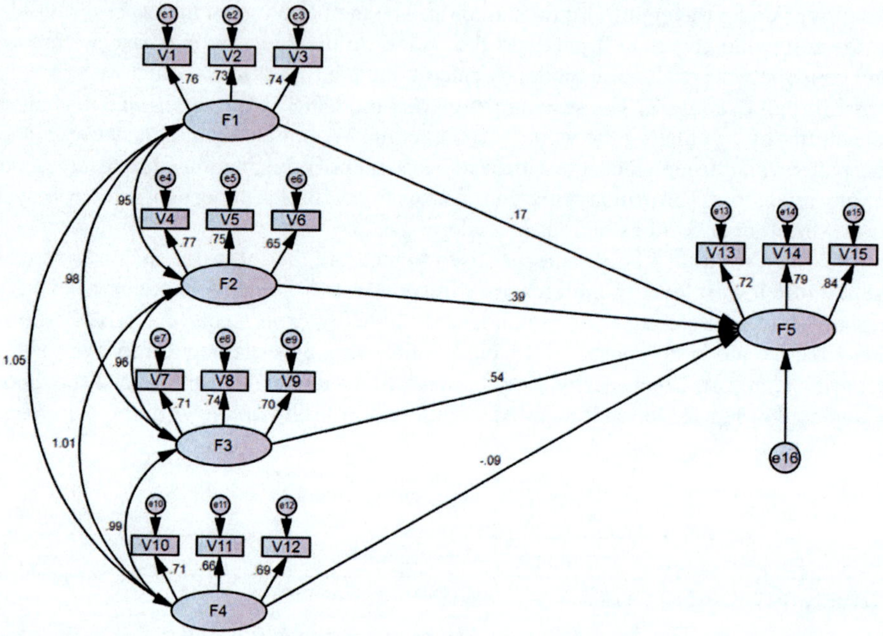

Fig. 2. Overview of standardized estimation of Structural Equation Model

In Fig. 2, F1 to F5 represent the latent variables and V1 to V15 represent the observation variables. The number over lines illustrate the regression weight of each variables, and e1 to e16 are the residuals. According to our calculation, the evaluation function of bikeability is:

$$Y_{(\text{bikeability})} = 0.167X_{1(\text{safety})} + 0.391X_{2(\text{comfort})} + 0.541X_{3(\text{connectivity})} \quad (1)$$

Detailed results for each variable are presented in Table 2. Obviously, connectivity and cormfort take high weight of the cycling satsfaction for shopping. By comparison, environment have little impact on satisfaction. Meanwhile, overpass and underpass will significantly distracts cyclists when going to shop, while the planting trees aroud the travel route get less concern of cyclists. Therefore, only three factors are chosen in our model.

Table 2. Standardized regression weights

Variables	Weights
F5 <--- F1	0.167
F5 <--- F2	0.391
F5 <--- F3	0.541
F5 <--- F4	−0.085
V1 <--- F1	0.765
V2 <--- F1	0.729
V3 <--- F1	0.739
V4 <--- F2	0.770
V5 <--- F2	0.747
V6 <--- F2	0.652
V7 <--- F3	0.708
V8 <--- F3	0.742
V9 <--- F3	0.701
V10 <--- F4	0.710
V11 <--- F4	0.657
V12 <--- F4	0.689
V13 <--- F5	0.715
V14 <--- F5	0.795
V15 <--- F5	0.841

In Table 3, it is known that this structural equation model has a chi-square (χ^2) of 112.3, degrees of freedom (DF) of 80 and a probability level (P) of 0.01. Besides, through confirmatory factor analysis, Good Fit Index (GFI), Number Fit Index (NFI), Incremental Fit Index (IFI), Non-Normed Fit Index are all above 0.90, Adjusted Good Fitting Index (AGFI) is greater than 0.08 and Root Mean Square Error Approximation (RMSEA) is under 0.08, which proves the validation of hypothesis for index significance. The indicators guarantee the credibility of our results.

Table 3. Notes of model for model fit summery

Index	Standard range	Model value
χ^2/DF	1.00 < NC < 3.00	1.486
GFI AGFI	>0.90 >0.80	0.902 0.881
RMSEA	<0.08	0.060
NFI	>0.90	0.919
IFI NNFI	>0.90 >0.90	0.972 0.969
P	<0.05	0.010

4 Conclusion and Discussion

This paper proposes a method for evaluating bikeability and analyzes the indicators that affect cycling under the purpose of shopping. First, the key factors affecting the cycling experiences are selected from previous literature. To systematically describe the characteristics of bikeability, these factors are divided into five main latent variables. Second, a questionnaire is conducted to obtain cyclists' satisfaction with various indicators reflecting the bikeability. Finally, the established structural equation model indicates that connectivity and comfort dominate the evaluation of bikeability.

The model shows that travelers place different values for distinct factors regarding to the satisfaction of cycling. For most cyclists, comfort and connectivity occupy the highest weight of cycling satisfaction for shopping. Intuitively, people pay more attention to convenience rather than time. It is, therefore, understandable that they would avoid overpass or underpass subconsciously for comfort. Furthermore, it is somehow surprising to see the natural environment and landscape around the cycling route are often ignored, namely the environment hardly affects the satisfaction of cycling experience. Compared with other aspects, cyclists are generally insensitive to the experience of the environment. Therefore, the authorities should lay more emphasis on connectivity of biking routes instead of landscape to promote bikeability.

There are some limitations and possible extensions. In this study, we only analyzed the results for riding for shopping but lacked other travel purposes, such as work commuting or leisure riding. Therefore, future researches can introduce other indicators of cycling bikeability that describe the individual psychology for a more robust evaluation of cycling bikeability. In addition, the bikeability model proposed in this paper could be integrated for cycling route planning under various travel purposes in the future study.

Acknowledgment. This study is supported by the Fundamental Research Funds for the Central Universities, Sun Yat-sen University (22qntd1701).

References

1. Porter, A.K., Kohl, H.W., Pérez, A., Reininger, B., Gabriel, K. P., Salvo, D.: Bikeability: assessing the objectively measured environment in relation to recreation and transportation bicycling. Environ. Behav. **52**(8), 861–894 (2020)
2. Puay Ping, K., Yiik Diew, W.: Influence of infrastructural compatibility factors on walking and cycling route choices. J. Environ. Psychol. **36**, 202–213 (2013)
3. Saghapour, T., Moridpour, S., Thompson, R.G.: Measuring cycling accessibility in metropolitan areas. Int. J. Sustain. Transp. **11**(5), 381–394 (2017)
4. Arellana, J., Saltarín, M., Larrañaga, A.M., González, V.I., Henao, C.A.: Developing an urban bikeability index for different types of cyclists as a tool to prioritise bicycle infrastructure investments. Transp. Res. Part A Policy Pract. **139**, 310–334 (2020)
5. Ehrgott, M., Wang, J.Y., Raith, A., Van Houtte, C.: A bi-objective cyclist route choice model. Transp. Res. Part A Policy Pract. **46**(4), 652–663 (2012)
6. Stinson, M.A., Bhat, C.R.: Commuter bicyclist route choice: Analysis using a stated preference survey. Transp. Res. Rec. **1828**(1), 107–115 (2003)
7. Bendik, B. et al.: Modelling Bikeability. In: Proceedings of the 11th International Space Syntax Symposium, pp. 3–7, Lisbon, Portugal (2017)
8. Krizek, K.J., Handy, S.L., Forsyth, A.: Explaining changes in walking and bicycling behavior: challenges for transportation research. Environ. Plann. B Plann. Des. **36**(4), 725–740 (2009)
9. Krenn, P.J., Oja, P., Titze, S.: Development of a bikeability index to assess the bicycle-friendliness of urban environments. Open J. Civ. Eng. **05**(04), 451 (2015)
10. Nathan, M.: Bikeability and the 20-min neighborhood: how infrastructure and destinations influence bicycle accessibility. Transp. Res. Rec. **2247**(1), 53–63 (2011)

Check for updates

Long Short-Term Memory Network and Ordinary Kriging Method for Prediction of PM2.5 Concentration

Junyou Liu, Bohong Zheng(✉), and Jinyu Fan

Central South University, Changsha 410083, Hunan, China
zhengbohong@csu.edu.cn

Abstract. When building a smart city, one must pay attention to the complex and diverse environmental problems in the city. Air quality prediction is an indispensable part of smart cities as it is conducive to guiding the development of air pollution prevention and control and alleviating urban air pollution. Based on the PM2.5 concentration data of 10 state-controlled air quality monitoring sites in Changsha City, Hunan Province, China from January 1, 2017 to December 30, 2021 and the meteorological data of Changsha city, a long short- term memory recurrent neural network model applicable to PM2.5 concentration prediction of Changsha city is formulated according to actual conditions with reference to related machine learning methods, and is used to predict the daily average concentration of PM2.5 at 10 air quality monitoring sites. Using the predicted PM2.5 concentration at the air quality monitoring sites in Changsha city, the spatial distribution simulation map of PM2.5 mass concentration within Changsha Third Ring Road is made by ordinary kriging method. Suggestions for PM2.5 pollution protection measures are put forward based on the predicted results. In addition, the interpolation analysis map of PM2.5 concentration distribution based on the measured value of each monitoring station is made. This is used to compare with the interpolation analysis map based on the predicted values so as to illustrate the rationality of the PM2.5 concentration prediction method.

Keywords: Smart city · PM2.5 · Long short- term memory network · Ordinary kriging method

1 Introduction

PM2.5, being a granular matter in the atmosphere, is an air pollutant widely concerned by people. It has a diameter of equal to or less than 2.5 microns, and can cause harm to human body when it enters the alveoli, inducing a series of respiratory diseases such as rhinitis, asthma, pharyngitis, and bronchitis as well as cardiovascular diseases such as hypertension, myocardial ischemia, and pneumonia. It is pointed out in studies that the risk of death due to cardiovascular disease increases by 12% when the concentration of PM2.5 in the environment increases by 10 μg per cubic meter. It is assumed by a report from the United Nations Environment Programme that about 340 thousand people

W. Guo and K. Qian (Eds.): GBCESC 2022, LNCE 211, pp. 1158–1169, 2023.
https://doi.org/10.1007/978-981-19-5217-3_119

die each year in China and India for every 20 μg per cubic meter increase in PM2.5 concentration in the environment. In addition, these fine particles are the main cause of haze weather, which lead to reduced atmospheric visibility and thus have a negative impact on urban traffic safety [1, 2].

Smart city is a space organically integrating digital city and physical city which form a space connecting virtuality and reality as well as provide various intelligent services by means of smart city technologies including cloud computing and artificial intelligence [3]. In terms of environmental impact, smart city is considered to be conducive to improving the level of science and technology and the efficiency of resource allocation to reduce environmental pollution [4]. Additionally, machine learning can be utilized to study computer simulation and realize human behaviors, which allows computers to predict the future based on past data, as humans do [5–7].

As one of the technical applications of artificial intelligence, machine learning has been applied by some researchers in urban PM2.5 prediction. Generally speaking, the prediction methods for air pollutant concentration are either simulation-based methods or data-mining-based methods. In simulation-based methods, physical and chemical models are often used to simulate pollutant emissions, transmissions, and chemical transformations. Whereas in data-mining-based methods, the correspondence between predicted values and dependent variables in time series data is predicted through statistical and machine learning technologies [5]. Some researchers have applied different machine learning methods to predict the concentration of PM2.5 in the environment. For instance, Qu et al. (2019) established three machine learning models concerning backpropagation (BP), convolutional neural network (CNN), and long and short-term memory (LSTM) models to predict the concentration of PM2.5 in Beijing. It is found that the prediction accuracy of the LSTM models was relatively high, at 81%, and that of BP and CNN were 48% [6]. However, Lin et al. (2020) proposed that a prediction model applicable in one city may not be applicable in another city due to different situations in different cities [7]. At present, some researchers are dedicated to improving existing prediction models for PM2.5. For example, Xu et al. (2018) conducted a case study in British Columbia to explore the application of 8 different machine learning models with aerosol optical thickness from satellite images to predict and map ground PM2.5 concentration [8]. It is found that the Cubist model performed the best in prediction. Meanwhile, some other researchers have also analyzed the harm of PM2.5 to the human body while predicting the PM2.5 concentration in some cities. For example, Goudarzi et al. (2021) predicted PM2.5 concentration in Ahvaz City of Iran through the artificial neural network method. In addition to that, they applied the World Health Organization's AirQ + software to evaluate the impact of different PM2.5 concentrations to human health. It is found that for a long time, the PM2.5 concentration in Ahvaz City had exceeded the standard limit set by Iran and some international standards. This long-term exposure to a PM2.5 polluted environment can result in a number of deadliest diseases, including Ischemic heart disease. However, as a data-mining-based method, machine learning can usually only be used to predict the average concentration of PM2.5 in measurement sites [9].

To study the distribution characteristics of PM2.5 quality concentration, some researchers deduced the distribution of the whole area by calculating the data of unknown points in the same area based on the concentration data of known points. For example, He et al. (2016) conducted an ordinary Kriging interpolation analysis based on the average value of PM2.5 quality concentration at 7 monitoring sites in Kunming, China in 2014, and obtained the predicted distribution map of PM2.5 quality concentration in urban Kunming [10]. In recent years, a growing number of researchers have applied simulation softwares such as CFD and ENVI-met to simulate and predict concentration of PM2.5 quality. For example, Wu et al. (2021) tested the ability of 23 common tree species in Shenzhen to reduce PM2.5 concentration through field investigation and wind tunnel test. The ENVI-met software was used to analyze the three green belt configurations (that is, trees, shrubs, and trees + shrubs) on the PM2.5 concentration of roads. Then, on this basis, the optimal road green belt configuration was proposed [11]. Meanwhile, some researchers have proposed different methods to analyze the PM2.5 concentration on the ground based on remote sensing images. For example, Xu and Zhang (2020) plotted the ground PM2.5 concentration map for Beijing from 2014 to 2016 using the MODIS aerosol optical thickness and modified regression model [12].

In this study, it is believed that there are few researches on the mapping of prediction distribution of PM2.5 concentration in the city according to the predicted value of PM2.5 concentration. We attempted to formulate a Long Short-Term Memory Recurrent Neural Network (LSTM-RNN) model applicable to PM2.5 concentration prediction in Changsha city according to actual conditions with reference to the related PM2.5 concentration prediction methods, and use it to predict the concentration of PM2.5 in Changsha City. In addition, on this basis, attempts are made in this study to draw a spatial distribution map of the predicted PM2.5 concentration at a certain time according to the PM2.5 concentration of each evaluating points as predicted. This paper will also include recommendations for people to carry out outdoor activities based on the prediction map produced.

2 Material and Method

2.1 Overview of the Study Area

Located between 27°53′ and 28°41′ north latitude and 111°53′ and 114°15′ east longitude, Changsha, the capital of Hunan Province in China, is an important central city in the middle reaches of the Yangtze River. It has a subtropical monsoon climate with varying temperature in spring, considerable amount of rainfall in early summer, constant high temperature in autumn, and less extreme weather in winter. In 2020, the total resident population of the city was 10.0479 million people [13]. According to the World Health Organization (WHO), the guideline annual average value for PM2.5 concentration is less than 5 $\mu g/m^3$ [14], and China's current annual average PM 2.5 concentration limit of the secondary standard is 35 $\mu g/m^3$ [15]. According to data from online monitoring and analysis platform of air quality in China, the annual average PM2.5 concentrations in Changsha City in 2018, 2019, and 2020 were 43.64, 47.17, and 40.82 $\mu g/m^3$, respectively [16]. The annual average value of PM2.5 concentration in Changsha City is higher than the safety value of PM2.5 concentration proposed by WHO and China's current annual average concentration limit of level 2 standard. This study predicts and analyzes

PM2.5 concentration in the area within Changsha Third Ring Road (Changsha Belt Highway).

2.2 Data Sources

Meteorological data and PM2.5 concentration data are studied in this paper. The historical meteorological data used (including minimum temperature, maximum temperature, wind strength, wind direction, and weather) are obtained from the large-scale weather forecasting website "Tianqi". PM2.5 concentration data are obtained from the real-time air quality index monitoring platform of China National Environmental Monitoring Center and zhenqi.tech. The experimental data will include historical meteorological data of Changsha City from January 1, 2017 to December 30, 2021 and the daily average PM2.5 concentration monitoring data at 10 state-controlled monitoring sites in Changsha.

2.3 Research Methods

In the experiment, under Windows operating system, Python was used as the programming language, and Jupyter notebook was used to build the PM2.5 concentration prediction model. Based on the daily average PM2.5 concentration monitoring data of 10 state-controlled monitoring sites in Changsha City from January 1, 2017 to December 30, 2021, we predicted the daily average PM2.5 concentration for 10 state-controlled monitoring points in Changsha City on December 31, 2021. In addition, in this paper, a spatial distribution map is drawn for predicted PM2.5 concentration for the area within the Third Ring Road of Changsha on that day of December 31, 2021 based on the predicted daily average PM2.5 concentration for 10 state-controlled monitoring sites in Changsha City on December 31, 2021 through the ordinary kriging method.

Data Coding. Meteorological data refers to the weather, wind direction, wind strength, maximum temperature, and minimum temperature, where weather and wind direction are text fields that need to be encoded. The principle of coding the weather and wind direction is based on the paper by Zheng and Zhu (2014) [17]. PM2.5 average values corresponding to different weather values were calculated and assigned values 1, 2…n according to the average values. The specific codes are shown in Table 1 and Table 2.

Table 1. List of codes for wind direction

Code	Wind direction
1	South wind
2	Southwest wind
3	Southeast wind

(continued)

Table 1. (*continued*)

Code	Wind direction
4	East wind
5	Northwest wind
6	Northeast wind
7	North wind
8	West wind

Table 2. List of codes for weather

Code	Weather	Code	Weather
1	Torrential rain	20	Thundershower
2	Windy to overcast	21	Overcast to cloudy
3	Light rain to torrential rain	22	Snowy
4	Moderate rain to rain	23	Light rain
5	Moderate rain to rainstorm	24	Light rain to overcast
6	Heavy rain	25	Cloudy
7	Light rain to clear	26	Cloudy to clear
8	Heavy rain to rainstorm	27	Shower
9	Overcast to light rain	28	Cloudy to overcast
10	Light rain to heavy rain	29	Overcast to rain
11	Moderate rain to heavy rain	30	Clear
12	Cloudy to light rain	31	Heavy rain to rain
13	Light rain to moderate rain	32	Overcast to light rain
14	Moderate rain	33	Overcast
15	Light rain to cloudy	34	Light rain to rain
16	Clear to rain	35	Haze to overcast
17	Cloudy to rain	36	Haze to cloudy
18	Rain	37	Haze
19	Clear to cloudy	38	Light snow

Data Normalization and Inverse Normalization Treatment. Normalization allows changing all the data to 0–1 to improve the accuracy of the model and increase the rate of convergence [9]. The maximum-minimum normalization treatment method was adopted in the research to normalize the meteorological data (weather, wind direction, wind strength, maximum temperature, and minimum temperature) and PM2.5 concentration. After that, inverse normalization is required to restore the results to the original dimensions. In the study, the meteorological data and the concentration of PM2.5 were normalized and inversely normalized according to Eqs. (1) and (2).

$$x_i' = (x_i - x_{min})/(x_{max} - x_{min}) \tag{1}$$

$$x_i = x_i'(x_i - x_{min}) + x_{min} \tag{2}$$

Detailed Description of the Prediction Method. In this study, the LSTM-RNN was used to build the model. The LSTM-RNN is good at dealing with continuous time series data. Because the LSTM-RNN can retain more long-term information compared to the general recurrent neural network model, the LSTM-RNN has the advantage of remembering information for a longer time [18]. The LSTM-RNN includes the input layer, output layer, and hidden layer (Fig. 1). The process is shown in Eqs. (3)–(7).

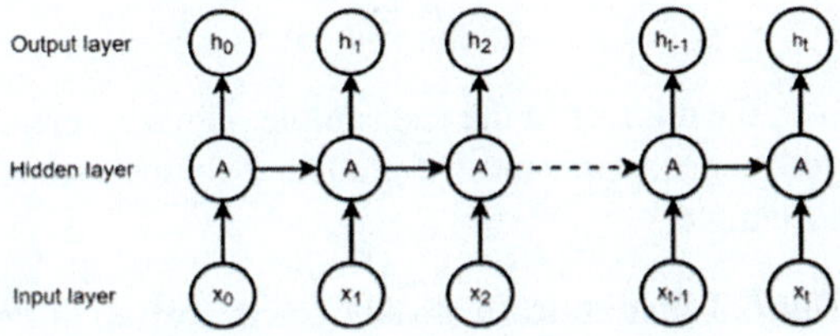

Fig. 1. Schematic diagram of LSTM network model

$$\text{Input gate} \qquad i_t = \sigma(W_{xi}x_i + W_{hi}h_{t-1} + b_i) \tag{3}$$

$$\text{Forget gate} \qquad f_t = \sigma\left(W_{xf}x_t + W_{hf}h_{t-1} + b_f\right) \tag{4}$$

$$\text{Neuron} \qquad c_t = i_t tanh(W_{xc}x_t + W_{hc}h_{t-1} + b_c) + c_{t-1}f_t \tag{5}$$

$$\text{Output gate} \qquad o_t = \sigma(W_{Xo}x_t + W_{ho}h_{t-1} + b_o) \tag{6}$$

$$\text{Output} \qquad h_t = o_t \tanh(c_t) \tag{7}$$

In which, c_t is the calculation rule of recurrent neural network cells of LSTM at Time t and h_t is the output of the calculation unit at Time t. W, U, and V are parameter matrices, b is the bias term, $\sigma(\bullet)$ is the activation function of Sigmoid, and $\tanh(\bullet)$ is the hyperbolic tangent activation function. $i_t, f_t,$ ando_t are the input gate, forget gate, and output gate at Time t [18].

In this study, the LSTM model provided by Tensor flow in the machine learning library was used to predict the concentration of PM2.5 in Changsha. 1825 pieces of original data were studied. After the elimination of outliers and null values and other treating processes, there were 1796 pieces of valid data. In this study, the first 1500 pieces of data will be used for training data, while the last 276 pieces of data will be used for test data. The number of samples per training (batch size) = 72. The number of rounds of training (epoch) = 400. In each round, the corresponding test data of the model was specified and each training record was output. The loss function was used as the mean square error and the adaptive moment was selected to estimate Adam as the optimizer. The data of the input layer includes historical daily average meteorological data and data of the concentration of PM2.5 of the station. The output layer is the predicted value of the mass concentration of PM2.5. That is, the related data of the n-1th day in history is used to predict the mass concentration value of PM2.5 of the nth day.

In this study, the root-mean-square error (RMSE) of the predicted values and real values of the concentration of PM2.5 was used to evaluate the model. In this study, the RMSE was calculated according to Formula (8).

$$RMSE = \sqrt{\frac{1}{n}\sum_{i=1}^{n}(y_i - y_i')^2} \tag{8}$$

In which, i refers to the number of the test sample points, n refers to the total number of days to be predicted; and y_i' represents the predicted values of the model. y_i represents the real values of the sample.

Predication of the Spatial Distribution of the Concentration of PM2.5. In this study, by using the ordinary Kriging interpolation analysis method, based on the predicted concentration of PM2.5 of 10 meteorological stations on January 1, 2022, the distribution diagram of the concentration of PM2.5 in the Third Ring Road of Changsha was drawn. With the ordinary Kriging interpolation method, we estimate the values of the unknown position based on the fitted variation function. The variation function can explain the space variation under the assumption of isotonicity and homogeneity. Some researchers describe the ordinary Kriging interpolation method with the following formula:

$$Z^*(x_o) = \sum_{i=1}^{n}\lambda_i Z(x_i) \tag{9}$$

In which, $Z^*(x_o)$ is the predicted value of the unmeasured position, $Z(x_i)$ is the measured value of the measured position, λ_i is the weighting coefficient from the measured position to the predicted position, and n is the number of known measuring points that can be searched in the field searching [19].

3 Results and Discussion

This study used the LSTM-RNN model to train and predict the PM2.5 concentration for each state-controlled air quality monitoring site according to the daily meteorological data of Changsha City from January 1, 2017 to December 30, 2021 and the daily average PM2.5 concentration data of the selected state-controlled air quality monitoring sites in Changsha City. The predicted results of the daily average PM2.5 concentration at 10 state-controlled air quality monitoring sites on December 31, 2021 are shown in Table 3. In this study, RMSE, a common evaluation indicator, is used to evaluate the model accuracy, as shown in Table 3. The LSTM-RNN model has produced good prediction results for each site in Changsha City, and the RMSE values are all below 20.

Table 3. Prediction effects and results for different sites

Site name	RMSE (μg/m^3)	Predicted daily average PM2.5 concentration on December 31, 2021 (μg/m^3)	Daily average PM 2.5 concentration measured at each station on December 31, 2021 (μg/m^3)
Mapoling	17.5	80	82
Hunan University of Chinese Medicine	16.3	80	81
New train station	17.3	90	98
Environmental Protection Bureau of Economic Development Zone	18.3	92	101
Environmental Protection Bureau of Tianxin District	15.9	80	93
Shaping	16.4	65	69
Wujialing	16.7	82	84
Environmental Protection Bureau of High and New Tech Development Zone	16.6	85	123
Environmental Protection Bureau of Yuhua District	17.6	86	90
Hunan Normal University	17.4	76	90

Based on the predicted data of daily average PM2.5 concentration in 10 state-controlled air quality monitoring sites in Changsha City on December 31, 2021, the Kriging model of the area within Changsha Third Ring Road was established according to Formula (9), and the obtained Kriging interpolation results are shown in Fig. 2. It can be seen from Fig. 2 that the predicted daily average PM2.5 concentration within Changsha Third Ring Road on December 31, 2021 shows a trend of high in the east and low in the west. The predicted daily average PM2.5 concentration in the region obviously exceeds the current first-class standard limit of daily average PM2.5 concentration in

China by 35 μg/m^3. On December 31, 2021, the predicted daily average PM2.5 concentration in most areas within the Changsha Third Ring Road exceeded the current second-class standard limit of daily average PM2.5 concentration in China by 75 μg/m^3 [15], and the predicted daily average PM2.5 concentration in certain areas in the north and southwest was lower than the second-class standard limit of daily average PM2.5 concentration.

According to the interpolation analysis map drawn for the predicted PM2.5 concentration at each station on December 31, 2021 (Fig. 2), 2021, residents (especially the elderly, children, pregnant women, and patients with cardiopulmonary diseases) should consider restricting travel to reduce exposure to PM2.5 pollution, lessen the times of window opening, and use air purifiers to purify indoor air. They should also consider taking public transport as much as possible if there is a need to travel, thereby reducing the use of bicycles and private cars. Areas with relatively low PM2.5 pollution concentrations in the north and southwest can be prioritized as travel destinations. In addition, the local government should attach great importance to PM2.5 pollution protection for workers who work outdoors for a long time on that day. A higher PM2.5 concentration may lead to a relatively large number of patients on that day. Hospitals should also take relevant measures in advance to cope with the possibility of a high number of patients on that day [1, 2].

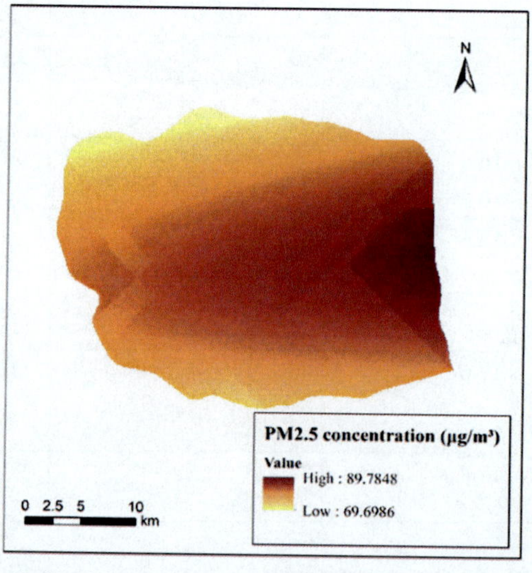

Fig. 2. Prediction map for PM2.5 concentration distribution within the Third Ring Road of Changsha drawn based on the predicted daily average PM2.5 concentration at each station on December 31, 2021

In this research, the distribution prediction map for the measured PM2.5 concentration within the Third Ring Road of Changsha (Fig. 3) was drawn based on the daily average PM2.5 concentration at each station on December 31, 2021 and through Kriging interpolation analysis. There will be errors in the predicted PM2.5 concentration distribution map whether based on the measured values or the predicted values. The interpolation results may be affected by the density, extent and attributes of sampling points. Since there are only ten nationally-controlled air-quality monitoring stations in and around this research region, there are only a few points available for interpolation analysis. This research found that, by comparing the predicted results in Figs. 2 and 3, there are some areas in the northern and southern parts of the Third Ring Road of Changsha City where PM2.5 concentration is lower than the national level 2 concentration limit of 75 $\mu g/m^3$. This also demonstrates that some suggestions made before the obtaining of the measured values and the interpolation analysis (for example, residents may give priority to the northern or southern suburb area within the Third Ring Road when outing) may be of practical significance. By comparing the measured and predicted daily average PM2.5 concentration values at each station on December 31, 2021 in Table 3, it can be found that the predicted and measured values for many stations are close to each other, with only differences of 1 to 2 $\mu g/m^3$. However, before the use of the measured PM2.5 concentration at each station in this research, it has been pointed out in this research that the overall PM2.5 concentration within the Third Ring Road of Changsha on December 31, 2021 was rather high, with that of most regions exceeding the level 2 concentration limit, 75 $\mu g/m^3$. So, a series of guiding opinions proposed to citizens in this research is also of immediate significance. In this research, the PM2.5 concentration at each monitoring station was predicted based on the LSTM-RNN model established in the research. There are only a few stations where the predicted values are quite different from the measured values, leading to the significant differences between the maximum PM2.5 concentration and PM2.5 concentration in some areas, as shown in Figs. 2 and 3. This also indicates that the overall interpolation results may not be significantly different even though there are only a few points whose values differ much. Adding interpolation points can, to some extent, make up for the obvious errors caused by a few interpolation points. To improve the accuracy of the interpolation analysis, researchers should continue to improve the prediction accuracy of PM2.5 concentration at each prediction point and make full use of the PM2.5 concentration prediction method with generally smaller errors to satisfy the needs of excessive reliance on the accuracy of each prediction point in the interpolation analysis. It is believed in this research that smart cities do not only present more and more unprecedented new technologies from the perspective of time but also make every corner of our cities benefit from these technologies from the perspective of space. Many studies have focused on the prediction of PM2.5 concentration hour by hour, with a relatively short time range of prediction, though. This paper argues that predicting the daily average PM2.5 concentration, increasing the range of prediction time, and predicting the situation of the whole city using interpolation analysis can help us early find problems throughout the city and take corresponding preventive measures. Additionally, we also found in our research that in our machine learning, it is rather difficult for the machine to take into account some factors such as holidays when predicting PM2.5 concentration, which may lead to significant changes in the daily average PM2.5 concentration. A machine is poor in prediction when the daily average PM2.5 concentration on a given day changes greatly compared with that of the previous days.

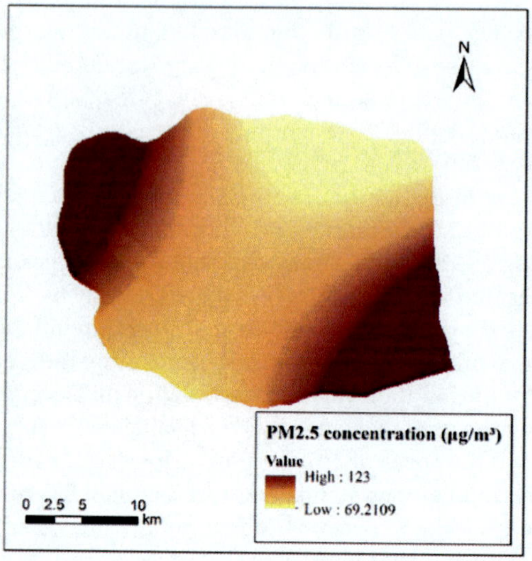

Fig. 3. Prediction map for PM2.5 concentration distribution within the Third Ring Road of Changsha drawn based on the measured daily average PM2.5 concentration at each station on December 31, 2021

4 Conclusion

Based on the meteorological data and PM2.5 concentration data in Changsha City, a LSTM-RNN model was established to train the meteorological data and pollution data of 10 state-controlled air quality monitoring sites to predict PM2.5 concentration of the next day. Based on the prediction data of PM2.5 concentration at state-controlled air quality monitoring sites, the spatial distribution of PM2.5 pollution was predicted by the ordinary Kriging method. According to the predicted values, some suggestions on the protective measures against PM2.5 on this day are put forward. The results show that the PM2.5 concentration prediction model based on the LSTM-RNN model and the ordinary Kriging method can predict the future PM2.5 concentration in time and broadly predict the PM2.5 value in a certain area.

References

1. Xinxing, Y., Lihua, F., Peng, W.: Air particulate matter PM2.5 in Beijing and its harm. Front. Sci. **6**(1), 22–31 (2012)
2. Qin, X., Yuting, L.: Application of the complex network in the PM 2.5 analysis of air pollution in regions. J. Technol. **19**(1), 77–83 (2019)
3. Chong, Y., Feng, Z., Guangliang, X.: Research progress and prospect of smart city in China. Hum. Geogr. **36**(5), 15–23 (2021)
4. Chu, Z., Cheng, M., Yu, N.N.: A smart city is a less polluted city. Technol. Forecast. Soc. Change **172**, 121037 (2021)
5. Karimian, H., et al.: Evaluation of different machine learning approaches to forecasting PM2.5 mass concentrations. Aerosol Air Qual. Res. **19**(6), 1400–1410 (2018)

6. Yue, Q., Xu, Q., Hongqing, S., Jie, H., Jianhui, L., Hao, X.: Machine-learning-based model and simulation analysis of PM2.5 concentration prediction in Beijing. Chin. J. Eng. **41**(3), 401–407 (2019)
7. Lin, L., Chen, C.Y., Yang, H.Y., Xu, Z., Fang, S.H.: Dynamic system approach for improved PM2.5 prediction in Taiwan. IEEE Access **8**, 210910–210921 (2020)
8. Xu, Y., et al.: Evaluation of machine learning techniques with multiple remote sensing datasets in estimating monthly concentrations of ground level PM2.5. Environ. Pollut. **242**(B), 1417–1426 (2018)
9. Gholamreza, G., Hopke, P.K., Mohsen, Y.: Forecasting PM2.5 concentration using artificial neural network and its health effects in Ahvaz, Iran. Chemosphere **283**, 131285 (2021)
10. Lu, H., Guangzhi, D., Zhenran, G.: Distribution characteristics of PM2.5 and PM10 mass concentration in Kunming main urban area. Environ. Eng. **34**(A1), 529–534 (2016)
11. Wu, J., Luo, K., Wang, Y., Wang, Z.: Urban road greenbelt configuration: the perspective of PM 2.5 removal and air quality regulation. Environ. Int. **157**, 106786 (2021)
12. Xu, X., Zhang, C.: Estimation of ground level PM2.5 concentration using MODIS AOD and corrected regression model over Beijing, China. PloS One **15**(10), e0240430 (2020)
13. Changsha overview. http://www.changsha.gov.cn/xfzs/zjmlzs/zsgl/200907/t20090727_5686409.html. Accessed 22 Jan 2022
14. Ambient (outdoor) air pollution. http://www.who.int/news-room/fact-sheets/detail/ambient-(outdoor)-air-quality-and-health. Accessed 22 Jan 2022
15. Ministry of environmental protection of the people's republic of China. National standard of the people's republic of China: ambient air quality standard (revised in 2012). Beijing: ministry of environmental protection of the people's republic of China (2012)
16. Monthly statistical historical data of Changsha air quality index. https://www.aqistudy.cn/historydata/monthdata.php?city=%E9%95%BF%E6%B2%99. Accessed 24 Jan 2022
17. Yi, Z., Chengzhang, Z.: A prediction method of atmospheric PM2.5 based on DBNs. J. Shandong Univ. (Eng. Sci.) **44**(6), 19–25 (2014)
18. Yang, G., Wang, Y., Li, X.: Prediction of the NOx emissions from thermal power plant using long-short term memory neural network. Energy **192**, 116597 (2020)
19. Gia Pham, T., Kappas, M., Van Huynh, C., Hoang Khanh Nguyen, L.: Application of ordinary kriging and regression kriging method for soil properties mapping in hilly region of central Vietnam. Int. J. Geo-Inf. **8**(3), 147–165 (2019)

Check for updates

The Influence of the Characteristics of the Elderly on the Landscape Design of Residential Area Suitable for Aging

Xiangyun Wang(✉) and Feng Wang

Shandong Youth Political College, Jingshi Dong Lu 31699, Jinan 250102, Shandong, China
wxy@sdyu.edu.cn

Abstract. With the continuous improvement of the age of the elderly, their psychological and physiological functions are weakening, and the demand for the use of landscape in residential areas has also changed greatly. This paper makes a detailed analysis and study of the psychological characteristics, physiological characteristics and behavioral activity rules of the elderly group, and concludes its influence on the overall planning, spatial layout, landscape facilities, construction technology and other aspects of the residential area suitable for aging garden landscape design principles and design strategies. It provides some theoretical support and design guidance for the construction of garden landscape space in line with the actual use needs of the elderly group.

Keywords: Physiological feature · Psychological feature · Behavior law · Behavioral activity needs · Design principles and strategies

1 Research Background

Under the background of the gradual acceleration of the proportion of aging, with the gradual growth of the elderly, the degradation of the elderly's comprehensive physical function, the physical and mental and pressure bearing capacity of the elderly and other actual situations, the requirements for the actual use of landscape for aging in residential areas have been increasing. At present, there is a serious imbalance in the supporting situation of most residential areas suitable for aging, which ignores the use needs of multi-layer suitable for aging garden space for healthy living activities of the elderly. Residential area landscape design should give top priority to optimum design of ageing, according to the old people's physical and mental characteristics, design for the elderly to accompany, trip, communication and simple rehabilitation exercise everyday activities such as ecological garden space, for our country the elderly health pension that occupy the home to provide comfortable and livable optimal aging landscape outdoor leisure communication.

W. Guo and K. Qian (Eds.): GBCESC 2022, LNCE 211, pp. 1170–1178, 2023.
https://doi.org/10.1007/978-981-19-5217-3_120

2 Research Contents

Refer to literature aging and residential area landscape design in the necessity of the selected topic research background, significance and research, based on the elderly outdoor landscape space that occupy the home use requirements, from the elderly physiological, psychological characteristics and behavioral activities from several aspects were analyzed influence of landscape design of residential area optimum aging element, The design principle and strategy of residential area suitable for aging garden landscape are summarized.

3 Research on the Characteristics of the Elderly

The landscape design and construction of residential area suitable for aging should fully consider the characteristics of the main users of landscape, namely, the elderly group at home. Based on the changes of the physiological, psychological and behavioral characteristics of the elderly, this paper analyzes and summarizes the changes of their special needs for the use of landscape. For the construction of humanized residential area suitable for aging landscape design to provide a premise guarantee.

3.1 Physiological Characteristics of the Elderly

As the growth of the age, the elderly perception, athletic ability and nervous system will gradually decline, the changes of these physical characteristics will directly influence the daily activities of the ability and the ways of participation, the landscape design of residential area optimum aging will try to make up for the elderly in the process of building physical characteristics.

Change in Perception. Perception mainly covers vision, hearing, taste, smell and touch, collectively known as the five senses of perception. Most people after 60 years old, visual acuity will appear obvious decline, as the age continues to grow to color perception and identification ability, visual distance, visual sensitivity will gradually decline. Old people's sense of smell, touch, hearing, taste, etc., will gradually deteriorate with the growth of age, and the decline of organs will lead to the decline of the elderly group space maintenance ability.

Decline in Motor Ability. As the elderly age, muscle mass decreases, leading to a decline in muscle strength and motor control. The bones of the body gradually become brittle and lead to osteoporosis, which is prone to fracture and ligament injury during exercise and has a long recovery cycle.

Deterioration of the Nervous System. The nervous system, respiratory system, metabolism and cardiovascular system of the elderly will gradually decline with the growth of age, resulting in the decline of behavioral response ability of the elderly, the decline of comprehensive physical fitness and other problems. Therefore, in the design of garden landscape suitable for aging, we should give full consideration to practical factors such as the activity ability and activity range of the elderly group, set up appropriate communication rest space, and pay attention to control the overall space layout and plant planting design and other related factors.

3.2 Psychological Characteristics of the Elderly

Mental Profile. With the decline of the physiological function of the elderly group, and the change of social school, social status, their psychological adaptability and psychological security weakened, bringing inferiority, loss, emptiness, loneliness and other negative psychological emotions. The sense of loss is the primary psychological feature of the elderly group. Withdrawing from the job makes the elderly unable to find a sense of personal value, making the extroverted elderly become stubborn and irritable, and the introverted elderly become withdrawn and depressed. Loneliness is also the main psychological characteristics of the elderly group, most of the elderly due to physical and age reasons gradually quit the work circle, return to home life, resulting in a tight social scope, the lack of family members to accompany loneliness and abandonment, resulting in the elderly spirit loss, reduced ability to live and other phenomena. There are anxiety is also an important psychological characteristics of the elderly group, along with the physical, sleep quality and social presence of the continuous decline, its risk resistance and psychological tolerance becomes fragile, easy to lead to depression and further insomnia [1]. The elderly need to get social attention and recognition like young people. In the process of landscape design, we should strengthen communication, exchange and experience space design, promote communication between the elderly and different groups, and reshape the life value of the elderly.

Psychological Needs of the Elderly. When the elderly leave work return to home, there will be a long time of adaptation in his heart, they will become more sensitive, loss, anxiety and other negative emotions, good landscape design of residential area to help the elderly to regain confidence, build good pension mentality, out of the heart trouble into healthy aging life. According to the psychological characteristics of elderly population, residential area landscape planning and design should be fully starting the elderly pension belonging needs that occupy the home, use security landscape tourist facilities, provide the communication of neighborhood contact space, communication, entertainment and chess, willingness to grow, sports, painting, music, walking, and so on self-realization of sexual activity, practical space requirements.

3.3 Behavior and Activity of the Elderly

The Time and Space Rule of the Behavior Activity of the Elderly. The elderly's daily activities mainly affected by the activity demand space range, and is limited by its own physiology, according to the research literature of jiang-ping wang and other scholars found that according to the division of the elderly daily activity space distance between the scope of activities can be divided into curtilage activities of activities and residential area, residential area in activity 3 types [2]. Due to the influence of physical function and other factors, indoor activity is the most frequent travel area for the elderly. According to relevant literature review and research, the time for outdoor activities of the elderly is relatively regular, with 7–10 am as the daily exercise stage. 3 to 5 PM is the stage of leisure activities, such as chess and cards, fishing, chatting, walking, exercise, etc. This period is an important communication time for the elderly group. From 7 to 9

PM in the evening, it is a healthy stage before going to bed. The main activities include brisk walking, square dancing and folk art. However, due to the difference in climate between north and South, the outdoor temperature in winter in north China is lower, so the number and frequency of evening travelers are significantly reduced.

Behavioral Preferences and Characteristics of the Elderly. With the decline of the elderly group's physical function, the elderly's desire to maintain physical health is increasingly strong, such as walking, tai chi, fitness equipment, square dancing and other outdoor space fitness anti-body shape activities are widely favored by the elderly group. At the same time, the elderly group lacks the company of their family members during the day, and recreational activities such as chess and cards, walking, chatting and fishing are also relatively important activities. Daily sports and hobby behavior activities are an important part of the daily life of the elderly group, which is highly valued, so it has long-term nature and regularity [3]. Under the background of the same life and cultural groups with common interests and common monitoring situation will take care of each other to attract each other and chess spontaneously gathered together, chat, a square dance, a fishing activities such as sex and gathering, but due to the lack of security, and most of the elderly they prefer gathered in half open and half illicit close sex of the space in daily communication activities. However, due to the restriction of the daily body conditions of the elderly group, most activities are concentrated in the garden space of the residential area, so their behavior activities have the characteristics of time domain and region.

4 Behavioral Needs and Planning and Design Principles

4.1 Behavioral Activity Needs

By the elderly physical, psychological and daily activities demand, the whole residential area optimum aging landscape planning and design layout, the elderly should be considered in the process of practical use, and in the landscape space layout, traffic system, the function partition, service facilities, such as the construction technology for optimal design of ageing, Based on the psychological needs of the elderly group, the layout of garden space should be graded, and the design of intersecting opening and semi-open space and private space of solitude regulation should be established [4]. Based on its physiological characteristics in the garden road, pavement, lighting, node design to consider accessibility, accessibility, skid resistance, lighting layout and other factors of the design. Based on the elderly health status, cultural level, habits, preference and heart factors should be considered in the overall layout of different requirements, pay attention to during the process of design activity zoning, spatial processing technique, the layout of the space to surround close and related infrastructure design, creating comfortable alzheimer's garden and sports communication space [5].

4.2 Principles of Landscape Planning and Design for Suitable Aging

Ecological health and environmental sustainability are the development themes of today's society. In the initial stage of residential area suitable for aging landscape design and construction, it is necessary to systematically explore the location environment around the residential area, climate characteristics, local culture and so on, and make comprehensive analysis, so as to achieve long-term ecological sustainability of residential area suitable for aging landscape [6]. With the continuous improvement of age, the ability of the elderly to adapt to the environment gradually decreases. The comfort and safety of landscape environment is the first principle of planning and design, and safety includes traffic safety and site safety. Comfort mainly includes environmental comfort, scale comfort and cognitive convenience. According to the actual use needs of the elderly group, the layout design of the overall garden space should consider the distribution of landscape levels and the rationality of the layout of function echo, and the layout of landscape function zoning should be coordinated with the spatial use distribution of the elderly group. For most of the elderly group, participation in community activities can improve communication opportunities, keep a happy mood, reduce inner loneliness, and restore their optimistic, open-minded and positive attitude. Benign community landscape space design will help the elderly health of body and mind development, based on the elderly body condition and the differences of physiological needs, in the process of planning and design should meet the demand of different levels of space as far as possible use demand, the pension for the elderly home health to provide benign promotion platform, so as to promote the benign development of the physiology and psychology.

4.3 Behavior Activity Preference and Type of the Elderly

With the decline of the elderly group's physical function, the elderly's desire to maintain physical health is increasingly strong, such as walking, tai chi, fitness equipment, square dancing and other outdoor space fitness anti-body shape activities are widely favored by the elderly group. At the same time, the elderly group lacks the company of their family members during the day, and recreational activities such as chess and cards, walking, chatting and fishing are also relatively important activities. The main types of behavioral activities for the elderly were leisure activities such as flower cultivation, planting, sitting, watching and gardening experience. In addition, the type of outdoor activities for the elderly is also affected by physical condition and gender.

4.4 Behavior and Activity Characteristics of the Elderly

Daily sports and hobby behavior activities are an important part of the daily life of the elderly group, which is highly valued, so it has long-term nature and regularity. Under the background of the same life and cultural groups with common interests and common monitoring situation will take care of each other to attract each other and chess spontaneously gathered together, chat, a square dance, a fishing activities such as sex and gathering, but due to the lack of security, and most of the elderly they prefer gathered in half open and half illicit close sex of the space in daily communication activities. Through the communication of daily activities, the elderly group will gain their sense of participation and regain their self-confidence. However, restricted by the daily body conditions of the elderly group, most activities are carried out in the garden space of the residential area, so their behavior activities have time domain and regional characteristics.

5 Planning and Design Strategy

5.1 Rational Planning of Landscape Spatial Layout

The design and layout of landscape space suitable for aging should conform to the behavioral activity characteristics and needs of the elderly, and follow the principles of reasonable distribution, safety and comfort. The spatial layout of residential gardens is mainly affected by the residential volume, topographic structure and distribution form, and usually has several common layout forms, such as block, center type and axis type [7]. According to the distance and location of landscape, residential landscape can be divided into three types of spatial hierarchical layout: house level, group level and community level. According to the needs of the elderly, the landscape space functions of the residential area suitable for aging can be divided into four types: recreational communication, recreation and viewing, health and fitness, gardening and planting. In the layout design process, reasonable classification planning should be carried out according to the functional needs of each type of space, use frequency, site interest and other factors (Table 1).

Table 1. Common health benefits of plants.

Health care effect	Floristics
Absorption of harmful substances	Poplar, elm, euonymus, crape myrtle, hibiscus
Reduce UV rays	Pewter, grape, eucommia, marigold
Sterilization, insect repellent	Cedar, mint, honeysuckle, citronella, tamex
Purified air	Begonia, chrysanthemum, pine, cypress, fir
Vision care	lidendron chinense, Acer chinense, iris chinensc
The sense of smell care	Cedar, magnolia, peach tree, cypress, lilac, rose
Tactile effects	Aloe, mimosa, Acacia, Magnolia grandiflora
Hearing care	Banana, lotus, cedar, bamboo
Edible health care	Rose, gingko, jasmine, chrysanthemum, plantain

5.2 Build a Healthy and Safe Road System

With the increasing age of the elderly group, the physical function gradually weakened, accompanied by weakened mobility, slow reaction ability and other problems, residential road traffic system is an important part of the elderly group in residential landscape space travel activities. A good road traffic system can organize all nodes, facilities and sites in the residential garden landscape space in an orderly way, so as to effectively guarantee the safe passage of people and vehicles (Fig. 1 and Table 2).

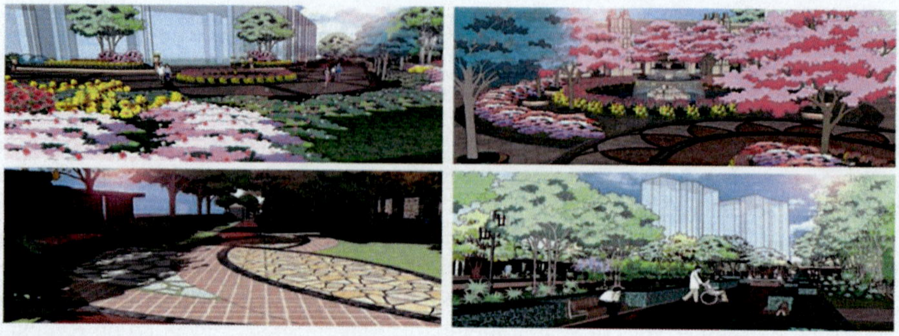

Fig. 1. In line with the needs of elderly behavior for aging flower belt, pavement, garden road, rest area design.

Table 2. The psychological influence of different colored plants on human beings.

Plant color	Dominant surface plant	Positive feelings	Health effects
Green	Willow, pine, cypress	Security, peace	Relieve visual fatigue
Red	Albizia, peony, Chinese rose	Enthusiastic, festive, excited	Promote blood circulation
Yellow	Ginkgo and Yuanbao maple	Warm, soft, calm	Stimulate the nervous system
Purple	Lavender, purple leaf plum	Elegant, mysterious	Calm down
Blue	Blue chamomile, hydrangea	Quiet, free, fresh	Calm spirit
White	Magnolia, jasmine, gardenia	Purity, peace	Relieve the pressure

5.3 Pay Attention to the Design of Elements Suitable for Aging Site

Terrain design is an important part of landscape design, landscape space terrain planning and design is the basis of all landscape elements layout, terrain ups and downs play a role in climate regulation, space enclosure, organizational axis and so on. The landscape space of residential area suitable for aging can be divided into three basic types: flat land, slope land and mountain land. Different terrain structure forms can guide the landscape visiting path, psychology and visiting behavior of the elderly group. Plant landscape design can beautify the environment, divide the space and so on, is an important element of residential landscape. Plant landscape design is both aesthetic and practical. In the process of landscape design for residential area suitable for aging, it should be selected for health care, rational space layout, suitable location of tree species and safe and comfortable under the premise of meeting the behavioral and activity needs of elderly groups [8]. Design of residential area landscape water landscape construction, can be adjusted to local air humidity and local microclimate, most people are on the water environment has the strong kindness, clever waterscape can make people relaxed mood, bring pleasant feeling, water landscape is divided into static and dynamic two kinds of design form, divided into dot, ribbon, planar layout of three forms.

5.4 Improve the Design of Barrier-Free Facilities

Due to the limitations of the physical conditions of the elderly, they cannot keep exercise for a long time, so they have to sit down for rest adjustment at certain intervals or distances. Rest and viewing activities are also important activities for the elderly. Good design of seat layout, scale and material in residential landscape space can better provide outdoor rest services for the elderly and help the elderly to recover their physical strength [9]. For the design of landscape lighting, not only the space beautification effect of lighting should be considered, but also according to the declining situation of vision and light sense of the elderly group, it is appropriate to enhance the lighting layout in landscape nodes, waterfront area, road space, gentle slope, steps and other dangerous locations to ensure the safety of the elderly group at night. Based on the elderly body status in the landscape space layout design should minimize the use of the steps, the gentle slope terrain ups and downs in the smaller area to replace, slope position can be adjusted to the appropriate terrain larger or use the font of ramp design, the location of the steps must be set to ramp beside the Settings for the action inconvenience of mobility, dielectric model use the old man, And set up a safety warning mark at the steps or adopt high identification of pavement to remind the change of height difference. Elderly affected by body function decline in its space recognition ability, memory will decline gradually, the suitable aging should be considered in the design of landscape space identity system of system design, assist the elderly cognitive enhancement of the landscape space identification and system, can through color, shape and material changes increase the identification of the elderly. To ensure the safe use of all kinds of elderly groups (Fig. 2).

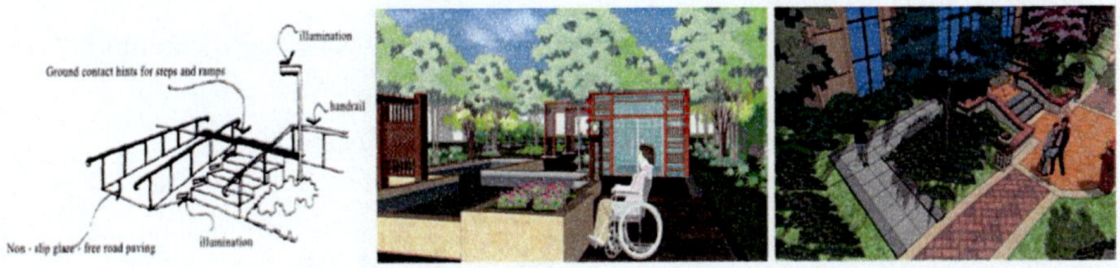

Fig. 2. Barrier-free facilities design.

6 Conclusions

Based on the changes of the physiological and psychological characteristics of the elderly group, through the study of the behavior and activity rules of the elderly group, activity needs, design principles and other aspects, the actual use needs of the elderly group for daily communication and activities should be fully considered in the landscape design of the aging residential area. This paper analyzes and discusses the planning and design strategy of residential area suitable for aging garden landscape from four aspects: landscape space layout, site design elements, transportation system design and barrier-free design. Based on the behavior and activity rules of the elderly, the planning and design process of residential garden landscape for aging should follow the principles of ecological health, comfort and safety, rational layout, participation and functional diversity. In the overall layout, dynamic and static zones are carried out. In the garden road design, pedestrians and vehicles are separated and pavement, lighting and barrier-free design are optimized. In addition, plant planting design is used to divide garden space and implant plant rehabilitation therapy design. Combined with the above elements, the residential area is designed to build a comfortable, safe, ecological and livable garden landscape space suitable for aging.

References

1. Shizhang, Y.: Urban Residential Environment Design. Chongqing University Press, Chongqing (1999)
2. Jiangping, W.: Planning and Design of Residential Environment for the Elderly. China Electric Power Press, Beijing (2009)
3. Havinghurst, R.: Mental life of the elderly. Landscape Architecture (2018)
4. Diane: Site Planning and Design for the Elderly. Van Nostrand Reinhold Company, New York (1985)
5. Xiaotong, C.: Application of rehabilitation landscape in residential areas for the elderly. Kunming University of Science and Technology (2018)
6. Liuyi, Z.: Research on landscape design method of community Park focusing on elderly group. East China University of Science and Technology (2017)
7. Yudai, Q.: Architectural Design Drawings for the Elderly. Science and Technology Press, Jinan (2004)
8. Xin, W.: Garden design of elderly community based on elderly behavior activities. Beijing Forestry University (2020)
9. Shufang, C., Jingjing, J.: Research on landscape facilities design of elderly apartment based on interactive concept. Packag. Eng. **37**(10), 85–89 (2016)

Travel Mode Recognition Using Mobile Phone Signaling Data

Fanghao Fu, Jiemin Xie(✉), Shuqi Zhong, and Ming Cai

Sun Yat-sen University, 51006 Guangzhou, China
xiejm28@mail.sysu.edu.cn

Abstract. Urban residents can choose different travel modes, including walking, cycling, taking public transportation, and driving, and the information about the residents' daily travel mode choices is important in the urban traffic planning and management. Therefore, the collection of such information is important. However, the traditional approaches are generally resource-consuming. Mobile phone signaling (MPS) data provides a chance to effectively collect residents' travel mode choices. MPS data are generated when mobile phones connect to their nearby base stations. With the widely use of mobile phones, millions of MPS records are generated every day. This study explores the application of MPS data in travel mode recognition. Based on the MPS data and navigation trajectory, the proposed algorithm has four main steps: a) invalid data cleaning, b) staying point recognition, c) road matching process based on the Hidden Markov Model, and d) trajectory similarity calculation based on the Dynamic Time Warping. An Experiment is preformed to test the applicability of the proposed algorithm.

Keywords: Travel mode recognition · Mobile phone signaling data · Road matching · Navigation trajectory

1 Introduction

The performance of the urban transport system is highly related to residents' travel mode preferences. When the most of residents choose to drive, the road network is generally congested. Thus, understanding residents' travel mode preferences is vital to optimize traffic planning and management. The traditional methods of collecting urban residents' travel mode data, such as resident travel questionnaire, are generally time-consuming and thus suffered from the low sampling rate. With the development of big data technology, there are increasing researches which explore new approaches to effectively collect the travel data. Vehicle GPS data is a relatively reliable data, but it cannot fully describe the residents' model choices. For example, it needs other methods to collect data of walking and cycling. Some researchers [1, 2] suggest that mobile phone signaling (MPS) data can be useful in the collection of the travel mode data.

W. Guo and K. Qian (Eds.): GBCESC 2022, LNCE 211, pp. 1179–1187, 2023.
https://doi.org/10.1007/978-981-19-5217-3_121

Mobile phones actively or passively connect to the communication network, these connections are recorded and formed MPS data. By March 2020, the number of mobile phone users in China reached 987 million [3]. Since the 4G era, the construction of base stations has been continuously promoted, and the coverage of base stations has been expanding year by year. By 2020, the total number of mobile communication base stations has reached 9.31 million nationwide [4]. A mobile phone continues generating interactive logs which record the information of its ID, the connection times, and the IDs of the connected base stations. As the locations of base stations are given,,the MPS data can be used in various cases, such as monitoring performance of urban road networks [5], analyzing people behaviors after the earthquake [6], and estimating travel demand patterns [7]. Also, the MPS data can be used to analyze mobile phone users' travel characteristics, such as identifying whether a mobile phone user is a resident or a tourist [8], recognizing residents' trip origin and destination (OD) [9], and distinguishing mobility phases from individual activities [10].

The existing studies on the travel mode identification using MPS data include the fuzzy identification method according to a prior knowledge [1, 2], the identification process based on a variety of machine learning models [11], and the method based on the combination of navigation trajectories and MPS trajectories [12]. However, the application of the fuzzy identification is limited because its core theory is based on empirical values. And the method with navigation trajectories does not consider residents' real travel trajectories and the measurements of identification are ambiguous.

This study proposes an algorithm to identify residents' choice of travel modes based on MPS data and navigation trajectory. The algorithm does not rely on empirical values, and it considers residents' real trajectories on the urban roads and clearly defines the measurement of identification. The main contributions of this study can be summarized as follows:

a) A systematic process of MPS data cleaning is established and applied to the real dataset.
b) A framework of the travel mode identification based on MPS data is formed.

2 Methodology

The framework of the travel mode identification is shown in Fig. 1.

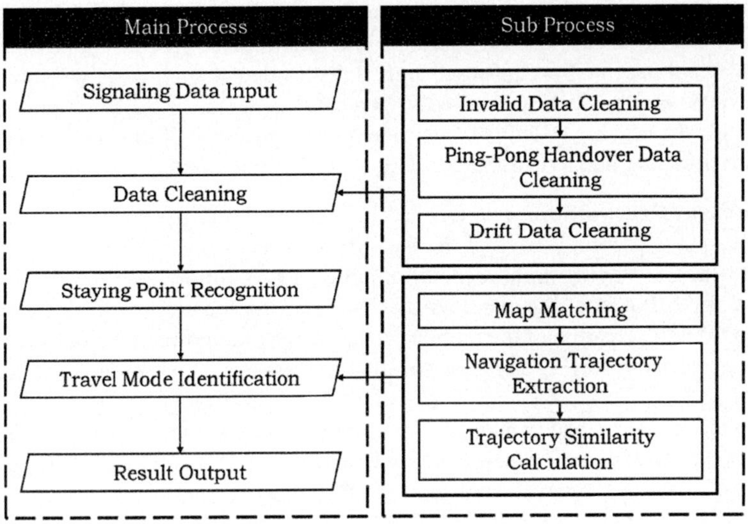

Fig. 1. Method framework.

After the MPS records are input, the data cleaning is performed, and then the staying point recognition is conducted. With the information of staying points, we match the trajectory of MPS data to the real road network by the Hidden Markov Model (HMM), and consider the result as the predicted trajectory. Based on the predicted trajectory's OD, the navigation trajectories of different travel modes are extracted. Finally, the Dynamic Time Warping (DTW) algorithm is used to identify resident's travel modes.

2.1 Data Preprocessing

Invalid Data Cleaning. Sometimes the signal is weak, so that the key information such as the base station's ID is missing, or the same record is repeatedly documented in a period, resulting in redundant records. The above invalid data should be detected and deleted.

Ping-Pong Handover Data Cleaning. When a mobile phone user is about to reach the radiation boundary between base stations A and B, a ping pong handover between these base stations may occur, including single-switch type "A-B-A" and multiple-switch type "A-B-B-A-B-A". In these types, the mobile phone communicates with base station A, but switches to base station B for a short time, and finally switches back to base station A. Besides, Ping Pong handover between three base stations of "B-C-D-B" type may occur. We call it as multi-base-station switch. Ping-pong handover data cleaning focuses on the above three types. A sequence of MPS records in which the first and the last records have the same base station is firstly found. If the recorded time length of connecting a base station is short and for two successive records which are for two different base stations, their time interval is short, these MPS records are ping-pong handover data and are reformed to be one record. This reformed record includes the ID of the base station in the first record and the connection times of the first and the last records.

Drift Data Cleaning. Drift data refers to the situation when the signal suddenly switches from the adjacent base station to the remote base station and switches back to the adjacent base station during a short time. To detect such data, we calculate the moving speed between two successive MPS records using the associated times and locations. When the moving speed is larger than the design speed of the urban arterial road, these data are identified as drift data and the record of the remote base station is deleted.

Staying Point Recognition. The OD can be extracted after the staying point recognition, and a staying point is used as the destination of the previous trip and the origin of the next trip. The staying point recognition process is as follows.

First, the MPS records are sorted by time, and the location of the base station is assumed to be the location of the mobile phone user at that moment. Besides, we set that t_{min} is the minimum staying duration and d_{max} is the maximum moving distance during stay.

Second, for a sequence of location points, we calculate the distance of every two adjacent points (D) and the corresponding interval time. If the distance D is larger than d_{max}, the previous points whose distance to the adjacent points are less than d_{max} are identified as staying points tentatively. Then, the associated staying duration is calculated and we check whether it is larger than t_{min}. If yes, these points are recognized as staying points. Otherwise, these points are not recognized as staying points.

When a mobile phone is not used for a long time, the signal frequency reduces and the time interval of MPS data thus increases. This situation usually occurs when the mobile phone user stays in the same location (e.g., sleeping at night), because the mobile phone user's movement will trigger the generation of MPS data. Hence, if the interval between two successive MPS data is over 30 min, the associated points are also identified as staying points.

2.2 Road Matching Based on the Hidden Markov Model (HMM)

This section focuses on the map of signaling trajectories to real road networks. The location information in MPS data is the location of base stations, so it cannot reflect the user's real trajectories in the road network. To handle this problem, we match the trajectories of MPS data to the real road network.

This study is based on the HMM [13] to match the MPS data to roads. We consider the order of the MPS data as the observation sequence, and the order of roads that mobile phone users have passed by is considered as the hidden sequence. As shown in Fig. 2, $p_1 \rightarrow p_2 \rightarrow p_3$ means that the mobile phone first connects to the base station at p_1, then that at p_2, and finally that at p_3. It is the observation sequence. A base station may serve more than one road. For example, the base station at p_1 serves r_1 and r_2. Thus, y be a variety of possibilities of the hidden sequence. For example, $r_1 \rightarrow r_2 \rightarrow r_3$, which means that the mobile phone user is first on r_1 and go to r_2 and then r_3. Also, $r_1 \rightarrow r_2 \rightarrow r_4$ is another possibility. Through the HMM, the hidden sequence is deduced from the observation sequence. The HMM process is as follows.

First, it selects the candidate roads within a certain range of each signaling point. The distance from the signaling point to all candidate roads d_{p1} should be lower than the specified threshold δ_m. For example, the candidate roads of p_1 is r_1 and r_2.

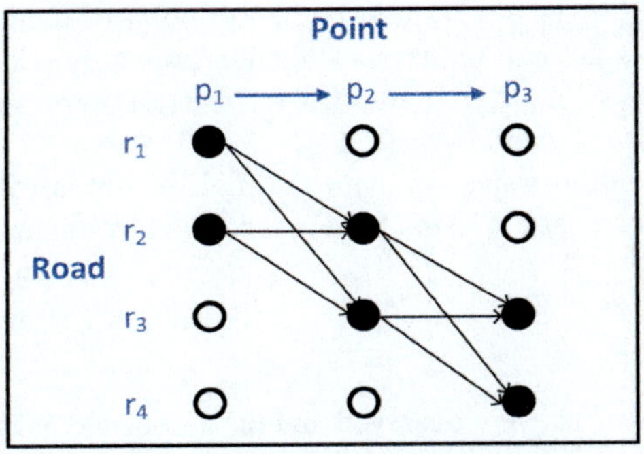

Fig. 2. Schematic diagram of the HMM.

Second, define the optimal transfer probability p which is divided into two parts p_{init} and p_r. p_{init} is the initial transfer probability between a signaling point and one of the corresponding candidate roads. It is calculated as shown in Eq. (1) and β is a constant parameter.

$$p_{init} = e^{-d_{p1}/\beta} \tag{1}$$

p_r is the road transfer probability among candidate roads, such as $p_{\mathrm{i}} : r_{\mathrm{i}} \rightarrow p_{\mathrm{j}} : r_{\mathrm{j}}$. It is calculated as shown in Eq. (2).

$$p_r = e^{-|d_{ij}-d_p|/\beta} \tag{2}$$

where d_{ij} is the shortest distance from r_i to r_{j} and this study uses Dijkstra algorithm to calculatedd_{ij}; d_p is the distance between and p_{j}.. With the transfer probabilities of p_{init} and p_r, we use Viterbi algorithm [13] to compute the best route which maximizes p_{init} and p_r. And the best route is considered as the predicted trajectory of the mobile phone user.

2.3 Navigation Track Extraction and Trajectory Similarity Calculation

After getting a predicted trajectory, with navigation trajectories of different modes, we use the DTW [14] to recognize travel modes.

First, with the OD of the predicted trajectory, we extract the navigation trajectories of each travel mode. This study adopts the navigation trajectory query method provided by Amap.

Second, the DTW is performed, and each navigation trajectory is compared with the predicted trajectory. The DTW uses Eq. (3) to calculate the similarity between each navigation trajectory and predicted trajectory.

$$DTW(X,Y) = \begin{cases} 0, \; if\, n = 0\; and\; m = 0 \\ \infty, \; if\, n = 0\; or\; m = 0 \\ D(Head(X), Head(Y)) + min\begin{cases} DTW(X, Rest(Y)) \\ DTW(Rest(X), Y) \\ DTW(Rest(X), Rest(Y)) \end{cases}, \; otherwise \end{cases} \tag{3}$$

In Eq. (3), $X = \{x_1, x_2, \ldots, x_n\}$, x_n is the nth point of predicted trajectory; $Y = \{y_1, y_2, \ldots, y_m\}$, y_m is the mth point of navigation trajectory; $Head(X)$ represents x_1, $Rest(X)$ represents $\{x_2, \ldots, x_n\}$, $D(Head(X), Head(Y))$ represents the geographical distance between $Head(X)$ and $Head(Y)$.

Finally, the navigation trajectory with the highest similarity is identified and its relative travel mode is regarded as the travel mode used by the mobile phone user.

3 Experiment

In this experiment, 67 trips were observed and the associated MPS data were provided by China Unicom.

First, the MPS data was preprocessed by invalid data cleaning, Ping Pong data cleaning, and drift data cleaning. After sensitivity tests, we found that when the minimum staying duration was set to be 14 min, the performance was better than those of other settings. Using the recognition method, 59 trips were correctly recognized, and the accuracy rate was about 88%. We compared the identified trips' length with the trip records. The results are shown in the Fig. 3. The differences are generally under 10 min, while a few differences are larger than 15 min when the mobile phone user travels within the area served by the same base station for a long time.

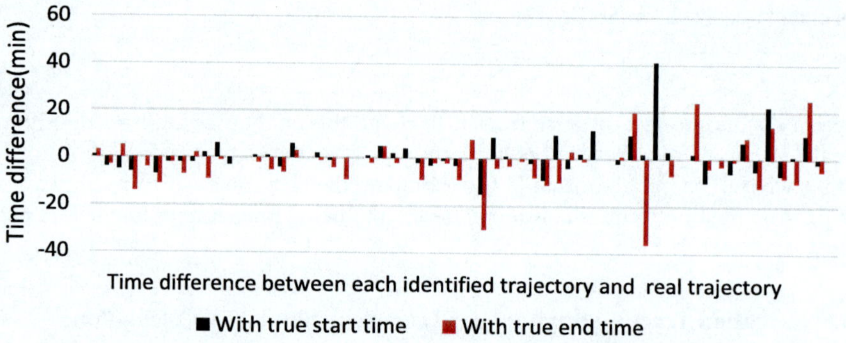

Fig. 3. Time difference between each identified trajectory and real trajectory.

Second, the signaling trajectory was matched to the real road as the predicted trajectory. The real MPS data is sparse, as shown in Fig. 4a, if the adjacent signaling points are far apart, the linear interpolation is used to overcome the problem. As shown in Fig. 4b, the black trajectory is the predicted trajectory, we take the starting point x_s and ending point x_d of the predicted trajectory (marked with red pentagrams) as the OD of the real trip, rather than the first and the last signaling points.

Third, the moving speed between OD points was calculated. Normally, the walking speed is 0–4 km/h. After experiment, we considered 0–3.6 km/h as the walking speed because the MPS data do not record the user's real travel trajectory. Therefore, the walking trips can be identified based on the moving speed.

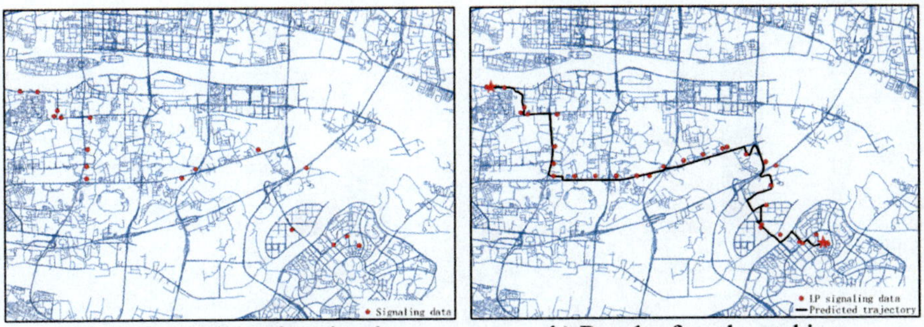

a) Signaling points after cleaning b) Result of road matching

Fig. 4. The process of one predicted trajectory generation.

Fourth, as shown in Fig. 5, x_s and x_d were used to request navigation data, including riding, public transportation under various strategies, and driving.

Finally, the navigation trajectory with the highest matching rate was identified, and the travel mode was determined as the mode of the identified navigation trajectory. For the trip shown in Fig. 5, driving was recognized as the travel mode.

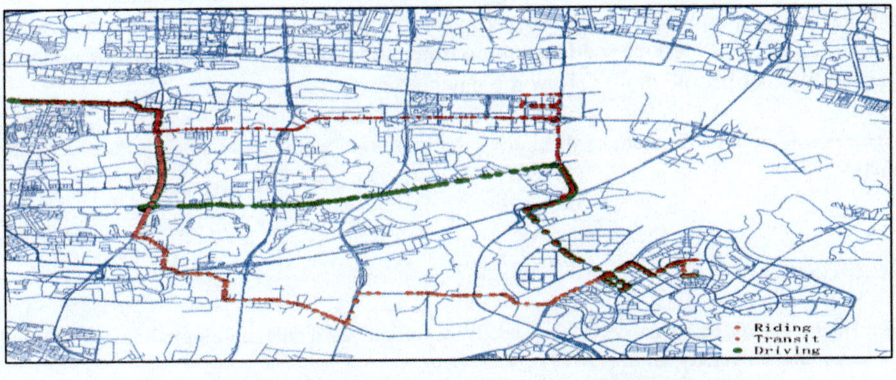

Fig. 5. The navigation trajectories in travel mode identification.

In the 67 trips, the real travel modes of 43 trips had been recorded. We extracted these 43 trips to verify our algorithm. 30 trips had been correctly identified, and the accuracy rate is approximately 69%. The result of each mode identification is shown in Fig. 6, and the slow traffic includes walking and riding.

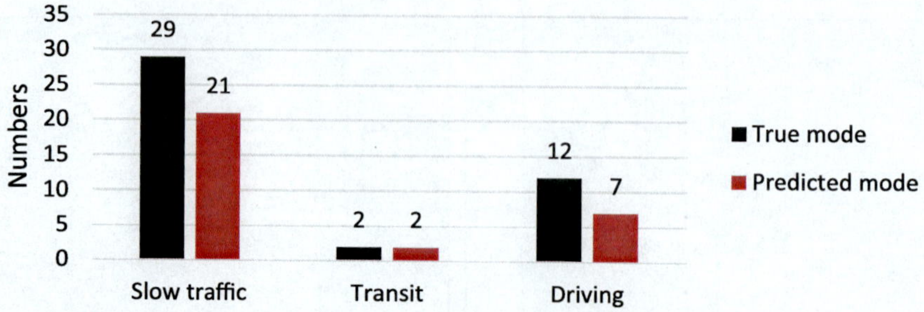

Fig. 6. The result of mode identification.

4 Discussing

In this paper, we use the MPS data to identify travel modes. The proposed algorithm mainly includes data cleaning, road matching, and travel mode analysis. The experiment showed that the identification accuracy of travel trajectories is about 88%, and the identification accuracy of travel modes is around 69%. This study has some limitations. First, the MPS data is sparse, so the accuracy of matching the signaling trajectory to the real travel road cannot be guaranteed. Second, the road matching process is time-consuming. Third, people may not follow the navigation. Hence, the future work includes: optimizing MPS data preprocess to improve the accuracy of signaling trajectory matching; classifying roads to reduce road-matching time; and designing methods to consider the probability that people don't follow the navigation.

Acknowledgement. The authors thank the support by National Natural Science Fund of China (62002398).

References

1. Huabing, L.: Urban transportation modes recognition based on mobile signaling data. Beijing Jiaotong University, China (2019)
2. Zhaohui, W., Yang, Y., Enjian, Y., Hekai, Z., Yigang, L.: Tourism travel mode identification based on cell phone signaling data. In: 2020 IEEE 5th International Conference on Intelligent Transportation Engineering, pp. 44–49. IEEE, China (2020)
3. Weihong, T., Shenghong, T., Canliang, L.: China's mobile internet entering the 5G era – release of China mobile internet development report (2020). Chinese newspaper industry, no. (17), pp. 32–35 (2020)
4. Statistical bulletin of communication industry in 2020. https://www.miit.gov.cn/gxsj/tjfx/txy/art/2021/art_057a331667154aaaa6767018dfd79a4f.html. Accessed 22 Jan 2021
5. Qiuyang, H., Yongjian, Y., Yuanbo, X., Funing, Y., Zhilu, Y., Yongxiong, S.: Citywide road-network traffic monitoring using large-scale mobile signaling data. Neurocomputing **444**, 136–146 (2021)
6. Kaixuan, D., Changxiu, C., Shi, S., Kai, S., Xingmei, Z., Ting, Z.: Postearthquake situational awareness based on mobile phone signaling data: an example from the 2017 Jiuzhaigou earthquake. Int. J. Disaster Risk Reduction **69**, 201736 (2022)

7. Mariem, F., et al.: Potential of cellular signaling data for time-of-day estimation and spatial classification of travel demand: a large-scale comparative study with travel survey and land use data. Transp. Lett. 1−19 (2022)
8. Haodong, S., Yanyan, C., Jianhui, L., Yang, W., Xiaoming, L.: Identifying tourists and locals by k-means clustering method from mobile phone signaling data. J. Transp. Eng. Part A Syst. **147**(10), 04021070 (2021)
9. Alexander, L., Jiang, S., Murga, M.: Origin–destination trips by purpose and time of day inferred from mobile phone data. Transp. Res. Part C Emerg. Technol. **58**, 240–250 (2015)
10. Loïc, B., et al.: TRANSIT: fine-grained human mobility trajectory inference at scale with mobile network signaling data. Transp. Res. Part C Emerg. Technol. **130**, 103257 (2021)
11. Juan, L.: Research on the traffic trip mode identification based on mobile phone data. Southeast University, China (2019)
12. Shuqi, Z., Rufeng, D., Hongping, D., Ming, C.: Recognition of traffic mode of mobile phone data based on the combination of point of interest data and navigation data. Acta Sci. Natur. Univ. Sunyatseni **59**(03), 87–96 (2020)
13. Newson, P., Krumm, J.: Hidden Markov map matching through noise and sparseness. In: 17th ACM SIGSPATIAL International Conference on Advances in Geographic Information Systems, pp. 336–343. Association for Computing Machinery, America (2009)
14. Dijkstra, W.: A note on two problems in connexion with graphs. Numberisch eMathematik **1**(1), 269–271 (1959)

Construction and Study of Automotive Working Conditions Based on Improved K_means Clustering

Wei Ding[1](✉) and Songwen Tian[2]

[1] Shangqiu Normal University, Shangqiu 476000, China
dingwei@sqnu.edu.cn
[2] Henan Polytechnic Institute, Nanyang 473000, China

Abstract. To address the problem that the random selection of the initial clustering centers when using K-means clustering for the construction of working conditions cannot guarantee the quality and stability of clustering, a method of selecting the initial clustering centers based on the data's own characteristics is proposed. Taking the actual operating conditions of passenger cars in Zhengzhou city as the research object, a processing method combining short-trip division of test data, principal component analysis (PCA) and improved K_means clustering is realized by using MATLAB programming to construct the cyclic working conditions of passenger cars in Zhengzhou city. The results of the comparison with the typical working conditions constructed by traditional K_means clustering show that the error of the characteristic parameters of the working conditions constructed by the improved K_means clustering has been reduced, and the average error of the joint velocity-acceleration distribution and the actual full working conditions has been reduced from 0.55 to 0.43, which is a 21.8% improvement and proves the effectiveness of the improvement, and the accuracy of the constructed working conditions is higher and more comprehensive to reflect the actual traffic conditions in Zhengzhou. The accuracy of the constructed conditions is higher, which can reflect the actual traffic conditions of Zhengzhou city.

Keywords: Passenger car driving conditions · Short trip classification · Principal component analysis · Improved K_means clustering

1 Introduction

The driving condition is the relationship curve of the speed of vehicle driving with time, and its main function is to detect the emission and fuel consumption of the vehicle, and provide an important basis for the development of new models, evaluation, and matching of the whole vehicle power system. In recent years, scholars in China have also conducted research in the construction of cyclic working conditions, and many regions have constructed cyclic working conditions that are compatible with the region [1–5].

W. Guo and K. Qian (Eds.): GBCESC 2022, LNCE 211, pp. 1188–1200, 2023.
https://doi.org/10.1007/978-981-19-5217-3_122

At this stage, most of the clustering methods used to construct working conditions are traditional K_means clustering, but their initial clustering centers are randomly selected, which cannot guarantee the quality and stability of clustering. In this study, we propose a method to determine the initial clustering centers based on the characteristics of the data samples themselves to construct the cyclic working conditions of passenger cars in Zhengzhou city, and the results of comparison and analysis with the traditional K_means clustering show that the improved K_means clustering can improve the accuracy of the constructed working conditions and has practical value.

2 Theoretical Basis of Work Condition Construction

2.1 Short Trip

To facilitate data processing, kinematic segments (short trips) are defined, and the collected working conditions are divided into short trips, i.e., the motion process of the vehicle from the beginning of one idle speed to the beginning of the next idle speed, as shown in Fig. 1.

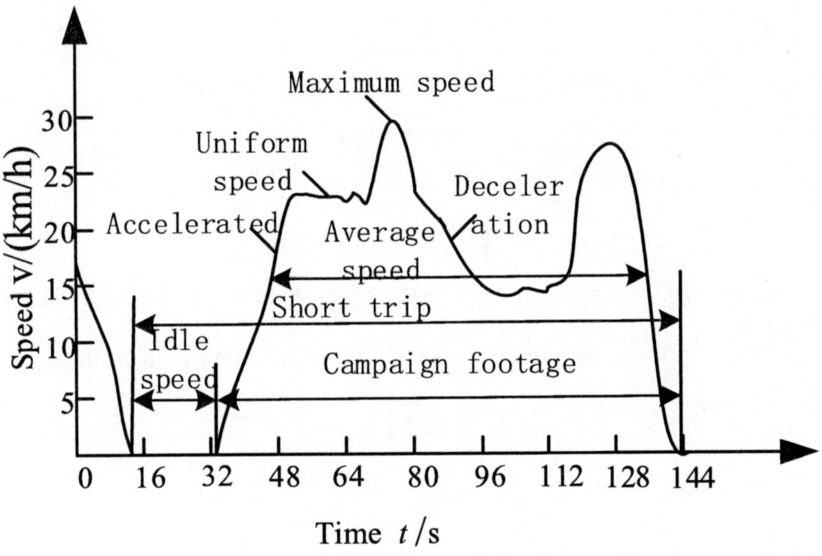

Fig. 1. Sketch map of short stroke

In this study, 12 characteristic parameters are defined to qualitatively describe the short stroke, as shown in Table 1.

Table 1. Short stroke characteristic value

Eigenvalue parameters	Meaning	Eigenvalue parameters	Meaning
T	Running time/s	V_{max}	Maximum speed/(km/h)
Ti	Idle time/s	V_{means}	Average speed/(km/h)
T_a	Acceleration time/s	A_{max}	Maximum acceleration/(m/s^2)
T_d	Deceleration time/s	A_{min}	Minimum acceleration/(m/s^2)
T_c	Homogenization time/s	$A_{a\text{-}means}$	Average acceleration of the acceleration section/(m/s^2)
S	Running distance/km	$A_{d\text{-}means}$	Average acceleration of deceleration section/(m/s^2)

2.2 Principal Component Analysis

The principal component analysis method [6] (PCA), representing all the characteristic parameters by a few composite variables, reduces the number of data sample dimensions and reduces the computational effort. The processing leads to a parameter matrix D with the number of short trips (rows) × characteristic parameters (columns).

$$D_{n\times m} = \begin{bmatrix} D_{11} & D_{12} & D_{13} & \cdots & D_{1m} \\ D_{21} & D_{22} & D_{23} & \cdots & D_{2m} \\ \vdots & \vdots & \vdots & \ddots & \vdots \\ D_{n1} & D_{n2} & D_{n3} & \cdots & D_{nm} \end{bmatrix} \tag{1}$$

where: D_{ij} (i = 1, 2, 3… n, j = 1, 2, 3… m) is the jth characteristic parameter in the i-th short stroke.

In order to unify the magnitudes, the matrix D is normalized to obtain the matrix X:

$$X = \begin{bmatrix} X_{11} & X_{12} & X_{13} & \cdots & X_{1m} \\ X_{21} & X_{22} & X_{23} & \cdots & X_{2m} \\ \vdots & \vdots & \vdots & \ddots & \vdots \\ X_{n1} & X_{n2} & X_{n3} & \cdots & X_{nm} \end{bmatrix} \tag{2}$$

where: $X_{ij} = \frac{(D_{ij} - D_{j-mean})}{S_j}$, D_{j-mean} is the mean of the jth column and S_j is the standard deviation of the jth column.

From X, we can obtain its covariance matrix T:

$$T = \begin{bmatrix} S_1^2 & Cov(1,2) & Cov(1,2) & \cdots & Cov(1,m) \\ Cov(2,1) & S_2^2 & Cov(2,3) & \cdots & Cov(2,m) \\ \vdots & \vdots & \vdots & \ddots & \vdots \\ Cov(n,1) & Cov(n,2) & Cov(n,3) & \cdots & S_m^2 \end{bmatrix} \tag{3}$$

where: $S_x^2 = Cov(x, x)$,$Cov(x, y) = Cov(y, x) = \frac{1}{n-1}\sum_{i=1}^{m}(x_i - x_{mean})(y_i - y_{meaan})$.

This leads to the correlation coefficient matrix R of the matrix D.

$$R = \begin{bmatrix} R_{11} & R_{12} & R_{13} & \cdots & R_{1m} \\ R_{21} & R_{22} & R_{23} & \cdots & R_{2m} \\ \vdots & \vdots & \vdots & \ddots & \vdots \\ R_{m1} & R_{m2} & R_{m3} & \cdots & R_{mm} \end{bmatrix} \tag{4}$$

where: $R_{11} = R_{22} = R_{33} = \cdots R_{mm} = 1,\ R_{xy} = \frac{Cov(x,y)}{S_x S_y}$.

According to the matrix R, we can find the characteristic parameters λi (i = 1, 2, 3…12) and the cumulative contribution of the first M principal components:

$$\psi = \frac{\sum_{i=1}^{M}\lambda_i}{\sum_{i=1}^{12}\lambda_i} \tag{5}$$

where λ_i denotes the eigenvalue of the ith principal component and the contribution margin indicates the ability of M_1, M_2, M_3…M_{12} to combine all the original variables, requiring a cumulative contribution margin greater than 80%.

2.3 K_means Clustering

It can be seen that the K-means clustering results are greatly influenced by the initial clustering center, and if the initial clustering center is not selected properly, the clustering results may fall into local optimal solutions and are more sensitive to isolated points, and these defects greatly limit its application scope and accuracy.

In order to ensure the quality and stability of clustering, this study proposes a method to determine the initial clustering centers according to the data samples' own characteristics, as follows.

(1) In the data set Q with n data samples, find the data object w whose distance from the origin is the closest.
(2) Calculate the distances Dis1, Dis2, Dis3 of all data objects to w in the data set Q one by one. Disn, and arrange all data samples in this distance order from smallest to largest.
(3) Specify respectively the
$$\begin{cases} 1 + (j-1) \times (n-2)/(k-1), \text{When n is an even number} \\ 1 + (j-1) \times (n-1)/(k-1), \text{When n is an odd number} \end{cases} \text{first data sample}$$
as the initial member of the corresponding group, respectively.
(4) Calculate the distance from the remaining data samples to all initial members in each grouping and assign them to the group with the smallest distance.

(5) After all data are grouped, the initial clustering center is determined according to the following formula The initial clustering center ci of cluster Ci is calculated by the formula.

$$c_i = \frac{1}{n_i}\sum_{x=1}^{n_i} R_x, R_x \in C_i \tag{6}$$

where R_x is the data samples belonging to cluster C_i and n_i is the number of all data samples in cluster C_i.

2.4 Data Acquisition

The experimental vehicle is equipped with the data acquisition terminal equipment developed by Beijing Boyang Company, as shown in Fig. 2, with a sampling frequency of 1 Hz. The equipment collects real-time vehicle driving condition data such as engine speed and fuel consumption through the vehicle CAN bus, and collects real-time vehicle driving data such as longitude, latitude, speed and mileage using GPS equipment, and realizes the collection and display of real-time vehicle data through GPRS network. Display through GPRS network.

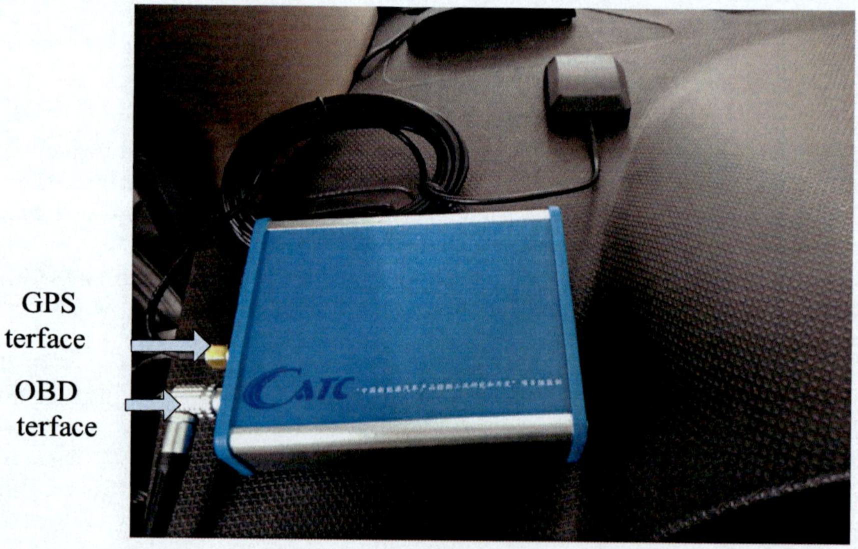

Fig. 2. Data acquisition terminal

The collected 613,792 effective driving conditions were divided into 5,555 short trips using MATLAB programming and solved for each short trip characteristic parameter separately to obtain a matrix with short trips as samples and the number of samples (rows) × characteristic parameters (columns), as shown in Table 2.

Table 2. Short stroke parameter library

Short trip number	S/km(10^{-3})	V_{means}/(km/h)	...	V_{max}/(km/h)	A_{max}/(m/s^2)	A_{min}/(m/s^2)
1	308.80	10.0069	...	26.10	1.38	−1.44
2	120.55	10.5854	...	22.60	2.08	−1.25
3	54.22	5.9152	...	11.30	2.77	−0.58
⋮	⋮	⋮	⋮	⋮	⋮	⋮
5553	291.11	14.3562	...	40.80	3.05	−4.22
5554	303.72	14.0179	...	27.30	1.77	−1.30
5555	101.36	8.2932	...	17.50	3.02	−0.66

2.5 Results of Principal Component Analysis

The authors prepared a program in MATLAB to perform PCA on the characteristic parameter matrix of Table 2 and obtained the contribution rates and cumulative contribution rates of the 12 principal components Mi (where i = 1, 2, 3... 12) as shown in Table 3.

Table 3. Contribution rate and cumulative contribution rate of each principal component.

Principal component number	Eigenvalue (λ)	Contribution rate/(%)	Cumulative contribution rate/(%)
M_1	6.0805	50.67	50.67
M_2	2.5113	20.93	71.60
M_3	1.0268	8.56	80.16
M_4	0.8926	7.43	87.59
M_5	0.4579	3.81	91.40
M_6	0.3900	3.26	94.66
M_7	0.2812	2.34	97.00
M_8	0.2170	1.81	98.81
M_9	0.0882	0.73	99.54
M_{10}	0.0347	0.29	99.83
M_{11}	0.0193	0.16	99.99
M_{12}	0.0012	0.01	100

As can be seen from Table 3, the cumulative contribution of the eigenvalues of the first three principal components is 81.06% and the eigenvalues of the first three principal components are greater than l. Therefore, the first three principal components were selected for further analysis [7].

2.6 Analysis of Clustering Results

An improved K-means clustering algorithm was written in MATLAB to cluster the short trips on the first 3 principal components of the scores for the work conditions with a number of clusters of 3 classes [8]. The clustering results are shown in Fig. 4, and the clustering differences of short trips are obvious and clearly classified by Fig. 3 and Fig. 4.

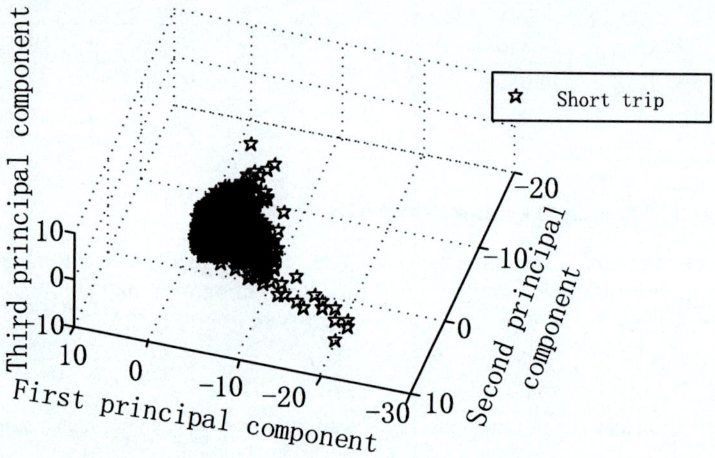

Fig. 3. Short stroke 3D scatter plot

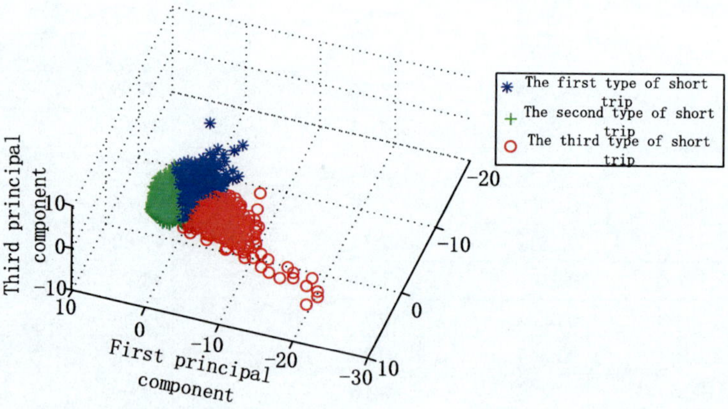

Fig. 4. Three dimensional scatter diagram of short stroke after clustering

2.7 Representative Working Condition Construction

Referring to the foreign standard working conditions and constant volume sampling system (CVS), the duration of the urban cycle working condition was taken to be about 1200 s [9, 10]. According to the proportion of the total time of all types of short trips to the total time of all collected conditions, the time taken by each type of short trips in the final constructed conditions can be determined, and the calculation formula is as follows.

$$T_c = \frac{\sum_{b=1}^{m_b} T_{c,b}}{T_{all}} T_{dc} \tag{7}$$

were T_c: the time of class c in the constructed working condition; $T_{c,b}$ is the duration of the bth short stroke in cluster c; the total number of short strokes in cluster c; the total time of all the collected working conditions; the total time of the final constructed representative working condition.

The representative working conditions of passenger cars in Zhengzhou city constructed by using traditional K-means clustering and improved K-means clustering according to the time proportion and the magnitude of correlation coefficient are shown in Figs. 5 and 6 respectively.

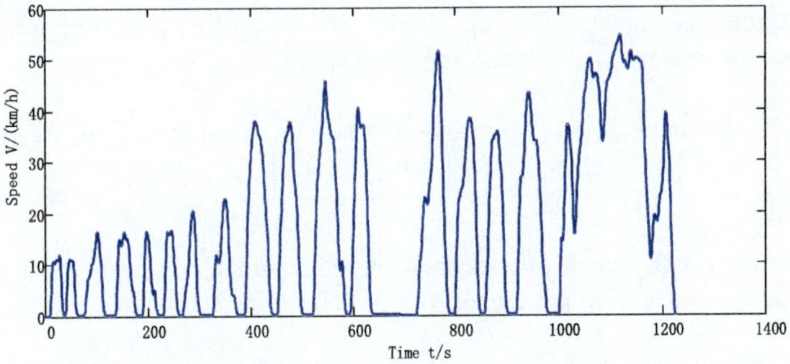

Fig. 5. Typical working conditions of traditional K-means clustering construction

By analyzing the data in Fig. 6, we found that the total running time of Zhengzhou passenger car cycle working condition constructed based on improved K-means clustering is 1240 s, the driving distance is 6.23 km, the maximum speed is 61.6 km/h, the average speed is 18.17 km/h, the average running speed is 25.46 km/h, the maximum acceleration is 2.89 m/s^2, the maximum deceleration is -1.75 m/s^2, the idling ratio was 28.63% s, the acceleration ratio was 27.18%, the deceleration ratio was 25.16%, and the uniform speed ratio was 19.03%.

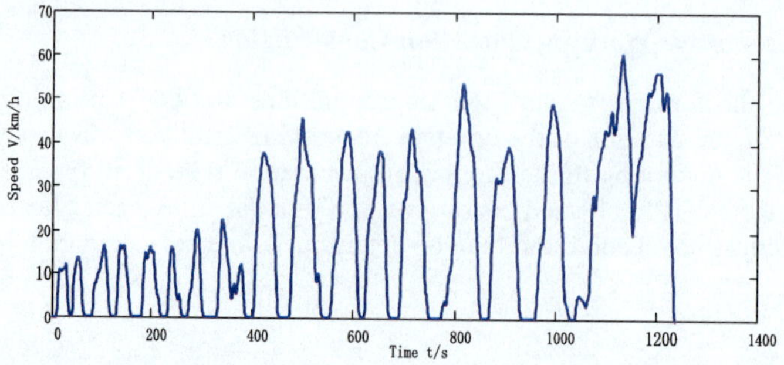

Fig. 6. Representative working conditions of improved k-means clustering construction

3 Analysis of the Accuracy of the Constructed Representative Working Conditions

3.1 Relative Error of Characteristic Parameters

Firstly, the characteristic parameters of all the working conditions collected from the base test were compared with working condition 1 and working condition 2, and the results are shown in Table 4. The Es is defined as the average value of the relative error between the constructed working condition characteristic parameters and all working condition characteristic parameters of the base test, and the accuracy of the constructed working condition is examined by the magnitude of Es. The calculation formula of Es is as follows (where Vyunxing is the operating speed).

$$E_s = \frac{\left[\begin{array}{c} \varepsilon_1|\Delta V_{\text{means}}| + \varepsilon_2\left|\Delta A_{\text{a\_ means}}\right| + \varepsilon_3\left|\Delta A_{\text{d\_ means}}\right| + \varepsilon_4\left|\Delta A_{\text{a\_ means}}\right| \\ + \varepsilon_5|\Delta P_{\text{a}}| + \varepsilon_6|\Delta P_{\text{d}}| + \varepsilon_7|\Delta P_{\text{i}}| + \varepsilon_8|\Delta P_{\text{c}}| \end{array}\right]}{8} \tag{8}$$

where: Δ is the relative error of the characteristic parameters; ε_i denotes the weighted value of each characteristic parameter, which is taken as 1 for the convenience of the study.

Table 4. Comparison of characteristic parameters between cycle conditions and all conditions

Characteristic parameters	Working condition 1	Working condition 2	All working conditions	Error 1/(%)	Error 1/(%)
V_{means}/(km/h)	18.17	17.04	18.61	2.3	8.4
$A_{\text{a-means}}$/(m/s^2)	0.56	0.61	0.58	3.4	5.2
$A_{\text{d-means}}$/(m/s^2)	−0.6	−0.63	−0.54	11.1	16.7
V_{yunxing}/(km/h)	25.46	25.43	26.06	2.3	2.4

(*continued*)

Table 4. *(continued)*

Characteristic parameters	Working condition 1	Working condition 2	All working conditions	Error 1/(%)	Error 1/(%)
P_a/(%)	27.18	24.57	26.01	4.5	5.5
P_d/(%)	25.16	23.67	24.10	4.4	1.8
P_i/(%)	28.63	32.98	30.00	4.6	9.9
P_c/(%)	19.03	18.78	19.89	4.3	5.6
E_S				4.6	6.9

Note: Error 1 indicates the relative error of working condition 1 and all working conditions, error 2 indicates the relative error of working condition 2 and all working conditions

By analyzing the errors in Table 4, it is found that only the error of Ad-mean is 11.1% for working condition 1 and all working conditions, and the errors of the rest of the characteristic parameters are less than 10%, and the errors of the rest of the characteristic parameters are reduced, among which ES is reduced from 6.9% to 4.6%.

3.2 Probability of Joint Velocity-Acceleration Distribution

In order to further describe the differences of the states in working condition 1, working condition 2, and all working conditions [6], the joint velocity-acceleration distribution probabilities of the three working conditions were calculated as shown in Fig. 7. From Fig. 7, it can be seen that the trends of working condition 1 and working condition 2 are basically consistent with those of all the working conditions of the basic experiment.

To quantitatively describe the differences, the error mean Ec of the probabilities of the velocity-acceleration interval distributions of the two was calculated, and Em was calculated as follows.

$$E_c = \frac{\sum_{i=1}^{10}\sum_{j=1}^{10}|\boldsymbol{A}-\boldsymbol{B}|}{100} \tag{10}$$

where: A is the constructed probability matrix of velocity-acceleration interval distribution for working conditions (working condition 1 and working condition 2), and B is the probability matrix of velocity-acceleration interval distribution for all working conditions. The mean error of the velocity-acceleration distribution probability of working condition 1 and working condition 2 is 0.43 and 0.55 respectively, which is 21.8% higher than that of all working conditions, indicating that the joint velocity-acceleration probability density distribution of working condition 1 is closer to the actual working condition.

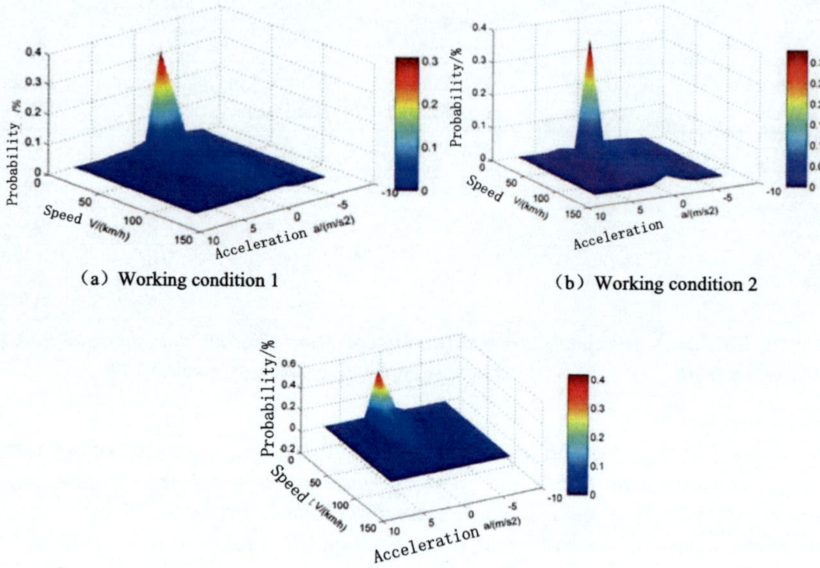

(a) Working condition 1

(b) Working condition 2

(c) All working conditions

Fig. 7. Velocity acceleration joint distribution under three working conditions

3.3 Simulation Verification

The whole vehicle model was built in AVL-Cruise, and the working condition 1 and working condition 2 were taken as the simulation conditions of the whole vehicle as shown in Fig. 8, and its 100 km fuel consumption and emission were compared with the actual ones as shown in Table 5.

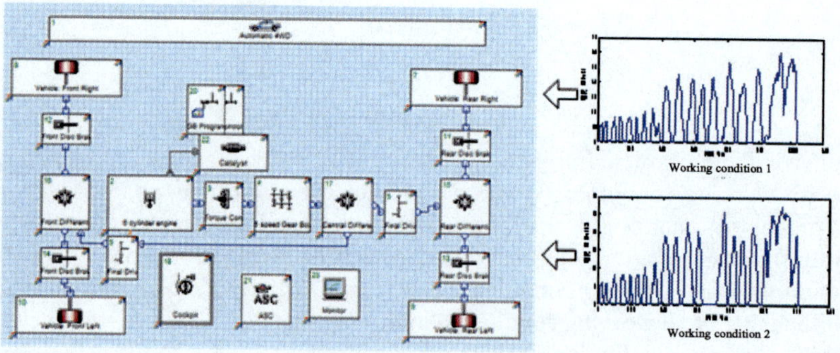

Fig. 8. Vehicle model

Table 5. Comparison between simulated fuel consumption and emission and actual

	Working condition 1	Working condition 2	Actual	Error 1/%	Error 2/%
Fuel consumption per 100 km/(L)	12.48	13.01	12.11	3.1	7.4
NO_X/g	8.25	8.7	7.83	5.3	11
CO/g	38.97	37.6	38.64	0.8	2.7
HC/g	7.17	6.99	7.12	0.7	1.8

Note: Error 1 indicates the relative error of working condition 1 and all working conditions, error 2 indicates the relative error of working condition 2 and all working conditions

From Table 5, it can be seen that the errors of fuel consumption and emission of working condition 1 with the actual are lower than those of working condition 2. In summary, the validation results from the three aspects of relative error of characteristic parameters, joint velocity-acceleration distribution and simulation validation can show that the working condition constructed by improved K_means is close to the actual working condition and is more representative.

4 Conclusion

In this study, the traditional K_means clustering was improved by the initial cluster center selection method with the goal of reducing the variability between data, and the cyclic working condition (working condition 1) of passenger cars in Zhengzhou City was constructed with the improved K_means clustering, and the cyclic working condition (working condition 2) constructed by the traditional K_means clustering was compared with the traditional K_means clustering in terms of the relative error of feature parameters, joint velocity-acceleration distribution, and The results show that the driving conditions constructed based on the improved K_means clustering are closer to the actual ones, which proves the effectiveness of the improvement. However, there are still errors with the actual working conditions, and further improvement is needed.

References

1. Zhen, W., Zhiyao, S.: Construction of vehicle driving conditions based on K-mean clustering algorithm. Journal **59**(6), 45–49 (2021)
2. Wenwu, L.: Construction and study of vehicle driving conditions in Fuzhou area. Journal **59**(01), 153–157 (2021)
3. Yao, M., Jianping, G., et al.: Multi-source information fusion for the construction of driving conditions of plug-in hybrid buses. Journal **42**(4), 40–41 (2021)
4. Mengliang, L., Xichan, Z., Jianwei, Z., et al.: Study on the driving cycle of typical urban vehicles. Journal **27**(5), 557–560 (2005)
5. Yuxi, Z., Xiaohui, S., et al.: Construction of vehicle driving conditions based on improved principal component and global k-means clustering. Journal **15**(11), 1253–1259 (2020)

6. Jiachen, G., Heng, J., et al.: Construction method of vehicle driving conditions on urban roads. Journal **20**(6), 197–209 (2020)
7. Jun, S., Tao, F., et al.: Construction of electric bus driving conditions in Hefei city based on improved k-mean clustering algorithm. Journal **15**(8), 56–62 (2020)
8. Yaohua, L., Zhongyu, L., et al.: Construction of urban bus line working conditions based on cluster analysis. Journal **38**(1), 83–89 (2019)
9. Rende, X.: Construction of vehicle driving conditions based on global K-means clustering algorithm. Journal **30**(1), 112–118 (2019)
10. Jianping, G., et al.: Improved fuzzy C-mean clustering method for the construction of actual driving conditions of vehicles. Journal **38**(6), 21–27 (2017)

Research on the Innovative Design Mode of Guangzhou Vernacular Architecture Under the Background of Carbon Neutrality

Qiang Tang[1], Yile Chen[2], Liang Zheng[2], and Junzhang Chen[3](✉)

[1] Shunde Polytechnic, Shunde District, Foshan City 528300, Guangdong, China
[2] Faculty of Humanities and Arts, Macau University of Science and Technology, Avenida Wai Long, Taipa, Macau 999078, China
[3] Faculty of Innovation and Design, City University of Macau, Avenida Padre Tomás Pereira Taipa, Macau 999078, China
U19091105192@cityu.mo

Abstract. Carbon neutrality is an important topic of environmental economics, and it is also an important strategy to promote the development of low-carbon buildings. The article takes an eco-leisure guesthouse next to the Art Town of Zinitang, Panyu, Guangzhou as a practical case, and studies the honeycomb paper calcium silicate board produced by construction users as the wall material, which is fully integrated into the design. Finally, it provides reference for the low-carbon design mode of modern vernacular architecture in Guangzhou.

Keywords: Carbon neutrality · Low-carbon buildings · Vernacular architecture

1 Introduction

At present, with the pressure and challenges of global environmental changes, mankind is increasingly aware of the huge impact of buildings and their operations on the environment. This set off a worldwide wave of "green building" development. At the same time, the intelligent systems created in the information and network era help humans build a comfortable, efficient, and safe living and working environment.

In March 2020, the European Commission announced the draft "European Climate Law" and decided to clarify in the form of legislation that Europe will achieve "carbon neutrality" by 2050, that is, net greenhouse gas emissions will be reduced to zero by 2050. The draft requires all EU institutions and member states to take necessary measures to achieve the above goals [1]. According to the "Green Agreement" announced in December 2019, the European Commission is working hard to achieve the EU's 2050 net zero emissions target. This long-term strategy was submitted to the United Nations in March 2020. In September 2020, Chinese leader Xi Jinping stated at the general debate of the 75th United Nations General Assembly that China will increase its nationally determined contribution, adopt more powerful policies and measures, and strive to reach its peak carbon dioxide emissions by 2030. Strive to achieve carbon neutrality by 2060 [2, 3]. This goal was later written into the 14th Five-Year Plan announced in 2021 [4].

W. Guo and K. Qian (Eds.): GBCESC 2022, LNCE 211, pp. 1201–1208, 2023.
https://doi.org/10.1007/978-981-19-5217-3_123

China is a country with dense cities, a large population, and rapid urbanization. However, there is a big gap between the level of construction technology, building energy consumption, and building housing—it is a big gap between China and developed countries. The construction life of China is only 30 years., Less than half of the world's average age. With the further deepening of world economic integration and the advent of new technological revolution, especially the bid for the 2008 Olympic Games with the theme of "green, technology, and humanities", it will have a huge boost to China's construction technology. Based on this background, the article takes an eco-leisure guesthouse next to the Art Town of Zinitang, Panyu, Guangzhou as a practical case, and studies the honeycomb paper calcium silicate board produced by construction users as the wall material, which is It is fully integrated into the design. It is hoped that through the introduction of internationally advanced crafts and technologies, my country will promote the development of vernacular architectural design in the context of carbon neutrality.

2 Literature Review

The undisclosed claim of natural resources by mankind has severely damaged the ecological environment of the earth. Various natural disasters show that mankind is in an unprecedented ecological crisis. The rapid development of modern cities has turned most cities in the world into reinforced concrete urban forests. Building energy consumption accounts for an important proportion of the total energy consumption of the entire society. The development of low-carbon ecological buildings is to adapt to global warming brought about by climate change and change buildings. Energy consumption characteristics, promote the sustainable development of the building type in the low-carbon era [5]. The era's call for ecological architecture is becoming more and more intense, and the development of ecological architecture has an important impact on the environmental protection of our country and the world. In 2015, Shigeru Ban, a Japanese architect who specializes in paper construction, won the Pritzker Prize, showing the world's high recognition of paper-based construction materials by the world's architecture circles; Simon Ville was hailed as the "Brave Colombian Bamboo Builder" and he used Bamboo creates a structure that subverts traditional building regulations, and the invented bolt cement connection method solves the problem of bamboo connection structure; [6] Vietnamese young architect Wu Chongyi skillfully uses bamboo to create multiple extraordinary spaces; Japanese architect Kuma Through bamboo, Kengo created a series of artistic conception spaces that meet the spiritual needs of Orientals. His representative work is the Bamboo House of the Commune by the Foot of the Great Wall, which ingeniously interprets the oriental artistic conception of Zen. At present, most bamboo buildings exist in a relatively single form, and buildings combining paper and bamboo are rarely heard. Eco-buildings should first focus on the selection of materials. Eco-buildings combined with paper and bamboo will greatly reduce resource loss and achieve optimal resource allocation through renewability.

3 Site and Design Positioning

This building is named "Paper Bamboo House" and is located next to the Art Town of Zinitang, Panyu, Guangzhou, with a construction area of 650 square meters. It is adjacent to the peaceful Shawan waterway, surrounded by farmland and canals, and is on an island surrounded by water. The base where the building is located was formerly agricultural land with high groundwater level and a lot of sandy soil. It is a weak foundation with weak structural performance. The building uses hollow square piles as the underground foundation, and the ground uses steel welded to form an integral frame structure, and all of them are of relatively high quality. The light honeycomb paper wallboard is used as the wall body, which effectively reduces the weight of the building to prevent the building from sinking (Fig. 1).

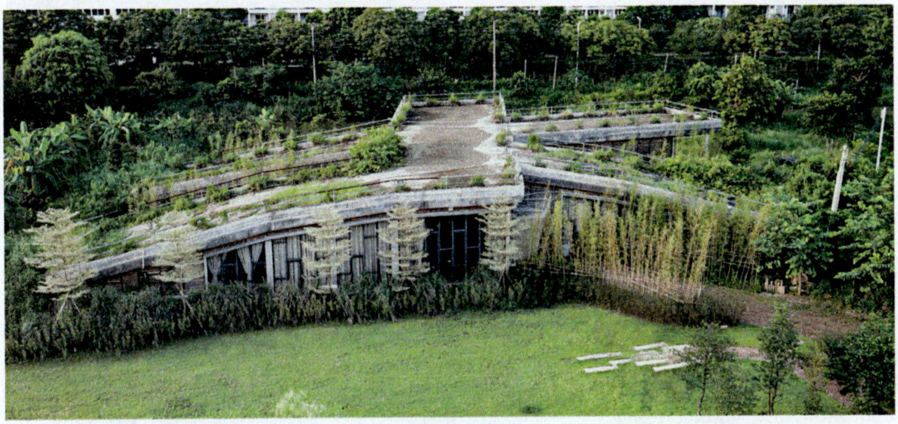

Fig. 1. Aerial view of the building (Source: photographed by the author)

Its design is inspired by the architect's nostalgia for bamboo, and the owner's hope that the architect will show the use of honeycomb cardboard in a physical space. Therefore, its functional positioning is an ecological club that includes leisure, negotiation, entertainment, living and other functions in one, and it is also a material display center for honeycomb cardboard. The building's pursuit of low-carbon environmental performance is fully demonstrated by the physical space of paper bamboo dwellings.

4 Low-Carbon Design Analysis

Paper Bamboo House perfectly combines the building energy-saving system with the smart home system. (1) Building energy-saving systems include: honeycomb panel wall system; solar collection system; intelligent shading system; ecological breathing curtain wall system; plant wall system; roof greening and vegetation system, etc. (2) The intelligent building system adopts the smart home IoT system (combo wireless IoT home system produced by Guangdong Kangbao Electric Co.) and the air purification system. Therefore, breaking the previous living mode, the whole building has the advantages of energy saving and intelligence. Making full use of the energy generated by the building itself for transformation, they can make the indoor space warm in winter and cool in summer, and air conditioning is almost unnecessary in summer. The integration of various energy-saving technologies can reduce the overall operating energy consumption of buildings by more than 50% (Fig. 2).

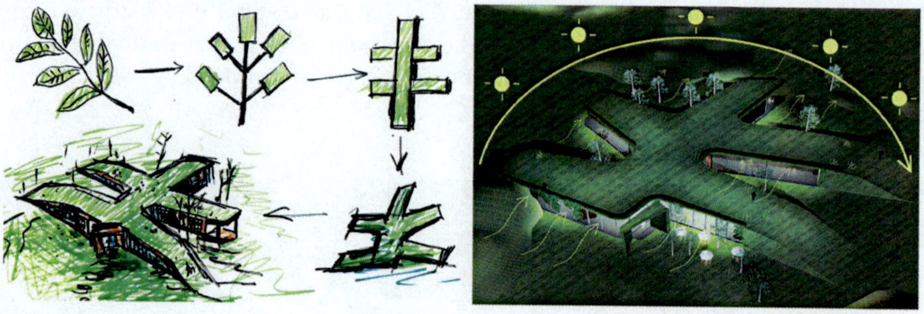

Fig. 2. Design concept and sunlight considerations (Source: Drawn by the author)

4.1 Honeycomb Panel Wall

Paper honeycomb panels are a new type of green material that has emerged in developed countries to save resources, protect the ecological environment, and have a low cost. It has many advantages such as lightness, strength, stability, heat preservation, heat insulation, and shock resistance. The formaldehyde content can reach E0 level (the national standard for furniture formaldehyde content E1 level). The walls of this building are all made of NAVIROTM, a new environmentally-friendly board with national patents (paper honeycomb panel in the middle) [7, 8].

Honeycomb paper often appears as a green packaging material with good cushioning properties and low prices. Honeycomb paperboard is commonplace in Europe as a furniture board. In recent years, honeycomb paper wallboards made of honeycomb paper combined with Eterboard have been used more and more widely in construction. Honeycomb paper wallboards have significant advantages in five aspects: (1) The internal honeycomb structure has light weight, high strength, and difficulty. The characteristics of deformation; (2) The honeycomb structure is full of air and has good evacuation and heat insulation; (3) It has good sound insulation and heat preservation performance [9]; (4) The formaldehyde content can reach E0 level, which is a real Environmentally friendly base material; (5) Both sides are Eterboard, with firm veneer, flexible process and convenient processing. The honeycomb paper wallboard is used as an external wall to solve the problem of difficulty in waterproofing. The construction personnel first hang a stainless steel wire mesh on the external wall of the Eterboard, and then use a 10 mm thick 1:2.5 cement mortar to shake and continue on the surface of the mortar. Make two exterior wall waterproof coatings, then use bamboo pieces of the same specification to pile up and decorate them from top to bottom, and finally use varnish on the surface of the bamboo pieces for protection (see Fig. 3). Once the waterproof problem of the honeycomb paper wallboard is solved, the use performance will be fully guaranteed. The “Paper Bamboo House” has been built and used for nearly 5 years, and it has stood still after many typhoons and rains in the south, verifying that the fragile materials can be built firmly through careful design The original intention of the space.

This technology has passed the evaluation of experimental testing, and was jointly prepared by the Guangdong Provincial Building Materials Research Institute, Foshan Dongnan Haiye Environmental Protection Material Co., Ltd., Guangzhou Wall Material Innovation and Building Energy Saving Office, and Guangdong Provincial Architectural Design and Research Institute. "Technical Specification for Lightweight Wall Engineering of Honeycomb Wallboard" was formally implemented on October 1, 2009. At present, "Paper Honeycomb Wallboard Lightweight Wall Engineering Technical Specifications" is the local standard of Guangdong Province, and the number is GBJ/T15–67-2009.

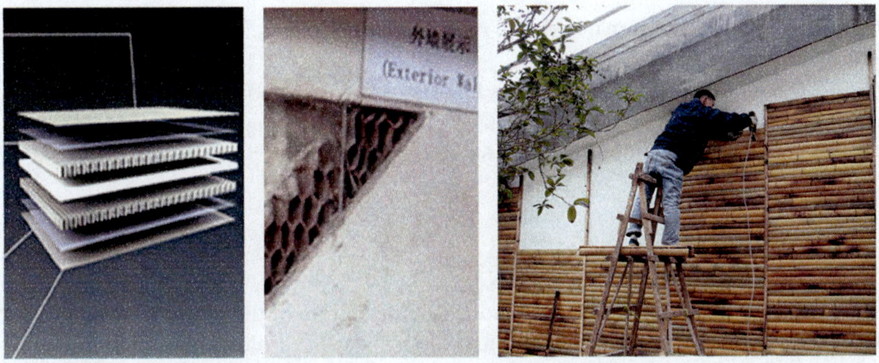

Fig. 3. The application of honeycomb paper in the construction process of the wall (Source: Photographed by the author)

4.2 Intelligent Shading System

Guangzhou belongs to the south subtropical monsoon climate zone. The summer is longer and the winter is shorter. The dominant wind directions in the summer are mostly southeast, south and southwest winds, and the winter half of the year is mostly northerly and northeasterly winds. Most of the doors and windows of the building are opened in the vertical direction of the dominant wind direction. The spatial form of north-south convection makes the room often have "through wind", and the wind can pass through the interior of the building to play the role of healthy ventilation, heat dissipation and dehumidification. In summer, the sun goes out from north to east in the morning, goes south from the center at noon, and sets from north to west in the evening. Based on the above factors, the main building of "Zhizhuju" is in the northeast and southwest directions. The sloped roofs on both sides of the east and west sides are covered by green vegetation to form a natural barrier, which can resist the strong direct sunlight in summer and autumn, and the sun exposure is not affected by the position of the sun. Affected by movement. It has obvious heat insulation and cooling effect on the buildings

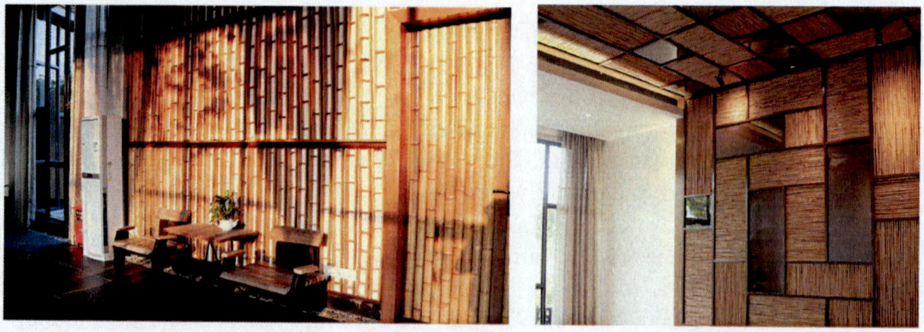

Fig. 4. Indoor light transmission (Source: Photographed by the author)

in the long-hot Lingnan area, and also has obvious effects on alleviating the urban heat island effect (Fig. 4).

In the indoor part, the building adopts an intelligent adjustable sunshade grille, which effectively reduces the sun's radiation in summer and improves the indoor lighting environment. At the same time, it plays a role in reconciling privacy and protecting the building.

4.3 Eco-Breathing Curtain Wall System

The system sets up additional long windows at the bottom of the floor between the inner ceiling of the outer glass curtain wall and the lower end of the curtain wall, forming a natural heat storage cavity to heat the air layer of the double-layer curtain wall, generating natural heat pressure and realizing natural ventilation Convection effectively achieves coolness in summer and warmness in winter.

4.4 Plant Wall System

The building plant wall system is the current mature vertical greening system, which can effectively improve the ecological environment and has the function of automatic irrigation. In summer, the building walls with creeping plants will reduce the absorption of solar radiant heat due to the shading of the leaves, thereby reducing the indoor temperature. In winter, it is good for the wall to fully receive the sunlight, which is very helpful for the insulation of the building.

At present, there are two types of eco-building technologies in the world: low-tech and high-tech. Low-tech eco-building refers to building based on local conditions and cultural customs, using traditional construction techniques and regional materials; while high-tech eco-building refers to the use of advanced Energy-saving technologies, including high-cost modern energy-saving equipment and new synthetic materials. "Paper Bamboo House" effectively combines the two based on regional ecological conditions. The honeycomb paper wallboard produced by modern technology and equipment represents a new type of synthetic material with high-tech strategies, and the simply processed moso bamboo represents an ecological material with low-tech strategies. Phyllostachys pubescens is the bamboo species with the highest industrial utilization value among the bamboo resources in the world. Huaiji County, Guangdong is rich in straight and long Phyllostachys pubescens. The Phyllostachys pubescens is boiled at high temperature-borax salt antiseptic-varnish finish treatment is tough and stable, and the bamboo is excellent. Characteristics. "Paper Bamboo House" creates a place with bamboo on all sides. Bamboos of different sizes are used in every corner of the building and presented in different forms in the space, giving full play to the rich decorative effects of ecological materials (Fig. 5).

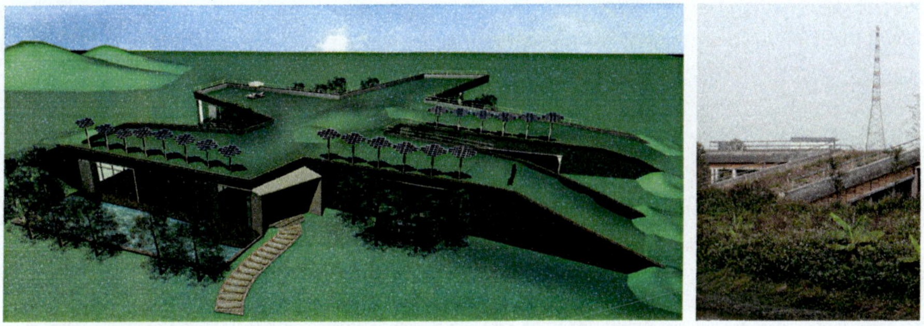

Fig. 5. Low-carbon roof design (Source: Drawn by the author)

4.5 Solar System

A tree-shaped smart solar panel is used to efficiently collect solar energy and save energy. Adopting high-efficiency and precise tracking solar photovoltaic panel power generation system to provide residential fan power, environmental protection lighting and other electricity. It can also be integrated into the intelligent control of the combo wireless IoT home system.

4.6 Green Roof System

The roof vegetation greening system blocks the solar radiation heat in summer, and the plants and the covering soil layer work together to increase the roof thermal resistance, reduce the roof heat transfer system, and play a role of heat insulation. In winter, it reduces the invasion of outdoor cold air to the roof panel, and indoor heat cannot be conducted to the outside, which plays a role of heat preservation; at the same time, it has the effect of water retention and slow drainage of heavy rain.

5 Conclusion

"Paper Bamboo House" establishes an emotional connection between space and people through ecological materials. The two materials of paper and bamboo are effectively combined to create a place with a special sense of experience, embodying the rhythm of architectural forms and materials, and advocating people's true perception of the place and the spirit of the place recognition. Practicing the practice of regionalization of buildings under the low-carbon ecological strategy, the design model of "Paper-Bamboo House" in Guangzhou's vernacular architecture will bring certain reference significance to the design of experiential low-carbon ecological buildings.

Acknowledgement. This article is based on the research staff from 2018 to 2020 from the 2020 Guangdong Province Ordinary Universities Characteristic Innovation Project (No. 2020WTSCX286); 2019 Guangdong Province Ordinary University Philosophy and Social Science Project (No.2019GXJK131); 2018 Guangdong Province Philosophy and Social Science Planning Discipline Co-construction with the result of the paper funded by the project (No. GD18XYS40).

References

1. Peeters, M., Stallworthy, M., de Larragán, J.D.C. (Eds.).: Climate law in EU member states: towards national legislation for climate protection. Edward Elgar Publishing (2012)
2. Zhao, X., Ma, X., Chen, B., Shang, Y., Song, M.: Challenges toward carbon neutrality in China: strategies and countermeasures. Resour. Conserv. Recycl. **176**, 105959 (2022)
3. Pollitt, H.: Analysis: going carbon neutral by 2060 'will make China richer'. Carbon Brief, 24 (2020)
4. Hepburn, C., Qi, Y., Stern, N., Ward, B., Xie, C., Zenghelis, D.: Towards carbon neutrality and China's 14th five-year plan: clean energy transition, sustainable urban development, and investment priorities. Environ. Sci. Ecotechnology **8**, 100130 (2021)
5. Zhaohui, Z.: Low-carbon ecological building design strategy and empirical analysis: a low-carbon exhibition hall as an example. Architecture, 9 (2011)
6. Yubo, L., Jia, L.: Colombian architect Simon Ville and his bamboo architecture. World Architect. **6**, 94–97 (2009)
7. Ladnykh, I., Ibragimov, A.: Low-rise house of plywood honeycomb panels. In: IOP Conference Series: Materials Science and Engineering, vol. 365, no. 6, p. 062028. IOP Publishing (2018)
8. Latka, J.F.: House of cards–design and implementation of a paper house prototype. In: Proceedings of IASS Annual Symposia, vol. 2017, no. 23, pp. 1–10. International Association for Shell and Spatial Structures (IASS) (2017)
9. Yingping, L., Bo, S., Yanfeng, G.: Research on the cushion coefficient and performance of honeycomb paperboard. J. Xi'an Inst. Technol. **23**(3), 194–197 (2003)

The Thermal Comfort Characteristics of PCS Applied in EVs

Jingjing Wu and Jingde Zhao(✉)

Donghua University, Shanghai 201620, China
zhaojingde@dhu.edu.cn

Abstract. In order to satisfy the needs of occupant's thermal comfort and reduction in HVAC energy consumption at the same time, based on the thermal comfort improvement measures in the building environment, this research applies the personal comfort system (PCS) to the electric vehicle cabin. By testing the physiological and psychological parameters of the occupants at different ambient temperatures, the effect of different local heating methods on the thermal comfort of the occupants is studied. The results show that PCS can effectively both increase the occupant's body surface temperature and the local and overall thermal sensation. In addition, the thermal sensation of the feet has the largest influence on the overall thermal sensation with the separate heating method which will provide a reference for the winter heating design of the EV in the future.

Keywords: Thermal comfort system · Thermal sensation · Electric vehicle

1 Introduction

Recently, with the aggravation of environmental problems and energy crisis, choosing electric vehicles (EVs) for travel has become a trend [1]. Electric vehicles are environmentally friendly and have the advantage of high energy efficiency. In addition, for the customers, people not only pay attention to the driving range of the vehicle, but also consider of the thermal environment in the vehicle cabin. Martinho [2] mentions that the human body thermal comfort in cabins is considered an important parameter in the design of vehicles. However, different from traditional fuel-powered vehicles which the engine generates enough waste heat to recover to heat the vehicle cabin, electric vehicle generates little heat by the battery, as a result, it needs to consume additional electric energy for space heating system which greatly reduces its driving range in winter. Lajunen [3] indicates that the vehicle driving range can be decreased more than 50% in extreme cold winter with the heating system on.

There are some studies have solved the contradiction between heating energy and driving range. Zhen Tian [4] found that with the PTC heater working, the EVs had 58.54% more driving range than under circumstance of the EVTMS off when the environment outside was −7 °C. Besides, personal comfort system (PCS) is a feasible way to solve these problems effectively in this study. PCS is generally used in buildings to satisfy occupants' thermal comfort. There are some related experiments to explore what impact

W. Guo and K. Qian (Eds.): GBCESC 2022, LNCE 211, pp. 1209–1217, 2023.
https://doi.org/10.1007/978-981-19-5217-3_124

the PCS can have on the energy consumption of the building's HVAC system. For example, the subjects who wear foot warmers can reduce the temperature setting point of the HVAC system from 21.1 °C to 18.9 °C so that the energy saving range can be from 38% to 75% in the office buildings [5]. Similarly, PCS can be used in EVs to save heating energy to increase the driving range. Jeffers found zonal heating strategies can increase the driving range by 6.9% to 18.7% [6].

However, the local thermal comfort has a great impact on the overall thermal comfort of the individual which has been improved. For example, Edward found that heating the local body parts will cause more intense thermal sensation than the thermal sensation of the overall body heating [7]. Heating different parts has different effects on other local thermal sensation and overall thermal sensation.

Currently there are some complete indices which applies to evaluate the building indoor thermal environment can be referred to, for example, predicted mean vote (PMV) and predicted percentage and dissatisfied (PPD) [8]. However, in the field of EVs, there are still few thermal comfort evaluations combined with PCS, and the degree of influence of local thermal comfort on the overall thermal comfort remains to be improved. At present, the most commonly used weighting factor of the local influence on the overall thermal comfort is the 9-point method in ISO [9] or an improvement on this basis [10]. The 9-point method is based on the ratio of the body surface area of each local part area to the total surface area. Therefore, this investigation mainly studies the influence weight of local thermal sensation on the overall thermal sensation with using local heating devices in an electric vehicle in winter.

2 Experiment

2.1 PCS Description

PCS is usually used in buildings to save energy consumption by heating the occupant's local body parts, as a result, the temperature setting point of the indoor air conditioning with the HVAC heating system can be reduced. Compared with traditional HVAC system, PCS is easier to satisfy the thermal comfort of different individuals in the same space by heating feet, seats, legs, back and so on. In this experiment, PCS is applied to EVs to save energy from HVAC system to increase the driving range in winter.

2.2 Experimental Setup

An experiment is conducted to verify the feasibility of the personalized heating methods and learn the performance and energy saving of PCS in EVs at low temperatures in winter. At the same time, the occupants' thermal sensation and thermal comfort can be satisfied.

Experimental Environment. An experiment was conducted in an open area on the campus of Donghua University from 18:00 to 22:00 from December 2018 to March 2019. The EV model used in the experiment is Roewe ERX5, and the car dimensions are 4554 * 1855 * 1719 (mm).

PCS Devices Description. In general, the knees and legs are more likely to be cold in cold winter. Drivers need to continue to operate with legs and feet in the process of driving. When these parts feel cold, it will decrease the flexibility of the two parts to affect the drivers' driving experience and safety. Besides, seat heating device has been confirmed that it has an obvious effect on improving the thermal comfort of human body. As a result, the PCS devices in this experiment mainly include seat heating device, leg heating device and foot heating device. The seat heating device mainly heats the back and buttocks. The carbon fiber heating pad (35 cm × 40 cm on the back and 40 cm × 40 cm on the buttocks) is covered on the seat, and the leg heating device (30 cm × 40 cm * 2) is worn on the subject's legs, and the foot heating device places the heating pad on the sole of the foot. Besides, 5 mm thick insulation cotton was laid between the heating pad and the tested part to reduce heat loss. Figure 1 is the physical picture of the foot heating and seat heating device used in the experiment.

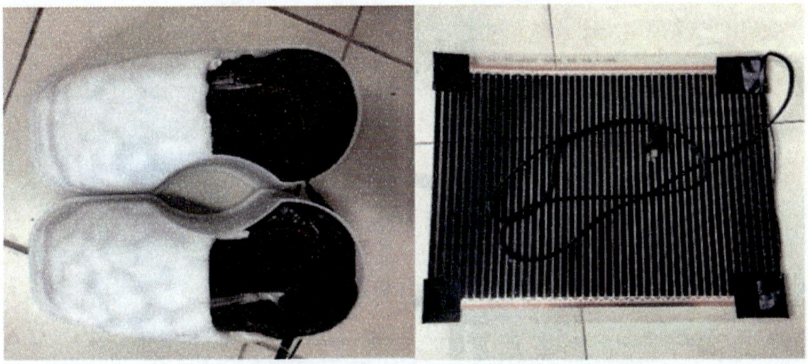

Fig. 1. Diagram of personalized heating equipment

Experimental Parameters and Instruments. The parameters should be measured in this experiment include the experimental environment parameters (air temperature of outdoor and cabin, velocity), the subject's body surface temperature, the subject's thermal sensation and thermal comfort. In order to test the above parameters, the instruments needed in the experiment are as Table 1, the layout points of human body temperature measurement are shown in Fig. 2:

Table 1. Experimental instruments and models

Parameter	Instrument	Model
Air temperature/relative humidity inside and outside the EV	Temperature and humidity meter thermistor	JXBS-3001-YJ OMEGA T
Body surface temperature	Thermistor	OMEGA T
Power	Regulator	YY-FTQS
	Ammeter	T-RS485

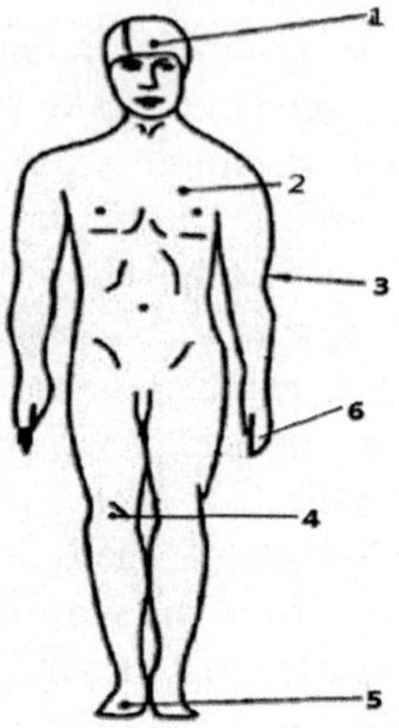

Fig. 2. Layout of measuring points on the human body

2.3 Subjective Questionnaire

In order to further verify that PCS can effectively improve the thermal sensation and thermal comfort of the human body, this subjective questionnaire is designed based on ASHRAE's 7-point model method which the thermal comfort is divided into 6 levels. As shown in Fig. 3, the scale ranges from 'cold' (−3), to 'cool' (−2), to 'slightly cool' (−1), to 'neutral' (0), to 'slightly warm' (+1), to 'warm' (+2), and to 'hot' (+3).

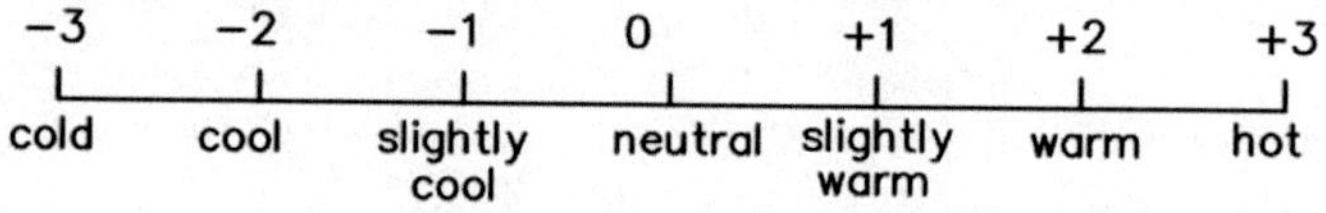

Fig. 3. 7-point model method

2.4 Subject

The experiment is officially approved by College of Environmental Science & Engineering, Donghua University. The experiment subjects are 12 healthy college students who were about 22 years old. The subjects' information is shown in Table 2. During the experiment, all subjects should wear uniform and typical winter clothing. In order to reduce the experimental error, the subject should be as close as possible to the PCS devices. In addition, the subjects should avoid strenuous exercise within 12 h before the experiment, and keep the body surface clean and dry to reduce the error caused by the thermistor temperature measurement.

Table 2. Basic information of subjects

Parameter	Age	Height (cm)	Weight (kg)	Body fat rate
Max	24	185	90	22
Min	21	170	58	11.6
Average	22.5 ± 1.5	177.5 ± 7.5	75 ± 15	16.8 ± 5.2

2.5 Experimental Working Situations

The experiment is carried out according to two variables: different outdoor temperatures and different personalized heating methods. Considering the actual outdoor temperature in winter in Shanghai and that temperature is the main factor affecting human thermal sensation and thermal comfort, 8 outdoor temperature conditions of 0, 2, 4, 6, 8, 10, 12, and 14 °C are selected for testing. For the heating methods, three heating methods including seat heating method, leg heating method and foot heating method are set up in the experiment to compare their effects on the human body's thermal sensation and thermal comfort under separate and combined effects. The experiment conditions information is shown in Table 3.

Table 3. Heating methods under different working conditions

Temperature	Heating methods	Serial number
0 °C	seat + leg + foot/no heating	1/2
2 °C	seat + legs + feet/no heating	3/4
4 °C	seat + legs + feet/no heating	5/6
6 °C	seat + legs + feet/no heating	7/8
8 °C	seat + legs + feet/seat/legs/feet/no heating	9/10/11/12/13
10 °C	seat + legs + feet/seat/legs/feet/no heating	14/15/16/17/18
12 °C	seat + legs + feet/seat/legs/feet/no heating	19/20/21/22/23
14 °C	seat + legs + feet/seat/legs/feet/no heating	24/25/26/27/28

3 Result

From the data analysis results, it can be seen that the local thermal sensation of various parts of the human body has different impact weight on the overall thermal sensation. This paper uses SPSS software and principal component analysis to process the subjective questionnaire data obtained under five heating methods, and verify the data by calculating the Bartlett sphericity test and the sampling appropriateness of Kaiser-Meyer-Olkin (KMO). When the significance level of the Bartlett sphericity test is $P < 0.05$, it indicates that there is a strong correlation between the original variables. When the KMO calculation result is close to 1.0, it indicates that it is suitable to use principal components component analysis.

Take the combined heating method condition as an example. The analysis results are shown in Table 4. From the table, it can be seen that the statistical significance level P of the sphericity test is less than 0.05, and the KMO measurement value is more than 0.5, this indicates that it is suitable to use the principal component analysis method to analyze the impact weight of each local thermal sensation on the overall thermal sensation with the combined heating method.

Table 4. KMO and Bartlett sphericity test coefficients with combined heating method

KMO sampling appropriateness quantity	0.807	
Bartlett sphericity test	Approximate chi-square	469.463
	Degree of freedom	36
	Significance	.000

The new variables obtained by standardization are selected for principal component analysis, and the characteristic roots and variance contributions obtained are shown in Table 5 and Table 6.

Table 5. Characteristic root and variance contribution table

Main ingredient	Initial eigenvalue			Extract the sum of squares of the load		
	Eigenvalues	Variance	Cumulative percentage of variance	Eigenvalues	Variance	Cumulative percentage of variance
1	4.304	47.818	47.818	4.304	47.818	47.818
2	1.751	19.452	67.270	1.751	19.452	67.270
3	.943	10.474	77.744			
4	.514	5.715	83.459			
5	.435	4.835	88.294			
6	.394	4.380	92.674			
7	.318	3.531	96.205			
8	.176	1.954	98.159			
9	.166	1.841	10.000			

Table 6. Principal component coefficient matrix with combined heating mode

	First principal component	Second principal component
Head	.776	−.127
Face	.760	−.216
Front chest	.779	.112
Back	.592	.726
Buttocks	.498	.781
Arm	.809	−.066
Finger	.682	−.437
Leg	.686	.062
Foot	.572	−.583

It can be seen from Table 5 that the eigenvalues of the first three principal components are all greater than 1, so three principal components are extracted. Combining the principal component coefficient matrix and the principal component eigenvalues in Table 3, 4 and 5, the linear combination of the three principal components can be obtained as shown in the following formula:

$$F_1 = 0.37X_{head} + 0.57X_{face} + 0.8X_{chest} + 0.83X_{back} + 0.76X_{buttocks} + 1.29X_{arm} + 1.21X_{finger} + 1.64X_{leg} + 1.4X_{foot} \quad (1)$$

$$F_2 = -0.06X_{head} - 0.16X_{face} + 0.12X_{chest} + 1.01X_{back} + 1.18X_{buttocks} - 0.11X_{arm} - 0.77X_{finger} + 0.15X_{leg} - 1.43X_{foot} \quad (2)$$

The percentage of the local thermal sensation obtained by the multiple regression under combined heating to the overall thermal sensation is shown in Table 7. It can be seen from the table that in the case of combined heating method, the head and face have little effect on the overall thermal sensation of the person, which weights are 0.03 and 0.05. Legs and feet have the highest influence weights on overall thermal sensation, both reaching 0.21.

Table 7. The influence weight of local thermal sensation on the overall thermal sensation

Head	Face	Front chest	Back	Buttocks	Arm	Finger	Leg	Foot
0.03	0.05	0.08	0.08	0.1	0.14	0.1	0.21	0.21

According to the above method, the three heating methods of seat heating, leg heating, and foot heating are respectively processed to obtain the impact weight of the local thermal sensation on the overall thermal sensation, the results as shown in Table 8.

Table 8. The influence weight of local thermal sensation on the overall thermal sensation under three heating methods

	Head	Face	Front chest	Back	Buttocks	Arm	Finger	Leg	Foot
Seat heating	0.04	0.08	0.11	0.09	0.1	0.12	0.09	0.11	0.26
Leg heating	0.04	0.04	0.06	0.11	0.14	0.03	0.14	0.17	0.27
Foot heating	0.02	0.07	0.07	0.13	0.13	0.02	0.14	0.18	0.24

From the data in Table 8, it can be seen that under the three local heating methods, the thermal sensation of the feet has the greatest impact on the overall thermal sensation, accounting for about a quarter; the thermal sensation of the legs is the second, about one-sixth of the overall thermal sensation.

4 Conclusion

In this study, PCS is used in EVs to improve the thermal comfort of occupants in winter and reduce energy consumption for heating. The interaction and influence among different working conditions, different personalized heating methods, local thermal sensation and overall thermal sensation, local thermal comfort and overall thermal comfort are investigated through the subjective questionnaire. Then the experiment data are analyzed by using principal component analysis method to obtain the weight coefficients of the influence of local thermal sensation on the overall thermal sensation. Through principal component analysis and the impact weight of each part under various heating methods, it can be seen that:

1. Under the same outdoor environment conditions, different heating methods have different effects on the thermal sensation of the human body. Under the condition of 0–6 °C, the improvement of the overall thermal sensation of the human is related to the improvement of the parts with the lowest thermal sensation among the local parts, which there is little correlation on lifting effect of body parts that are in a better thermal sensation condition. However, it has little relation with the improvement of thermal sensation of body parts under the condition of good thermal sensation.
2. By comparing the weight of each local part's thermal sensation to the overall thermal sensation with single heating method, the thermal sensation of the feet always accounts for the largest proportion, accounting for about a quarter; the thermal sensation of the legs accounts for about one-sixth of the overall thermal sensation.

References

1. Harvey, L.: Rethinking electric vehicle subsidies, rediscovering energy efficiency. Energy Policy **146**, 111760 (2020)
2. Martinho, G., Silva, M.G., Ramos, J.A.E.: Evaluation of thermal comfort in a vehicle cabin. Proc. Inst. Mech. Eng. Part D: J. Automob. Eng. **218**(2) (20014)

3. Lajunen, A.: Energy efficiency and performance of cabin thermal management in electric vehicles. SAE Int. (2017). https://doi.org/10.4271/2017-01-0192
4. Zhen, T., Wei, G., Zhang, X., et al.: Investigation on an integrated thermal management system with battery cooling and motor waste heat recovery for electric vehicle. Appl. Therm. Eng. **136**, 16–27 (2018)
5. Hui, Z., Edward, A., Mallory, T., et al.: Using footwarmers in offices for thermal comfort and energy savings. Energy Build. **104**, 233–243 (2015)
6. Jeffers, M.A., Larry, C., Rugh, J.P.: Climate control load reduction strategies for electric drive vehicles in cold weather. SAE Int. J. Passeng. Cars Mech. Syst. **9**(1), 75–82 (2016)
7. Edward, A., Hui, Z., Charlie, H.: Partial- and whole-body thermal sensation and comfort—Part I: Uniform environmental conditions. J. Therm. Biol **31**(1–2), 53–59 (2006)
8. Fanger, P.O.: Thermal Comfort Analysis and Applications in Environmental Engineering (1972)
9. ISO: ISO 9920: 2007 Ergonomics of the thermal environment—Estimation of thermal insulation and water vapour resistance of a clothing ensemble (2007). (in Genevan)
10. Weiwei, L., Zhiwei, L., Qihong, D., Yuanmou, L.: Evaluation of calculation methods of mean skin temperature for use in thermal comfort study. Build. Environ. **46**(2), 478–488 (2011)

Mode Choice Behavior of Ride-Sourcing Passengers Under Travel Time Variability

Aitong Pang[1], Xin Zhang[2], and Gege Jiang[1](✉)

[1] Sun Yat-sen University, Shenzhen, China
jianggg@mail.sysu.edu.cn
[2] Hohai University, Nanjing, China

Abstract. Travel time variability has a great impact on travel behavior. In this paper, we conduct a survey which shows that travel time variability is a vital element of mode choice in ride-sourcing services. According to the survey results, we address some key factors that influence travelers' mode choices, including travel time, travel time variability, monetary cost and comfort. Then we propose a theoretical framework to investigate the mode choice behavior under travel time variability. In addition, the pricing strategy of the ride-sourcing company is formulated as an optimization problem. Our study shows that considering the impact of travel time variability is of vital importance both in modeling the travel behavior of travelers and the pricing strategy of the ride-sourcing market.

Keywords: Ride-sourcing · Travel time variability · Mode choice

1 Introduction

With the rapid development of electronic communication technologies, app-based, on-demand ride-sourcing services (RS) are flourishing and becoming increasingly popular. Travelers can match with vehicles and drivers through the ride-sourcing platform to obtain door-to-door travel services. Transportation network companies (TNCs) provide diversified services, including Express, Premier, and Ride-splitting. Premier service mainly serves the upmarket, while Express service has a large customer overlap with taxis [1]. As for the term ride-splitting, it means a ride-sourcing service in which travelers can split both a ride and the fare [2]. Jiayi Joey et al. noted that in extreme cases, ride-sourcing services might drive the traditional taxi industry out of the market [3], which aroused heated discussion among the transportation industry and researchers.

With ride-sourcing services emerging as a new travel option, travelers have more travel modes to choose from. Plenty of studies focused on the factors that influence ride-sourcing behavior, such as price, convenience, and accessibility [4–6]. In addition, unpredictable travel time, which is called travel time variability, also has a great impact on travelers' mode choice. Empirical study has proved that the variations in travel time can cause substantial early or late penalties [7]. With increasing attention paid to travel time variability, studies have taken it into account in travel choice analysis [8–10]. Early researches mainly concentrated on demonstrating the influence of variability via stated

W. Guo and K. Qian (Eds.): GBCESC 2022, LNCE 211, pp. 1218–1225, 2023.
https://doi.org/10.1007/978-981-19-5217-3_125

preference (SP) survey and revealed preference (RP) study [11, 12]. Recently, Jiancheng et al. developed a stochastic ride-sharing system considering travel time variability [13]. Gege and Hong Kam designed pricing strategies for a ride-sourcing service under travel time variability [14].

Although travel time variability has a substantial influence on travel behavior, only a handful of studies considered it in the mode choice problem, especially under the ride-sourcing context, let alone pricing strategies under travel time variability for the TNCs. To fill this gap, this paper makes the following contribution. First, through empirical investigation, we discover the key factors influencing travelers' choice when choosing ride-sourcing services. Second, based on the survey results, we select four factors to be considered in the mode choice problem, namely comfort, travel time, travel time variability and monetary cost. By combing the theoretical framework with empirical research, this study can shed insights on understanding the motivation of travelers to join ride-sourcing services. Third, we propose a fixed pricing strategy for TNCs company to maximize its revenue.

The remainder of the paper is as follows. Section 2 analyzes the results of an empirical survey. Section 3 constructs the theoretical framework. The numerical studies are presented in Sect. 4. Section 5 gives the conclusion and future research directions.

2 Empirical Survey

We conducted an online questionnaire to identify the critical factors that impact travelers' mode choice behavior regarding to ride-sourcing services, which provides an empirical basis for subsequent modeling. A total of 650 valid questionnaires were collected through the Internet.

According to our survey results, nearly 60% of respondents prefer to travel by express. Despite the lower price, less than 40% choose ride-sharing services due to the high travel time variability. Furthermore, 73.38% of respondents said they would attach great importance to the impact of late penalties caused by travel time variability.

Combined with the key elements of travelers' mode choices and the characteristics of ride-sourcing, some specific factors that affect travelers' ride-sourcing behavior are listed in the survey. Respondents were asked to choose and rank the five factors they cared about most of ride-splitting service or express/premier service. According to the ranking results, the average score of each factor is obtained, as shown in Fig. 1.

The higher the score is, the more attention people pay to the factor. It implies that time and monetary cost are always the primary concerns. Besides, the travel time variability also has a high score, surpassing the scores for travel cost variability, comfort, service quality and traffic condition. Therefore, it is quite necessary to introduce travel time variability into the mode choice problem.

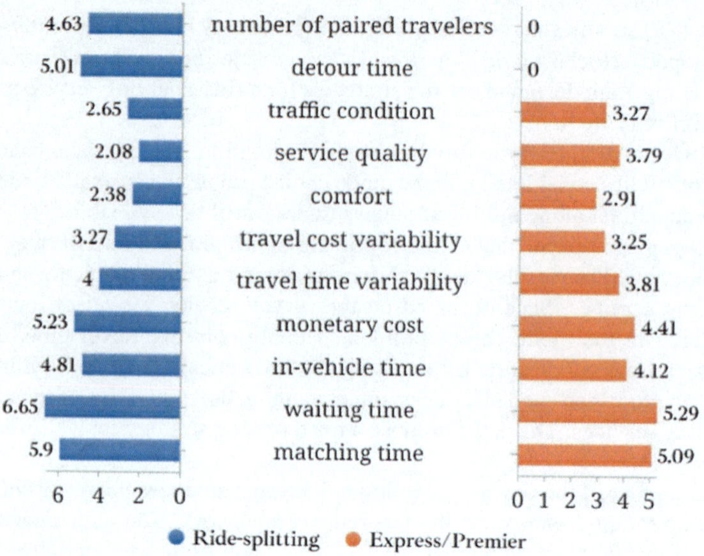

Fig. 1. The score of each factor under different ride-sourcing services

3 Problem Definition

Research shows that the TNCs not only influence the taxi industry but also compete with public transit. Public transit has a predictable travel time, with little variability and lower fare. However, travelers taking the public transit have to endure a long and crowded trip. In contrast, the RS charges cheaper than conventional taxis and runs faster than public transit. Moreover, it has an unpredictable fluctuation in travel time due to degradable road capacity. Thus, the problem is defined for *N* travelers facing choices considering two travel modes: taking the subway versus taking the ride-sourcing services provided by TNCs. All travelers know the distribution of the capacity in advance. Travelers can make decisions by comparing the pros and cons between ride-sourcing service and public transit service. For the sake of analytical purpose, we only consider time, monetary cost, travel time variability and comfort in the modelling framework to draw some general conclusions.

3.1 Modeling of Travel Time Variability

Travel time variability stems from the uncertainty of demand and supply. The demand-side variation refers to factors like varying reactions to traffic information can cause the fluctuation of travel demand. The supply-side variation concludes weather condition, accidents and other factors influencing the road capacity. This study considers travel time variability is endogenously affected by the traffic flow and the degradable capacity.

Thus, the travel time for RS is assumed to be stochastic, and the bureau of public roads (BPR) link performance function is employed:

$$t_r(N_r, C) = t_0\left[1 + \gamma\left(\frac{N_r}{C}\right)^q\right] \tag{1}$$

where t_0 is the deterministic free-flow travel time on the road. N_r is the number of travelers for RS. γ and q are constant coefficients of the function. C is the realized capacity of the road. We assume C follows a uniform distribution with the lower bound $\theta\overline{C}$ and the upper bound $\overline{C}$, and θ is a parameter that indicates the variability of capacity. The probability density function (PDF) $f(C)$ and the cumulative density function (CDF) $F(C)$ of the road capacity are:

$$f(c) = \frac{1}{\overline{C} - \theta\overline{C}} \tag{2}$$

$$F(C) = \int_{\theta\overline{C}}^{C} \frac{1}{\overline{C} - \theta\overline{C}} dC \tag{3}$$

Following the derivation in Barbara and Hong Kam [15], the mean and standard deviation of travel time are:

$$\bar{t}_r = t_0 + \gamma t_0 N_r^q E\left(\frac{1}{C^q}\right) = t_0 + \gamma t_0 N_r^q \frac{1 - \theta^{1-q}}{\overline{C}^q(1-\theta)(1-q)} \tag{4}$$

$$\sigma_r = \gamma t_0 N_r^q \operatorname{var}\left(\frac{1}{C^q}\right) = \gamma t_0 N_r^q \sqrt{\frac{1 - \theta^{1-2q}}{\overline{C}^{2q}(1-\theta)(1-2q)} - \left[\frac{1 - \theta^{1-q}}{\overline{C}^q(1-\theta)(1-q)}\right]^2} \tag{5}$$

We can simplify Eq. (4) and (5) as:

$$\bar{t}_r = t_0 + aN_r^q \tag{6}$$

$$\sigma_r = bN_r^q \tag{7}$$

where $a = \gamma t_0 \frac{1-\theta^{1-q}}{\overline{C}^q(1-\theta)(1-q)}$, $b = \gamma t_0 \sqrt{\frac{1-\theta^{1-2q}}{\overline{C}^{2q}(1-\theta)(1-2q)} - \left[\frac{1-\theta^{1-q}}{\overline{C}^q(1-\theta)(1-q)}\right]^2}$, both of them are positive.

3.2 Disutility Function

Based on the survey results, the four factors that travelers most concerned are comfort, travel time, travel time variability and monetary cost. For simplicity, we assume that the travel time variability for the subway and the discomfort of taking RS are zero. Thus, the disutility function of taking subway and taking ride-sourcing service are respectively expressed as below:

$$-U_s = \alpha_1 N_s + \alpha_2 \bar{t}_s + g(\tau_s) \tag{8}$$

$$-U_r = \alpha_2 \bar{t}_r + \alpha_3 \sigma_r + g(\tau_r) \tag{9}$$

N_s is the number of travelers taking subway. $\bar{t}_s$ is the mean travel time of taking subway. α_1 is the inconvenience parameter of taking subway and the discomfort of the subway is linear to N_s. α_2 and α_3 are the value of time (VOT) and the value of reliability (VOR), representing the sensitivity of mean travel time and standard deviation, respectively. τ_i is a fixed monetary cost of taking subway or RS, and $g(x)$ is the disutility function of monetary cost with $g(x) > 0$.

3.3 Maximum Profit Function

The pricing strategy can be modeled by an optimization problem. The goal is to maximize the revenue of ride-sourcing company with the constraint of user-equilibrium. In equilibrium, all travelers have a same disutility and no travelers can get a lower cost by changing their mode choice unilaterally, so $U_s = U_r$. Suppose the operating cost per unit time for ride-sourcing services is β, the objective function of profit maximization can be formulated as:

$$\max_{\tau_r} Z = N_r(\tau_r - \beta \bar{t}_t) \tag{10}$$

$$\text{s.t.} \quad \begin{array}{l} N_s + N_r = N \\ (-U_s + U_r)N_r \geq 0 \\ (-U_s + U_r)N_s \leq 0 \\ N_s, N_r > 0 \end{array}$$

Since the origin and destination are fixed, the ride-sourcing service charges a fixed fare τ_r every day regardless of the actual travel time on a specific day. To capture the relationship between $g(\tau_r)$ and N_r, combining (8) and (9), we have:

$$g(\tau_r) = f(N_r) = c - \left[(\alpha_2 a + \alpha_3 b)N_r^q + \alpha_1 N_r\right] \tag{11}$$

where $c = \alpha_2(\bar{t}_s - t_0) + g(\tau_s) + \alpha_1 N$. Define $\tau_r = h(N_r) = g^{-1}(f(N_r))$, we can rewrite Eq. (10) as:

$$\max_{N_r} Z = N_r\left[h(N_r) - \beta \bar{t}_r)\right] \tag{12}$$

Suppose the optimal solution is N_r^* and $g(\tau_r^*) = f(N_r^*)$. The corresponding maximum profit is Z_*.

4 Numerical Studies

In order to illustrate the relationship between pricing strategy, disutility and travel demand under travel time variability, an example of the fixed pricing was provided by numerical analysis.

Consider $N = 100$ travelers facing choices considering two travel modes: taking the subway versus taking the ride-sourcing services provided by TNCs. The price of taking subway is $\tau_s = 5$ \$/min. The operating cost per unit time for RS is $\beta = 0.1$ \$/min. As the price of taking RS is usually higher than taking the subway, we suppose $\tau_r \geq \tau_s$. The free-flow travel time $t_0 = 15$ min. The mean travel time for subway $\overline{t}_s = 30$ min. The upper bound of the capacity $\overline{C}$ is 30, and the degradation level of capacity θ is 0.7. $\alpha_1 = 0.07$, $\alpha_2 = 0.2$ \$/min, $\alpha_3 = 0.3$ \$/min are the inconvenience parameter, VOT, and VOR respectively. Other constant coefficients are $\gamma = 0.13$, $q = 4$.

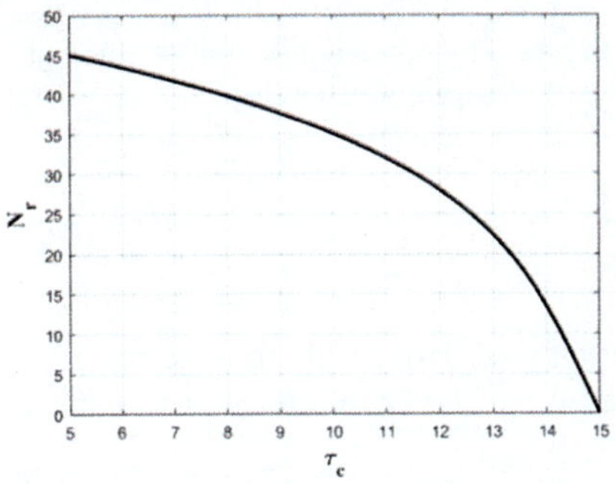

Fig. 2. The number of travelers choosing RS under different prices

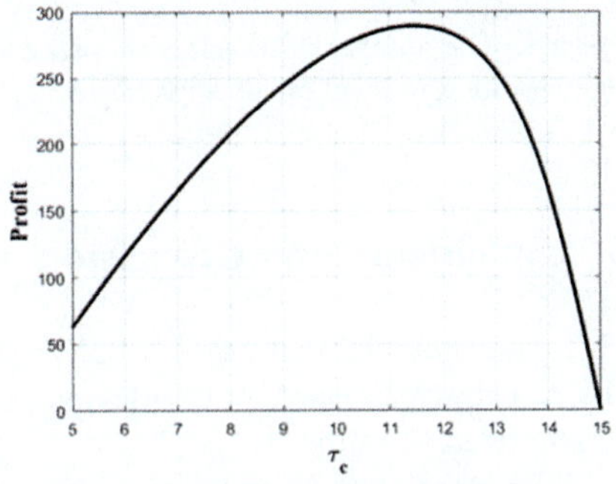

Fig. 3. The profit of the ride-sourcing company under different prices

When travelers have a same value of travel time and value of variability, the disutility for them increases with the price. As a result, high travel cost distracts travelers taking the ride-sourcing services, so the ride-sourcing company gradually lose their market share, as shown in Fig. 2.

The profit of the ride-sourcing company under different prices is illustrated in Fig. 3. The profit increases first and then decreases. Although facing travel time variability, the lower price attracts more travelers choosing RS, which reduces their travel time and inconvenience. When the price is 11.4, the ride-sourcing company gets a maximum profit $Z^* = 289.24$, after that the profit drops. The ride-sourcing company trade between the market share and the revenue they obtain.

The degradation level of capacity θ refers to the fluctuation in travel time. Larger θ means less mean and standard deviation of travel time under same demand. So, when θ increases, the ride-sourcing company can serve more travelers with a large decrease in the mean travel time and travel time variability (see Fig. 4). As a result, the optimal profit increases with the level of capacity variability (see Fig. 5). Significantly, the profit at $\theta = 0.7$ is lower than $\theta = 1$, which can be seen from Fig. 4. It can be concluded that the ride-sourcing company will overestimate their revenue if travel time variability is ignored.

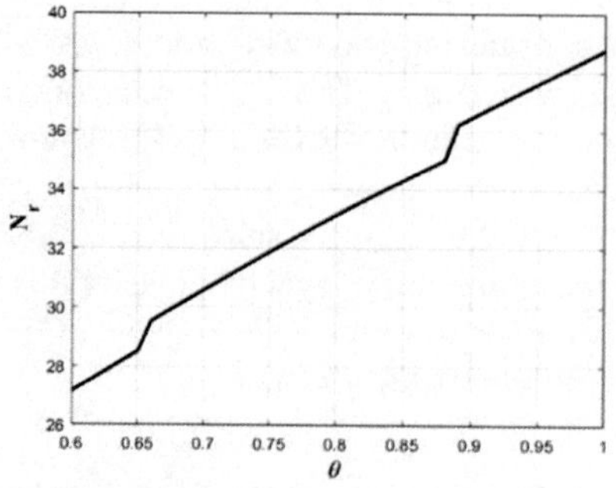

Fig. 4. The number of travelers choosing RS under different level of capacity variability

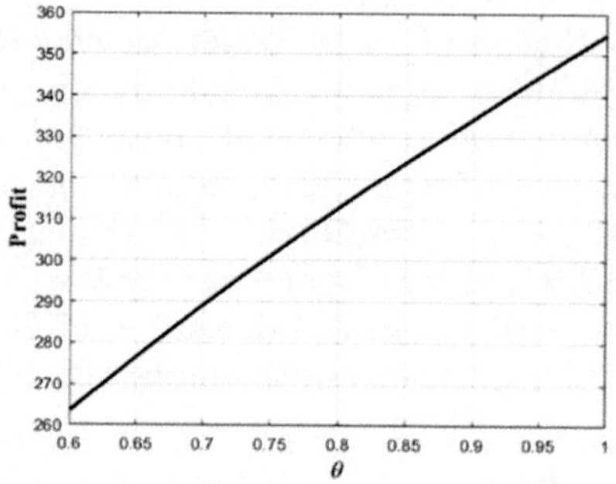

Fig. 5. The optimal profit under different level of capacity variability

5 Conclusion and Discussion

This paper showed that travel time variability plays a critical role in passengers' mode choice behavior regarding to ride-sourcing services. The key findings of this study are summarized as follows:

First, according to our survey results, more than 70% of participants highly value the travel time variability. Nearly 60% of them prefer to travel by express, which has less fluctuation in travel time. Thus, excluding travel time and monetary cost, variability of travel time also affects travelers' mode choices. Secondly, a theoretical model considering the impact of travel time variability on modal choice behavior is proposed. We introduce an endogenous travel time variability coming from capacity degradation into the utility function. When the degradation level of capacity increases, people are more willing to travel by ride-sourcing service. We also study a fixed pricing strategy and explore the maximum profit function for the ride-sourcing services. The disutility increases with the fixed price, thus distracting travelers taking the ride-sourcing services. A reasonable pricing strategy can maximize revenue for TNCs. The ride-sourcing company are less profitable at a low level of capacity variability than at a high level. They will overestimate the revenue when travel time variability is ignored. This phenomenon is well demonstrated in the numerical example.

There are some potential extensions in the future. First, our study is limited to homogenous travelers with same VOT and VOR. Travelers' heterogeneity and heterogeneous monetary disutility function can be studied in this modeling framework as extensions. Besides, we can investigate more pricing strategies under travel time variability.

Acknowledgment. This study is supported by the Fundamental Research Funds for the Central Universities, Sun Yat-sen University (22qntd1701).

References

1. Xiaolei, W., Wei, L., Hai, Y., Dan, W., Jieping, Y.: Customer behavioural modelling of order cancellation in coupled ride-sourcing and taxi markets. Transp. Res. Part B: Methodol. **132**, 358–378 (2020)
2. Hai, W., Hai, Y.: Ridesourcing systems: a framework and review. Transp. Res. Part B: Methodol. **129**, 122–155 (2019)
3. Jiayi Joey, Y., Christopher, S., Zuo-Jun Max, S., Xiqun, C.: Should On-Demand Ride Services Be Regulated? An Analytical Evaluation of Chinese Government Policies. Social Science Electronic Publishing (2017)
4. Lisa, R., Danielle, D., Nelson, C., Robert, C., Susan, S.: Just a better taxi? A survey-based comparison of taxis, transit, and ridesourcing services in San Francisco. Transp. Policy **45**, 168–178 (2016)
5. Dias, F.F., Lavieri, P.S., Garikapati, V.M., Astroza, S., Pendyala, R.M., Bhat, C.R.: A behavioral choice model of the use of car-sharing and ride-sourcing services. Transp. **44**(6), 1307–1323 (2017)
6. Roger, F.T.: Carpooling: who, how and why. Transp. Res. Part A: General **21**(3), 203–214 (1987)
7. Richard, B.J., Nicolás, I.: Randomness in preference orderings, outcomes and attribute tastes: an application to journey time risk. J. Choice Model. **5**(3), 157–175 (2012)
8. Siu, B.W., Lo, H.K.: Equilibrium trip scheduling in congested traffic under uncertainty. In: Lam, W., Wong, S., Lo, H. (eds.) Transportation and Traffic Theory 2009: Golden Jubilee, pp. 147–167. Springer, Boston (2009). https://doi.org/10.1007/978-1-4419-0820-9_8
9. Mogens, F., Anders, K.: The value of reliability. Transp. Res. Part B: Methodol. **44**, 38–49 (2010)
10. Mogens, F., Leonid, E.: The value of travel time variance. Transp. Res. Part B: Methodol. **45**(1), 1–8 (2011)
11. John, B., John, P., Peter, J., Andrew, C.: The valuation of reliability for personal travel. Transp. Res. Part E: Logist. Transp. Rev. **37**, 191–229 (2001)
12. David, B., Kenneth, A.S.: Valuing time and reliability: assessing the evidence from road pricing demonstrations. Transp. Res. Part A: Policy Pract. **39**(4), 279–293 (2005)
13. Jiancheng, L., Weimin, T., Szeto, W.Y., Yao, L.: Ride-sharing with travel time uncertainty. Transp. Res. Part B: Methodol. **118**, 143–171 (2018)
14. Gege, J., Hong Kam, L.: Pricing strategy for a ridesourcing service under travel time variability. In: Transportation Systems in the Connected Era-Proceedings of the 23rd International Conference of Hong Kong Society for Transportation Studies, HKSTS 2018, p. 219 (2018)
15. Barbara, W.Y.S., Hong Kam, L.: Doubly uncertain transportation network: degradable capacity and stochastic demand. Eur. J. Oper. Res. **191**(1), 166–181 (2008)

Check for updates

Intelligent Control of Shield Tunneling from the Perspective of Complex Network

Xiaojing Zhou, Yue Pan, and Jianjun Qin(✉)

Shanghai Key Laboratory for Digital Maintenance of Buildings and Infrastructure, School of Naval Architecture, Ocean and Civil Engineering, Shanghai Jiao Tong University, Shanghai 200240, China
jianjunqin@sjtu.edu.cn

Abstract. In order to more comprehensively analyze the influence of shield tunneling parameters on shield machine performance, this study develops a complex network (CN) -enabled framework to explore the shield tunneling performance. In the first step, Principal Component Analysis (PCA) is used to observe time series data. In the second step, the adjacency matrix is obtained by calculating the correlation between parameters of any two rings and setting appropriate threshold values. Each node of the complex network stands for different rings, and the links between nodes (i.e. edges) represent the nonlinear relations between the rings. Finally, Leiden, as an excellent community discovery algorithm, aggregates nodes with similar tunneling status to form clusters of different sizes in the network, and different clusters show different tunneling performance in terms of over excavation rate. We verify the feasibility of complex network theory in the case study about the Contract T308 along the Eastern Regional Line (ERL) in Singapore.

Keywords: Shield tunneling · Complex network · Community detection · Data mining

1 Introduction

In the tunneling process of tunnel boring machines (TBMs), excavation rate [1] is an output parameter to measure the penetration and advance rate of a TBM. Many previous studies have shown that it is mainly affected by both geological and mechanical parameters [2]. By observing these parameters, engineers can understand the operation of the TBM, further assess the risks and costs of construction. But considering the site monitoring data collected often reflects high-dimensional, mixed and highly nonlinear characteristics, engineers are always subjective to parameter setting, at the same time hard to juggle all the parameters (the types of parameters are even more than 200 [3]). Therefore, to develop a comprehensive analysis method of shield performance based on monitoring parameters is of great significance to realize intelligent control of tunnel construction.

In order to explore TBM performance optimization based on multidimensional shield parameters, many scholars have used the complex network (CN) theory, which has been

W. Guo and K. Qian (Eds.): GBCESC 2022, LNCE 211, pp. 1226–1233, 2023.
https://doi.org/10.1007/978-981-19-5217-3_126

gradually applied in construction industry [4] recently. However, in the existing studies, the original monitoring data tend to set the average value as the representative of each ring value. Obviously, although the average value reduces the complexity of the original data set, it is susceptible to the influence of individual extreme values in the database, affecting the accuracy of the final performance analysis results.

Principal Component Analysis (PCA), as a broad algorithm to reduce data dimension[5], realizes feature extraction and compression of the whole time series database and offers even greater advantages in data mining (DM) with the combination of CN. For example, Pan, Y., et al. [6] adopt an PCA-CN method to fuse observed time-series data about the evolution of the COVID-19 pandemic in some parts of the United States (US), which builds a complex network based on a low-dimensional representation of a large data set and can serve as rich scientific evidence to systematically understand the dynamic behavior of COVID-19 from a new point.

In our study, the time series data composed of high-dimensional shield parameters is taken as the research object, and new data mining methods such as PCA, CN and community detection (CD) are innovatively introduced to show the difference of tunneling performance in tunnel loop number. Different from previous studies, feature variables extracted after PCA are used as node information for subsequent similarity calculation instead of the mean value, which would greatly reduce information loss. Finally, the case of Contract T308 along the Eastern Regional Line (ERL) in Singapore proves the feasibility of complex network theory.

2 Construction of Complex Network

In our study, each node of the complex network corresponds to different rings of tunnel engineering, and the links between nodes (edges in other words) represent the nonlinear relations between rings. Our work in Fig. 1 is divided into three steps: 1) Dimension reduction by PCA;2) Network construction, mainly including similar matrix and adjacent matrix; 3) Network topology features and community detection. We respectively describe these methods below.

2.1 PCA

Before PCA, it is first to set normalized processing of the original data. After normalization, a matrix $\mathbb{R}^{M\times N}$ composed of shield parameters of each tunnel segment is obtained, that is, each parameter has a time series $X^{i} = \{X_1^{i}, X_2^{i}, X_3^{i}, \ldots, X_M^{i}\}$ in a fixed acquisition period T, and the number of shield parameters $X_j = \left\{X_j^1, X_j^2, X_j^3, \ldots, X_j^M\right\}$ is N. Singular value decomposition is performed on the derived covariance matrix (Eq. (1)) [7]

$$\Sigma = \frac{1}{N-1}XX^T = P\Lambda P^T \tag{1}$$

Here $\Lambda = diag(\lambda_1, \lambda_2, \lambda_3, \ldots, \lambda_M)$ and $\lambda_1 \geq \lambda_2 \geq \lambda_3 \geq \ldots \geq \lambda_M \geq 0$. In PCA, the load matrix $P\in \mathbb{R}^{M\times M}$ is divided into $P = \left[P_{\text{pc}}\ P_{\text{res}}\right]$ and Λ is divided

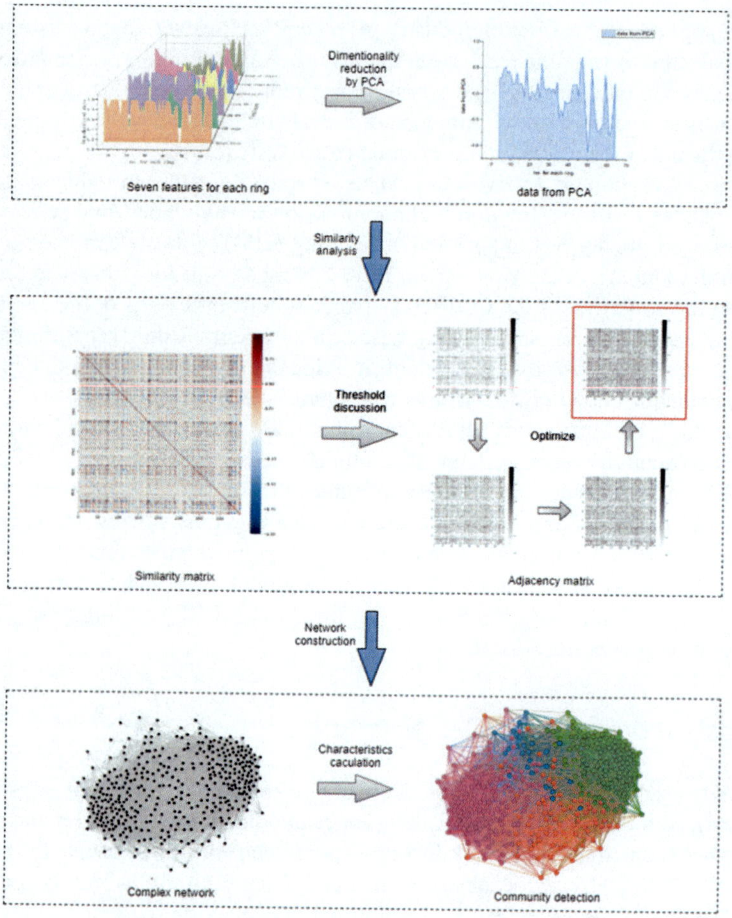

Fig. 1. The construction process of complex network

into $\Lambda = \begin{bmatrix} \Lambda_{pc} & 0 \\ 0 & \Lambda_{res} \end{bmatrix}$. Given a sample data $\mathrm{x} \in \mathbb{R}^{M \times 1}$, there are the following two parameters T^2 (Eq. (2)) and Q (Eq. (3))

$$T^2 = \mathrm{x}^{\mathrm{T}} \mathrm{P}_{\mathrm{pc}} \Lambda_{pc}^{-1} P_{pc}^{T} x \tag{2}$$

$$Q = \left\| (I - P_{\mathrm{pc}} \mathrm{P}_{\mathrm{pc}}^{\mathrm{T}}) \mathrm{x} \right\|_E^2 = \mathrm{x}T(I - P_{pc} P_{pc}^{T})x \tag{3}$$

Here $I \in \mathbb{R}^{M \times M}$ is the identity matrix.

2.2 Similarity Matrix and Adjacent Matrix

In this complex network, the interaction between nodes is determined by the similarity between two time series, which is measured by Pearson correlation coefficient [8], and then similarity matrix is created. The complex network model based on the similarity matrix is fully connected, no matter whether the link is strong or weak, which will lead to information redundancy. Therefore, the threshold value ε is introduced to retain the strong link in the network.

2.3 Network Topology Features and Community Detection

At the same time, it is necessary to analyze the network topology structure in order to further explore the collective behavior and other emergent phenomena of tunnel segmental ring in tunneling performance. Therefore, we introduce four typical indicators (including diameter, density, clustering and length) of the network [9] in the selection of threshold. Among them, we believe that the average path length is a sufficient condition to ensure small-world property of complex network, which is also the basis for network optimization with different thresholds.

Further, by observing the similarity matrix, we believe that there are some community structures in the shield system, and the process of discovering community structures is community discovery (CN). Using 10 known community discovery algorithms [9] (including AGDL, DER, Greedy_Modularity, Leiden, Louvain, etc.), the indicators for clustering effects (including erdos_renyi_modularity, newman_girvan_modularity and z_modularity) are compared. The Leiden algorithm [10] is determined as the CD algorithm with best performance. On this basis, further excavation can be carried out for each community, and centralized parameter regulation and optimization can be done for tunnel segments with similar tunneling performance.

3 Case Study

This paper is based on the E1002 contract for the Eastern Regional Line (ERL) in Singapore. We take the statistical data of 400 rings (the serial number is 1–400 rings) as the research object. Each ring pair should have 61 groups (the serial number is 4–64) of shield parameters, and these data constitute a generalized time series. The seven parameters of TBM we selected here are Thrust Force (X1), cutter head (CHD) Torque (X2), CHD rotation speed (RPM) (X3), SCREW RPM (X4), and penetration rate (X5). On the other hand, geological conditions along the tunnel project are considered, and the main parameters are Average chamber pressure (X6) and Average soil pressure (X7).

After the processing of the previous two steps, the network topology analysis under different thresholds is shown in Table 1. After observing the visualized complex network in Fig. 2, it is easy to find that with the increase of threshold, the network is connected more closely. According to the precondition of average path length, 0.20 is determined as a reasonable threshold.

Table 1. Characteristics of complex network topology under different thresholds.

Characteristics	Threshold		
	0.10	0.15	0.20
Number of nodes	400	400	400
Number of links	8200	12200	16200
Graph density	0.103	0.153	0.203
Network diameter	5	4	3
Average clustering coefficient	0.575	0.547	0.541
Average path length	2.531	2.143	1.912

Subsequently, based on the adjacency matrix under the threshold, 10 community detection algorithms have been used to explore the possible community structure in the network. Table 2 compares the clustering performance using three evaluations indicators. It is can be seen that the Leiden algorithm can show the best performance in community detection, whose result is visualized in Fig. 3.

Table 2. Community detection algorithm comparison.

Algorithm	Evaluation indicator		
	erdos_renyi_modularity	newman_girvan_modularity	z_modularity
agdl	0.2683	0.2695	0.5409
Fluid	0.4561	0.3937	0.8323
DER	0.3961	0.3926	0.7852
eigenvector	0.3933	0.3924	0.7847
greedy_modularity	0.3937	0.3851	0.7702
Leiden	**0.4493**	**0.4007**	**0.8244**
louvain	0.3964	0.3933	0.7867
significance_communities	0.4176	0.3170	0.7455
surprise_communities	0.4950	0.3770	0.7964
walktrap	0.4470	0.3944	0.8058

After all tunnel segmentation rings are divided into 4 communities with the help of Leiden algorithm, statistical analysis is made to explore the over excavation rate of each ring in each community, the result of which is shown in Fig. 4. The lower the overcut rate is, the more accurate the performance of shield machine is and the higher construction requirements are met. It can be observed that the over excavation ratio in cluster 4 is significantly larger than the other three clusters. That is to say, the rings grouped into cluster 4 are more likely to suffer from a relatively higher over excavation ratio. It

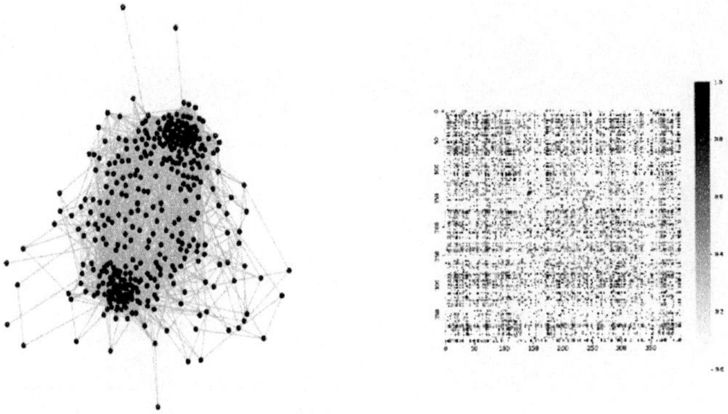

(a) Threshold=0.10

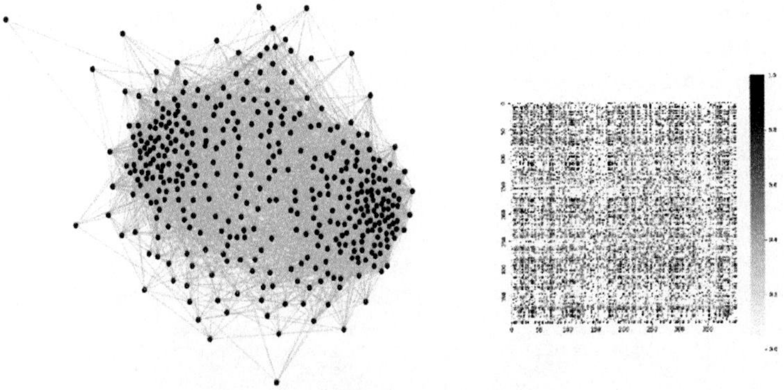

(b) Threshold=0.15

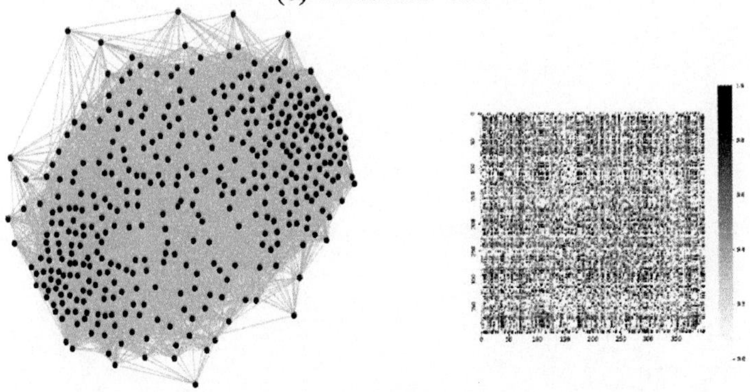

(c)Threshold=0.20

Fig. 2. Adjacent matrix and their corresponding CN with different thresholds

is essential to pay more attention to properly adjust machine parameters according to surrounding rock geological conditions in cluster 4 and timely survey excavation section and send feedbacks to engineers during excavation. On the contrary, cluster 1 contains

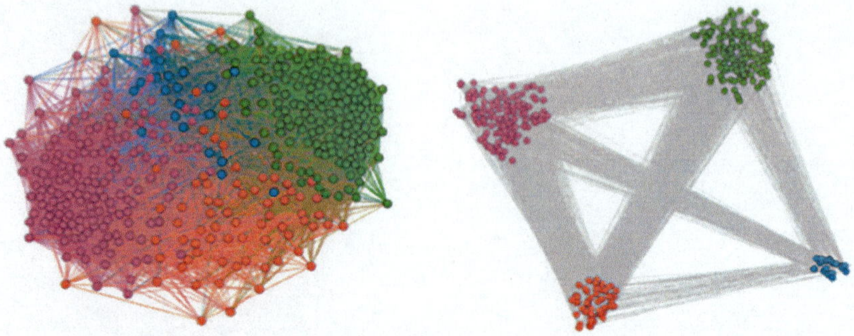

Fig. 3. Community detection results based on the Leiden algorithm

more segment rings with lower over excavation rate than other clusters detected, suggesting more accurate tunneling performance and possibly providing reference information for parameter settings of other clusters. It is desirable to follow such a variation pattern in cluster 1.

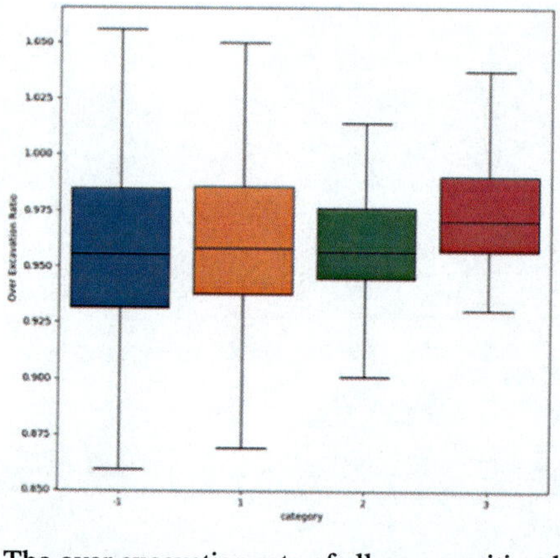

Fig. 4. The over excavation rate of all communities detected

4 Conclusion

In our study, we develop a shield machine working performance analysis method, with the help of a complex network theory. To deal with the parameters of time series data of the shield machine, the PCA, as a common data dimension reduction method, is innovatively performed to automatically extract effective information from the original data set, which reduce information loss more effectively than simply using the mean value. By using the CD algorithms to mine the hidden information in internal community structure, the clustering results show that some clusters are generally higher in over excavation rate, while others are more precise. These findings provide decision-making information for tunnel construction and contribute to a centralized control of shield parameters with partial rings in spatial distribution.

However, there are still many in our study: 1) Lack of correlation comparison between the real tag set of tunnel segmentation rings and the predicted results, which reduces the credibility of the clustering performance. 2) Our current research only focuses on rings with similar performance in spatial distribution, without further analysis of their corresponding shield parameters. In the future, it is expected to solve the problem of how to adjust the input parameters of the communities detected in detail.

Acknowledgement. The research work described in this paper was supported by a grant from the Shanghai Sailing Program, China (No: 22YF1419100).

References

1. Chen, Z., et al.: Topological mapping and assessment of multiple settlement time series in deep excavation: a complex network perspective. Adv. Eng. Inform. **36**, 1–19 (2018)
2. Khalid, E., et al.: Prediction model of shield performance during Tunneling via incorporating improved particle swarm optimization into ANFIS. IEEE Access **8**, 39659–39671 (2020)
3. Chen, Z., et al.: Unsupervised spectral clustering for shield tunneling machine monitoring data with complex network theory. Autom. Constr. **107**, 102924 (2019)
4. Yue, P., et al.: Clustering of designers based on building information modeling event logs. Comput.-Aided Civ. Infrastruct. Eng. **35**(7), 701–718 (2020)
5. Rasmus, B., Smilde, A.K.: Principal component analysis. Anal. Methods **6**(9), 2812–2831 (2014)
6. Yue, P., et al.: Discovering spatial-temporal patterns via complex networks in investigating COVID-19 pandemic in the United States. Sustain. Cities Soc. **77**, 103508 (2022)
7. Yining, D., Joe, Q.: A novel dynamic PCA algorithm for dynamic data modeling and process monitoring. J. Process Control **67**, 1–11 (2018)
8. Patrick, S., et al.: Correlation coefficients: appropriate use and interpretation. Anesth. Analg. **126**(5), 1763–1768 (2018)
9. Giulio, R., et al.: CDLIB: a python library to extract, compare and evaluate communities from complex networks. Appl. Network Sci. **4**(1), 52 (2019). https://doi.org/10.1007/s41109-019-0165-9
10. Traag, V.A., et al.: From Louvain to Leiden: guaranteeing well-connected communities. Sci. Rep. **9**, 1–2 (2019)

A Systematic Review of Urban Design and Computer Modelling Methods to Support Smart City Development in a Post-COVID Era

Liu Yang[1](✉), Michiyo Iwami[2], Yishan Chen[3], Mingbo Wu[4,5], and Koen H. van Dam[6]

[1] School of Architecture, Southeast University, Nanjing, China
yangliu2020@seu.edu.cn
[2] Department of Infectious Disease, Faculty of Medicine, Imperial College London, London, UK
[3] Architecture and Urban Design Research Center, China IPPR International Engineering Co., Ltd, Beijing, China
[4] University of Chinese Academy of Sciences, Beijing, China
[5] State Key Laboratory of Resources and Environmental Information System, Institute of Geographic Sciences and Natural Resources Research, Chinese Academy of Sciences, Beijing, China
[6] Centre for Process Systems Engineering, Department of Chemical Engineering, Imperial College London, London, UK

Abstract. The COVID-19 pandemic emphasised the need for decision-support tools to assist urban designers in building resilient and smart cities. Therefore, a multi-disciplinary systematic review was conducted following the PRISMA guideline to identify papers relevant for selecting appropriate methodologies that can be applied to build decision-support tools for resilient cities. This paper presents a list of 109 key references, selected from 8,737 records found from the searches, and identified major research themes, fundamental design interventions, and computer modelling techniques. We extracted six groups of interventions categorised by different scales of action: from an individual, crowds (social distancing and travel-related interventions), to a building, a neighbourhood/district, and a city. In addition, there are three sorts of computational modelling approaches, i.e., computer simulation, statistical models, and AI algorithms. Most of the studies developed models for predictive purposes, and 28% of the modelling studies built models for descriptive purposes. This work intends to empower urban designers and planners to overcome and get prepared for unpredictable disasters in pursuit of resilient and smart cities, particularly in the post-pandemic world. This review enables them to quickly find relevant papers as well as suitable methodologies and tools for a particular research purpose.

Keywords: Urban design · Computer model · Decision-support tool · Resilience

W. Guo and K. Qian (Eds.): GBCESC 2022, LNCE 211, pp. 1234–1246, 2023.
https://doi.org/10.1007/978-981-19-5217-3_127

1 Introduction

Cities are intertwined with public health issues. The emergence of the Coronavirus Disease 2019 (COVID-19) pandemic was a severe threat to public health, especially in cities with high population density and high levels of connectivity. To tackle COVID-19 and its impact in different urban contexts, urban design and planning methods play a role in the mitigation and suppression of the virus in the short term, as well as resilience planning in the long term. The research argued the importance of urban design strategies in addressing the negative impacts and the spread of diseases, e.g., building ventilation and green spaces design [1]. Experts called for an "anti-virus built environment", resulting in designs that are resilient to infectious diseases and can play a role in minimising future outbreaks [2, 3].

Increasing attention has been paid to combining physical space design methods with quantitative computer modelling technologies under the normalisation and adjustment stage of COVID-19 [1]. Studies focused on modelling virus transmission and evaluating various design scenarios using statistical models, Artificial Intelligence (AI) algorithms, and computer simulation [4]. Previous literature reviews in this space focused on general guidelines for responding to pandemics and COVID-19 in particular [5]. With consideration of the impact of pandemics on city planning, reviews on lessons learned for the built environment were previously published [2, 3]. Despite growing interest in the use of computer-aided tools, few literature reviews were conducted in a way that urban designers can use them to select relevant urban design and computer modelling methods.

To bridge this gap, this paper aims to: 1) list key references for designers to more easily find relevant work, 2) extract major themes in the literature, and 3) identify fundamental design interventions and computer modelling tools that can be applied in this space. We conducted a systematic review from a multi-disciplinary perspective, including planning, architecture, infectious diseases, urban systems, and computer science and AI. The novelty of this paper lies in proposing research themes, design interventions and suggesting computer modelling tools for computer-aided urban design during the post-pandemic era. Besides, this work longlists the aims of the modelling studies against COVID-19.

2 Method

We conducted this systematic review following the Preferred Reporting Item for Systematic Reviews and Meta-Analyses (PRISMA) 2020 recommendations [6] and Cochrane Handbook. In November 2020 and May 2021, this study searched six databases (Web of Science, Scopus, IEEE Xplore, and Ovid MEDLINE/Embase/PsycInfo), combined with a grey literature search and hand-searching of references of the papers that were found eligible under our inclusion/exclusion criteria. All publication years, geographical regions/areas, and languages were considered. We used the Covidence platform for screening titles/abstracts, full-text review, and data extraction. The risk of bias assessment was assessed using a modified version of existing tools. The final included papers were evaluated, and the themes of each article were analysed using the CiteSpace software. A comprehensive description of the methodology was presented in our recent publication (in press) [7].

3 Results

A total of 8,737 records were found and imported to the Covidence platform. After manual selection based on inclusion criteria related to the interface between public health, urban design, and computer-aided decision support, 109 distinctive eligible papers were included for analysis. Among them, 84 articles investigated COVID-19, others examined SARS (4), MERS (2), H1N1 influenza (5), H5N1 influenza (4), other influenza (7), and infectious disease in general (3). Table 1 presents identified aims of the 84 COVID-related studies, while a complete list of the references, the case study country and research type were archived at http://doi.org/10.17632/h43fhkvjjg.1.

Table 1. Characteristics of the studies included in the review which investigated COVID-19.

Author (Year)	Aim of the study
Adiga et al. (2020)	To highlight some essential computational models used to support the COVID-19 pandemic response
Adly et al. (2020)	To discuss the applicability of using AI to control COVID-19 by analysing relevant literature
Badr et al. (2020)	Using mobility data and the number of reported cases to empirically estimate the relationship between decreased mobility and reduced transmission of COVID-19 in the United States
Balocco & Leoncini (2020)	To introduce and compare school retrofit operations in building-plant system retrofits and air purification; to correct for IAQ ventilation rates and ventilation efficiencies
Blečić et al. (2020)	A high-fidelity evaluation model and corresponding UPGS metrics are proposed to evaluate pedestrian accessibility of UPGS on a high-granularity spatial scale of street network nodes
Block et al. (2020)	Evaluating the effectiveness of three different distancing strategies to keep the infection curve flat and helping compliance in a post-lockdown world using a social networking approach
Campisi et al. (2020)	To explore changes in travel patterns during COVID-19; to predict walking and cycling changes; to determine the impact of stress and anxiety levels on public transport use

(continued)

Table 1. (*continued*)

Author (Year)	Aim of the study
Cartenì et al. (2021)	To support policymakers in developing strategies to respond to COVID-19; investigate the link between the number of cases in an area and its accessibility to transportation
Dabachine et al. (2020)	To propose a simulation framework to manage passenger flows integrating quality of service standards and emerging health regulations for airports
Di Mascio et al. (2020)	Implementing computational tools to validate airport terminal service levels to study and evaluate airport development plans
Dragoicea et al. (2020)	To broaden the discussion on resilience for the design of public services with a "Systems Thinking" lens
Silva et al. (2020)	To present a framework to simulate a multi-level virtual urban environment, considering the environment structure and the population
Grantz et al. (2020)	To review the use of mobile phone signalling data in assessing the response to COVID-19 and their relevance to studying disease spread and control
Wei & Chen (2020)	To explore the internal mechanism of subgroups' flocking problem, which can be regarded as collision avoidance between neighbouring agents or obstacle avoidance between agents and static/dynamic obstacles
Hamidi et al. (2020)	Investigating direct and indirect effects of population density on COVID-19 infection and mortality figures in metropolitan counties in the USA
Hamidi & Zandiatashbar (2021)	Exploring the effectiveness of social distancing in the United States; examining the impact of compact development on compliance with social distancing measures
Hao et al. (2020)	Understanding the spread of COVID-19 within an urban environment through mobility and population distribution

(*continued*)

Table 1. (*continued*)

Author (Year)	Aim of the study
Hernandez-Mejia & Hernandez-Vargas (2020)	Testing the effect of different population sizes and distance rules on potential new infections
Jorritsma et al. (2020)	Simulation of the spread of epidemics (such as COVID-19) finite-space and non-spatial network models incorporating temporary immunity
Kain et al. (2021)	To build a model that uses mobility, death, and case data for estimation, incorporating heterogeneity in transmission rates, to be applied to real-world scenarios
Kaplan (2020)	Recommend population size limitations, hospital surgery planning, timing decisions, and scenario analysis; identify issues faced and provide advice during the real-time response to COVID-19 at the local level
Kierzkowski & Kisiel (2020)	To examine the impact of social distancing on safety control channel performance
Klompmaker et al. (2020)	To assess whether access to green space is directly associated with morbidity and mortality risk for COVID-19 in the United States
Kudela (2020)	To evaluate the best use of available space in a certain location place, while social distancing requirements, formulated as a p-dispersion problem, as satisfied
Leiva et al. (2020)	To understand the impact of urban structure and population movement on social distancing and the spread of COVID-19
Leng et al. (2020)	Numerical strategies were developed to optimise the physical environment of a courtyard, including the distribution of air pollutants, drought perception and infection risk
Xin Li et al. (2020)	To increase understanding of COVID-19 and its impact at the city level by validating urban factors and identifying relevant areas
Maghdid & Ghafoor (2020)	To support contact tracing of confirmed COVID-19 cases by developing a smartphone-based method

(*continued*)

Table 1. (*continued*)

Author (Year)	Aim of the study
Gharakhanlou & Hooshangi (2020)	To develop an ABM that simulates the spatio-temporal outbreak of COVID-19
Miller et al. (2020)	To develop strategies to increase surgical capacity and expand self-isolation capacity over the COVID-19 pandemic within buildings that offer collective living; to minimise disease transmission within the facility
C.T. Nguyen et al. (2020)	Providing a survey of how emerging technologies, such as wireless and networking, are evolving to help fight the pandemic
Perlman & Yechiali (2020)	To optimise queue sizes and customer waiting times to ensure public safety by constructing and analysing two models designed
Q.C. Nguyen et al. (2020)	To investigate the relationship of neighbourhood built environment factors with COVID-19 cases
Rader et al. (2020)	To investigate the role of climate, urbanisation, and appropriate interventions by analysing spatial and case study data in cities
M.R. Rahman et al. (2020)	Assessing the spread of COVID-19 in Bangladesh concerning implemented interventions; understanding epidemic spread patterns and identifying hotspots; generating vulnerability zoning maps; and developing management strategies
M.M. Rahman et al. (2020)	An empirical study of the relationships between the severity of the COVID-19 pandemic and changes in mobility, including retail and leisure, public transport stations, workplaces, residential places, and related lockdown measures
Ramchandani et al. (2020)	Developing a model to predict the spread of diseases that can take various relevant factors and their possible interactions into account for supporting pandemic surveillance and associated policymaking
Renardy et al. (2020)	To predict local trends of the COVID-19 pandemic in Washtenaw County of US

(*continued*)

Table 1. (*continued*)

Author (Year)	Aim of the study
Rezaei & Azarmi (2020)	To support reducing the spread of the coronavirus and its economic costs by providing artificial intelligence solutions to monitor and detect social distancing violations by individuals automatically
Ronchi & Lovreglio (2020)	To assess occupant exposure in confined spaces using crowd modelling
Scarpone et al. (2020)	To investigate the associations between COVID-19 incidence and socioeconomic, infrastructural, and built environment characteristics
Shuvo et al. (2020)	To clarify the importance of social distancing and hospital bed counts to shorten outbreak times and reduce infection rates and total deaths
Small & Cavanagh (2020)	A model of contact network topology under external control conditions is proposed
C. Sun & Zhai (2020)	To answer the questions "what is the safe distance" and "what is sufficient ventilation" to minimise infections
QC. Sun et al. (2020)	Presenting a new method to determine where to allocate limited resources to mitigate heat, resulting in the best outcomes for the most vulnerable communities
Swinarski (2020)	To predict elevator flow under social distancing restrictions in a university building and study the effect of four interventions on improving this flow
Ban et al. (2020)	To present a detailed simulation for the COVID-19 pandemic using the GAMA platform
Tang et al. (2020)	To review the key models that have been used to study infectious diseases and their dynamics and introduce the basics of infectious disease modelling
Domino (2021)	To establish a methodology to link high-fidelity and dose-response modelling to model droplet evaporation physics, deposition, and possible persistence of aerosols

(*continued*)

Table 1. (*continued*)

Author (Year)	Aim of the study
Bazant & Bush (2021)	To derive an indoor safety guideline based on airborne disease transmission models; to demonstrate how this guideline depends on the face mask use and respiratory activity, breathing rate, room dimensions, ventilation and air filtration, and infectiousness of the respiratory aerosols
Elavarasan et al. (2021)	To highlight the typical pandemic characteristics over time and their associated public health measures that were demonstrated to support keeping the pandemic under control
Subahi (2021)	To design a model transformation and full weighted graph-based IoT edge system for detecting violations of social distancing measures in indoor public spaces
Harweg et al. (2021)	To develop a pedestrian dynamics simulation model to assess pedestrian behaviour with respect to contact transmission and gather insights on the overall effectiveness of exposure time and distance measures
H. Li et al. (2021)	To simulate droplet spread in cough and study the risk of transmission associated with SARS-CoV-2
Ugail et al. (2020)	To automate the process of generating the optimal arrangement of public spaces to safely accommodate individuals with given social distancing requirements in a given area
Farthing & Lanzas (2021)	To support the use of simulation techniques to determine the most effective strategies to limit the spread of SARS-CoV-2 in indoor settings
Patil et al. (2021)	To understand the response of urban travel patterns to partial lockdowns or the creation of containment zones; to explore the relationship between intra-city mobility and SARS-CoV-2 spread in different metropolitan areas
Pang et al. (2021)	To predict epidemic trends of a new infectious disease, overcoming the challenges of uncertain influence, cold start problem, and privacy protection; to predict infection trends after developing a response plan

(*continued*)

Table 1. (*continued*)

Author (Year)	Aim of the study
Yamakawa et al. (2021)	A detailed study of the long-term spread of novel coronavirus-carrying droplets in the air using numerical simulations
Komperda et al. (2021)	Assessing the risk of SARS-CoV-2 virus infection in large enclosed spaces
Mohammadi et al. (2021)	To explore the implications of physical distancing requirements for pedestrians on urban sidewalks
Z. Li et al. (2021)	To analyse the dispersion of droplets produced by passengers coughing on an escalator by numerical simulation, focusing on the influence of the slope and speed of the escalator on the diffusion of droplets
Tupper et al. (2020)	To provide a quantitative framework to identify the settings where specific interventions have the most significant impact
Selvakarthi et al. (2021)	To detect the number of passengers inside the bus using an image processing camera to facilitate and monitor social distancing protocols
Hong et al. (2021)	To quantify neighbourhood activity behaviours at high spatio-temporal resolutions and test whether behavioural responses to social-distancing policies vary with socioeconomic and demographic characteristics
Silalahi et al. (2020)	To introduce GIS-based distribution of COVID-19; to recommend nearest medical institutions based on service area and origin-destination cost matrix; to create standard deviation ellipses for the study area
Megahed & Ghoneim (2021)	To draw architects' attention to airborne disease risk by providing up-to-date information and solutions to understand better environmental and health concerns surrounding COVID-19
Gomez et al. (2021)	To simulate the transmission of infectious diseases under social distancing with computer modelling, considering a complex network of individuals and places

(*continued*)

Table 1. (*continued*)

Author (Year)	Aim of the study
Xiang Li (2020)	To find out if the policies of extensive gatherings prohibition, schools and non-essential businesses closures, and stay-at-home orders are effective controlling interventions for COVID-19 and compare with alternative solutions
Klinker et al. (2021)	To propose a method to dynamically guide passengers around dangerously congested stations better distribute passenger loads across an urban public transport network
Roy et al. (2021)	To guide the movement of people and limit exposure of susceptible and infected individuals
Fang et al. (2020)	To develop a disease dynamics model considering influences from people's surroundings
Q. Xu & Chraibi (2020)	To simulate the effect of limiting the number of customers, maintaining social distancing, and using shopping cart entry in reducing contact between customers
Gkiotsalitis & Cats (2021)	To support public transport operators by providing simulation models that can be used to evaluate the impact of various social distancing policies on indicators, including operational and passenger-related costs
Pavon et al. (2020)	Detecting crowd hot spots using mathematical models and BIM shows the best time to move around the campus
Kudryashova et al. (2021)	To describe the indoor transmission of aerosols as potential virus-carrier; to demonstrate that large numbers of fine droplets released from infected individuals travel rapidly and over long distances over obstacles or stairs
Qiu et al. (2020)	To reveal the spatial response law of urban vulnerability in the context of disease-causing disasters; to build a research framework for planning responses based on major epidemics and explore planning technologies

(*continued*)

Table 1. (*continued*)

Author (Year)	Aim of the study
W. Xu et al. (2021)	To identify the factors that influence community resilience and analyse their causation toward helping manage the risks associated with the COVID-19 pandemic
Shorfuzzaman et al. (2021)	To provide an active surveillance system that enforces social distancing to slow the spread of the virus effectively
Y. Yang et al. (2021)	To examine the changes in recreational physical activity during the pandemic; to test whether the urban greenery of neighbourhoods is associated with this change
Ye and Qiu (2021)	To determine the role of population density, identify spatial heterogeneity; to identify critical components of key landscape indicators and develop models to assess the risk of infection
Sen & Singh (2021)	To simulate how coughing droplets move around a door and quantify the proportion of droplets entering through an open door
UN-Habitat China et al. (2020)	To demonstrate how to use big data and QR apps to trace the path of the disease transmission, understand the trend of the pandemic situation, project pandemic transmission, and assist in prevention and control
UN-Habitat & WHO (2020)	To outline available tools to assist with the appraisal of health and health equity

The keywords of the included 109 papers are classified into eleven clusters and mapped by the clusters over time, as presented in Fig. 1. Coronavirus-laden droplet/aerosol, agent-based modelling, spatial-temporal transmission, hospital ward, individual protective behaviour, operational use, using GPS technology, effective response, and COVID-19 scratch model are frequently identified themes in the multi-disciplinary research field covered here.

By referring to the literature, we grouped the design interventions into six categories: social distancing, travel-related, individual-level, building-level, neighbourhood-level, and city-level design interventions. Among the computational modelling tools, computer simulation ranked first (60% papers used), and statistical models ranked second (18% papers used), followed by AI algorithms (see Fig. 2). Most of the studies (70%) built models for predictive purposes, while 28% for descriptive goals.

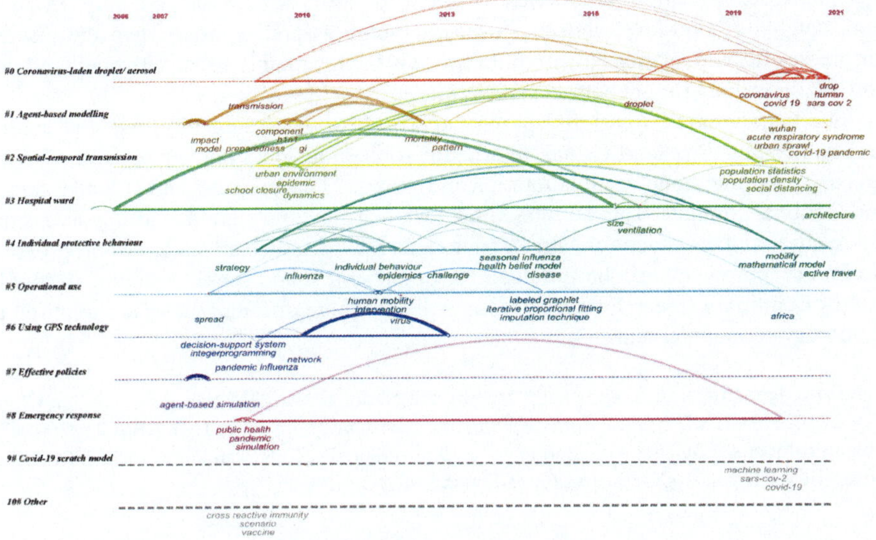

Fig. 1. Clustering of the keywords of the included 109 research papers over time.

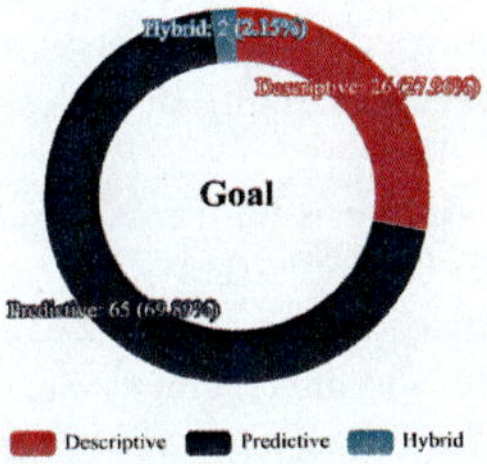

Fig. 2. Goals of the included models: number of papers (percentage in the modelling papers).

4 Discussions and Conclusion

We conducted a multi-disciplinary systematic review to address the lack of literature review that can support urban designers in selecting the relevant urban design and computer modelling methods in building smart and resilient cities during the post-pandemic era. As a result, we identified major research themes and fundamental design interventions in line with computer modelling tools. Furthermore, the aims of 84 papers that investigated COVID-19 were presented to allow the reader to more easily identify relevant work that sits at the interface of public health, urban planning and design, and computational modelling.

This work also discovered that after an infectious disease spreads, social distancing and building-level interventions, travel-related and individual-level interventions can prevent the infection inflation at an early stage. To support the implementation of these methods, a computer simulation built for predictive purposes, focusing on mechanism examination, and driven by theories, were utilised substantially. In addition, neighbourhood-level and city-level interventions are required when the disease progresses from local areas to communities. This stage mainly used statistical methods and AI algorithms for phenomenon description.

Our collaborative, multi-disciplinary approach to the systematic review is unique and transferable. We held virtual, international meetings on a regular basis, led by the first author, throughout the collaboration and discussed the progress and task allocations and resolved conflicts. This facilitated the task clarification, shared learning across the disciplines, and co-generating strategies to address the issues arisen as we progressed. A similar approach could be followed for other review and scoping studies in highly multi-disciplinary research fields and for projects where different views are required to support systemic interventions.

Acknowledgement. Liu Yang is supported by the National Natural Science Foundation China (No. 5210081743), the Natural Science Foundation of Jiangsu Province (No. BK20210260), the China Postdoctoral Science Foundation (No. 2021M690612). Koen van Dam works on Climate Compatible Growth (CCG) project funded by the FCDO.

References

1. Wang, J.: Vision of China's future urban construction reform: in the perspective of comprehensive prevention and control for multi disasters. Sustain. Cities Soc. **64**, 102511 (2021)
2. Megahed, N.A., Ghoneim, E.M.: Antivirus-built environment: Lessons learned from Covid-19 pandemic. Sustain. Cities Soc. **61**, 102350 (2020)
3. Sharifi, A., Khavarian-Garmsir, A.R.: The COVID-19 pandemic: impacts on cities and major lessons for urban planning, design, and management. Sci. Total Environ. **749**, 142391 (2020)
4. De Las, A., Heras, A.L.-S., Zamora-Polo, F.: Machine learning technologies for sustainability in smart cities in the post-COVID era. Sustainability **12**(22), 9320 (2020)
5. Kakodkar, P., Kaka, N., Baig, M.: A comprehensive literature review on the clinical presentation, and management of the pandemic coronavirus disease 2019 (COVID-19). Cureus **12**, e7560 (2020)
6. Page, M.J., et al.: The PRISMA 2020 statement: an updated guideline for reporting systematic reviews. BMJ **2021**, 372 (2021)
7. Yang, L., Iwami, M., Chen, Y., Wu, M., van Dam, K.H.: Computational decision-support tools for urban design to improve resilience against COVID-19 and other infectious diseases: a systematic review. Prog. Plann. 100657 (2022)

Impact Estimation of Traffic Accident Duration Based on Survival Analysis by Using Field Urban Traffic Condition

Duanya Lyu[1] and Yongjie Lin[2(✉)]

[1] School of Civil Engineering and Transportation, South China University of Technology, Guangzhou 510641, Guangdong, China
[2] Guangzhou Institute of Modern Industrial Technology, Guangzhou 511458, Guangdong, China
linyjscut@scut.edu.cn

Abstract. Analyzing and predicting accident duration is important in the field of Traffic Incident Management (TIM). This study imports survival analysis to identify the influencing factors for the duration of urban arterial-way accidents based on field accident data and traffic condition data. First, the Kaplan-Meier method is used to compare the accident duration distributions under different attribute values. And then, the Cox proportional hazard model is developed to investigate the relationship between accident duration and different attributes. Finally, the analysis result shows that the impacted duration of nighttime accidents and off-rush time accidents is significantly longer than that of daytime accidents and rush-hour accidents, and the arrival time of TIM staff and accident location significantly influence the risk of accident ending. The proposed model can help accident clearance operators investigate urban traffic accident duration and make real-time management strategies.

Keywords: Traffic Incident Management · Accident Duration Estimation · Survival analysis

1 Introduction and Literature Review

Traffic accident remains a problem in modern society. According to the National Bureau of Statistics of China [1], there were 211,074 vehicle traffic accidents in China in 2020 with 55,950 deaths and 214,442 injuries. Meanwhile, traffic accidents also cause a reduction in traffic efficiency, bringing considerable loss. Wu et al. [2] reported that the annual economic loss due to congestion alone in Beijing amounted to 105.6 billion yuan in 2010 with the time delay loss accounting for 76.7%.

To understand the duration of traffic accidents, some studies have been conducted and related analysis methods have been used including data-driven methods [3–5] and statistical methods [6–8]. Existing researches mainly focus on expressway accidents and less on urban road accidents. However, urban road traffic has higher heterogeneity due to transverse traffic and signal control which makes the analysis of urban traffic accidents complicated and requires more information. In 2018, Li et al. [9] made a comprehensive

W. Guo and K. Qian (Eds.): GBCESC 2022, LNCE 211, pp. 1247–1254, 2023.
https://doi.org/10.1007/978-981-19-5217-3_128

overview of the studies on incident duration and proposed three types of data that should be focused on: 1) accurate event timing obtained through intelligent vehicle systems, 2) traffic flow data obtained through floating cars or road detectors or simulations, and 3) more complete descriptive traffic event information obtained through social networks. Thus, this study aims to analyze the impact of traffic accidents on urban arterial roads based on accident record data provided by the local traffic incident management (TIM) department and traffic condition data provided by Amap (Gaode).

2 Problem Statement

TIM departments are dedicated to analyzing and reducing the impact of accidents on road traffic. One of their main focuses is reducing accident duration via many measures, such as legislating drivers' post-accident acts, encouraging online negotiation between drivers. These measures are based on the consideration of the safety of drivers and traffic efficiency of the flow. Unfortunately, as traffic volume increases, road condition changes more frequently and rapidly, the accident clearing measures need to be modified and redesigned in practice. Nowadays, apart from the local traffic management departments, navigation service providers can also accurately estimate real-time traffic conditions based on the crowdsourcing data from their expanding user base. Thus, this research aspires to study the relationship between accident duration and traffic condition to support incident prevention and management by integrating incident record data from TIM departments and road traffic condition data from navigation service providers.

3 Methodology

See the Fig. 1.

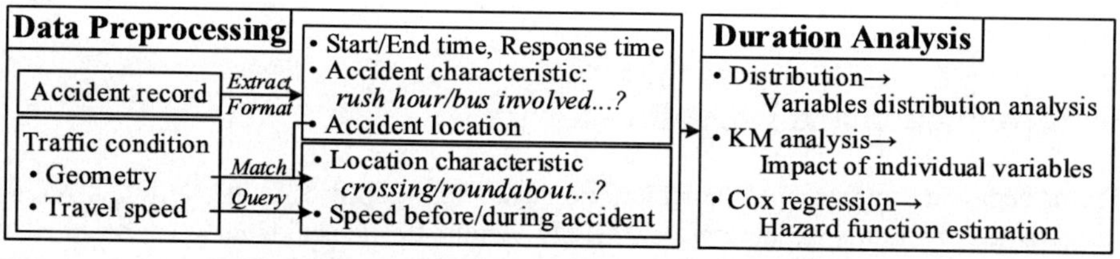

Fig. 1. Overall framework of the proposed method

3.1 Data Preprocessing

The available data resources are the accident records from the local TIM agency and the road condition data from the navigation service provider. This study focuses on urban arterial road accidents, and the target area is the rectangular area enclosed by Fenjiang Road, Wenhua Road, Lvjing Road, and Jianxin-Zhaoxiang Road in the city of Foshan.

Accident Data. The accident records include event start/happening time, event end time, TIM staff arrival time, location, and related accident description. The location information and accident description are kept in natural language. Vehicle-related accidents on urban arterial roads are focused, and 49 records were sampled.

Traffic Condition Data. Road travel speed with the interval of 10 min is used in this study. In the dataset, the whole roads are split into many segments according to vehicle interweaving or conflicting locations (e.g., intersections, zebra crossing) and a unique id is assigned to each segment. Each segment includes the static geometry data (e.g., segment location, name, type, and the number of lanes) and dynamic travel speed.

Data Fusion. The road traffic condition data are linked to the records.
Since the original records were kept in natural language, the characteristics were manually extracted and formatted. Then, the accident locations were linked to segments, so that the records were fused with the traffic condition data. After data fusion, there are 13 fields in the modified record data including 5 time-related fields (workday indicator, rush-hour indicator, nighttime indicator, time till TIM staff arrival, accident duration), 3 location-related fields (segment id, segment type, number of lanes), 3 type-related fields (number of vehicles involved, bus/truck involvement indicator, pedestrian/cyclist involvement indicator) and 2 condition-related fields (segment travel speed before and during the accident). The segment travel speed *before* the accident is set to the average speed in the nearest 3 time-bins (30 min) before the accident, and the speed *during* the accident is set to the average speed in the time-bins that are between the accident occurrence time and the TIM staff arrival time. These 13 fields were used as input for the analysis model.

3.2 Accident Duration Analysis

Survival Analysis. Survival analysis is a probabilistic model that studies the duration of events and it has strong capabilities in processing censored data and analyzing impact variables [10]. Survival analysis has been applied in the field of transportation [11–13], especially accident duration analysis [14–16].

In survival analysis, the survival time is focused, and two functions are modeled: the survival function $S(t)$ and the hazard function $h(t)$.

The survival function, or the cumulative survival function, measures the probability that the research object survives longer than a specific time:

$$S(t) = P(T > t) = \int_t^{\infty} f(x)dx = 1 - F(t) \tag{1}$$

where T is the duration time, and $f(t)$ and $F(x)$ are the probability density function and the cumulative distribution function respectively.

The hazard function, or the conditional death function, measures the probability that the research object has lasted to a specific period and then ends:

$$h(t) = \lim_{\Delta t \to 0} \frac{P(t \leq T < t + \Delta t | T \geq t)}{\Delta t} = \frac{f(t)}{S(t)} - \frac{d}{dt} S(t) \tag{2}$$

Kaplan–Meier Analysis. As a non-parametric method, Kaplan–Meier analysis (KM analysis) can be used to estimate the survival function $S(t)$. For n events within time T, it estimates the survival function $S(t)$ with $\widehat{S}(t)$:

$$\hat{S}(t) = \prod\nolimits_{T_i^{total} \leq t} \frac{n-i}{n-i+1} \quad (3)$$

where *total* represents total events, and i is the i-th uncensored event after sorting. All events are uncensored in this study (all accidents are with start and end time).

KM analysis is often used to compare the duration distributions under different variable values within a variable group. The Log-rank test (LR test) is used to determine whether the difference is significant.

Cox Model. Cox model, also called the proportional hazards model, studies the ratio of the hazard function h to the reference hazard function h_0. It is a semi-parametric regression model and be expressed as:

$$\frac{h(t,x)}{h_0(t)} = \exp(\beta_1 x_1 + \beta_2 x_2 + \cdots + \beta_i x_i) \quad (4)$$

where t is the duration, x_i is the dependent variable, β_i is the coefficient.

Positive coefficient ($\beta_i > 0$) indicates that the variable will terminate the record, or end the event in the study of traffic accidents; on the contrary, $\beta_i < 0$ will prolong the event; $\beta_i = 0$ indicates that the duration is independent of this variable.

4 Results and Discussions

4.1 Distribution of Variables

Dependent Variable. The distribution of the dependent variable, duration, is shown in Fig. 2. It shows that, as duration increases, the number of records decreases, and half of the accidents lasted less than 50 min.

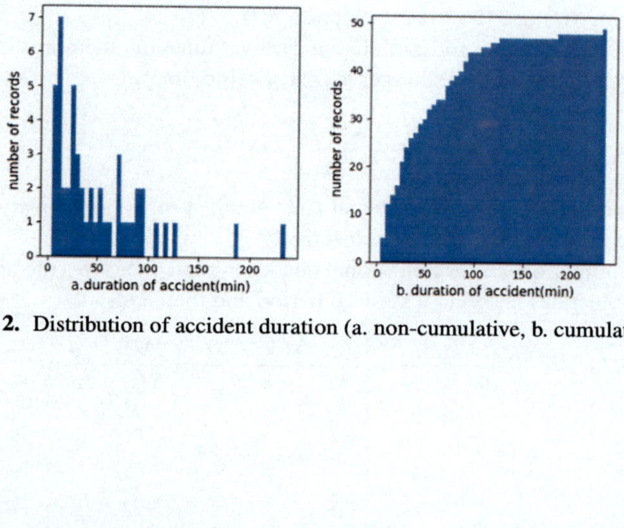

Fig. 2. Distribution of accident duration (a. non-cumulative, b. cumulative)

Independent Variables. The distribution of the categorical input variables is shown in Fig. 3 and the distribution of numerical variables is shown in Fig. 4. As shown in Fig. 3, most of the accidents are car-collision accidents, workday accidents, general-segment or intersection accidents; few accidents involve buses/trucks, pedestrians/cyclists; only three accidents happened on special road segments (roundabout, pedestrian crossing, sideroad). As shown in Fig. 4-a, for most accidents, TIM staff arrived within 10 min. As shown in Fig. 4-b and 3-c, the travel speed approximately follows a normal distribution. The average speed before and during the accident are 31.70 km/h and 31.11 km/h respectively.

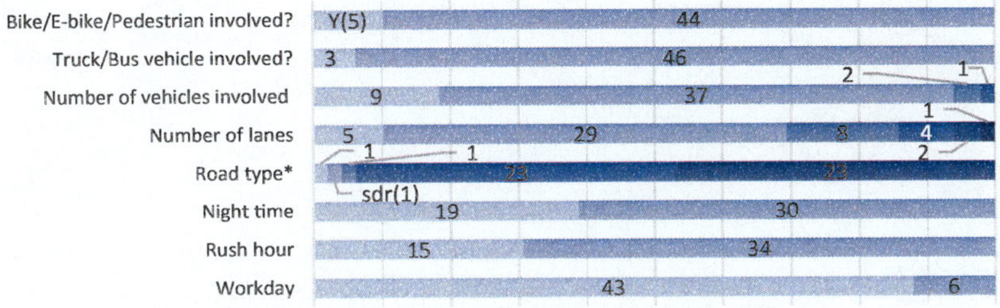

* pcr-pedestrian crossing; sdr-side road; rda-roundabout.
* The figure in brackets is the number of happened events under the corresponding category

Fig. 3. Distribution of categorical variables

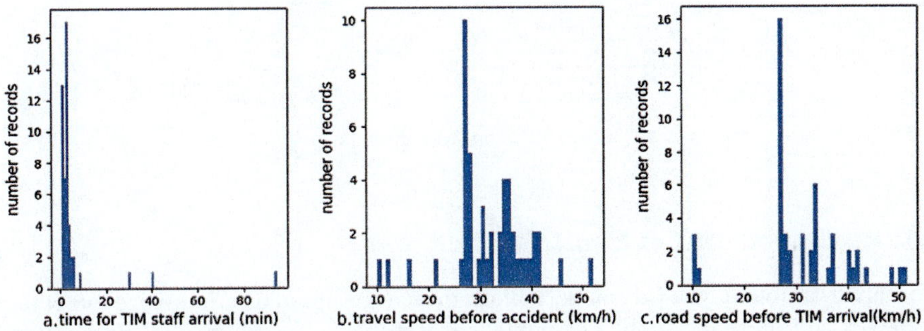

Fig. 4. Distribution of numerical variables (a. time for TIM staff to arrive, b. travel speed before the accident, c. travel speed before TIM staff arrived)

4.2 Results from the Kaplan-Meier Analysis

Based on the discussion in Sect. 4.1, five groups of variables with high differentiation were selected for the KM analysis, including the nighttime indicator, the rush-hour indicator, the number of vehicles involved, the number of lanes, and the location of the accident. As shown in Fig. 5, Nighttime and daytime are proved to have significantly different influences on duration at 0.05 level, and the rush and off-rush hour group gives the second-smallest p-value of 0.12. Previous studies on expressway incidents have also found that more accidents occur during daytime and have a shorter duration [6, 14]. However, rush-hour expressway accidents are proved to last longer [6] or have insignificant differences [4].

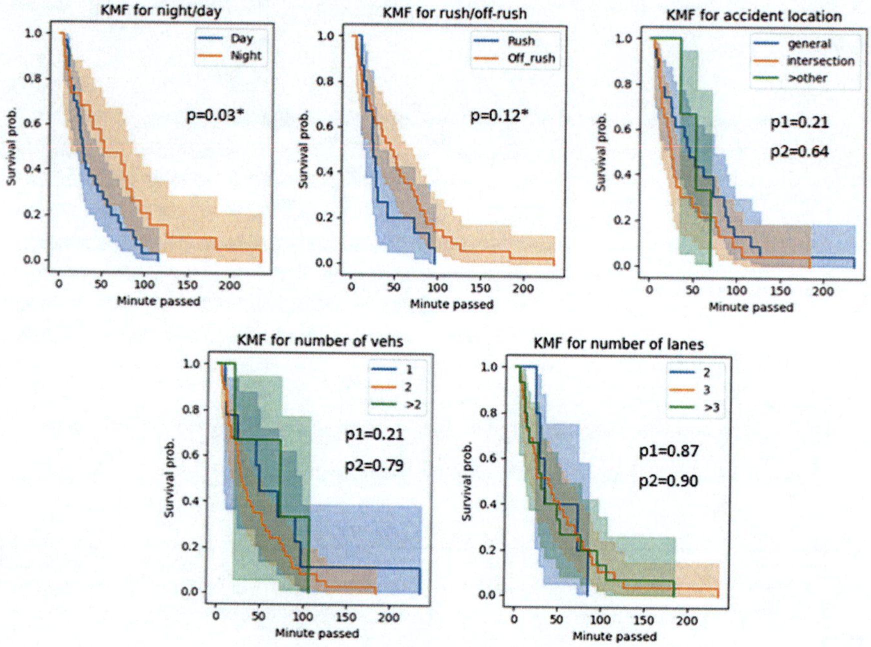

Fig. 5. Kaplan-Meier analysis

4.3 Results from the Cox Model

The input variables of the Cox model include the five groups of discriminative categorical variables and the three groups of numerical variables. The estimation results of the significant variables are listed in Table 1.

Table 1. Parameter estimation from cox model.

	B	exp(B) HR	se(B)	B-95%		exp(B)-95%		z	p	−log2(p)
				lower	upper	lower	upper			
tim_tm	−0.02	0.98	0.01	−0.04	0.00	0.96	1	−1.81	0.07*	3.84
location	0.58	1.79	0.28	0.03	1.13	1.03	3.11	2.06	0.04*	4.67

The time till TIM staff arrival (*tim_tm*) and the accident location (*location*) are considered significant (at 0.1 level). For the two significant variables, their influence can be analyzed with their HR values.

The p-value for *tim_tm* is 0.07 and the HR value is 0.98 < 1, indicating a significant negative relationship between *tim_tm* and the risk of the event ending. This means that, when holding other covariates constant, accidents with longer TIM staff arrival time have a lower risk of ending. This can be explained by the fact that accidents with lower severity tend to have longer TIM reaction time and shorter duration.

The p-value for location is 0.04 and the HR value is 1.79 > 1, indicating a significant positive relationship between the complexity of location and the risk of the event ending (since general segments were coded as 1, intersections as 2, others as 3 in the model). This means that, when holding other covariates constant, accidents with higher location complexity have a higher risk of ending. This could be explained by the fact that vehicles involved in accidents that happened on complex roads usually leave the scene more quickly to avoid causing severe congestion.

In previous studies on expressway accidents, researchers have found out that the increase in TIM response time will lead to an increase in accident clearing time [4, 6] and road complexity has little impact [5, 7] on time spent for accident clearance. The urban transportation system is more comprehensive than expressways on road configuration and traffic flow behavior because vehicles on urban roads have more available routes to detour.

5 Conclusions

This study investigated the duration of urban arterial way traffic accidents. The accident records were fused with traffic condition data. Survival analysis was applied where the Kaplan-Meier analysis was used to compare the duration under different time periods, road alignments, and road traffic conditions separately, and the Cox model was employed to find the quantitative relationship between duration and these variables. The main findings are as follows:

1) Data fusing results show that the speed before and during the accidents are approximately normally distributed and there is a drop in the average speed after the accidents happened.
2) Kaplan-Meier analysis results show that nighttime accidents last significantly longer than daytime accidents, and the accidents that happened during off-rush hours last longer than those that happened during rush hours.
3) The Cox model analysis results show that TIM staff site arrival time has a significantly negative impact on the accident ending risk, and location complexity has a significantly positive impact on the accident ending risk.

However, this study focuses only on accidents that happened on urban arterial roads. For future work, one can extend to study the traffic accidents on minor arterial roads and collector roads, and discuss the difference of accidents on different types of roads (or on intersections formed by different types of roads). Moreover, in the fusing of traffic condition data, this study only considers the traffic condition of the accident road segments and it can be extended to adjacent road segments to consider the influence of road structure on accident duration.

Acknowledgments. This research was funded by the Nansha District Science and Technology Foundation of Guangzhou (2021MS018), and the authors appreciated data support from the Amap and department of traffic incident management, Foshan Chancheng District.

References

1. Basic statistics on traffic accidents. https://data.stats.gov.cn/english/easyquery.htm?cn=C01. Accessed 10 Jan 2021
2. Dongdong, W., Yi, S., et al.: Assessment of ecological-economic loss caused by traffic congestion in Beijing. Ecol. Econ. **4**, 75–79 (2013)
3. Yuexu, Z., Wei, D.: Prediction in traffic accident duration based on heterogeneous ensemble learning. Appl. Artif. Intell. 1–24 (2022)
4. Jinjun, T., Lanlan, Z., et al.: Traffic incident clearance time prediction and influencing factor analysis using extreme gradient boosting model. J. Adv. Transp. **1**, 1–12 (2020)
5. Jie, Z., Junhua, W., Shou'en, F.: Prediction of urban expressway total traffic accident duration based on multiple linear regression and artificial neural network. In: ICTIS 2019, pp. 503–510. IEEE (2019)
6. Junhua, W., Haozhe, C., Shi, Q.: Estimating freeway incident duration using accelerated failure time modeling. Saf. Sci. **54**, 43–50 (2013)
7. Ying, L., Chienhung, W., Kaichon, C.: Non-parametric machine learning methods for evaluating the effects of traffic accident duration on freeways. Arch. Transp. **43** (2017)
8. Xiaobing, L., Jun, L., et al.: Sequential prediction for large-scale traffic incident duration: application and comparison of survival models. Transp. Res. Rec. **2674**(1), 79–93 (2020)
9. Li, R., Pereira, F.C., Ben-Akiva, M.E.: Overview of traffic incident duration analysis and prediction. Eur. Trans. Res. Rev. **10**(2), 1–13 (2018). https://doi.org/10.1186/s12544-018-0300-1
10. Simon, W., Matthew. K., et al.: Statistical and Econometric Methods for Transportation Data Analysis, 2nd edn. CRC Press, Boca Raton (2010)
11. Xing, G., Jing, Z., Hang, G.: Red-light running behavior of delivery-service E-cyclists based on survival analysis. Traffic Inj. Prev. **21**(8), 558–562 (2020)
12. Yang, L., Linbo, L., et al.: Comprehensive survival analysis of lane-changing duration. Measurement **182**, 109707 (2021)
13. Yao, W., Yanyong, G., Wei, Y.: Real time safety model for pedestrian red-light running at signalized intersections in China. Sustainability **13**(4), 1695 (2021)
14. Hongke, X., Qilin, Y., et al: Analysis on the factors of expressway traffic incident duration based on survival analysis. In: CICTP 2021, pp. 1421–1430. American Society of Civil Engineers (2021)
15. Zongni, G., Ilir, B., Binbin, P.: Exploring characteristics and influencing factors of crash duration on urban arterials and collectors. J. Transp. Saf. Secur. 1–20 (2021)
16. Chung, Y., Walubita, L.F., Choi, K.: Modeling accident duration and its mitigation strategies on South Korean freeway systems. Transp. Res. Rec. **2178**(1), 49–57 (2010)

An Integrated Framework for Population Synthesis at Fine-Grained Spatial Scales

Yinni Huang, Meng Zhou(✉), Runqin Deng, Zeteng Huang, and Linlin You

School of Intelligent Systems Engineering, Sun Yat-Sen University, Shenzhen 518107, Guangdong, China
zhoum89@mail.sysu.edu.cn

Abstract. With the conception and development of Smart City and its applications, Digital Twins and other related technologies have been growing rapidly during the past years. Simultaneously, in the fields of transportation and urban analytics, agent-based models (ABMs) and microsimulations have also gained popularity, partly echoing the Digital Twins initiatives in Smart City applications, and is also owing to the sharply rising utilizability of high-performance computation resources and rapidly boosting high scale data storage. This study mainly developed a population synthesis pipeline to generate spatially detailed populations in urban environments and combined probabilistic models and re-weighting techniques for agent synthesis to cope with microdata on heterogeneity and marginal control matching. We also structured the built environment with higher spatial resolution than previous studies. We then created a realistic and spatially fine-grained synthetic population by the assignment of agents to the spatial units via rule-based and behavioral modeling. The results show good performance in terms of the accuracies over different dimensions and the proposed pipeline can facilitate the use of ABMs in Smart City projects.

Keywords: Smart city · Agent-based modeling · Population synthesis · Probabilistic modeling · Data fusion

1 Introduction

To adapt to the growing complexity and worldwide sustainability concerns, numerous urban communities are carrying out innovative measures in their tasks towards more efficient operations and management, changing into smart urban communities that follow a smart development plan. However, the achievement of such change depends on the ability to comprehend and address new difficulties that arise in reality [1].

In this way, city administrators attempt to consolidate information and communication technologies (ICTs) with physical entities to shape a virtual space where reality can be perceived automatically and in real time. Some argue that a smart city is characterized by the use of data and technological innovation to work on the viability of government, business administrations, and resident government assistance [2] or that

W. Guo and K. Qian (Eds.): GBCESC 2022, LNCE 211, pp. 1255–1262, 2023.
https://doi.org/10.1007/978-981-19-5217-3_129

the idea of 'Smart City' involves the automatic collection of environmental and natural information, leading to the proficient administration of the urban regions [3]. We, however, adopt the notion that, instead of a conceptual substance, a purpose, or a state, a smart city means an iterative process where cities become smarter as time passes and the varying definitions share the similar conception of a comprehensive vision of subareas [4]. In a nutshell, the concept of Smart City is an important and significant subject today.

In many instances, the smart city initiatives ask for realistic digital replicas of the real-world urban systems. The rise of Digital Twins adds to the establishment for cognitive development of Smart City. This advanced reproduction of the city is continuously informed by the genuine city through ongoing spatiotemporal information from the foundation and human frameworks [1]. Its aim, therefore, is to collect and transmit data from the physical world in real time to the virtual space and to replicate living and non-living entities from reality to the digital world.

In the fields of transportation and urban analytics, agent-based modeling (ABM) is a popular analytical framework that seeks to mimic the functioning of urban systems [5, 6]. Echoing the philosophy of the Digital Twins in Smart City applications, ABMs require the microscopic representation of the urban environments and the population residing within. This process is often termed population synthesis and it is becoming increasingly detailed both semantically and spatially.

This study is aimed at developing a population synthesis pipeline to generate spatially-detailed populations in urban environments. We mainly propose a structure that joined the Bayesian Network and Generalized Raking to represent the special blend of micro data heterogeneity and match marginal controls. We likewise establish the built environment at a higher spatial resolution compared with the past studies. By linking probabilistically sampled synthetic agents to detailed spatial units, we are able to create realistic and fine-grained synthetic urban environments for microsimulations while preserving the privacy to the original data. Through a detailed representation of the built environment in the synthesis process, the synthetic population created in this study mimics the real-world urban environment at a greater spatial granularity and thus could support more detailed and in-depth analysis of urban systems.

2 Methods

2.1 Study Area

In this study, the synthetic population of Singapore is taken as the case study. Singapore covers an area of about 719 square kilometers in total. There is around 5.61 million population in total as of 2016, among which 3.93 million are Citizens and Permanent Residents, appertaining to 1.26 million occupant families. As shown in Fig. 1, it is separated into 6 planning regions, which can be subdivided into 55 planning areas in total, with around 126,000 postcodes in use.

We note that postcode commonly refers to a specific building or a less-developed area in most areas of Singapore, which provides higher spatial resolution to the portrayal of civic systems. Thus, this study adopts the postcodes to represent detailed spatial units.

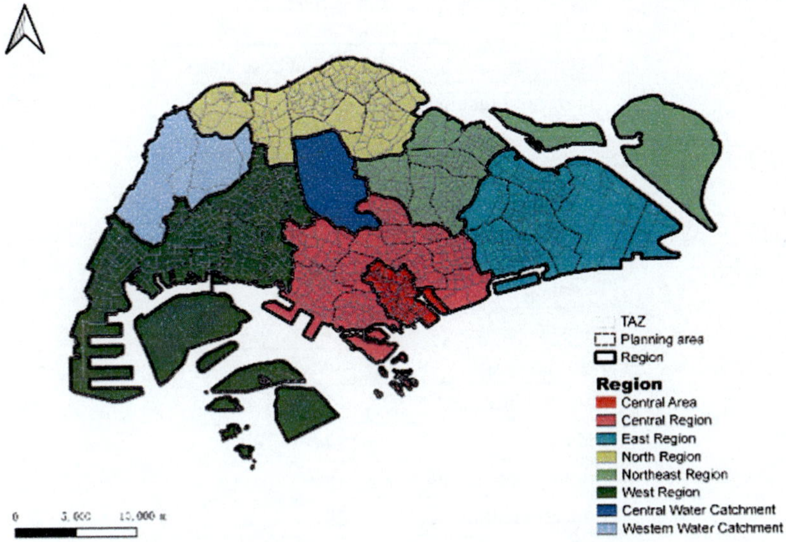

Fig. 1. The specific geography distribution in Singapore.

2.2 Population Synthesis Approach

As mentioned, we utilize the Bayesian Network (BN) for the synthesis of agents to catch heterogeneous multivariate joint distributions and Generalized Raking (GR) to match marginal controls at both the household and individual levels.

BN is a graphical model for learning and representing complicated relationships between a group of variables. Nodes refer to variables and edges represent the correlation between variables. Every node owns a conditional probability distribution (CPD), defining the probability of observing specific values. We trained different BNs according to the well-defined household types to reflect different family structures. We then applied a data-driven approach to select a suitable model structure on the base of AIC metric and constructed a BN at the household level for different household types. Greedy hill-climbing algorithm is used to learn the model structure and maximum likelihood estimation is used to learn the model parameters [7]. Household and individual models were developed and applied separately in similar manners with important household-level attributes included in the individual-level models. A total of 9 variables (i.e., nodes) in household-level networks and 6 additional individual-level variables in individual-level networks. They are listed in Table 1 below with the number of states for each variable. The BN models were trained with data sample from a recent travel survey in the study area.

So as to alter the BN-produced samples, we utilized the Generalized Raking for both household and individual ponds to concurrently make them possibly correspond to a bunch of chosen multivariate marginal controls in several key spatial and demographic dimensions [8]. The GR procedure was conducted with public statistics. For

Table 1. List of variables

Name	Number of states
Household income	9
Number of workers	4
Number of cars	4
Age of head	3
Ethnicity of head	4
Distance to public transit	3
Gender	2
Age	4
Personal income	7
Industry	14
Employment status	8
Education attainment	10

every individual and every household after convergence, the GR procedure created fractional weight independently. The truncate, replicate, sample (TRS) method [9] was then used for converting fractional weight to integers. The GR process maintains the total population size.

Next, various spatial objects in different scales hierarchical structures were utilized to represent the civic built environment. These levels may vary in different scenarios while the generic idea of the spatial hierarchy should work for all cases. We specifically constructed several spatial objects to depict the occupant and working space for every household and worker.

We adopted an ontology-based approach for the built environment with multi-sourced urban data in this study. Readers are referred to [10] for further details of this approach. This process comprises cleansing geometric information and estimating sundry building semantic information. We first acquired sites and geometry of buildings directly from tracking information with possibly less processing. Next, building features (e.g., building type, height, and space) were estimated with third-party open data. We also identified specific types and subtypes for dwelling buildings. The number of stories was searched out from the obtained information or evaluated by the height of buildings. The space of different uses was evaluated by floor number and the area searched from the track information.

According to the synthetic buildings and using kinds, we then, at that point, made different kinds of housing units, various industry sectors of establishments and companies. The quantity of these objects was acquired straightforwardly by accessing given the space of the particular uses type or retrieving from the accessible information. We assessed other attribute like covered areas, building ages, and occupations counts (for establishments) based on the correlative features of the buildings. In addition, land parcels were generated by available land-use dataset and accessibility indexes from buildings to the roads were calculated by data of transport infrastructures, shopping centers and freeway entrances, etc. The framework for the built environment synthesis is as shown below in Fig. 2.

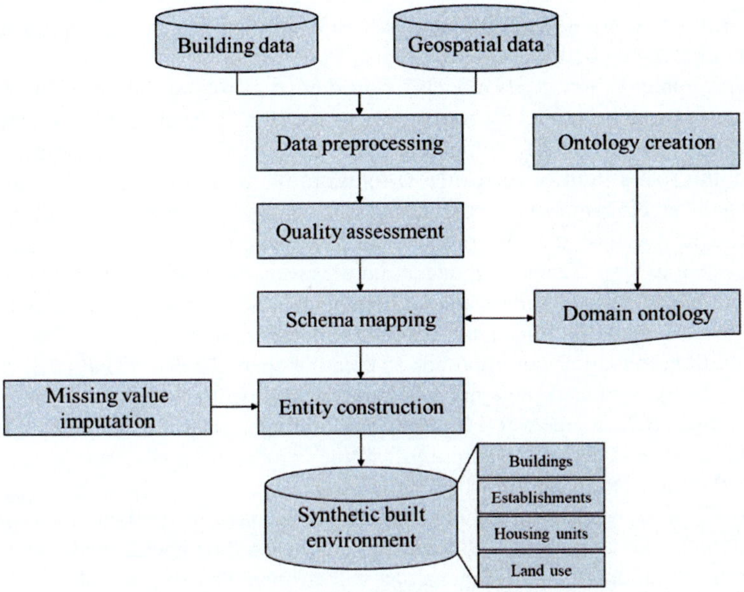

Fig. 2. Framework of built environment synthesis

After the synthesizing, we matched relevant spatial units to synthetic agents to achieve the detailed representation of the synthetic population. Firstly, bins of a preset rate were kept out for each resident type in each sector to reflect the vacancy rate. After that, housing units were relegated to households at random, and bigger units more probably were relegated to bigger and richer households. Adjustments were made in the rare cases where units in specific combinations ran out before all households were settled down. We then used a destination chosen model which is evaluated by the travel survey data to assign job slots to worker agents.

3 Results

To assess our population synthesis approach, we compared the accuracies of synthetic populations which were separately produced by using the approach proposed in this study, the BN-only approach and iterative proportional updating (IPU), and the similarities of three aforementioned approaches to the joint distribution of the weighted micro-sample and the marginal controlling distributions measured by the standard root mean square error (SRMSE). Not surprisingly, the BN-only approach has the minimum SRMSE of 22.34 since it attempts to directly capture the joint distribution of samples. However, after introducing GR in our approach resulted in a small increase in the error from the micro sample with an SRMSE of 23.66. IPU produced the lowest similarities among the three with an SRMSE of 98.96. On the other hand, our approach outperforms the BN-only method in terms of the similarity to the marginal distribution and their spatial variations. It, along with IPU, has near-perfect matches to various marginal control distributions (detailed results omitted for space), due to the explicit fitting mechanism to align with the marginals. Combined, assessments suggest that our approach is more capable of balancing between the two goals.

We formulated different BN models for different types of households in the study area and obtained separate modeling results. There are some commonalities shared among different sub-models and also some differences in the structure and the parameters. For instance, the spatial unit of residence is found to be the root node of all sub-models, reflecting the significance of the spatial aspect in synthesizing population. The observed differences in structure and parameter values justify the use of separate models for different household types. Detailed model structures were omitted due to space limitations. As a result, we generated a detailed and heterogeneous synthetic population, including 1.26 million occupant families and over 3 million individuals.

In addition, to exhibit our approach to creating spatially-detailed synthetic populations, this study produced the synthetic built environment containing different spatial objects, among which are over 116 thousand buildings. As shown in Fig. 3, dwelling buildings are spread over the island, most of the commerce architecture concentrate in the center of the island, and industrial buildings are located mostly in the rural regions. Besides, we made over 1 million housing units on the basis of synthetic dwelling buildings and generated around 200 thousand synthetic establishments with the synthetic commercial and industrial buildings, along with around 3.42 million jobs.

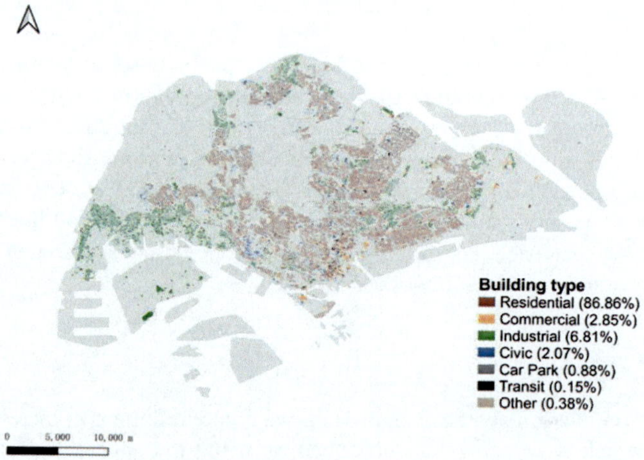

Fig. 3. Spatial distribution of synthetic buildings.

With synthetic agents and built environment generated, the last step is to coordinate the agents with the spatial objects, assigning housing units for households and jobs for laborers. The rule-based heuristics produced good results in the matching between households and housing units with more than nearly 100% of households able to find suitable units within their categories of zone-type combination. Figure 4a shows dwelling locations distribution of all households. Subsequently, we assigned job slots to all resident workers according to their industries. Figure 4b shows job locations distribution. The resultant median job-housing distance is around 8.44 km, compared to 7.93 in the micro sample, indicating similar and realistic representation of our synthetic population.

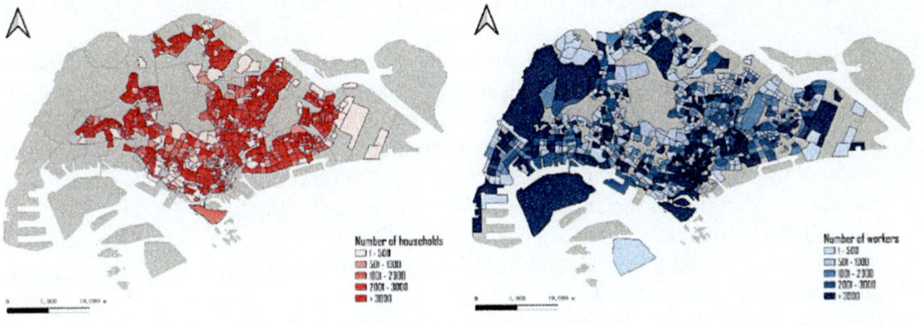

a b

Fig. 4. Spatial distribution of (a) dwelling locations of households and (b) job locations of laborers.

4 Conclusion

ABMs have filled in fame because of the conception and development of the Digital Twins in Smart City applications and the convenience of superior computational resources as well as huge information stockpiling abilities. This study aims at developing a population synthesis pipeline for the creation of spatially detailed populations in urban environments with the BN-GR model for agent synthesis and an ontology-based data integration effort for built environment synthesis. We created heterogenous and fine-grained synthetic populations with realistic and representative distribution by joining synthetic agents with the detailed spatial unit through probabilistic methods. In future studies, we would like to extend the present study by developing a more integrated framework of the synthesis of the two sides and testing the applicability of recent deep generative models to the synthesis of high-dimension and detailed populations.

Acknowledgement. This research was supported in part by the National Natural Science Fund of China (62002398) and Guangdong Basic and Applied Basic Research Foundation (2021A1515110049).

References

1. Mohammadi, N., Taylor, J.E.: Smart city digital twins. In: 2017 IEEE Symposium Series on Computational Intelligence (SSCI), pp. 1–5. IEEE, United States (2017)
2. Melakessou, F., Kugener, P., Alnaffakh, N., Faye, S., Khadraoui, D.: Heterogeneous sensing data analysis for commercial waste collection. Sensors **20**(4), 978 (2020)
3. Fortes, S., et al.: The campus as a smart city: university of Málaga environmental. Learn. Res. Approaches Sens. **19**(6), 1349 (2019)
4. Andrés, C., Enrique, A.: Smart city and information technology: a review. Cities **93**, 84–94 (2019)
5. Daniel, J.F., Kara, M.K.: The travel and environmental implications of shared autonomous vehicles, using agent-based model scenarios. Transp. Res. Part C Emerg. Technol. **40**, 1–13 (2014)
6. Salvini, P., Miller, E.J.: ILUTE: an operational prototype of a comprehensive microsimulation model of urban systems. Netw. Spat. Econ. **5**, 217–234 (2005)
7. Marco, S.: Learning Bayesian networks with the bnlearn R package. J. Stat. Softw. **35**(3), 1–22 (2010)
8. Iterative proportional fitting algorithms for nested structures (2021). https://mlfit.github.io/mlfit/, https://github.com/mlfit/mlfit
9. Lovelace, R., Ballas, D.: 'Truncate, replicate, sample': a method for creating integer weights for spatial microsimulation. Comput. Environ. Urban Syst. **41**, 1–11 (2013)
10. Geertman, S., Ferreira, J., Goodspeed, R., Stillwell, J. (eds.): Planning support systems and smart cities. LNGC, Springer, Cham (2015). https://doi.org/10.1007/978-3-319-18368-8

Author Index

B
Bai, Jiulin, 14
Bai, Zhiyu, 984
Ban, Xinlin, 1091
Bao, Chengyuan, 78

C
Cai, Jing, 562
Cai, Ming, 1179
Candeias, António, 465
Cao, Junrui, 356
Cao, Shiyuan, 32
Cao, Xiaocheng, 716
Cao, Xuyang, 869
Cao, Yunzhong, 741
Che, Weixian, 734
Che, Yujun, 542
Chen, Baikun, 974
Chen, Fei, 893
Chen, Feixiang, 445
Chen, Huan, 32, 65, 151
Chen, Huiguo, 1125
Chen, Huiming, 14
Chen, Jianjun, 176
Chen, Jin, 436
Chen, Jinjian, 884
Chen, Junting, 239
Chen, Junzhang, 319, 1201
Chen, Kang, 918
Chen, Ke, 1039
Chen, Minghao, 1139
Chen, Shizhi, 869
Chen, Xi, 276
Chen, Xuhao, 57
Chen, Yangqing, 176
Chen, Yile, 319, 1201
Chen, Yishan, 1234
Chen, Yuchen, 311
Chen, Zhexuan, 239
Cheng, Baoquan, 141
Cui, Weijiu, 679

D
Dai, Hao, 1062
Dai, Kao-shan, 579
Dai, Xueyan, 542
Deng, Runqin, 1255
Deng, Xiaofang, 415, 762, 910
Deng, Xueyuan, 141
Ding, Wei, 1188
Ding, Xuanming, 519, 851
Dong, Hantuo, 734
Dong, Jiangfeng, 901
Dong, Zhensheng, 504
Du, Shenghua, 816
Du, Zhiqiang, 562
Duan, Shuyi, 616
Duan, Xinhui, 255

F
Faber, Michael Havbro, 591
Fan, Jinyu, 1158
Fan, Lulu, 672
Fan, Shuanglong, 427
Fang, Pailin, 779
Fang, Zirui, 78
Fu, Bo, 436

W. Guo and K. Qian (Eds.): GBCESC 2022, LNCE 211, pp. 1263–1267, 2023.
https://doi.org/10.1007/978-981-19-5217-3

Fu, Fanghao, 1179
Fu, Helin, 750, 960

G
Gao, Fukang, 3
Gao, Yirong, 836
Ge, Hongjing, 288
Guo, Hongyu, 750, 960
Guo, Jichao, 1091
Guo, Jingen, 734
Guo, Xianqing, 571
Guo, Yichenfei, 356

H
He, Bo, 211
He, Jiang, 288
He, Junfeng, 1045
He, Minjuan, 893
He, Zhiqi, 1099
Hou, Guanyu, 934
Hu, Gui, 816
Hu, Hao, 579
Hu, Jianjun, 816
Hu, Song, 826
Hu, Yixi, 629
Huang, Gege, 536
Huang, Hao, 649
Huang, Hengchao, 176
Huang, Jing, 696
Huang, Kaiqian, 474
Huang, Keke, 176
Huang, Mei, 276
Huang, Min, 1091
Huang, Qinshou, 337, 373
Huang, Wei, 951
Huang, Weizhuo, 415
Huang, Yinni, 1255
Huang, Zeteng, 1255
Huang, Zheng, 255

I
Iwami, Michiyo, 1234

J
Ji, Yanqiang, 616
Jiang, Chunyong, 519
Jiang, Gege, 1151, 1218
Jiang, Xin, 1091
Jiang, Yunmu, 1022
Jiang, Zhilong, 629
Jin, Jian, 816
Jin, Longmei, 762
Jin, Tongxin, 166

K
Kang, Xueyun, 455
Kang, Zhiqiang, 166, 211
Kazadi, Billy, 394

L
Lai, Zimeng, 1030
Lei, Yixin, 1125
Li, Bing, 92
Li, Binhua, 22
Li, Chaoyang, 484
Li, Chengyu, 1030
Li, Gang, 445, 455
Li, Guangtao, 415
Li, Guoxing, 672
Li, Jing, 629
Li, Jinzhou, 552
Li, Jixiang, 934
Li, Linfeng, 607
Li, Mengyu, 672
Li, Mingguang, 884
Li, Mingjun, 255
Li, Nan, 57
Li, Ning, 724
Li, Qi, 934
Li, Sheng, 876
Li, Shihong, 1053
Li, Shuaibo, 263
Li, Shunuo, 385
Li, Tengfei, 943
Li, Weihong, 57
Li, Xiao, 484
Li, Yang, 708
Li, Yiming, 78
Li, Yue, 562
Li, Zheng, 893
Li, Zhenqiang, 92
Li, Zhi, 415, 474, 762, 910
Liang, Yanfang, 1009
Liang, Zhen, 108
Liao, Gang, 133
Lin, Hongjie, 536, 689
Lin, Wei, 629
Lin, Yongjie, 1247
Lin, Yuhan, 552
Liu, Bing, 836, 974
Liu, Feng-Tao, 599
Liu, Haodong, 3
Liu, Jiankun, 689
Liu, Jie, 347, 1107
Liu, Junyou, 1158
Liu, Min, 591
Liu, Ming, 248
Liu, Penghui, 504
Liu, Qizhou, 552, 779

Liu, Shi, 771
Liu, Shuai, 696
Liu, Wenhao, 365
Liu, Xiaobei, 696
Liu, Xiongfei, 934
Liu, Yang, 579, 660
Liu, Yijun, 1151
Liu, Zhangjun, 1022
Liu, Zhenqing, 427, 771
Liu, Zixin, 1022
Lourenço, Sérgio D. N., 1039, 1084
Lu, Dagang, 591
Lu, Dan, 779
Lu, Kun, 141
Luo, Guijun, 750, 960
Luo, Yanwen, 22
Lyu, Duanya, 1247

M
Ma, Chi, 427
Ma, Chunyin, 672
Ma, Zichang, 496
Magar, Sandeep Gharti, 860
Mao, Yishan, 44, 190
Mei, Zhu, 579, 660
Meng, Fantao, 992
Meng, Xin, 504
Min, Xin, 445

N
Ning, Xizhan, 951

P
Pan, Baichong, 734
Pan, Dan, 779
Pan, Yue, 202, 884, 1226
Pang, Aitong, 1151, 1218

Q
Qi, Yuliang, 176
Qi, Yunyang, 649
Qian, Xudong, 221
Qin, Jianjun, 591, 607, 1226
Qu, Chengping, 679
Quan, Rongsong, 92

R
Ruan, Xingqun, 992
Ruan, Xinxin, 1022

S
Shen, Huijun, 455, 1062
Shen, Jing Hua, 465
Shen, Jinghua, 329
Shen, Yuxuan, 202
Shen, Zhonghui, 984
Shi, Aofeng, 230
Shi, Xuhua, 311
Shu, Zhan, 365, 893
Song, Guohua, 860
Su, Jian, 860
Sun, Ranran, 185
Sun, Xiaofeng, 893

T
Tan, Xiaojun, 1139
Tang, Liang, 1009
Tang, Mingyi, 926
Tang, Qiang, 319, 1201
Tang, Xiaoxue, 385
Tang, Yi, 716
Tang, Zhenyun, 3
Tao, Ran, 221
Tian, Fei, 1062
Tian, Songwen, 1188
Tian, Xinchen, 542
Tian, Zhen, 1009
Tong, Longyong, 789

V
van Dam, Koen H., 1234

W
Wang, Bingyan, 403
Wang, Biyan, 32, 65, 151
Wang, Chao, 708
Wang, Chengyuan, 571
Wang, Chunyan, 851
Wang, Feng, 1107, 1170
Wang, Haidong, 716
Wang, Hanxue, 562
Wang, Huan, 542
Wang, Hui, 92, 616
Wang, Jianlin, 1099
Wang, Li, 394, 934
Wang, Nuo, 1001
Wang, Peng, 1091
Wang, Qingyuan, 901
Wang, Rong, 221
Wang, Shan, 65
Wang, Shangzhang, 1009
Wang, Sheng, 679
Wang, Shuai, 1139
Wang, Tao, 484
Wang, Wei, 504
Wang, Weiwu, 32, 65, 151
Wang, Xiangyun, 1107, 1170
Wang, Xiaolei, 1070
Wang, Xinghua, 734
Wang, Xiong, 884

Wang, Yanfeng, 734
Wang, Yigan, 504
Wang, Yize, 771
Wang, Yunfei, 724
Wang, Zengliang, 789
Wang, Zhen, 1009
Wang, Zhiqiang, 1099
Wang, Zhongxing, 672
Wei, Huixian, 542
Wei, Kai, 984
Wei, Yan, 255
Wen, Caifeng, 741
Wen, Zhengmin, 92
Weng, Yunhao, 1053
Wenqi, E, 926
Wu, Bin, 1009
Wu, Hongju, 571
Wu, Jianmei, 78
Wu, Jie, 329, 385, 465
Wu, Jingjing, 1209
Wu, Junhua, 1001
Wu, Mingbo, 1234
Wu, Wenjing, 741
Wu, Yao, 44, 122, 190
Wu, Yonglan, 44, 190

X

Xi, Changqin, 484
Xia, Xiaodong, 211
Xian, Jiahua, 1045
Xiang, Yuanzhi, 741
Xiao, Jiajun, 1009
Xie, Jiemin, 1179
Xie, Peixing, 629
Xing, Xin, 1084
Xiong, Feng, 660
Xu, Chengguang, 221
Xu, Hongpeng, 78
Xu, Jianqiang, 108
Xu, Jigang, 869
Xu, Liang, 465
Xu, Ruijie, 385
Xu, Yi, 901
Xu, Yifeng, 826
Xue, Huiqing, 1116
Xue, Tianqi, 910, 1053
Xue, Zhaoyang, 329

Y

Yan, Qianqian, 385
Yang, Bo, 918
Yang, Chen, 876
Yang, Gang, 519
Yang, Ge, 1009
Yang, Huashan, 542
Yang, Liu, 1234
Yang, Mingming, 860
Yang, Ronghui, 445
Yang, Ronglu, 151
Yang, Tong, 347
Yang, Xiaorui, 1125
Yang, Yuxin, 151
Yao, Liang, 394
Yao, Wu, 133
Yi, Lu, 798
Yin, Chao, 826
Yin, Jing, 504
Yin, Sai, 299
You, Linlin, 1255
You, Qi, 455
Yu, Guoqing, 230
Yu, Qi, 679
Yu, Sen, 716
Yu, Zhen, 527
Yuan, Xiaolan, 415, 474, 836, 910, 974, 1053
Yuan, Yufang, 918

Z

Zeng, Cong, 926
Zeng, Yixin, 629
Zhang, Baoxi, 166
Zhang, Dan, 724
Zhang, Dong-Dong, 660
Zhang, Futai, 436
Zhang, Hongwei, 750, 960
Zhang, Jiazhong, 1139
Zhang, Jie, 329
Zhang, Jing, 809
Zhang, Jixun, 496
Zhang, Junfei, 724
Zhang, Kaiyuan, 750, 960
Zhang, Min, 552, 779
Zhang, Mo, 724
Zhang, Wei, 465
Zhang, Weiheng, 591
Zhang, Xin, 910, 1218
Zhang, Xinyue, 679
Zhang, Yan, 599
Zhang, Yi, 484, 836, 974
Zhang, Yuxian, 496
Zhang, Zhouyu, 1091
Zhao, Jianfeng, 992
Zhao, Jiayue, 741
Zhao, Jingde, 1209
Zhao, Xuexiu, 22
Zhao, Zixu, 1070
Zheng, Bohong, 1158
Zheng, Hehui, 1062
Zheng, Liang, 319, 1201
Zheng, Yan, 696

Zhong, Minting, 1045
Zhong, Shuqi, 1179
Zhong, Xi, 984
Zhou, Hang, 789
Zhou, Ji, 934
Zhou, Li, 826
Zhou, Longlong, 1125
Zhou, Meng, 1255
Zhou, Xiaojing, 1226
Zhou, Xi-Wen, 599
Zhou, Yingying, 263
Zhou, Ziye, 689
Zhu, Han, 809
Zhu, Hui, 122
Zhu, Liujie, 496
Zhu, Ruiguang, 1070
Zhu, Shaocai, 1045
Zhu, Wenwei, 734
Zhu, Youqing, 385
Zou, Qingyu, 826